Lecture Notes in Mathematics 2125

More information about this series at
http://www.springer.com/series/304

Takuro Mochizuki

Mixed Twistor $\mathscr{D}$-modules

Takuro Mochizuki
Research Institute for Mathematical
Sciences (RIMS)
Kyoto University
Kyoto, Japan

ISSN 0075-8434 ISSN 1617-9692 (electronic)
Lecture Notes in Mathematics
ISBN 978-3-319-10087-6 ISBN 978-3-319-10088-3 (eBook)
DOI 10.1007/978-3-319-10088-3

Library of Congress Control Number: 2015943589

Mathematics Subject Classification (2010): 32C38, 14F10

Springer Cham Heidelberg New York Dordrecht London

Printed on acid-free paper

Springer International Publishing AG Switzerland is part of Springer Science+Business Media (www.springer.com)

Dedicated to
Professor Akira Kono and
Professor Mikiya Masuda
with appreciation to their supports
in my younger days

Preface

Let us begin with the most basic concept of twistor structure, which was introduced by Simpson [82] as a generalization of the concept of Hodge structure. A twistor structure is a holomorphic vector bundle on $\mathbb{P}^1$, and a (complex) Hodge structure is a $\mathbb{C}$-vector space with two filtrations. The standard Rees construction allows us to obtain an equivalence of Hodge structure and twistor structure with a $\mathbb{C}^*$-action.

Ideally, any notion for Hodge structure can be translated to those for twistor structure. Indeed, by applying the Rees construction, we obtain something on $\mathbb{P}^1$ equipped with a $\mathbb{C}^*$-action. Then, by forgetting the $\mathbb{C}^*$-action, we obtain the counterpart of the notion in the context of twistor structure.

For instance, a twistor structure is called pure of weight m if it is isomorphic to a direct sum of $\mathcal{O}_{\mathbb{P}^1}(m)$. A mixed twistor structure is defined to be a twistor structure V with an increasing filtration W indexed by integers such that $\mathrm{Gr}_m^W(V)$ are pure of weight m. These generalize the notions of pure and mixed Hodge structure. We also have the twistor version of polarizations and Tate twists. They share important properties with their counterparts in the Hodge context. For example, the categories of mixed Hodge structures and mixed twistor structures are abelian. We can compare it with the fact that the category of vector spaces with a filtration is not abelian.

On the basis of the fact that many important features of Hodge structures already appear in the level of twistor structures, Simpson proposed a principle, called Simpson's Meta Theorem. It roughly says that most objects and most theorems in Hodge theory should have their counterparts in the context of twistor structures. This principle leads us to a promising and interesting project, which we call "from Hodge toward twistor". I cannot exaggerate the philosophical significance of his principle.

Let us recall that one of the important branches in the Hodge theory is the study of the functoriality of Hodge structure. Later in Introduction, we will briefly review it. Here, we just remind that the functoriality of Hodge structure was thoroughly established by the extremely deep theory of mixed Hodge modules [69, 73] due to Morihiko Saito. Roughly saying, mixed Hodge modules are regular holonomic $\mathcal{D}$-modules with mixed Hodge structure. Saito's theory ensures that mixed Hodge structures are functorial with respect to standard operations for regular holonomic $\mathscr{D}$-modules.

According to Simpson's Meta theorem, we should have the twistor version of mixed Hodge modules. That is the concept of mixed twistor $\mathscr{D}$-modules which we shall investigate in this monograph. We shall establish their fundamental properties, in particular, the functoriality with respect to the standard operations. That is the goal of this study. Indeed, it is the ultimate goal for me in the project "from Hodge toward twistor".

We should remark that Claude Sabbah introduced the concept of pure twistor $\mathscr{D}$-modules [66, 67], which is a twistor version of pure Hodge modules. The theory of pure twistor $\mathscr{D}$-modules was further studied by Sabbah and myself [52, 55]. We could regard mixed twistor $\mathscr{D}$-modules as the mixed version of pure twistor $\mathscr{D}$-modules. We note that the ingredients for twistor $\mathscr{D}$-modules are not the same as those for mixed Hodge modules. We also note that there are some phenomena which do not appear in the context of Hodge modules. So, although we owe much to the fundamental strategy of Saito in the Hodge case, we also have some additional issues to deal with. It is contrast to the fact that generalization of mixed Hodge structure to mixed twistor structure is technically rather straightforward.

I should mention why it is interesting to have a twistor version of mixed Hodge modules. It is the most important reason that we can apply the theory of mixed twistor $\mathscr{D}$-modules to a wider class of holonomic $\mathscr{D}$-modules possibly with irregular singularities.

Indeed, any algebraic semisimple holonomic $\mathscr{D}$-module underlies a pure twistor $\mathscr{D}$-module [55]. It implies the Hard Lefschetz theorem for algebraic semisimple holonomic $\mathscr{D}$-modules with respect to the push-forward by projective morphisms, which is one of the most interesting results in the study of pure twistor $\mathscr{D}$-modules.

As for the mixed case, we have the mixed twistor $\mathscr{D}$-modules associated with meromorphic functions. Namely, let X be a complex manifold with a hypersurface H. Let f be a meromorphic function on X whose poles are contained in H. We have the $\mathscr{D}_X$-module $L_*(f,H)$ obtained as the $\mathcal{O}_X$-module $\mathcal{O}_X(*H)$ with the flat connection $d+df$. We have the natural mixed twistor $\mathscr{D}$-module over $L_*(f,H)$. By applying the standard operations to such mixed twistor $\mathscr{D}$-modules, we can observe that many important $\mathscr{D}$-modules naturally underlie mixed twistor $\mathscr{D}$-modules.

For instance, some type of GKZ-hypergeometric systems are naturally enriched to mixed twistor $\mathscr{D}$-modules, which also naturally appear in the study of the Landau-Ginzburg models in the toric mirror symmetry. Recently, we applied the degeneration of the mixed twistor $\mathscr{D}$-modules over the GKZ-hypergeometric systems to the study of local mirror symmetry in [60]. See also [59] for an application to the study of Kontsevich complexes.

I hope that our study would be a part of the foundation for the further study on the generalized Hodge theory of holonomic $\mathscr{D}$-modules possibly with irregular singularity. I also hope that it would be a help for readers to get into a technical part of the deep theory due to Saito.

This study grew out of my attempt to understand the works due to Beilinson [2, 3], Kashiwara [32], and Saito [69–71, 73]. The readers can find most essential ideas in their papers. I thank Morihiko Saito for some discussions. I thank Claude Sabbah for numerous discussions on many occasions. I deeply thank Carlos Simpson. It is impossible for me to mention what I owe to him. I just note here that his most fundamental principle (Simpson's Meta-Theorem) invited me to this study. Special thanks go to Mark de Cataldo, Pierre Deligne, Kenji Fukaya, William Fulton, David Gieseker, Akira Kono, Mikiya Masuda, Atsushi Moriwaki, Masa-Hiko Saito, Tomohide Terasoma, Michael Thaddeus, and Kari Vilonen. I would like to express my gratitude to Yves André, Philip Boalch, Helene Esnault, Claus Hertling, Maxim Kontsevich, Thomas Reichelt, Kyoji Saito, Christian Schnell, and Christian Sevenheck, for some discussions. I am grateful to Indranil Biswas for his excellent hospitality during my stay at the Tata Institute of Fundamental Research. I thank Akira Ishii and Yoshifumi Tsuchimoto for their constant encouragement. I appreciate the referees for their valuable comments to improve this monograph.

I studied harmonic bundles and twistor $\mathscr{D}$-modules in the Department of Mathematics at Osaka City University, the Institute for Advanced Study, the Max-Planck Institute for Mathematics, l'Institut des Hautes Études Scientifique, the Department of Mathematics at Kyoto University, and the Research Institute for Mathematical Sciences at Kyoto University. I thank the colleagues and the staff of the institutions for their excellent support. I thank the Tata Institute of Fundamental Research for the excellent hospitality during my stay where I wrote a part of the final manuscript of this book. I thank the organizer of the conference "International Conference on Noncommutative Geometry and Physics" in which I gave a talk on this topic.

This work was partially supported by the Grant-in-Aid for Scientific Research (S) (No. 24224001), the Grant-in-Aid for Scientific Research (A) (No. 22244003), the Grant-in-Aid for Scientific Research (C) (No. 22540078), Japan Society for the Promotion of Science.

Kyoto, Japan Takuro Mochizuki

Contents

Chapter 1
Introduction

1.1 Mixed Hodge Modules

Let us briefly review a stream in the Hodge theory. It is concerned with the functoriality with respect to various operations, and it was finally accomplished with the most great generality by the theory of mixed Hodge modules due to M. Saito. The author regrets that this review is quite restricted by his personal interest, and that it is not exhaustive. The readers can find more thorough reviews in [21, 62] and the recent book "Hodge Theory" edited by E. Cattani, F. El Zein, P.A. Griffiths, and L.D. Tràng (Princeton University Press). We also refer to [72, 74, 75] for the introduction to mixed Hodge modules.

A variation of Hodge structure on a complex manifold X is a pair of $\mathbb{Q}$-local system $L_{\mathbb{Q}}$ of finite rank and a Hodge filtration F. Here, a Hodge filtration is a decreasing filtration of holomorphic subbundles of $L_{\mathbb{Q}} \otimes_{\mathbb{Q}} \mathcal{O}_X$ indexed by integers satisfying the Griffiths transversality. Although we may replace $\mathbb{Q}$ with other algebras such as $\mathbb{Z}$ and $\mathbb{R}$, we omit such details here. A variation of Hodge structure is called pure of weight w if each restriction $(L_{\mathbb{Q}}, F)_{|Q}$ $(Q \in X)$ is a pure Hodge structure of weight w. It is called polarizable if moreover it admits a polarization, i.e., there exists a flat $(-1)^w$-symmetric pairing S of $L_{\mathbb{Q}}$ such that each $S_{|Q}$ $(Q \in X)$ is a polarization of the pure Hodge structure $(L_{\mathbb{Q}}, F)_{|Q}$. A variation of mixed Hodge structure is a variation of Hodge structure $(L_{\mathbb{Q}}, F)$ with a weight filtration W of $L_{\mathbb{Q}}$ which is an increasing filtration indexed by integers, such that $\mathrm{Gr}^W_w(L_{\mathbb{Q}})$ $(w \in \mathbb{Z})$ with the induced Hodge filtration F are pure of weight w. The variation of mixed Hodge structure is called graded polarizable if each $\mathrm{Gr}^W_w(L_{\mathbb{Q}}, F)$ is polarizable. In this paper, we almost always impose the polarizability (resp. the graded polarizability) to variations of pure (resp. mixed) Hodge structure. So, we often omit the adjective "graded polarizable".

The notion of polarized variation of Hodge structure was originally discovered by P. A. Griffiths as *something* on the Gauss-Manin connections associated to smooth projective families of varieties. This can already be regarded as one of the most basic

T. Mochizuki, *Mixed Twistor $\mathcal{D}$-modules*, Lecture Notes in Mathematics 2125,
DOI 10.1007/978-3-319-10088-3_1

and interesting cases of the functoriality of Hodge structure for the push-forward by any smooth projective morphism. The seminal work of Griffiths opened several interesting research projects, for example, the study of polarized variation of Hodge structure with singularity, which we will return later.

Inspired by the dream of motives, P. Deligne discovered the notion of mixed Hodge structure, and he proved a deep theorem which ensures that the cohomology group of any complex algebraic variety is naturally equipped with a mixed Hodge structure. This can be regarded as one of the most important cases of the functoriality of mixed Hodge structure. He also proved the functoriality in various cases. For example, if we are given a smooth projective family of varieties $f : \mathcal{Y} \longrightarrow S$ and a graded polarizable variation of mixed Hodge structure $(L_{\mathbb{Q}}, F, W)$ on $\mathcal{Y}$, then it was proved that the local system $R^i f_*(L_{\mathbb{Q}})$ is equipped with the naturally induced Hodge filtration F and weight filtration W and that $(R^i f_* L_{\mathbb{Q}}, F, W)$ is graded polarizable variation of mixed Hodge structure. He also observed crucial properties of the induced variation of mixed Hodge structure, including the Hard Lefschetz Theorem in the pure case. His insight has been quite influential on the subsequent works. (See [11, 14–16] for instance for more details on his work).

It is natural to ask what happens in the other more general cases. For example, if we are given a polarizable variation of Hodge structure $(L_{\mathbb{Q}}, F)$ on a quasi projective variety Y which is not extendable on any projective completion of Y, it is asked whether the cohomology group $H^p(Y, L_{\mathbb{Q}})$ or its variant may have mixed or pure Hodge structure. It is also natural to ask what happens for the singular family of varieties. Finally, all of these questions were answered by the functoriality of mixed Hodge modules. But, historically, it was first investigated with the L^2-method on the basis of the study of the asymptotic behaviour of polarized variations of pure Hodge structure and admissible variations of mixed Hodge structure. The study of asymptotic behaviour is also fundamental as the foundation of the theory of mixed Hodge modules.

As mentioned, the work of Griffiths naturally lead to the study of polarized variations of pure Hodge structure with singularity. An extremely important contribution was done by Schmid [77]. Let $X := \{(z_1, \ldots, z_n) \mid |z_i| < 1\}$ and $D := \bigcup_{i=1}^{\ell} \{z_i = 0\}$. Let $(L_{\mathbb{Q}}, F)$ be a polarizable variation of pure Hodge structure on $X \setminus D$. For simplicity, we assume that the local monodromy automorphisms around any irreducible components of D are unipotent. The nilpotent orbit theorem of Schmid ensures that, around any $P \in D$, the polarized variation of Hodge structure can be approximated by an easier one called a nilpotent orbit. It is not only interesting itself but also the most important foundation for the further investigation. One of the important consequences is that $(L_{\mathbb{Q}}, F)$ given on $X \setminus D$ naturally induces a nice object on X. Namely, let (V, ∇) be the Deligne extension of $L_{\mathbb{Q}} \otimes_{\mathbb{Q}} \mathcal{O}_{X \setminus D}$ on X, i.e., V is the locally free $\mathcal{O}_X$-module with a logarithmic connection ∇ such that (1) $(V, \nabla)_{|X \setminus D} = L_{\mathbb{Q}} \otimes_{\mathbb{Q}} \mathcal{O}_{X \setminus D}$, (2) the residues of ∇ are nilpotent. (See [12] for more details on the Deligne extension). Then, the nilpotent orbit theorem implies that F is extended to a filtration of V by holomorphic subbundles. In the case $n = \ell = 1$, the study of singular polarized variation of pure Hodge structure was accomplished by his SL(2)-orbit theorem, which ensures that the polarized variation of Hodge structure can be approximated by an easier one called an SL(2)-orbit. As a consequence, in the one variable case, he obtained that the weight filtration of the nilpotent part of the local monodromy controls the growth order of the norms of flat sections with respect to the Hermitian metric associated to the polarization. He also obtained the polarized mixed Hodge structure from the asymptotic data around the singularity, which is called the limit mixed Hodge structure. Note that, for the polarized variation of pure Hodge structure associated to a degenerating family of smooth projective varieties, the asymptotic behaviour was intensively studied by Steenbrink with a different method [86].

The higher dimensional case was accomplished by the definitive works by Cattani and Kaplan [7], Cattani et al. [8, 9], Kashiwara [31] and Kashiwara and Kawai [35]. In the above situation, for each point $P \in D$, we obtain the limit mixed Hodge structure with the induced bi-linear form from the asymptotic data around P, which is a polarized mixed Hodge structure in several variables. It turned out that the limit mixed Hodge structure controls the behaviour of $(L_{\mathbb{Q}}, F)$ around P. They obtained a generalization of the norm estimate. They also obtained a rather strong constraint on the nilpotent parts of the local monodromy along the loops around $\{z_i = 0\}$ $(i = 1, \ldots, \ell)$. Moreover, they proved various interesting properties of polarized mixed Hodge structures, which are significant for their study on the L^2-cohomology group associated to any polarized variation of Hodge structure. Although it requires much more preparation to describe their results precisely, which we do not intend here, they are quite impressive.

As for singular graded polarizable variations of mixed Hodge structure, it was one of the main issues to clarify what condition should be imposed at the boundary. Thanks to the studies of Kashiwara [32], Steenbrink and Zucker [87] and Zucker [89], it turned out that the admissibility condition is the most appropriate one. Let us recall it in the case that $X = \{(z_1, \ldots, z_n) \mid |z_i| < 1\}$ and $D = \bigcup_{i=1}^{\ell} \{z_i = 0\}$ as above. Let $(L_{\mathbb{Q}}, F, W)$ be a graded polarizable variation of mixed Hodge structure on $X \setminus D$. For simplicity, suppose that the monodromy g_i along the loops around $\{z_i = 0\}$ are unipotent. Let $N_i := \log g_i$. Let (V, ∇) be the Deligne extension of $L_{\mathbb{Q}} \otimes_{\mathbb{Q}} \mathcal{O}_{X \setminus D}$, which is naturally equipped with the flat filtration W. We should impose that the filtration F is extended to a filtration of V by holomorphic subbundles such that $\mathrm{Gr}^F \mathrm{Gr}^W(V)$ is a locally free $\mathcal{O}_X$-module. We should also impose the existence of a relative monodromy weight filtration $M(N_i; W)$ of N_i with respect to the induced filtration W on the space of the multi-valued flat sections of $L_{\mathbb{Q}}$. It was introduced by Steenbrink-Zucker in the one variable case, by Kashiwara in the higher dimensional case. Note that Kashiwara clarified many issues to ensure that the condition is good. (The condition here is not equal but equivalent to that in [32], by results in [32].) Moreover, Kashiwara introduced and studied *infinitesimal mixed Hodge modules*, which is the "mixed version" of polarizable mixed Hodge structures. He constructed some natural filtrations, as a generalization of some filtrations in [87], which are crucial in the study on mixed Hodge modules.

As mentioned, one of the motivations in the study of singular variations of Hodge structure was to establish the functoriality of Hodge structure, as a generalization of the results of Deligne. Let X be a smooth projective variety with a normal crossing hypersurface D. Let $(L_{\mathbb{Q}}, F)$ be any polarizable variation of pure Hodge structure on $X \setminus D$. One of the issues in those days was to show that there exists a natural pure Hodge structure on the intersection cohomology group of $L_{\mathbb{Q}}$. If $(L_{\mathbb{Q}}, F)$ has no singularity at D, then it follows from the result of Deligne. In the singular case, the contribution of Zucker [88] is quite important. He studied the issue in the case $\dim X = 1$, and he proved that the intersection cohomology group is isomorphic to the L^2-cohomology group if $X \setminus D$ is equipped with a Poincaré like metric. He also developed the L^2-harmonic theory for singular polarized variation of pure Hodge structure on projective curves. As a result, he obtained a naturally induced pure Hodge structure on the intersection cohomology of $L_{\mathbb{Q}}$. In the higher dimensional case, Cattani-Kaplan-Schmid and Kashiwara-Kawai established it by making good use of their results on the asymptotic behaviour of polarized variation of Hodge structure, and by generalizing the method of Zucker. As for the mixed case, for an admissible variation of mixed Hodge structure on curves, Steenbrink-Zucker proved that the various naturally associated cohomology groups have mixed Hodge structure, based on their results on the asymptotic behaviour.

This stream of research for functoriality was eventually accomplished with much more great generality by the theory of mixed Hodge modules due to M. Saito. A cohomology theory can be regarded as a part of the theory of six functors on the derived categories of some type of sheaves. Briefly, the theory of mixed Hodge modules ensures that the derived functors for $\mathbb{Q}$-perverse sheaves on complex algebraic varieties can be enriched by mixed Hodge structures. (In this introduction, we consider only polarizable pure Hodge modules and graded polarizable mixed Hodge modules, we omit the adjectives "polarizable" or "graded polarizable".)

Very roughly, a Hodge module on a complex manifold X consists of a $\mathbb{Q}$-perverse sheaf $P_{\mathbb{Q}}$ with a Hodge filtration F on the regular holonomic $\mathscr{D}_X$-module M corresponding to $P_{\mathbb{Q}}$, i.e., (1) $\mathrm{DR}_X(M) \simeq P_{\mathbb{Q}} \otimes_{\mathbb{Q}} \mathbb{C}$, (2) F is an increasing filtration of M by coherent $\mathcal{O}_X$-modules indexed by integers such that $F_j(\mathscr{D}_X) \cdot F_i(M) \subset F_{i+j}(M)$, where $\mathscr{D}_X$ denotes the sheaf of holomorphic differential operators on X with the filtration F by the order of operators. In his highly original and genius work, Saito invented the appropriate definitions of pure and mixed conditions for such filtered $\mathscr{D}$-modules, and he established their fundamental properties. The most important theorems in the theory are the functoriality with respect to six operations, and the description of pure and mixed Hodge modules.

For the functoriality in the pure case, he proved the Hard Lefschetz Theorem. Namely, let $f : X \longrightarrow Y$ be a projective morphism of smooth projective varieties. Let $(P_{\mathbb{Q}}, F)$ be any polarizable pure Hodge module of weight w on X. Let $f_\dagger^i P_{\mathbb{Q}}$ denote the i-th cohomology perverse sheaf of the push-forward of $P_{\mathbb{Q}}$ with respect to f. Then, $f_\dagger^i P$ is equipped with a naturally induced Hodge filtration $f_\dagger^i F$, so that $f_\dagger^i(P_{\mathbb{Q}}, F)$ is a polarizable pure Hodge module of weight $w + i$. Moreover, for the morphisms $L : f_\dagger^i P_{\mathbb{Q}} \longrightarrow f_\dagger^{i+2} P_{\mathbb{Q}}$ $(i \in \mathbb{Z})$ induced by the first Chern class of a relatively ample line bundle, the morphisms $L^i : f_\dagger^{-i} P_{\mathbb{Q}} \longrightarrow f_\dagger^i P_{\mathbb{Q}}$ $(i \geq 0)$ are isomorphisms. This theorem is a generalization of the classical and important theorem of Beilinson-Bernstein-Deligne-Gabber on perverse sheaves of geometric origin [4].

As for the functoriality in the mixed case, he constructed the six operations together with the nearby and vanishing cycle functors for the derived category of mixed Hodge modules on algebraic varieties, which are compatible with those for the derived category of perverse sheaves.

Because the definitions of pure and mixed Hodge modules are complicated, it is important to know what objects are contained in the categories. Saito proved that a polarizable (resp. graded polarizable) variation of pure (resp. mixed) Hodge structure on X naturally gives a pure (mixed) Hodge module on X, as expected. Hence, the simplest variation of pure Hodge structure $\mathbb{Q}_X$ naturally gives a pure Hodge module. (But, note that while the variation of pure Hodge structure is of weight 0, the pure Hodge module is of weight $\dim X$.) Therefore, if a perverse sheaf $P_{\mathbb{Q}}$ on X is obtained from $\mathbb{Q}_Y$ on a complex algebraic manifold Y by successive use of six functors, it naturally underlies a mixed Hodge module. If a perverse sheaf is of geometric origin, then it naturally underlies a pure Hodge module. In particular, the category of mixed Hodge modules contain many objects. Moreover, he proved the more general results for the description. In the pure case, he proved the following.

- Let $Z \subset X$ be a closed irreducible complex analytic subvariety. Let $U \subset Z$ be a complement of a closed analytic subset, such that U is smooth. Let $\iota : U \longrightarrow X$ be the inclusion, and set $d_U := \dim U$. Let $(L_{\mathbb{Q}}, F)$ be any polarizable variation of Hodge structure on U. Then, the perverse sheaf $\iota_{*!}L_{\mathbb{Q}}[d_U]$, which is the minimal extension of $L_{\mathbb{Q}}[d_U]$ on X, is naturally equipped with the Hodge filtration F so that $(\iota_{*!}L_{\mathbb{Q}}[d_U], F)$ is a polarizable pure Hodge module.
- Conversely, any polarizable pure Hodge module is the direct sum of such minimal extensions.

Hence, for example, suppose that we are given a polarizable variation of Hodge structure $(L_{\mathbb{Q}}, F)$ on $X \setminus D$, where X is a complex manifold, and D is a normal crossing hypersurface. We obtain the pure Hodge module $(P_{\mathbb{Q}}, F)$ on X, obtained as the minimal extension of $(L_{\mathbb{Q}}, F)$, as in the above description. For the canonical map a_X of X to a point, the i-th cohomology of the push-forward $a^i_{X\dagger}(P_{\mathbb{Q}})$ is naturally equipped with the Hodge filtration by the functoriality of the pure Hodge modules. It means that the intersection cohomology of $L_{\mathbb{Q}}$ is equipped with a naturally induced pure Hodge structure, which implies the theorem of Zucker, Cattani-Kaplan-Schmid and Kashiwara-Kawai.

In the mixed case, Saito established the following:

- Let X, Z, U, ι and d_U be as above. Let $(L_{\mathbb{Q}}, F, W)$ be an admissible variation of mixed Hodge structure. Then, the perverse sheaves $\iota_* L_{\mathbb{Q}}[d_U]$ and $\iota_! L_{\mathbb{Q}}[d_U]$ are naturally equipped with the Hodge filtrations $\tilde{F}$ and the weight filtrations $\tilde{W}$ such that $(\iota_\star L_{\mathbb{Q}}[d_U], \tilde{F}, \tilde{W})$ $(\star = *, !)$ are mixed Hodge modules.
- Conversely, any mixed Hodge modules on X are locally obtained as the "gluing" of admissible variation of mixed Hodge structures.

It implies that, for example, we have a natural mixed Hodge structure on various cohomology groups associated to an admissible variation of mixed Hodge structure.

Remark 1.1.1 The theory of pure and mixed Hodge modules can be regarded as a counterpart of the theory of pure and mixed ℓ-adic sheaves on algebraic varieties over finite fields [4], which has been influential in various fields of mathematics including number theory and representation theory. See a very nice book [27] for more details on the philosophical background of Hodge modules, and for applications of the theory of Hodge modules to representation theory. □

Remark 1.1.2 See also a more recent work [10] for another approach to the functoriality of Hodge structures.

1.2 From Hodge Toward Twistor

As mentioned, it is our purpose in this monograph to study a twistor version of mixed Hodge modules. It is C. Simpson who introduced the notion of twistor structure as an underlying structure of Hodge structure. He proposed a principle called Simpson's Meta Theorem, which says that stories of Hodge structures should be generalized to stories of twistor structures.

When he introduced the concept of twistor structure, he was motivated to understand harmonic bundles in a deeper way. Let (V, ∇) be a flat bundle on a complex manifold X with a Hermitian metric h. We have a unique decomposition $\nabla = \nabla^u + \Phi$ into a unitary connection and a self-adjoint section of $\mathrm{End}(V) \otimes \Omega^1$. We have the decompositions into $\nabla^u = \overline{\partial}_V + \partial_V$ and $\Phi = \theta^\dagger + \theta$ into the $(0, 1)$-part and the $(1, 0)$-part. Then, (V, ∇, h) is called a harmonic bundle, if $(V, \overline{\partial}_V, \theta)$ is a Higgs bundle in the sense that $\overline{\partial}_V$ is a holomorphic structure of V and θ is a Higgs field of $(V, \overline{\partial}_V)$. In that case, the metric h is called pluri-harmonic. The concept was discovered by N. Hitchin [26] in the one dimensional case, and by Simpson [78–80] in the higher dimensional case.

One of the most important classes of harmonic bundles is given by polarized variations of Hodge structure. The Hermitian metrics induced by polarizations of polarizable variations of Hodge structure are pluri-harmonic. From the beginning of his study, Simpson was motivated by the investigation of polarized variations of Hodge structure. (See [78–83]). He gave a method to construct polarized variation of Hodge structure by using the Kobayashi-Hitchin correspondence for harmonic bundles. He observed that various properties of polarized variation of Hodge structure are naturally generalized to those for harmonic bundles. For example, he developed the harmonic theory for harmonic bundles as a generalization of that for polarized variation of Hodge structure, and he proved that the push-forward of any harmonic bundle of any smooth projective family of varieties is naturally a harmonic bundle.

To pursue this analogy in a deeper level, he introduced the concept of twistor structure, and observed that harmonic bundles can be regarded as *polarized variations of pure twistor structure [82]*. Thus, he established the analogy between polarized variations of Hodge structure and harmonic bundles in the level of the definitions. This important idea enables us to consider a twistor version of various objects appeared in the Hodge theory.

This is quite efficient in the study of the asymptotic behaviour of singular harmonic bundles, which was studied by Simpson himself and the author. (See [79, 82], [50, 52, 55]). The twistor viewpoint suggests us how to formulate generalizations of results of Cattani, Kaplan, Kashiwara, Kawai and Schmid. Indeed, we obtain a nice object on X from a harmonic bundle on $X \setminus D$, and we also obtain the limit mixed twistor structure, which is quite useful to control the nilpotent part of the residues. However, we would like to mention that there are also some phenomena which do not appear for polarized variation of Hodge structure, such as KMS-structure and Stokes structure, and that the proofs are not necessarily given in parallel ways.

It is also suggested by Simpson's Meta Theorem that we should have a twistor version of the theory of pure and mixed Hodge modules. In the pure case, it was pursued by C. Sabbah and the author. Sabbah prepared the notion of $\mathcal{R}$-triples as an ingredient to define twistor $\mathscr{D}$-modules, which can be regarded as a counterpart of pairs of $\mathbb{R}$-perverse sheaf and filtered $\mathscr{D}$-module. They are suitable even in the case that the underlying $\mathscr{D}$-modules are irregular. Based on Saito's strategy, he gave the appropriate definition of pure twistor $\mathscr{D}$-modules and the framework to prove the Hard Lefschetz Theorem, i.e., the functoriality for projective push-forward. The correspondence between tame harmonic bundles and regular pure twistor $\mathscr{D}$-modules was established in [51–53]. In the wild case, the basic properties were established in [55].

It is interesting to have the correspondence between semisimple holonomic $\mathscr{D}$-modules and pure twistor $\mathscr{D}$-modules on projective varieties, which does not appear in the theory of pure Hodge modules. As a result, we obtain that semisimplicity of algebraic holonomic $\mathscr{D}$-modules is preserved by projective push-forward.

In this monograph, we introduce mixed twistor $\mathscr{D}$-modules, and prove the fundamental properties. For the author, it is one of the ultimate goals of the research for years, driven by Simpson's Meta Theorem.

There are various intermediate objects between twistor structure and Hodge structure such as integrable twistor structure and TERP-structure (see [22–25, 56, 66]). So, we could have variants of mixed twistor $\mathscr{D}$-modules by considering additional structures. Because the twistor structure could be most basic among them, the author hopes that mixed twistor $\mathscr{D}$-modules would play a basic role in the study of generalized Hodge structure on holonomic $\mathscr{D}$-modules.

Remark 1.2.1 See [17] for a philosophical background toward a generalized Hodge theory in the context of irregular singularities. See [37] for a generalized Hodge theoretic aspect of the mirror symmetry.

Remark 1.2.2 See [81] for the functoriality of harmonic bundles with respect to smooth projective morphisms. See also [84] for a twistor structure on the cohomology group of orbifolds.

1.3 Mixed Twistor $\mathscr{D}$-Modules

In the rest of this introduction, let us briefly review the theory of pure twistor $\mathscr{D}$-modules, and explain our issues in the study of mixed twistor $\mathscr{D}$-modules.

1.3.1 Pure Twistor $\mathscr{D}$-Modules

For any complex manifold X, the product $\mathbb{C}_\lambda \times X$ is denoted by $\mathcal{X}$. Let $p : \mathcal{X} \longrightarrow X$ be the projection. Let Θ_X be the tangent sheaf of X. We have the sheaf of holomorphic differential operators $\mathscr{D}_{\mathcal{X}}$ on $\mathcal{X}$. Then, $\mathcal{R}_X$ is the sheaf of subalgebras in $\mathscr{D}_{\mathcal{X}}$ generated by $\lambda p^*\Theta_X$ over $\mathcal{O}_{\mathcal{X}}$.

We have two basic conditions on $\mathcal{R}_X$-modules. One is strictness, i.e., flat over $\mathcal{O}_{\mathbb{C}_\lambda}$. The other is holonomicity. Namely, the characteristic variety of any coherent $\mathcal{R}_X$-module $\mathcal{M}$ is defined as in the case of $\mathscr{D}$-modules, denoted by $Ch(\mathcal{M})$. It is a subvariety in $\mathbb{C}_\lambda \times T^*X$. If $Ch(\mathcal{M})$ is contained in the product of $\mathbb{C}_\lambda$ and a Lagrangian subvariety in T^*X, the $\mathcal{R}_X$-module $\mathcal{M}$ is called holonomic.

An $\mathcal{R}_X$-triple is a tuple of $\mathcal{R}_X$-modules $\mathcal{M}_i$ $(i = 1, 2)$ with a sesqui-linear pairing C. To explain what is sesqui-linear paring, we need a preparation. Let $\boldsymbol{S}$ denote the circle $\{\lambda \in \mathbb{C}_\lambda \mid |\lambda| = 1\}$. Let $\sigma : \boldsymbol{S} \longrightarrow \boldsymbol{S}$ be given by $\sigma(\lambda) = -\lambda = -\overline{\lambda}^{-1}$. The induced involution $\boldsymbol{S} \times X \longrightarrow \boldsymbol{S} \times X$ is also denoted by σ.

Let $\mathfrak{Db}_{\boldsymbol{S}\times X/\boldsymbol{S}}$ denote the sheaf of distributions on $\boldsymbol{S} \times X$ which are continuous in the $\boldsymbol{S}$-direction in an appropriate sense. This sheaf is naturally a module over $\mathcal{R}_{X|\boldsymbol{S}\times X} \otimes \sigma^*\mathcal{R}_{X|\boldsymbol{S}\times X}$. Then, a sesqui-linear pairing of $\mathcal{R}_X$-modules $\mathcal{M}_i$ $(i = 1, 2)$ is an $\mathcal{R}_{X|\boldsymbol{S}\times X} \otimes \sigma^*\mathcal{R}_{X|\boldsymbol{S}\times X}$-homomorphism $\mathcal{M}_{1|\boldsymbol{S}\times X} \otimes \sigma^*\mathcal{M}_{2|\boldsymbol{S}\times X} \longrightarrow \mathfrak{Db}_{\boldsymbol{S}\times X/\boldsymbol{S}}$. An $\mathcal{R}_X$-triple is called strict (resp. holonomic), if the underlying $\mathcal{R}$-modules are strict (resp. holonomic). The category of pure twistor $\mathscr{D}$-module is constructed as a full subcategory of strict holonomic $\mathcal{R}$-triples.

Let us recall how to impose some conditions on strict holonomic $\mathcal{R}$-triples. In the case of a variation of Hodge structure which is a $\mathbb{Q}$-local system with a Hodge filtration, it is defined to be pure, if its restriction to the fiber over each point is pure. But, for $\mathcal{R}$-triples or even for $\mathscr{D}$-modules, the restriction to a point is not so well behaved. Instead, for holonomic $\mathscr{D}$-modules, there is a nice theory for restriction to hypersurfaces. Namely, we have the nearby and vanishing cycle functors, which describe the behaviour of the holonomic $\mathscr{D}$-modules around the hypersurface in some degree. To define some condition for holonomic $\mathscr{D}$-modules, it seems natural to consider the conditions on nearby and vanishing cycle sheaves inductively, instead of the restriction to a point. Similarly, to define some condition for $\mathcal{R}$-triples, we would like to consider the condition on the appropriately defined nearby and vanishing cycle functors for $\mathcal{R}$-triples. This is a basic strategy due to Saito, and it may lead us to inductive definitions of pure and mixed twistor $\mathscr{D}$-modules, as a variant of pure and mixed Hodge modules.

A strict holonomic $\mathcal{R}$-triple $\mathcal{T}$ is called pure of weight w if the following holds. First, we impose that, for any open subset $U \subset X$ with a holomorphic function g, $\mathcal{T}_{|U}$ is strictly S-decomposable along g. It implies that we have the decomposition $\mathcal{T} = \bigoplus \mathcal{T}_Z$ by strict support, where Z runs through closed irreducible subsets of X. Then, we impose the conditions on each $\mathcal{T}_Z$. If Z is a point, $\mathcal{T}_Z$ is supposed to be the push-forward of a pure twistor structure of weight w by the inclusion of Z into X. In the positive dimensional case, for any open subset of X with a holomorphic function if we take the grading of the weight filtration of the naturally induced nilpotent morphism on the nearby cycle functor along the function, the m-th graded pieces are pure of weight $w+m$. Then, inductively, the notion of pure twistor $\mathscr{D}$-module is defined. Precisely, we should consider the polarizable object. A polarization of $\mathcal{T}$ is defined as a Hermitian sesqui-linear duality of weight w satisfying some condition on positivity, which is also given in an inductive way using the nearby cycle functor.

Let $\mathrm{MT}(X,w)$ denote the category of polarizable pure twistor $\mathscr{D}$-modules of weight w on X. Let us recall some of their fundamental properties; (1) The category $\mathrm{MT}(X,w)$ is abelian and semisimple; (2) For objects $\mathcal{T}_i \in \mathrm{MT}(X,w_i)$ with a morphism $f : \mathcal{T}_1 \longrightarrow \mathcal{T}_2$ as $\mathcal{R}$-triples such that $w_1 > w_2$, we have $f = 0$; (3) For any projective morphism $f : X \longrightarrow Y$, and for any $\mathcal{T} \in \mathrm{MT}(X,w)$, the i-th cohomology of the push-forward $f_\dagger^i\mathcal{T}$ is an object in $\mathrm{MT}(Y,w+i)$. Moreover, $f_\dagger\mathcal{T} \simeq \bigoplus f_\dagger^i\mathcal{T}[-i]$ in the derived category of $\mathcal{R}$-triples; (4) Let $Z \subset X$ be a closed complex analytic subset. Let $Z_0 \subset Z$ be a closed complex analytic subset such that $Z \setminus Z_0$ is smooth. Then, a wild harmonic bundle on (Z, Z_0) is naturally extended to a pure twistor $\mathscr{D}$-module on X; (5) Conversely, any pure twistor $\mathscr{D}$-modules are the direct sum of such minimal extensions of wild harmonic bundles; (6) Any semisimple algebraic holonomic $\mathscr{D}$-module naturally underlies a polarizable wild pure twistor $\mathscr{D}$-module.

1.3.2 Mixed Twistor $\mathscr{D}$-Modules

To define mixed twistor $\mathscr{D}$-modules, we first consider filtered $\mathcal{R}$-triples $(\mathcal{T}, W)$ such that $\mathrm{Gr}^W_w(\mathcal{T})$ are pure of weight w, where W are locally finite increasing complete exhaustive filtrations indexed by integers. They are too naive, and they play only auxiliary roles. Tentatively, they are called pre-mixed twistor $\mathscr{D}$-modules in this monograph. They have nice functoriality for the push-forward by projective morphisms. However, we need to impose additional conditions for other standard functoriality such as push-forward for open embeddings and pull back. Very briefly, to define mixed twistor $\mathscr{D}$-module, we impose (1) the filtered strict compatibility of W and the V-filtrations, (2) the existence of relative monodromy filtrations on the nearby and vanishing cycle sheaves, (3) the relative monodromy filtrations give the weight filtrations of mixed twistor $\mathscr{D}$-modules with smaller supports. (It will be explained in Chap. 7.)

It is easy to show that mixed twistor $\mathscr{D}$-modules have nice functorial property for nearby and vanishing cycle functors and projective push-forward. However, it is not so easy to show the other functorial properties, for example, the localization

$M \longmapsto M(*H)$ for a hypersurface H. To establish more detailed property, we need a concrete description of mixed twistor $\mathscr{D}$-modules as the gluing of admissible variations of mixed twistor structure.

1.3.3 Gluing Procedure

For perverse sheaves and holonomic $\mathscr{D}$-modules, there are well established theories to glue objects on $\{f = 0\}$ and objects on $\{f \neq 0\}$ [3, 40, 90]. We need such gluing procedure in the context of $\mathcal{R}$-triples. Because of the difference of ingredients, it is not easy to generalize the method of gluing prepared in [73] for Hodge modules to the case of $\mathcal{R}$-triples. Instead, we adopt the excellent method of Beilinson in [3], which reduces the issue to the construction of canonical prolongations $\mathcal{T}[\star t]$ ($\star = *, !$). (See Chaps. 3–4.)

1.3.4 Admissible Variation of Mixed Twistor Structure

We prepare a general theory for admissible variations of mixed twistor structure (Chaps. 8–9), which is a natural generalization of the theory of admissible variations of mixed Hodge structure. Let X be a complex manifold with a simple normal crossing hypersurface D. Very briefly, it is a filtered $\mathcal{R}$-triple $(\mathcal{V}, W)$ on X with poles along D. We impose the conditions (1) each $\mathrm{Gr}^W_w(\mathcal{V})$ comes from a good wild harmonic bundle, (2) $\mathcal{V}$ has good-KMS structure along D compatible with W, (3) the residues along the divisors have relative monodromy filtrations. It is important to understand the specialization of admissible variations of mixed twistor structure along the divisors. For that purpose, it is essential to study the relative monodromy filtrations and their compatibility. So, as in [32], we study in Chap. 8 the infinitesimal version of admissible variations of mixed twistor structure, called infinitesimal mixed twistor modules. We can show that they have nice properties as in the Hodge case.

Then, we study the canonical prolongation of admissible mixed twistor structure $(\mathcal{V}, W)$ on (X, D) to pre-mixed twistor $\mathscr{D}$-modules on X. Let $D = D^{(1)} \cup D^{(2)}$ be a decomposition. Recall that any good meromorphic flat bundle V on (X, D) is extended to a $\mathscr{D}$-module $V[*D^{(1)}!D^{(2)}]$ on X. We prepare a similar procedure to make a pre-mixed twistor $\mathscr{D}$-module $(\mathcal{V}, W)[*D^{(1)}!D^{(2)}]$ from $(\mathcal{V}, W)$. First, we construct the underlying $\mathcal{R}$-triple $\mathcal{V}[*D^{(1)}!D^{(2)}]$ in Chap. 5. One of the main tasks is to construct a correct weight filtration W on $\mathcal{V}[*D^{(1)}!D^{(2)}]$. By applying the procedure in Chap. 5 to each $W_j\mathcal{V}$, we obtain a naively induced filtration L on $\mathcal{V}[*D^{(1)}!D^{(2)}]$, i.e., $L_j(\mathcal{V}[*D^{(1)}!D^{(2)}]) = W_j(\mathcal{V})[*D^{(1)}!D^{(2)}]$. But, this is not the correct filtration. We need much more considerations for the construction of the correct weight filtration W. It is essentially contained in [32, 73], but we shall give rather details, which is one of the main themes in Chap. 6 and Chaps. 8–10.

Once we obtain canonical prolongations of admissible variations of mixed twistor structures across normal crossing hypersurfaces, we can glue them to obtain pre-mixed twistor $\mathscr{D}$-modules, which are called good pre-mixed twistor $\mathscr{D}$-modules. It is one of the main theorems to show that any good pre-mixed twistor $\mathscr{D}$-module is a mixed twistor $\mathscr{D}$-module (Theorem 10.3.1). Then, by a rather formal argument, we can show that any mixed twistor $\mathscr{D}$-module can be expressed as gluing of admissible variation of mixed twistor $\mathscr{D}$-modules as in Sect. 11.1. We can deduce some basic functorial properties by using this description. See Sects. 11.2–11.4.

Remark 1.3.1 In this monograph, we will often omit "variation of" just for simplification. For example, an admissible variation of mixed twistor structure is often called an admissible mixed twistor structure. □

1.3.5 Duality and Real Structure

1.3.5.1 Duality

In the study of $\mathscr{D}$-modules, the duality functor is fundamental. To define the duality functor for mixed twistor $\mathscr{D}$-modules there are two issues which we should address. Let X be any complex manifold. Let $d_X := \dim X$. Let ω_X be the sheaf of holomorphic d_X-forms. We put $\omega_{\mathcal{X}} := \lambda^{-d_X} p^* \omega_X$. As given in Sect. 13.1, the dual of any coherent $\mathcal{R}_X$-module $\mathcal{M}$ is defined as follows in the derived category of $\mathcal{R}_X$-modules:

$$\boldsymbol{D}\mathcal{M} := \lambda^{d_X} R\mathcal{H}om_{\mathcal{R}_X}\big(\mathcal{M}, \mathcal{R}_X \otimes \omega_{\mathcal{X}}^{-1}\big)[d_X] \tag{1.1}$$

The dual $\boldsymbol{D}M$ of any holonomic $\mathscr{D}$-module M in the derived category is also a holonomic $\mathscr{D}$-module, i.e., the j-th cohomology sheaf of $\boldsymbol{D}M$ vanishes unless $j = 0$. We cannot expect such a property for general holonomic $\mathcal{R}_X$-modules. We need to prove that if $\mathcal{M}$ is an $\mathcal{R}_X$-module underlying a mixed twistor $\mathscr{D}$-module then $\boldsymbol{D}\mathcal{M}$ is also a strict holonomic $\mathscr{D}$-module. This issue already appeared in the Hodge case, and solved by Saito. Even in the twistor case, we can apply Saito's method in a rather straightforward way. (See Sect. 13.2.)

The other issue is the construction of a sesqui-linear pairing for the dual. Let $\mathcal{T} = (\mathcal{M}_1, \mathcal{M}_2, C)$ be the $\mathcal{R}$-triple underlying a mixed twistor $\mathscr{D}$-module on X. We need to construct a sesqui-linear pairing $\boldsymbol{D}C$ of $\boldsymbol{D}\mathcal{M}_1$ and $\boldsymbol{D}\mathcal{M}_2$. This issue did not appear in the Hodge case. It is non-trivial even for sesqui-linear pairings of holonomic $\mathscr{D}$-modules. Let M_i $(i = 1, 2)$ be holonomic $\mathscr{D}$-modules. Let $C : M_1 \otimes \overline{M_2} \longrightarrow \mathfrak{Db}_X$ be a $\mathscr{D}_X \otimes \mathscr{D}_{\overline{X}}$-homomorphism. Here, $\mathfrak{Db}_X$ denotes the sheaf of distributions on X. We need to construct an induced sesqui-linear pairing $\boldsymbol{D}C$ of $\boldsymbol{D}M_1$ and $\boldsymbol{D}M_2$. We obviously have such a pairing, if M_i are smooth, i.e., flat bundles. It is not difficult to construct it in the case of regular holonomic $\mathscr{D}$-modules, thanks to the Riemann-Hilbert correspondence [28, 30, 46–48]. But, at this moment, some additional arguments are required in the irregular case, which will be given in Chap. 12. Once we have such a pairing in the case of $\mathcal{D}$-modules, it is rather formal to construct it in the context of mixed twistor $\mathcal{D}$-modules. (See Sect. 13.3.)

1.3.5.2 Real Structure

Combining the duality functor with some other functors, we can introduce the concept of real structures on mixed twistor $\mathcal{D}$-modules.

We have a contravariant auto equivalence on the category of mixed twistor $\mathcal{D}$-modules, called the Hermitian adjoint. For any $\mathcal{R}_X$-triple $\mathcal{T} = (\mathcal{M}_1, \mathcal{M}_2, C)$, we have the associated $\mathcal{R}_X$-triple $\mathcal{T}^* = (\mathcal{M}_2, \mathcal{M}_1, C^*)$, where $C^*(a, \sigma^* b) := \overline{\sigma^* C(b, \sigma^* a)}$. If $\mathcal{T}$ is equipped with a filtration W, we set $W_j(\mathcal{T}^*)$ as the image of $(\mathcal{T}/W_{-j-1})^* \longrightarrow \mathcal{T}^*$. If $(\mathcal{T}, W)$ is a mixed twistor $\mathcal{D}$-module, then $(\mathcal{T}, W)^*$ is also a mixed twistor $\mathcal{D}$-module. This can be regarded as the enhancement of the operation $M \longmapsto \overline{\mathcal{H}om_{\mathcal{D}_X}(M, \mathfrak{Db}_X)}$ for holonomic $\mathcal{D}_X$-modules, where $\mathfrak{Db}_X$ denotes the sheaf of distributions on X. We also denote $\mathcal{T}^*$ by $\boldsymbol{D}^{\mathrm{herm}}(\mathcal{T})$.

Let $j : \mathcal{X} \longrightarrow \mathcal{X}$ be given by $j(\lambda, Q) = (-\lambda, Q)$. For any $\mathcal{R}_X$-triple $\mathcal{T} = (\mathcal{M}_1, \mathcal{M}_2, C)$, the $\mathcal{R}_X$-triple $j^*\mathcal{T} = (j^*\mathcal{M}_1, j^*\mathcal{M}_2, j^*C)$ is naturally defined. If $(\mathcal{T}, W)$ is a mixed twistor $\mathcal{D}_X$-module, then $j^*(\mathcal{T}, W)$ is naturally a mixed twistor $\mathcal{D}_X$-module.

We define the functor $\tilde{\gamma}^*$ on the category of mixed twistor $\mathcal{D}_X$-modules by

$$\tilde{\gamma}^*(\mathcal{T}, W) = j^* \circ \boldsymbol{D} \circ \boldsymbol{D}^{\mathrm{herm}}(\mathcal{T}, W).$$

Namely, for $\mathcal{T} = (\mathcal{M}_1, \mathcal{M}_2, C)$, we set $\tilde{\gamma}^*(\mathcal{T}) = (j^*\boldsymbol{D}\mathcal{M}_2, j^*\boldsymbol{D}\mathcal{M}_1, j^*\boldsymbol{D}C^*)$. Then, a real structure on a mixed twistor $\mathcal{D}$-module $(\mathcal{T}, W)$ is defined to be an isomorphism $\kappa : \tilde{\gamma}^*(\mathcal{T}, W) \simeq (\mathcal{T}, W)$ such that $\tilde{\gamma}^*(\kappa) \circ \kappa = \mathrm{id}$. We show the functorial property of such real structure in Sect. 13.4.

1.3.5.3 Relation with Mixed Hodge Modules

We shall see in Sect. 13.5 that integrable mixed twistor $\mathcal{D}$-modules with real structure are closely related with mixed Hodge modules. Let (P, F, W) be a mixed Hodge module on X. Let M be a regular holonomic $\mathcal{D}_X$-module with an isomorphism $\mathrm{DR}_X M \simeq P \otimes_{\mathbb{Q}} \mathbb{C}$ on which F is defined. It is also naturally equipped with the weight filtration W. The real structure of $\mathrm{DR}_X M$ given by $P \otimes \mathbb{R}$ naturally induces a sesqui-linear pairing $\boldsymbol{D}M \times \overline{M} \longrightarrow \mathfrak{Db}_X$. Let $\mathcal{M}$ be the analytification of the Rees module of (M, F). Then, we have the induced sesqui-linear pairing C of $j^*\boldsymbol{D}\mathcal{M}$ and $\mathcal{M}$. It turns out that the $\mathcal{R}_X$-triple $\mathcal{T} = (j^*\boldsymbol{D}\mathcal{M}, \mathcal{M}, C)$ with the induced filtration W is a mixed twistor $\mathcal{D}$-module. It is equipped with the real structure given by $\kappa = (\mathrm{id}, \mathrm{id}) : \tilde{\gamma}^*\mathcal{T} \simeq \mathcal{T}$. If (P, F, W) is pure of weight w, then the associated $(\mathcal{T}, W)$ is also pure of weight w. Moreover, $(\mathcal{T}, W, \kappa)$ is naturally integrable. So, we obtain a functor from the category of mixed Hodge modules to the category of integrable mixed twistor $\mathscr{D}$-modules with real structure.

We shall give more details on this functor in Sect. 13.5, and observe that it is naturally compatible with various operations. Briefly, it is compatible with the push-forward by projective morphisms, the localizations and the duality, by construction of the functors. As a consequence, the functor is compatible with the six operations in the algebraic setting. Note that the pull back functors for a closed immersion are described in terms of the localizations and the Kashiwara equivalence. We also check in Sect. 13.5.2 the coincidence of the concepts of polarizations for pure Hodge modules and pure twistor $\mathscr{D}$-modules. So, we follow the same rules for signatures and weights in the Hodge context and the twistor context.

Let us look at the simplest variation of pure Hodge structure $(\mathbb{R}_X[d_X], F)$ of weight 0 on a complex manifold X, where F is the Hodge filtration on $\mathcal{O}_X$ given by $F_0(\mathcal{O}_X) = \mathcal{O}_X$ and $F_{-1}(\mathcal{O}_X) = 0$. It naturally gives a pure Hodge module of weight d_X. The analytification of the Rees module of $(\mathcal{O}_X, F)$ is $\mathcal{O}_{\mathcal{X}}$. We have a natural isomorphism $j^*\boldsymbol{D}\mathcal{O}_{\mathcal{X}} \simeq \lambda^{d_X}\mathcal{O}_{\mathcal{X}}$. It turns out that $\mathcal{T}_X := (j^*\boldsymbol{D}\mathcal{O}_{\mathcal{X}}, \mathcal{O}_{\mathcal{X}}, C)$ is isomorphic to the pure twistor $\mathscr{D}$-module $(\lambda^{d_X}\mathcal{O}_{\mathcal{X}}, \mathcal{O}_{\mathcal{X}}, C_0)$ of weight d_X, where $C_0(f, \sigma^*g) = f \cdot \overline{\sigma^*(g)}$. A natural polarization of $\mathcal{T}_X$ is induced by the pairing $\mathbb{R}[d_X] \otimes \mathbb{R}[d_X] \longrightarrow \mathbb{R}[d_X]$ given by $(a, b) \longmapsto (-1)^{d_X(d_X-1)/2}ab$. Let H be a smooth hypersurface of X. Let $\iota_H : H \longrightarrow X$ be the inclusion. We have the exact sequences $0 \longrightarrow \iota_{H\dagger}\mathcal{T}_H \longrightarrow \mathcal{T}_X[!H] \longrightarrow \mathcal{T}_X \longrightarrow 0$ and $0 \longrightarrow \mathcal{T}_X \longrightarrow \mathcal{T}_X[*H] \longrightarrow \iota_{H\dagger}\mathcal{T}_H \otimes \boldsymbol{T}(-1) \longrightarrow 0$, where $\boldsymbol{T}(-1)$ is the (-1)-th Tate twist. In the algebraic setting, we have $\iota_H^*\mathcal{T}_X = \mathcal{T}_H[1]$ and $\iota_H^!\mathcal{T}_X = \mathcal{T}_H[-1] \otimes \boldsymbol{T}(-1)$.

Part I
Gluing and Specialization of $\mathcal{R}$-Triples

Chapter 2
Preliminary

We shall begin with a review of the basis on $\mathcal{R}$-triples and variants in Sect. 2.1. In particular, we recall the strictly specializability along holomorphic functions.

The concept of $\mathcal{R}$-triples was introduced by Sabbah in [66], and it is the most basic ingredient when we consider twistor structures on holonomic $\mathscr{D}$-modules which may admit irregular singularities. When we impose additional conditions to $\mathcal{R}$-triples, they are given in terms of V-filtrations, the nearby cycle functors, and the vanishing cycle functors which are defined for $\mathcal{R}$-triples satisfying the strictly specializability condition. This is the basic strategy of Saito [69, 73] in his theory of Hodge modules.

In the context of $\mathscr{D}$-modules, the concepts of V-filtration, the nearby cycle functor and the vanishing cycle functor along any function f for $\mathscr{D}$-modules are due to Kashiwara and Malgrange [29, 42]. It might be instructive to recall them in this situation. We fix a total order on $\mathbb{C}/\mathbb{Z}$ and a section $\sigma : \mathbb{C}/\mathbb{Z} \longrightarrow \mathbb{C}$. The section gives a bijection $\mathbb{C} \simeq \mathbb{Z}\times\mathbb{C}/\mathbb{Z}$, and we obtain the total order $\leq_{\mathbb{C}}$ on $\mathbb{C}$ corresponding to the lexicographic order on $\mathbb{Z} \times \mathbb{C}/\mathbb{Z}$.

First, let us consider the case that X is an open subset in the product of a complex manifold X_0 and $\mathbb{C}$. Let t be the standard coordinate function on $\mathbb{C}$. Let $V\mathscr{D}_X \subset \mathscr{D}_X$ be the sheaf of subalgebras generated by the pull back of the tangent sheaf of X_0 and $t\partial_t$ over $\mathcal{O}_X$. Let M be any holonomic $\mathscr{D}$-module on X. Then, a V-filtration of M along t is a unique filtration by coherent $V\mathscr{D}_X$-submodules $V_\bullet(M)$ indexed by $(\mathbb{C}, \leq_{\mathbb{C}})$ such that (i) $V_\alpha M = \bigcap_{\beta >_{\mathbb{C}} \alpha} V_\beta M$, (ii) $tV_\alpha(M) \subset V_{\alpha-1}(M)$ and $\partial_t V_\alpha(M) \subset V_{\alpha+1}(M)$, (ii) $\partial_t t + \alpha$ is locally nilpotent on $\psi_{t,\alpha}(M) := \mathrm{Gr}^V_\alpha(M)$. We obtain the $\mathscr{D}_{X_0}$-modules $\psi_{t,\alpha}(M)$. We have the natural maps $t : \psi_{t,0}(M) \longrightarrow \psi_{t,-1}(M)$ and $\partial_t : \psi_{t,-1}(M) \longrightarrow \psi_{t,0}(M)$.

For any general holomorphic function f on any complex manifold X, let $\iota_f : X \longrightarrow X \times \mathbb{C}$ be the graph. For any holonomic $\mathscr{D}$-module M on X, we consider the induced $\mathscr{D}$-module $\iota_{f\dagger}M$ on $\mathbb{C} \times X$. We have the V-filtration of $\iota_{f\dagger}M$, and we obtain the $\mathscr{D}$-modules $\psi_{f,\alpha}(M) := \psi_{t,\alpha}(\iota_{f\dagger}M)$ on X, which are equipped with the nilpotent maps. We also have the morphisms $\psi_{f,-1}(M) \longrightarrow \psi_{f,0}(M)$ and

T. Mochizuki, *Mixed Twistor $\mathscr{D}$-modules*, Lecture Notes in Mathematics 2125,
DOI 10.1007/978-3-319-10088-3_2

$\psi_{f,0}(M) \longrightarrow \psi_{f,-1}(M)$. They contain much information of the $\mathscr{D}$-module M along f. For instance, we can recover M from $M(*f)$ and the morphisms $\psi_{f,-1}(M) \longrightarrow \psi_{f,0}(M) \longrightarrow \psi_{f,-1}(M)$.

We mention some easy examples. Set $X := \mathbb{C}$, and let t be the standard coordinate on X. For a complex number $\alpha \in \mathbb{C}$ and a positive integer ℓ, let $M := \mathscr{D}_X/(t\partial_t + \alpha)^\ell$. We may naturally regard M as a meromorphic flat bundle $\bigoplus_{i=1}^{\ell} \mathscr{O}_X(*t)e_i$ with the flat connection ∇ determined by $\nabla e_i = (-\alpha e_i + e_{i-1})dt/t$. We set $M_0 := \bigoplus_{i=1}^{\ell} \mathscr{O}_X e_i$. Then, the V-filtration of M along t is determined by $V_{\alpha+n}(M) = t^{-n-1}M_0$ $(n \in \mathbb{Z})$ and $\mathrm{Gr}^V_\beta(M) = 0$ $(\beta - \alpha \notin \mathbb{Z})$. We have $\psi_{t,\alpha}(M) \simeq \mathbb{C}^\ell$. When $\alpha \in \mathbb{Z}$, $\psi_{t,0}(M) \longrightarrow \psi_{t,-1}(M)$ is an isomorphism, and the rank of $\psi_{-1}(M) \longrightarrow \psi_0(M)$ is $\ell - 1$. In this case, M can be recovered from $\psi_{t,\alpha}(M)$ with the endomorphisms induced by $t\partial_t$. In general, we could say that the nearby cycle sheaves and the vanishing cycle sheaves contain much information on the regular part of the $\mathscr{D}$-modules.

We mention other easy examples. Let X and t be as above. We set $Q := \mathscr{D}_X/(t^2\partial_t+1)^\ell$. We regard it as $\bigoplus_{i=1}^{\ell} \mathscr{O}_X(*t)\, e_i$ with the connection ∇ determined by $\nabla e_i = e_i d(t^{-1}) + e_{i-1}dt/t$. In this case, Q is $V\mathscr{D}_X$-coherent. So, we have $V_\alpha(Q) = Q$ and $\psi_\alpha(Q) = 0$ for any $\alpha \in \mathbb{C}$. As shown by this example, the V-filtration, the nearby cycle sheaf and the vanishing cycle sheaf do not contain any information on "the irregular part" of the $\mathscr{D}$-module.

As shown by the above examples, the V-filtration and the nearby cycle sheaf and the vanishing cycle sheaf of a $\mathscr{D}$-module M contain only "the regular part" of M. To improve this, P. Deligne proposed to consider the V-filtrations, the nearby cycle sheaves of the ramified exponential twist of the $\mathscr{D}$-modules. (We shall review it in Sect. 2.1.9.1 in the context of $\mathcal{R}$-triples. Note that Sabbah introduced the concept of ramified exponential twist in the context of $\mathcal{R}$-triples [67].)

It is M. Saito who studied the V-filtrations, the nearby cycle functors, and the vanishing cycle functors in the context of $\mathscr{D}$-modules with a good filtration. It is one of the most fundamental basis in his theory of Hodge modules. Although any holonomic $\mathscr{D}$-module M is equipped with a V-filtration $V(M)$ along any holomorphic function f, we need to impose some compatibility conditions for the filtrations $V(M)$ and $F(M)$ when we consider a filtered $\mathscr{D}$-module (M, F). When we consider mixed Hodge modules, we have an additional weight filtration W. So, we need to impose compatibility conditions on V, F and W. Saito introduced most important compatibility conditions, and studied their functoriality.

As in the case of filtered $\mathscr{D}$-modules, a holonomic $\mathcal{R}$-module does not necessarily have a V-filtration with a good property along a holomorphic function. The existence of such a V-filtration in the context of $\mathcal{R}$-modules is called strictly specializability condition which is a generalization of the compatibility condition of V and F. In Sect. 4.4 later, we shall introduce a generalization of the compatibility condition of V, F and W in the context of $\mathcal{R}$-modules and $\mathcal{R}$-triples.

In the rest of Sect. 2.1, we also review the basis on smooth $\mathcal{R}$-triples, variation of twistor structure, Tate triples, etc.

In Sect. 2.2, we give some procedure to construct an $\mathcal{R}$-triple from a given $\mathcal{R}$-triple with a commuting tuple of nilpotent morphisms. It is a reformulation of the construction of twistor nilpotent orbits in [52, 56], and it has the origin in the study of nilpotent orbit in the Hodge theory [77]. It will be useful to deduce some properties of polarized mixed twistor structures and infinitesimal mixed twistor modules in Chap. 8.

In Sect. 2.3, we introduce the concept of Beilinson triple. It is the preparation for the reformulation of the nearby and the vanishing cycle functors studied in Chap. 4. It will be useful in the gluing procedure of $\mathcal{R}$-triples.

2.1 $\mathcal{R}$-Triples

2.1.1 $\mathcal{R}$-Modules

Let X be a complex manifold. We set $\mathcal{X} := \mathbb{C}_\lambda \times X$. The projection $\mathcal{X} \longrightarrow X$ is denoted by p_λ. Let Θ_X denote the tangent sheaf of X. Let $\mathcal{D}_\mathcal{X}$ denote the sheaf of differential operators on $\mathcal{X}$. Recall that $\mathcal{R}_X$ denote the sheaf of subalgebras in $\mathcal{D}_\mathcal{X}$ generated by $\lambda\, p_\lambda^* \Theta_X$ over $\mathcal{O}_\mathcal{X}$.

Let H be a hypersurface of X. We set $\mathcal{H} := \mathbb{C}_\lambda \times H$. For any $\mathcal{O}_\mathcal{X}$-module $\mathcal{M}$, we define $\mathcal{M}(*H) := \mathcal{M} \otimes_{\mathcal{O}_\mathcal{X}} \mathcal{O}_\mathcal{X}(*\mathcal{H})$, where $\mathcal{O}_\mathcal{X}(*\mathcal{H})$ be the sheaf of meromorphic functions on $\mathcal{X}$ whose poles are contained in $\mathcal{H}$.

We put $\mathcal{R}_{X(*H)} := \mathcal{R}_X(*H)$. The sheaf of algebras $\mathcal{R}_{X(*H)}$ is Noetherian. The notions of left and right $\mathcal{R}_{X(*H)}$-modules are naturally defined. They are exchanged by a standard formalism for $\mathcal{D}$-modules or $\mathcal{R}$-modules. Let ω_X denote the sheaf of holomorphic n-forms, where $n := \dim X$. We set $\omega_\mathcal{X} := \lambda^{-n} p_\lambda^* \omega_X$ as the subsheaf of $p_\lambda^* \omega_X \otimes \mathcal{O}_\mathcal{X}\big(*(\{0\} \times X)\big)$. Then, for any left $\mathcal{R}_{X(*H)}$-module $\mathcal{M}^\ell$, we have the natural right $\mathcal{R}_{X(*H)}$-module structure on $\mathcal{M}^r := \omega_\mathcal{X} \otimes_{\mathcal{O}_\mathcal{X}} \mathcal{M}^\ell$. For any right $\mathcal{R}_{X(*H)}$-module $\mathcal{N}^r$, we have the natural left $\mathcal{R}_{X(*H)}$-module structure on $\mathcal{N}^\ell := \mathcal{N}^r \otimes \omega_\mathcal{X}^{-1}$. We consider left $\mathcal{R}_{X(*H)}$-modules in this paper, unless otherwise specified.

A left $\mathcal{R}_{X(*H)}$-module $\mathcal{M}$ is naturally regarded as an $\mathcal{O}_\mathcal{X}(*H)$-module with a family of flat λ-connections. Namely, $\mathcal{M}$ is equipped with a differential operator $\mathbb{D} : \mathcal{M} \longrightarrow \mathcal{M} \otimes p_\lambda^* \Omega_X^1$ such that (i) $\mathbb{D}(f\,s) = \lambda d(f)\, s + f\, \mathbb{D}(s)$ for $f \in \mathcal{O}_\mathcal{X}(*H)$ and $s \in \mathcal{M}$, (ii) $\mathbb{D} \circ \mathbb{D} = 0$. Let $\Omega_\mathcal{X}^1 := \lambda^{-1} p_\lambda^* \Omega_X^1$. We set $\mathbb{D}^f = \lambda^{-1}\mathbb{D}$, then we have $\mathbb{D}^f : \mathcal{M} \longrightarrow \mathcal{M} \otimes \Omega_\mathcal{X}^1$, and it is a meromorphic family of flat connections in the sense (i) $\mathbb{D}^f(f\,s) = d(f)\, s + f\, \mathbb{D}^f(s)$ for $f \in \mathcal{O}_\mathcal{X}(*H)$ and $s \in \mathcal{M}$, (ii) $\mathbb{D}^f \circ \mathbb{D}^f = 0$.

An $\mathcal{R}_{X(*H)}$-module is called strict if it is $\mathcal{O}_{\mathbb{C}_\lambda}$-flat. Let $\mathcal{M}$ be a strict $\mathcal{R}_{X(*H)}$-module. For any $\lambda_0 \in \mathbb{C}$, let $i_{\lambda_0} : X \longrightarrow \mathcal{X}$ be given by $i_{\lambda_0}(x) = (\lambda_0, x)$. We set $\mathcal{M}^{\lambda_0} := i_{\lambda_0}^{-1}(\mathcal{M}/(\lambda - \lambda_0)\mathcal{M})$. It is naturally a module over $\mathcal{R}_{X(*H)}^{\lambda_0} := i_{\lambda_0}^{-1}\big(\mathcal{R}_{X(*H)}/(\lambda - \lambda_0)\mathcal{R}_{X(*H)}\big)$. Note that $\mathcal{R}_{X(*H)}^{\lambda_0} \simeq \mathcal{D}_X(*H)$ if $\lambda_0 \neq 0$, and $\mathcal{R}_{X(*H)}^0 \simeq \operatorname{Sym} \Theta_X(*H)$.

If an $\mathcal{R}_{X(*H)}$-module $\mathcal{M}$ is pseudo-coherent as an $\mathcal{O}_\mathcal{X}(*H)$-module, and locally finitely generated as an $\mathcal{R}_{X(*H)}$-module, then it is a coherent $\mathcal{R}_{X(*H)}$-module. The sheaf of algebras $\mathcal{R}_{X(*H)}$ is naturally filtered by the order of differential operators, and we have the notion of coherent filtration for $\mathcal{R}_{X(*H)}$-modules as in the case of $\mathcal{D}$-modules. Let $\mathcal{M}$ be an $\mathcal{R}_{X(*H)}$-module on an open set $U \subset \mathcal{X}$. We say that $\mathcal{M}$ is *good* if for any compact subset $K \subset U$ we have a neighbourhood U' of K in U and a finite filtration F of $\mathcal{M}_{|U'}$ by $\mathcal{R}_{X(*H)}$-submodules such that $\mathrm{Gr}^F(\mathcal{M}_{|U'})$ has a coherent filtration.

The push forward and the pull back of $\mathcal{R}_{X(*H)}$-modules are defined by the formula of those for $\mathcal{R}_X$-modules in [66]. Let $f : X' \longrightarrow X$ be a morphism of complex manifolds. Let H be a hypersurface of X, and we put $H' := f^{-1}(H)$. Let $\mathcal{R}_{X'\to X} := \mathcal{O}_{\mathcal{X}'} \otimes_{f^{-1}\mathcal{O}_{\mathcal{X}}} f^{-1}\mathcal{R}_X$, which is naturally an $(\mathcal{R}_{X'}, f^{-1}\mathcal{R}_X)$-module. We set $\mathcal{R}_{X\leftarrow X'} := \omega_{\mathcal{X}'} \otimes \mathcal{R}_{X'\to X} \otimes f^{-1}\omega_{\mathcal{X}}^{-1}$. For any $\mathcal{R}_{X'(*H')}$-module $\mathcal{M}'$, we have

$$f_\dagger(\mathcal{M}') := Rf_*\big(\mathcal{R}_{X\leftarrow X'} \otimes^L_{\mathcal{R}_{X'}} \mathcal{M}'\big)$$

in $D^b(\mathcal{R}_{X(*H)})$. Note that $\mathcal{R}_{X\leftarrow X'} \otimes^L_{\mathcal{R}_{X'}} \mathcal{M}' \simeq \mathcal{R}_{X\leftarrow X'}(*H') \otimes^L_{\mathcal{R}_{X'(*H')}} \mathcal{M}'$, and that an $f^{-1}\mathcal{R}_{X(*H)}$-injective resolution of $\mathcal{R}_{X\leftarrow X'} \otimes^L_{\mathcal{R}_{X'}} \mathcal{M}'$ is naturally an $f^{-1}\mathcal{R}_X$-injective resolution. For an $\mathcal{R}_{X(*H)}$-module $\mathcal{N}$, we set $f^\dagger\mathcal{N} := \mathcal{R}_{X'\to X} \otimes^L_{f^{-1}\mathcal{R}_X} f^{-1}\mathcal{N}$ in $D^b(\mathcal{R}_{X'(*H')})$. If $\mathcal{M}'$ is good, $f_\dagger\mathcal{M}'$ is cohomologically good, which can be proved by the argument in the case of $\mathcal{D}$-modules.

Lemma 2.1.1 *Assume that f is proper and birational, and it induces an isomorphism $X'\setminus H' \simeq X\setminus H$. Then, we have a natural isomorphism $f_\dagger\mathcal{M}' \simeq f_*\mathcal{M}'$ for any good $\mathcal{R}_{X'(*H')}$-module $\mathcal{M}'$, and an isomorphism $f^\dagger\mathcal{M} \simeq f^*\mathcal{M} := \mathcal{O}_{\mathcal{X}'} \otimes_{\mathcal{O}_{\mathcal{X}}} f^{-1}\mathcal{M}$ for any good $\mathcal{R}_{X(*H)}$-module $\mathcal{M}$. They give an equivalence of the categories of good $\mathcal{R}_{X(*H)}$-modules and good $\mathcal{R}_{X'(*H')}$-modules.*

Proof Because $\mathcal{R}_{X'\to X}(*H') \simeq \mathcal{R}_{X'(*H')}$, we obtain the first claim. For an $\mathcal{R}_{X(*H)}$-module $\mathcal{M}$ of the form $\mathcal{R}_{X(*H)} \otimes_{\mathcal{O}_{\mathcal{X}}} M$, where M is an $\mathcal{O}_{\mathcal{X}}$-coherent module, we have $f^\dagger\mathcal{M} \simeq \mathcal{R}_{X'(*H')} \otimes_{\mathcal{O}_{\mathcal{X}'}} f^*M$. For an $\mathcal{R}_{X'(*H')}$-module $\mathcal{M}'$ of the form $\mathcal{R}_{X'(*H')} \otimes_{\mathcal{O}_{\mathcal{X}'}} M'$, where M' is an $\mathcal{O}_{\mathcal{X}'}$-coherent module, we have $f_\dagger\mathcal{M}' \simeq \mathcal{R}_{X(*D)} \otimes_{\mathcal{O}_{\mathcal{X}}} f_*(M')$. Then, the second claim is clear. □

2.1.2 Strict Specializability for $\mathcal{R}$-Modules

First we recall the strictly specializability along a coordinate function. Let $\mathbb{C}_t$ be a complex line with a coordinate function t. Let X_0 be a complex manifold. We put $X := X_0 \times \mathbb{C}_t$. We identify X_0 and $X_0 \times \{0\}$. Let H be a hypersurface of X. The notion of strict specializability for $\mathcal{R}_{X(*H)}$-module is defined as in the case of $\mathcal{R}_X$-modules [66]. (See also Sect. 3.1.1.)

Let $\Theta_X(\log X_0)$ denote the sheaf of vector fields on X which are logarithmic along X_0. We recall that $V_0\mathcal{R}_X$ denotes the sheaf of subalgebras in $\mathcal{R}_X$ generated by $\lambda p_\lambda^*\Theta_X(\log X_0)$. Note that it depends only on t, i.e., it is independent of the choice of a decomposition into the product $X = X_0 \times \mathbb{C}_t$.

2.1.2.1 The Case $X_0 \not\subset H$

We set $H_0 := X_0 \cap H$. We put $V_0\mathcal{R}_{X(*H)} := V_0\mathcal{R}_X(*H)$. For any point $\lambda_0 \in \mathbb{C}_\lambda$, let $\mathcal{X}^{(\lambda_0)}$ denote a small neighbourhood of $\{\lambda_0\} \times X$. We use the symbol $\mathcal{X}_0^{(\lambda_0)}$ with a similar meaning. Let $\mathcal{M}$ be a coherent $\mathcal{R}_{X(*H)}$-module on $\mathcal{X}$. It is called strictly specializable along t at λ_0 if $\mathcal{M}_{|\mathcal{X}^{(\lambda_0)}}$ is equipped with an increasing and exhaustive filtration $V^{(\lambda_0)} = \big(V_a^{(\lambda_0)}(\mathcal{M}) \,\big|\, a \in \mathbb{R}\big)$ by coherent $V_0\mathcal{R}_{X(*H)}$-modules satisfying the following conditions (Conditions 22.3.1 and 22.3.2 in [55]).

(i) For any $a \in \mathbb{R}$ and $P \in \mathcal{X}_0^{(\lambda_0)}$, there exists $\epsilon > 0$ such that $V_a^{(\lambda_0)}(\mathcal{M}) = V_{a+\epsilon}^{(\lambda_0)}(\mathcal{M})$ on a neighbourhood of P. Set $V_{<a}^{(\lambda_0)}(\mathcal{M}) := \bigcup_{b<a} V_b^{(\lambda_0)}(\mathcal{M})$ and $\mathrm{Gr}_a^{V^{(\lambda_0)}}(\mathcal{M}) := V_a^{(\lambda_0)}(\mathcal{M})/V_{<a}^{(\lambda_0)}(\mathcal{M})$.
(ii) $tV_a^{(\lambda_0)}(\mathcal{M}) \subset V_{a-1}^{(\lambda_0)}(\mathcal{M})$ for any $a \in \mathbb{R}$. Moreover, $tV_a^{(\lambda_0)}(\mathcal{M}) = V_{a-1}^{(\lambda_0)}(\mathcal{M})$ if $a < 0$.
(iii) Set $\eth_t := \lambda\partial_t$. Then, $\eth_t V_a^{(\lambda_0)}(\mathcal{M}) \subset V_{a+1}^{(\lambda_0)}(\mathcal{M})$ for any $a \in \mathbb{R}$. Moreover, the induced morphisms $\eth_t : \mathrm{Gr}_a^{V^{(\lambda_0)}}(\mathcal{M}) \longrightarrow \mathrm{Gr}_{a+1}^{V^{(\lambda_0)}}(\mathcal{M})$ are isomorphisms for $a > -1$.
(iv) The $\mathcal{R}_{X_0(*H_0)}$-modules $\mathrm{Gr}_a^{V^{(\lambda_0)}}(\mathcal{M})$ are strict for any $a \in \mathbb{R}$.
(v) For any $a \in \mathbb{R}$ and $P \in \mathcal{X}_0^{(\lambda_0)}$, there exists a finite subset

$$\mathcal{K}(a, \lambda_0, P) \subset \big\{u \in \mathbb{R} \times \mathbb{C} \,\big|\, \mathfrak{p}(\lambda_0, u) = a\big\}$$

such that $\prod_{u \in \mathcal{K}(a,\lambda_0,P)}\big(-\eth_t t + \mathfrak{e}(\lambda, u)\big)$ is nilpotent on $\mathrm{Gr}_a^{V^{(\lambda_0)}}(\mathcal{M})$. Here,

$$\mathfrak{p}(\lambda, (b, \beta)) = b + 2\,\mathrm{Re}(\lambda\overline{\beta}), \qquad \mathfrak{e}(\lambda, (b, \beta)) = \beta - b\lambda - \overline{\beta}\lambda^2.$$

We implicitly assume that $\mathcal{K}(a, \lambda_0, P)$ is minimal among such sets.

Such a filtration is unique, if it exists. Note that the condition is independent of the choice of a decomposition into the product $X = X_0 \times \mathbb{C}_t$. We say that $\mathcal{M}$ is called strictly specializable along t if it is strictly specializable along t at any λ_0.

Suppose that a coherent $\mathcal{R}_{X(*H)}$-module $\mathcal{M}$ is strictly specializable along t. For any $P \in \mathcal{X}_0^{(\lambda_0)}$ and any $u \in \mathcal{K}(a, \lambda_0, P)$, we define $\psi_{t,u}^{(\lambda_0)}(\mathcal{M}) \subset \mathrm{Gr}_a^{V^{(\lambda_0)}}(\mathcal{M})$ as

$$\psi_{t,u}^{(\lambda_0)}(\mathcal{M}) = \bigcup_N \mathrm{Ker}\Big(\big(-\eth_t t + \mathfrak{e}(\lambda, u)\big)^N\Big)$$

on a neighbourhood of P. We have $\mathrm{Gr}_a^{V^{(\lambda_0)}}(\mathcal{M}) = \bigoplus_{\mathfrak{p}(\lambda_0,u)=a} \psi_{t,u}^{(\lambda_0)}(\mathcal{M})$. For a fixed $u \in \mathbb{R} \times \mathbb{C}$, by varying λ_0 and P, and by gluing $\psi_u^{(\lambda_0)}(\mathcal{M})$, we obtain an $\mathcal{R}_{X_0(*H_0)}$-module $\psi_{t,u}(\mathcal{M})$ as in the case of $\mathcal{R}$-modules. It is also denoted as $\psi_u(\mathcal{M})$ if there is no risk of confusion.

We have the naturally defined morphisms $\eth_t : \psi_{t,u}(\mathcal{M}) \longrightarrow \psi_{t,u+\delta}(\mathcal{M})$, and $t : \psi_{t,u}(\mathcal{M}) \longrightarrow \psi_{t,u-\delta}(\mathcal{M})$, where $\delta = (1,0) \in \mathbb{R} \times \mathbb{C}$. We set $\tilde{\psi}_{t,u}(\mathcal{M}) := \varinjlim \psi_{t,u-N\delta}(\mathcal{M})$ for $u \in \mathbb{R} \times \mathbb{C}$. It is isomorphic to $\psi_{t,u}(\mathcal{M}(*t))$ below.

If $\mathcal{M}_i$ $(i = 1,2)$ are strictly specializable along t with a morphism $F : \mathcal{M}_1 \longrightarrow \mathcal{M}_2$ of $\mathcal{R}_{X(*H)}$-modules, F is compatible with the V-filtrations. We have an induced morphism $\psi_{t,u}(F) : \psi_{t,u}(\mathcal{M}_1) \longrightarrow \psi_{t,u}(\mathcal{M}_2)$. If the cokernel of $\psi_{t,u}(F)$ is strict, the morphism F is called strictly specializable. We use the following easy lemma implicitly. (See [52, 66] for the proof.)

Lemma 2.1.2 *If F is strictly specializable, F is strictly compatible with the V-filtrations. Hence,* $\mathrm{Ker}(F)$, $\mathrm{Im}(F)$ *and* $\mathrm{Cok}(F)$ *are also strictly specializable. Moreover, we have $\psi_{t,u}(\mathrm{Im}\,F) \simeq \mathrm{Im}\,\psi_{t,u}(F)$, $\psi_{t,u}(\mathrm{Ker}\,F) \simeq \mathrm{Ker}\,\psi_{t,u}(F)$ and $\psi_{t,u}(\mathrm{Cok}\,F) \simeq \mathrm{Cok}\,\psi_{t,u}(F)$.* □

2.1.2.2 The Case $X_0 \subset H$

If $X_0 \subset H$, we decompose $H = X_0 \cup H'$, where $X_0 \cap H'$ is codimension one in X_0. We put $H_0' := H' \cap X_0$. Let $\mathcal{M}$ be a coherent $\mathcal{R}_{X(*H)} = \mathcal{R}_{X(*H')}(*t)$-module. It is called strictly specializable along t at λ_0 if $\mathcal{M}_{|\mathcal{X}^{(\lambda_0)}}$ is equipped with an increasing and exhaustive filtration $V^{(\lambda_0)}(\mathcal{M}) = \big(V_a^{(\lambda_0)}(\mathcal{M}) \,\big|\, a \in \mathbb{R}\big)$ by coherent $V_0\mathcal{R}_{X(*H')}$-modules indexed by $\mathbb{R}$, satisfying the following conditions in addition to (i), (iv), (v) (Definition 22.4.1 in [55]):

(ii') $tV_a^{(\lambda_0)}(\mathcal{M}) = V_{a-1}^{(\lambda_0)}(\mathcal{M})$ for any $a \in \mathbb{R}$.
(iii') $\eth_t V_a^{(\lambda_0)}(\mathcal{M}) \subset V_{a+1}^{(\lambda_0)}(\mathcal{M})$.

As before, we obtain $\mathcal{R}_{X_0(*H_0')}$-modules $\psi_{t,u}(\mathcal{M})$ for any $u \in \mathbb{R} \times \mathbb{C}$. The induced morphisms $t : \psi_{t,u}(\mathcal{M}) \longrightarrow \psi_{t,u-\delta}(\mathcal{M})$ are isomorphisms. It is also denoted by $\tilde{\psi}_{t,u}(\mathcal{M})$.

2.1.2.3 Strictly Specializability Along Holomorphic Functions

Let us recall the strict specializability along any holomorphic function g on X. We set $X_1 := X \times \mathbb{C}_t$ and $H_1 := H \times \mathbb{C}_t$. We have the map $\iota_g : X \longrightarrow X \times \mathbb{C}_t$ given by $\iota_g(x) = (x, g(x))$. For any $\mathcal{O}_{\mathcal{X}}$-module M, we set $M(*g) := M \otimes \mathcal{O}_{\mathcal{X}}(*g)$.

Let $\mathcal{M}$ be a coherent $\mathcal{R}_{X(*H)}$-module. We obtain the $\mathcal{R}_{X_1(*H_1)}$-module $\iota_{g\dagger}\mathcal{M}$. We say that $\mathcal{M}$ is strictly specializable along g if $\iota_{g\dagger}\mathcal{M}$ is strictly specializable along t. In that case, by applying the procedure in Sect. 2.1.2.1, we obtain $\mathcal{R}_{X(*H)}$-modules $\psi_{g,u}(\mathcal{M}) := \psi_{t,u}(\iota_{g\dagger}\mathcal{M})$ and $\tilde{\psi}_{g,u}(\mathcal{M}) := \tilde{\psi}_{t,u}(\iota_{g\dagger}\mathcal{M})$.

Let $\mathcal{M}$ be a coherent $\mathcal{R}_{X(*H)}(*g)$-module. We obtain the $\mathcal{R}_{X_1(*H_1)}(*t)$-module $\iota_{g\dagger}\mathcal{M}$. We say that $\mathcal{M}$ is strictly specializable along g if $\iota_{g\dagger}\mathcal{M}$ is strictly specializable along t. In that case, by applying the procedure in Sect. 2.1.2.2, we obtain $\mathcal{R}_{X(*H)}$-modules $\psi_{g,u}(\mathcal{M}) := \psi_{t,u}(\iota_{g\dagger}\mathcal{M})$. It is also denoted by $\tilde{\psi}_{g,u}(\mathcal{M})$.

2.1.2.4 Genericity

Let g be a holomorphic function on X. Let $\mathcal{M}$ be a coherent $\mathcal{R}_{X(*H)}$-module or a coherent $\mathcal{R}_{X(*H)}(*g)$-module which is strictly specializable along g. We denote by $\mathrm{KMS}(\mathcal{M},g)$ the set of $u \in \mathbb{R}\times\mathbb{C}$ such that $\psi_{g,u}(\mathcal{M}) \neq 0$.

We say that a non-zero complex number λ_0 is generic for $(\mathcal{M},g)$ if the map

$$\mathfrak{e}(\lambda_0) : \mathrm{KMS}(\mathcal{M},g) \cup (\mathbb{Z}\times\{0\}) \longrightarrow \mathbb{C}$$

is injective. Suppose that λ_0 is generic. Set $\mathcal{K}_b := \{u \in \mathrm{KMS}(\mathcal{M},g) \,|\, \mathfrak{p}(\lambda_0,u) = b\}$ for any $b \in \mathbb{R}$. We take total orders on the sets $\mathcal{K}_b$ for each $-1 \leq b \leq 0$. By the natural bijections $\mathcal{K}_b \simeq \mathcal{K}_{b+n}$ $(n \in \mathbb{Z})$, we obtain total orders on $\mathcal{K}_b$ for any $b \in \mathbb{R}$. Together with the natural order on $\mathbb{R}$, we obtain a total order $\leq$ on $\mathrm{KMS}(\mathcal{M},g)$. Let $V^{(\lambda_0)}$ be the V-filtration of $\iota_{g\dagger}\mathcal{M}$ along g at λ_0 on a neighbourhood $\mathcal{X}^{(\lambda_0)}$. We have the filtration $\tilde{V}^{(\lambda_0)}$ of $\iota_{g\dagger}\mathcal{M}$ indexed by $(\mathrm{KMS}(\mathcal{M},g),\leq)$ given as follows. For any $b \in \mathbb{R}$, let $\pi_b : V_b^{(\lambda_0)}(\iota_{g\dagger}\mathcal{M}) \longrightarrow \mathrm{Gr}_b^{V^{(\lambda_0)}}(\iota_{g\dagger}\mathcal{M})$ be the projection. We set

$$\tilde{V}_u^{(\lambda_0)}(\iota_{g\dagger}\mathcal{M}) := \pi_{\mathfrak{p}(\lambda_0,u)}^{-1}\Big(\bigoplus_{u_1\leq u}\psi_{u_1}^{(\lambda_0)}(\mathcal{M})\Big)$$

We have the induced filtration $\tilde{V}$ on $\iota_{g\dagger}\mathcal{M}^{\lambda_0}$. We shall use the following obvious lemma implicitly.

Lemma 2.1.3 *$\tilde{V}(\mathcal{M}^{\lambda_0})$ is a V-filtration of the $\mathcal{D}_X$-module $\iota_{g\dagger}\mathcal{M}^{\lambda_0}$.* □

2.1.3 Some Sheaves

We recall some sheaves, following Sabbah in [65, 67], to which we refer for more details and precision. We refer to the nice textbooks [49, 63] for distributions. Let X be an n-dimensional complex manifold. Let $\Omega_X^{n,n}$ denote the bundle of differential (n,n)-forms on X. Let T be a real C^∞-manifold. Let $\Omega_{X\times T/T}^{n,n}$ denote the pull back of $\Omega_X^{n,n}$ by $X\times T \longrightarrow X$. For any open subset V of $X\times T$, let $\mathcal{E}_{X\times T/T,c}^{(n,n)}(V)$ denote the space of C^∞-sections of $\Omega_{X\times T/T}^{n,n}$ on V with compact supports. Let $\mathrm{Diff}_{X\times T/T}(V)$ denote the space of C^∞-differential operators on V relative to T, i.e., we consider only differentials in the X-direction on V. For any compact subset $K \subset V$ and $P \in \mathrm{Diff}_{X\times T/T}(V)$, we have the semi-norm $\|\varphi\|_{P,K} = \sup_K |P\varphi|$. For any closed subset $Z \subset X$, let $\mathcal{E}_{X\times T/T,c}^{<Z,(n,n)}(V)$ denote the subspace of $\mathcal{E}_{X\times T/T,c}^{(n,n)}(V)$, which consists of the sections φ such that $(P\varphi)_{|(Z\times T)\cap V} = 0$ for any $P \in \mathrm{Diff}_{X\times T/T}(V)$. We have the induced semi-norms $\|\cdot\|_{P,K}$ on the space $\mathcal{E}_{X\times T/T,c}^{<Z,(n,n)}(V)$. By the semi-norms, the spaces $\mathcal{E}_{X\times T/T,c}^{(n,n)}(V)$ and $\mathcal{E}_{X\times T/T,c}^{<Z,(n,n)}(V)$ are locally convex topological spaces. Let $C_c^0(T)$ denote the space of continuous functions on T with compact supports. It is a normed vector space with the sup norms.

The space of continuous $C^\infty(T)$-linear maps $\mathcal{E}^{(n,n)}_{X\times T/T,c}(V) \longrightarrow C^0_c(T)$ is denoted by $\mathfrak{Db}_{X\times T/T}(V)$. Let H be a hypersurface of X. Let $\mathfrak{Db}^{\mathrm{mod}\,H}_{X\times T/T}(V)$ denote the space of continuous $C^\infty(T)$-linear maps from $\mathcal{E}^{<H,(n,n)}_{X\times T/T,c}(V)$ to $C^0_c(T)$. Any elements of $\mathfrak{Db}^{\mathrm{mod}\,H}_{X\times T/T}(V)$ are called distributions with moderate growth along H. Let $\mathfrak{Db}_{X\times T/T,H}(V)$ be the space of continuous C^∞-linear maps $\Phi : \mathcal{E}^{(n,n)}_{X\times T/T,c}(V) \longrightarrow C^0_c(T)$ whose supports are contained in $H\times T$, i.e., $\Phi(\varphi) = 0$ if $\varphi = 0$ on some neighbourhood of $H\times T$. They give the sheaves $\mathfrak{Db}_{X\times T/T}$, $\mathfrak{Db}^{\mathrm{mod}\,H}_{X\times T/T}$ and $\mathfrak{Db}_{X\times T/T,H}$ on $X\times T$. We have the following lemma as in [41, 55, 65, 66].

Lemma 2.1.4 *We have the exact sequence*

$$0 \longrightarrow \mathfrak{Db}_{X\times T/T,H} \longrightarrow \mathfrak{Db}_{X\times T/T} \longrightarrow \mathfrak{Db}^{\mathrm{mod}\,H}_{X\times T/T} \longrightarrow 0.$$

In particular, $\mathfrak{Db}^{\mathrm{mod}\,H}_{X\times T/T}$ is isomorphic to the image of the natural morphism $\mathfrak{Db}_{X\times T/T} \longrightarrow j_*j^{-1}\mathfrak{Db}_{X\times T/T}$, where $j : X\setminus H \longrightarrow X$. □

Let V be any open subset of X with a real coordinate $(x_1,\ldots,x_{2n})$. For a given $J = (j_1,\ldots,j_{2n}) \in \mathbb{Z}^{2n}_{\geq 0}$, we set $\partial^J := \prod_{i=1}^{2n} \partial^{j_i}_{x_i}$ and $|J| = \sum j_i$. The following lemma can be proved by using a standard argument in the theory of distributions.

Lemma 2.1.5 *For any $\Phi \in \mathfrak{Db}^{\mathrm{mod}\,H}_{X\times T/T}(V)$ and any compact subset K in V, we have $m \in \mathbb{Z}_{>0}$ and $C > 0$ such that $\sup_T\bigl|\Phi(\varphi)\bigr| \leq C \sup_{|J|\leq m} |\partial^J\varphi|_K$ for any $\varphi \in \mathcal{E}^{<H\,(n,n)}_{X\times T/T,c}(V)$ with $\mathrm{Supp}(\varphi) \subset K$.* □

Corollary 2.1.6 *Let g be a holomorphic function such that $g^{-1}(0) \subset H$. Let H_1 be a hypersurface of X such that $g^{-1}(0) \cup H_1 = H$. Let $\Phi \in \mathfrak{Db}^{\mathrm{mod}\,H}_{X\times T/T}(V)$. For any compact subset K in V, there exists $C_K > 0$ such that $\Phi\bigl(|g|^s\varphi\bigr) \in C^0_c(T)$ is well defined for $\varphi \in \mathcal{E}^{<H_1\,(n,n)}_{X\times T/T,c}(V)$ with $\mathrm{Supp}(\varphi) \subset K$ and for $\mathrm{Re}(s) > C_K$. As a function of s, $\Phi\bigl(|g|^s\varphi\bigr)$ is holomorphic.* □

Let $C^\infty_{X\times T}$ be the sheaf of C^∞-functions on $X\times T$. Let H be a hypersurface of X. Let $U \subset X\times T$ be an open subset. Let φ be a C^∞-function on an open subset $U\setminus(H\times T)$. We say that φ is a C^∞-function of moderate growth along H if the following holds for any $(Q_1,Q_2) \in U\cap(H\times T)$.

- For any $P \in \mathrm{Diff}_{X\times T/T}(U)$, there exists $N > 0$ such that $\bigl|P\varphi\bigr| = O\bigl(|g|^{-N}\bigr)$ around (Q_1,Q_2).

Let $C^{\infty\,\mathrm{mod}\,H}_{X\times T}$ denote the sheaf of C^∞-functions on $X\times T$ with moderate growth along H. It is standard to prove that the sheaf is c-soft.

Let $f : X' \longrightarrow X$ be a morphism of complex manifolds. Let H be a hypersurface of X. We put $H' := f^{-1}(H)$.

Lemma 2.1.7 *Assume that f is proper and induces $X'\setminus H' \simeq X\setminus H$. We naturally have an isomorphism $a_1 : f_!\mathfrak{Db}^{\mathrm{mod}\,H'}_{X'\times T/T} \simeq \mathfrak{Db}^{\mathrm{mod}\,H}_{X\times T/T}$ and an epimorphism*

$a_2 : f^{-1}\mathcal{C}^{\infty \bmod H}_{X\times T} \longrightarrow \mathcal{C}^{\infty \bmod H'}_{X'\times T}$. *We also naturally have a morphism* $f^{-1}\mathfrak{Db}^{\bmod H}_{X\times T/T} \longrightarrow \mathfrak{Db}^{\bmod H'}_{X'\times T/T}$ *and an isomorphism* $\mathcal{C}^{\infty \bmod H}_{X\times T} \simeq f_*\mathcal{C}^{\infty \bmod H'}_{X'\times T}$.

Proof The natural morphism $\mathcal{C}^{\infty \bmod H}(X\times T) \longrightarrow \mathcal{C}^{\infty \bmod H'}(X'\times T)$ is bijective. Because $\mathcal{C}^{\infty \bmod H'}_{X'\times T}$ is c-soft, we obtain that a_2 is an epimorphism. It is easy to prove that the induced morphism $\mathcal{C}^{\infty \bmod H}_{X\times T} \longrightarrow f_*\mathcal{C}^{\infty \bmod H'}_{X'\times T}$ is an isomorphism.

Let V be any open subset in $X \times T$, and $V' := f^{-1}(V)$. The natural continuous morphism $\mathcal{E}^{<H\,(n,n)}_{X\times T/T,c}(V) \longrightarrow \mathcal{E}^{<H\,(n,n)}_{X'\times T/T,c}(V')$ is bijective. It is a homeomorphism, which follows from Lemma 2.1.8 below. Then, we obtain the claims for $\mathfrak{Db}^{\bmod H'}_{X'\times T/T}$ and $\mathfrak{Db}^{\bmod H}_{X\times T/T}$. □

Lemma 2.1.8 *Let Y be any complex manifold with a hypersurface H_1. Let g be any holomorphic function on Y such that $g^{-1}(0) \subset H_1$. Let $K \subset Y$ be any compact subset. Let $P \in \mathrm{Diff}_{Y\times T/T}$ and $N \in \mathbb{Z}_{\geq 0}$. Then, there exist $C > 0$ and $P_i \in \mathrm{Diff}_{Y\times T/T}$ $(i = 1,\ldots,M)$ such that $\sup_K |g^{-N}P\varphi| \leq C\sum_i \sup_K |P_i\varphi|$ for any $\varphi \in \mathcal{E}^{<H_1\,(n,n)}_{Y\times T/T,c}(K\times T)$.*

Proof We give only a sketch of the proof. We have only to consider the case $P = 1$. We need estimates only around any point of $K \times T$.

First, let us consider the case dg is nowhere vanishing on X. We may assume that X is equipped with a holomorphic coordinate system $(z_1,\ldots,z_n)$ and $z_1 = g$. Let $z_i = x_i + \sqrt{-1}y_i = r_i e^{\sqrt{-1}\theta_i}$. By using the relation $r_1^L\partial_{r_1}^L = \sum_{q=0}^{L}((L,q))\, x_1^q y_1^{L-q}\, \partial_{x_1}^q \partial_{y_1}^{L-q}$, where $((L,q))$ are binomial coefficients, we can easily deduce the desired estimate for $|g^{-N}\varphi|$. For example, in the case of one variable and $N = 1$, we have

$$\left|z^{-1}f(z)\right| \leq \int_0^1 \left|(\partial_r f)(sz)\right| ds \leq \sup\left|\partial_r f\right| \leq \sup\left|\partial_x f\right| + \sup\left|\partial_y f\right|.$$

Let us consider the case $g = \prod_{i=1}^{\ell} z_i^{m_i}$. We set $g_j = \prod_{i=1}^{j} z_i^{m_i}$. Then, we can deduce $|g^{-N}\varphi| \leq C\sum_i \sup\left|g_{j-1}^{-N}Q_{j,i}\varphi\right|$ by an easy descending induction on j.

Let us consider the general case. We take a projective birational morphism $F : X_1 \longrightarrow X$ such that $F^{-1}(H)$ is normal crossing. By the above consideration, we have an estimate $\left|F^{-1}(g^{-N}\varphi)\right| \leq C\sum \sup\left|P_i F^{-1}(\varphi)\right|$ on $X_1 \times T$ for some $P_i \in \mathrm{Diff}_{X_1\times T/T}$. We can find $\tilde{P}_{i,j} \in \mathrm{Diff}_{X\times T/T}$ $(j = 1,\ldots,M_i)$ such that $\left|P_i F^{-1}(\varphi)\right| \leq \sum \sup\left|F^{-1}(\tilde{P}_{i,j}\varphi)\right|$ for any φ. Thus, we obtain the estimate on $X \times T$. □

Remark 2.1.9 $\mathfrak{Db}^{\bmod H}_{X\times T/T}$ is also denoted by $\mathfrak{Db}_{X(*H)\times T/T}$ or $\mathfrak{Db}_{X\times T/T}(*H)$ in the following. □

Suppose that $X = X_0 \times \mathbb{C}_t$, and $X_0 \times \{0\} \not\subset H$. We put $H_0 := (X_0 \times \{0\}) \cap H$. For an open subset $V \subset X \times T$, we set $V_0 := V \cap (X_0 \times \{0\} \times T)$. We have the naturally defined map given by the restriction:

$$\mathcal{E}^{<H\,(n,n)}_{X\times T/T,c}(V) \longrightarrow \mathcal{E}^{<H_0\,(n-1,n-1)}_{X_0\times T/T,c}(V_0) \otimes (\mathbb{C}\,dt\,d\bar{t}) \tag{2.1}$$

Lemma 2.1.10 *The map (2.1) is surjective.*

Proof We set $H_1 := (X_0 \times \{0\}) \cup H$. We take a projective birational map $F : X' \longrightarrow X$ such that (i) $H_1' := F^{-1}(H_1)$ is normal crossing, (ii) $X' \setminus H' \simeq X \setminus H$, where $H' := F^{-1}(H)$. Let $X_1 \subset X'$ be the strict transform of $(X_0 \times \{0\})$. Let $V' := F^{-1}(V)$. Let V_0' be the closure of $V_0 \setminus H_0$ in X'. Then, by a direct construction, it is easy to see that

$$\mathcal{E}^{<H'\,(n,n)}_{X'\times T/T,c}(V') \longrightarrow \mathcal{E}^{<H_0'\,(n-1,n-1)}_{X_1\times T/T,c}(V_0') \otimes (\mathbb{C}F^*(dt\,d\overline{t}))$$

is surjective. Then, we obtain that (2.1) is surjective. □

2.1.4 $\mathcal{R}$-Triple

Let us recall the concept of $\mathcal{R}$-triples very briefly. We refer [66] for details.

Let X be a complex manifold, and let H be a hypersurface of X. Let $\sigma : \mathbb{C}_\lambda^* \longrightarrow \mathbb{C}_\lambda^*$ be given by $\sigma(\lambda) = -\overline{\lambda}^{-1}$. We set $\boldsymbol{S} = \{\lambda \in \mathbb{C} \,\big|\, |\lambda| = 1\}$. If $\lambda \in \boldsymbol{S}$, we have $\sigma(\lambda) = -\lambda$. The induced maps $\mathbb{C}_\lambda^* \times X \longrightarrow \mathbb{C}_\lambda^* \times X$ and $\boldsymbol{S} \times X \longrightarrow \boldsymbol{S} \times X$ are also denoted by σ.

We have the natural action of $\mathcal{R}_{X(*H)|\boldsymbol{S}\times X}$ on $\mathfrak{Db}^{\mathrm{mod}\,H}_{\boldsymbol{S}\times X/\boldsymbol{S}}$. We also have the action of $\sigma^*\mathcal{R}_{X(*H)|\boldsymbol{S}\times X}$ on $\mathfrak{Db}^{\mathrm{mod}\,H}_{\boldsymbol{S}\times X/\boldsymbol{S}}$ by $\sigma^*(f) \bullet F = \overline{\sigma^*(f)} \cdot F$ for any local section f of $\mathcal{O}_\mathcal{X}(*H)$, and $\sigma^*(\lambda v) \bullet F = -\lambda^{-1}\overline{v}F$ for any holomorphic vector field v on X.

Let $\mathcal{M}_i$ $(i = 1, 2)$ be $\mathcal{R}_{X(*H)}$-modules. A sesqui-linear pairing of $\mathcal{M}_1$ and $\mathcal{M}_2$ is an $\mathcal{R}_{X(*H)|\boldsymbol{S}\times X} \otimes \sigma^*\mathcal{R}_{X(*H)|\boldsymbol{S}\times X}$-morphism $C : \mathcal{M}_{1|\boldsymbol{S}\times X} \otimes \sigma^*\mathcal{M}_{2|\boldsymbol{S}\times X} \longrightarrow \mathfrak{Db}^{\mathrm{mod}\,H}_{\boldsymbol{S}\times X/\boldsymbol{S}}$. Such a tuple $(\mathcal{M}_1, \mathcal{M}_2, C)$ is called an $\mathcal{R}_{X(*H)}$-triple. It is called good (coherent, strict, etc.), if the underlying $\mathcal{R}_{X(*H)}$-modules are good (coherent, strict, etc.). A morphism $(\mathcal{M}_1, \mathcal{M}_2, C) \longrightarrow (\mathcal{M}_1', \mathcal{M}_2', C')$ is a pair of morphisms $\varphi_1 : \mathcal{M}_1' \longrightarrow \mathcal{M}_1$ and $\varphi_2 : \mathcal{M}_2 \longrightarrow \mathcal{M}_2'$ such that $C\big(\varphi_1(m_1'), \sigma^*(m_2)\big) = C'\big(m_1', \sigma^*\varphi_2(m_2)\big)$.

The category of $\mathcal{R}_{X(*H)}$-triples is abelian, and denoted by $\mathcal{R}\text{-Tri}(X, H)$. For $\mathcal{T} = (\mathcal{M}_1, \mathcal{M}_2, C) \in \mathcal{R}\text{-Tri}(X, H)$, we have the Hermitian adjoint $\mathcal{T}^* := (\mathcal{M}_2, \mathcal{M}_1, C^*)$, where $C^*(x, \sigma^*y) := \overline{\sigma^*C(y, \sigma^*x)}$.

The push-forward for $\mathcal{R}_{X(*H)}$-triples is defined as in the case of $\mathcal{R}$-triples [66]. Let $f : X' \longrightarrow X$ be a morphism of complex manifolds. Let H be a hypersurface, and set $H' := f^{-1}(H)$. Let $\mathcal{T} = (\mathcal{M}_1, \mathcal{M}_2, C)$ be a coherent $\mathcal{R}_{X'(*H')}$-triple. We have the $\mathcal{R}_{X(*H)}$-triples $f_\dagger^i(\mathcal{T}) = \big(f_\dagger^{-i}\mathcal{M}_1, f_\dagger^i\mathcal{M}_2, f_\dagger^i C\big)$. We recall the formulas from [66] for the convenience to us in the cases (i) f is a closed immersion, (ii) f is the projection.

Let us see the case (i). It is enough to consider the case where X' and X are open subsets in $\mathbb{C}^\ell$ and $\mathbb{C}^n$, and f is given by $\tilde{f}(z_1, \ldots, z_\ell) = (z_1, \ldots, z_\ell, 0, \ldots, 0)$. Let $\eth_i := \lambda(\partial/\partial_{z_i})$. We have $f_\dagger\mathcal{M}_i = f_*\big(\mathcal{M}_i \otimes (\omega_{\mathcal{X}'} \otimes \omega_\mathcal{X}^{-1})\big)[\eth_{\ell+1}, \ldots, \eth_n]$. We set

$\tau := (dz_{\ell+1}/\lambda) \wedge \cdots \wedge (dz_n/\lambda)$. It is enough to describe $f_\dagger C\big(m_1 \otimes \tau^{-1}, \sigma^*(m_2 \otimes \tau^{-1})\big)$ for local sections m_i of $\mathcal{M}_{i|\mathbf{S}\times X'}$. For a section $\phi \cdot \prod_{i=1}^n (dz_i \wedge d\bar{z}_i)$ of $\mathcal{E}^{<H,(n,n)}_{\mathbf{S}\times X/\mathbf{S},c}$, we have

$$\big\langle f_\dagger C\big(m_1 \otimes \tau, \sigma^*(m_2 \otimes \tau)\big), \varphi\big\rangle = (-2\pi\sqrt{-1})^{n-\ell}\Big\langle C(m_1, \sigma^* m_2), f^*(\phi)\prod_{i=1}^{\ell}(dz_i \wedge d\bar{z}_i)\Big\rangle \tag{2.2}$$

Let us see the case (ii), i.e., $X' = Z \times X$, $H' = Z \times H$, and f is the projection. Let $\mathcal{E}^{a,b}_Z$ be the sheaf of C^∞ (a,b)-forms on Z. Let $q : \mathcal{X}' \longrightarrow Z$ be the projection. We set $\mathcal{E}^{a,b}_{\mathcal{X}'/\mathcal{X}} := \lambda^{-a}\mathcal{O}_{\mathcal{X}'} \otimes_{q^{-1}\mathcal{O}_Z} q^{-1}(\mathcal{E}^{a,b}_Z)$. Let $d_Z := \dim Z$. We have $f_\dagger \mathcal{M}_i = f_!\big(\mathrm{Tot}\,\mathcal{E}^{\bullet,\bullet} \otimes_{\mathcal{O}_{\mathcal{X}'}} \mathcal{M}_i[d_Z]\big)$. It is enough to describe $f_\dagger C\big(\eta_1^{d_Z-j} \otimes m_1, \sigma^*(\eta_2^{d_Z+j} \otimes m_2)\big)$, where η_i^k be sections of $\bigoplus_{a+b=k}\mathcal{E}^{a,b}_{\mathcal{X}'/\mathcal{X}}$, and m_i be C^∞-sections of $\mathcal{M}_i$. It is given as follows:

$$f_\dagger C\big(\eta_1^{d_Z-j} \otimes m_1, \sigma^*(\eta_2^{d_Z+j} \otimes m_2)\big) = \frac{(-1)^{(d_Z+j)(d_Z+j-1)/2}}{(2\pi\sqrt{-1})^{d_Z}} \int_f C(m_1, \sigma^* m_2)\,\eta_1^{d_Z-j}\,\overline{\sigma^*(\eta_2^{d_Z+j})} \tag{2.3}$$

We shall give a different expression in Sect. 12.1.4.2.

The following lemma follows from Lemmas 2.1.1 and 2.1.7.

Lemma 2.1.11 *Let $f : X' \longrightarrow X$ be a proper birational morphism. Let $H \subset X$ be a hypersurface, and we put $H' := f^{-1}(H)$. Assume that f gives an isomorphism $X' \setminus H' \simeq X \setminus H$. Then, $f_\dagger$ induces an equivalence of the categories of good $\mathcal{R}_{X'(*H')}$-triples and good $\mathcal{R}_{X(*H)}$-triples.* □

2.1.4.1 Strict Specializability

First, let us recall the specialization in the case $X = X_0 \times \mathbb{C}_t$. We set $H_0 := H \cap (X_0 \times \{0\})$.

A coherent $\mathcal{R}_{X(*H)}(*t)$-triple $\mathcal{T} = (\mathcal{M}', \mathcal{M}'', C)$ is called strictly specializable along t if $\mathcal{M}'$ and $\mathcal{M}''$ are strictly specializable along t. If $\mathcal{T}$ is strictly specializable, we have the induced $\mathcal{R}_{X_0(*H_0)}$-triple $\psi_{t,u}(\mathcal{T})$ for $u \in \mathbb{R} \times \mathbb{C}$ as in the case of $\mathcal{R}$-triples. It consists of $\mathcal{R}_{X_0(*H_0)}$-modules $\psi_{t,u}(\mathcal{M}')$ and $\psi_{t,u}(\mathcal{M}'')$ with the induced sesqui-linear pairing $\psi_{t,u}(C)$. (See [66] for the construction of $\psi_{t,u}(C)$. Note Lemma 2.1.10.) The induced pairing is independent of the choice of a decomposition into the product $X = X_0 \times \mathbb{C}_t$. It is also denoted by $\tilde{\psi}_{t,u}(\mathcal{T})$. The pair of morphisms (t^{-1}, t) gives an isomorphism $\psi_{t,u}(\mathcal{T}) \simeq \psi_{t,u-\boldsymbol{\delta}}(\mathcal{T})$. We have the induced morphisms $-\eth_t t : \psi_{t,u}(\mathcal{M}')\,\lambda \longrightarrow \psi_{t,u}(\mathcal{M}')$ and $-\eth_t t : \psi_{t,u}(\mathcal{M}'') \longrightarrow \psi_{t,u}(\mathcal{M}'')\,\lambda^{-1}$. The nilpotent part is denoted by N' and N''. Then, we obtain

$\mathcal{N} := (N', N'') : \psi_{t,u}(\mathcal{T}) \longrightarrow \psi_{t,u}(\mathcal{T}) \otimes \boldsymbol{T}(-1)$, where $\boldsymbol{T}(-1) = \big(\mathcal{O}_{\mathcal{X}}\,\lambda,\ \mathcal{O}_{\mathcal{X}}\,\lambda^{-1},\ C_0\big)$ and $C_0(s_1, \sigma^* s_2) = s_1 \cdot \overline{\sigma^* s_2}$.

Remark 2.1.12 In [52, 56, 66], we preferred to use $-\mathcal{N}$ as the canonically defined nilpotent map. See Remark 2.1.26. □

A coherent $\mathcal{R}_{X(*H)}$-triple $\mathcal{T}$ is called strictly specializable along t if the underlying $\mathcal{R}_{X(*H)}$-modules are strictly specializable along t. In that case, we naturally obtain a coherent $\mathcal{R}_{X(*H)}(*t)$-triple $\mathcal{T}(*t)$, which is strictly specializable along t. We define $\tilde{\psi}_{t,u}(\mathcal{T}) := \psi_{t,u}(\mathcal{T}(*t))$ as in the case of $\mathcal{R}$-triples.

Let g be any holomorphic function on X. We set $X_1 := X \times \mathbb{C}_t$ and $H_1 := H \times \mathbb{C}_t$. Let $\iota_g : X \longrightarrow X_1$ be given by $\iota_g(x) = (x, g(x))$. A coherent $\mathcal{R}_{X(*H)}(*g)$-triple (resp. $\mathcal{R}_{X(*H)}$-triple) $\mathcal{T}$ is called strictly specializable along g if the coherent $\mathcal{R}_{X_1(*H_1)}$-triple (resp. $\mathcal{R}_{X_1(*H_1)}(*t)$-triple) $\iota_{g\dagger}\mathcal{T}$ is strictly specializable along t. We obtain the $\mathcal{R}_{X(*H)}$-triple $\tilde{\psi}_{g,u}(\mathcal{T}) := \psi_{t,u}\big(\iota_{g\dagger}\mathcal{T}(*t)\big)$.

Let $\mathcal{KMS}(\mathcal{T}, g)$ denote the set of $u \in \mathbb{R} \times \mathbb{C}$ such that $\psi_{g,u}(\mathcal{T}) \neq 0$. We say that a non-zero complex number λ_0 is generic to $(\mathcal{T}, g)$ if $\mathfrak{e}(\lambda_0) : \mathcal{KMS}(\mathcal{T}, g) \cup (\mathbb{Z} \times \{0\}) \longrightarrow \mathbb{C}$ is injective.

2.1.4.2 Functor j^*

Let $j : \mathbb{C}_\lambda \longrightarrow \mathbb{C}_\lambda$ be given by $j(\lambda) = -\lambda$. The induced morphism $\mathcal{X} \longrightarrow \mathcal{X}$ is also denoted by j. Let $\mathcal{T} = (\mathcal{M}_1, \mathcal{M}_2, C) \in \mathcal{R}\text{-Tri}(X, H)$. We have the naturally defined pairing $j^*C : j^*\mathcal{M}_{1|\boldsymbol{S}\times X} \times \sigma^* j^* \mathcal{M}_{2|\boldsymbol{S}\times X} \longrightarrow \mathfrak{Db}^{\mathrm{mod}\,H}_{\boldsymbol{S}\times X/\boldsymbol{S}}$ by which we have $j^*\mathcal{T} := (j^*\mathcal{M}_1, j^*\mathcal{M}_2, j^*C) \in \mathcal{R}\text{-Tri}(X, H)$. We have a natural identification $j^*(\mathcal{T}^*) = (j^*\mathcal{T})^*$.

2.1.5 *Integrable $\mathcal{R}$-Triple*

Let us recall the concepts of integrable $\mathcal{R}$-module and integrable $\mathcal{R}$-triple, introduced in Sect. 7 of [66]. Let X be any complex manifold with a hypersurface H. We set $\mathcal{X}^\lambda := \{\lambda\} \times X \subset \mathcal{X}$ for any $\lambda \in \mathbb{C}$. Let $\Theta_{\mathcal{X}}(\log \mathcal{X}^0)$ denote the sheaf of holomorphic vector fields on $\mathcal{X}$ logarithmic along $\mathcal{X}^0$. Let $\tilde{\mathcal{R}}_X$ denote the sheaf of subalgebras in $\mathscr{D}_{\mathcal{X}}$ generated by $\lambda\Theta_{\mathcal{X}}(\log \mathcal{X}^0)$ over $\mathcal{O}_{\mathcal{X}}$. It is equal to the sheaf of subalgebras in $\mathscr{D}_{\mathcal{X}}$ generated by $\lambda^2\partial_\lambda$ over $\mathcal{R}_X$. We set $\tilde{\mathcal{R}}_{X(*H)} := \tilde{\mathcal{R}}_X(*H)$. A module over $\tilde{\mathcal{R}}_{X(*H)}$ is also called an integrable $\mathcal{R}_{X(*H)}$-module. For $\tilde{\mathcal{R}}_{X(*H)}$-modules $\mathcal{M}_i$ $(i = 1, 2)$, an $\mathcal{R}_{X(*H)}$-homomorphism $\mathcal{M}_1 \longrightarrow \mathcal{M}_2$ is called integrable if it gives an $\tilde{\mathcal{R}}_{X(*H)}$-homomorphism.

Let $\tilde{\Omega}^1_{\mathcal{X}}(\log \mathcal{X}^0)$ be the sheaf of meromorphic 1-forms on $\mathcal{X}$ which are logarithmic along $\mathcal{X}^0 = \{0\} \times X$. Then, $\tilde{\mathcal{R}}_{X(*H)}$-module is equivalent to an $\mathcal{O}_{\mathcal{X}}(*H)$-module $\mathcal{M}$ with a meromorphic flat connection $\nabla : \mathcal{M} \longrightarrow \mathcal{M} \otimes \tilde{\Omega}^1_{\mathcal{X}}(\log \mathcal{X}^0) \otimes \mathcal{O}_{\mathcal{X}}(\mathcal{X}^0)$.

Let $\mathcal{M}_i$ $(i = 1,2)$ be $\widetilde{\mathcal{R}}_{X(*H)}$-modules. Let θ be the polar coordinate on $\boldsymbol{S}$. A sesqui-linear pairing C of $\mathcal{R}_{X(*H)}$-modules $\mathcal{M}_i$ is called integrable if the following holds:

$$\partial_\theta C\big(m_1, \overline{\sigma^* m_2}\big) = C\big(\partial_\theta m_1, \overline{\sigma^* m_2}\big) + C\big(m_1, \overline{\sigma^*(\partial_\theta m_2)}\big)$$

Here, $\partial_\theta m_i := \sqrt{-1}\lambda\partial_\lambda m_i$, which can be regarded as $\big(\sqrt{-1}\lambda\partial_\lambda - \sqrt{-1}\overline{\lambda}\partial_{\overline{\lambda}}\big)m_i$. If C is extended to a pairing $\mathcal{M}_{1|\mathbb{C}_\lambda^*\times X}\times\sigma^*\mathcal{M}_{2|\mathbb{C}_\lambda^*\times X}\longrightarrow\mathfrak{Db}^{an\ \mathrm{mod}\,H}_{\mathbb{C}_\lambda^*\times X/\mathbb{C}_\lambda^*}$, where $\mathfrak{Db}^{an,\ \mathrm{mod}\,H}_{\mathbb{C}_\lambda^*\times X/\mathbb{C}_\lambda^*}$ is the sheaf of λ-holomorphic distributions on $\mathbb{C}_\lambda^*\times X$ with moderate growth along H, then we have

$$\lambda\partial_\lambda C\big(m_1, \overline{\sigma^*(m_2)}\big) = C\big(\lambda\partial_\lambda m_1, \overline{\sigma^*(m_2)}\big) - C\big(m_1, \overline{\sigma^*(\lambda\partial_\lambda m_2)}\big).$$

A tuple of $\tilde{\mathcal{R}}_{X(*H)}$-modules and an integrable sesqui-linear pairing is called an $\tilde{\mathcal{R}}_{X(*H)}$-triple or an integrable $\mathcal{R}_{X(*H)}$-triple. If $\mathcal{T}$ is an integrable $\mathcal{R}_{X(*H)}$-triple, the induced $\mathcal{R}_{X(*H)}$-triples $\mathcal{T}^*$ and $j^*\mathcal{T}$ are also naturally integrable. For $\tilde{\mathcal{R}}_{X(*H)}$-triples $\mathcal{T}_i$ $(i = 1,2)$, a morphism $\mathcal{T}_1\longrightarrow\mathcal{T}_2$ in $\mathcal{R}$-Tri(X,H) is called integrable, if the underlying morphisms of $\mathcal{R}_{X(*H)}$-modules are integrable.

Let $f : X'\longrightarrow X$ be a morphism of complex manifolds. Let H be a hypersurface of X, and we put $H' := f^{-1}(H)$. Let $\mathcal{T}'$ be a good integrable $\mathcal{R}_{X'(*H')}$-module. We have the push-forward $f^i_\dagger\mathcal{T}'$ as $\mathcal{R}_{X(*H)}$-triple. As observed in [66], $f^i_\dagger\mathcal{T}'$ is naturally an integrable $\mathcal{R}_{X(*H)}$-triple. Similarly, for an integrable $\mathcal{R}_{X(*H)}$-triple $\mathcal{T}$ on X, the pull back $f^\dagger\mathcal{T}$ is naturally integrable. We have the integrable variant of Lemma 2.1.11.

Lemma 2.1.13 *Assume that f is proper and induces an isomorphism $X'\setminus H'\simeq X\setminus H$. Then, $f_\dagger$ induces an equivalence of the categories of good integrable $\mathcal{R}_{X'(*H')}$-triples and good integrable $\mathcal{R}_{X(*H)}$-triples.* □

Let g be a holomorphic function on X. We set $X_1 := X\times\mathbb{C}_t$ and $H_1 := H\times\mathbb{C}_t$. Let $\iota_g : X\longrightarrow X_1$ be the graph. We recall the following in [66].

Lemma 2.1.14
- *Let $\mathcal{M}$ be a coherent $\mathcal{R}_{X(*H)}$-module or $\mathcal{R}_{X(*H)}(*g)$-module, which is strictly specializable along g. If $\mathcal{M}$ is integrable, the V-filtrations of $\iota_{g\dagger}\mathcal{M}$ are preserved by $\lambda^2\partial_\lambda$. In particular, the nearby cycle sheaves $\psi_{g,u}(\mathcal{M})$ are naturally integrable.*
- *Let $\mathcal{T}$ be a coherent $\mathcal{R}_{X(*H)}$-triple or a coherent $\mathcal{R}_{X(*H)}(*g)$-triple, which is strictly specializable along g. Then, $\psi_{g,u}(\mathcal{T})$ are also integrable.* □

Note that if an $\mathcal{R}_{X(*H)}$-module $\mathcal{M}$ is integrable and strictly specializable along g, we have $\psi_{g,u}(\mathcal{M}) = 0$ unless $u\in\mathbb{R}\times\{0\}$. It easily follows from the commutativity of $t\partial_t$ and $\lambda^2\partial_\lambda$ on $\mathbb{C}_\lambda\times X\times\mathbb{C}_t$. Hence, we have $\mathcal{KMS}(\mathcal{M},g)\subset\mathbb{R}\times\{0\}$. Similarly, if an $\mathcal{R}_{X(*H)}$-triple $\mathcal{T}$ is integrable and strictly specializable along g, we have $\mathcal{KMS}(\mathcal{T},g)\subset\mathbb{R}\times\{0\}$.

2.1.6 Smooth $\mathcal{R}$-triples and Some Functorial Properties

Let X be a complex manifold with a normal crossing hypersurface H. A coherent $\mathcal{R}_{X(*H)}$-module $\mathcal{M}$ is called smooth if $\mathcal{M}_{|X\setminus H}$ is a locally free $\mathcal{O}_{\mathcal{X}\setminus\mathcal{H}}$-module. A smooth sesqui-linear pairing of smooth $\mathcal{R}_X$-modules $\mathcal{M}_i$ ($i=1,2$) is an $\mathcal{R}_{X(*H)|\mathbf{S}\times X}\otimes\sigma^*\mathcal{R}_{X(*H)|\mathbf{S}\times X}$-homomorphism $C:\mathcal{M}_{1|\mathbf{S}\times X}\otimes\sigma^*\mathcal{M}_{2|\mathbf{S}\times X}\longrightarrow \mathcal{C}^{\infty\ \mathrm{mod}\ H}_{\mathbf{S}\times X}$ such that its restriction to $\mathbf{S}\times(X\setminus H)$ is extended to a pairing of $\mathcal{M}_{1|\mathbb{C}^*_\lambda\times(X\setminus H)}$ and $\sigma^*\mathcal{M}_{1|\mathbb{C}^*_\lambda\times(X\setminus H)}$ to the sheaf of C^∞-functions on $\mathbb{C}^*_\lambda\times(X\setminus H)$ which are λ-holomorphic. Let $\mathcal{R}\text{-Tri}_{\mathrm{sm}}(X,H)$ denote the full subcategory of smooth $\mathcal{R}_{X(*H)}$-triples in $\mathcal{R}\text{-Tri}(X,H)$.

2.1.6.1 Tensor Product

We consider $\mathcal{T}=(\mathcal{M}_1,\mathcal{M}_2,C)\in\mathcal{R}\text{-Tri}_{\mathrm{sm}}(X,H)$ and $\mathcal{T}'=(\mathcal{M}'_1,\mathcal{M}'_2,C')\in\mathcal{R}\text{-Tri}(X,H)$. We set $\mathcal{M}''_i:=\mathcal{M}_i\otimes_{\mathcal{O}_\mathcal{X}}\mathcal{M}'_i$. We have the naturally defined pairing C'' of $\mathcal{M}''_i$ ($i=1,2$), given as follows:

$$C''\big(m_1\otimes m'_1,\sigma^*(m_2\otimes m'_2)\big)=C(m_1,\sigma^*m_2)\cdot C(m'_1,\sigma^*m'_2)$$

The pairing C'' is also denoted by $C\otimes C'$. The $\mathcal{R}_{X(*H)}$-triple $(\mathcal{M}''_1,\mathcal{M}''_2,C'')$ is denoted by $\mathcal{T}\otimes\mathcal{T}'$. If $\mathcal{T}'$ is smooth, $\mathcal{T}\otimes\mathcal{T}'$ is also smooth. If $\mathcal{T}$ and $\mathcal{T}'$ are integrable, $\mathcal{T}\otimes\mathcal{T}'$ is also integrable.

2.1.6.2 Pull Back

Let $f:X_1\longrightarrow X_2$ be a morphism of complex manifolds. Let H_i be hypersurfaces of X_i such that $H_1=f^{-1}(H_2)$. Let $\mathcal{T}=(\mathcal{M}_1,\mathcal{M}_2,C)\in\mathcal{R}\text{-Tri}_{\mathrm{sm}}(X_2,H_2)$. We set $f^*\mathcal{M}_i:=\mathcal{O}_{\mathcal{X}_2}\otimes_{f^{-1}\mathcal{O}_{\mathcal{X}_1}}f^{-1}\mathcal{M}_i$ which are naturally smooth $\mathcal{R}_{X_1(*H_1)}$-modules. We have a smooth sesqui-linear pairing $f^{-1}C$ of $f^{-1}\mathcal{M}_1$ and $f^{-1}\mathcal{M}_2$ obtained as the composite of the following:

$$f^{-1}(\mathcal{M}_1)_{|\mathbf{S}\times X_1}\times\sigma^*f^{-1}(\mathcal{M}_2)_{|\mathbf{S}\times X_1}\longrightarrow f^{-1}\mathcal{C}^{\infty\ \mathrm{mod}\,H_2}_{\mathbf{S}\times X_2}\longrightarrow\mathcal{C}^{\infty\ \mathrm{mod}\,H_1}_{\mathbf{S}\times X_1}$$

It induces a smooth sesqui-linear pairing f^*C of $f^*\mathcal{M}_1$ and $f^*\mathcal{M}_2$. Thus, we obtain a smooth $\mathcal{R}_{X_1(*H_1)}$-triple $f^*\mathcal{T}:=(f^*\mathcal{M}_1,f^*\mathcal{M}_2,f^*C)$. If $\mathcal{T}$ is integrable, $f^*\mathcal{T}$ is also integrable.

Lemma 2.1.15 *Assume that f is a proper morphism, and that it gives an isomorphism $X_1\setminus H_1\simeq X_2\setminus H_2$. Then, f^* gives an equivalence of the categories $\mathcal{R}\text{-Tri}_{\mathrm{sm}}(X_i,H_i)$. It also gives an equivalence of the categories of smooth integrable $\mathcal{R}_{X_i(*H_i)}$-triples.* □

2.1.6.3 A Projection Formula

Let $f : X_1 \longrightarrow X_2$ be as above. Let $\mathcal{T}'$ be any good $\mathcal{R}_{X_1(*H_1)}$-triple. Assume that the support of $\mathcal{T}'$ is proper over X_2. Let $\mathcal{T} \in \mathcal{R}\text{-Tri}_{sm}(X_2, H_2)$.

Lemma 2.1.16 *We have a natural isomorphism of* $\mathcal{R}_{X_2(*H_2)}$*-triples*

$$f_\dagger\big(f^*(\mathcal{T}) \otimes \mathcal{T}'\big) \simeq \mathcal{T} \otimes f_\dagger \mathcal{T}'.$$

If $\mathcal{T}$ *and* $\mathcal{T}'$ *are integrable, the isomorphism is also integrable.*

Proof Let us construct an isomorphism in the level of $\mathcal{R}$-modules. Let $\Omega^p_{\mathcal{X}} := \bigwedge^p \Omega^1_{\mathcal{X}}$. Let $d := \dim X_1$. We have the resolution $\mathcal{R}_{X_1} \otimes \Omega^\bullet_{\mathcal{X}_1}[d]$ of $\omega_{\mathcal{X}_1}$. Hence, $\mathcal{R}_{X_2 \leftarrow X_1} \simeq \mathcal{R}_{X_1} \otimes \Omega^\bullet_{\mathcal{X}_1}[d] \otimes_{f^{-1}\mathcal{O}_{\mathcal{X}_2}} f^{-1}\big(\mathcal{R}_{X_2} \otimes \omega^{-1}_{\mathcal{X}_2}\big)$. Let $\mathcal{M}'$ be a good $\mathcal{R}_{X_1(*H_1)}$-module, and let $\mathcal{M}$ be a smooth $\mathcal{R}_{X_2(*H_2)}$-module. With the notation in Sect. 2.1.6.4, we have

$$\begin{aligned} &\mathcal{R}_{X_2 \leftarrow X_1} \otimes^L_{\mathcal{R}_{X_1}} (\mathcal{M}' \otimes f^*\mathcal{M}) \\ &\quad \simeq \big(\Omega^\bullet_{\mathcal{X}_1}[d] \otimes \mathcal{M}'\big) \otimes^\ell_{f^{-1}\mathcal{O}_{\mathcal{X}_2}} f^{-1}\big((\mathcal{R}_{X_2} \otimes \omega^{-1}_{\mathcal{X}_2})^\ell \otimes_{\mathcal{O}_{\mathcal{X}_2}} \mathcal{M}\big) \\ &\quad \simeq \big(\Omega^\bullet_{\mathcal{X}_1}[d] \otimes \mathcal{M}'\big) \otimes^\ell_{f^{-1}\mathcal{O}_{\mathcal{X}_2}} f^{-1}\big((\mathcal{R}_{X_2} \otimes \omega^{-1}_{\mathcal{X}_2})^r \otimes_{\mathcal{O}_{\mathcal{X}_2}} \mathcal{M}\big). \end{aligned} \tag{2.4}$$

The second isomorphism is due to Lemma 2.1.17. Here, "$\otimes^\ell_{f^{-1}\mathcal{O}_{\mathcal{X}_2}}$" means that we use the $f^{-1}\mathcal{O}_{\mathcal{X}_2}$-action ℓ on $f^{-1}\big((\mathcal{R}_{X_2} \otimes \omega^{-1}_{\mathcal{X}_2})^\kappa \otimes_{\mathcal{O}_{\mathcal{X}_2}} \mathcal{N}\big)$ $(\kappa = \ell, r)$ for the tensor products. Hence, we obtain the desired isomorphism as follows:

$$\begin{aligned} f_\dagger(\mathcal{M}' \otimes f^*\mathcal{M}) &\simeq Rf_*\Big(\mathcal{R}_{X_2 \leftarrow X_1} \otimes^L_{\mathcal{R}_X} (\mathcal{M}' \otimes f^*\mathcal{M})\Big) \\ &\simeq Rf_*\Big(\big(\mathcal{R}_{X_2 \leftarrow X_1} \otimes^L_{\mathcal{R}_X} \mathcal{M}'\big) \otimes_{f^{-1}\mathcal{O}_{\mathcal{X}_2}} f^{-1}\mathcal{M}\Big) \\ &\simeq Rf_*\Big(\big(\mathcal{R}_{X_2 \leftarrow X_1} \otimes^L_{\mathcal{R}_X} \mathcal{M}'\big)\Big) \otimes_{\mathcal{O}_{\mathcal{X}_2}} \mathcal{M} \simeq f_\dagger(\mathcal{M}') \otimes_{\mathcal{O}_{\mathcal{X}_2}} \mathcal{M} \end{aligned} \tag{2.5}$$

If $\mathcal{M}$ and $\mathcal{M}'$ are integrable, the isomorphism (2.5) is also integrable by construction.

If $\mathcal{M}$ is isomorphic to $\mathcal{O}^{\oplus r}_{\mathcal{X}_2}$ as an $\mathcal{R}_{X_2}$-module, then the isomorphism is equal to the natural one $f_\dagger\big(\mathcal{M}'^{\oplus r}\big) \simeq f_\dagger(\mathcal{M}')^{\oplus r}$. We can find such an isomorphism $\mathcal{M}_2 \simeq \mathcal{O}^{\oplus r}_{\mathcal{X}_2}$ locally around any point of $\boldsymbol{S} \times (X_2 \setminus H_2)$. Then, we can easily compare the pairings $f_\dagger\big(f^*(C) \otimes C'\big)$ and $C \otimes f_\dagger C'$. □

2.1.6.4 Appendix

Let X be a complex manifold with a hypersurface H. We have two $\mathcal{R}_X$-actions on $\mathcal{R}_X \otimes \omega_{\mathcal{X}}^{-1}$. Let ℓ denote the left multiplication, and let r denote the $\mathcal{R}_X$-action induced by the right multiplication.

Let $\mathcal{N}$ be an $\mathcal{R}_{X(*H)}$-module. We consider

$$\big(\mathcal{R}_X \otimes \omega_{\mathcal{X}}^{-1}\big)^{\ell} \otimes_{\mathcal{O}_{\mathcal{X}}} \mathcal{N} \simeq \big(\mathcal{R}_{X(*H)} \otimes \omega_{\mathcal{X}}^{-1}\big)^{\ell} \otimes_{\mathcal{O}_{\mathcal{X}}} \mathcal{N} \tag{2.6}$$

where "${}^{\ell}\otimes$" means that we use the $\mathcal{O}_{\mathcal{X}}$-action on $\mathcal{R}_X \otimes \omega_{\mathcal{X}}^{-1}$ induced by ℓ for the tensor product. It is naturally an $\mathcal{R}_{X(*H)}$-bi-module. Let r denote the $\mathcal{R}_{X(*H)}$-action on (2.6) induced by the $\mathcal{R}_{X(*H)}$-action r on $\mathcal{R}_{X(*H)} \otimes \omega_{\mathcal{X}}^{-1}$. Let ℓ denote the $\mathcal{R}_{X(*H)}$-action on (2.6) induced by the $\mathcal{R}_{X(*H)}$-action ℓ on $\mathcal{R}_{X(*H)} \otimes \omega_{\mathcal{X}}$ and the $\mathcal{R}_{X(*H)}$-action on $\mathcal{N}$. We also consider

$$\big(\mathcal{R}_X \otimes \omega_{\mathcal{X}}^{-1}\big)^{r} \otimes_{\mathcal{O}_{\mathcal{X}}} \mathcal{N} \simeq \big(\mathcal{R}_{X(*H)} \otimes \omega_{\mathcal{X}}^{-1}\big)^{r} \otimes_{\mathcal{O}_{\mathcal{X}}} \mathcal{N} \tag{2.7}$$

It is naturally an $\mathcal{R}_{X(*H)}$-bi-module. Let ℓ denote the $\mathcal{R}_{X(*H)}$-action on (2.7) induced by the $\mathcal{R}_{X(*H)}$-action ℓ on $\mathcal{R}_{X(*H)} \otimes \omega_{\mathcal{X}}^{-1}$. Let r denote the $\mathcal{R}_{X(*H)}$-action on (2.7) induced by the $\mathcal{R}_{X(*H)}$-action r on $\mathcal{R}_{X(*H)} \otimes \omega_{\mathcal{X}}$ and the $\mathcal{R}_{X(*H)}$-action on $\mathcal{N}$. Let us recall the following from [66, 69].

Lemma 2.1.17 *We have a unique isomorphism of $\mathcal{R}_{X(*H)}$-bi-modules*

$$\Phi : \Big(\big(\mathcal{R}_X \otimes \omega_{\mathcal{X}}^{-1}\big)^{r} \otimes_{\mathcal{O}_{\mathcal{X}}} \mathcal{N}, \ell, r\Big) \longrightarrow \Big(\big(\mathcal{R}_X \otimes \omega_{\mathcal{X}}^{-1}\big)^{\ell} \otimes_{\mathcal{O}_{\mathcal{X}}} \mathcal{N}, \ell, r\Big) \tag{2.8}$$

such that it induces the identity on $\omega_{\mathcal{X}}^{-1} \otimes \mathcal{N}$.

Proof The inclusion $\omega_{\mathcal{X}}^{-1} \otimes \mathcal{N} \longrightarrow \big(\mathcal{R}_X \otimes \omega_{\mathcal{X}}^{-1}\big)^{r} \otimes_{\mathcal{O}_{\mathcal{X}}} \mathcal{N}$ is uniquely extended to an isomorphism of $\mathcal{R}_{X(*H)}$-modules

$$\Phi : \Big(\big(\mathcal{R}_X \otimes \omega_{\mathcal{X}}^{-1}\big)^{r} \otimes_{\mathcal{O}_{\mathcal{X}}} \mathcal{N}, \ell\Big) \longrightarrow \Big(\big(\mathcal{R}_X \otimes \omega_{\mathcal{X}}^{-1}\big)^{\ell} \otimes_{\mathcal{O}_{\mathcal{X}}} \mathcal{N}, \ell\Big).$$

Let us check that Φ is compatible with the $\mathcal{R}_{X(*H)}$-actions ℓ. It is enough to check locally around any point of X. We may assume to have a holomorphic coordinate system $(z_1, \dots, z_n)$ on X. Let $\tau = \lambda^{-n} dz_1 \cdots dz_n$. Then, for $P \in \mathcal{R}_X$ and $m \in \mathcal{N}$, we have $\Phi(P \otimes \tau \otimes m) = \ell(P)(\tau \otimes m)$ in (2.6). For any $g \in \mathcal{O}_{\mathcal{X}}(*H)$, we have

$$\begin{aligned}
\Phi \circ r(g)(P \otimes \tau \otimes m)\Phi(P \otimes \tau \otimes gm) &= \ell(P)(1 \otimes \tau \otimes gm) \\
&= \ell(P)r(g)(1 \otimes \tau \otimes m) \\
&= r(g)\ell(P)(1 \otimes \tau \otimes m) \\
&= r(g) \circ \Phi(P \otimes \tau \otimes m).
\end{aligned} \tag{2.9}$$

Let $\eth_i := \lambda \partial_i$. We have

$$\begin{aligned}\Phi \circ r(\eth_i)(P \otimes \tau \otimes m) &= \Phi\big(-P\eth_i \otimes \tau \otimes m\big) + \Phi\big(P \otimes \tau \otimes \eth_i m\big)\\ &= -\ell(P\eth_i)(1 \otimes \tau \otimes m) + \ell(P)(1 \otimes \tau \otimes \eth_i m)\\ &= -\ell(P)(\eth_i \otimes \tau \otimes m) = \ell(P) r(\eth_i)(1 \otimes \tau \otimes m)\\ &= r(\eth_i) \circ \Phi(P \otimes \tau \otimes m). \end{aligned} \tag{2.10}$$

Hence, we obtain that Φ is compatible with the $\mathcal{R}_{X(*H)}$-actions r. □

2.1.7 Variation of Twistor Structure

A variation of twistor structure on (X, H) (or simply a twistor structure on (X, H)) is a smooth $\mathcal{R}_{X(*H)}$-triple $\mathcal{T} = (\mathcal{M}_1, \mathcal{M}_2, C)$ satisfying the following conditions:

- If X is a point and $H = \emptyset$, the extended pairing $\mathcal{M}'_{|\mathbb{C}^*_\lambda} \otimes \sigma^* \mathcal{M}''_{|\mathbb{C}^*_\lambda} \longrightarrow \mathcal{O}_{\mathbb{C}^*_\lambda}$ is perfect, i.e., the induced morphism
$$\mathcal{M}'_{|\mathbb{C}^*_\lambda} \longrightarrow \mathcal{H}om_{\mathcal{O}_{\mathbb{C}^*_\lambda}}\big(\sigma^*\mathcal{M}''_{|\mathbb{C}^*_\lambda}, \mathcal{O}_{\mathbb{C}^*_\lambda}\big)$$
is an isomorphism.
- In the general case, $\iota_P^* C$ is a twistor structure for any point $\iota_P : \{P\} \longrightarrow X \setminus H$.

Let $\mathrm{TS}(X, H) \subset \mathcal{R}\text{-Tri}_{\mathrm{sm}}(X, H)$ denote the full subcategory of twistor structures on (X, H). For $\mathcal{T}_i \in \mathrm{TS}(X, H)$, we have $\mathcal{T}_1 \otimes \mathcal{T}_2 \in \mathrm{TS}(X, H)$. For any morphism $f : (X_1, H_1) \longrightarrow (X_2, H_2)$, we have the naturally defined pull back $f^* : \mathrm{TS}(X_2, H_2) \longrightarrow \mathrm{TS}(X_1, H_1)$.

2.1.7.1 Duality

Let $\mathcal{T} = (\mathcal{M}_1, \mathcal{M}_2, C)$ be a variation of twistor structure on (X, H). We set $\mathcal{M}_i^\vee := \mathcal{H}om_{\mathcal{O}_{\mathcal{X}(*H)}}\big(\mathcal{M}_i, \mathcal{O}_{\mathcal{X}}(*H)\big)$. We shall observe that we have a naturally induced pairing

$$C^\vee : \mathcal{M}^\vee_{1|\boldsymbol{S}\times X} \times \sigma^* \mathcal{M}^\vee_{2|\boldsymbol{S}\times X} \longrightarrow \mathcal{C}^{\infty \bmod H}_{\boldsymbol{S}\times X},$$

and we shall set $\mathcal{T}^\vee := (\mathcal{M}_1^\vee, \mathcal{M}_2^\vee, C^\vee)$, which is called the dual of $\mathcal{T}$. For a neighbourhood $U(\lambda_0)$ of λ_0 in $\mathbb{C}_\lambda$, let $\boldsymbol{I}(\lambda_0) := \boldsymbol{S} \cap U(\lambda_0)$.

First, let us consider the case $H = \emptyset$. We set $\mathcal{M}'_{C^\infty, \boldsymbol{S}\times X} := \mathcal{M}' \otimes \mathcal{C}^\infty_{\boldsymbol{S}\times X}$ and $\mathcal{M}''_{C^\infty, \boldsymbol{S}\times X} := \mathcal{M}'' \otimes \mathcal{C}^\infty_{\boldsymbol{S}\times X}$ on $\boldsymbol{S} \times X$. Then, $\mathcal{M}'_{C^\infty, \boldsymbol{S}\times X}$ and $\mathcal{M}''_{C^\infty, \boldsymbol{S}\times X}$ are naturally $\mathrm{Diff}_{\boldsymbol{S}\times X/\boldsymbol{S}}$-modules. For a section $P \in \mathrm{Diff}_{\boldsymbol{S}\times X/\boldsymbol{S}}$ on $\boldsymbol{I}(\lambda_0) \times X$ and $m'' \in \mathcal{M}''_{C^\infty, \boldsymbol{S}\times X}$ on $\boldsymbol{I}(-\lambda_0) \times X$, we set $P \cdot \sigma^* m'' := \sigma^*\big(\overline{\sigma^*(P)} \cdot m''\big)$. Thus, $\sigma^*(\mathcal{M}''_{C^\infty, \boldsymbol{S}\times X})$ is a $\mathrm{Diff}_{\boldsymbol{S}\times X/\boldsymbol{S}}$-module.

We have the induced pairing of $\mathcal{C}^{\infty}_{S\times X}$-modules

$$C : \mathcal{M}'_{C^{\infty},S\times X} \times \sigma^* \mathcal{M}''_{C^{\infty},S\times X} \longrightarrow \mathcal{C}^{\infty}_{S\times X}. \tag{2.11}$$

We have induced morphisms of $\mathcal{C}^{\infty}_{S\times X}$-modules

$$\Psi_C : \mathcal{M}'_{C^{\infty},S\times X} \longrightarrow \sigma^* \mathcal{M}''^{\vee}_{C^{\infty},S\times X}, \qquad \Phi_C : \sigma^* \mathcal{M}''_{C^{\infty},S\times X} \longrightarrow \mathcal{M}'^{\vee}_{C^{\infty},S\times X}$$

given as follows:

$$\big\langle \Psi_C(m'), \sigma^*(m'')\big\rangle := C(m', \sigma^* m''), \quad \big\langle \Phi_C(\sigma^* m''), m'\big\rangle := C(m', \sigma^* m'') \tag{2.12}$$

Note that we use the identification $\sigma^*(\mathcal{M}''^{\vee}) \simeq \sigma^*(\mathcal{M}'')^{\vee}$ given by the pairing $\sigma^* \mathcal{M}''^{\vee}_{C^{\infty},S\times X} \times \sigma^* \mathcal{M}''_{C^{\infty},S\times X} \longrightarrow \mathcal{C}^{\infty}_{S\times X}$:

$$\langle \sigma^* n'', \sigma^* m''\rangle = \overline{\sigma^* \langle n'', m''\rangle}.$$

We can check that Ψ_C and Φ_C are $\mathrm{Diff}_{S\times X/S}$-homomorphisms by direct computations. Because $\mathcal{T} \in \mathrm{TS}(X, D)$, Ψ_C and Φ_C are isomorphisms. Hence, we have the induced $\mathrm{Diff}_{S\times X/S}$-morphism: $C^{\vee} : \mathcal{M}'^{\vee}_{C^{\infty},S\times X} \otimes \sigma^* \mathcal{M}''^{\vee}_{C^{\infty},S\times X} \longrightarrow \mathcal{C}^{\infty}_{S\times X}$:

$$C^{\vee}(n', \sigma^* n'') := C\big(\Psi_C^{-1}(\sigma^* n''), \Phi_C^{-1}(n')\big)$$

The composition $\mathcal{M}'^{\vee}_{|S\times X} \times \sigma^* \mathcal{M}''^{\vee}_{|S\times X} \longrightarrow \mathcal{M}'^{\vee}_{C^{\infty},S\times X} \times \sigma^* \mathcal{M}''^{\vee}_{C^{\infty},S\times X} \longrightarrow \mathcal{C}^{\infty}_{S\times X}$ is also denoted by $C^{\vee}$, which is the desired smooth sesqui-linear pairing.

Let us consider the case that H is not necessarily empty. We have the smooth pairing $C_0^{\vee} : \mathcal{M}'_{|S\times(X\setminus H)} \times \sigma^* \mathcal{M}''_{|S\times(X\setminus H)} \longrightarrow \mathcal{C}^{\infty}_{S\times(X\setminus H)}$ by the previous consideration.

Lemma 2.1.18 *$C_0^{\vee}$ is extended to a pairing $C^{\vee} : \mathcal{M}'_{|S\times X} \times \sigma^* \mathcal{M}''_{|S\times X} \longrightarrow \mathcal{C}^{\infty\,\mathrm{mod}\,H}_{S\times X}$.*

Proof By using a resolution and Lemma 2.1.15, we may assume that $\mathcal{M}'$ and $\mathcal{M}''$ are locally free $\mathcal{O}_{\mathcal{X}}(*H)$-modules, and that H is normal crossing. Let us consider the case that $\mathcal{M}'$ and $\mathcal{M}''$ are of rank 1. Let $P \in H$. We will shrink X around P. We may assume that the monodromy of $\mathcal{M}'$ and $\mathcal{M}''$ are trivial. Let $\lambda_0 \in \boldsymbol{S}$. Let $U(\lambda_0)$ be a small neighbourhood of λ_0 in $\mathbb{C}$, and let $U(-\lambda_0) := \sigma(U(\lambda_0))$. We can find a frame m' of $\mathcal{M}'$ on $U(\lambda_0) \times X$, and a frame m'' of $\mathcal{M}''$ on $U(-\lambda_0) \times X$. We have meromorphic functions $\mathfrak{a}'$ and $\mathfrak{a}''$ such that (i) the poles are contained in $\mathcal{H}$, (ii) $\mathbb{D}m' = m'\, d\mathfrak{a}'$ and $\mathbb{D}m'' = m''\, d\mathfrak{a}''$.

Lemma 2.1.19 *We have $\mathfrak{a}' = \sigma^* \mathfrak{a}''$ modulo holomorphic functions.*

Proof Set $F := C(m', \sigma^* m'')$. Then, we have a holomorphic function $G(\lambda)$ so that $F = G(\lambda) \exp\Big(2\sqrt{-1}\,\mathrm{Im}(\lambda^{-1}\mathfrak{a}') + \lambda\overline{(\mathfrak{a}' - \sigma^* \mathfrak{a}'')}\Big)$. Because F is of moderate growth, we obtain the claim of Lemma 2.1.19. □

In particular, $\mathfrak{a}'$ and $\mathfrak{a}''$ are independent of λ. By considering the tensor product with the rank one object, we may assume that $\mathfrak{a}' = \mathfrak{a}'' = 0$. Then, $C_0^\vee$ is clearly extended. Moreover, we have a bound of $C(m', \sigma^* m'')$ from below. Then, we can deduce the claim of Lemma 2.1.18 in the general rank case by using the exterior product. □

The following lemma is clear by construction.

Lemma 2.1.20 *If $\mathcal{T} \in \mathrm{TS}(X, H)$ is integrable, $\mathcal{T}^\vee$ is naturally integrable.* □

2.1.7.2 Real Structure

Let $\mathcal{T}$ be a variation of twistor structure on (X, H). We have natural identifications $(\mathcal{T}^*)^\vee = (\mathcal{T}^\vee)^*$ and $j^*(\mathcal{T}^\vee) = (j^*\mathcal{T})^\vee$. We formally set

$$\tilde{\gamma}_{\mathrm{sm}}^* \mathcal{T} := j^*(\mathcal{T}^\vee)^*.$$

We naturally have $\tilde{\gamma}_{\mathrm{sm}}^* \circ \tilde{\gamma}_{\mathrm{sm}}^*(\mathcal{T}) \simeq \mathcal{T}$. A real structure of a variation of twistor structure $\mathcal{T}$ is defined to be an isomorphism $\kappa : \tilde{\gamma}_{\mathrm{sm}}^* \mathcal{T} \longrightarrow \mathcal{T}$ such that $\kappa \circ \tilde{\gamma}_{\mathrm{sm}}^* \kappa = \mathrm{id}$.

Remark 2.1.21 $\tilde{\gamma}_{\mathrm{sm}}^*$ is the counterpart of the pull back of vector bundles on $\mathbb{P}^1$ by $\gamma : \mathbb{P}^1 \longrightarrow \mathbb{P}^1$, where $\gamma([z_0 : z_1]) = [\bar{z}_1 : \bar{z}_0]$. □

Let $f : X_1 \longrightarrow X_2$ be any morphism of complex manifolds. Let H_i $(i = 1, 2)$ be hypersurfaces of X_i such that $H_1 = f^{-1}(H_2)$. Let $\mathcal{T} \in \mathrm{TS}(X_2, H_2)$. We have natural isomorphisms $f^*(\mathcal{T}^*) = (f^*\mathcal{T})^*, f^*(\mathcal{T}^\vee) \simeq f^*(\mathcal{T})^\vee, f^*j^*(\mathcal{T}) \simeq j^*f^*(\mathcal{T})$. A real structure is functorial with respect to the pull back.

Remark 2.1.22 Let $\mathcal{T} \in \mathrm{TS}(X, H)$. Let $\mathcal{M}_i$ be the underlying $\mathcal{R}_{X(*H)}$-modules. If $\mathcal{T} \in \mathrm{TS}(X, H)$ is equipped with a real structure, each $\mathcal{M}_{i|P}^\lambda$ for $(\lambda, P) \in \boldsymbol{S} \times (X \setminus H)$ has an induced real structure. If moreover $\mathcal{T}$ is equipped with an integrable structure, the flat bundle $\mathcal{M}_{i|\mathbb{C}_\lambda^* \times (X \setminus H)}$ has an induced real structure. □

2.1.8 Tate Object

We have integrable $\mathcal{R}$-triples $\boldsymbol{T}(w) := \big(\mathcal{O}_{\mathbb{C}_\lambda}\lambda^{-w}, \mathcal{O}_{\mathbb{C}_\lambda}\lambda^{w}, C_0\big)$ for $w \in \mathbb{Z}$, where C_0 is the natural pairing:

$$C_0(f, \sigma^* g) = f \cdot \overline{\sigma^* g}.$$

We naturally have $\boldsymbol{T}(w)^* = \big(\mathcal{O}_{\mathbb{C}_\lambda}\lambda^{w}, \mathcal{O}_{\mathbb{C}_\lambda}\lambda^{-w}, C_0\big)$. We shall use the twisted identification $c_w : \boldsymbol{T}(w)^* \simeq \boldsymbol{T}(-w)$ given by $\big((-1)^w, (-1)^w\big)$. We shall naturally identify $\boldsymbol{T}(a) \otimes \boldsymbol{T}(b)$ with $\boldsymbol{T}(a+b)$.

We shall also use the natural identifications $d_w : \boldsymbol{T}(w)^\vee \simeq \boldsymbol{T}(-w)$ and $j^*\boldsymbol{T}(w) \simeq \boldsymbol{T}(w)$. Then, as the composition of the above morphisms, we obtain an induced real structure $\kappa : \tilde{\gamma}_{\mathrm{sm}}^* \boldsymbol{T}(w) \simeq \boldsymbol{T}(w)$ given by $\big((-1)^w, (-1)^w\big)$.

Remark 2.1.23 The real part of the fiber $\mathcal{O}_{\mathbb{C}_\lambda|1}$ is given by $(\sqrt{-1})^w\mathbb{R}$. □

Let $\mathbb{T}^S(w)$ be the smooth $\mathcal{R}$-triple given by $\big(\mathcal{O}_{\mathbb{C}_\lambda}, \mathcal{O}_{\mathbb{C}_\lambda}, (\sqrt{-1}\lambda)^{-2w}\big)$. Note that we have an isomorphism $\Psi_w : \boldsymbol{T}(w) \simeq \mathbb{T}^S(w)$ given by the pair of the following morphisms:

$$\mathcal{O}_{\mathbb{C}_\lambda}\lambda^{-w} \xleftarrow{(\sqrt{-1}\lambda)^{-w}} \mathcal{O}_{\mathbb{C}_\lambda}, \qquad \mathcal{O}_{\mathbb{C}_\lambda}\lambda^{w} \xrightarrow{(-\sqrt{-1}\lambda)^{-w}} \mathcal{O}_{\mathbb{C}_\lambda}$$

Then, the following diagram is commutative:

$$\begin{array}{ccc} \boldsymbol{T}(w)^* & \xrightarrow{c_w} & \boldsymbol{T}(-w) \\ {\scriptstyle \Psi_w^*}\big\uparrow & & \big\downarrow{\scriptstyle \Psi_{-w}} \\ \mathbb{T}^S(w)^* & \xrightarrow{c_w^S:=(\mathrm{id},\mathrm{id})} & \mathbb{T}^S(-w) \end{array}$$

Hence, we may replace the family $\big(\mathbb{T}^S(w), c_w^S \,\big|\, w \in \mathbb{Z}\big)$ with $\big(\boldsymbol{T}(w), c_w \,\big|\, w \in \mathbb{Z}\big)$.

Remark 2.1.24 Let X be any complex manifold. For the canonical morphism a_X of X to a point, the pull back $a_X^*\boldsymbol{T}(w)$ is denoted by $\boldsymbol{T}(w)_X$, or just by $\boldsymbol{T}(w)$, when there is no risk of confusion. □

Remark 2.1.25 Let X be any complex manifold. Let $\mathcal{T} = (\mathcal{M}', \mathcal{M}'', C)$ be an $\mathcal{R}_X$-triple. Recall that a morphism $\mathcal{S} : \mathcal{T} \longrightarrow \mathcal{T}^* \otimes \boldsymbol{T}(-w)$ is called Hermitian sesqui-linear duality of weight w if $\mathcal{S}^* = (-1)^w\mathcal{S}$. Here, $\mathcal{S}^*$ is the composite of the following morphisms:

$$\mathcal{T} = (\mathcal{T}^*)^* \simeq \big(\mathcal{T}^* \otimes \boldsymbol{T}(-w)\big)^* \otimes \boldsymbol{T}(-w) \longrightarrow \mathcal{T}^* \otimes \boldsymbol{T}(-w)$$

Here, we use the above identification $\boldsymbol{T}(-w)^* \simeq \boldsymbol{T}(w)$. The morphism $\mathcal{S}$ is expressed as a pair of morphisms $\mathcal{S}', \mathcal{S}'' : \mathcal{M}'' \longrightarrow \mathcal{M}'$. The condition is equivalent to $\mathcal{S}' = \mathcal{S}''$. □

Remark 2.1.26 Let $X = X_0 \times \mathbb{C}_t$. Let $\mathcal{T} = (\mathcal{M}', \mathcal{M}'', C)$ be any $\mathcal{R}_X$-triple which is strictly specializable along t. Under the above identification $\mathbb{T}^S(-1) \simeq \boldsymbol{T}(-1)$, the morphism $\mathcal{N} : \psi_{t,u}(\mathcal{T}) \longrightarrow \psi_{t,u}(\mathcal{T}) \otimes \boldsymbol{T}(-1)$ in Sect. 2.1.4 is identified with $-\mathcal{N}$ in Sect. 3.6.1 of [66]. Note that $\eth_t t = (\sqrt{-1}\lambda)^{-1}(\sqrt{-1}\eth_t t) = (-\sqrt{-1}\lambda)^{-1}(-\sqrt{-1}\eth_t t)$. □

Remark 2.1.27 Let $f : X \longrightarrow Y$ be a proper morphism. For any coherent $\mathcal{R}_X$-triple $\mathcal{T}$, we have the $\mathcal{R}$-triples $f_\dagger^i\mathcal{T}$ on Y. For a real $(1,1)$-form ω on X, the morphism $L_\omega^S : f_\dagger^i\mathcal{T} \longrightarrow f_\dagger^i\mathcal{T} \otimes \mathbb{T}^S(1)$ is given in [66] as the pair $(-L_\omega, L_\omega)$, where L_ω be the morphism induced by the multiplication of $\lambda^{-1}\omega$. Under the isomorphism

$\mathbb{T}^S(1) \simeq \boldsymbol{T}(1)$, it is identified with the morphism $f_\dagger^i(\mathcal{T}) \longrightarrow f_\dagger^{i+2}(\mathcal{T}) \otimes \boldsymbol{T}(1)$ given as the pair $(\varphi_\omega, \varphi_\omega)$, where φ_ω is the multiplication of $-\sqrt{-1}\omega$. □

2.1.8.1 Variant

For $(p,q) \in \mathbb{Z}^2$, we set let $\mathcal{U}(p,q) := \big(\mathcal{O}_{\mathcal{X}}\lambda^p, \mathcal{O}_{\mathcal{X}}\lambda^q, C_0\big)$. We have an isomorphism $\Psi : \mathbb{T}^S\big((q-p)/2\big) \longrightarrow \mathcal{U}(p,q)$ given by $\big((\sqrt{-1}\lambda)^{-p}, (-\sqrt{-1}\lambda)^q\big)$. Then, the following diagram is commutative:

$$\begin{array}{ccc}
\mathbb{T}^S\big((q-p)/2\big) & \longrightarrow & \mathbb{T}^S\big((q-p)/2\big)^* \otimes \mathbb{T}^S\big((q-p)\big) \\
\Psi \big\downarrow & & \big\uparrow \Psi^* \otimes \Phi \\
\mathcal{U}(p,q) & \xrightarrow{((-1)^p,(-1)^p)} & \mathcal{U}(p,q)^* \otimes \boldsymbol{T}\big(-(p-q)\big)
\end{array}$$

Hence, we use the polarization of $\mathcal{U}(p,q)$ given by $\big((-1)^p, (-1)^p\big)$.

2.1.9 *Other Basic Examples of Smooth $\mathcal{R}$-Triples of Rank One*

Let $\mathfrak{a}$ be a section of $\mathcal{O}_X(*H)$. Let $L(\mathfrak{a}) = \mathcal{O}_{X(*H)}\, e$ with $h(e,e) = 1$ and $\theta = d\mathfrak{a}$, which gives a wild harmonic bundle on (X,H). We have $\overline{\partial} e = 0$, $\partial e = 0$ and $\theta^\dagger = d\overline{\mathfrak{a}}$. The associated family of λ-flat bundles $\mathcal{L}(\mathfrak{a})$ has a frame $v = e\,\exp(-\lambda\overline{\mathfrak{a}})$. We have the pairing $C(v, \sigma^* v) = h(v, \sigma^* v) = \exp(-\lambda\overline{\mathfrak{a}} + \lambda^{-1}\mathfrak{a})$ on $\boldsymbol{S} \times X$, which takes values in $C^{\infty \bmod H}_{\boldsymbol{S}\times X}$. We obtain a variation of twistor structure $\mathcal{T}_\mathfrak{a} = \big(\mathcal{L}(\mathfrak{a}), \mathcal{L}(\mathfrak{a}), C\big)$ on (X,H).

We have the natural isomorphism $\mathcal{T}_\mathfrak{a}^* \simeq \mathcal{T}_\mathfrak{a}$ given by $(\mathrm{id},\mathrm{id})$. We also have natural isomorphisms $j^*\mathcal{T}_\mathfrak{a} \simeq \mathcal{T}_{-\mathfrak{a}}$ and $\mathcal{T}_\mathfrak{a}^\vee \simeq \mathcal{T}_{-\mathfrak{a}}$. Hence, we have a natural isomorphism $\tilde{\gamma}_{\mathrm{sm}}^* \mathcal{T}_\mathfrak{a} \simeq \mathcal{T}_\mathfrak{a}$, which is a real structure of the variation of twistor structure $\mathcal{T}_\mathfrak{a}$. The variation of twistor structure is naturally integrable. The natural meromorphic flat connection ∇ of $\mathcal{L}(\mathfrak{a})$ is given by $\nabla v = v\,d(\lambda^{-1}\mathfrak{a})$.

Let $a \in \mathbb{R}$. We consider a harmonic bundle $L(a) = \mathcal{O}_{\mathbb{C}_z^*}\, e$ with $h(e,e) = |z|^{-2a}$ and $\theta = 0$. We have $\overline{\partial} e = 0$, $\partial e = e\,(-a)\,dz/z$ and $\theta^\dagger = 0$. We have $(\overline{\partial} + \lambda\theta^\dagger)e = 0$, $\mathbb{D}e = e(-\lambda\, a)\,dz/z$ and $\lambda^2 \partial_\lambda\, e = 0$. We have the pairing $C(e, \sigma^* e) = |z|^{-2a}$. We obtain the variation of twistor structure $\mathcal{T}_a = \big(\mathcal{L}(a), \mathcal{L}(a), C\big)$.

We have the natural isomorphism $\mathcal{T}_a^* \simeq \mathcal{T}_a$ given by $(\mathrm{id},\mathrm{id})$. The restriction to $X \setminus H$ is a polarization of $\mathcal{T}_{a|X\setminus H}$. We have a natural isomorphism $j^*\mathcal{T}_a \simeq \mathcal{T}_a$. We have a natural isomorphism $\mathcal{T}_a^\vee \simeq \mathcal{T}_{-a}$. Hence, for $0 < a < 1/2$, we have a real structure of $\mathcal{T}_a \oplus \mathcal{T}_{-a}$, given as follows:

$$\tilde{\gamma}_{\mathrm{sm}}^* \mathcal{T}_a \oplus \tilde{\gamma}_{\mathrm{sm}}^* \mathcal{T}_{-a} \simeq \mathcal{T}_{-a} \oplus \mathcal{T}_a \simeq \mathcal{T}_a \oplus \mathcal{T}_{-a},$$

where the second isomorphism is the exchange of the components. If $a = 1/2$, we have a real structure of $\mathcal{T}_{1/2}$ given by $\tilde{\gamma}_{\mathrm{sm}}^{*}\mathcal{T}_{1/2} \simeq \mathcal{T}_{-1/2} \simeq \mathcal{T}_{1/2}$. The variation of twistor structure $\mathcal{T}_a$ is naturally integrable. The meromorphic flat connection of $\mathcal{L}(a)$ is given by $\nabla e = e(-adz/z)$.

Let $\alpha \in \mathbb{C}$. We have a harmonic bundle $L(\alpha) = \mathcal{O}_{\mathbb{C}_z^*}\, e$ with $h(e,e) = 1$ and $\theta\, e = e\, \alpha\, dz/z$. We put $v = e\, \exp(-\lambda\overline{\alpha}\, \log|z|^2)$, which gives a frame of $\mathcal{L}(\alpha)$ on $\mathbb{C}_\lambda \times \mathbb{C}_z^*$. We have $z\partial_z v = v\,(\lambda^{-1}\alpha - \lambda\overline{\alpha})$. We have the pairing $C(v, \sigma^* v) = |z|^{2(-\lambda\overline{\alpha}+\lambda^{-1}\alpha)}$. We obtain an $\mathcal{R}$-triple $\mathcal{T}_\alpha = \big(\mathcal{L}(\alpha), \mathcal{L}(\alpha), C_\alpha\big)$.

We have natural isomorphisms $\mathcal{T}_\alpha^* \simeq \mathcal{T}_\alpha$ and $\mathcal{T}_\alpha^\vee \simeq \mathcal{T}_{-\alpha}$ and $j^*\mathcal{T}_\alpha \simeq \mathcal{T}_{-\alpha}$. Hence, we have a real structure $\tilde{\gamma}_{\mathrm{sm}}^{*}\mathcal{T}_\alpha \simeq \mathcal{T}_\alpha$. But, it is easy to observe that $\mathcal{T}_\alpha$ $(\alpha \neq 0)$ cannot be integrable, i.e., $\mathcal{L}(\alpha)$ cannot underlie an $\tilde{\mathcal{R}}_{\mathbb{C}}$-module. Indeed, assume that an $\tilde{\mathcal{R}}_{\mathbb{C}}$-module. We have a meromorphic function A whose poles is contained in $\mathbb{C}_\lambda^* \times \{0\}$ such that $\partial_\lambda v = v\, A$ on $\mathbb{C}_\lambda^* \times X$. We obtain the relation $z\partial_z A = -\overline{\alpha} - \lambda^2\alpha$, which contradicts with the condition on A.

2.1.9.1 Ramified Exponential Twist

Let $\varphi_n : \mathbb{C}_s \longrightarrow \mathbb{C}_t$ be given by $\varphi_n(s) = s^n$. We set $X^{(n)} := X_0 \times \mathbb{C}_s$. The induced map $X^{(n)} \longrightarrow X$ is also denoted by φ_n. We give a complement to [55, Sects. 22.4.2 and 22.11.2]. (See also [66, 67].) Let $\mathcal{M}$ be a strict coherent $\mathcal{R}_{X^{(n)}}(*s)$-module. We have $\varphi_{n\dagger}\mathcal{M} = \varphi_{n*}\mathcal{M}$.

Lemma 2.1.28 *$\mathcal{M}$ is strictly specializable along s if and only if $\varphi_{n\dagger}\mathcal{M}$ is strictly specializable along t. In that case, we have natural isomorphisms $\tilde{\psi}_{-\boldsymbol{\delta}+u}(\mathcal{M}) \simeq \tilde{\psi}_{-\boldsymbol{\delta}+u/n}(\varphi_{n\dagger}\mathcal{M})$ for any $u \in \mathbb{R}\times\mathbb{C}$.*

Proof Suppose that $\mathcal{M}$ is strictly specializable along s. Then, $\varphi_{n\dagger}\mathcal{M} = \varphi_{n*}\mathcal{M}$ is strictly specializable along t. Indeed, the V-filtration at λ_0 is given by

$$V^{(\lambda_0)}_{-1+a}(\varphi_{n\dagger}\mathcal{M}) = \varphi_{n*}\big(V^{(\lambda_0)}_{-1+a/n}\mathcal{M}\big).$$

Note the equality $-\eth_s s + e(\lambda, -\boldsymbol{\delta}+u) = n(-\eth_t t + \mathfrak{e}(\lambda, -\boldsymbol{\delta}+u/n))$. In particular, we have a natural isomorphism $\tilde{\psi}_{-\boldsymbol{\delta}+u}(\mathcal{M}) \simeq \tilde{\psi}_{-\boldsymbol{\delta}+u/n}(\varphi_{n\dagger}\mathcal{M})$. Conversely, suppose that $\varphi_{n\dagger}(\mathcal{M})$ is strictly specializable along t. As remarked in [55, Lemma 22.4.7], $\varphi_n^\dagger\varphi_{n\dagger}(\mathcal{M})$ is strictly specializable along s. Because $\mathcal{M}$ is a direct summand of $\varphi_n^\dagger\varphi_{n\dagger}(\mathcal{M})$, we obtain that $\mathcal{M}$ is strictly specializable along s. □

Corollary 2.1.29 *A strict coherent $\mathcal{R}_{X^{(n)}}(*s)$-triple $\mathcal{T}_1$ is strictly specializable along s if and only if $\varphi_{n\dagger}\mathcal{T}_1$ is strictly specializable along t. In that case, we have a natural isomorphism:*

$$\tilde{\psi}_{-\boldsymbol{\delta}+u}(\mathcal{T}_1) \simeq \tilde{\psi}_{-\boldsymbol{\delta}+u/n}\big(\varphi_{n\dagger}\mathcal{T}_1\big).$$

Proof The comparison of the induced sesqui-linear pairings can be done by an easy computation. □

Let $\mathfrak{a} \in s^{-1}\mathbb{C}[s^{-1}]$. We have the $\mathcal{R}_{\mathbb{C}_s}(*s)$-triple $\mathcal{T}_{-\mathfrak{a}}$. Via the pull back by the projection $X^{(n)} \longrightarrow \mathbb{C}_s$, it induces a smooth $\mathcal{R}_{X^{(n)}}(*s)$-triple, which is also denoted by $\mathcal{T}_{-\mathfrak{a}}$. Let $\mathcal{T}$ be a strict coherent $\mathcal{R}_X(*t)$-triple. Recall that, if $\varphi_n^{\dagger}(\mathcal{T}) \otimes \mathcal{T}_{-\mathfrak{a}}$ is strictly specializable along s, we set

$$\tilde{\psi}_{t,\mathfrak{a},u}(\mathcal{T}) := \tilde{\psi}_{s,u}\big(\varphi_n^{\dagger}(\mathcal{T}) \otimes \mathcal{T}_{-\mathfrak{a}}\big).$$

We have a smooth $\mathcal{R}_X(*t)$-triple $\varphi_{n\dagger}(\mathcal{T}_{-\mathfrak{a}})$. It is easy to observe that $\varphi_n^{\dagger}\mathcal{T} \otimes \mathcal{T}_{-\mathfrak{a}}$ is strictly specializable along s if and only if $\mathcal{T} \otimes \varphi_{n\dagger}\mathcal{T}_{-\mathfrak{a}}$ is strictly specializable along t. By the above consideration, we have natural isomorphism

$$\tilde{\psi}_{t,\mathfrak{a},-\boldsymbol{\delta}+u}(\mathcal{T}) \simeq \tilde{\psi}_{t,-\boldsymbol{\delta}+u/n}\big(\mathcal{T} \otimes \varphi_{n\dagger}\mathcal{T}_{-\mathfrak{a}}\big). \tag{2.13}$$

2.2 Deformation Associated to Nilpotent Morphisms

An $\mathcal{R}$-triple with a tuple of nilpotent morphisms has a natural deformation. Such a deformation was used in the contexts of variations of twistor structure in [52, 56], which originated from the procedure to construct nilpotent orbits in the Hodge theory.

2.2.1 *Twistor Nilpotent Orbit in $\mathcal{R}$-Triple*

Let X be a complex manifold, and let H be a hypersurface. Let Λ be a finite set. Let Λ-$\mathcal{R}$-Tri(X,H) be the category of $\mathcal{R}_{X(*H)}$-triple $\mathcal{T}$ with a tuple $\boldsymbol{\mathcal{N}} = (\mathcal{N}_i \mid i \in \Lambda)$ of mutually commuting morphisms $\mathcal{N}_i : \mathcal{T} \longrightarrow \mathcal{T} \otimes \boldsymbol{T}(-1)$. A morphism $F : (\mathcal{T}_1,\boldsymbol{\mathcal{N}}) \longrightarrow (\mathcal{T}_2,\boldsymbol{\mathcal{N}})$ in Λ-$\mathcal{R}$-Tri(X,H) is defined to be a morphism F in $\mathcal{R}$-Tri(X,H) such that $F \circ \mathcal{N}_i = \mathcal{N}_i \circ F$ for $i \in \Lambda$.

For $I \subset \Lambda$, we put $X\{I\} := X \times \mathbb{C}^I$ and $H\{I\} := \big(H \times \mathbb{C}^I\big) \cup \bigcup_{i\in I}\{z_i = 0\}$. We shall construct a functor

$$\mathrm{TNIL}_I : \Lambda\text{-}\mathcal{R}\text{-Tri}(X,H) \longrightarrow \Lambda\text{-}\mathcal{R}\text{-Tri}\big(X\{I\},H\{I\}\big) \tag{2.14}$$

Let $(\mathcal{T},\boldsymbol{\mathcal{N}}) \in \Lambda$-$\mathcal{R}$-Tri$(X,H)$. Let $\mathcal{T} = (\mathcal{M}',\mathcal{M}'',C)$. Recall that morphisms $\mathcal{N}_i$ are given as pairs $(\mathcal{N}_i',\mathcal{N}_i'')$, where

$$\mathcal{M}' \xleftarrow{\mathcal{N}_i'} \mathcal{M}' \cdot \lambda, \qquad \mathcal{M}'' \xrightarrow{\mathcal{N}_i''} \mathcal{M}'' \cdot \lambda^{-1}.$$

We set $\eth_i := \lambda \partial_{z_i}$. The underlying $\mathcal{R}_{X\{I\}(*H\{I\})}$-modules of $\mathrm{TNIL}_I(\mathcal{T}, \boldsymbol{N})$ are given by the natural actions of $\mathcal{R}_X$ and the actions of $z_i\eth_i$ as follows:

$$\mathcal{M}' \otimes \mathcal{O}_{\mathcal{X}\{I\}}(*H\{I\}), \quad z_i\eth_i m' = -\lambda\, \mathcal{N}_i' m',$$
$$\mathcal{M}'' \otimes \mathcal{O}_{\mathcal{X}\{I\}}(*H\{I\}), \quad z_i\eth_i m'' = -\lambda\, \mathcal{N}_i'' m''.$$

Here, m' and m'' denote the sections $m' \otimes 1$ and $m'' \otimes 1$. We would like to define the pairing $\tilde{C}$ of $\mathrm{TNIL}_I(\mathcal{T}, \boldsymbol{N})$ by the following formula:

$$\tilde{C}(m', \sigma^* m'') = C\Big(\exp\Big(\sum_{i\in I} -\mathcal{N}_i' \log |z_i|^2\Big) m',\ \sigma^* m''\Big) \tag{2.15}$$

Because $C\big(\mathcal{N}_i' m', \sigma^* m''\big) = C\big(m', \sigma^*(\mathcal{N}_i'' m'')\big)$, the right hand side of (2.15) is equal to the following:

$$C\Big(m',\ \sigma^*\Big(\exp\Big(\sum_{i\in I} -\mathcal{N}_i'' \log |z_i|^2\Big) m''\Big)\Big)$$

Let us prove that $\tilde{C}$ is an $\mathcal{R}_{X\{I\}(*H\{I\})} \times \sigma^* \mathcal{R}_{X\{I\}(*H\{I\})}$-homomorphism. For that purpose, it is enough to prove the following equalities:

$$z_i\partial_i \tilde{C}(m', \sigma^* m'') = \tilde{C}(z_i\partial_i m', \sigma^* m'')$$
$$\overline{z}_i\partial_{\overline{z}_i} \tilde{C}(m', \sigma^* m'') = \tilde{C}\big(m', \sigma^*(z_i\partial_i m'')\big) \tag{2.16}$$

We have the following:

$$\begin{aligned} z_i\partial_i \tilde{C}(m', \sigma^* m'') &= z_i\partial_i C\Big(\exp\Big(-\sum_{j\in I} \mathcal{N}_j' \log |z_j|^2\Big) m', \sigma^* m''\Big) \\ &= C\Big(-\exp\Big(-\sum_{j\in I} \mathcal{N}_j' \log |z_j|^2\Big)\, \mathcal{N}_i' m',\ \sigma^* m''\Big) \\ &= \tilde{C}(z_i\partial_i m', \sigma^* m'') \end{aligned} \tag{2.17}$$

We also have the following:

$$\begin{aligned} \overline{z}_i\partial_{\overline{z}_i} \tilde{C}(m', \sigma^* m'') &= \overline{z}_i\partial_{\overline{z}_i} C\Big(m', \sigma^*\Big(\exp\Big(-\sum_{j\in I} \mathcal{N}_j'' \log |z_i|^2\Big) m''\Big)\Big) \\ &= C\Big(m', \sigma^*\Big(-\exp\Big(-\sum_{j\in I} \mathcal{N}_j'' \log |z_j|^2\Big)\, \mathcal{N}_i'' m''\Big)\Big) \\ &= \tilde{C}\big(m', \sigma^*(z_i\partial_i m'')\big) \end{aligned} \tag{2.18}$$

Hence, we obtain (2.16), and $\tilde{C}$ gives a sesqui-linear pairing of $\mathcal{M}'$ and $\mathcal{M}''$.

By using the commutativity of $\mathcal{N}_j$ and other $\mathcal{N}_i$ ($i \in I$), we obtain $\tilde{C}(\mathcal{N}_j' m', \sigma^* m'') = \tilde{C}(m', \sigma^* \mathcal{N}_j'' m'')$. Hence, we have morphisms

$$\mathrm{TNIL}_I(\mathcal{N}_j) : \mathrm{TNIL}_I(\mathcal{T}, \boldsymbol{\mathcal{N}}) \longrightarrow \mathrm{TNIL}_I(\mathcal{T}, \boldsymbol{\mathcal{N}}) \otimes \boldsymbol{T}(-1)$$

for $j \in \Lambda$. In particular, an $\mathcal{R}$-triple $\mathrm{TNIL}_I(\mathcal{T}, \boldsymbol{\mathcal{N}})$ with $\mathrm{TNIL}_I(\boldsymbol{\mathcal{N}}) = \big(\mathrm{TNIL}_I(\mathcal{N}_i) \,|\, i \in \Lambda\big)$ is an object of Λ-$\mathcal{R}$-$\mathrm{Tri}\big(X\{I\}, H\{I\}\big)$. Thus, we obtain the functor (2.14).

It is easy to see that $\mathrm{TNIL}_I(\mathcal{T}, \boldsymbol{\mathcal{N}})$ is strictly specializable along z_i ($i \in I$), and we have

$$\tilde{\psi}_{z_i, -\boldsymbol{\delta}}\, \mathrm{TNIL}_I(\mathcal{T}, \boldsymbol{\mathcal{N}}) = \mathrm{TNIL}_{I \setminus i}(\mathcal{T}, \boldsymbol{\mathcal{N}}).$$

The following lemma is easy to see.

Lemma 2.2.1 *If $(\mathcal{T}, \boldsymbol{\mathcal{N}})$ is integrable, $\mathrm{TNIL}_I(\mathcal{T}, \boldsymbol{\mathcal{N}})$ is naturally integrable.* □

2.2.1.1 Hermitian Adjoint

For $(\mathcal{T}, \boldsymbol{\mathcal{N}}) \in \Lambda$-$\mathcal{R}$-$\mathrm{Tri}(X, H)$, we set $(\mathcal{T}, \boldsymbol{\mathcal{N}})^* := (\mathcal{T}^*, -\boldsymbol{\mathcal{N}}^*)$, which gives a contravariant functor on Λ-$\mathcal{R}$-$\mathrm{Tri}(X, H)$.

Lemma 2.2.2 $\big(\mathrm{TNIL}_I(\mathcal{T}, \boldsymbol{\mathcal{N}})\big)^*$ *is naturally identified with* $\mathrm{TNIL}_I\big((\mathcal{T}, \boldsymbol{\mathcal{N}})^*\big)$.

Proof Let $\mathcal{T} = (\mathcal{M}', \mathcal{M}'', C)$ and $\mathcal{N}_i = (\mathcal{N}_i', \mathcal{N}_i'')$. Because the identification $c_{-1} : \boldsymbol{T}(-1)^* \simeq \boldsymbol{T}(1)$ is given by $(-1, -1)$, we have $-\mathcal{N}_i^* = (\mathcal{N}_i'', \mathcal{N}_i')$. Hence, the underlying $\mathcal{R}$-modules are naturally identified.

Let us compare the pairings. Let $\tilde{C}^*$ be the pairing for $\big(\mathrm{TNIL}_I(\mathcal{T}, \boldsymbol{\mathcal{N}})\big)^*$. We have

$$\begin{aligned}\tilde{C}^*(m'', \sigma^* m') &= \overline{\sigma^* \tilde{C}(m', \sigma^* m'')} = \overline{\sigma^* C\Big(\exp\Big(-\sum_{i \in I} \mathcal{N}_i' \log |z_i|^2\Big) m',\, \sigma^* m''\Big)} \\ &= C^*\Big(m'', \sigma^*\Big(\exp\Big(-\sum_{i \in I} \mathcal{N}_i' \log |z_i|^2\Big) m'\Big)\Big) \qquad (2.19)\end{aligned}$$

This is exactly the pairing for $\mathrm{TNIL}_I\big((\mathcal{T}, \boldsymbol{\mathcal{N}})^*\big)$. □

2.2.1.2 Duality

If $\mathcal{T} \in \mathrm{TS}(X, H)$, we set $(\mathcal{T}, \boldsymbol{\mathcal{N}})^\vee := (\mathcal{T}^\vee, -\boldsymbol{\mathcal{N}}^\vee)$.

Lemma 2.2.3 $\big(\mathrm{TNIL}_I(\mathcal{T}, \boldsymbol{\mathcal{N}})\big)^\vee$ *is identified with* $\mathrm{TNIL}_I\big((\mathcal{T}, \boldsymbol{\mathcal{N}})^\vee\big)$ *naturally.*

Proof The underlying $\mathcal{R}_X$-modules are naturally identified. Let us compare the pairings. For $\mathcal{T} = (\mathcal{M}', \mathcal{M}'', C)$, let Ψ_C and Φ_C be defined as in (2.12). Similarly, we obtain $\Psi_{\tilde{C}}$ and $\Phi_{\tilde{C}}$ from the pairing $\tilde{C}$. Note that we have $\Psi_C \circ \mathcal{N}_i' = \sigma^*(\mathcal{N}_i'')^\vee \circ \Psi_C$. Indeed,

$$\begin{aligned}\big\langle \Psi_C(\mathcal{N}_i' m'), \sigma^* m''\big\rangle &= C(\mathcal{N}_i' m', \sigma^* m'') = C\big(m', \sigma^*(\mathcal{N}_i'' m'')\big) \\ &= \big\langle \Psi_C(m'),\, \sigma^*(\mathcal{N}_i'' m'')\big\rangle = \big\langle \sigma^*(\mathcal{N}_i'')^\vee \Psi_C(m'),\, \sigma^* m''\big\rangle\end{aligned} \tag{2.20}$$

Similarly, we have $\Phi_C \circ \sigma^* \mathcal{N}_i'' = (\mathcal{N}_i')^\vee \circ \Phi_C$. We have

$$\Psi_{\tilde{C}} = \Psi_C \circ \exp\Big(-\sum_i \log|z_i|^2 \mathcal{N}_i'\Big) = \exp\Big(-\sum_i \log|z_i|^2 \sigma^*(\mathcal{N}_i'')^\vee\Big) \circ \Psi_C$$

Indeed, we have the equalities:

$$\begin{aligned}\langle \Psi_{\tilde{C}}(m'), \sigma^* m''\rangle &= \tilde{C}(m', \sigma^* m'') = C\Big(\exp\Big(-\sum_i \log|z_i|^2 \mathcal{N}_i'\Big) m',\, \sigma^* m''\Big) \\ &= \Big\langle \Psi_C\Big(\exp\Big(-\sum_i \log|z_i|^2 \mathcal{N}_i'\Big) m'\Big),\, \sigma^* m''\Big\rangle\end{aligned} \tag{2.21}$$

Similarly, $\Phi_{\tilde{C}} = \Phi_C \circ \exp\Big(-\sum_i \log|z_i|^2 \sigma^* \mathcal{N}_i''\Big) = \exp\Big(-\sum_i \log|z_i|^2 (\mathcal{N}_i')^\vee\Big) \circ \Phi_C$. Then, we obtain the following:

$$\begin{aligned}(\tilde{C})^\vee(n', \sigma^* n'') &= C\Big(\exp\Big(-\sum_i \log|z_i|^2 \mathcal{N}_i'\Big) \Psi_{\tilde{C}}^{-1}(\sigma^* n''),\, \Phi_{\tilde{C}}^{-1}(n')\Big) \\ &= C\Big(\Psi_C^{-1}(\sigma^* n''),\, \Phi_C^{-1}\Big(\exp\Big(\sum_i \log|z_i|^2 (\mathcal{N}_i')^\vee\Big) n'\Big)\Big) \\ &= C^\vee\Big(\exp\Big(\sum_i \log|z_i|^2 (\mathcal{N}_i')^\vee\Big) n',\, \sigma^* n''\Big) = \widetilde{(C^\vee)}(n', \sigma^* n'')\end{aligned} \tag{2.22}$$

Thus, we are done. □

2.2.1.3 Real Structure

We also have a natural isomorphism $j^* \operatorname{TNIL}_I(\mathcal{T}, \boldsymbol{\mathcal{N}}) \simeq \operatorname{TNIL}_I j^*(\mathcal{T}, \boldsymbol{\mathcal{N}})$. Hence, we have a natural isomorphism

$$\tilde{\gamma}_{\mathrm{sm}}^* \operatorname{TNIL}_I(\mathcal{T}, \boldsymbol{\mathcal{N}}) \simeq \operatorname{TNIL}_I \tilde{\gamma}_{\mathrm{sm}}^*(\mathcal{T}, \boldsymbol{\mathcal{N}}).$$

A real structure of an object $(\mathcal{T},\boldsymbol{\mathcal{N}})$ in Λ-$\mathcal{R}$-Tri(X,H) is an isomorphism κ : $\tilde{\gamma}^*_{\rm sm}(\mathcal{T},\boldsymbol{\mathcal{N}}) \simeq (\mathcal{T},\boldsymbol{\mathcal{N}})$ such that $\kappa \circ \tilde{\gamma}^*_{\rm sm}\kappa = \mathrm{id}$. A real structure of $(\mathcal{T},\boldsymbol{\mathcal{N}})$ naturally induces a real structure of $\mathrm{TNIL}_I(\mathcal{T},\boldsymbol{\mathcal{N}})$.

2.2.2 Variant

Let $(\mathcal{T},\boldsymbol{\mathcal{N}}) \in \Lambda$-$\mathcal{R}$-Tri$(X,H)$. Let $\boldsymbol{\varphi} = (\varphi_i \,|\, i \in \Lambda)$ be a tuple of meromorphic functions whose poles are contained in H. We shall construct an $\mathcal{R}_{X(*H)}$-triple $\mathrm{Def}_{\boldsymbol{\varphi}}(\mathcal{T},\boldsymbol{\mathcal{N}})$. We define a new $\mathcal{R}$-action on $\mathcal{M}'$ by $f \bullet m' = f m'$ for $f \in \mathcal{O}_{\mathcal{X}}$ and $v \bullet m' = v\, m' - \sum_i \mathcal{N}_i'(m')\, v(\varphi_i)$ for $v \in \lambda\, \Theta_X$. We define a new $\mathcal{R}$-action on $\mathcal{M}''$ in the same way. The $\mathcal{R}_{X(*H)}$-modules are denoted by $\mathcal{M}'_{(\boldsymbol{\varphi},\boldsymbol{\mathcal{N}})}$ and $\mathcal{M}''_{(\boldsymbol{\varphi},\boldsymbol{\mathcal{N}})}$. We define a sesqui-linear pairing $C_{(\boldsymbol{\varphi},\boldsymbol{\mathcal{N}})}$ of $\mathcal{M}'_{(\boldsymbol{\varphi},\boldsymbol{\mathcal{N}})}$ and $\mathcal{M}''_{(\boldsymbol{\varphi},\boldsymbol{\mathcal{N}})}$ as follows:

$$C_{(\boldsymbol{\varphi},\boldsymbol{\mathcal{N}})} := C\Big(\exp(-\boldsymbol{\varphi}\cdot\boldsymbol{\mathcal{N}}')\, m',\, \sigma^*\big(\exp(-\boldsymbol{\varphi}\cdot\boldsymbol{\mathcal{N}}'')\, m''\big)\Big)$$

Here, $\boldsymbol{\varphi}\cdot\boldsymbol{\mathcal{N}}' := \sum_{i\in\Lambda} \varphi_i \mathcal{N}_i'$ and $\boldsymbol{\varphi}\cdot\boldsymbol{\mathcal{N}}'' := \sum_{i\in\Lambda} \varphi_i \mathcal{N}_i''$. The triple is denoted by $\mathrm{Def}_{\boldsymbol{\varphi}}(\mathcal{T},\boldsymbol{\mathcal{N}})$. The following lemma can be checked as in Sect. 2.2.1.

Lemma 2.2.4 *We have natural isomorphisms*

$$\mathrm{Def}_{\boldsymbol{\varphi}}\big((\mathcal{T},\boldsymbol{\mathcal{N}})\big)^* \simeq \mathrm{Def}_{\boldsymbol{\varphi}}\big((\mathcal{T},\boldsymbol{\mathcal{N}})^*\big),$$

$$\mathrm{Def}_{\boldsymbol{\varphi}}\big((\mathcal{T},\boldsymbol{\mathcal{N}})\big)^\vee \simeq \mathrm{Def}_{\boldsymbol{\varphi}}\big((\mathcal{T},\boldsymbol{\mathcal{N}})^\vee\big),$$

$$j^*\,\mathrm{Def}_{\boldsymbol{\varphi}}(\mathcal{T},\boldsymbol{\mathcal{N}}) \simeq \mathrm{Def}_{\boldsymbol{\varphi}}\big(j^*(\mathcal{T},\boldsymbol{\mathcal{N}})\big).$$

In particular, we have $\tilde{\gamma}^*_{\rm sm}\,\mathrm{Def}_{\boldsymbol{\varphi}}(\mathcal{T},\boldsymbol{\mathcal{N}}) \simeq \mathrm{Def}_{\boldsymbol{\varphi}}\big(\tilde{\gamma}^*_{\rm sm}(\mathcal{T},\boldsymbol{\mathcal{N}})\big)$. □

A real structure of $(\mathcal{T},\boldsymbol{\mathcal{N}})$ naturally induces a real structure of $\mathrm{Def}_{\boldsymbol{\varphi}}(\mathcal{T},\boldsymbol{\mathcal{N}})$. If $(\mathcal{T},\boldsymbol{\mathcal{N}})$ is integrable, $\mathrm{Def}_{\boldsymbol{\varphi}}(\mathcal{T},\boldsymbol{\mathcal{N}})$ is also integrable. The following lemma can be checked directly.

Lemma 2.2.5 $\mathrm{Def}_{\boldsymbol{\varphi}'}\big(\mathrm{Def}_{\boldsymbol{\varphi}}(\mathcal{T},\boldsymbol{\mathcal{N}})\big) \simeq \mathrm{Def}_{\boldsymbol{\varphi}'+\boldsymbol{\varphi}}(\mathcal{T},\boldsymbol{\mathcal{N}})$ *naturally.* □

Let Y be a complex manifold with a hypersurface H_Y. Let

$$\iota_j : (Y, H_Y) \subset (X\{I\}, H\{I\}) \qquad (j = 1, 2)$$

be open embeddings such that (i) the compositions $Y \xrightarrow{\iota_j} X\{I\} \longrightarrow X$ are equal, where the latter is the natural projection, (ii) $\iota_1^*(z_i) = e^{\varphi_i}\iota_2^*(z_i)$ for $i \in I$, where φ_i are holomorphic functions on Y. The following lemma can be checked directly.

Lemma 2.2.6 $\iota_1^*\,\mathrm{TNIL}_I(\mathcal{T},\boldsymbol{\mathcal{N}}) \simeq \mathrm{Def}_{\boldsymbol{\varphi}}\Big(\iota_2^*\,\mathrm{TNIL}_I(\mathcal{T},\boldsymbol{\mathcal{N}})\Big)$ *naturally.* □

2.3 Beilinson Triples

We introduce a special type of smooth $\mathcal{R}$-triples, which we call Beilinson triples. It will be used in the construction of Beilinson functor in Chap. 4.

2.3.1 *Triples on a Point*

We put $A := \mathcal{O}_{\mathbb{C}_\lambda}[\lambda s, (\lambda s)^{-1}] = \bigoplus_n \mathcal{O}_{\mathbb{C}_\lambda} \lambda^n s^n$. We set $A^a := (\lambda s)^a \, \mathcal{O}[\lambda s] \subset A$. For $a \leq b$, we put

$$\mathbb{I}_1^{a,b} := A^{-b+1}/A^{-a+1} \simeq \bigoplus_{a \leq i < b} \mathcal{O}_{\mathbb{C}_\lambda} \cdot (\lambda s)^{-i}, \quad \mathbb{I}_2^{a,b} := A^a/A^b \simeq \bigoplus_{a \leq i < b} \mathcal{O}_{\mathbb{C}_\lambda} \cdot (\lambda s)^i.$$

Let $C_0^{(i)}$ be the sesqui-linear pairing of $\mathcal{O}_{\mathbb{C}_\lambda} \cdot (\lambda s)^{-i}$ and $\mathcal{O}_{\mathbb{C}_\lambda} \cdot (\lambda s)^i$ given by $C_0^{(i)}(f\, s^{-i}, \sigma^* g\, s^i) = f \cdot \overline{\sigma^* g}$. They induce a sesqui-linear pairing $C_{\mathbb{I}}$ of $\mathbb{I}_1^{a,b}$ and $\mathbb{I}_2^{a,b}$. The integrable $\mathcal{R}$-triple is denoted by $\mathbb{I}^{a,b}$. We have a natural identification $\mathbb{I}^{a,b} = \oplus_{a \leq i < b} \mathbb{I}^{i,i+1}$. We shall also use the identification $\mathbb{I}^{i,i+1} \simeq \boldsymbol{T}(i)$ given by $s^j \longleftrightarrow 1$ $(j = i, -i)$. We have natural isomorphisms $\Upsilon_j : \mathbb{I}^{a,b} \otimes \boldsymbol{T}(j) \simeq \mathbb{I}^{a+j,b+j}$ given by the multiplication of s^j.

The multiplication of $-s$ induces $\mathcal{N}_{\mathbb{I}} = (\mathcal{N}_{\mathbb{I},1}, \mathcal{N}_{\mathbb{I},2}) : \mathbb{I}^{a,b} \longrightarrow \mathbb{I}^{a,b} \otimes \boldsymbol{T}(-1)$:

$$\mathbb{I}_1^{a,b} \xleftarrow{\mathcal{N}_{\mathbb{I},1}} \mathbb{I}_1^{a,b} \otimes \mathcal{O}_{\mathbb{C}_\lambda} \lambda, \qquad \mathbb{I}_2^{a,b} \xrightarrow{\mathcal{N}_{\mathbb{I},2}} \mathbb{I}_2^{a,b} \otimes \mathcal{O}_{\mathbb{C}_\lambda} \lambda^{-1}.$$

2.3.1.1 Hermitian Adjoint

We shall use the identifications

$$\mathcal{S}^{a,b} : (\mathbb{I}^{a,b})^* \simeq \mathbb{I}^{-b+1,-a+1}$$

given by the natural correspondence $s^i \longmapsto s^i$. In other words, they are given by $\bigoplus_{a \leq i < b} (-1)^i c_i$, where $c_i : (\mathbb{I}^{i,i+1})^* \simeq \mathbb{I}^{-i,-i+1}$ are induced by $c_i : \boldsymbol{T}(i)^* \simeq \boldsymbol{T}(-i)$ given in Sect. 2.1.8. They give isomorphisms $\mathcal{S}^{a,b} : (\mathbb{I}^{a,b}, \mathcal{N}_{\mathbb{I}})^* \simeq (\mathbb{I}^{-b+1,-a+1}, \mathcal{N}_{\mathbb{I}})$. Note that $-\mathcal{N}_{\mathbb{I}}^* = (\mathcal{N}_{\mathbb{I},2}, \mathcal{N}_{\mathbb{I},1})$. Under the identifications, we have $\Upsilon_j^* = (-1)^j \Upsilon_{-j}$.

2.3.1.2 Duality

We shall use the identifications $(\mathbb{I}^{a,b})^\vee \simeq \mathbb{I}^{-b+1,-a+1}$ given by $\bigoplus_{a \leq i < b} (-1)^i d_i$, where $d_i : (\mathbb{I}^{i,i+1})^\vee \simeq \mathbb{I}^{-i,-i+1}$ are natural isomorphisms. Then, we have

$$(\mathbb{I}^{a,b}, \mathcal{N}_{\mathbb{I}})^\vee \simeq (\mathbb{I}^{-b+1,-a+1}, \mathcal{N}_{\mathbb{I}}).$$

Under the identifications, we have $\Upsilon_j^\vee = (-1)^j \Upsilon_{-j}$.

2.3.1.3 Real Structure

We naturally have $j^*(\mathbb{I}^{a,b}, \mathcal{N}_\mathbb{I}) \simeq (\mathbb{I}^{a,b}, \mathcal{N}_\mathbb{I})$. By composition of the above isomorphisms we obtain a real structure $\kappa_\mathbb{I} : \tilde{\gamma}_{\rm sm}^*(\mathbb{I}^{a,b}, \mathcal{N}_\mathbb{I}) \simeq (\mathbb{I}^{a,b}, \mathcal{N}_\mathbb{I})$. The real structure is the same as the direct sum $\bigoplus_{a \le i < b} \boldsymbol{T}(i)$, under the natural isomorphism $\mathbb{I}^{a,b} \simeq \bigoplus_{a \le i < b} \boldsymbol{T}(i)$.

2.3.2 *The Associated Twistor Nilpotent Orbit*

We obtain integrable $\mathcal{R}_{\mathbb{C}_t}(*t)$-triples $\tilde{\mathbb{I}}^{a,b} := \mathrm{TNIL}(\mathbb{I}^{a,b}, \mathcal{N}_\mathbb{I}) = \big(\tilde{\mathbb{I}}_1^{a,b}, \tilde{\mathbb{I}}_2^{a,b}, \tilde{C}_\mathbb{I}\big)$. The underlying $\mathcal{R}$-modules are

$$\tilde{\mathbb{I}}_1^{a,b} = \bigoplus_{a \le i < b} \mathcal{O}_{\mathbb{C}^2_{\lambda,t}}(*t) \cdot (\lambda\, s)^{-i}, \quad \text{and} \quad \tilde{\mathbb{I}}_2^{a,b} = \bigoplus_{a \le i < b} \mathcal{O}_{\mathbb{C}^2_{\lambda,t}}(*t) \cdot (\lambda\, s)^{i}$$

with the action $t\eth_t (\lambda\, s)^i = (\lambda\, s)^{i+1}$. The pairing $\tilde{C}_\mathbb{I}$ is given as follows:

$$\tilde{C}_\mathbb{I}\big((\lambda\, s)^i, \sigma^*(\lambda\, s)^j\big) = \frac{(\log |t|^2)^{-i-j}}{(-i-j)!} (-1)^j \lambda^{i-j} \cdot \chi_{i+j \le 0} \tag{2.23}$$

Here, $\chi_{i+j\le 0}$ is 1 if $i + j \le 0$, and 0 if $i + j > 0$. Indeed, we have the following equalities:

$$\begin{aligned}
\tilde{C}_\mathbb{I}\Big((\lambda\, s)^i,\, \sigma^*\big((\lambda\, s)^j\big)\Big) &= C_\mathbb{I}\Big(\exp(-\mathcal{N}_{\mathbb{I},1} \log |t|^2)\, (\lambda\, s)^i,\, \sigma^*\big((\lambda\, s)^j\big)\Big) \\
&= \sum_{k=0}^{\infty} \frac{(\log |t|^2)^k}{k!} C_\mathbb{I}\big((\lambda\, s)^i s^k,\, \sigma^*\big((\lambda\, s)^j\big)\big) \\
&= \sum_{k=0}^{\infty} \frac{(\log |t|^2)^k}{k!} \lambda^i (-\lambda)^{-j} \delta_{i+k+j,0}
\end{aligned} \tag{2.24}$$

Here, $\delta_{m,0}$ is 0 if $m \ne 0$ and 1 if $m = 0$. Then, we obtain (2.23).

We have the induced isomorphisms

$$(\tilde{\mathbb{I}}^{a,b})^* \simeq \tilde{\mathbb{I}}^{-b+1,-a+1}, \quad j^*\tilde{\mathbb{I}}^{a,b} \simeq \tilde{\mathbb{I}}^{a,b}, \quad (\tilde{\mathbb{I}}^{a,b})^\vee \simeq \tilde{\mathbb{I}}^{-b+1,-a+1}.$$

It is equipped with the induced real structure $\kappa : \tilde{\gamma}_{\rm sm}^{*}\tilde{\mathbb{I}}^{a,b} \simeq \tilde{\mathbb{I}}^{a,b}$.

2.3.2.1 Pull Back

Let X be a complex manifold. Let f be a holomorphic function on X. We obtain a smooth $\mathcal{R}(*f)$-triple $\tilde{\mathbb{I}}_f^{a,b} := f^{*}\tilde{\mathbb{I}}^{a,b}$. The following lemma is clear by construction.

Lemma 2.3.1 *Let $g = e^{\varphi} \cdot f$, where φ is holomorphic. We have a natural isomorphism $\tilde{\mathbb{I}}_g^{a,b} \simeq \mathrm{Def}_{\varphi}(\tilde{\mathbb{I}}_f^{a,b}, \mathcal{N})$.* □

2.3.3 *Appendix*

Let $\mathcal{M}$ be an $\mathcal{R}_X$-module with a nilpotent morphism $\mathcal{N} : \mathcal{M} \longrightarrow \mathcal{M}$ such that $\mathcal{N}^L = 0$ for some $L \in \mathbb{Z}_{>0}$. We put $\mathcal{N}''_{\mathbb{I}} := \lambda\, \mathcal{N}_{\mathbb{I},2}$. For integers $M_1 \leq M_2$, we consider the morphism $\mathcal{N} \otimes \mathrm{id} + \mathrm{id} \otimes \mathcal{N}''_{\mathbb{I}} : \mathcal{M} \otimes \mathbb{I}_2^{M_1,M_2} \longrightarrow \mathcal{M} \otimes \mathbb{I}_2^{M_1,M_2}$. The inclusion $\mathbb{I}_2^{M_1,M_1+1} \longrightarrow \mathbb{I}_2^{M_1,M_2}$ induces the following morphism:

$$\mathcal{M} \otimes \mathbb{I}_2^{M_1,M_1+1} \longrightarrow \mathrm{Cok}\Bigl(\mathcal{M} \otimes \mathbb{I}_2^{M_1,M_2} \xrightarrow{\mathcal{N}\otimes\mathrm{id} + \mathrm{id}\otimes\mathcal{N}''_{\mathbb{I}}} \mathcal{M} \otimes \mathbb{I}_2^{M_1,M_2}\Bigr) \tag{2.25}$$

The projection $\mathbb{I}_2^{M_1,M_2} \longrightarrow \mathbb{I}_2^{M_2-1,M_2}$ induces the following morphism:

$$\mathrm{Ker}\Bigl(\mathcal{M} \otimes \mathbb{I}_2^{M_1,M_2} \xrightarrow{\mathcal{N}\otimes\mathrm{id} + \mathrm{id}\otimes\mathcal{N}''_{\mathbb{I}}} \mathcal{M} \otimes \mathbb{I}_2^{M_1,M_2}\Bigr) \longrightarrow \mathcal{M} \otimes \mathbb{I}_2^{M_2-1,M_2} \tag{2.26}$$

The following lemma can be checked by direct computations.

Lemma 2.3.2 *If $M_2 - M_1 > 2L$, then the morphisms (2.25) and (2.26) are isomorphisms.* □

Let $a \leq b$. Let $M > \max\{|a|, |b|\} + 2L$. We have the following natural commutative diagrams:

$$\begin{array}{ccccccc}
\mathrm{Ker}_1 & \longrightarrow & \mathcal{M}\otimes \mathbb{I}_2^{b,M+1} & \xrightarrow{\mathcal{N}\otimes \mathrm{id}+\mathrm{id}\otimes \mathcal{N}_{\mathbb{I}}''} & \mathcal{M}\otimes \mathbb{I}_2^{a,M+1} & \longrightarrow & \mathrm{Cok}_1\\
\varphi_1\downarrow & & \downarrow & & \downarrow & & \varphi_2\downarrow\\
\mathrm{Ker}_2 & \longrightarrow & \mathcal{M}\otimes \mathbb{I}_2^{b,M} & \xrightarrow{\mathcal{N}\otimes \mathrm{id}+\mathrm{id}\otimes \mathcal{N}_{\mathbb{I}}''} & \mathcal{M}\otimes \mathbb{I}_2^{a,M} & \longrightarrow & \mathrm{Cok}_2
\end{array} \tag{2.27}$$

$$\begin{array}{ccccccc}
\mathrm{Ker}_3 & \longrightarrow & \mathcal{M}\otimes \mathbb{I}_2^{-M,b} & \xrightarrow{\mathcal{N}\otimes \mathrm{id}+\mathrm{id}\otimes \mathcal{N}_{\mathbb{I}}''} & \mathcal{M}\otimes \mathbb{I}_2^{-M,a} & \longrightarrow & \mathrm{Cok}_3\\
\varphi_3\downarrow & & \downarrow & & \downarrow & & \varphi_4\downarrow\\
\mathrm{Ker}_4 & \longrightarrow & \mathcal{M}\otimes \mathbb{I}_2^{-M-1,b} & \xrightarrow{\mathcal{N}\otimes \mathrm{id}+\mathrm{id}\otimes \mathcal{N}_{\mathbb{I}}''} & \mathcal{M}\otimes \mathbb{I}_2^{-M-1,a} & \longrightarrow & \mathrm{Cok}_4
\end{array} \tag{2.28}$$

We easily obtain the following lemma.

Lemma 2.3.3 *The morphisms φ_i $(i = 2, 3)$ are isomorphisms.* □

We mention variants of Lemma 2.3.2. Let $M > \max\{|a|, |b|\} + 2L$. Let ψ_1 be the composite of the following morphisms:

$$\mathcal{M}\otimes \mathbb{I}_2^{-M,b} \xrightarrow{\text{projection}} \mathcal{M}\otimes \mathbb{I}_2^{-M,a} \xrightarrow{\mathcal{N}\otimes \mathrm{id}+\mathrm{id}\otimes \mathcal{N}_{\mathbb{I}}''} \mathcal{M}\otimes \mathbb{I}_2^{-M,a}$$

Let ψ_2 be the composite of the following morphisms:

$$\mathcal{M}\otimes \mathbb{I}_2^{b,M} \xrightarrow{\mathcal{N}\otimes \mathrm{id}+\mathrm{id}\otimes \mathcal{N}_{\mathbb{I}}''} \mathcal{M}\otimes \mathbb{I}_2^{b,M} \xrightarrow{\text{inclusion}} \mathcal{M}\otimes \mathbb{I}_2^{a,M}$$

The inclusion $\mathcal{M}\otimes\mathbb{I}_2^{-M,-M+1} \longrightarrow \mathcal{M}\otimes\mathbb{I}_2^{-M,a}$ induces $\mathcal{M}\otimes\mathbb{I}_2^{-M,-M+1} \longrightarrow \mathrm{Cok}\,\psi_1$. The natural projection $\mathcal{M}\otimes\mathbb{I}_2^{b,M} \longrightarrow \mathcal{M}\otimes\mathbb{I}_2^{M-1,M}$ induces $\mathrm{Ker}\,\psi_2 \longrightarrow \mathcal{M}\otimes\mathbb{I}_2^{M-1,M}$. The following lemma can be checked easily.

Lemma 2.3.4 *The morphisms $\mathcal{M}\otimes\mathbb{I}_2^{-M,-M+1} \longrightarrow \mathrm{Cok}\,\psi_1$ and $\mathrm{Ker}\,\psi_2 \longrightarrow \mathcal{M}\otimes\mathbb{I}_2^{M-1,M}$ are isomorphisms.* □

Chapter 3
Canonical Prolongations

Let X be a complex manifold with a hypersurface D. Let M be a holonomic $\mathscr{D}$-module on a complex manifold X. We have the canonically defined $\mathscr{D}$-modules $M[*D] := M \otimes \mathcal{O}_X(*D)$ and $M[!D] := \boldsymbol{D}_X\big((\boldsymbol{D}_X M)(*D)\big)$, where $\boldsymbol{D}_X$ denotes the duality functor of $\mathscr{D}_X$-modules. They are also denoted as $M(\star D)$ $(\star = *, !)$.

As we will recall in Chap. 4, the functors $[\star D]$ $(\star = *, !)$ are the most basic in various operations along the hypersurface D. Namely, we may construct the maximal functor, the nearby cycle functor, the vanishing cycle functor, the gluing procedure by using the functors $[\star D]$ within the abelian category of holonomic $\mathscr{D}$-modules.

It is fundamental for us to construct such functors in the context of $\mathcal{R}$-triples. In this chapter, we mainly consider the case that $D = \{t = 0\}$ for a coordinate function t on X, i.e., dt is nowhere vanishing. In this case, we can construct $\mathcal{T}[\star t]$ $(\star = *, !)$ for any $\mathcal{R}_X$-triples $\mathcal{T}$ which are strictly specializable along t. We study the issue for $\mathcal{R}_X$-modules in Sect. 3.1, and we study the issue for $\mathcal{R}_X$-triples in Sect. 3.2.

Note that the construction of $\mathcal{M}[\star t]$ for an $\mathcal{R}_X$-module $\mathcal{M}$ are not given as in the case of holonomic $\mathscr{D}$-modules. Indeed, $\mathcal{M}[*t]$ is not necessarily equal to $\mathcal{M}(*t) = \mathcal{M} \otimes_{\mathcal{O}_{\mathcal{X}}} \mathcal{O}_{\mathcal{X}}(*t)$ because $\mathcal{M}(*t)$ is not $\mathcal{R}_X$-coherent in general. We cannot use the duality in the construction of $\mathcal{M}[!t]$ at this stage, although we will later see $\mathcal{M}[!t] = \boldsymbol{D}\big((\boldsymbol{D}\mathcal{M})[*t]\big)$ for any $\mathcal{R}_X$-module $\mathcal{M}$ underlying a mixed twistor $\mathscr{D}$-module. (See Chap. 13.) Instead, we construct $\mathcal{M}[\star t]$ $(\star = *, !)$ directly in terms of V-filtrations.

In Sect. 3.3, we study the functors $[\star D]$ for $\mathcal{R}$-triples in more general cases. When $D = \{g = 0\}$ for a general holomorphic function g on X, it is standard to consider the graph $\iota_g : X \longrightarrow X \times \mathbb{C}$. We have $(\iota_{g\dagger}\mathcal{T})[\star t]$ $(\star = *, !)$, where t is the standard coordinate of $\mathbb{C}$. If we have $\mathcal{R}_X$-triples $\mathcal{T}'_\star$ such that $\iota_{g\dagger}\mathcal{T}'_\star = (\iota_{g\dagger}\mathcal{T})[\star t]$, it is appropriate to set $\mathcal{T}[\star g] := \mathcal{T}'_\star$. Later, we shall prove that such $\mathcal{T}'_\star$ always exist if the $\mathcal{R}_X$-triple $\mathcal{T}$ underlies a mixed twistor $\mathscr{D}$-module. (See Sect. 11.2.) We do not study whether such $\mathcal{T}'_\star$ always exist. Instead, we regard it as a condition for $\mathcal{R}$-triples along g. It is naturally generalized to a condition along any effective divisor.

T. Mochizuki, *Mixed Twistor $\mathscr{D}$-modules*, Lecture Notes in Mathematics 2125,
DOI 10.1007/978-3-319-10088-3_3

3.1 Canonical Prolongations of $\mathcal{R}(*t)$-Modules

First, we study the canonical prolongations of $\mathcal{R}(*t)$-modules, for a coordinate function t, i.e., t is a holomorphic function whose derivative dt is nowhere vanishing. Then, the canonical prolongations of $\mathcal{R}$-modules $\mathcal{M}$ are given as the canonical prolongation of $\mathcal{M}(*t)$, as in Sect. 3.1.6.

3.1.1 *Strictly Specializable $\mathcal{R}(*t)$-Modules*

Let X_0 be a complex manifold, and let X be an open subset of $X_0 \times \mathbb{C}_t$. We shall often identify X_0 and $X_0 \times \{0\}$. Let H be a hypersurface of X such that $(X_0 \times \{0\}) \not\subset H$. Let $\mathcal{M}$ be a coherent $\mathcal{R}_{X(*H)}(*t)$-module which is strictly specializable along t. Let $V^{(\lambda_0)} = \bigl(V_a^{(\lambda_0)} \,\big|\, a \in \mathbb{R}\bigr)$ be the V-filtration of $\mathcal{M}$ at λ_0 on a neighbourhood $\mathcal{X}^{(\lambda_0)}$ of $\{\lambda_0\} \times X$, i.e., (i) each $V_a^{(\lambda_0)}(\mathcal{M})$ is $V_0\mathcal{R}_{X(*H)}$-coherent, (ii) we have $V_a^{(\lambda_0)}(\mathcal{M}) \subset V_b^{(\lambda_0)}(\mathcal{M})$ for $a < b$, $\bigcup_{a\in\mathbb{R}} V_a^{(\lambda_0)}(\mathcal{M}) = \mathcal{M}$, and $V_a^{(\lambda_0)}(\mathcal{M}) = \bigcap_{a<b} V_b^{(\lambda_0)}(\mathcal{M})$, (iii) each $\mathrm{Gr}_a^{V^{(\lambda_0)}}(\mathcal{M}) = V_a^{(\lambda_0)}(\mathcal{M})/V_{<a}^{(\lambda_0)}(\mathcal{M})$ is strict, (iv) $t\, V_a^{(\lambda_0)}(\mathcal{M}) = V_{a-1}^{(\lambda_0)}(\mathcal{M})$, $\eth_t V_a^{(\lambda_0)}(\mathcal{M}) \subset V_{a+1}^{(\lambda_0)}(\mathcal{M})$, (v) for each $P \in X_0$, there exists a discrete subset $\mathcal{S} \subset \mathbb{R} \times \mathbb{C}$ such that

$$\prod_{\substack{u\in\mathcal{S} \\ \mathfrak{p}(\lambda_0,u)=a}} \bigl(-\eth_t t + \mathfrak{e}(\lambda, u)\bigr)$$

is nilpotent on $\mathrm{Gr}_a^{V^{(\lambda_0)}}(\mathcal{M})$ around P. Let $\mathcal{KMS}(\mathcal{M}, P)$ denote the minimum among such $\mathcal{S}$. Note that $\mathcal{KMS}(\mathcal{M}, P) \subset \mathbb{R} \times \mathbb{C}$ is invariant under the natural action of $\mathbb{Z} \times \{0\}$ on $\mathbb{R} \times \mathbb{C}$. Let $\mathcal{KMS}(\mathcal{M}, t) := \bigcup_{P\in X_0} \mathcal{KMS}(\mathcal{M}, P)$.

We have the decomposition

$$\mathrm{Gr}_a^{V^{(\lambda_0)}}(\mathcal{M}) = \bigoplus_{\substack{u\in\mathbb{R}\times\mathbb{C} \\ \mathfrak{p}(\lambda_0,u)=a}} \psi_{t,u}^{(\lambda_0)}(\mathcal{M}),$$

such that $-\eth_t t + \mathfrak{e}(\lambda, u)$ is locally nilpotent on $\psi_{t,u}^{(\lambda_0)}(\mathcal{M})$. For a section $f \in \mathcal{M}$, we have $\inf\bigl\{a \in \mathbb{R} \,\big|\, f \in V_a^{(\lambda_0)}(\mathcal{M})\bigr\}$ in $\mathbb{R} \cup \{-\infty\}$, which is denoted by $\deg^{V^{(\lambda_0)}}(f)$. It is also denoted by $\deg(f)$, if there is no risk of confusion.

3.1.2 *The $\mathcal{R}$-Module $\mathcal{M}[*t]$*

Let $\mathcal{M}^{(\lambda_0)}[*t]$ be the $\mathcal{R}_{X(*H)}$-submodule of $\mathcal{M}_{|\mathcal{X}^{(\lambda_0)}}$ generated by $V_0^{(\lambda_0)}\mathcal{M}$.

Lemma 3.1.1

- *$\mathcal{M}^{(\lambda_0)}[*t]$ is $\mathcal{R}_{X(*H)}$-coherent and strictly specializable. The V-filtration is given by*

$$V_a^{(\lambda_0)}(\mathcal{M}^{(\lambda_0)}[*t]) = \mathcal{M}^{(\lambda_0)}[*t] \cap V_a^{(\lambda_0)}(\mathcal{M}). \tag{3.1}$$

- *$V_a^{(\lambda_0)}$ are described as follows:*

$$V_a^{(\lambda_0)}(\mathcal{M}^{(\lambda_0)}[*t]) = \begin{cases} V_a^{(\lambda_0)}(\mathcal{M}) & (a \leq 0) \\ \eth_t V_{a-1}^{(\lambda_0)}(\mathcal{M}^{(\lambda_0)}[*t]) + V_{<a}^{(\lambda_0)}(\mathcal{M}^{(\lambda_0)}[*t]) & (a > 0) \end{cases}$$

In particular, we have

$$\eth_t : \psi_u^{(\lambda_0)}(\mathcal{M}^{(\lambda_0)}[*t]) \xrightarrow{\simeq} \psi_{u+\delta}^{(\lambda_0)}(\mathcal{M}^{(\lambda_0)}[*t]), \qquad \mathfrak{p}(\lambda_0, u) > -1 \tag{3.2}$$

$$t : \psi_u^{(\lambda_0)}(\mathcal{M}^{(\lambda_0)}[*t]) \xrightarrow{\simeq} \psi_{u-\delta}^{(\lambda_0)}(\mathcal{M}^{(\lambda_0)}[*t]), \qquad \mathfrak{p}(\lambda_0, u) \leq 0 \tag{3.3}$$

Proof It is clear that $\mathcal{M}^{(\lambda_0)}[*t]$ is $\mathcal{R}_{X(*H)}$-coherent. Let $V^{(\lambda_0)}(\mathcal{M}^{(\lambda_0)}[*t])$ be the filtration given by (3.1). Then, we clearly have $tV_a^{(\lambda_0)}(\mathcal{M}^{(\lambda_0)}[*t]) \subset V_{a-1}^{(\lambda_0)}(\mathcal{M}^{(\lambda_0)}[*t])$ and $\eth_t V_a^{(\lambda_0)}(\mathcal{M}^{(\lambda_0)}[*t]) \subset V_{a+1}^{(\lambda_0)}(\mathcal{M}^{(\lambda_0)}[*t])$, and the submodule $\mathrm{Gr}_a^{V^{(\lambda_0)}}(\mathcal{M}^{(\lambda_0)}[*t]) \subset \mathrm{Gr}_a^{V^{(\lambda_0)}}(\mathcal{M})$ is strict. We have the decomposition

$$\mathrm{Gr}_a^{V^{(\lambda_0)}}(\mathcal{M}^{(\lambda_0)}[*t]) = \bigoplus_{\substack{u \in \mathbb{R}\times\mathbb{C} \\ \mathfrak{p}(\lambda_0,u)=a}} \psi_u^{(\lambda_0)}(\mathcal{M}^{(\lambda_0)}[*t])$$

such that $-\eth_t t + \mathfrak{e}(\lambda, u)$ are nilpotent on $\psi_u^{(\lambda_0)}(\mathcal{M}^{(\lambda_0)}[*t])$. For a fixed point $P \in X$, we have $\psi_u^{(\lambda_0)}(\mathcal{M}^{(\lambda_0)}[*t]) \neq 0$ around P if and only if $u \in \mathrm{KMS}(\mathcal{M}, P)$.

Let $a \leq 0$. Because $V_a^{(\lambda_0)}(\mathcal{M}) \subset \mathcal{M}^{(\lambda_0)}[*t]$, we have $V_a^{(\lambda_0)}(\mathcal{M}^{(\lambda_0)}[*t]) = V_a^{(\lambda_0)}(\mathcal{M})$. They are $V_0\mathcal{R}_{X(*H)}$-coherent, and we have $t \cdot V_a^{(\lambda_0)}(\mathcal{M}^{(\lambda_0)}[*t]) = V_{a-1}^{(\lambda_0)}(\mathcal{M}^{(\lambda_0)}[*t])$ for $a \leq 0$.

Take $u \in \mathrm{KMS}(\mathcal{M})$ such that $\mathfrak{e}(\lambda_0, u) \neq 0$. The induced action of $-\eth_t t$ on $\psi_u^{(\lambda_0)}(\mathcal{M}^{(\lambda_0)}[*t])$ is invertible, and hence $t : \psi_u^{(\lambda_0)}(\mathcal{M}^{(\lambda_0)}[*t]) \longrightarrow \psi_{u-\delta}^{(\lambda_0)}(\mathcal{M}^{(\lambda_0)}[*t])$ is injective, and $\eth_t : \psi_{u-\delta}^{(\lambda_0)}(\mathcal{M}^{(\lambda_0)}[*t]) \longrightarrow \psi_u^{(\lambda_0)}(\mathcal{M}^{(\lambda_0)}[*t])$ is surjective.

If $\mathfrak{p}(\lambda_0, u) > -1$, the function $\mathfrak{e}(\lambda, u + \delta)$ of λ is not constantly 0. The action of $-t\eth_t + \mathfrak{e}(\lambda, u + \delta)$ on $\psi_u^{(\lambda_0)}(\mathcal{M})$ is locally nilpotent. Hence, the action of $-t\eth_t$ on $\psi_u^{(\lambda_0)}(\mathcal{M})$ is injective. Therefore, $\eth_t : \psi_u^{(\lambda_0)}(\mathcal{M}) \longrightarrow \psi_{u+\delta}^{(\lambda_0)}(\mathcal{M})$ is injective. It implies that $\eth_t : \mathrm{Gr}_a^{V^{(\lambda_0)}}(\mathcal{M}) \longrightarrow \mathrm{Gr}_{a+1}^{V^{(\lambda_0)}}(\mathcal{M})$ is injective for any $a > -1$. Hence, if $[g] \neq 0$ in $\mathrm{Gr}_a^{V^{(\lambda_0)}}(\mathcal{M})$ for some $-1 < a \leq 0$, then $[\eth_t^N g] \neq 0$ in $\mathrm{Gr}_{a+N}^{V^{(\lambda_0)}}(\mathcal{M})$.

For any $f \in \mathcal{M}^{(\lambda_0)}[*t]$, we have an expression $f = \sum_{j=0}^{N} \eth_t^j f_j$, where $f_j \in V_0^{(\lambda_0)} \mathcal{M}^{(\lambda_0)}[*t]$. We can assume $-1 < \deg(f_j) \leq 0$ for $j \geq 1$. If f is not contained in $V_0^{(\lambda_0)}$, we have $j + \deg(f_j) < N + \deg(f_N)$ for any $j < N$.

Let us prove $V_a^{(\lambda_0)} \mathcal{M}^{(\lambda_0)}[*t] = V_{<a}^{(\lambda_0)} \mathcal{M}^{(\lambda_0)}[*t] + \eth_t V_{a-1}^{(\lambda_0)} \mathcal{M}^{(\lambda_0)}[*t]$ for $a > 0$. Take $f \in V_a^{(\lambda_0)} \mathcal{M}^{(\lambda_0)}[*t]$ such that the induced section $[f]$ of $\mathrm{Gr}_a^{V^{(\lambda_0)}} \big(\mathcal{M}^{(\lambda_0)}[*t]\big)$ is non-zero. Then, we have $N + \deg(f_N) = a$, and $[f] = [\eth_t^N f_N]$ in $\mathrm{Gr}_a^{V^{(\lambda_0)}} \mathcal{M}^{(\lambda_0)}[*t]$. It follows that $V_a^{(\lambda_0)} \mathcal{M}^{(\lambda_0)}[*t] = V_{<a}^{(\lambda_0)} \mathcal{M}[*t] + \eth_t V_{a-1}^{(\lambda_0)} \mathcal{M}^{(\lambda_0)}[*t]$. In particular, we obtain that $V_a^{(\lambda_0)} \mathcal{M}[*t]$ $(a > 0)$ are $V_0 \mathcal{R}_{X(*H)}$-coherent. Hence the filtration $V^{(\lambda_0)}$ gives a V-filtration. □

Lemma 3.1.2 *$\mathcal{R}_X \otimes_{V_0 \mathcal{R}_X} V_0^{(\lambda_0)} \mathcal{M} \simeq \mathcal{M}^{(\lambda_0)}[*t]$ naturally around any point of $\{\lambda_0\} \times X$.*

Proof We put $\mathcal{M}_1 := \mathcal{R}_X \otimes_{V_0 \mathcal{R}_X} V_0^{(\lambda_0)} \mathcal{M}$. We have a naturally defined surjection $\mathcal{M}_1 \longrightarrow \mathcal{M}^{(\lambda_0)}[*t]$. The composite of the morphisms $V_0^{(\lambda_0)} \mathcal{M} \longrightarrow \mathcal{M}_1 \longrightarrow \mathcal{M}$ is injective, where the first morphism is given by $m \longmapsto 1 \otimes m$. For $a \leq 0$, let $V_a^{(\lambda_0)}(\mathcal{M}_1)$ denote the image of $V_a^{(\lambda_0)} \mathcal{M}$. For $a > 0$, we define $V_a^{(\lambda_0)}(\mathcal{M}_1) := \sum_{\substack{b+n \leq a \\ b \leq 0, n \in \mathbb{Z}_{\geq 0}}} \eth_t^n V_b^{(\lambda_0)}(\mathcal{M}_1)$. For $a > 0$, we have $-1 < a_0 \leq 0$ such that $n := a - a_0 \in \mathbb{Z}$. Then, the following diagram is commutative:

$$\begin{array}{ccc}
\mathrm{Gr}_{a_0}^{V^{(\lambda_0)}} \mathcal{M}_1 & \xrightarrow{\simeq} & \mathrm{Gr}_{a_0}^{V^{(\lambda_0)}} \mathcal{M}[*t] \\
{\scriptstyle \eth_t^n} \downarrow & & {\scriptstyle \eth_t^n} \downarrow \simeq \\
\mathrm{Gr}_{a}^{V^{(\lambda_0)}} \mathcal{M}_1 & \longrightarrow & \mathrm{Gr}_{a}^{V^{(\lambda_0)}} \mathcal{M}[*t]
\end{array}$$

Hence, we obtain $\mathrm{Gr}_{a_0}^{V^{(\lambda_0)}} \mathcal{M}_1 \longrightarrow \mathrm{Gr}_{a}^{V^{(\lambda_0)}} \mathcal{M}_1$ is injective. It is also surjective by construction. Hence, $\mathrm{Gr}_{a_0}^{V^{(\lambda_0)}} \mathcal{M}_1 \longrightarrow \mathrm{Gr}_{a}^{V^{(\lambda_0)}} \mathcal{M}_1$ and $\mathrm{Gr}_{a}^{V^{(\lambda_0)}} \mathcal{M}_1 \longrightarrow \mathrm{Gr}_{a}^{V^{(\lambda_0)}} \mathcal{M}[*t]$ are isomorphisms. Then, we obtain that $\mathcal{M}_1 \longrightarrow \mathcal{M}[*t]$ is an isomorphism. □

The following obvious lemma will be used implicitly.

Lemma 3.1.3 *Suppose $\mathfrak{e}(\lambda_0, u) \neq 0$ for any $u \in \mathcal{KMS}(\mathcal{M}, t) \setminus \{(0,0)\}$. Then, we have $\mathcal{M}^{(\lambda_0)}[*t] = \mathcal{M}_{|\mathcal{X}^{(\lambda_0)}}$.* □

We obtain a globally defined $\mathcal{R}$-module.

Lemma 3.1.4 *Let λ_1 be sufficiently close to λ_0, and let $\mathcal{X}^{(\lambda_1)} \subset \mathcal{X}^{(\lambda_0)}$ be a neighbourhood of $\{\lambda_1\} \times X$. Then, we have $\mathcal{M}^{(\lambda_1)}[*t] = \mathcal{M}^{(\lambda_0)}[*t]_{|\mathcal{X}^{(\lambda_1)}}$. Therefore, we have a globally defined $\mathcal{R}_{X(*H)}$-module $\mathcal{M}[*t]$.*

Proof Take a sufficiently small $\epsilon > 0$. Then, $\mathcal{M}^{(\lambda_0)}[*t]$ is generated by $V_\epsilon^{(\lambda_0)}(\mathcal{M})$ over $\mathcal{R}_{X(*H)}$. If λ_1 is sufficiently close to λ_0, then we have $V_\epsilon^{(\lambda_0)}(\mathcal{M})_{|\mathcal{X}^{(\lambda_1)}} = V_\epsilon^{(\lambda_1)}(\mathcal{M})$. Then, we obtain the desired equality. □

The module $\mathcal{M}[*t]$ has the following universal property.

Lemma 3.1.5 *Let $\mathcal{M}_1$ be a coherent $\mathcal{R}_{X(*H)}$-module such that (i) $\mathcal{M}_1(*t) = \mathcal{M}$, (ii) $\mathcal{M}_1$ is strictly specializable along t. Then, the natural morphism $\mathcal{M}_1 \longrightarrow \mathcal{M}_1(*t) = \mathcal{M}$ factors through $\mathcal{M}[*t]$.*

Proof We have only to argue around $\{\lambda_0\} \times X$ for any λ_0. By a standard argument, we can prove that the image of $V_0^{(\lambda_0)}\mathcal{M}_1$ in $\mathcal{M}$ is contained in $V_0^{(\lambda_0)}\mathcal{M}$. Because $\mathcal{M}_1$ is generated by $V_0^{(\lambda_0)}\mathcal{M}_1$ over $\mathcal{R}_{X(*H)}$, the image of $\mathcal{M}_1$ in $\mathcal{M}$ is contained in $\mathcal{M}[*t]$. □

Corollary 3.1.6 *$\mathcal{M}[*t]$ is independent of the choice of a decomposition $X = X_0 \times \mathbb{C}_t$. Moreover, for any nowhere vanishing function A, we naturally have $\mathcal{M}[*t] \simeq \mathcal{M}[*(At)]$.* □

The following lemma is clear by construction.

Lemma 3.1.7 *If $\mathcal{M}$ is integrable, $\mathcal{M}[*t]$ is also naturally integrable.* □

Lemma 3.1.8 *If $\mathcal{M}$ is strict, then $\mathcal{M}[*t]$ is also strict.*

Proof Because $\mathcal{M}[*t] \subset \mathcal{M}(*t)$, the claim is clear. □

The following lemma is also clear by construction.

Lemma 3.1.9 *If $\lambda_0 \neq 0$ is generic for $(\mathcal{M}, t)$, then $(\mathcal{M}[*t])^{\lambda_0} = \mathcal{M}^{\lambda_0}(*t)$.* □

Indeed, the same assumption in Lemma 3.1.3 is enough.

3.1.3 The $\mathcal{R}$-Module $\mathcal{M}[!t]$

Let $\mathcal{M}^{(\lambda_0)}[!t]$ be the $\mathcal{R}_{X(*H)}$-module on $\mathcal{X}^{(\lambda_0)}$ defined as follows:

$$\mathcal{M}^{(\lambda_0)}[!t] := \mathcal{R}_{X(*H)} \otimes_{V_0\mathcal{R}_{X(*H)}} V_{<0}^{(\lambda_0)}\mathcal{M}$$

Because the composite of the natural morphisms $V_{<0}^{(\lambda_0)}\mathcal{M} \longrightarrow \mathcal{M}^{(\lambda_0)}[!t] \longrightarrow \mathcal{M}$ is injective, we can naturally regard $V_{<0}^{(\lambda_0)}\mathcal{M}$ as a $V_0\mathcal{R}_{X(*H)}$-submodule of $\mathcal{M}^{(\lambda_0)}[!t]$, which will be used implicitly.

Lemma 3.1.10

- *$\mathcal{M}^{(\lambda_0)}[!t]$ is $\mathcal{R}_{X(*H)}$-coherent and strictly specializable along t.*
- *We have a natural isomorphism $V_a^{(\lambda_0)}\big(\mathcal{M}^{(\lambda_0)}[!t]\big) \simeq V_a^{(\lambda_0)}(\mathcal{M})$ for any $a < 0$.*

- *The natural morphism* $\eth_t : \psi_{-\delta}^{(\lambda_0)}(\mathcal{M}^{(\lambda_0)}[!t]) \longrightarrow \psi_0^{(\lambda_0)}(\mathcal{M}^{(\lambda_0)}[!t])$ *is an isomorphism.*
- *We have a globally defined* $\mathcal{R}_{X(*H)}$*-module* $\mathcal{M}[!t]$.

Proof We have only to consider the issues locally on X, which we will implicitly use in the following argument. We consider the following $V_0\mathcal{R}_{X(*H)}$-submodules:

$$V_a^{(\lambda_0)}\mathcal{M}[!t] := \begin{cases} V_a^{(\lambda_0)}\mathcal{M} & (a<0) \\ \eth_t V_{a-1}^{(\lambda_0)}\mathcal{M}[!t] + V_{<a}^{(\lambda_0)}\mathcal{M}[!t] & (a \geq 0) \end{cases} \tag{3.4}$$

We clearly have (i) $tV_a^{(\lambda_0)} \subset V_{a-1}^{(\lambda_0)}$, (ii) $\eth_t V_a^{(\lambda_0)} \subset V_{a+1}^{(\lambda_0)}$, (iii) $\mathrm{Gr}_a^{V^{(\lambda_0)}}\mathcal{M}[!t] \simeq \mathrm{Gr}_a^{V^{(\lambda_0)}}\mathcal{M}$ for $a<0$, (iv) the natural morphism $\eth_t : \mathrm{Gr}_a^{V^{(\lambda_0)}}\mathcal{M}[!t] \longrightarrow \mathrm{Gr}_{a+1}^{V^{(\lambda_0)}}\mathcal{M}[!t]$ is onto for $a \geq -1$. Let us prove the claims (A1) $\mathrm{Gr}_a^{V^{(\lambda_0)}}(\mathcal{M}[!t])$ is strict, (A2) the induced action of

$$\prod_{\substack{u\in \mathcal{KMS}(\mathcal{M})\\ \mathfrak{p}(\lambda_0,u)=a}} (-\eth t + \mathfrak{e}(\lambda,u))$$

is nilpotent on $\mathrm{Gr}_a^{V^{(\lambda_0)}}(\mathcal{M}[!t])$. If $a<0$, they are clear by the construction.

Let us consider the case $a=0$. If $f \in V_0^{(\lambda_0)}\mathcal{M}[!t]$, we have $f = f_0 + \eth_t \otimes f_1$, where $f_0 \in V_{<0}^{(\lambda_0)}$ and $f_1 \in V_{-1}^{(\lambda_0)}$. For $\lambda \neq 0$, we put $\mathcal{M}^\lambda := \mathcal{M}_{|\{\lambda\}\times X}$, which can naturally be regarded as a $\mathscr{D}_X$-module. For general λ, we naturally have $\mathcal{M}^\lambda[!t] \simeq D_X \otimes_{V_0D_X} V_{<0}\mathcal{M}^\lambda$ for $\mathscr{D}$-modules. (See Appendix 3.1.3.1 below.) We have the following commutative diagram for any generic λ:

$$\begin{array}{ccc} V_0^{(\lambda_0)}\mathcal{M}[!t] & \xrightarrow{|\lambda} & V_0(\mathcal{M}^\lambda[!t]) \\ \eth_t\uparrow & & \eth_t\uparrow \\ V_{-1}^{(\lambda_0)}\mathcal{M} & \xrightarrow[|\lambda]{} & V_{-1}\mathcal{M}^\lambda \end{array}$$

If $\eth_t(f_{1|\lambda}) \in V_{<0}\mathcal{M}^\lambda[!t]$, we obtain $f_{1|\lambda} \in V_{<-1}\mathcal{M}^\lambda$. Hence, if $\eth_t \otimes f_1 \in V_{<0}^{(\lambda_0)}\mathcal{M}[!t]$, then we have $f_{1|\lambda} \in V_{<-1}\mathcal{M}^\lambda$ for any generic $\lambda \neq 0$. We obtain $f_1 \in V_{<-1}^{(\lambda_0)}\mathcal{M}$, because $\mathrm{Gr}_{-1}^{V^{(\lambda_0)}}\mathcal{M}$ is strict. Hence, the map $\eth_t : \mathrm{Gr}_{-1}^{V^{(\lambda_0)}}\mathcal{M}[!t] \longrightarrow \mathrm{Gr}_0^{V^{(\lambda_0)}}\mathcal{M}[!t]$ is an isomorphism. It implies (A1) and (A2) for $\mathrm{Gr}_0^{V^{(\lambda_0)}}(\mathcal{M}[!t])$.

Let us consider the case $a>0$. For $0<a\leq 1$, we obtain the injectivity of $t : \mathrm{Gr}_a^{V^{(\lambda_0)}}\mathcal{M}[!t] \longrightarrow \mathrm{Gr}_{a-1}^{V^{(\lambda_0)}}\mathcal{M}[!t]$ from the injectivity of $t\eth_t$ on $\mathrm{Gr}_{a-1}^{V^{(\lambda_0)}}\mathcal{M}[!t]$. Then, we obtain (A1) and (A2) for $\mathrm{Gr}_a^{V^{(\lambda_0)}}\mathcal{M}[!t]$. By an easy inductive argument, we obtain (A1) and (A2) for any $a \geq 0$. We also obtain that $\eth_t : \mathrm{Gr}_a^{V^{(\lambda_0)}}(\mathcal{M}) \longrightarrow \mathrm{Gr}_{a+1}^{V^{(\lambda_0)}}(\mathcal{M})$ is an isomorphism for $a>-1$.

Because $V^{(\lambda_0)}_{<0}\mathcal{M}$ is $V_0\mathcal{R}_{X(*H)}$-coherent, we obtain that $\mathcal{M}[!t]$ is $\mathcal{R}_{X(*H)}$-coherent. Note that $V_0\mathcal{R}_{X(*H)}$ and $\mathcal{R}_{X(*H)}$ are Noetherian. Because $V_a(\mathcal{M}[!t])$ are $\mathcal{O}_X(*H)$-pseudo-coherent and locally $V_0\mathcal{R}_{X(*H)}$-finitely generated, it is $V_0\mathcal{R}_{X(*H)}$-coherent. Thus, we obtain that $\mathcal{M}[!t]$ is strictly specializable along t, and the V-filtration is given by $V^{(\lambda_0)}$.

If $\epsilon > 0$ is sufficiently small, a natural morphism $\mathcal{R}_{X(*H)} \otimes_{V_0\mathcal{R}_{X(*H)}} V^{(\lambda_0)}_{<-\epsilon}\mathcal{M} \longrightarrow \mathcal{M}^{(\lambda_0)}[!t]$ is an isomorphism. If λ_1 is sufficiently close to λ_0, we have $V^{(\lambda_0)}_{<-\epsilon|\mathcal{X}^{(\lambda_1)}} = V^{(\lambda_1)}_{<-\epsilon}$. Hence, we obtain the global well definedness of $\mathcal{M}[!t]$. □

From the proof, we also obtain the following corollary.

Corollary 3.1.11 *If λ is generic for $(\mathcal{M}, t)$, the specialization $(\mathcal{M}[!t])^{\lambda}$ is naturally isomorphic to $\mathcal{M}^{\lambda}[!t]$.* □

We have the following universality.

Lemma 3.1.12 *Let $\mathcal{M}_1$ be a coherent strict $\mathcal{R}_{X(*H)}$-module such that (i) $\mathcal{M}_1$ is strictly specializable along t, (ii) $\mathcal{M}_1(*t) = \mathcal{M}$. Then, we have a uniquely defined morphism $\mathcal{M}[!t] \longrightarrow \mathcal{M}_1$.*

Proof We have only to consider the issue around $\{\lambda_0\} \times X$. We have $V^{(\lambda_0)}_a(\mathcal{M}_1) = V^{(\lambda_0)}_a(\mathcal{M})$ for any $a < 0$. In particular, $V^{(\lambda_0)}_{<0}\mathcal{M} \subset \mathcal{M}_1$. Hence, we obtain the uniquely induced morphism $\mathcal{M}[!t] = \mathcal{R}_{X(*H)} \otimes_{V_0\mathcal{R}_{X(*H)}} V^{(\lambda_0)}_{<0}\mathcal{M} \longrightarrow \mathcal{M}_1$. □

Corollary 3.1.13 *$\mathcal{M}[!t]$ is independent of the choice of a decomposition $X = X_0 \times \mathbb{C}_t$. Moreover, for any nowhere vanishing function A, we naturally have $\mathcal{M}[!t] \simeq \mathcal{M}[!(At)]$.* □

The following lemma is also clear by construction.

Lemma 3.1.14 *If $\mathcal{M}$ is integrable, $\mathcal{M}[!t]$ is also integrable.* □

Lemma 3.1.15 *If $\mathcal{M}$ is strict, $\mathcal{M}[!t]$ is also strict.*

Proof We have only to argue the issue locally around any point $(\lambda, P) \in \mathbb{C}_\lambda \times \{t = 0\}$. Because $V^{(\lambda_0)}_{<0}(\mathcal{M}) \subset \mathcal{M}(*t)$, it is strict. By construction, each $\psi_{t,u}(\mathcal{M}[!t])$ is strict. Hence, we obtain that $\mathcal{M}[!t]$ is strict. □

3.1.3.1 Appendix: The Case of $\mathscr{D}$-Modules

Let M be a coherent $\mathscr{D}_X$-module with a V-filtration along t. (We fix an appropriate total order $\leq_{\mathbb{C}}$ on $\mathbb{C}$.) For simplicity, we assume $M = M(*t)$. Assume that we are given morphisms of $\mathscr{D}_{X_0}$-modules $\psi_{-1}M \xrightarrow{u} Q \xrightarrow{v} \psi_0 M$ such that $v \circ u$ is equal to the induced map $\eth_t : \psi_{-1}M \longrightarrow \psi_0 M$. Although the following lemma is known, we give a construction by hand for our understanding.

Lemma 3.1.16 *We have a coherent $\mathscr{D}$-module $\tilde{M}$ with a V-filtration along t such that (i) $\tilde{M}(*t) = M$, (ii) we have $\psi_0(\tilde{M}) \simeq Q$, and the following commutative diagram:*

$$\begin{array}{ccccc} \psi_{-1}(\widetilde{M}) & \xrightarrow{\partial_t} & \psi_0(\widetilde{M}) & \xrightarrow{t} & \psi_{-1}(\widetilde{M}) \\ {\scriptstyle =}\downarrow & & {\scriptstyle \simeq}\downarrow & & {\scriptstyle =}\downarrow \\ \psi_{-1}(M) & \xrightarrow{u} & Q & \xrightarrow{t \circ v} & \psi_{-1}(M) \end{array}$$

Proof In general, let N be a nilpotent endomorphism of a $\mathscr{D}_{X_0}$-module Q. Let $\iota : X_0 \times \{0\} \longrightarrow X$ be the inclusion. We set $G(Q,N) := \iota_* Q \otimes_{\mathbb{C}} \mathbb{C}[\partial_t]$ as a sheaf on X. We define the action of t on $G(Q,N)$ by $t(m\,\partial_t^j) = (-j-N)m\,\partial_t^{j-1}$. Because it is nilpotent, we obtain an $\mathcal{O}_X$-action on $G(Q,N)$. We define an action of $\partial_t t$ on $G(Q,N)$ by $(\partial_t t)(m\,\partial_t^j) = (-j-N)m\,\partial_t^j$. For $f \in \mathcal{O}_X$, let $\mu_f : G(Q,N) \longrightarrow G(Q,N)$ be the multiplication of f. We can check that $[(\partial_t t), \mu_f] = \mu_{[\partial_t t, f]}$. Hence, with the above actions, $G(Q,N)$ is a $V_0 D_X$-module. Let $\partial_t : G(Q,N) \longrightarrow G(Q,N)$ be given by $\partial_t(m\,\partial_t^j) = m\,\partial_t^{j+1}$. Then, we have $\mu_t \circ \partial_t = (\partial_t t) - \mathrm{id}$. We also have $\partial_t \circ \mu_t = (\partial_t t) + N\pi_0$, where π_0 is the projection of $G(Q,N)$ onto the 0-th degree part Q. Hence, we have $[\partial_t, \mu_t] = \mathrm{id} + N \circ \pi_0$.

Let N on Q be the composite of the morphisms $Q \xrightarrow{v} \psi_0(M) \xrightarrow{-t} \psi_{-1}(M) \xrightarrow{u} Q$. Let N on $\psi_0(M)$ be $-\partial_t t$. Because $N \circ v = v \circ N$, we naturally have $V_0 D_X$-homomorphism $G(Q,N) \longrightarrow G(\psi_0(M), N)$, denoted by ρ_1. Note that $G(\psi_0(M), N)$ is naturally a direct summand of $M/V_{<0}M$ as $V_0 D_X$-module. Let ρ_2 be the composite of the $V_0 D_X$-homomorphisms $M \longrightarrow M/V_{<0}M \longrightarrow G(\psi_0(M), N)$. We consider the $V_0 D_X$-submodule $\tilde{M}$ of $G(Q,N) \oplus M$ given as follows:

$$\tilde{M} = \big\{(m_1, m_2) \in G(Q,N) \oplus M \,\big|\, \rho_1(m_1) = \rho_2(m_2)\big\}$$

We shall make a ∂_t-action on $\tilde{M}$. Let $-L <_{\mathbb{C}} -1$. We have the decomposition $M/V_{-L}M \simeq \bigoplus_{-L <_{\mathbb{C}} b} \psi_b M$ of the $p^{-1}D_{X_0}$-module compatible with the action of $t\partial_t$, where $p : X \longrightarrow X_0$ be the projection. Let π_{-1} denote the projection $M \longrightarrow \psi_{-1}(M)$, which is independent of a choice of L. Then, let $\partial_t : \tilde{M} \longrightarrow \tilde{M}$ be the $\mathbb{C}$-linear morphism given as follows:

$$\partial_t(m_1, m_2) = \big(\partial_t m_1 + u(\pi_{-1}(m_2)),\ \partial_t m_2\big)$$

By construction, we have $[\partial_t, P] = 0$ for $P \in p^{-1}\mathcal{O}_{X_0}$. For $f \in \mathcal{O}_X$, let $\mu_f : \tilde{M} \longrightarrow \tilde{M}$ be the multiplication of f. We can check $\partial_t \circ \mu_t = \partial_t t$ and $[\partial_t, \mu_f] = \mu_{\partial_t f}$ by a direct computation. Hence we obtain the structure of an $\mathscr{D}_X$-module on $\tilde{M}$. It is easy to observe that $\tilde{M}$ is the desired one. □

Applying the above construction, in the case $Q = \psi_{-1}M$, $u = \mathrm{id}$ and $v = \partial_t$, we obtain the following.

Corollary 3.1.17 *We have a coherent $\mathscr{D}_X$-module $M[!t]$ with a V-filtration along t, such that $\eth_t : \psi_{-1}(M[!t]) \longrightarrow \psi_0(M[!t])$ is an isomorphism.* □

We have a natural isomorphism $\mathscr{D}_X \otimes_{V_0\mathscr{D}_X} V_{<0}M \simeq M[!t]$. Indeed, the natural map $\mathscr{D}_X \otimes_{V_0\mathscr{D}_X} V_{<0}M \longrightarrow M[!t]$ is surjective by the construction. We define a filtration V of $\mathscr{D}_X \otimes_{V_0\mathscr{D}_X} V_{<0}M$ as in (3.4). The natural map preserves the filtration V. We have $V_a\big(D_X \otimes_{V_0D_X} V_{<0}M\big) \simeq V_a(M[!t]) \simeq V_a(M)$ for $a < 0$. We can prove that the induced map on Gr_a^V is an isomorphism for each $a \geq 0$, by using the argument in Lemma 3.1.10. Similarly, we obtain a natural isomorphism $\mathscr{D}_X \otimes_{V_0D_X} V_0M \simeq M = M(*t)$.

3.1.4 Characterization

Let $\mathcal{M}$ be any coherent $\mathcal{R}_{X(*H)}(*t)$-module which is strictly specializable along t.

Lemma 3.1.18 *Let $\mathcal{Q}$ be a coherent $\mathcal{R}_{X(*H)}$-module which is strictly specializable along t such that $\mathcal{Q}(*t) \simeq \mathcal{M}$.*

- *If $t : \psi_0(\mathcal{Q}) \simeq \psi_{-\delta}(\mathcal{Q})$, we naturally have $\mathcal{Q} \simeq \mathcal{M}[*t]$.*
- *If $\eth_t : \psi_{-\delta}(\mathcal{Q}) \simeq \psi_0(\mathcal{Q})$, we naturally have $\mathcal{Q} \simeq \mathcal{M}[!t]$.*

Proof Let us prove the first claim. Both $\mathcal{Q}$ and $\mathcal{M}[*t]$ are naturally $\mathcal{R}_{X(*H)}$-submodules of $\mathcal{M}$. It is easy to observe the coincidence of $V_0^{(\lambda_0)}$ for any λ_0. Because they are locally generated by $V_0^{(\lambda_0)}$, they are the same.

Let us consider the second claim. By the universal property, we have the naturally induced morphism $\mathcal{M}[!t] \longrightarrow \mathcal{Q}$. We have only to check that it is an isomorphism locally around each $(\lambda_0, P) \in \mathcal{X}$. We have $V_{<0}^{(\lambda_0)}\mathcal{M}[!t] \simeq V_{<0}^{(\lambda_0)}\mathcal{Q}$. By the condition, we obtain that $\psi_0^{(\lambda_0)}(\mathcal{M}[!t]) \longrightarrow \psi_0^{(\lambda_0)}(\mathcal{Q})$ is an isomorphism. Then, we can check that $\psi_u^{(\lambda_0)}(\mathcal{M}[!t]) \longrightarrow \psi_u^{(\lambda_0)}(\mathcal{Q})$ is an isomorphism for each $u \in \mathbb{R} \times \mathbb{C}$. Then, the second claim follows. □

3.1.5 Morphisms

Let $\mathcal{M}_i$ $(i = 1, 2)$ be $\mathcal{R}_{X(*H)}(*t)$-modules which are strictly specializable along t. For $\star = *, !$, we have a natural map given by the localization

$$\mathrm{Hom}_{\mathcal{R}_{X(*H)}}\big(\mathcal{M}_1[\star t], \mathcal{M}_2[\star t]\big) \longrightarrow \mathrm{Hom}_{\mathcal{R}_{X(*H)}(*t)}\big(\mathcal{M}_1, \mathcal{M}_2\big). \tag{3.5}$$

Lemma 3.1.19 *The map (3.5) is an isomorphism. If $\mathcal{M}_i$ are integrable, we have a bijection between integrable homomorphisms.*

Proof Any $\mathcal{R}_{X(*H)}(*t)$-morphism $f:\mathcal{M}_1\longrightarrow\mathcal{M}_2$ induces $V^{(\lambda_0)}_{<0}\mathcal{M}_1\longrightarrow V^{(\lambda_0)}_{<0}\mathcal{M}_2$ and $V^{(\lambda_0)}_0\mathcal{M}_1\longrightarrow V^{(\lambda_0)}_0\mathcal{M}_2$ for each λ_0. We obtain $\mathcal{M}_1[!t]\longrightarrow\mathcal{M}_2[!t]$ and $\mathcal{M}_1[*t]\longrightarrow\mathcal{M}_2[*t]$, respectively. It gives the converse of (3.5). □

Let $f:\mathcal{M}_1\longrightarrow\mathcal{M}_2$ be a morphism of $\mathcal{R}_{X(*H)}(*t)$-modules. We have the induced morphisms $\psi_u(f):\psi_u(\mathcal{M}_1)\longrightarrow\psi_u(\mathcal{M}_2)$ for any $u\in\mathbb{R}\times\mathbb{C}$. Recall the following lemma.

Lemma 3.1.20 *Assume that f is strictly specializable, i.e.,* $\operatorname{Cok}\psi_u(f)$ *are strict. Then, the following holds:*

- *f is strict with respect to the filtrations $V^{(\lambda_0)}$ for each λ_0.*
- $\operatorname{Ker}(f)$, $\operatorname{Im}(f)$ *and* $\operatorname{Cok}(f)$ *are strictly specializable along t. The V-filtrations are the same as the naturally induced filtrations $V^{(\lambda_0)}$.*
- $\psi_u\operatorname{Ker}(f)\simeq\operatorname{Ker}\psi_u(f)$, $\psi_u\operatorname{Im}(f)\simeq\operatorname{Im}\psi_u(f)$ *and* $\psi_u\operatorname{Cok}(f)\simeq\operatorname{Cok}\psi_u(f)$ *naturally.* □

As we have already remarked, we have the induced morphism $f[\star t]:\mathcal{M}_1[\star t]\longrightarrow\mathcal{M}_2[\star t]$.

Lemma 3.1.21 *Under the assumption of Lemma 3.1.20, we have the following natural isomorphisms:*

$$(\operatorname{Ker} f)[\star t]\simeq\operatorname{Ker}\bigl(f[\star t]\bigr),\quad(\operatorname{Im} f)[\star t]\simeq\operatorname{Im}\bigl(f[\star t]\bigr),\quad(\operatorname{Cok} f)[\star t]\simeq\operatorname{Cok}\bigl(f[\star t]\bigr)$$

Proof We naturally have $(\operatorname{Ker} f)[\star t](*t)\simeq(\operatorname{Ker} f)(*t)\simeq\operatorname{Ker}\bigl(f[\star t]\bigr)(*t)$. Then, we obtain the first isomorphism by the characterization. The others are obtained similarly. □

Let $\mathcal{M}^\bullet$ be a bounded complex of $\mathcal{R}_{X(*H)}(*t)$-modules such that each $\mathcal{M}^\bullet$ is strict and strictly specializable along t.

Corollary 3.1.22 *Assume that $\mathcal{H}^\bullet(\psi_u\mathcal{M}^\bullet)$ are strict for any $u\in\mathbb{R}\times\mathbb{C}$. Then, the following holds:*

- *The differential of $\mathcal{M}^\bullet$ is strict with respect to the V-filtrations.*
- *$\mathcal{H}^\bullet(\mathcal{M}^\bullet)$ are strictly specializable along t. The V-filtrations are the same as the naturally induced filtrations $V^{(\lambda_0)}$.*
- *$\psi_u\mathcal{H}^\bullet\simeq\mathcal{H}^\bullet\psi_u$ and $\mathcal{H}^\bullet\bigl(\mathcal{M}^\bullet[\star t]\bigr)\simeq\mathcal{H}^\bullet\bigl(\mathcal{M}^\bullet\bigr)[\star t]$ naturally.* □

3.1.6 Canonical Prolongations of $\mathcal{R}$-Modules

Let $\mathcal{M}$ be any coherent $\mathcal{R}_{X(*H)}$-module which is strictly specializable along t. Note that $\mathcal{M}(*t)$ is a coherent $\mathcal{R}_{X(*H)}(*t)$-module, and it is strictly specializable along t. For $\star=*,!$, we define $\mathcal{M}[\star t]:=\bigl(\mathcal{M}(*t)\bigr)[\star t]$. If $\mathcal{M}$ is integrable, $\mathcal{M}[\star t]$ are naturally also integrable.

Lemma 3.1.23 *We have a natural morphism $\iota : \mathcal{M} \longrightarrow \mathcal{M}[*t]$. We have the following naturally defined isomorphisms:*

$$\mathrm{Ker}(\iota) \simeq \mathrm{Ker}\Big(\psi_0(\mathcal{M}) \xrightarrow{t} \psi_{-\boldsymbol{\delta}}(\mathcal{M})\Big)[\eth_t]$$

$$\mathrm{Cok}(\iota) \simeq \mathrm{Cok}\Big(\psi_0(\mathcal{M}) \xrightarrow{t} \psi_{-\boldsymbol{\delta}}(\mathcal{M})\Big)[\eth_t] \tag{3.6}$$

For any $u \notin \mathbb{Z}_{\geq 0} \times \{0\}$, we have a natural isomorphism $\psi_u^{(\lambda_0)}(\iota) : \psi_u^{(\lambda_0)}(\mathcal{M}) \simeq \psi_u^{(\lambda_0)}(\mathcal{M}[*t])$.

Proof We have a naturally defined morphism $\mathcal{M} \longrightarrow \mathcal{M}(*t)$, for which the image of $V_0^{(\lambda_0)}(\mathcal{M})$ is contained in $V_0^{(\lambda_0)}\big(\mathcal{M}(*t)\big) = V_0^{(\lambda_0)}(\mathcal{M}[*t])$. Because $\mathcal{M}_{|\mathcal{X}^{(\lambda_0)}}$ is generated by $V_0^{(\lambda_0)}(\mathcal{M})$ over $\mathcal{R}_{X(*H)}$, we obtain $\mathcal{M} \longrightarrow \mathcal{M}[*t]$. Let us consider $\psi_u^{(\lambda_0)}(\iota) : \psi_u^{(\lambda_0)}(\mathcal{M}) \longrightarrow \psi_u^{(\lambda_0)}(\mathcal{M}[*t])$. It is an isomorphism if $\mathfrak{p}(\lambda_0, u) < 0$. If $u \notin \mathbb{Z}_{\geq 0} \times \{0\}$, we obtain $\psi_u(\iota)$ is an isomorphism by using an easy induction and isomorphisms (3.2).

The kernel and the cokernel of $\mathcal{M} \longrightarrow \mathcal{M}[*t]$ are naturally isomorphic to those of

$$\mathcal{M}/V_{<0}^{(\lambda_0)}\mathcal{M} \longrightarrow \mathcal{M}[*t]/V_{<0}^{(\lambda_0)}\mathcal{M}[*t],$$

which are naturally isomorphic to those for $\psi_0^{(\lambda_0)}(\mathcal{M})[\eth_t] \longrightarrow \psi_0^{(\lambda_0)}(\mathcal{M}[*t])[\eth_t]$ by the above consideration. Note that the induced morphism $\psi_0^{(\lambda_0)}(\iota)$ is naturally identified with $t : \psi_0^{(\lambda_0)}(\mathcal{M}) \longrightarrow \psi_{-\boldsymbol{\delta}}^{(\lambda_0)}(\mathcal{M})$ under the natural identification $t : \psi_0^{(\lambda_0)}(\mathcal{M}[*t]) \simeq \psi_{-\boldsymbol{\delta}}^{(\lambda_0)}(\mathcal{M}[*t]) = \psi_{-\boldsymbol{\delta}}^{(\lambda_0)}(\mathcal{M})$. Then, we obtain (3.6). □

Similarly, we have the following lemma.

Lemma 3.1.24 *We have a naturally defined morphism $\iota : \mathcal{M}[!t] \longrightarrow \mathcal{M}$. The induced morphism $\psi_u(\iota)$ is an isomorphism unless $u \in \mathbb{Z}_{\geq 0} \times \{0\}$. We have natural isomorphisms*

$$\mathrm{Ker}(\iota) \simeq \mathrm{Ker}\Big(\psi_{-\boldsymbol{\delta}}(\mathcal{M}) \xrightarrow{\eth_t} \psi_0(\mathcal{M})\Big)[\eth_t],$$

$$\mathrm{Cok}(\iota) \simeq \mathrm{Cok}\Big(\psi_{-\boldsymbol{\delta}}(\mathcal{M}) \xrightarrow{\eth_t} \psi_0(\mathcal{M})\Big)[\eth_t].$$

Proof It can be proved as in the case of $\mathcal{M}[*t]$. □

Let $\mathcal{M}_i$ $(i = 1, 2)$ be strict coherent $\mathcal{R}_{X(*H)}$-modules, which are strictly specializable along t. We obtain the following lemma from Lemma 3.1.19.

Lemma 3.1.25 *We have the following natural morphisms for $\star = *, !$:*

$$\mathrm{Hom}_{\mathcal{R}_{X(*H)}}(\mathcal{M}_1, \mathcal{M}_2) \longrightarrow \mathrm{Hom}_{\mathcal{R}_{X(*H)}}(\mathcal{M}_1[\star t], \mathcal{M}_2[\star t]).$$

If $\mathcal{M}_i$ are integrable, we have the morphism between spaces of integrable homomorphisms. □

We also obtain the following corollary.

Corollary 3.1.26 *We have natural bijections:*

$$\mathrm{Hom}_{\mathcal{R}_{X(*H)}}\big(\mathcal{M}_1[*t], \mathcal{M}_2[*t]\big) \overset{a_1}{\simeq} \mathrm{Hom}_{\mathcal{R}_{X(*H)}}\big(\mathcal{M}_1, \mathcal{M}_2[*t]\big)$$

$$\mathrm{Hom}_{\mathcal{R}_{X(*H)}}\big(\mathcal{M}_1[!t], \mathcal{M}_2[!t]\big) \overset{a_2}{\simeq} \mathrm{Hom}_{\mathcal{R}_{X(*H)}}\big(\mathcal{M}_2[!t], \mathcal{M}_2\big).$$

If $\mathcal{M}_i$ are integrable, we have the bijections of integrable homomorphisms.

Proof The morphisms $\mathcal{M}_1 \longrightarrow \mathcal{M}_1[*t]$ and $\mathcal{M}_2[!t] \longrightarrow \mathcal{M}_2$ induce a_i. Note that the morphism $\mathcal{M}_1 \longrightarrow \mathcal{M}_2[*t]$ is uniquely determined by the induced morphism $\mathcal{M}_1(*t) \longrightarrow \mathcal{M}_2[*t]$. We can construct the inverse by using Lemma 3.1.19. Note that $\mathcal{M}_1 \longrightarrow \mathcal{M}_2[*t]$ and $\mathcal{M}_1[!t] \longrightarrow \mathcal{M}_2$ are determined by $V_{<0}\mathcal{M}_1 \longrightarrow V_{<0}\mathcal{M}_2$, and which are uniquely determined by the induced morphism $\mathcal{M}_1(*t) \longrightarrow \mathcal{M}_2(*t)$. Hence, we obtain that the morphisms a_i are bijective. □

Lemma 3.1.27 *Let $\mathcal{M}$ be any coherent $\mathcal{R}_{X(*H)}$-module which is strictly specializable along t. Then, $\mathcal{M}[\star t]$ is independent of the choice of a decomposition into the product $X_0 \times \mathbb{C}_t$. If A is a nowhere vanishing holomorphic function on X, we have $\mathcal{M}[\star(At)] \simeq \mathcal{M}[\star t]$ naturally.*

Proof It follows from Corollaries 3.1.6 and 3.1.13. □

We obtain the following from Lemma 3.1.9 and Corollary 3.1.11.

Lemma 3.1.28 *Let $\mathcal{M}$ be any coherent strict $\mathcal{R}_X$-module which is strictly specializable along t. If a non-zero complex number λ_0 is generic for $(\mathcal{M}, t)$ then $\mathcal{M}[\star t]^{\lambda_0}$ is naturally isomorphic to $\mathcal{M}^{\lambda_0}[\star t]$.* □

3.2 Canonical Prolongations of $\mathcal{R}$-Triples

3.2.1 Canonical Prolongations of $\mathcal{R}(*t)$-Triples

We use the setting in Sect. 3.1.1. Let $\mathcal{T} = (\mathcal{M}', \mathcal{M}'', C)$ be an $\mathcal{R}_{X(*H)}(*t)$-triple which is strictly specializable along t, i.e., $\mathcal{M}'$ and $\mathcal{M}''$ are $\mathcal{R}_{X(*H)}(*t)$-modules which are strictly specializable along t, and C is a sesqui-linear pairing of $\mathcal{M}'$ and $\mathcal{M}''$, which is an $\mathcal{R}_{X(*H)} \otimes \sigma^*\mathcal{R}_{X(*H)}$-homomorphism

$$C : \mathcal{M}'_{|\mathbf{S}\times X} \otimes \sigma^*\mathcal{M}''_{|\mathbf{S}\times X} \longrightarrow \mathfrak{Db}^{\mathrm{mod}\,H}_{\mathbf{S}\times X/\mathbf{S}}(*t).$$

Proposition 3.2.1 *We have sesqui-linear pairings*

$$C[!t] : \mathcal{M}'[*t]_{|S\times X} \otimes \sigma^*\mathcal{M}''[!t]_{|S\times X} \longrightarrow \mathfrak{Db}^{\mathrm{mod}\,H}_{S\times X/S}$$

$$C[*t] : \mathcal{M}'[!t]_{|S\times X} \otimes \sigma^*\mathcal{M}''[*t]_{|S\times X} \longrightarrow \mathfrak{Db}^{\mathrm{mod}\,H}_{S\times X/S}$$

such that $C[\star t]_{|S\times(X\setminus\{t=0\})} = C_{|S\times(X\setminus\{t=0\})}$. *They are determined uniquely by the conditions. In particular, we obtain the following* $\mathcal{R}$*-triples:*

$$\mathcal{T}[!t] := \big(\mathcal{M}'[*t], \mathcal{M}''[!t], C[!t]\big), \qquad \mathcal{T}[*t] := \big(\mathcal{M}'[!t], \mathcal{M}''[*t], C[*t]\big)$$

Proof By the uniqueness, we have only to consider it locally. We have the given pairing

$$V_{<0}\mathcal{M}'_{|S\times X} \otimes \sigma^* V_{<0}\mathcal{M}''_{|S\times X} \longrightarrow \mathfrak{Db}^{\mathrm{mod}\,H}_{S\times X/S}(*t).$$

Let us observe that it is extended to the following pairing:

$$V_0\mathcal{M}'_{|S\times X} \otimes \sigma^* V_{<0}\mathcal{M}''_{|S\times X} \longrightarrow \mathfrak{Db}^{\mathrm{mod}\,H}_{S\times X/S} \tag{3.7}$$

Let $X_1 \subset X$ be open. Let $U(\lambda_0)$ be a neighbourhood of λ_0 in $\mathbb{C}_\lambda$. Let u and v be local sections of $\mathcal{M}'$ and $\mathcal{M}''$ on $U(\lambda_0)\times X_1$ and $U(-\lambda_0)\times X_1$ respectively. We put $\boldsymbol{I}(\lambda_0) := U(\lambda_0)\cap \boldsymbol{S}$. Let ϕ be a C^∞-section of $p_\lambda^*\Omega_X^{n,n}$ on $\boldsymbol{I}(\lambda_0)\times X_1$ with compact support such that $\phi_{|\boldsymbol{I}(\lambda_0)\times\hat{H}} = 0$. Note that $\big\langle C(u,v), |t|^{2s}\phi\big\rangle$ is well defined for $s \in \mathcal{H} := \big\{s\in\mathbb{C}\,\big|\,\mathrm{Re}\,s >> 0\big\}$, and holomorphic for s and continuous for $\lambda \in \boldsymbol{S}$. By a standard argument, if $u \in V_0^{(\lambda_0)}\mathcal{M}'$ and $v \in V_{<0}^{(-\lambda_0)}\mathcal{M}''$, it is extended to a function on $\boldsymbol{S}\times\big\{s\in\mathbb{C}\,\big|\,\mathrm{Re}(s) > -\epsilon\big\}$ for some $\epsilon > 0$. (See the proof of Lemma 20.10.9 of [55], for example.) We can take the value at $s = 0$, which defines $\tilde{C}(u,v)$. It gives a section of $\mathfrak{Db}^{\mathrm{mod}\,H}_{S\times X/S}$. Thus, we obtain (3.7).

Lemma 3.2.2 *Let* $\lambda_0 \in \boldsymbol{S}$. *Assume* $\sum_{i=0}^N \eth_t^i f_i = 0$ *in* $\mathcal{M}'[*t] = 0$ *for* $f_i \in V_0^{(\lambda_0)}\mathcal{M}'[*t]$. *Then, for any* $g \in V_{<0}^{(-\lambda_0)}\mathcal{M}''$ *we have*

$$\sum_{i=0}^N \eth_t^i \tilde{C}(f_i, \sigma^* g) = 0. \tag{3.8}$$

Assume $\sum_{j=0}^N \eth_t^j g_j = 0$ *in* $\mathcal{M}''[!t]$ *for some* $g_j \in V_{<0}^{(-\lambda_0)}\mathcal{M}''$. *Then, for any* $f \in V_0^{(\lambda_0)}\mathcal{M}'[*t]$, *we have*

$$\sum_{j=0}^N \overline{\eth}_t^j \tilde{C}(f, \sigma^* g_j) = 0.$$

Proof Let us prove (3.8). The other can be proved in a similar way. We use an induction on N. In the case $N = 0$, the claim is trivial. If $\sum_{i=0}^{N} \partial_t^i f_i = 0$, we have $f_N \in V_{-1}^{(\lambda_0)}\mathcal{M}[*t]$, and hence $\partial_t f_N \in V_0^{(\lambda_0)}\mathcal{M}[*t]$. By using the hypothesis of the induction, we obtain

$$\sum_{i=0}^{N-2} \partial_t^i \tilde{C}(f_i, \sigma^* g) + \partial_t^{N-1} \tilde{C}\big(f_{N-1} + \partial_t f_N, \sigma^* g\big) = 0.$$

So, we have only to prove that $\tilde{C}(\partial_t f, \sigma^* g) = \partial_t \tilde{C}(f, \sigma^* g)$ for $f \in V_{-1}^{(\lambda_0)}(\mathcal{M}')$. For a test form ϕ, we have

$$\begin{aligned} \big\langle \tilde{C}(\partial_t f, \sigma^* g),\, \phi \big\rangle &= \big\langle C(\partial_t f, \sigma^* g),\, |t|^{2s}\phi \big\rangle_{|s=0} \\ &= -\big\langle C(f, \sigma^* g),\, |t|^{2s}\, \partial_t \phi \big\rangle_{|s=0} - \Big(s \big\langle C(f, \sigma^* g),\, |t|^{2s-2} \bar{t}\, \phi \big\rangle \Big)_{|s=0} \\ &= \big\langle \partial_t \tilde{C}(f, \sigma^* g),\, \phi \big\rangle - \Big(s \big\langle C(f, \sigma^* g),\, |t|^{2s-2} \bar{t}\, \phi \big\rangle \Big)_{|s=0} \end{aligned} \tag{3.9}$$

Because $f \in V_{-1}^{(\lambda_0)}\mathcal{M}'$ and $g \in V_{<0}^{(-\lambda_0)}\mathcal{M}''$, $\big\langle C(f, \sigma^* g),\, |t|^{2s-2}\bar{t}\phi \big\rangle$ is holomorphic at $s = 0$. Hence, we obtain $\tilde{C}(\partial_t f, \sigma^* g) = \partial_t \tilde{C}(f, \sigma^* g)$ for $f \in V_{-1}^{(\lambda_0)}(\mathcal{M}')$. □

Then, we can naturally extend $\tilde{C}$ to the pairing $C[!t]$ of $\mathcal{M}'[*t]$ and $\mathcal{M}''[!t]$. Let C' be another pairing of $\mathcal{M}'[*t]$ and $\mathcal{M}''[!t]$ whose restriction to $\boldsymbol{S} \times (X \setminus \{t = 0\})$ is equal to C. Then, we can prove that $C_1 := C[!t] - C'$ is 0 on $V_0^{(\lambda_0)}\mathcal{M}' \otimes \sigma^* V_{<0}^{(-\lambda_0)}\mathcal{M}''$ by a standard argument. (See the proof of Lemma 22.10.8 of [55], for example.) Then, we obtain that $C_1 = 0$ on $\mathcal{M}'[*t] \otimes \sigma^*\mathcal{M}''[!t]$. Thus, the proof of Proposition 3.2.1 is finished. □

The following lemma is obvious.

Lemma 3.2.3 *We have* $\big(\mathcal{T} \otimes \boldsymbol{T}(w)\big)[\star t] = \mathcal{T}[\star t] \otimes \boldsymbol{T}(w)$ *and* $\mathcal{T}^*[!t] = \big(\mathcal{T}[*t]\big)^*$. □

Lemma 3.2.4 *If $\mathcal{T}$ is integrable, then $\mathcal{T}[\star t]$ are also integrable.*

Proof Let us consider $C[!t]$. We set

$$C_0(m', \sigma^* m'') := \partial_\theta C(m', \sigma^* m'') - C(\partial_\theta m', \sigma^* m'') - C(m', \sigma^* \partial_\theta m'').$$

It is 0 outside $\{t = 0\}$. It is standard to prove the vanishing of C_0 on $V_0 \otimes \sigma^* V_{<0}$. (See the proof of Lemma 22.10.8 of [55], for example.) Then, we obtain the vanishing on $\mathcal{M}'[*t] \otimes \sigma^*\mathcal{M}''[!t]$. □

3.2.2 Morphisms

Let $\mathcal{T}_i$ $(i = 1, 2)$ be coherent $\mathcal{R}_{X(*H)}(*t)$-triples, which are strictly specializable along t.

Lemma 3.2.5 *Let $\star = *, !$. Morphisms $\mathcal{T}_1 \longrightarrow \mathcal{T}_2$ of $\mathcal{R}_{X(*H)}(*t)$-triples bijectively correspond to morphisms $\mathcal{T}_1[\star t] \longrightarrow \mathcal{T}_2[\star t]$ of $\mathcal{R}_{X(*H)}$-triples. If $\mathcal{T}_i$ are integrable, we have such bijections for integrable morphisms.*

Proof A morphism $\mathcal{T}_1[\star t] \longrightarrow \mathcal{T}_2[\star t]$ naturally induces $\mathcal{T}_1 \longrightarrow \mathcal{T}_2$. Let $\mathcal{T}_1 \longrightarrow \mathcal{T}_2$ be a morphism of $\mathcal{R}_{X(*H)}(*t)$-triples. Let us observe that we have an induced morphism $\mathcal{T}_1[\star t] \longrightarrow \mathcal{T}_2[\star t]$. By Lemma 3.1.19, we have the morphisms of the underlying $\mathcal{R}_{X(*H)}$-modules. We have only to prove the compatibility of Hermitian sesqui-linear pairings. By construction, they are compatible on $V_0^{(\lambda_0)} \otimes \sigma^* V_{<0}^{(\lambda_0)}$ or $V_{<0}^{(\lambda_0)} \otimes \sigma^* V_0^{(\lambda_0)}$. Then, the claim is easy to see. □

Let $f : \mathcal{T}_1 \longrightarrow \mathcal{T}_2$ be a morphism. We have the induced morphism $f[\star t] : \mathcal{T}_1[\star t] \longrightarrow \mathcal{T}_2[\star t]$. We obtain the following lemma from Lemma 3.1.21.

Lemma 3.2.6 *Suppose that f is strictly specializable. Then, we naturally have* $\mathrm{Ker}(f[\star t]) \simeq \mathrm{Ker}(f)[\star t]$, $\mathrm{Im}(f[\star t]) \simeq \mathrm{Im}(f)[\star t]$, *and* $\mathrm{Cok}(f[\star t]) \simeq \mathrm{Cok}(f)[\star t]$. □

Let $\mathcal{T}^\bullet$ be a bounded complex of $\mathcal{R}_{X(*H)}(*t)$-triples such that each $\mathcal{T}^p$ is strictly specializable along t. We obtain the following from Corollary 3.1.22.

Lemma 3.2.7 *Assume that $\mathcal{H}^\bullet(\psi_u \mathcal{T}^\bullet)$ are strict for any $u \in \mathbb{R} \times \mathbb{C}$. Then, the following holds:*

- *$\mathcal{H}^\bullet(\mathcal{T}^\bullet)$ are strictly specializable along t.*
- *We have natural isomorphisms $\psi_u \mathcal{H}^\bullet(\mathcal{T}^\bullet) \simeq \mathcal{H}^\bullet \psi_u(\mathcal{T}^\bullet)$ and $\mathcal{H}^\bullet(\mathcal{T}^\bullet[\star t]) \simeq \mathcal{H}^\bullet(\mathcal{T}^\bullet)[\star t]$.* □

3.2.3 Canonical Prolongations of $\mathcal{R}$-Triples

Let $\mathcal{T}$ be an $\mathcal{R}_{X(*H)}$-triple which is strictly specializable along t. By applying the previous construction to the $\mathcal{R}_{X(*H)}(*t)$-triple $\mathcal{T}(*t)$, we obtain $\mathcal{R}_{X(*H)}$-triples $\mathcal{T}[\star t]$ for $\star = *, !$. If $\mathcal{T}$ is integrable, $\mathcal{T}[\star t]$ are also integrable.

Lemma 3.2.8 *We have natural morphisms $\mathcal{T}[!t] \longrightarrow \mathcal{T} \longrightarrow \mathcal{T}[*t]$.*

Proof We consider only the morphism $\mathcal{T}[!t] \longrightarrow \mathcal{T}$. The other can be proved similarly. We have the morphisms $\mathcal{M}''[!t] \longrightarrow \mathcal{M}''$ and $\mathcal{M}' \longrightarrow \mathcal{M}'[*t]$. Let us check that they are compatible with pairings. Let $f \in V_0^{(\lambda_0)}\mathcal{M}'$ and $g \in V_{<0}^{(\lambda_0)}\mathcal{M}''$. We have only to prove $C[!t](f, \sigma^* g) = C(f, \sigma^* g)$. As in the proof of Proposition 3.2.1, $\langle C(f, \sigma^* g), |t|^{2s}\phi \rangle$ is extended to a function on $\boldsymbol{S} \times \{s \in \mathbb{C} \mid \mathrm{Re}(s) > -\epsilon\}$ for some

$\epsilon > 0$. Moreover, we have $\big\langle C(f,\sigma^*g), |t|^{2s}\phi\big\rangle_{s=0} = \big\langle C(f,\sigma^*g),\phi\big\rangle$. (See the proof of Lemma 20.10.9 of [55], for example.) Then, we obtain the desired compatibility of the pairings. □

Let $\mathcal{T}_i$ $(i = 1,2)$ be $\mathcal{R}_{X(*H)}$-triples which are strictly specializable along t. We obtain the following lemma from Lemma 3.2.5.

Lemma 3.2.9 *We have the following natural morphisms for* $\star = *, !$*:*

$$\mathrm{Hom}_{\mathcal{R}\text{-}Tri(X,H)}(\mathcal{T}_1,\mathcal{T}_2) \longrightarrow \mathrm{Hom}_{\mathcal{R}\text{-}Tri(X,H)}(\mathcal{T}_1[\star t],\mathcal{T}_2[\star t])$$

If $\mathcal{T}_i$ are integrable, we have the morphisms between the spaces of integrable morphisms. □

Corollary 3.2.10 *We have the following natural bijections:*

$$\mathrm{Hom}_{\mathcal{R}\text{-}Tri(X,H)}(\mathcal{T}_1,\mathcal{T}_2[*t]) \overset{a_1}{\simeq} \mathrm{Hom}_{\mathcal{R}\text{-}Tri(X,H)}(\mathcal{T}_1[*t],\mathcal{T}_2[*t])$$

$$\mathrm{Hom}_{\mathcal{R}\text{-}Tri(X,H)}(\mathcal{T}_1[!t],\mathcal{T}_2[!t]) \overset{a_2}{\simeq} \mathrm{Hom}_{\mathcal{R}\text{-}Tri(X,H)}(\mathcal{T}_1[!t],\mathcal{T}_2)$$

The morphisms a_i are induced by the morphisms in Lemma 3.2.8. If $\mathcal{T}_i$ are integrable, we have the bijections of the spaces of integrable morphisms.

Proof We can construct the inverse of a_i by using Lemma 3.2.9. By using Corollary 3.1.26, we obtain that the morphisms a_i are bijective. □

Proposition 3.2.11 *$\mathcal{T}[\star t]$ is independent of the choice of a decomposition into the product $X_0 \times \mathbb{C}_t$. For any nowhere vanishing holomorphic function A on X, we have natural isomorphisms $\mathcal{T}[\star(At)] \simeq \mathcal{T}[\star t]$.*

Proof We have natural morphisms $\mathcal{T}[\star(At)] \longrightarrow \mathcal{T}[\star t]$. The underlying $\mathcal{R}_{X(*H)}$-modules are isomorphic by Lemma 3.1.27 □

3.2.4 Compatibility of Canonical Prolongation with Push-Forward

Let $F_0 : X_0 \longrightarrow Y_0$ be a morphism of complex manifolds. Let $F : X \longrightarrow Y$ be $F_0 \times \mathrm{id}$, where $X = X_0 \times \mathbb{C}_t$ and $Y = Y_0 \times \mathbb{C}_t$. Let H_Y be a hypersurface of Y such that $\{t = 0\} \not\subset H_Y$. We put $H := F^{-1}(H_Y)$. Let $\mathcal{M}$ be a coherent $\mathcal{R}_{X(*H)}(*t)$-module, which is strictly specializable along t. Assume the following:

- $F^i_\dagger\psi_u(\mathcal{M})$ are strict for any $u \in \mathbb{R}\times\mathbb{C}$.

According to [66], $F^i_\dagger\mathcal{M}$ are strictly specializable $\mathcal{R}_{Y(*H_Y)}(*t)$-module, and we naturally have $\psi_u F^i_\dagger\mathcal{M} \simeq F^i_\dagger\psi_u\mathcal{M}$.

Lemma 3.2.12 *Under the assumption, we have* $F^i_\dagger(\mathcal{M}[\star t]) \simeq \big(F^i_\dagger \mathcal{M}\big)[\star t]$ *naturally.*

Proof It easily follows from the characterization in Lemma 3.1.18. □

Corollary 3.2.13 *Let* $\mathcal{T}$ *be a coherent* $\mathcal{R}_{X(*H)}(*t)$*-triple, which is strictly specializable along* t. *Assume that* $F^i_\dagger \psi_u \mathcal{T}$ *is strict. Then, we naturally have* $F^i_\dagger\big(\mathcal{T}[\star t]\big) \simeq \big(F^i_\dagger \mathcal{T}\big)[\star t]$. □

3.3 Canonical Prolongations Across Hypersurfaces

3.3.1 Canonical Prolongations Across Holomorphic Functions

3.3.1.1 $\mathcal{R}_X(*g)$-Modules and $\mathcal{R}_X(*g)$-Triples

Let g be a holomorphic function on a complex manifold X. Let $\iota_g : X \longrightarrow X \times \mathbb{C}_t$ denote the embedding $\iota_g(x) = (x, g(x))$.

Definition 3.3.1 Let $\mathcal{M}$ be a coherent $\mathcal{R}_X(*g)$-module which is strictly specializable along g. It is called localizable along g, if there exist $\mathcal{R}_X$-modules $\mathcal{M}[\star g]$ $(\star = *, !)$ such that $\iota_{g\dagger}(\mathcal{M}[\star g]) \simeq \big(\iota_{g\dagger}(\mathcal{M})\big)[\star t]$. □

Note that such $\mathcal{R}_X$-modules $\mathcal{M}[\star g]$ are uniquely determined up to canonical isomorphisms because they are recovered as the kernel of the multiplication of $t - g$ on $\iota_{g\dagger}(\mathcal{M})[\star t]$. But, in general, it is not clear whether $\iota_{g\dagger}(\mathcal{M})[\star t]$ are strictly specializable along $t - g$. The notation is verified by the following lemma.

Lemma 3.3.2 *If* g *is a coordinate function, then* $\mathcal{M}[\star g]$ *given in Sect. 3.1 satisfy* $\iota_{g\dagger}(\mathcal{M}[\star g]) \simeq (\iota_{g\dagger}\mathcal{M})[\star t]$.

Proof We may assume that X is an open subset of $X_1 \times \mathbb{C}_g$. We have the bi-holomorphic map $F : \mathbb{C}_y \times \mathbb{C}_s \longrightarrow \mathbb{C}_g \times \mathbb{C}_t$ given by $F(y, s) = (y + s, y)$. It induces a bi-holomorphic map $X_1 \times \mathbb{C}_y \times \mathbb{C}_s \longrightarrow X_1 \times \mathbb{C}_g \times \mathbb{C}_t$, which is also denoted by F. Let $\iota'_g : X \longrightarrow X_1 \times \mathbb{C}_y \times \mathbb{C}_s$ be given by $\iota'_g(x, g) = (x, g, 0)$. We have $F \circ \iota'_g = \iota_g$. We also have $t \circ F = y$. Then, $\iota'_{g\dagger}\mathcal{M}$ is clearly strictly specializable along $y = 0$, and we have $\iota'_{g\dagger}(\mathcal{M}[\star g]) \simeq \iota'_{g\dagger}(\mathcal{M})[\star y]$. By using the bi-holomorphic map, we obtain the claim of the lemma. □

Lemma 3.3.3 *If* $\mathcal{M}$ *is integrable and localizable along* g, *then* $\mathcal{M}[\star g]$ *is also integrable.*

Proof Because $\mathcal{M}[\star g]$ is obtained as the kernel of the multiplication of $t - g$ on the integrable $(\iota_{g\dagger}\mathcal{M})[\star t]$, it is integrable. □

We say that a non-zero complex number λ_0 is generic with respect to $(\mathcal{M}, g)$ if it is generic with respect to $(\iota_{g\dagger}\mathcal{M}, t)$. The following lemma is clear by Lemma 3.1.9 and Corollary 3.1.11.

Lemma 3.3.4 *Let $\mathcal{M}$ be a strict coherent $\mathcal{R}_X(*g)$-module which is localizable along g. If λ_0 is generic for $(\mathcal{M}, t)$, we have $(\mathcal{M}[\star g])^{\lambda_0} \simeq \mathcal{M}^{\lambda_0}[\star g]$.* □

Lemma 3.3.5 *Let A be a nowhere vanishing holomorphic function. We set $g_1 := A\, g$. Then, $\mathcal{M}$ is strictly specializable along g if and only if it is strictly specializable along g_1. Moreover, it is localizable along g if and only if it is localizable along g_1, and we have $\mathcal{M}[\star g] = \mathcal{M}[\star g_1]$ in that case.*

Proof Let $F : X \times \mathbb{C}_t \longrightarrow X \times \mathbb{C}_s$ be given by $F(x,t) = \big(x, A(x)t\big)$. We have $\iota_{g_1} = F \circ \iota_g$. Then, the claim is clear. □

Similarly, for $\mathcal{R}_X(*g)$-triples, we use the following.

Definition 3.3.6 Let $\mathcal{T}$ be a coherent $\mathcal{R}_X(*g)$-triple which is strictly specializable along g. It is called localizable along g, if there exist $\mathcal{R}_X$-triple $\mathcal{T}[\star g]$ for $\star = *, !$ with isomorphisms $\iota_{g\dagger}(\mathcal{T}[\star g]) \simeq (\iota_{g\dagger}\mathcal{T})[\star t]$. □

Such $\mathcal{R}_X$-triples are uniquely determined up to canonical isomorphisms if they exist. The following lemma is clear.

Lemma 3.3.7 *Suppose that a coherent $\mathcal{R}_X(*g)$-triple $\mathcal{T}$ is localizable along g.*

- *If g is a coordinate function, then $\mathcal{T}[\star g]$ are equal to those in Sect. 3.2.*
- *If $\mathcal{T}$ is integrable, then $\mathcal{T}[\star g]$ are also integrable.*
- *$\mathcal{T}[\star g]$ are independent of the choice of a decomposition into the product $X_0 \times \mathbb{C}_t$. If A is a nowhere vanishing holomorphic function, then $\mathcal{T}[\star(Ag)] \simeq \mathcal{T}[\star g]$ naturally.* □

3.3.1.2 Morphisms

Let $\mathcal{T}_i$ $(i = 1, 2)$ be coherent $\mathcal{R}_X(*g)$-triples which are localizable along g. We obtain the following lemma from Lemma 3.2.5.

Lemma 3.3.8 *Let $\star = *, !$. Morphisms $\mathcal{T}_1 \longrightarrow \mathcal{T}_2$ of $\mathcal{R}_X(*g)$-triples bijectively correspond to morphisms $\mathcal{T}_1[\star g] \longrightarrow \mathcal{T}_2[\star g]$ of $\mathcal{R}_X$-triples. If $\mathcal{T}_i$ are integrable, we have such bijections for integrable morphisms. Similar claims hold for $\mathcal{R}_X(*g)$-modules.* □

Let $f : \mathcal{T}_1 \longrightarrow \mathcal{T}_2$ be a morphism. We have the induced morphism $f[\star g] : \mathcal{T}_1[\star g] \longrightarrow \mathcal{T}_2[\star g]$. We obtain the following lemma from Lemma 3.2.6.

Lemma 3.3.9 *Suppose that f is strictly specializable along g, i.e., the induced morphism $\iota_{g\dagger}\mathcal{T}_1 \longrightarrow \iota_{g\dagger}\mathcal{T}_2$ is strictly specializable along t. Then, $\mathrm{Ker} f$, $\mathrm{Im} f$ and $\mathrm{Cok} f$ are strictly specializable and localizable along g, and we naturally have*

$$\mathrm{Ker}(f[\star g]) \simeq \mathrm{Ker}(f)[\star g], \quad \mathrm{Im}(f[\star g]) \simeq \mathrm{Im}(f)[\star g], \quad \mathrm{Cok}(f[\star g]) \simeq \mathrm{Cok}(f)[\star g].$$

□

Let $\mathcal{T}^\bullet$ be a bounded complex of $\mathcal{R}_X(*g)$-triples such that each $\mathcal{T}^p$ is localizable along g. We obtain the following from Lemma 3.2.7.

Lemma 3.3.10 *Assume moreover that $\mathcal{H}^\bullet(\psi_{g,u}\mathcal{T}^\bullet)$ are strict for any $u \in \mathbb{R} \times \mathbb{C}$. Then, the following holds:*

- *$\mathcal{H}^\bullet(\mathcal{T}^\bullet)$ are strictly specializable and localizable along g.*
- *We have isomorphisms $\psi_{g,u}\mathcal{H}^\bullet(\mathcal{T}^\bullet) \simeq \mathcal{H}^\bullet\psi_{g,u}(\mathcal{T}^\bullet)$ and $\mathcal{H}^\bullet(\mathcal{T}^\bullet[\star g]) \simeq \mathcal{H}^\bullet(\mathcal{T}^\bullet)[\star g]$ naturally.* □

3.3.1.3 Compatibility with the Push-Forward

Let $F : X \longrightarrow Y$ be a morphism of complex manifolds. Let g_Y be a holomorphic function on Y, and we set $g_X := g_Y \circ F$. Let $\mathcal{M}$ be a coherent $\mathcal{R}_X(*g_X)$-module which is strictly specializable along g_X. Assume the following:

- The support of $\mathcal{M}$ is proper over Y.
- $F^i_\dagger \tilde{\psi}_{g_X,u}(\mathcal{M})$ are strict for any $u \in \mathbb{R} \times \mathbb{C}$.

According to [66], $F^i_\dagger \mathcal{M}$ are strictly specializable $\mathcal{R}_Y(*g_Y)$-modules.

Lemma 3.3.11 *Assume moreover that $\mathcal{M}$ is localizable along g_X. Then, $F^i_\dagger\mathcal{M}$ are also localizable along g_Y, and we have natural isomorphisms $F^i_\dagger(\mathcal{M}[\star g_X]) \simeq (F^i_\dagger\mathcal{M})[\star g_Y]$.*

Proof We naturally have $F^i_\dagger(\iota_{g_X\dagger}\mathcal{M}[\star g_X]) \simeq \iota_{g_Y\dagger}\big(F^i_\dagger(\mathcal{M}[\star g_X])\big)$. We also have $F^i_\dagger(\iota_{g_X\dagger}\mathcal{M}[\star g_X]) \simeq F^i_\dagger\big((\iota_{g_X\dagger}\mathcal{M})[\star t]\big) \simeq F^i_\dagger(\iota_{g_X\dagger}\mathcal{M})[\star t]$. Then, the claim follows. □

Let $\mathcal{T}$ be a coherent $\mathcal{R}_X(*g_X)$-triple which is strictly specializable along g_X. We suppose the following.

- The support of $\mathcal{T}$ is proper over Y.
- (ii) $F^i_\dagger\psi_{g_X,u}\mathcal{T}$ are strict for any $i \in \mathbb{Z}$ and $u \in \mathbb{R} \times \mathbb{C}$.

Corollary 3.3.12 *Assume moreover that $\mathcal{T}$ is localizable along g_X. Then, $F^i_\dagger(\mathcal{T})$ are also localizable along g_Y, and we have natural isomorphisms $F^i_\dagger(\mathcal{T})[\star g_Y] \simeq F^i_\dagger\big(\mathcal{T}[\star g_X]\big)$.* □

3.3.1.4 $\mathcal{R}$-Modules and $\mathcal{R}$-Triples

Definition 3.3.13 Let $\mathcal{M}$ (resp. $\mathcal{T}$) be a coherent $\mathcal{R}_X$-module (resp. $\mathcal{R}_X$-triple) which is strictly specializable along g. It is called localizable along g if $\mathcal{M}(*g)$ (resp. $\mathcal{T}(*g)$) is localizable along g. □

In that case, we set $\mathcal{M}[\star g] := \big(\mathcal{M}(*g)\big)[\star g]$ and $\mathcal{T}[\star g] := \big(\mathcal{T}(*g)\big)[\star g]$.

3.3.2 Canonical Prolongations Across Hypersurfaces

Let D be an effective divisor of X.

Definition 3.3.14 A coherent $\mathcal{R}_{X(*D)}$-module $\mathcal{M}$ (resp. $\mathcal{R}_{X(*D)}$-triple $\mathcal{T}$) is called strictly specializable along D if the following holds:

- Let U be any open subset of X with a generator g_U of $\mathcal{O}(-D)_{|U}$. Then, $\mathcal{M}_{|U}$ (resp. $\mathcal{T}_{|U}$) is strictly specializable along g_U.

We say that $\mathcal{M}$ (resp. $\mathcal{T}$) is localizable along D if the following holds:

- Let U and g_U be as above. Then, $\mathcal{M}_{|U}$ (resp. $\mathcal{T}_{|U}$) is localizable along g_U. □

When $\mathcal{M}$ is strictly specializable along D, we set

$$\mathcal{KMS}(\mathcal{M},D) := \bigcup_{(U,g)} \mathcal{KMS}(\mathcal{M}_{|U},g),$$

where (U,g) runs through a pair of open subsets $U \subset X$ with a generator g of $\mathcal{O}(-D)_{|U}$. A non-zero complex number λ_0 is called generic for $(\mathcal{M},D)$ if $\mathfrak{e}(\lambda_0) : \mathcal{KMS}(\mathcal{M},D) \cup (\mathbb{Z} \times \{0\}) \longrightarrow \mathbb{C}$ is injective.

If $\mathcal{M}$ is localizable along D, let $\mathcal{M}[\star D]$ denote an $\mathcal{R}_X$-module with an isomorphism $\big(\mathcal{M}[\star D]\big)(*D) \simeq \mathcal{M}$ determined by the following condition:

- Let U and g_U be as above. Then, $\mathcal{M}[\star D]_{|U} \simeq \mathcal{M}_{|U}[\star g_U]$.

We use the symbols $\mathcal{KMS}(\mathcal{T},D)$ and $\mathcal{T}[\star D]$ with similar meanings.

The following lemma is clear by Lemma 3.3.4.

Lemma 3.3.15 *Suppose that a coherent strict $\mathcal{R}_X(*D)$-module $\mathcal{M}$ is localizable along D. If a non-zero complex number λ_0 is generic for $(\mathcal{M},D)$ then $\mathcal{M}[\star D]^{\lambda_0}$ is naturally isomorphic to $\mathcal{M}^{\lambda_0}[\star D]$.* □

3.3.2.1 Morphisms

Let $\mathcal{T}_i$ $(i = 1,2)$ be coherent $\mathcal{R}_X(*D)$-triples which are localizable along D. We obtain the following lemma from Lemma 3.3.8.

Lemma 3.3.16 *Let $\star = *,!$. Morphisms $\mathcal{T}_1 \longrightarrow \mathcal{T}_2$ of $\mathcal{R}_X(*D)$-triples bijectively correspond to morphisms $\mathcal{T}_1[\star D] \longrightarrow \mathcal{T}_2[\star D]$ of $\mathcal{R}_X$-triples. If $\mathcal{T}_i$ are integrable, we have such bijections for integrable morphisms. Similar claims hold for $\mathcal{R}_X(*D)$-modules.* □

Let $f : \mathcal{T}_1 \longrightarrow \mathcal{T}_2$ be a morphism. We have the induced morphism $f[\star D] : \mathcal{T}_1[\star D] \longrightarrow \mathcal{T}_2[\star D]$. We obtain the following lemma from Lemma 3.3.9.

Lemma 3.3.17 *Suppose that f is strictly specializable along D, i.e., for any open subset $U \subset X$ with a generator g_U of $\mathcal{O}(-D)_{|U}$, $f_{|U}$ is strictly specializable along g_U. Then,* $\mathrm{Ker} f$, $\mathrm{Im} f$ *and* $\mathrm{Cok} f$ *are strictly specializable and localizable along D, and we naturally have* $\mathrm{Ker}(f[\star D]) \simeq \mathrm{Ker}(f)[\star D]$, $\mathrm{Im}(f[\star D]) \simeq \mathrm{Im}(f)[\star D]$, *and* $\mathrm{Cok}(f[\star D]) \simeq \mathrm{Cok}(f)[\star D]$. □

Let $\mathcal{T}^{\bullet}$ be a bounded complex of $\mathcal{R}_{X(*D)}$-triples such that each $\mathcal{T}^p$ is localizable along D. We obtain the following from Lemma 3.3.10.

Lemma 3.3.18 *Assume that the complex $\mathcal{T}^{\bullet}$ is strictly specializable in the following sense:*

- *For any open subset $U \subset X$ with a generator g_U of $\mathcal{O}(-D)_{|U}$, $\mathcal{H}^{\bullet}(\psi_{g_U,u}\mathcal{T}^{\bullet})$ are strict for any $u \in \mathbb{R} \times \mathbb{C}$.*

Then, the following holds:

- *$\mathcal{H}^{\bullet}(\mathcal{T}^{\bullet})$ are strictly specializable and localizable along D.*
- *We have natural isomorphisms $\mathcal{H}^{\bullet}\big(\mathcal{T}^{\bullet}[\star D]\big) \simeq \mathcal{H}^{\bullet}\big(\mathcal{T}^{\bullet}\big)[\star D]$.* □

3.3.2.2 Compatibility with Push-Forward

Let $F : X \longrightarrow Y$ be a morphism of complex manifolds. Let D_Y be an effective divisor of Y, and let $D_X := F^*(D_Y)$ be the pull back of the divisor. Let $\mathcal{T}$ be a coherent $\mathcal{R}_X$-triple which is strictly specializable along D_X. Assume the following:

- The support of $\mathcal{T}$ is proper over Y.
- Let $U_Y \subset Y$ be open with a generator g_{U_Y} of $\mathcal{O}(-D_Y)_{|U}$. We set $U_X := F^{-1}(U_Y)$ and $g_{U_X} := g_{U_Y} \circ F_{|U_X}$. We suppose that $F^i_{\dagger}\tilde{\psi}_{g_{U_X},u}(\mathcal{T}_{|U_X})$ are strict for any $u \in \mathbb{R} \times \mathbb{C}$.

Lemma 3.3.19 *If moreover $\mathcal{T}$ is localizable along D_X, then $F^i_{\dagger}\mathcal{T}$ is localizable along D_Y, and we have a natural isomorphism $F^i_{\dagger}(\mathcal{T}[\star D_X]) \simeq (F^i_{\dagger}\mathcal{T})[\star D_Y]$. A similar claim holds for $\mathcal{R}_X$-modules.* □

3.3.2.3 $\mathcal{R}$-Modules and $\mathcal{R}$-Triples

Definition 3.3.20 Let $\mathcal{M}$ (resp. $\mathcal{T}$) be a coherent $\mathcal{R}_X$-module (resp. $\mathcal{R}_X$-triple) which is strictly specializable along D. It is called localizable along D, if and only if $\mathcal{M}(*D)$ (resp. $\mathcal{T}(*D)$) is localizable along D. □

In that case, we set $\mathcal{M}[\star D] := \big(\mathcal{M}(*D)\big)[\star D]$ and $\mathcal{T}[\star D] := \big(\mathcal{T}(*D)\big)[\star D]$. We use the notation $\mathrm{KMS}(\mathcal{M}, D)$ and $\mathrm{KMS}(\mathcal{T}, D)$ as in the case of $\mathcal{R}_{X(*D)}$-modules and $\mathcal{R}_{X(*D)}$-triples.

Chapter 4
Gluing and Specialization of $\mathcal{R}$-Triples

Let us recall the excellent formalism of Beilinson [3] for the specialization and the gluing of holonomic $\mathscr{D}$-modules along holomorphic functions. Let g be a holomorphic function on X. We set $H := g^{-1}(0)$. We put $A^{a,b} := s^a\mathbb{C}[s]/s^b\mathbb{C}[s]$. We have locally free $\mathcal{O}_X(*H)$-module $\tilde{I}_g^{a,b} := A^{a,b} \otimes_{\mathbb{C}} \mathcal{O}[*H]$ with the flat connection $\nabla = d + s \cdot dg/g$. Here, $s\cdot$ means the multiplication of s. For any holonomic $\mathscr{D}$-module M on X, we set $\Pi_{g\star}^{a,b}(M) := (M \otimes \tilde{I}_g^{a,b})[\star H]$ ($\star = *, !$). Beilinson introduced the following:

$$\Pi_{g,*!}^{a,b}(M) := \varprojlim_{N\to\infty} \operatorname{Cok}\bigl(\Pi_{g,!}^{b,N}(M) \longrightarrow \Pi_{g,*}^{a,N}(M)\bigr)$$

In particular, we obtain the nearby cycle sheaf $\psi_g^{(a)}(M) := \Pi_{g,*!}^{a,a}(M)$ and the maximal sheaf $\Xi_g^{(a)}(M) := \Pi_{g,*!}^{a,a+1}(M)$. The vanishing cycle sheaf $\phi_g^{(0)}(M)$ is defined as the cohomology of the following natural complex:

$$M(!H) \longrightarrow \Xi_g^{(0)}(M) \oplus M \longrightarrow M(*H)$$

Beilinson found that M is reconstructed as the cohomology of the following natural complex:

$$\psi_g^{(1)}(M) \longrightarrow \Xi_g^{(0)}(M) \oplus M \longrightarrow \psi_g^{(0)}(M)$$

The gluing of $\mathscr{D}$-modules along H is given as follows. Let N_1 be a holonomic $\mathscr{D}_X(*H)$-module, and let N_2 be a holonomic $\mathscr{D}_X$-module whose support is contained in H. Suppose that we are given morphisms $\psi_g^{(1)}(N_1) \xrightarrow{u} N_2 \xrightarrow{v} \psi_g^{(0)}(N_1)$ such that the composite is equal to the natural morphism $\psi_g^{(1)}(N_1) \longrightarrow \psi_g^{(0)}(N_1)$.

T. Mochizuki, *Mixed Twistor $\mathscr{D}$-modules*, Lecture Notes in Mathematics 2125,
DOI 10.1007/978-3-319-10088-3_4

Then, we have the $\mathscr{D}_X$-module $\mathrm{Glue}(N_1, N_2, u, v)$ obtained as the cohomology of the following complex:

$$\psi_g^{(1)}(N_1) \longrightarrow \Xi_g^{(0)}(N_1) \oplus N_2 \longrightarrow \psi_g^{(0)}(N_1)$$

We naturally have $\phi_g^{(0)}\,\mathrm{Glue}(N_1, N_2, u, v) \simeq N_2$ and $\mathrm{Glue}(N_1, N_2, u, v)(*H) = N_1$. In this sense, $\mathrm{Glue}(N_1, N_2, u, v)$ gives the gluing of N_1 and N_2 along H.

Although there are several ways for gluing of $\mathscr{D}$-modules or perverse sheaves, the formalism of Beilinson is most convenient for our purpose. Once we obtain the functors $[*H]$, the others are obtained by general procedures in an abelian category.

We introduce the counterparts of the functors $\Pi^{a,b}_{g*!}$, $\Xi_g^{(a)}$, $\psi_g^{(a)}$ and $\phi_g^{(0)}$ in the context of $\mathcal{R}$-triples. We explain more details on the constructions in Sects. 4.1–4.2.

As studied in Chap. 3, we have the localization functors $[\star t]$ for any coordinate function t. So, we can obtain the functors $\Pi^{a,b}_{t*!}$, $\Xi_t^{(a)}$, $\psi_t^{(a)}$ and $\phi_t^{(0)}$ by the above procedure in the abelian category of $\mathcal{R}$-modules or $\mathcal{R}$-triples. The gluing procedure along t is also given in the same way.

For more general functions, in this stage, we can obtain the functors $\psi_g^{(a)}$ and $\phi_g^{(a)}$ for any $\mathcal{R}$-triples or $\mathcal{R}$-modules which are strictly specializable along g. For the maximal functor $\Xi_g^{(a)}$, we have the issue as in the construction of $[\star g]$ mentioned in Chap. 3. Later, in Sect. 11.3, we will see that $\Xi_g^{(a)}(\mathcal{T})$ exist for any holomorphic function g if the $\mathcal{R}_X$-triple $\mathcal{T}$ underlies a mixed twistor $\mathscr{D}$-module.

As recalled in Sect. 2.1, in the context of strictly specializable $\mathcal{R}$-triples, Sabbah defined the nearby cycle functor $\psi_{g,-\boldsymbol{\delta}}$ [66]. We compare $\psi_g^{(a)}$ and $\psi_{g,-\boldsymbol{\delta}}$ in Proposition 4.3.1. See also Proposition 4.3.2. Both $\psi_g^{(a)}$ and $\psi_{g,-\boldsymbol{\delta}}$ have their advantages. For example, $\psi_g^{(a)}$ is more useful when we study the gluing of $\mathcal{R}$-triples, and $\psi_{g,-\boldsymbol{\delta}}$ is more useful for computations.

In Sect. 4.4, we introduce the admissible specializability condition for filtered $\mathcal{R}$-triples. It is used in the definition of mixed twistor $\mathscr{D}$-modules in Chap. 7.

4.1 Beilinson Functors for $\mathcal{R}$-Modules

4.1.1 The Functors $\Pi^{a,b}$, $\Pi^{a,b}_!$, $\Pi^{a,b}_*$ and $\Pi^{a,b}_{*!}$ for $\mathcal{R}$-Modules

We continue to use the setting in Sect. 3.1. For any $u \in \mathbb{R} \times \mathbb{C}$, we denote $\psi_{t,u}$ by ψ_u for simplicity if there is no risk of confusion. Let $a \leq b$ be integers. For any $\mathcal{R}_{X(*H)}(*t)$-module $\mathcal{M}$, we set $\Pi^{a,b}\mathcal{M} := \mathcal{M} \otimes \tilde{\mathbb{I}}_2^{a,b}$. (See Sect. 2.3 for $\tilde{\mathbb{I}}_2^{a,b}$.) We have naturally defined morphisms $\Pi^{a,b}\mathcal{M} \longrightarrow \Pi^{c,d}\mathcal{M}$ for $a > c$ and $b > d$. If $\mathcal{M}$ is coherent and strictly specializable along t, so is $\Pi^{a,b}\mathcal{M}$. In that case, we set

$\Pi^{a,b}_{\star}\mathcal{M} := \Pi^{a,b}\mathcal{M}[\star t]$. We define

$$\Pi^{a,b}_{*!}\mathcal{M} := \varprojlim_{N\to\infty} \operatorname{Cok}\Big(\Pi^{b,N}_{!}\mathcal{M} \longrightarrow \Pi^{a,N}_{*}\mathcal{M}\Big)$$

We also denote them by $\Pi^{a,b}_{t}\mathcal{M}$, $\Pi^{a,b}_{t\star}\mathcal{M}$ ($\star = *, !$) and $\Pi^{a,b}_{t*!}\mathcal{M}$ when we emphasize the dependence on t.

Lemma 4.1.1 *Let $P \in \mathcal{X}$. There exists $N(P) > 0$ such that, for any $N > N(P)$, on a neighbourhood $\mathcal{X}_P$ of P, the morphism $\Pi^{b,N}_{!}\mathcal{M} \longrightarrow \Pi^{a,N}_{*}\mathcal{M}$ is strictly specializable along t, and its cokernel is independent of N in the sense that the naturally defined morphism*

$$\operatorname{Cok}\big(\Pi^{b,N+1}_{!}\mathcal{M} \longrightarrow \Pi^{a,N+1}_{*}\mathcal{M}\big) \longrightarrow \operatorname{Cok}\big(\Pi^{b,N}_{!}\mathcal{M} \longrightarrow \Pi^{a,N}_{*}\mathcal{M}\big) \tag{4.1}$$

is an isomorphism on $\mathcal{X}_P$.

Proof In this proof, N is a sufficiently large number, and we omit to distinguish a neighbourhood $\mathcal{X}_P$. If $u \notin \mathbb{Z}_{\geq 0} \times \{0\}$, we have isomorphisms $\psi_u\big(\Pi^{a,N}_{\star}\mathcal{M}\big) \simeq \psi_u(\mathcal{M}) \otimes \mathbb{I}^{a,N}_{2}$ and the following commutative diagram:

$$\begin{array}{ccc}
\psi_u\big(\Pi^{b,N}_{!}\mathcal{M}\big) & \longrightarrow & \psi_u\big(\Pi^{a,N}_{*}\mathcal{M}\big) \\
\simeq\downarrow & & \simeq\downarrow \\
\psi_u(\mathcal{M}) \otimes \mathbb{I}^{b,N}_{2} & \xrightarrow{\mathrm{id}\otimes\iota} & \psi_u(\mathcal{M}) \otimes \mathbb{I}^{a,N}_{2}
\end{array} \tag{4.2}$$

Here ι is a natural inclusion $\mathbb{I}^{b,N}_{2} \longrightarrow \mathbb{I}^{a,N}_{2}$. Hence, $\psi_u\big(\Pi^{b,N}_{!}\mathcal{M}\big) \longrightarrow \psi_u\big(\Pi^{a,N}_{*}\mathcal{M}\big)$ is strict in this case.

We have the following isomorphisms:

$$\psi_0\big(\Pi^{a,N}_{!}(\mathcal{M})\big) \xleftarrow[\simeq]{-\eth_t} \lambda\psi_{-\boldsymbol{\delta}}\big(\Pi^{a,N}_{!}\mathcal{M}\big) \simeq \lambda\psi_{-\boldsymbol{\delta}}(\mathcal{M}) \otimes \mathbb{I}^{a,N}_{2}$$

$$\psi_0\big(\Pi^{a,N}_{*}(\mathcal{M})\big) \xrightarrow[\simeq]{t} \psi_{-\boldsymbol{\delta}}\big(\Pi^{a,N}_{*}\mathcal{M}\big) \simeq \psi_{-\boldsymbol{\delta}}(\mathcal{M}) \otimes \mathbb{I}^{a,N}_{2}$$

We have the following commutative diagram:

$$\begin{array}{ccccc}
\psi_0\big(\Pi^{b,N}_{!}\mathcal{M}\big) & \longrightarrow & \psi_0\big(\Pi^{b,N}_{*}\mathcal{M}\big) & \longrightarrow & \psi_0\big(\Pi^{a,N}_{*}\mathcal{M}\big) \\
\simeq\uparrow & & \simeq\downarrow & & \simeq\downarrow \\
\lambda\psi_{-\boldsymbol{\delta}}(\mathcal{M}) \otimes \mathbb{I}^{b,N}_{2} & \xrightarrow{\mathcal{N}_{\mathcal{M}}\otimes\mathrm{id}+\mathrm{id}\otimes\mathcal{N}_{\mathbb{I},2}} & \psi_{-\boldsymbol{\delta}}(\mathcal{M}) \otimes \mathbb{I}^{b,N}_{2} & \xrightarrow{\mathrm{id}\otimes\iota} & \psi_{-\boldsymbol{\delta}}(\mathcal{M}) \otimes \mathbb{I}^{a,N}_{2}
\end{array} \tag{4.3}$$

Here, $\mathcal{N}_{\mathcal{M}}$ is induced by $-t\partial_t$. Then, it is easy to check that $\psi_0(\Pi_!^{b,N}\mathcal{M}) \longrightarrow \psi_0(\Pi_*^{a,N}\mathcal{M})$ is strict. The case $u \in \mathbb{Z}_{\geq 0} \times \{0\}$ also follows. Hence, the natural morphism $\Pi_!^{b,N}\mathcal{M} \longrightarrow \Pi_*^{a,N}\mathcal{M}$ is strict.

The following morphism of $\mathcal{R}_{X(*H)}(*t)$-modules is an isomorphism:

$$\mathrm{Cok}(\Pi^{b,N+1}\mathcal{M} \longrightarrow \Pi^{a,N+1}\mathcal{M}) \longrightarrow \mathrm{Cok}(\Pi^{b,N}\mathcal{M} \longrightarrow \Pi^{a,N}\mathcal{M})$$

We have the following identifications:

$$\begin{array}{ccc} \psi_u \mathrm{Cok}(\Pi_!^{b,N+1}\mathcal{M} \to \Pi_*^{a,N+1}\mathcal{M}) & \longrightarrow & \psi_u \mathrm{Cok}(\Pi_!^{b,N}\mathcal{M} \to \Pi_*^{a,N}\mathcal{M}) \\ \simeq\downarrow & & \simeq\downarrow \\ \mathrm{Cok}\Big(\psi_u(\Pi_!^{b,N+1}\mathcal{M}) \to \psi_u(\Pi_*^{a,N+1}\mathcal{M})\Big) & \longrightarrow & \mathrm{Cok}(\psi_u(\Pi_!^{b,N}\mathcal{M}) \to \psi_u(\Pi_*^{a,N}\mathcal{M})) \end{array} \tag{4.4}$$

By using the identification (4.2), we obtain that the horizontal arrows in (4.4) are isomorphisms in the case $u \notin \mathbb{Z}_{\geq 0} \times \{0\}$. By using Lemma 2.3.3, with an easy diagram chasing, we obtain that the horizontal arrows in (4.4) are isomorphisms in the case $u = (0,0)$. It also follows that they are isomorphisms in the case $u \in \mathbb{Z}_{>0} \times \{0\}$. Thus, (4.1) is an isomorphism. □

Corollary 4.1.2 *The $\mathcal{R}_{X(*H)}$-modules $\Pi_{*!}^{a,b}\mathcal{M}$ are strictly specializable along t.* □

We obtain the following lemma by a similar argument.

Lemma 4.1.3 *Let $P \in \mathcal{X}$. There exists $N(P) > 0$ such that, for any $N > N(P)$, on a neighbourhood $\mathcal{X}_P$ of P, the morphism $\Pi_!^{-N,b}\mathcal{M} \longrightarrow \Pi_*^{-N,a}\mathcal{M}$ is strictly specializable along t, and its kernel is independent of the choice of N in the sense that the following naturally defined morphism*

$$\mathrm{Ker}\Big(\Pi_!^{-N,b}\mathcal{M} \longrightarrow \Pi_*^{-N,a}\mathcal{M}\Big) \longrightarrow \mathrm{Ker}\Big(\Pi_!^{-N-1,b}\mathcal{M} \longrightarrow \Pi_*^{-N-1,a}\mathcal{M}\Big)$$

is an isomorphism. □

The following lemma is clear by construction.

Lemma 4.1.4 *If $\mathcal{M}$ is integrable, $\Pi_\star^{a,b}\mathcal{M}$ ($\star = *, !$) are naturally integrable.* □

4.1.2 Another Description

We have another description of the functor $\Pi_{*!}^{a,b}$.

Lemma 4.1.5 *We have the following natural isomorphism:*

$$\Pi^{a,b}_{*!}(\mathcal{M}) \simeq \varinjlim_{N\to\infty} \operatorname{Ker}\Big(\Pi_!^{-N,b}\mathcal{M} \longrightarrow \Pi_*^{-N,a}\mathcal{M}\Big) \tag{4.5}$$

Proof First, let us construct such an isomorphism on a neighbourhood $\mathcal{X}_P$ of P. We have the following natural commutative diagram:

$$\begin{array}{ccccc}
\Pi_!^{b,N}\mathcal{M} & \longrightarrow & \Pi_!^{-N,N}\mathcal{M} & \longrightarrow & \Pi_!^{-N,b}\mathcal{M} \\
\downarrow & & \downarrow & & \downarrow \\
\Pi_*^{a,N}\mathcal{M} & \longrightarrow & \Pi_*^{-N,N}\mathcal{M} & \longrightarrow & \Pi_*^{-N,a}\mathcal{M}
\end{array}$$

Hence, we have only to prove the following morphisms are isomorphisms on $\mathcal{X}_P$, if N is sufficiently large:

$$\operatorname{Cok}\Big(\Pi_!^{-N,N}\mathcal{M} \longrightarrow \Pi_*^{-N,N}\mathcal{M}\Big) \longrightarrow \operatorname{Cok}\Big(\Pi_!^{-N,b}\mathcal{M} \longrightarrow \Pi_*^{-N,a}\mathcal{M}\Big) \tag{4.6}$$

$$\operatorname{Ker}\Big(\Pi_!^{b,N}\mathcal{M} \longrightarrow \Pi_*^{a,N}\mathcal{M}\Big) \longrightarrow \operatorname{Ker}\Big(\Pi_!^{-N,N}\mathcal{M} \longrightarrow \Pi_*^{-N,N}\mathcal{M}\Big) \tag{4.7}$$

We have $\operatorname{Cok}\Big(\Pi^{-N,N}\mathcal{M} \longrightarrow \Pi^{-N,N}\mathcal{M}\Big) \simeq \operatorname{Cok}\Big(\Pi^{-N,b}\mathcal{M} \longrightarrow \Pi^{-N,a}\mathcal{M}\Big) \simeq 0$. Let us consider the following morphism:

$$\begin{aligned}
&\operatorname{Cok}\Big(\psi_u(\Pi_!^{-N,N}\mathcal{M}) \longrightarrow \psi_u(\Pi_*^{-N,N}\mathcal{M})\Big) \\
&\quad \longrightarrow \operatorname{Cok}\Big(\psi_u(\Pi_!^{-N,b}\mathcal{M}) \longrightarrow \psi_u(\Pi_*^{-N,a}\mathcal{M})\Big)
\end{aligned} \tag{4.8}$$

If $u \notin \mathbb{Z}_{\geq 0} \times \{0\}$, the both sides are 0. For $u = (0,0)$, we obtain that (4.8) is an isomorphism from Lemma 2.3.4. Hence, (4.8) is an isomorphism for each u. Then, we can deduce (4.6) is an isomorphism by using an argument in the proof of Lemma 4.1.1. We obtain that (4.7) is an isomorphism by a similar argument. Thus, we obtain the isomorphism (4.5) on a neighbourhood of $\mathcal{X}_P$. By varying P, we obtain the isomorphism globally. □

4.1.3 The Induced Morphism

Let $\mathcal{M}_i$ $(i = 1, 2)$ be coherent $\mathcal{R}_{X(*H)}(*t)$-modules which are strictly specializable along t. For any morphism $f : \mathcal{M}_1 \longrightarrow \mathcal{M}_2$, we have the induced morphisms $\Pi^{a,b}_{\star}(f) : \Pi^{a,b}_{\star}(\mathcal{M}_1) \longrightarrow \Pi^{a,b}_{\star}(\mathcal{M}_2)$ and $\Pi^{a,b}_{*!}(f) : \Pi^{a,b}_{*!}(\mathcal{M}_1) \longrightarrow \Pi^{a,b}_{*!}(\mathcal{M}_2)$.

Lemma 4.1.6 *If f is strictly specializable, we have the natural isomorphisms* $\mathrm{Ker}(\Pi^{a,b}_{*!}f) \simeq \Pi^{a,b}_{*!}\,\mathrm{Ker} f$, $\mathrm{Im}(\Pi^{a,b}_{*!}f) \simeq \Pi^{a,b}_{*!}\,\mathrm{Im} f$ *and* $\mathrm{Cok}(\Pi^{a,b}_{*!}f) \simeq \Pi^{a,b}_{*!}\,\mathrm{Cok} f$.

Proof We obtain the claim for Ker from the following commutative diagram:

$$\begin{array}{ccccc} \Pi^{a,b}_{*!}\mathcal{M}_1 & \longrightarrow & \Pi^{-N,b}_{!}\mathcal{M}_1 & \longrightarrow & \Pi^{-N,a}_{*}\mathcal{M}_1 \\ \downarrow & & \downarrow & & \downarrow \\ \Pi^{a,b}_{*!}\mathcal{M}_2 & \longrightarrow & \Pi^{-N,b}_{!}\mathcal{M}_2 & \longrightarrow & \Pi^{-N,a}_{*}\mathcal{M}_2 \end{array}$$

We obtain the claim for Cok from the following commutative diagram:

$$\begin{array}{ccccc} \Pi^{b,N}_{!}\mathcal{M}_1 & \longrightarrow & \Pi^{a,N}_{*}\mathcal{M}_1 & \longrightarrow & \Pi^{a,b}_{*!}\mathcal{M}_1 \\ \downarrow & & \downarrow & & \downarrow \\ \Pi^{b,N}_{!}\mathcal{M}_2 & \longrightarrow & \Pi^{a,N}_{*}\mathcal{M}_2 & \longrightarrow & \Pi^{a,b}_{*!}\mathcal{M}_2 \end{array}$$

The claim for the image follows from the claims for the kernel and the cokernel with an easy diagram chase. □

Corollary 4.1.7 *Let $\mathcal{M}^{\bullet}$ be as in Corollary* 3.1.22. *We have natural isomorphisms* $\Pi^{a,b}_{*!}\mathcal{H}^{\bullet}\mathcal{M}^{\bullet} \simeq \mathcal{H}^{\bullet}\Pi^{a,b}_{*!}\mathcal{M}^{\bullet}$. □

4.1.4 Compatibility with the Push-Forward

Let us consider the compatibility of the functors $\Pi^{a,b}_{\star}$ and $\Pi^{a,b}_{*!}$ with the push-forward. We use the notation in Sect. 3.2.4. Let $\mathcal{M}$ be a good coherent $\mathcal{R}_{X(*H)}(*t)$-module which is strictly specializable along t. Assume the following:

- The support of $\mathcal{M}$ is proper over Y with respect to F.
- $F^i_{\dagger}\tilde{\psi}_u\mathcal{M}$ is strict for any $u \in \mathbb{R}\times\mathbb{C}$ and $i \in \mathbb{Z}$.

We have a natural isomorphism $F^i_{\dagger}(\Pi^{a,b}\mathcal{M}) \simeq \Pi^{a,b}F^i_{\dagger}(\mathcal{M})$ of coherent $\mathcal{R}_{Y(*H_Y)}(*t)$-modules. By the second assumption, $F^i_{\dagger}(\Pi^{a,b}\mathcal{M})$ are strictly specializable along t, and we have $F^i_{\dagger}(\Pi^{a,b}_{\star}\mathcal{M}) \simeq \Pi^{a,b}_{\star}F^i_{\dagger}\mathcal{M}$ by Lemma 3.2.12. We have the following naturally defined morphism of $\mathcal{R}_{Y(*H_Y)}$-modules:

$$\Pi^{a,b}_{*!}F^i_{\dagger}\mathcal{M} \longrightarrow F^i_{\dagger}\Pi^{a,b}_{*!}\mathcal{M}. \tag{4.9}$$

Proposition 4.1.8 *The morphism* (4.9) *is an isomorphism.*

Proof It is enough to consider the issue locally around any point $P \in \mathcal{Y}$. Let $\mathcal{K}_N$ be the kernel of $\Pi^{b,N}_{!}\mathcal{M} \longrightarrow \Pi^{a,N}_{*}\mathcal{M}$. If N is sufficiently large, we have the following exact sequence:

$$0 \longrightarrow \psi_u \mathcal{K}_N \longrightarrow \psi_u(\Pi_!^{b,N}\mathcal{M}) \longrightarrow \psi_u(\Pi_*^{a,N}\mathcal{M}) \longrightarrow \psi_u \Pi_{*!}^{a,b}\mathcal{M} \longrightarrow 0$$

It can be rewritten as follows:

$$0 \longrightarrow \psi_u \mathcal{K}_N \longrightarrow \psi_u \mathcal{M} \otimes \mathbb{I}^{b,N} \longrightarrow \psi_u \mathcal{M} \otimes \mathbb{I}^{a,N} \longrightarrow \psi_u \Pi_{*!}^{a,b}\mathcal{M} \longrightarrow 0 \qquad (4.10)$$

Note that, as an $\mathcal{R}_{X_0(*H_0)}$-complex, the exact sequence (4.10) has a splitting, and as $\mathcal{R}_{X_0(*H_0)}$-modules, $\psi_u \mathcal{K}_N$ and $\psi_u \Pi_{*!}^{a,b}\mathcal{M}$ are direct sums of some copies of $\psi_u \mathcal{M}$. Hence, $F_\dagger^i \psi_u \Pi_{*!}^{a,b}\mathcal{M}$ and $F_\dagger^i \psi_u \mathcal{K}_N$ are strict, and the induced sequence

$$0 \longrightarrow F_\dagger^i \psi_u \mathcal{K}_N \longrightarrow F_\dagger^i \psi_u \mathcal{M} \otimes \mathbb{I}^{b,N} \longrightarrow F_\dagger^i \psi_u \mathcal{M} \otimes \mathbb{I}^{a,N} \longrightarrow F_\dagger^i \psi_u \Pi_{*!}^{a,b}\mathcal{M} \longrightarrow 0$$

is exact. The following complex of $\mathcal{R}_{Y(*H_Y)}(*t)$-modules is clearly exact:

$$\begin{aligned} 0 \longrightarrow F_\dagger^i \mathcal{K}_N(*t) \longrightarrow F_\dagger^i \mathcal{M} \otimes \tilde{\mathbb{I}}^{b,N} \longrightarrow F_\dagger^i \mathcal{M} \otimes \tilde{\mathbb{I}}^{a,N} \\ \longrightarrow F_\dagger^i \Pi_{*!}^{a,b}\mathcal{M}(*t) \longrightarrow 0 \qquad (4.11) \end{aligned}$$

So, $0 \longrightarrow F_\dagger^i \mathcal{K}_N \longrightarrow F_\dagger^i \Pi_!^{b,N}\mathcal{M} \longrightarrow F_\dagger^i \Pi_*^{a,N}\mathcal{M} \longrightarrow F_\dagger^i \Pi_{*!}^{a,b}\mathcal{M} \longrightarrow 0$ is exact. In particular, we obtain that (4.9) is an isomorphism. □

4.1.5 The Functors $\psi^{(a)}$ and $\Xi^{(a)}$ for $\mathcal{R}(*t)$-Modules

Let $\mathcal{M}$ be any coherent $\mathcal{R}_{X(*H)}(*t)$-module which is strictly specializable along t. We define

$$\psi^{(a)}(\mathcal{M}) := \Pi_{*!}^{a,a}(\mathcal{M}), \quad \Xi^{(a)}(\mathcal{M}) := \Pi_{*!}^{a,a+1}(\mathcal{M}).$$

By Corollary 4.1.2, they are strictly specializable along t. We also denote them by $\psi_t^{(a)}(\mathcal{M})$ and $\Xi_t^{(a)}(\mathcal{M})$ when we emphasize the dependence on t. We have naturally defined exact sequences:

$$0 \longrightarrow \mathcal{M}[!t] \otimes s^a \xrightarrow{\alpha_a} \Xi^{(a)}(\mathcal{M}) \xrightarrow{\beta_a} \psi^{(a)}(\mathcal{M}) \longrightarrow 0$$

$$0 \longrightarrow \psi^{(a+1)}(\mathcal{M}) \xrightarrow{\gamma_a} \Xi^{(a)}(\mathcal{M}) \xrightarrow{\delta_a} \mathcal{M}[*t] \otimes s^a \longrightarrow 0$$

According to Corollary 4.1.7, we have natural isomorphisms

$$\psi^{(a)}\mathcal{H}^\bullet(\mathcal{M}^\bullet) \simeq \mathcal{H}^\bullet\psi^{(a)}(\mathcal{M}^\bullet), \qquad \Xi^{(a)}\mathcal{H}^\bullet(\mathcal{M}^\bullet) \simeq \mathcal{H}^\bullet\Xi^{(a)}(\mathcal{M}^\bullet)$$

for any complex $\mathcal{M}^\bullet$ as in Corollary 3.1.22. In the situation of Sect. 4.1.4, we have natural isomorphisms

$$F^i_\dagger\psi^{(a)}(\mathcal{M}) \simeq \psi^{(a)}F^i_\dagger(\mathcal{M}), \qquad F^i_\dagger\Xi^{(a)}(\mathcal{M}) \simeq \Xi^{(a)}F^i_\dagger(\mathcal{M}).$$

4.1.6 Beilinson Functors for $\mathcal{R}$-Modules

Let $\mathcal{M}$ be a coherent $\mathcal{R}_{X(*H)}$-module which is strictly specializable along t. We obtain a coherent $\mathcal{R}_{X(*H)}(*t)$-module $\tilde{\mathcal{M}} := \mathcal{M}(*t)$, which is strictly specializable along t. We define $\Pi^{a,b}(\mathcal{M}) := \Pi^{a,b}(\tilde{\mathcal{M}})$, $\Pi^{a,b}_\star(\mathcal{M}) := \Pi^{a,b}_\star(\tilde{\mathcal{M}})$ and $\Pi^{a,b}_{*!}(\mathcal{M}) := \Pi^{a,b}_{*!}(\tilde{\mathcal{M}})$. We also define $\psi^{(a)}(\mathcal{M}) := \psi^{(a)}(\tilde{\mathcal{M}})$ and $\Xi^{(a)}(\mathcal{M}) := \Xi^{(a)}(\tilde{\mathcal{M}})$. As in the case of $\mathcal{R}(*t)$-modules, we also denote them by $\Pi^{a,b}_t(\mathcal{M})$, etc., when we emphasize the dependence on t.

Recall we have the morphisms $\mathcal{M}[!t] \xrightarrow{\iota_1} \mathcal{M} \xrightarrow{\iota_2} \mathcal{M}[*t]$. Let $\phi^{(a)}(\mathcal{M})$ be defined as the cohomology of the following complex:

$$\mathcal{M}[!t]\otimes s^a \xrightarrow{\alpha_a+\iota_1} \Xi^{(a)}(\mathcal{M})\oplus\big(\mathcal{M}\otimes s^a\big) \xrightarrow{\delta_1-\iota_2} \mathcal{M}[*t]\otimes s^a$$

We also denote them by $\phi^{(a)}_t(\mathcal{M})$ when we emphasize the dependence on t.

Lemma 4.1.9 *The $\mathcal{R}_{X(*H)}$-modules $\psi^{(a)}(\mathcal{M})$, $\Xi^{(a)}(\mathcal{M})$ and $\phi^{(a)}(\mathcal{M})$ are strictly specializable along t. We have the following natural isomorphisms:*

$$\psi_0\big(\psi^{(a)}(\mathcal{M})\big) \simeq \psi_{-\delta}(\mathcal{M})\,s^a,$$
$$\psi_0\big(\Xi^{(a)}(\mathcal{M})\big) \simeq \psi_{-\delta}(\mathcal{M})\,s^a \oplus \psi_{-\delta}(\mathcal{M})\,s^{a+1},$$
$$\psi_0\big(\phi^{(a)}(\mathcal{M})\big) \simeq \psi_0(\mathcal{M})\,s^a$$

Proof By Corollary 4.1.2, $\psi^{(a)}(\mathcal{M})$ and $\Xi^{(a)}(\mathcal{M})$ are strictly specializable along t. We have the following isomorphism:

$$\begin{aligned}\psi_0(\Pi^{a,b}_{*!}(\mathcal{M})) &\simeq \mathrm{Cok}\Big(\psi_0(\Pi^{b,N}_!\mathcal{M}) \longrightarrow \psi_0(\Pi^{a,N}_*\mathcal{M})\Big)\\ &\simeq \mathrm{Cok}\Big(\psi_{-\delta}(\mathcal{M})\otimes\mathbb{I}_2^{b,N} \longrightarrow \psi_{-\delta}(\mathcal{M})\otimes\mathbb{I}_2^{a,N}\Big)\end{aligned} \tag{4.12}$$

We obtain the first two isomorphisms. Let us consider $\phi^{(a)}$ in the case $a=0$. Let us consider the following induced complex:

$$\psi_{-\delta}(\mathcal{M}) \longrightarrow \Big(\psi_{-\delta}(\mathcal{M})\,s\oplus\psi_{-\delta}(\mathcal{M})\Big)\oplus\psi_0(\mathcal{M}) \longrightarrow \psi_{-\delta}(\mathcal{M})$$

The morphism $\psi_{-\boldsymbol{\delta}}(\mathcal{M}) \longrightarrow \psi_{-\boldsymbol{\delta}}(\mathcal{M})\, s \oplus \psi_{-\boldsymbol{\delta}}(\mathcal{M})$ is given by (s,N). The morphism $\psi_{-\boldsymbol{\delta}}(\mathcal{M})\, s \oplus \psi_{-\boldsymbol{\delta}}(\mathcal{M}) \longrightarrow \psi_{-\boldsymbol{\delta}}(\mathcal{M})$ is the projection. Then, it is easy to check that the cohomology is naturally isomorphic to $\psi_0(\mathcal{M})$. In particular, it is strict. Hence, we obtain that $\phi^{(0)}(\mathcal{M})$ is strictly specializable along t, and $\psi_0(\phi^{(0)}(\mathcal{M})) \simeq \psi_0(\mathcal{M})$. □

We can reconstruct $\mathcal{M}$ as the cohomology of the following, as in [3]:

$$\psi^{(1)}(\mathcal{M}) \longrightarrow \Xi^{(0)}(\mathcal{M}) \oplus \phi^{(0)}(\mathcal{M}) \longrightarrow \psi^{(0)}(\mathcal{M})$$

Remark 4.1.10 $\phi^{(0)}$ will often be denoted by ϕ. □

The following lemma is clear by Lemma 3.3.4.

Lemma 4.1.11 *Let $\mathcal{M}$ be a coherent strict $\mathcal{R}_X$-module which is strictly specializable along t. Let λ_0 be a non-zero complex number which is generic for $(\mathcal{M},t)$. Then, we have natural isomorphisms $\Pi^{a,b}_{*!}(\mathcal{M})^{\lambda_0} \simeq \Pi^{a,b}_{*!}(\mathcal{M}^{\lambda_0})$. In particular, we have $\psi^{(a)}(\mathcal{M})^{\lambda_0} \simeq \psi^{(a)}(\mathcal{M}^{\lambda_0})$ and $\Xi^{(a)}(\mathcal{M})^{\lambda_0} \simeq \Xi^{(a)}(\mathcal{M}^{\lambda_0})$. We also have $\phi^{(a)}(\mathcal{M})^{\lambda_0} \simeq \phi^{(a)}(\mathcal{M}^{\lambda_0})$.* □

See [3, 57] for Beilinson functors for $\mathscr{D}$-modules.

4.1.7 Beilinson Functors Along General Holomorphic Functions

Let X be a complex manifold. Let g be a holomorphic function on X. Let $\iota_g : X \longrightarrow X \times \mathbb{C}_t$ be the graph. Let $\iota_0 : X \longrightarrow X \times \mathbb{C}_t$ be given by $\iota_0(x) = (x,0)$. Let $\mathcal{M}$ be a coherent $\mathcal{R}_X$-module which is strictly specializable along g. The following is clear from Lemma 4.1.9.

Lemma 4.1.12 *We have coherent strict $\mathcal{R}_X$-modules $\psi_g^{(a)}(\mathcal{M})$ and $\phi_g^{(a)}(\mathcal{M})$ such that $\iota_{0\dagger}\psi_g^{(a)}(\mathcal{M}) \simeq \psi^{(a)}\big(\iota_{g\dagger}\mathcal{M}\big)$ and $\iota_{0\dagger}\phi_g^{(a)}(\mathcal{M}) := \phi^{(a)}\big(\iota_{g\dagger}\mathcal{M}\big)$.*

If $\mathcal{M}$ is integrable, they are also naturally integrable. They are uniquely determined up to canonical isomorphisms. □

We have natural isomorphisms $\psi_g^{(a)}(\mathcal{M}) \simeq \psi_{g,-\boldsymbol{\delta}}(\mathcal{M})s^a$ and $\phi_g^{(a)}(\mathcal{M}) \simeq \psi_{g,0}(\mathcal{M})s^a$. We shall often denote $\phi_g^{(0)}$ by ϕ_g.

Assume that $\mathcal{M}$ is strict for simplicity. If there exists an $\mathcal{R}_X$-module $\mathcal{M}'$ such that $\iota_{g\dagger}\mathcal{M}' = \Pi^{a,b}_!(\iota_{g\dagger}\mathcal{M})$, it is uniquely determined up to isomorphisms, and denoted by $\Pi^{a,b}_{g!}(\mathcal{M})$. We use the notation $\Pi^{a,b}_{g*}(\mathcal{M})$, $\Pi^{a,b}_{*!}(\iota_{g\dagger}\mathcal{M})$ and $\Xi_g^{(a)}(\mathcal{M})$ with similar meaning. If $\mathcal{M}$ is integrable, they are naturally integrable.

The following lemma is clear by Lemma 3.3.4.

Lemma 4.1.13 *Let $\mathcal{M}$ be a coherent strict $\mathcal{R}_X$-module which is strictly specializable along g. Suppose that a non-zero complex number λ_0 is generic for $(\mathcal{M}, g)$.*

- *We have natural isomorphisms $\psi_g^{(a)}(\mathcal{M})^{\lambda_0} \simeq \psi_g^{(a)}(\mathcal{M}^{\lambda_0})$ and $\phi_g^{(a)}(\mathcal{M})^{\lambda_0} \simeq \phi_g^{(a)}(\mathcal{M}^{\lambda_0})$.*
- *If $\Pi_{g!}^{a,b}(\mathcal{M})$ exists, we naturally have $\Pi_{g!}^{a,b}(\mathcal{M})^{\lambda_0} \simeq \Pi_{g!}^{a,b}(\mathcal{M}^{\lambda_0})$. We have similar claims for $\Pi_{g*}^{a,b}$, $\Pi_{g*!}^{a,b}$ and $\Xi_g^{(a)}$.* □

4.2 Beilinson Functors for $\mathcal{R}$-Triples

4.2.1 Functors $\Pi^{a,b}$, $\Pi_*^{a,b}$ and $\Pi_!^{a,b}$ for $\mathcal{R}(*t)$-Triple

We continue to use the setting in Sect. 3.1. Let $\mathcal{T} = (\mathcal{M}', \mathcal{M}'', C)$ be a strictly specializable $\mathcal{R}_{X(*H)}(*t)$-triple. As in Sect. 2.1.6, we obtain the following $\mathcal{R}_{X(*H)}(*t)$-triple:

$$\Pi^{a,b}\mathcal{T} := \mathcal{T} \otimes \tilde{\mathbb{I}}^{a,b} = \left(\mathcal{M}' \otimes \tilde{\mathbb{I}}_1^{a,b}, \mathcal{M}'' \otimes \tilde{\mathbb{I}}_2^{a,b}, C \otimes C_{\tilde{\mathbb{I}}}\right)$$

Here, $C \otimes C_{\tilde{\mathbb{I}}}$ is given as follows:

$$\begin{aligned}\left(C \otimes C_{\tilde{\mathbb{I}}}\right)\left(u \otimes (\lambda s)^i, \sigma^*(v \otimes (\lambda s)^j)\right) &= C(u, \sigma^* v) \cdot C_{\tilde{\mathbb{I}}}\left((\lambda s)^i, \sigma^*(\lambda s)^j\right) \\ &= C(u, \sigma^* v) \frac{(\log |t|^2)^{-i-j}}{(-i-j)!}(-1)^j \lambda^{i-j} \chi_{i+j \le 0}\end{aligned} \tag{4.13}$$

(See Sect. 2.3.2.) The $\mathcal{R}_{X(*H)}(*t)$-triple $\Pi^{a,b}\mathcal{T}$ is strictly specializable along t. Then, we obtain the following $\mathcal{R}_{X(*H)}$-triples:

$$\Pi_*^{a,b}\mathcal{T} := \Pi^{a,b}\mathcal{T}[*t], \qquad \Pi_!^{a,b}\mathcal{T} := \Pi^{a,b}\mathcal{T}[!t].$$

We also denote them by $\Pi_t^{a,b}\mathcal{T}$ and $\Pi_{t\star}^{a,b}\mathcal{T}$ ($\star = *, !$) when we emphasize the dependence on t. If $\mathcal{T}$ is integrable, $\Pi^{a,b}\mathcal{T}$ and $\Pi_\star^{a,b}\mathcal{T}$ ($\star = *, !$) are also naturally integrable.

Lemma 4.2.1 *Let $u \in (\mathbb{R} \times \mathbb{C}) \setminus (\mathbb{Z}_{\ge 0} \times \{0\})$. We have*

$$\psi_u\left(\Pi^{a,b}\mathcal{T}\right) = \psi_u(\mathcal{T}) \otimes \mathbb{I}^{a,b} \simeq \left(\psi_u(\mathcal{M}') \otimes \mathbb{I}_1^{a,b}, \psi_u(\mathcal{M}'') \otimes \mathbb{I}_2^{a,b}, \psi_u C \otimes C_{\mathbb{I}}\right).$$

Proof We have only to check the claim for the pairing. We have the following equality for any positive integer M:

$$\operatorname*{Res}_{s+\lambda^{-1}\mathfrak{e}(\lambda,u)} \int \big\langle C(u,\sigma^* v),\, \phi\big\rangle (\log|t|^2)^M\, |t|^{2s}\, dt\, d\bar{t}$$
$$= \operatorname*{Res}_{s+\lambda^{-1}\mathfrak{e}(\lambda,u)} \frac{d^M}{ds^M} \int \big\langle C(u,\sigma^* v),\, \phi\big\rangle |t|^{2s}\, dt\, d\bar{t} = 0 \tag{4.14}$$

Then, the claim follows. □

We have $\Pi^{a,a+1}\mathcal{T} \simeq \mathcal{T} \otimes \boldsymbol{T}(a)$, and hence

$$\mathcal{T}[\star t] \otimes \boldsymbol{T}(a) \simeq \operatorname{Cok}\Big(\Pi_\star^{a+1,N}\mathcal{T} \longrightarrow \Pi_\star^{a,N}\mathcal{T}\Big) \quad (N > a),$$
$$\mathcal{T}[\star t] \otimes \boldsymbol{T}(b) \simeq \operatorname{Ker}\Big(\Pi_\star^{N,b+1}\mathcal{T} \longrightarrow \Pi_\star^{N,b}\mathcal{T}\Big) \quad (N < b).$$

4.2.1.1 Relation with the Hermitian Adjoint

Note that $\mathcal{S}^{a,b}$ (Sect. 2.3.1) induces an isomorphism

$$\Pi^{a,b}(\mathcal{T}^*) \simeq (\Pi^{-b+1,-a+1}\mathcal{T})^*.$$

If $b = a+1$, it is equal to the canonical isomorphism $\mathcal{T}^* \otimes \boldsymbol{T}(a) \simeq \big(\mathcal{T} \otimes \boldsymbol{T}(-a)\big)^*$ multiplied by $(-1)^a$. We have the induced isomorphisms

$$\Pi_!^{a,b}(\mathcal{T}^*) \simeq (\Pi_*^{-b+1,-a+1}\mathcal{T})^*, \qquad \Pi_*^{a,b}(\mathcal{T}^*) \simeq (\Pi_!^{-b+1,-a+1}\mathcal{T})^*.$$

For $a \geq a'$ and $b \geq b'$, the following natural diagrams are commutative:

$$\begin{array}{ccc} \Pi_!^{a,b}(\mathcal{T}^*) & \xrightarrow{\simeq} & (\Pi_*^{-b+1,-a+1}\mathcal{T})^* \\ \downarrow & & \downarrow \\ \Pi_!^{a',b'}(\mathcal{T}^*) & \xrightarrow{\simeq} & (\Pi_*^{-b'+1,-a'+1}\mathcal{T})^* \end{array} \qquad \begin{array}{ccc} \Pi_*^{a,b}(\mathcal{T}^*) & \xrightarrow{\simeq} & (\Pi_!^{-b+1,-a+1}\mathcal{T})^* \\ \downarrow & & \downarrow \\ \Pi_*^{a',b'}(\mathcal{T}^*) & \xrightarrow{\simeq} & (\Pi_!^{-b'+1,-a'+1}\mathcal{T})^* \end{array}$$

If $b = a+1$, they are equal to the canonical isomorphisms multiplied by $(-1)^a$.

$$\mathcal{T}^*[!t] \otimes \boldsymbol{T}(a) \simeq (\mathcal{T}[*t])^* \otimes \boldsymbol{T}(a) \simeq \big(\mathcal{T}[*t] \otimes \boldsymbol{T}(-a)\big)^*$$
$$\mathcal{T}^*[*t] \otimes \boldsymbol{T}(a) \simeq (\mathcal{T}[!t])^* \otimes \boldsymbol{T}(a) \simeq \big(\mathcal{T}[!t] \otimes \boldsymbol{T}(-a)\big)^*$$

Let $\mathcal{S} : \mathcal{T} \longrightarrow \mathcal{T}^* \otimes \boldsymbol{T}(-w)$ be a Hermitian sesqui-linear duality of weight w. We obtain the induced morphism:

$$\Pi^{a,b}\mathcal{S} : \Pi^{a,b}\mathcal{T} \longrightarrow \Pi^{a,b}\big(\mathcal{T}^* \otimes \boldsymbol{T}(-w)\big) \simeq \big(\Pi^{-b+1,-a+1}\mathcal{T}\big)^* \otimes \boldsymbol{T}(-w)$$

For $u \in (\mathbb{R} \times \mathbb{C}) \setminus (\mathbb{Z}_{\geq 0} \times \{0\})$, we have the following commutativity:

$$\begin{array}{ccc}
\psi_u(\Pi^{a,b}\mathcal{T}) & \xrightarrow{\psi_u(\Pi^{a,b}\mathcal{S})} & \psi_u\Big((\Pi^{-b+1,-a+1}\mathcal{T})^* \otimes \boldsymbol{T}(-w)\Big) \\
\downarrow & & \downarrow \\
\psi_u(\mathcal{T}) \otimes \mathbb{I}^{a,b} & \xrightarrow{\psi_u(\mathcal{S})\otimes\mathcal{S}^{a,b}} & \psi_u(\mathcal{T}^*) \otimes (\mathbb{I}^{-b+1,-a+1})^* \otimes \boldsymbol{T}(-w)
\end{array}$$

4.2.2 Functors $\Pi^{a,b}_{*!}$, $\psi^{(a)}$ and $\Xi^{(a)}$

Let $\mathcal{T}$ be any $\mathcal{R}_{X(*H)}(*t)$-triple which is strictly specializable along t.

Lemma 4.2.2 *Let $P \in \mathcal{X}$. There exists a large number $N(P) > 0$ such that, for any $N > N(P)$, on a small neighbourhood $\mathcal{X}_P$ of P, the following naturally defined morphisms*

$$\mathrm{Cok}\big(\Pi_!^{b,N+1}\mathcal{T} \longrightarrow \Pi_*^{a,N+1}\mathcal{T}\big) \longrightarrow \mathrm{Cok}\big(\Pi_!^{b,N}\mathcal{T} \longrightarrow \Pi_*^{a,N}\mathcal{T}\big)$$
$$\mathrm{Ker}\big(\Pi_!^{-N,b}\mathcal{T} \longrightarrow \Pi_*^{-N,a}\mathcal{T}\big) \longrightarrow \mathrm{Cok}\big(\Pi_!^{-N-1,b}\mathcal{T} \longrightarrow \Pi_*^{-N-1,a}\mathcal{T}\big)$$

are isomorphisms. In this sense, $C_N := \mathrm{Cok}\big(\Pi_!^{b,N}\mathcal{T} \longrightarrow \Pi_*^{a,N}\mathcal{T}\big)$ and $K_N := \mathrm{Ker}\big(\Pi_!^{-N,b}\mathcal{T} \longrightarrow \Pi_*^{-N,a}\mathcal{T}\big)$ are independent of $N > N(P)$. Moreover, C_N and K_N are naturally isomorphic.

Proof We obtain the first claim from Lemma 4.1.1 and Lemma 4.1.3. We have the following natural commutative diagram:

$$\begin{array}{ccccc}
\Pi_!^{b,N}\mathcal{T} & \longrightarrow & \Pi_!^{-N,N}\mathcal{T} & \longrightarrow & \Pi_!^{-N,b}\mathcal{T} \\
\downarrow & & \downarrow & & \downarrow \\
\Pi_*^{a,N}\mathcal{T} & \longrightarrow & \Pi_*^{-N,N}\mathcal{T} & \longrightarrow & \Pi_!^{-N,a}\mathcal{T}
\end{array}$$

Then, by using Lemma 4.1.5, we obtain the second claim. □

We define

$$\Pi_{*!}^{a,b}\mathcal{T} := \varprojlim_{N\to\infty} \mathrm{Cok}\big(\Pi_!^{b,N}\mathcal{T} \longrightarrow \Pi_*^{a,N}\mathcal{T}\big) \simeq \varinjlim_{N\to\infty} \mathrm{Ker}\big(\Pi_!^{-N,b}\mathcal{T} \longrightarrow \Pi_*^{-N,a}\mathcal{T}\big) \tag{4.15}$$

We also denote it by $\Pi_{t*!}^{a,b}\mathcal{T}$ when we emphasize the dependence on t. If $\mathcal{T}$ is integrable, $\Pi_{*!}^{a,b}\mathcal{T}$ are also naturally integrable. Let $\mathcal{T}_i$ $(i = 1,2)$ be coherent

$\mathcal{R}_{X(*H)}(*t)$-triples, which are strictly specializable along t. For any morphism $f : \mathcal{T}_1 \longrightarrow \mathcal{T}_2$, we have the induced morphisms $\Pi_{\star}^{a,b}(f) : \Pi_{\star}^{a,b}(\mathcal{T}_1) \longrightarrow \Pi_{\star}^{a,b}(\mathcal{T}_2)$ and $\Pi_{*!}^{a,b}(f) : \Pi_{*!}^{a,b}(\mathcal{T}_1) \longrightarrow \Pi_{*!}^{a,b}(\mathcal{T}_2)$. We obtain the following lemma from Lemma 4.1.6.

Lemma 4.2.3 *If f is strictly specializable, we have natural isomorphisms* $\mathrm{Ker}\big(\Pi_{*!}^{a,b}f\big) \simeq \Pi_{*!}^{a,b}\,\mathrm{Ker} f$, $\mathrm{Im}\big(\Pi_{*!}^{a,b}f\big) \simeq \Pi_{*!}^{a,b}\,\mathrm{Im} f$ *and* $\mathrm{Cok}\big(\Pi_{*!}^{a,b}f\big) \simeq \Pi_{*!}^{a,b}\,\mathrm{Cok} f$. □

Corollary 4.2.4 *Let $\mathcal{T}^{\bullet}$ be as in Lemma* 3.2.7. *We have natural isomorphisms* $\Pi_{*!}^{a,b}\mathcal{H}^{\bullet}\mathcal{M}^{\bullet} \simeq \mathcal{H}^{\bullet}\Pi_{*!}^{a,b}\mathcal{M}^{\bullet}$. □

4.2.2.1 Functors $\psi^{(a)}$ and $\Xi^{(a)}$

In particular, for any $a \in \mathbb{Z}$, we define

$$\psi^{(a)}\mathcal{T} := \Pi_{*!}^{a,a}\mathcal{T}, \quad \Xi^{(a)}\mathcal{T} := \Pi_{*!}^{a,a+1}\mathcal{T}.$$

We denote them by $\psi_t^{(a)}\mathcal{T}$ and $\Xi_t^{(a)}\mathcal{T}$ when we emphasize the dependence on t. If $\mathcal{T}$ is integrable, they have naturally induced integrable structure. We naturally have the following exact sequences:

$$0 \longrightarrow \mathcal{T}[!t] \otimes \boldsymbol{T}(a) \xrightarrow{\alpha_a} \Xi^{(a)}\mathcal{T} \xrightarrow{\beta_a} \psi^{(a)}\mathcal{T} \longrightarrow 0$$

$$0 \longrightarrow \psi^{(a+1)}\mathcal{T} \xrightarrow{\gamma_a} \Xi^{(a)}\mathcal{T} \xrightarrow{\delta_a} \mathcal{T}[*t] \otimes \boldsymbol{T}(a) \longrightarrow 0$$

We also have natural identifications

$$\psi^{(a+1)}\mathcal{T} \simeq \psi^{(a)}\mathcal{T} \otimes \boldsymbol{T}(1), \quad \Xi^{(a+1)}\mathcal{T} \simeq \Xi^{(a)}\mathcal{T} \otimes \boldsymbol{T}(1).$$

The composite $\beta_a \circ \gamma_a : \psi^{(a+1)}\mathcal{T} \longrightarrow \psi^{(a)}\mathcal{T} \simeq \psi^{(a+1)}\mathcal{T} \otimes \boldsymbol{T}(-1)$ is induced by the natural morphisms $\Pi_{\star}^{a+1+i,N}\mathcal{T} \longrightarrow \Pi_{\star}^{a+i,N}\mathcal{T}$.

4.2.2.2 Relation with Hermitian Adjoint

Lemma 4.2.5 *We have natural isomorphisms* $\Pi_{*!}^{a,b}(\mathcal{T}^*) \simeq \big(\Pi_{*!}^{-b+1,-a+1}\mathcal{T}\big)^*$. *In particular, we naturally have*

$$\big(\psi^{(a)}\mathcal{T}\big)^* \simeq \psi^{(-a+1)}\big(\mathcal{T}^*\big), \qquad \big(\Xi^{(a)}\mathcal{T}\big)^* \simeq \Xi^{(-a)}\big(\mathcal{T}^*\big).$$

Proof We have the following natural isomorphisms:

$$\begin{aligned}\Pi_{*!}^{a,b}(\mathcal{T}^*) &= \mathrm{Cok}\Big(\Pi_!^{b,N}(\mathcal{T}^*) \longrightarrow \Pi_*^{a,N}(\mathcal{T}^*)\Big) \\ &\simeq \mathrm{Cok}\Big(\big(\Pi_*^{-N+1,-b+1}\mathcal{T}\big)^* \longrightarrow \big(\Pi_!^{-N+1,-a+1}\mathcal{T}\big)^*\Big) \\ &= \mathrm{Ker}\Big(\Pi_*^{-N+1,-b+1}\mathcal{T} \longleftarrow \Pi_!^{-N+1,-a+1}\mathcal{T}\Big)^* \simeq \Pi_{*!}^{-b+1,-a+1}(\mathcal{T})^*\end{aligned} \tag{4.16}$$

Thus, we are done. □

For $a \geq a'$ and $b \geq b'$, the following diagram is commutative:

$$\begin{array}{ccc} \Pi_{*!}^{a,b}(\mathcal{T}^*) & \longrightarrow & \Pi_{*!}^{-b+1,-a+1}(\mathcal{T})^* \\ \downarrow & & \downarrow \\ \Pi_{*!}^{a',b'}(\mathcal{T}^*) & \longrightarrow & \Pi_{*!}^{-b'+1,-a'+1}(\mathcal{T})^* \end{array}$$

In particular, the following diagram is commutative:

$$\begin{array}{ccccc} \psi^{(a+1)}(\mathcal{T}^*) & \xrightarrow{\gamma_a(\mathcal{T}^*)} & \Xi^{(a)}(\mathcal{T}^*) & \xrightarrow{\beta_a(\mathcal{T}^*)} & \psi^{(a)}(\mathcal{T}^*) \\ \downarrow & & \downarrow & & \downarrow \\ \psi^{(-a)}(\mathcal{T})^* & \xrightarrow{\beta_{-a}(\mathcal{T})^*} & \Xi^{(-a)}(\mathcal{T})^* & \xrightarrow{\gamma_{-a}(\mathcal{T})^*} & \psi^{(-a+1)}(\mathcal{T})^* \end{array}$$

We have the isomorphisms $\Upsilon_{a,\mathcal{T}} : \psi^{(a+1)}(\mathcal{T}) \simeq \psi^{(a)}(\mathcal{T}) \otimes \boldsymbol{T}(1)$ and $\Upsilon_{-a,\mathcal{T}^*} : \psi^{(-a+1)}(\mathcal{T}^*) \simeq \psi^{(-a)}(\mathcal{T}^*) \otimes \boldsymbol{T}(1)$. The following diagram is commutative:

$$\begin{array}{ccc} \psi^{(a)}(\mathcal{T})^* & \xrightarrow[-(\Upsilon_{a,\mathcal{T}})^*]{\simeq} & \psi^{(a+1)}(\mathcal{T})^* \otimes \boldsymbol{T}(1) \\ \simeq\big\downarrow g_1 & & \simeq\big\downarrow g_2 \\ \psi^{(-a+1)}(\mathcal{T}^*) & \xrightarrow[\Upsilon_{-a,\mathcal{T}^*}]{\simeq} & \psi^{(-a)}(\mathcal{T}^*) \otimes \boldsymbol{T}(1) \end{array}$$

Here, the vertical isomorphisms are as in Lemma 4.2.5.

4.2.3 Vanishing Cycle Functor for $\mathcal{R}$-Triple

Let $\mathcal{T}$ be an $\mathcal{R}_{X(*H)}$-triple which is strictly specializable along t. Then, applying the above construction to $\tilde{\mathcal{T}} := \mathcal{T}(*t)$, we define

$$\Xi^{(a)}\mathcal{T} := \Xi^{(a)}\tilde{\mathcal{T}}, \qquad \psi^{(a)}\mathcal{T} := \psi^{(a)}\tilde{\mathcal{T}}.$$

We shall introduce the vanishing cycle functor $\phi^{(a)}$. We have the canonical morphisms $\mathcal{T}[!t] \xrightarrow{\iota_1} \mathcal{T} \xrightarrow{\iota_2} \mathcal{T}[*t]$. We have $\iota_2 \circ \iota_1 = \delta_a \circ \alpha_a$, because the restrictions to $X \setminus \{t = 0\}$ are equal. Then, we define $\phi^{(a)}(\mathcal{T})$ as the cohomology of the following complex:

$$\mathcal{T}[!t] \otimes \boldsymbol{T}(a) \xrightarrow{\alpha_a + \iota_1} \Xi^{(a)}\mathcal{T} \oplus \big(\mathcal{T} \otimes \boldsymbol{T}(a)\big) \xrightarrow{\delta_a - \iota_2} \mathcal{T}[*t] \otimes \boldsymbol{T}(a)$$

We naturally have $\phi^{(a)}(\mathcal{T}) \simeq \phi^{(a+1)}(\mathcal{T}) \otimes \boldsymbol{T}(-1)$. In particular, we set $\phi(\mathcal{T}) := \phi^{(0)}(\mathcal{T})$. If $\mathcal{T}$ is integrable, $\phi^{(a)}(\mathcal{T})$ has a naturally induced integrable structure.

The morphisms β_a and γ_a induce the following morphisms:

$$\psi^{(a+1)}\mathcal{T} \xrightarrow{\mathrm{can}^{(a)}} \phi^{(a)}\mathcal{T} \xrightarrow{\mathrm{var}^{(a)}} \psi^{(a)}\mathcal{T}$$

As in [3], we can reconstruct $\mathcal{T} \otimes \boldsymbol{T}(a)$ as the cohomology of

$$\psi^{(a+1)}\mathcal{T} \xrightarrow{\gamma_a + \mathrm{can}^{(a)}} \Xi^{(a)}\mathcal{T} \oplus \phi^{(a)}\mathcal{T} \xrightarrow{\delta_a - \mathrm{var}^{(a)}} \psi^{(a)}\mathcal{T}$$

We have natural isomorphisms $\psi^{(a)}(\mathcal{T})^* \simeq \psi^{(-a+1)}(\mathcal{T}^*)$ and $\Xi^{(a)}(\mathcal{T})^* \simeq \Xi^{(-a)}(\mathcal{T}^*)$. We also obtain an induced isomorphism $\phi^{(a)}(\mathcal{T})^* \simeq \phi^{(-a)}(\mathcal{T}^*)$.

4.2.4 Gluing of $\mathcal{R}$-Triples

Let $\mathcal{T}$ be a coherent $\mathcal{R}_{X(*H)}(*t)$-triple which is strictly specializable along t. Let $\mathcal{Q}$ be a coherent strict coherent $\mathcal{R}_{X(*H)}$-triple with morphisms

$$\psi^{(1)}\mathcal{T} \xrightarrow{u} \mathcal{Q} \xrightarrow{v} \psi^{(0)}\mathcal{T},$$

such that (i) $v \circ u = \delta_0 \circ \gamma_0$, (ii) $\mathrm{Supp}\,\mathcal{Q} \subset \{t = 0\}$. Then, we obtain an $\mathcal{R}$-triple $\mathrm{Glue}(\mathcal{T}, \mathcal{Q}, u, v)$ as the cohomology of the following complex:

$$\psi^{(1)}\mathcal{T} \xrightarrow{\gamma_0 + u} \Xi^{(0)}\mathcal{T} \oplus \mathcal{Q} \xrightarrow{\delta_0 - v} \psi^{(0)}\mathcal{T}$$

We naturally have $\mathrm{Glue}(\mathcal{T}, \mathcal{Q}, u, v)(*t) \simeq \mathcal{T}$ and $\phi^{(0)}\big(\mathrm{Glue}(\mathcal{T}, \mathcal{Q}, u, v)\big) \simeq \mathcal{Q}$. Under the isomorphisms, we have $\mathrm{can}^{(0)} = u$ and $\mathrm{var}^{(0)} = v$. We naturally have $\mathrm{Glue}(\mathcal{T}, \mathcal{Q}, u, v)^* \simeq \mathrm{Glue}(\mathcal{T}^*, \mathcal{Q}^*, -v^*, -u^*)$. If $\mathcal{T}$, $\mathcal{Q}$, u and v are integrable, the object $\mathrm{Glue}(\mathcal{T}, \mathcal{Q}, u, v)$ is also naturally integrable.

We give a remark for Lemma 4.2.9 below. We have the naturally induced morphisms

$$\Xi^{(0)}(\mathcal{T}) \longrightarrow \Xi^{(-1)}(\mathcal{T}) \simeq \Xi^{(0)}(\mathcal{T}) \otimes \boldsymbol{T}(-1),$$
$$\psi^{(a)}(\mathcal{T}) \longrightarrow \psi^{(a-1)}(\mathcal{T}) \simeq \psi^{(a)}(\mathcal{T}) \otimes \boldsymbol{T}(-1).$$

We also have $\mathcal{Q} \longrightarrow \psi^{(0)}(\mathcal{T}) \simeq \psi^{(1)}(\mathcal{T}) \otimes \boldsymbol{T}(-1) \longrightarrow \mathcal{Q} \otimes \boldsymbol{T}(-1)$. The morphisms induce $N : \mathrm{Glue}(\mathcal{T}, \mathcal{Q}, u, v) \longrightarrow \mathrm{Glue}(\mathcal{T}, \mathcal{Q}, u, v) \otimes \boldsymbol{T}(-1)$.

Lemma 4.2.6 *The morphism N is* 0. □

Proof The map $\mathrm{Ker}\big(\Xi^{(0)}\mathcal{T} \oplus \mathcal{Q} \stackrel{\delta_0 - v}{\longrightarrow} \psi^{(0)}\mathcal{T}\big) \longrightarrow \big(\Xi^{(0)}\mathcal{T} \oplus \mathcal{Q}\big) \otimes \boldsymbol{T}(-1)$ factors through $\psi^{(0)}\mathcal{T}$ by construction, which implies the claim of the lemma. □

4.2.5 Dependence on the Function *t*

To distinguish the dependence of $\Pi^{a,b}_{\star}$ on t, we use the symbol $\Pi^{a,b}_{t\star}$. We use the symbols $\Pi^{a,b}_{t,*!}$, $\Xi^{(a)}_t$, $\psi^{(a)}_t$ and $\phi^{(a)}_t$ in similar meanings. We have the following morphism denoted by $\mathcal{N}_t$:

$$\Pi^{a,b}_{t\star}(\mathcal{T}) \longrightarrow \Pi^{a-1,b-1}_{t\star}(\mathcal{T}) \simeq \Pi^{a,b}_{t\star}(\mathcal{T}) \otimes \boldsymbol{T}(-1)$$

The induced morphisms for $\Pi^{a,b}_{t,*!}(\mathcal{T})$, $\Xi^{(a)}_t(\mathcal{T})$, $\psi^{(a)}_t(\mathcal{T})$ and $\phi^{(a)}_t(\mathcal{T})$ are also denoted by $\mathcal{N}_t$.

Let φ be a holomorphic function. Let $s = e^{\varphi}\, t$. Let us compare the functors for t and s.

Lemma 4.2.7 *We naturally have* $\big(\Pi^{a,b}_{s,\star}(\mathcal{T}), \mathcal{N}_s\big) \simeq \mathrm{Def}_{\varphi}\big(\Pi^{a,b}_{t,\star}(\mathcal{T}), \mathcal{N}_t\big)$.

Proof We have $\big(\Pi^{a,b}_{s}(\mathcal{T}), \mathcal{N}_s\big) \simeq \mathrm{Def}_{\varphi}\big(\Pi^{a,b}_{t}(\mathcal{T}), \mathcal{N}_t\big)$ from Lemma 2.3.1. Then, the claim of the lemma immediately follows. □

Corollary 4.2.8 *We naturally have* $\big(\Pi^{a,b}_{s,*!}(\mathcal{T}), \mathcal{N}_s\big) \simeq \mathrm{Def}_{\varphi}\big(\Pi^{a,b}_{t,*!}(\mathcal{T}), \mathcal{N}_t\big)$. *In particular,*

$$\big(\psi^{(a)}_s(\mathcal{T}), \mathcal{N}_s\big) \simeq \mathrm{Def}_{\varphi}\big(\psi^{(a)}_t(\mathcal{T}), \mathcal{N}_t\big), \quad \big(\Xi^{(a)}_s(\mathcal{T}), \mathcal{N}_s\big) \simeq \mathrm{Def}_{\varphi}\big(\Xi^{(a)}_t(\mathcal{T}), \mathcal{N}_t\big).$$

We also have $\big(\phi^{(a)}_s(\mathcal{T}), \mathcal{N}_s\big) \simeq \mathrm{Def}_{\varphi}\big(\phi^{(a)}_t(\mathcal{T}), \mathcal{N}_t\big)$. *In particular,*

$$\big(\mathrm{Gr}^{W(\mathcal{N}_s)}\, \psi^{(a)}_s(\mathcal{T}), \mathcal{N}^{(0)}_s\big) \simeq \big(\mathrm{Gr}^{W(\mathcal{N}_t)}\, \psi^{(a)}_t(\mathcal{T}), \mathcal{N}^{(0)}_t\big),$$
$$\big(\mathrm{Gr}^{W(\mathcal{N}_s)}\, \phi^{(a)}_s(\mathcal{T}), \mathcal{N}^{(0)}_s\big) \simeq \big(\mathrm{Gr}^{W(\mathcal{N}_t)}\, \phi^{(a)}_t(\mathcal{T}), \mathcal{N}^{(0)}_t\big),$$

where $W(\mathcal{N}_\kappa)$ ($\kappa = s, t$) denote the weight filtrations of $\mathcal{N}_\kappa$, and $\mathcal{N}_\kappa^{(0)}$ are induced nilpotent morphisms. □

Let $\mathcal{T}'$ be an $\mathcal{R}_X(*t)$-module, which is strictly specializable along t. Let $\mathcal{Q}$ be a strict $\mathcal{R}_{X_0}$-module with morphisms

$$\psi_t^{(1)}(\mathcal{T}') \xrightarrow{\ u\ } \mathcal{Q} \xrightarrow{\ v\ } \psi_t^{(0)}(\mathcal{T}')$$

such that $v \circ u = \mathcal{N}_t$. Then, we have an $\mathcal{R}_X$-module $\mathrm{Glue}_t(\mathcal{T}', \mathcal{Q}, u, v)$.

We have $\mathcal{N}_t : \mathcal{Q} \longrightarrow \mathcal{Q} \otimes \boldsymbol{T}(-1)$ induced by $\mathcal{Q} \longrightarrow \psi_t^{(0)}(\mathcal{T}') \simeq \psi_t^{(1)}(\mathcal{T}') \otimes \boldsymbol{T}(-1) \longrightarrow \mathcal{Q} \otimes \boldsymbol{T}(-1)$. We put $(\tilde{\mathcal{Q}}, \mathcal{N}_s) := \mathrm{Def}_\varphi(\mathcal{Q}, \mathcal{N}_t)$. By Corollary 4.2.8, we have the following naturally induced morphisms:

$$\psi_s^{(1)}(\mathcal{T}') \xrightarrow{\ \tilde{u}\ } \tilde{\mathcal{Q}} \xrightarrow{\ \tilde{v}\ } \psi_s^{(0)}(\mathcal{T}')$$

We have $\tilde{v} \circ \tilde{u} = \mathcal{N}_s$. We obtain an $\mathcal{R}$-triple $\mathrm{Glue}_s(\mathcal{T}', \tilde{\mathcal{Q}}, \tilde{u}, \tilde{v})$.

Lemma 4.2.9 $\mathrm{Glue}_t(\mathcal{T}', \mathcal{Q}, u, v) \simeq \mathrm{Glue}_s(\mathcal{T}', \tilde{\mathcal{Q}}, \tilde{u}, \tilde{v})$ *naturally.*

Proof We have the induced morphism

$$\mathcal{N}_t : \mathrm{Glue}(\mathcal{T}', \mathcal{Q}, u, v) \longrightarrow \mathrm{Glue}(\mathcal{T}', \mathcal{Q}, u, v) \otimes \boldsymbol{T}(-1),$$

and by construction, we have a natural isomorphism

$$\mathrm{Def}_\varphi\big(\mathrm{Glue}(\mathcal{T}', \mathcal{Q}, u, v), \mathcal{N}_t\big) \simeq \big(\mathrm{Glue}(\mathcal{T}', \tilde{\mathcal{Q}}, \tilde{u}, \tilde{v}), \mathcal{N}_s\big).$$

Because $\mathcal{N}_t = 0$ on $\mathrm{Glue}(\mathcal{T}', \mathcal{Q}, u, v)$ as remarked in Lemma 4.2.6, we obtain the desired isomorphism. □

4.2.6 Compatibility with Push-Forward

Let us consider the compatibility of the functors $\Pi_{!*}^{a,b}$ with the push-forward. We use the notation in Sect. 3.2.4. Let $\mathcal{T}$ be a good $\mathcal{R}_{X(*H)}(*t)$-triple which is strictly specializable along t. Assume the following:

- The support of $\mathcal{T}$ are proper over Y with respect to F.
- $F_\dagger^i \tilde{\psi}_u \mathcal{T}$ are strict for any $u \in \mathbb{R} \times \mathbb{C}$ and $i \in \mathbb{Z}$.

We naturally have $F_\dagger\big(\Pi^{a,b}\mathcal{T}\big) \simeq \Pi^{a,b} F_\dagger(\mathcal{T})$ of good $\mathcal{R}_{Y(*H_Y)}(*t)$-triples. By the second assumption, $F_\dagger^i\big(\Pi^{a,b}\mathcal{T}\big)$ are strictly specializable along t, and we have $F_\dagger^i\big(\Pi_\star^{a,b}\mathcal{T}\big) \simeq \Pi_\star^{a,b} F_\dagger^i \mathcal{T}$ according to Corollary 3.2.13. Then, we obtain the following morphism:

$$\Pi^{a,b}_{*!} F^i_{\dagger} \mathcal{T} \longrightarrow F^i_{\dagger} \Pi^{a,b}_{*!} \mathcal{T} \tag{4.17}$$

Proposition 4.2.10 *The morphism* (4.17) *is an isomorphism. In particular, we have natural isomorphisms*

$$F^i_{\dagger} \psi^{(a)} \mathcal{T} \simeq \psi^{(a)} F^i_{\dagger} \mathcal{T}, \quad F^i_{\dagger} \Xi^{(a)} \mathcal{T} \simeq \Xi^{(a)} F^i_{\dagger} \mathcal{T}.$$

Proof It follows from Proposition 4.1.8. □

Let $\mathcal{T}$ be an $\mathcal{R}$-triple which is strictly specializable along t. Assume that $F^i_{\dagger} \tilde{\psi}_u \mathcal{T}$ ($u \in \mathbb{R} \times \mathbb{C}$) and $F^i_{\dagger} \phi^{(0)} \mathcal{T}$ are strict. According to [66], $F^i_{\dagger} \mathcal{T}$ are strictly specializable along t. The following lemma is proved in [66] with a different method.

Corollary 4.2.11 *We have a natural isomorphism* $F^i_{\dagger} \phi^{(0)}(\mathcal{T}) \simeq \phi^{(0)} F^i_{\dagger} \mathcal{T}$.

Proof We have the following descriptions:

$$\phi^{(0)}(\mathcal{T}) = H^1\Big(\mathcal{T}[!t] \longrightarrow \Xi^{(0)}(\mathcal{T}) \oplus \mathcal{T} \longrightarrow \mathcal{T}[*t]\Big)$$

$$\phi^{(0)}(F^i_{\dagger} \mathcal{T}) = H^1\big((F^i_{\dagger} \mathcal{T})[!t] \longrightarrow \Xi^{(0)} F^i_{\dagger} \mathcal{T} \oplus F^i_{\dagger} \mathcal{T} \longrightarrow (F^i_{\dagger} \mathcal{T})[*t]\big)$$

Then, the claim of the lemma follows from Proposition 4.2.10. □

4.2.7 Beilinson Functors Along General Holomorphic Functions

Let X be a complex manifold. Let g be a holomorphic function on X. Let $\iota_g : X \longrightarrow X \times \mathbb{C}_t$ be the graph. Let $\mathcal{T}$ be a coherent $\mathcal{R}_X$-triple, which is strictly specializable along g. The following is clear by Lemma 4.1.12.

Lemma 4.2.12 *There exist $\mathcal{R}_X$-triples $\psi^{(a)}_g(\mathcal{T})$ and $\phi^{(a)}_g(\mathcal{T})$ such that*

$$\iota_{g\dagger} \psi^{(a)}_g(\mathcal{T}) \simeq \psi^{(a)}\big(\iota_{g\dagger} \mathcal{T}\big), \qquad \iota_{g\dagger} \phi^{(a)}_g(\mathcal{T}) := \phi^{(a)}\big(\iota_{g\dagger} \mathcal{T}\big)$$

They are uniquely determined up to canonical isomorphisms. If $\mathcal{T}$ is integrable, they are also naturally integrable. □

Assume that $\mathcal{T}$ is strict for simplicity. If there exists an $\mathcal{R}_X$-triples $\mathcal{T}'$ such that $\iota_{g\dagger} \mathcal{T}' = \Pi^{a,b}_!(\iota_{g\dagger} \mathcal{T})$, it is uniquely determined up to isomorphisms, and denoted by $\Pi^{a,b}_{g!}(\mathcal{T})$. We use the notation $\Pi^{a,b}_{g*}(\mathcal{T})$, $\Pi^{a,b}_{*!}(\iota_{g\dagger} \mathcal{T})$ and $\Xi^{(a)}_g(\mathcal{T})$ with similar meaning. If $\mathcal{T}$ is integrable, they are naturally integrable.

4.3 Comparison of the Nearby Cycle Functors

4.3.1 Statements

Let $\mathcal{T}$ be an $\mathcal{R}_{X(*H)}(*t)$-triple which is strictly specializable along t. Let ι denote the inclusion $X_0 \times \{0\} \longrightarrow X$. In this subsection, we distinguish the $\mathcal{R}_{X_0}$-triple $\tilde{\psi}_{-\boldsymbol{\delta}}(\mathcal{T})$ and the $\mathcal{R}_X$-triple $\iota_\dagger\tilde{\psi}_{-\boldsymbol{\delta}}(\mathcal{T})$. We shall compare the $\mathcal{R}_X$-triples $\iota_\dagger\tilde{\psi}_{-\boldsymbol{\delta}}(\mathcal{T})$ and $\psi^{(1)}(\mathcal{T})$. We shall prove the following proposition in Sects. 4.3.2–4.3.4.

Proposition 4.3.1 *We have a natural isomorphism*

$$\Psi : \iota_\dagger\tilde{\psi}_{-\boldsymbol{\delta}}(\mathcal{T}) \simeq \psi^{(1)}(\mathcal{T}) \otimes \mathcal{U}(1,0)$$

with the following property:

- *Let* $\iota_\dagger\tilde{\psi}_{-\boldsymbol{\delta}}(\mathcal{T}) \simeq \psi^{(0)}(\mathcal{T}) \otimes \mathcal{U}(0,1)$ *be the induced isomorphism. The following diagram is commutative:*

$$\begin{array}{ccc} \psi^{(1)}(\mathcal{T}) & \xrightarrow{\delta_0 \circ \gamma_0} & \psi^{(0)}(\mathcal{T}) \\ \simeq\downarrow & & \simeq\downarrow \\ \iota_\dagger\tilde{\psi}_{-\boldsymbol{\delta}}(\mathcal{T}) \otimes \mathcal{U}(-1,0) & \xrightarrow{\mathcal{N}} & \iota_\dagger\tilde{\psi}_{-\boldsymbol{\delta}}(\mathcal{T}) \otimes \mathcal{U}(0,-1) \end{array} \tag{4.18}$$

  Here, $\mathcal{N} = (N', N'')$ *is as in* Sect. 2.1.4, *i.e.,* N' *and* N'' *are the nilpotent part of* $-t\partial_t$. *See* Sect. 2.1.8.1 *for the smooth integrable* $\mathcal{R}$*-triples* $\mathcal{U}(p,q)$.

If $\mathcal{T}$ *is integrable, the isomorphism* Ψ *is also integrable.*

Let $\mathcal{S} : \mathcal{T} \longrightarrow \mathcal{T}^* \otimes \boldsymbol{T}(-w)$ be a Hermitian sesqui-linear duality. We have the induced morphism

$$\Pi^{1,N}\mathcal{T} \longrightarrow (\Pi^{-N+1,0}\mathcal{T})^* \otimes \boldsymbol{T}(-w).$$

Hence, we have the following naturally induced morphisms:

$$\begin{aligned} \psi^{(1)}(\mathcal{T}) \longrightarrow \psi^{(1)}(\mathcal{T}^*) \otimes \boldsymbol{T}(-w) &\simeq \psi^{(0)}(\mathcal{T}^*) \otimes \boldsymbol{T}(-w+1) \\ &\simeq \psi^{(1)}(\mathcal{T})^* \otimes \boldsymbol{T}(-w+1) \end{aligned} \tag{4.19}$$

The composite is denoted by $\psi^{(1)}(\mathcal{S})$. We shall prove the following proposition in Sect. 4.3.5.

Proposition 4.3.2 *The following diagram is commutative:*

$$\begin{array}{ccc} \psi^{(1)}(\mathcal{T}) & \xrightarrow{\psi^{(1)}(\mathcal{S})} & \psi^{(1)}(\mathcal{T})^* \otimes \boldsymbol{T}(-w+1) \\ \simeq\downarrow & & \simeq\downarrow \\ \iota_\dagger \widetilde{\psi}_{-\boldsymbol{\delta}}(\mathcal{T}) \otimes \mathcal{U}(-1,0) & \longrightarrow & \big(\iota_\dagger \widetilde{\psi}_{-\boldsymbol{\delta}}(\mathcal{T}) \otimes \mathcal{U}(-1,0)\big)^* \otimes \boldsymbol{T}(-w+1) \end{array}$$

Here, the vertical arrows are induced by the isomorphism in Proposition 4.3.1, *and the lower horizontal arrow is induced by* $\tilde{\psi}_{-\boldsymbol{\delta}}(\mathcal{S})$ *and the polarization* $(-1,-1)$ *of* $\mathcal{U}(-1,0)$, *given in* Sect. 2.1.8.1.

4.3.2 Preliminary (1)

Let $\mathcal{M}$ be an $\mathcal{R}(*t)$-module strictly specializable along t. Assume that the cokernel of the nilpotent part of $-\eth_t t$ on $\psi_{-\boldsymbol{\delta}}(\mathcal{M})$ is strict. We have the following commutative diagram:

$$\begin{array}{ccc} \psi_0(\mathcal{M}[!t]) & \xrightarrow{\psi_0(\varphi)} & \psi_0(\mathcal{M}[*t]) \\ -\eth_t \uparrow \simeq & & t \downarrow \simeq \\ \psi_{-\boldsymbol{\delta}}(\mathcal{M}) & \xrightarrow{-t\eth_t} & \psi_{-\boldsymbol{\delta}}(\mathcal{M}) \end{array}$$

Hence, the cokernel of $\psi_u(\varphi)$ are strict for any u.

Let $\mathcal{K}(\mathcal{M})$ and $\mathcal{Q}(\mathcal{M})$ be the kernel and the cokernel of $\varphi : \mathcal{M}[!t] \longrightarrow \mathcal{M}[*t]$. We have the exact sequence:

$$0 \longrightarrow \psi_0\mathcal{K}(\mathcal{M}) \longrightarrow \psi_0(\mathcal{M}[!t]) \xrightarrow{\psi_0(\varphi)} \psi_0(\mathcal{M}[*t]) \longrightarrow \psi_0\mathcal{Q}(\mathcal{M}) \longrightarrow 0$$

Let $N : \tilde{\psi}_{-\boldsymbol{\delta}}(\mathcal{M}) \longrightarrow \tilde{\psi}_{-\boldsymbol{\delta}}(\mathcal{M})\lambda^{-1}$ be induced by $-t\partial_t$. We obtain the following commutative diagram:

$$\begin{array}{ccccccc} \psi_0\mathcal{K}(\mathcal{M}) & \longrightarrow & \psi_0(\mathcal{M}[!t]) & \xrightarrow{\psi_0(\varphi)} & \psi_0(\mathcal{M}[*t]) & \longrightarrow & \psi_0\mathcal{Q}(\mathcal{M}) \\ \mu'_{\mathcal{K}(\mathcal{M})}\uparrow\simeq & & -\lambda\partial_t\uparrow\simeq & & \lambda^{-1}t\downarrow\simeq & & \mu'_{\mathcal{Q}(\mathcal{M})}\downarrow\simeq \\ \operatorname{Ker} N & \longrightarrow & \psi_{-\boldsymbol{\delta}}(\mathcal{M}) & \xrightarrow{N} & \psi_{-\boldsymbol{\delta}}(\mathcal{M})\,\lambda^{-1} & \longrightarrow & \operatorname{Cok} N \end{array}$$

We obtain the isomorphism $\mu_{\mathcal{K}(\mathcal{M})} : \iota_{\dagger}\operatorname{Ker} N \longrightarrow \mathcal{K}(\mathcal{M})$ given as follows:

$$\iota_{\dagger}(\operatorname{Ker} N) = \bigoplus_{n=0}^{\infty} \operatorname{Ker} N \cdot (dt/\lambda)^{-1} \cdot \eth_t^n \longrightarrow \mathcal{K}(\mathcal{M})$$

$$= \bigoplus_{n=0}^{\infty} \psi_0(\mathcal{K}(\mathcal{M}))\, \eth_t^n$$

$$\mu_{\mathcal{K}(\mathcal{M})}\Big(\sum a_n \cdot (dt/\lambda)^{-1}\, \eth_t^n\Big) := \sum \mu'_{\mathcal{K}(\mathcal{M})}(a_n)\, \eth_t^n$$

We also obtain the following isomorphism $\mathcal{Q}(\mathcal{M}) \longrightarrow \iota_{\dagger}(\operatorname{Cok} N)$:

$$\mathcal{Q}(\mathcal{M}) = \bigoplus_{n=0}^{\infty} \psi_0(\mathcal{Q}(\mathcal{M}))\, \eth_t^n \longrightarrow \iota_{\dagger}(\operatorname{Cok} N)$$

$$= \bigoplus_{n=0}^{\infty} \operatorname{Cok} N \cdot (dt/\lambda)^{-1} \cdot \eth_t^n$$

$$\mu_{\mathcal{Q}(\mathcal{M})}\Big(\sum b_n\, \eth_t^n\Big) := \sum \mu'_{\mathcal{Q}(\mathcal{M})}(b_n) \cdot (dt/\lambda)^{-1}\, \eth_t^n$$

4.3.3 Preliminary (2)

Let $\mathcal{T} = (\mathcal{M}', \mathcal{M}'', C)$ be a coherent $\mathcal{R}_{X(*H)}(*t)$-triple which is strictly specializable along t. We have the induced morphism $\mathcal{N} : \tilde{\psi}_{-\boldsymbol{\delta}}(\mathcal{T}) \longrightarrow \tilde{\psi}_{-\boldsymbol{\delta}}(\mathcal{T}) \otimes \boldsymbol{T}(-1)$. We have $\operatorname{Ker} \mathcal{N} = \big(\lambda \cdot \operatorname{Cok} N', \operatorname{Ker} N'', C_1\big)$ and $\operatorname{Cok} \mathcal{N} = \big(\lambda \cdot \operatorname{Ker} N', \operatorname{Cok} N'', C_2\big)$, where C_i are naturally induced pairings. Let $\mathcal{K}$ and $\mathcal{Q}$ be the kernel and the cokernel of $\mathcal{T}[!t] \longrightarrow \mathcal{T}[*t]$.

Lemma 4.3.3 *The pair of morphisms* $(\mu_{\mathcal{K}(\mathcal{M}')}, \mu_{\mathcal{Q}(\mathcal{M}'')})$ *gives an isomorphism* $\mathcal{Q} \xrightarrow{\simeq} \iota_{\dagger} \operatorname{Cok} \mathcal{N} \otimes \mathcal{U}(-1,0)$. *Similarly, the pair* $(\mu_{\mathcal{Q}(\mathcal{M}')}, \mu_{\mathcal{K}(\mathcal{M}'')})$ *gives an isomorphism* $\iota_{\dagger} \operatorname{Ker} \mathcal{N} \otimes \mathcal{U}(-1,0) \xrightarrow{\simeq} \mathcal{K}$.

Proof We have only to check the compatibility of pairings. Let us check the first claim. Let $\lambda\, t^{-1} f \in V_0(\mathcal{Q}(\mathcal{M}''))$ be lifted to $V_0\mathcal{M}''[*t]$, and $-\eth_t \otimes g \in V_0(\mathcal{K}(\mathcal{M}'))$ be mapped to $V_0\mathcal{M}'[!t]$. Note $t\eth_t g = 0$ in $V_{<0}\mathcal{M}'$. Let $\phi = (\sqrt{-1}/2\pi) \cdot \chi \cdot \varphi \cdot dt\, d\bar{t}$, where φ is a C^{∞}-top form on X_0 with compact support, and χ is a cut function on

$\mathbb{C}_t$ around $t = 0$. We have

$$\begin{aligned}\Big\langle C\big(-\eth_t \otimes g, \sigma^*(\lambda t^{-1} f)\big), \phi\Big\rangle &= \big\langle C\big(\partial_t \otimes g, \sigma^*(t^{-1} f)\big), \phi\big\rangle \\ &= -\big\langle C(g, \sigma^* f), \overline{t}^{-1} \partial_t \phi\big\rangle \\ &= -\big\langle C(g, \sigma^* f),\ \overline{t}^{-1} |t|^{2s} \partial_t \phi\big\rangle_{|s=0} \\ &= -\big\langle C(g, \sigma^* f),\ \partial_t(\overline{t}^{-1} |t|^{2s} \phi)\big\rangle_{|s=0} \\ &\quad + \big\langle C(g, \sigma^* f),\ s\,|t|^{2(s-1)} \phi\big\rangle_{|s=0} \\ &= \big\langle \psi_{-\delta} C([g], \sigma^*[f]),\ \varphi\big\rangle \\ &= \big\langle \iota_\dagger \psi_{-\delta} C\big([g] \cdot (dt/\lambda)^{-1},\ \sigma^*\big([f] \cdot (dt/\lambda)^{-1}\big)\big), \phi\big\rangle \end{aligned} \tag{4.20}$$

Let us check the second claim. Let $-\eth_t \otimes g \in V_0(\mathcal{K}(\mathcal{M}''))$ be lifted to $V_0\mathcal{M}''[!t]$, and let $\lambda\, t^{-1} f \in V_0(\mathcal{Q}(\mathcal{M}'))$ be lifted to $V_0\mathcal{M}'[*t]$. We have the following:

$$\begin{aligned}\big\langle C\big(\lambda\, t^{-1} f, \sigma^*(-\eth_t \otimes g)\big),\ \phi\big\rangle &= \big\langle C\big(t^{-1} f, \sigma^*(\partial_t \otimes g)\big),\ \phi\big\rangle \\ &= -\big\langle C(f, \sigma^* g),\ t^{-1} \overline{\partial}_t \phi\big\rangle \\ &= -\big\langle C(f, \sigma^* g),\ t^{-1} |t|^{2s} \overline{\partial}_t \phi\big\rangle_{|s=0} \\ &= -\big\langle C(f, \sigma^* g),\ \overline{\partial}_t\big(t^{-1} |t|^{2s} \phi\big)\big\rangle_{|s=0} \\ &\quad + \big\langle C(f, \sigma^* g),\ s\,|t|^{2(s-1)} \phi\big\rangle_{|s=0} \\ &= \big\langle \tilde{\psi}_{-\delta} C([f], \sigma^*[g]),\ \varphi\big\rangle \\ &= \big\langle \iota_\dagger \tilde{\psi}_{-\delta} C\big([f]\,(dt/\lambda)^{-1},\ \sigma^*([g]\,(dt/\lambda)^{-1})\big),\ \phi\big\rangle \end{aligned} \tag{4.21}$$

Thus, we are done. □

4.3.4 Construction of Isomorphisms

We set $\mathbb{I}_0^{a,b} := \mathbb{I}_2^{a,b} = \mathbb{I}_1^{-b+1,-a+1}$. We have the isomorphism

$$\tilde{\psi}_{-\delta}(\mathcal{T}) \xrightarrow{\simeq} \operatorname{Cok}\Big(\mathcal{N} : \tilde{\psi}_{-\delta}(\Pi^{1,N}\mathcal{T}) \longrightarrow \tilde{\psi}_{-\delta}(\Pi^{1,N}\mathcal{T}) \otimes \boldsymbol{T}(-1)\Big)$$

given as follows:

- We use $\tilde{\psi}_{-\boldsymbol{\delta}}(\mathcal{M}') \stackrel{\simeq}{\longleftarrow} \operatorname{Ker}\Bigl(\tilde{\psi}_{-\boldsymbol{\delta}}(\mathcal{M}') \otimes \mathbb{I}_0^{-N+1,0} \longleftarrow \tilde{\psi}_{-\boldsymbol{\delta}}(\mathcal{M}') \otimes \mathbb{I}_0^{-N+1,0} \cdot \lambda\Bigr)$ induced by the projection together with $s \longmapsto -s$:

$$\mathbb{I}_0^{-N+1,0}\lambda \longrightarrow \mathcal{O}_{\mathbb{C}_\lambda}\,\lambda; \quad \sum_{j=1}^{N-1} a_j\,(\lambda s)^{-j}\,\lambda \longmapsto -a_1.$$

- We use $\tilde{\psi}_{-\boldsymbol{\delta}}(\mathcal{M}'') \stackrel{\simeq}{\longrightarrow} \operatorname{Cok}\Bigl(\tilde{\psi}_{-\boldsymbol{\delta}}(\mathcal{M}'') \otimes \mathbb{I}_0^{1,N} \longrightarrow \tilde{\psi}_{-\boldsymbol{\delta}}(\mathcal{M}'') \otimes \mathbb{I}_0^{1,N}\,\lambda^{-1}\Bigr)$ induced by the inclusion together with $s \longmapsto -s$:

$$\mathcal{O}_{\mathbb{C}_\lambda} \longrightarrow \mathbb{I}_0^{1,N}\,\lambda^{-1}; \quad a \longmapsto -a\,s.$$

We have the isomorphism

$$\tilde{\psi}_{-\boldsymbol{\delta}}(\mathcal{T}) \stackrel{\simeq}{\longleftarrow} \operatorname{Ker}\Bigl(\mathcal{N} : \tilde{\psi}_{-\boldsymbol{\delta}}(\Pi^{-N,1}\mathcal{T}) \longrightarrow \tilde{\psi}_{-\boldsymbol{\delta}}(\Pi^{-N,1}\mathcal{T}) \otimes \boldsymbol{T}(-1)\Bigr)$$

given as follows:

- We use $\tilde{\psi}_{-\boldsymbol{\delta}}(\mathcal{M}') \stackrel{\simeq}{\longrightarrow} \operatorname{Cok}\Bigl(\tilde{\psi}_{-\boldsymbol{\delta}}(\mathcal{M}') \otimes \mathbb{I}_0^{0,N+1} \longleftarrow \tilde{\psi}_{-\boldsymbol{\delta}}(\mathcal{M}') \otimes \mathbb{I}_0^{0,N+1}\,\lambda\Bigr)$ given by the inclusion $\mathcal{O}_{\mathbb{C}_\lambda} \longrightarrow \mathbb{I}_0^{0,N+1}$; $a \longmapsto a$.
- We use $\tilde{\psi}_{-\boldsymbol{\delta}}(\mathcal{M}'') \stackrel{\simeq}{\longleftarrow} \operatorname{Ker}\Bigl(\tilde{\psi}_{-\boldsymbol{\delta}}(\mathcal{M}'') \otimes \mathbb{I}_0^{-N,1} \longrightarrow \tilde{\psi}_{-\boldsymbol{\delta}}(\mathcal{M}'') \otimes \mathbb{I}_0^{-N,1}\,\lambda^{-1}\Bigr)$ induced by the projection $\mathbb{I}_0^{-N,1} \longrightarrow \mathcal{O}_{\mathbb{C}_\lambda}$; $\sum_{j=0}^{N} a_j\,(\lambda s)^{-j} \longmapsto a_0$.

  Then, we obtain the following isomorphisms:

$$\Psi_1 : \psi^{(1)}(\mathcal{T}) \simeq \iota_\dagger \operatorname{Cok}(\mathcal{N}) \otimes \mathcal{U}(-1,0) \simeq \iota_\dagger \tilde{\psi}_{-\boldsymbol{\delta}}(\mathcal{T}) \otimes \mathcal{U}(-1,0)$$
$$\Psi_2 : \psi^{(1)}(\mathcal{T}) \simeq \iota_\dagger \operatorname{Ker}(\mathcal{N}) \otimes \mathcal{U}(-1,0) \simeq \iota_\dagger \tilde{\psi}_{-\boldsymbol{\delta}}(\mathcal{T}) \otimes \mathcal{U}(-1,0)$$

Lemma 4.3.4 *We have $\Psi_1 = \Psi_2$. They are denoted by Ψ.*

Proof It can be checked by a direct computation. We will give an indication. The first component of $\Psi_2 \circ \Psi_1^{-1}$ is the composition of the following:

$$\begin{aligned}
\iota_\dagger \tilde{\psi}_{-\boldsymbol{\delta}}(\mathcal{M}')\,\lambda^{-1} &\simeq \iota_\dagger \operatorname{Cok}\Bigl(\tilde{\psi}_{-\boldsymbol{\delta}}(\mathcal{M}') \otimes \mathbb{I}_0^{0,N+1}\,\lambda^{-1} \longleftarrow \tilde{\psi}_{-\boldsymbol{\delta}}(\mathcal{M}') \otimes \mathbb{I}_0^{0,N+1}\Bigr) \\
&\simeq \mathcal{Q}(\Pi^{0,N+1}\mathcal{M}') \stackrel{\Phi}{\simeq} \mathcal{K}(\Pi^{-N+1,0}\mathcal{M}') \\
&\simeq \iota_\dagger \operatorname{Ker}\Bigl(\tilde{\psi}_{-\boldsymbol{\delta}}(\mathcal{M}') \otimes \mathbb{I}_0^{-N+1,0}\,\lambda^{-1} \longleftarrow \tilde{\psi}_{-\boldsymbol{\delta}}(\mathcal{M}') \otimes \mathbb{I}_0^{-N+1,0}\Bigr) \\
&\simeq \iota_\dagger \tilde{\psi}_{-\boldsymbol{\delta}}(\mathcal{M}')\,\lambda^{-1}
\end{aligned} \tag{4.22}$$

The morphism Φ is induced by the following:

$$\begin{array}{ccccc}
\mathrm{Ker} & \xrightarrow{=} & \mathrm{Ker} & \longrightarrow & \mathcal{K}(\Pi^{-N+1,0}\mathcal{M}') \\
\downarrow & & \downarrow & & \downarrow \\
\Pi_!^{0,N+1}\mathcal{M}' & \longrightarrow & \Pi_!^{-N+1,N+1}\mathcal{M}' & \longrightarrow & \Pi_!^{-N+1,0}\mathcal{M}' \\
\downarrow & & \downarrow & & \downarrow \\
\Pi_*^{0,N+1}\mathcal{M}' & \longrightarrow & \Pi_*^{-N+1,N+1}\mathcal{M}' & \longrightarrow & \Pi_*^{-N+1,0}\mathcal{M}' \\
\downarrow & & \downarrow & & \downarrow \\
\mathcal{Q}(\Pi^{0,N}\mathcal{M}') & \longrightarrow & \mathrm{Cok} & \xrightarrow{=} & \mathrm{Cok}
\end{array}$$

Hence, the composite of the morphisms in (4.22) is induced by the following diagram:

$$\begin{array}{ccccccc}
 & & \mathcal{P}'\otimes\mathbb{I}_0^{0,N_+} & \to & \mathcal{P}'\otimes\mathbb{I}_0^{-N_-,N_+} & \to & \mathcal{P}'\otimes\mathbb{I}_0^{-N_-,0} \to \mathcal{P}'\lambda^{-1} \\
 & & {\scriptstyle -t\partial_t}\downarrow & & \downarrow & & \downarrow \\
\mathcal{P}'\lambda^{-1} & \to & \mathcal{P}'\otimes\mathbb{I}_0^{0,N_+}\lambda^{-1} & \to & \mathcal{P}'\otimes\mathbb{I}_0^{-N_-,N_+}\lambda^{-1} & \to & \mathcal{P}'\otimes\mathbb{I}_0^{-N_-,0}\lambda^{-1}
\end{array}$$

Here, $N_+ = N+1$, $N_- = N-1$ and $\mathcal{P}'$ denotes $\tilde{\psi}_{-\boldsymbol{\delta}}(\mathcal{M}')$ Hence, the composite is the identity. Let us look at the second component of $\Psi_2^{-1}\circ\Psi_1$. It is the composite of the following morphisms:

$$\begin{aligned}
\iota_\dagger\tilde{\psi}_{-\boldsymbol{\delta}}(\mathcal{M}'') &\simeq \iota_\dagger\,\mathrm{Cok}\Big(\tilde{\psi}_{-\boldsymbol{\delta}}(\mathcal{M}'')\otimes\mathbb{I}_0^{1,N}\longrightarrow\tilde{\psi}_{-\boldsymbol{\delta}}(\mathcal{M}'')\otimes\mathbb{I}_0^{1,N}\lambda^{-1}\Big) \\
&\simeq \mathcal{Q}(\Pi^{1,N}\mathcal{M}'')\simeq\mathcal{K}(\Pi^{-N,1}\mathcal{M}'') \\
&\simeq \iota_\dagger\,\mathrm{Ker}\Big(\tilde{\psi}_{-\boldsymbol{\delta}}(\mathcal{M}'')\otimes\mathbb{I}_0^{-N,1}\longrightarrow\tilde{\psi}_{-\boldsymbol{\delta}}(\mathcal{M}'')\otimes\mathbb{I}_0^{-N,1}\lambda^{-1}\Big) \\
&\simeq \iota_\dagger\tilde{\psi}_{-\boldsymbol{\delta}}(\mathcal{M}'')
\end{aligned} \tag{4.23}$$

The composite of the morphisms is obtained from the following diagram:

$$\begin{array}{ccccccc}
 & & \mathcal{P}''\otimes\mathbb{I}_0^{1,N} & \to & \mathcal{P}''\otimes\mathbb{I}_0^{-N,1} & \to & \mathcal{P}''\otimes\mathbb{I}_0^{-N,0} \to \mathcal{P}'' \\
 & & {\scriptstyle -t\partial_t}\downarrow & & \downarrow & & \downarrow \\
\mathcal{P}'' & \to & \mathcal{P}''\otimes\mathbb{I}_0^{1,N}\lambda^{-1} & \to & \mathcal{P}''\otimes\mathbb{I}_0^{-N,N}\lambda^{-1} & \to & \mathcal{P}''\otimes\mathbb{I}_0^{-N,1}\lambda^{-1}
\end{array}$$

Here, $\mathcal{P}''$ denotes $\tilde{\psi}_{-\boldsymbol{\delta}}(\mathcal{M}'')$. Then, we can check that the composite is the identity. □

It is easy to check that the diagram (4.18) is commutative, because we have $t\partial_t = -N' \otimes \mathrm{id} + \mathrm{id} \otimes s$ on $\tilde{\psi}_{-\delta}(\mathcal{M}') \otimes \mathbb{I}_0^{p,q}$ and $t\partial_t = -N'' \otimes \mathrm{id} + \mathrm{id} \otimes s$ on $\tilde{\psi}_{-\delta}(\mathcal{M}'') \otimes \mathbb{I}_0^{p,q}$. Thus, we obtain Proposition 4.3.1.

4.3.5 Hermitian Adjoint

Let $\mathcal{S} : \mathcal{T} \longrightarrow \mathcal{T}^* \otimes \boldsymbol{T}(-w)$ be a Hermitian sesqui-linear duality of weight w. Let us check Proposition 4.3.2. Under the isomorphism Ψ in Sect. 4.3.4, $\psi^{(1)}(\mathcal{S})'' : \psi^{(1)}(\mathcal{M}'') \longrightarrow \psi^{(1)}(\mathcal{M}')\,\lambda^{-w+1}$ is induced by

$$
\begin{aligned}
\tilde{\psi}_{-\delta}(\mathcal{M}'') &\overset{b_1}{\simeq} \mathrm{Cok}\Big(\mathcal{N}'' : \tilde{\psi}_{-\delta}(\mathcal{M}'') \otimes \mathbb{I}^{1,N} \longrightarrow \tilde{\psi}_{-\delta}(\mathcal{M}'') \otimes \mathbb{I}^{1,N}\,\lambda^{-1}\Big) \\
&\xrightarrow{f_1} \mathrm{Cok}\Big(\mathcal{N}' : \tilde{\psi}_{-\delta}(\mathcal{M}') \otimes \mathbb{I}^{1,N} \longrightarrow \tilde{\psi}_{-\delta}(\mathcal{M}') \otimes \mathbb{I}^{1,N}\,\lambda^{-1}\Big) \cdot \lambda^{-w} \\
&\xrightarrow{s^{-1}} \mathrm{Cok}\Big(\mathcal{N}' : \tilde{\psi}_{-\delta}(\mathcal{M}') \otimes \mathbb{I}^{0,N-1} \longrightarrow \tilde{\psi}_{-\delta}(\mathcal{M}') \otimes \mathbb{I}^{0,N-1}\,\lambda^{-1}\Big) \cdot \lambda^{-w+1} \\
&\xrightarrow{g_1} \mathrm{Cok}\Big(\mathcal{N}' : \tilde{\psi}_{-\delta}(\mathcal{M}') \otimes \mathbb{I}^{0,N-1} \longrightarrow \tilde{\psi}_{-\delta}(\mathcal{M}') \otimes \mathbb{I}^{0,N-1}\,\lambda^{-1}\Big) \cdot \lambda^{-w+1} \\
&\overset{b_2}{\simeq} \big(\tilde{\psi}_{-\delta}(\mathcal{M}')\,\lambda^{-1}\big) \cdot \lambda^{-w+1}
\end{aligned}
\tag{4.24}
$$

Here, b_i are isomorphisms in Sect. 4.3.4, f_1 is induced by $\mathcal{S}''$, and g_1 is induced by $\mathcal{S}^{a,b}$. The composite of the morphisms is $-\tilde{\psi}_{-\delta}(\mathcal{S}'')$.

Let us look at the following morphisms:

$$
\begin{aligned}
\lambda^{-1}\,\tilde{\psi}_{-\delta}(\mathcal{M}') &\simeq \mathrm{Ker}\Big(\tilde{\psi}_{-\delta}(\mathcal{M}') \otimes \mathbb{I}^{-N+1,0}\,\lambda^{-1} \longleftarrow \tilde{\psi}_{-\delta}(\mathcal{M}') \otimes \mathbb{I}^{-N+1,0}\Big) \\
&\xleftarrow{f_2} \mathrm{Ker}\Big(\tilde{\psi}_{-\delta}(\mathcal{M}'') \otimes \mathbb{I}^{-N+1,0}\,\lambda^{-1} \longleftarrow \tilde{\psi}_{-\delta}(\mathcal{M}'') \otimes \mathbb{I}^{-N+1,0}\Big)\,\lambda^{w} \\
&\xleftarrow{s} \mathrm{Ker}\Big(\tilde{\psi}_{-\delta}(\mathcal{M}'') \otimes \mathbb{I}^{-N+2,1}\,\lambda^{-1} \longleftarrow \tilde{\psi}_{-\delta}(\mathcal{M}'') \otimes \mathbb{I}^{-N+2,1}\Big)\,\lambda^{w-1} \\
&\xleftarrow{g_2} \mathrm{Ker}\Big(\tilde{\psi}_{-\delta}(\mathcal{M}'') \otimes \mathbb{I}^{-N+2,1}\,\lambda^{-1} \longleftarrow \tilde{\psi}_{-\delta}(\mathcal{M}'') \otimes \mathbb{I}^{-N+2,1}\Big)\,\lambda^{w-1} \\
&\simeq \tilde{\psi}_{-\delta}(\mathcal{M}'') \cdot \lambda^{w-1}
\end{aligned}
\tag{4.25}
$$

Here, f_2 is induced by $\mathcal{S}'$, and g_2 is induced by $\mathcal{S}^{a,b}$. Hence, the composite is $-\psi_{-\delta}(\mathcal{S}')$. Thus, we obtain Proposition 4.3.2. □

4.4 Admissible Specializability

4.4.1 Filtered $\mathcal{R}$-Modules

4.4.1.1 Filtered Strictly Specializability

In this paper, a (coherent) filtered $\mathcal{R}_X$-module means an $\mathcal{R}_X$-module with a locally finite increasing exhaustive $\mathbb{Z}$-indexed filtration in the category of (coherent) $\mathcal{R}_X$-modules.

Let $(\mathcal{M}, L)$ be a coherent filtered $\mathcal{R}_X$-module. Let g be any holomorphic function on X.

Definition 4.4.1 $(\mathcal{M}, L)$ is called filtered strictly specializable along g if the following holds:

- Each $L_j\mathcal{M}$ is strictly specializable along g with any ramified exponential twist.
- The cokernel of $\tilde{\psi}_{g,\mathfrak{a},u}(L_j\mathcal{M}) \longrightarrow \tilde{\psi}_{g,\mathfrak{a},u}(\mathcal{M})$ are strict for any $u \in \mathbb{R}\times\mathbb{C}$ and $\mathfrak{a} \in \mathbb{C}[t_n^{-1}]$. The cokernel of $\phi_g(L_j\mathcal{M}) \longrightarrow \phi_g(\mathcal{M})$ are also strict. □

Lemma 4.4.2 *Suppose that $(\mathcal{M}, L)$ is filtered strictly specializable. Then, each $\big(L_j\mathcal{M}/L_k\mathcal{M}, L\big)$ is also filtered strictly specializable.*

Proof In this case, each $\mathcal{M}/L_j\mathcal{M}$ is also strictly specializable along g with any ramified exponential twist. The morphisms $L_j\mathcal{M} \longrightarrow \mathcal{M} \longrightarrow \mathcal{M}/L_j\mathcal{M}$ are strict with respect to the V-filtrations for any ramified exponential twist. We have the exact sequence $0 \longrightarrow \tilde{\psi}_{g,\mathfrak{a},u}(L_j\mathcal{M}) \longrightarrow \tilde{\psi}_{g,\mathfrak{a},u}\mathcal{M} \longrightarrow \tilde{\psi}_{g,\mathfrak{a},u}(\mathcal{M}/L_j\mathcal{M}) \longrightarrow 0$. We have

$$\mathrm{Cok}\Big(\tilde{\psi}_{g,\mathfrak{a},u}(L_m\mathcal{M}) \to \tilde{\psi}_{g,\mathfrak{a},u}(L_j\mathcal{M})\Big) \subset \mathrm{Cok}\Big(\tilde{\psi}_{g,\mathfrak{a},u}(L_m\mathcal{M}) \to \tilde{\psi}_{g,\mathfrak{a},u}(\mathcal{M})\Big),$$

$$\mathrm{Cok}\Big(\phi_g(L_m\mathcal{M}) \to \phi_g(L_j\mathcal{M})\Big) \subset \mathrm{Cok}\Big(\phi_g(L_m\mathcal{M}) \to \phi_g(\mathcal{M})\Big).$$

We also have the following isomorphisms:

$$\mathrm{Cok}\Big(\tilde{\psi}_{g,\mathfrak{a},u}(L_j\mathcal{M}/L_m\mathcal{M}) \longrightarrow \tilde{\psi}_{g,\mathfrak{a},u}(\mathcal{M}/L_m\mathcal{M})\Big) \simeq \tilde{\psi}_{g,\mathfrak{a},u}(\mathcal{M}/L_j\mathcal{M})$$

$$\mathrm{Cok}\Big(\phi_g(L_j\mathcal{M}/L_m\mathcal{M}) \longrightarrow \phi_g(\mathcal{M}/L_m\mathcal{M})\Big) \simeq \phi_g(\mathcal{M}/L_j\mathcal{M})$$

Then, the claim of Lemma 4.4.2 follows. □

From the proof of Lemma 4.4.2, we obtain the naturally induced filtrations L on $\tilde{\psi}_{g,\mathfrak{a},u}(\mathcal{M})$ and $\phi_g(\mathcal{M})$:

$$L_j\tilde{\psi}_{g,\mathfrak{a},u}(\mathcal{M}) := \tilde{\psi}_{g,\mathfrak{a},u}(L_j\mathcal{M}), \qquad L_j\phi_g(\mathcal{M}) := \phi_g(L_j\mathcal{M})$$

The induced filtered $\mathcal{R}_X$-modules $\big(\tilde{\psi}_{g,\mathfrak{a},u}(\mathcal{M}), L\big)$ and $\big(\phi_g(\mathcal{M}), L\big)$ are also denoted by $\tilde{\psi}_{g,\mathfrak{a},u}(\mathcal{M}, L)$ and $\phi_g(\mathcal{M}, L)$. From the proof of Lemma 4.4.2, we also obtain the following.

Lemma 4.4.3 $\mathrm{Gr}^L\,\tilde{\psi}_{g,\mathfrak{a},u}(\mathcal{M})$ *and* $\mathrm{Gr}^L\,\phi_g(\mathcal{M})$ *are strict* $\mathcal{R}_X$*-modules.* □

Let $(\mathcal{M}_i, L)$ $(i = 1, 2)$ be coherent filtered $\mathcal{R}_X$-modules which are filtered strictly specializable along g. Let $f : (\mathcal{M}_1, L) \longrightarrow (\mathcal{M}_2, L)$ be a morphism of filtered $\mathcal{R}_X$-modules. Then, we have the induced morphisms of coherent filtered $\mathcal{R}_X$-modules:

$$\tilde{\psi}_{g,\mathfrak{a},u}(f) : \tilde{\psi}_{g,\mathfrak{a},u}(\mathcal{M}_1, L) \longrightarrow \tilde{\psi}_{g,\mathfrak{a},u}(\mathcal{M}_2, L)$$
$$\phi_g(f) : \phi_g(\mathcal{M}_1, L) \longrightarrow \phi_g(\mathcal{M}_2, L)$$

4.4.1.2 Admissible Specializability

Definition 4.4.4 Let $(\mathcal{M}, L)$ be a filtered $\mathcal{R}_X$-module, which is filtered strictly specializable along g. It is called admissibly specializable along g if the following holds:

(P1) $\big(\tilde{\psi}_{g,\mathfrak{a},u}(\mathcal{M}, L), \mathcal{N}\big)$ has a relative monodromy filtration for each $u \in \mathbb{R} \times \mathbb{C}$ and $\mathfrak{a} \in \mathbb{C}[t_n^{-1}]$.

(P2) $\big(\phi_g(\mathcal{M}, L), \mathcal{N}\big)$ has a relative monodromy filtration.

(P3) For the morphisms $\psi_{g,-\boldsymbol{\delta}}(\mathcal{M}) \xrightarrow{u} \phi_g(\mathcal{M}) \xrightarrow{v} \psi_{g,-\boldsymbol{\delta}}(\mathcal{M})$, where u and v are induced by $\eth_t$ and t, we have $u \cdot M_k(N; L) \subset M_{k-1}(N; L)$ and $v \cdot M_k(N; L) \subset M_{k-1}(N; L)$.

(We shall give a review on relative monodromy filtration in Sect. 6.1.1 below.) □

The following lemma is clear.

Lemma 4.4.5 *If* $(\mathcal{M}, L)$ *is admissibly specializable along a holomorphic function* g*, any* $\big(L_j\mathcal{M}/L_k\mathcal{M}, L\big)$ *are also admissibly specializable along* g*.* □

A filtered $\mathcal{R}_X$-module $(\mathcal{M}, L)$ is called integrable if each $L_j\mathcal{M}$ is integrable. The following lemma is clear by Deligne's formula for relative monodromy filtration (6.2), (6.3) below.

Lemma 4.4.6 *If* $(\mathcal{M}, L)$ *is integrable and admissibly specializable along* g*, the relative monodromy filtrations are also integrable.* □

The following lemma is also clear by Deligne's formula.

Lemma 4.4.7 *Let* $(\mathcal{M}_i, L)$ $(i = 1, 2)$ *be admissibly specializable along* g*. Let* $f : (\mathcal{M}_1, L) \longrightarrow (\mathcal{M}_2, L)$ *be any morphism. Then, the induced morphisms* $\tilde{\psi}_{g,\mathfrak{a},u}(f)$ *and* $\phi_g(f)$ *are compatible with* $M(N; L)$*.* □

Let $\iota : Y \longrightarrow X$ be a closed immersion of complex manifolds. Let $(\mathcal{M}, L)$ be a coherent filtered $\mathcal{R}_Y$-module. Let g be a holomorphic function on X.

Lemma 4.4.8 *$(\mathcal{M}, L)$ is filtered strictly specializable (resp. admissibly) along $g_Y := g_{|Y}$, if and only if $\iota_\dagger(\mathcal{M}, L)$ is filtered strictly (resp. admissibly) specializable along g.*

Proof Assume that $(\mathcal{M}, L)$ is filtered strictly specializable along g_Y. Because $\iota_\dagger\tilde{\psi}_{g_Y,\mathfrak{a},u}(L_j\mathcal{M})$ is strict, according to the compatibility of the push-forward and the strict specializability, $\iota_\dagger L_j\mathcal{M}$ is strictly specializable along g with any ramified exponential twist, and $\tilde{\psi}_{g,\mathfrak{a},u}\bigl(\iota_\dagger L_j\mathcal{M}\bigr) \simeq \iota_\dagger\tilde{\psi}_{g_Y,\mathfrak{a},u}(L_j\mathcal{M})$. Because the cokernel of $\tilde{\psi}_{g_Y,\mathfrak{a},u}(L_j\mathcal{M}) \longrightarrow \tilde{\psi}_{g_Y,\mathfrak{a},u}(\mathcal{M})$ are strict, the cokernel of $\tilde{\psi}_{g,\mathfrak{a},u}(\iota_\dagger L_j\mathcal{M}) \longrightarrow \tilde{\psi}_{g,\mathfrak{a},u}(\iota_\dagger\mathcal{M})$ are strict. A similar claim holds for ϕ_g. Hence, $\iota_\dagger(\mathcal{M}, L)$ is filtered strictly specializable along g.

Suppose that $\iota_\dagger(\mathcal{M}, L)$ is filtered strictly specializable along g. Each $\iota_\dagger(L_j\mathcal{M})$ is strictly specializable along g with any ramified exponential twist. By Lemma 4.4.9 below, each $L_j\mathcal{M}$ is strictly specializable along g_Y, and hence we have $\tilde{\psi}_{g,\mathfrak{a},u}\bigl(\iota_\dagger L_j\mathcal{M}\bigr) \simeq \iota_\dagger\tilde{\psi}_{g_Y,\mathfrak{a},u}(L_j\mathcal{M})$. We have a similar isomorphism for ϕ_g. Then, we obtain that $(\mathcal{M}, L)$ is filtered strictly specializable along g.

If $(\mathcal{M}, L)$ is filtered specializable along g_Y, $\iota_\dagger\tilde{\psi}_{g_Y,\mathfrak{a},u}(\mathcal{M})$ and $\tilde{\psi}_{g,\mathfrak{a},u}(\iota_\dagger\mathcal{M})$ are naturally isomorphic, which is compatible with the induced filtrations and the nilpotent maps. A similar claim holds for ϕ_g. Hence, $(\mathcal{M}, L)$ is admissibly specializable along g_Y if and only if $\iota_\dagger(\mathcal{M}, L)$ is admissibly specializable along g. □

Lemma 4.4.9 *Let $\mathcal{M}$ be a coherent strict $\mathcal{R}_Y$-module. If $\iota_\dagger\mathcal{M}$ is strictly specializable along g, then $\mathcal{M}$ is strictly specializable along g_Y.*

Proof We have only to consider the case $\operatorname{codim} Y = 1$. It is enough to consider the issue locally around any point of Y. We will shrink X and Y without mention. By the graph construction, we replace X and Y with $X \times \mathbb{C}_t$ and $Y \times \mathbb{C}_t$, respectively, and the functions g and g_Y are replaced with t. Then, we may assume that g is a coordinate function, and $g^{-1}(0)$ and Y are transversal. Then, after a bi-holomorphic transform, we may assume that X is an open subset of $Z \times \mathbb{C}_s \times \mathbb{C}_t$, and $Y = \{s = 0\}$ and $g = t$. We have the grading $\iota_\dagger\mathcal{M} = \bigoplus_j \iota_*\mathcal{M}\,\eth_s^j$. Let $V^{(\lambda_0)}$ be the V-filtration of $\iota_\dagger\mathcal{M}$ at λ_0. We have only to prove that it is compatible with the grading, i.e., for a given $f = \sum_{j=0}^N f_j\eth_s^j \in V_a^{(\lambda_0)}(\iota_\dagger\mathcal{M})$, we have $f_j\,\eth_s^j \in V_a^{(\lambda_0)}(\iota_\dagger\mathcal{M})$ for each j. We use an induction on N, and so it is enough to prove $f_N\eth_s^N \in V_a^{(\lambda_0)}(\iota_\dagger\mathcal{M})$. We have $s^N f \in V_a^{(\lambda_0)}(\iota_\dagger(\mathcal{M}))$. We obtain $\lambda^N f_N \in V_a^{(\lambda_0)}(\iota_\dagger\mathcal{M})$. By using the strictness of $\psi_b^{(\lambda_0)}$ for any $b \in \mathbb{R}$, we obtain that $f_N \in V_a^{(\lambda_0)}$, and hence $f_N\eth_s^N \in V_a^{(\lambda_0)}$. □

4.4.2 Filtered $\mathcal{R}$-Triples

In this paper, a (coherent) filtered $\mathcal{R}_X$-triple means an $\mathcal{R}_X$-triple with a locally finite increasing exhaustive $\mathbb{Z}$-indexed filtration in the category of (coherent) $\mathcal{R}_X$-modules.

Let $(\mathcal{T}, L)$ be a filtered $\mathcal{R}_X$-triple. The $\mathcal{R}_X$-triple $\mathcal{T}$ consists of $\mathcal{R}_X$-modules $\mathcal{M}_i$ $(i = 1, 2)$ and a sesqui-linear pairing C. The $\mathcal{R}_X$-triple $L_j\mathcal{T}$ consists of $\mathcal{R}_X$-modules $\mathcal{M}_{1,j}$ and $\mathcal{M}_{2,j}$ and a sesqui-linear pairing C_j. For $j \geq k$, we have the monomorphism $\mathcal{M}_{2,k} \subset \mathcal{M}_{2,j}$ and the epimorphism $\mathcal{M}_{1,j} \longrightarrow \mathcal{M}_{1,k}$. For any j, we set $L_j(\mathcal{M}_2) := \mathcal{M}_{2,j}$ and $L_j(\mathcal{M}_1) := \operatorname{Ker}\bigl(\mathcal{M}_1 \longrightarrow \mathcal{M}_{1,-j-1}\bigr)$. Then, we obtain filtered $\mathcal{R}_X$-modules $(\mathcal{M}_i, L)$ $(i = 1, 2)$. They are called filtered $\mathcal{R}_X$-modules underlying $\mathcal{T}$. For $k \leq j$, the $\mathcal{R}_X$-triple $\mathcal{T}_j/\mathcal{T}_k$ consists of the $\mathcal{R}_X$-modules $L_{-k-1}\mathcal{M}_1/L_{-j-1}\mathcal{M}_1$ and $L_j\mathcal{M}_2/L_k\mathcal{M}_2$ with the induced sesqui-linear pairing.

Definition 4.4.10 A coherent filtered $\mathcal{R}_X$-triple $(\mathcal{T}, L)$ is called filtered strictly specializable along g if the underlying filtered $\mathcal{R}$-modules are filtered strictly specializable along g. □

Lemma 4.4.11 *Assume that $(\mathcal{T}, L)$ is filtered strictly specializable along g.*

- *Each $(L_j\mathcal{T}/L_k\mathcal{T}, L)$ is filtered strictly specializable along g.*
- *We have the exact sequence*

$$0 \longrightarrow \tilde{\psi}_{g,\mathfrak{a},u}(L_j\mathcal{T}) \longrightarrow \tilde{\psi}_{g,\mathfrak{a},u}\mathcal{T} \longrightarrow \tilde{\psi}_{g,\mathfrak{a},u}(\mathcal{T}/L_j\mathcal{T}) \longrightarrow 0$$

  for $\mathfrak{a} \in \mathbb{C}[t_n^{-1}]$ and $u \in \mathbb{R} \times \mathbb{C}$.
- *We have the exact sequence $0 \longrightarrow \phi_g(L_j\mathcal{T}) \longrightarrow \phi_g\mathcal{T} \longrightarrow \phi_g(\mathcal{T}/L_j\mathcal{T}) \longrightarrow 0.$* □

We set $L_j\tilde{\psi}_{g,\mathfrak{a},u}(\mathcal{T}) := \tilde{\psi}_{g,\mathfrak{a},u}(L_j\mathcal{T})$ and $L_j\phi_g(\mathcal{T}) := \phi_g(L_j\mathcal{T})$. We obtain filtered $\mathcal{R}$-triples $\tilde{\psi}_{g,\mathfrak{a},u}(\mathcal{T}, L) = (\tilde{\psi}_{g,\mathfrak{a},u}(\mathcal{T}), L)$ and $\phi_g(\mathcal{T}, L) = (\phi_g(\mathcal{T}), L)$ such that Gr^L are strict.

Definition 4.4.12 Let $(\mathcal{T}, L)$ be a filtered $\mathcal{R}$-triple, which is filtered strictly specializable along g. It is called admissibly specializable along g if the underlying $\mathcal{R}_X$-modules are admissibly specializable along g. □

Lemma 4.4.13 *If $(\mathcal{T}, L)$ is admissibly specializable along g, then the tuples $(\tilde{\psi}_{g,\mathfrak{a},u}(\mathcal{T}, L), \mathcal{N})$ and $(\phi_g(\mathcal{T}, L), \mathcal{N})$ have relative monodromy filtrations $M(\mathcal{N}; L)$ in the category of $\mathcal{R}$-triples.*

Proof It follows from the following general Lemma 4.4.14, which is a consequence of Deligne's inductive formula (6.2), (6.3) below. □

Lemma 4.4.14 *Let $\mathcal{T} = (\mathcal{M}_1, \mathcal{M}_2, C)$ be an $\mathcal{R}_X$-triple with a filtration L. Let $\mathcal{N} : \mathcal{T} \longrightarrow \mathcal{T} \otimes \boldsymbol{T}(-1)$ be a morphism such that $\mathcal{N} : L_j\mathcal{T} \longrightarrow L_j\mathcal{T} \otimes \boldsymbol{T}(-1)$.*

Assume that $\mathcal{M}_i$ $(i = 1, 2)$ have relative monodromy filtrations $\tilde{L} = M(N; L)$. Then, we have a filtration $\tilde{L}$ of $\mathcal{T}$ which is a relative monodromy filtration $M(\mathcal{N}; L)$. □

We obtain the following lemma from Lemma 4.4.7.

Lemma 4.4.15 *Let $(\mathcal{T}_i, L)$ $(i = 1, 2)$ be admissibly specializable along g. Let $f : (\mathcal{T}_1, L) \longrightarrow (\mathcal{T}_2, L)$ be any morphism. Then, the induced morphisms $\tilde{\psi}_{g,\mathfrak{a},u}(f)$ and $\phi_g(f)$ are compatible with the filtrations $M(\mathcal{N}; L)$.* □

Lemma 4.4.16 *Suppose that $(\mathcal{T}, L)$ is admissibly specializable along g. Let A be a nowhere vanishing holomorphic function on X. Then, $(\mathcal{T}, L)$ is admissibly specializable along $g_1 := Ag$, and we have the following natural isomorphisms:*

$$\big(\mathrm{Gr}^{M(\mathcal{N};L)}\, \tilde{\psi}_{g,\mathfrak{a},u}(\mathcal{T}), \mathcal{N}^{(0)}\big) \simeq \big(\mathrm{Gr}^{M(\mathcal{N};L)}\, \tilde{\psi}_{g_1,\mathfrak{a},u}(\mathcal{T}), \mathcal{N}^{(0)}\big)$$
$$\big(\mathrm{Gr}^{M(\mathcal{N};L)}\, \phi_g(\mathcal{T}), \mathcal{N}^{(0)}\big) \simeq \big(\mathrm{Gr}^{M(\mathcal{N};L)}\, \phi_{g_1}(\mathcal{T}), \mathcal{N}^{(0)}\big)$$

They preserve the canonical splitting of the induced filtration L.

Proof It follows from Corollary 4.2.8. □

We obtain the following lemma from Lemma 4.4.8.

Lemma 4.4.17 *Let $\iota : Y \longrightarrow X$ be a closed embedding of complex manifolds. Let $(\mathcal{T}, L)$ be a coherent strict filtered $\mathcal{R}_Y$-triple. Let g be a holomorphic function on X such that $Y \not\subset g^{-1}(0)$. Then, $(\mathcal{T}, L)$ is filtered strictly specializable along $g_{|Y}$, if and only if $\iota_\dagger(\mathcal{T}, L)$ is filtered strictly specializable along g.* □

Definition 4.4.18 Let $(\mathcal{T}, L)$ be a filtered $\mathcal{R}_X$-triple. It is called admissibly (resp. filtered strictly) specializable if the following holds

- Let $U \subset X$ be any open subset with a holomorphic function g. Then, $(\mathcal{T}, L)_{|U}$ is admissibly (resp. filtered strictly) specializable along g. □

4.4.3 Admissible Specializability Along Hypersurfaces

Let H be an effective divisor of X.

Definition 4.4.19 A filtered $\mathcal{R}_X$-triple $(\mathcal{T}, L)$ is called admissibly (resp. filtered strictly) specializable along H if the following holds:

- Let U be any open subset of X with a generator g_U of $\mathcal{O}(-H)_{|U}$. Then, $(\mathcal{T}, L)_{|U}$ is admissibly (resp. filtered strictly) specializable along g_U.

Admissible specializability and filtered strict specializability along H for $\mathcal{R}_X$-modules are defined in similar ways. □

The following lemma is easy to see.

Lemma 4.4.20 *If $(\mathcal{T}, L)$ is admissibly (resp. filtered strictly) specializable along H, each $L_j\mathcal{T}/L_k\mathcal{T}$ is admissibly (resp. filtered strictly) specializable along H.* □

4.4.3.1 Induced Graded $\mathcal{R}$-Triples

Let $U \subset X$ be any open subset with a generator g_U of $\mathcal{O}(-H)_{|U}$. Let L be the naively induced filtrations on $\phi_{g_U}^{(a)}(\mathcal{T}_{|U})$ and $\psi_{g_U}^{(a)}(\mathcal{T}_{|U})$, i.e., $L_j\phi_{g_U}^{(a)}(\mathcal{T}_{|U}) = \phi_{g_U}^{(a)}(L_j\mathcal{T}_{|U})$ and $L_j\psi_{g_U}^{(a)}(\mathcal{T}_{|U}) = \psi_{g_U}^{(a)}(L_j\mathcal{T}_{|U})$. Set $W := M(\mathcal{N}; L)[-2a]$ on $\phi_{g_U}^{(a)}(\mathcal{T}_{|U})$, and $W := M(\mathcal{N}; L)[-2a+1]$ on $\psi_{g_U}^{(a)}(\mathcal{T}_{|U})$. We have the $\mathbb{Z}$-graded $\mathcal{R}$-triples

$$\operatorname{Gr}^W \psi_{g_U}^{(a)}(\mathcal{T}_{|U}), \quad \operatorname{Gr}^W \phi_{g_U}^{(a)}(\mathcal{T}_{|U})$$

on U with the endomorphisms

$$\mathcal{N} : \operatorname{Gr}_k^W \psi_{g_U}^{(a)}(\mathcal{T}_{|U}) \longrightarrow \operatorname{Gr}_{k-2}^W \psi_{g_U}^{(a)}(\mathcal{T}_{|U}) \otimes \boldsymbol{T}(-1),$$
$$\mathcal{N} : \operatorname{Gr}_k^W \phi_{g_U}^{(a)}(\mathcal{T}_{|U}) \longrightarrow \operatorname{Gr}_{k-2}^W \phi_{g_U}^{(a)}(\mathcal{T}_{|U}) \otimes \boldsymbol{T}(-1).$$

According to Lemma 4.4.16, they are independent of the choice of a generator g_U. Hence, by gluing them for varied (U, g_U), we obtain graded $\mathcal{R}_X$-triples with morphisms, denoted by

$$\Big((\operatorname{Gr}^W \psi_H^{(a)})(\mathcal{T}), \mathcal{N}\Big), \quad \Big((\operatorname{Gr}^W \phi_H^{(a)})(\mathcal{T}), \mathcal{N}\Big).$$

Similarly, we obtain a graded $\mathcal{R}$-triple $\operatorname{Gr}^W \psi_{H,-\boldsymbol{\delta}}(\mathcal{T})$ with morphisms

$$\mathcal{N} : \operatorname{Gr}_k^W \psi_{H,-\boldsymbol{\delta}}(\mathcal{T}) \longrightarrow \operatorname{Gr}_{k-2}^W \psi_{H,-\boldsymbol{\delta}}(\mathcal{T}) \otimes \boldsymbol{T}(-1)$$

as the gluing of $(\operatorname{Gr}^W \psi_{g_U,-\boldsymbol{\delta}}(\mathcal{T}_{|U}), \mathcal{N})$. According to Proposition 4.3.1, we have following natural isomorphisms:

$$\Big((\operatorname{Gr}^W \psi_{H,-\boldsymbol{\delta}}(\mathcal{T})), \mathcal{N}\Big) \simeq \Big((\operatorname{Gr}^W \psi_H^{(1)})(\mathcal{T}), \mathcal{N}\Big) \otimes \mathcal{U}(1,0)$$
$$\simeq \Big((\operatorname{Gr}^W \psi_H^{(0)})(\mathcal{T}), \mathcal{N}\Big) \otimes \mathcal{U}(0,1) \tag{4.26}$$

Remark 4.4.21 Although $\big(\operatorname{Gr}^W \psi_H^{(a)}\big)(\mathcal{T})$ is defined, we do not necessarily have "$\psi_H^{(a)}(\mathcal{T})$" as an $\mathcal{R}$-triple. We have similar remarks for $\big(\operatorname{Gr}^W \phi_H^{(a)}\big)(\mathcal{T})$ and $\big(\operatorname{Gr}^W \psi_{H,-\boldsymbol{\delta}}(\mathcal{T}), \mathcal{N}\big)$. □

The filtration L induces filtrations on $\big(\operatorname{Gr}^W \psi_H^{(a)}\big)(\mathcal{T})$ and $\big(\operatorname{Gr}^W \phi_H^{(a)}\big)(\mathcal{T})$, which are preserved by $\mathcal{N}$. By gluing the canonical decomposition of Kashiwara (see Sect. 6.1.2 below), we obtain the following isomorphisms, which are splittings of L:

$$\big(\mathrm{Gr}^W \psi_H^{(a)}\big)(\mathcal{T}) \simeq \big(\mathrm{Gr}^W \psi_H^{(a)}\big)(\mathrm{Gr}^L \mathcal{T}) \tag{4.27}$$

$$\big(\mathrm{Gr}^W \phi_H^{(a)}\big)(\mathcal{T}) \simeq \big(\mathrm{Gr}^W \phi_H^{(a)}\big)(\mathrm{Gr}^L \mathcal{T}) \tag{4.28}$$

By the condition of the admissible specializability, we have the following induced morphisms:

$$\big(\mathrm{Gr}^W \psi_H^{(a+1)}\big)(\mathcal{T}) \xrightarrow{\alpha} \big(\mathrm{Gr}^W \phi_H^{(a)}\big)(\mathcal{T}) \xrightarrow{\beta} \big(\mathrm{Gr}^W \psi_H^{(a)}\big)(\mathcal{T})$$

The morphisms $\beta \circ \alpha$ and $\alpha \circ \beta$ are identified with the morphisms $\mathcal{N}$ on $(\mathrm{Gr}^W \psi_H^{(a)})(\mathcal{T})$ and $(\mathrm{Gr}^W \phi_H^{(a)})(\mathcal{T})$ up to constants.

Let $(\mathcal{T}_i, L)$ $(i = 1, 2)$ be admissibly specializable along H. Let $f : (\mathcal{T}_1, L) \longrightarrow (\mathcal{T}_2, L)$ be any morphism. According to Lemma 4.4.15, we have the naturally induced morphisms:

$$\big(\mathrm{Gr}^W \psi_H^{(a)}\big)(f) : \big(\mathrm{Gr}^W \psi_H^{(a)}\big)(\mathcal{T}_1) \longrightarrow \big(\mathrm{Gr}^W \psi_H^{(a)}\big)(\mathcal{T}_2)$$

$$\big(\mathrm{Gr}^W \phi_H^{(a)}\big)(f) : \big(\mathrm{Gr}^W \phi_H^{(a)}\big)(\mathcal{T}_1) \longrightarrow \big(\mathrm{Gr}^W \phi_H^{(a)}\big)(\mathcal{T}_2)$$

Lemma 4.4.22 *If $(\mathcal{T}, L)$ is integrable, the induced $\mathcal{R}_X$-triples $\big(\mathrm{Gr}^W \psi_H^{(a)}\big)(\mathcal{T})$ and $\big(\mathrm{Gr}^W \phi_H^{(a)}\big)(\mathcal{T})$ are also integrable.* □

Chapter 5
Gluing of Good-KMS Smooth $\mathcal{R}$-Triples

Let $X := \{(z_1, \ldots, z_n) \in \mathbb{C}^n \mid |z_i| < 1\}$, $D_i := \{z_i = 0\} \cap X$ and $D := \bigcup_{i=1}^{\ell} D_i$. For any $I \subset \underline{\ell} = \{1, \ldots, \ell\}$, we put $D_I := \bigcap_{i \in I} D_i$ and $\partial D_I := D_I \cap \bigcup_{i \in \underline{\ell} \setminus I} D_i$. For $J \subset \underline{\ell} \setminus I$, we set $D_I(J) := D_I \cap \bigcup_{i \in J} D_i$. Let V_I $(I \subset \underline{\ell})$ be regular singular meromorphic flat bundles on $(D_I, \partial D_I)$. With some additional data, we may consider the $\mathscr{D}$-module obtained as the gluing of V_I $(I \subset \underline{\ell})$. Namely, we can successively apply the formalism of Beilinson along z_i to obtain a $\mathscr{D}$-module. In the construction of the gluing of V_I $(I \subset \underline{\ell})$, we naturally encounter $\mathscr{D}$-modules of the form $V_I'[*D_I(J)][!D_I(K)]$ $(I \sqcup J \sqcup K = \underline{\ell})$ where V_I' are regular singular meromorphic flat bundles on $(D_I, \partial D_I)$.

We also have the gluing procedure of meromorphic flat bundles V_I on $(D_I, \partial D_I)$ which are not necessarily regular singular. In that case, we impose that they are *good*, for which we have $V_I[*z_i][!z_j] = V_I[!z_j][*z_i]$ $(i, j \in \underline{\ell} \setminus I)$. So, again the $\mathscr{D}$-modules $V_I'[*D_I(J)][!D_I(K)]$ naturally appear, where V_I' are good meromorphic flat bundles on $(D_I, \partial D_I)$.

In this chapter, we consider such gluing procedure in the context of $\mathcal{R}$-triples. Namely, we study the gluing of smooth $\mathcal{R}_{D_I(*\partial D_I)}$-triples $\mathcal{T}_I$ by the successive use of the formalism of Beilinson along z_i. We should impose a suitable condition to $\mathcal{T}_I$ along ∂D_I, that is the good-KMS condition. The condition was already appeared in [55]. We review it in Sect. 5.1. Indeed, we need to consider the gluing of filtered smooth $\mathcal{R}_{D_I(*\partial D_I)}$-triples. So, it will be useful to clarify the relation of the filtration and good-KMS structure, which is studied in Sect. 5.2.

In Sect. 5.3, we shall construct the counterpart of $V[!D_I(J)][*D_I(K)]$ in the context of $\mathcal{R}$-modules. In Sect. 5.4, we study the strictly specializability and the localizability of such $\mathcal{R}$-modules along functions g such that $g^{-1}(0) \subset D$. Recall that we did not study in Chap. 4 the localizability of $\mathcal{R}$-modules along general holomorphic functions. The results in Sects. 5.3–5.4 are useful in our proof of the localizability of mixed twistor $\mathscr{D}$-modules. (See Sect. 11.2.)

We study $\mathcal{R}_{D_I(*\partial D_I)}$-triples satisfying the good-KMS condition in Sect. 5.5, and their gluing in Sect. 5.6. We shall use the category of the filtered $\mathcal{R}$-modules

T. Mochizuki, *Mixed Twistor $\mathscr{D}$-modules*, Lecture Notes in Mathematics 2125,
DOI 10.1007/978-3-319-10088-3_5

obtained as the gluing of filtered good-KMS smooth $\mathcal{R}_{D_I(*\partial D_I)}$-triples in Chap. 10. It might be useful to see the more classical theory of $\mathscr{D}$-modules on curves, for which we refer to [59], for example.

5.1 Good-KMS Smooth $\mathcal{R}$-Modules

5.1.1 Good-KMS Meromorphic Prolongment

Let X be any complex manifold with a simply normal crossing hypersurface D. Let $\mathcal{M}$ be a smooth $\mathcal{R}_{X(*D)}$-module. The natural family of flat λ-connections on $\mathcal{M}$ is denoted by $\mathbb{D}$. In this paper, we say that $\mathcal{M}$ is unramifiedly good, if the following holds:

- For each point $P \in D$, there exists a good set of irregular values $\mathrm{Irr}(\mathcal{M}, P) \subset \mathcal{O}_X(*D)_P/\mathcal{O}_{X,P}$, such that the formal completion of $\mathcal{M}$ at (λ, P) is isomorphic to a direct sum

$$\bigoplus_{\mathfrak{a}\in \mathrm{Irr}(\mathcal{M},P)} (\mathcal{M}_{\mathfrak{a}}, \mathbb{D}_{\mathfrak{a}}) \tag{5.1}$$

  where each $(\mathcal{M}_{\mathfrak{a}}, \mathbb{D}_{\mathfrak{a}} - d\tilde{\mathfrak{a}})$ is regular, i.e., it has a logarithmic lattice. Here, $\tilde{\mathfrak{a}} \in \mathcal{O}_X(*D)_P$ is a lift of $\mathfrak{a}$. (See §2.1 of [55] for the notion of good set of irregular values.) The decomposition (5.1) is called the Hukuhara-Levelt-Turrittin type decomposition.

It is called unramifiedly good-KMS if moreover it has the KMS-structure (see §2.8 of [55]). It is called good-KMS (resp. good) if it is locally the descent of an unramifiedly good-KMS (resp. good) smooth $\mathcal{R}_{X(*D)}$-module. It is called regular-KMS if moreover $\mathcal{M}$ is a regular along D. For a decomposition $D = D^{(1)} \cup D^{(2)}$, a good-KMS smooth $\mathcal{R}_{X(*D)}$-module $\mathcal{M}$ is called regular along $D^{(1)}$ if $\mathcal{M}_{|\mathcal{X}\setminus\mathcal{D}^{(2)}}$ is regular-KMS.

Let $D = \bigcup_{i\in\Lambda} D_i$ be the irreducible decomposition. In the following, for $I \subset \Lambda$, we set $D_I := \bigcap_{i\in I} D_i$, $D(I) := \bigcup_{i\in I} D_i$ and $\partial D_I := \bigcup_{j\notin I}(D_I \cap D_j)$. We put $D_I^\circ := D_I \setminus \partial D_I$. We set $I^c := \Lambda \setminus I$. If $I = \emptyset$, we put $D_\emptyset := X$.

5.1.2 Induced Bundles on the Intersection of Divisors

Let X and D be as above. For $\lambda_0 \in \mathbb{C}_\lambda$, let $\mathcal{X}^{(\lambda_0)}$ be a neighbourhood of $\{\lambda_0\}\times X$ in $\mathcal{X}$. We use the notation $\mathcal{D}_i^{(\lambda_0)}$, etc., with a similar meaning. Recall that, for $u = (a, \alpha)$ and for $\lambda \in \mathbb{C}$, we set $\mathfrak{p}(\lambda, u) := a + 2\,\mathrm{Re}(\lambda\overline{\alpha}) \in \mathbb{R}$ and $\mathfrak{e}(\lambda, u) := \alpha - a\lambda - \overline{\alpha}\lambda^2 \in \mathbb{C}$. We also set $\mathfrak{k}(\lambda, u) = \big(\mathfrak{p}(\lambda, u), \mathfrak{e}(\lambda, u)\big) \in \mathbb{R}\times\mathbb{C}$. The induced maps $(\mathbb{R}\times\mathbb{C})^I \longrightarrow \mathbb{R}^I$, $(\mathbb{R}\times\mathbb{C})^I \longrightarrow \mathbb{C}^I$ and $(\mathbb{R}\times\mathbb{C})^I \longrightarrow (\mathbb{R}\times\mathbb{C})^I$ are also denoted by the same notation.

Let $\mathcal{M}$ be a good-KMS smooth $\mathcal{R}_{X(*D)}$-module. We have the associated family of good filtered λ-flat bundles $\mathcal{Q}_*^{(\lambda_0)}\mathcal{M}$ on $(\mathcal{X}^{(\lambda_0)},\mathcal{D}^{(\lambda_0)})$ indexed by $\mathbb{R}^\Lambda$. We have the induced filtration ${}^iF^{(\lambda_0)}$ of $\mathcal{Q}_{\boldsymbol{a}}^{(\lambda_0)}\mathcal{M}_{|\mathcal{D}_i^{(\lambda_0)}}$ given by

$$
{}^iF_b^{(\lambda_0)}\big(\mathcal{Q}_{\boldsymbol{a}}^{(\lambda_0)}\mathcal{M}_{|\mathcal{D}_i^{(\lambda_0)}}\big) := \operatorname{Im}\Big(\mathcal{Q}_{\boldsymbol{a}'}^{(\lambda_0)}\mathcal{M}_{|\mathcal{D}_i^{(\lambda_0)}} \longrightarrow \mathcal{Q}_{\boldsymbol{a}}^{(\lambda_0)}\mathcal{M}_{|\mathcal{D}_i^{(\lambda_0)}}\Big),
$$

where the i-th component of $\boldsymbol{a}'$ is b, and the other components of $\boldsymbol{a}'$ are equal to those of $\boldsymbol{a}$. The induced filtrations ${}^iF^{(\lambda_0)}$ of $\mathcal{Q}_{\boldsymbol{a}}^{(\lambda_0)}\mathcal{M}_{|\mathcal{D}_I^{(\lambda_0)}}$ $(i \in I)$ are compatible. For $\boldsymbol{b} \in \mathbb{R}^I$, we put ${}^IF_{\boldsymbol{b}}^{(\lambda_0)} := \bigcap_{i\in I} {}^iF_{b_i}^{(\lambda_0)}$ on $\mathcal{D}_I^{(\lambda_0)}$, and

$$
{}^I\mathrm{Gr}_{\boldsymbol{b}}^{F^{(\lambda_0)}}\,\mathcal{Q}_{\boldsymbol{a}}^{(\lambda_0)}\mathcal{M} := {}^IF_{\boldsymbol{b}}^{(\lambda_0)}\Big/\sum_{\boldsymbol{c}\lneq\boldsymbol{b}} {}^IF_{\boldsymbol{c}}^{(\lambda_0)}.
$$

On ${}^I\mathrm{Gr}_{\boldsymbol{b}}^{F^{(\lambda_0)}}\,\mathcal{Q}_{\boldsymbol{a}}^{(\lambda_0)}\mathcal{M}$, we have the induced endomorphisms $\mathrm{Res}_i(\mathbb{D})$ $(i \in I)$. We have the decomposition

$$
{}^I\mathrm{Gr}_{\boldsymbol{b}}^{F^{(\lambda_0)}}\,\mathcal{Q}_{\boldsymbol{a}}^{(\lambda_0)}\mathcal{M} = \bigoplus_{\substack{\boldsymbol{u}\in(\mathbb{R}\times\mathbb{C})^I\\ \mathfrak{p}(\lambda_0,\boldsymbol{u})=\boldsymbol{b}}} {}^I\mathcal{G}_{\boldsymbol{u}}^{(\lambda_0)}\mathcal{Q}_{\boldsymbol{a}}^{(\lambda_0)}\mathcal{M},
$$

such that (1) it is preserved by $\mathrm{Res}_i(\mathbb{D})$ $(i \in I)$, (2) the restriction of $\mathrm{Res}_i(\mathbb{D}) - \mathfrak{e}(\lambda,u_i)$ to ${}^I\mathcal{G}_{\boldsymbol{u}}^{(\lambda_0)}\mathcal{Q}_{\boldsymbol{a}}^{(\lambda_0)}\mathcal{M}$ are nilpotent, where u_i is the i-th component of $\boldsymbol{u}$.

For $I \subset \Lambda$ and $\boldsymbol{b} \in \mathbb{R}^I$, we take $\boldsymbol{a} \in \mathbb{R}^\Lambda$ which is mapped to $\boldsymbol{b}$ by the projection $\mathbb{R}^\Lambda \longrightarrow \mathbb{R}^I$, and we put ${}^I\mathcal{Q}_{\boldsymbol{b}}^{(\lambda_0)}\mathcal{M} := \mathcal{Q}_{\boldsymbol{a}}^{(\lambda_0)}\mathcal{M}\otimes\mathcal{O}\big(*\mathcal{D}(I^c)\big)$. It is independent of the choice of $\boldsymbol{a}$. We obtain ${}^I\mathcal{G}_{\boldsymbol{u}}\mathcal{Q}_{\boldsymbol{b}}\mathcal{M}$ as above. In particular, for $\boldsymbol{u} \in (\mathbb{R}\times\mathbb{C})^I$, we obtain

$$
{}^I\mathcal{G}_{\boldsymbol{u}}^{(\lambda_0)}\mathcal{M} := {}^I\mathcal{G}_{\boldsymbol{u}}^{(\lambda_0)}\,{}^I\mathcal{Q}_{\mathfrak{p}(\lambda_0,\boldsymbol{u})}^{(\lambda_0)}\mathcal{M}.
$$

If λ_1 is sufficiently close to λ_0 and if $\mathcal{X}^{(\lambda_1)} \subset \mathcal{X}^{(\lambda_0)}$, we naturally have ${}^I\mathcal{G}_{\boldsymbol{u}}^{(\lambda_0)}(\mathcal{M})_{|\mathcal{D}_I^{(\lambda_1)}} = {}^I\mathcal{G}_{\boldsymbol{u}}^{(\lambda_1)}(\mathcal{M})$. Hence, we can glue ${}^I\mathcal{G}_{\boldsymbol{u}}^{(\lambda_0)}\mathcal{M}$ for varied $\lambda_0 \in \mathbb{C}_\lambda$, and we obtain an $\mathcal{O}_{\mathcal{D}_I}(*\partial\mathcal{D}_I)$-module ${}^I\mathcal{G}_{\boldsymbol{u}}\mathcal{M}$ on $\mathcal{D}_I$.

Let $\mathrm{KMS}(\mathcal{M},I) \subset (\mathbb{R}\times\mathbb{C})^I$ denote the set of $\boldsymbol{u}$ such that ${}^I\mathcal{G}_{\boldsymbol{u}}\mathcal{M} \neq 0$. The set is called the KMS-spectrum of $\mathcal{M}$.

5.1.3 *Hukuhara-Levelt-Turrittin Type Decomposition*

Let Δ denote the disc $\{z \in \mathbb{C} \,\big|\, |z| < 1\}$. Let us consider the case $X = \Delta^n$ and $D = \bigcup_{i=1}^{\ell}\{z_i = 0\}$. We set $D_i = \{z_i = 0\}$. For $J \subset \underline{\ell} := \{1,\dots,\ell\}$, let $M(X,D(J))$ be the space of global sections of $\mathcal{O}_X(*D(J))$. Let $H(X)$ be the space of global sections of $\mathcal{O}_X$. Let $\mathcal{M}$ be an unramifiedly good-KMS smooth

$\mathcal{R}_{X(*D)}$-module. After shrinking X, we have the good set of irregular values $\mathrm{Irr}(\mathcal{M})$ in $M(X,D)/H(X)$. For $I \subset \Lambda$, let $\mathrm{Irr}(\mathcal{M},I)$ denote the image of $\mathrm{Irr}(\mathcal{M})$ by $M(X,D)/H(X) \longrightarrow M(X,D)/M(X,D(I^c))$. We have the Hukuhara-Levelt-Turrittin type decomposition (see §2.4.2 of [55]):

$$(\mathcal{M},\mathbb{D})_{|\hat{\mathcal{D}}_I} = \bigoplus_{\mathfrak{a}\in\mathrm{Irr}(\mathcal{M},I)} \big(\mathcal{M}_{\mathfrak{a},\hat{\mathcal{D}}_I},\hat{\mathbb{D}}_{\mathfrak{a}}\big)$$

Here, $\hat{\mathcal{D}}_I$ denotes the completion of $\mathcal{X}$ along $\mathcal{D}_I$. (See [1, 5], for example.) On $\mathcal{X}^{(\lambda_0)}$, the decomposition is compatible with the KMS-structure, i.e., for $\boldsymbol{a} \in \mathbb{R}^\ell$, we have the induced decomposition of the family of filtered λ-flat bundles

$$\big(\mathcal{Q}^{(\lambda_0)}_{\boldsymbol{a}}\mathcal{M},\mathbb{D}\big)_{|\hat{\mathcal{D}}_I^{(\lambda_0)}} = \bigoplus_{\mathfrak{a}\in\mathrm{Irr}(\mathcal{M},I)} \big(\mathcal{Q}^{(\lambda_0)}_{\boldsymbol{a}}\mathcal{M}_{\mathfrak{a},\hat{\mathcal{D}}_I},\hat{\mathbb{D}}_{\mathfrak{a}}\big),$$

and $\hat{\mathbb{D}}_{\mathfrak{a}} - d\mathfrak{a}$ are logarithmic along $\mathcal{D}(I)$ in the following sense:

$$\begin{aligned}&\big(\hat{\mathbb{D}}_{\mathfrak{a}} - d\mathfrak{a}\big)\mathcal{Q}^{(\lambda_0)}_{\boldsymbol{a}}\mathcal{M}_{\mathfrak{a},\hat{\mathcal{D}}_I}\\ &\quad\subset \mathcal{Q}^{(\lambda_0)}_{\boldsymbol{a}}\mathcal{M}_{\mathfrak{a},\hat{\mathcal{D}}_I} \otimes \Big(\Omega^1_{\mathcal{X}^{(\lambda_0)}/\mathbb{C}_\lambda}\big(\log \hat{\mathcal{D}}^{(\lambda_0)}(I)\big) + \Omega^1_{\mathcal{X}^{(\lambda_0)}/\mathbb{C}_\lambda}\big(*\hat{\mathcal{D}}^{(\lambda_0)}(I^c)\big)\Big)\end{aligned} \tag{5.2}$$

Here, $\Omega^1_{\mathcal{X}^{(\lambda_0)}/\mathbb{C}_\lambda}$ denote the sheaf of relative one forms on $\mathcal{X}^{(\lambda_0)}/\mathbb{C}_\lambda$. We set $\mathcal{M}^{(\mathrm{reg})}_{\widehat{\mathcal{D}}_I} := \mathcal{M}_{0,\widehat{\mathcal{D}}_I}$, and $\mathcal{M}^{(\mathrm{irr})}_{\hat{\mathcal{D}}_I} := \bigoplus_{\mathfrak{a}\neq 0}\mathcal{M}_{\mathfrak{a},\hat{\mathcal{D}}_I}$.

Let us consider the case that $\mathcal{M}$ is good-KMS but is not necessarily unramified. Let $\varphi : (X',D') \longrightarrow (X,D)$ be a ramified covering such that $\mathcal{M}' = \varphi^*\mathcal{M}$ is unramified. Applying the above construction, we obtain the decomposition $\mathcal{M}'_{|\hat{\mathcal{D}}'_I} = \mathcal{M}'^{(\mathrm{reg})}_{\hat{\mathcal{D}}'_I} \oplus \mathcal{M}'^{(\mathrm{irr})}_{\hat{\mathcal{D}}'_I}$, which is preserved by the action of the Galois group of the ramified covering. We obtain a decomposition $\mathcal{M}_{|\hat{\mathcal{D}}_I} = \mathcal{M}^{(\mathrm{reg})}_{\hat{\mathcal{D}}_I} \oplus \mathcal{M}^{(\mathrm{irr})}_{\hat{\mathcal{D}}_I}$. It is independent of the choice of a ramified covering and a choice of the coordinate. It is compatible with the KMS-structure, i.e, we have the induced decomposition $\mathcal{Q}^{(\lambda_0)}_*\mathcal{M}_{|\hat{\mathcal{D}}_I} = \mathcal{Q}^{(\lambda_0)}_*\mathcal{M}^{(\mathrm{reg})}_{\hat{\mathcal{D}}_I} \oplus \mathcal{Q}^{(\lambda_0)}_*\mathcal{M}^{(\mathrm{irr})}_{\hat{\mathcal{D}}_I}$.

5.1.4 Specialization

Let X and D be as in Sect. 5.1.3. Let $\mathcal{M}$ be a good-KMS $\mathcal{R}_{X(*D)}$-module. We set $\boldsymbol{\delta}_I := (\boldsymbol{\delta},\ldots,\boldsymbol{\delta}) \in (\mathbb{R}\times\mathbb{C})^I$. For $\boldsymbol{u} \in (\mathbb{R}\times\mathbb{C})^I$, we put $(\boldsymbol{b},\boldsymbol{\beta}) := \mathfrak{k}(\lambda_0,\boldsymbol{u}+\boldsymbol{\delta}_I)$, and we obtain the following on $\mathcal{D}_I$:

$$^{I}\tilde{\psi}_{\boldsymbol{u}}(\mathcal{M}) := {}^{I}\mathcal{G}_{\boldsymbol{u}+\delta_I}\mathcal{M}^{(\mathrm{reg})}_{\hat{\mathcal{D}}_I}$$

It is naturally a good-KMS smooth $\mathcal{R}_{D_I(*\partial D_I)}$-module, although it depends on the choice of a coordinate system. Let φ_I denote the natural map $\mathrm{Irr}(\mathcal{M}) \longrightarrow M(X,D)/M(X,\mathcal{D}(I^c))$. If $\mathcal{M}$ is unramifiedly good, then $\mathrm{Irr}({}^{I}\tilde{\psi}_{\boldsymbol{u}}\mathcal{M})$ is $\{\mathfrak{a}_{|D_I} \mid \mathfrak{a} \in \varphi_I^{-1}(0)\}$. The bundle ${}^{I}\tilde{\psi}_{\boldsymbol{u}}(\mathcal{M})$ is also equipped with the induced endomorphisms $\mathrm{Res}_i(\mathbb{D})$ ($i \in I$), which are independent of the choice of a coordinate system. By construction, for $I_1 \sqcup I_2 \subset \underline{\ell}$, we have a natural isomorphism ${}^{I_1}\tilde{\psi}_{\boldsymbol{u}_1}\left({}^{I_2}\tilde{\psi}_{\boldsymbol{u}_2}(\mathcal{M})\right) \simeq {}^{I_1\sqcup I_2}\tilde{\psi}_{(\boldsymbol{u}_1,\boldsymbol{u}_2)}(\mathcal{M})$.

It is easy to see that $\mathcal{M}$ is strictly specializable along z_i as an $\mathcal{R}_{X(*D)}$-module (Sect. 2.1.2.2), and that we have ${}^{i}\tilde{\psi}_u(\mathcal{M}) \simeq \tilde{\psi}_{z_i,u}(\mathcal{M})$. So, ${}^{I}\tilde{\psi}_{\boldsymbol{u}}$ is obtained as the composition $\tilde{\psi}_{z_{i_1},u_{i_1}} \circ \cdots \circ \tilde{\psi}_{z_{i_m},u_{i_m}}$.

5.1.5 Reduction with Respect to Stokes Structure

Let X, D and $\mathcal{M}$ be as in Sect. 5.1.4. We recall the procedure of the reduction of $\mathcal{M}$ with respect to the Stokes structure. We will shrink X around the origin without mention.

If $\mathcal{M}$ is unramifiedly good-KMS, we have the reduction with respect to the Stokes structure along $D(I)$, and we obtain a graded good-KMS smooth $\mathcal{R}_{X(*D)}$-module

$$^{I}\mathrm{Gr}^{\mathrm{St}}(\mathcal{M}) = \bigoplus_{\mathfrak{a}\in\mathrm{Irr}(\mathcal{M},I)} {}^{I}\mathrm{Gr}^{\mathrm{St}}_{\mathfrak{a}}(\mathcal{M}).$$

Indeed, we can choose an auxiliary sequence $\boldsymbol{m}(0), \boldsymbol{m}(1), \ldots, \boldsymbol{m}(N)$ for $\mathrm{Irr}(\mathcal{M})$ as in §2.1.2 of [55]. By successive use of the reduction with respect to the Stokes structure in the level $\boldsymbol{m}(j)$ in §3.3 of [55], we obtain a sequence of unramifiedly good smooth $\mathcal{R}_{X(*D)}$-modules $\mathrm{Gr}^{\boldsymbol{m}(j)}(\mathcal{M})$. Let j_0 be the minimum among j such that the p-th components of $\boldsymbol{m}(j)$ are 0 for any $p \in I$. We define ${}^{I}\mathrm{Gr}^{\mathrm{St}}(\mathcal{M}) := \mathrm{Gr}^{\boldsymbol{m}(j_0-1)}(\mathcal{M})$. We also have filtered bundles $\mathcal{Q}_*^{(\lambda_0)}{}^{I}\mathrm{Gr}^{\mathrm{St}}(\mathcal{M}) := {}^{I}\mathrm{Gr}^{\mathrm{St}}(\mathcal{Q}_*^{(\lambda_0)}\mathcal{M})$ with which ${}^{I}\mathrm{Gr}^{\mathrm{St}}(\mathcal{M})$ is good-KMS.

For each $\mathfrak{a}$, we take an appropriate lift $\tilde{\mathfrak{a}}$ to $M(X,D)$, and then the tensor product ${}^{I}\mathrm{Gr}^{\mathrm{St}}_{\mathfrak{a}}(\mathcal{M}) \otimes L(-\tilde{\mathfrak{a}})$ is regular along $D(I)$. We have a natural isomorphism ${}^{I}\mathrm{Gr}^{\mathrm{St}}(\mathcal{M})_{|\hat{\mathcal{D}}_I} \simeq \mathcal{M}_{|\hat{\mathcal{D}}_I}$. So, we naturally have

$$^{I}\tilde{\psi}_{\boldsymbol{u}}(\mathcal{M}) \simeq {}^{I}\tilde{\psi}_{\boldsymbol{u}}\left({}^{I}\mathrm{Gr}^{\mathrm{St}}(\mathcal{M})\right) \simeq {}^{I}\tilde{\psi}_{\boldsymbol{u}}\left({}^{I}\mathrm{Gr}^{\mathrm{St}}_0(\mathcal{M})\right).$$

We put ${}^{I}\mathrm{Gr}^{\mathrm{St,reg}}(\mathcal{M}) := {}^{I}\mathrm{Gr}^{\mathrm{St}}_0(\mathcal{M})$ and ${}^{I}\mathrm{Gr}^{\mathrm{St,irr}}(\mathcal{M}) := \bigoplus_{\mathfrak{a}\neq 0} {}^{I}\mathrm{Gr}^{\mathrm{St}}_{\mathfrak{a}}(\mathcal{M})$.

Suppose that $\mathcal{M}$ is good-KMS but not necessarily unramified. We take a ramified covering $\varphi : (X',D') \longrightarrow (X,D)$ such that $\mathcal{M}' := \varphi^*\mathcal{M}$ is unramifiedly good-

KMS. By applying the above procedure, we obtain the reduction ${}^I\mathrm{Gr}^{\mathrm{St}}(\mathcal{M}') = {}^I\mathrm{Gr}^{\mathrm{St,reg}}(\mathcal{M}') \oplus {}^I\mathrm{Gr}^{\mathrm{St,irr}}(\mathcal{M}')$, on which the Galois group of the covering naturally acts. As the descent, we obtain a good-KMS smooth $\mathcal{R}_{X(*D)}$-module

$$ {}^I\mathrm{Gr}^{\mathrm{St}}(\mathcal{M}) = {}^I\mathrm{Gr}^{\mathrm{St,reg}}(\mathcal{M}) \oplus {}^I\mathrm{Gr}^{\mathrm{St,irr}}(\mathcal{M}) $$

on $(\mathcal{X}, \mathcal{D})$. We have natural isomorphisms

$$ {}^I\mathrm{Gr}^{\mathrm{St,reg}}(\mathcal{M})_{|\hat{\mathcal{D}}_I} \simeq \mathcal{M}^{(\mathrm{reg})}_{|\hat{\mathcal{D}}_I}, \quad {}^I\mathrm{Gr}^{\mathrm{St,irr}}(\mathcal{M})_{|\hat{\mathcal{D}}_I} \simeq \mathcal{M}^{(\mathrm{irr})}_{|\hat{\mathcal{D}}_I}. $$

We also have ${}^I\tilde{\psi}_{\boldsymbol{u}}\Big({}^I\mathrm{Gr}^{\mathrm{St}}(\mathcal{M})\Big) \simeq {}^I\tilde{\psi}_{\boldsymbol{u}}\Big({}^I\mathrm{Gr}^{\mathrm{St}}_0(\mathcal{M})\Big) \simeq {}^I\tilde{\psi}_{\boldsymbol{u}}(\mathcal{M})$ naturally.

5.2 Compatibility of Filtrations

5.2.1 *Compatibility with Hukuhara-Levelt-Turrittin Type Decomposition*

Let X be any complex manifold, and let D be a simple normal crossing hypersurface. Let $\mathcal{M}$ be a good smooth $\mathcal{R}_{X(*D)}$-module. Let L be a filtration of $\mathcal{M}$ in the category of smooth $\mathcal{R}_{X(*D)}$-modules, i.e., a filtration in the category of $\mathcal{R}_{X(*D)}$-modules such that $\mathrm{Gr}^L(\mathcal{M})$ is also a smooth $\mathcal{R}_{X(*D)}$-module.

Lemma 5.2.1 *Let $P \in D$. Suppose that $\mathcal{M}$ is unramified around P. Then, the filtration L is compatible with the formal decomposition of $\mathcal{M}$ at $(\lambda, P) \in \mathcal{D}$ as in* (5.1)*:*

$$ \mathcal{M}_{|\widehat{(\lambda,P)}} = \bigoplus_{\mathfrak{a} \in \mathrm{Irr}(\mathcal{M},P)} \hat{\mathcal{M}}_{\mathfrak{a},(\lambda,P)}. \tag{5.3} $$

Proof Recall a standard and easy result for $\mathbb{C}((t))$-differential module. Let M be a $\mathbb{C}((t))$-differential module of finite rank with a decomposition $M = \bigoplus_{\mathfrak{a}\in t^{-1}\mathbb{C}[t^{-1}]} M_{\mathfrak{a}}$ such that $\partial_t - \partial_t(\mathfrak{a})$ are regular singular. Then, any differential submodule $M' \subset M$ is compatible with the decomposition.

Similarly, let M be a finite dimensional $\mathbb{C}((t))$-vector space with an $\mathbb{C}((t))$-endomorphism f, with a decomposition $(M, f) = \bigoplus_{\mathfrak{a}\in t^{-1}\mathbb{C}[t^{-1}]}(M_{\mathfrak{a}}, f_{\mathfrak{a}})$ such that $M_{\mathfrak{a}}$ has a lattice $L_{\mathfrak{a}}$ preserved by $f_{\mathfrak{a}} - \partial_t\mathfrak{a}$. If $M' \subset M$ is preserved by f, then M' is compatible with the decomposition, which can be proved in a similar way.

Let $\varphi : \Delta \longrightarrow X$ be any morphism such that $\varphi(0) = P$ and that $\varphi(C) \not\subset D$. If φ is general, the induced map $\mathrm{Irr}(\mathcal{M}, P) \longrightarrow \mathcal{O}_\Delta(*0)_0/\mathcal{O}_{\Delta,0}$ is injective. By applying the above result, we obtain that φ^*L is compatible with the decomposition obtained as the pull back of (5.3). It follows that we obtain that L is compatible with (5.3). □

We obtain the following lemma from the previous lemma.

Lemma 5.2.2 *Each $L_j\mathcal{M}$ is a good smooth $\mathcal{R}_{X(*D)}$-module.* □

Remark 5.2.3 The Stokes structure of $\mathcal{M}$ is compatible with the filtration L. It follows from the characterization of the Stokes filtration in terms of the growth order of flat sections. (See §3.2.3 [55], for example.) □

5.2.2 Extension of Good-KMS Smooth $\mathcal{R}$-Modules

Let $\mathcal{M}$ be a good smooth $\mathcal{R}_{X(*D)}$-module given on $\mathcal{X}^{(\lambda_0)}$. Let L be a filtration of $\mathcal{M}$ in the category of $\mathcal{R}_{X(*D)}$-modules.

Proposition 5.2.4 *Assume that (i) $\mathrm{Gr}^L\,\mathcal{M}$ is good-KMS, (ii) for any smooth point P of D, the restriction of $\mathcal{M}$ to a small neighbourhood of P is good-KMS. Then, $\mathcal{M}$ is good-KMS.*

Proof We may assume that $X = \Delta^n$, $D_i = \{z_i = 0\}$ and $D = \bigcup_{i=1}^{\ell} D_i$. We set $D_{[2]} := \bigcup_{i\neq j} D_i \cap D_j$. We may also assume that $\mathcal{M}$ is unramifiedly good. For any λ_0, by the assumption, we have the lattices $\mathcal{Q}^{(\lambda_0)\prime}_{\boldsymbol{a}}\mathcal{M}$ ($\boldsymbol{a} \in \mathbb{R}^{\ell}$) of $\mathcal{M}_{|\mathcal{X}^{(\lambda_0)}\setminus\mathcal{D}^{(\lambda_0)}_{[2]}}$.

Let us consider the case that $\lambda_0 \neq 0$ is generic with respect to $\mathrm{Gr}^L\,\mathcal{M}$, i.e., the map $\mathfrak{e}(\lambda_0) : \mathcal{KMS}(\mathcal{M}, i) \longrightarrow \mathbb{C}$ is injective for any i. The reduction $\mathrm{Gr}^{\mathrm{St}}(\mathcal{M})$ is also equipped with the induced filtration L, and $\mathrm{Gr}^L\,\mathrm{Gr}^{\mathrm{St}}(\mathcal{M})$ is unramifiedly good-KMS. Moreover, for any smooth point P of D, the restriction of $\mathrm{Gr}^{\mathrm{St}}\,\mathcal{M}$ on a neighbourhood of P has KMS structure. Because λ_0 is generic, $\mathrm{Gr}^{\mathrm{St}}\,\mathcal{M}$ has KMS structure at λ_0. Indeed, $\mathrm{Gr}^{\mathrm{St}}(\mathcal{M})$ has a decomposition $\bigoplus \mathrm{Gr}^{\mathrm{St}}_{\mathfrak{a}}(\mathcal{M})$, and $\mathrm{Gr}^{\mathrm{St}}_{\mathfrak{a}}(\mathcal{M}) \otimes \mathcal{L}(-\mathfrak{a})$ can be regarded as a family of regular singular meromorphic flat bundles. Hence, it is easy to construct the family of lattices with the desired property in a direct way. By applying the Riemann-Hilbert-Birkhoff correspondence (§4 of [55]) with the lattices $\mathcal{Q}^{(\lambda_0)}_{\boldsymbol{a}}\,\mathrm{Gr}^{\mathrm{St}}\,\mathcal{M}$, we obtain a lattice $\mathcal{Q}^{(\lambda_0)}_{\boldsymbol{a}}\mathcal{M}$ such that (i) it induces $\mathcal{Q}^{(\lambda_0)}_{\boldsymbol{a}}\,\mathrm{Gr}^{\mathrm{St}}\,\mathcal{M}$, (ii) $\mathcal{Q}^{(\lambda_0)}_{\boldsymbol{a}}\mathcal{M}_{|\mathcal{X}^{(\lambda_0)}\setminus\mathcal{D}^{(\lambda_0)}_{[2]}} \simeq \mathcal{Q}^{(\lambda_0)\prime}_{\boldsymbol{a}}\mathcal{M}$. The second condition implies $\mathrm{Gr}^L\,\mathcal{Q}^{(\lambda_0)}_{\boldsymbol{a}}\mathcal{M} \simeq \mathcal{Q}^{(\lambda_0)}_{\boldsymbol{a}}\,\mathrm{Gr}^L\,\mathcal{M}$. Then, it is easy to see that $\mathcal{M}$ is unramifiedly good-KMS around λ_0.

Let us consider the case that $\lambda_0 \neq 0$ is not necessarily generic. By the assumption and the consideration in the generic case, the restriction of $\mathcal{M}$ to $\mathcal{X}^{(\lambda_0)} \setminus \mathcal{D}^{\lambda_0}_{[2]}$ is good-KMS, i.e., we have vector bundles $\mathcal{Q}^{(\lambda_0)\prime}_{\boldsymbol{a}}\mathcal{M}$ on $\mathcal{X}^{(\lambda_0)} \setminus \mathcal{D}^{\lambda_0}_{[2]}$, which induce $\mathcal{Q}^{(\lambda_0)}_{\boldsymbol{a}}\,\mathrm{Gr}^L\,\mathcal{M}_{|\mathcal{X}\setminus\mathcal{D}^{\lambda_0}_{[2]}}$. We have only to prove that $\mathcal{Q}^{(\lambda_0)\prime}_{\boldsymbol{a}}\mathcal{M}$ is uniquely extended to vector bundles on $\mathcal{X}$. Because $\mathcal{D}^{\lambda_0}_{[2]}$ is of codimension 3 in $\mathcal{X}$, it follows from Lemma 5.2.5 below. The case $\lambda_0 = 0$ can be argued similarly. Thus, we obtain Proposition 5.2.4. □

Lemma 5.2.5 *Let Y be a complex manifold. Let $Z \subset Y$ be an analytic closed subset of codimension 3. Let E_i be vector bundles on Y. Let E_0' be a vector bundle on $Y - Z$ with an exact sequence $0 \longrightarrow E_{1|Y\setminus Z} \longrightarrow E_0' \longrightarrow E_{2|Y\setminus Z} \longrightarrow 0$. Then, E_0' is uniquely extended to a vector bundle on Y with an exact sequence $0 \longrightarrow E_1 \longrightarrow E_0 \longrightarrow E_2 \longrightarrow 0$.*

Proof Let $j : Y \setminus Z \longrightarrow Y$ be the open immersion. The claim of the lemma follows from $R^1 j_* \mathcal{O}_{Y\setminus Z} = 0$, which holds because $\operatorname{codim} Z \geq 3$. (See Lemma 5 of [18], for example.) □

5.2.3 Compatibility with KMS Structure

Definition 5.2.6 Let $\mathcal{M}$ be a good-KMS smooth $\mathcal{R}_{X(*D)}$-module. We say that a filtration L of $\mathcal{M}$ is compatible with the KMS-structure, if the induced increasing sequence $\mathrm{Gr}^L(\mathcal{Q}_*^{(\lambda_0)}\mathcal{M})$ gives a KMS-structure of $\mathrm{Gr}^L(\mathcal{M})$. □

We will give an example of a filtration below, which is not compatible with a KMS-structure.

We obtain the following lemma from Proposition 5.2.4.

Lemma 5.2.7 *Let $\mathcal{M}$ be a good $\mathcal{R}_{X(*D)}$-module. Let L be a filtration of $\mathcal{M}$ in the category of smooth $\mathcal{R}_{X(*D)}$-modules. Assume the following:*

- *$\mathrm{Gr}^L(\mathcal{M})$ has a KMS-structure.*
- *For any smooth point P of D, there exists a neighbourhood X_P of P such that $\mathcal{M}_{|\mathcal{X}_P}$ has a KMS-structure with which $L_{|\mathcal{X}_P}$ is compatible.*

Then, $\mathcal{M}$ has a unique KMS-structure with which L is compatible. □

We state it in a slightly different way.

Corollary 5.2.8 *Let $\mathcal{M}$ be a good-KMS smooth $\mathcal{R}_{X(*D)}$-module. Let L be a filtration of $\mathcal{M}$ in the category of smooth $\mathcal{R}_{X(*D)}$-modules. Assume the following:*

- *$\mathrm{Gr}^L(\mathcal{M})$ has a good-KMS structure.*
- *For any smooth point P of D, there exists a small neighbourhood X_P of P such that $L_{|\mathcal{X}_P}$ is compatible with the KMS-structure of $\mathcal{M}_{|\mathcal{X}_P}$.*

Then, L is compatible with the KMS-structure of $\mathcal{M}$. □

5.2.3.1 Example

Let $X = \mathbb{C}_z$. Let $a, b \in \mathbb{R}$ such that $a \neq b$. We consider the $\mathcal{R}_X$-module $\mathcal{M} := \bigoplus_{i=1,2} \mathcal{O}_{\mathcal{X}}(*z)\, e_i$ with $z\eth_z e_1 = -\lambda\, a\, e_1$ and $z\eth_z e_2 = e_1 - \lambda\, b\, e_2$. We put $v_2 := e_1 + \lambda\,(a-b)\, e_2$. Then, we have $z\eth_z v_2 = -\lambda\, b\, v_2$.

Assume $a < b$. For $c = a, b$, let $\mathcal{M}(c)$ be $\mathcal{O}_{\mathcal{X}}(*z)f$ with $z\eth f = \lambda\, c f$. We have the morphism $\varphi : \mathcal{M}(b) \longrightarrow \mathcal{M}$ given by $f \longmapsto v$. By a direct computation, we can check that the cokernel of $\tilde{\psi}_b(\varphi) : \tilde{\psi}_b(\mathcal{M}(b)) \longrightarrow \tilde{\psi}_b(\mathcal{M})$ is not strict. The cokernel of φ is generated by $[e_2]$, and it is naturally isomorphic to $\mathcal{M}(a)$. We can check that the morphism $\mathcal{M} \longrightarrow \mathcal{M}(a)$ is not strict with respect to the V-filtrations $V^{(0)}$.

5.2.4 *Curve Test*

Let $\mathcal{M}$ be a good smooth $\mathcal{R}_{X(*D)}$-module. Let L be a filtration in the category of $\mathcal{R}_{X(*)}$-modules.

Proposition 5.2.9 *We suppose the following:*

- $\mathrm{Gr}^L(\mathcal{M})$ *is good-KMS.*
- *Let C be any curve in X which intersects with the smooth part of D transversally. Then, $\mathcal{M}_{|C}$ has a KMS-structure with which $L_{|C}$ is compatible.*

Then, $\mathcal{M}$ has a KMS-structure with which L is compatible.

Proof By Lemma 5.2.7, we may assume $X = \Delta^n$ and $D = \{z_1 = 0\}$. As in the proof of Proposition 5.2.4, $\mathcal{M}$ has KMS-structure at generic λ_0. (This case is easier. Indeed, because the divisor is smooth, we have only to consider the lattice in the formal completion.)

Let us consider the case that $\lambda_0 \neq 0$ is not necessarily generic. Let $U(\lambda_0) \subset \mathbb{C}_\lambda$ be a small neighbourhood of λ_0. For any generic $\lambda_1 \in U(\lambda_0)$, we take a small neighbourhood $U(\lambda_1) \subset U(\lambda_0)$ such that any $\lambda \in U(\lambda_1)$ is generic. We can construct $\mathcal{Q}^{(\lambda_0)}(\mathcal{M}_{|U(\lambda_1)\times X})$ from $\mathcal{Q}^{(\lambda_1)}(\mathcal{M}_{|U(\lambda_1)\times X})$ by the relation of $\mathcal{Q}^{(\lambda_0)}$ and $\mathcal{Q}^{(\lambda_1)}$, which is compatible with L. (See §2.8.4 of [55], for example.) By applying Lemma 5.2.10 below, we obtain that it is extended to the KMS-structure at λ_0 which is compatible with L. The case $\lambda_0 = 0$ can be argued similarly. □

Lemma 5.2.10 *Let Z be a complex manifold. We set $W := \{(z_1, z_2) \mid |z_i| < 1\}$ and $W^* := W \setminus \{(0,0)\}$. Let E_i $(i = 1, 2)$ be locally free sheaves on $Z \times W$. We consider an extension of $\mathcal{O}_{Z\times W^*}$-modules*

$$0 \longrightarrow E_{1|Z\times W^*} \longrightarrow E' \longrightarrow E_{2|Z\times W^*} \longrightarrow 0 \tag{5.4}$$

We suppose the following condition:

- *For any $P \in Z$, the specialization of* (5.4) *to $\{P\} \times W^*$ is prolonged to*

$$0 \longrightarrow E_{1|\{P\}\times W} \longrightarrow E'_{P\times W} \longrightarrow E_{2|\{P\}\times W} \longrightarrow 0.$$

Then, we have a unique extension $0 \longrightarrow E_1 \longrightarrow E \longrightarrow E_2 \longrightarrow 0$ on $Z \times W$ whose restriction to $Z \times W^*$ is (5.4).

Proof We may assume that $E_2 = \mathcal{O}_{Z\times W}$. By using Theorem 6.10 in [85], we obtain a reflexive coherent $\mathcal{O}_{Z\times W}$-module E_3 such that $E_{3|Z\times W^*} = E'$. We naturally have the exact sequence $0 \longrightarrow E_1 \longrightarrow E_3 \longrightarrow E_2$. Let us prove that $\kappa : E_3 \longrightarrow E_2$ is surjective. Fix a point $P \in Z$, and we will shrink Z around a fixed point P. Let $\mathfrak{m}$ denote the ideal sheaf of $Z\times\{(0,0)\}$. If we take a sufficiently large N, the composite of $\mathfrak{m}^N \longrightarrow \mathcal{O}_X \longrightarrow \mathrm{Cok}(\kappa)$ is 0. We have the induced extension

$$0 \longrightarrow E_1 \longrightarrow E_4 \longrightarrow \mathfrak{m}^N \longrightarrow 0. \tag{5.5}$$

Let us observe that (5.5) has a splitting. Let $\mathfrak{c}$ be the extension class of (5.5), which is a section of $\mathcal{E}xt^1(\mathfrak{m}^N, E_1)$. Note that $\mathfrak{m}^N$ is flat over $\mathcal{O}_Z$. Let $\mathfrak{m}_0 \subset \mathcal{O}_W$ be the ideal sheaf of $(0,0)$. For any $Q \in Z$, we have

$$\mathcal{E}xt^1(\mathfrak{m}^N, E_1) \otimes \mathcal{O}_{\{Q\}\times W} \simeq \mathcal{E}xt^1\big(\mathfrak{m}_0^N, E_{1|\{Q\}\times W}\big) \tag{5.6}$$

By the assumption, the image of $\mathfrak{c}$ in (5.6) are 0 for any Q. By the coherence of $\mathcal{E}xt^1(\mathfrak{m}^N, E_1)$, we obtain that $\mathfrak{c}$ is 0, i.e., (5.5) has a splitting.

We have the natural inclusion $E_4 \subset E_3$ and an isomorphism $E_4 \simeq E_1 \oplus \mathfrak{m}^N$. Hence, we have $E_3 \simeq E_1 \oplus \mathcal{O}$, and $E_3 \longrightarrow E_1$ is equal to the projection $E_3 \longrightarrow \mathcal{O}$. Thus, we obtain Lemma 5.2.10. □

5.3 Canonical Prolongations of Good-KMS Smooth $\mathcal{R}$-Modules

We shall study the canonical prolongations of good-KMS smooth $\mathcal{R}$-modules across normal crossing hypersurfaces.

5.3.1 Goal

Let X be any complex manifold with a simple normal crossing hypersurface D. Let $D = \bigcup_{i\in\Lambda} D_i$ be the irreducible decomposition. We set $(\mathcal{X},\mathcal{D}) := \mathbb{C}_\lambda \times (X,D)$. Let $\mathcal{M}$ be a good-KMS smooth $\mathcal{R}_{X(*D)}$-module. An $\mathcal{R}_X$-module $\tilde{\mathcal{M}}$ with an isomorphism $\rho : \tilde{\mathcal{M}} \otimes \mathcal{O}_{\mathcal{X}}(*\mathcal{D}) \simeq \mathcal{M}$ is called a prolongment of $\mathcal{M}$. We say prolongments $(\tilde{\mathcal{M}}_i, \rho_i)$ $(i = 1,2)$ are isomorphic, if there exists an isomorphism of $\mathcal{R}_X$-modules $F : \tilde{\mathcal{M}}_1 \longrightarrow \tilde{\mathcal{M}}_2$ such that the following diagram is commutative:

$$\begin{array}{ccc}
\widetilde{\mathcal{M}}_1 \otimes \mathcal{O}_{\mathcal{X}}(*\mathcal{D}) & \xrightarrow{F} & \widetilde{\mathcal{M}}_2 \otimes \mathcal{O}_{\mathcal{X}}(*\mathcal{D}) \\
{\scriptstyle \rho_1}\downarrow & & {\scriptstyle \rho_2}\downarrow \\
\mathcal{M} & \xrightarrow{=} & \mathcal{M}
\end{array}$$

A prolongment $(\tilde{\mathcal{M}}, \rho)$ is often denoted just by $\tilde{\mathcal{M}}$ in the following. We shall prove the following proposition in Sects. 5.3.2–5.3.6.

Proposition 5.3.1 *For any decomposition $\Lambda = I \sqcup J$, there exists a prolongment $\mathcal{M}[*I!J]$ of $\mathcal{M}$ with the following property, which is unique up to isomorphisms:*

(P1) *$\mathcal{M}[*I!J]$ is $\mathcal{R}_X$-coherent, holonomic and strict.*

(P2) *For any $P \in D$, take a small coordinate neighbourhood $(X_P; z_1, \dots, z_n)$ around P such that for each $i \in \Lambda$, we have $D_i \cap X_P = \emptyset$ or $D_i \cap X_P = \{z_{k(i)} = 0\}$ for some $k(i)$. Then, $\mathcal{M}[*I!J]_P := \mathcal{M}[*I!J]_{|X_P}$ is strictly specializable along z_k for any k, and we have*

$$\mathcal{M}[*I!J]_P[*z_{k(i)}] = \mathcal{M}[*I!J]_P \quad (i \in I),$$
$$\mathcal{M}[*I!J]_P[!z_{k(i)}] = \mathcal{M}[*I!J]_P \quad (i \in J).$$

We shall also prove the following lemmas.

Lemma 5.3.2 *Let $\varphi : \mathcal{M}[*I!J] \simeq \mathcal{M}[*I!J]$ be an isomorphism as prolongments of $\mathcal{M}$. Then, φ is the identity.*

For any $I \subset \Lambda$ and $i \in \Lambda$, we put $I_{\cup i} := I \cup \{i\}$ and $I_{\setminus i} := I \setminus \{i\}$.

Lemma 5.3.3 *We have isomorphisms of the following prolongments*

$$\big(\mathcal{M}[*I!J]_P\big)[*z_{k(i)}] \simeq \mathcal{M}[*I_{\cup i}!J_{\setminus i}]_P, \qquad \big(\mathcal{M}[*I!J]_P\big)[!z_{k(i)}] \simeq \mathcal{M}[*I_{\setminus i}!J_{\cup i}]_P.$$

5.3.2 Uniqueness and Lemma 5.3.2

Let $\mathcal{M}[*I!J]_\kappa$ $(\kappa = 1, 2)$ be prolongments of $\mathcal{M}$ satisfying the conditions (P1) and (P2). We prove that there uniquely exists an isomorphism $\mathcal{M}[*I!J]_1 \simeq \mathcal{M}[*I!J]_2$.

By the uniqueness in the claim, we have only to consider the case $X = \Delta^n$ and $D = \bigcup_{i=1}^{\ell}\{z_i = 0\}$. For $L \subset \underline{\ell}$, we denote "$\otimes\mathcal{O}(*\mathcal{D}(L))$" by "$(*L)$" for simplicity. We also denote $\mathcal{R}_{X(*D_L)}$ by $\mathcal{R}_X(*L)$. An $\mathcal{R}_X(*L)$-module $\tilde{\mathcal{M}}$ with an isomorphism $\rho : \tilde{\mathcal{M}} \otimes \mathcal{O}_{\mathcal{X}}(*\mathcal{D}) \simeq \mathcal{M}$ is called a prolongment of $\mathcal{M}$. An isomorphism of prolongments as $\mathcal{R}_X(*L)$-modules is defined as in the case of prolongments as $\mathcal{R}_X$-modules.

We consider the $\mathcal{R}_X(*L)$-modules $\mathcal{M}[*I!J]_\kappa(*L)$. They are strictly specializable along z_i as $\mathcal{R}_X(*L)$-modules. We have the following natural isomorphisms as $\mathcal{R}_X(*L)$-modules for any $i \notin L$:

$$\mathcal{M}[*I!J]_\kappa(*L)[*z_i] \simeq \mathcal{M}[*I!J]_\kappa(*L) \quad (i \in I)$$
$$\mathcal{M}[*I!J]_\kappa(*L)[!z_i] \simeq \mathcal{M}[*I!J]_\kappa(*L) \quad (i \in J)$$

Hence, we have the following natural isomorphisms as $\mathcal{R}_X(*L_{\setminus i})$-modules for any $i \in L$:

$$\mathcal{M}[*I!J]_\kappa(*L)[*z_i] \simeq \mathcal{M}[*I!J]_\kappa(*L_{\setminus i}) \quad (i \in I) \tag{5.7}$$

$$\mathcal{M}[*I!J]_\kappa(*L)[!z_i] \simeq \mathcal{M}[*I!J]_\kappa(*L_{\setminus i}) \quad (i \in J) \tag{5.8}$$

We have the unique isomorphism $\varphi : \mathcal{M}[*I!J]_1(*\underline{\ell}) \simeq \mathcal{M}[*I!J]_2(*\underline{\ell})$. By using (5.7) and (5.8), we obtain unique isomorphisms $\mathcal{M}[*I!J]_1(*L) \simeq \mathcal{M}[*I!J]_2(*L)$ for any $L \subset \underline{\ell}$ as prolongments with a descending induction on $|L|$. □

5.3.3 Local Construction

Let $X := \Delta^n$ and $D := \bigcup_{i=1}^{\ell}\{z_i = 0\}$. Set $\eth_i := \lambda\partial_i$. In this subsection, let $V_0\mathcal{R}_X \subset \mathcal{R}_X$ be generated by $z_i\eth_i$ $(i \leq \ell)$ and $\eth_i$ $(i > \ell)$ over $\mathcal{O}_{\mathcal{X}}$. Let $\mathcal{K} \subset \mathbb{C}$ be a small neighbourhood of λ_0. We set $(\mathcal{X}^{(\lambda_0)}, \mathcal{D}^{(\lambda_0)}) := \mathcal{K} \times (X, D)$. Let $\mathcal{M}^{(\lambda_0)}$ be a good-KMS smooth $\mathcal{R}_{X(*D)}$-module given on $\mathcal{X}^{(\lambda_0)}$. Let $(\mathcal{Q}_*^{(\lambda_0)}\mathcal{M}, \mathbb{D})$ be the associated filtered λ-flat bundles at λ_0 on $(\mathcal{X}^{(\lambda_0)}, \mathcal{D}^{(\lambda_0)})$.

Let $\underline{\ell} = I \sqcup J$. We put $\boldsymbol{a}(I,J) := \boldsymbol{\delta}_I + (1-\epsilon)\boldsymbol{\delta}_J$ for some sufficiently small $\epsilon > 0$. We obtain a coherent $V_0\mathcal{R}_X$-module $V_0\mathcal{R}_X \cdot \mathcal{Q}^{(\lambda_0)}_{\boldsymbol{a}(I,J)}\mathcal{M} \subset \mathcal{Q}^{(\lambda_0)}\mathcal{M}$. Then, we obtain a coherent $\mathcal{R}_X$-module

$$\mathcal{M}^{(\lambda_0)}[*I!J] := \mathcal{R}_X \otimes_{V_0\mathcal{R}_X} \big(V_0\mathcal{R}_X \cdot \mathcal{Q}^{(\lambda_0)}_{\boldsymbol{a}(I,J)}\mathcal{M}\big).$$

We shall prove that $\mathcal{M}^{(\lambda_0)}[*I!J]$ has the desired property.

Remark 5.3.4 For any good meromorphic flat bundle $\mathcal{V}$ on (X, D), we have a similar description of $\mathcal{V}[*I!J]$. □

Lemma 5.3.5 *The $\mathcal{R}_X$-module $\mathcal{M}^{(\lambda_0)}[*I!J]$ is holonomic, and its characteristic variety* $\mathrm{Ch}(\mathcal{M}^{(\lambda_0)}[*I!J])$ *is contained in* $\mathcal{S} = \bigcup_{L\subset\underline{\ell}} \mathcal{K} \times T^*_{D_L}X$.

Proof Let F_0 be the image of the naturally defined morphism $\mathcal{Q}^{(\lambda_0)}_{\boldsymbol{a}(I,J)}\mathcal{M} \longrightarrow \mathcal{M}^{(\lambda_0)}[*I!J]$. By the construction, $\mathcal{M}^{(\lambda_0)}[*I!J]$ is generated by F_0 over $\mathcal{R}_X$. For $\boldsymbol{p} \in \mathbb{Z}^n_{\geq 0}$, we put $\eth^{\boldsymbol{p}} := \prod \eth_i^{p_i}$. We set $F_m := \sum_{|\boldsymbol{p}|\leq m} \eth^{\boldsymbol{p}} F_0$. Then, $\{F_m\}$ is a coherent filtration of $\mathcal{M}^{(\lambda_0)}[*I!J]$. Let us prove that the support of $\mathrm{Gr}^F \mathcal{M}^{(\lambda_0)}[*I!J]$ is contained in $\mathcal{S}$. Let $\pi : \mathcal{K} \times T^*X \longrightarrow X$.

First, let us consider the case $\mathcal{M}$ is regular-KMS. Let $Q \in D_K^\circ$ for some $K \subset \underline{\ell}$. For any $j \notin K$, we have $\eth_j F_0 \subset F_0$ around Q. Hence, we obtain $\eth_j F_m \subset F_m$ for any $m \geq 0$. Then, the action of $\eth_j$ on $\mathrm{Gr}^F \mathcal{M}^{(\lambda_0)}[*I!J]$ is 0 around Q. Then, we obtain that $\mathrm{Ch}(\mathcal{M}^{(\lambda_0)}[*I!J]) \cap \pi^{-1}(Q) \subset \mathcal{K} \times (T^*_{D_K}X)_Q$.

We put $\mathfrak{a} := \boldsymbol{z}^{\boldsymbol{m}}$ for some $\boldsymbol{m} \in \mathbb{Z}^{p}_{<0}$, where $1 \leq p \leq \ell$. Let $L(\mathfrak{a}) = \mathcal{O}_{\mathcal{X}^{(\lambda_0)}}\, e$ with $\mathbb{D}e = e\, d\mathfrak{a}$. Let $\pi : \mathcal{X}^{(\lambda_0)} \longrightarrow \mathcal{X}^{(\lambda_0)}$ be the ramified covering, given by $z_i^{q_i}$ for $1 \leq i \leq p$ and z_i for $p+1 \leq i \leq n$. Let us consider the case $\mathcal{M}$ is the tensor product of $\pi_* L(\mathfrak{a})$ and a regular-KMS smooth $\mathcal{R}_{X(*D)}$-module $\mathcal{M}'$. Let $Q \in D_K^{\circ}$ for some $K \subset \underline{\ell}$. If $K \cap \underline{p} = \emptyset$, we have $\mathrm{Ch}(\mathcal{M}^{(\lambda_0)}[*I!J]) \subset \mathcal{S}$ around Q by the consideration in the regular singular case. Let us consider the case $K \cap \underline{p} \neq \emptyset$. Take $i \in K \cap \underline{p}$. For $j \in \underline{p} \setminus K$, we put $v_j := m_j^{-1} z_j \eth_j - m_i^{-1} z_i \eth_i$. Note we have $v_j e = 0$. Hence, we have $v_j F_m \subset F_m$ around Q. For $j \in \underline{\ell} \setminus (\underline{p} \cup K)$, we have $\eth_j F_0 \subset F_0$, and hence $\eth_j F_m \subset F_m$. Thus, we obtain $\mathrm{Ch}(\mathcal{M}[*I!J])_{|\pi^{-1}(Q)} \subset \mathcal{K} \times (T^*_{D_K} X)_Q$.

The general case can be reduced to the above cases, by using the formal decomposition. □

5.3.4 *Preliminary*

We have the formal decomposition into the regular part and the irregular part:

$$\mathcal{Q}_*^{(\lambda_0)} \mathcal{M}_{|\hat{\mathcal{D}}_i} = \mathcal{Q}_*^{(\lambda_0)} \mathcal{M}^{(\mathrm{reg})}_{\hat{\mathcal{D}}_i} \oplus \mathcal{Q}_*^{(\lambda_0)} \mathcal{M}^{(\mathrm{irr})}_{\hat{\mathcal{D}}_i}$$

(See Sect. 5.1.3.) Let $u \in \mathbb{R} \times \mathbb{C}$. Take $\boldsymbol{a} \in (\mathbb{R} \times \mathbb{C})^I$ such that the i-th component is $(\mathfrak{p}(\lambda_0, u), \mathfrak{e}(\lambda_0, u))$. Then, we set

$${}^{i}\tilde{\psi}_u^{(\lambda_0)}(\mathcal{Q}_{\boldsymbol{a}}^{(\lambda_0)} \mathcal{M}) := {}^{i}\mathcal{G}_{u+\boldsymbol{\delta}}^{(\lambda_0)}(\mathcal{Q}_{\boldsymbol{a}}^{(\lambda_0)} \mathcal{M}^{(\mathrm{reg})}_{\hat{\mathcal{D}}_i})$$

Here, $\boldsymbol{\delta} = (1, 0) \in \mathbb{R} \times \mathbb{C}$. We obtain a good-KMS filtered λ-flat bundle ${}^{i}\tilde{\psi}_u^{(\lambda_0)}(\mathcal{Q}_*^{(\lambda_0)} \mathcal{M})$ on $(\mathcal{D}^{(\lambda_0)}, \partial \mathcal{D}^{(\lambda_0)})$.

Let $I \sqcup J \sqcup \{i\} = \underline{\ell}$ be a decomposition. Let q_j be the projection of $\mathbb{R}^{\ell}$ to the j-th component. Let $\boldsymbol{a}(b) \in \mathbb{R}^{\ell}$ be determined by $q_j(\boldsymbol{a}(b)) = 1$ for $j \in I$, $q_j(\boldsymbol{a}(b)) = 1 - \epsilon$ for $j \in J$, and $q_i(\boldsymbol{a}(b)) = b + 1$, where $\epsilon > 0$ is any sufficiently small number.

Lemma 5.3.6 *We have the following natural isomorphism:*

$$\frac{V_0 \mathcal{R}_X \cdot \mathcal{Q}_{\boldsymbol{a}(b)}^{(\lambda_0)} \mathcal{M}}{V_0 \mathcal{R}_X \cdot \mathcal{Q}_{\boldsymbol{a}(b-\epsilon)}^{(\lambda_0)} \mathcal{M}} \simeq \bigoplus_{\mathfrak{p}(\lambda_0, u) = b} V_0 \mathcal{R}_{D_i} \cdot {}^{i}\tilde{\psi}_u^{(\lambda_0)}(\mathcal{Q}_{\boldsymbol{a}(b)}^{(\lambda_0)} \mathcal{M}) \tag{5.9}$$

Proof We have $V_0 \mathcal{R}_X \cdot \mathcal{Q}_{\boldsymbol{a}(b)}^{(\lambda_0)} \mathcal{M}_{|\hat{\mathcal{D}}_i} = V_0 \mathcal{R}_X \cdot \mathcal{Q}_{\boldsymbol{a}(b)}^{(\lambda_0)} \mathcal{M}^{(\mathrm{reg})}_{\hat{\mathcal{D}}_i} \oplus V_0 \mathcal{R}_X \cdot \mathcal{Q}_{\boldsymbol{a}(b)}^{(\lambda_0)} \mathcal{M}^{(\mathrm{irr})}_{\hat{\mathcal{D}}_i}$. Let $V_0 \mathcal{R}_{X, \setminus i} \subset \mathcal{R}_X$ be generated by $z_j \eth_j$ ($j \in \underline{\ell} \setminus \{i\}$) and $\eth_j$ ($j \in \underline{n} \setminus \underline{\ell}$) over $\mathcal{O}_{\mathcal{X}}$. The following natural morphisms are isomorphisms:

$$V_0 \mathcal{R}_{X, \setminus i} \cdot \mathcal{Q}_{\boldsymbol{a}(b)}^{(\lambda_0)} \mathcal{M}^{(\mathrm{reg})}_{\hat{\mathcal{D}}_i} \longrightarrow V_0 \mathcal{R}_X \cdot \mathcal{Q}_{\boldsymbol{a}(b)}^{(\lambda_0)} \mathcal{M}^{(\mathrm{reg})}_{\hat{\mathcal{D}}_i} \tag{5.10}$$

$$V_0 \mathcal{R}_X \cdot \mathcal{Q}_{\boldsymbol{a}(b-\epsilon)}^{(\lambda_0)} \mathcal{M}^{(\mathrm{irr})}_{\hat{\mathcal{D}}_i} \longrightarrow V_0 \mathcal{R}_X \cdot \mathcal{Q}_{\boldsymbol{a}(b)}^{(\lambda_0)} \mathcal{M}^{(\mathrm{irr})}_{\hat{\mathcal{D}}_i}. \tag{5.11}$$

Indeed, it can be reduced to the case that $\mathcal{Q}\mathcal{M}$ is unramified. We have only to prove that their formal completions along $\mathcal{K}\times P$ are isomorphisms for each $P\in D_i$. Then, the claim can be checked by a direct computation.

We obtain the following:

$$\frac{V_0\mathcal{R}_X\cdot\mathcal{Q}^{(\lambda_0)}_{\boldsymbol{a}(b)}\mathcal{M}}{V_0\mathcal{R}_X\cdot\mathcal{Q}^{(\lambda_0)}_{\boldsymbol{a}(b-\epsilon)}\mathcal{M}}\simeq\frac{V_0\mathcal{R}_X\cdot\mathcal{Q}^{(\lambda_0)}_{\boldsymbol{a}(b)}\mathcal{M}_{|\hat{\mathcal{D}}_i}}{V_0\mathcal{R}_X\cdot\mathcal{Q}^{(\lambda_0)}_{\boldsymbol{a}(b-\epsilon)}\mathcal{M}_{|\hat{\mathcal{D}}_i}}\simeq\bigoplus_{\mathfrak{p}(\lambda_0,u)=b}V_0\mathcal{R}_{D_i}\cdot{}^i\tilde{\psi}^{(\lambda_0)}_u\big(\mathcal{Q}^{(\lambda_0)}_{\boldsymbol{a}(b)}\mathcal{M}\big)$$

Thus, we are done. □

Let ${}^iV_0\mathcal{R}_X\subset\mathcal{R}_X$ be generated by $\eth_j$ $(j\neq i)$ and $z_i\eth_i$ over $\mathcal{O}_{\mathcal{X}}$.

Corollary 5.3.7 *We have the following natural isomorphism:*

$$\frac{{}^iV_0\mathcal{R}_X\otimes_{V_0\mathcal{R}_X}\big(V_0\mathcal{R}_X\cdot\mathcal{Q}^{(\lambda_0)}_{\boldsymbol{a}(b)}\mathcal{M}\big)}{{}^iV_0\mathcal{R}_X\otimes_{V_0\mathcal{R}_X}\big(V_0\mathcal{R}_X\cdot\mathcal{Q}^{(\lambda_0)}_{\boldsymbol{a}(b-\epsilon)}\mathcal{M}\big)}$$
$$\simeq\bigoplus_{\mathfrak{p}(\lambda_0,u)=b}\mathcal{R}_{D_i}\otimes_{V_0\mathcal{R}_{D_i}}\Big(V_0\mathcal{R}_{D_i}\cdot{}^i\tilde{\psi}^{(\lambda_0)}_u\big(\mathcal{Q}^{(\lambda_0)}_{\boldsymbol{a}(b)}\mathcal{M}\big)\Big)\tag{5.12}$$

Proof By Lemma 5.3.6, we obtain the following:

$$\frac{{}^iV_0\mathcal{R}_X\otimes_{V_0\mathcal{R}_X}\big(V_0\mathcal{R}_X\cdot\mathcal{Q}^{(\lambda_0)}_{\boldsymbol{a}(b)}\mathcal{M}\big)}{{}^iV_0\mathcal{R}_X\otimes_{V_0\mathcal{R}_X}\big(V_0\mathcal{R}_X\cdot\mathcal{Q}^{(\lambda_0)}_{\boldsymbol{a}(b-\epsilon)}\mathcal{M}\big)}$$
$$\simeq\bigoplus_{\mathfrak{p}(\lambda_0,u)=b}{}^iV_0\mathcal{R}_X\otimes_{V_0\mathcal{R}_X}\Big(V_0\mathcal{R}_{D_i}\cdot{}^i\tilde{\psi}^{(\lambda_0)}_u\big(\mathcal{Q}^{(\lambda_0)}_{\boldsymbol{a}}\mathcal{M}\big)\Big)\tag{5.13}$$

It is easy to observe that the right hand side of (5.13) is naturally isomorphic to the right hand side of (5.12). □

5.3.5 *Some Filtrations*

For any subset $K\subset\underline{\ell}$, let ${}^KV_0\mathcal{R}_X\subset\mathcal{R}_X$ be generated by $\eth_i$ $(i\in\underline{n}\setminus K)$ and $z_i\eth_i$ $(i\in K)$ over $\mathcal{O}_{\mathcal{X}}$. Let $\underline{\ell}=I\sqcup J\sqcup K$ be a decomposition. We put $\boldsymbol{a}(I,J):=\boldsymbol{\delta}_I+(1-\epsilon)\boldsymbol{\delta}_J\in\mathbb{R}^{I\sqcup J}$. Let $\mathcal{Q}^{(\lambda_0)}_{\boldsymbol{a}(I,J)}\mathcal{M}$ mean

$$\mathcal{Q}^{(\lambda_0)}_{\boldsymbol{a}(I,J)+\boldsymbol{c}}\mathcal{M}\otimes\mathcal{O}_{\mathcal{X}^{(\lambda_0)}}(*\mathcal{D}^{(\lambda_0)}(K))$$

for any $\boldsymbol{c} \in \mathbb{R}^K$. We consider the following coherent $\mathcal{R}_X(*\mathcal{D}^{(\lambda_0)}(K))$-module:

$$\mathcal{M}^{(\lambda_0)}[*I!J] := {}^K V_0 \mathcal{R}_X \otimes_{V_0\mathcal{R}_X} \big(V_0\mathcal{R}_X \cdot \mathcal{Q}^{(\lambda_0)}_{\boldsymbol{a}(I,J)}\mathcal{M}\big)$$
$$\simeq \mathcal{R}_X \otimes_{V_0\mathcal{R}_X} \big(V_0\mathcal{R}_X \cdot \mathcal{Q}^{(\lambda_0)}_{\boldsymbol{a}(I,J)}\mathcal{M}\big) \tag{5.14}$$

For $\boldsymbol{b} \in \mathbb{R}^K$, we consider the following ${}^K V_0\mathcal{R}_X$-module:

$${}^K V^{(\lambda_0)}_{\boldsymbol{b}} \mathcal{M}^{(\lambda_0)}[*I!J] := {}^K V_0 \mathcal{R}_X \otimes_{V_0\mathcal{R}_X} \big(V_0\mathcal{R}_X \cdot \mathcal{Q}^{(\lambda_0)}_{\boldsymbol{a}(I,J)+\boldsymbol{b}+\boldsymbol{\delta}_K}\mathcal{M}\big)$$

For $i \in I$ and $b \leq 0$, or for $i \in J$ and $b < 0$, let ${}^{K,i}V^{(\lambda_0)}_{\boldsymbol{b},b}\mathcal{M}^{(\lambda_0)}[*I!J]$ be the image of the following morphism:

$${}^{K\sqcup\{i\}}V_0\mathcal{R}_X \otimes_{V_0\mathcal{R}_X} \big(V_0\mathcal{R}_X \cdot \mathcal{Q}^{(\lambda_0)}_{\boldsymbol{a}'+\boldsymbol{b}+\boldsymbol{\delta}_K}\mathcal{M}\big) \longrightarrow {}^K V^{(\lambda_0)}_{\boldsymbol{b}}\mathcal{M}^{(\lambda_0)}[*I!J]. \tag{5.15}$$

Here, $\boldsymbol{a}' \in \mathbb{R}^{I\sqcup J}$ is determined such that $q_j(\boldsymbol{a}') = q_j(\boldsymbol{a}(I,J))$ if $j \neq i$, and $q_i(\boldsymbol{a}') = b + 1$.

For $i \in I$ and $b > 0$, or for $i \in J$ and $b \geq 0$, we set

$${}^{K,i}V^{(\lambda_0)}_{\boldsymbol{b},b}\big(\mathcal{M}^{(\lambda_0)}[*I!J]\big) = \sum_{(c,p)\in\mathcal{U}(b,i)} \eth_i^p\big({}^{K,i}V^{(\lambda_0)}_{\boldsymbol{b},c}(\mathcal{M}^{(\lambda_0)}[*I!J])\big)$$

Here, $\mathcal{U}(b,i)$ denotes the set $\big\{(c,p) \in \mathbb{R}_{\leq 0} \times \mathbb{Z}_{\geq 0} \,\big|\, c+p \leq b\big\}$ if $i \in I$, or $\big\{(c,p) \in \mathbb{R}_{<0} \times \mathbb{Z}_{\geq 0} \,\big|\, c+p \leq b\big\}$ if $i \in J$.

Lemma 5.3.8 *The following holds for $m = |I \sqcup J|$:*

$P(m)$: *The morphisms* (5.15) *are injective.*

$Q(m)$: *For any $\boldsymbol{c},\boldsymbol{d} \in \mathbb{R}^K$ with $\boldsymbol{c} \leq \boldsymbol{d}$, the naturally defined morphisms ${}^K V^{(\lambda_0)}_{\boldsymbol{c}}\mathcal{M}^{(\lambda_0)}[*I!J] \longrightarrow {}^K V^{(\lambda_0)}_{\boldsymbol{d}}\mathcal{M}^{(\lambda_0)}[*I!J]$ are injective. In particular, we have the injectivity of the morphism ${}^K V^{(\lambda_0)}_{\boldsymbol{c}}\mathcal{M}^{(\lambda_0)}[*I!J] \longrightarrow \mathcal{M}^{(\lambda_0)}[*I!J]$.*

$R(m)$: *${}^K V^{(\lambda_0)}_{\boldsymbol{c}}\mathcal{M}^{(\lambda_0)}[*I!J]$ are strict.*

Proof We shall prove the claims by an induction on $m = |I \sqcup J|$ and $\dim X$. If $\dim X = 0$, the claim is trivial. The claim $P(0)$ is trivial, and the claims $Q(0)$ and $R(0)$ are obvious. We obtain $P(m)$ from $Q(m-1)$, by considering the composition of the morphism (5.15) and the natural one

$${}^K V^{(\lambda_0)}_{\boldsymbol{b}}\mathcal{M}^{(\lambda_0)}[*I!J] \longrightarrow {}^K V^{(\lambda_0)}_{\boldsymbol{b}}\mathcal{M}^{(\lambda_0)}[*I!J](*i).$$

Assume $P(m)$ and the claims in the strictly lower dimensional case. For $i \in I$ and $b \leq 0$, or for $i \in J$ and $b < 0$, we obtain the following by using Lemma 5.3.6:

$${}^i\mathrm{Gr}^{V^{(\lambda_0)}}_b\, {}^K V^{(\lambda_0)}_{\boldsymbol{b}}\mathcal{M}^{(\lambda_0)}[*I!J] \simeq \bigoplus_{\mathfrak{p}(\lambda_0,u)=b} {}^K V_0\mathcal{R}_{D_i} \otimes_{V_0\mathcal{R}_{D_i}} \Big(V_0\mathcal{R}_{D_i} \cdot {}^i\tilde{\psi}_u\big(\mathcal{Q}^{(\lambda_0)}_{\boldsymbol{a}'+\boldsymbol{b}}\mathcal{M}\big)\Big)$$
$$= \bigoplus_{\mathfrak{p}(\lambda_0,u)=b} {}^K V^{(\lambda_0)}_{\boldsymbol{b}}\big({}^i\tilde{\psi}_u(\mathcal{M})^{(\lambda_0)}[*I_{\setminus i}\,!J_{\setminus i}]\big) \tag{5.16}$$

For $i \in I$ and $d > 0$, (resp. $i \in J$ and $d \geq 0$), we take $p \in \mathbb{Z}_{>0}$ such that $-1 < d-p \leq 0$ (resp. $-1 \leq d-p < 0$). We consider the following surjection:

$$\eth_i^p : {}^i\mathrm{Gr}_{d-p}^{V^{(\lambda_0)}}\,{}^K V_{\boldsymbol{b}}^{(\lambda_0)} \mathcal{M}^{(\lambda_0)}[*I!J] \longrightarrow {}^i\mathrm{Gr}_{d}^{V^{(\lambda_0)}}\,{}^K V_{\boldsymbol{b}}^{(\lambda_0)} \mathcal{M}^{(\lambda_0)}[*I!J] \tag{5.17}$$

If $i \in I$, the morphism $z_i^p \eth_i^p$ on ${}^i\mathrm{Gr}_{d-p}^{V^{(\lambda_0)}}\,{}^K V_{\boldsymbol{b}}^{(\lambda_0)} \mathcal{M}^{(\lambda_0)}[*I!J]$ is injective by the assumption of the induction on the base space. Hence, we obtain that (5.17) is an isomorphism. If $i \in J$ and $d = 0$, we can prove that the restriction of (5.17) to $\mathcal{D}_i^{(\lambda_0)} \setminus \bigcup_{j \neq i} \mathcal{D}_j^{(\lambda_0)}$ is an isomorphism, by using Lemma 3.1.10. Then, by using the description (5.16) and the hypothesis of the induction on $\dim X$, we obtain that (5.17) is isomorphism. In the case $d > 0$, we can check that (5.17) is an isomorphism by using the argument in the case $i \in I$.

Now, assume $P(m)$, $R(m-1)$, $Q(m-1)$ and the claims in the strictly lower dimensional case. We obtain that

$${}^i\mathrm{Gr}_{d}^{V^{(\lambda_0)}}\,{}^K V_{\boldsymbol{b}}^{(\lambda_0)} \mathcal{M}^{(\lambda_0)}[*I!J] \longrightarrow {}^i\mathrm{Gr}_{d}^{V^{(\lambda_0)}}\,{}^K V_{\boldsymbol{c}}^{(\lambda_0)} \mathcal{M}^{(\lambda_0)}[*I!J]$$

is injective for each d by using the isomorphisms (5.16) and (5.17). Then, $Q(m)$ follows. We also obtain $R(m)$ from $R(m-1)$ and the strictness in the lower dimensional case. Thus, the proof of Lemma 5.3.8 is finished. □

Corollary 5.3.9 *Let $I \sqcup J \subset \underline{\ell}$, and $K := \underline{\ell} \setminus (I \sqcup J)$.*

- *$\mathcal{M}^{(\lambda_0)}[*I!J]$ is a coherent, holonomic and strict $\mathcal{R}_X(*K)$-module.*
- *It is strictly specializable along z_i with the V-filtration ${}^iV^{(\lambda_0)}$. For any $i \in K$ and any $u \in \mathbb{R} \times \mathbb{C}$, or for any $i \in I \sqcup J$ and any $u \in (\mathbb{R} \times \mathbb{C}) \setminus (\mathbb{Z}_{\geq 0} \times \{0\})$, we have the following natural isomorphisms:*

$${}^i\tilde{\psi}_u^{(\lambda_0)}(\mathcal{M}^{(\lambda_0)}[*I!J]) \simeq {}^i\tilde{\psi}_u^{(\lambda_0)}(\mathcal{M}^{(\lambda_0)})[*I_{\setminus i}!J_{\setminus i}]. \tag{5.18}$$

- *The following morphisms are isomorphisms:*

$$\eth_i : {}^i\psi_{-\boldsymbol{\delta}}^{(\lambda_0)}(\mathcal{M}^{(\lambda_0)}[*I!J]) \longrightarrow {}^i\psi_0^{(\lambda_0)}(\mathcal{M}^{(\lambda_0)}[*I!J]) \qquad (i \in J)$$

$$z_i : {}^i\psi_0^{(\lambda_0)}(\mathcal{M}^{(\lambda_0)}[*I!J]) \longrightarrow {}^i\psi_{-\boldsymbol{\delta}}^{(\lambda_0)}(\mathcal{M}^{(\lambda_0)}[*I!J]) \qquad (i \in I)$$

In particular, we have isomorphisms $\mathcal{M}^{(\lambda_0)}[*I!J][*z_i] \simeq \mathcal{M}^{(\lambda_0)}[*I_{\cup i}!J_{\setminus i}]$ and $\mathcal{M}^{(\lambda_0)}[*I!J][!z_i] \simeq \mathcal{M}^{(\lambda_0)}[*I_{\setminus i}!J_{\cup i}]$. □

5.3.6 Globalization

Let us return to the situation in Sect. 5.3.1. Let $I \sqcup J = \Lambda$ be a decomposition. Let $P \in D$. We take a small coordinate neighbourhood $(X_P; z_1, \ldots, z_n)$ of X around P such that $D_P := D \cap X_P = \bigcup_{i=1}^{\ell} \{z_i = 0\}$. Let $U(\lambda_0)$ be a sufficiently small neighbourhood of λ_0. We set $(\mathcal{X}_P^{(\lambda_0)}, \mathcal{D}_P^{(\lambda_0)}) := U(\lambda_0) \times (X_P, X_P \cap D)$. By applying the procedure in Sect. 5.3.3 to $(\mathcal{Q}_*^{(\lambda_0)}\mathcal{M}_P, \mathbb{D}) := (\mathcal{Q}_*^{(\lambda_0)}\mathcal{M}, \mathbb{D})_{|\mathcal{X}_P^{(\lambda_0)}}$, we obtain an $\mathcal{R}_{X_P}$-module $\mathcal{M}_P^{(\lambda_0)}[*I_P!J_P]$ on $\mathcal{X}_P^{(\lambda_0)}$ for any decomposition $\underline{\ell} = I_P \sqcup J_P$. According to Corollary 5.3.9, they satisfy the conditions (P1) and (P2), and the claim in Lemma 5.3.3.

By using the uniqueness and Lemma 5.3.2, we obtain an $\mathcal{R}_X$-module $\mathcal{M}[*I!J]$ by gluing $\mathcal{M}_P^{(\lambda_0)}[*I_P!J_P]$ for varied $(\lambda_0, P) \in \mathbb{C}_\lambda \times D$, where $I_P \sqcup J_P = \underline{\ell}$ is the induced decomposition induced by $I \sqcup J = \Lambda$. By construction, the $\mathcal{R}_X$-modules $\mathcal{M}[*I!J]$ $(I \sqcup J = \underline{\ell})$ satisfy the conditions (P1) and (P2), and the claim in Lemma 5.3.3. Thus, the proof of Proposition 5.3.1 and Lemma 5.3.3 are finished. □

5.3.7 *Ramified Covering*

We give a remark on the functoriality with respect to a ramified covering. Let $X = \Delta^n$ and $D = \bigcup_{i=1}^{\ell}\{z_i = 0\}$. Let $\varphi : (X', D') \longrightarrow (X, D)$ be a ramified covering along (X, D). Namely, $X' = \Delta^n$, $D' = \bigcup_{i=1}^{\ell}\{w_i = 0\}$, and $\varphi(w_1, \dots, w_n) = (w_1^{a_1}, \dots, w_\ell^{a_\ell}, w_{\ell+1}, \dots, w_n)$. Let $\mathcal{M}'$ be a good-KMS smooth $\mathcal{R}_{X'(*D')}$-module. We naturally obtain the good-KMS smooth $\mathcal{R}_{X(*D)}$-module $\varphi_*\mathcal{M}'$.

Proposition 5.3.10 *For any decomposition $I \sqcup J = \underline{\ell}$, we naturally have*

$$(\varphi_*\mathcal{M}')[*I!J] \simeq \varphi_\dagger\big(\mathcal{M}'[*I!J]\big). \tag{5.19}$$

Proof By using the induction on the dimension and Corollary 5.3.9, we can check that the assumption in Lemma 3.2.12 is satisfied for $\mathcal{M}'[*I!J]$ and φ along z_i $(i = 1, \dots, \ell)$. Then, by the lemma and the characterization of $(\varphi_*\mathcal{M}')[*I!J]$, we obtain (5.19). □

Let $\mathcal{M}$ be a good-KMS smooth $\mathcal{R}_{X(*D)}$-module. We obtain the good-KMS smooth $\mathcal{R}_{X'(*D')}$-module $\varphi^*\mathcal{M}$.

Corollary 5.3.11 *$\mathcal{M}[*I!J]$ is a direct summand of $\varphi_\dagger\big(\varphi^*\mathcal{M}[*I!J]\big)$. It is the invariant part with respect to the action of the Galois group of φ.* □

5.4 Strict Specializability Along Monomial Functions

5.4.1 *Statement*

Let X be a complex manifold with a simple normal crossing hypersurface D. Let $D = \bigcup_{i\in\Lambda} D_i$ be the irreducible decomposition. Let $\mathcal{M}$ be a good-KMS smooth $\mathcal{R}_{X(*D)}$-module. For simplicity, we assume the following:

(A) $\mathcal{M}$ is equipped with a filtration L in the category of smooth $\mathcal{R}_{X(*D)}$-modules, such that $\mathrm{Gr}^L(\mathcal{M})$ is the canonical prolongment of a good wild harmonic bundle around any $P \in D$. (See §11.1 of [55] for the canonical prolongment of wild harmonic bundles.)

Let $\Lambda = I \sqcup J$ be a decomposition. Let g be a holomorphic function on X such that $g^{-1}(0) = \bigcup_{i \in K} D_i$ for some $K \subset \Lambda$.

Proposition 5.4.1
- *$\mathcal{M}[*I!J](*g)$ is strictly specializable along g.*
- *For $\star\ =!, *$, there exist $\big(\mathcal{M}[*I!J]\big)[\star g]$, and we have the following natural isomorphisms:*

$$\big(\mathcal{M}[*I!J]\big)[!g] \simeq \mathcal{M}\big[*(I \setminus K)!(J \cup K)\big],$$
$$\big(\mathcal{M}[*I!J]\big)[*g] \simeq \mathcal{M}\big[*(I \cup K)!(J \setminus K)\big].$$

In particular, if we have $K \subset I$ or $K \subset J$, then $\mathcal{M}[*I!J]$ is strictly specializable along g.

Remark 5.4.2 Because $\mathcal{M}[*I!J]$ underlies a mixed twistor $\mathcal{D}$-module, we eventually have that $\mathcal{M}[*I!J]$ is strictly specializable along g without the condition $K \subset I$ or $K \subset J$. □

5.4.2 Refinement

We give a refined claim in the local case. Let $X = \Delta^n$ and $D = \bigcup_{i=1}^{\ell}\{z_i = 0\}$. Let g be a monomial function $g = \boldsymbol{z}^{\boldsymbol{p}}$, where $\boldsymbol{p} \in \mathbb{Z}_{>0}^{K}$ and $K \subset \underline{\ell}$. Let $i_g : X \longrightarrow X \times \mathbb{C}_t$. Let $\mathcal{K}$ be a small neighbourhood of λ_0 in $\mathbb{C}_\lambda$. We set $(\mathcal{X}^{(\lambda_0)}, \mathcal{D}^{(\lambda_0)}) := \mathcal{K} \times (X, D)$. Let $\mathcal{Q}_*^{(\lambda_0)}\mathcal{M}$ be a good-KMS family of filtered λ-flat bundles. For a decomposition $\underline{\ell} = I \sqcup J \sqcup K$, let us consider $\iota_{g\dagger}\mathcal{M}[*I!J \star K] = \iota_{g*}\mathcal{M}[*I!J \star K] \otimes \mathbb{C}[\eth_t]$ for $\star = *, !$, where $\eth_t = \lambda\partial_t$. Let $\mathcal{R}_{X,K} \subset \mathcal{R}_X$ be generated by $\eth_i = \lambda\partial_i$ $(i \in K)$ over $\mathcal{O}_{\mathcal{X}}$. The following proposition implies Proposition 5.4.1.

Proposition 5.4.3 *Assume that $\mathcal{M}$ is a good-KMS smooth $\mathcal{R}_X(*D)$-module satisfying the condition* **(A)** *in* Sect. 5.4.1.

- *$\iota_{g\dagger}\mathcal{M}[*I!J \star K]$ are strictly specializable along t, and we have*

$$\iota_{g\dagger}\mathcal{M}[*I!J \star K] \simeq \big(\iota_{g\dagger}\mathcal{M}[*I!J \star K]\big)[\star t].$$

- *The V-filtration $U^{(\lambda_0)}$ of $\iota_{g\dagger}\mathcal{M}[*I!J * K]$ is given as follows: For $b \le 0$, we have*

$$U_b^{(\lambda_0)}\big(\iota_{g\dagger}\mathcal{M}[*I!J * K]\big) = \mathcal{R}_{X,K}\Big({}^{K}V_{b\boldsymbol{p}}^{(\lambda_0)}\mathcal{M}[*I!J] \otimes 1\Big). \tag{5.20}$$

For $b > 0$, *we have* $U_b^{(\lambda_0)} = \sum_{c,j} \eth_t^j U_c^{(\lambda_0)}$, *where* (c, j) *runs through* $\mathbb{R}_{\leq 0} \times \mathbb{Z}_{\geq 0}$ *satisfying* $c + j \leq b$. *(See* Sect. 5.3.5 *for the filtration* ${}^K V^{(\lambda_0)}$.*)*

- *The V-filtration* $U^{(\lambda_0)}$ *of* $\iota_{g\dagger}\mathcal{M}^{(\lambda_0)}[*I!J!K]$ *is given as follows: For* $b < 0$, *we have*

$$U_b^{(\lambda_0)}\Big(\iota_{g\dagger}\mathcal{M}[*I!J!K]\Big) = \mathcal{R}_{X,K}\Big({}^K V_{b\boldsymbol{p}}^{(\lambda_0)}\mathcal{M}[*I!J] \otimes 1\Big). \tag{5.21}$$

For $b \geq 0$, *we have* $U_b^{(\lambda_0)} = \sum_{c,j} \eth_t^j U_c^{(\lambda_0)}$, *where* (c, j) *runs through* $\mathbb{R}_{<0} \times \mathbb{Z}_{\geq 0}$ *satisfying* $c + j \leq b$.

5.4.3 Preliminary

Recall that for an $\mathcal{R}_X$-module $\mathcal{N}$ on $\mathcal{X}^{(\lambda_0)}$, the push-forward $i_{g\dagger}\mathcal{N}$ is naturally isomorphic to $i_{g*}\mathcal{N}[\eth_t]$, where the action of $\mathcal{R}_{X\times\mathbb{C}_t}$ is given as follows:

$$a \cdot \big(u \otimes \eth_t^j\big) = au \otimes \eth_t^j \quad (a \in \mathcal{O}_X),$$

$$\eth_i\big(u \otimes \eth_t^j\big) = (\eth_i u) \otimes \eth_t^j - (\partial_i g \cdot u) \otimes \eth_t^{j+1} \tag{5.22}$$

$$t \cdot \big(u \otimes \eth_t^j\big) = (g \cdot u) \otimes \eth_t^j - j\lambda u \otimes \eth_t^{j-1}, \quad \eth_t\big(u \otimes \eth_t^j\big) = u \otimes \eth_t^{j+1} \tag{5.23}$$

In particular, we have

$$(p_i \eth_t t + \eth_i z_i)(u \otimes \eth_t^j) = -p_i u \otimes j\lambda \eth_t^j + \big(\eth_i(z_i u)\big) \otimes \eth_t^j. \tag{5.24}$$

Let ${}^{\prime}V_0\mathcal{R}_{X\times\mathbb{C}_t} \subset \mathcal{R}_{X\times\mathbb{C}_t}$ be the sheaf of subalgebras generated by $t\eth_t$ and $\mathcal{R}_X$ over $\mathcal{O}_{\mathbb{C}_\lambda\times X\times\mathbb{C}_t}$.

Lemma 5.4.4 $U_b^{(\lambda_0)}(i_{g\dagger}\mathcal{M}[*I!J \star K])$ *are* ${}^{\prime}V_0\mathcal{R}_{X\times\mathbb{C}_t}$*-coherent modules. We have* $\bigcup_{b\in\mathbb{R}} U_b^{(\lambda_0)}(i_{g\dagger}\mathcal{M}[*I!J \star K]) = i_{g\dagger}\mathcal{M}[*I!J \star K]$.

Proof Let us prove the first claim. We have only to consider the cases that $U_b^{(\lambda_0)}$ are expressed as (5.20) or (5.21). By using the relation (5.24), we can check that $U_b^{(\lambda_0)}(i_{g\dagger}\mathcal{M}[*I!J \star K])$ are ${}^{\prime}V_0\mathcal{R}_{X\times\mathbb{C}_t}$-modules. Let $\boldsymbol{a}(I, J)$ be as in Sect. 5.3. We have naturally defined morphisms:

$$\mathcal{Q}_{\boldsymbol{a}(I,J)+b\boldsymbol{p}+\boldsymbol{\delta}_K}^{(\lambda_0)}\mathcal{M} \longrightarrow {}^K V_{b\boldsymbol{p}}^{(\lambda_0)}\mathcal{M}[*I!J] \longrightarrow U_b^{(\lambda_0)}(i_{g\dagger}\mathcal{M}[*I!J * K]) \tag{5.25}$$

We can check that the image of (5.25) generates $U_b^{(\lambda_0)}(i_{g\dagger}\mathcal{M}[*I!J * K])$ over ${}^{\prime}V_0\mathcal{R}_{X\times\mathbb{C}_t}$ by using (5.24). Then, we can deduce the coherence. (See the last argument in the proof of Proposition 12.3.3 of [55], for example.)

Let us prove the second claim in the case $\star = *$. Put $\mathcal{P} := \bigcup_{b\in\mathbb{R}} U_b^{(\lambda_0)}$. By construction, it is an $\mathcal{R}_{X\times\mathbb{C}_t}$-submodule of $i_{g\dagger}\mathcal{M}[*I!J * K]$. We have $\mathcal{P} \supset {}^K V_0^{(\lambda_0)}\mathcal{M}[*I!J] \otimes 1$ by the assumption. Suppose $u \otimes \eth_t^j \in \mathcal{P}$. By the first in (5.22), we have $(\partial_i g)u \otimes \eth_t^j \in \mathcal{P}$. By the second formulas in (5.22) and (5.23), we obtain $(\eth_i u) \otimes \eth_t^j \in \mathcal{P}$. Because ${}^K V_0^{(\lambda_0)}\mathcal{M}[*I!J]$ generates $\mathcal{M}[*I!J]$ over $\mathcal{R}_{X,K}$, we obtain that $\mathcal{M}[*I!J * K] \otimes 1 \subset \mathcal{P}$, which implies $\mathcal{P} = i_{g\dagger}\mathcal{M}[*I!J * K]$. The case $\star =!$ can be argued similarly. □

We set $\mathrm{KMS}(\mathcal{M}, i) := \{u \in \mathbb{R}\times\mathbb{C} \mid {}^i\mathcal{G}_u(\mathcal{M}) \neq 0\}$. For $b \in \mathbb{R}$, we set

$$\mathcal{K}(b, \lambda_0) := \bigcup_{i\in K} \{v \in \mathbb{R}\times\mathbb{C} \mid p_i v \in \mathrm{KMS}(\mathcal{M}, i),\ \mathfrak{p}(\lambda_0, v) = b\}.$$

Lemma 5.4.5 *The filtration $U^{(\lambda_0)}$ is monodromic. Namely, the induced endomorphism $\prod_{u\in\mathcal{K}(b,\lambda_0)}(-\eth_t t + \mathfrak{e}(\lambda, u))$ is nilpotent on $U_b^{(\lambda_0)}/U_{<b}^{(\lambda_0)}$.*

Proof (See §16.1 of [52]) By the relation (5.24), for $s \in V_{b\boldsymbol{p}}^{(\lambda_0)}\mathcal{M}[*I!J]$ and for $i \in K$, we have

$$\begin{aligned}\big(-\eth_t t + \mathfrak{e}(\lambda, u)\big)(s \otimes 1) - p_i^{-1}\big((-\eth_i z_i + \mathfrak{e}(\lambda, p_i u))s\big) \otimes 1 \\ = p_i^{-1}\eth_i\big((z_i s) \otimes 1\big) \in U_{<b}^{(\lambda_0)}.\end{aligned} \quad (5.26)$$

Then, we can check the claim easily. □

By construction, we have $t \cdot U_b^{(\lambda_0)} = U_{b-1}^{(\lambda_0)}$ for $b < 0$, and $\eth_t : \mathrm{Gr}_c^{U^{(\lambda_0)}} \longrightarrow \mathrm{Gr}_{c+1}^{U^{(\lambda_0)}}$ is surjective for $c > -1$. If $\star = *$, we also have $t \cdot U_0^{(\lambda_0)} = U_{-1}^{(\lambda_0)}$. Hence, we have only to prove that (1) $\mathrm{Gr}_b^{U^{(\lambda_0)}}$ are strict for $b < 0$, (2) $\eth_t : \mathrm{Gr}_{-1}^{U^{(\lambda_0)}} \longrightarrow \mathrm{Gr}_0^{U^{(\lambda_0)}}$ is injective in the case $\star =!$, which we shall consider in the following.

5.4.4 Regular and Pure Case

Let us consider the regular and pure case, i.e., $\mathcal{M}$ comes from a tame harmonic bundle. Let $I \sqcup J \sqcup K \subset \underline{\ell}$. For $b \in \mathbb{R}$ and $\boldsymbol{c} \in \mathbb{R}^I$, we consider

$$\begin{aligned}{}^I V_{\boldsymbol{c}}^{(\lambda_0)} U_b^{(\lambda_0)}\big(\iota_{g\dagger}\mathcal{M}[!J]\big) &:= \mathcal{R}_{X,K}\Big({}^I V_{\boldsymbol{c}}^{(\lambda_0)}\, {}^K V_{b\boldsymbol{p}}^{(\lambda_0)}\mathcal{M}[!J] \otimes 1\Big) \\ &= \mathcal{R}_{X,K} \cdot \Big\{\Big[\mathcal{R}_{X,J} \otimes_{V_0\mathcal{R}_{X,J}} {}^I V_{\boldsymbol{c}}^{(\lambda_0)}\, {}^J V_{<0}^{(\lambda_0)}\, {}^K V_{b\boldsymbol{p}}^{(\lambda_0)}(\mathcal{M})\Big] \otimes 1\Big\} \\ &= \mathcal{R}_{X,J} \otimes_{V_0\mathcal{R}_{X,J}} \Big\{\mathcal{R}_{X,K}\Big[{}^I V_{\boldsymbol{c}}^{(\lambda_0)}\, {}^J V_{<0}^{(\lambda_0)}\, {}^K V_{b\boldsymbol{p}}^{(\lambda_0)}(\mathcal{M})\Big] \otimes 1\Big\} \\ &=: \mathcal{R}_{X,J} \otimes_{V_0\mathcal{R}_{X,J}} \Big({}^I V_{\boldsymbol{c}}^{(\lambda_0)}\, {}^J V_{<0}^{(\lambda_0)}\, U_b^{(\lambda_0)}(\iota_{g\dagger}\mathcal{M})\Big).\end{aligned} \quad (5.27)$$

Here, ${}^{I}V^{(\lambda_0)}_{\boldsymbol{c}}{}^{K}V^{(\lambda_0)}_{\boldsymbol{bp}}\mathcal{M}[!J]$ is as in Sect. 5.3.5. We obtain the following isomorphism:

$$ {}^{I}V^{(\lambda_0)}_{\boldsymbol{c}}\,\mathrm{Gr}^{U(\lambda_0)}_{b}\big(\iota_{g\dagger}\mathcal{M}[!J]\big)\simeq\mathcal{R}_{X,J}\otimes_{V_0\mathcal{R}_{X,J}}\Big({}^{I}V^{(\lambda_0)}_{\boldsymbol{c}}{}^{J}V^{(\lambda_0)}_{<0}\,\mathrm{Gr}^{U(\lambda_0)}_{b}\big(\iota_{g\dagger}\mathcal{M}\big)\Big) $$

Let $\mathcal{M}_J$ denote the right hand side.

Lemma 5.4.6 *$\mathcal{M}_J$ is strict.*

Proof We use an induction on $|J|$. The claim follows from Corollary 16.45 of [52] in the case $|J|=0$, $\boldsymbol{c}\in\mathbb{R}^{I}_{<0}$ and $b<0$. (Note that the minimal extension of $\mathcal{M}$ is studied in §16 of [52], which is the image of $\mathcal{M}[!\underline{\ell}]\longrightarrow\mathcal{M}[*\underline{\ell}]$ in the terminology of this paper.) By the isomorphism given by the multiplication of t and z_i $(i\in I)$, we obtain the claim in the case $|J|=0$.

Let $J=J_0\sqcup\{j\}$. By the assumption of the induction, we have the strictness of $\mathcal{R}_{X,J_0}\otimes_{V_0\mathcal{R}_{X,J_0}}{}^{I}V^{(\lambda_0)}_{\boldsymbol{c}}{}^{J}V^{(\lambda_0)}_{<0}\,\mathrm{Gr}^{U(\lambda_0)}_{b}\,\iota_{g\dagger}\mathcal{M}$. We have

$$ \begin{aligned} &\mathcal{R}_{X,J_0}\otimes_{V_0\mathcal{R}_{X,J_0}}\Big({}^{I}V^{(\lambda_0)}_{\boldsymbol{c}}{}^{J_0}V^{(\lambda_0)}_{<0}{}^{j}\mathrm{Gr}^{V(\lambda_0)}_{d}\,\mathrm{Gr}^{U(\lambda_0)}_{b}\big(\iota_{g\dagger}\mathcal{M}\big)\Big)\\ &\simeq\mathcal{R}_{X,J_0}\otimes_{V_0\mathcal{R}_{X,J_0}}\Big({}^{I}V^{(\lambda_0)}_{\boldsymbol{c}}{}^{J_0}V^{(\lambda_0)}_{<0}\,\mathrm{Gr}^{U(\lambda_0)}_{b}\big(\iota_{g\dagger}{}^{j}\mathrm{Gr}^{V(\lambda_0)}_{d}\,\mathcal{M}\big)\Big), \end{aligned} \tag{5.28} $$

which is strict by the assumption of the induction.

Let us consider the filtration ${}^{j}V^{(\lambda_0)}$ of $\mathcal{M}_J$ given as follows. For $d<0$, we put

$$ {}^{j}V^{(\lambda_0)}_{d}\mathcal{M}_J:=\mathcal{R}_{X,J_0}\otimes_{V_0\mathcal{R}_{X,J_0}}{}^{I}V^{(\lambda_0)}_{\boldsymbol{c}}{}^{j}V^{(\lambda_0)}_{d}{}^{J_0}V^{(\lambda_0)}_{<0}\,\mathrm{Gr}^{U(\lambda_0)}_{b}\big(\iota_{g\dagger}\mathcal{M}\big) $$

For $d\geq 0$, we put ${}^{j}V^{(\lambda_0)}_{d}(\mathcal{M}_J):=\sum_{c,n}\eth_j^n\big({}^{j}V^{(\lambda_0)}_{c}(\mathcal{M}_J)\big)$, where (c,n) runs through $\mathbb{R}_{<0}\times\mathbb{Z}_{\geq 0}$ such that $c+n\leq b$. Note that ${}^{j}V^{(\lambda_0)}_{<0}\mathcal{M}_J$ is strict, and that ${}^{j}\mathrm{Gr}^{V(\lambda_0)}_{c}(\mathcal{M}_J)$ is strict for $c<0$, by the considerations in the previous paragraph. Let us consider the morphism

$$ {}^{j}\mathrm{Gr}^{V(\lambda_0)}_{-1}(\mathcal{M}_J)\xrightarrow{\eth_j}{}^{j}\mathrm{Gr}^{V(\lambda_0)}_{0}(\mathcal{M}_J). \tag{5.29} $$

Lemma 5.4.7 *The specializations of* (5.29) *to any generic λ are isomorphisms.*

Proof The surjectivity is clear by construction. Let $\pi:X\longrightarrow D_I$ be the projection. For $P\in D_I^{\circ}$, we can naturally regard $\mathcal{M}^{\lambda}_{J,P}:=\mathcal{M}_J\otimes\mathcal{O}_{\{\lambda\}\times\pi^{-1}(P)}$ as a $\mathscr{D}$-module. If λ is generic, we have $\mathcal{M}^{\lambda}_{J,P}[!z_j]=\mathcal{M}^{\lambda}_{J,P}$, and the specialization of ${}^{j}V^{(\lambda_0)}$ gives a V-filtration along z_j, which can be checked by a direct computation. Hence the specialization of (5.29) at a generic λ is injective. □

By Lemma 5.4.7 and the strictness of ${}^{j}\mathrm{Gr}^{V(\lambda_0)}_{-1}\mathcal{M}_J$, we obtain that (5.29) is injective, and hence an isomorphism. In particular, ${}^{j}\mathrm{Gr}^{V(\lambda_0)}_{0}(\mathcal{M}_J)$ is strict. For $b>-1$, the morphism $\eth_j:{}^{j}\mathrm{Gr}^{V(\lambda_0)}_{b}(\mathcal{M}_J)\longrightarrow{}^{j}\mathrm{Gr}^{V(\lambda_0)}_{b+1}(\mathcal{M}_J)$ is surjective, and

$z_j\eth_j$ on ${}^j\mathrm{Gr}_b^{V^{(\lambda_0)}}(\mathcal{M}_J)$ is injective. Hence, $\eth_j$ is an isomorphism. We also have the strictness of ${}^j\mathrm{Gr}_b^{V^{(\lambda_0)}}(\mathcal{M}_J)$. Thus, we obtain the claim in the case of J, and the proof of Lemma 5.4.6 is finished. □

Corollary 5.4.8 *Suppose that $\mathcal{M}$ comes from a tame harmonic bundle.*

- *For any $J \subset \underline{\ell} \setminus K$, $\mathrm{Gr}_b^{U^{(\lambda_0)}}(\iota_{g\dagger}\mathcal{M}[!J])$ is strict, i.e., $\mathcal{M}[!J]$ is strictly specializable along g. The V-filtration is given in the standard way as in Proposition 5.4.3.*
- *For $I \sqcup J \subset \underline{\ell} \setminus K$, $\mathcal{M}[*I!J]$ is strictly specializable along g. The V-filtration is given in the standard way as in Proposition 5.4.3. (Note that $\mathcal{M}[*I!J] \subset \mathcal{M}[!J]$.)*

□

Let $I \sqcup J \sqcup K = \underline{\ell}$, and we prove the claim of Proposition 5.4.3 in the case that $\mathcal{M}$ comes from a tame harmonic bundle. Let us consider $\iota_{g\dagger}\mathcal{M}[*I!J*K]$. We have already known the strictness of $\mathrm{Gr}_b^{U^{(\lambda_0)}}$ for $b<0$. Because $t:\mathrm{Gr}_0^{U^{(\lambda_0)}} \longrightarrow \mathrm{Gr}_{-1}^{U^{(\lambda_0)}}$ is an isomorphism, $\mathrm{Gr}_0^{U^{(\lambda_0)}}$ is also strict. By the standard argument, we obtain the strictness of $\mathrm{Gr}_b^{U^{(\lambda_0)}}$ for $b>0$. Hence, we obtain the claim for $\iota_{g\dagger}\mathcal{M}[*I!J*K]$. Let us consider $\iota_{g\dagger}\mathcal{M}[*I!J!K]$. For $b<0$, we have already known that $\mathrm{Gr}_b^{U^{(\lambda_0)}}$ are strict. Let us consider the specialization to any generic λ. We have

$$\big(\mathcal{M}^\lambda[*I!J!K]\big)[!g] = \mathcal{M}^\lambda[*I!J!K],$$

and the specialization of $U^{(\lambda_0)}$ gives a V-filtration. Hence, we obtain that $\eth_t:\mathrm{Gr}_{-1|\lambda}^{U^{(\lambda_0)}} \longrightarrow \mathrm{Gr}_{0|\lambda}^{U^{(\lambda_0)}}$ are isomorphisms for generic λ. Because $\mathrm{Gr}_{-1}^{U^{(\lambda_0)}}$ is strict, we obtain that $\eth_t:\mathrm{Gr}_{-1}^{U^{(\lambda_0)}} \longrightarrow \mathrm{Gr}_0^{U^{(\lambda_0)}}$ is an isomorphism. In particular, $\mathrm{Gr}_0^{U^{(\lambda_0)}}$ is strict. Then, by the standard argument, we obtain that $\mathrm{Gr}_b^{U^{(\lambda_0)}}$ are strict for any b. Hence, we obtain the claim for $\mathcal{M}[*I!J!K]$.

5.4.5 Regular and Filtered Case

Let L be the filtration of $\mathcal{M}$ as in the condition (A) in Sect. 5.4.1. Let us consider the case that $\mathrm{Gr}^L\mathcal{M}$ comes from a tame harmonic bundle. In this case, we obtain the claim of Proposition 5.4.3 from the following general lemma with an easy induction.

Lemma 5.4.9 *Let $0 \longrightarrow \mathcal{N}_1 \longrightarrow \mathcal{N}_2 \longrightarrow \mathcal{N}_3 \longrightarrow 0$ be an exact sequence. They are equipped with monodromic filtrations $U^{(\lambda_0)}$, which are preserved by morphisms. Assume the following.*

- *$\mathcal{N}_i$ $(i=1,3)$ are strictly specializable with $U^{(\lambda_0)}(\mathcal{N}_i)$.*
- *$\mathcal{N}_2 \longrightarrow \mathcal{N}_3$ is strict with respect to $U^{(\lambda_0)}$.*

Then, $\mathcal{N}_2$ is also strictly specializable with $U^{(\lambda_0)}$.

Proof We have only to prove that $\mathcal{N}_1 \longrightarrow \mathcal{N}_2$ is strict with respect to $U^{(\lambda_0)}$. Because the restriction of $U^{(\lambda_0)}(\mathcal{N}_2)$ to $\mathcal{N}_1$ is also monodromic, we obtain $U_b^{(\lambda_0)}(\mathcal{N}_1) \supset U_b^{(\lambda_0)}(\mathcal{N}_2) \cap \mathcal{N}_1$ by a general result (see Lemma 14.23 of [52]). Thus, we obtain Lemma 5.4.9. □

5.4.6 Good Irregular Case with Unique Irregular Value

We consider a ramified covering $\pi : (X', D') \longrightarrow (X, D)$ given by

$$\pi(z_1, \ldots, z_n) = (z_1^{e_1}, \ldots, z_k^{e_k}, z_{k+1}, \ldots, z_n).$$

Let $\mathfrak{a}$ be a meromorphic function on (X', D') such that $\mathfrak{a} = \boldsymbol{z}^{\boldsymbol{m}}\mathfrak{a}_1$ for some holomorphic function $\mathfrak{a}_1$ with $\mathfrak{a}_1(O) \neq 0$ and $\boldsymbol{m} \in \mathbb{Z}^k_{<0}$. Let $\mathcal{M}'$ be a good-KMS smooth $\mathcal{R}_{X(*D)}$-module satisfying the condition in Sect. 5.4.5. Let us consider the case that $\mathcal{M}$ is obtained as $\mathcal{M}' \otimes \pi_*\mathcal{L}(\mathfrak{a})$. By using the characterization of $\mathcal{M}[*I!J \star K]$ in Proposition 5.3.1, we have a natural isomorphism $\mathcal{M}[*I!J \star K] \simeq \mathcal{M}'[*I!J \star K] \otimes \pi_*\mathcal{L}(\mathfrak{a})$. Let us observe

$$^{K}V^{(\lambda_0)}_{b\boldsymbol{p}}\big(\mathcal{M}[*I!J \star K]\big) = {}^{K}V^{(\lambda_0)}_{b\boldsymbol{p}}(\mathcal{M}'[*I!J \star K]) \otimes \pi_*\mathcal{L}(\mathfrak{a}). \tag{5.30}$$

Both of them are $^{K}V_0\mathcal{R}_X$-submodules, and contains $\mathcal{Q}_{\boldsymbol{a}(I,J)+b\boldsymbol{p}+\boldsymbol{\delta}_K}$. The left hand side is generated by $\mathcal{Q}_{\boldsymbol{a}(I,J)+b\boldsymbol{p}+\boldsymbol{\delta}_K}$. Hence, the right hand side contains the left hand side. Let $\mathcal{N} \subset \mathcal{M}'[*I!J \star K]$ be a $V_0\mathcal{R}_X$-submodule. Suppose that $\mathcal{N} \otimes 1$ is contained in the left hand side. By considering the action of $z_i\eth_i$ $(1 \leq i \leq k)$, we obtain that $\mathcal{N} \otimes \pi_*\mathcal{L}(\mathfrak{a})$ is also contained in the left hand side. Let $F_*(^{K}V_0\mathcal{R}_X)$ be the filtration given by the order of differential operators. We set $\mathcal{N}_m := F_m(^{K}V_0\mathcal{R}_X)\mathcal{Q}_{\boldsymbol{a}(I,J)+b\boldsymbol{p}+\boldsymbol{\delta}_K}$ in $\mathcal{M}'[*I!J \star K]$, which are $V_0\mathcal{R}_X$-submodules. If $\mathcal{N}_m$ is contained in the left hand side, we obtain that $\mathcal{N}_m \otimes \pi_*\mathcal{L}(\mathfrak{a})$ is also contained in the left hand side as remarked above. Then, by a formal computation, we obtain that $\mathcal{N}_{m+1}$ is contained in the left hand side. Hence, by using an induction, we obtain that $\big(\bigcup_m \mathcal{N}_m\big) \otimes \pi_*\mathcal{L}(\mathfrak{a})$ is contained in the left hand side. Because $\bigcup_m \mathcal{N}_m = {}^{K}V^{(\lambda_0)}_{b\boldsymbol{p}}(\mathcal{M}'[*I!J \star K])$, we obtain (5.30).

Let $q : X \times \mathbb{C}_t \longrightarrow X$ be the projection. We have a natural isomorphism

$$i_{g\dagger}\mathcal{M}[*I!J \star K] \simeq i_{g\dagger}\mathcal{M}'[*I!J \star K] \otimes q^*\big(\pi_*\mathcal{L}(\mathfrak{a})\big).$$

Let us observe

$$U_b^{(\lambda_0)}\big(i_{g\dagger}\mathcal{M}[*I!J \star K]\big) \simeq U_b^{(\lambda_0)}\big(i_{g\dagger}\mathcal{M}'[*I!J \star K]\big) \otimes q^*\big(\pi_*\mathcal{L}(\mathfrak{a})\big). \tag{5.31}$$

Both of them are ${}^{t}V_0\mathcal{R}_{X\times\mathbb{C}_t}$-modules, and contain $i_{g*}\big({}^{K}V^{(\lambda_0)}_{\boldsymbol{bp}}(\mathcal{M}[*I!J\star K])\big)$. Because the left hand side is generated by $i_{g*}\big({}^{K}V^{(\lambda_0)}_{\boldsymbol{bp}}(\mathcal{M}[*I!J\star K])\big)$, it is contained in the right hand side. Let $\mathcal{N}\subset U^{(\lambda_0)}_{b}\big(i_{g\dagger}\mathcal{M}'[*I!J\star K]\big)$ be a $V_0\mathcal{R}_X$-submodule. Suppose that $\mathcal{N}$ is contained in the left hand side. By considering the action of $z_i\eth_i$ $(i=1,\dots,k)$, we obtain that $\mathcal{N}\otimes q^*\pi_*(\mathcal{L})$ is contained in the left hand side. Let $F_*(\mathcal{R}_{X,K})$ be the filtration given by the order of differential operators. We set $\mathcal{N}_m:=F_m(\mathcal{R}_{X,K})\cdot i_{g*}\big({}^{K}V^{(\lambda_0)}_{\boldsymbol{bp}}(\mathcal{M}'[*I!J\star K])\big)$ in $U^{(\lambda_0)}_{b}\big(i_{g\dagger}\mathcal{M}'[*I!J\star K]\big)$. We have $\mathcal{N}_0\otimes\pi^*\mathcal{L}(\mathfrak{a})\simeq i_{g*}\big({}^{K}V^{(\lambda_0)}_{\boldsymbol{bp}}(\mathcal{M}[*I!J\star K])\big)$, which is contained in the left hand side. Then, by an easy induction, we obtain that $\mathcal{N}_m\otimes q^*\pi_*\mathcal{L}(\mathfrak{a})$ is contained in the left hand side. Because $\bigcup_m\mathcal{N}_m=U^{(\lambda_0)}_{b}\big(i_{g\dagger}\mathcal{M}'[*I!J\star K]\big)$, we obtain (5.31).

The claim of Proposition 5.4.3 in this case immediately follows from (5.31).

5.4.7 End of the Proof of Proposition 5.4.3

By using the formal completion as in §12.4 of [55], we obtain the claims of Proposition 5.4.3 in the general case. □

5.5 Good-KMS Smooth $\mathcal{R}$-Triple

Let X be a complex manifold with a simple normal crossing hypersurface D. Let $\mathcal{T}=(\mathcal{M}_1,\mathcal{M}_2,C)$ be a smooth $\mathcal{R}_{X(*D)}$-triple. It is called good(-KMS) if $\mathcal{M}_i$ are good(-KMS). We use the other adjectives "unramifiedly good-KMS", "regular-KMS" with similar meanings. For simplicity, we assume the following:

- Let X_P be any small neighbourhood of $P\in X$ with a ramified covering $\varphi_P:(X'_P,D'_P)\longrightarrow(X_P,D\cap X_P)$ such that $\varphi_P^*(\mathcal{T})$ is unramified. Then, $\mathrm{Irr}(\mathcal{T},P):=\mathrm{Irr}(\mathcal{M}_1,P)\cup\mathrm{Irr}(\mathcal{M}_2,P)$ is a good set of irregular values.

Let $D=\bigcup_{i\in\Lambda}D_i$ be the irreducible decomposition. For $I\subset\Lambda$, we set $\mathrm{KMS}(\mathcal{T},I):=\mathrm{KMS}(\mathcal{M}_1,I)\cup\mathrm{KMS}(\mathcal{M}_2,I)$ which is called the KMS-spectrum of $\mathcal{T}$ along D_I.

5.5.1 Reduction with Respect to Stokes Structure

Let $X=\Delta^n$ and $D=\bigcup_{i=1}^{\ell}\{z_i=0\}$. Let $\mathcal{T}$ be a good-KMS smooth $\mathcal{R}_{X(*D)}$-triple. Let $I\subset\underline{\ell}$. We have the induced good-KMS smooth $\mathcal{R}_{X(*D)}$-modules ${}^{I}\mathrm{Gr}^{\mathrm{St}}(\mathcal{M}_i)$ as in Sect. 5.1.5. Let us observe that we have the induced sesqui-linear pairing ${}^{I}\mathrm{Gr}^{\mathrm{St}}(C)$

of ${}^{I}\mathrm{Gr}^{\mathrm{St}}(\mathcal{M}_1)$ and ${}^{I}\mathrm{Gr}^{\mathrm{St}}(\mathcal{M}_2)$. We will shrink X around the origin in the following argument.

We give a preliminary. Let $\pi : \tilde{X}(D) \longrightarrow X$ be the real blow up of X along D. The induced morphism $\mathbb{C}_\lambda \times \tilde{X}(D) \longrightarrow \mathbb{C}_\lambda \times X$ is also denoted by π. Let $U \subset \boldsymbol{S} \times \tilde{X}(D)$ be any open subset. Let f be a C^∞-function on $U \setminus \big(\boldsymbol{S} \times \pi^{-1}(D)\big)$. We say that f is a C^∞-function on U with moderate growth along D if for any real differential operator P on U there exists $N > 0$ such that $|Pf| = O\big(\prod_{i=1}^{\ell} |z_i|^{-N}\big)$ around any point of $U \cap \big(\boldsymbol{S} \times \pi^{-1}(D)\big)$. Let $\mathcal{C}^{\infty \bmod D}_{\boldsymbol{S}\times\tilde{X}(D)}(U)$ denote the space of such C^∞-functions on U with moderate growth along D. We obtain a sheaf $\mathcal{C}^{\infty \bmod D}_{\boldsymbol{S}\times\tilde{X}(D)}$ on $\boldsymbol{S} \times X$.

The projection $\mathbb{C}_\lambda \times \tilde{X}(D) \longrightarrow \mathcal{X}$ is also denoted by π. Let U be any open subset in $\mathbb{C}_\lambda \times \tilde{X}(D)$. A C^∞-function g on U is called holomorphic if $g_{|U\setminus\pi^{-1}(\mathcal{D})}$is holomorphic. Let $\mathcal{O}_{\tilde{X}(D)}$ denote the sheaf of holomorphic functions on $\tilde{X}(D)$. For any $\mathcal{O}_{\mathcal{X}}$-module $\mathcal{N}$, let $\pi^*\mathcal{N} := \mathcal{O}_{\mathbb{C}_\lambda\times\tilde{X}(D)} \otimes_{\pi^{-1}\mathcal{O}_{\mathcal{X}}} \pi^{-1}\mathcal{N}$.

Suppose that $\mathcal{T}$ is unramifiedly good-KMS. We have the induced pairing

$$\tilde{C} : \big(\pi^*\mathcal{M}_1\big)_{|\boldsymbol{S}\times\tilde{X}(D)} \times \big(\sigma^*\pi^*\mathcal{M}_2\big)_{|\boldsymbol{S}\times\tilde{X}(D)} \longrightarrow \mathcal{C}^{\infty \bmod D}_{\boldsymbol{S}\times\tilde{X}(D)}.$$

Let $\lambda_0 \in \boldsymbol{S}$ and $Q \in \pi^{-1}(D)$. We have the full Stokes filtration $\tilde{\mathcal{F}}^{(\lambda_0,Q)}$ (resp. $\tilde{\mathcal{F}}^{(-\lambda_0,Q)}$) of $\pi^*\mathcal{M}_1$ (resp. $\pi^*\mathcal{M}_2$) at (λ_0, Q) (resp. $(-\lambda_0, Q)$). We obtain the following lemma, by considering the growth order.

Lemma 5.5.1 *The restriction of $\tilde{C}$ to $\tilde{\mathcal{F}}^{(\lambda_0,Q)}_{\mathfrak{a}} \times \sigma^*\tilde{\mathcal{F}}^{(-\lambda_0,Q)}_{\mathfrak{b}}$ vanishes, unless* $\mathrm{Re}(\mathfrak{a}/\lambda_0) - \mathrm{Re}(\mathfrak{b}/\lambda_0) \leq 0$. □

By shrinking X around the origin O, we obtain the pairing

$${}^{I}\mathrm{Gr}^{\mathrm{St}}(C) : \pi^*\big({}^{I}\mathrm{Gr}^{\mathrm{St}}(\mathcal{M}_1)\big)_{|\boldsymbol{S}\times\tilde{X}(D)} \times \sigma^*\pi^*\big({}^{I}\mathrm{Gr}^{\mathrm{St}}(\mathcal{M}_2)\big)_{|\boldsymbol{S}\times\tilde{X}(D)} \to \mathcal{C}^{\infty \bmod D}_{\boldsymbol{S}\times\tilde{X}(D)}.$$

We obtain the induced pairing ${}^{I}\mathrm{Gr}^{\mathrm{St}}(C)$ of ${}^{I}\mathrm{Gr}^{\mathrm{St}}(\mathcal{M}_i)$ $(i = 1,2)$. It is easy to observe that ${}^{I}\mathrm{Gr}^{\mathrm{St}}(C)_{|\boldsymbol{S}\times(X\setminus D)}$ is extended to a pairing on $\mathbb{C}^*_\lambda \times (X - D)$ which is holomorphic with respect to λ. The tuple of ${}^{I}\mathrm{Gr}^{\mathrm{St}}(\mathcal{M}_i)$ $(i = 1,2)$ with ${}^{I}\mathrm{Gr}^{\mathrm{St}}(C)$ is denoted by ${}^{I}\mathrm{Gr}^{\mathrm{St}}(\mathcal{T})$, which is an unramifiedly good-KMS smooth $\mathcal{R}_{X(*D)}$-triple. We have the natural grading ${}^{I}\mathrm{Gr}^{\mathrm{St}}(\mathcal{T}) = \bigoplus_{\mathfrak{a}\in\mathrm{Irr}(\mathcal{T},I)} {}^{I}\mathrm{Gr}^{\mathrm{St}}_{\mathfrak{a}}(\mathcal{T})$, and we put ${}^{I}\mathrm{Gr}^{\mathrm{St,reg}}(\mathcal{T}) = {}^{I}\mathrm{Gr}^{\mathrm{St}}_0(\mathcal{T})$ and ${}^{I}\mathrm{Gr}^{\mathrm{St,irr}}(\mathcal{T}) = \bigoplus_{\mathfrak{a}\neq 0} {}^{I}\mathrm{Gr}^{\mathrm{St}}_{\mathfrak{a}}(\mathcal{T})$.

Let us consider the case that $\mathcal{T}$ is not necessarily unramified. We take a ramified covering $\varphi : (X', D') \longrightarrow (X, D)$ such that $\varphi^*\mathcal{T}$ is unramified. By applying the above procedure to each good summand, we obtain the decomposition ${}^{I}\mathrm{Gr}^{\mathrm{St}}(\mathcal{T}') = {}^{I}\mathrm{Gr}^{\mathrm{St,reg}}(\mathcal{T}') \oplus {}^{I}\mathrm{Gr}^{\mathrm{St,irr}}(\mathcal{T}')$, on which the Galois group of the ramified covering naturally acts. As the descent, we obtain a good-KMS smooth $\mathcal{R}_{X(*D)}$-triple with the decomposition:

$${}^{I}\mathrm{Gr}^{\mathrm{St}}(\mathcal{T}) = {}^{I}\mathrm{Gr}^{\mathrm{St,reg}}(\mathcal{T}) \oplus {}^{I}\mathrm{Gr}^{\mathrm{St,irr}}(\mathcal{T})$$

5.5.2 Specialization

Let X and D be as in Sect. 5.5.1. Let $\mathcal{T}$ be a good-KMS smooth $\mathcal{R}_{X(*D)}$-triple. Because $\mathcal{T}$ is strictly specializable along z_i as an $\mathcal{R}_{X(*D)}$-triple, we obtain an $\mathcal{R}_{D_i(*\partial D_i)}$-triple ${}^i\tilde{\psi}_u(\mathcal{T}) := \tilde{\psi}_{z_i,u}(\mathcal{T})$ for any $u \in \mathbb{R}\times\mathbb{C}$ (Sect. 2.1.4).

Lemma 5.5.2 *${}^i\tilde{\psi}_u(\mathcal{T})$ are good-KMS smooth $\mathcal{R}_{D_i(*\partial D_i)}$-triples. We have natural isomorphisms ${}^i\tilde{\psi}_u{}^I\mathrm{Gr}^{\mathrm{St}}(\mathcal{T}) \simeq {}^{I\setminus\{i\}}\mathrm{Gr}^{\mathrm{St}}\,{}^i\tilde{\psi}_u(\mathcal{T})$.*

Proof By using Lemma 22.11.2 of [55], we can reduce the issue to the unramified case. Let $\mathrm{Irr}(\mathcal{M}_k)$ be the set of irregular values of $\mathcal{M}_k$. For each $\mathfrak{a} \in \mathrm{Irr}(\mathcal{M}_k)$, we have $\boldsymbol{m}(\mathfrak{a}) \in \mathbb{Z}^{\ell}_{\leq 0}$ such that we have $\mathfrak{a} = \boldsymbol{z}^{\boldsymbol{m}(\mathfrak{a})}\mathfrak{a}_1$ for a holomorphic function $\mathfrak{a}_1$ such that $\mathfrak{a}_1(0) \neq 0$. Here, for $\boldsymbol{m} = (m_1,\dots,m_\ell) \in \mathbb{Z}^\ell$, we set $\boldsymbol{z}^{\boldsymbol{m}} = \prod_{i=1}^{\ell} z_i^{m_i}$. For each $\mathfrak{a} \in \mathrm{Irr}(\mathcal{M}_k)$, set $\mathfrak{s}(\mathfrak{a}) := \bigl\{p \bigm| m_p(\mathfrak{a}) \neq 0\bigr\}$. Because $\mathrm{Irr}(\mathcal{M}_1) \cup \mathrm{Irr}(\mathcal{M}_2)$ is a good set of irregular values, we may assume that for each $\mathfrak{a}$ there exists $p(\mathfrak{a})$ such that $\mathfrak{s}(\mathfrak{a}) = \{1,\dots,p(\mathfrak{a})\}$.

Let $j_0 := \min I$. If $j_0 \geq i$, we have natural isomorphisms ${}^i\tilde{\psi}_u{}^I\mathrm{Gr}^{\mathrm{St}}(\mathcal{M}_k) \simeq {}^i\tilde{\psi}_u(\mathcal{M}_k) \simeq {}^{I\setminus\{i\}}\mathrm{Gr}^{\mathrm{St}}\,{}^i\tilde{\psi}_u(\mathcal{M}_k)$. We can compare the induced pairings ${}^i\tilde{\psi}_u(C)$ and ${}^i\tilde{\psi}_u{}^I\mathrm{Gr}^{\mathrm{St}}(C)$ by using the argument in Proposition 12.7.1 of [55]. Let us consider the case $j_0 < i$. By using the claim in the case $j_0 \geq i$, we may assume that $\mathcal{M}_k$ are regular along D_i. By using the compatibility of the KMS-structure along D_i and the Stokes structure, which follows from Proposition 3.3.10 of [55], we obtain natural isomorphisms ${}^i\tilde{\psi}_u{}^I\mathrm{Gr}^{\mathrm{St}}(\mathcal{M}_k) \simeq {}^{I\setminus\{i\}}\mathrm{Gr}^{\mathrm{St}}\,{}^i\tilde{\psi}_u(\mathcal{M}_k)$. The comparison of ${}^i\tilde{\psi}_u{}^I\mathrm{Gr}^{\mathrm{St}}(C)$ and ${}^{I\setminus\{i\}}\mathrm{Gr}^{\mathrm{St}}\,{}^i\tilde{\psi}_u(C)$ can be done directly from the construction. Thus, we obtain the second claim.

Let us prove the first claim. By using the second claim with $I = \{i\}$, we may assume that $\mathcal{T}$ is regular along D_i. We set $i^c := \{i\}^c$. We consider the issue on $X \setminus D(i^c)$ for a while. By the regularity along D_i, we obtain the pairing C of $\mathcal{M}_{1|\mathbb{C}^*_\lambda\times(X\setminus D(i^c))}$ and $\sigma^*\mathcal{M}_{2|\mathbb{C}^*_\lambda\times(X\setminus D(i^c))}$ taking values in the sheaf of C^∞-functions of moderate growth along D_i on $\mathbb{C}^*_\lambda\times(X\setminus D(i^c))$ which are holomorphic with respect to λ. Hence, we obtain that ${}^i\tilde{\psi}_u(\mathcal{T})_{|D_i\setminus\partial D_i}$ is a smooth $\mathcal{R}_{D_i(*\partial D_i)}$-triple. The underlying $\mathcal{R}_{D_i(*\partial D_i)}$-modules are good-KMS smooth. We obtain the growth estimate around ∂D_i directly by construction, or by using Proposition 5.5.7 below. □

Lemma 5.5.3 *We have a natural isomorphism ${}^i\tilde{\psi}_{u_i}{}^j\tilde{\psi}_{u_j}(\mathcal{T}) \simeq {}^j\tilde{\psi}_{u_j}{}^i\tilde{\psi}_{u_i}(\mathcal{T})$.*

Proof By using the previous lemma, we can reduce the issue to the case that $\mathcal{T}$ is regular along $D_i \cup D_j$. We have only to compare ${}^i\tilde{\psi}_{u_i}{}^j\tilde{\psi}_{u_j}C$ and ${}^j\tilde{\psi}_{u_j}{}^i\tilde{\psi}_{u_i}C$ around generic λ_0. It can be done by a direct computation. Indeed, by using the regularity along $D_i \cup D_j$ and the genericity of λ_0, we have an expression

$$C(m_1,\sigma^*m_2) = \sum a_{u_i,u_j,k_i,k_j}(\lambda,z_1,\dots,z_n)|z_i|^{\mathfrak{e}(\lambda,u_i)}|z_j|^{\mathfrak{e}(\lambda,u_j)}(\log|z_i|^2)^{k_i}(\log|z_j|^2)^{k_j} \tag{5.32}$$

Here, a_{u_i,u_j,k_i,k_j} are C^∞ with respect to (z_i, z_j). Both ${}^i\tilde{\psi}_{u_i}{}^j\tilde{\psi}_{u_j}C([m_1],[m_2])$ and ${}^j\tilde{\psi}_{u_j}{}^i\tilde{\psi}_{u_i}C([m_1],[m_2])$ are equal to $\big(a_{-\boldsymbol{\delta},-\boldsymbol{\delta},0,0}\big)_{|(z_i,z_j)=(0,0)}$ up to a constant term. □

Remark 5.5.4 We can also obtain the commutativity in Lemma 5.5.3 by using the Beilinson construction. (See Sect. 5.5.4 below.) □

For any $I \subset \underline{\ell}$ and $\boldsymbol{u} \in (\mathbb{R}\times\mathbb{C})^I$, we obtain ${}^I\tilde{\psi}_{\boldsymbol{u}}(\mathcal{T}) := {}^{i_1}\tilde{\psi}_{u_1} \circ \cdots \circ {}^{i_m}\tilde{\psi}_{u_m}(\mathcal{T})$. It is equipped with nilpotent morphisms $\mathcal{N}_j : {}^I\tilde{\psi}_{\boldsymbol{u}}(\mathcal{T}) \longrightarrow {}^I\tilde{\psi}_{\boldsymbol{u}}(\mathcal{T}) \otimes \boldsymbol{T}(-1)$ $(j \in I)$. We have a natural isomorphism ${}^I\tilde{\psi}_{\boldsymbol{u}}(\mathcal{T}) \simeq {}^I\tilde{\psi}_{\boldsymbol{u}}{}^I\mathrm{Gr}^{\mathrm{St}}(\mathcal{T}) \simeq {}^I\tilde{\psi}_{\boldsymbol{u}}{}^I\mathrm{Gr}^{\mathrm{St}}_0(\mathcal{T})$. If $\mathcal{T}$ is unramified, for each $\mathfrak{a} \in \mathrm{Irr}(\mathcal{T}, I)$, we also obtain a good-KMS smooth $\mathcal{R}_{D_I(*\partial D_I)}$-triple ${}^I\tilde{\psi}_{\mathfrak{a},\boldsymbol{u}}(\mathcal{T}) := {}^I\tilde{\psi}_{\boldsymbol{u}}\big(\mathcal{T} \otimes L(-\tilde{\mathfrak{a}})\big)$. Note that it depends on the choice of a lift $\tilde{\mathfrak{a}} \in M(X,D)$ of $\mathfrak{a} \in M(X,D)/H(X)$.

5.5.3 *Canonical Prolongations*

Let X and D be as in Sect. 5.3.1. Let $\mathcal{T} = (\mathcal{M}_1, \mathcal{M}_2, C)$ be a good-KMS smooth $\mathcal{R}_{X(*D)}$-triple. By applying the procedure in Sect. 3.2 inductively, we obtain a uniquely determined pairing $C[*I!J]$ of $\mathcal{M}_1[!I * J]$ and $\mathcal{M}_2[*I!J]$. Thus, we obtain an $\mathcal{R}$-triple

$$\mathcal{T}[*I!J] := \big(\mathcal{M}_1[!I * J],\ \mathcal{M}_2[*I!J],\ C[*I!J]\big).$$

We obtain the following lemma by the property of the canonical prolongments.

Lemma 5.5.5 *Let X and D be as in Sect. 5.5.1. Recall that we set* $I_{\cup i} := I \cup \{i\}$ *and* $I_{\setminus i} := I \setminus \{i\}$.

- $\mathcal{T}[*I!J]$ *is strictly specializable along* z_i*, and we have the following:*

$$\mathcal{T}[*I!J][*z_i] \simeq \mathcal{T}[*I_{\cup i}!J_{\setminus i}], \quad \mathcal{T}[*I!J][!z_i] \simeq \mathcal{T}[*I_{\setminus i}!J_{\cup i}].$$

- *The following canonical morphisms are isomorphisms:*

$$\phi^{(0)}_{z_i}\mathcal{T}[*I!J] \longrightarrow \psi^{(0)}_{z_i}\mathcal{T}[*I!J], \quad (i \in I),$$
$$\psi^{(1)}_{z_i}\mathcal{T}[*I!J] \longrightarrow \phi^{(0)}_{z_i}\mathcal{T}[*I!J], \quad (i \in J).$$

- *For* $u \in (\mathbb{R}\times\mathbb{C}) \setminus (\mathbb{Z}_{\geq 0} \times \{0\})$*, we naturally have* ${}^i\tilde{\psi}_u\big(\mathcal{T}[*I!J]\big) \simeq \big({}^i\tilde{\psi}_u(\mathcal{T})\big)[*I_{\setminus i}!J_{\setminus i}]$. □

5.5.4 Variant of Beilinson Functors

Let $X = \Delta^n$, $D_i := \{z_i = 0\}$ and $D = \bigcup_{i=1}^{\ell} D_i$. We take $K \sqcup J \sqcup I \sqcup A \sqcup B = L \subset \underline{\ell}$. For any function $f : K \sqcup J \longrightarrow \{0,1\}$, we put $K_i(f) := f^{-1}(i) \cap K$ for $i = 0, 1$. Let $\mathcal{T}_I$ be a good-KMS smooth $\mathcal{R}_{D_I(*\partial D_I)}$-triple. We set $\mathbb{I}_g^{-\infty,a} := \varinjlim \mathbb{I}_g^{-N,a}$. For $\boldsymbol{a} = (\boldsymbol{a}_K, \boldsymbol{a}_J) \in \mathbb{R}^{K \sqcup J}$, we put

$$\mathcal{C}_{f,\boldsymbol{a}}(J,K,\mathcal{T}_I)[*A!B] := \Big(\mathcal{T}_I \otimes \bigotimes_{k \in K_0(f)} \mathbb{I}_{z_k}^{-\infty,a_k+1} \otimes \bigotimes_{k \in J \sqcup K_1(f)} \mathbb{I}_{z_k}^{-\infty,a_k}\Big)\Big[!(f^{-1}(0) \sqcup B) * (K \sqcup J \sqcup A \setminus f^{-1}(0))\Big] \tag{5.33}$$

Let $\mathbf{0}$ and χ_i be functions $K \sqcup J \longrightarrow \{0,1\}$ given by $\mathbf{0}(j) = 0$ for any $j \in K \sqcup J$, and $\chi_i(j) = 1$ $(j = i)$ or $\chi_i(j) = 0$ $(j \neq i)$. Then, let $\Xi_K^{(\boldsymbol{a}_K)} \psi_J^{(\boldsymbol{a}_J)} \mathcal{T}_I[*A!B]$ be the $\mathcal{R}_X(*D(\underline{\ell} \setminus L))$-triple obtained as the kernel of the following morphism:

$$\mathcal{C}_{\mathbf{0},\boldsymbol{a}}(J,K,\mathcal{T}_I)[*A!B] \longrightarrow \bigoplus_{i \in K \sqcup J} \mathcal{C}_{\chi_i,\boldsymbol{a}}(J,K,\mathcal{T}_I)[*A!B] \tag{5.34}$$

Then, $\Xi_K^{(\boldsymbol{a}_K)} \psi_J^{(\boldsymbol{a}_J)} \mathcal{T}_I[*A!B](*\partial D_{I \sqcup J})$ are good-KMS smooth $\mathcal{R}_{D_{I \sqcup J}(*\partial D_{I \sqcup J})}$-triple.

Lemma 5.5.6 *$\Xi_K^{(\boldsymbol{a}_K)} \psi_J^{(\boldsymbol{a}_J)} \mathcal{T}_I[*A!B]$ is strict, and strictly specializable along any z_j $(j \in \underline{\ell})$. Moreover, for $j \in \underline{\ell} \setminus (I \sqcup J)$ and for $u \in (\mathbb{R} \times \mathbb{C}) \setminus (\mathbb{Z}_{\geq 0} \times \{0\})$, we have natural isomorphisms*

$$\tilde{\psi}_{z_j,u} \Xi_K^{(\boldsymbol{a}_K)} \psi_J^{(\boldsymbol{a}_J)} \mathcal{T}_I[*A!B] \simeq \Xi_{K \setminus j}^{(\boldsymbol{a}_{K \setminus j})} \psi_J^{(\boldsymbol{a}_J)} \tilde{\psi}_{z_j,u} \mathcal{T}_I[*A_{\setminus j}!B_{\setminus j}]. \tag{5.35}$$

We also have the following for $j \in \underline{\ell} \setminus (I \sqcup J)$:

$$\Xi_j^{(a_j)} \Xi_K^{(\boldsymbol{a}_K)} \psi_J^{(\boldsymbol{a}_J)} \mathcal{T}_I[*A!B] \simeq \Xi_{Kj}^{(\boldsymbol{a}_{Kj})} \psi_J^{(\boldsymbol{a}_J)} \mathcal{T}_I[*A_{\setminus j}!B_{\setminus j}] \tag{5.36}$$

Proof We have only to consider the case $I = J = \emptyset$. We use an induction on $|K|$. If $|K| = 0$, the claim follows from Proposition 5.4.1. We shall prove the claim in the case $|K| = m$, by assuming the claim in the case $|K| < m$.

Let $j \in K$. By construction, we naturally have $\Xi_K^{(\boldsymbol{a}_K)} \mathcal{T}[*A!B](*z_j) \simeq \Xi_{K \setminus j}^{(\boldsymbol{a}_{K \setminus j})} \mathcal{T}[*A!B]$. By the assumption of the induction, it is strictly specializable along z_i $(i \in \underline{\ell})$. We obtain (5.36) in this case by construction, which implies $\Xi_K^{(\boldsymbol{a}_K)} \mathcal{T}[*A!B]$ is strictly specializable along z_j. After applying $\tilde{\psi}_{z_j,u}$ to (5.34), the morphism is strict. Hence, we obtain (5.35) in this case.

Let $j \in \underline{\ell} \setminus K$. Take any $k \in K$. Recall that $\Xi_K^{(\boldsymbol{a}_K)} \mathcal{T}[*A!B]$ is the kernel of the following morphism:

$$\varphi : \Pi_{k!}^{-N,a_k+1} \Xi_{K \setminus k}^{(\boldsymbol{a}_{K \setminus k})}(\mathcal{T}[*A!B]) \longrightarrow \Pi_{k*}^{-N,a_k} \Xi_{K \setminus k}^{(\boldsymbol{a}_{K \setminus k})}(\mathcal{T}[*A!B])$$

For any $u \in \mathbb{R}\times\mathbb{C}$, let ${}^j\psi_u := \psi_{z_j,u}$. (See Sect. 2.1.2 for $\psi_{z_j,u}$.) We have the following commutative diagram:

$$\begin{array}{ccc} {}^j\psi_u\Big(\Pi_{k!}^{-N,a_k+1}\Xi_{K\setminus k}^{(\boldsymbol{a}_{K\setminus k})}\big(\mathcal{T}[*A!B]\big)\Big) & \xrightarrow{{}^j\psi_u(\varphi)} & {}^j\psi_u\Big(\Pi_{k*}^{-N,a_k}\Xi_{K\setminus k}^{(\boldsymbol{a}_{K\setminus k})}\big(\mathcal{T}[*A!B]\big)\Big) \\ \simeq\downarrow & & \simeq\downarrow \\ \Pi_{k!}^{-N,a_k+1}\Xi_{K\setminus k}^{(\boldsymbol{a}_{K\setminus k})}\big({}^j\psi_u\mathcal{T}[*A!B]\big) & \longrightarrow & \Pi_{k*}^{-N,a_k}\Xi_{K\setminus k}^{(\boldsymbol{a}_{K\setminus k})}\big({}^j\psi_u\mathcal{T}[*A!B]\big) \end{array}$$

We also have a similar diagram for ${}^j\tilde{\psi}_u$. Hence, the cokernel of ${}^j\psi_u(\varphi)$ is naturally isomorphic to $\Xi_{K\setminus k}^{(\boldsymbol{a}_{K\setminus k})}{}^k\psi_{-\boldsymbol{\delta}}^{(-N)}{}^j\psi_u\mathcal{T}[*A!B]$, which is strict by the assumption of the induction. We obtain that $\Xi_K^{(\boldsymbol{a}_K)}\mathcal{T}[*A!B]$ is strictly specializable along z_j. We also have the following natural isomorphisms for $u \in (\mathbb{R}\times\mathbb{C}) \setminus (\mathbb{Z}_{\geq 0}\times\{0\})$:

$${}^j\psi_u\Xi_K^{(\boldsymbol{a}_K)}\big(\mathcal{T}[*A!B]\big) \simeq \Xi_k^{(a_k)}\Xi_{K\setminus k}^{(\boldsymbol{a}_{K\setminus k})}{}^j\psi_u\mathcal{T}[*A!B] \simeq \Xi_K^{(\boldsymbol{a}_K)}\big({}^j\psi_u\mathcal{T}[*A!B]\big) \tag{5.37}$$

We obtain (5.35) from (5.37). We obtain (5.36) by construction. Thus, the induction can proceed. □

5.5.5 *Growth Order and the Compatibility of Stokes Filtrations*

Although a condition is imposed on sesqui-linear pairings of smooth $\mathcal{R}$-triples, we do not have to impose the assumption in the good case. Namely, the following proposition holds.

Proposition 5.5.7 *Let $\mathcal{T}$ be an $\mathcal{R}_{X(*D)}$-triple such that the underlying $\mathcal{R}_{X(*D)}$-modules are smooth and good. Then, $\mathcal{T}$ is smooth.*

Proof We have only to consider the case that the underlying smooth $\mathcal{R}_{X(*D)}$-modules $\mathcal{M}_i$ $(i = 1, 2)$ are unramifiedly good. We use the notation in Sect. 5.5.1. Let C_0 denote the restriction of C to $\boldsymbol{S}\times(X\setminus D)$. Let $\lambda_0 \in \boldsymbol{S}$ and $Q \in \pi^{-1}(D)$. Let $I(\lambda_0) \subset \boldsymbol{S}$ and $U_Q \subset \tilde{X}(D)$ be neighbourhoods of λ_0 and Q, respectively. We consider the restriction of C_0 to $\big(I(\lambda_0)\times U_Q\big)\setminus\pi^{-1}(D)$, denoted by C_Q. We have only to prove that the restriction of C_Q to $\tilde{\mathcal{F}}_{\mathfrak{a}}^{(\lambda_0,Q)} \otimes \sigma^*\tilde{\mathcal{F}}_{\mathfrak{b}}^{(-\lambda_0,Q)}$ is 0, unless $\mathrm{Re}(\mathfrak{a}/\lambda) - \mathrm{Re}(\mathfrak{b}/\lambda) \leq 0$. By considering the specialization to curves which are transversal with the smooth part of D, we can reduce the issue to the one dimensional case. Then, the claim follows from Lemma 12.6.10 in [55]. □

5.5.6 $\mathcal{I}$-Good-KMS Smooth $\mathcal{R}$-Triples

We have a refined notion.

Definition 5.5.8 Let $\mathcal{I} = (\mathcal{I}_P \mid P \in D)$ be a good system of ramified irregular values on (X, D). (See Sect. 15.1 below.) A good(-KMS) smooth $\mathcal{R}_{X(*D)}$-module $\mathcal{M}$ is called $\mathcal{I}$-good(-KMS), if $\mathrm{Irr}(\mathcal{M}, P) \subset \mathcal{I}_P$ for any $P \in D$. A good(-KMS) smooth $\mathcal{R}_{X(*D)}$-triple is called $\mathcal{I}$-good(-KMS), if the underlying $\mathcal{R}_{X(*D)}$-modules are $\mathcal{I}$-good(-KMS). □

A direct sum of $\mathcal{I}$-good(-KMS) smooth $\mathcal{R}_{X(*D)}$-triples is $\mathcal{I}$-good(-KMS). Let $F : X' \longrightarrow X$ be a morphism of complex manifolds. If $\mathcal{T}$ is an $\mathcal{I}$-good(-KMS) smooth $\mathcal{R}_{X(*D)}$-triple, then $F^*\mathcal{T}$ is an $F^{-1}(\mathcal{I})$-good(-KMS) smooth $\mathcal{R}_{X'(*D')}$-triple. Let $P \in D_I$ be any point. If we take a holomorphic coordinate system around P, we obtain the $\mathcal{I}(I)_{|D_I}$-good(-KMS) smooth $\mathcal{R}_{D_I(*\partial D_I)}$-triple ${}^I\tilde{\psi}_{\boldsymbol{u}}(\mathcal{T})$.

5.6 Gluing of Good-KMS Smooth $\mathcal{R}$-Triples on the Intersections

Let X be an open subset of $\mathbb{C}^n$. Let $D_i := \{z_i = 0\} \cap X$ and $D = \bigcup_{i=1}^{\ell} D_i$. We shall introduce a procedure to glue good-KMS smooth $\mathcal{R}$-triples given on the intersections D_I ($I \subset \underline{\ell}$). For $I \subset \underline{\ell}$ and $i_1, i_2, \ldots, i_m \in \underline{\ell} \setminus I$, we set $Ii_1i_2\cdots i_m := I \sqcup \{i_1, \ldots, i_m\}$.

Let $I \sqcup \{i\} \subset \underline{\ell}$. For any good-KMS smooth $\mathcal{R}_{D_I(*\partial D_I)}$-triple $\mathcal{T}$, we obtain good-KMS smooth $\mathcal{R}_{D_{Ii}(*\partial D_{Ii})}$-triples

$$\overline{\psi}_i^{(1)}(\mathcal{T}) := \psi_{z_i,-\boldsymbol{\delta}}(\mathcal{T}) \otimes \mathcal{U}(-1,0), \quad \overline{\psi}_i^{(0)}(\mathcal{T}) := \psi_{z_i,-\boldsymbol{\delta}}(\mathcal{T}) \otimes \mathcal{U}(0,-1).$$

Let $\iota_{I,i} : D_{Ii} \longrightarrow D_I$ be the inclusion. According to Proposition 4.3.1, we have the following commutative diagram, in which the morphisms are given as in Proposition 4.3.1.

$$\begin{array}{ccc} \iota_{I,i\dagger}\overline{\psi}_i^{(1)}(\mathcal{T}) & \longrightarrow & \iota_{I,i\dagger}\overline{\psi}_i^{(0)}(\mathcal{T}) \\ \simeq\downarrow & & \simeq\downarrow \\ \psi_i^{(1)}(\mathcal{T}) & \longrightarrow & \psi_i^{(0)}(\mathcal{T}) \end{array}$$

5.6.1 A Category

Let $\boldsymbol{C}(X,D)$ be the category of tuples:

$$\boldsymbol{\mathcal{T}}=\big(\mathcal{T}_I,\ (I\subset\underline{\ell});\, f_{I,i},\, g_{I,i}\ (I\subset\underline{\ell},\ i\in\underline{\ell}\setminus I)\big)$$

- $\mathcal{T}_I$ are good-KMS $\mathcal{R}_{D_I(*\partial D_I)}$-triples.
- $f_{I,i}$ and $g_{I,i}$ are morphisms of $\mathcal{R}$-triples

$$\overline{\psi}_i^{(1)}\mathcal{T}_I\xrightarrow{g_{I,i}}\mathcal{T}_{Ii}\xrightarrow{f_{I,i}}\overline{\psi}_i^{(0)}\mathcal{T}_I$$

such that $f_{I,i}\circ g_{I,i}$ is equal to the canonical morphism $\overline{\psi}_i^{(1)}\mathcal{T}_I\longrightarrow\overline{\psi}_i^{(0)}\mathcal{T}_I$. We impose the commutativity of the following diagrams for $j,k\in\underline{\ell}\setminus I$ with $j\neq k$:

$$\begin{array}{ccc}
\mathcal{T}_{Ikj} & \xrightarrow{f_{Ij,k}} & \overline{\psi}_k^{(0)}(\mathcal{T}_{Ij})\\
{\scriptstyle f_{Ik,j}}\big\downarrow & & \big\downarrow{\scriptstyle \overline{\psi}_k^{(0)}(f_{I,j})}\\
\overline{\psi}_j^{(0)}(\mathcal{T}_{Ik}) & \xrightarrow{\overline{\psi}_j^{(0)}(f_{I,k})} & \overline{\psi}_k^{(0)}\overline{\psi}_j^{(0)}(\mathcal{T}_I)
\end{array}\tag{5.38}$$

$$\begin{array}{ccc}
\overline{\psi}_k^{(1)}\overline{\psi}_j^{(1)}(\mathcal{T}_I) & \xrightarrow{\overline{\psi}_j^{(1)}(g_{I,k})} & \overline{\psi}_j^{(1)}(\mathcal{T}_{Ik})\\
\big\downarrow{\scriptstyle \overline{\psi}_k^{(1)}(g_{I,j})} & & {\scriptstyle g_{Ik,j}}\big\downarrow\\
\overline{\psi}_k^{(1)}(\mathcal{T}_{Ij}) & \xrightarrow{g_{Ij,k}} & \mathcal{T}_{Ikj}
\end{array}\tag{5.39}$$

$$\begin{array}{ccc}
\overline{\psi}_j^{(1)}(\mathcal{T}_{Ik}) & \xrightarrow{\overline{\psi}_j^{(1)}(f_{I,k})} & \overline{\psi}_k^{(0)}\overline{\psi}_j^{(1)}(\mathcal{T}_I)\\
{\scriptstyle g_{Ik,j}}\big\downarrow & & \big\downarrow{\scriptstyle \overline{\psi}_k^{(0)}(g_{I,j})}\\
\mathcal{T}_{Ikj} & \xrightarrow{f_{Ij,k}} & \overline{\psi}_k^{(0)}(\mathcal{T}_{Ij})
\end{array}\tag{5.40}$$

A morphism $\boldsymbol{\mathcal{T}}^{(1)}\longrightarrow\boldsymbol{\mathcal{T}}^{(2)}$ in $\boldsymbol{C}(X,D)$ is a tuple of morphisms $F_I:\mathcal{T}_I^{(1)}\longrightarrow\mathcal{T}_I^{(2)}$ such that the following diagram is commutative:

$$\begin{array}{ccccc}
\overline{\psi}_i^{(1)}\mathcal{T}_I^{(1)} & \xrightarrow{g_{I,i}^{(1)}} & \mathcal{T}_{Ii}^{(1)} & \xrightarrow{f_{I,i}^{(1)}} & \overline{\psi}_i^{(0)}\mathcal{T}_I^{(1)}\\
{\scriptstyle \overline{\psi}_i^{(1)}F_I}\big\downarrow & & {\scriptstyle F_{Ii}}\big\downarrow & & {\scriptstyle \overline{\psi}_i^{(0)}F_I}\big\downarrow\\
\overline{\psi}_i^{(1)}\mathcal{T}_I^{(2)} & \xrightarrow{g_{I,i}^{(2)}} & \mathcal{T}_{Ii}^{(2)} & \xrightarrow{f_{I,i}^{(2)}} & \overline{\psi}_i^{(0)}\mathcal{T}_I^{(2)}
\end{array}$$

For $p \in \underline{\ell}$ and $u \in \mathbb{R} \times \mathbb{C}$, we have the functors ${}^{P}\tilde{\psi}_u$ and ϕ_p from $\boldsymbol{C}(X, D)$ to $\boldsymbol{C}(D_p, \partial D_p)$, given as follows:

$$ {}^{P}\tilde{\psi}_u(\boldsymbol{\mathcal{T}}) := \Bigl({}^{P}\tilde{\psi}_u(\mathcal{T}_I),\ (I \subset \underline{\ell} \setminus p);\ {}^{P}\tilde{\psi}_u(f_{I,j}),\ {}^{P}\tilde{\psi}_u(g_{I,j})\ \bigl(I \subset \underline{\ell} \setminus p,\ j \in \underline{\ell} \setminus Ip\bigr)\Bigr) $$

$$ \phi_p^{(m)}(\boldsymbol{\mathcal{T}}) := \Bigl(\mathcal{T}_{Ip} \otimes \boldsymbol{T}(m),\ (I \subset \underline{\ell} \setminus p);\ f_{Ip,j},\ g_{Ip,j}\ \bigl(I \subset \underline{\ell} \setminus p,\ j \in \underline{\ell} \setminus Ip\bigr)\Bigr) $$

We shall construct a fully faithful functor Ψ_X from $\boldsymbol{C}(X, D)$ to the category of $\mathcal{R}_X$-triples.

5.6.2 Construction of the Functor

For a non-empty finite subset I, we take a $2|I|$-dimensional Hermitian vector space (V, h) with an orthonormal base u_k, v_k $(k \in I)$. For $\boldsymbol{i} = (i_k | k \in I) \in \{0, 1\}^I$ and $\boldsymbol{j} = (j_k | k \in I) \in \{0, 1\}^I$, let $E(\boldsymbol{i}, \boldsymbol{j})$ denote the one dimensional subspace of $\bigwedge^{\bullet} V$ generated by $\bigwedge_{k \in I} u_k^{i_k} \wedge \bigwedge_{k \in I} v_k^{j_k}$, where we set $u_k^0 = v_k^0 = 0$. It is naturally equipped with the Hermitian metric denoted by h. We obtain a smooth $\mathcal{R}$-triple $\tilde{E}(\boldsymbol{i}, \boldsymbol{j}) := \bigl(\mathcal{O}_{\mathbb{C}_\lambda} \otimes E(\boldsymbol{i}, \boldsymbol{j}), \mathcal{O}_{\mathbb{C}_\lambda} \otimes E(\boldsymbol{i}, \boldsymbol{j}), C_h\bigr)$, where C_h is the sesqui-linear pairing induced by h. If $i'_p = i_p + 1$, $i'_k = i_k$ $(k \neq p)$ and $j'_k = j_k$ for any k, we have a map $\boldsymbol{a}_p^{(1)} : \tilde{E}(\boldsymbol{i}, \boldsymbol{j}) \longrightarrow \tilde{E}(\boldsymbol{i}', \boldsymbol{j}')$ given by the inner product of u_p on the first component and the exterior product of u_p on the second component. If $j'_p = j_p + 1$, $j'_k = j_k$ $(k \neq p)$ and $i'_k = i_k$ for any k, we have a map $\boldsymbol{a}_p^{(2)} : \tilde{E}(\boldsymbol{i}, \boldsymbol{j}) \longrightarrow \tilde{E}(\boldsymbol{i}', \boldsymbol{j}')$ given by the inner product of v_p on the first component and the exterior product of v_p on the second component. We have $\boldsymbol{a}_{p_1}^{(\kappa_1)} \circ \boldsymbol{a}_{p_2}^{(\kappa_2)} = -\boldsymbol{a}_{p_2}^{(\kappa_2)} \circ \boldsymbol{a}_{p_1}^{(\kappa_1)}$.

Let Γ be the category given by the following commutative diagram:

$$ \begin{array}{ccc} (0,0) & \xrightarrow{\ a\ } & (0,1) \\ {\scriptstyle b}\big\downarrow & & \big\downarrow{\scriptstyle c} \\ (1,0) & \xrightarrow{\ d\ } & (1,1) \end{array} $$

For a non-empty finite set I, let Γ^I denote the product of the categories of the I-tuples of objects in Γ. An object $\bigl((i_k, j_k) \bigm| k \in I\bigr)$ is denoted as the pair of $\boldsymbol{i} = (i_k)$ and $\boldsymbol{j} = (j_k)$. For such an object, we set $|\boldsymbol{i}| := \sum i_k$ and $|\boldsymbol{j}| := \sum j_k$.

Let $J \subset \underline{\ell}$, and let F be a functor from Γ^I to the category of $\mathcal{R}_{X(*D(J))}$-triples. For $n \in \mathbb{Z}$, we define

$$ \bigl(\pi_I F\bigr)^n := \bigoplus_{|\boldsymbol{i}| + |\boldsymbol{j}| = n + |I|} F(\boldsymbol{i}, \boldsymbol{j}) \otimes \tilde{E}(\boldsymbol{i}, \boldsymbol{j}). $$

The morphisms $F(\boldsymbol{i},\boldsymbol{j}) \longrightarrow F(\boldsymbol{i}',\boldsymbol{j}')$ and $a_p^{(\kappa)}$ naturally give a differential. The complex of $\mathcal{R}_{X(*D(J))}$-triples is denoted by $\pi_I F$.

For $1 \leq m \leq \ell$, we set $\underline{m} = \{1,\ldots,m\}$. Let $\Gamma^m := \Gamma^{\underline{m}}$. For $(\boldsymbol{i},\boldsymbol{j}) \in \Gamma^m$, we put

$$I(\boldsymbol{i},\boldsymbol{j}) := \bigl\{k \in \underline{m} \,\big|\, (i_k,j_k) = (0,1)\bigr\}, \quad K(\boldsymbol{i},\boldsymbol{j}) := \bigl\{k \in \underline{m} \,\big|\, (i_k,j_k) = (1,0)\bigr\},$$
$$J_1(\boldsymbol{i},\boldsymbol{j}) := \bigl\{k \in \underline{m} \,\big|\, (i_k,j_k) = (0,0)\bigr\}, \quad J_0(\boldsymbol{i},\boldsymbol{j}) := \bigl\{k \in \underline{m} \,\big|\, (i_k,j_k) = (1,1)\bigr\}.$$

For $\mathcal{T} \in \boldsymbol{C}(X,D)$ and $\boldsymbol{a}_{\underline{m}} \in \mathbb{Z}^m$, we define a functor $\mathcal{Q}_X^{\underline{m}}(\mathcal{T},\boldsymbol{a}_{\underline{m}})$ from Γ^m to the category of $\mathcal{R}_{X(*D(\underline{\ell}-\underline{m}))}$-triples given as follows. We set

$$\mathcal{Q}_X^{\underline{m}}\bigl(\mathcal{T},(\boldsymbol{i},\boldsymbol{j}),\boldsymbol{a}_{\underline{m}}\bigr) := \iota_{K(\boldsymbol{i},\boldsymbol{j})\dagger}\Bigl(\Xi_{I(\boldsymbol{i},\boldsymbol{j})}^{(\boldsymbol{a}_{I(\boldsymbol{i},\boldsymbol{j})})}\psi_{J_1(\boldsymbol{i},\boldsymbol{j})}^{(\boldsymbol{a}_{J_1(\boldsymbol{i},\boldsymbol{j})}+\boldsymbol{\delta}_{J_1(\boldsymbol{i},\boldsymbol{j})})}\psi_{J_0(\boldsymbol{i},\boldsymbol{j})}^{(\boldsymbol{a}_{J_0(\boldsymbol{i},\boldsymbol{j})})}\bigl(\mathcal{T}_{K(\boldsymbol{i},\boldsymbol{j})}\bigr)\Bigr)$$

Here, we set $a_L := (a_i \mid i \in L)$, and let $\iota_{K(\boldsymbol{i},\boldsymbol{j})} : D_{K(\boldsymbol{i},\boldsymbol{j})} \longrightarrow X$ denote the inclusion. Let $(\boldsymbol{i}_0,\boldsymbol{j}_0)$ be an object in $\Gamma^{\underline{m}\setminus k}$. Let $I := I(\boldsymbol{i}_0,\boldsymbol{j}_0)$, $J_i := J_i(\boldsymbol{i}_0,\boldsymbol{j}_0)$ and $K := K(\boldsymbol{i}_0,\boldsymbol{j}_0)$. We have the following diagram:

$$\begin{array}{ccc}
\psi_k^{(a_k+1)}\Xi_I^{(\boldsymbol{a}_I)}\psi_{J_1}^{(\boldsymbol{a}_{J_1}+\boldsymbol{\delta}_{J_1})}\psi_{J_0}^{(\boldsymbol{a}_{J_0})}(\mathcal{T}_K) & \xrightarrow{\alpha_1} & \Xi_k^{(a_k)}\Xi_I^{(\boldsymbol{a}_I)}\psi_{J_1}^{(\boldsymbol{a}_{J_1}+\boldsymbol{\delta}_{J_1})}\psi_{J_0}^{(\boldsymbol{a}_{J_0})}(\mathcal{T}_K) \\
{\scriptstyle\alpha_2}\big\downarrow & & {\scriptstyle\alpha_3}\big\downarrow \\
\iota_{K,k\dagger}\Xi_I^{(\boldsymbol{a}_I)}\psi_{J_1}^{(\boldsymbol{a}_{J_1}+\boldsymbol{\delta}_{J_1})}\psi_{J_0}^{(\boldsymbol{a}_{J_0})}(\mathcal{T}_{Kk}) & \xrightarrow{\alpha_4} & \psi_k^{(a_k)}\Xi_I^{(\boldsymbol{a}_I)}\psi_{J_1}^{(\boldsymbol{a}_{J_1}+\boldsymbol{\delta}_{J_1})}\psi_{J_0}^{(\boldsymbol{a}_{J_0})}(\mathcal{T}_K)
\end{array} \tag{5.41}$$

Here, α_1 and α_3 are the natural morphisms, and α_2 and α_4 are induced by $g_{K,k}$ and $f_{K,k}$ respectively. Let $(\boldsymbol{i}_1,\boldsymbol{j}_1)$, $(\boldsymbol{i}_2,\boldsymbol{j}_2)$, $(\boldsymbol{i}_3,\boldsymbol{j}_3)$ and $(\boldsymbol{i}_4,\boldsymbol{j}_4)$ be objects in $\Gamma^{\underline{m}}$ induced by $(0,0)$, $(0,1)$, $(1,0)$, $(1,1)$ with $(\boldsymbol{i}_0,\boldsymbol{j}_0)$. Then, the morphisms

$$\begin{array}{ccc}
\mathcal{Q}_X^{\underline{m}}(\mathcal{T},(\boldsymbol{i}_1,\boldsymbol{j}_1),\boldsymbol{a}_{\underline{m}}) & \longrightarrow & \mathcal{Q}_X^{\underline{m}}(\mathcal{T},(\boldsymbol{i}_2,\boldsymbol{j}_2),\boldsymbol{a}_{\underline{m}}) \\
\big\downarrow & & \big\downarrow \\
\mathcal{Q}_X^{\underline{m}}(\mathcal{T},(\boldsymbol{i}_3,\boldsymbol{j}_3),\boldsymbol{a}_{\underline{m}}) & \longrightarrow & \mathcal{Q}_X^{\underline{m}}(\mathcal{T},(\boldsymbol{i}_4,\boldsymbol{j}_4),\boldsymbol{a}_{\underline{m}})
\end{array}$$

are induced by the morphisms in (5.41). By using the construction of the morphisms and the conditions (5.38)–(5.40), we can check that $\mathcal{Q}_X^{\underline{m}}(\mathcal{T},\boldsymbol{a}_{\underline{m}})$ gives a functor from $\Gamma^{\underline{m}}$ to the category of $\mathcal{R}_{X(*D(\underline{\ell}-\underline{m}))}$-triples.

We obtain a complex of $\mathcal{R}_{X(*D(\underline{\ell}-\underline{m}))}$-triples $\pi_{\underline{m}}\mathcal{Q}_X^{\underline{m}}(\mathcal{T},\boldsymbol{a}_{\underline{m}})$. We formally set $\pi_{\underline{0}}\mathcal{Q}_X^{\underline{0}}(\mathcal{T},\boldsymbol{a}_{\underline{0}}) := \mathcal{T}_{\emptyset}$.

Lemma 5.6.1

- $\mathcal{H}^p\pi_{\underline{m}}\mathcal{Q}_X^{\underline{m}}(\mathcal{T},\boldsymbol{a}_{\underline{m}}) = 0$ *unless* $p \neq 0$.
- $\mathcal{H}^0\pi_{\underline{m}}\mathcal{Q}_X^{\underline{m}}(\mathcal{T},\boldsymbol{a}_{\underline{m}})$ *is strict, and strictly specializable along any* z_i $(i \in \underline{\ell})$.
- *For any* $p \in \underline{\ell} \setminus \underline{m}$ *and* $u \in \mathbb{R}\times\mathbb{C}$, *we have*

$$^{P}\tilde{\psi}_u\mathcal{H}^0\pi_{\underline{m}}\mathcal{Q}_X^{\underline{m}}(\mathcal{T},\boldsymbol{a}_{\underline{m}}) \simeq \mathcal{H}^0\pi_{\underline{m}}\mathcal{Q}_{D_p}^{\underline{m}}\bigl({}^{P}\tilde{\psi}_u\mathcal{T},\boldsymbol{a}_{\underline{m}}\bigr)$$

For any $p \in \underline{m}$ *and* $u \in \mathbb{R} \times \mathbb{C}$, *we have*

$$^{p}\tilde{\psi}_u \mathcal{H}^0 \pi_{\underline{m}} \mathcal{Q}_X^{\underline{m}}(\boldsymbol{\mathcal{T}}, \boldsymbol{a}_{\underline{m}}) \simeq \mathcal{H}^0 \pi_{\underline{m} \setminus p} \mathcal{Q}_{D_p}^{\underline{m} \setminus p}({}^{p}\tilde{\psi}_u \boldsymbol{\mathcal{T}}, \boldsymbol{a}_{\underline{m} \setminus p})$$

- *For any* $p \in \underline{m}$, *we have*

$$\phi_p^{(a_p)} \mathcal{H}^0 \pi_{\underline{m}} \mathcal{Q}_X^{\underline{m}}(\boldsymbol{\mathcal{T}}, \boldsymbol{a}_{\underline{m}}) \simeq \iota_{p\dagger} \mathcal{H}^0 \pi_{\underline{m} \setminus p} \mathcal{Q}_{D_p}^{\underline{m} \setminus p}(\phi_p^{(a_p)} \boldsymbol{\mathcal{T}}, \boldsymbol{a}_{\underline{m} \setminus p})$$

- *For any* $p \in \underline{\ell}$, *we have*

$$\Xi_p^{(a_p)} \mathcal{H}^0 \pi_{\underline{m}} \mathcal{Q}_X^{\underline{m}}(\boldsymbol{\mathcal{T}}, \boldsymbol{a}_{\underline{m}}) \simeq \mathcal{H}^0 \Xi_p^{(a_p)} \pi_{\underline{m}} \mathcal{Q}_X^{\underline{m}}(\boldsymbol{\mathcal{T}}, \boldsymbol{a}_{\underline{m}})$$

Proof We use an induction on m. If $m = 0$, the claim is clear. We put $\mathcal{T}_m := \mathcal{H}^0 \pi_{\underline{m}} \mathcal{Q}_X^{\underline{m}}(\boldsymbol{\mathcal{T}}, \boldsymbol{a}_{\underline{m}})$. We consider the following complexes:

$$C_{0,0}^{\bullet} := \psi_m^{(a_m+1)} \pi_{\underline{m-1}} \mathcal{Q}_X^{\underline{m-1}}(\boldsymbol{\mathcal{T}}, \boldsymbol{a}_{\underline{m-1}}), \quad C_{0,1}^{\bullet} := \Xi_m^{(a_m)} \pi_{\underline{m-1}} \mathcal{Q}_X^{\underline{m-1}}(\boldsymbol{\mathcal{T}}, \boldsymbol{a}_{\underline{m-1}}),$$
$$C_{1,0}^{\bullet} := \pi_{\underline{m-1}} \mathcal{Q}_{D_m}^{\underline{m-1}}(\phi_m^{(a_m)} \boldsymbol{\mathcal{T}}, \boldsymbol{a}_{\underline{m-1}}), \quad C_{1,1}^{\bullet} := \psi_m^{(a_m)} \pi_{\underline{m-1}} \mathcal{Q}_X^{\underline{m-1}}(\boldsymbol{\mathcal{T}}, \boldsymbol{a}_{\underline{m-1}}).$$

The complex $\pi_{\underline{m}} \mathcal{Q}_X^{\underline{m}}(\boldsymbol{\mathcal{T}}, \boldsymbol{a}_{\underline{m}})$ is associated to the following naturally obtained triple complex:

$$\begin{array}{ccc} C_{0,0}^{\bullet} & \longrightarrow & C_{0,1}^{\bullet} \\ \downarrow & & \downarrow \\ C_{1,0}^{\bullet} & \longrightarrow & C_{1,1}^{\bullet} \end{array} \tag{5.42}$$

Let us consider the double complex obtained as $\mathcal{H}^0$ of (5.42).

Lemma 5.6.2 *The complex associated to the double complex may have non-trivial cohomology only in the degree* 0, *i.e., the first claim holds in the case* m.

Proof By the assumption in the case $m-1$, $\mathcal{T}_{m-1}$ is strictly specializable along z_m. We also have $\mathcal{H}^p(C_{i,j}) = 0$ unless $p \neq 0$, and we have natural isomorphisms:

$$\mathcal{H}^0(C_{0,0}) \simeq \psi_m^{(a_m+1)} \mathcal{T}_{m-1}, \ \mathcal{H}^0(C_{0,1}) \simeq \Xi_m^{(a_m)} \mathcal{T}_{m-1}, \ \mathcal{H}^0(C_{1,1}) \simeq \psi_m^{(a_m)} \mathcal{T}_{m-1}.$$

Hence, $\rho_1 : \mathcal{H}^0(C_{0,0}^{\bullet}) \longrightarrow \mathcal{H}^0(C_{0,1}^{\bullet})$ is a monomorphism, and $\rho_2 : \mathcal{H}^0(C_{0,1}^{\bullet}) \longrightarrow \mathcal{H}^0(C_{1,1}^{\bullet})$ is an epimorphism. Then the claim of Lemma 5.6.2 follows. □

We have the exact sequence

$$0 \longrightarrow \mathcal{H}^0(C_{1,0}^{\bullet}) \longrightarrow \Big(\mathcal{H}^0(C_{1,0}^{\bullet}) \oplus \mathcal{H}^0(C_{0,1}^{\bullet})\Big) \Big/ \mathcal{H}^0(C_{0,0})$$
$$\longrightarrow \mathcal{H}^0(C_{0,1}^{\bullet}) / \mathcal{H}^0(C_{0,0}) \longrightarrow 0. \tag{5.43}$$

We obtain that the middle term of (5.43) is strict. Because $\mathcal{T}_m$ is a subobject of the middle term, we obtain that the second $\mathcal{R}$-module of $\mathcal{T}_m$ is strict. We can also prove that the other underlying $\mathcal{R}$-module is also strict by using the Hermitian adjoint. Hence, we obtain that $\mathcal{T}_m$ is strict.

Because $\mathcal{T}_m(*z_m) = \mathcal{T}_{m-1}$, we have ${}^m\tilde{\psi}_u\mathcal{T}_m = {}^m\tilde{\psi}_u\mathcal{T}_{m-1}$. Hence, we obtain the desired formula for ${}^m\tilde{\psi}_u\mathcal{T}_m$ from the formula for ${}^m\tilde{\psi}_u\mathcal{T}_{m-1}$. We also have $\Xi_m^{(a_m)}\mathcal{T}_m = \Xi_m^{(a_m)}\mathcal{T}_{m-1}$. Because $\Xi_m^{(a_m)}\mathcal{H}^0C_{i,j} = 0$ unless $(i,j) \neq (0,1)$, we obtain the desired formula for $\Xi_m^{(a_m)}\mathcal{T}_m$. Because $\mathcal{T}_m$ is obtained as the gluing of $\mathcal{T}_m(*z_m)$ and $\mathcal{H}^0\big(C^\bullet_{1,0}\big)$, we obtain that $\mathcal{T}_m$ is strictly specializable along z_m. We also obtain the formula for $\phi_m^{(a_m)}\mathcal{T}_m$. By exchanging the roles, we obtain that, for $p \leq m$, $\mathcal{T}_m$ is strictly specializable along z_p, and the desired formulas for $\Xi_p^{(a_m)}\mathcal{T}_m$, ${}^p\tilde{\psi}_u\mathcal{T}_m$ and $\phi_p^{(a_m)}\mathcal{T}_m$.

Let $p > m$. We have

$$
\begin{aligned}
{}^p\tilde{\psi}_u C^\bullet_{0,0} &= \psi_m^{(a_m+1)}\pi_{\underline{m-1}}\mathcal{Q}_{D_p}^{\underline{m-1}}({}^p\tilde{\psi}_u\boldsymbol{\mathcal{T}},\boldsymbol{a}_{\underline{m-1}}),\\
{}^p\tilde{\psi}_u C^\bullet_{0,1} &= \Xi_m^{(a_m)}\pi_{\underline{m-1}}\mathcal{Q}_{D_p}^{\underline{m-1}}({}^p\tilde{\psi}_u\boldsymbol{\mathcal{T}},\boldsymbol{a}_{\underline{m-1}}),\\
{}^p\tilde{\psi}_u C^\bullet_{1,0} &= \pi_{\underline{m-1}}\mathcal{Q}_{D_p}^{\underline{m-1}}\big(\phi_m^{(a_m)}{}^p\tilde{\psi}_u\boldsymbol{\mathcal{T}},\boldsymbol{a}_{\underline{m-1}}\big),\\
{}^p\tilde{\psi}_u C^\bullet_{1,1} &= \psi_m^{(a_m)}\pi_{\underline{m-1}}\mathcal{Q}_{D_p}^{\underline{m-1}}({}^p\tilde{\psi}_u\boldsymbol{\mathcal{T}},\boldsymbol{a}_{\underline{m-1}}).
\end{aligned}
$$

The complex ${}^p\tilde{\psi}_u\pi_{\underline{m}}\mathcal{Q}_X^{\underline{m}}(\boldsymbol{\mathcal{T}},\boldsymbol{a}_{\underline{m}})$ is associated to the triple complex:

$$
\begin{array}{ccc}
{}^p\tilde{\psi}_u C^\bullet_{0,0} & \longrightarrow & {}^p\tilde{\psi}_u C^\bullet_{0,1}\\
\downarrow & & \downarrow\\
{}^p\tilde{\psi}_u C^\bullet_{1,0} & \longrightarrow & {}^p\tilde{\psi}_u C^\bullet_{1,1}
\end{array}
$$

Hence, by using the previous result, we obtain that ${}^p\tilde{\psi}_u\pi_{\underline{m}}\mathcal{Q}_X^{\underline{m}}(\boldsymbol{\mathcal{T}},\boldsymbol{a}_{\underline{m}})$ may have non-trivial cohomology only in the degree 0, and the 0-th cohomology sheaf is strict. It implies that the morphisms in the complex $\pi_{\underline{m}}\mathcal{Q}_X^{\underline{m}}(\boldsymbol{\mathcal{T}},\boldsymbol{a}_{\underline{m}})$ is strict with respect to the V-filtration along z_p. Therefore, we obtain that $\mathcal{T}_m$ is strictly specializable along z_p, and we obtain the formula for $\Xi_p^{(a_p)}\mathcal{T}_m$. We also have the following natural isomorphisms for ${}^p\tilde{\psi}_u\mathcal{T}_m$:

$$
{}^p\tilde{\psi}_u\mathcal{T}_m \simeq \mathcal{H}^0\,{}^p\tilde{\psi}_u\pi_{\underline{m}}\mathcal{Q}_X^{\underline{m}}(\boldsymbol{\mathcal{T}},\boldsymbol{a}_{\underline{m}}) \simeq \mathcal{H}^0\pi_{\underline{m}}\mathcal{Q}_{D_p}^{\underline{m}}({}^p\tilde{\psi}_u\boldsymbol{\mathcal{T}},\boldsymbol{a}_{\underline{m}})
$$

Thus, the induction can proceed, and the proof of Lemma 5.6.1 is finished. □

By Lemma 5.6.1, we obtain the functor $\Psi_X := \mathcal{H}^0\pi_{\underline{\ell}}\mathcal{Q}_X^{\underline{\ell}}$ from the category $\boldsymbol{C}(X,D)$ to the category of $\mathcal{R}_X$-triples. It is independent of the choice of the Hermitian vector space V with an orthonormal basis, up to canonical isomorphisms.

5.6.3 Some Properties

By Lemma 5.6.1, $\Psi_X(\mathcal{T})$ is strictly specializable along z_i ($i = 1, \ldots, \ell$), and we have the following natural isomorphisms:

$$^P\tilde{\psi}_u \Psi_X(\mathcal{T}) \simeq \Psi_{D_p}(^P\tilde{\psi}_u \mathcal{T}), \quad \phi_p^{(a)} \Psi_X(\mathcal{T}) \simeq \iota_{p\dagger} \Psi_{D_p}(\phi_p^{(a)} \mathcal{T}).$$

We have the following natural isomorphisms for $p \neq q$:

$$^P\tilde{\psi}_{u_1} {}^q\tilde{\psi}_{u_2} \Psi_X(\mathcal{T}) \simeq {}^q\tilde{\psi}_{u_2} {}^P\tilde{\psi}_{u_1} \Psi_X(\mathcal{T}) \quad \phi_p^{(a_p)} \phi_q^{(a_q)} \Psi_X(\mathcal{T}) \simeq \phi_q^{(a_q)} \phi_p^{(a_p)} \Psi_X(\mathcal{T}).$$

We denote $\phi_{i_1}^{(a_{i_1})} \circ \cdots \circ \phi_{i_m}^{(a_{i_m})} \Psi_X(\mathcal{T})$ by $\phi_I^{(\boldsymbol{a})} \Psi_X(\mathcal{T})$.

Lemma 5.6.3 *The functor Ψ_X is fully faithful.*

Proof We use an induction on $\dim X$. Let $\boldsymbol{F} = (F_I \mid I \subset \underline{\ell}) : \mathcal{T}_1 \longrightarrow \mathcal{T}_2$ be a morphism in $\boldsymbol{C}(X, D)$ such that $\Psi_X(\boldsymbol{F}) = 0$. Then, we have $F_\emptyset = 0$. For each $p \in \underline{\ell}$, we have $\phi_p(\boldsymbol{F}) = 0$. It implies $F_{Ip} = 0$ for any $I \subset \underline{\ell} \setminus p$ by the assumption of the induction. Hence, we obtain $\boldsymbol{F} = 0$.

Let $G : \Psi_X(\mathcal{T}_1) \longrightarrow \Psi_X(\mathcal{T}_2)$ be a morphism of $\mathcal{R}$-triples. We have the induced morphism $G_\emptyset : \mathcal{T}_{1,\emptyset} \longrightarrow \mathcal{T}_{2,\emptyset}$. By applying ϕ_I, we obtain $G_I : \mathcal{T}_{1,I} \longrightarrow \mathcal{T}_{2,I}$. By using an induction, we can prove that $\Psi_X\big((G_I)\big) = G$. □

Lemma 5.6.4 *Let $\mathcal{T}$ be an object in $\boldsymbol{C}(X, D)$. Fix $i \in \underline{\ell}$.*

- *If $f_{I,i}$ are isomorphisms for any $I \subset \underline{\ell} \setminus \{i\}$, then $\Psi_X(\mathcal{T}) \simeq \Psi_X(\mathcal{T})[*z_i]$.*
- *If $g_{I,i}$ are isomorphisms for any $I \subset \underline{\ell} \setminus \{i\}$, then $\Psi_X(\mathcal{T}) \simeq \Psi_X(\mathcal{T})[!z_i]$.*

Proof Suppose that $f_{I,i}$ are isomorphisms for any $I \subset \underline{\ell} \setminus \{i\}$. Then, by Lemma 5.6.1, the canonical morphism $\phi_i^{(0)} \Psi_X(\mathcal{T}) \longrightarrow \psi_i^{(0)} \Psi_X(\mathcal{T})$ is an isomorphism. Hence, we obtain that $\Psi_X(\mathcal{T}) \simeq \Psi_X(\mathcal{T})[*z_i]$. The second claim can be shown similarly. □

Let $X' \overset{\iota}{\subset} X$ be any open subset. We put $D' := D \cap X'$. We have a natural functor $\iota^*; \boldsymbol{C}(X, D) \longrightarrow \boldsymbol{C}(X', D')$ given by the restriction. It is easy to observe that $\Psi_{X'}\big(\iota^*(\mathcal{T})\big)$ is naturally isomorphic to $\iota^* \Psi_X(\mathcal{T})$.

5.6.4 Dependence on Coordinate Systems

The category $\boldsymbol{C}(X, D)$ and the functor Ψ_X depend on the coordinate system $\boldsymbol{z} = (z_1, \ldots, z_n)$. To emphasize it, we use the symbols $\boldsymbol{C}(X, D, \boldsymbol{z})$ and $\Psi_{X,\boldsymbol{z}}$. We shall study their dependence on $\boldsymbol{z}$.

Let $\boldsymbol{w} = (w_1, \ldots, w_n)$ be another holomorphic coordinate system such that $w_i = e^{\varphi_i} \cdot z_i$ ($i = 1, \ldots, \ell$) for holomorphic functions φ_i on X. We obtain a category $\boldsymbol{C}(X, D, \boldsymbol{w})$ and a functor $\Psi_{X,\boldsymbol{w}}$.

Lemma 5.6.5 *We have an equivalence* $\mathrm{Def}_{\varphi}:\boldsymbol{C}(X,D,\boldsymbol{z})\longrightarrow\boldsymbol{C}(X,D,\boldsymbol{w})$ *such that* $\Psi_{X,\boldsymbol{z}}\simeq\Psi_{X,\boldsymbol{w}}\circ\mathrm{Def}_{\varphi}$.

Proof Let $\mathcal{T}\in\boldsymbol{C}(X,D,\boldsymbol{z})$. For each $I\subset\underline{\ell}$, $\mathcal{T}_I$ is equipped with a tuple $\boldsymbol{\mathcal{N}}_I=(\mathcal{N}_i\,|\,i\in I)$ of morphisms $\mathcal{N}_i:\mathcal{T}_I\longrightarrow\mathcal{T}_I\otimes\boldsymbol{T}(-1)$ induced as follows:

$$\mathcal{T}_I\xrightarrow{f_{I\setminus i,i}}\overline{\psi}^{(0)}_{z_i}\mathcal{T}_{I\setminus i}\simeq\overline{\psi}^{(1)}_{z_i}\mathcal{T}_{I\setminus i}\otimes\boldsymbol{T}(-1)\xrightarrow{g_{I\setminus i,i}}\mathcal{T}_I\otimes\boldsymbol{T}(-1)$$

Let $\boldsymbol{\varphi}_I:=(\varphi_i\,|\,i\in I)$. We put $\tilde{\mathcal{T}}_I:=\mathrm{Def}_{\boldsymbol{\varphi}_I}(\mathcal{T}_I,\boldsymbol{\mathcal{N}}_I)$.

For $i\in I$ and $j\in I\setminus i$, we have the induced morphisms $\mathcal{N}_j:\overline{\psi}^{(a)}_{z_i}(\mathcal{T}_{I\setminus i})\longrightarrow\overline{\psi}^{(a)}_{z_i}(\mathcal{T}_{I\setminus i})\otimes\boldsymbol{T}(-1)$. We also have $\mathcal{N}_i:\overline{\psi}^{(a)}_{z_i}(\mathcal{T}_{I\setminus i})\longrightarrow\overline{\psi}^{(a)}_{z_i}(\mathcal{T}_{I\setminus i})\otimes\boldsymbol{T}(-1)$. Let $\boldsymbol{\mathcal{N}}_I:=(\mathcal{N}_i\,|\,i\in I)$. Then, we have natural isomorphisms

$$\overline{\psi}^{(a)}_{w_i}(\tilde{\mathcal{T}}_{I\setminus i})\simeq\mathrm{Def}_{\varphi_I}\big(\overline{\psi}^{(a)}_{z_i}(\mathcal{T}_{I\setminus i}),\boldsymbol{\mathcal{N}}\big).$$

Hence, we have the following naturally induced morphisms:

$$\overline{\psi}^{(1)}_{w_i}(\tilde{\mathcal{T}}_I)\xrightarrow{\tilde{f}_{I,i}}\tilde{\mathcal{T}}_{Ii}\xrightarrow{\tilde{g}_{I,i}}\overline{\psi}^{(0)}_{w_i}(\tilde{\mathcal{T}}_I)$$

They satisfy the conditions (5.38)–(5.40). We put $\mathrm{Def}_{\varphi}(\mathcal{T}):=\big(\tilde{\mathcal{T}}_I\,;\tilde{f}_{I,i},\tilde{g}_{I,i}\big)$, and thus we obtain the functor $\mathrm{Def}_{\varphi}:\boldsymbol{C}(X,D,\boldsymbol{z})\longrightarrow\boldsymbol{C}(X,D,\boldsymbol{w})$. By using Lemma 4.2.9 inductively, we obtain that $\Psi_{X,\boldsymbol{w}}\circ\mathrm{Def}_{\varphi}(\mathcal{T})$ is naturally isomorphic to $\Psi_{X,\boldsymbol{z}}(\mathcal{T})$. □

Part II
Mixed Twistor $\mathscr{D}$-Modules

Chapter 6
Preliminary for Relative Monodromy Filtrations

Let us recall the degeneration phenomena of polarizable variations of pure Hodge structure on a punctured disk Δ^*, as a motivation to consider (relative) monodromy filtrations. Let $(\mathcal{V}_{\mathbb{Q}}, F)$ be a polarizable variation of Hodge structure of weight w on Δ^*, where $\mathcal{V}_{\mathbb{Q}}$ is a $\mathbb{Q}$-local system on Δ^*, and F is a Hodge filtration of the vector bundle $V = \mathcal{O}_{\Delta^*} \otimes_{\mathbb{Q}} \mathcal{V}_{\mathbb{Q}}$. We have the induced connection ∇ on V. For simplicity, we assume that the local monodromy automorphism is unipotent. Let $\tilde{V}$ be the Deligne extension of V on the disk Δ. Then, F is extended to a filtration of $\tilde{V}$ by subbundles. Moreover, we have an isomorphism of the fiber $\tilde{V}_{|0}$ and the space of multi-valued flat sections of V, which induces a $\mathbb{Q}$-structure on $\tilde{V}_{|0}$, i.e., we obtain a $\mathbb{Q}$-vector space $H_{\mathbb{Q}}$ with an isomorphism $H_{\mathbb{Q}} \otimes_{\mathbb{Q}} \mathbb{C} \simeq \tilde{V}_{|0}$. In general, $(H_{\mathbb{Q}}, \tilde{F}_{|0})$ is not a pure Hodge structure of weight w. Instead, we should consider the weight filtration induced by the residue $N = \mathrm{Res}(\nabla)$ on $\tilde{V}_{|0}$. Namely, let $W(N)$ denote the monodromy weight filtration of N on $\tilde{V}_{|0}$ i.e., $W(N)$ be the increasing filtration indexed by integers such that (1) $W_j(N) = 0$ $(j << 0)$, $W_j(N) = H_{\mathbb{Q}}$ $(j >> 0)$, (2) $N \cdot W_j(N) \subset W_{j-2}(N)$ $(j \in \mathbb{Z})$ (3) the induced morphisms $N^j : \mathrm{Gr}_j^{W(N)}(\tilde{V}_{|0}) \longrightarrow \mathrm{Gr}_{-j}^{W(N)}(\tilde{V}_{|0})$ are isomorphisms for any $j \geq 0$. It turns out that $W(N)$ is defined over $H_{\mathbb{Q}}$. Let W be the filtration on $H_{\mathbb{Q}}$ determined by $W_j = W_{j-w}(N)$. Then, $(H_{\mathbb{Q}}, W, F)$ is a polarizable mixed Hodge structure.

In other words, roughly speaking, when a pure object of weight w has singularity, it degenerates to a mixed object, and the weight filtration of the mixed object is the monodromy weight filtration of the induced nilpotent morphism up to the shift depending on w. When we consider the degeneration of a mixed object, *the relative monodromy filtration* should control the weight filtration on the degenerated object.

The concept of relative monodromy filtration is crucial in the study of variation of mixed Hodge structure and mixed Hodge modules. They are clearly inevitable even in the twistor setting. In Sect. 6.1, we recall some basic facts on relative monodromy filtrations by following [32, 87], and especially [73].

One of the first issues in the study of mixed twistor $\mathscr{D}$-modules is to obtain the correct weight filtration on the object obtained by a gluing procedure. Roughly, it

T. Mochizuki, *Mixed Twistor $\mathscr{D}$-modules*, Lecture Notes in Mathematics 2125,
DOI 10.1007/978-3-319-10088-3_6

is explained as follows. Let f be a holomorphic function on a complex manifold X. Let M_1 and M_2 be holonomic $\mathscr{D}$-modules on X such that $M_1(*f) = M_1$ and that $\mathrm{Supp}(M_2) \subset \{f = 0\}$. Suppose that we are given morphisms $\psi_f^{(1)}(M_1) \xrightarrow{u} M_2 \xrightarrow{v} \psi_f^{(0)}(M_1)$. We have the $\mathscr{D}$-module $\mathrm{Glue}(M_1, M_2, u, v)$ obtained as the cohomology of the complex

$$\psi_f^{(1)}(M_1) \longrightarrow \Xi_f^{(0)}(M_1) \oplus M_2 \longrightarrow \psi_f^{(0)}(M_1) \tag{6.1}$$

We would like to construct a "weight filtration" on $\mathrm{Glue}(M_1, M_2, u, v)$ from "weight filtrations" W on M_i. We have the induced filtrations L on $\psi_f^{(a)}(M_1)$ and $\Xi_f^{(0)}(M_1)$ naively induced by W:

$$L_j\psi_f^{(a)}(M_1) := \psi_f^{(a)}(W_jM_1), \quad L_j\Xi_f^{(0)}(M_1) := \Xi_f^{(0)}(W_jM_1)$$

As mentioned above, on the degenerated object such as $\psi_f^{(a)}(M_1)$ and $\Xi_f^{(0)}(M_1)$, the naively induced filtrations are not "the weight filtration", in general. In this situation, the morphisms u and v are not compatible with filtrations L on $\psi_f^{(a)}(M_1)$ and W on M_2. Instead, it turns out that u and v are compatible with W on M_2 and the shift of the relative monodromy filtration of N on $(\psi_f^{(a)}(M_1), L)$, where N is the natural nilpotent morphism. Namely, let $M\big(N, L(\psi_f^{(a)}(M_1))\big)$ denote the relative monodromy filtration. We set $W_j(\psi_f^{(a)}(M_1)) := M_{j-1+2a}\big(N, L(\psi_f^{(a)}(M_1))\big)$. Then, u and v are compatible with the "weight filtrations" W on $\psi_f^{(a)}(M_1)$ and M_2. For the construction of a filtration on $\mathrm{Glue}(M_1, M_2, u, v)$, we need a filtration L on M_2 such that the morphisms in (6.1) are compatible with the filtrations L on each term. Saito [73] showed that there is a unique choice of a filtration L on M_2 with the desired property. (Some special cases appeared in [32, 87].) We will review it Sect. 6.2. By applying such "transfer of filtration", we could obtain a correct weight filtration on $\mathrm{Glue}(M_1, M_2, u, v)$.

In Sect. 6.3, we consider the successive use of the procedure. It is suitable to impose some good properties which are summarized as **(P0–3)** and **(M0–3)**. The conditions naturally appeared in the study of polarizable mixed Hodge structures and infinitesimal mixed Hodge modules [7–9, 32, 35].

This chapter is independent of the previous chapters.

6.1 Relative Monodromy Filtrations

6.1.1 Definition and Basic Properties

Let $\mathcal{A}$ be an abelian category with a family of additive auto equivalences $\boldsymbol{\Sigma} = \big(\Sigma^{p,q} \,\big|\, p, q \in \mathbb{Z}\big)$ such that $\Sigma^{p,q} \circ \Sigma^{r,s} = \Sigma^{p+r,q+s}$. Let $C \in \mathcal{A}$. Let L be a finite increasing exhaustive complete filtration of C in $\mathcal{A}$ indexed by $\mathbb{Z}$, i.e., L is a tuple of subobjects $\big(L_k(C) \subset C \,\big|\, k \in \mathbb{Z}\big)$ such that (1) $L_k(C) \subset L_{k+1}(C)$, (2) $L_k(C) = C$ for any sufficiently large k, (3) $L_k(C) = 0$ for any sufficiently small k. In this monograph, such a pair (C, L) is called a filtered object in $\mathcal{A}$. A morphism of filtered objects $F : \big(C, L(C)\big) \longrightarrow \big(C', L(C')\big)$ is defined to be a morphism $F : C \longrightarrow C'$ preserving L, i.e., $F \cdot L_k(C) \subset L_k(C')$. The category of filtered objects in $\mathcal{A}$ is denoted by $\mathcal{A}^{\mathrm{fil}}$. It is an additive category. Let $(C, L) \in \mathcal{A}^{\mathrm{fil}}$. We consider two naturally induced filtrations $L^{(i)}$ $(i = 1, 2)$ on $\Sigma^{p,q}C$ given by $L^{(1)}_k \Sigma^{p,q}C = \Sigma^{p,q}(L_k C)$ and $L^{(2)}_k \Sigma^{p,q}C = \Sigma^{p,q}(L_{k+p+q}C)$. The object $(\Sigma^{p,q}C, L^{(1)})$ is denoted by $(\Sigma^{p,q}C, L)$, and $(\Sigma^{p,q}C, L^{(2)})$ is denoted by $\Sigma^{p,q}(C, L)$.

For $(C, L) \in \mathcal{A}^{\mathrm{fil}}$, put $\mathrm{Gr}^L_k(C) := L_k(C)/L_{k-1}(C) \in \mathcal{A}$. We have a natural isomorphism $\mathrm{Gr}^L_k \Sigma^{p,q}C \simeq \Sigma^{p,q}\,\mathrm{Gr}^L_k C$. If C is equipped with a filtration W, $\mathrm{Gr}^L_k(C)$ has the induced filtration given by $W_m \mathrm{Gr}^L_k(C) := \mathrm{Im}\big(W_m \cap L_k(C) \longrightarrow \mathrm{Gr}^L_k(C)\big)$.

We put $\boldsymbol{T} := \Sigma^{1,1}$. Let $\mathcal{A}^{\mathrm{nil}}$ be the category of objects $C \in \mathcal{A}$ equipped with a nilpotent endomorphism $N : C \longrightarrow \boldsymbol{T}^{-1}C$. A morphism $F : (C, N) \longrightarrow (C', N')$ in $\mathcal{A}^{\mathrm{nil}}$ is a morphism $F : C \longrightarrow C'$ such that $F \circ N = N' \circ F$. The category $\mathcal{A}^{\mathrm{nil}}$ is abelian. We set $\mathcal{A}^{\mathrm{fil,nil}} := (\mathcal{A}^{\mathrm{nil}})^{\mathrm{fil}}$. For $(C, L, N) \in \mathcal{A}^{\mathrm{fil,nil}}$, let $\mathrm{Gr}^L_k(N)$ denote the induced nilpotent endomorphism of $\mathrm{Gr}^L_k(C)$.

Let us recall the notion of relative monodromy filtration.

Definition 6.1.1 Let $(C, L, N) \in \mathcal{A}^{\mathrm{fil,nil}}$. A filtration W of C in $\mathcal{A}$ is called a relative monodromy filtration of N with respect to L if the following holds:

- $N \cdot W_i(C) \subset \boldsymbol{T}^{-1} W_{i-2}(C)$, i.e., N induces $(C, W, L) \longrightarrow \big(\boldsymbol{T}^{-1}(C, W), L\big)$.
- The induced morphisms

$$\mathrm{Gr}^W \mathrm{Gr}^L(N)^i : \mathrm{Gr}^W_{k+i} \mathrm{Gr}^L_k(C) \longrightarrow \boldsymbol{T}^{-i}\, \mathrm{Gr}^W_{k-i} \mathrm{Gr}^L_k(C)$$

are isomorphisms.

The relative monodromy filtration is often denoted by $M(N; L)$ in this paper. In this situation, we say that (C, L, N) has a relative monodromy filtration. □

According to §1.1 of [73] (see also [87]), a relative monodromy filtration is uniquely determined by Deligne's inductive formula, if it exists:

$$W_{-i+k}L_kC = W_{-i+k}L_{k-1}C + N^i \boldsymbol{T}^i\big(W_{i+k}L_kC\big) \qquad (i > 0) \tag{6.2}$$

$$W_{i+k}L_kC = \mathrm{Ker}\Big(N^{i+1} : L_kC \longrightarrow \boldsymbol{T}^{-i}\big(L_kC/W_{-i-2+k}L_kC\big)\Big) \qquad (i \geq 0) \tag{6.3}$$

We should recall that a relative monodromy filtration does not necessarily exist. If there exists a $k \in \mathbb{Z}$ such that $\mathrm{Gr}^L_m = 0$ unless $m = k$, then a relative monodromy filtration is the weight filtration of the nilpotent morphism up to a shift of the degree, and it always exists. By Deligne's inductive formula, relative monodromy filtrations are functorial in the following sense.

Lemma 6.1.2 *Assume that* $(C^{(i)}, L, N^{(i)}) \in \mathcal{A}^{\mathrm{fil,nil}}$ $(i = 1,2)$ *have relative monodromy filtrations* $M(N^{(i)}; L)$. *Let* $F : (C^{(1)}, L, N^{(1)}) \longrightarrow (C^{(2)}, L, N^{(2)})$ *be a morphism in* $\mathcal{A}^{\mathrm{fil,nil}}$. *Then,* $F \cdot M_k(N^{(1)}; L) \subset M_k(N^{(2)}; L)$. □

Let $\mathcal{A}^{\mathrm{RMF}}$ be the full subcategory of $\mathcal{A}^{\mathrm{fil,nil}}$, whose objects have relative monodromy filtrations. According to Lemma 6.1.2, the correspondence $(C, L, N) \longmapsto (C, M(N; L))$ gives a functor $\Phi_1 : \mathcal{A}^{\mathrm{RMF}} \longrightarrow \mathcal{A}^{\mathrm{fil}}$. We have $\Phi_1 \circ \Sigma^{p,q}(C, L, N) = \Sigma^{p,q}(C, M(N; L))$.

6.1.2 Canonical Decomposition

Let us recall the notion of canonical decomposition due to Kashiwara [32], with a generalization in [73], which is one of the most fundamental in the study of relative monodromy filtrations. Let $(C, L, N) \in \mathcal{A}^{\mathrm{RMF}}$. Let M denote the relative monodromy filtration for (C, L, N). We put $C^{(0)} := \mathrm{Gr}^M(C)$. It is equipped with a filtration induced by L, which is also denoted by L. Then, there exists a canonical splitting of the induced filtration $L(C^{(0)})$ in $\mathcal{A}$:

$$C^{(0)} = \bigoplus C_i^{(0)}, \quad \text{such that } L_j C^{(0)} = \bigoplus_{i \le j} C_i^{(0)} \tag{6.4}$$

(See [32, 73] for the explicit construction of the splitting.) In particular, we have $C_i^{(0)} \simeq \mathrm{Gr}_i^L \mathrm{Gr}^M(C) \simeq \mathrm{Gr}^M \mathrm{Gr}_i^L(C)$ in $\mathcal{A}$. It is functorial in the following sense, which is clear by the construction in [32, 73].

Lemma 6.1.3 *Let* $F : (C^{(1)}, L, N) \longrightarrow (C^{(2)}, L, N)$ *be a morphism in* $\mathcal{A}^{\mathrm{RMF}}$. *Let* $M = M(N; L)$ *on* $C^{(i)}$. *Then, the induced morphism* $\mathrm{Gr}^M(F) : \mathrm{Gr}^M(C^{(1)}) \longrightarrow \mathrm{Gr}^M(C^{(2)})$ *preserves the canonical splittings* (6.4). □

6.1.3 A Criterion

Let us recall a condition for the existence of relative monodromy filtration in [87]. Let $\mathcal{A}$ be an abelian category. Let $(C, L, N) \in \mathcal{A}^{\mathrm{nil,fil}}$ such that (1) $L_k(C) = L_{k'}(C)$ for any $k' \ge k$, (2) $(L_{k-1}(C), L, N) \in \mathcal{A}^{\mathrm{RMF}}$. Let $M\big(L_{k-1}(C)\big)$ denote the relative monodromy filtration of $(L_{k-1}(C), L, N)$. We have a naturally defined morphism

$$\mathrm{Ker}\big(N^\ell : \boldsymbol{T}^\ell \,\mathrm{Gr}_k^L \longrightarrow \mathrm{Gr}_k^L\big) \xrightarrow{\ a\ } \frac{L_{k-1}}{N^\ell \boldsymbol{T}^\ell L_{k-1} + M_{k-\ell-1}(L_{k-1}(C))} \tag{6.5}$$

Here, a is induced by N^ℓ.

Proposition 6.1.4 ([87], See Also [73]) *We have* $(C, L, N) \in \mathcal{A}^{\mathrm{RMF}}$ *if and only if the morphisms* (6.5) *vanish for all integers* $\ell > 0$. □

6.1.4 Functoriality for Tensor Product and Duality

Let us consider the case that $\mathcal{A}$ is the category Vect_K of finite dimensional vector spaces over a ground field K, and $\Sigma^{p,q} = \mathrm{id}$. Recall that we have tensor product and inner homomorphism in the category $\mathrm{Vect}_K^{\mathrm{fil,nil}}$, given in the standard manner.

Let $(V^{(i)}, L, N^{(i)}) \in \mathrm{Vect}_K^{\mathrm{fil,nil}}$ $(i = 1, 2)$. The tensor product $V^{(1)} \otimes V^{(2)}$ has the induced endomorphism $N_{V^{(1)} \otimes V^{(2)}} := N^{(1)} \otimes 1 + 1 \otimes N^{(2)}$, and the filtration of $V^{(1)} \otimes V^{(2)}$ given by $L_k(V^{(1)} \otimes V^{(2)}) = \sum_{p+q \leq k} L_p(V^{(1)}) \otimes L_q(V^{(2)})$. The tuple $(V^{(1)} \otimes V^{(2)}, L, N)$ is also denoted by $(V^{(1)}, L, N^{(1)}) \otimes (V^{(2)}, L, N^{(2)})$. The space $\mathrm{Hom}(V^{(1)}, V^{(2)})$ has the induced endomorphism $N_{\mathrm{Hom}(V^{(1)},V^{(2)})}(f) := N^{(2)} \circ f - f \circ N^{(1)}$ and the filtration L given by

$$L_k \mathrm{Hom}(V^{(1)}, V^{(2)}) := \bigl\{ f \in \mathrm{Hom}(V^{(1)}, V^{(2)}) \,\big|\, f(L_j V^{(1)}) \subset L_{j+k} V^{(2)} \ \forall j \bigr\}.$$

The tuple $\bigl(\mathrm{Hom}(V^{(1)}, V^{(2)}), L, N\bigr)$ is also denoted by

$$\mathrm{Hom}\bigl((V^{(1)}, L, N^{(1)}), (V^{(2)}, L, N^{(2)})\bigr).$$

The following proposition is due to Steenbrink and Zucker [87].

Proposition 6.1.5 ([87]) $\mathrm{Vect}_K^{\mathrm{RMF}}$ *is closed under tensor product and inner homomorphism in* $\mathrm{Vect}_K^{\mathrm{fil,nil}}$. *Moreover, the relative monodromy filtrations for*

$$(V^{(1)}, L, N^{(1)}) \otimes (V^{(2)}, L, N^{(2)}) \quad \text{and} \quad \mathrm{Hom}\bigl((V^{(1)}, L, N^{(1)}), (V^{(2)}, L, N^{(2)})\bigr)$$

are naturally induced by $M(N^{(i)}; L)$ $(i = 1, 2)$. □

In particular, if $(V, L, N) \in \mathrm{Vect}_K^{\mathrm{RMF}}$, its dual object is also an object in $\mathrm{Vect}_K^{\mathrm{RMF}}$, and the relative monodromy filtration is naturally induced by $M(N; L)$.

6.2 Transfer of Filtrations

In some cases, a relative monodromy filtration or a base filtration is inherited. A fundamental general result is due to Saito [73], which we will review in Sect. 6.2.3.

6.2.1 Gluing Data

We introduce some terminology for our argument. Let $\mathcal{A}$ be an abelian category with additive auto equivalences $\boldsymbol{\Sigma}$.

Definition 6.2.1 A tuple $(C, C'; u, v)$ of morphisms in $\mathcal{A}$

$$\Sigma^{1,0} C \xrightarrow{\ u\ } C' \xrightarrow{\ v\ } \Sigma^{0,-1} C$$

is called a gluing datum in $(\mathcal{A}, \boldsymbol{\Sigma})$. We define a morphism of gluing data $(C_1, C_1'; u_1, v_1) \longrightarrow (C_2, C_2'; u_2, v_2)$ as a commutative diagram:

$$\begin{array}{ccccc} \Sigma^{1,0}C_1 & \xrightarrow{u_1} & C_1' & \xrightarrow{v_1} & \Sigma^{0,-1}C_1 \\ {\scriptstyle F}\downarrow & & {\scriptstyle F'}\downarrow & & {\scriptstyle F}\downarrow \\ \Sigma^{1,0}C_2 & \xrightarrow{u_2} & C_2' & \xrightarrow{v_2} & \Sigma^{0,-1}C_2 \end{array}$$

The category of gluing data in $(\mathcal{A}, \boldsymbol{\Sigma})$ is denoted by $\mathrm{Glu}(\mathcal{A}, \boldsymbol{\Sigma})$. □

An object of $\mathrm{Glu}(\mathcal{A}, \boldsymbol{\Sigma})^{\mathrm{fil}}$ is often denoted by $(C, C'; u, v; L)$, which means an object $(C, C'; u, v)$ with a filtration L in $\mathrm{Glu}(\mathcal{A}, \boldsymbol{\Sigma})$.

Definition 6.2.2 • $(C, C'; u, v) \in \mathrm{Glu}(\mathcal{A}, \boldsymbol{\Sigma})$ is called S-decomposable if $C' = \mathrm{Im}\, u \oplus \mathrm{Ker}\, v$.

- $(C, C'; u, v; L) \in \mathrm{Glu}(\mathcal{A}, \boldsymbol{\Sigma})^{\mathrm{fil}}$ is called filtered S-decomposable if the associated graded object $\mathrm{Gr}^L(C, C'; u, v)$ is S-decomposable. □

In the following, when we say that $(C, C'; u, v; L)$ is filtered S-decomposable for given $(C, L), (C', L) \in \mathcal{A}^{\mathrm{fil}}$ with morphisms $u : \Sigma^{1,0}C \longrightarrow C'$ and $v : C' \longrightarrow \Sigma^{0,-1}C$, we implicitly imply that u and v preserve the filtration L, i.e., $u : (\Sigma^{1,0}C, L) \longrightarrow (C', L)$ and $v : (C', L) \longrightarrow (\Sigma^{0,-1}C, L)$. We remark that this kind of condition appeared in the study on mixed Hodge structure in [32, 73]. The terminology "S-decomposable" is taken from [66].

Let $\mathrm{Glu}(\mathcal{A}^{\mathrm{fil}}, \boldsymbol{\Sigma})$ denote the category of gluing data in $\mathcal{A}^{\mathrm{fil}}$, where the action of $\Sigma^{p,q}$ on $\mathcal{A}^{\mathrm{fil}}$ is given by $(C, L) \longmapsto \Sigma^{p,q}(C, L)$. An object of $\mathrm{Glu}(\mathcal{A}^{\mathrm{fil}}, \boldsymbol{\Sigma})$ is often denoted by $(C, C', \tilde{L}; u, v)$, i.e., a pair of objects $(C, \tilde{L})$ and $(C', \tilde{L})$ with morphisms in $\mathcal{A}^{\mathrm{fil}}$:

$$\Sigma^{1,0}(C, \widetilde{L}) \xrightarrow{u} (C', \widetilde{L}) \xrightarrow{v} \Sigma^{0,-1}(C, \widetilde{L})$$

A morphism $(C_1, C_1', \tilde{L}; u_1, v_1) \longrightarrow (C_2, C_2', \tilde{L}; u_2, v_2)$ in $\mathrm{Glu}(\mathcal{A}^{\mathrm{fil}}, \boldsymbol{\Sigma})$ is a commutative diagram:

$$\begin{array}{ccccc} \Sigma^{1,0}(C_1, \widetilde{L}) & \xrightarrow{u_1} & (C_1', \widetilde{L}) & \xrightarrow{v_1} & \Sigma^{0,-1}(C_1, \widetilde{L}) \\ \downarrow & & \downarrow & & \downarrow \\ \Sigma^{1,0}(C_2, \widetilde{L}) & \xrightarrow{u_2} & (C_2', \widetilde{L}) & \xrightarrow{v_2} & \Sigma^{0,-1}(C_2, \widetilde{L}) \end{array}$$

6.2.2 Inheritance of Relative Monodromy Filtration

Let us recall a lemma due to Kashiwara. Let $(C, C'; u, v; L) \in \mathrm{Glu}(\mathcal{A}, \boldsymbol{\Sigma})^{\mathrm{fil}}$. We put $N := v \circ u$ and $N' := u \circ v$.

Proposition 6.2.3 ([32, 73]) *Assume that (i) the object $(C, C'; u, v; L)$ is filtered S-decomposable, (ii) N' is nilpotent, and $(C', L, N') \in \mathcal{A}^{\mathrm{RMF}}$. Then, we also have $(C, L, N) \in \mathcal{A}^{\mathrm{RMF}}$. Moreover, u and v give the following morphisms in $\mathcal{A}^{\mathrm{fil}}$:*

$$\Sigma^{1,0}(C, M(N; L)) \xrightarrow{u} (C', M(N'; L)) \xrightarrow{v} \Sigma^{0,-1}(C, M(N; L))$$

i.e., $u \cdot M_k(N; LC) \subset M_{k-1}(N'; LC')$ and $v \cdot M_k(N'; LC') \subset M_{k-1}(N; LC)$.

Proof We give only a remark. In [32], the case $\Sigma^{p,q} = \mathrm{id}$ is proved. The key is Lemma 3.32 in [32], which was generalized in Corollary 1.7 [73]. Then, we can deduce the claim in the general case by using the argument in [32]. □

We have an obvious reformulation. Let $\mathcal{C}_1$ be the full subcategory of $\mathrm{Glu}(\mathcal{A}, \boldsymbol{\Sigma})^{\mathrm{fil}}$ whose objects $(C, C'; u, v; L)$ are filtered S-decomposable and satisfy the following:

- We put $N := v \circ u$ and $N' := u \circ v$, and then (C, L, N) and (C', L, N') are objects in $\mathcal{A}^{\mathrm{RMF}}$.
- We put $\tilde{L}(C) := M\big(N; L(C)\big)$ and $\tilde{L}(C') := M\big(N'; L(C')\big)$, and then we have $(C, C', \tilde{L}; u, v) \in \mathrm{Glu}(\mathcal{A}^{\mathrm{fil}}, \boldsymbol{\Sigma})$.

Let $\mathcal{C}_2$ be the full subcategory of $\mathrm{Glu}(\mathcal{A}, \boldsymbol{\Sigma})^{\mathrm{fil}}$ whose objects $(C, C'; u, v; L)$ are filtered S-decomposable and satisfy the following:

- We put $N' := u \circ v$, and then (C', L, N') is an object in $\mathcal{A}^{\mathrm{RMF}}$.

According to Proposition 6.2.3, we have $\mathcal{C}_1 = \mathcal{C}_2$.

6.2.3 *Transfer of Filtration*

We shall recall a fundamental result due to Saito in [73]. Let $(C, C'; u, v) \in \mathrm{Glu}(\mathcal{A})$. We set $N := v \circ u$ and $N' := u \circ v$. Assume that C and C' are equipped with two filtrations L and $\tilde{L}$ such that $(C, C'; u, v; L) \in \mathrm{Glu}(\mathcal{A}, \boldsymbol{\Sigma})^{\mathrm{fil}}$ and $(C, C', \tilde{L}; u, v) \in \mathrm{Glu}(\mathcal{A}^{\mathrm{fil}}, \boldsymbol{\Sigma})$, i.e., u and v give the following morphisms in $\mathcal{A}^{\mathrm{fil}}$:

$$(\Sigma^{1,0}C, L) \xrightarrow{u} (C', L) \xrightarrow{v} (\Sigma^{0,-1}C, L) \tag{6.6}$$

$$\Sigma^{1,0}(C, \tilde{L}) \xrightarrow{u} (C', \tilde{L}) \xrightarrow{v} \Sigma^{0,-1}(C, \tilde{L}) \tag{6.7}$$

Proposition 6.2.4 (Corollary 1.9 of [73]) *We assume $(C, L, N) \in \mathcal{A}^{\mathrm{RMF}}$ and $\tilde{L}(C) = M(N; LC)$. Then the following conditions are equivalent.*

(A1) $(C, C'; u, v; L)$ *is filtered S-decomposable, and we have* $(C', L, N') \in \mathcal{A}^{\mathrm{RMF}}$ *and* $M(N'; LC') = \tilde{L}(C')$.

(A2) $L_k C' = u\big(\Sigma^{1,0} L_k C\big) + \Big(v^{-1}(\Sigma^{0,-1} L_k C) \cap \tilde{L}_k C'\Big)$ *for each k.*

(A2') $L_k C' = v^{-1}\big(\Sigma^{0,-1} L_k C\big) \cap \Big(u(\Sigma^{1,0} L_k C) + \tilde{L}_k C'\Big)$ *for each k.* □

Let us mention some immediate consequences of Proposition 6.2.4. Let $(C, C', \tilde{L}; u, v) \in \mathrm{Glu}(\mathcal{A}^{\mathrm{fil}}, \boldsymbol{\Sigma})$. Let $N := v \circ u$ and $N' := u \circ v$. Note that they are nilpotent. Let $L(C)$ be a filtration of C such that $(C, L, N) \in \mathcal{A}^{\mathrm{RMF}}$ and $\tilde{L}(C) = M\big(N; L(C)\big)$.

Corollary 6.2.5 *Under the situation, there exists a unique filtration* $L(C')$ *such that (i)* $(C, C'; u, v; L) \in \mathrm{Glu}(\mathcal{A}, \boldsymbol{\Sigma})^{\mathrm{fil}}$ *is filtered S-decomposable, (ii)* $M\big(N'; L(C')\big) = \tilde{L}(C')$.

Proof The filtration $L(C')$ is given by Saito's formula in Proposition 6.2.4. □

Definition 6.2.6 The filtration $L(C')$ in Corollary 6.2.5 is called the transfer of $L(C)$ by (u, v) in this paper. □

By Proposition 6.2.4, the transfer is functorial. It is reformulated as follows. Let $\mathcal{C}_3$ be the category of objects $\big((C, C', \tilde{L}; u, v), L(C)\big)$, where (1) $(C, C', \tilde{L}; u, v) \in \mathrm{Glu}(\mathcal{A}^{\mathrm{fil}}, \boldsymbol{\Sigma})$, (2) $L(C)$ is an exhaustive complete finite increasing filtration of C in $\mathcal{A}$, satisfying the following:

- We put $N := v \circ u$, and then $(C, L, N) \in \mathcal{A}^{\mathrm{RMF}}$ and $\tilde{L}(C) = M\big(N; L(C)\big)$.

A morphism $\big((C_1, C_1', \tilde{L}; u_1, v_1), L(C_1)\big) \longrightarrow \big((C_2, C_2', \tilde{L}; u_2, v_2), L(C_2)\big)$ in $\mathcal{C}_3$ is a pair (F, F') of morphisms

$$F : C_1 \longrightarrow C_2, \qquad F' : C_1' \longrightarrow C_2'$$

such that (1) (F, F') gives a morphism $(C_1, C_1', \tilde{L}; u_1, v_1) \longrightarrow (C_2, C_2', \tilde{L}; u_2, v_2)$ in $\mathrm{Glu}(\mathcal{A}^{\mathrm{fil}}, \boldsymbol{\Sigma})$, (2) F gives a morphism $\big(C_1, L(C_1)\big) \longrightarrow \big(C_2, L(C_2)\big)$ in $\mathcal{A}^{\mathrm{fil}}$.

Let $\mathcal{C}_1$ be the category considered in Sect. 6.2.2. We have the functor $\Psi : \mathcal{C}_1 \longrightarrow \mathcal{C}_3$ given by

$$\Psi(C, C'; u, v; L) = \big((C, C', \tilde{L}; u, v), L(C)\big),$$

where $\tilde{L}(C) := M\big(N; L(C)\big)$ and $\tilde{L}(C') := M\big(N; L(C')\big)$.

Corollary 6.2.7 *The functor* Ψ *is an equivalence.*

Proof Essential surjectivity is already stated in Corollary 6.2.5. The functor is clearly faithful. It is full by Saito's formula in Proposition 6.2.4. □

6.2.4 Special Case

As remarked in [73], Proposition 6.2.4 can be regarded as a generalization of the filtrations introduced in [32, 87], which we recall here.

Let $(C_1, L, N) \in \mathcal{A}^{\mathrm{RMF}}$, and let $W := M(N; L)$. We have the induced filtration $N_* L$ on C_1, obtained as the transfer in the following case:

$$(C, L, \tilde{L}) = \Sigma^{0,1}(C_1, L, W), \quad (C', \tilde{L}) = (C_1, W), \quad u = N, \quad v = \mathrm{id}.$$

We also have the filtration $N_! L$ on C_1 obtained as the transfer in the following case:

$$(C, L, \tilde{L}) = \Sigma^{-1,0}(C_1, L, W), \quad (C', \tilde{L}) = (C_1, W), \quad u = \mathrm{id}, \quad v = N.$$

Indeed, the explicit formulas are given in [32, 87] (we omit $\boldsymbol{\Sigma}$):

$$(N_* L)_k = NL_{k+1} + M_k(N; L) \cap L_k = NL_{k+1} + M_k(N; L) \cap L_{k+1}$$
$$(N_! L)_k = L_{k-1} + M_k(N; L) \cap N^{-1} L_{k-1} = L_{k-1} + M_k(N; L) \cap N^{-1} L_{k-2}$$

The morphism $N : (C_1, W) \longrightarrow \boldsymbol{T}^{-1}(C_1, W)$ induces $(C_1, N_! L) \longrightarrow \boldsymbol{T}^{-1}(C_1, N_* L)$, which follows from the commutativity of the following diagram:

$$\begin{array}{ccccc}
(C_1, W) & \xrightarrow{\mathrm{id}} & (C_1, W) & \xrightarrow{N} & \boldsymbol{T}^{-1}(C_1, W) \\
\mathrm{id}\big\downarrow & & N\big\downarrow & & \mathrm{id}\big\downarrow \\
(C_1, W) & \xrightarrow{N} & \boldsymbol{T}^{-1}(C_1, W) & \xrightarrow{\mathrm{id}} & \boldsymbol{T}^{-1}(C_1, W)
\end{array}$$

We define filtrations $\hat{N}_* L$ on $\Sigma^{0,-1} C_1$, and $\hat{N}_! L$ on $\Sigma^{1,0} C_1$ as follows:

$$(\Sigma^{0,-1} C_1, \hat{N}_* L) = \Sigma^{0,-1}(C_1, N_* L), \qquad (\Sigma^{1,0} C_1, \hat{N}_! L) = \Sigma^{1,0}(C_1, N_! L)$$

6.2.5 Duality and Tensor Product

Let us consider the case that $\mathcal{A}$ is the category of finite dimensional vector spaces with $\Sigma^{p,q} = \mathrm{id}$. We have the functoriality of transfer for duality. Let $(C, C'; u, v; L) \in \mathrm{Glu}(\mathcal{A}, \boldsymbol{\Sigma})^{\mathrm{fil}}$. Then, we have the induced morphisms of dual filtered vector spaces

$$(C, L)^\vee \xrightarrow{v^\vee} (C', L)^\vee \xrightarrow{u^\vee} (C, L)^\vee.$$

Set $(C, C'; u, v; L)^\vee := \big(C^\vee, (C')^\vee; v^\vee, u^\vee; L\big) \in \mathrm{Glu}(\mathcal{A}, \boldsymbol{\Sigma})^{\mathrm{fil}}$. If $(C, C'; u, v; L)$ is a filtered S-decomposable, $(C, C'; u, v; L)^\vee$ is also filtered S-decomposable.

By the functoriality of relative monodromy filtration with respect to duality, the correspondence induces contravariant functors $\vee_i$ on the category $\mathcal{C}_i$ $(i = 1, 2)$.

An object $(C, C', W; u, v) \in \mathrm{Glu}(\mathcal{A}^{\mathrm{fil}}, \boldsymbol{\Sigma})$ has its dual object

$$(C, C', W; u, v)^{\vee} := \big(C^{\vee}, (C')^{\vee}, W; v^{\vee}, u^{\vee}\big) \in \mathrm{Glu}(\mathcal{A}^{\mathrm{fil}}, \boldsymbol{\Sigma})$$

If $(C, C', W; u, v)$ is equipped with a filtration L of C, $(C, C', W; u, v)^{\vee}$ is also equipped with an induced filtration L of $C^{\vee}$. By the functoriality of relative monodromy filtration with respect to duality, the correspondence of objects $\big((C, C', W; u, v), L(C)\big)$ and $\big((C, C', W; u, v)^{\vee}, L(C^{\vee})\big)$ gives a contravariant functor $\vee_3$ on $\mathcal{C}_3$. The following lemma is clear by construction.

Lemma 6.2.8 *We have $\Psi \circ \vee_1 = \vee_3 \circ \Psi$ as functors from $\mathcal{C}_1$ to $\mathcal{C}_3$.* □

Let us reword the claim of the lemma. Let $\big((C, C', W; u, v), L(C)\big) \in \mathcal{C}_3$. We have a filtration $L(C')$ obtained as the transfer of $L(C)$ by (u, v). We also have $\big((C, C', W; u, v)^{\vee}, L(C^{\vee})\big) \in \mathcal{C}_3$, and we obtain a filtration $L(C'^{\vee})$ obtained as the transfer of $L(C^{\vee})$. Then, according to Lemma 6.2.8, $L(C'^{\vee})$ is the same as the filtration obtained as the dual filtration of $L(C')$.

Remark 6.2.9 Let $(C, L, N) \in \mathcal{A}^{\mathrm{RMF}}$. Let $N_! L$ and $N_* L$ be the filtrations in Sect. 6.2.4. Then, as remarked in [32], $N_! L(C^{\vee})$ is obtained as the dual filtration of $N_* L(C)$, i.e., we have $(N_! L)_k(C^{\vee}) = (N_* L)_{-k-1}(C)^{\perp}$. It also follows from the above functoriality of transfer with respect to duality. □

Let $(U, L) \in \mathcal{A}^{\mathrm{fil}}$. For any $(C, C', L; u, v) \in \mathrm{Glu}(\mathcal{A}, \boldsymbol{\Sigma})^{\mathrm{fil}}$, we have the naturally induced object $(C, C', L; u, v) \otimes U \in \mathrm{Glu}(\mathcal{A}, \boldsymbol{\Sigma})^{\mathrm{fil}}$, where the filtrations of $C \otimes U$ and $C' \otimes U$ are induced by L. If $(C, C', L; u, v)$ is filtered S-decomposable, $(C, C', L; u, v) \otimes U$ is also filtered S-decomposable. By the functoriality of relative monodromy filtrations with respect to tensor products, it induces functors $\otimes U$ on $\mathcal{C}_i$ $(i = 1, 2)$.

For $(C, C', W; u, v) \in \mathrm{Glu}(\mathcal{A}^{\mathrm{fil}}, \boldsymbol{\Sigma})$, we have the naturally induced object

$$(C, C', W; u, v) \otimes U \in \mathrm{Glu}(\mathcal{A}^{\mathrm{fil}}, \boldsymbol{\Sigma}),$$

where the filtrations of $C \otimes U$ and $C' \otimes U$ are induced by $W(C)$, $W(C')$ and $L(U)$. If $(C, C', W; u, v)$ is equipped with a filtration L of C, then $(C, C', W; u, v) \otimes U$ is equipped with a filtration L of $C \otimes U$ induced by the filtrations L of C and U. By the functoriality of relative monodromy filtrations, the correspondence of $\big((C, C', W; u, v), L(C)\big)$ and $\big((C, C', W; u, v) \otimes U, L(C \otimes U)\big)$ induces a functor $\otimes U$ on $\mathcal{C}_3$. The following lemma is clear by construction.

Lemma 6.2.10 *We have $\Psi \circ (\otimes U) = (\otimes U) \circ \Psi$.* □

Let us reword the lemma. Let $(U, L) \in \mathcal{A}^{\mathrm{fil}}$ and $\big((C, C', W; u, v), L(C)\big) \in \mathcal{C}_3$. We have a filtration $L(C')$ obtained as the transfer of $L(C)$ by (u, v). We have $\big((C, C', W; u, v) \otimes U, L(C \otimes U)\big) \in \mathcal{C}_3$, and we obtain a filtration $L(C' \otimes U)$ obtained as the transfer of $L(C \otimes U)$. Then, according to Lemma 6.2.10, $L(C' \otimes U)$ is the same as the filtration induced by the filtrations $L(C')$ and $L(U)$.

6.3 Pure and Mixed Objects

This subsection is a preparation for gluing of admissible variations of mixed twistor structure (Chaps. 8–10). The contents of this subsection are essentially contained in [32, 73].

6.3.1 Setting

Let $\mathcal{A}$ be an abelian category with a family of additive auto equivalences $\boldsymbol{\Sigma}$ as in Sect. 6.1.1. For any finite set Λ, we shall consider the category $\mathcal{A}(\Lambda)$ of objects C in $\mathcal{A}$ with a commuting Λ-tuple of nilpotent morphisms $N_i : C \longrightarrow \boldsymbol{T}^{-1}C$ $(i \in \Lambda)$. A morphism $(C_1,\boldsymbol{N}) \longrightarrow (C_2,\boldsymbol{N})$ in $\mathcal{A}(\Lambda)$ is a morphism $F : C_1 \longrightarrow C_2$ such that $N_j \circ F = F \circ N_j$. The category $\mathcal{A}(\Lambda)$ is abelian. It is equipped with auto equivalences given by $(C,\boldsymbol{N}) \longmapsto (\Sigma^{p,q}C,\boldsymbol{N})$, which is also denoted by $\Sigma^{p,q}$. When we are given $(C,\boldsymbol{N}) \in \mathcal{A}(\Lambda)$, for any subset $\Lambda_0 \subset \Lambda$, we set $N(\Lambda_0) := \sum_{i\in\Lambda_0} N_i$, and the weight filtration $M\big(N(\Lambda_0)\big)$ is denoted by $M(\Lambda_0)$.

Let $\Phi : \Lambda_1 \longrightarrow \Lambda_2$ be a map. For $(C_1,\boldsymbol{N}_{\Lambda_1}) \in \mathcal{A}(\Lambda_1)$, we have a naturally induced object $(C_1,\boldsymbol{N}_{\Lambda_2}) \in \mathcal{A}(\Lambda_2)$, where $N_j := \sum_{i\in\Phi^{-1}(j)} N_i$ for $j \in \operatorname{Im}\Phi$, and $N_j := 0$ for $j \notin \operatorname{Im}\Phi$. It gives a functor $\Phi_* : \mathcal{A}(\Lambda_1) \longrightarrow \mathcal{A}(\Lambda_2)$. For $(C_2,\boldsymbol{N}_{\Lambda_2}) \in \mathcal{A}(\Lambda_2)$, we have a naturally induced object $(C_1,\boldsymbol{N}_{\Lambda_1}) \in \mathcal{A}(\Lambda_1)$, where $N_j := N_{\Phi(j)}$ for $j \in \Lambda_1$. It gives a functor $\Phi^* : \mathcal{A}(\Lambda_2) \longrightarrow \mathcal{A}(\Lambda_1)$. If Φ is a bijection, both Φ_* and Φ^* are equivalences.

Assume that we are given saturated full subcategories $\mathcal{P}_w(\Lambda) \subset \mathcal{A}(\Lambda)$ $(w \in \mathbb{Z})$ for any finite sets Λ with the following property:

(P0) $\mathcal{P}_w(\Lambda)$ are abelian subcategories of $\mathcal{A}(\Lambda)$, and they are semisimple. Any injection $\Phi : \Lambda_1 \longrightarrow \Lambda_2$ induces a functor $\Phi_* : \mathcal{P}_w(\Lambda_1) \longrightarrow \mathcal{P}_w(\Lambda_2)$, i.e., for $(C,\boldsymbol{N}) \in \mathcal{P}_w(\Lambda_1)$, we have $\Phi_*(C,\boldsymbol{N}) \in \mathcal{P}_w(\Lambda_2)$.
(P1) $\Sigma^{p,q}$ induces an equivalence $\mathcal{P}_w(\Lambda) \simeq \mathcal{P}_{w-p-q}(\Lambda)$.
(P2) For $(C_i,\boldsymbol{N}) \in \mathcal{P}_{w_i}(\Lambda)$ with $w_1 > w_2$, we have

$$\operatorname{Hom}_{\mathcal{A}(\Lambda)}\big((C_1,\boldsymbol{N}),(C_2,\boldsymbol{N})\big) = 0.$$

(P3) Let $(C,\boldsymbol{N}) \in \mathcal{P}_w(\Lambda)$.

(P3.1) For any $\Lambda_2 \subset \Lambda_1 \subset \Lambda$, there exists a relative monodromy filtration $M\big(N(\Lambda_1);M(\Lambda_2)\big)$, and it is equal to $M(\Lambda_1)$.
(P3.2) $\big(\operatorname{Gr}^{M(\Lambda_1)}_{w_1}(C),\boldsymbol{N}_{\Lambda_1^c}\big)$ is an object in $\mathcal{P}_{w+w_1}(\Lambda_1^c)$, where $\Lambda_1^c := \Lambda \setminus \Lambda_1$.

(P3.3) For $\bullet \in \Lambda$, $\big(\operatorname{Im} N_\bullet, \boldsymbol{N}\big)$ is an object in $\mathcal{P}_{w+1}(\Lambda)$. Moreover, for $* \in \Lambda \setminus \bullet$, $\big(C, \Sigma^{1,0} \operatorname{Im} N_\bullet; u, v; M(*)\big)$ is filtered S-decomposable, where $u : \Sigma^{1,0} C \longrightarrow \Sigma^{1,0} \operatorname{Im} N_\bullet$ and $v : \Sigma^{1,0} \operatorname{Im} N_\bullet \longrightarrow \Sigma^{0,-1} C$ are induced by $N_\bullet$ and the canonical morphism $\operatorname{Im} N_\bullet \longrightarrow \boldsymbol{T}^{-1} C$, respectively. Namely,

$$\Sigma^{1,0} \operatorname{Gr}^{M(*)} C \longrightarrow \Sigma^{1,0} \operatorname{Gr}^{M(*)} \operatorname{Im} N_\bullet \longrightarrow \Sigma^{0,-1} \operatorname{Gr}^{M(*)} C$$

is S-decomposable.

We have the category $\mathcal{A}(\Lambda)^{\mathrm{fil}}$ of the filtered objects in $\mathcal{A}(\Lambda)$, on which $\Sigma^{p,q}$ naturally acts by $\Sigma^{p,q}(C, L, \boldsymbol{N}) = \big(\Sigma^{p,q}(C, L), \boldsymbol{N}\big)$. Let $\mathcal{M}'(\Lambda) \subset \mathcal{A}(\Lambda)$ be the saturated full subcategory of the objects $(C, L, \boldsymbol{N})$ such that $\operatorname{Gr}^L_w(C, \boldsymbol{N}) \in \mathcal{P}_w(\Lambda)$ for any $w \in \mathbb{Z}$.

Lemma 6.3.1 *Let $F : (C_1, L, \boldsymbol{N}) \longrightarrow (C_2, L, \boldsymbol{N})$ be a morphism in $\mathcal{M}'(\Lambda)$.*

- *F is strict with respect to L.*
- *Let $(\operatorname{Ker} F, \boldsymbol{N})$, $(\operatorname{Im} F, \boldsymbol{N})$ and $(\operatorname{Cok} F, \boldsymbol{N})$ be the kernel, the image and the cokernel in $\mathcal{A}(\Lambda)$. They are equipped with the naturally induced filtrations L, with which they are objects in $\mathcal{M}'(\Lambda)$.*
- *In particular, $\mathcal{M}'(\Lambda)$ is abelian, and the forgetful functor $\mathcal{M}'(\Lambda) \longrightarrow \mathcal{A}(\Lambda)$ is exact.*

Proof The first claim follows from **P0** and **P2**. Then, the second claim follows from **P0** and the first claim. □

Corollary 6.3.2 *A sequence*

$$0 \longrightarrow (C_1, L, \boldsymbol{N}_1) \longrightarrow (C_2, L, \boldsymbol{N}_2) \longrightarrow (C_3, L, \boldsymbol{N}_3) \longrightarrow 0$$

in $\mathcal{M}'(\Lambda)$ is exact if and only if

$$0 \longrightarrow (C_1, \boldsymbol{N}_1) \longrightarrow (C_2, \boldsymbol{N}_2) \longrightarrow (C_3, \boldsymbol{N}_3) \longrightarrow 0$$

is exact in $\mathcal{A}(\Lambda)$. □

Assume that we are given a saturated full subcategory $\mathcal{M}(\Lambda) \subset \mathcal{M}'(\Lambda)$ with the following property:

(M0) Any injection $\Phi : \Lambda \longrightarrow \Lambda_1$ induces a functor $\Phi_* : \mathcal{M}(\Lambda) \longrightarrow \mathcal{M}(\Lambda_1)$. The categories $\mathcal{M}(\Lambda)$ are abelian subcategories of $\mathcal{M}'(\Lambda)$. We naturally have $\mathcal{P}_w(\Lambda) \subset \mathcal{M}(\Lambda)$ for any $w \in \mathbb{Z}$, i.e., if $(C, L) \in \mathcal{M}'(\Lambda)$ satisfies $\operatorname{Gr}^L_j(C) = 0$ unless $j \neq w$, then $(C, L) \in \mathcal{M}(\Lambda)$.

(M1) If $(C, L, \boldsymbol{N}) \in \mathcal{M}(\Lambda)$, then $\Sigma^{p,q}(C, L, \boldsymbol{N}) \in \mathcal{M}(\Lambda)$.

(M2) Let $(C, L, \boldsymbol{N}) \in \mathcal{M}(\Lambda)$.

(M2.1) $\operatorname{Gr}^L_w(C, \boldsymbol{N}) \in \mathcal{P}_w(\Lambda)$.

(M2.2) For any decomposition $\Lambda = \Lambda_0 \sqcup \Lambda_1$, there exists a relative monodromy filtration $M\big(N(\Lambda_1); L\big) =: M(\Lambda_1; L)$, and

$$\mathrm{res}^{\Lambda}_{\Lambda_0}(C, L, \boldsymbol{N}) := (C, M(\Lambda_1; L), \boldsymbol{N}_{\Lambda_0})$$

is an object in $\mathcal{M}(\Lambda_0)$.

(M3) Let $\bullet \in \Lambda$, and put $\Lambda_0 := \Lambda \setminus \bullet$. Let us consider $(C, L, \boldsymbol{N}_\Lambda) \in \mathcal{M}(\Lambda)$ and $(C', \tilde{L}, \boldsymbol{N}'_{\Lambda_0}) \in \mathcal{M}(\Lambda_0)$ with morphisms

$$\Sigma^{1,0}\,\mathrm{res}^{\Lambda}_{\Lambda_0}(C, L, \boldsymbol{N}_\Lambda) \xrightarrow{\;u\;} (C', \widetilde{L}, \boldsymbol{N}'_{\Lambda_0}) \xrightarrow{\;v\;} \Sigma^{0,-1}\,\mathrm{res}^{\Lambda}_{\Lambda_0}(C, L, \boldsymbol{N}_\Lambda)$$

in $\mathcal{M}(\Lambda_0)$ such that $v \circ u = N_\bullet$. Let $L(C')$ be the filtration in the category of $\mathcal{M}(\Lambda_0)$ obtained as the transfer of $L(C)$, and put $N'_\bullet := u \circ v$. Then, $(C', L, \boldsymbol{N}'_\Lambda) \in \mathcal{M}(\Lambda)$.

The following lemma clearly follows from **M0**.

Lemma 6.3.3 *Let $(C, L, \boldsymbol{N}) \in \mathcal{M}(\Lambda)$. Then, $(L_j C, L, \boldsymbol{N}) \in \mathcal{M}(\Lambda)$ for each $j \in \mathbb{Z}$.* □

Corollary 6.3.4 *In (M2.2), L gives a filtration of* $\mathrm{res}^{\Lambda}_{\Lambda_0}(C, L, \boldsymbol{N})$ *in $\mathcal{M}(\Lambda_0)$.* □

Lemma 6.3.5 *Let $F : (C, L, \boldsymbol{N}) \longrightarrow (C', L, \boldsymbol{N}')$ be a morphism in $\mathcal{M}(\Lambda)$. Then, F is strict with respect to the filtrations $M(\Lambda_1; L)$ for any $\Lambda_1 \subset \Lambda$.*

Proof It follows from the conditions **P2** and **M2**. □

Lemma 6.3.6 *The functor* $\mathrm{res}^{\Lambda}_{\Lambda_0} : \mathcal{M}(\Lambda) \longrightarrow \mathcal{M}(\Lambda_0)$ *is exact.*

Proof It follows from Corollary 6.3.2 and Lemma 6.3.5. □

Lemma 6.3.7 *Let $(C, L, \boldsymbol{N}) \in \mathcal{M}(\Lambda)$. For any $\Lambda_2 \subset \Lambda_1 \subset \Lambda$, we have*

$$M\big(N(\Lambda_1); M(\Lambda_2; L)\big) = M(\Lambda_1; L). \tag{6.8}$$

Proof We use an induction on the length of the filtration L. If L is pure, it follows from **P3.1**. Assume that (i) $L_k(C) = C$, (ii) the claim holds for $(L_{k-1}(C), L, \boldsymbol{N})$. We have the exact sequence in $\mathcal{M}(\Lambda)$:

$$0 \longrightarrow (L_{k-1}C, L, \boldsymbol{N}) \longrightarrow (C, L, \boldsymbol{N}) \longrightarrow (\mathrm{Gr}^L_k C, \boldsymbol{N}) \longrightarrow 0$$

We obtain the exact sequence in $\mathcal{M}(\Lambda_2^c)$:

$$0 \longrightarrow \mathrm{res}^{\Lambda}_{\Lambda_2^c}(L_{k-1}C, L, \boldsymbol{N}) \longrightarrow \mathrm{res}^{\Lambda}_{\Lambda_2^c}(C, L, \boldsymbol{N}) \longrightarrow \mathrm{res}^{\Lambda}_{\Lambda_2^c}(\mathrm{Gr}^L_k C, \boldsymbol{N}) \longrightarrow 0$$

Because it is strict with respect to $M\big(N(\Lambda_1); M(\Lambda_2; L)\big)$, we obtain (6.8). □

6.3.2 A Category $L\mathcal{A}(\Lambda)$

We introduce a category $L\mathcal{A}(\Lambda)$ as follows. An object of $L\mathcal{A}(\Lambda)$ is a tuple $\bigl(C_I \in \mathcal{A} \bigm| I \subset \Lambda\bigr)$, with morphisms for $I \subset J \subset \Lambda$

$$\Sigma^{|J\setminus I|,0} C_I \xrightarrow{g_{JI}} C_J \xrightarrow{f_{IJ}} \Sigma^{0,-|J\setminus I|} C_I$$

satisfying the compatibility conditions for $I \subset J \subset K \subset \Lambda$:

$$f_{IJ} \circ f_{JK} = f_{IK}, \qquad g_{KJ} \circ g_{JI} = g_{KI}, \qquad g_{J,I\cap J} \circ f_{I\cap J,I} = f_{J\,I\cup J} \circ g_{I\cup J,I}.$$

Such a tuple $\bigl(C_I; f_{IJ}, g_{JI}\bigr)$ is often denoted by $\mathcal{T}$. A morphism $\mathcal{T}_1 \longrightarrow \mathcal{T}_2$ is a tuple of morphisms $F_I : C_{1,I} \longrightarrow C_{2,I}$ such that the following diagrams are commutative:

$$\begin{array}{ccccc}
\Sigma^{|J\setminus I|,0} C_{1,I} & \xrightarrow{g_{1,JI}} & C_{1,J} & \xrightarrow{f_{1,IJ}} & \Sigma^{0,-|J\setminus I|} C_{1,I} \\
{\scriptstyle F_I}\big\downarrow & & {\scriptstyle F_J}\big\downarrow & & {\scriptstyle F_I}\big\downarrow \\
\Sigma^{|J\setminus I|,0} C_{2,I} & \xrightarrow{g_{2,JI}} & C_{2,J} & \xrightarrow{f_{2,IJ}} & \Sigma^{0,-|J\setminus I|} C_{2,I}
\end{array}$$

The category $L\mathcal{A}(\Lambda)$ is abelian, and equipped with naturally induced auto equivalences $\Sigma^{p,q}$.

For any object $\mathcal{T} = (C_I, f_{IJ}, g_{JI})$, each C_I is equipped with a tuple of morphisms $N_i : C_I \longrightarrow \boldsymbol{T}^{-1} C_I$ $(i \in \Lambda)$ given as follows:

$$N_i = f_{I,I\cup\{i\}} \circ g_{I\cup\{i\},I} \quad (i \notin I), \qquad N_i = g_{I,I\setminus i} \circ f_{I\setminus i,I} \quad (i \in I).$$

The tuple is denoted by $\boldsymbol{N}_I$.

6.3.3 S-Decomposability and Strict Support

An object $\mathcal{T} \in L\mathcal{A}(\Lambda)$ is called S-decomposable, if the following holds:

- For $J = I \sqcup \{i\}$, we have $C_J = \operatorname{Im} g_{JI} \oplus \operatorname{Ker} f_{IJ}$, i.e., the tuple $(C_I, C_J, g_{JI}, f_{IJ})$ is S-decomposable.

Let $\mathcal{T} \in L\mathcal{A}(\Lambda)$. We say that $\mathcal{T}$ has strict support I if the following holds:

- We have $C_J = 0$ unless $J \supset I$. The morphisms g_{JI} (resp. f_{IJ}) are epimorphisms (resp. monomorphisms) for any $J \supset I$. In particular, $\mathcal{T}$ is S-decomposable.

The 0 object has strict support I for any $I \subset \Lambda$. If a non-zero object $\mathcal{T}$ has a strict support, the strict support is uniquely determined for $\mathcal{T}$.

The following lemma is clear.

Lemma 6.3.8 *Let $\mathcal{T} = \mathcal{T}_1 \oplus \mathcal{T}_2$ be an object in $L\mathcal{A}(\Lambda)$. Then, $\mathcal{T}$ is S-decomposable if and only if $\mathcal{T}_i$ $(i = 1, 2)$ are S-decomposable.* □

Lemma 6.3.9 *If $\mathcal{T}_i \in L\mathcal{A}(\Lambda)$ $(i = 1, 2)$ have strict supports I_i with $I_1 \neq I_2$, then any morphism $\mathcal{T}_1 \longrightarrow \mathcal{T}_2$ is 0.*

Proof Let $F : \mathcal{T}_1 \longrightarrow \mathcal{T}_2$ be a morphism. If $I_1 \supsetneq I_2$, we have the following:

$$\begin{array}{ccc} C_{1,I_1} & \longrightarrow & C_{1,I_2} = 0 \\ {\scriptstyle F_{I_1}}\downarrow & & {\scriptstyle F_{I_2}}\downarrow \\ C_{2,I_1} & \xrightarrow{\text{mono}} & C_{2,I_2} \end{array}$$

Hence, we have $F_{I_1} = 0$. For any $J \supset I_1$, we have the commutative diagram:

$$\begin{array}{ccc} C_{1,I_1} & \xrightarrow{\text{epi}} & C_{1,J} \\ {\scriptstyle F_{I_1}}\downarrow & & {\scriptstyle F_{J}}\downarrow \\ C_{2,I_1} & \xrightarrow{g_{J,I_1}} & C_{2,J} \end{array}$$

We obtain $F_J = 0$ for any $J \supset I$, and hence $F = 0$. If $I_1 \not\supset I_2$, we have $C_{2,I_1} = 0$ and hence $F_{I_1} : C_{1,I_1} \longrightarrow C_{2,I_1}$ is 0, which implies $F = 0$ as above. □

Lemma 6.3.10 *Let $\mathcal{T} \in L\mathcal{A}(\Lambda)$ be S-decomposable. Then, we have a unique decomposition $\mathcal{T} = \bigoplus_{I \subset \Lambda} \mathcal{T}_I$ such that each $\mathcal{T}_I$ has strict support I.*

Proof The uniqueness follows from Lemma 6.3.9. Let us prove the existence by using an induction on $|\Lambda|$. If $|\Lambda| = 0$, the claim is clear. For any $I \subset \Lambda$ and $i \in \Lambda \setminus I$, let $Ii := I \cup \{i\}$.

Let $\mathcal{T} \in L\mathcal{A}(\Lambda)$ be S-decomposable. Assume that we have $\Lambda_0 \subset \Lambda$ such that $g_{Ii,I}$ are epimorphisms for any $i \in \Lambda_0$ and $I \subset \Lambda \setminus \{i\}$. If $\Lambda = \Lambda_0$, $\mathcal{T}$ has strict support $\emptyset$. Let us consider the case $\Lambda \neq \Lambda_0$. Fix $j \in \Lambda \setminus \Lambda_0$. We put

$$C'_I := \begin{cases} C_I & (j \notin I) \\ \operatorname{Im} g_{I,I \setminus j} & (j \in I), \end{cases} \qquad C''_I := \begin{cases} 0 & (j \notin I) \\ \operatorname{Ker} f_{I \setminus j,I} & (j \in I). \end{cases}$$

We obtain a decomposition $C_I = C'_I \oplus C''_I$ for any $I \subset \Lambda$, which induces a decomposition $\mathcal{T} = \mathcal{T}' \oplus \mathcal{T}''$ in $L\mathcal{A}(\Lambda)$. By Lemma 6.3.8, both $\mathcal{T}'$ and $\mathcal{T}''$ are S-decomposable. We may apply the hypothesis of the induction to $\mathcal{T}''$. We have the surjectivity of $g_{Ii,I}$ for $\mathcal{T}'$ if $i \in \Lambda_0 \sqcup \{j\}$ and $i \notin I$. Hence, we obtain a decomposition by strict supports by an easy induction. □

6.3.4 A Category $\mathcal{LA}(\Lambda_1, \Lambda_2)$

Let Λ_i $(i = 1, 2)$ be finite sets. We consider the category $\mathcal{LA}(\Lambda_1, \Lambda_2) := \big(\mathcal{LA}(\Lambda_1)\big)(\Lambda_2)$, i.e., let $\mathcal{LA}(\Lambda_1, \Lambda_2)$ be the category of objects $\mathcal{T}$ in $\mathcal{LA}(\Lambda_1)$ with a commuting Λ_2-tuple of morphisms $\mathcal{N}_j : \mathcal{T} \longrightarrow \boldsymbol{T}^{-1}\mathcal{T}$ $(j \in \Lambda_2)$. It is an abelian category equipped with naturally induced auto equivalences $\Sigma^{p,q}$. An object $(\mathcal{T}, \boldsymbol{\mathcal{N}}) \in \mathcal{LA}(\Lambda_1, \Lambda_2)$ is called S-decomposable if $\mathcal{T}$ is S-decomposable. In the case, $\boldsymbol{\mathcal{N}}$ preserves the decomposition in Lemma 6.3.10.

Let $(\mathcal{T}, \boldsymbol{\mathcal{N}}) \in \mathcal{LA}(\Lambda_1, \Lambda_2)$. Each underlying C_I is equipped with the morphisms N_i $(i \in \Lambda_1)$ and $\mathcal{N}_j$ $(j \in \Lambda_2)$. We set $\boldsymbol{N}_I^{(\Lambda_2)} := \big(N_i \,\big|\, i \in \Lambda_1\big) \sqcup \big(\mathcal{N}_j \,\big|\, j \in \Lambda_2\big)$.

Remark 6.3.11 $(\mathcal{T}, \boldsymbol{\mathcal{N}})$ will often be denoted by $\mathcal{T}$ if there is no risk of confusion.

□

6.3.5 Pure Objects in $\mathcal{LA}(\Lambda_1, \Lambda_2)$

An object $(\mathcal{T}, \boldsymbol{\mathcal{N}}) \in \mathcal{LA}(\Lambda_1, \Lambda_2)$ is called pure of weight w if the following holds:

(LP1) $(\mathcal{T}, \boldsymbol{\mathcal{N}})$ is S-decomposable.
(LP2) Each $\big(C_I, \boldsymbol{N}_I^{(\Lambda_2)}\big)$ is an object in $\mathcal{P}_w(\Lambda_1 \sqcup \Lambda_2)$.

Let $P_w\mathcal{LA}(\Lambda_1, \Lambda_2) \subset \mathcal{LA}(\Lambda_1, \Lambda_2)$ denote the full subcategory of pure objects of weight w. If $\Lambda_2 = \emptyset$, it is denoted by $P_w\mathcal{LA}(\Lambda_1)$.

Proposition 6.3.12 *The family* $\big\{P_w\mathcal{LA}(\Lambda_1, \Lambda_2)\big\}$ *satisfies the conditions* **P0–P3**.

Proof The condition **P2** for $P_w\mathcal{LA}(\Lambda_1, \Lambda_2)$ follows from that for $\mathcal{P}_w(\Lambda_1 \sqcup \Lambda_2)$. The condition **P1** is clearly satisfied. Any injection $\Lambda_2 \longrightarrow \Lambda_2'$ clearly induces a functor $P_w\mathcal{LA}(\Lambda_1, \Lambda_2) \longrightarrow P_w\mathcal{LA}(\Lambda_1, \Lambda_2')$. Let us prove that $P_w\mathcal{LA}(\Lambda_1, \Lambda_2)$ is abelian and semisimple. The following lemma is clear.

Lemma 6.3.13 *Let* $\mathcal{T}_i \in \mathcal{LA}(\Lambda_1, \Lambda_2)$ $(i = 1, 2)$*. Then,* $\mathcal{T}_1 \oplus \mathcal{T}_2$ *is an object in* $P_w\mathcal{LA}(\Lambda_1, \Lambda_2)$ *if and only if* $\mathcal{T}_i$ *are objects in* $P_w\mathcal{LA}(\Lambda_1, \Lambda_2)$ *for* $i = 1, 2$. □

We obtain the following lemma from Lemmas 6.3.10 and 6.3.13.

Lemma 6.3.14 *For any* $\mathcal{T} \in P_w\mathcal{LA}(\Lambda_1, \Lambda_2)$*, we have the decomposition* $\mathcal{T} = \bigoplus_{I \subset \Lambda_1} \mathcal{T}_I$ *in* $P_w\mathcal{LA}(\Lambda_1, \Lambda_2)$*, where each* $\mathcal{T}_I$ *has strict support* I*. It is uniquely determined.* □

Lemma 6.3.15 *Let* $\mathcal{T}, \mathcal{T}' \in P_w\mathcal{LA}(\Lambda_1, \Lambda_2)$*. Assume that they have strict support* I*. Let* $F : \mathcal{T} \longrightarrow \mathcal{T}'$ *be a morphism in* $P_w\mathcal{LA}(\Lambda_1, \Lambda_2)$*. Then, we have*

$$\operatorname{Ker} F, \operatorname{Im} F, \operatorname{Cok} F \in P_w\mathcal{LA}(\Lambda_1, \Lambda_2),$$

and they have strict support I*.*

Proof It is enough to prove that $\operatorname{Ker} F$, $\operatorname{Im} F$ and $\operatorname{Cok} F$ have strict supports I. Let $F_I : C_I \longrightarrow C'_I$ ($I \subset \Lambda_1$) denote the underlying morphisms. Let $I \subset J$. We omit to denote the shift by $\Sigma^{p,q}$. The induced morphisms $g_{J,I} : \operatorname{Im}(F_I) \longrightarrow \operatorname{Im}(F_J)$ and $g_{J,I} : \operatorname{Cok}(F_I) \longrightarrow \operatorname{Cok}(F_J)$ are clearly epimorphisms. The induced morphisms $f_{I,J} : \operatorname{Im}(F_J) \longrightarrow \operatorname{Im}(F_I)$ and $f_{I,J} : \operatorname{Ker}(F_J) \longrightarrow \operatorname{Ker}(F_I)$ are clearly monomorphisms. Hence, we have only to prove that (i) $g_{J,I} : \operatorname{Ker}(F_I) \longrightarrow \operatorname{Ker}(F_J)$ are epimorphisms, (ii) $f_{I,J} : \operatorname{Cok}(F_J) \longrightarrow \operatorname{Cok}(F_I)$ are monomorphisms. We have only to consider the case $J = I \sqcup \{i\}$.

Because the category $\mathcal{P}_w(\Lambda_1 \sqcup \Lambda_2)$ is semisimple, we can take a decomposition $C_I = \operatorname{Ker} F_I \oplus C$ compatible with N_k ($k \in \Lambda_1 \sqcup \Lambda_2$). Then, we have $\operatorname{Im} N_i = \big(\operatorname{Im} N_i \cap \operatorname{Ker} F_I\big) \oplus \big(\operatorname{Im} N_i \cap C\big) = N_i(\operatorname{Ker} F_I) \oplus N_i(C)$. We have $N_i \operatorname{Ker}(F_I) \subset f_{IJ}(\operatorname{Ker} F_J) \subset \operatorname{Im} N_i \cap \operatorname{Ker} F_I$. Then, we obtain (i). The claim (ii) can be proved similarly. □

By **P3.3** for the categories $\mathcal{P}(\Lambda)$, the decompositions by strict support induce an equivalence of the categories:

$$P_w L\mathcal{A}(\Lambda_1, \Lambda_2) \simeq \bigoplus_{I \subset \Lambda_1} \mathcal{P}_w(I \sqcup \Lambda_2)$$

In particular, the category $P_w L\mathcal{A}(\Lambda_1, \Lambda_2)$ is abelian and semisimple, and we obtain that $P_w L\mathcal{A}(\Lambda_1, \Lambda_2)$ satisfies the condition **P0**.

Let us consider **P3** for $P_w L\mathcal{A}(\Lambda_1, \Lambda_2)$. Let $(\mathcal{T}, \boldsymbol{\mathcal{N}}) \in P_w L\mathcal{A}(\Lambda_1, \Lambda_2)$. The condition **P3.1** follows from that of objects in $\mathcal{P}_w(\Lambda_1 \sqcup \Lambda_2)$.

Let $\Lambda_2 = \Lambda_{2,0} \sqcup \Lambda_{2,1}$ be a decomposition. We have the induced object

$$\big(\operatorname{Gr}^{M(\Lambda_{2,1})}_{w'} \mathcal{T}, \boldsymbol{\mathcal{N}}_{\Lambda_{2,0}}\big) \in L\mathcal{A}(\Lambda_1, \Lambda_{2,0}).$$

To prove that it is pure of weight $w + w'$, we have only to consider the case in which $\Lambda_{2,1}$ consists of a unique element $*$, according to **P3.1** for $P_w L\mathcal{A}(\Lambda_1, \Lambda_2)$. It satisfies **LP2**, due to the condition **P3.2** for objects in $\mathcal{P}_w(\Lambda_1 \sqcup \Lambda_2)$. To prove **LP1** for $\big(\operatorname{Gr}^{M(*)} \mathcal{T}, \boldsymbol{N}_{\Lambda_{2,0}}\big)$, we may assume that $(\mathcal{T}, \boldsymbol{\mathcal{N}})$ has strict support I. Then, it follows from **P3.3** for objects in $\mathcal{P}_w(\Lambda_1 \sqcup \Lambda_2)$. Thus, we obtain **P3.2** for $P_w L\mathcal{A}(\Lambda_1, \Lambda_2)$.

Let $\bullet \in \Lambda_2$. We obtain an object $\big(\operatorname{Im} \mathcal{N}_\bullet, \boldsymbol{\mathcal{N}}\big)$ in $L\mathcal{A}(\Lambda_1, \Lambda_2)$. We shall prove that the object $\big(\operatorname{Im} \mathcal{N}_\bullet, \boldsymbol{\mathcal{N}}\big)$ is pure of weight $w + 1$. We have only to consider the case in which $\mathcal{T}$ has strict support I. It is clear that **LP1** is satisfied. The condition **LP2** follows from **P3.3** for $\mathcal{P}_w(\Lambda_1 \sqcup \Lambda_2)$.

Let $* \in \Lambda_2 \setminus \bullet$. Let us look at the induced morphisms:

$$\Sigma^{1,0} \operatorname{Gr}^{M(*)} \mathcal{T} \longrightarrow \Sigma^{1,0} \operatorname{Gr}^{M(*)} \operatorname{Im} \mathcal{N}_\bullet \longrightarrow \Sigma^{0,-1} \operatorname{Gr}^{M(*)} \mathcal{T} \tag{6.9}$$

It is a tuple of the following morphisms:

$$\Sigma^{1,0} \operatorname{Gr}^{M(*)} C_I \longrightarrow \Sigma^{1,0} \operatorname{Gr}^{M(*)} \operatorname{Im} \mathcal{N}_{\bullet,I} \longrightarrow \Sigma^{0,-1} \operatorname{Gr}^{M(*)} C_I$$

It is S-decomposable by **P3.3** for objects in $\mathcal{P}_w(\Lambda)$. Hence, (6.9) is S-decomposable. Thus, the proof of Proposition 6.3.12 is finished. □

6.3.6 Mixed Objects in $L\mathcal{A}(\Lambda_1, \Lambda_2)$

A filtered object $(\mathcal{T},\boldsymbol{\mathcal{N}},\mathcal{L}) \in L\mathcal{A}(\Lambda_1,\Lambda_2)^{\mathrm{fil}}$ is called mixed, if the following holds:

(LM1) $\mathrm{Gr}^{\mathcal{L}}_w(\mathcal{T},\boldsymbol{\mathcal{N}}) \in P_wL\mathcal{A}(\Lambda_1,\Lambda_2)$.
(LM2) Each $(C_I,\mathcal{L},\boldsymbol{N}^{(\Lambda_2)})$ is an object in $\mathcal{M}(\Lambda_1 \sqcup \Lambda_2)$ for $I \subset \Lambda_1$.

Let $ML\mathcal{A}(\Lambda_1,\Lambda_2) \subset \mathcal{A}(\Lambda_1,\Lambda_2)^{\mathrm{fil}}$ denote the full subcategory of mixed objects. If $\Lambda_2 = \emptyset$, it is denoted by $ML\mathcal{A}(\Lambda_1)$.

Proposition 6.3.16 *The family $\{ML\mathcal{A}(\Lambda_1,\Lambda_2)\}$ satisfies the conditions* **M0–M3**.

Proof The condition **M1** is clearly satisfied. Let $F : (\mathcal{T}_1,\mathcal{L},\boldsymbol{\mathcal{N}}) \longrightarrow (\mathcal{T}_2,\mathcal{L},\boldsymbol{\mathcal{N}})$ be a morphism in $ML\mathcal{A}(\Lambda_1,\Lambda_2)$. By **P2** for $P_wL\mathcal{A}(\Lambda_1,\Lambda_2)$, we have the strictness of F with respect to $\mathcal{L}$. In particular, $\mathrm{Ker}\,F$, $\mathrm{Cok}\,F$ and $\mathrm{Im}\,F$ are naturally equipped with filtrations $\mathcal{L}$, and we have natural isomorphisms $\mathrm{Gr}^{\mathcal{L}}\,\mathrm{Ker}\,F \simeq \mathrm{Ker}\,\mathrm{Gr}^{\mathcal{L}}\,F$, $\mathrm{Gr}^{\mathcal{L}}\,\mathrm{Im}\,F \simeq \mathrm{Im}\,\mathrm{Gr}^{\mathcal{L}}\,F$ and $\mathrm{Gr}^{\mathcal{L}}\,\mathrm{Cok}\,F \simeq \mathrm{Cok}\,\mathrm{Gr}^{\mathcal{L}}\,F$. Hence, **LM1** holds for $\mathrm{Ker}\,F$, $\mathrm{Cok}\,F$ and $\mathrm{Im}\,F$ by Proposition 6.3.12. By the condition **M0** for $\mathcal{M}(\Lambda_1 \sqcup \Lambda_2)$, the condition **LM2** also holds for $\mathrm{Ker}\,F$, $\mathrm{Im}\,F$ and $\mathrm{Cok}\,F$. Hence, $ML\mathcal{A}(\Lambda_1,\Lambda_2)$ is abelian. Clearly, any injection $\Lambda_2' \longrightarrow \Lambda_2$ induces a functor $ML\mathcal{A}(\Lambda_1,\Lambda_2') \longrightarrow ML\mathcal{A}(\Lambda_1,\Lambda_2)$. Hence, **M0** is satisfied for $ML\mathcal{A}(\Lambda_1,\Lambda_2)$.

Let us consider **M2**. For $(\mathcal{T},\mathcal{L},\boldsymbol{\mathcal{N}}) \in ML\mathcal{A}(\Lambda_1,\Lambda_2)$, the condition **M2.1** is satisfied by definition. Let $\Lambda_2 = \Lambda_{2,0} \sqcup \Lambda_{2,1}$ be a decomposition. Each underlying C_I has $\tilde{\mathcal{L}} := M\big(\mathcal{N}(\Lambda_{2,1}),\mathcal{L}(C_I)\big)$ by the condition **M2.2** for objects in $\mathcal{M}(\Lambda_1 \sqcup \Lambda_2)$. It induces a filtration $\tilde{\mathcal{L}}$ of $(\mathcal{T},\boldsymbol{\mathcal{N}}_{\Lambda_{2,0}})$ in $L\mathcal{A}(\Lambda_1,\Lambda_{2,0})$, which is a relative monodromy filtration $M(\mathcal{N}(\Lambda_{2,1}),\mathcal{L})$:

$$\mathrm{res}^{\Lambda_2}_{\Lambda_{2,0}}(\mathcal{T},\mathcal{L},\boldsymbol{\mathcal{N}}) := (\mathcal{T},\tilde{\mathcal{L}},\boldsymbol{\mathcal{N}}_{\Lambda_{2,0}})$$

The condition **LM2** for $\mathrm{res}^{\Lambda_2}_{\Lambda_{2,0}}(\mathcal{T},\mathcal{L},\boldsymbol{\mathcal{N}})$ follows from **M2.2** for objects in $\mathcal{M}(\Lambda_1 \sqcup \Lambda_2)$. By the canonical decomposition (Sect. 6.1.2), we have an isomorphism

$$\mathrm{Gr}^{\tilde{\mathcal{L}}}(\mathcal{T},\boldsymbol{\mathcal{N}}_{\Lambda_{2,0}}) \simeq \mathrm{Gr}^{\tilde{\mathcal{L}}}\,\mathrm{Gr}^{\mathcal{L}}(\mathcal{T},\boldsymbol{\mathcal{N}}_{\Lambda_{2,0}}).$$

Hence, the condition **LM1** for $\mathrm{res}^{\Lambda_2}_{\Lambda_{2,0}}(\mathcal{T},\mathcal{L},\boldsymbol{\mathcal{N}})$ follows from **P3.2** for objects in $P_wL\mathcal{A}(\Lambda_1,\Lambda_2)$. Thus, we obtain that $\mathrm{res}^{\Lambda_2}_{\Lambda_{2,0}}(\mathcal{T},\mathcal{L},\boldsymbol{\mathcal{N}})$ is an object in $ML\mathcal{A}(\Lambda_1,\Lambda_2)$, i.e., **M2** holds for $ML\mathcal{A}(\Lambda_1,\Lambda_2)$.

Let us consider **M3** for $ML\mathcal{A}(\Lambda_1,\Lambda_2)$. Let $\bullet \in \Lambda_2$, and we put $\Lambda_{2,0} := \Lambda_2 \setminus \bullet$. We consider $(\mathcal{T},\mathcal{L},\boldsymbol{\mathcal{N}}_{\Lambda_2}) \in ML\mathcal{A}(\Lambda_1,\Lambda_2)$ and $(\mathcal{T}',\tilde{\mathcal{L}},\boldsymbol{\mathcal{N}}'_{\Lambda_{2,0}}) \in ML\mathcal{A}(\Lambda_1,\Lambda_{2,0})$ with morphisms

$$\Sigma^{1,0}\,\mathrm{res}^{\Lambda_2}_{\Lambda_{2,0}}(\mathcal{T},\mathcal{L},\boldsymbol{\mathcal{N}}_{\Lambda_2}) \xrightarrow{u} (\mathcal{T}',\tilde{\mathcal{L}},\boldsymbol{\mathcal{N}}'_{\Lambda_{2,0}}) \xrightarrow{v} \Sigma^{0,-1}\,\mathrm{res}^{\Lambda_2}_{\Lambda_{2,0}}(\mathcal{T},\mathcal{L},\boldsymbol{\mathcal{N}}_{\Lambda_2})$$

in $\mathit{MLA}(\Lambda_1,\Lambda_{2,0})$ such that $v\circ u=\mathcal{N}_\bullet$. We have the filtration $\mathcal{L}(\mathcal{T}')$ obtained as the transfer of $\mathcal{L}(\mathcal{T})$. We put $\mathcal{N}'_\bullet := u\circ v$. Let us prove that $(\mathcal{T}',\mathcal{L},\boldsymbol{\mathcal{N}}'_{\Lambda_2})$ is an object in $\mathit{MLA}(\Lambda_1,\Lambda_2)$. The condition **LM2** for $\mathit{MLA}(\Lambda_1,\Lambda_2)$ follows from **M3** for objects in $\mathcal{M}(\Lambda_1\sqcup\Lambda_2)$. By the construction of $\mathcal{L}(\mathcal{T}')$, we have the decomposition

$$\mathrm{Gr}^{\mathcal{L}}_w(\mathcal{T}')=\mathrm{Im}\,\mathrm{Gr}^{\mathcal{L}}_w(u)\oplus\mathrm{Ker}\,\mathrm{Gr}^{\mathcal{L}}_w(v). \tag{6.10}$$

The decomposition is compatible with the action of $\mathcal{N}_\bullet$, and the restriction of $\mathcal{N}_\bullet$ on $\mathrm{Ker}\,\mathrm{Gr}^{\mathcal{L}}_w(v)$ is 0. Because $\tilde{\mathcal{L}}=M(\mathcal{N}_\bullet,\mathcal{L})$, it is pure of weight w on $\mathrm{Ker}\,\mathrm{Gr}^{\mathcal{L}}_w(v)$, i.e., we have a natural isomorphism

$$\mathrm{Gr}^{\tilde{\mathcal{L}}}_w\,\mathrm{Ker}\,\mathrm{Gr}^{\mathcal{L}}_w(v)\simeq\mathrm{Ker}\,\mathrm{Gr}^{\mathcal{L}}_w(v).$$

Hence, $\mathrm{Ker}\,\mathrm{Gr}^{\mathcal{L}}_w(v)$ with the induced tuple of morphisms $\mathrm{Gr}^{\mathcal{L}}_w(\boldsymbol{\mathcal{N}}_{\Lambda_{2,0}})$ is isomorphic to a direct summand of

$$\mathrm{Gr}^{\tilde{\mathcal{L}}}_w\,\mathrm{Gr}^{\mathcal{L}}_w(\mathcal{T}',\boldsymbol{\mathcal{N}}_{\Lambda_{2,0}})\simeq\mathrm{Gr}^{\tilde{\mathcal{L}}}_w(\mathcal{T}',\boldsymbol{\mathcal{N}}_{\Lambda_{2,0}})\in P_w\mathit{LA}(\Lambda_1,\Lambda_{2,0})\subset P_w\mathit{LA}(\Lambda_1,\Lambda_2)$$

By **P0** for $P_w\mathit{LA}(\Lambda_1,\Lambda_2)$, we obtain that $\big(\mathrm{Ker}\,\mathrm{Gr}^{\mathcal{L}}(v),\mathrm{Gr}^{\mathcal{L}}(\boldsymbol{\mathcal{N}}_{\Lambda_{2,0}})\big)$ is an object in $P_w\mathit{LA}(\Lambda_1,\Lambda_2)$. We have the natural isomorphism

$$\Sigma^{1,0}\big(\mathrm{Im}\,\mathrm{Gr}^{\mathcal{L}}\mathcal{N}_\bullet,\mathrm{Gr}^{\mathcal{L}}\boldsymbol{\mathcal{N}}_{\Lambda_2}\big)\simeq\big(\mathrm{Im}\,\mathrm{Gr}^{\mathcal{L}}u,\mathrm{Gr}^{\mathcal{L}}\boldsymbol{\mathcal{N}}_{\Lambda_2}\big).$$

Hence, we obtain that $\big(\mathrm{Im}\,\mathrm{Gr}^{\mathcal{L}}u,\mathrm{Gr}^{\mathcal{L}}\boldsymbol{\mathcal{N}}_{\Lambda_2}\big)$ is an object in $P_w\mathit{LA}(\Lambda_1,\Lambda_2)$ by **P3.3** in Proposition 6.3.12. Hence, $(\mathcal{T}',\mathcal{L},\boldsymbol{\mathcal{N}}'_{\Lambda_2})$ is an object in $\mathit{MLA}(\Lambda_1,\Lambda_2)$, and the condition **M3** holds for $\mathit{MLA}(\Lambda_1,\Lambda_2)$. Thus, the proof of Proposition 6.3.16 is finished. □

6.3.7 Some Functors

For any $K\subset\Lambda_1$, we have the functors

$$\psi_K,\phi_K:\mathit{LA}(\Lambda_1,\Lambda_2)\longrightarrow\mathit{LA}(\Lambda_1\setminus K,\Lambda_2\sqcup K)$$

given as follows. Let $\mathcal{T}\in\mathit{LA}(\Lambda_1,\Lambda_2)$. For $I\subset\Lambda_1\setminus K$, we set $\psi_K(\mathcal{T})_I:=C_I$ and $\phi_K(\mathcal{T})_I:=C_{I\sqcup K}$. For $I\subset J$, we have the naturally induced morphisms

$$\Sigma^{|J\setminus I|,0}\psi_K(\mathcal{T})_I\longrightarrow\psi_K(\mathcal{T})_J\longrightarrow\Sigma^{0,-|J\setminus I|}\psi_K(\mathcal{T})_I$$
$$\Sigma^{|J\setminus I|,0}\phi_K(\mathcal{T})_I\longrightarrow\phi_K(\mathcal{T})_J\longrightarrow\Sigma^{0,-|J\setminus I|}\phi_K(\mathcal{T})_I$$

with which they are objects in $\mathit{LA}(\Lambda_1\setminus K)$. They are equipped with the naturally induced morphisms $\boldsymbol{\mathcal{N}}_{\Lambda_2}$. Moreover, $\psi_K(\mathcal{T})$ and $\phi_K(\mathcal{T})$ are equipped with naturally induced morphisms $\mathcal{N}_K := (\mathcal{N}_i \mid i \in K)$. We set $\mathcal{N}_{\Lambda_2 \sqcup K} = \mathcal{N}_{\Lambda_2} \sqcup \mathcal{N}_K$. Hence, we obtain $(\phi_K(\mathcal{T}), \mathcal{N}_{\Lambda_2 \sqcup K})$ and $(\psi_K(\mathcal{T}), \mathcal{N}_{\Lambda_2 \sqcup K})$. They naturally induce the functors

$$\psi_K, \phi_K : LA(\Lambda_1, \Lambda_2)^{\mathrm{fil}} \longrightarrow LA(\Lambda_1 \setminus K, \Lambda_2 \sqcup K)^{\mathrm{fil}}.$$

They naturally induce

$$\psi_K, \phi_K : MLA(\Lambda_1, \Lambda_2) \longrightarrow MLA(\Lambda_1 \setminus K, \Lambda_2 \sqcup K)$$

$$\psi_K, \phi_K : P_w LA(\Lambda_1, \Lambda_2) \longrightarrow P_w LA(\Lambda_1 \setminus K, \Lambda_2 \sqcup K).$$

The induced functor $\mathrm{res}^{\Lambda_2 \sqcup K}_{\Lambda_2} \phi_K : MLA(\Lambda_1, \Lambda_2) \longrightarrow MLA(\Lambda_1 \setminus K, \Lambda_2)$ is also denoted by ϕ_K, if there is no risk of confusion.

6.3.8 Gluing

Fix an element $\bullet \in \Lambda_1$. We consider the category $\mathrm{Glue}(\Lambda_1, \Lambda_2, \bullet)$ given as follows. An object in $\mathrm{Glue}(\Lambda_1, \Lambda_2, \bullet)$ is a tuple of $(\mathcal{T}', \mathcal{L}, \mathcal{N}') \in MLA(\Lambda_1 \setminus \bullet, \Lambda_2 \sqcup \bullet)$ and $(\mathcal{T}'', \tilde{\mathcal{L}}, \mathcal{N}'') \in MLA(\Lambda_1 \setminus \bullet, \Lambda_2)$ with morphisms $\mathcal{U}$ and $\mathcal{V}$ in $MLA(\Lambda_1 \setminus \bullet, \Lambda_2)$,

$$\Sigma^{1,0}\,\mathrm{res}^{\Lambda_2 \sqcup \bullet}_{\Lambda_2}(\mathcal{T}', \mathcal{L}, \mathcal{N}') \xrightarrow{\mathcal{U}} (\mathcal{T}'', \tilde{\mathcal{L}}, \mathcal{N}'') \xrightarrow{\mathcal{V}} \Sigma^{0,-1}\,\mathrm{res}^{\Lambda_2 \sqcup \bullet}_{\Lambda_2}(\mathcal{T}', \mathcal{L}, \mathcal{N}')$$

such that $\mathcal{N}'_{\bullet} = \mathcal{V} \circ \mathcal{U}$. A morphism in $\mathrm{Glue}(\Lambda_1, \Lambda_2, \bullet)$ is a tuple of morphisms

$$F' : (\mathcal{T}'_1, \mathcal{L}, \mathcal{N}') \longrightarrow (\mathcal{T}'_2, \mathcal{L}, \mathcal{N}') \quad \text{in } MLA(\Lambda_1 \setminus \bullet, \Lambda_2 \sqcup \bullet)$$

$$F'' : (\mathcal{T}''_1, \tilde{\mathcal{L}}, \mathcal{N}'') \longrightarrow (\mathcal{T}''_2, \tilde{\mathcal{L}}, \mathcal{N}'') \quad \text{in } MLA(\Lambda_1 \setminus \bullet, \Lambda_2)$$

such that the following diagram is commutative:

$$\begin{array}{ccccc}
\Sigma^{1,0}\,\mathrm{res}^{\Lambda_2 \sqcup \bullet}_{\Lambda_2}(\mathcal{T}'_1, \mathcal{L}, \mathcal{N}') & \longrightarrow & (\mathcal{T}''_1, \tilde{\mathcal{L}}, \mathcal{N}'') & \longrightarrow & \Sigma^{0,-1}\,\mathrm{res}^{\Lambda_2 \sqcup \bullet}_{\Lambda_2}(\mathcal{T}'_1, \mathcal{L}, \mathcal{N}') \\
{\scriptstyle F'}\downarrow & & {\scriptstyle F''}\downarrow & & {\scriptstyle F'}\downarrow \\
\Sigma^{1,0}\,\mathrm{res}^{\Lambda_2 \sqcup \bullet}_{\Lambda_2}(\mathcal{T}'_2, \mathcal{L}, \mathcal{N}') & \longrightarrow & (\mathcal{T}''_2, \tilde{\mathcal{L}}, \mathcal{N}'') & \longrightarrow & \Sigma^{0,-1}\,\mathrm{res}^{\Lambda_2 \sqcup \bullet}_{\Lambda_2}(\mathcal{T}'_2, \mathcal{L}, \mathcal{N}')
\end{array}$$

For $(\mathcal{T}, \mathcal{L}, \mathcal{N}) \in MLA(\Lambda_1, \Lambda_2)$, we have $\Gamma(\mathcal{T}, \mathcal{L}, \mathcal{N})$ in $\mathrm{Glue}(\Lambda_1, \Lambda_2, \bullet)$ which is the tuple of the objects

$$\psi_{\bullet}(\mathcal{T}, \mathcal{L}, \mathcal{N}) \in MLA(\Lambda_1 \setminus \bullet, \Lambda_2 \sqcup \bullet),$$

$$\mathrm{res}^{\Lambda_2 \sqcup \bullet}_{\Lambda_2} \phi_{\bullet}(\mathcal{T}, \mathcal{L}, \mathcal{N}) \in MLA(\Lambda_1 \setminus \bullet, \Lambda_2),$$

with naturally induced morphisms $\mathcal{U}$ and $\mathcal{V}$:

$$\Sigma^{1,0}\operatorname{res}^{\Lambda_2\sqcup\bullet}_{\Lambda_2}\psi_\bullet(\mathcal{T},\mathcal{L},\mathcal{N})\xrightarrow{\mathcal{U}}\operatorname{res}^{\Lambda_2\sqcup\bullet}_{\Lambda_2}\phi_\bullet(\mathcal{T},\mathcal{L},\mathcal{N})\xrightarrow{\mathcal{V}}\Sigma^{0,-1}\operatorname{res}^{\Lambda_2\sqcup\bullet}_{\Lambda_2}\psi_\bullet(\mathcal{T},\mathcal{L},\mathcal{N})$$

Thus, we obtain a functor $\Gamma : MLA(\Lambda_1,\Lambda_2)\longrightarrow \mathrm{Glue}(\Lambda_1,\Lambda_2,\bullet)$.

Corollary 6.3.17 *The functor Γ is an equivalence.*

Proof Let $(\mathcal{T}',\mathcal{L},\mathcal{N}')$, $(\mathcal{T}'',\tilde{\mathcal{L}},\mathcal{N}'')$, $\mathcal{U}$ and $\mathcal{V}$ be as above. We have the filtration $\mathcal{L}$ of $(\mathcal{T}'',\mathcal{N}'')$ obtained as the transfer. By **M3** in Proposition 6.3.16, $(\mathcal{T}'',\mathcal{L},\mathcal{N}'')$ is an object in $MLA(\Lambda_1\setminus\bullet,\Lambda_2\sqcup\bullet)$. We put $C_I := C'_I$ if $\bullet\notin I$ and $C_I := C''_{I\setminus\bullet}$ if $\bullet\in I$. The tuple $(C_I\,|\,I\subset\Lambda_1)$ with the induced morphisms naturally gives an object in $LA(\Lambda_1,\Lambda_2)$. Then, it is easy to check **LM1–2** for this object, and we obtain the essential surjectivity of Γ. It is obvious that Γ is faithful. It is full thanks to the functoriality of the transfer. □

6.3.9 *Another Description of $MLA(\Lambda_1,\Lambda_2)$*

We shall introduce a category $M'LA(\Lambda_1,\Lambda_2)$. An object of $M'LA(\Lambda_1,\Lambda_2)$ is $(\mathcal{T},\boldsymbol{\mathcal{N}}_{\Lambda_2})\in LA(\Lambda_1,\Lambda_2)$ with a tuple of filtrations $\boldsymbol{L}=\bigl(L_I\,\big|\,I\subset\Lambda_1\bigr)$ of the underlying objects C_I such that the following holds:

- Each $\mathcal{U}_I=(C_I,L_I,\boldsymbol{N}^{(\Lambda_2)}_{\Lambda_1\setminus I})$ is an object in $\mathcal{M}(\Lambda_1\sqcup\Lambda_2\setminus I)$, where $\boldsymbol{N}^{(\Lambda_2)}_{\Lambda_1\setminus I}=\bigl(N_j\,\big|\,j\in\Lambda_1\setminus I\bigr)\sqcup\boldsymbol{\mathcal{N}}_{\Lambda_2}$.
- g_{JI} and f_{IJ} give the following morphisms in $\mathcal{M}\bigl(\Lambda_1\sqcup\Lambda_2\setminus J\bigr)$:

$$\Sigma^{|J\setminus I|,0}\operatorname{res}^{\Lambda_1\sqcup\Lambda_2\setminus I}_{\Lambda_1\sqcup\Lambda_2\setminus J}\bigl(\mathcal{U}_I\bigr)\xrightarrow{g_{JI}}\mathcal{U}_J\xrightarrow{f_{IJ}}\Sigma^{0,-|J\setminus I|}\operatorname{res}^{\Lambda_1\sqcup\Lambda_2\setminus I}_{\Lambda_1\sqcup\Lambda_2\setminus J}\bigl(\mathcal{U}_I\bigr)$$

A morphism $(\mathcal{T}_1,\boldsymbol{\mathcal{N}}_{\Lambda_2},\boldsymbol{L})\longrightarrow(\mathcal{T}_2,\boldsymbol{\mathcal{N}}_{\Lambda_2},\boldsymbol{L})$ in $M'LA(\Lambda_1,\Lambda_2)$ is a morphism $F:(\mathcal{T}_1,\boldsymbol{\mathcal{N}}_{\Lambda_2})\longrightarrow(\mathcal{T}_2,\boldsymbol{\mathcal{N}}_{\Lambda_2})$ in $LA(\Lambda_1,\Lambda_2)$ such that each $F_I : C_{1,I}\longrightarrow C_{2,I}$ is compatible with the filtrations L_I.

We have the functor $\Phi_{\Lambda_1,\Lambda_2} : MLA(\Lambda_1,\Lambda_2)\longrightarrow M'LA(\Lambda_1,\Lambda_2)$ defined as follows. For $(\mathcal{T},\mathcal{L},\boldsymbol{\mathcal{N}}_{\Lambda_2})$, we set $L_I(C_I) := M\Bigl(N(I);\mathcal{L}(C_I)\Bigr)$. Then, $(\mathcal{T},\boldsymbol{\mathcal{N}}_{\Lambda_2})$ with $\boldsymbol{L}=(L_I\,|\,I\subset\Lambda_1)$ is an object in $M'LA(\Lambda_1,\Lambda_2)$, which is defined to be $\Phi_{\Lambda_1,\Lambda_2}(\mathcal{T},\mathcal{L},\boldsymbol{\mathcal{N}}_{\Lambda_2})$.

Theorem 6.3.18 *The above functors $\Phi_{\Lambda_1,\Lambda_2}$ are equivalent.*

Proof The claim is clear in the case $|\Lambda_1|=0$. We use an induction on $\bigl(|\Lambda_1|+|\Lambda_2|,|\Lambda_1|\bigr)$. Take $\bullet\in\Lambda_1$. We assume that $\Phi_{\Lambda_1\setminus\bullet,\Lambda_2\sqcup\bullet}$ and $\Phi_{\Lambda_1\setminus\bullet,\Lambda_2}$ are equivalent, and we will prove that $\Phi_{\Lambda_1,\Lambda_2}$ is equivalent.

Let $(\mathcal{T},\boldsymbol{\mathcal{N}}_{\Lambda_2},\boldsymbol{L}) \in ML\mathcal{A}'(\Lambda_1,\Lambda_2)$. Each C_I is equipped with a filtration $\tilde{L}_I := M(\mathcal{N}_\bullet, L_I(C_I))$. Thus, we obtain a functor

$$\mathrm{res}^{\Lambda_2}_{\Lambda_2\setminus\bullet} : M'L\mathcal{A}(\Lambda_1,\Lambda_2) \longrightarrow M'L\mathcal{A}(\Lambda_1,\Lambda_2\setminus\bullet)$$

given by $\mathrm{res}^{\Lambda_2}_{\Lambda_2\setminus\bullet}(\mathcal{T},\boldsymbol{\mathcal{N}}_{\Lambda_2},\boldsymbol{L}) = \big(\mathcal{T},\boldsymbol{\mathcal{N}}_{\Lambda_2\setminus\bullet},\tilde{\boldsymbol{L}}\big)$. We also have the functor

$$\mathrm{res}^{\Lambda_2}_{\Lambda_2\setminus\bullet} : ML\mathcal{A}(\Lambda_1,\Lambda_2) \longrightarrow ML\mathcal{A}(\Lambda_1,\Lambda_2\setminus\bullet).$$

Lemma 6.3.19 *We have* $\mathrm{res}^{\Lambda_2}_{\Lambda_2\setminus\bullet}\circ\Phi_{\Lambda_1,\Lambda_2} = \Phi_{\Lambda_1,\Lambda_2\setminus\bullet}\circ\mathrm{res}^{\Lambda_2}_{\Lambda_2\setminus\bullet}$.

Proof We have

$$M\Big(N(I);M(\mathcal{N}_\bullet;\mathcal{L})\Big) = M\Big(\mathcal{N}_\bullet + N(I);\mathcal{L}\Big) = M\Big(\mathcal{N}_\bullet;M\Big(N(I);\mathcal{L}\Big)\Big).$$

Then, we obtain the claim of the lemma. □

We consider a category $\mathrm{Glue}'(\Lambda_1,\Lambda_2,\bullet)$ given as follows. An object is a tuple of

$$\big(\mathcal{T}',\boldsymbol{\mathcal{N}}',\boldsymbol{L}\big) \in M'L\mathcal{A}(\Lambda_1\setminus\bullet,\Lambda_2\sqcup\bullet), \qquad \big(\mathcal{T}'',\boldsymbol{\mathcal{N}}'',\tilde{\boldsymbol{L}}\big) \in M'L\mathcal{A}(\Lambda_1\setminus\bullet,\Lambda_2) \tag{6.11}$$

with morphisms in $M'L\mathcal{A}(\Lambda_1\setminus\bullet,\Lambda_2)$

$$\Sigma^{1,0}\,\mathrm{res}^{\Lambda_2\sqcup\bullet}_{\Lambda_2}\big(\mathcal{T}',\boldsymbol{\mathcal{N}}',\boldsymbol{L}\big) \xrightarrow{\mathcal{U}} \big(\mathcal{T}'',\boldsymbol{\mathcal{N}}'',\tilde{\boldsymbol{L}}\big) \xrightarrow{\mathcal{V}} \Sigma^{0,-1}\,\mathrm{res}^{\Lambda_2\sqcup\bullet}_{\Lambda_2}\big(\mathcal{T}',\boldsymbol{\mathcal{N}}',\boldsymbol{L}\big) \tag{6.12}$$

such that $\mathcal{V}\circ\mathcal{U} = \mathcal{N}_\bullet$. Morphisms in $\mathrm{Glue}'(\Lambda_1,\Lambda_2,\bullet)$ are naturally defined.

By Lemma 6.3.19, we naturally obtain a functor $\Phi_0 : \mathrm{Glue}(\Lambda_1,\Lambda_2,\bullet) \longrightarrow \mathrm{Glue}'(\Lambda_1,\Lambda_2,\bullet)$.

Lemma 6.3.20 $\Phi_0 : \mathrm{Glue}(\Lambda_1,\Lambda_2,\bullet) \longrightarrow \mathrm{Glue}'(\Lambda_1,\Lambda_2,\bullet)$ *is equivalent.*

Proof It is easy to observe that the functor Φ_0 is fully faithful. Let us prove the essential surjectivity. Take an object in $\mathrm{Glue}'(\Lambda_1,\Lambda_2,\bullet)$, i.e., a tuple as in (6.11) with morphisms as in (6.12). By using the hypothesis of the induction, we obtain the following:

- A filtration $\mathcal{L}$ of $(\mathcal{T}',\boldsymbol{\mathcal{N}}')$ such that $(\mathcal{T}',\mathcal{L},\boldsymbol{\mathcal{N}}') \in ML\mathcal{A}(\Lambda_1\setminus\bullet,\Lambda_2\sqcup\bullet)$ and $\Phi_{\Lambda_1\setminus\bullet,\Lambda_2\sqcup\bullet}(\mathcal{T}',\mathcal{L},\boldsymbol{\mathcal{N}}') = (\mathcal{T}',\boldsymbol{\mathcal{N}}',\boldsymbol{L})$.
- A filtration $\tilde{\mathcal{L}}$ of $(\mathcal{T}'',\boldsymbol{\mathcal{N}}'')$ such that $(\mathcal{T}'',\tilde{\mathcal{L}},\boldsymbol{\mathcal{N}}'') \in ML\mathcal{A}(\Lambda_1\setminus\bullet,\Lambda_2)$ and

$$\Phi_{\Lambda_1\setminus\bullet,\Lambda_2}(\mathcal{T}'',\tilde{\mathcal{L}},\boldsymbol{\mathcal{N}}'') = (\mathcal{T}'',\boldsymbol{\mathcal{N}}'',\boldsymbol{L}).$$

- A filtration $\tilde{\mathcal{L}}$ of $(\mathcal{T}', \boldsymbol{\mathcal{N}}'_{\Lambda_2})$ such that $(\mathcal{T}', \tilde{\mathcal{L}}, \boldsymbol{\mathcal{N}}'_{\Lambda_2}) \in MLA(\Lambda_1 \setminus \bullet, \Lambda_2)$ and
$$\Phi_{\Lambda_1\setminus\bullet,\Lambda_2}(\mathcal{T}', \tilde{\mathcal{L}}, \boldsymbol{\mathcal{N}}'_{\Lambda_2}) = \mathrm{res}^{\Lambda_2\sqcup\bullet}_{\Lambda_2}(\mathcal{T}', \boldsymbol{\mathcal{N}}', \boldsymbol{L}),$$
where $\boldsymbol{\mathcal{N}}'_{\Lambda_2} := \big(\mathcal{N}'_j \,\big|\, j \in \Lambda_2\big)$.

Because $\Phi_{\Lambda_1\setminus\bullet,\Lambda_2\sqcup\bullet}$ is equivalent, $\mathrm{res}^{\Lambda_2\sqcup\bullet}_{\Lambda_2}(\mathcal{T}', \mathcal{L}, \boldsymbol{\mathcal{N}}')$ is equal to $(\mathcal{T}', \tilde{\mathcal{L}}, \boldsymbol{\mathcal{N}}'_{\Lambda_2})$, and $\mathcal{U}$ and $\mathcal{V}$ give morphisms in $MLA(\Lambda_1 \setminus \bullet, \Lambda_2)$. Hence, $(\mathcal{T}', \mathcal{L}, \boldsymbol{\mathcal{N}}')$, $(\mathcal{T}'', \tilde{\mathcal{L}}, \boldsymbol{\mathcal{N}}'')$ with $(\mathcal{U}, \mathcal{V})$ naturally gives an object in $\mathrm{Glue}(\Lambda_1, \Lambda_2)$, and we can deduce that Φ_0 is essentially surjective. □

We prepare some functors. We have the naturally induced functors
$$\phi_\bullet : M'LA(\Lambda_1, \Lambda_2) \longrightarrow M'LA(\Lambda_1 \setminus \bullet, \Lambda_2),$$
$$\psi_\bullet : M'LA(\Lambda_1, \Lambda_2) \longrightarrow M'LA(\Lambda_1 \setminus \bullet, \Lambda_2 \sqcup \bullet).$$
For $(\mathcal{T}, \mathcal{L}, \boldsymbol{\mathcal{N}}) \in M'LA(\Lambda_1, \Lambda_2)$, we have the induced morphisms
$$\Sigma^{1,0}\psi(\mathcal{T}) \xrightarrow{\mathcal{U}} \phi(\mathcal{T}) \xrightarrow{\mathcal{V}} \Sigma^{0,-1}\psi(\mathcal{T})$$
induced by $g_{I,I\setminus\bullet}$ and $f_{I\setminus\bullet,I}$. They give morphisms in $M'LA(\Lambda_1 \setminus \bullet, \Lambda_2)$:
$$\Sigma^{1,0}\,\mathrm{res}^{\Lambda_2\sqcup\bullet}_{\Lambda_2}\,\psi_\bullet(\mathcal{T}, \boldsymbol{\mathcal{N}}, \boldsymbol{L}) \longrightarrow \phi_\bullet(\mathcal{T}, \boldsymbol{\mathcal{N}}, \boldsymbol{L}) \longrightarrow \Sigma^{0,-1}\,\mathrm{res}^{\Lambda_2\sqcup\bullet}_{\Lambda_2}\,\psi_\bullet(\mathcal{T}, \boldsymbol{\mathcal{N}}, \boldsymbol{L})$$
Hence, we obtain a functor $\Gamma' : M'LA(\Lambda, \Lambda_2) \longrightarrow \mathrm{Glue}'(\Lambda_1, \Lambda_2, \bullet)$. Because $\Gamma' \circ \Phi_{\Lambda_1,\Lambda_2} = \Phi_0 \circ \Gamma$ by construction, we have only to prove that Γ' is an equivalence. But, it is clear by construction. Thus, the proof of Theorem 6.3.18 is finished. □

For any $K \subset \Lambda_1$, we naturally have a functor $\psi_K : M'LA(\Lambda_1, \Lambda_2) \longrightarrow M'LA(\Lambda_1 \setminus K, \Lambda_2 \sqcup K)$, and the following is commutative:
$$\begin{array}{ccc} MLA(\Lambda_1, \Lambda_2) & \xrightarrow{\psi_K} & MLA(\Lambda_1 \setminus K, \Lambda_2 \sqcup K) \\ \simeq\big\downarrow & & \simeq\big\downarrow \\ M'LA(\Lambda_1, \Lambda_2) & \xrightarrow{\psi_K} & M'LA(\Lambda_1 \setminus K, \Lambda_2 \sqcup K) \end{array} \tag{6.13}$$

6.3.10 Commutativity of the Transfer

Let $V = (C, L, \boldsymbol{N}) \in \mathcal{M}(\underline{2})$, where $\underline{2} := \{1, 2\}$. We obtain the following object $\mathcal{T} =: V[!1 * 2]$ in $M'LA(\underline{2})$. We put
$$(\mathcal{T}_\emptyset, L_\emptyset) := (C, L), \quad (\mathcal{T}_1, L_1) := \Sigma^{1,0}\big(C, M(N_1; L)\big),$$
$$(\mathcal{T}_2, L_2) := \Sigma^{0,-1}\big(C, M(N_2; L)\big), \quad (\mathcal{T}_{\underline{2}}, L_{\underline{2}}) := \Sigma^{1,-1}\big(C, M(N_1 + N_2; L)\big).$$

The morphisms f_{IJ} and g_{JI} are given as follows:

$$\Sigma^{1,0}(\mathcal{T}_{\emptyset}, M(N_1; L_{\emptyset})) \xrightarrow{\text{id}} (\mathcal{T}_1, L_1) \xrightarrow{N_1} \Sigma^{0,-1}(\mathcal{T}_{\emptyset}, M(N_1; L_{\emptyset}))$$

$$\Sigma^{1,0}(\mathcal{T}_{\emptyset}, M(N_2; L_{\emptyset})) \xrightarrow{N_2} (\mathcal{T}_2, L_2) \xrightarrow{\text{id}} \Sigma^{0,-1}(\mathcal{T}_{\emptyset}, M(N_2; L_{\emptyset}))$$

$$\Sigma^{1,0}(\mathcal{T}_1, M(N_2; L_1)) \xrightarrow{N_2} (\mathcal{T}_{\underline{2}}, L_{\underline{2}}) \xrightarrow{\text{id}} \Sigma^{0,-1}(\mathcal{T}_1, M(N_2; L_1))$$

$$\Sigma^{1,0}(\mathcal{T}_2, M(N_1; L_2)) \xrightarrow{\text{id}} (\mathcal{T}_{\underline{2}}, L_{\underline{2}}) \xrightarrow{N_1} \Sigma^{0,-1}(\mathcal{T}_2, M(N_1; L_2))$$

By Theorem 6.3.18, we obtain the filtration $\mathcal{L}$ of $\mathcal{T} = V[!1 * 2]$ such that $(\mathcal{T}, \mathcal{L}) \in MLA(\underline{2})$. It is easy to check the following:

$$\mathcal{L}_{\emptyset}(C) = L, \quad \mathcal{L}_1(\Sigma^{1,0}C) = \hat{N}_{1!}L,$$
$$\mathcal{L}_2(\Sigma^{0,-1}C) = \hat{N}_{2*}L, \quad \mathcal{L}_{\underline{2}}(\Sigma^{1,-1}C) = \hat{N}_{2*}\hat{N}_{1!}L = \hat{N}_{1!}\hat{N}_{2*}L.$$

In particular, we obtain the following commutativity.

Lemma 6.3.21 $N_{2*}N_{1!}L = N_{1!}N_{2*}L$. □

Similarly, we obtain the following lemma.

Lemma 6.3.22 $N_{2\star}N_{1\star}L = N_{1\star}N_{2\star}L$ *for* $\star = *, !$. □

Let Λ be a finite set.

Proposition 6.3.23 *Let* $(V, L, \boldsymbol{N})$ *be an object in* $\mathcal{M}(\Lambda)$. *Let* $i, j \in \Lambda$ *be distinct elements. Let* $\star_i, \star_j \in \{*, !\}$. *Then, we have* $N_{i\star_i}(N_{j\star_j}L) = N_{j\star_j}(N_{i\star_i}L)$.

Proof Let $\Lambda_1 := \Lambda \setminus \{i, j\}$ and $\Lambda_2 := \{i, j\}$. We set $\tilde{\mathcal{A}} := \mathcal{A}(\Lambda_1)$, $\tilde{\mathcal{P}}_w(\Lambda') := \mathcal{P}_w(\Lambda_1 \sqcup \Lambda')$ and $\tilde{\mathcal{M}}(\Lambda') = \mathcal{M}(\Lambda_1 \sqcup \Lambda')$. Then, they satisfy the conditions in Sect. 6.3.1. Then, the claim follows from Lemmas 6.3.21 and 6.3.22. □

Let $(C, L, \boldsymbol{N})$ be an object in $\mathcal{M}(\Lambda)$. Let $J \subset \Lambda$. We define a filtration $\hat{\boldsymbol{N}}_{J*}L$ of $\Sigma^{0,-|J|}C$ inductively; take $j \in J$, put $J_0 := J \setminus \{j\}$, and define $\hat{\boldsymbol{N}}_{J*}L := \hat{N}_{j*}\hat{\boldsymbol{N}}_{J_0*}L$. By Proposition 6.3.23, it is independent of the choice of j. Similarly, we define $\hat{\boldsymbol{N}}_{J!}L$ of $\Sigma^{|J|,0}C$. We obtain $\boldsymbol{N}_{J*}L$ and $\boldsymbol{N}_{J!}L$ on C by $(C, \boldsymbol{N}_{J*}L) := \Sigma^{0,|J|}(C, \hat{\boldsymbol{N}}_{J*}L)$ and $(C, \boldsymbol{N}_{J!}L) := \Sigma^{-|J|,0}(C, \hat{\boldsymbol{N}}_{J!}L)$.

For $J_i \subset \Lambda$ $(i = 1, 2)$ with $J_1 \cap J_2 = \emptyset$, we obtain a filtration $\boldsymbol{N}_{J_1!}\boldsymbol{N}_{J_2*}L = \boldsymbol{N}_{J_2*}\boldsymbol{N}_{J_1!}L$ on C.

6.3.11 Canonical Prolongations

Let $V = (C, L, \boldsymbol{N})$ be an object in $\mathcal{M}(\Lambda)$. Let $\Lambda_1 \sqcup \Lambda_2 = \Lambda$ be a decomposition. We obtain an object $\mathcal{T} =: V[*K_1!K_2] \in MLA(\Lambda)$ given as follows. For $I \subset \Lambda$, we set $I_j := I \cap \Lambda_j$, and $\mathcal{T}_I := \Sigma^{|I_2|,-|I_1|}C$. It is equipped with a filtration $\hat{\boldsymbol{N}}_{I_1 !}\hat{\boldsymbol{N}}_{I_2 *}L$. For $I \sqcup \{i\} \subset \Lambda$, morphisms $g_{Ii,I}$ and $f_{I,Ii}$ are given as follows:

$$g_{Ii,I} := \begin{cases} \mathrm{id} & (i \in \Lambda_2) \\ N_i & (i \in \Lambda_1) \end{cases} \qquad f_{I,Ii} := \begin{cases} N_i & (i \in \Lambda_2) \\ \mathrm{id} & (i \in \Lambda_1) \end{cases}$$

Chapter 7
Mixed Twistor $\mathcal{D}$-Modules

We shall introduce the concept of mixed twistor $\mathcal{D}$-module (Definition 7.2.1). The definition is not completely parallel to that for mixed Hodge modules in [73]. It is partially because we are concerned with only graded polarizable objects. Since we shall later prove that any mixed twistor $\mathcal{D}$-module is expressed as a gluing of some admissible variations of mixed twistor structures (Sects. 10.3 and 11.1), mixed twistor $\mathcal{D}$-modules can be regarded as a twistor version of mixed Hodge modules. We shall also construct a natural functor from the category of mixed Hodge modules to the category of mixed twistor $\mathcal{D}$-modules in Chap. 13.

We briefly mention how mixed twistor $\mathcal{D}$-module is defined. Let X be a complex manifold. First it is natural to consider filtered $\mathcal{R}_X$-triples $(\mathcal{T}, W)$ such that each $\mathrm{Gr}_j^W \mathcal{T}$ is a polarizable pure twistor $\mathcal{D}$-module of weight j on X. Such filtered $\mathcal{R}_X$-triples are called pre-mixed twistor $\mathcal{D}$-modules in this paper. They already have some nice properties. For example, the category $\mathrm{MTW}(X)$ of pre-mixed twistor $\mathcal{D}_X$-modules is naturally an abelian category. For any projective morphism of complex manifolds $f : X \longrightarrow Y$, we have a naturally defined cohomological functor $f_\dagger^i$: $\mathrm{MTW}(X) \longrightarrow \mathrm{MTW}(Y)$ ($i \in \mathbb{Z}$).

But, for the definition of mixed twistor $\mathcal{D}$-modules, we have to impose additional conditions along any holomorphic functions. A pre-mixed twistor $\mathcal{D}$-module $(\mathcal{T}, W)$ is a mixed twistor $\mathcal{D}$-module if the following holds: Let $U \subset X$ be any open subset with a holomorphic function g. Then, $(\mathcal{T}, W)_{|U}$ is admissibly specializable along g in the sense of Sect. 4.4, and $(\tilde{\psi}_{g,\mathfrak{a},u}(\mathcal{T}), W)$ ($\mathfrak{a} \in \mathbb{C}[t^{-1/n}]$, $u \in \mathbb{R} \times \mathbb{C}$) and $(\phi_g^{(0)}(\mathcal{T}), W)$ are mixed twistor $\mathcal{D}$-modules with smaller supports. Here, W is the filtration obtained as the relative monodromy filtration of the canonically induced nilpotent map with respect to the naively induced filtration.

In Sect. 7.1, we study the admissible specializability condition along any holomorphic functions and the localizability along any effective divisor for pre-mixed twistor $\mathcal{D}$-modules. We check that the properties are preserved by various operations. Moreover, we check that for any $(\mathcal{T}, W) \in \mathrm{MTW}(X)$ which is localizable along an effective divisor H, we have $(\mathcal{T}[\star H], W) \in \mathrm{MTW}(X)$ ($\star = *, !$). We

T. Mochizuki, *Mixed Twistor $\mathcal{D}$-modules*, Lecture Notes in Mathematics 2125,
DOI 10.1007/978-3-319-10088-3_7

note that we shall care about the existence of polarizations on the graded pieces $\mathrm{Gr}_j^W \mathcal{T}[\star H]$.

In Sect. 7.2, we introduce the concept of mixed twistor $\mathscr{D}$-modules as explained above, and we show some of basic properties which follow from the definition rather easily. We shall return to the study of basic properties of mixed twistor $\mathscr{D}$-modules in Chap. 11 after the study of admissible variations of mixed twistor structure and their gluing.

7.1 Admissible Specializability of Pre-mixed Twistor $\mathscr{D}$-Modules

7.1.1 Pre-mixed Twistor $\mathscr{D}$-Modules

Let $\mathrm{MT}(X, w)$ denote the category of polarizable pure twistor $\mathscr{D}$-modules of weight w on X. (We omit the adjective "wild".) Recall that $\mathrm{MT}(X, w)$ is abelian and semisimple. (See [52, 55, 66, 67].)

In this monograph, a filtered $\mathcal{R}_X$-triple means an $\mathcal{R}_X$-triple $\mathcal{T}$ with an increasing filtration W in the category of $\mathcal{R}_X$-triples indexed by integers such that $W_j = 0$ for $j << 0$ and $W_j(\mathcal{T}) = \mathcal{T}$ for $j >> 0$ locally on X. A filtered $\mathcal{R}_X$-triple $(\mathcal{T}, W)$ is called a pre-mixed twistor $\mathscr{D}$-module if $\mathrm{Gr}_w^W(\mathcal{T}) \in \mathrm{MT}(X, w)$ for each w.

Lemma 7.1.1 *If $(\mathcal{T}, W)$ is a pre-mixed twistor $\mathscr{D}$-module, the underlying $\mathcal{R}_X$-modules $\mathcal{M}_i$ are coherent. Moreover, for any $\lambda_0 \in \mathbb{C}_\lambda$ and for any relatively compact subset $X_1 \subset X$, there exists a small neighbourhood $U(\lambda_0)$ in $\mathbb{C}_\lambda$ such that $\mathcal{M}_{i|U(\lambda_0)\times X_1}$ is good, i.e., it is a sum of $\mathcal{O}_{U(\lambda_0)\times X_1}$-coherent subsheaves.*

Proof If W is pure, it follows from the description of polarizable pure twistor $\mathscr{D}$-modules as the minimal extension of wild variation of pure twistor structure in [55]. The coherence of $\mathcal{R}$-modules is preserved by extensions. The goodness is also preserved by extensions (Proposition 4.23 of [34]). Hence, we obtain the claim of the lemma. □

Let $\mathrm{MTW}(X)$ denote the full subcategory of pre-mixed twistor $\mathscr{D}$-modules in the category of filtered $\mathcal{R}$-triples. It is equipped with auto equivalences $\Sigma^{p,q}(\mathcal{T}, W) := (\mathcal{T}, W) \otimes \big(\mathcal{U}(-p, q), W\big)$, where W on $\mathcal{U}(-p, q)$ is pure of weight $-(p+q)$. For any $(\mathcal{T}, W) \in \mathrm{MTW}(X)$, we set $W_k(\mathcal{T}^*) := (\mathcal{T}/W_{-k-1}\mathcal{T})^*$ and $W_k(j^*\mathcal{T}) = j^*W_k(\mathcal{T})$. Then, we obtain pre-mixed twistor $\mathscr{D}$-modules $(\mathcal{T}, W)^* := (\mathcal{T}^*, W)$ and $j^*(\mathcal{T}, W) := (j^*\mathcal{T}, W)$. Recall the following lemma in [55, 66, 67].

Lemma 7.1.2 *The category* $\mathrm{MTW}(X)$ *is abelian, and any morphism $f : (\mathcal{T}_1, W) \longrightarrow (\mathcal{T}_2, W)$ is strict with respect to the weight filtration.* □

Corollary 7.1.3 *Let $f : (\mathcal{T}_1, W) \longrightarrow (\mathcal{T}_2, W)$ be a morphism in* $\mathrm{MTW}(X)$. *If the morphism of $\mathcal{R}_X$-triples $f : \mathcal{T}_1 \longrightarrow \mathcal{T}_2$ is an isomorphism, f is an isomorphism of pre-mixed twistor $\mathscr{D}$-modules.* □

7.1.1.1 Some Abelian Categories

We introduce some auxiliary categories as in Chap. 6. Let $\mathrm{MTW}(X,\bullet)$ denote the category of objects $(\mathcal{T},W)\in\mathrm{MTW}(X)$ equipped with a morphism $N:(\mathcal{T},W)\longrightarrow(\mathcal{T},W)\otimes\boldsymbol{T}(-1)$. Morphisms in the category are naturally defined. It is clearly an abelian category. For $w\in\mathbb{Z}$, let $\mathrm{MTN}(X,w)\subset\mathrm{MTW}(X,\bullet)$ be the full subcategory of objects $(\mathcal{T},W,N)$ such that $W=M(N)[w]$, i.e., $W_j=M(N)_{j-w}$.

Lemma 7.1.4 *The category* $\mathrm{MTN}(X,w)$ *is abelian. Let* $(\mathcal{T}_i,W,N)$ *be objects in* $\mathrm{MTN}(X,w_i)$ *with* $w_1>w_2$*, and let* $F:(\mathcal{T}_1,W,N)\longrightarrow(\mathcal{T}_2,W,N)$ *be a morphism in* $\mathrm{MTW}(X,\bullet)$*. Then,* $F=0$*.*

Proof It is easy to deduce the first claim from Lemma 7.1.2. As for the second claim, because the induced morphism $\mathrm{Gr}^W(F)$ is 0, we obtain $F=0$ by Lemma 7.1.2. □

Let $\mathrm{MTW}(X,\bullet)^{\mathrm{RMF}}\subset\mathrm{MTW}(X,\bullet)^{\mathrm{fil}}$ denote the full subcategory of the objects $(\mathcal{T},W,L,N)$ such that $\mathrm{Gr}^L_w(\mathcal{T},W,N)\in\mathrm{MTN}(X,w)$. We obtain the following lemma from Lemma 7.1.4.

Lemma 7.1.5 $\mathrm{MTW}(X,\bullet)^{\mathrm{RMF}}$ *is an abelian category. Any morphism in* $\mathrm{MTW}(X,\bullet)^{\mathrm{RMF}}$ *is strict with respect to the filtration* L*.* □

7.1.2 Push-Forward by Projective Morphisms

Let $F:X\longrightarrow Y$ be a projective morphism. Let $(\mathcal{T},W)\in\mathrm{MTW}(X)$. The $\mathcal{R}$-triples $F^i_\dagger\mathcal{T}$ are equipped with the filtration W given as follows: $W_k(F^i_\dagger\mathcal{T})$ are the image of $F^i_\dagger(W_{k-i}\mathcal{T})\longrightarrow F^i_\dagger(\mathcal{T})$. The following proposition can be found in [73] essentially.

Proposition 7.1.6 $F^i_\dagger(\mathcal{T},W):=(F^i_\dagger\mathcal{T},W)$ *is a pre-mixed twistor* $\mathscr{D}_Y$*-module.*

Proof If $\mathcal{T}\in\mathrm{MT}(X,w)$, we know that $F^i_\dagger\mathcal{T}\in\mathrm{MT}(Y,w+i)$ [55, 66, 69]. According to [66], we have a spectral sequence $E_1^{-i,i+j}=F^j_\dagger\,\mathrm{Gr}^W_i(\mathcal{T})\Longrightarrow F^j_\dagger\mathcal{T}$. Because $F^j_\dagger\,\mathrm{Gr}^W_i(\mathcal{T})\in\mathrm{MT}(Y,i+j)$, we have $E_2^{-i,i+j}\in\mathrm{MT}(Y,i+j)$, and the spectral sequence degenerates. So, we have $\mathrm{Gr}^W_{i+j}F^j_\dagger(\mathcal{T})\simeq E_2^{-i,i+j}\in\mathrm{MT}(Y,i+j)$. □

Remark 7.1.7 The claim of the proposition holds if $F^i_\dagger\mathcal{T}\in\mathrm{MT}(Y,w+i)$ for any $\mathcal{T}\in\mathrm{MT}(X,w)$. □

From the proof, we have a description of $\mathrm{Gr}^W_j F^i_\dagger(\mathcal{T})$. We have the exact sequences $0\longrightarrow\mathrm{Gr}^W_{k-1}(\mathcal{T})\longrightarrow W_k(\mathcal{T})/W_{k-2}(\mathcal{T})\longrightarrow\mathrm{Gr}^W_k(\mathcal{T})\longrightarrow 0$ for any $k\in\mathbb{Z}$. They induce morphisms $a_{j,k}:F^j_\dagger\,\mathrm{Gr}^W_k(\mathcal{T})\longrightarrow F^{j+1}_\dagger\,\mathrm{Gr}^W_{k-1}(\mathcal{T})$. We obtain the following complex:

$$F^{j-1}_\dagger\,\mathrm{Gr}^W_{k+1}(\mathcal{T})\xrightarrow{a_{j-1,k+1}}F^j_\dagger\,\mathrm{Gr}^W_k(\mathcal{T})\xrightarrow{a_{j,k}}F^{j+1}_\dagger\,\mathrm{Gr}^W_{k-1}(\mathcal{T})\tag{7.1}$$

Corollary 7.1.8 *$\mathrm{Gr}^W_{k+j} F^j_{\dagger}(\mathcal{T})$ is isomorphic to* $\mathrm{Ker}\, a_{j,k} / \mathrm{Im}\, a_{j-1,k+1}$ *naturally.* □

Let $0 \longrightarrow (\mathcal{T}_1, W) \longrightarrow (\mathcal{T}_2, W) \longrightarrow (\mathcal{T}_3, W) \longrightarrow 0$ be an exact sequence in $\mathrm{MTW}(X)$. We have the exact sequence of $\mathcal{R}_Y$-triples:

$$\cdots \longrightarrow F^i_{\dagger}(\mathcal{T}_1) \xrightarrow{a_1} F^i_{\dagger}(\mathcal{T}_2) \xrightarrow{a_2} F^i_{\dagger}(\mathcal{T}_3) \xrightarrow{a_3} F^{i+1}_{\dagger}(\mathcal{T}_1) \longrightarrow \cdots$$

Lemma 7.1.9 *a_i are compatible with the weight filtrations.*

Proof The claims for a_1 and a_2 are clear by the construction of the weight filtration. The exact sequence $0 \longrightarrow \mathrm{Gr}^W_w(\mathcal{T}_1) \longrightarrow \mathrm{Gr}^W_w(\mathcal{T}_2) \longrightarrow \mathrm{Gr}^W_w(\mathcal{T}_3) \longrightarrow 0$ splits because the category of polarizable pure twistor $\mathscr{D}$-modules is semisimple, So, the induced morphism $F^i_{\dagger}\, \mathrm{Gr}^W_w(\mathcal{T}_3) \longrightarrow F^{i+1}_{\dagger}\, \mathrm{Gr}^W_w(\mathcal{T}_1)$ is 0. Note that $\mathrm{Gr}^W_{w+k} F^k_{\dagger} W_w \mathcal{T}_j \longrightarrow F^k_{\dagger}\, \mathrm{Gr}^W_w \mathcal{T}_j$ are monomorphisms. Hence, $F^i_{\dagger}(W_w \mathcal{T}_3) \longrightarrow F^{i+1}_{\dagger}(W_w \mathcal{T}_1)$ factor through $W_{w+i} F^{i+1}_{\dagger} W_w \mathcal{T}_1$. Then, the claim is clear by the construction of W on $F^i_{\dagger}(\mathcal{T}_j)$. □

Corollary 7.1.10 *We have a naturally induced cohomological functor $F^i_{\dagger}$:* $\mathrm{MTW}(X) \longrightarrow \mathrm{MTW}(Y)$. □

7.1.2.1 Composition of Projective Morphisms

Let $F : X \longrightarrow Y$ and $G : Y \longrightarrow Z$ be projective morphisms of complex manifolds. Let $(\mathcal{T}, W) \in \mathrm{MTW}(X)$. Because $(G \circ F)_{\dagger} = G_{\dagger} \circ F_{\dagger}$ as functors on complexes of $\mathcal{R}$-triples (see [66]), we have the Leray spectral sequence of $\mathcal{R}$-triples $E_2^{p,q}(\mathcal{T}) = G^p_{\dagger} \circ F^q_{\dagger} \mathcal{T} \Longrightarrow (G \circ F)^j_{\dagger} \mathcal{T}$. It gives a filtration $\mathcal{L}$ on $(G \circ F)^j_{\dagger} \mathcal{T}$ such that $\mathrm{Gr}^{\mathcal{L}}_p (G \circ F)^{p+q}_{\dagger}(\mathcal{T}) \simeq E_{\infty}^{p,q}(\mathcal{T})$.

Proposition 7.1.11 *For any $r \geq 2$ and any $p, q \in \mathbb{Z}$, the terms $E_r^{p,q}(\mathcal{T})$ are naturally equipped with weight filtrations W so that the morphisms $d_r : E_r^{p,q}(\mathcal{T}) \longrightarrow E_r^{p+r,q-r+1}(\mathcal{T})$ are morphisms in* $\mathrm{MTW}(Z)$, *and that $(E_{r+1}^{p,q}, W)$ are obtained as the cohomology of $(E_r^{p-r,q+r-1}, W) \longrightarrow (E_r^{p,q}, W) \longrightarrow (E_r^{p+r,q-r+1}, W)$ in* $\mathrm{MTW}(Z)$. *Moreover, $\mathcal{L}$ is a filtration in* $\mathrm{MTW}(Z)$, *and $\mathrm{Gr}^{\mathcal{L}}_p\big((G \circ F)^{p+q}_{\dagger} \mathcal{T}, W\big)$ is isomorphic to $(E_{\infty}^{p,q}(\mathcal{T}), W)$ in* $\mathrm{MTW}(Z)$.

Proof We have the weight filtration W on $E_2^{p,q}(\mathcal{T})$ obtained as $G^p_{\dagger}(F^q_{\dagger}\mathcal{T}, W)$. If $\mathcal{T}$ is pure of weight w, $E_2^{p,q}(\mathcal{T})$ are pure of weight $w + p + q$, and the spectral sequence degenerates by the decomposition theorem.

Let $\mathcal{K}_{k,j}$ be the kernel of the epimorphism $F^j_{\dagger} W_{k-j} \mathcal{T} \longrightarrow W_k F^j_{\dagger} \mathcal{T}$.

Lemma 7.1.12 *$\mathcal{K}_{k,j}$ are pure of weight k, and the exact sequence $0 \longrightarrow \mathcal{K}_{k,j} \longrightarrow F^j_{\dagger} W_{k-j} \mathcal{T} \longrightarrow W_k F^j_{\dagger} \mathcal{T} \longrightarrow 0$ admits a splitting.*

Proof By definition, $\mathcal{K}_{k,j}$ is equal to the kernel of $F_{\dagger}^{j} W_{k-j}\mathcal{T} \longrightarrow F_{\dagger}^{j}\mathcal{T}$. Hence, $\mathcal{K}_{k,j}$ is the image of $F_{\dagger}^{j-1}(\mathcal{T}/W_{k-j}\mathcal{T}) \longrightarrow F_{\dagger}^{j} W_{k-j}\mathcal{T}$ in $\mathrm{MTW}(X)$. Because $\mathrm{Gr}_m^W F_{\dagger}^{j-1}(\mathcal{T}/W_{k-j}\mathcal{T}) = 0$ $(m < k)$ and $\mathrm{Gr}_m^W F_{\dagger}^{j}(W_{k-j}\mathcal{T}) = 0$ $(m > k)$, we obtain that $\mathcal{K}_{k,j}$ is pure of weight k. The composite of the morphisms $\mathcal{K}_{k,j} \longrightarrow F_{\dagger}^{j} W_{k-j}\mathcal{T} \longrightarrow \mathrm{Gr}_k^W F_{\dagger}^{j} W_{k-j}\mathcal{T}$ is a monomorphism. Because $\mathrm{MT}(X,k)$ is semisimple, we have a splitting $\mathrm{Gr}_k^W F_{\dagger}^{j} W_{k-j}\mathcal{T} \longrightarrow \mathcal{K}_{k,j}$. Hence, we have a splitting of the exact sequence in the lemma. □

Lemma 7.1.13 *The following holds:*

(A_2) *The morphisms* $E_2^{p,q}(W_m\mathcal{T}) \longrightarrow W_{m+p+q}E_2^{p,q}(\mathcal{T})$ *are epimorphisms, and the kernel are pure of weight* $m+p+q$.
(B_2) *The morphisms* $W_{m+p+q-1}E_2^{p,q}(W_m\mathcal{T}) \longrightarrow W_{m+p+q-1}E_2^{p,q}(\mathcal{T})$ *are isomorphisms.*
(C_2) *The induced morphisms* $\mathrm{Gr}^W_{p+q+m} E_2^{p,q} W_m\mathcal{T} \longrightarrow E_2^{p,q}\,\mathrm{Gr}_m^W \mathcal{T}$ *are monomorphisms.*
(D_2) *We have* $\mathrm{Gr}_j^W E_2^{p,q} W_m\mathcal{T} = 0$ *unless* $j \leq m+p+q$.

Proof The property (D_2) is clear by construction. We have $G_{\dagger}^{p}F_{\dagger}^{q}W_m\mathcal{T} \xrightarrow{\alpha_1} G_{\dagger}^{p}W_{m+p}F_{\dagger}^{p}\mathcal{T} \xrightarrow{\alpha_2} W_{m+p+q}G_{\dagger}^{p}F_{\dagger}^{q}\mathcal{T}$. By definition α_2 is an epimorphism. By Lemma 7.1.12, α_1 is an epimorphism. Hence, their composite is an epimorphism. By the construction, α_i are morphisms in $\mathrm{MTW}(Z)$. Hence, $\mathrm{Ker}\,\alpha_2\circ\alpha_1$ is an object in $\mathrm{MTW}(Z)$. By Lemma 7.1.12, $\mathrm{Ker}\,\alpha_2$ is pure of weight $m+p+q$. By the construction $\mathrm{Ker}\,\alpha_1$ is pure of weight $m+p+q$. Hence, $\mathrm{Ker}\,\alpha_2\circ\alpha_1$ is also pure of weight $m+p+q$, and we obtain (A_2). The property (B_2) follows from (A_2).

We have the exact sequence $F_{\dagger}^{q}W_{m-1}\mathcal{T} \xrightarrow{a_1} F_{\dagger}^{q}W_m\mathcal{T} \xrightarrow{a_2} F_{\dagger}^{q}\,\mathrm{Gr}_m^W\mathcal{T}$. We have that $\mathrm{Gr}_j^W(\mathrm{Im}\,a_1) = 0$ unless $j \leq m-1+q$, and that $\mathrm{Im}\,a_2$ is a direct summand of $F_{\dagger}^{q}\,\mathrm{Gr}_m^W\mathcal{T}$. Hence, $G_{\dagger}^{p}\,\mathrm{Im}\,a_1 \longrightarrow G_{\dagger}^{p}F_{\dagger}^{q}W_m\mathcal{T} \longrightarrow G_{\dagger}^{p}F_{\dagger}^{q}\,\mathrm{Gr}_m^W\mathcal{T}$ is an exact sequence. Then, (C_2) is clear. □

Lemma 7.1.14 *The morphisms* $d_2 : E_2^{p,q}(W_m\mathcal{T}) \longrightarrow E_2^{p+2,q-1}(W_m\mathcal{T})$ *factors through* $W_{m+p+q}E_2^{p+2,q-1}(W_m\mathcal{T})$.

Proof The morphisms $d_2 : E_2^{p,q}\,\mathrm{Gr}_m^W\mathcal{T} \longrightarrow E_2^{p+2,q-1}\,\mathrm{Gr}_m^W\mathcal{T}$ are 0. Then, the claim follows from (C_2). □

Corollary 7.1.15 $d_2 : E_2^{p,q}(\mathcal{T}) \longrightarrow E_2^{p+2,q-1}(\mathcal{T})$ *are morphisms in* $\mathrm{MTW}(Z)$. *In particular,* $E_3^{p,q}(\mathcal{T})$ *are equipped with the naturally induced weight filtration* W.

Proof It follows from Lemma 7.1.14 and (A_2). □

Lemma 7.1.16 *We have the following exact sequence which admits a splitting:*

$$0 \longrightarrow \mathrm{Ker}(E_2^{p,q}W_m\mathcal{T} \longrightarrow W_{m+p+q}E_2^{p,q}\mathcal{T}) \longrightarrow \mathrm{Ker}(E_2^{p,q}W_m\mathcal{T} \longrightarrow E_2^{p+2,q-1}W_m\mathcal{T})$$
$$\longrightarrow \mathrm{Ker}(W_{m+p+q}E_2^{p,q}\mathcal{T} \longrightarrow W_{m+p+q}E_2^{p+2,q-1}\mathcal{T}) \longrightarrow 0 \qquad (7.2)$$

Proof It follows from (A_2) and (B_2). □

We have the following naturally defined morphism in MTW(Z).

$$\mathrm{Im}(E_2^{p-2,q+1}W_m\mathcal{T} \longrightarrow E_2^{p,q}W_m\mathcal{T}) \longrightarrow \mathrm{Im}(E_2^{p-2,q+1}\mathcal{T} \longrightarrow E_2^{p,q}\mathcal{T}).$$

Lemma 7.1.17 *The induced morphism* $W_{m+p+q-1}\,\mathrm{Im}(E_2^{p-2,q+1}W_m\mathcal{T} \longrightarrow E_2^{p,q}W_m\mathcal{T}) \longrightarrow W_{m+p+q-1}\,\mathrm{Im}(E_2^{p-2,q+1}\mathcal{T} \longrightarrow E_2^{p,q}\mathcal{T})$ *is an isomorphism.*

Proof By the strictness of the morphisms in MTW(Z) with respect to the weight filtration, we have $W_{m+p+q-1}\,\mathrm{Im}(E_2^{p-2,q+1}\mathcal{T} \longrightarrow E_2^{p,q}\mathcal{T})$ is equal to $\mathrm{Im}(W_{m+p+q-1}E_2^{p-2,q+1}\mathcal{T} \longrightarrow E_2^{p,q}\mathcal{T})$. By (A_2), it is equal to the image of $E_2^{p-2,q+1}W_m\mathcal{T} \longrightarrow E_2^{p,q}\mathcal{T}$. Then, the claim of the lemma follows from (B_2). □

Lemma 7.1.18 *The following holds for* $(E_r^{p,q}, W)$ *with* $r = 3$.

(A_r) *The morphisms* $E_r^{p,q}(W_m\mathcal{T}) \longrightarrow W_{m+p+q}E_r^{p,q}(\mathcal{T})$ *are epimorphisms, and the kernel are pure of weight* $m + p + q$.
(B_r) *The morphisms* $W_{m+p+q-1}E_r^{p,q}(W_m\mathcal{T}) \longrightarrow W_{m+p+q-1}E_r^{p,q}(\mathcal{T})$ *are isomorphisms.*
(C_r) *The induced morphisms* $\mathrm{Gr}^W_{p+q+m} E_r^{p,q}W_m\mathcal{T} \longrightarrow E_r^{p,q}\,\mathrm{Gr}^W_m\mathcal{T}$ *are monomorphisms.*
(D_r) *We have* $\mathrm{Gr}^W_j E_r^{p,q}W_m\mathcal{T} = 0$ *unless* $j \leq m + p + q$.

Proof The property (D_3) is clear by the construction. By Lemma 7.1.16, The morphisms $E_3^{p,q}(W_m\mathcal{T}) \longrightarrow W_{m+p+q}E_3^{p,q}(\mathcal{T})$ are epimorphisms. By Lemma 7.1.17, we obtain (B_3). We obtain that the kernel of $E_3^{p,q}(W_m\mathcal{T}) \longrightarrow W_{m+p+q}E_3^{p,q}(\mathcal{T})$ is pure of weight $m + p + q$, and (A_3) is obtained. The property (C_3) follows from Lemma 7.1.14. □

By an inductive argument on r, we obtain the weight filtrations W on $E_r^{p,q}(W_m\mathcal{T})$ satisfying the conditions (A_r), (B_r), (C_r) and (D_r). There exists r_0 such that $E_\infty^{p,q}(\mathcal{T}) = E_r^{p,q}(\mathcal{T})$ for any $r \geq r_0$. We obtain the filtrations W on $E_\infty^{p,q}$ satisfying (A_∞), (B_∞), (C_∞) and (D_∞).

Fix $j \in \mathbb{Z}$. Let us prove that $\mathcal{L}$ on $(G \circ F)^j_\dagger(\mathcal{T})$ is a filtration in MTW(Z). If m is sufficiently small, then $\mathcal{L}_m = 0$, and the claim is trivial. Suppose that we have already known that $(\mathcal{L}_{m-1}, W)$ is a subobject of $((G \circ F)^j_\dagger\mathcal{T}, W)$ in MTW(Z). Then, we have $\big((G\circ F)^j_\dagger\mathcal{T}/\mathcal{L}_{m-1}, W\big)$ in MTW(Z). We have the morphism of $\mathcal{R}$-triples $\rho_m : E_\infty^{m,j-m}(\mathcal{T}) \longrightarrow (G \circ F)^j_\dagger\mathcal{T}/\mathcal{L}_{m-1}$. By (A_∞), ρ_m is a morphism in MTW(Z), which is strict with respect to the weight filtration W. Hence, we have $0 \longrightarrow W_k\mathcal{L}_{m-1} \longrightarrow W_k\mathcal{L}_m \longrightarrow W_k\,\mathrm{Gr}^{\mathcal{L}}_m\mathcal{T} \longrightarrow 0$ for any k. We obtain $\mathrm{Gr}^{\mathcal{L}}_m\,\mathrm{Gr}^W_k(\mathcal{T}) \simeq \mathrm{Gr}^W_k\,\mathrm{Gr}^{\mathcal{L}}_m(\mathcal{T}) \in \mathrm{MT}(Z,k)$. We also have $\mathrm{Gr}^W_k(\mathcal{T}) \in \mathrm{MT}(Z,k)$ and $\mathcal{L}_{m-1}\,\mathrm{Gr}^W_k(\mathcal{T}) \in \mathrm{MT}(Z,k)$. By using the semisimplicity of MT(Z, k), we obtain that $\mathcal{L}_m\,\mathrm{Gr}^W_k\mathcal{T} \in \mathrm{MT}(Z,k)$. Thus, we obtain Proposition 7.1.11. □

Remark 7.1.19 In Proposition 7.1.11, we can replace the projectivity of G with the assumption that $G^j_\dagger \mathcal{T} \in \mathrm{MT}(Z, w+j)$ for any $\mathcal{T} \in \mathrm{MT}(Z, w)$. As for the morphism F, we need that the decomposition theorem holds for F, not only that $F_\dagger$ preserves the purity and the weight. □

Corollary 7.1.20 *Let $F : X \longrightarrow Y$ and $G : Y \longrightarrow Z$ be projective morphisms of complex manifolds. Let $(\mathcal{T}, W) \in \mathrm{MTW}(X)$. Suppose that $F^j_\dagger(\mathcal{T}) = 0$ $(j \neq 0)$. Then, the isomorphisms of $\mathcal{R}$-triples $(G \circ F)^j_\dagger(\mathcal{T}) \simeq G^j_\dagger(F^0_\dagger \mathcal{T})$ are isomorphisms in* $\mathrm{MTW}(Z)$.

Proof Although this is a consequence of Proposition 7.1.11, we give a direct argument in this easier situation. By Lemma 7.1.12 and the assumption $F^i_\dagger(\mathcal{T}) = 0$ $(i \neq 0)$, we have that $F^i_\dagger(W_{m-i}\mathcal{T})$ $(i \neq 0)$ are pure of weight m. Hence, the morphisms $F^i_\dagger W_{m-i}(\mathcal{T}) \longrightarrow F^i_\dagger \mathrm{Gr}^W_{m-i}(\mathcal{T})$ are monomorphisms if $i \neq 0$. Let $\mathcal{I}_{m,i}$ denote the image. Because of the semisimplicity of the category $\mathrm{MT}(X, m)$, $\mathcal{I}_{m,i}$ is a direct summand of $F^i_\dagger \mathrm{Gr}^W_{m-i}(\mathcal{T})$. Hence, the morphisms $G^j_\dagger F^i_\dagger W_{m-i}(\mathcal{T}) \longrightarrow G^j_\dagger F^i_\dagger \mathrm{Gr}^W_{m-i}(\mathcal{T})$ are monomorphisms if $i \neq 0$.

We have the spectral sequence $E_2^{p,q}(\mathcal{T}) = G^p_\dagger \circ F^q_\dagger(\mathcal{T}) \Longrightarrow (G \circ F)^j_\dagger \mathcal{T}$. If $\mathcal{T}$ is pure of weight w, it degenerates by the Hard Lefschetz Theorem.

The morphisms $E_2^{p,q}(W_k\mathcal{T}) \longrightarrow E_2^{p,q}(\mathrm{Gr}^W_k(\mathcal{T}))$ are monomorphisms if $q \neq 0$. The morphisms $d_2 : E_2^{p,q}(\mathrm{Gr}^W_k \mathcal{T}) \longrightarrow E_2^{p+2,q-1}(\mathrm{Gr}^W_k \mathcal{T})$ are 0. Then, we obtain that $d_2 : E_2^{p,0}(W_k\mathcal{T}) \longrightarrow E_2^{p+2,-1}(W_k\mathcal{T})$ are 0 for any p. We also have $E_3^{p,q} W_k\mathcal{T} \longrightarrow E_3^{p,q} \mathrm{Gr}^W_k(\mathcal{T})$ are monomorphisms for any $q \neq 0$ and any p. Then, by an induction on r, we obtain that (i) $d_r : E_r^{p,0} W_k\mathcal{T} \longrightarrow E_r^{p+r,-r+1} W_k\mathcal{T}$ are 0 for any p, (ii) $E_r^{p,q} W_k\mathcal{T} \longrightarrow E_r^{p,q} \mathrm{Gr}^W_k \mathcal{T}$ are monomorphisms for any $q \neq 0$ and any p. Then, we obtain that the image of $G^j_\dagger W_{m-j} F^0_\dagger \mathcal{T} \longrightarrow G^j_\dagger F^0_\dagger \mathcal{T}$ are contained in the image of $G^j_\dagger F^0_\dagger W_{m-j}\mathcal{T} \longrightarrow G^j_\dagger F^0_\dagger \mathcal{T}$. Then, the claim of the corollary follows. □

7.1.3 Admissible Specializability for Pre-mixed Twistor $\mathscr{D}$-Modules

For a holomorphic function g on X, let $\mathrm{MTW}^{\mathrm{sp}}(X, g) \subset \mathrm{MTW}(X)$ denote the full subcategory of $(\mathcal{T}, W) \in \mathrm{MTW}(X)$ which are admissibly specializable along g. We obtain the following lemma from Proposition 6.2.3 and the strictly S-decomposability condition for pure twistor $\mathscr{D}$-modules.

Lemma 7.1.21 *A pre-mixed twistor $\mathscr{D}$-module $(\mathcal{T}, W)$ is admissibly specializable along g if only if the underlying $\mathcal{R}$-modules are filtered strictly specializable and satisfy the conditions* **(P1–2)** *in Definition 4.4.4. In other words,* **(P3)** *is automatically satisfied.* □

For $(\mathcal{T}, W) \in \mathrm{MTW}^{\mathrm{sp}}(X, g)$, we have the filtrations L of $\tilde{\psi}_{g,\mathfrak{a},u}(\mathcal{T})$ and $\phi_g(\mathcal{T})$ which are naively induced by W:

$$L_j\tilde{\psi}_{g,\mathfrak{a},u}(\mathcal{T}) := \tilde{\psi}_{g,\mathfrak{a},u}(W_j\mathcal{T}), \qquad L_j\phi_g(\mathcal{T}) := \phi_g(W_j\mathcal{T})$$

Let W denote the filtrations of $\tilde{\psi}_{g,\mathfrak{a},u}(\mathcal{T})$ and $\phi_g(\mathcal{T})$ obtained as the relative monodromy filtration of $\mathcal{N}$ with respect to L.

Lemma 7.1.22 $\big(\tilde{\psi}_{g,\mathfrak{a},u}(\mathcal{T}), W, L, \mathcal{N}\big)$ *and* $\big(\phi_g(\mathcal{T}), W, L, \mathcal{N}\big)$ *are objects in the category* $\mathrm{MTW}(X, \bullet)^{\mathrm{RMF}}$.

Proof We have only to prove that $\big(\tilde{\psi}_{g,\mathfrak{a},u}(\mathcal{T}), W\big)$ and $\big(\phi_g(\mathcal{T}), W\big)$ are objects in $\mathrm{MTW}(X)$, which follows from the canonical decomposition $\mathrm{Gr}^W \simeq \mathrm{Gr}^W \mathrm{Gr}^L$. (See Sect. 6.1.2.) □

We shall often denote pre-mixed twistor $\mathscr{D}$-modules $\big(\tilde{\psi}_{g,\mathfrak{a},u}(\mathcal{T}), W\big)$ and $\big(\phi_g(\mathcal{T}), W\big)$ by $\tilde{\psi}_{g,\mathfrak{a},u}(\mathcal{T}, W)$ and $\phi_g(\mathcal{T}, W)$. We should remark that the filtrations are changed to the relative monodromy filtration from the naively induced filtrations. (Compare $\tilde{\psi}_{g,\mathfrak{a},u}(\mathcal{T}, L)$ and $\phi_g(\mathcal{T}, L)$ in Sect. 4.4.2.)

Remark 7.1.23 If we would like to obtain the correct weight filtration of $\psi_g^{(1)}(\mathcal{T})$ as the relative monodromy filtration of $\mathcal{N}$ with respect to a filtration L', then L' is given as $L'_j\psi_g^{(1)}(\mathcal{T}) = \psi_g^{(1)}(W_{j+1}\mathcal{T})$. Recall the relation in Proposition 4.3.1.

However, we prefer to use the filtration L on $\psi_g^{(a)}(\mathcal{T})$ naively induced by W, i.e., $L_j\psi_g^{(a)}(\mathcal{T}) = \psi_g^{(a)}(W_j\mathcal{T})$, for which the isomorphisms in Proposition 4.3.1 preserve the filtrations L. □

7.1.3.1 Sub-quotients

Following M. Saito [73], we prove that the admissible specializability is stable for sub-quotients in $\mathrm{MTW}(X)$. We consider an exact sequence in $\mathrm{MTW}(X)$:

$$0 \longrightarrow (\mathcal{T}_1, W) \longrightarrow (\mathcal{T}_2, W) \longrightarrow (\mathcal{T}_3, W) \longrightarrow 0$$

Proposition 7.1.24 *Let g be any holomorphic function on X. If $(\mathcal{T}_2, W)$ is admissibly (resp. filtered strictly) specializable along g, $(\mathcal{T}_i, W)$ $(i = 1, 3)$ are also admissibly (resp. filtered strictly) specializable along g.*

Proof Let $\mathcal{T}_i = (\mathcal{M}'_i, \mathcal{M}''_i, C_i)$. We may assume that $X = X_0 \times \mathbb{C}_t$ and $g = t$. Suppose that $(\mathcal{T}_2, W)$ is filtered strictly specializable along t. Let us prove that $(\mathcal{T}_i, W)$ $(i = 1, 3)$ are also filtered strictly specializable along t, together with the following claim.

(C) For any $\lambda_0 \in \mathbb{C}$, the followings complexes are exact:

$$0 \longrightarrow V_a^{(\lambda_0)}(\mathcal{M}_3') \longrightarrow V_a^{(\lambda_0)}(\mathcal{M}_2') \longrightarrow V_a^{(\lambda_0)}(\mathcal{M}_1') \longrightarrow 0$$
$$0 \longrightarrow V_a^{(\lambda_0)}(\mathcal{M}_1'') \longrightarrow V_a^{(\lambda_0)}(\mathcal{M}_2'') \longrightarrow V_a^{(\lambda_0)}(\mathcal{M}_3'') \longrightarrow 0$$

We shall use the induction on the length of W. If W is pure, the claim of Proposition 7.1.24 is trivial. Because the category of polarizable pure twistor $\mathscr{D}$-modules are semisimple, the claim (C) also holds.

We may assume that $W_0(\mathcal{T}_i) = \mathcal{T}_i$, and that the claims hold for $W_{-1}(\mathcal{T}_i)$. Suppose that $(\mathcal{T}_2, W)$ is filtered strictly specializable along t. Let $\lambda_0 \in \mathbb{C}$. We define $V_a^{(\lambda_0)}(W_j\mathcal{M}_3'')$ as the image of $V_a^{(\lambda_0)}(W_j\mathcal{M}_2'') \longrightarrow W_j\mathcal{M}_3''$. By the claim (**C**) for W_{-1}, the filtrations $V_\bullet^{(\lambda_0)}(W_j\mathcal{M}_3'')$ $(j < 0)$ are equal to the V-filtrations of $W_j\mathcal{M}_3''$. We have the following natural epimorphisms:

$$V_a^{(\lambda_0)}(\mathcal{M}_2'') \longrightarrow V_a^{(\lambda_0)} \mathrm{Gr}_0^W(\mathcal{M}_2'') \longrightarrow V_a^{(\lambda_0)} \mathrm{Gr}_0^W(\mathcal{M}_3'')$$

Hence, we have the natural surjection $V_a^{(\lambda_0)}(\mathcal{M}_3'') \longrightarrow V_a^{(\lambda_0)} \mathrm{Gr}_0^W(\mathcal{M}_3'')$. We set $V_a'^{(\lambda_0)}(W_{-1}\mathcal{M}_3'') := V_a^{(\lambda_0)}(\mathcal{M}_3'') \cap W_{-1}\mathcal{M}_3''$. By construction, we have $V_a^{(\lambda_0)}(W_{-1}\mathcal{M}_3'') \subset V_a'^{(\lambda_0)}(W_{-1}\mathcal{M}_3'')$. Because $V'^{(\lambda_0)}$ is a monodromic filtration by $V_0\mathcal{R}_{X_0\times\mathbb{C}_t}$-coherent subsheaves, we have $V_a^{(\lambda_0)}(W_{-1}\mathcal{M}_3'') \supset V_a'^{(\lambda_0)}(W_{-1}\mathcal{M}_3'')$. Hence,

$$0 \longrightarrow V_a^{(\lambda_0)}(W_{-1}\mathcal{M}_3'') \longrightarrow V_a^{(\lambda_0)}(\mathcal{M}_3'') \longrightarrow V_a^{(\lambda_0)}(\mathrm{Gr}_0^W \mathcal{M}_3'') \longrightarrow 0$$

is exact for any a. By considering any ramified exponential twist, we obtain that $(\mathcal{M}_3'', W)$ is filtered strictly specializable along t.

We define $V_a^{(\lambda_0)}(\mathcal{M}_1'') := \mathcal{M}_1'' \cap V_a^{(\lambda_0)}(\mathcal{M}_2'')$. By an easy diagram chasing, we obtain that $0 \longrightarrow V_a^{(\lambda_0)}(W_{-1}\mathcal{M}_1'') \longrightarrow V_a^{(\lambda_0)}(\mathcal{M}_1'') \longrightarrow V_a^{(\lambda_0)}(\mathrm{Gr}_0^W \mathcal{M}_1'') \longrightarrow 0$ is exact for any a. Hence, by considering any ramified exponential twist, we obtain that $(\mathcal{M}_1'', W)$ is also filtered strictly specializable along t. We can prove the filtered strictly specializability along t for $(\mathcal{M}_i', W)$ $(i = 1, 3)$ in a similar way. The claim (**C**) is also proved.

Suppose that $(\mathcal{T}_2, W)$ is admissibly specializable along t. We have already known that $(\mathcal{T}_i, W)$ $(i = 1, 3)$ are filtered strictly specializable along t. Let us prove that $(\mathcal{T}_3, W)$ is admissibly specializable along t by using Proposition 6.1.4. For $\mathfrak{a}, u$, we set $\mathfrak{T}_i := \tilde{\psi}_{t,\mathfrak{a},u}(\mathcal{T}_i)$ which we naturally regard as $\mathcal{R}_X$-triples. We have the following commutative diagram:

$$\begin{array}{ccc}
\mathrm{Ker}(\mathcal{N}^\ell : \mathrm{Gr}_0^L \mathfrak{T}_2 \otimes \boldsymbol{T}(\ell) \to \mathrm{Gr}_0^L \mathfrak{T}_2) & \xrightarrow{\gamma_2} & \dfrac{L_{-1}\mathfrak{T}_2}{\mathcal{N}^\ell L_{-1}\mathfrak{T}_2 \otimes \boldsymbol{T}(\ell) + W_{-\ell-1}L_{-1}\mathfrak{T}_2} \\
\gamma_1 \big\downarrow & & \big\downarrow \\
\mathrm{Ker}(\mathcal{N}^\ell : \mathrm{Gr}_0^L \mathfrak{T}_3 \otimes \boldsymbol{T}(\ell) \to \mathrm{Gr}_0^L \mathfrak{T}_3) & \xrightarrow{\gamma_3} & \dfrac{L_{-1}\mathfrak{T}_3}{\mathcal{N}^\ell L_{-1}\mathfrak{T}_3 \otimes \boldsymbol{T}(\ell) + W_{-\ell-1}L_{-1}\mathfrak{T}_3}
\end{array}$$

The vertical arrows are natural morphisms, and the horizontal arrows are induced by $\mathcal{N}^{\ell}$. By Proposition 6.1.4, we have $\gamma_2 = 0$. By the semisimplicity of polarizable pure twistor $\mathscr{D}$-modules, γ_1 is an epimorphism. Hence, we obtain that $\gamma_3 = 0$. By Proposition 6.1.4, we obtain the existence of a relative monodromy filtration of $\mathcal{N}$ on $(\tilde{\psi}_{t,\mathfrak{a},u}(\mathcal{T}_3), L)$. Similarly, we obtain the existence of a relative monodromy filtration of $\mathcal{N}$ on $(\phi_t(\mathcal{T}_3), L)$. Thus, we obtain that $(\mathcal{T}_3, W)$ is admissibly specializable along t. By using the Hermitian adjoint, we obtain that $(\mathcal{T}_1, W)$ is also admissibly specializable along t. □

As an immediate corollary, we obtain the following proposition.

Proposition 7.1.25 $\mathrm{MTW}^{\mathrm{sp}}(X, g)$ *is an abelian subcategory of* $\mathrm{MTW}(X)$. □

7.1.3.2 The Associated Gluing Data

Let $(\mathcal{T}, W) \in \mathrm{MTW}^{\mathrm{sp}}(X, g)$. Recall Proposition 4.3.1. We also use the notation in Sects. 6.2.1 and 7.1.3. We have the following object in $\mathrm{Glu}(\mathrm{MTW}(X), \boldsymbol{\Sigma})^{\mathrm{fil}}$, which is filtered S-decomposable:

$$\left(\Sigma^{1,0}\tilde{\psi}_{g,-\boldsymbol{\delta}}(\mathcal{T}), L\right) \longrightarrow \left(\phi_g^{(0)}(\mathcal{T}), L\right) \longrightarrow \left(\Sigma^{0,-1}\tilde{\psi}_{g,-\boldsymbol{\delta}}(\mathcal{T}), L\right)$$

We also have the object in $\mathrm{Glu}\bigl(\mathrm{MTW}(X)^{\mathrm{fil}}, \boldsymbol{\Sigma}\bigr)$:

$$\Sigma^{1,0}\bigl(\tilde{\psi}_{g,-\boldsymbol{\delta}}(\mathcal{T}), W\bigr) \longrightarrow \bigl(\phi_g^{(0)}(\mathcal{T}), W\bigr) \longrightarrow \Sigma^{0,-1}\bigl(\tilde{\psi}_{g,-\boldsymbol{\delta}}(\mathcal{T}), W\bigr)$$

We remark that the filtration $L\phi_g^{(0)}(\mathcal{T})$ is the transfer of $L\tilde{\psi}_{g,-\boldsymbol{\delta}}(\mathcal{T})$.

7.1.3.3 Independence from Functions

Lemma 7.1.26 *Let A be a nowhere vanishing holomorphic function. We set $g_1 := Ag$.*

- *We have* $\mathrm{MTW}^{\mathrm{sp}}(X, g_1) = \mathrm{MTW}^{\mathrm{sp}}(X, g)$.
- *For any $(\mathcal{T}, W) \in \mathrm{MTW}^{\mathrm{sp}}(X, g)$, we have natural isomorphisms*

$$\mathrm{Gr}^W \tilde{\psi}_{g,-\boldsymbol{\delta}}(\mathcal{T}) \simeq \mathrm{Gr}^W \tilde{\psi}_{g_1,-\boldsymbol{\delta}}(\mathcal{T}), \quad \mathrm{Gr}^W \phi_g(\mathcal{T}) \simeq \mathrm{Gr}^W \phi_{g_1}(\mathcal{T}), \tag{7.3}$$

which preserve the natural nilpotent morphisms.

Proof It follows from Lemma 4.4.16. □

Let $(\mathcal{T}, W) \in \mathrm{MTW}^{\mathrm{sp}}(X, g)$. Let $\mathcal{S} = (\mathcal{S}_k \mid k \in \mathbb{Z})$ be a tuple of polarizations on $\mathrm{Gr}_k^W(\mathcal{T})$. We have the natural polarizations on

$$\mathrm{Gr}_k^W \psi_{g,-\boldsymbol{\delta}}(\mathcal{T}) \simeq \mathrm{Gr}_k^W \mathrm{Gr}^L \psi_{g,-\boldsymbol{\delta}}(\mathcal{T}) = \mathrm{Gr}_k^W \psi_{g,-\boldsymbol{\delta}}(\mathrm{Gr}^W \mathcal{T}),$$
$$\mathrm{Gr}_k^W \phi_g(\mathcal{T}) \simeq \mathrm{Gr}_k^W \mathrm{Gr}^L \phi_g(\mathcal{T}) = \mathrm{Gr}_k^W \phi_g(\mathrm{Gr}^W \mathcal{T}),$$

induced by $\mathcal{S}$, the natural nilpotent maps $\mathcal{N}$ on $\tilde{\psi}_{g,-\boldsymbol{\delta}}(\mathrm{Gr}^W \mathcal{T})$ and $\phi_g(\mathrm{Gr}^W \mathcal{T})$, and the primitive decompositions. The following corollary is clear by construction.

Corollary 7.1.27 *The induced polarizations are preserved by the isomorphisms (7.3).* □

7.1.3.4 Admissible Specializability Along Hypersurfaces

Let H be an effective divisor of X. Let $\mathrm{MTW}^{\mathrm{sp}}(X, H)$ denote the full subcategory of $(\mathcal{T}, W) \in \mathrm{MTW}(X)$ which are admissibly specializable along H (Definition 4.4.19). We obtain the following from Proposition 7.1.25.

Proposition 7.1.28 $\mathrm{MTW}^{\mathrm{sp}}(X, H)$ *is an abelian subcategory of* $\mathrm{MTW}(X)$. □

As in Sect. 4.4.3.1, we obtain graded $\mathcal{R}_X$-triples:

$$\big(\mathrm{Gr}^W \tilde{\psi}_{H,-\boldsymbol{\delta}}\big)(\mathcal{T}), \quad \big(\mathrm{Gr}^W \phi_H\big)(\mathcal{T}).$$

They are equipped with the nilpotent maps

$$\mathcal{N} : \big(\mathrm{Gr}_k^W \tilde{\psi}_{H,-\boldsymbol{\delta}}\big)(\mathcal{T}) \longrightarrow \big(\mathrm{Gr}_{k-2}^W \tilde{\psi}_{H,-\boldsymbol{\delta}}\big)(\mathcal{T}) \otimes \boldsymbol{T}(-1),$$
$$\mathcal{N} : \big(\mathrm{Gr}_k^W \phi_H\big)(\mathcal{T}) \longrightarrow \big(\mathrm{Gr}_{k-2}^W \phi_H\big)(\mathcal{T}) \otimes \boldsymbol{T}(-1).$$

They are also equipped with the filtration L naively induced by the weight filtration W on $\mathcal{T}$ and the canonical splitting:

$$\big(\mathrm{Gr}^W \tilde{\psi}_{H,-\boldsymbol{\delta}}\big)(\mathcal{T}) \simeq \mathrm{Gr}^L\big(\mathrm{Gr}^W \tilde{\psi}_{H,-\boldsymbol{\delta}}\big)(\mathcal{T}) \simeq \big(\mathrm{Gr}^W \tilde{\psi}_{H,-\boldsymbol{\delta}}\big)(\mathrm{Gr}^W \mathcal{T}),$$
$$\big(\mathrm{Gr}^W \phi_H\big)(\mathcal{T}) \simeq \mathrm{Gr}^L\big(\mathrm{Gr}^W \phi_H\big)(\mathcal{T}) \simeq \big(\mathrm{Gr}^W \phi_H\big)(\mathrm{Gr}^W \mathcal{T}).$$

A graded polarization $\mathcal{S} = \bigoplus \mathcal{S}_k$ on $\mathrm{Gr}^W(\mathcal{T})$ with $\mathcal{N}$ canonically induces Hermitian sesqui-linear dualities $\mathcal{S} \circ (-\mathcal{N})^m$ on $\mathrm{Gr}_m^W \tilde{\psi}_{H,-\boldsymbol{\delta}} \mathrm{Gr}_k^W \mathcal{T}$ and $\mathrm{Gr}_m^W \phi_H \mathrm{Gr}_k^W \mathcal{T}$.

Proposition 7.1.29 $\big(\mathrm{Gr}_m^W \tilde{\psi}_{H,-\boldsymbol{\delta}}\big)\big(\mathrm{Gr}_k^W \mathcal{T}\big)$ *and* $\big(\mathrm{Gr}_m^W \phi_H\big)\big(\mathrm{Gr}_k^W \mathcal{T}\big)$ *are polarizable pure twistor* $\mathscr{D}$*-modules of weight* m.

Proof We can check that $\mathrm{Gr}_m^W \tilde{\psi}_{H,-\boldsymbol{\delta}} \mathrm{Gr}_k^W \mathcal{T}$ and $\mathrm{Gr}_m^W \phi_H \mathrm{Gr}_k^W \mathcal{T}$ are pure twistor $\mathscr{D}$-modules of weight m by using Lemma 7.1.22. We can construct their polarizations by using the above Hermitian sesqui-linear dualities and the primitive decompositions with respect to the action of $\mathcal{N}$. Hence, we obtain that $\mathrm{Gr}^W_m \tilde{\psi}_{H,-\delta} \mathrm{Gr}^W_k \mathcal{T}$ and $\mathrm{Gr}^W_m \phi_H \mathrm{Gr}^W_k \mathcal{T}$ are polarizable pure twistor $\mathscr{D}$-modules. □

Corollary 7.1.30 *$\big(\mathrm{Gr}^W_m \tilde{\psi}_{H,-\delta}\big)(\mathcal{T})$ and $\big(\mathrm{Gr}^W_m \phi_H\big)(\mathcal{T})$ are polarizable wild pure twistor $\mathscr{D}$-modules of weight m.* □

7.1.4 Admissible Specializability and Push-Forward

Let $F : X \longrightarrow Y$ be a projective morphism. Let g_Y be a holomorphic function on Y. We put $g_X := F^{-1}(g_Y)$. Let $(\mathcal{T}, W) \in \mathrm{MTW}^{\mathrm{sp}}(X, g_X)$. We obtain $F^j_\dagger(\mathcal{T}, W)$ in $\mathrm{MTW}(Y)$ as in Sect. 7.1.2.

Lemma 7.1.31 *We have $F^j_\dagger(\mathcal{T}, W) \in \mathrm{MTW}^{\mathrm{sp}}(Y, g_Y)$.*

Proof We have $\big(\tilde{\psi}_{g_X,\mathfrak{a},u}(\mathcal{T}), W\big)$ and $\big(\phi_{g_X}(\mathcal{T}), W\big)$ in $\mathrm{MTW}(X)$. We obtain the push-forward $\big(F^j_\dagger\tilde{\psi}_{g_X,\mathfrak{a},u}(\mathcal{T}), W\big)$ and $\big(F^j_\dagger\phi_{g_X}(\mathcal{T}), W\big)$ in $\mathrm{MTW}(Y)$. In particular, they are strict. Hence, we obtain that $F^j_\dagger(\mathcal{T})$ is strictly specializable along g_Y with any ramified exponential twist, and we have $\tilde{\psi}_{g_Y,\mathfrak{a},u}F^j_\dagger\mathcal{T} \simeq F^j_\dagger\tilde{\psi}_{g_X,\mathfrak{a},u}\mathcal{T}$ and $\phi_{g_Y}F^j_\dagger\mathcal{T} \simeq F^j_\dagger\phi_{g_X}\mathcal{T}$. Because the complex

$$F^j_\dagger\tilde{\psi}_{g_X,\mathfrak{a},u}\big(W_k\mathcal{T}\big) \longrightarrow F^j_\dagger\tilde{\psi}_{g_X,\mathfrak{a},u}(W_\ell\mathcal{T}) \longrightarrow F^j_\dagger\tilde{\psi}_{g_X,\mathfrak{a},u}\big(W_\ell\mathcal{T}/W_k\mathcal{T}\big)$$

is exact, we obtain that $F^j_\dagger(\mathcal{T})$ is filtered strictly specializable by using Lemma 2.1.2, and that $L_k\tilde{\psi}_{g_Y,\mathfrak{a},u}F^j_\dagger(\mathcal{T})$ is isomorphic to the image of the natural morphism $F^j_\dagger L_{j-k}\tilde{\psi}_{g_X,\mathfrak{a},u}\mathcal{T} \longrightarrow F^j_\dagger\tilde{\psi}_{g_X,\mathfrak{a},u}\mathcal{T}$.

We have the epimorphism $F^j_\dagger W_{k-j}\mathcal{T} \longrightarrow \mathrm{Gr}^W_k F^j_\dagger\mathcal{T}$. It is factorized as follows:

$$F^j_\dagger W_{k-j}\mathcal{T} \longrightarrow \mathrm{Ker}\big(F^j_\dagger \mathrm{Gr}^W_{k-j}\mathcal{T} \longrightarrow F^{j+1}_\dagger \mathrm{Gr}^W_{k-j-1}\mathcal{T}\big) \longrightarrow \mathrm{Gr}^W_k F^j_\dagger\mathcal{T}$$

Note that $\tilde{\psi}_{g_Y,\mathfrak{a},u}F^j_\dagger \mathrm{Gr}^W_{k-j}\mathcal{T}$, $\tilde{\psi}_{g_Y,\mathfrak{a},u}F^{j+1}_\dagger \mathrm{Gr}^W_{k-j-1}\mathcal{T}$ and $\tilde{\psi}_{g_Y,\mathfrak{a},u}\mathrm{Gr}^W_k F^j_\dagger\mathcal{T}$ have the filtration W obtained as $M(\mathcal{N})[k]$, the shift of the monodromy filtration of $\mathcal{N}$, with which they are pre-mixed twistor $\mathscr{D}_Y$-modules. Note also that the filtrations W on $\tilde{\psi}_{g_Y,\mathfrak{a},u}F^j_\dagger \mathrm{Gr}^W_{k-j}\mathcal{T}$ and $\tilde{\psi}_{g_Y,\mathfrak{a},u}F^{j+1}_\dagger \mathrm{Gr}^W_{k-j-1}\mathcal{T}$ are equal to the induced weight filtrations of $F^j_\dagger(\tilde{\psi}_{g_X,\mathfrak{a},u}\mathrm{Gr}^W_{k-j}\mathcal{T}, W)$ and $F^{j+1}_\dagger(\tilde{\psi}_{g_X,\mathfrak{a},u}\mathrm{Gr}^W_{k-j-1}\mathcal{T}, W)$. Hence, the morphism $\tilde{\psi}_{g_Y,\mathfrak{a},u}(F^j_\dagger W_{k-j}\mathcal{T}) \longrightarrow \tilde{\psi}_{g_Y,\mathfrak{a},u}(\mathrm{Gr}^W_k F^j_\dagger\mathcal{T})$ is compatible with the weight filtrations, and it is a morphism in $\mathrm{MTW}(Y)$. We obtain that the morphism $L_k\big(\tilde{\psi}_{g_Y,\mathfrak{a},u}F^j_\dagger(\mathcal{T}), W\big) \longrightarrow \big(\tilde{\psi}_{g_Y,\mathfrak{a},u}\mathrm{Gr}^L_k F^j_\dagger(\mathcal{T}), W\big)$ is a morphism in $\mathrm{MTW}(Y)$. In particular, it is strict with respect to W. It implies that the filtration W on $\tilde{\psi}_{g,\mathfrak{a},u}F^j_\dagger(\mathcal{T})$ is equal to the relative monodromy filtration of $\mathcal{N}$ with respect to the filtration L.

Similarly, we obtain that the weight filtration W of $F^j_\dagger\phi_{g_X}(\mathcal{T}) \simeq \phi_{g_Y}F^j_\dagger\mathcal{T}$ is the relative monodromy filtration of $\mathcal{N}$ with respect to L. Therefore, $F^j_\dagger(\mathcal{T},W)$ is admissibly specializable. □

In the proof, we also obtain the following which relates the relative monodromy filtrations for $(\tilde{\psi}_{g_X,\mathfrak{a},u}(\mathcal{T}),\mathcal{N},L)$ (resp. $(\phi_{g_X}(\mathcal{T}),\mathcal{N},L)$) and $(\psi_{g_Y,\mathfrak{a},u}F^j_\dagger(\mathcal{T}),\mathcal{N},L)$ (resp. $(\phi_{g_Y}F^j_\dagger(\mathcal{T}),\mathcal{N},L)$).

Corollary 7.1.32 *We have natural isomorphisms*

$$F^j_\dagger\tilde{\psi}_{g_X,\mathfrak{a},u}(\mathcal{T},W) \simeq \tilde{\psi}_{g_Y,\mathfrak{a},u}F^j_\dagger(\mathcal{T},W), \quad F^j_\dagger\phi_{g_X}(\mathcal{T},W) \simeq \phi_{g_Y}F^j_\dagger(\mathcal{T},W)$$

in $\mathrm{MTW}(Y)$. □

Corollary 7.1.33 *Let H_Y be a hypersurface of Y. We put $H_X := F^{-1}(H_Y)$. For any $(\mathcal{T},W) \in \mathrm{MTW}^{\mathrm{sp}}(X,H_X)$, we have $F^j_\dagger(\mathcal{T},W) \in \mathrm{MTW}^{\mathrm{sp}}(Y,H_Y)$.* □

7.1.5 Gluing Along a Coordinate Function

Let t be a coordinate function, i.e., it is a holomorphic function such that dt is nowhere vanishing. Let $(\mathcal{T},W) \in \mathrm{MTW}^{\mathrm{sp}}(X,t)$. We have $\big(\tilde{\psi}_{t,-\boldsymbol{\delta}}(\mathcal{T}),W,L,\mathcal{N}\big)$ in $\mathrm{MTW}(X,\bullet)^{\mathrm{RMF}}$. Let $(\mathcal{Q},W) \in \mathrm{MTW}(X)$ with morphisms

$$\Sigma^{1,0}(\tilde{\psi}_{t,-\boldsymbol{\delta}}(\mathcal{T}),W) \xrightarrow{\ u\ } (\mathcal{Q},W) \xrightarrow{\ v\ } \Sigma^{0,-1}(\tilde{\psi}_{t,-\boldsymbol{\delta}}(\mathcal{T}),W)$$

such that (1) $\mathcal{N} = v\circ u$, (2) $\mathrm{Supp}\,\mathcal{Q} \subset \{t=0\}$. We obtain an $\mathcal{R}_X$-triple $\mathcal{T}' := \mathrm{Glue}(\mathcal{T},\mathcal{Q},u,v)$, as explained in Sect. 4.2.4. Let us observe that it is naturally equipped with a filtration W such that $(\mathcal{T}',W) \in \mathrm{MTW}(X)$. We have the filtration L on $\mathcal{Q}$ obtained as the transfer of $L\tilde{\psi}_{t,-\boldsymbol{\delta}}(\mathcal{T})$ with respect to (u,v). Note that $\Xi^{(0)}(\mathcal{T})$ has the filtration L naively induced by W, i.e., $L_j\Xi^{(0)}(\mathcal{T}) = \Xi^{(0)}(W_j\mathcal{T})$. The natural morphisms $\big(\psi_t^{(1)}(\mathcal{T}),L\big) \longrightarrow \big(\Xi_t^{(0)}(\mathcal{T}),L\big) \longrightarrow \big(\psi_t^{(0)}(\mathcal{T}),L\big)$ are strict with respect to L. Then, we obtain a filtered $\mathcal{R}$-triple $(\mathcal{T}',W)$ as the cohomology of the following complex:

$$\big(\psi_t^{(1)}(\mathcal{T}),L\big) \longrightarrow \big(\Xi_t^{(0)}(\mathcal{T}),L\big)\oplus(\mathcal{Q},L) \longrightarrow \big(\psi_t^{(0)}(\mathcal{T}),L\big)$$

By construction, $\mathrm{Gr}^W_w(\mathcal{T}')$ are strictly S-decomposable along t. The components whose supports are contained in $\{t=0\}$, are isomorphic to direct summands of $\mathrm{Gr}^W_w(\mathcal{Q})$. The components whose supports are not contained in $\{t=0\}$, are isomorphic to direct summands of $\mathrm{Gr}^W_w(\mathcal{T})$. Hence, they are objects in $\mathrm{MT}(X,w)$, i.e., $(\mathcal{T}',W) \in \mathrm{MTW}(X)$. Moreover, by construction, it is easy to observe that $(\mathcal{T}',W) \in \mathrm{MTW}^{\mathrm{sp}}(X,t)$. We have $\mathcal{Q} \simeq \phi_t^{(0)}(\mathcal{T}')$ and $L_j\mathcal{Q} \simeq \phi_t^{(0)}W_j\mathcal{T}'$.

In particular, $(\mathcal{T}, W)$ is reconstructed as the cohomology of the following complex in filtered $\mathcal{R}$-triples:

$$(\psi_t^{(1)}(\mathcal{T}), L) \longrightarrow (\phi_t^{(0)}(\mathcal{T}), L) \oplus (\Xi_t^{(0)}(\mathcal{T}), L) \longrightarrow (\psi_t^{(0)}(\mathcal{T}), L) \tag{7.4}$$

7.1.6 Localization

7.1.6.1 The Case of Coordinate Functions

Let t be a coordinate function. Let $(\mathcal{T}, W) \in \mathrm{MTW}(X, t)^{\mathrm{sp}}$. We apply the gluing construction in Sect. 7.1.5 to some special cases.

We have $(\tilde{\psi}_{t,-\boldsymbol{\delta}}(\mathcal{T}), W, L, \mathcal{N})$ in $\mathrm{MTW}(X, \bullet)^{\mathrm{RMF}}$. We obtain the filtrations $\hat{N}_! L$ on $\Sigma^{1,0}\tilde{\psi}_{t,-\boldsymbol{\delta}}(\mathcal{T}) = \psi_t^{(1)}(\mathcal{T})$ and $\hat{N}_* L$ on $\Sigma^{0,-1}\tilde{\psi}_{t,-\boldsymbol{\delta}}(\mathcal{T}) = \psi_t^{(0)}(\mathcal{T})$ by applying the procedure in Sect. 6.2.4. We obtain the filtered $\mathcal{R}$-triple $(\mathcal{T}, W)[!t] := (\mathcal{T}[!t], W)$ as the cohomology of the following:

$$(\psi_t^{(1)}(\mathcal{T}), L) \longrightarrow (\Xi_t^{(0)}(\mathcal{T}), L) \oplus (\psi_t^{(1)}(\mathcal{T}), \widehat{N}_! L) \longrightarrow (\psi_t^{(0)}(\mathcal{T}), L)$$

We obtain the filtered $\mathcal{R}$-triple $(\mathcal{T}, W)[*t] := (\mathcal{T}[*t], W)$ as the cohomology of the following complex:

$$(\psi_t^{(1)}(\mathcal{T}), L) \longrightarrow (\Xi_t^{(0)}(\mathcal{T}), L) \oplus (\psi_t^{(0)}(\mathcal{T}), \widehat{N}_* L) \longrightarrow (\psi_t^{(0)}(\mathcal{T}), L)$$

By Lemma 7.1.22, we have $(\mathcal{T}, W)[\star t] \in \mathrm{MTW}(X, t)^{\mathrm{sp}}$ for $\star = *, !$.

Lemma 7.1.34 *We have natural morphisms* $(\mathcal{T}, W)[!t] \longrightarrow (\mathcal{T}, W) \longrightarrow (\mathcal{T}, W)[\star t]$ *in* $\mathrm{MTW}(X, t)^{\mathrm{sp}}$.

Proof We consider the pre-mixed twistor $\mathscr{D}_X$-modules

$$(\psi_t^{(1)}(\mathcal{T}), W) = \Sigma^{1,0}(\tilde{\psi}_{t,-\boldsymbol{\delta}}(\mathcal{T}), W), \quad (\psi_t^{(0)}(\mathcal{T}), W) = \Sigma^{0,-1}(\tilde{\psi}_{t,-\boldsymbol{\delta}}(\mathcal{T}), W).$$

We have the following commutative diagram:

$$\begin{array}{ccccc}
(\psi_t^{(1)}(\mathcal{T}), W) & \xrightarrow{=} & (\psi_t^{(1)}(\mathcal{T}), W) & \xrightarrow{v\circ u} & (\psi_t^{(0)}(\mathcal{T}), W) \\
{=}\downarrow & & u\downarrow & & {=}\downarrow \\
(\psi_t^{(1)}(\mathcal{T}), W) & \xrightarrow{u} & (\phi_t^{(0)}(\mathcal{T}), W) & \xrightarrow{v} & (\psi_t^{(0)}(\mathcal{T}), W) \\
{=}\downarrow & & v\downarrow & & {=}\downarrow \\
(\psi_t^{(1)}(\mathcal{T}), W) & \xrightarrow{v\circ u} & (\psi_t^{(0)}(\mathcal{T}), W) & \xrightarrow{=} & (\psi_t^{(0)}(\mathcal{T}), W)
\end{array}$$

Here, v and u are the natural morphisms. By the functoriality of the transfer in Sect. 6.2.3, we have the following commutative diagram:

$$\begin{array}{ccccc}
(\psi_t^{(1)}(\mathcal{T}),L) & \xrightarrow{=} & (\psi_t^{(1)}(\mathcal{T}),\widehat{N}_!L) & \longrightarrow & (\psi_t^{(0)}(\mathcal{T}),L) \\
=\downarrow & & \downarrow & & =\downarrow \\
(\psi_t^{(1)}(\mathcal{T}),L) & \longrightarrow & (\phi_t^{(0)}(\mathcal{T}),L) & \longrightarrow & (\psi_t^{(0)}(\mathcal{T}),L) \\
=\downarrow & & \downarrow & & =\downarrow \\
(\psi_t^{(1)}(\mathcal{T}),L) & \longrightarrow & (\psi_t^{(0)}(\mathcal{T}),\widehat{N}_*L) & \xrightarrow{=} & (\psi_t^{(0)}(\mathcal{T}),L)
\end{array}$$

Then, we obtain the claim of Lemma 7.1.34. □

We have the following characterization.

Lemma 7.1.35 *Let* $\star = *$ *or* $!$. *Let* $(\mathcal{T},W) \in \mathrm{MTW}^{\mathrm{sp}}(X,t)$. *Assume that* $\mathcal{T}[\star t] = \mathcal{T}$ *as an* $\mathcal{R}_X$*-triple. Then, we have* $(\mathcal{T},W) \simeq (\mathcal{T},W)[\star t]$ *in* $\mathrm{MTW}(X)$.

Proof Let us consider the case $\star = !$. The other case can be argued similarly. By the assumption, we have $\psi_t^{(1)}(\mathcal{T}) \simeq \phi_t^{(0)}(\mathcal{T})$ as $\mathcal{R}$-triples. Hence, $\big(\psi_t^{(1)}(\mathcal{T}),W\big) \longrightarrow \big(\phi_t^{(0)}(\mathcal{T}),W\big)$ is an isomorphism in $\mathrm{MTW}(X)$. We obtain that the naively induced filtration of $\phi_t^{(0)}(\mathcal{T})$ is the same as $\hat{N}_!L$ under the isomorphism. Recall that $(\mathcal{T},W)$ is reconstructed as the cohomology of (7.4). Hence, we obtain that $(\mathcal{T},W) \simeq (\mathcal{T},W)[!t]$. □

Similarly, we obtain the following lemma.

Lemma 7.1.36 *Let* $\star = *$ *or* $!$. *Let* $(\mathcal{T}_i,W) \in \mathrm{MTW}^{\mathrm{sp}}(X,t)$ $(i = 1,2)$. *We have a natural bijective correspondence between morphisms of filtered* $\mathcal{R}_X(*t)$*-triples* $(\mathcal{T}_1,W)(*t) \longrightarrow (\mathcal{T}_2,W)(*t)$, *and morphisms* $(\mathcal{T}_1,W)[\star t] \longrightarrow (\mathcal{T}_2,W)[\star t]$ *in* $\mathrm{MTW}(X)$.

Proof We may assume $(\mathcal{T}_i,W) \simeq (\mathcal{T}_i,W)[\star t]$ from the beginning. Let $F : (\mathcal{T}_1,W)(*t) \longrightarrow (\mathcal{T}_2,W)(*t)$ be a morphism. It naturally induces a morphism $\big(\tilde{\psi}_{t,-\delta}(\mathcal{T}_1),L\big) \longrightarrow \big(\tilde{\psi}_{t,-\delta}(\mathcal{T}_2),L\big)$. It gives a morphism $(\tilde{\psi}_{t,-\delta}(\mathcal{T}_1),W) \longrightarrow (\tilde{\psi}_{t,-\delta}(\mathcal{T}_2),W)$ in $\mathrm{MTW}(X)$. By the assumption, we obtain $(\phi_t(\mathcal{T}_1),W) \longrightarrow (\phi_t(\mathcal{T}_2),W)$ in $\mathrm{MTW}(X)$. Because the naively induced filtrations L on $\phi_t(\mathcal{T}_i)$ are obtained as the transfer of L of $\tilde{\psi}_{t,-\delta}(\mathcal{T}_i)$, we obtain $\big(\phi_t(\mathcal{T}_1),L\big) \longrightarrow \big(\phi_t(\mathcal{T}_2),L\big)$. Hence, we have $F : (\mathcal{T}_1,W) \longrightarrow (\mathcal{T}_2,W)$. □

Corollary 7.1.37 *Let* $(\mathcal{T}_i,W) \in \mathrm{MTW}^{\mathrm{sp}}(X,t)$ $(i = 1,2)$. *We have natural bijections:*

$$\mathrm{Hom}_{\mathrm{MTW}(X)}\big((\mathcal{T}_1,W)[*t],\ (\mathcal{T}_2,W)[*t]\big) \overset{a_i}{\simeq} \mathrm{Hom}_{\mathrm{MTW}(X)}\big((\mathcal{T}_1,W),\ (\mathcal{T}_2,W)[*t]\big)$$

$$\mathrm{Hom}_{\mathrm{MTW}(X)}\big((\mathcal{T}_1,W)[!t],\ (\mathcal{T}_2,W)[!t]\big) \overset{a_2}{\simeq} \mathrm{Hom}_{\mathrm{MTW}(X)}\big((\mathcal{T}_1,W)[!t],\ (\mathcal{T}_2,W)\big)$$

The morphisms are induced by those in Lemma 7.1.34.

Proof We can construct the inverse of the morphisms a_i by using Lemma 7.1.36. We obtain that they are bijective by Corollary 3.2.10. □

Lemma 7.1.38 *Let $f : (\mathcal{T}_1, W) \longrightarrow (\mathcal{T}_2, W)$ be a morphism in $\mathrm{MTW}^{\mathrm{sp}}(X,t)$. For the induced morphism $f[\star t] : (\mathcal{T}_1, W)[\star t] \longrightarrow (\mathcal{T}_2, W)[\star t]$, we have the following natural isomorphisms in $\mathrm{MTW}^{\mathrm{sp}}(X,t)$:*

$$\mathrm{Ker}(f)[\star t] \simeq \mathrm{Ker}(f[\star t]), \quad \mathrm{Im}(f)[\star t] \simeq \mathrm{Im}(f[\star t]), \quad \mathrm{Cok}(f)[\star t] \simeq \mathrm{Cok}(f[\star t]).$$

Proof By the construction, we have the following morphisms in $\mathrm{MTW}(X)$:

$$\mathrm{Ker}(f)[\star t] \longrightarrow \mathcal{T}_1[\star t] \longrightarrow \mathrm{Im}(f)[\star t] \longrightarrow \mathcal{T}_2[\star t] \longrightarrow \mathrm{Cok}(f)[\star t]$$

We obtain the naturally defined morphisms $\mathrm{Ker}(f)[\star t] \longrightarrow \mathrm{Ker}(f[\star t])$ and $\mathrm{Cok}(f[\star t]) \longrightarrow \mathrm{Cok}(f)[\star t]$. Because they give isomorphisms of $\mathcal{R}$-triples, and because any morphism in $\mathrm{MTW}(X)$ is strict with respect to the weight filtration, they are isomorphisms. Then, we also obtain that $\mathrm{Im}(f[\star t]) \simeq \mathrm{Im}(f)[\star t]$. □

Let $(\mathcal{T}, W) \in \mathrm{MTW}^{\mathrm{sp}}(X,t)$. We obtain the graded polarizable pure twistor $\mathscr{D}$-module $\mathrm{Gr}^W(\mathcal{T}) = \bigoplus \mathrm{Gr}^W_k(\mathcal{T})$. We have the decomposition

$$\mathrm{Gr}^W_k(\mathcal{T}) = \mathcal{P}_{k0} \oplus \mathcal{P}_{k1},$$

such that the support of $\mathcal{P}_{k1}$ is contained in $\{t = 0\}$, and that the support of any direct summand of $\mathcal{P}_{k0}$ is not contained in $\{t = 0\}$.

Lemma 7.1.39 *We have the following natural isomorphisms:*

$$\mathrm{Gr}^W_k(\mathcal{T}[*t]) = \mathcal{P}_{k0} \oplus \mathrm{Ker}\big(\mathrm{Gr}^W_k \mathrm{Gr}^{\hat{N}_* L}_k \psi_t^{(0)}(\mathcal{T}) \longrightarrow \mathrm{Gr}^W_k \mathrm{Gr}^L_k \psi_t^{(0)}(\mathcal{T})\big) \quad (7.5)$$

$$\mathrm{Gr}^W_k(\mathcal{T}[!t]) = \mathcal{P}_{k0} \oplus \mathrm{Ker}\big(\mathrm{Gr}^W_k \mathrm{Gr}^{\hat{N}_! L}_k \psi_t^{(1)}(\mathcal{T}) \longrightarrow \mathrm{Gr}^W_k \mathrm{Gr}^L_k \psi_t^{(0)}(\mathcal{T})\big) \quad (7.6)$$

Proof By the construction, we have the natural isomorphism:

$$\mathrm{Gr}^W_k(\mathcal{T}[*t]) = \mathcal{P}_{k0} \oplus \mathrm{Ker}\big(\mathrm{Gr}^{\hat{N}_* L}_k \psi_t^{(0)}(\mathcal{T}) \longrightarrow \mathrm{Gr}^L_k \psi_t^{(0)}(\mathcal{T})\big)$$

By the construction, the morphism $\mathrm{Gr}^{\hat{N}_* L}_k \psi_t^{(0)}(\mathcal{T}) \longrightarrow \mathrm{Gr}^L_k \psi_t^{(0)}(\mathcal{T})$ is a morphism in $\mathrm{MTW}(X)$, and $\mathrm{Ker}\big(\mathrm{Gr}^{\hat{N}_* L}_k \psi_t^{(0)}(\mathcal{T}) \longrightarrow \mathrm{Gr}^L_k \psi_t^{(0)}(\mathcal{T})\big)$ is pure of weight k. Hence, we obtain (7.5). We obtain (7.6) in a similar way. □

7.1.6.2 The Case of General Holomorphic Functions

Let g be a holomorphic function. Let $\iota_g : X \longrightarrow X \times \mathbb{C}_t$ be the graph of g. Let $(\mathcal{T}, W)$ in $\mathrm{MTW}^{\mathrm{sp}}(X,g)$. We obtain $\iota_{g\dagger}(\mathcal{T}, W)[\star t]$ in $\mathrm{MTW}^{\mathrm{sp}}(X \times \mathbb{C}_t, t)$.

Definition 7.1.40 We say that $(\mathcal{T}, W) \in \mathrm{MTW}^{\mathrm{sp}}(X, g)$ is localizable along g if there exist objects $(\mathcal{T}[\star g], W) \in \mathrm{MTW}^{\mathrm{sp}}(X, g)$ $(\star = *, !)$ satisfying

$$\iota_{g\dagger}(\mathcal{T}[\star g], W) \simeq \iota_{g\dagger}(\mathcal{T}, W)[\star t]$$

in $\mathrm{MTW}^{\mathrm{sp}}(X \times \mathbb{C}_t, t)$. Such $(\mathcal{T}[\star g], W)$ is uniquely determined up to canonical isomorphisms if it exists. □

Let $\mathrm{MTW}^{\mathrm{loc}}(X, g) \subset \mathrm{MTW}^{\mathrm{sp}}(X, g)$ denote the full subcategory of $(\mathcal{T}, W)$ which are localizable along g. We obtain the following from Lemma 7.1.38.

Lemma 7.1.41 *Let* $f : (\mathcal{T}_1, W) \to (\mathcal{T}_2, W)$ *be a morphism in* $\mathrm{MTW}^{\mathrm{loc}}(X, g)$*. Then,* $\mathrm{Ker}(f)$*,* $\mathrm{Im}(f)$ *and* $\mathrm{Cok}(f)$ *are also objects in* $\mathrm{MTW}^{\mathrm{loc}}(X, g)$*. For the induced morphism* $f[\star g] : (\mathcal{T}_1, W)[\star g] \longrightarrow (\mathcal{T}_2, W)[\star g]$*, we have the following natural isomorphisms in* $\mathrm{MTW}^{\mathrm{loc}}(X, g)$*:*

$$\mathrm{Ker}(f)[\star g] \simeq \mathrm{Ker}(f[\star g]), \quad \mathrm{Im}(f)[\star g] \simeq \mathrm{Im}(f[\star g]), \quad \mathrm{Cok}(f)[\star g] \simeq \mathrm{Cok}(f[\star g]).$$

□

Clearly the following holds.

Lemma 7.1.42 $\mathrm{MTW}^{\mathrm{loc}}(X, g)$ *is an abelian subcategory of* $\mathrm{MTW}(X)$. □

We obtain the following from Lemma 7.1.36.

Lemma 7.1.43 *Let* $\star = *$ *or* $!$*. Let* $(\mathcal{T}_i, W) \in \mathrm{MTW}^{\mathrm{loc}}(X, g)$ $(i = 1, 2)$ *such that* $\mathcal{T}_i[\star g] = \mathcal{T}_i$*. Then, we have a natural bijective correspondence between morphisms* $(\mathcal{T}_1, W)(*g) \longrightarrow (\mathcal{T}_2, W)(*g)$ *as filtered* $\mathcal{R}_X(*g)$*-triples, and morphisms* $(\mathcal{T}_1, W) \longrightarrow (\mathcal{T}_2, W)$ *in* $\mathrm{MTW}(X)$*. In particular, if moreover* $\mathcal{T}_1(*g) \simeq \mathcal{T}_2(*g)$*, the isomorphism is extended to* $\mathcal{T}_1 \simeq \mathcal{T}_2$*.* □

Lemma 7.1.44 *For any* $(\mathcal{T}, W) \in \mathrm{MTW}^{\mathrm{loc}}(X, g)$*, we have naturally defined morphisms* $(\mathcal{T}, W)[!g] \longrightarrow (\mathcal{T}, W) \longrightarrow (\mathcal{T}, W)[*g]$ *in* $\mathrm{MTW}^{\mathrm{loc}}(X, g)$*.* □

We also have the following.

Corollary 7.1.45 *Let* $(\mathcal{T}_i, W) \in \mathrm{MTW}^{\mathrm{loc}}(X, g)$ $(i = 1, 2)$*. We have natural bijections:*

$$\mathrm{Hom}_{\mathrm{MTW}(X)}\big((\mathcal{T}_1, W)[*g], (\mathcal{T}_2, W)[*g]\big) \simeq \mathrm{Hom}_{\mathrm{MTW}(X)}\big((\mathcal{T}_1, W), (\mathcal{T}_2, W)[*g]\big)$$
$$\mathrm{Hom}_{\mathrm{MTW}(X)}\big((\mathcal{T}_1, W)[!g], (\mathcal{T}_2, W)[!g]\big) \simeq \mathrm{Hom}_{\mathrm{MTW}(X)}\big((\mathcal{T}_1, W)[!g], (\mathcal{T}_2, W)\big)$$

□

Let $\mathcal{T} \in \mathrm{MTW}^{\mathrm{loc}}(X, g)$. We have the decomposition $\mathrm{Gr}_k^W(\mathcal{T}) = \mathcal{P}_{k0} \oplus \mathcal{P}_{k1}$, where $\mathcal{P}_{k0}$ (resp. $\mathcal{P}_{k1}$) is the direct sum of pure twistor $\mathscr{D}$-modules whose strict supports are not contained (resp. contained) in $\{g = 0\}$. We obtain the following isomorphisms from Lemma 7.1.39.

$$\mathrm{Gr}_k^W(\mathcal{T}[*g]) = \mathcal{P}_{k0} \oplus \mathrm{Ker}\bigl(\mathrm{Gr}_k^W \mathrm{Gr}_k^{\hat{N}_*L} \psi_g^{(0)}(\mathcal{T}) \longrightarrow \mathrm{Gr}_k^W \mathrm{Gr}_k^L \psi_g^{(0)}(\mathcal{T})\bigr) \quad (7.7)$$

$$\mathrm{Gr}_k^W(\mathcal{T}[!g]) = \mathcal{P}_{k0} \oplus \mathrm{Ker}\bigl(\mathrm{Gr}_k^W \mathrm{Gr}_k^{\hat{N}_!L} \psi_g^{(1)}(\mathcal{T}) \longrightarrow \mathrm{Gr}_k^W \mathrm{Gr}_k^L \psi_g^{(0)}(\mathcal{T})\bigr) \quad (7.8)$$

Lemma 7.1.46 *Let $\mathcal{T} \in \mathrm{MTW}(X)$. Let $\{U_i \mid i \in \Lambda\}$ be an open covering of X. Then, $\mathcal{T}$ is localizable along g if and only if $\mathcal{T}_{|U_i}$ are localizable along $g_{|U_i}$ for any $i \in \Lambda$. Similar claims hold for the filtered specializability and admissible specializability.*

Proof The "only if" part is clear. Let us look at the "if" part. We construct the filtered $\mathcal{R}$-triples $(\mathcal{T}[\star g], W)$ by gluing $(\mathcal{T}_{|U_i}[\star g_{|U_i}], W)$. By (7.7) and (7.8), we can also construct polarizations on $\mathrm{Gr}_k^W(\mathcal{T}[\star g])$ by gluing. □

The following lemmas are easy to see.

Lemma 7.1.47 *Let A be a nowhere vanishing function on X. We set $g_1 := Ag$.*

- $\mathrm{MTW}^{\mathrm{loc}}(X, g) = \mathrm{MTW}^{\mathrm{loc}}(X, g_1)$.
- *For any $(\mathcal{T}, W) \in \mathrm{MTW}^{\mathrm{loc}}(X, g)$, we have isomorphisms $(\mathcal{T}, W)[\star g] \simeq (\mathcal{T}, W)[\star g_1]$ naturally.*
- *The isomorphisms* (7.7) *and* (7.8) *for g and g_1 are equal under the isomorphisms in Lemma* 4.4.16. □

7.1.6.3 The Case of Hypersurfaces

Let H be an effective divisor of X.

Definition 7.1.48 $(\mathcal{T}, W) \in \mathrm{MTW}(X)$ is called admissibly specializable (resp. localizable) along H if the following holds:

- Let $U \subset X$ be any open subset with a generator g_U of $\mathcal{O}(-H)_{|U}$. Then, $(\mathcal{T}, W)_{|U}$ is admissibly specializable (resp. localizable) along g_U. □

Let $\mathrm{MTW}^{\mathrm{sp}}(X, H)$ (resp. $\mathrm{MTW}^{\mathrm{loc}}(X, H)$) denote the full subcategory of $(\mathcal{T}, W) \in \mathrm{MTW}(X)$ which are admissibly specializable (resp. localizable) along H. We obtain the following from Lemma 7.1.41.

Lemma 7.1.49 *Let $f : (\mathcal{T}_1, W) \to (\mathcal{T}_2, W)$ be a morphism in $\mathrm{MTW}^{\mathrm{loc}}(X, H)$. Then, $\mathrm{Ker}(f)$, $\mathrm{Im}(f)$ and $\mathrm{Cok}(f)$ are also objects in $\mathrm{MTW}^{\mathrm{loc}}(X, H)$. For the induced morphism $f[\star H] : (\mathcal{T}_1, W)[\star H] \longrightarrow (\mathcal{T}_2, W)[\star H]$, we have the natural isomorphisms $\mathrm{Ker}(f)[\star H] \simeq \mathrm{Ker}(f[\star H])$, $\mathrm{Im}(f)[\star H] \simeq \mathrm{Im}(f[\star H])$ and $\mathrm{Cok}(f)[\star H] \simeq \mathrm{Cok}(f[\star H])$ in $\mathrm{MTW}^{\mathrm{loc}}(X, H)$.* □

Clearly the following holds.

Lemma 7.1.50 $\mathrm{MTW}^{\mathrm{sp}}(X,H)$ *and* $\mathrm{MTW}^{\mathrm{loc}}(X,H)$ *are abelian subcategories of* $\mathrm{MTW}(X)$. □

For $(\mathcal{T}, W) \in \mathrm{MTW}^{\mathrm{loc}}(X, H)$, we obtain a filtered $\mathcal{R}_X$-triple $(\mathcal{T}, W)[\star H]$ with an isomorphism $(\mathcal{T}, W)[\star H](*H) \simeq (\mathcal{T}, W)(*H)$ determined by the following condition:

- Let (U, g_U) be as above. Then, we have $(\mathcal{T}, W)[\star H]_{|U} \simeq (\mathcal{T}, W)_{|U}[\star g_U]$ which induces $(\mathcal{T}, W)[\star H](*H)_{|U} \simeq (\mathcal{T}, W)(*H)_{|U}$.

Proposition 7.1.51 $(\mathcal{T}, W)[\star H] \in \mathrm{MTW}^{\mathrm{loc}}(X, H)$.

Proof We have the decomposition $\mathrm{Gr}_k^W(\mathcal{T}) = \mathcal{P}_{k0} \oplus \mathcal{P}_{k1}$, where the support of $\mathcal{P}_{k1}$ is contained in H, and the support of any sub-$\mathcal{R}_X$-triple of $\mathcal{P}_{k0}$ is not contained in H. By Lemma 7.1.47, we obtain the following isomorphisms:

$$\mathrm{Gr}_k^W(\mathcal{T}[*H]) = \mathcal{P}_{k0} \oplus \mathrm{Ker}\Big(\mathrm{Gr}_k^{\hat{N}_* L}(\mathrm{Gr}_k^W \psi_H^{(0)})(\mathcal{T}) \longrightarrow \mathrm{Gr}_k^L(\mathrm{Gr}_k^W \psi_H^{(0)})(\mathcal{T})\Big) \quad (7.9)$$

$$\mathrm{Gr}_k^W(\mathcal{T}[!H]) = \mathcal{P}_{k0} \oplus \mathrm{Ker}\Big(\mathrm{Gr}_k^{\hat{N}_! L}(\mathrm{Gr}_k^W \psi_H^{(1)})(\mathcal{T}) \longrightarrow \mathrm{Gr}_k^L(\mathrm{Gr}_k^W \psi_H^{(0)})(\mathcal{T})\Big) \quad (7.10)$$

The Ker-part in (7.9) and (7.10) are direct summands of polarizable pure twistor $\mathscr{D}$-modules $(\mathrm{Gr}_k^W \psi_H^{(a)})(\mathcal{T})$ of weight k. Note the semisimplicity of the category of polarizable pure twistor $\mathscr{D}$-modules. Then, the claim of the proposition follows from Proposition 7.1.29. □

Lemma 7.1.52 *Let* $(\mathcal{T}_i, W) \in \mathrm{MTW}^{\mathrm{loc}}(X, H)$ $(i = 1, 2)$ *such that* $\mathcal{T}_i[*H] \simeq \mathcal{T}_i$ *as* $\mathcal{R}_X$*-triples. Then, we have a natural bijective correspondence of morphisms* $(\mathcal{T}_1, W)(*H) \longrightarrow (\mathcal{T}_2, W)(*H)$ *as filtered* $\mathcal{R}_X(*H)$*-triples and morphisms* $(\mathcal{T}_1, W) \longrightarrow (\mathcal{T}_2, W)$ *in* $\mathrm{MTW}(X)$*. In particular, if moreover we have* $\mathcal{T}_1(*H) \simeq \mathcal{T}_2(*H)$*, the isomorphism is extended to* $(\mathcal{T}_1, W) \simeq (\mathcal{T}_2, W)$*.* □

Lemma 7.1.53 *For any* $(\mathcal{T}, W) \in \mathrm{MTW}^{\mathrm{loc}}(X, H)$*, we have naturally defined morphisms* $(\mathcal{T}, W)[!H] \longrightarrow (\mathcal{T}, W) \longrightarrow (\mathcal{T}, W)[*H]$ *in* $\mathrm{MTW}(X)$*.* □

Corollary 7.1.54 *For* $(\mathcal{T}_i, W) \in \mathrm{MTW}^{\mathrm{loc}}(X, H)$ $(i = 1, 2)$*, we have natural bijections:*

$$\mathrm{Hom}_{\mathrm{MTW}(X)}\big((\mathcal{T}_1, W)[*H], (\mathcal{T}_2, W)[*H]\big) \simeq \mathrm{Hom}_{\mathrm{MTW}(X)}\big((\mathcal{T}_1, W), (\mathcal{T}_2, W)[*H]\big)$$

$$\mathrm{Hom}_{\mathrm{MTW}(X)}\big((\mathcal{T}_1, W)[!H], (\mathcal{T}_2, W)[!H]\big) \simeq \mathrm{Hom}_{\mathrm{MTW}(X)}\big((\mathcal{T}_1, W)[!H], (\mathcal{T}_2, W)\big)$$

□

7.1.7 Integrable Case

An integrable pure twistor $\mathscr{D}$-module of weight w [66] is an integrable $\mathcal{R}$-triple $\mathcal{T}$ such that (1) it is a pure twistor $\mathscr{D}$-module of weight w, (2) it has a polarization which is integrable as a morphism $\mathcal{T} \longrightarrow \mathcal{T}^* \otimes \boldsymbol{T}(-w)$. A morphism of integrable pure twistor $\mathscr{D}$-modules of weight w is defined to be an integrable morphism for integrable $\mathcal{R}$-triples. Let $\mathrm{MT}^{\mathrm{int}}(X,w)$ denote the category of integrable pure twistor $\mathscr{D}$-modules of weight w.

An integrable filtered $\mathcal{R}_X$-triple $(\mathcal{T},W)$ is called an integrable pre-mixed twistor $\mathscr{D}$-module if $\mathrm{Gr}^W_w(\mathcal{T}) \in \mathrm{MT}^{\mathrm{int}}(X,w)$ for each w. A morphism is defined to be an integrable morphism of the underlying filtered $\mathcal{R}$-triples. Let $\mathrm{MTW}^{\mathrm{int}}(X)$ denote the category of integrable pre-mixed twistor $\mathscr{D}$-modules. The following proposition is easy to see.

Proposition 7.1.55 $\mathrm{MT}^{\mathrm{int}}(X)$ *is abelian and semisimple, and* $\mathrm{MTW}^{\mathrm{int}}(X)$ *is an abelian category.*

Proof The first claim can be proved as in the ordinary case. The second claim immediately follows from the first one. □

The following proposition is an analogue of Proposition 7.1.6.

Proposition 7.1.56 *Let* $F : X \longrightarrow Y$ *be a projective morphism. For any object* $\mathcal{T} \in \mathrm{MT}^{\mathrm{int}}(X,w)$, *we have* $F^i_\dagger\mathcal{T} \in \mathrm{MT}^{\mathrm{int}}(Y,w+i)$. *For* $(\mathcal{T},W) \in \mathrm{MTW}^{\mathrm{int}}(X)$, *we have* $(F^i_\dagger\mathcal{T},W) \in \mathrm{MTW}^{\mathrm{int}}(Y)$, *where the weight filtration* W *of* $F^i_\dagger\mathcal{T}$ *is defined as in* Sect. 7.1.2. □

7.1.7.1 Admissible Specializability, Gluing and Localizability

For a given holomorphic function g on X, let $\mathrm{MTW}^{\mathrm{int\,sp}}(X,g) \subset \mathrm{MTW}^{\mathrm{int}}(X)$ denote the full subcategory of $(\mathcal{T},W) \in \mathrm{MTW}^{\mathrm{int}}(X)$ which are admissibly specializable along g. As in Proposition 7.1.25, $\mathrm{MTW}^{\mathrm{int\,sp}}(X,g)$ is an abelian category. We remark that the naively induced filtrations on $\tilde{\psi}_{g,\mathfrak{a},u}(\mathcal{T})$ and $\phi_g(\mathcal{T})$ are integrable. The induced nilpotent morphisms are integrable. Hence, the relative monodromy filtrations are also integrable.

Let t be a coordinate function. Let $(\mathcal{T},W) \in \mathrm{MTW}^{\mathrm{int\,sp}}(X,t)$. Then, as in the ordinary case, we have $\big(\tilde{\psi}_{t,-\boldsymbol{\delta}}(\mathcal{T}),W\big)$ in $\mathrm{MTW}^{\mathrm{int}}(X)$, which is equipped with the filtration L naively induced by W, and the integrable morphism $\mathcal{N}$: $\big(\tilde{\psi}_{t,-\boldsymbol{\delta}}(\mathcal{T}),W\big) \longrightarrow \big(\tilde{\psi}_{t,-\boldsymbol{\delta}}(\mathcal{T}),W\big)\otimes\boldsymbol{T}(-1)$. We have $W = M(\mathcal{N};L)$. Let $(\mathcal{Q},W) \in \mathrm{MTW}^{\mathrm{int}}(X)$ with integrable morphisms

$$\Sigma^{1,0}(\widetilde{\psi}_{t,-\boldsymbol{\delta}}(\mathcal{T}),W) \xrightarrow{\ u\ } (\mathcal{Q},W) \xrightarrow{\ v\ } \Sigma^{0,-1}(\widetilde{\psi}_{t,-\boldsymbol{\delta}}(\mathcal{T}),W)$$

such that (1) $\mathcal{N} = v \circ u$, (2) $\mathrm{Supp}\,\mathcal{Q} \subset \{t=0\}$. As explained in Sect. 7.1.5, we have constructed $\mathrm{Glue}(\mathcal{T},\mathcal{Q},u,v)$ with the weight filtration W in $\mathrm{MTW}^{\mathrm{sp}}(X,g)$. Note that the naively induced filtrations L of $\psi^{(a)}(\mathcal{T})$ and its transfer L to $\mathcal{Q}$ are integrable. The naively induced filtration L of $\Xi^{(a)}(\mathcal{T})$ is also integrable. Hence, we have $\big(\mathrm{Glue}(\mathcal{T},\mathcal{Q},u,v),W\big) \in \mathrm{MTW}^{\mathrm{int}}(X)$. In particular, we have the following.

Lemma 7.1.57 *For $(\mathcal{T}, W) \in \mathrm{MTW}^{\mathrm{int\,sp}}(X,t)$ we naturally have the objects $(\mathcal{T}, W)[\star t]$ ($\star = *, !$) in $\mathrm{MTW}^{\mathrm{int}}(X)$ with the morphisms $(\mathcal{T}, W)[!t] \longrightarrow (\mathcal{T}, W) \longrightarrow (\mathcal{T}, W)[*t]$ in $\mathrm{MTW}^{\mathrm{int}}(X)$.* □

Let H be an effective divisor of X.

Lemma 7.1.58 *Suppose that $(\mathcal{T}, W) \in \mathrm{MTW}^{\mathrm{int}}(X)$ is localizable along H. Then, $(\mathcal{T}, W)[\star H] \in \mathrm{MTW}(X)$ are also integrable.*

Proof By the construction, the weight filtration W on $\mathcal{T}[\star H]$ is integrable. In the proof of Proposition 7.1.51, an integrable graded polarization of $\mathrm{Gr}^W \mathcal{T}$ induces integrable polarizations of $(\mathrm{Gr}_k^W \psi_H^{(a)})(\mathcal{T})$. In (7.9) and (7.10), the Ker-parts are integrable. Hence, we obtain that $\mathrm{Gr}_k^W(\mathcal{T}[\star H])$ have integrable polarizations. □

Lemma 7.1.59 *Suppose that $(\mathcal{T}_i, W) \in \mathrm{MTW}^{\mathrm{int}}(X)$ $(i = 1, 2)$ are localizable along H such that $\mathcal{T}_i \simeq \mathcal{T}_i[\star H]$. We have a natural bijective correspondence of morphisms $(\mathcal{T}_1, W)(*H) \longrightarrow (\mathcal{T}_2, W)(*H)$ as filtered integrable $\mathcal{R}_X(*H)$-triples, with morphisms $(\mathcal{T}_1, W) \longrightarrow (\mathcal{T}_2, W)$ in $\mathrm{MTW}^{\mathrm{int}}(X)$. In particular, if moreover $\mathcal{T}_1(*H) \simeq \mathcal{T}_2(*H)$, the isomorphism is extended to $(\mathcal{T}_1, W) \simeq (\mathcal{T}_2, W)$.* □

Lemma 7.1.60 *Suppose that $(\mathcal{T}, W) \in \mathrm{MTW}^{\mathrm{int}}(X)$ is localizable along H. Then, we have natural morphisms $(\mathcal{T}, W)[!H] \longrightarrow (\mathcal{T}, W) \longrightarrow (\mathcal{T}, W)[*H]$ in $\mathrm{MTW}^{\mathrm{int}}(X)$.* □

The following is an analogue of Corollary 7.1.45.

Lemma 7.1.61 *Let $(\mathcal{T}_i, W) \in \mathrm{MTW}^{\mathrm{int}}(X)$ $(i = 1, 2)$ be localizable along H. We have natural bijections:*

$$\mathrm{Hom}_{\mathrm{MTW}^{\mathrm{int}}(X)}\big((\mathcal{T}_1, W)[*H], (\mathcal{T}_2, W)[*H]\big) \simeq \mathrm{Hom}_{\mathrm{MTW}^{\mathrm{int}}(X)}\big((\mathcal{T}_1, W), (\mathcal{T}_2, W)[*H]\big)$$

$$\mathrm{Hom}_{\mathrm{MTW}^{\mathrm{int}}(X)}\big((\mathcal{T}_1, W)[!H], (\mathcal{T}_2, W)[!H]\big) \simeq \mathrm{Hom}_{\mathrm{MTW}^{\mathrm{int}}(X)}\big((\mathcal{T}_1, W)[!H], (\mathcal{T}_2, W)\big)$$

□

7.1.8 Restriction of KMS-Spectrum

Let $\mathcal{A}$ be a $\mathbb{Q}$-vector space in $\mathbb{R} \times \mathbb{C}$ such that $\mathbb{Q} \times \{0\} \subset \mathcal{A}$. We recall the concept of $\mathcal{A}$-pure twistor $\mathscr{D}$-module in [55].

Let $\mathrm{MT}_{\leq n}(X, w)$ be the full subcategory of $\mathrm{MT}(X, w)$ of the objects $\mathcal{T}$ such that $\dim \mathrm{Supp}(\mathcal{T}) \leq n$. For any closed irreducible complex subvariety $Z \subset X$, let $\mathrm{MT}_Z(X, w)$ be the full subcategory of $\mathrm{MT}(X, w)$ of the objects whose strict support is Z. We include $0 \in \mathrm{MT}_Z(X, w)$. We define saturated full subcategories $\mathrm{MT}_{\leq n}(X, w)_{\mathcal{A}} \subset \mathrm{MT}_{\leq n}(X, w)$ and $\mathrm{MT}_Z(X, w)_{\mathcal{A}} \subset \mathrm{MT}_Z(X, w)$. If Z is a point in X, we set $\mathrm{MT}_Z(X, w)_{\mathcal{A}} := \mathrm{MT}_Z(X, w)$ for any $w \in \mathbb{Z}$. Suppose that we have already defined $\mathrm{MT}_Z(X, w)_{\mathcal{A}} \subset \mathrm{MT}_Z(X, w)$ for any complex manifold X, any $w \in \mathbb{Z}$ and any closed irreducible complex subvariety $Z \subset X$ with $\dim Z \leq n$. Then, we define

$\mathrm{MT}_{\leq n}(X,w)_{\mathcal{A}} \subset \mathrm{MT}_{\leq n}(X,w)$ as the full subcategory of the objects $\mathcal{T}$ satisfying the following condition.

- Let U be any relatively compact open subset in X. Note that we have the decomposition $\mathcal{T}_{|U} = \bigoplus \mathcal{T}_Z$ by the strict support, where Z runs through closed irreducible subvariety in U with $\dim Z \leq n$, and the strict supports of $\mathcal{T}_Z$ are Z. Then, we impose that $\mathcal{T}_Z \in \mathrm{MT}_Z(X,w)_{\mathcal{A}}$.

For any closed subvariety $Z \subset X$ with $\dim Z = n+1$, let $\mathrm{MT}_Z(X,w)_{\mathcal{A}} \subset \mathrm{MT}_Z(X,w)$ be the full subcategory of the objects $\mathcal{T} \in \mathrm{MT}_Z(X,w)$ satisfying the following conditions.

- Let $U \subset X$ be any open subset with a holomorphic function g on U such that $\dim g^{-1}(0) \cap Z < n+1$. Then, for any $\mathfrak{a} \in \mathbb{C}[t_m^{-1}]$, we impose that $\tilde{\psi}_{g,\mathfrak{a},u}(\mathcal{T}_{|U}) = 0$ unless $u \in \mathcal{A}$. We also impose $\mathrm{Gr}_j^{M(\mathcal{N})} \tilde{\psi}_{g,\mathfrak{a},u}(\mathcal{T}_{|U}) \in \mathrm{MT}_{\leq n}(U, w+j)_{\mathcal{A}}$ and $\mathrm{Gr}_j^{M(\mathcal{N})} \phi_g^{(0)}(\mathcal{T}_{|U}) \in \mathrm{MT}_{\leq n}(U, w+j)_{\mathcal{A}}$, where $M(\mathcal{N})$ denote the weight filtration of the canonical nilpotent morphism on $\tilde{\psi}_{g,\mathfrak{a},u}(\mathcal{T}_{|U})$.

We set $\mathrm{MT}(X,w)_{\mathcal{A}} := \mathrm{MT}_X(X,w)_{\mathcal{A}}$. It is an abelian full subcategory in $\mathrm{MT}(X,w)$. Any direct summand of an object in $\mathrm{MT}(X,w)_{\mathcal{A}}$ is also an object in $\mathrm{MT}(X,w)_{\mathcal{A}}$. If $\mathcal{T} \in \mathrm{MT}(X,w)_{\mathcal{A}}$, then we have $\mathcal{T}^* \in \mathrm{MT}(X,w)_{\mathcal{A}}$. We also have $j^*\mathcal{T} \in \mathrm{MT}(X,w)_{j^*\mathcal{A}}$, where $j^*\mathcal{A} = \{(a,-\alpha) \mid (a,\alpha) \in \mathcal{A}\}$.

We define the full subcategory $\mathrm{MTW}(X)_{\mathcal{A}} \subset \mathrm{MTW}(X)$ as the full subcategory of the objects $(\mathcal{T},W) \in \mathrm{MTW}(X)$ such that $\mathrm{Gr}_w^W(\mathcal{T}) \in \mathrm{MT}(X,w)_{\mathcal{A}}$. It is an abelian subcategory of $\mathrm{MTW}(X)_{\mathcal{A}}$. For $(\mathcal{T},W) \in \mathrm{MTW}(X)_{\mathcal{A}}$, we have $j^*(\mathcal{T},W) \in \mathrm{MTW}(X)_{j^*\mathcal{A}}$ and $(\mathcal{T},W)^* \in \mathrm{MTW}(X)_{\mathcal{A}}$.

It is easy to see the following.

Lemma 7.1.62 *Let $f : X \longrightarrow Y$ be any projective morphism of complex manifolds. For any $\mathcal{T} \in \mathrm{MT}(X,w)_{\mathcal{A}}$, we have $f_\dagger^i \mathcal{T} \in \mathrm{MT}(X,w+i)_{\mathcal{A}}$. For any $(\mathcal{T},W) \in \mathrm{MTW}(X)_{\mathcal{A}}$, we have $f_\dagger^i(\mathcal{T},W) \in \mathrm{MTW}(X)_{\mathcal{A}}$.* □

Let g be any holomorphic function on X. We define $\mathrm{MTW}^{\mathrm{sp}}(X,g)_{\mathcal{A}} \subset \mathrm{MTW}^{sp}(X,g)$ as the full subcategory of the objects $(\mathcal{T},W) \in \mathrm{MTW}^{\mathrm{sp}}(X,g)$ such that $\mathrm{Gr}_w^W(\mathcal{T}) \in \mathrm{MT}(X,w)_{\mathcal{A}}$. The following is clear.

Lemma 7.1.63 *For any $(\mathcal{T},W) \in \mathrm{MTW}^{\mathrm{sp}}(X,g)_{\mathcal{A}}$, the pre-mixed twistor $\mathscr{D}$-modules $(\tilde{\psi}_{t,\mathfrak{a},u}(\mathcal{T}),W)$ and $(\phi_g(\mathcal{T}),W)$ are also objects in $\mathrm{MTW}(X)_{\mathcal{A}}$.* □

Let H be any hypersurface of X. Let $\mathrm{MTW}(X,H)_{\mathcal{A}}$ be the full subcategory $\mathrm{MTW}(X,H) \cap \mathrm{MTW}(X)_{\mathcal{A}}$ in $\mathrm{MTW}(X)$. For any $\mathcal{T} \in \mathrm{MTW}^{\mathrm{sp}}(X,H)$, we have the polarizable pure twistor $\mathscr{D}$-modules $\mathrm{Gr}_m^W \tilde{\psi}_{H,-\boldsymbol{\delta}}(\mathcal{T})$ and $\mathrm{Gr}_m^W \phi_H(\mathcal{T})$ in Corollary 7.1.30. The following is easy to see.

Lemma 7.1.64 *If $(\mathcal{T},W) \in \mathrm{MTW}^{\mathrm{sp}}(X,H)_{\mathcal{A}}$, then the pure twistor $\mathscr{D}$-modules $\mathrm{Gr}_m^W \tilde{\psi}_{H,-\boldsymbol{\delta}}(\mathcal{T})$ and $\mathrm{Gr}_m^W \phi_H(\mathcal{T})$ are objects in $\mathrm{MT}(X,m)_{\mathcal{A}}$.* □

The following lemma is easy to see.

Lemma 7.1.65 *In the setting of* Sect. 7.1.5, *if* $(\mathcal{T}, W)$ *and* $(\mathcal{Q}, W)$ *are objects in* $\mathrm{MTW}(X)_{\mathcal{A}}$, *then* $(\mathcal{T}', W)$ *is also an object in* $\mathrm{MTW}(X)_{\mathcal{A}}$. □

For any hypersurface $H \subset X$, we set $\mathrm{MTW}^{\mathrm{loc}}(X, H)_{\mathcal{A}} := \mathrm{MTW}^{\mathrm{loc}}(X, H) \cap \mathrm{MTW}(X)_{\mathcal{A}}$. For any $(\mathcal{T}, W) \in \mathrm{MTW}^{\mathrm{loc}}(X, H)$ we have $(\mathcal{T}, W)[\star H] \in \mathrm{MTW}^{\mathrm{loc}}(X, H)$ in Sect. 7.1.6.3. The following lemma is easy to see.

Lemma 7.1.66 *If* $(\mathcal{T}, W) \in \mathrm{MTW}^{\mathrm{loc}}(X, H)_{\mathcal{A}}$, *then* $(\mathcal{T}, W)[\star H]$ *is an object in* $\mathrm{MTW}^{\mathrm{loc}}(X, H)_{\mathcal{A}}$.

Proof It follows from the description of $\mathrm{Gr}^W(\mathcal{T}[\star H])$ in the proof of Proposition 7.1.51. □

We remark the following.

Proposition 7.1.67 *Let* $(\mathcal{T}, W) \in \mathrm{MTW}(X)$. *For any* $P \in X$, *there exist a neighbourhood* X_P *of* P *in* X *and a finite dimensional* $\mathbb{Q}$*-vector subspace* $\mathcal{A} \subset \mathbb{R} \times \mathbb{C}$ *such that (i)* $\mathbb{Q} \times \{0\} \subset \mathcal{A}$, *(ii)* $(\mathcal{T}, W)_{|X_P} \in \mathrm{MTW}(X_P)_{\mathcal{A}}$.

Proof It is enough to consider the case where $\mathcal{T}$ is pure of weight w. Take a small neighbourhood of X_P of P in X. We can take a complex manifold Z with a simple normal crossing hypersurface H_Z and a projective morphism $\varphi : Z \longrightarrow X_P$, and a polarizable wild variation of pure twistor structure $\mathcal{V}$ of weight w on (Z, H_Z), such that $\mathcal{T}$ is a direct summand of $\varphi^0_\dagger(\mathcal{T}_1)$, where $\mathcal{T}_1$ is a pure twistor $\mathscr{D}$-module of weight w on Z obtained as the minimal extension of $\mathcal{V}$. We can find a finite dimensional $\mathbb{Q}$-vector subspace $\mathcal{A} \subset \mathbb{R} \times \mathbb{C}$ such that (i) $\mathbb{Q} \times \{0\} \subset \mathcal{A}$, (ii) the union of the KMS-spectra of $\mathcal{V}$ along the irreducible components of D_Z is contained in $\mathcal{A}$. A shown in [55], $\varphi^0_\dagger \mathcal{T}_1 \in \mathrm{MT}(X_P, w)_{\mathcal{A}}$. Hence, $\mathcal{T} \in \mathrm{MT}(X_P, w)_{\mathcal{A}}$. □

We have the following as mentioned in Sect. 2.1.5.

Lemma 7.1.68 *If* $\mathcal{T} \in \mathrm{MTW}(X)$ *is integrable, then* $\mathcal{T} \in \mathrm{MTW}(X)_{\mathbb{R} \times \{0\}}$. □

7.2 Mixed Twistor $\mathscr{D}$-Modules

7.2.1 Definition

Let X be a complex manifold. We define mixed twistor $\mathscr{D}$-modules on X by using a Noetherian induction on the support.

Definition 7.2.1 A pre-mixed twistor $\mathscr{D}$-module $(\mathcal{T}, W)$ is called a mixed twistor $\mathscr{D}$-module if the following holds for any open subset $U \subset X$ with a holomorphic function g:

- $(\mathcal{T}, W)_{|U}$ is admissibly specializable along g.
- $\big(\tilde{\psi}_{g,\mathfrak{a},u}(\mathcal{T}), W\big)$ $(\mathfrak{a} \in \mathbb{C}[t_n^{-1}], u \in \mathbb{R} \times \mathbb{C})$ and $\big(\phi_g(\mathcal{T}), W\big)$ are mixed twistor $\mathscr{D}$-modules with strictly smaller supports if $\mathrm{Supp}\, \mathcal{T}_{|U} \not\subset g^{-1}(0)$. Here, W on

$\tilde{\psi}_{g,\mathfrak{a},u}(\mathcal{T})$ and $\phi_g(\mathcal{T})$ are the relative monodromy filtrations of the canonical nilpotent morphism $\mathcal{N}$ with respect to the naively induced filtration. (See Sect. 7.1.3.)

(We use a Noetherian induction on the dimensions and the number of the components of the maximal dimension.) □

Remark 7.2.2 In this paper, we consider only graded polarizable mixed objects. We shall often omit to denote the weight filtration W if there is no risk of confusion. □

7.2.2 Some Basic Properties

Let $\mathrm{MTM}(X) \subset \mathrm{MTW}(X)$ denote the full subcategory of mixed twistor $\mathscr{D}$-modules on X.

Proposition 7.2.3 *Let* $0 \longrightarrow (\mathcal{T}_1, W) \longrightarrow (\mathcal{T}_2, W) \longrightarrow (\mathcal{T}_3, W) \longrightarrow 0$ *be an exact sequence in* $\mathrm{MTW}(X)$. *If* $(\mathcal{T}_2, W) \in \mathrm{MTM}(X)$, *then* $(\mathcal{T}_i, W)$ $(i = 1, 3)$ *are also mixed twistor* $\mathscr{D}$*-modules on* X.

Proof Let $\mathrm{Map}^*(\mathbb{Z}_{\geq 0}, \mathbb{Z}_{\geq 0})$ be the set of maps $\rho : \mathbb{Z}_{\geq 0} \longrightarrow \mathbb{Z}_{\geq 0}$ such that $\rho(n) = 0$ except for finitely many n. We use the order on $\mathrm{Map}^*(\mathbb{Z}_{\geq 0}, \mathbb{Z}_{\geq 0})$ given by $\rho_1 < \rho_2$ if and only if $\rho_1(i_0) < \rho_2(i_0)$ for $i_0 = \max\{i \mid \rho_1(i) \neq \rho_2(i)\}$.

Let $P \in X$. Let X_P be any small neighbourhood of P. For any closed analytic subset $\Gamma \subset X_P$, let $\rho(\Gamma, i)$ be the number of the i-dimensional irreducible components. We obtain $\rho(\Gamma) \in \mathrm{Map}^*(\mathbb{Z}_{\geq 0}, \mathbb{Z}_{\geq 0})$. We use an induction on $\rho(\Gamma)$.

Suppose that $\mathrm{Supp}(\mathcal{T}_2) = \Gamma$. By Proposition 7.1.24, $\mathcal{T}_i$ $(i = 1, 3)$ are admissibly specializable along any holomorphic function g on X_P. Suppose $\Gamma \not\subset g^{-1}(0)$. We obtain the exact sequence in $\mathrm{MTW}(X_P)$ for any $\mathfrak{a} \in t_n^{-1}\mathbb{C}[t_n^{-1}]$ and $u \in \mathbb{R} \times \mathbb{C}$:

$$0 \longrightarrow \tilde{\psi}_{g,\mathfrak{a},u}(\mathcal{T}_1, W) \longrightarrow \tilde{\psi}_{g,\mathfrak{a},u}(\mathcal{T}_2, W) \longrightarrow \tilde{\psi}_{g,\mathfrak{a},u}(\mathcal{T}_3, W) \longrightarrow 0$$

By definition, $\tilde{\psi}_{g,\mathfrak{a},u}(\mathcal{T}_2, W) \in \mathrm{MTM}(X_P)$. By the hypothesis of the induction, we obtain that $\tilde{\psi}_{g,\mathfrak{a},u}(\mathcal{T}_i, W) \in \mathrm{MTM}(X_P)$ $(i = 1, 3)$. Similarly, we obtain that $\phi_g(\mathcal{T}_i, W) \in \mathrm{MTM}(X)$ $(i = 1, 3)$. Hence, we obtain that $(\mathcal{T}_i, W) \in \mathrm{MTM}(X)$ $(i = 1, 3)$. □

Remark 7.2.4 It is M. Saito who emphasized the usefulness of the fact that any subquotients of mixed Hodge modules in the category of weakly mixed Hodge modules are mixed Hodge modules. □

As an immediate corollary, we obtain the following proposition.

Proposition 7.2.5 $\mathrm{MTM}(X)$ *is an abelian category.* □

Proposition 7.2.6 *$(\mathcal{T}, W)^*$ and $j^*(\mathcal{T}, W)$ are mixed twistor $\mathscr{D}$-modules for any $(\mathcal{T}, W) \in \mathrm{MTM}(X)$.*

Proof We have only to use a Noetherian induction. Note the compatibility of Hermitian adjoint and Beilinson's formalism. □

Proposition 7.2.7 *Let $F : X \longrightarrow Y$ be any projective morphism. For any $(\mathcal{T}, W) \in \mathrm{MTM}(X)$, the pre-mixed twistor $\mathscr{D}_Y$-modules $F_\dagger^i(\mathcal{T}, W)$ $(i \in \mathbb{Z})$ are mixed twistor $\mathscr{D}_Y$-modules. So, we obtain a cohomological functor $F_\dagger^i : \mathrm{MTM}(X) \longrightarrow \mathrm{MTM}(Y)$ $(i \in \mathbb{Z})$.*

Proof By Lemma 7.1.31, $F_\dagger^i(\mathcal{T}, W)$ is admissibly specializable. Let $P \in Y$. Let g_Y be any holomorphic function on Y, such that $\mathrm{Supp}\, F_\dagger^i \mathcal{T} \not\subset g_Y^{-1}(0)$. We put $g_X := F^{-1}(g_Y)$. Let $(\mathcal{T}, W) \in \mathrm{MTM}(X)$. By applying the hypothesis of the induction to $\tilde{\psi}_{g_X,\mathfrak{a},u}(\mathcal{T}, W)$ and $\phi_{g_X}(\mathcal{T}, W)$ together with Corollary 7.1.32, we obtain that $\tilde{\psi}_{g_Y,\mathfrak{a},u} F_\dagger^i(\mathcal{T}, W)$ and $\phi_{g_Y} F_\dagger^i(\mathcal{T}, W)$ are objects in $\mathrm{MTM}(Y)$. Hence, we obtain $F_\dagger^i(\mathcal{T}, W) \in \mathrm{MTM}(Y)$. □

Let $\iota : Y \subset X$ be a closed immersion of complex manifolds. Let $\mathrm{MTM}_Y(X) \subset \mathrm{MTM}(X)$ be the full subcategory of $(\mathcal{T}, W) \in \mathrm{MTM}(X)$ such that $\mathrm{Supp}\, \mathcal{T} \subset Y$. We obtain a functor $\iota_\dagger : \mathrm{MTM}(Y) \longrightarrow \mathrm{MTM}_Y(X)$. We have the following version of Kashiwara equivalence.

Proposition 7.2.8 *The functor $\iota_\dagger : \mathrm{MTM}(Y) \longrightarrow \mathrm{MTM}_Y(X)$ is an equivalence.*

Proof We may assume that $X = \Delta^n$ and $Y = \{z_1 = 0\}$. We use a Noetherian induction on the support. Let $(\mathcal{T}, W) \in \mathrm{MTM}_Y(X)$. Because $(\mathcal{T}, W)$ is filtered strictly specializable along z_1, we have a filtered $\mathcal{R}_Y$-triple $(\mathcal{T}_0, W)$ such that $(\mathcal{T}, W) = \iota_\dagger(\mathcal{T}_0, W)$. By Lemma 4.4.8, $(\mathcal{T}_0, W)$ is admissibly specializable. Let g be a holomorphic function on Y such that $\mathrm{Supp}\, \mathcal{T}_0 \not\subset g^{-1}(0)$. We extend g to a holomorphic function on X by the pull back via the projection $X \longrightarrow Y$. Then, we have

$$\iota_\dagger \psi_{g,\mathfrak{a},u}(\mathcal{T}, W) \simeq \psi_{g,\mathfrak{a},u} \iota_\dagger(\mathcal{T}, W), \qquad \iota_\dagger \phi_g(\mathcal{T}, W) \simeq \phi_g \iota_\dagger(\mathcal{T}, W).$$

By definition, $\psi_{g,\mathfrak{a},u}\iota_\dagger(\mathcal{T}, W)$ and $\phi_g \iota_\dagger(\mathcal{T}, W)$ are objects in $\mathrm{MTM}(X)$. Hence, by the hypothesis of the induction, $\psi_{g,\mathfrak{a},u}(\mathcal{T}, W)$ and $\phi_g(\mathcal{T}, W)$ are objects in $\mathrm{MTM}(Y)$. Thus, we are done. □

Remark 7.2.9 Let $\mathcal{T} \in \mathrm{MTM}(X)$. Let $(\mathcal{M}_1, \mathcal{M}_2, C)$ be the underlying $\mathcal{R}_X$-triple of $\mathcal{T}$. We define $\Xi_{\mathrm{DR}}(\mathcal{T}) := \mathcal{M}_2^1$. It gives a functor Ξ_{DR} from $\mathrm{MTM}(X)$ to the category of holonomic $\mathscr{D}$-modules. We can observe that it is faithful. Indeed, let $F : (\mathcal{T}_1, L) \longrightarrow (\mathcal{T}_2, L)$ be a morphism in $\mathrm{MTM}(X)$ such that $\Xi_{\mathrm{DR}}(F) = 0$. By a result in the pure case, we obtain $\mathrm{Gr}^W(F) = 0$. By the strictness, we obtain $F = 0$. □

Let $\mathcal{A}$ be a $\mathbb{Q}$-vector subspace in $\mathbb{R} \times \mathbb{C}$ such that $\mathbb{Q} \times \{0\} \subset \mathcal{A}$. We set $\mathrm{MTM}(X)_{\mathcal{A}} := \mathrm{MTM}(X) \cap \mathrm{MTW}(X)_{\mathcal{A}}$ in $\mathrm{MTW}(X)_{\mathcal{A}}$. Any object in $\mathrm{MTM}(X)_{\mathcal{A}}$ is called $\mathcal{A}$-mixed twistor $\mathscr{D}$-module on X. The following is clear.

Proposition 7.2.10

- *In the setting of Proposition* 7.2.3, *if* $(\mathcal{T}_2, W) \in \mathrm{MTM}(X)_{\mathcal{A}}$ *then we have* $(\mathcal{T}_i, W) \in \mathrm{MTM}(X)_{\mathcal{A}}$ $(i = 1, 3)$.
- $\mathrm{MTM}(X)_{\mathcal{A}}$ *is an abelian subcategory in* $\mathrm{MTM}(X)$.
- *If* $(\mathcal{T}, W)$ *is an object in* $\mathrm{MTM}(X)_{\mathcal{A}}$ *then we have* $j^*(\mathcal{T}, W) \in \mathrm{MTM}(X)_{j^*\mathcal{A}}$ *and* $(\mathcal{T}, W)^* \in \mathrm{MTM}(X)_{\mathcal{A}}$.
- *In the setting of Proposition* 7.2.7, *if* $(\mathcal{T}, W) \in \mathrm{MTM}(X)_{\mathcal{A}}$ *then* $F^i_{\dagger}(\mathcal{T}, W)$ *are objects in* $\mathrm{MTM}(Y)_{\mathcal{A}}$.
- *In the setting of Proposition* 7.2.8, $\iota_{\dagger}$ *induces an equivalence* $\mathrm{MTM}(Y)_{\mathcal{A}} \simeq \mathrm{MTM}_Y(X)_{\mathcal{A}}$. □

7.2.3 Integrable Case

An integrable pre-mixed twistor $\mathscr{D}$-module is called integrable mixed twistor $\mathscr{D}$-module if the underlying pre-mixed twistor $\mathscr{D}$-module is a mixed twistor $\mathscr{D}$-module. Let $\mathrm{MTM}^{\mathrm{int}}(X) \subset \mathrm{MTW}^{\mathrm{int}}(X)$ denote the full subcategory of integrable mixed twistor $\mathscr{D}$-modules. The following proposition can be proved by a Noetherian induction as in the case of Proposition 7.2.5.

Proposition 7.2.11 $\mathrm{MTM}^{\mathrm{int}}(X)$ *is closed for sub-quotients in* $\mathrm{MTW}^{\mathrm{int}}(X)$. *In particular,* $\mathrm{MTM}^{\mathrm{int}}(X)$ *is abelian.* □

We mention some basic property.

Proposition 7.2.12 *Let* $(\mathcal{T}, W) \in \mathrm{MTM}^{\mathrm{int}}(X)$.

- *For any open* $U \subset X$ *with a holomorphic function* g, *the mixed twistor* $\mathscr{D}$*-modules* $\tilde{\psi}_{g,\mathfrak{a},u}(\mathcal{T}, W)$ *and* $\phi_g(\mathcal{T}, W)$ *are naturally integrable.*
- $(\mathcal{T}, W)^*$ *and* $j^*(\mathcal{T}, W)$ *are objects in* $\mathrm{MTM}^{\mathrm{int}}(X)$.
- *For any projective morphism* $F : X \longrightarrow Y$, *we have* $F^i_{\dagger}(\mathcal{T}, W) \in \mathrm{MTM}^{\mathrm{int}}(Y)$. □

Let $\iota : Y \subset X$ be a closed immersion. Let $\mathrm{MTM}^{\mathrm{int}}_Y(X) \subset \mathrm{MTM}^{\mathrm{int}}(X)$ be the full subcategory of $(\mathcal{T}, W) \in \mathrm{MTM}^{\mathrm{int}}(X)$ such that $\mathrm{Supp}\,\mathcal{T} \subset Y$. We obtain a functor $\iota_{\dagger} : \mathrm{MTM}^{\mathrm{int}}(Y) \longrightarrow \mathrm{MTM}^{\mathrm{int}}_Y(X)$.

Proposition 7.2.13 *The functor* $\iota_{\dagger} : \mathrm{MTM}^{\mathrm{int}}(Y) \longrightarrow \mathrm{MTM}^{\mathrm{int}}_Y(X)$ *is an equivalence.* □

Chapter 8
Infinitesimal Mixed Twistor Modules

Any holonomic $\mathscr{D}$-module is locally expressed as the gluing of some meromorphic flat bundles on subvarieties. In this sense, meromorphic flat bundles are building blocks of holonomic $\mathscr{D}$-modules. In a similar sense, *admissible variations of mixed twistor structure* are building blocks of mixed twistor $\mathscr{D}$-modules. We shall study admissible variations of mixed twistor structure in Chap. 9. In this chapter, as a preliminary, we shall study the linear version of admissible variation of mixed Hodge structure, that is *infinitesimal mixed twistor module*. Infinitesimal mixed twistor structures naturally appear at the singularities of admissible variations of mixed twistor structure. Moreover, they contain much important information on the behaviour of the variations.

In the Hodge case, the concept of infinitesimal mixed Hodge modules was introduced and studied in [32]. The concept of infinitesimal mixed twistor module is obviously its twistor version.

One of the main purpose in this chapter is to prove that the category of infinitesimal mixed twistor modules has the property **(M0–3)** given in Sect. 6.3.1. Note that pure objects in this situation are polarizable mixed twistor structures which was previously studied in [52, 56], on the basis of the work of Cattani and Kaplan [7], Cattani, Kaplan and Schmid [8, 9], Kashiwara [31] and Kashiwara and Kawai [35]. As explained in Sect. 8.2, the results in [52, 56] imply that the category of polarizable mixed twistor structures has the property **(P0–3)** (Proposition 8.2.1). Then, in Sect. 8.3, we explain that the category of infinitesimal mixed twistor modules has the property **(M0–3)** (Theorem 8.3.6), on the basis of the work of Kashiwara [32] and Saito [73].

The results will be useful in the gluing of admissible variations of mixed twistor structure in Chap. 10. Let $X := \Delta^n$, $D_i := \{z_i = 0\} \cap X$ and $D := \bigcup_{i=1}^{\ell} D_i$. For $I \subset \underline{\ell} = \{1, \dots, \ell\}$, we set $D_I := \bigcap_{i \in I} D_i$ and $\partial D_I := D_I \cap \bigcup_{j \in \underline{\ell} \setminus I} D_j$. We shall study mixed twistor $\mathscr{D}$-modules obtained as the gluing of admissible variations of mixed

T. Mochizuki, *Mixed Twistor $\mathscr{D}$-modules*, Lecture Notes in Mathematics 2125,
DOI 10.1007/978-3-319-10088-3_8

twistor structure (V_I, W) on $(D_I, \partial D_I)$ $(I \subset \underline{\ell})$. We naturally encounter diagrams of mixed twistor structures as in Sect. 6.3. We shall efficiently use Proposition 8.2.1 and Theorem 8.3.6 together with Theorem 6.3.18 in the study of such objects.

8.1 Preliminary

8.1.1 Pure Twistor Structure

Let us recall the concept of polarizable pure twistor structure [82]. Let X be a complex manifold. Let $\mathcal{T} = (\mathcal{M}', \mathcal{M}'', C)$ be a variation of twistor structure on X (Sect. 2.1.7). For any point $P \in X$, we have the isomorphism $\mathcal{M}'^{\vee}_{|\mathbb{C}^*_\lambda \times \{P\}}$ and $\sigma^* \mathcal{M}''_{|\mathbb{C}^*_\lambda \times \{P\}}$ induced by the sesqui-linear pairing C. Hence, we obtain a vector bundle $V_P(\mathcal{T})$. Recall that $\mathcal{T}$ is called pure of weight w if $V_P(\mathcal{T})$ is isomorphic to a direct sum of $\mathcal{O}_{\mathbb{P}^1}(w)$. In this paper, if there is no risk of confusion, a variation of pure twistor structure of weight w on X will be often called pure twistor structure of weight w on X, i.e., we shall omit "variation of".

Let us recall the notion of polarization of pure twistor structure. First, let us consider the case where X is a point. Let $\mathcal{T}$ be a pure twistor structure of weight w. Let $\mathcal{S}$ be a Hermitian sesqui-linear duality of weight w on $\mathcal{T}$. (See Remark 2.1.25 for Hermitian sesqui-linear duality.)

(Case $w = 0$) We have the induced Hermitian pairing on $H^0(\mathbb{P}^1, V(\mathcal{T}))$ induced by $\mathcal{S}$. If the Hermitian pairing is positive definite, $\mathcal{S}$ is called a polarization of $\mathcal{T}$.

(General case) Let $\mathcal{S}'$ be the induced Hermitian sesqui-linear duality of weight 0 on $\mathcal{T} \otimes \mathcal{U}(0, w)$. Here, we use the isomorphism $\mathcal{U}(p, q) \simeq \mathcal{U}(p, q)^* \otimes \boldsymbol{T}(q - p)$ given by $((-1)^p, (-1)^p)$ given in Sect. 2.1.8. If the Hermitian sesqui-linear duality $\mathcal{S}'$ is a polarization of $\mathcal{T} \otimes \mathcal{U}(0, w)$, then $\mathcal{S}$ is called a polarization of $\mathcal{T}$. (In the following, the induced Hermitian sesqui-linear duality $\mathcal{S}'$ is also denoted by $\mathcal{S}$.)

Note that the isomorphisms $\mathcal{U}(p, q) \simeq \mathcal{U}(p, q)^* \otimes \boldsymbol{T}(q - p)$ are polarizations.

Let $\mathcal{T}$ be a pure twistor structure of weight w on X. Let $\mathcal{S}$ be a Hermitian sesqui-linear duality of weight w on $\mathcal{T}$. Let $P \in X$ be any point, and let $\iota_P : \{P\} \longrightarrow X$ be the inclusion. We have the induced Hermitian sesqui-linear duality $\iota_P^* \mathcal{S}$ of weight w on $\iota^* \mathcal{T}$. If $\iota_P^* \mathcal{S}$ are polarizations of $\iota_P^* \mathcal{T}$ for any $P \in X$, then $\mathcal{S}$ is called a polarization on $\mathcal{T}$. If $\mathcal{T}$ admits a polarization, $\mathcal{T}$ is called a polarizable pure twistor structure. Let $\mathrm{PTS}(X, w) \subset \mathrm{TS}(X)$ denote the full subcategory of polarizable pure twistor structure of weight w on X.

8.1.1.1 Some Operations

Let $\mathcal{T}$ be a pure twistor structure of weight w on X with a polarization $\mathcal{S}$. We have the naturally induced Hermitian sesqui-linear duality of weight $-w$ on $\mathcal{T}^*$:

$$\mathcal{S}^{-1} : \mathcal{T}^* \longrightarrow (\mathcal{T}^*)^* \otimes \boldsymbol{T}(w).$$

We also have the naturally induced Hermitian sesqui-linear duality $j^*\mathcal{S}$ of weight w on $j^*\mathcal{T}$, where $j^*\boldsymbol{T}(-w) \simeq \boldsymbol{T}(-w)$ is given as in Sect. 2.1.8.

Lemma 8.1.1 *$\mathcal{S}^{-1}$ and $j^*\mathcal{S}$ are polarizations of $\mathcal{T}^*$ and $j^*\mathcal{T}$, respectively. As a result, we naturally have the Hermitian adjoint functor and the functor j^* on* $\mathrm{PTS}(X, w)$.

Proof It is enough to consider the case where X is a point. In that case, $\mathcal{T}$ is isomorphic to $\mathcal{U}(0, w)^{\oplus r}$, and $\mathcal{S}$ is induced by the isomorphism $\mathcal{U}(0, w) \simeq \mathcal{U}(0, w)^* \otimes \boldsymbol{T}(w)$ and a Hermitian metric of a vector space. Then, we can check the claim directly. □

Let $\mathcal{T}^{(i)}$ $(i = 1, 2)$ be pure twistor structure of weight w_i on X. Let $\mathcal{S}^{(i)}$ be Hermitian sesqui-linear dualities of $\mathcal{T}^{(i)}$. We have the naturally induced Hermitian sesqui-linear dualities $\mathcal{S}^{(1)} \otimes \mathcal{S}^{(2)}$ of weight $w_1 + w_2$ on $\mathcal{T}^{(1)} \otimes \mathcal{T}^{(2)}$. We also have the induced Hermitian sesqui-linear duality $\hat{\mathcal{S}}$ of weight $w_2 - w_1$ on $\mathcal{H}om(\mathcal{T}^{(1)}, \mathcal{T}^{(2)})$. The following lemma can be also checked by the argument in Lemma 8.1.1.

Lemma 8.1.2 *$\mathcal{S}^{(1)} \otimes \mathcal{S}^{(2)}$ and $\hat{\mathcal{S}}$ are polarizations.* □

Let $\mathcal{T}$ be a pure twistor structure of weight w on X with a polarization $\mathcal{S}$. By Lemma 8.1.2, we have the induced polarization $\mathcal{S}^\vee$ on $\mathcal{T}^\vee$. Together with Lemma 8.1.1, we also have the induced polarization $\tilde{\gamma}_{sm}^*\mathcal{S}$ on $\tilde{\gamma}_{sm}^*\mathcal{T}$. (See Sect. 2.1.7 for $\tilde{\gamma}_{sm}^*$.)

8.1.1.2 Basic Facts

We recall some basic facts on polarizable pure twistor structure.

Lemma 8.1.3 *Let $\mathcal{T}$ be a smooth $\mathcal{R}_X$-triple with a decomposition $\mathcal{T} = \mathcal{T}_1 \oplus \mathcal{T}_2$. Then, $\mathcal{T} \in$ $\mathrm{PTS}(X, w)$ if and only if $\mathcal{T}_i \in$ $\mathrm{PTS}(X, w)$ $(i = 1, 2)$.*

Proof The claim is clear. We only remark that if $\mathcal{S}$ is a polarization of a pure twistor structure $\mathcal{T} = \mathcal{T}_1 \oplus \mathcal{T}_2$, then the induced Hermitian sesqui-linear dualities

$$\mathcal{S}_i : \mathcal{T}_i \longrightarrow \mathcal{T} \longrightarrow \mathcal{T}^* \otimes \boldsymbol{T}(-w) \longrightarrow \mathcal{T}_i^* \otimes \boldsymbol{T}(-w)$$

are polarizations. They are called the restriction of $\mathcal{S}$ to $\mathcal{T}_i$. □

Lemma 8.1.4 *Let $\mathcal{T}_i \in$ $\mathrm{PTS}(X, w_i)$ with $w_1 > w_2$. Then, any morphism $F : \mathcal{T}_1 \longrightarrow \mathcal{T}_2$ in $\mathcal{R}$-$Tri(X)$ is* 0.

Proof It is reduced to the case where X is a point. Then, it follows from the fact $\mathrm{Hom}_{\mathcal{O}}(\mathcal{O}_{\mathbb{P}^1}(w_1), \mathcal{O}_{\mathbb{P}^1}(w_2)) = 0$ if $w_1 > w_2$. □

Lemma 8.1.5 *Let $\mathcal{T} \in \mathrm{PTS}(X, w)$. Let $\mathcal{T}_1 \subset \mathcal{T}$ be a sub-$\mathcal{R}_X$-triple which is also a pure twistor structure of weight w.*

- *$\mathcal{T}_1$ is also a polarizable pure twistor structure on X.*
- *Let $\mathcal{S}$ be a polarization on $\mathcal{T}$. Let $\mathcal{T}_2$ be the kernel of*

$$\mathcal{T} \simeq \mathcal{T}^* \otimes \boldsymbol{T}(-w) \longrightarrow \mathcal{T}_1^* \otimes \boldsymbol{T}(-w)$$

Then, we have $\mathcal{T} = \mathcal{T}_1 \oplus \mathcal{T}_2$.

Proof The first claim follows from the second claim and Lemma 8.1.3. The second is reduced to the case where X is a point and $w = 0$. Then, $\mathcal{T}_2$ is the orthogonal complement of $\mathcal{T}_1$ with respect to the Hermitian metric, and we obtain the decomposition. □

Corollary 8.1.6 *The category* $\mathrm{PTS}(X, w)$ *is abelian and semisimple.*

Proof Let $F : \mathcal{T}_1 \longrightarrow \mathcal{T}_2$ be a morphism in $\mathrm{PTS}(X, w)$. We can easily check that $\mathrm{Im}(F)$, $\mathrm{Ker}(F)$ and $\mathrm{Cok}(F)$ are also pure twistor structure of weight w. Then, by using Lemma 8.1.5, we obtain that they are polarizable. Hence, $\mathrm{PTS}(X, w)$ is an abelian category. We also obtain the semisimplicity from Lemma 8.1.5. □

8.1.2 Mixed Twistor Structure

Let $(\mathcal{T}, W)$ be a filtered object in the category of variations of twistor structure on X (Sect. 2.1.7), where W is a finite complete exhaustive increasing filtration indexed by $\mathbb{Z}$. It is called a variation of mixed twistor structure on X if each $\mathrm{Gr}^W_w(\mathcal{T})$ is a polarizable variation of pure twistor structure of weight w in the sense of Sect. 8.1.1. The condition is equivalent to that each $\mathrm{Gr}^W_w(\mathcal{T})$ is a polarizable pure twistor $\mathscr{D}$-module of weight w, i.e., $(\mathcal{T}, W)$ is a pre-mixed twistor $\mathscr{D}$-module [66, 69]. In this paper, if there is no risk of confusion, we shall often call it a mixed twistor structure on X, i.e., we omit "variation of". It is called pure of weight w if $\mathrm{Gr}^L_m = 0$ unless $m = w$.

Let $\mathrm{MTS}(X) \subset \mathrm{TS}(X)^{\mathrm{fil}}$ denote the category of mixed twistor structure on X. It is an abelian category. For any morphism $F : (\mathcal{T}_1, W) \longrightarrow (\mathcal{T}_2, W)$ in $\mathrm{MTS}(X)$, F is strictly compatible with the filtration W.

Remark 8.1.7 It might be more appropriate that the above object is called graded polarizable variation of mixed twistor structure. We omit to distinguish the graded polarizability because we consider only graded polarizable ones. □

For any mixed twistor structure $(\mathcal{T}, W)$ on X, we define the filtration W on $\mathcal{T}^*$ by $W_j(\mathcal{T}^*) := (\mathcal{T}/W_{-j-1})^*$. Because $\mathrm{Gr}_j^W(\mathcal{T}^*) \simeq \big(\mathrm{Gr}_{-j}^W(\mathcal{T})\big)^*$, $(\mathcal{T}, W)^* = (\mathcal{T}^*, W)$ is a mixed twistor structure. It gives the Hermitian adjoint functor on $\mathrm{MTS}(X)$.

The category $\mathrm{MTS}(X)$ is naturally equipped with tensor product and inner homomorphism by Lemma 8.1.2. The induced weight filtrations are given as in Sect. 6.1.4. In particular, it is equipped with the natural duality functor. It is also equipped with additive auto equivalences $\Sigma^{p,q}$ given by the tensor product of $\mathcal{U}(-p,q)$. We have the naturally defined functors j^* and $\tilde{\gamma}_{sm}^*$ on $\mathrm{MTS}(X)$.

Remark 8.1.8 A mixed twistor structure $(\mathcal{T}, W)$ is often denoted just by $\mathcal{T}$ if there is no risk of confusion. For a filtration L of a mixed twistor structure $\mathcal{T}$ in the category $\mathrm{MTS}(X)$, the filtered objects $\Sigma^{p,q}(\mathcal{T}, L)$ and $\big(\Sigma^{p,q}(\mathcal{T}), L\big)$ are also denoted by $(\mathcal{T}, L) \otimes \mathcal{U}(-p,q)$ and $(\mathcal{T} \otimes \mathcal{U}(-p,q), L)$, respectively. If $p = q$, they are also denoted by $(\mathcal{T}, L) \otimes \boldsymbol{T}(q)$ and $\big(\mathcal{T} \otimes \boldsymbol{T}(q), L\big)$, or $\boldsymbol{T}^q(\mathcal{T}, L)$ and $\big(\boldsymbol{T}^q\mathcal{T}, L\big)$, respectively. □

Lemma 8.1.9 *Let $(\mathcal{T}, W)$ be a mixed twistor $\mathscr{D}$-module on X. Assume that the underlying $\mathscr{D}$-module is a flat bundle. Then, $(\mathcal{T}, W)$ comes from a variation of mixed twistor structure.*

Proof In the pure case, it is easy to prove the claim by using the correspondence between pure twistor $\mathscr{D}$-modules and wild harmonic bundles. Let us consider the mixed case. We may assume $X = \Delta^n$. Let $\mathcal{T} = (\mathcal{M}_1, \mathcal{M}_2, C)$. By using the result in the pure case, we obtain that the $\mathcal{R}_X$-modules $\mathcal{M}_i$ are smooth. By using flat frames, we can check that the pairing takes values in the sheaf of continuous functions on $\boldsymbol{S} \times X$ which are C^∞ in the X-direction. By successive use of $\psi_{z_i,-\boldsymbol{\delta}}$, we obtain that its restriction to $\boldsymbol{S} \times \{P\}$ can be extended to holomorphic function on $\mathbb{C}_\lambda^*$. Hence, we obtain a pairing of $\mathcal{M}_{1|\mathbb{C}_\lambda^* \times X}$ and $\sigma^*\mathcal{M}_{2|\mathbb{C}_\lambda^* \times X}$ valued in the sheaf of C^∞-functions on $\mathbb{C}_\lambda^* \times X$ which are holomorphic in the $\mathbb{C}_\lambda^*$-direction. Hence $(\mathcal{T}, W)$ comes from a smooth $\mathcal{R}$-triple. Because the w-th graded piece corresponds to the pure twistor $\mathscr{D}$-module of weight w, we obtain that $(\mathcal{T}, W)$ comes from a variation of mixed twistor structure. □

8.1.3 Reduction

We shall use the following lemma implicitly.

Lemma 8.1.10 *Let $(\mathcal{T}, W) \in \mathrm{MTS}(X)$ with subobjects $(\mathcal{T}_i, W) \subset (\mathcal{T}, W)$ $(i = 1,2)$. If $\mathrm{Gr}^W(\mathcal{T}_1) = \mathrm{Gr}^W(\mathcal{T}_2)$ in $\mathrm{Gr}^W(\mathcal{T})$, then we have $\mathcal{T}_1 = \mathcal{T}_2$.*

Proof Consider $F : (\mathcal{T}_1, W) \longrightarrow (\mathcal{T}, W)/(\mathcal{T}_2, W)$. If $\mathrm{Gr}^W(F) = 0$, we have $F = 0$, which implies the claim of the lemma. □

Let $\big((\mathcal{T}, W), L, N\big) \in \mathrm{MTS}(X)^{\mathrm{fil,nil}}$. We put $\mathcal{T}^{(0)} := \mathrm{Gr}^W(\mathcal{T})$. It is equipped with naturally induced filtrations $W^{(0)}$ and $L^{(0)}$. We also have an induced map $N^{(0)}$:

$(\mathcal{T}^{(0)}, W^{(0)}) \longrightarrow (\mathcal{T}^{(0)}, W^{(0)}) \otimes \boldsymbol{T}(-1)$. Thus, we obtain $\big((\mathcal{T}^{(0)}, W^{(0)}), L^{(0)}, N^{(0)}\big)$ in $\mathrm{MTS}(X)^{\mathrm{fil,nil}}$.

Lemma 8.1.11 *$\big((\mathcal{T}, W), L, N\big)$ has a relative monodromy filtration if and only if $\big((\mathcal{T}^{(0)}, W^{(0)}), L^{(0)}, N^{(0)}\big)$ has a relative monodromy filtration.*

Proof Let M' be the filtration of $(\mathcal{T}, W)$ in $\mathrm{MTS}(X)$, given by Deligne's inductive formula for N. The induced filtration $M'^{(0)}$ on $(\mathcal{T}^{(0)}, W^{(0)})$ satisfies Deligne's inductive formula for $N^{(0)}$. Note $\mathrm{Gr}^L(M')^{(0)} \simeq \mathrm{Gr}^{L^{(0)}}(M'^{(0)})$. Thus, we are done. □

Let $\mathrm{MTS}(X)^{\mathrm{RMF}} \subset \mathrm{MTS}(X)^{\mathrm{fil,nil}}$ denote the full subcategory of the objects which have relative monodromy filtrations.

Lemma 8.1.12 *The category $\mathrm{MTS}(X)^{\mathrm{RMF}}$ is equipped with tensor product and inner homomorphism as in Proposition 6.1.5.*

Proof By using Proposition 6.1.4, it can be reduced to the case $\mathcal{A} = \mathrm{Vect}_{\mathbb{C}}$. □

8.1.4 *Some Conditions for the Existence of Relative Monodromy Filtration*

Let $(\mathcal{T}, L)$ be a filtered smooth $\mathcal{R}_X$-triple. Let $N : \mathcal{T} \longrightarrow \mathcal{T} \otimes \boldsymbol{T}(-1)$ be a morphism such that $N \cdot L_k(\mathcal{T}) \subset L_k(\mathcal{T}) \otimes \boldsymbol{T}(-1)$. Because $\mathrm{Gr}^L(\mathcal{T}, L, N)$ is graded, it has a relative monodromy filtration $W\big(\mathrm{Gr}^L(\mathcal{T})\big)$ in the category of $\mathcal{R}_X$-triples. Assume the following:

- $(\mathrm{Gr}^L(\mathcal{T}), W)$ is a mixed twistor structure on X.

The underlying $\mathcal{R}_X$-modules of $\mathcal{T}$ are denoted by $\mathcal{M}_1$ and $\mathcal{M}_2$.

Proposition 8.1.13 *Assume that, for each $P \in X$, there exists a subset $U_P \subset \mathbb{C}_\lambda$ such that (i) $|U_P| = \infty$, (ii) $(\mathcal{M}_1, L, N)_{|(\lambda,P)} \in \mathrm{Vect}_{\mathbb{C}}^{\mathrm{RMF}}$ for any $\lambda \in U_P$. Then, $\mathcal{T}$ is equipped with a filtration $W(\mathcal{T})$ such that (i) $(\mathcal{T}, W)$ is a mixed twistor structure, (ii) W is a relative monodromy filtration of $(\mathcal{T}, L, N)$ in $\mathrm{MTS}(X)$.*

Proof We have only to consider the case that X is a point $\{P\}$. The set U_P is denoted just by U. Because $\mathrm{Gr}^L(\mathcal{T})$ is a mixed twistor structure, the pairing C of $\mathcal{M}_1$ and $\mathcal{M}_2$ is non-degenerate. Hence, we can regard $\mathcal{T}$ as a vector bundle V on $\mathbb{P}^1$, obtained as the gluing of $\mathcal{M}_1^\vee$ and $\sigma^*\mathcal{M}_2$. It is equipped with a filtration L and a nilpotent morphism $N : V \longrightarrow V \otimes \mathcal{O}_{\mathbb{P}^1}(2)$ preserving L. The relative monodromy filtration W on $\mathrm{Gr}^L(V)$ gives a mixed twistor structure.

Lemma 8.1.14 *If $(V, L, N)_{|(\lambda,P)} \in \mathrm{Vect}_{\mathbb{C}}^{\mathrm{RMF}}$ for any $\lambda \in \mathbb{P}^1$, then V is equipped with a filtration $W(V)$ such that (i) (V, W) is a mixed twistor structure, (ii) W is a relative monodromy filtration of (V, L, N).*

Proof Let $M(N_{|\lambda}; L_{|\lambda})$ be the relative monodromy filtration of $(V, L, N)_{|\lambda}$. For each $k \in \mathbb{Z}$, the rank of $M_k(N_{|\lambda}; L_{|\lambda})$ is independent of $\lambda \in \mathbb{P}^1$. Then, we obtain that it depends on λ continuously, from Deligne's inductive formula (6.2) and (6.3). The property (i) follows from the canonical decomposition $\mathrm{Gr}^W \simeq \mathrm{Gr}^W \mathrm{Gr}^L$. Thus, we obtain Lemma 8.1.14. □

According to the lemma, we have only to prove that $(V, L, N)_{|\lambda} \in \mathrm{Vect}_{\mathbb{C}}^{\mathrm{RMF}}$ for any $\lambda \in \mathbb{P}^1$. We use an induction on the length of the filtration L. We assume that (i) $V = L_k(V)$, (ii) the claim holds for $L_{k-1}(V)$, and we shall prove that the claim holds for V. We consider the morphisms (6.5) for $(V, L, N)_{|\lambda}$ $(\lambda \in \mathbb{P}^1)$. The assumption implies that (6.5) for $\lambda \in U$ vanishes. Note that the both hand sides for $\lambda \in \mathbb{P}^1$ in (6.5) give vector bundles on $\mathbb{P}^1$. Hence, by the continuity, we obtain the vanishing of (6.5) for any $\lambda \in \mathbb{P}^1$. Thus, Proposition 8.1.13 is proved. □

Corollary 8.1.15 *Suppose that X is connected. Assume that, for a point $P \in X$, there exists a subset $U_P \subset \mathbb{C}_\lambda$ such that (i) $|U_P| = \infty$, (ii) $(\mathcal{M}_1, L, N)_{|(\lambda,P)} \in$* $\mathrm{Vect}_{\mathbb{C}}^{\mathrm{RMF}}$ *for any $\lambda \in U_P$. Then, $\mathcal{T}$ is equipped with a filtration $W(\mathcal{T})$ such that (i) $(\mathcal{T}, W)$ is a mixed twistor structure, (ii) W is a relative monodromy filtration of $\big((\mathcal{T}, W), L, N\big)$ in* MTS(X). □

The following lemma can be proved similarly and more easily.

Lemma 8.1.16 *Let $\big((\mathcal{T}, W), L, N\big) \in$* $\mathrm{MTS}(X)^{\mathrm{fil,nil}}$. *Assume that, for each $P \in X$, there exists $U_P \in \mathbb{C}_\lambda$ such that (i) $|U_P| = \infty$, (ii) $(\mathcal{M}_1, L, N)_{|(\lambda,P)} \in$* $\mathrm{Vect}_{\mathbb{C}}^{\mathrm{RMF}}$. *Then, there exists a relative monodromy filtration $M(N; L)$ in* MTS(X). □

8.2 Polarizable Mixed Twistor Structure

8.2.1 Statements

Let X be a complex manifold. We consider an abelian category $\mathcal{A} = \mathrm{MTS}(X)$ with additive auto equivalences $\Sigma^{p,q}(\mathcal{T}) = \mathcal{T} \otimes \mathcal{U}(-p, q)$. Then, for any finite set Λ, we obtain the abelian category $\mathrm{MTS}(X, \Lambda) := \mathrm{MTS}(X)(\Lambda)$ as in Sect. 6.1.1. For an object $(\mathcal{T}, W, \boldsymbol{N}) \in \mathrm{MTS}(X, \Lambda)$, we set $N(\Lambda) := \sum_{j \in \Lambda} N_j$.

An object $(\mathcal{T}, W, \boldsymbol{N}) \in \mathrm{MTS}(X, \Lambda)$ is called a (w, Λ)-polarizable mixed twistor structure, if (1) $W = M\big(N(\Lambda)\big)[w]$, (2) there exists a Hermitian sesqui-linear duality $\mathcal{S}$ of weight w on $\mathcal{T}$ such that (1) $\mathcal{S} \circ N_i = -N_i^* \circ \mathcal{S}$, (2) $(N(\Lambda)^*)^\ell \circ \mathcal{S}$ induces a polarization of $P\,\mathrm{Gr}_\ell^W(\mathcal{T})$. Such $\mathcal{S}$ is called a polarization of $(\mathcal{T}, W, \boldsymbol{N})$. Let $\mathcal{P}(X, w, \Lambda) \subset \mathrm{MTS}(X, \Lambda)$ denote the full subcategory of (w, Λ)-polarizable mixed twistor structure on X. The following proposition is essentially proved in [52], on the basis of the results in the Hodge case in [7–9, 35]. We will give an indication in Sect. 8.2.2.

Proposition 8.2.1 *The family of the categories $\mathcal{P}(X, w, \Lambda)$ satisfies the property* **P0–3** *in* Sect. 6.3.1.

8.2.1.1 Decomposition

We state some complementary property. First, we give remarks on the ambiguity of polarizations.

Lemma 8.2.2 *Let $(\mathcal{T}, W, \boldsymbol{N}) \in \mathcal{P}(X, w, \Lambda)$. If it is simple, i.e., there is no non-trivial subobject, then a polarization of $(\mathcal{T}, W, \boldsymbol{N})$ is unique up to constant multiplication.*

Proof Let S_i $(i = 1, 2)$ be polarizations of $(\mathcal{T}, W, \boldsymbol{N})$. They induce an endomorphism of $(\mathcal{T}, W, \boldsymbol{N})$ in $\mathcal{P}(X, w, \Lambda)$. Because $(\mathcal{T}, W, \boldsymbol{N})$ is simple, it is a scalar multiplication, which implies $S_1 = \alpha S_2$ for some $\alpha \in \mathbb{C}$. Because they are polarizations, we obtain α is a positive number. □

According to Proposition 8.2.1, $(\mathcal{T}, W, \boldsymbol{N}) \in \mathcal{P}(X, w, \Lambda)$ has a canonical decomposition

$$(\mathcal{T}, W, \boldsymbol{N}) \simeq \bigoplus_i (\mathcal{T}_i, W_i, \boldsymbol{N}_i) \otimes U_i, \tag{8.1}$$

where (1) $(\mathcal{T}_i, W_i, \boldsymbol{N}_i) \not\simeq (\mathcal{T}_j, W_j, \boldsymbol{N}_j)$ for $i \neq j$, (2) each $(\mathcal{T}_i, W_i, \boldsymbol{N}_i)$ is irreducible, (3) U_i are vector spaces. (We regard a vector space as a constant pure twistor structure of weight 0 on X.) We take a polarization S_i of each $(\mathcal{T}_i, W_i, \boldsymbol{N}_i)$, which is unique up to positive multiplication. We argue the following proposition in Sect. 8.2.3.

Proposition 8.2.3 *Any polarization of $(\mathcal{T}, W, \boldsymbol{N})$ is of the form $\bigoplus S_i \otimes h_i$, where h_i are Hermitian metrics of U_i.*

8.2.1.2 Some Operations

Let $(\mathcal{T}^{(i)}, W^{(i)}, \boldsymbol{N}^{(i)}) \in \mathcal{P}(X, w_i, \Lambda)$ $(i = 1, 2)$. We have the induced filtration $\tilde{W}$ on $\tilde{\mathcal{T}} := \mathcal{T}^{(1)} \otimes \mathcal{T}^{(2)}$, and $(\tilde{\mathcal{T}}, \tilde{W})$ is a mixed twistor structure given as in Sect. 6.1.4. We have the induced morphisms $\tilde{N}_j := N_j^{(1)} \otimes \mathrm{id} + \mathrm{id} \otimes N_j^{(2)}$ for $j \in \Lambda$. Let $\mathcal{S}^{(i)}$ be polarizations of $(\mathcal{T}^{(i)}, W^{(i)}, \boldsymbol{N}^{(i)})$. We have the induced Hermitian sesqui-linear duality $\mathcal{S}^{(1)} \otimes \mathcal{S}^{(2)}$ on $\mathcal{T}^{(1)} \otimes \mathcal{T}^{(2)}$. We prove the following proposition in Sect. 8.2.4.

Lemma 8.2.4 *$\mathcal{S}^{(1)} \otimes \mathcal{S}^{(2)}$ is a polarization of*

$$(\mathcal{T}^{(1)}, W^{(1)}, \boldsymbol{N}^{(1)}) \otimes (\mathcal{T}^{(2)}, W^{(2)}, \boldsymbol{N}^{(2)}) := (\tilde{\mathcal{T}}, \tilde{W}, \widetilde{\boldsymbol{N}}).$$

In particular, $(\mathcal{T}^{(1)}, W^{(1)}, \boldsymbol{N}^{(1)}) \otimes (\mathcal{T}^{(2)}, W^{(2)}, \boldsymbol{N}^{(2)})$ is an object in $\mathcal{P}(X, w_1 + w_2, \Lambda)$.

We also have the induced filtration $\hat{W}$ on $\hat{\mathcal{T}} = \mathcal{H}om(\mathcal{T}_1, \mathcal{T}_2)$, and $(\hat{\mathcal{T}}, \hat{W})$ is a mixed twistor structure. We have the induced morphisms $\hat{N}_j$ $(j \in \Lambda)$ given by $\hat{N}_j(f) = N_j^{(2)} \circ f - f \circ N_j^{(1)}$. We have the Hermitian sesqui-linear duality $\widehat{\mathcal{S}}$ of weight

$w_2 - w_1$ on $\mathcal{H}om(\mathcal{T}^{(1)}, \mathcal{T}^{(2)})$ induced by $\mathcal{S}^{(i)}$. We prove the following proposition in Sect. 8.2.4.

Lemma 8.2.5 *$\widehat{\mathcal{S}}$ is a polarization of*

$$\mathcal{H}om\big((\mathcal{T}^{(1)}, W^{(1)}, \boldsymbol{N}^{(1)}),\ (\mathcal{T}^{(2)}, W^{(2)}, \boldsymbol{N}^{(2)})\big) := (\hat{\mathcal{T}}, \hat{W}, \widehat{\boldsymbol{N}}).$$

In particular, $\mathcal{H}om\big((\mathcal{T}^{(1)}, W^{(1)}, \boldsymbol{N}^{(1)}),\ (\mathcal{T}^{(2)}, W^{(2)}, \boldsymbol{N}^{(2)})\big)$ is an object in $\mathcal{P}(X, w_2 - w_1, \Lambda)$.

In particular, for $(\mathcal{T}, W, \boldsymbol{N}) \in \mathcal{P}(X, w, \Lambda)$, we have the induced filtration $\overline{W}$ on $\overline{\mathcal{T}} := \mathcal{T}^{\vee}$, and the induced morphisms $\overline{N}_j := -N_j^{\vee}$. Then, $(\overline{\mathcal{T}}, \overline{W}, \overline{\boldsymbol{N}}) \in \mathcal{P}(X, -w, \Lambda)$. It is denoted by $(\mathcal{T}, W, \boldsymbol{N})^{\vee}$.

The operations j^* and $\tilde{\gamma}_{sm}^*$ on $\mathrm{TS}(X)$ naturally induce operations on $\mathrm{MTS}(X, \Lambda)$, and they preserve the subcategories $\mathcal{P}(X, \Lambda, w)$.

8.2.2 *Proof of Proposition 8.2.1*

The property **P1** clearly holds. Let us prove **P2**. We have only to consider the case $|\Lambda| = 1$ and $X = \{P\}$. Let $(\mathcal{T}_i, W, N) \in \mathcal{P}(w_i, \Lambda)$ with $w_1 > w_2$. A morphism $F : (\mathcal{T}_1, W, N) \longrightarrow (\mathcal{T}_2, W, N)$ induces $\mathrm{TNIL}(F) : \mathrm{TNIL}(\mathcal{T}_1, N) \longrightarrow \mathrm{TNIL}(\mathcal{T}_2, N)$ on $\Delta^*(R)$ for any $R > 0$. (See Sect. 2.2.1 for TNIL.) If R is sufficiently small, $\mathrm{TNIL}(\mathcal{T}_i, N)$ are pure of weight w_i (see [56]), and hence we have $\mathrm{TNIL}(F) = 0$. It implies $F = 0$. Hence, **P2** holds.

8.2.2.1 The Property P0

We prepare some lemmas.

Lemma 8.2.6 *Let* $(\mathcal{T}_i, W, \boldsymbol{N}) \in \mathrm{MTS}(X, \Lambda)$ $(i = 1, 2)$. *The direct sum $\bigoplus_{i=1,2}(\mathcal{T}_i, W, \boldsymbol{N})$ is an object in $\mathcal{P}(X, w, \Lambda)$ if and only if both $(\mathcal{T}_i, W, \boldsymbol{N})$ $(i = 1, 2)$ are objects in $\mathcal{P}(X, w, \Lambda)$*

Proof If S_i are polarizations of $(\mathcal{T}_i, W, \boldsymbol{N})$, $S_1 \oplus S_2$ is a polarization of $\bigoplus(\mathcal{T}_i, W, \boldsymbol{N})$. If S is a polarization of $\bigoplus(\mathcal{T}_i, W, \boldsymbol{N})$, we have the induced Hermitian sesqui-linear duality S_i of $\mathcal{T}_i$ $(i = 1, 2)$, given by $\mathcal{T}_i \longrightarrow \bigoplus_{j=1,2} \mathcal{T}_j \longrightarrow \bigoplus \mathcal{T}_j^* \otimes \boldsymbol{T}(-w) \longrightarrow \mathcal{T}_i^* \otimes \boldsymbol{T}(-w)$. It is easy to check that S_i are polarizations of $(\mathcal{T}_i, W, \boldsymbol{N})$. □

Let $(\mathcal{T}, W, \boldsymbol{N}, S)$ be a (w, Λ)-polarized mixed twistor structure. Suppose that we are given a subobject $(\mathcal{T}', W', \boldsymbol{N}') \subset (\mathcal{T}, W, \boldsymbol{N})$ in the category $\mathrm{MTS}(X, \Lambda)$. We assume that the monodromy weight filtration $M\big(N'(\Lambda)\big)$ on $\mathcal{T}'$ satisfies $W' =$

$M\big(N'(\Lambda)\big)[w]$. Let $\mathcal{T}''$ be the kernel of the composite of the following morphisms:

$$\mathcal{T} \xrightarrow{\ S\ } \mathcal{T}^* \otimes \boldsymbol{T}(-w) \longrightarrow (\mathcal{T}')^* \otimes \boldsymbol{T}(-w)$$

It induces a subobject $(\mathcal{T}'', W'', \boldsymbol{N}) \subset (\mathcal{T}, W, \boldsymbol{N})$ in the category $\mathrm{MTS}(X, \Lambda)$.

Lemma 8.2.7 *We have $\mathcal{T}' \cap \mathcal{T}'' = 0$. Namely, we have a decomposition which is orthogonal with respect to S:*

$$(\mathcal{T}, W, \boldsymbol{N}) = (\mathcal{T}', W', \boldsymbol{N}') \oplus (\mathcal{T}'', W'', \boldsymbol{N}'')$$

In particular, $(\mathcal{T}', W', \boldsymbol{N}')$ and $(\mathcal{T}'', W'', \boldsymbol{N}'')$ are polarizable.

Proof We have only to consider the case $|\Lambda| = 1$ and $X = \{P\}$. We have the induced morphism $\mathrm{TNIL}(\mathcal{T}', N') \subset \mathrm{TNIL}(\mathcal{T}, N)$ on $\Delta^*(R)$. If R is sufficiently small, both $\mathrm{TNIL}(\mathcal{T}, N)$ and $\mathrm{TNIL}(\mathcal{T}', N')$ are pure of weight w, and the pairing S induces a polarization of $\mathrm{TNIL}(\mathcal{T}, N)$. We have the orthogonal decomposition $\mathrm{TNIL}(\mathcal{T}, N) = \mathrm{TNIL}(\mathcal{T}', N') \oplus \mathrm{TNIL}(\mathcal{T}', N')^{\perp}$ as in Lemma 8.1.5. We obtain $\mathcal{T}' \cap \mathcal{T}'' = 0$. □

Lemma 8.2.8 *The category $\mathcal{P}(X, w, \Lambda)$ is abelian and semisimple, i.e.,* **P0** *holds.*

Proof Let $F : (\mathcal{T}_1, W, \boldsymbol{N}) \longrightarrow (\mathcal{T}_2, W, \boldsymbol{N})$ be a morphism in $\mathcal{P}(X, w, \Lambda)$. We have the kernel, the image and the cokernel in $\mathrm{MTS}(X, \Lambda)$, denoted by $(\mathrm{Ker}\, F, W, \boldsymbol{N})$, $(\mathrm{Im}\, F, W, \boldsymbol{N})$, and $(\mathrm{Cok}\, F, W, \boldsymbol{N})$. Let us prove that they are (w, Λ)-polarizable mixed twistor structure. It is easy to prove that the filtrations W on $\mathrm{Ker}\, F$, $\mathrm{Im}\, F$ and $\mathrm{Cok}\, F$ are equal to $M\big(N(\Lambda)\big)[w]$. By Lemma 8.2.7, they are (w, Λ)-polarizable mixed twistor structures. Hence, $\mathcal{P}(X, w, \Lambda)$ is abelian.

Let $(\mathcal{T}, W, \boldsymbol{N}) \in \mathcal{P}(X, w, \Lambda)$, and let $(\mathcal{T}', W', \boldsymbol{N}') \subset (\mathcal{T}, W, \boldsymbol{N})$ be a subobject in $\mathcal{P}(X, w, \Lambda)$. By Lemma 8.2.7, we have a decomposition $(\mathcal{T}, W, \boldsymbol{N}) = (\mathcal{T}', W', \boldsymbol{N}') \oplus (\mathcal{T}'', W'', \boldsymbol{N}'')$ in $\mathcal{P}(X, w, \Lambda)$. Hence, $\mathcal{P}(X, w, \Lambda)$ is semisimple. □

8.2.2.2 Property P3

Let $(\mathcal{T}, W, \boldsymbol{N}) \in \mathcal{P}(X, w, \Lambda)$. For $I \subset \Lambda$ and $\boldsymbol{a} \in \mathbb{R}^I_{>0}$, we put $N(\boldsymbol{a}) := \sum_{i \in I} a_i N_i$.

Lemma 8.2.9

- *The filtrations $M\big(N(\boldsymbol{a})\big)$ are independent of $\boldsymbol{a} \in \mathbb{R}^I_{>0}$. It is denoted by $M(I)$.*
- *Let $I \subset J \subset \Lambda$. For any $\boldsymbol{a} \in \mathbb{R}^J_{>0}$, the relative monodromy filtration $M\big(N(\boldsymbol{a}); M(I)\big)$ exists, and it is equal to $M(J)$.*

In particular, **P3.1** *holds.*

Proof We have only to consider the case that X is a point. The claims are known in the Hodge case [7]. The twistor case can be easily reduced to the Hodge case. (See §3 of [52].) □

Take $\bullet \in \Lambda$, and put $\Lambda_0 := \Lambda \setminus \bullet$. We put $(\mathcal{T}_k^{(1)}, W^{(1)}) := \mathrm{Gr}_k^{M(N_\bullet)}(\mathcal{T}, W)$. Let $\boldsymbol{N}^{(1)}$ denote the tuple of morphisms $N_j^{(1)} : \mathcal{T}_k^{(1)} \longrightarrow \mathcal{T}_k^{(1)} \otimes \boldsymbol{T}(-1)$ induced by N_j $(j \in \Lambda \setminus \bullet)$.

Lemma 8.2.10 $\left(\mathcal{T}_k^{(1)}, W^{(1)}, \boldsymbol{N}^{(1)}\right) \in \mathcal{P}(X, w+k, \Lambda_0)$. *A polarization is naturally induced by* $N_\bullet$ *and a polarization* $\mathcal{S}$ *of* $(\mathcal{T}, W, \boldsymbol{N})$. *On the primitive part, it is induced as* $(N_\bullet^*)^k \mathcal{S}$.

Proof We have only to consider the case that X is a point. By considering Gr^W, we can reduce the issue to the Hodge case, where the claim is known by the work due to Cattani-Kaplan-Schmid and Kashiwara-Kawai. (See [8, 9, 35].) It is also easy to apply their argument in our case. □

Let $(\mathcal{T}^{(2)}, W^{(2)}, \boldsymbol{N}^{(2)})$ denote the image of $N_\bullet : (\mathcal{T}, W, \boldsymbol{N}) \longrightarrow (\mathcal{T}, W, \boldsymbol{N}) \otimes \boldsymbol{T}(-1)$ in the category MTS(X, Λ). Let $\mathcal{S}$ be a polarization of $(\mathcal{T}, W, \boldsymbol{N})$. It is easy to observe that the composite $\mathrm{Im}(N_\bullet) \longrightarrow \mathcal{T} \otimes \boldsymbol{T}(-1) \xrightarrow{\mathcal{S}} \mathcal{T}^* \otimes \boldsymbol{T}(-w-1)$ factors through $\mathrm{Im}(N_\bullet)^* \otimes \boldsymbol{T}(-w-1)$, by using $N_\bullet \circ \mathcal{S} = -N_\bullet^* \circ \mathcal{S}$. Namely, $\mathcal{S}$ and $N_\bullet$ induce a sesqui-linear duality $\tilde{\mathcal{S}}$ of $\mathcal{T}^{(2)}$.

Lemma 8.2.11 (Proposition 3.126 [52]) $(\mathcal{T}^{(2)}, W^{(2)}, \boldsymbol{N}^{(2)}) \in \mathcal{P}(X, w+1, \Lambda)$, *and* $\tilde{\mathcal{S}}$ *is a polarization of* $(\mathcal{T}^{(2)}, W^{(2)}, \boldsymbol{N}^{(2)})$. □

We set $\mathcal{T}^{(3)} := \Sigma^{1,0}\mathcal{T}^{(2)}$. We have naturally induced morphisms

$$\Sigma^{1,0}(\mathcal{T}) \xrightarrow{\ u\ } \mathcal{T}^{(3)} \xrightarrow{\ v\ } \Sigma^{0,-1}(\mathcal{T})$$

Lemma 8.2.12 *Let* $* \in \Lambda_0$. *We have the decomposition*

$$\mathrm{Gr}^{M(N_*)}(\mathcal{T}^{(3)}) = \mathrm{Im}\,\mathrm{Gr}^{M(N_*)}(u) \oplus \mathrm{Ker}\,\mathrm{Gr}^{M(N_*)}(v).$$

Proof It follows from Proposition 3.134 of [52]. □

We obtain **P3.2** by an inductive use of Lemma 8.2.10 with Lemma 8.2.9. We obtain **P3.3** from Lemmas 8.2.11 and 8.2.12. Thus, the proof of Proposition 8.2.1 is finished. □

8.2.3 *Proof of Proposition 8.2.3*

The restriction of S to $(\mathcal{T}_i \otimes U_i) \otimes \sigma^*\left(\mathcal{T}_j \otimes U_j\right)$ induces a morphism

$$(\mathcal{T}_i, W_i, \boldsymbol{N}_i) \otimes U_i \longrightarrow (\mathcal{T}_j, W_j, \boldsymbol{N}_j) \otimes U_j$$

in $\mathcal{P}(w, \Lambda)$. It has to be 0 if $i \neq j$. Hence, (8.1) is orthogonal with respect to S. By using Lemma 8.2.7, we obtain that a polarization S of $(\mathcal{T}, W, \boldsymbol{N})$ is of the form $S = \bigoplus S_i \otimes h_i$. □

8.2.4 Proof of Lemmas 8.2.4 and 8.2.5

It is enough to consider the case $X = \{P\}$ and $|\Lambda| = 1$. Let $(\mathcal{T}_i, W, N) \in \mathcal{P}(w_i, \Lambda)$ $(i = 1, 2)$. We have the following natural isomorphisms of smooth $\mathcal{R}_{\mathbb{C}^*}$-triples:

$$\mathrm{TNIL}\big((\mathcal{T}_1, N) \otimes (\mathcal{T}_2, N)\big) \simeq \mathrm{TNIL}(\mathcal{T}_1, N) \otimes \mathrm{TNIL}(\mathcal{T}_2, N) \tag{8.2}$$

$$\mathrm{TNIL}\Big(\mathcal{H}om((\mathcal{T}_1, N), (\mathcal{T}_2, N))\Big) \simeq \mathcal{H}om\Big(\mathrm{TNIL}(\mathcal{T}_1, N), \mathrm{TNIL}(\mathcal{T}_2, N)\Big) \tag{8.3}$$

Let $R > 0$ be sufficiently small. On $\Delta^*(R) = \big\{z \in \mathbb{C} \,\big|\, 0 < |z| < R\big\}$, $\mathrm{TNIL}(\mathcal{T}_i, N)$ is a polarizable variation of pure twistor structure, and the Hermitian sesqui-linear dualities induced by $\mathcal{S}_i$ are polarizations. Hence, we obtain the claim of the lemmas by Theorem 12.1 of [52]. □

8.3 Infinitesimal Mixed Twistor Modules

8.3.1 Definition

We have the category of filtered objects in $\mathrm{MTS}(X, \Lambda)$ which is denoted by $\mathrm{MTS}(X, \Lambda)^{\mathrm{fil}}$. We consider a twistor version of infinitesimal mixed Hodge modules introduced by Kashiwara [32].

Definition 8.3.1 Let $(\mathcal{T}, W, L, \boldsymbol{N}) \in \mathrm{MTS}(X, \Lambda)^{\mathrm{fil}}$.

- It is called a variation of Λ-pre-infinitesimal mixed twistor module on X, or simply a Λ-pre-IMTM on X, if $\mathrm{Gr}^L_w(\mathcal{T}, W, \boldsymbol{N})$ is a (w, Λ)-polarizable mixed twistor structure on X.
- It is called a variation of Λ-infinitesimal mixed twistor module on X, or simply Λ-IMTM on X, if moreover there exists a relative monodromy filtration $M(N_j; L)$ for any $j \in \Lambda$. □

If we do not have to distinguish Λ, we use "IMTM" instead of "Λ-IMTM". The full subcategory of Λ-IMTM (resp. Λ-pre-IMTM) in $\mathrm{MTS}(X, \Lambda)^{\mathrm{fil}}$ is denoted by $\mathcal{M}(X, \Lambda)$ (resp. $\mathcal{M}^{\mathrm{pre}}(X, \Lambda)$). Note the following lemma, which follows from Lemma 6.3.1 and Proposition 8.2.1.

Lemma 8.3.2 *$\mathcal{M}^{\rm pre}(X,\Lambda)$ is abelian. Any morphism in $\mathcal{M}^{\rm pre}(X,\Lambda)$ is strict with respect to the filtration L.* □

Remark 8.3.3 The definitions of IMTM and pre-IMTM are not given in a parallel way to those of infinitesimal mixed Hodge module (IMHM) and pre-IMHM in [32]. For pre-IMHM, the weight filtration W is given only for $\mathrm{Gr}^L(\mathcal{T})$. For IMHM, the existence of relative monodromy filtration $M\big(N(J);L\big)$ is assumed for each $J\subset\Lambda$. But, it was proved in Theorem 4.4.1 of [32] that, for a given pre-IMHM, if $M\big(N_j;L\big)$ exists for each $j\in\Lambda$, then $M\big(N(J);L\big)$ exists for each $J\subset\Lambda$. □

The following lemma is a weaker version of Proposition 8.5.1 below. For simplicity, assume that X is connected.

Lemma 8.3.4 *Let $(\mathcal{T},L,\boldsymbol{N})\in\mathrm{TS}(X,\Lambda)^{\rm fil}$, i.e., $(\mathcal{T},L,\boldsymbol{N})$ is a filtered object in $\mathrm{TS}(X,\Lambda)$. Assume the following:*

- $\mathrm{Gr}^L_w(\mathcal{T},\boldsymbol{N})\in\mathcal{P}(X,\Lambda,w)$.
- *For a point $P\in X$, there exists $U_P\subset\mathbb{C}_\lambda$ such that (i) $|U_P|=\infty$, (ii) for any $\lambda\in U_P$, $(\mathcal{M}_1,N(\Lambda),L)_{|(\lambda,P)}\in\mathrm{Vect}^{\rm RMF}_{\mathbb{C}}$ and $(\mathcal{M}_1,N_i,L)_{|(\lambda,P)}\in\mathrm{Vect}^{\rm RMF}_{\mathbb{C}}$ $(i\in\Lambda)$. Here, $\mathcal{M}_1$ is one of the underlying $\mathcal{R}_X$-modules.*

Then, there exists a relative monodromy filtration $W=M(N(\Lambda);L)$, and $(\mathcal{T},W,L,\boldsymbol{N})$ is an object in $\mathcal{M}(X,\Lambda)$.

Proof Applying Proposition 8.1.13 to $(\mathcal{T},L,N(\Lambda))$, we obtain the existence of $W=M\big(N(\Lambda);L\big)$, and we have $(\mathcal{T},W,\boldsymbol{N})\in\mathrm{MTS}(X,\Lambda)$. By applying Lemma 8.1.16 to $\big((\mathcal{T},W),N_i,L\big)$ $(i\in\Lambda)$, we obtain the existence of $M(N_i;L)$. Hence, $(\mathcal{T},W,L,\boldsymbol{N})\in\mathcal{M}(X,\Lambda)$. □

Corollary 8.3.5 *Let $(\mathcal{T},W,L,\boldsymbol{N})\in\mathrm{MTS}(X,\Lambda)^{\rm fil}$. Assume the following:*

- *For a point $P\in X$, there exists $U_P\subset\mathbb{C}_\lambda$ such that (i) $|U_P|=\infty$, (ii) for any $\lambda\in U_P$, $(\mathcal{M}_1,N_i,L)_{|(\lambda,P)}\in\mathrm{Vect}^{\rm RMF}_{\mathbb{C}}$ $(i\in\Lambda)$.*

Then, $(\mathcal{T},W,L,\boldsymbol{N})$ is an object in $\mathcal{M}(X,\Lambda)$. □

8.3.2 Statements

We state some basic property of IMTM. We will prove the following theorem in Sects. 8.3.4–8.3.9.

Theorem 8.3.6 *The categories $\mathcal{M}(X,\Lambda)$ have the property* **M0–3** *in* Sect. 6.3.1.

We give some complement on operations on the categories. Let us consider $(\mathcal{T}^{(i)},W,L,\boldsymbol{N}^{(i)})\in\mathrm{MTS}(X,\Lambda)^{\rm fil}$ $(i=1,2)$. We have the naturally induced filtrations W and L, and tuples of morphisms $\boldsymbol{N}$ on $\mathcal{T}^{(1)}\otimes\mathcal{T}^{(2)}$ and $\mathcal{H}om(\mathcal{T}^{(1)},\mathcal{T}^{(2)})$, given as in Sect. 6.1.4. The induced tuples are denoted by

$$\begin{gathered}\big(\mathcal{T}^{(1)},W,L,\boldsymbol{N}^{(1)}\big)\otimes\big(\mathcal{T}^{(2)},W,L,\boldsymbol{N}^{(2)}\big),\\ \mathcal{H}om\Big(\big(\mathcal{T}^{(1)},W,L,\boldsymbol{N}^{(1)}\big),\big(\mathcal{T}^{(2)},W,L,\boldsymbol{N}^{(2)}\big)\Big).\end{gathered}\tag{8.4}$$

They are objects in $\mathrm{MTS}(X,\Lambda)^{\mathrm{fil}}$.

Proposition 8.3.7 *Suppose that* $(\mathcal{T}^{(1)},W,L,\boldsymbol{N}^{(i)})$ *are objects in* $\mathcal{M}(X,\Lambda)$ *(resp.* $\mathcal{M}^{\mathrm{pre}}(X,\Lambda)$*). Then, the objects in* (8.4) *are also in* $\mathcal{M}(X,\Lambda)$ *(resp.* $\mathcal{M}^{\mathrm{pre}}(X,\Lambda)$*). In particular, we have the duality functor on* $\mathcal{M}(X,\Lambda)$ *and* $\mathcal{M}^{\mathrm{pre}}(X,\Lambda)$.

Proof The claims for Λ-pre-IMTM are clear from Lemmas 8.2.4 and 8.2.5. To prove the claims for Λ-IMTM, we have only to care the existence of relative monodromy filtrations. It follows from a result due to Steenbrink-Zucker (see Proposition 6.1.5). □

Let $(\mathcal{T},W,L,\boldsymbol{N})\in\mathrm{MTS}(X,\Lambda)^{\mathrm{fil}}$. The dual $\mathcal{T}^{\vee}$ is naturally equipped with the filtrations W and L and tuple of nilpotent morphisms $\boldsymbol{N}$. The induced object in $\mathrm{MTS}(X,\Lambda)^{\mathrm{fil}}$ is denoted by $(\mathcal{T},W,L,\boldsymbol{N})^{\vee}$. We define $W_j(\mathcal{T}^*):=(\mathcal{T}/W_{-j-1}\mathcal{T})^*$ and $L_j(\mathcal{T}^*):=(\mathcal{T}/L_{-j-1}\mathcal{T})^*$. We have the naturally induced tuple of morphisms $\boldsymbol{N}$ on $\mathcal{T}^*$. The induced tuple in $\mathrm{MTS}(X,\Lambda)^{\mathrm{fil}}$ is denoted by $(\mathcal{T},W,L,\boldsymbol{N})^*$. We also naturally have $j^*(\mathcal{T},W,L,\boldsymbol{N})$ in $\mathrm{MTS}(X,\Lambda)^{\mathrm{fil}}$. By the composition, we obtain $\tilde{\gamma}^*_{sm}(\mathcal{T},W,L,\boldsymbol{N})$ in $\mathrm{MTS}(X,\Lambda)^{\mathrm{fil}}$. They give the functors the dual $\vee$, the Hermitian adjoint, j^*, and $\tilde{\gamma}^*_{sm}$ on $\mathrm{MTS}(X,\Lambda)^{\mathrm{fil}}$. The following proposition is clear.

Proposition 8.3.8 *The functors preserve* $\mathcal{M}(X,\Lambda)$ *and* $\mathcal{M}^{\mathrm{pre}}(X,\Lambda)$. □

8.3.3 Canonical Filtrations

We reword the construction and the results in Sects. 6.2.4 and 6.3.10. We consider $(\mathcal{T},W,L,\boldsymbol{N}_{\Lambda})\in\mathcal{M}(X,\Lambda)$. Take an element $\bullet\in\Lambda$, and put $\Lambda_0:=\Lambda\setminus\bullet$. Let $\tilde{L}:=M(N_{\bullet};L)$. By considering the morphisms

$$\Sigma^{1,0}(\mathcal{T},W,\tilde{L},\boldsymbol{N}_{\Lambda_0})\xrightarrow{N_\bullet}\Sigma^{0,-1}(\mathcal{T},W,\tilde{L},\boldsymbol{N}_{\Lambda_0})\xrightarrow{\mathrm{id}}\Sigma^{0,-1}(\mathcal{T},W,\tilde{L},\boldsymbol{N}_{\Lambda_0})$$

we obtain the filtration $\hat{N}_{\bullet *}L$ of $\Sigma^{0,-1}(\mathcal{T},W,\tilde{L},\boldsymbol{N}_{\Lambda_0})$ in the category of $\mathcal{M}(X,\Lambda_0)$. Similarly, by considering the morphisms

$$\Sigma^{1,0}(\mathcal{T},W,\tilde{L},\boldsymbol{N}_{\Lambda_0})\xrightarrow{\mathrm{id}}\Sigma^{1,0}(\mathcal{T},W,\tilde{L},\boldsymbol{N}_{\Lambda_0})\xrightarrow{N_\bullet}\Sigma^{0,-1}(\mathcal{T},W,\tilde{L},\boldsymbol{N}_{\Lambda_0})$$

we obtain the filtration $\hat{N}_{\bullet !}L$ of $\Sigma^{1,0}(\mathcal{T},W,\tilde{L},\boldsymbol{N}_{\Lambda_0})$ in the category of $\mathcal{M}(X,\Lambda_0)$. By **M3** of Theorem 8.3.6,

$$\big(\Sigma^{0,-1}(\mathcal{T},W),\hat{N}_{\bullet *}L,\boldsymbol{N}_{\Lambda}\big)\ \text{and}\ \big(\Sigma^{1,0}(\mathcal{T},W),\hat{N}_{\bullet !}L,\boldsymbol{N}_{\Lambda}\big)$$

are Λ-IMTM. In particular,

$$(\mathcal{T}, W, N_{\bullet *}L, \boldsymbol{N}_\Lambda) := \Sigma^{0,1}\big(\Sigma^{0,-1}(\mathcal{T}, W), \hat{N}_{\bullet *}L, \boldsymbol{N}_\Lambda\big)$$
$$(\mathcal{T}, W, N_{\bullet !}L, \boldsymbol{N}_\Lambda) := \Sigma^{-1,0}\big(\Sigma^{1,0}(\mathcal{T}, W), \hat{N}_{\bullet !}L, \boldsymbol{N}_\Lambda\big)$$

are Λ-IMTM.

Remark 8.3.9 We can also deduce that they are Λ-IMTM by using the reduction to the Hodge case using Gr^W. The Hodge case was proved in [32]. □

We obtain the following as a special case of Proposition 6.3.23.

Proposition 8.3.10 *Let $(\mathcal{T}, W, L, \boldsymbol{N}) \in \mathcal{M}(X, \Lambda)$. Let $i, j \in \Lambda$ be distinct elements. Let $\star(i), \star(j) \in \{*, !\}$. Then, we have $N_{i\,\star(i)}(N_{j\,\star(j)}L) = N_{j\,\star(j)}(N_{i\,\star(i)}L)$.* □

We use the notation $\boldsymbol{N}_{J*}\boldsymbol{N}_{I!}L$ ($I \cap J = \emptyset$) in the meaning as in Sect. 6.3.10.

8.3.3.1 Canonical Prolongations

We repeat the construction in Sect. 6.3.11 for this situation. We have the category $ML\big(\mathrm{MTS}(X, \Lambda)\big)$ as in Sect. 6.3.6. Let $(\mathcal{T}, W, L, \boldsymbol{N}) \in \mathcal{M}(X, \Lambda)$. For a decomposition $K_1 \sqcup K_2 = \Lambda$, we have an object $\mathcal{T}[*K_1!K_2] \in ML\big(\mathrm{MTS}(X, \Lambda)\big)$ given as follows. For $I \subset \Lambda$, we set $I_j := I \cap K_j$. Then, we put $\mathcal{T}[*K_1!K_2]_I := \Sigma^{|I_2|,-|I_1|}\mathcal{T}$. It is equipped with a filtration $\hat{\boldsymbol{N}}_{I_1!}\hat{\boldsymbol{N}}_{I_2*}L$. For $I \sqcup \{i\} \subset \Lambda$, morphisms $g_{Ii,I}$ and $f_{I,Ii}$ are given as follows:

$$g_{Ii,I} := \begin{cases} \mathrm{id} & (i \in \Lambda_2) \\ N_i & (i \in \Lambda_1) \end{cases} \qquad f_{I,Ii} := \begin{cases} N_i & (i \in \Lambda_2) \\ \mathrm{id} & (i \in \Lambda_1) \end{cases}$$

8.3.4 Property M2.2

The claims for **M1** and **M2.1** are clear by definition. Let us consider **M2.2**.

Proposition 8.3.11 *Let $(\mathcal{T}, W, L, \boldsymbol{N}) \in \mathcal{M}(X, \Lambda)$. For any subset $J \subset \Lambda$, we put $N(J) := \sum_{j \in J} N_j$.*

- *There exists a relative monodromy filtration $M\big(N(J); L\big)$. We denote it by $M(J; L)$.*
- *Let $I \subset J \subset \Lambda$. Then, $M\big(J; L\big) = M\Big(N(J); M\big(I; L\big)\Big)$.*

Note that $M(I; L)$ is also a relative monodromy filtration of $\sum_{i \in I} a_i N_i$ ($\boldsymbol{a} \in \mathbb{R}^I_{>0}$).

Proof In the Hodge case, it was proved in [32]. The twistor case can be reduced to the Hodge case, by considering $\mathrm{Gr}^W(\mathcal{T})$ and using Lemma 8.1.11. □

We take $\bullet \in \Lambda$, and put $\Lambda_0 := \Lambda \setminus \bullet$.

Lemma 8.3.12 *Let $(\mathcal{T}, W, L, \boldsymbol{N})$ be a Λ-pre-IMTM. Assume that there exists a relative monodromy filtration $M = M(N_\bullet; L)$. Then, $\big(\mathcal{T}, W, M, \boldsymbol{N}_{\Lambda_0}\big)$ is Λ_0-pre-IMTM, where $\boldsymbol{N}_{\Lambda_0} = \big(N_i \,\big|\, i \in \Lambda_0\big)$.*

If $(\mathcal{T}, W, L, \boldsymbol{N})$ is a Λ-IMTM, then $(\mathcal{T}, W, M, \boldsymbol{N}_{\Lambda_0})$ is a Λ_0-IMTM.

Proof The first claim follows from the canonical splitting $\mathrm{Gr}^M \simeq \mathrm{Gr}^M \mathrm{Gr}^L$ and Lemma 8.2.10. The second claim follows from Proposition 8.3.11. □

Let $(\mathcal{T}, W, L, \boldsymbol{N}) \in \mathcal{M}(X, \Lambda)$. For a decomposition $\Lambda = \Lambda_0 \sqcup \Lambda_1$, we obtain an object in $\mathrm{MTS}(X, \Lambda_0)^{\mathrm{fil}}$:

$$\mathrm{res}^{\Lambda}_{\Lambda_0}\big(\mathcal{T}, W, L, \boldsymbol{N}\big) := \big(\mathcal{T}, W, M(\Lambda_1; L), \boldsymbol{N}_{\Lambda_0}\big)$$

Here, $\boldsymbol{N}_{\Lambda_0} := \big(N_j \,\big|\, j \in \Lambda_0\big)$. We obtain the following corollary by an inductive use of Lemma 8.3.12 with Proposition 8.3.11.

Corollary 8.3.13 $\mathrm{res}^{\Lambda}_{\Lambda_0}(\mathcal{T}, W, L, \boldsymbol{N})$ *is a Λ_0-IMTM on X.* □

Thus, we have proved the claim for **M2.2**. For any $I \subset \Lambda$, we obtain a functor $\mathrm{res}^{\Lambda}_I : \mathcal{M}(X, \Lambda) \longrightarrow \mathcal{M}(X, I)$. We naturally have $\mathrm{res}^I_J \circ \mathrm{res}^{\Lambda}_I = \mathrm{res}^{\Lambda}_J$.

8.3.5 Property M0

We consider **M0** for $\mathcal{M}(X, \Lambda)$. It is clear that (1) any injection $\Phi : \Lambda \longrightarrow \Lambda_1$ induces $\mathcal{M}(X, \Lambda) \longrightarrow \mathcal{M}(X, \Lambda_1)$, (2) we naturally have $\mathcal{P}(X, w, \Lambda) \subset \mathcal{M}(X, \Lambda)$.

Let us prove that $\mathcal{M}(X, \Lambda)$ is an abelian category. Let $F : (\mathcal{T}, W, L, \boldsymbol{N}) \longrightarrow (\mathcal{T}', W', L', \boldsymbol{N}')$ be a morphism in $\mathcal{M}(X, \Lambda)$. According to Lemma 8.3.2, we have $(\mathrm{Ker}\, F, W, L, \boldsymbol{N})$, $(\mathrm{Im}\, F, W, L, \boldsymbol{N})$, $(\mathrm{Cok}\, F, W, L, \boldsymbol{N})$ in $\mathcal{M}^{\mathrm{pre}}(\Lambda)$. It remains to prove that there exist relative monodromy filtrations $M(N_j; L)$ on them for any $j \in \Lambda$. For that purpose, we have only to prove that F is strict with respect to $M(N_j; L)$. Fix $j \in \Lambda$, and we put $\Lambda_0 := \Lambda \setminus \{j\}$. Let $\tilde{\boldsymbol{N}} = (N_i \mid i \in \Lambda_0)$ and $\tilde{\boldsymbol{N}}' := (N'_i \mid i \in \Lambda_0)$. Because

$$\big(\mathcal{T}, W, M(N_j; L), \tilde{\boldsymbol{N}}\big) \longrightarrow \big(\mathcal{T}', W, M(N'_j; L'), \tilde{\boldsymbol{N}}'\big)$$

is a morphism in $\mathcal{M}^{\mathrm{pre}}(X, \Lambda_0)$, we have the desired strictness. Thus, we proved the claim for **M0**.

8.3.6 Property M3

We state the property **M3** in this situation. We will prove it in Sects. 8.3.7–8.3.9. Let Λ be a finite set. Fix an element $\bullet \in \Lambda$, and we put $\Lambda_0 := \Lambda \setminus \bullet$. Let $(\mathcal{T}, W, L, \boldsymbol{N}_\Lambda) \in \mathcal{M}(X, \Lambda)$. We have the induced object $(\mathcal{T}, W, \tilde{L}, \boldsymbol{N}_{\Lambda_0}) := \mathrm{res}^{\Lambda}_{\Lambda_0}(\mathcal{T}, W, L, \boldsymbol{N}_\Lambda)$ in $\mathcal{M}(X, \Lambda_0)$. We consider $(\mathcal{T}', W, \tilde{L}, \boldsymbol{N}'_{\Lambda_0}) \in \mathcal{M}(X, \Lambda_0)$ with the following morphisms

$$\Sigma^{1,0}(\mathcal{T}, W, \widetilde{L}, \boldsymbol{N}_{\Lambda_0}) \xrightarrow{\ u\ } (\mathcal{T}', W, \widetilde{L}, \boldsymbol{N}'_{\Lambda_0}) \xrightarrow{\ v\ } \Sigma^{0,-1}(\mathcal{T}, W, \widetilde{L}, \boldsymbol{N}_{\Lambda_0})$$

in $\mathcal{M}(X, \Lambda_0)$, such that $v \circ u = N_\bullet$. We set $N'_\bullet := u \circ v$, and the induced tuple $\boldsymbol{N}'_{\Lambda_0} \sqcup \{N'_\bullet\}$ is denoted by $\boldsymbol{N}'_\Lambda$. According to Corollary 6.2.7, we have a filtration L of $\mathcal{T}'$ in $\mathcal{M}(X, \Lambda_0)$, obtained as the transfer of $L(\mathcal{T})$ by (u, v). We will prove the following proposition.

Proposition 8.3.14 *$(\mathcal{T}', W, L, \boldsymbol{N}'_\Lambda)$ is a Λ-IMTM, i.e., $\mathcal{M}(X, \Lambda)$ have the property **M3**. The relative monodromy filtrations $M(N'_j; L)$ $(j \in \Lambda_0)$ are obtained as the transfer of $M(N_j; L)$ by (u, v).*

8.3.7 Transfer for Pre-IMTM

As a preparation, let us address a similar issue for pre-IMTM. We consider objects $(\mathcal{T}, W, L, \boldsymbol{N}_\Lambda) \in \mathcal{M}^{\mathrm{pre}}(X, \Lambda)$ and $(\mathcal{T}', W, \tilde{L}, \boldsymbol{N}'_{\Lambda_0}) \in \mathcal{M}^{\mathrm{pre}}(X, \Lambda_0)$ with morphisms in $\mathcal{M}^{\mathrm{pre}}(X.\Lambda_0)$,

$$\Sigma^{1,0}(\mathcal{T}, W, \widetilde{L}, \boldsymbol{N}_{\Lambda_0}) \xrightarrow{\ u\ } (\mathcal{T}', W, \widetilde{L}, \boldsymbol{N}'_{\Lambda_0}) \xrightarrow{\ v\ } \Sigma^{0,-1}(\mathcal{T}, W, \widetilde{L}, \boldsymbol{N}_{\Lambda_0}) \tag{8.5}$$

such that $v \circ u = N_\bullet$. We have a unique filtration L of $(\mathcal{T}', W, \tilde{L}, \boldsymbol{N}'_{\Lambda_0})$ in $\mathcal{M}^{\mathrm{pre}}(X, \Lambda_0)$ obtained as the transfer of $L(\mathcal{T})$ by (u, v). We set $N'_\bullet := u \circ v$, and the tuple $\boldsymbol{N}'_{\Lambda_0} \sqcup \{N'_\bullet\}$ is denoted by $\boldsymbol{N}'_\Lambda$.

Lemma 8.3.15 $(\mathcal{T}', W, L, \boldsymbol{N}'_\Lambda) \in \mathcal{M}^{\mathrm{pre}}(X, \Lambda)$.

Proof Because $\mathrm{Gr}^L_k(N'_\bullet) = 0$ on the direct summand $\mathrm{Ker}\,\mathrm{Gr}^L_k(v)$, the induced filtration $\tilde{L}$ on $\mathrm{Ker}\,\mathrm{Gr}^L_k(v)$ is pure of weight k, i.e., $\mathrm{Gr}^{\tilde{L}}_j = 0$ unless $j = k$. Hence, $(\mathrm{Ker}\,\mathrm{Gr}^L_k(v), W, \boldsymbol{N}'_{\Lambda_0})$ is isomorphic to a direct summand of $\mathrm{Gr}^{\tilde{L}}_k \mathrm{Gr}^L_k(\mathcal{T}', W, \boldsymbol{N}'_{\Lambda_0})$. Thanks to the canonical splitting in Sect. 6.1.2, $\mathrm{Gr}^{\tilde{L}}_k \mathrm{Gr}^L_k(\mathcal{T}', W, \boldsymbol{N}'_{\Lambda_0})$ is isomorphic to a direct summand of $\mathrm{Gr}^{\tilde{L}}_k(\mathcal{T}', W, \boldsymbol{N}'_{\Lambda_0})$ naturally. Hence, $(\mathrm{Ker}\,\mathrm{Gr}^L_k(v), W, \boldsymbol{N}'_{\Lambda_0})$ is an object in $\mathcal{P}(X, k, \Lambda_0)$.

For the morphism $N_\bullet : \mathcal{T} \longrightarrow \boldsymbol{T}^{-1}\mathcal{T}$, $\bigl(\operatorname{Im}\operatorname{Gr}^L_k(N_\bullet), W, \operatorname{Gr}^L_k \boldsymbol{N}_\Lambda\bigr)$ is an object in $\mathcal{P}(X, k+1, \Lambda)$, according to **P3.3** in Proposition 8.2.1. Because $\operatorname{Gr}^L_k(v)$ induces the following isomorphism

$$\bigl(\operatorname{Im}\operatorname{Gr}^L_k(u),\ W,\ \operatorname{Gr}^L_k \boldsymbol{N}'_\Lambda\bigr) \simeq \Sigma^{1,0}\bigl(\operatorname{Im}\operatorname{Gr}^L_k(N_\bullet),\ W,\ \operatorname{Gr}^L_k \boldsymbol{N}_\Lambda\bigr),$$

we have $\bigl(\operatorname{Im}\operatorname{Gr}^L_k(u),\ W,\ \operatorname{Gr}^L_k \boldsymbol{N}'_\Lambda\bigr) \in \mathcal{P}(X, k, \Lambda)$. Hence, $(\mathcal{T}', W, L, \boldsymbol{N}'_\Lambda)$ is an object in $\mathcal{M}^{\mathrm{pre}}(X, \Lambda)$. □

For the proof of Proposition 8.3.14, it remains to prove the existence of relative monodromy filtrations $M\bigl(N'_j; L(\mathcal{T}')\bigr)$ for $j \in \Lambda_0$.

8.3.8 Existence of Relative Monodromy Filtration in a Special Case

Let $\underline{2} := \{1, 2\}$. Let $(\mathcal{T}', W, L, \boldsymbol{N}') \in \mathcal{M}^{\mathrm{pre}}(X, \underline{2})$ and $(\mathcal{T}, W, L, \boldsymbol{N}) \in \mathcal{M}(X, \underline{2})$. Assume that we are given morphisms in $\mathrm{MTS}(X)^{\mathrm{fil,nil}}$

$$\bigl(\Sigma^{1,0}(\mathcal{T}, W), L, N_1\bigr) \xrightarrow{\ u\ } (\mathcal{T}', W, L, N'_1) \xrightarrow{\ v\ } \bigl(\Sigma^{0,-1}(\mathcal{T}, W), L, N_1\bigr)$$

such that (i) $v \circ u = N_2$ and $u \circ v = N'_2$, (ii) $\bigl(\mathcal{T}, \mathcal{T}'; u, v; L\bigr)$ is filtered S-decomposable, (iii) $(\mathcal{T}', L, N'_2) \in \mathrm{MTS}(X)^{\mathrm{RMF}}$. By Proposition 6.2.3, u and v give

$$\Sigma^{1,0}\bigl(\mathcal{T}, W, M(N_2; L)\bigr) \longrightarrow \bigl(\mathcal{T}', W, M(N'_2; L)\bigr) \longrightarrow \Sigma^{0,-1}\bigl(\mathcal{T}, W, M(N_2; L)\bigr).$$

The following lemma is based on an argument in [73].

Lemma 8.3.16 *Under the assumption, $(\mathcal{T}', W, L, \boldsymbol{N}')$ is a $\underline{2}$-IMTM, namely, there exists a relative monodromy filtration $M(N'_1; L)$. Moreover, $M(N'_1; L)$ is obtained as the transfer of $M(N_1; L)$ by (u, v).*

Proof We put $W^{(1)} := M(N_1; L)$. Then, we have $(\mathcal{T}, W, W^{(1)}, N_2) \in \mathcal{M}(X, 1)$. We have a unique filtration $W^{(1)}$ of $\mathcal{T}'$ obtained as the transfer of $W^{(1)}(\mathcal{T})$ with respect to (u, v). It satisfies the following conditions:

(A1) $W(\mathcal{T}') = M(N'_2; W^{(1)})$.
(A2) $\bigl(\mathcal{T}, \mathcal{T}'; u, v; W^{(1)}\bigr)$ is filtered S-decomposable.

We shall prove that $W^{(1)}(\mathcal{T}')$ gives a relative monodromy filtration $M(N'_1; L)$.

Let us prove that $(\mathcal{T}', W, W^{(1)}, N'_2)$ is a 1-IMTM, i.e., $\bigl(\operatorname{Gr}^{W^{(1)}}_w \mathcal{T}', W, N'_2\bigr)$ is an object in $\mathcal{P}(X, w, 1)$. On $\operatorname{Ker}\operatorname{Gr}^{W^{(1)}}_w v$, the filtration W is pure of weight w by construction. According to **P3.3** in Proposition 8.2.1, $\operatorname{Im}\operatorname{Gr}^{W^{(1)}}_w u$ is a polarizable mixed twistor structure of weight w. Hence, $(\mathcal{T}', W^{(1)}, N'_2)$ is 1-IMTM.

For $m \leq k$, we put $(\mathcal{T}_{k,m}, W) := L_k(\mathcal{T}, W)/L_m(\mathcal{T}, W)$ and $(\mathcal{T}'_{k,m}, W) := L_k(\mathcal{T}', W)/L_m(\mathcal{T}', W)$ in the category MTS(X). They are equipped with the induced filtrations L and the induced tuple of morphisms $\boldsymbol{N}$ and $\boldsymbol{N}'$. Then, we have $(\mathcal{T}_{k,m}, W, L, \boldsymbol{N}) \in \mathcal{M}(X, \underline{2})$ and $(\mathcal{T}'_{k,m}, W, L, \boldsymbol{N}') \in \mathcal{M}^{\mathrm{pre}}(X, \underline{2})$. The induced morphisms $u : \Sigma^{1,0}\mathcal{T}_{k,m} \longrightarrow \mathcal{T}'_{k,m}$ and $v : \mathcal{T}'_{k,m} \longrightarrow \Sigma^{0,-1}\mathcal{T}_{k,m}$ satisfy the assumptions (i) and (ii). Hence, by the above argument, we obtain a filtration $W^{(1)}(\mathcal{T}'_{k,m})$ with which $(\mathcal{T}'_{k,m}, W, W^{(1)}, N'_2) \in \mathcal{M}(X, 1)$.

Let us consider the exact sequence

$$0 \longrightarrow L_{k-1}(\mathcal{T}', W) \longrightarrow L_k(\mathcal{T}', W) \longrightarrow \mathrm{Gr}^L_k(\mathcal{T}', W) \longrightarrow 0.$$

The arrows are morphisms in $\mathcal{M}(1)$. In particular, they are strict with respect to $W^{(1)}$.

Note that $M(N'_1)[k]$ on $\mathrm{Gr}^L_k(\mathcal{T}')$ satisfies the condition (A1) for $W^{(1)}$. It also satisfies (A2) according to **P3.3** in Proposition 8.2.1. Hence, the induced filtration $W^{(1)}$ on $\mathrm{Gr}^L_k(\mathcal{T}')$ is the same as $M(N'_1)[k]$. Then, we can conclude that $W^{(1)}$ gives $M(N_1; L)$ on $\mathcal{T}$. □

Remark 8.3.17 For any 2-IMTM $(\mathcal{T}, W, L, \boldsymbol{N})$, we have

$$N_{2\star}M(N_1; L) = M(N_1; N_{2\star}L)$$

for $\star = *, !$. Indeed, we can deduce it in our situation from Lemma 8.3.16. Alternatively, by considering Gr^W, we can also reduce it to the Hodge case proved in [32]. □

8.3.9 End of the Proof of Proposition 8.3.14

Let us return to the situation in Sect. 8.3.6. According to Lemma 8.3.15, we have only to prove the existence of $M(N'_j; L\mathcal{T}')$ for $j \in \Lambda_0$. We have only to consider the case that X is a point. Put $\Lambda_1 := \Lambda_0 \setminus \{j\}$. We set $\boldsymbol{N}_{\Lambda_1} := \big(N_i \,|\, i \in \Lambda_1\big)$ and $\boldsymbol{N}'_{\Lambda_1} := \big(N'_i \,\big|\, i \in \Lambda_1\big)$. We obtain smooth $\mathcal{R}$-triples $\mathrm{TNIL}_{\Lambda_1}(\mathcal{T}, \boldsymbol{N}_{\Lambda_1})$ and $\mathrm{TNIL}_{\Lambda_1}(\mathcal{T}', \boldsymbol{N}'_{\Lambda_1})$ on $(\mathbb{C}^*)^{\Lambda_1}$ with filtrations L. We take a point $\iota_P : \{P\} \longrightarrow (\mathbb{C}^*)^{\Lambda_1}$, which is sufficiently close to the origin $(0, \ldots, 0)$. We set $\mathcal{T}_P := \iota_P^*\,\mathrm{TNIL}_{\Lambda_1}(\mathcal{T}, \boldsymbol{N}_{\Lambda_1})$, which is equipped with the induced filtration L and the morphisms $N_j, N_\bullet : \mathcal{T}_P \longrightarrow \boldsymbol{T}^{-1}\mathcal{T}_P$. Similarly, we put $\mathcal{T}'_P := \iota_P^*\,\mathrm{TNIL}_{\Lambda_1}(\mathcal{T}', \boldsymbol{N}'_{\Lambda_1})$ equipped with the induced filtration L and the morphisms $N'_j, N'_\bullet : \mathcal{T}'_P \longrightarrow \boldsymbol{T}^{-1}\mathcal{T}'_P$. We can apply Lemma 8.3.16 to $(\mathcal{T}_P, L_P, N_j, N_\bullet)$ and $(\mathcal{T}'_P, L_P, N'_j, N'_\bullet)$, and we obtain a relative monodromy filtration $M\big(N'_j; L(\mathcal{T}'_P)\big)$. By construction of TNIL, we obtain a relative monodromy filtration $M\big(N'_j; L(\mathcal{T}')\big)$. Thus, the proof of Proposition 8.3.14 is finished. □

8.4 Nearby Cycle Functor Along a Monomial Function

8.4.1 Beilinson IMTM and Its Deformation

Recall the Beilinson triple in Sect. 2.3. We use the same symbol to denote the pull back via a morphism from X to a point. It is naturally equipped with the weight filtration W given by $W_k(\mathbb{I}^{a,b}) = \bigoplus_{-2i\leq k} \boldsymbol{T}(i)$, and $(\mathbb{I}^{a,b}, W)$ is a mixed twistor structure on X. The tuple $(\mathbb{I}^{a,b}, W, L, N_{\mathbb{I}})$ is a 1-IMTM on X, where $L = W$. For any $\boldsymbol{c} \in \mathbb{R}^{\Lambda}$, let $\boldsymbol{c} N_{\mathbb{I}}$ denote the tuple $\bigl(c_i N_{\mathbb{I}} \bigm| i \in \Lambda\bigr)$. We obtain $\bigl(\mathbb{I}^{a,b}, W, L, \boldsymbol{c} N_{\mathbb{I}}\bigr) \in \mathcal{M}(X, \Lambda)$. Let $\boldsymbol{\varphi} = (\varphi_i \,|\, i \in \Lambda)$ be a tuple of holomorphic functions on X. We set $\mathbb{I}^{a,b}_{\boldsymbol{c},\boldsymbol{\varphi}} := \mathrm{Def}_{\boldsymbol{\varphi}}\bigl(\mathbb{I}^{a,b}, W, L, \boldsymbol{c} N_{\mathbb{I}}\bigr)$ in $\mathcal{M}(X, \Lambda)$.

8.4.2 Statement

Note that we will omit to denote the weight filtration W of mixed twistor structure in this section. We consider $(\mathcal{T}, L, \boldsymbol{N}) \in \mathcal{M}(X, \Lambda)$ and $\boldsymbol{m} \in \mathbb{Z}^{\Lambda}_{>0}$. We obtain the following Λ-IMTM on X:

$$\bigl(\Pi^{a,b}_{\boldsymbol{m},\boldsymbol{\varphi}} \mathcal{T}, L, \tilde{\boldsymbol{N}}\bigr) := \bigl(\mathcal{T}, L, \boldsymbol{N}\bigr) \otimes \mathbb{I}^{a,b}_{\boldsymbol{m},\boldsymbol{\varphi}}$$

For any subset $I \subset \Lambda$, we put $\tilde{N}_I := \prod_{i\in I} \tilde{N}_i$, i.e., the composite of the morphisms $\tilde{N}_i$ $(i \in I)$. For $\star = *, !$, we have the filtrations $\tilde{\boldsymbol{N}}_{I\star} L$ on $\Pi^{a,b}_{\boldsymbol{m},\boldsymbol{\varphi}} \mathcal{T}$. (See Sect. 8.3.3.) Let M be a sufficiently large integer. We have an induced morphism of Λ-IMTM:

$$\tilde{N}_I : \Sigma^{|I|,0}\bigl(\Pi^{0,M}_{\boldsymbol{m},\boldsymbol{\varphi}} \mathcal{T},\, \tilde{\boldsymbol{N}}_{I!} L,\, \tilde{\boldsymbol{N}}\bigr) \longrightarrow \Sigma^{0,-|I|}\bigl(\Pi^{0,M}_{\boldsymbol{m},\boldsymbol{\varphi}} \mathcal{T},\, \tilde{\boldsymbol{N}}_{I*} L,\, \tilde{\boldsymbol{N}}\bigr)$$

The cokernel in $\mathcal{M}(X, \Lambda)$ is denoted by $\bigl(\psi^{(0)}_{\boldsymbol{m},\boldsymbol{\varphi}}(\mathcal{T})_I, \hat{L}, \tilde{\boldsymbol{N}}\bigr)$.

On the other hand, the filtration L of $(\mathcal{T}, \boldsymbol{N})$ naively induces a filtration on $\Pi^{a,b}_{\boldsymbol{m},\boldsymbol{\varphi}} \mathcal{T}$ given by $L_k\bigl(\Pi^{a,b}_{\boldsymbol{m},\boldsymbol{\varphi}} \mathcal{T}\bigr) = \Pi^{a,b}_{\boldsymbol{m},\boldsymbol{\varphi}} L_k(\mathcal{T})$ in $\mathcal{M}(X, \Lambda)$. It induces a filtration of $\psi^{(0)}_{\boldsymbol{m},\boldsymbol{\varphi}}(\mathcal{T})_I$ in $\mathcal{M}(X, \Lambda)$. They are also denoted by L.

The morphism $N_{\mathbb{I}} : \mathbb{I}^{a,b} \longrightarrow \boldsymbol{T}^{-1}\mathbb{I}^{a,b}$ naturally induces a morphism $\psi^{(0)}_{\boldsymbol{m},\boldsymbol{\varphi}}(\mathcal{T})_I \longrightarrow \boldsymbol{T}^{-1}\psi^{(0)}_{\boldsymbol{m},\boldsymbol{\varphi}}(\mathcal{T})_I$, which is also denoted by $N_{\mathbb{I}}$. We shall prove the following theorem.

Theorem 8.4.1 *We have $\hat{L} = M(N_{\mathbb{I}}; L)[1]$ on $\psi^{(0)}_{\boldsymbol{m},\boldsymbol{\varphi}}(\mathcal{T})_I$.*

Here, for a filtration F, we set $F[a]_j := F_{j-a}$. The Hodge version of this theorem appeared in [73]. We obtain the following as a special case.

Corollary 8.4.2 *We have $\hat{L} = M(N_{\mathbb{I}}; L)[1]$ on $\psi^{(0)}_{\boldsymbol{m},\boldsymbol{\varphi}}(\mathcal{T})_{\Lambda}$.* □

8.4.3 Variant

Let $K \subset \Lambda$ be any non-empty subset. Take $\boldsymbol{m} \in \mathbb{Z}^K_{>0}$. (Note that we considered $\boldsymbol{m} \in \mathbb{Z}^{\Lambda}_{>0}$ in Sect. 8.4.2.) We obtain $\big(\Pi^{a,b}_{\boldsymbol{m},\varphi}\mathcal{T}, L, \tilde{\boldsymbol{N}}\big) := \big(\mathcal{T}, L, \boldsymbol{N}\big) \otimes \mathbb{I}^{a,b}_{\boldsymbol{m},\varphi}$ in $\mathcal{M}(X,\Lambda)$. We have the induced morphism in $\mathcal{M}(X,\Lambda)$:

$$\tilde{N}_K : \Sigma^{|K|,0}\big(\Pi^{0,M}_{\boldsymbol{m},\varphi}\mathcal{T}, \tilde{\boldsymbol{N}}_{K!}L, \tilde{\boldsymbol{N}}\big) \longrightarrow \Sigma^{0,-|K|}\big(\Pi^{0,M}_{\boldsymbol{m},\varphi}\mathcal{T}, \tilde{\boldsymbol{N}}_{K*}L, \tilde{\boldsymbol{N}}\big)$$

On the cokernel $\psi^{(0)}_{\boldsymbol{m},\varphi}(\mathcal{T})_K$, we have the induced filtration $\hat{L}$ and the tuple $\tilde{\boldsymbol{N}}$ so that $\big(\psi^{(0)}_{\boldsymbol{m},\varphi}(\mathcal{T})_K, \hat{L}, \tilde{\boldsymbol{N}}\big)$ is a Λ-IMTM. It is also equipped with the naively induced filtration L. We have a naturally induced morphism $N_{\mathbb{I}} : \psi^{(0)}_{\boldsymbol{m},\varphi}(\mathcal{T})_K \longrightarrow \psi^{(0)}_{\boldsymbol{m},\varphi}(\mathcal{T})_K \otimes \boldsymbol{T}(-1)$ in $\mathcal{M}(X,\Lambda)$.

Corollary 8.4.3 *We have* $\hat{L} = M(N_{\mathbb{I}}; L)[1]$.

Proof By using TNIL_{K^c}, we can reduce it to Corollary 8.4.2. □

8.4.4 Reformulation

Let $I' := I \sqcup \{i\}$. We have the following commutative diagram:

$$\begin{array}{ccccc}
\Sigma^{|I|+1,0}\Pi^{a,b}_{\boldsymbol{m},\varphi}\mathcal{T} & \xrightarrow{=} & \Sigma^{|I'|,0}\Pi^{a,b}_{\boldsymbol{m},\varphi}\mathcal{T} & \xrightarrow{\widetilde{N}_i} & \Sigma^{|I|,-1}\Pi^{a,b}_{\boldsymbol{m},\varphi}\mathcal{T} \\
\widetilde{N}_I\big\downarrow & & \widetilde{N}_{I'}\big\downarrow & & \widetilde{N}_I\big\downarrow \\
\Sigma^{1,-|I|}\Pi^{a,b}_{\boldsymbol{m},\varphi}\mathcal{T} & \xrightarrow{\widetilde{N}_i} & \Sigma^{0,-|I'|}\Pi^{a,b}_{\boldsymbol{m},\varphi}\mathcal{T} & \xrightarrow{=} & \Sigma^{0,-|I|-1}\Pi^{a,b}_{\boldsymbol{m},\varphi}\mathcal{T}
\end{array}$$

Hence, we obtain naturally induced morphisms:

$$\mathrm{can} : \Sigma^{1,0}\psi^{(0)}_{\boldsymbol{m},\varphi}(\mathcal{T})_I \longrightarrow \psi^{(0)}_{\boldsymbol{m},\varphi}(\mathcal{T})_{I'} \qquad \mathrm{var} : \psi^{(0)}_{\boldsymbol{m},\varphi}(\mathcal{T})_{I'} \longrightarrow \Sigma^{0,-1}\psi^{(0)}_{\boldsymbol{m},\varphi}(\mathcal{T})_I$$

The tuple $\psi^{(0)}_{\boldsymbol{m},\varphi}(\mathcal{T}) := \big(\psi^{(0)}_{\boldsymbol{m},\varphi}(\mathcal{T})_I \,\big|\, I \subset \underline{\ell}\big)$ with the filtration $\hat{L}$ is an object in $ML\big(\mathrm{MTS}(X,\Lambda)\big)$. (See Sect. 6.3.6 for the category $ML\big(\mathrm{MTS}(X,\Lambda)\big)$.) Indeed, it is obtained as the cokernel of $(\Pi^{0,N}_{\boldsymbol{m},\varphi}\mathcal{T})_! \longrightarrow (\Pi^{0,N}_{\boldsymbol{m},\varphi}\mathcal{T})_*$ in $ML\big(\mathrm{MTS}(X,\Lambda)\big)$.

On the other hand, the filtration L naively induces a filtration of $\psi^{(0)}_{\boldsymbol{m},\varphi}(\mathcal{T})$ in $ML\big(\mathrm{MTS}(X,\Lambda)\big)$, which is also denoted by L. We have $N_{\mathbb{I}} : \psi^{(0)}_{\boldsymbol{m},\varphi}(\mathcal{T}) \longrightarrow \boldsymbol{T}^{-1}\psi^{(0)}_{\boldsymbol{m},\varphi}(\mathcal{T})$. The previous theorem can be reformulated as follows.

Theorem 8.4.4 *We have* $\hat{L} = M(N_{\mathbb{I}}; L)[1]$ *on* $\psi^{(0)}_{\boldsymbol{m},\varphi}(\mathcal{T})$.

8.4.5 Proof

We have only to consider the case that L is pure of weight w, i.e., $(\mathcal{T}, W, \boldsymbol{N}) \in \mathcal{P}(X, w, \Lambda)$. Moreover, we may assume $w = 0$. We have the weight filtration $M(N_{\mathbb{I}})$ on each $\psi^{(0)}_{\boldsymbol{m},\boldsymbol{\varphi}}(\mathcal{T})_I$, which is preserved by can and var. Hence, we obtain a filtration $M(N_{\mathbb{I}})$ on $\psi^{(0)}_{\boldsymbol{m},\boldsymbol{\varphi}}(\mathcal{T})$. We have only to prove that $M(N_{\mathbb{I}})[1] = \hat{L}$. We may assume that X is a point. We have only to consider the case that $\varphi_i = 0$ $(i \in \Lambda)$.

According to Corollary 3.132 of [52], $\big(\psi^{(0)}_{\boldsymbol{m}}(\mathcal{T})_I, (\boldsymbol{N}, N_{\mathbb{I}})\big)$ is a $(1, \Lambda \sqcup \{\bullet\})$-polarizable mixed twistor structure. Hence, $\big(\psi^{(0)}_{\boldsymbol{m}}(\mathcal{T})_I, M(N_{\mathbb{I}})[1], \tilde{\boldsymbol{N}}\big)$ is a Λ-IMTM. For $I' = I \sqcup \{i\}$, according to Lemma 8.2.12, the following is S-decomposable:

$$\Sigma^{1,0}\operatorname{Gr}^{M(N_{\mathbb{I}})}\psi^{(0)}_{\boldsymbol{m}}(\mathcal{T})_I \xrightarrow{u} \operatorname{Gr}^{M(N_{\mathbb{I}})}\psi^{(0)}_{\boldsymbol{m}}(\mathcal{T})_{I'} \xrightarrow{v} \Sigma^{0,-1}\operatorname{Gr}^{M(N_{\mathbb{I}})}\psi^{(0)}_{\boldsymbol{m}}(\mathcal{T})_I$$

Namely, $\big(\psi^{(0)}_{\boldsymbol{m}}(\mathcal{T}), M(N_{\mathbb{I}})[1]\big)$ is an object in $ML\big(\mathrm{MTS}(\Lambda)\big)$.

Let $I \subset \Lambda$. We have the filtrations $M\big(\tilde{N}(I); M(N_{\mathbb{I}})[1]\big)$ and $M(\tilde{N}(I); \hat{L})$ on $\psi^{(0)}_{\boldsymbol{m}}(\mathcal{T})_I$. To prove $M(N_{\mathbb{I}})[1] = \hat{L}$, we have only to prove $M\big(\tilde{N}(I); M(N_{\mathbb{I}})[1]\big) = M(\tilde{N}(I); \hat{L})$ on $\psi^{(0)}_{\boldsymbol{m}}(\mathcal{T})_I$ for any $I \subset \Lambda$, according to Theorem 6.3.18. If $I = \Lambda$, both the filtrations $M\big(\tilde{N}(\Lambda); M(N_{\mathbb{I}})[1]\big)$ and $M(\tilde{N}(\Lambda); \hat{L})$ are the weight filtration of the mixed twistor structure. In particular, they are the same. Let us consider the case $J := \Lambda \setminus I \neq \emptyset$. We take $Q \in (\mathbb{C}^*)^J$, which is sufficiently close to the origin. We have a $(0, I)$-polarizable mixed twistor structure $\mathrm{TNIL}_{J,Q}(\mathcal{T})$ with $\boldsymbol{N}_I := (N_i \in I)$. Let $\boldsymbol{m}_I := (m_i \,|\, i \in I)$. On $\psi^{(0)}_{\boldsymbol{m}_I}\big(\mathrm{TNIL}_{J,Q}(\mathcal{T})\big)_I$, we have $M\big(\tilde{N}(I); M(N_{\mathbb{I}})[1]\big) = M\big(\tilde{N}(I); \hat{L}\big)$. The underlying $\mathcal{R}$-modules of $\psi^{(0)}_{\boldsymbol{m}}(\mathcal{T})_I$ and $\psi^{(0)}_{\boldsymbol{m}_I}\big(\mathrm{TNIL}_{J,Q}(\mathcal{T})_I\big)$ are naturally isomorphic. Therefore, we have

$$M\big(\tilde{N}(I); M(N_{\mathbb{I}})[1]\big) = M(\tilde{N}(I); \hat{L})$$

on $\psi^{(0)}_{\boldsymbol{m}}(\mathcal{T})_I$, and the proof of Theorem 8.4.4 is finished. □

8.5 Twistor Version of a Theorem of Kashiwara

Let $(\mathcal{T}, L, \boldsymbol{N}) \in \mathrm{TS}(X, \Lambda)^{\mathrm{fil}}$. Let $\mathcal{M}_i$ $(i = 1, 2)$ denote the underlying $\mathcal{R}_X$-modules. We assume the following:

- For each w, we have $\operatorname{Gr}^L_w(\mathcal{T}, L, \boldsymbol{N}) \in \mathcal{P}(X, w, \Lambda)$.
- For each $P \in X$, there exists a subset $U_P \subset \mathbb{C}_\lambda$ such that (a) $|U_P| = \infty$, (b) for any $\lambda \in U_P$ and for $i \in \Lambda$, we have $(\mathcal{M}_1, L, N_i)_{|P} \in \mathrm{Vect}^{\mathrm{RMF}}_{\mathbb{C}}$.

The following proposition is a twistor version of Theorem 4.4.1 of [32].

Proposition 8.5.1 *$\big(\mathcal{T}, L, N(\Lambda)\big)$ has a relative monodromy filtration W, and $(\mathcal{T}, W, L, \boldsymbol{N})$ is a Λ-IMTM.*

8.5.1 A Purity Theorem (Special Case)

Let $(\mathcal{T}, W, L, \boldsymbol{N})$ be a $\underline{2}$-IMTM. For $i = 1, 2$, we have the morphisms

$$N_i : (\Sigma^{1,0}\mathcal{T}, L) \longrightarrow \Sigma^{0,-1}(\mathcal{T}, N_{i*}L).$$

They give morphisms $N_i : \Sigma^{1,0}L_k(\mathcal{T}) \longrightarrow \Sigma^{0,-1}(N_{i*}L_{k-1}\mathcal{T})$, namely $N_i : L_k(\mathcal{T}) \longrightarrow \boldsymbol{T}^{-1}N_{i*}L_{k-1}(\mathcal{T})$. Similarly, we have the morphisms $N_i : N_{j*}L_k(\mathcal{T}) \longrightarrow \boldsymbol{T}^{-1}N_{j*}N_{i*}L_{k-1}(\mathcal{T})$. Then, we obtain the following filtered complex Π in MTS(X):

$$L_{-1}\mathcal{T} \xrightarrow{N_1 \oplus N_2} \boldsymbol{T}^{-1}(N_{1*}L)_{-2}\mathcal{T} \oplus \boldsymbol{T}^{-1}(N_{2*}L)_{-2}\mathcal{T} \xrightarrow{N_2 - N_1} \boldsymbol{T}^{-2}(N_{1*}N_{2*}L)_{-3}\mathcal{T}$$

The cohomology group $H^i(\Pi)$ with the induced weight filtration W is a mixed twistor structure.

Lemma 8.5.2 $\mathrm{Gr}_j^W H^i(\Pi) = 0$ *unless* $j \leq i - 1$.

Proof We have $\mathrm{Gr}_j^W H^i(\Pi) \simeq H^i \mathrm{Gr}_j^W \Pi$. Then, the claim can be reduced to the Hodge case in [32]. □

Remark 8.5.3 The purity theorem can be proved in a more general situation as in the Hodge case. □

8.5.2 Proof of Proposition 8.5.1

We have only to consider the case that X is a point $\{P\}$. By using TNIL, we can reduce the issue to the case $\Lambda = \{1, 2\}$. We regard $(\mathcal{T}, L)$ as a filtered vector bundle on $\mathbb{P}^1$, denoted by (V, L). We may assume $V = L_0(V)$. By an inductive argument, we may assume that the claim holds for $(L_{-1}V, L, \boldsymbol{N})$. According to Lemma 8.3.4, we have only to prove the existence of relative monodromy filtration for $N_{1|\lambda} + N_{2|\lambda}$ on $(V, L)_{|\lambda}$ for any $\lambda \in U_P$. Then, we have only to apply the argument in Sect. 6 of [32], with Lemma 8.5.2 above. □

8.6 Integrable Case

8.6.1 Integrable Mixed Twistor Structure

Let X be a complex manifold. Let $(\mathcal{T}, W)$ be a filtered object in the category of integrable smooth $\mathcal{R}_X$-triples. It is called an integrable variation of mixed twistor structure on X, (or simply an integrable mixed twistor structure on X) if each $\mathrm{Gr}_w^W(\mathcal{T})$ is integrable polarizable pure twistor $\mathscr{D}$-module of weight w. It is called pure of weight w if $\mathrm{Gr}^L_m = 0$ unless $m = w$. The category of integrable mixed twistor structure on X is denoted by $\mathrm{MTS}^{\mathrm{int}}(X)$. It is an abelian category. It is equipped with tensor product and inner homomorphism. It is also equipped with additive auto equivalences $\Sigma^{p,q}$ given by the tensor product with $\mathcal{U}(-p,q)$.

8.6.2 *Integrable Polarizable Mixed Twistor Structure*

We consider an abelian category $\mathcal{A} = \mathrm{MTS}^{\mathrm{int}}(X)$ with additive auto equivalences $\Sigma^{p,q}(\mathcal{T}) = \mathcal{T} \otimes \mathcal{U}(-p,q)$. Then, for any finite set Λ, we obtain the abelian category $\mathrm{MTS}^{\mathrm{int}}(X,\Lambda) := \mathrm{MTS}^{\mathrm{int}}(X)(\Lambda)$ as in Sect. 6.1.1. An object $(\mathcal{T}, W, \boldsymbol{N}) \in \mathrm{MTS}^{\mathrm{int}}(X,\Lambda)$ is called an integrable (w,Λ)-polarizable mixed twistor structure if the following holds:

- We have $W = M\big(N(\Lambda)\big)[w]$.
- There exists an integrable Hermitian sesqui-linear duality $\mathcal{S} : \mathcal{T} \longrightarrow \mathcal{T}^* \otimes \boldsymbol{T}(-w)$ of weight w which gives a polarization of the underlying $(\mathcal{T}, W, \boldsymbol{N}) \in \mathrm{MTS}(X,\Lambda)$.

The full subcategory of integrable (w,Λ)-polarizable mixed twistor structure is denoted by $\mathcal{P}^{\mathrm{int}}(X,w,\Lambda)$. The following is an analogue of Proposition 6.1.4.

Proposition 8.6.1 *The categories* $\mathcal{P}^{\mathrm{int}}(X,w,\Lambda)$ *satisfy the property* **P0–3** *in* Sect. 6.3.1.

Proof As for **P0**, we have only to repeat the argument in the ordinary case. The other property in the integrable case can be reduced to those in the ordinary case. □

As in the case of $\mathcal{P}(X,\Lambda,w)$, any object in $\mathcal{P}^{\mathrm{int}}(X,\Lambda,w)$ has the canonical decomposition, and its polarization is unique up to automorphisms. The category $\mathcal{P}^{\mathrm{int}}(X,\Lambda,w)$ is equipped with tensor products, duality, j^* and $\tilde{\gamma}^*_{\mathrm{sm}}$.

8.6.3 *Infinitesimal Mixed Twistor Module*

We consider the category of filtered objects in $\mathrm{MTS}^{\mathrm{int}}(X,\Lambda)$, denoted by $\mathrm{MTS}^{\mathrm{int}}(X,\Lambda)^{\mathrm{fil}}$. Let $(\mathcal{T}, W, L, \boldsymbol{N}) \in \mathrm{MTS}^{\mathrm{int}}(X,\Lambda)^{\mathrm{fil}}$.

- It is called an integrable Λ-pre-IMTM on X if $\mathrm{Gr}^L_w(\mathcal{T}, W, \boldsymbol{N})$ is an integrable (w,Λ)-polarizable mixed twistor structure on X.
- It is called an integrable Λ-IMTM on X if moreover there exists a relative monodromy filtration $M(N_j; L)$ for any $j \in \Lambda$. In other words, $(\mathcal{T}, W, L, \boldsymbol{N})$ is an integrable Λ-IMTM if (a) it is an integrable pre-Λ-IMTM, (b) it is a Λ-IMTM.

Let $\mathcal{M}^{\mathrm{int}}(X,\Lambda)$ (resp. $\mathcal{M}^{\mathrm{int\,pre}}(X,\Lambda)$) denote the full subcategory of integrable Λ-IMTM (resp. integrable Λ-pre-IMTM) in $\mathrm{MTS}^{\mathrm{int}}(X,\Lambda)^{\mathrm{fil}}$. The following lemma is obvious.

Lemma 8.6.2 *$\mathcal{M}^{\mathrm{int\,pre}}(\Lambda)$ is abelian. Any morphism in $\mathcal{M}^{\mathrm{int\,pre}}(\Lambda)$ is strict with respect to the filtration L.* □

Proposition 8.6.3 *The family of the categories $\mathcal{M}(X,\Lambda)$ has the property* **M0–3** *in* Sect. 6.3.1.

Proof We give only an indication. Let us consider **M0**. It is clear that (i) any injection $\Phi:\Lambda\longrightarrow\Lambda_1$ induces $\mathcal{M}^{\mathrm{int}}(X,\Lambda)\longrightarrow\mathcal{M}^{\mathrm{int}}(X,\Lambda_1)$, (ii) we naturally have $\mathcal{P}^{\mathrm{int}}(X,w,\Lambda)\subset\mathcal{M}^{\mathrm{int}}(X,\Lambda)$. Let us prove that $\mathcal{M}^{\mathrm{int}}(X,\Lambda)$ is abelian. Let $F:(\mathcal{T},W,L,\boldsymbol{N})\longrightarrow(\mathcal{T}',W',L',\boldsymbol{N}')$ be a morphism in $\mathcal{M}^{\mathrm{int}}(X,\Lambda)$. We have the objects $(\operatorname{Ker}F,W,L,\boldsymbol{N})$, $(\operatorname{Im}F,W,L,\boldsymbol{N})$, $(\operatorname{Cok}F,W,L,\boldsymbol{N})$ in $\mathcal{M}(X,\Lambda)$. They are naturally integrable smooth filtered $\mathcal{R}_X$-triples, and objects in $\mathcal{M}^{\mathrm{int\,pre}}(X,\Lambda)$. Hence, they are naturally objects in $\mathcal{M}^{\mathrm{int}}(X,\Lambda)$. The claims for **M1** and **M2.1** are clear by definition. Let us consider **M2.2**. Let $(\mathcal{T},W,L,\boldsymbol{N})\in\mathcal{M}^{\mathrm{int}}(X,\Lambda)$. By **M2.2** for $\mathcal{M}(X,\Lambda)$, we have the filtration $M(\Lambda_1;L)$ and the object $\mathrm{res}^{\Lambda}_{\Lambda_0}(\mathcal{T},W,L,\boldsymbol{N})$ in $\mathcal{M}(X,\Lambda_0)$. We can prove that it is naturally an object in $\mathcal{M}^{\mathrm{int}}(X,\Lambda_0)$ by using Deligne's formula for relative monodromy filtration and Kashiwara's canonical decomposition, and we obtain **M2.2**. To argue **M3**, let us consider the situation in Sect. 8.3.6 with the integrability condition. Let $(\mathcal{T},W,L,\boldsymbol{N}_\Lambda)\in\mathcal{M}^{\mathrm{int}}(\Lambda)$. We have the induced object $\mathrm{res}^{\Lambda}_{\Lambda_0}(\mathcal{T},W,L,\boldsymbol{N}_\Lambda)=:(\mathcal{T},W,\tilde{L},\boldsymbol{N}_{\Lambda_0})$ in $\mathcal{M}^{\mathrm{int}}(X,\Lambda_0)$. Let $(\mathcal{T}',W,\tilde{L},\boldsymbol{N}'_{\Lambda_0})\in\mathcal{M}^{\mathrm{int}}(X,\Lambda_0)$ with morphisms as in (8.5) in $\mathcal{M}^{\mathrm{int}}(X,\Lambda_0)$, such that $v\circ u=N_\bullet$. We set $N'_\bullet:=u\circ v$, and the induced tuple $\boldsymbol{N}'_{\Lambda_0}\sqcup\{N'_\bullet\}$ is denoted by $\boldsymbol{N}'_\Lambda$. We have a filtration L of $\mathcal{T}'$ in $\mathcal{M}^{\mathrm{int}}(X,\Lambda_0)$, obtained as the transfer of $L(\mathcal{T})$ by (u,v). By **M3** for $\mathcal{M}(X,\Lambda)$, $(\mathcal{T}',W,L,\boldsymbol{N}'_\Lambda)$ is a Λ-IMTM. By repeating the argument for Lemma 8.3.15, we obtain that it is an integrable Λ-pre-IMTM, and hence it is an integrable Λ-IMTM. □

Remark 8.6.4 Although we do not give the statements of the integral version of Theorem 8.4.1 and Proposition 8.5.1, they can easily be reduced to the ordinary case. □

Chapter 9
Admissible Mixed Twistor Structures and Their Variants

In this chapter, we study admissible variations of mixed twistor structure and their variant. Let X be a complex manifold with a normal crossing hypersurface D. Let $\mathcal{V}$ be a smooth $\mathcal{R}_{X(*D)}$-triple with a filtration W such that $(\mathcal{V}, W)_{|X\setminus D}$ is a variation of mixed twistor structure. We are interested in the degeneration at any points of D. The relative monodromy filtration should control the weight filtration of the degenerated objects. If it is satisfied, $(\mathcal{V}, W)$ is called an admissible variation of mixed twistor structure.

As already mentioned, admissible variations of mixed twistor structure are building blocks of mixed twistor $\mathscr{D}$-modules. Many of basic properties of mixed twistor $\mathscr{D}$-modules are eventually reduced to the properties of admissible variations of mixed twistor structure.

The concept of admissible variation of mixed twistor structure is a twistor version of the concept of admissible variation of mixed Hodge structure, which was introduced by Steenbrink and Zucker [87] in the one dimensional case, and by Kashiwara [32] in the higher dimensional case.

We explain basic results on admissible variations of mixed twistor structure in Sect. 9.1.

We shall also consider variants of admissible variation of mixed twistor structure. Indeed, we study admissible variation of polarizable mixed twistor structure in Sect. 9.2, and admissible variation of infinitesimal mixed twistor module in Sect. 9.3. But, we should emphasize that they are just rather auxiliary objects for the argument in Chap. 10.

Let us mention why we discuss such objects. For simplicity, let us consider the case $X = \Delta^n$, $D_i = \{z_i = 0\}$ and $D = \bigcup_{i=1}^{\ell} D_i$. If we are given an admissible variation of mixed twistor structure on (X, D), we naturally obtain admissible variations of infinitesimal mixed twistor module on $(D_I, \partial D_I)$ for any $I \subset \underline{\ell}$. If the admissible variation of mixed twistor structure is pure, we obtain admissible variation of polarized mixed twistor structure. So, they are useful as intermediate objects when we consider gluing of admissible variation of mixed twistor structure on $(D_I, \partial D_I)$ $(I \subset \underline{\ell})$.

T. Mochizuki, *Mixed Twistor $\mathscr{D}$-modules*, Lecture Notes in Mathematics 2125,
DOI 10.1007/978-3-319-10088-3_9

9.1 Admissible Mixed Twistor Structure

9.1.1 *Mixed Twistor Structure on* (X, D)

Let X be a complex manifold, and D be a simply normal crossing hypersurface of X with the irreducible decomposition $D = \bigcup_{i \in \Lambda} D_i$. For $I \subset \Lambda$, we set $D_I := \bigcap_{i \in I} D_i$, $\partial D_I := \bigcup_{j \notin I}(D_I \cap D_j)$ and $D_I^\circ := D_I \setminus \partial D_I$. Let $\mathcal{I}$ be a good system of ramified irregular values on (X, D) (see Sect. 15.1). The tuple $(X, D, \mathcal{I})$ is denoted by $\boldsymbol{X}$.

An object $(\mathcal{T}, W) \in \mathrm{TS}(X, D)^{\mathrm{fil}}$ is called a variation of mixed twistor structure on (X, D) if the restriction $(\mathcal{T}, W)_{|X \setminus D}$ is a mixed twistor structure on $X \setminus D$. The full subcategory $\mathrm{MTS}(X, D) \subset \mathrm{TS}(X, D)^{\mathrm{fil}}$ is abelian. An object $(\mathcal{T}, W) \in \mathrm{MTS}(X, D)$ is called a variation of mixed twistor structure on $\boldsymbol{X}$ if $\mathcal{T}$ is $\mathcal{I}$-good. Let $\mathrm{MTS}(\boldsymbol{X}) \subset \mathrm{TS}(X, D)^{\mathrm{fil}}$ denote the full subcategory of variation of mixed twistor structure on $\boldsymbol{X}$. It is an abelian subcategory. We shall often omit "variation of".

9.1.2 *Pre-admissibility*

A mixed twistor structure $(\mathcal{T}, W)$ on $\boldsymbol{X}$ is called pre-admissible if the following holds:

(Adm0) For each w, $\mathrm{Gr}^W_w(\mathcal{T})$ is the canonical prolongment of an $\mathcal{I}$-good wild polarizable variation of pure twistor structure of weight w. (See §11.1 of [55].)

(Adm1) $\mathcal{T}$ is a $\mathcal{I}$-good-KMS smooth $\mathcal{R}_{X(*D)}$-triple, and the filtration W is compatible with the KMS-structure. (See Sect. 5.2 for the compatibility condition.)

We shall impose additional conditions for admissibility.

9.1.2.1 Specialization of Pre-admissible Mixed Twistor Structure

We give a remark on a specialization to the intersections D_I°. First let us consider the case $X = \Delta^n$, $D = \bigcup_{i=1}^{\ell}\{z_i = 0\}$. Let $(\mathcal{T}, W) \in \mathrm{MTS}(\boldsymbol{X})$ be pre-admissible. For $\boldsymbol{u} \in (\mathbb{R} \times \mathbb{C})^\ell$, we obtain a smooth $\mathcal{R}_{D_{\underline{\ell}}}$-triple ${}^{\underline{\ell}}\tilde{\psi}_{\boldsymbol{u}}(\mathcal{T})$ which is equipped with a tuple of morphisms $\boldsymbol{N}_{\underline{\ell}} = (N_i \,|\, i = 1, \ldots, \ell)$. (See Sect. 5.5.2 for ${}^{\underline{\ell}}\tilde{\psi}_{\boldsymbol{u}}(\mathcal{T})$.) Here

$$N_i : {}^{I}\tilde{\psi}_{\boldsymbol{u}}(\mathcal{T}) \longrightarrow {}^{I}\tilde{\psi}_{\boldsymbol{u}}(\mathcal{T}) \otimes \boldsymbol{T}(-1)$$

denote the canonically induced morphisms. We define

$$L_j\big({}^{\underline{\ell}}\tilde{\psi}_{\boldsymbol{u}}(\mathcal{T})\big) := {}^{\underline{\ell}}\tilde{\psi}_{\boldsymbol{u}}(W_j\mathcal{T}). \tag{9.1}$$

They give a filtration L of ${}^{\underline{\ell}}\tilde{\psi}_{\boldsymbol{u}}(\mathcal{T})$ in the category of $\mathrm{TS}(D_{\underline{\ell}})$ by **(Adm1)**. The filtration is preserved by each N_i. Thus, we obtain $\big({}^{\underline{\ell}}\tilde{\psi}_{\boldsymbol{u}}(\mathcal{T}), L, \boldsymbol{N}_{\underline{\ell}}\big)$ in $\mathrm{TS}(D_{\underline{\ell}}, \underline{\ell})^{\mathrm{fil}}$.

We have $\mathrm{Gr}^L_w \,{}^{\underline{\ell}}\tilde{\psi}_{\boldsymbol{u}}(\mathcal{T}) \simeq {}^{\underline{\ell}}\tilde{\psi}_{\boldsymbol{u}}\,\mathrm{Gr}^W_w(\mathcal{T})$. If $\mathcal{I}$ is unramified, we also have the induced object $\big({}^{\underline{\ell}}\tilde{\psi}_{\mathfrak{a},\boldsymbol{u}}(\mathcal{T}), L, \boldsymbol{N}_{\underline{\ell}}\big)$ in $\mathrm{TS}(D_{\underline{\ell}}, \underline{\ell})^{\mathrm{fil}}$ for $\boldsymbol{u} \in (\mathbb{R}\times\mathbb{C})^{\ell}$ and $\mathfrak{a} \in \mathrm{Irr}(\mathcal{T}, \underline{\ell})$, where L is given as in (9.1).

Let us consider the case in which $\boldsymbol{X} = (X, D, \mathcal{I})$ is general. Take $I \subset \Lambda$ and $P \in D_I^{\circ}$. We take a holomorphic coordinate neighbourhood $(X_P, z_1, \ldots, z_n)$ around P such that $X_P \cap D = \bigcup_{j=1}^{|I|}\{z_j = 0\}$. On $D_{I,P} := X_P \cap D_I$, for any $\boldsymbol{u} \in (\mathbb{R}\times\mathbb{C})^{|I|}$, we obtain objects $\big({}^{\underline{|I|}}\tilde{\psi}_{\boldsymbol{u}}(\mathcal{T}), L, \boldsymbol{N}_{\underline{|I|}}\big)$ in $\mathrm{TS}(D_{I,P}, I)^{\mathrm{fil}}$. If $\mathcal{I}$ is unramified around P, we have $\big({}^{\underline{|I|}}\tilde{\psi}_{\mathfrak{a},\boldsymbol{u}}(\mathcal{T}), L, \boldsymbol{N}_{\underline{|I|}}\big)$ for any $\boldsymbol{u} \in (\mathbb{R}\times\mathbb{C})^{I}$ and $\mathfrak{a} \in \mathcal{I}_P$. We denote them by $\big({}^{I}\tilde{\psi}_{\boldsymbol{u}}(\mathcal{T}), L, \boldsymbol{N}_I\big)_P$ and $\big({}^{I}\tilde{\psi}_{\mathfrak{a},\boldsymbol{u}}(\mathcal{T}), L, \boldsymbol{N}_I\big)_P$. For any $J \subset I$, we set $N(J) := \sum_{j\in J} N_j$.

9.1.3 Admissibility in the Smooth Divisor Case

Let us consider the case that D is smooth. Let $(\mathcal{T}, W) \in \mathrm{MTS}(\boldsymbol{X})$ be pre-admissible. Let $\mathcal{M}_i$ $(i = 1, 2)$ be the underlying $\mathcal{R}_{X(*D)}$-modules of $\mathcal{T}$. We have the induced bundles $\mathcal{G}_u\mathcal{M}_i$ $(u \in \mathbb{R}\times\mathbb{C})$ on $\mathcal{D}$ by taking Gr with respect to the KMS-structure. (See Sect. 5.1.2 for $\mathcal{G}_u(\mathcal{M}_i)$.) It is equipped with the endomorphism $\mathrm{Res}(\mathbb{D})$. Let N denote the nilpotent part. We define

$$L_j\mathcal{G}_u(\mathcal{M}_i) := \mathcal{G}_u(W_j\mathcal{M}_i).$$

They give a filtration L on $\mathcal{G}_u(\mathcal{M}_i)$ by **(Adm1)**. We say that $\mathcal{T}$ is admissible if moreover the following holds:

(Adm2) The nilpotent part of the residue N_1 on $\big(\mathcal{G}_u(\mathcal{M}_i), L\big)$ has a relative monodromy filtration.

The following lemma is a special case of Proposition 9.1.5 below.

Lemma 9.1.1 *Let $(\mathcal{T}, W) \in \mathrm{MTS}(\boldsymbol{X})$ be admissible. Assume $X = \Delta^n$ and $D = \{z_1 = 0\}$.*

- *For each $u \in \mathbb{R}\times\mathbb{C}$, N_1 on $(\tilde{\psi}_{z_1,u}(\mathcal{T}), L)$ has a relative monodromy filtration W, and $(\tilde{\psi}_{z_1,u}(\mathcal{T}), W, L, N_1)$ is a 1-IMTM on D.*
- *If $\mathcal{I}$ is unramified, $(\tilde{\psi}_{z_1,\mathfrak{a},u}(\mathcal{T}), L, N_1)$ has a relative monodromy filtration W for each $u \in \mathbb{R}\times\mathbb{C}$ and $\mathfrak{a} \in \mathcal{I}$, so that $\big(\tilde{\psi}_{z_1,\mathfrak{a},u}(\mathcal{T}), W, L, N_1\big)$ is a 1-IMTM on D.*

Proof We have only to consider the unramified case. By the compatibility of the filtration W and the KMS-structure, $\big(\mathrm{Gr}^L_w\,\tilde{\psi}_{z_1,\mathfrak{a},u}(\mathcal{T}), N_1\big)$ comes from $\mathrm{Gr}^W_w(\mathcal{T})$. Hence, it is a polarizable pure twistor structure. □

Lemma 9.1.2 *We can replace the condition* **(Adm2)** *with the following condition:*

(Adm2') *There exists a subset $U \subset \mathbb{C}_\lambda$ with $|U| = \infty$ such that each $\big(\mathcal{G}_u(\mathcal{M}_1), L, N\big)_{|(\lambda,P)}$ is an object in $\mathrm{Vect}^{\mathrm{RMF}}_{\mathbb{C}}$ for any $(\lambda, P) \in U \times D$.*

Proof It follows from Proposition 8.1.13. □

9.1.4 Admissibility in the Normal Crossing Case

Let X be a complex manifold, and let D be a simple normal crossing hypersurface of X.

Definition 9.1.3 Let $(\mathcal{T}, W) \in \mathrm{MTS}(\boldsymbol{X})$ be pre-admissible. It is called admissible if for any smooth point $P \in D$ there exists a small neighbourhood X_P of P such that $\mathcal{T}_{|X_P}$ is admissible in the sense of the smooth divisor case in Sect. 9.1.3. □

The following lemma is clear by definition.

Lemma 9.1.4 *If $\mathcal{T} \in \mathrm{TS}(X, D)$ comes from a good wild polarizable variation of pure twistor structure of weight w, it is naturally an admissible mixed twistor structure on (X, D) with the weight filtration pure of weight w.* □

We have the following proposition on the specialization of admissible mixed twistor structure. (See Sect. 9.1.2.1 for the specialization of pre-admissible mixed twistor structure.)

Proposition 9.1.5 *Let $(\mathcal{T}, W) \in \mathrm{MTS}(\boldsymbol{X})$ be admissible. Let $P \in D_I^{\circ}$. We take a holomorphic coordinate neighbourhood $(X_P; z_1, \dots, z_n)$ around P such that $D = \bigcup_{i=1}^{|I|}\{z_i = 0\}$.*

- *For any $\boldsymbol{u} \in (\mathbb{R}\times\mathbb{C})^I$, the object $\big({}^I\tilde{\psi}_{\boldsymbol{u}}(\mathcal{T}), L, N(I)\big)_P$ in $\mathrm{TS}(D_{I,P})^{\mathrm{fil,nil}}$ has a relative monodromy filtration W, and $\big({}^I\tilde{\psi}_{\boldsymbol{u}}(\mathcal{T}), W, L, \boldsymbol{N}_I\big)_P$ is an I-IMTM.*
- *Suppose that $\mathcal{T}$ is unramified around P. For any $\mathfrak{a} \in \mathrm{Irr}(\mathcal{T}, P)$ and for any $\boldsymbol{u} \in (\mathbb{R} \times \mathbb{C})^I$, the object $\big({}^I\tilde{\psi}_{\mathfrak{a},\boldsymbol{u}}(\mathcal{T}), L, N(I)\big)_P$ in $\mathrm{TS}(D_{I,P})^{\mathrm{fil,nil}}$ has a relative monodromy filtration W, and $\big({}^I\tilde{\psi}_{\mathfrak{a},\boldsymbol{u}}(\mathcal{T}), W, L, \boldsymbol{N}_I\big)_P$ is an I-IMTM.*

The objects $\big({}^I\tilde{\psi}_{\boldsymbol{u}}(\mathcal{T}), W, L, \boldsymbol{N}_I\big)_P$ and $\big({}^I\tilde{\psi}_{\mathfrak{a},\boldsymbol{u}}(\mathcal{T}), W, L, \boldsymbol{N}_I\big)_P$ in $\mathcal{M}(D_{I,P}, I)$ are denoted by ${}^I\tilde{\psi}_{\boldsymbol{u}}(\mathcal{T}, W)$ and ${}^I\tilde{\psi}_{\mathfrak{a},\boldsymbol{u}}(\mathcal{T}, W)$ if there is no risk of confusion.

Proof Let $\mathcal{M}$ denote an underlying $\mathcal{R}_{X(*D)}$-module. It is enough to prove the first claim in the case $X = \Delta^n$, $D = \bigcup_{i=1}^{\ell}\{z_i = 0\}$ and $I = \underline{\ell}$. We may assume that $\mathcal{I}$ is unramified. Moreover, we have only to consider the case $\mathfrak{a} = 0$. Let $\lambda_0 \in \mathbb{C}$ be generic, i.e., $\mathfrak{e}(\lambda_0) : \mathrm{KMS}(\mathcal{M}, j) \longrightarrow \mathbb{C}$ are injective for any j. Let $\boldsymbol{a} = \mathfrak{p}(\lambda_0, \boldsymbol{u})$. On $\mathcal{D}_i^{(\lambda_0)}$, we have the bundle ${}^i\mathcal{G}_{u_i}\mathcal{Q}_{\boldsymbol{a}}^{(\lambda_0)}(\mathcal{M})$. It is equipped with the induced filtration L given by $L_j{}^i\mathcal{G}_{u_i}\mathcal{Q}_{\boldsymbol{a}}^{(\lambda_0)}(\mathcal{M}) = {}^i\mathcal{G}_{u_i}\mathcal{Q}_{\boldsymbol{a}}^{(\lambda_0)}(W_j\mathcal{M})$, and the nilpotent endomorphism N_i obtained as the nilpotent part of $\mathrm{Res}_i(\mathbb{D})$. Let ∂D_i be the union of $D_i \cap D_j$ ($1 \leq j \leq \ell$, $j \neq i$). The induced morphism N_i' on ${}^i\mathcal{G}_{u_i}\mathcal{Q}_{\boldsymbol{a}}^{(\lambda_0)}(\mathcal{M})(*\partial\mathcal{D}_i^{(\lambda_0)})$ has a relative monodromy filtration with respect to the induced filtration L.

Let $\hat{\mathcal{D}}_{\underline{\ell}}^{(\lambda_0)} \cap \mathcal{D}_i^{(\lambda_0)} \subset \hat{\mathcal{D}}_{\underline{\ell}}^{(\lambda_0)}$ be determined by $z_i = 0$. We have the decomposition

$$
{}^i\mathcal{G}_{u_i}\mathcal{Q}_{\boldsymbol{a}}^{(\lambda_0)}(\mathcal{M})_{|\hat{\mathcal{D}}_{\underline{\ell}}^{(\lambda_0)}\cap\mathcal{D}_i^{(\lambda_0)}} = {}^i\mathcal{G}_{u_i}\mathcal{Q}_{\boldsymbol{a}}^{(\lambda_0)}(\mathcal{M}_{\mathcal{D}_{\underline{\ell}}}^{(\mathrm{reg})}) \oplus {}^i\mathcal{G}_{u_i}\mathcal{Q}_{\boldsymbol{a}}^{(\lambda_0)}(\mathcal{M}_{\mathcal{D}_{\underline{\ell}}}^{(\mathrm{irr})})
$$

as in Sect. 5.1.3. It is compatible with N_i and L. Hence, the induced endomorphism N_i on ${}^{i}\mathcal{G}_{u_i}\mathcal{Q}_{\boldsymbol{a}}^{(\lambda_0)}(\mathcal{M}_{\mathcal{D}_{\underline{\ell}}}^{(\text{reg})})(*\partial\mathcal{D}_i^{(\lambda_0)})$ has a relative monodromy filtration with respect to the induced filtration L.

The bundle ${}^{i}\mathcal{G}_{u_i}\mathcal{Q}_{\boldsymbol{a}}^{(\lambda_0)}(\mathcal{M}_{\mathcal{D}_{\underline{\ell}}}^{(\text{reg})})$ is naturally equipped with an induced family of meromorphic flat λ-connections ${}^{i}\mathbb{D}$ which is logarithmic with respect to $\partial\mathcal{D}_i^{(\lambda_0)}$, and N_i and L are ${}^{i}\mathbb{D}$-flat. Because λ_0 is generic, we obtain that N_i on ${}^{i}\mathcal{G}_{u_i}\mathcal{Q}_{\boldsymbol{a}}^{(\lambda_0)}(\mathcal{M}_{\mathcal{D}_{\underline{\ell}}}^{(\text{reg})})$ has a relative monodromy filtration with respect to L, and the filtration is given as an increasing sequence of subbundles.

We have the decomposition

$$
{}^{i}\mathcal{G}_{u_i}\mathcal{Q}_{\boldsymbol{a}}^{(\lambda_0)}(\mathcal{M}_{\mathcal{D}_{\underline{\ell}}}^{(\text{reg})})_{|(\lambda_0,P)} = \bigoplus_{\substack{\mathfrak{p}(\lambda_0,\boldsymbol{u}')=\boldsymbol{a}\\ u_i'=u_i}} {}^{\underline{\ell}}\psi_{\boldsymbol{u}'}(\mathcal{M})_{|(\lambda_0,P)},
$$

which is compatible with N_i and L. Hence, we obtain that N_i on ${}^{\underline{\ell}}\psi_{\boldsymbol{u}}(\mathcal{M})_{|(\lambda_0,P)}$ has a relative monodromy filtration with respect to L. Then, the claim of Proposition 9.1.5 follows from Proposition 8.5.1. □

9.1.4.1 Pull Back

We give a consequence on the pull back. Let $f : X' \longrightarrow X$ be a morphism of complex manifolds such that $D' := f^{-1}(D)$ is normal crossing. Let $f^{-1}\mathcal{I}$ denote a good system of ramified irregular values on (X', D') obtained as the pull back of $\mathcal{I}$. Let $\boldsymbol{X}' := (X', D', f^{-1}\mathcal{I})$.

Corollary 9.1.6 *For any admissible variation of mixed twistor structure* $(\mathcal{T}, W)$ *on* $\boldsymbol{X}$*, the pull back* $f^*(\mathcal{T}, W)$ *is an admissible variation of mixed twistor structure on* $\boldsymbol{X}'$.

Proof By using the functoriality of the deformation associated to a variation of irregular values with respect to the pull back (Sect. 4.5 of [55]), it is easy to see that $f^*(\mathcal{T}, W)$ satisfies the condition **(Adm0)**. We can check the other conditions by using Propositions 9.1.5 and 8.3.11. □

9.1.5 Category of Admissible MTS

Let $\mathrm{MTS}^{\mathrm{adm}}(\boldsymbol{X}) \subset \mathrm{MTS}(\boldsymbol{X})$ be the full subcategory of admissible mixed twistor structure on $\boldsymbol{X}$. It is equipped with additive auto equivalences $\Sigma^{p,q}$ given by the tensor product with $\mathcal{U}(-p, q)$.

Proposition 9.1.7 $\mathrm{MTS}^{\mathrm{adm}}(\boldsymbol{X})$ *is an abelian subcategory of* $\mathrm{MTS}(\boldsymbol{X})$.

Proof Let $F : (\mathcal{T}_1, W) \longrightarrow (\mathcal{T}_2, W)$ be any morphism in $\mathrm{MTS}^{\mathrm{adm}}(\boldsymbol{X})$. Note that F is strict with respect to W. We obtain the filtered $\mathcal{R}_{X(*D)}$-triples $(\mathrm{Ker}\, F, W)$, $(\mathrm{Im}\, F, W)$ and $(\mathrm{Cok}\, F, W)$. Let us prove that they are objects in $\mathrm{MTS}^{\mathrm{adm}}(\boldsymbol{X})$. If W is pure of weight w, it follows from the theory of polarizable wild pure twistor $\mathscr{D}$-modules. Indeed, we have the corresponding polarizable pure twistor $\mathscr{D}$-modules $\mathfrak{T}_i$ of weight w with morphisms $\tilde{F} : \mathfrak{T}_1 \longrightarrow \mathfrak{T}_2$. We have the polarizable pure twistor $\mathscr{D}$-modules $\mathrm{Ker}\, \tilde{F}$, $\mathrm{Im}\, \tilde{F}$ and $\mathrm{Cok}\, \tilde{F}$, which are smooth on $X \setminus D$. By the semisimplicity of the category of polarizable pure twistor $\mathscr{D}$-modules, $\mathrm{Ker}\, \tilde{F}$ and $\mathrm{Cok}\, \tilde{F}$ are direct summands of $\mathfrak{T}_1$ and $\mathfrak{T}_2$, respectively, and $\mathrm{Im}\, \tilde{F}$ is a direct summand of both of $\mathfrak{T}_i$ $(i = 1, 2)$. Because $\mathrm{Ker}\, F = \mathrm{Ker}\, \tilde{F}(*D)$, $\mathrm{Cok}\, F = \mathrm{Cok}\, \tilde{F}(*D)$, and $\mathrm{Im}\, F = \mathrm{Im}\, \tilde{F}(*D)$, they satisfy **(Adm0)** and **(Adm1)**. The condition **(Adm2)** is trivial.

Let us consider the mixed case. Because F is strict with respect to W, we have $\mathrm{Gr}^W \mathrm{Ker}\, F \simeq \mathrm{Ker}\, \mathrm{Gr}^W F$, $\mathrm{Gr}^W \mathrm{Im}\, F \simeq \mathrm{Im}\, \mathrm{Gr}^W F$ and $\mathrm{Gr}^W \mathrm{Cok}\, F \simeq \mathrm{Cok}\, \mathrm{Gr}^W F$. Hence, **(Adm0)** is satisfied for $\mathrm{Ker}\, F$, $\mathrm{Im}\, F$ and $\mathrm{Cok}\, F$. Let us check **(Adm1–2)**. By **(Adm0)**, we have already known that $\mathrm{Ker}\, F$, $\mathrm{Im}\, F$ and $\mathrm{Cok}\, F$ are smooth good $\mathcal{R}_{X(*D)}$-triples.

Let us check that they are $\boldsymbol{\mathcal{I}}$-good $\mathcal{R}_{X(*D)}$-triples. We have only to consider the unramified case. Let $\mathcal{M}_{i,c}$ $(c = 1, 2)$ be the underlying $\mathcal{R}_{X(*D)}$-modules of $\mathcal{T}_i$. We have the irregular decomposition

$$\mathcal{M}_{i,c|\widehat{(\lambda,P)}} = \bigoplus_{\mathfrak{a} \in \mathrm{Irr}(\mathcal{T}_i)} \hat{\mathcal{M}}_{i,c,\mathfrak{a}}.$$

Let $F_1 : \mathcal{M}_{2,1} \longrightarrow \mathcal{M}_{1,1}$ and $F_2 : \mathcal{M}_{1,2} \longrightarrow \mathcal{M}_{2,2}$ be the underlying morphisms of F. It is easy to see that $F_{c|\hat{\mathcal{D}}}$ is compatible with the decompositions. We have natural isomorphisms $\mathrm{Ker}(F_c)_{|\widehat{(\lambda,P)}} = \mathrm{Ker}(F_{c|\widehat{(\lambda,P)}})$, $\mathrm{Im}(F_c)_{|\widehat{(\lambda,P)}} = \mathrm{Im}(F_{c|\widehat{(\lambda,P)}})$ and $\mathrm{Cok}(F_c)_{|\widehat{(\lambda,P)}} = \mathrm{Cok}(F_{c|\widehat{(\lambda,P)}})$. Hence, $\mathrm{Ker}(F_c)$, $\mathrm{Im}(F_c)$ and $\mathrm{Cok}(F_c)$ are good smooth $\mathcal{R}_{X(*D)}$-modules, and the set of irregular values are contained in $\mathcal{I}_P$.

To check the remaining claims, we may and will assume that D is smooth. (See Corollary 5.2.8.) Moreover, we may assume that $X = \Delta^n$ and $D = \{z_1 = 0\}$.

Let us consider the regular singular case. For any $u \in \mathbb{R} \times \mathbb{C}$, we have the following morphisms in $\mathcal{M}(D, 1)$:

$$\tilde{\psi}_{z_1,u}(F) : \big(\tilde{\psi}_{z_1,u}(\mathcal{T}_1), W, L, N_1\big) \longrightarrow \big(\tilde{\psi}_{z_1,u}(\mathcal{T}_2), W, L, N_1\big)$$

Hence, in particular, the cokernel is strict with respect to λ. Then, we obtain that F is strict with respect to the KMS-structure, and $\mathrm{Ker}\, F$, $\mathrm{Im}\, F$ and $\mathrm{Cok}\, F$ are equipped with the induced KMS-structure. We obtain the existence of relative monodromy filtrations, because the category of 1-IMTM is abelian. Thus, we are done in the regular singular case.

Let us consider the good irregular case. By the reduction with respect to the Stokes structure in Sect. 5.5.1, we obtain $\mathrm{Gr}^{\mathrm{St}}(F) : \mathrm{Gr}^{\mathrm{St}}(\mathcal{T}_1) \longrightarrow \mathrm{Gr}^{\mathrm{St}}(\mathcal{T}_2)$. By using the result in the regular singular case, we obtain that the cokernel of $\tilde{\psi}_{z_1,\mathfrak{a},u}(F)$:

$\tilde{\psi}_{z_1,\mathfrak{a},u}(\mathcal{T}_1) \longrightarrow \tilde{\psi}_{z_1,\mathfrak{a},u}(\mathcal{T}_2)$ are strict. Hence, we obtain that F is strict with respect to the KMS structure. We also obtain the existence of the relative monodromy filtration of the nilpotent part of the residues. □

Proposition 9.1.8 *Let $F : \mathcal{T}_1 \longrightarrow \mathcal{T}_2$ be a morphism in $\mathrm{MTS}^{\mathrm{adm}}(\boldsymbol{X})$. Then, F is strictly compatible with the KMS-structure. Namely, for the $\mathcal{R}_{X(*D)}$-triples $\mathcal{T}_i = (\mathcal{M}_{i,1}, \mathcal{M}_{i,2}, C_i)$, we have*

$$F(\mathcal{Q}^{(\lambda_0)}_{\boldsymbol{a}}\mathcal{M}_{1,2}) = \mathcal{Q}^{(\lambda_0)}_{\boldsymbol{a}}(\mathcal{M}_{2,2}) \cap \operatorname{Im} F, \tag{9.2}$$

$$F(\mathcal{Q}^{(\lambda_0)}_{\boldsymbol{a}}\mathcal{M}_{2,1}) = \mathcal{Q}^{(\lambda_0)}_{\boldsymbol{a}}(\mathcal{M}_{1,1}) \cap \operatorname{Im} F, \tag{9.3}$$

for any $\lambda_0 \in \mathbb{C}$ and $\boldsymbol{a} \in \mathbb{R}^{\Lambda}$, where Λ is the set of irreducible components of D.

Proof If D is smooth, the claim has already been proved in the proof of Proposition 9.1.7. If L is pure, the claim follows from that $\operatorname{Im} F$ is a direct summand of both of $\mathcal{T}_i$, as remarked in the proof of Proposition 9.1.7.

We consider the general case. By Proposition 9.1.7, and by considering the Hermitian adjoint, we have only to consider the case that F is an epimorphism. We obtain (9.3) from the smooth case and the Hartogs property. We obtain (9.2) from the pure case by using an easy induction on the length of L. □

Let $X = \Delta^n$ and $D = \bigcup_{i=1}^{\ell}\{z_i = 0\}$. As noted in Proposition 9.1.5, we have a functor

$${}^{\underline{\ell}}\tilde{\psi}_{\boldsymbol{u}} : \mathrm{MTS}^{\mathrm{adm}}(\boldsymbol{X}) \longrightarrow \mathcal{M}(D_{\underline{\ell}}, \underline{\ell}).$$

We obtain the following corollary from Proposition 9.1.8.

Corollary 9.1.9 *The functor ${}^{\underline{\ell}}\tilde{\psi}_{\boldsymbol{u}}$ is exact.* □

Let $N : (\mathcal{T}, W) \longrightarrow (\mathcal{T}, W) \otimes \boldsymbol{T}(-1)$ be a morphism in $\mathrm{MTS}^{\mathrm{adm}}(\boldsymbol{X})$. The monodromy weight filtration of N is a filtration in the category $\mathrm{MTS}^{\mathrm{adm}}(\boldsymbol{X})$. It induces a filtration of ${}^{\underline{\ell}}\tilde{\psi}_{\boldsymbol{u}}(\mathcal{T}, W)$ in $\mathcal{M}(D_{\underline{\ell}}, \underline{\ell})$, denoted by ${}^{\underline{\ell}}\tilde{\psi}_{\boldsymbol{u}}M(N)$. We also have the induced morphism ${}^{\underline{\ell}}\tilde{\psi}_{\boldsymbol{u}}(N) : {}^{\underline{\ell}}\tilde{\psi}_{\boldsymbol{u}}(\mathcal{T}, W) \longrightarrow {}^{\underline{\ell}}\tilde{\psi}_{\boldsymbol{u}}(\mathcal{T}, W) \otimes \boldsymbol{T}(-1)$ in $\mathcal{M}(D_{\underline{\ell}}, \underline{\ell})$. Because ${}^{\underline{\ell}}\tilde{\psi}_{\boldsymbol{u}}$ is exact, we obtain the following corollary by Deligne's inductive formula for monodromy weight filtrations.

Corollary 9.1.10 *We have ${}^{\underline{\ell}}\tilde{\psi}_{\boldsymbol{u}}M(N) = M\big({}^{\underline{\ell}}\tilde{\psi}_{\boldsymbol{u}}(N)\big)$.* □

9.1.6 Some Operations

Let us use the setting in Sect. 9.1.1. Let $(\mathcal{T}, W)$ be a mixed twistor structure on (X, D). We have naturally defined objects $(\mathcal{T}, W)^{\vee}$, $j^*(\mathcal{T}, W)$, $(\mathcal{T}, W)^*$, and $\tilde{\gamma}^*_{sm}(\mathcal{T}, W)$ in $\mathrm{MTS}(X, D)$. The following lemma is easy to see.

Lemma 9.1.11 *If $(\mathcal{T},W)$ is $\boldsymbol{\mathcal{I}}$-good, then $(\mathcal{T},W)^{\vee}$ and $j^*(\mathcal{T},W)$ are $-\boldsymbol{\mathcal{I}}$-good, and $(\mathcal{T},W)^*$ and $\tilde{\gamma}^*_{sm}(\mathcal{T},W)$ are $\boldsymbol{\mathcal{I}}$-good.* □

Lemma 9.1.12 *Suppose that $(\mathcal{T},W) \in \mathrm{MTS}(\boldsymbol{X})$ is admissible. Then, the mixed twistor structures $j^*(\mathcal{T},W)$, $(\mathcal{T},W)^{\vee}$, $(\mathcal{T},W)^*$ and $\tilde{\gamma}^*(\mathcal{T},W)$ are also admissible.*

Proof Let us consider **(Adm0)**. It is enough to consider the case that W is pure of weight 0. Let $(E,\overline{\partial}_E,\theta,h)$ be a $\boldsymbol{\mathcal{I}}$-good wild harmonic bundle on (X,D) corresponding to $\mathcal{T}$. Then, we can check that $j^*(\mathcal{T})$, $\mathcal{T}^{\vee}$, $\mathcal{T}^*$ and $\tilde{\gamma}^*_{sm}\mathcal{T}$ correspond to $(E,\overline{\partial}_E,-\theta,h)$, $(E^{\vee},\overline{\partial}_{E^{\vee}},-\theta^{\vee},h)$, $(E,\overline{\partial}_E,\theta,h)$ and $(E^{\vee},\overline{\partial}_{E^{\vee}},\theta^{\vee},h)$ respectively. Here, h denotes the induced pluri-harmonic metrics. Thus, the condition **(Adm0)** is satisfied. The condition **(Adm1)** is clearly satisfied. The condition **(Adm2)** is clearly satisfied for $j^*(\mathcal{T})$ and $\mathcal{T}^*$. We can use Lemma 8.1.12 to check **(Adm2)** for $\mathcal{T}^{\vee}$ and $\tilde{\gamma}^*_{sm}\mathcal{T}$. □

For $u=(a,\alpha)\in\mathbb{R}\times\mathbb{C}$, we set $-u:=(-a,-\alpha)$, $j^*u:=(a,-\alpha)$, and $\widetilde{\gamma}^*(a,\alpha)=(-a,\alpha)$. The induced maps on $(\mathbb{R}\times\mathbb{C})^{\ell}$ are denoted by the same notation.

Lemma 9.1.13 *Let $X=\Delta^n$, $D_i=\{z_i=0\}$ and $D=\bigcup_{i=1}^{\ell}\{z_i=0\}$. Let $(\mathcal{T},W)\in \mathrm{MTS}^{\mathrm{adm}}(\boldsymbol{X})$. Then, we have the following isomorphisms in $\mathcal{M}(D_{\underline{\ell}},\underline{\ell})$:*

$$\tilde{\psi}_{j^*(\boldsymbol{u})}\big(j^*(\mathcal{T},W)\big)\simeq j^*\tilde{\psi}_{\boldsymbol{u}}(\mathcal{T},W),\qquad \tilde{\psi}_{-\boldsymbol{u}}\big((\mathcal{T},W)^{\vee}\big)\simeq \tilde{\psi}_{\boldsymbol{u}}\big((\mathcal{T},W)\big)^{\vee},$$

$$\tilde{\psi}_{\boldsymbol{u}}\big((\mathcal{T},W)^*\big)\simeq \tilde{\psi}_{\boldsymbol{u}}\big((\mathcal{T},W)\big)^*,\qquad \tilde{\psi}_{\widetilde{\gamma}^*(\boldsymbol{u})}\big(\widetilde{\gamma}^*_{sm}(\mathcal{T},W)\big)\simeq \widetilde{\gamma}^*_{sm}\tilde{\psi}_{\boldsymbol{u}}(\mathcal{T},W).$$

Proof The claims for $j^*(\mathcal{T},W)$ and $(\mathcal{T},W)^*$ are clear. Let us see the claim for $(\mathcal{T},W)^{\vee}$. By using the compatibility of the duality and the reduction with respect to the Stokes structure (see [55]), we may assume that $\mathcal{T}$ is regular. Then, the claim for $(\mathcal{T},W)^{\vee}$ can be checked directly. The claim for $\tilde{\gamma}^*_{sm}(\mathcal{T},W)$ follows from those for others. □

Let $\mathbf{0}$ denote the trivial good set of ramified irregular values (see Sect. 15.1).

Lemma 9.1.14 *Let us consider $(\mathcal{T}_1,W)\in\mathrm{MTS}^{\mathrm{adm}}(\boldsymbol{X})$ and $(\mathcal{T}_2,W)\in\mathrm{MTS}^{\mathrm{adm}}(X,D,\mathbf{0})$. Then, we have $(\mathcal{T}_1,W)\otimes(\mathcal{T}_2,W)\in\mathrm{MTS}^{\mathrm{adm}}(\boldsymbol{X})$. If $X=\Delta^n$ and $D=\bigcup_{i=1}^{\ell}\{z_i=0\}$, we have the following natural isomorphisms for $\boldsymbol{u}\in(\mathbb{R}\times\mathbb{C})^{\ell}$:*

$$^{\underline{\ell}}\tilde{\psi}_{\boldsymbol{u}}\big((\mathcal{T}_1,W)\otimes(\mathcal{T}_2,W)\big)\simeq\bigoplus_{(\boldsymbol{u}_1,\boldsymbol{u}_2)\in\mathcal{S}(\boldsymbol{u})}{}^{\underline{\ell}}\tilde{\psi}_{\boldsymbol{u}_1}(\mathcal{T}_1,W)\otimes{}^{\underline{\ell}}\tilde{\psi}_{\boldsymbol{u}_2}(\mathcal{T}_2,W)$$

Here, $\mathcal{S}(\boldsymbol{u})$ is the set of $(\boldsymbol{u}_1,\boldsymbol{u}_2)\in\big(]-1,0]\times\mathbb{C}\big)^{\ell}\times\big(\mathbb{R}\times\mathbb{C}\big)^{\ell}$ such that $\boldsymbol{u}_1+\boldsymbol{u}_2=\boldsymbol{u}$.

Proof We obtain isomorphisms

$$\big({}^{\underline{\ell}}\tilde{\psi}_{\boldsymbol{u}}(\mathcal{T}_1\otimes\mathcal{T}_2),L,\boldsymbol{N}_{\underline{\ell}}\big)\simeq\bigoplus_{(\boldsymbol{u}_1,\boldsymbol{u}_2)\in\mathcal{S}(\boldsymbol{u})}\big({}^{\underline{\ell}}\tilde{\psi}_{\boldsymbol{u}_1}(\mathcal{T}_1),L,\boldsymbol{N}_{\underline{\ell}}\big)\otimes\big({}^{\underline{\ell}}\tilde{\psi}_{\boldsymbol{u}_2}(\mathcal{T}_2),L,\boldsymbol{N}_{\underline{\ell}}\big)$$

in $\mathrm{TS}(D_{\underline{\ell}},\underline{\ell})^{\mathrm{fil}}$. Hence, Then, we obtain the claim by Proposition 8.3.7. □

9.1.7 *Curve Test*

Let us use the setting in Sect. 9.1.1. Let $(\mathcal{T}, W) \in \mathrm{MTS}(\boldsymbol{X})$ satisfying **(Adm0)**.

Proposition 9.1.15 *Suppose the following condition:*

- *Let C be any curve contained in X which transversally intersects with the smooth part of D. Then, $(\mathcal{T}, W)_{|C}$ is an admissible mixed twistor structure on $(C, C \cap D)$.*

Then, $(\mathcal{T}, W)$ is admissible.

Proof By Proposition 5.2.9, $(\mathcal{T}, W)$ satisfies the condition **(Adm1)**. It is easy to check **(Adm2)**. □

9.1.8 *Tensor Products*

We give a complement on the functoriality with respect to the tensor product. Let $\mathcal{I}_i$ $(i = 1, 2)$ be good systems of ramified irregular values on (X, D). We set $\boldsymbol{X}_i := (X, D, \mathcal{I}_i)$. We obtain a system $\mathcal{I}_1 \otimes \mathcal{I}_2$ on (X, D), where $(\mathcal{I}_1 \otimes \mathcal{I}_2)_P = \{\mathfrak{a}_1 + \mathfrak{a}_2 \,\big|\, \mathfrak{a}_i \in \mathcal{I}_{i,P}\}$ for $P \in D$. By Proposition 15.1.5, we take a projective morphism of complex manifolds $\varphi : X' \longrightarrow X$ such that (1) $D' = \varphi^{-1}(D)$ is normal crossing, (2) φ induces an isomorphism $X' \setminus D' \simeq X \setminus D$, (3) $\mathcal{I}' := \varphi^{-1}(\mathcal{I}_1 \otimes \mathcal{I}_2)$ is a good system of ramified irregular vales. We set $\boldsymbol{X}' := (X', D', \mathcal{I}')$.

Proposition 9.1.16 *For admissible mixed twistor structures $(\mathcal{T}_i, W)$ $(i = 1, 2)$ on $\boldsymbol{X}_i$, the pull back $\varphi^*\big((\mathcal{T}_1, W) \otimes (\mathcal{T}_2, W)\big)$ is an admissible mixed twistor structure on $\boldsymbol{X}'$.*

Proof As remarked in Corollary 9.1.6, $\varphi^*(\mathcal{T}_i, W)$ are admissible mixed twistor structures on $(X', D', \varphi^*\mathcal{I}_i)$. Hence, it is enough to consider the case where $\mathcal{I}_1 \otimes \mathcal{I}_2$ is a good system of ramified irregular values. Then, the claim follows from Proposition 6.1.5. □

9.2 Admissible Polarizable Mixed Twistor Structure

9.2.1 *Definition*

Let us use the setting in Sect. 9.1.1. For any finite set Λ, we put $\mathrm{MTS}^{\mathrm{adm}}(\boldsymbol{X}, \Lambda) := \mathrm{MTS}^{\mathrm{adm}}(\boldsymbol{X})(\Lambda)$. We begin with a preliminary on the specializations of objects in $\mathrm{MTS}^{\mathrm{adm}}(\boldsymbol{X}, \Lambda)$.

Let $(\mathcal{T}, W, \boldsymbol{N}_\Lambda)$ be an object in $\mathrm{MTS}^{\mathrm{adm}}(\boldsymbol{X}, \Lambda)$. Let P be any point of D_I°. We take a holomorphic coordinate neighbourhood $(X_P; z_1, \ldots, z_n)$ around P such that $D = \bigcup_{i=1}^{|I|}\{z_i = 0\}$. For any $\boldsymbol{u} \in (\mathbb{R} \times \mathbb{C})^I$, we have the objects $\big({}^I\tilde{\psi}_{\boldsymbol{u}}(\mathcal{T}), L, \boldsymbol{N}_I\big)$ in

$\mathrm{TS}(D_{I,P}, I)^{\mathrm{fil}}$ as mentioned in Sect. 9.1.2.1. We also have the induced morphisms

$$N_j : ({}^I\tilde{\psi}_{\boldsymbol{u}}(\mathcal{T}), \boldsymbol{N}_I) \longrightarrow ({}^I\tilde{\psi}_{\boldsymbol{u}}(\mathcal{T}), \boldsymbol{N}_I) \otimes \boldsymbol{T}(-1) \quad (j \in \Lambda)$$

in $\mathrm{TS}(D_{I,P}, I)$. We set $\boldsymbol{N}_{I\sqcup\Lambda} := \big(N_i \,\big|\, i \in I\big) \sqcup \big(N_j \,\big|\, j \in \Lambda\big)$. Thus, we obtain objects $\big({}^I\tilde{\psi}_{\boldsymbol{u}}(\mathcal{T}), \boldsymbol{N}_{I\sqcup\Lambda}\big)$ in $\mathrm{TS}(D_{I,P}, I \sqcup \Lambda)$. If $\boldsymbol{\mathcal{I}}$ is unramified around P, we also obtain $\big({}^I\tilde{\psi}_{\mathfrak{a},\boldsymbol{u}}(\mathcal{T}), \boldsymbol{N}_{I\sqcup\Lambda}\big)$ in $\mathrm{TS}(D_{I,P}, I\sqcup\Lambda)$ for $\mathfrak{a} \in \mathcal{I}_P$ and $\boldsymbol{u} \in (\mathbb{R}\times\mathbb{C})^I$. We set $N(I\sqcup\Lambda) := \sum_{i\in I} N_i + \sum_{j\in\Lambda} N_j$.

9.2.1.1 Unramified Case

Let us consider the case where $\boldsymbol{\mathcal{I}}$ is unramified. An object $(\mathcal{T}, W, \boldsymbol{N}_\Lambda)$ in $\mathrm{MTS}^{\mathrm{adm}}(\boldsymbol{X}, \Lambda)$ is called (w, Λ)-polarizable if the following holds.

(R1) There exists a Hermitian sesqui-linear duality

$$S : (\mathcal{T}, W, \boldsymbol{N}_\Lambda) \longrightarrow (\mathcal{T}, W, \boldsymbol{N}_\Lambda)^* \otimes \boldsymbol{T}(-w)$$

of weight w such that $(\mathcal{T}, W, \boldsymbol{N}_\Lambda, S)_{|X\setminus D} \in \mathcal{P}(X \setminus D, \Lambda, w)$. Such an S is called a polarization of $(\mathcal{T}, W, \boldsymbol{N}_\Lambda)$.

(R2) For any $P \in D_I^\circ$, we take a coordinate neighbourhood $(X_P; z_1, \ldots, z_n)$ such that $X_P \cap D = \bigcup_{i=1}^{|I|}\{z_i = 0\}$. Set $W := M\big(N(I \sqcup \Lambda)\big)[w]$ on ${}^I\tilde{\psi}_{\mathfrak{a},\boldsymbol{u}}(\mathcal{T})$ for any $\mathfrak{a} \in \mathcal{I}_P$ and $\boldsymbol{u} \in (\mathbb{R} \times \mathbb{C})^I$. Then, $\big({}^I\tilde{\psi}_{\mathfrak{a},\boldsymbol{u}}(\mathcal{T}), W, \boldsymbol{N}_{I\sqcup\Lambda}\big)_P$ with ${}^I\tilde{\psi}_{\mathfrak{a},\boldsymbol{u}}(S)$ is a $(w, \Lambda \sqcup I)$-polarized mixed twistor structure, where ${}^I\tilde{\psi}_{\mathfrak{a},\boldsymbol{u}}(S)$ is induced by S in **(R1)**.

Lemma 9.2.1 *The second condition is independent of the choice of S in the first condition.*

Proof Let S_i $(i = 1, 2)$ be polarizations of $(\mathcal{T}, W, \boldsymbol{N}_\Lambda)$ as in the first condition. Assume that the second condition is satisfied for S_1. Let us prove that it is also satisfied for S_2. There exists a decomposition $(\mathcal{T}, \boldsymbol{N}_\Lambda)_{|X\setminus D} = \bigoplus(\mathcal{T}_j, \boldsymbol{N}_\Lambda)$ which are orthogonal with respect to both S_i, and $S_{1|\mathcal{T}_j} = a_j \cdot S_{2|\mathcal{T}_j}$ for some $a_j > 0$. We have the endomorphism F of $\mathcal{T}$ induced by S_1 and S_2. We have $\mathcal{T}_j = \mathrm{Ker}(F_{|X\setminus D} - a_j)$. Let $P \in D_I^\circ$. We have the induced endomorphisms ${}^I\tilde{\psi}_{\mathfrak{a},\boldsymbol{u}}(F)$ on ${}^I\tilde{\psi}_{\mathfrak{a},\boldsymbol{u}}(\mathcal{T})_{|P}$. It is a morphism in the category of polarizable mixed twistor structures. Hence, we obtain that $F - \alpha$ are strict for any $\alpha \in \mathbb{C}$. Then, the decomposition $\mathcal{T}_{|X\setminus D} = \bigoplus \mathcal{T}_j$ is extended to $\mathcal{T} = \bigoplus \mathcal{T}_j'$, and it is compatible with the KMS-structures. Then, the claim is clear. □

9.2.1.2 Ramified Case

Let us consider the case where $\mathcal{I}$ is not necessarily unramified. Let $(\mathcal{T}, W, \boldsymbol{N}_\Lambda)$ be an object in $\mathrm{MTS}^{\mathrm{adm}}(\boldsymbol{X}, \Lambda)$, which is not necessarily unramified. It is called (w, Λ)-polarizable if the following holds:

- It is locally the descent of an unramified admissible (w, Λ)-polarizable mixed twistor structure.
- There exists a morphism $S : (\mathcal{T}, W, \boldsymbol{N}_\Lambda) \longrightarrow (\mathcal{T}, W, \boldsymbol{N}_\Lambda)^* \otimes \boldsymbol{T}(-w)$ such that $(\mathcal{T}, W, \boldsymbol{N}_\Lambda, S)_{|X \setminus D}$ is a polarized mixed twistor structure on $X \setminus D$.

9.2.2 *Category of Admissible (w, Λ)-Polarizable Mixed Twistor Structure*

Let $\mathcal{P}(\boldsymbol{X}, \Lambda, w) \subset \mathrm{MTS}^{\mathrm{adm}}(\boldsymbol{X}, \Lambda)$ denote the full subcategory of admissible (w, Λ)-polarizable mixed twistor structure on $\boldsymbol{X}$.

Proposition 9.2.2 *The family of the subcategories $\mathcal{P}(\boldsymbol{X}, \Lambda, w)$ has the property* **P0–3** *in Sect. 6.3.1.*

Proof The claims for **P1–2** are clear. Let us study the property **P0**. Let $F : (\mathcal{T}_1, W, \boldsymbol{N}_\Lambda) \longrightarrow (\mathcal{T}_2, W, \boldsymbol{N}_\Lambda)$ be a morphism in $\mathcal{P}(\boldsymbol{X}, \Lambda, w)$. We have the objects $(\mathrm{Ker}\, F, W, \boldsymbol{N}_\Lambda)$, $(\mathrm{Im}\, F, W, \boldsymbol{N}_\Lambda)$ and $(\mathrm{Cok}\, F, W, \boldsymbol{N}_\Lambda)$ in $\mathrm{MTS}^{\mathrm{adm}}(\boldsymbol{X}, \Lambda)$. We choose polarizations S_i for $(\mathcal{T}_i, W, \boldsymbol{N}_\Lambda)$. We have the adjoint $F^\vee := S_1^{-1} \circ F^* \circ S_2 : (\mathcal{T}_2, W, \boldsymbol{N}_\Lambda) \longrightarrow (\mathcal{T}_1, W, \boldsymbol{N}_\Lambda)$, giving splittings $\mathcal{T}_1 = \mathrm{Ker}\, F \oplus \mathrm{Im}\, F^\vee$ and $\mathcal{T}_2 = \mathrm{Im}\, F \oplus \mathrm{Ker}\, F^\vee$ in $\mathrm{MTS}^{\mathrm{adm}}(\boldsymbol{X}, \Lambda)$, as in Lemma 8.2.7. The decompositions are compatible with the filtrations W and the tuple of morphisms $\boldsymbol{N}_\Lambda$. The decompositions are also compatible with ${}^I\tilde{\psi}_{\boldsymbol{u}}(\cdot)_P$ and ${}^I\tilde{\psi}_{\mathfrak{a},\boldsymbol{u}}(\cdot)_P$ for any $P \in D_I^\circ$. The restriction of the polarizations to direct summands are also polarizations. Hence, we obtain that $\mathcal{P}(\boldsymbol{X}, \Lambda, w)$ is abelian and semisimple.

Let us prove **P3.1**. For $\Lambda_1 \subset \Lambda$, we have the filtration $M(\Lambda_1) := M\big(N(\Lambda_1)\big)$ of $(\mathcal{T}, W)$ in the category $\mathrm{MTS}^{\mathrm{adm}}(\boldsymbol{X})$. Let $\Lambda_2 \subset \Lambda_1$. The restriction $M(\Lambda_1)_{|X \setminus D}$ is a relative monodromy filtration of $N(\Lambda_1)_{|X \setminus D}$ with respect to $M(\Lambda_2)_{|X \setminus D}$. Hence, it is easy to observe that $M(\Lambda_1)$ is a relative monodromy filtration of $N(\Lambda_1)$ with respect to $M(\Lambda_2)$.

Let us consider **P3.2**. Let $\Lambda = \Lambda_0 \sqcup \Lambda_1$ be a decomposition. We consider the filtration $\tilde{L} := M(\Lambda_1)$ on $(\mathcal{T}, W, \boldsymbol{N}_{\Lambda_0})$ in $\mathrm{MTS}^{\mathrm{adm}}(\boldsymbol{X}, \Lambda_0)$.

Lemma 9.2.3 $\mathrm{Gr}_k^{\tilde{L}}(\mathcal{T}, W, \boldsymbol{N}_{\Lambda_0})$ *is an object in $\mathcal{P}(\boldsymbol{X}, \Lambda_0, w + k)$.*

Proof Let S be a polarization of $(\mathcal{T}, W, \boldsymbol{N}_\Lambda)$. We obtain a Hermitian sesqui-linear duality S_k of weight $w + k$ on the primitive part $P\,\mathrm{Gr}_k^{\tilde{L}}(\mathcal{T}, W, \boldsymbol{N}_{\Lambda_0})$ induced by S and $N(\Lambda_1)$. The restriction of $S_{k|X \setminus D}$ is a polarization of $P\,\mathrm{Gr}_k^{\tilde{L}}(\mathcal{T}, W, \boldsymbol{N}_{\Lambda_0})_{|X \setminus D}$. For the remaining conditions, it is enough to consider the case that $\mathcal{I}$ is unramified.

Let $P \in D_I^\circ$. We take a coordinate neighbourhood $(X_P; z_1, \ldots, z_n)$ around P such that $X_P \cap D = \bigcup_{i=1}^{|I|}\{z_i = 0\}$. We define a filtration $\tilde{L}$ on ${}^I\tilde{\psi}_{\mathfrak{a},\boldsymbol{u}}(\mathcal{T})$ by $\tilde{L}_j{}^I\tilde{\psi}_{\mathfrak{a},\boldsymbol{u}}(\mathcal{T}) := {}^I\tilde{\psi}_{\mathfrak{a},\boldsymbol{u}}(L_j\mathcal{T})$. By the exactness of the functor ${}^I\tilde{\psi}_{\mathfrak{a},\boldsymbol{u}}$, it is equal to $M(N(\Lambda_1))$ on ${}^I\tilde{\psi}_{\mathfrak{a},\boldsymbol{u}}(\mathcal{T})$. We have natural isomorphisms

$$\big({}^I\tilde{\psi}_{\mathfrak{a},\boldsymbol{u}} P\,\mathrm{Gr}_k^{\tilde{L}}(\mathcal{T}), \boldsymbol{N}_{I\sqcup\Lambda_0}\big)_P \simeq \big(P\,\mathrm{Gr}_k^{\tilde{L}}\,{}^I\tilde{\psi}_{\mathfrak{a},\boldsymbol{u}}(\mathcal{T}), \boldsymbol{N}_{I\sqcup\Lambda_0}\big)_P$$

The right hand side is equipped with the polarization induced by S and $N(\Lambda_1)$ (Lemma 8.2.10). By the construction, it is also induced by S_k. Hence, we obtain that $P\,\mathrm{Gr}_k^{\tilde{L}}(\mathcal{T}, L, \boldsymbol{N}_{\Lambda_0})$ is an object in $\mathcal{P}(\boldsymbol{X}, \Lambda_0, w+k)$. □

Let us prove **P3.3**. Let $(\mathcal{T}, W, \boldsymbol{N})$ be an object in $\mathcal{P}(\boldsymbol{X}, \Lambda, w)$. Let $\bullet \in \Lambda$. We have the induced object $(\mathrm{Im}\,N_\bullet, W, \boldsymbol{N}_\Lambda)$ in $\mathrm{MTS}^{\mathrm{adm}}(\boldsymbol{X}, \Lambda)$.

Lemma 9.2.4 $(\mathrm{Im}\,N_\bullet, W, \boldsymbol{N}_\Lambda)$ *is an object in* $\mathcal{P}(\boldsymbol{X}, \Lambda, w+1)$.

Proof Let S be a polarization of $(\mathcal{T}, W, \boldsymbol{N}_\Lambda)$. Let S' be a Hermitian sesqui-linear duality of weight $w+1$ on $\mathrm{Im}\,N_\bullet$ induced by S and $N_\bullet$. The restriction $S'_{|X\setminus D}$ gives a polarization of $(\mathrm{Im}\,N_\bullet, W, \boldsymbol{N}_\Lambda)_{|X\setminus D}$. To check the remaining conditions, we may assume that $\mathcal{I}$ is unramified. Let $P \in D_I^\circ$. We take a holomorphic coordinate neighbourhood $(X_P; z_1, \ldots, z_n)$ such that $X_P \cap D = \bigcup_{i=1}^{|I|}\{z_i = 0\}$. We have the following natural isomorphisms:

$$\big({}^I\tilde{\psi}_{\mathfrak{a},u}(\mathrm{Im}\,N_\bullet), \boldsymbol{N}_{\Lambda\sqcup I}\big)_P \simeq \big(\mathrm{Im}\big({}^I\tilde{\psi}_{\mathfrak{a},u}(N_\bullet)\big), \boldsymbol{N}_{\Lambda\sqcup I}\big)_P.$$

Hence, the claim follows from Lemma 8.2.11. □

Let $* \in \Lambda \setminus \bullet$. We have the following induced morphisms in $\mathrm{MTS}^{\mathrm{adm}}(\boldsymbol{X}, \Lambda \setminus *)$:

$$\Sigma^{1,0}\,\mathrm{Gr}^{M(N_*)}\,\mathcal{T} \longrightarrow \Sigma^{1,0}\,\mathrm{Gr}^{M(N_*)}\,\mathrm{Im}\,N_\bullet \longrightarrow \Sigma^{0,-1}\,\mathrm{Gr}^{M(N_*)}\,\mathcal{T} \tag{9.4}$$

The restriction of (9.4) to $X \setminus D$ is S-decomposable by Lemma 8.2.12. Hence, we obtain that (9.4) is S-decomposable. Thus, the proof of Proposition 9.2.2 is finished. □

9.2.3 An Equivalent Condition

Let $X = \Delta^n$, $D_i = \{z_i = 0\}$ and $D = \bigcup_{i=1}^{\ell} D_i$. Let $(\mathcal{T}, \boldsymbol{N}) \in \mathrm{TS}(X, D)(\Lambda)$. Let $W := M\big(N(\Lambda)\big)[w]$. We assume the following:

- $\mathcal{T}$ is an $\mathcal{I}$-good-KMS smooth $\mathcal{R}_{X(*D)}$-triple.
- **Adm0** holds for $(\mathcal{T}, W)$.

- Take a ramified covering $\varphi_m : (X,D) \longrightarrow (X,D)$ given by $\varphi_m(z_1,\ldots,z_n) = (z_1^m,\ldots,z_\ell^m,z_{\ell+1},\ldots,z_n)$ such that $\varphi_m^*\mathcal{I}$ is unramified. Then, $\varphi_m^*(\mathcal{T},\boldsymbol{N})$ satisfies **R1–2** with w.

Note that W is not assumed to be compatible with the KMS-structure. The following proposition implies that the compatibility is automatically satisfied under the above assumption.

Proposition 9.2.5 *We have $(\mathcal{T},W,\boldsymbol{N}) \in \mathcal{P}(\boldsymbol{X},w,\Lambda)$ if the above conditions are satisfied.*

Proof It is enough to consider the case where $\mathcal{I}$ is unramified. For any $\boldsymbol{a} \in \mathbb{C}^{\Lambda}$, we put $N(\boldsymbol{a}) := \sum a_i\,\lambda N_i$, which gives an endomorphism of $\mathcal{Q}_{\boldsymbol{b}}^{(\lambda_0)}\mathcal{M}_i$.

Lemma 9.2.6 *The conjugacy classes of $N(\boldsymbol{a})_{|(\lambda,P)}$ are independent of the choice of $(\lambda,P) \in \mathbb{C}_\lambda \times X$.*

Proof If we fix a point $P \in X \setminus D$, they are independent of λ, because of the mixed twistor property. If we fix $\lambda \neq 0$, they are independent of $P \in X \setminus D$ because of the flatness. Hence, we obtain that they are independent of $(\lambda,P) \in \mathbb{C}_\lambda \times (X \setminus D)$.

Let us fix $P \in D_I^{\circ}$. We use the notation in Sect. 5.1.2. The conjugacy classes of $N(\boldsymbol{a})_{|(\lambda,P)}$ on ${}^{I}\mathcal{G}^{(\lambda_0)}\mathcal{Q}_{\boldsymbol{b}}^{(\lambda_0)}(\mathcal{M}_i)_{|(\lambda,P)}$ are independent of λ, because of **(R2)** and the mixed twistor property. If λ is generic, the filtrations ${}^{i}F^{(\lambda_0)}$ have the splitting around λ given by the generalized eigen decomposition of $\mathrm{Res}_i(\mathbb{D})$. Then, it is easy to observe that the conjugacy classes of $N(\boldsymbol{a})_{|(\lambda,P)}$ are independent around λ_0.

Let us consider the regular singular case. If we fix a generic λ, they are independent of $P \in X$. Hence, we are done.

Let us consider the unramifiedly good irregular case. We have $\mathrm{Gr}^{\mathrm{St}}(\mathcal{Q}_{\boldsymbol{b}}\mathcal{M}_i)$ with $\mathrm{Gr}^{St}(N(\boldsymbol{a}))$. By using the previous consideration, for generic λ, we obtain that the conjugacy classes of $\mathrm{Gr}^{\mathrm{St}}(N(\boldsymbol{a}))_{|(\lambda,P)}$ are independent of $P \in X$. By considering the completion, we obtain that the conjugacy classes of $\mathrm{Gr}^{\mathrm{St}}(N(\boldsymbol{a}))_{|(\lambda,P)}$ and $N(\boldsymbol{a})_{|(\lambda,P)}$ are the same. Thus, we are done. □

We also obtain the following lemma.

Lemma 9.2.7 *The weight filtration $M(\boldsymbol{a})$ of $N(\boldsymbol{a})$ is a filtration by subbundles of $\mathcal{Q}_{\boldsymbol{b}}^{(\lambda_0)}\mathcal{M}_i$. Moreover, $M(\boldsymbol{a})$ is compatible with the KMS-structure.*

Proof The first claim follows from the previous lemma. To prove the compatibility, according to Proposition 5.2.9, we have only to consider the case $n = \ell = 1$. On $\mathcal{Q}_b^{(\lambda_0)}(\mathcal{M}_i)_{|\mathcal{D}^{(\lambda_0)}}$, we have the parabolic filtration $F^{(\lambda_0)}$. By using Lemma 5.2 of [52], we obtain that the weight filtration $M(\boldsymbol{a})$ of $N(\boldsymbol{a})$ on $\mathcal{Q}_b^{(\lambda_0)}(\mathcal{M}_i)_{|\mathcal{D}^{(\lambda_0)}}$ induces the weight filtration of $M(\boldsymbol{a})$ of $N(\boldsymbol{a})$ on $\mathrm{Gr}^{F^{(\lambda_0)}}(\mathcal{Q}_b^{(\lambda_0)}\mathcal{M}_i)$. In particular, $\mathrm{Gr}^{M(\boldsymbol{a})}\mathrm{Gr}^{F^{(\lambda_0)}}(\mathcal{Q}_b^{(\lambda_0)}\mathcal{M})$ is a locally free $\mathcal{O}_{\mathcal{D}^{(\lambda_0)}}$-module. Then, we obtain that $\mathrm{Gr}^{M(\boldsymbol{a})}(\mathcal{Q}_*^{(\lambda_0)}\mathcal{M}_i)$ is a KMS-structure of $\mathrm{Gr}^{M(\boldsymbol{a})}\mathcal{M}_i$ around λ_0, i.e., $M(\boldsymbol{a})$ is compatible with the KMS-structure. □

In particular, $W = M\big(N(\Lambda)\big)[w]$ is compatible with the KMS-structure. The existence of relative monodromy filtrations follows from the property of polarizable mixed twistor structures. Thus, Proposition 9.2.5 is proved. □

9.2.4 Specialization

Let $X = \Delta^n$, $D_i = \{z_i = 0\}$ and $D = \bigcup_{i=1}^{\ell} D_i$. Recall that we have the specialization of the good system of ramified irregular values $\boldsymbol{\mathcal{I}}$ to D_I, denoted by $\boldsymbol{\mathcal{I}}(I)_{|D_I}$. (See Sect. 15.1.3.) The induced tuple $(D_I, \partial D_I, \boldsymbol{\mathcal{I}}(I)_{|D_I})$ is denote by $\boldsymbol{D}_I$.

Let $(\mathcal{T}, W, \boldsymbol{N}_\Lambda) \in \mathcal{P}(\boldsymbol{X}, \Lambda, w)$. We obtain ${}^{I}\tilde{\psi}_{\boldsymbol{u}}(\mathcal{T}) \in \mathrm{TS}(\boldsymbol{D}_I)$ with the induced morphisms $\boldsymbol{N}_\Lambda$. We also have the naturally induced morphisms $\boldsymbol{N}_I = (N_i \mid i \in I)$. We set $\boldsymbol{N}_{\Lambda\sqcup I} := \boldsymbol{N}_\Lambda \sqcup \boldsymbol{N}_I$. We set $W({}^{I}\tilde{\psi}_{\boldsymbol{u}}(\mathcal{T})) := M\big(N(\Lambda \sqcup I)\big)[w]$.

Proposition 9.2.8 *We have* $\big({}^{I}\tilde{\psi}_{\boldsymbol{u}}(\mathcal{T}), W, \boldsymbol{N}_{\Lambda\sqcup I}\big) \in \mathcal{P}(\boldsymbol{D}_I, \Lambda \sqcup I, w)$. *The object* $\big({}^{I}\tilde{\psi}_{\boldsymbol{u}}(\mathcal{T}), W, \boldsymbol{N}_{\Lambda\sqcup I}\big)$ *is denoted by* ${}^{I}\tilde{\psi}_{\boldsymbol{u}}(\mathcal{T}, W, \boldsymbol{N}_\Lambda)$ *if there is no risk of confusion. In this way, we obtain an exact functor* ${}^{I}\tilde{\psi}_{\boldsymbol{u}} : \mathcal{P}(\boldsymbol{X}, \Lambda, w) \longrightarrow \mathcal{P}(\boldsymbol{D}_I, \Lambda \sqcup I, w)$.

Proof By an inductive argument, we have only to consider the case $I = \{i\}$. We define the filtration L on ${}^{i}\tilde{\psi}_u(\mathcal{T})$ by $L_k{}^{i}\tilde{\psi}_u(\mathcal{T}) := {}^{i}\tilde{\psi}_u(W_k\mathcal{T})$. By Corollary 9.1.10, it is equal to $M(N(\Lambda))[w]$. Because W is the relative monodromy filtration of $N(\Lambda \sqcup \{i\})$ with respect to L, we have natural isomorphisms

$$\mathrm{Gr}^W\big({}^{i}\tilde{\psi}_u(\mathcal{T})\big) \simeq \mathrm{Gr}^W \mathrm{Gr}^L\big({}^{i}\tilde{\psi}_u(\mathcal{T})\big) \simeq \mathrm{Gr}^W\big({}^{i}\tilde{\psi}_u(\mathrm{Gr}^W \mathcal{T})\big).$$

Hence, according to a special case of §12.7 of [55], we obtain that $\big({}^{i}\tilde{\psi}_u(\mathcal{T}), W\big)$ satisfies **Adm0**. Then, it is easy to check $\big({}^{i}\tilde{\psi}_u(\mathcal{T}), W, \boldsymbol{N}_{\Lambda\sqcup\{i\}}\big)$ satisfies the assumptions for Proposition 9.2.5 for $\boldsymbol{D}_i$. Hence, $\big({}^{i}\tilde{\psi}_u(\mathcal{T}), W, \boldsymbol{N}_{\Lambda\sqcup\{i\}}\big)$ is an object in $\mathcal{P}(\boldsymbol{D}_i, \Lambda \sqcup \{i\}, w)$. □

We give a variant. Suppose that $\boldsymbol{\mathcal{I}}$ is unramified. We obtain $\boldsymbol{\mathcal{I}}(-\mathfrak{a}, I)_{|D_I}$ as in Sect. 15.1.3. We set $\boldsymbol{D}_I(-\mathfrak{a}) := (D_I, \partial D_I, \boldsymbol{\mathcal{I}}(-\mathfrak{a}, I)_{|D_I})$.

Corollary 9.2.9 *Let* $(\mathcal{T}, W, \boldsymbol{N}_\Lambda) \in \mathcal{P}(\boldsymbol{X}, \Lambda, w)$. *For any* $\boldsymbol{u} \in (\mathbb{R} \times \mathbb{C})^I$, *we have the induced objects* $\big({}^{I}\tilde{\psi}_{\mathfrak{a},\boldsymbol{u}}(\mathcal{T}), \boldsymbol{N}_{\Lambda\sqcup I}\big)$ *in* $\mathcal{P}(\boldsymbol{D}_I(-\mathfrak{a}), \Lambda \sqcup I, w)$. □

9.2.5 Some Operations

We have a complement. We set $\boldsymbol{X}^- := (X, D, -\boldsymbol{\mathcal{I}})$. Let $\boldsymbol{0}$ denote the trivial good system of ramified irregular values. The following lemma easily follows from Lemmas 9.1.12, 9.1.13 and 9.1.14.

Lemma 9.2.10

- *Let* $(\mathcal{T}, W, \boldsymbol{N}_\Lambda) \in \mathcal{P}(\boldsymbol{X}, \Lambda, w)$. *Then, we have*

$$(\mathcal{T}, W, \boldsymbol{N}_\Lambda)^* \in \mathcal{P}(\boldsymbol{X}, \Lambda, -w), \quad (\mathcal{T}, W, \boldsymbol{N}_\Lambda)^\vee \in \mathcal{P}(\boldsymbol{X}^-, \Lambda, -w),$$

$$j^*(\mathcal{T}, W, \boldsymbol{N}_\Lambda) \in \mathcal{P}(\boldsymbol{X}^-, \Lambda, w), \quad \tilde{\gamma}^*_{\mathrm{sm}}(\mathcal{T}, W, \boldsymbol{N}_\Lambda) \in \mathcal{P}(\boldsymbol{X}, \Lambda, w).$$

- *For* $(\mathcal{T}_1, W, \boldsymbol{N}_\Lambda) \in \mathcal{P}(\boldsymbol{X}, \Lambda, w_1)$ *and* $(\mathcal{T}_2, W, \boldsymbol{N}_\Lambda) \in \mathcal{P}((X, D, \boldsymbol{0}), \Lambda, w_2)$, *we have* $(\mathcal{T}_1, W, \boldsymbol{N}_\Lambda) \otimes (\mathcal{T}_2, W, \boldsymbol{N}_\Lambda) \in \mathcal{P}(\boldsymbol{X}, \Lambda, w_1 + w_2)$. □

Let us consider the case $X = \Delta^n$, $D_i = \{z_i = 0\}$ and $D = \bigcup_{i=1}^{\ell} D_i$. We also obtain the following.

Lemma 9.2.11

- *Let* $(\mathcal{T}, W, \boldsymbol{N}_\Lambda) \in \mathcal{P}(\boldsymbol{X}, \Lambda, w)$. *Then, for any* $I \subset \underline{\ell}$, *we have the following natural isomorphisms:*

$$^I\tilde{\psi}_{-\boldsymbol{u}}\big((\mathcal{T}, W, \boldsymbol{N}_\Lambda)^\vee\big) \simeq \big(^I\tilde{\psi}_{\boldsymbol{u}}(\mathcal{T}, W, \boldsymbol{N}_\Lambda)\big)^\vee$$

$$^I\tilde{\psi}_{\boldsymbol{u}}\big((\mathcal{T}, W, \boldsymbol{N}_\Lambda)^*\big) \simeq \big(^I\tilde{\psi}_{\boldsymbol{u}}(\mathcal{T}, W, \boldsymbol{N}_\Lambda)\big)^*$$

$$^I\tilde{\psi}_{j^*(\boldsymbol{u})}\big(j^*(\mathcal{T}, W, \boldsymbol{N}_\Lambda)\big) \simeq j^*\big(^I\tilde{\psi}_{\boldsymbol{u}}(\mathcal{T}, W, \boldsymbol{N}_\Lambda)\big)$$

$$^I\tilde{\psi}_{\widetilde{\gamma}^*(\boldsymbol{u})}\big(\widetilde{\gamma}^*_{sm}(\mathcal{T}, W, \boldsymbol{N}_\Lambda)\big) \simeq \widetilde{\gamma}^*_{sm}\big(^I\tilde{\psi}_{\boldsymbol{u}}(\mathcal{T}, W, \boldsymbol{N}_\Lambda)\big)$$

- *For* $(\mathcal{T}_1, W, \boldsymbol{N}_\Lambda) \in \mathcal{P}(\boldsymbol{X}, \Lambda, w_1)$ *and* $(\mathcal{T}_2, W, \boldsymbol{N}_\Lambda) \in \mathcal{P}((X, D, \boldsymbol{0}), \Lambda, w_2)$, *we have the following natural isomorphisms:*

$$\begin{aligned} &^I\tilde{\psi}_{\boldsymbol{u}}\big((\mathcal{T}_1, W, \boldsymbol{N}_\Lambda) \otimes (\mathcal{T}_2, W, \boldsymbol{N}_\Lambda)\big) \\ &\simeq \bigoplus_{(\boldsymbol{u}_1, \boldsymbol{u}_2) \in \mathcal{S}(\boldsymbol{u})} {}^I\tilde{\psi}_{\boldsymbol{u}_1}(\mathcal{T}_1, W, \boldsymbol{N}_\Lambda) \otimes {}^I\tilde{\psi}_{\boldsymbol{u}_2}(\mathcal{T}_2, W, \boldsymbol{N}_\Lambda) \end{aligned} \tag{9.5}$$

Here, $\mathcal{S}(\boldsymbol{u})$ *are as in Lemma 9.1.14.* □

9.3 Admissible IMTM

9.3.1 Definitions

An object $(\mathcal{T}, W, L, \boldsymbol{N}) \in \mathrm{MTS}^{\mathrm{adm}}(\boldsymbol{X}, \Lambda)^{\mathrm{fil}}$ is called pre-admissible Λ-IMTM on $\boldsymbol{X}$ if the following holds:

(R3) $(\mathcal{T}, W, L, \boldsymbol{N})_{|X \setminus D} \in \mathcal{M}(X \setminus D, \Lambda)$, and $\mathrm{Gr}^L_w(\mathcal{T}, W, \boldsymbol{N}) \in \mathcal{P}(\boldsymbol{X}, \Lambda, w)$.

Note that $(\mathcal{T}, L, N_i)$ $(i \in \Lambda)$ have relative monodromy filtrations. We shall impose an additional condition to define the admissibility condition.

9.3.1.1 The Smooth Divisor Case

Let us consider the case that D is smooth. Let $(\mathcal{T}, W, L, \boldsymbol{N}_\Lambda)$ be a pre-admissible Λ-IMTM on $\boldsymbol{X}$. Let $\mathcal{M}_i$ be the underlying smooth $\mathcal{R}_{X(*D)}$-modules, which is equipped with the KMS-structure. For $u \in \mathbb{R} \times \mathbb{C}$, we obtain the vector bundles $\mathcal{G}_u(\mathcal{M}_i)$ on $\mathcal{D}$ equipped with the induced morphism N obtained as the nilpotent part of the residue $\mathrm{Res}(\mathbb{D})$. It is also equipped with the filtration L defined by $L_j\mathcal{G}_u(\mathcal{M}_i) := \mathcal{G}_u(L_j\mathcal{M}_i)$.

The object $(\mathcal{T}, W, L, \boldsymbol{N})$ is called admissible if the following holds:

(R4) $\big(\mathcal{G}_u(\mathcal{M}_i), L, N\big)$ $(i = 1, 2)$ have relative monodromy filtrations.

Let us consider the case Let $X = \Delta^n$ and $D = \{z_1 = 0\}$. Let L be the filtration on $\tilde{\psi}_{z_1,u}(\mathcal{T})$ defined by $L_j\tilde{\psi}_{z_1,u}(\mathcal{T}) := \tilde{\psi}_{z_1,u}(L_j\mathcal{T})$. We obtain the object $(\tilde{\psi}_{z_1,u}(\mathcal{T}), L, \boldsymbol{N}_\Lambda)$ in $\mathrm{TS}(D_1, \Lambda)^{\mathrm{fil}}$. Let N_1 denote the induced nilpotent morphism on $(\tilde{\psi}_{z_1,u}(\mathcal{T}), L, \boldsymbol{N}_\Lambda)$. We set $\boldsymbol{N}_{\Lambda\cup\{1\}} := \boldsymbol{N}_\Lambda \sqcup \{N_1\}$. We set $N(\Lambda \sqcup \{1\}) := N(\Lambda) + N_1$.

Lemma 9.3.1

- *For $u \in \mathbb{R} \times \mathbb{C}$, $\big(\tilde{\psi}_{z_1,u}(\mathcal{T}), L, N(\Lambda \sqcup \{1\})\big)$ has a relative monodromy filtration W, and $\big(\tilde{\psi}_{z_1,u}(\mathcal{T}), W, L, \boldsymbol{N}_{\Lambda\sqcup\{1\}}\big)$ is $(\Lambda \sqcup \{1\})$-IMTM on D.*
- *Suppose that $\mathcal{I}$ is unramified. Then, for each $\mathfrak{a} \in \mathcal{I}$ and $u \in \mathbb{R} \times \mathbb{C}$, the tuple $\big(\tilde{\psi}_{z_1,\mathfrak{a},u}(\mathcal{T}, L), \boldsymbol{N}_{\Lambda\sqcup\{1\}}\big)$ is $(\Lambda \sqcup \{1\})$-IMTM on D.*

Proof We have only to consider the unramified case. By the compatibility of L with the KMS-structure, $\mathrm{Gr}^L_w\big(\tilde{\psi}_{z_1,\mathfrak{a},u}(\mathcal{T}), \boldsymbol{N}_{\Lambda\sqcup\{1\}}\big)$ comes from $\mathrm{Gr}^L_w(\mathcal{T}, \boldsymbol{N})$. Hence, it is a $(w, \Lambda \sqcup \{1\})$-polarizable mixed twistor structure. By the assumption, N_1 has a relative monodromy filtration. For generic λ, $N_{i|\lambda}$ $(i \in \Lambda)$ have relative monodromy filtrations. Then, the claim follows from Proposition 8.5.1. (We can also deduce it directly from Proposition 8.1.13 and Lemma 8.1.16.) □

We can also deduce the following lemma from Proposition 8.5.1 (or easier Proposition 8.1.13 and Lemma 8.1.16).

Lemma 9.3.2 *We can replace the above admissibility condition with the following:*

- *There exists $U \subset \mathbb{C}_\lambda$ with $|U| = \infty$ such that $\big(\mathcal{G}_u(\mathcal{M}_1), L, N\big)_{(\lambda,P)} \in \mathrm{Vect}^{\mathrm{RMF}}_{\mathbb{C}}$ for any $(\lambda, P) \in U \times D$.* □

9.3.1.2 The Normal Crossing Case

Let us consider the case that D is normal crossing. A pre-admissible Λ-IMTM $(\mathcal{T}, W, L, \boldsymbol{N}_\Lambda)$ on $\boldsymbol{X}$ is called admissible, if the following holds:

- For any smooth point $P \in D$, there exists a small neighbourhood X_P of P such that $(\mathcal{T}, W, L, \boldsymbol{N}_\Lambda)_{|X_P}$ is admissible in the sense of the smooth divisor case.

Let $P \in D_I^\circ$ be any point. We take a holomorphic coordinate neighbourhood $(X_P; z_1, \dots, z_n)$ such that $X_P \cap D = \bigcup_{i=1}^{|I|}\{z_i = 0\}$. We define the filtration L on ${}^I\tilde{\psi}_{\boldsymbol{u}}(\mathcal{T})$ by $L_j({}^I\tilde{\psi}_{\boldsymbol{u}}\mathcal{T}) := {}^I\tilde{\psi}_{\boldsymbol{u}}(L_j\mathcal{T})$. We obtain the induced objects $\big({}^I\tilde{\psi}_{\boldsymbol{u}}(\mathcal{T}), L, \boldsymbol{N}_\Lambda\big)$ in $\mathrm{TS}(D_{I,P}, \Lambda)^{\mathrm{fil}}$. We have the induced morphisms N_i $(i \in I)$ on $\big({}^I\tilde{\psi}_{\boldsymbol{u}}(\mathcal{T}), L, \boldsymbol{N}_\Lambda\big)_P$. We set $\boldsymbol{N}_{\Lambda \sqcup I} := \boldsymbol{N}_\Lambda \sqcup \big(N_i \,\big|\, i \in I\big)$. Thus, we obtain objects $\big({}^I\tilde{\psi}_{\boldsymbol{u}}(\mathcal{T}), L, \boldsymbol{N}_{\Lambda \sqcup I}\big)_P$ in $\mathrm{TS}(D_{I,P}, \Lambda \sqcup I)$. If $\mathcal{I}$ is unramified at P, we obtain objects $\big({}^I\tilde{\psi}_{\mathfrak{a},\boldsymbol{u}}(\mathcal{T}), L, \boldsymbol{N}_{\Lambda \sqcup I}\big)$ in $\mathrm{TS}(D_{I,P}, \Lambda \sqcup I)$ for each $\mathfrak{a} \in \mathcal{I}_P$. We set $N(\Lambda \sqcup I) = N(I) + \sum_{j \in I} N_j$.

The following proposition is an analogue of Proposition 9.1.5.

Proposition 9.3.3 *$\big({}^I\tilde{\psi}_{\boldsymbol{u}}(\mathcal{T}), L, N(\Lambda \sqcup I)\big)_P$ has a relative monodromy filtration W, and $\big({}^I\tilde{\psi}_{\boldsymbol{u}}(\mathcal{T}), W, L, \boldsymbol{N}_{\Lambda \sqcup I}\big)_P$ is a $(\Lambda \sqcup I)$-IMTM on $D_{I,P}$.*

If $\mathcal{I}$ is unramified at P, $\big({}^I\tilde{\psi}_{\mathfrak{a},\boldsymbol{u}}(\mathcal{T}), L, N(\Lambda \sqcup I)\big)_P$ has a relative monodromy filtration W, and $\big({}^I\tilde{\psi}_{\mathfrak{a},\boldsymbol{u}}(\mathcal{T}), W, L, \boldsymbol{N}_{\Lambda \sqcup I}\big)_P$ is a $(\Lambda \sqcup I)$-IMTM on $D_{I,P}$. □

The objects $\big({}^I\tilde{\psi}_{\boldsymbol{u}}(\mathcal{T}), W, L, \boldsymbol{N}_{\Lambda \sqcup I}\big)_P$ and $\big({}^I\tilde{\psi}_{\mathfrak{a},\boldsymbol{u}}(\mathcal{T}), W, L, \boldsymbol{N}_{\Lambda \sqcup I}\big)_P$ are denoted by ${}^I\tilde{\psi}_{\boldsymbol{u}}(\mathcal{T}, W, L, \boldsymbol{N}_\Lambda)_P$ and ${}^I\tilde{\psi}_{\mathfrak{a},\boldsymbol{u}}(\mathcal{T}, W, L, \boldsymbol{N}_\Lambda)_P$ if there is no risk of confusion. They are objects in $\mathcal{M}(D_{I,P}, \Lambda \sqcup I)$.

The following lemma is obvious by definition.

Lemma 9.3.4 *An admissible mixed twistor structure is equivalent to an admissible Λ-IMTM with trivial morphisms. An admissible (w, Λ)-polarizable mixed twistor structure is equivalent to an admissible Λ-IMTM whose weight filtration is pure of weight w.* □

9.3.2 Category of Admissible IMTM

Let $\mathcal{M}(\boldsymbol{X}, \Lambda)$ denote the full subcategory of admissible Λ-IMTM on $\boldsymbol{X}$ in $\mathrm{MTS}^{\mathrm{adm}}(\boldsymbol{X}, \Lambda)^{\mathrm{fil}}$.

Proposition 9.3.5 *The family of the subcategories $\mathcal{M}(\boldsymbol{X}, \Lambda)$ has the property* **M0–3**.

Proof The claim for **M1** is clear. Let us consider the property **M0**. Let $F : (\mathcal{T}_1, W, L, \boldsymbol{N}_\Lambda) \longrightarrow (\mathcal{T}_2, W, L, \boldsymbol{N}_\Lambda)$ be a morphism in $\mathcal{M}(\boldsymbol{X}, \Lambda)$. We have the objects $(\operatorname{Ker} F, W, \boldsymbol{N}_\Lambda)$, $(\operatorname{Im} F, W, \boldsymbol{N}_\Lambda)$ and $(\operatorname{Cok} F, W, \boldsymbol{N}_\Lambda)$ in $\mathrm{MTS}^{\mathrm{adm}}(\boldsymbol{X}, \Lambda)$. Because F is strict with respect to L, they are equipped with naturally induced filtrations L in the category $\mathrm{MTS}^{\mathrm{adm}}(\boldsymbol{X}, \Lambda)$. We also obtain $\operatorname{Ker} \mathrm{Gr}^L F = \mathrm{Gr}^L \operatorname{Ker} F$, $\operatorname{Im} \mathrm{Gr}^L F = \mathrm{Gr}^L \operatorname{Im} F$ and $\operatorname{Cok} \mathrm{Gr}^L F = \mathrm{Gr}^L \operatorname{Cok} F$. Hence, we obtain that $(\operatorname{Ker} F, W, L, \boldsymbol{N}_\Lambda)$, $(\operatorname{Im} F, W, L, \boldsymbol{N}_\Lambda)$ and $(\operatorname{Cok} F, W, L, \boldsymbol{N}_\Lambda)$ are pre-admissible. To check the admissibility, we may assume that $X = \Delta^n$, $D = \{z_1 = 0\}$ and that $\mathcal{I}$ is unramified. Then, we have the natural isomorphisms $\tilde{\psi}_{\mathfrak{a},u}(\operatorname{Ker} F) \simeq \operatorname{Ker} \tilde{\psi}_{\mathfrak{a},u}(F)$, $\tilde{\psi}_{\mathfrak{a},u}(\operatorname{Im} F) \simeq$

$\operatorname{Im}\tilde{\psi}_{\mathfrak{a},u}(F)$, $\tilde{\psi}_{\mathfrak{a},u}(\operatorname{Cok}F) \simeq \operatorname{Cok}\tilde{\psi}_{\mathfrak{a},u}(F)$. The isomorphisms are compatible with the filtrations L and a tuple of morphisms $\boldsymbol{N}_\Lambda$, and the nilpotent morphism N_1. Hence, the isomorphisms are also compatible with $W = M(N(\Lambda \sqcup \{1\}); L)$. Because $\big(\operatorname{Ker}\tilde{\psi}_{\mathfrak{a},u}(F), W, L, \boldsymbol{N}_{\Lambda\sqcup I}\big) \in \mathcal{M}(D, \Lambda \sqcup I)$, we have a relative monodromy filtration of $\big(\operatorname{Ker}\tilde{\psi}_{\mathfrak{a},u}(F), L, N_1\big) \simeq \big(\tilde{\psi}_{\mathfrak{a},u}(\operatorname{Ker}F), L, N_1\big)$. Similar claims hold for $\operatorname{Im}F$ and $\operatorname{Cok}F$. Thus, we obtain that $\mathcal{M}(\boldsymbol{X}, \Lambda)$ is abelian.

Let us consider **M2**. The claim for **M2.1** is clear. We check **M2.2**. Let $\Lambda = \Lambda_0 \sqcup \Lambda_1$ be a decomposition. Let $(\mathcal{T}, W, L, \boldsymbol{N}_\Lambda) \in \mathcal{M}(\boldsymbol{X}, \Lambda)$. Let us prove that the existence of a relative monodromy filtration of $N(\Lambda_1)$ with respect to L. We assume the induction on the length of L. If L is pure, there is nothing to prove. Assume that $\mathcal{T} = L_0\mathcal{T}$, and that the claim holds for $L_{-1}\mathcal{T}$. We construct a filtration $\tilde{L}$ of $\mathcal{T}$ in the category $\mathrm{MTS}^{\mathrm{adm}}(\boldsymbol{X})$ by Deligne's inductive formula. Its restriction to $X - D$ is the relative monodromy filtration of $N(\Lambda_1)_{|X-D}$ with respect to $L_{|X-D}$. Then, we obtain that $\tilde{L}$ is a relative monodromy filtration of $N(\Lambda_1)$ with respect to L. Let us check that $\operatorname{res}^{\Lambda}_{\Lambda_0}(\mathcal{T}, W, L, \boldsymbol{N}_\Lambda) := (\mathcal{T}, W, \tilde{L}, \boldsymbol{N}_{\Lambda_0})$ is admissible Λ_0-IMTM. Because $\operatorname{Gr}^{\tilde{L}}(\mathcal{T}, W, \boldsymbol{N}_{\Lambda_0}) \simeq \operatorname{Gr}^{\tilde{L}}\operatorname{Gr}^{L}(\mathcal{T}, W, \boldsymbol{N}_{\Lambda_0})$, $(\mathcal{T}, W, \tilde{L}, \boldsymbol{N}_{\Lambda_0})$ is pre-admissible. To check the admissibility of $(\mathcal{T}, W, \tilde{L}, \boldsymbol{N}_{\Lambda_0})$, we may assume that $X = \Delta^n$ and $D = \{z_1 = 0\}$, and that $\boldsymbol{\mathcal{I}}$ is unramified. We define the filtration $\tilde{L}$ on $\tilde{\psi}_{z_1,\mathfrak{a},u}(\mathcal{T})$ by $\tilde{L}_j\tilde{\psi}_{z_1,\mathfrak{a},u}(\mathcal{T}) := \tilde{\psi}_{z_1,\mathfrak{a},u}(\tilde{L}_j\mathcal{T})$. Then, we can observe that $\tilde{L} = M(N(\Lambda_1); L)$ on $\tilde{\psi}_{z_1,\mathfrak{a},u}(\mathcal{T})$. Because $\tilde{\psi}_{z_1,\mathfrak{a},u}(\mathcal{T}, W, L, \boldsymbol{N}_\Lambda) \in \mathcal{M}(D, \Lambda \sqcup \{1\})$, we have a relative monodromy filtration of $\big(\tilde{\psi}_{z_1,\mathfrak{a},u}(\mathcal{T}), \tilde{L}, N_1\big)$. Hence, $(\mathcal{T}, W, \tilde{L}, \boldsymbol{N}_{\Lambda_0})$ is admissible.

Let us consider **M3**. Let $\bullet \in \Lambda$, and put $\Lambda_0 := \Lambda \setminus \bullet$. Let $(\mathcal{T}, W, L, \boldsymbol{N}_\Lambda)$ be an object in $\mathcal{M}(\boldsymbol{X}, \Lambda)$. Let $(\mathcal{T}', W, \tilde{L}, \boldsymbol{N}_{\Lambda_0})$ be an object in $\mathcal{M}(\boldsymbol{X}, \Lambda_0)$ with morphisms

$$\Sigma^{1,0}\operatorname{res}^{\Lambda}_{\Lambda_0}(\mathcal{T}, W, L, \boldsymbol{N}_\Lambda) \stackrel{u}{\longrightarrow} (\mathcal{T}', W, \tilde{L}, \boldsymbol{N}_{\Lambda_0}) \stackrel{v}{\longrightarrow} \Sigma^{0,-1}\operatorname{res}^{\Lambda}_{\Lambda_0}(\mathcal{T}, W, L, \boldsymbol{N}_\Lambda)$$

in $\mathcal{M}(\boldsymbol{X}, \Lambda_0)$ such that $v \circ u = N_\bullet$. We obtain the filtration $L(\mathcal{T}')$ of the object $(\mathcal{T}', W, \tilde{L}, \boldsymbol{N}_{\Lambda_0})$ in the category $\mathcal{M}(\boldsymbol{X}, \Lambda_0)$, as the transfer of $L(\mathcal{T})$. We set $N_\bullet := u \circ v$ on $\mathcal{T}'$. The tuple $\boldsymbol{N}_{\Lambda_0} \sqcup \{N_\bullet\}$ on $\mathcal{T}'$ is also denoted by $\boldsymbol{N}_\Lambda$.

Lemma 9.3.6 *$(\mathcal{T}', W, L, \boldsymbol{N}_\Lambda)$ is an object in $\mathcal{M}(\boldsymbol{X}, \Lambda)$.*

Proof We have the decomposition $\operatorname{Gr}^L_w\mathcal{T}' = \operatorname{Im}\operatorname{Gr}^L_w u \oplus \operatorname{Ker}\operatorname{Gr}^L_w v$. Because

$$\operatorname{Im}\operatorname{Gr}^L_w u \simeq \Sigma^{1,0}\operatorname{Im}\operatorname{Gr}^L_w N_\bullet,$$

it is an object in $\mathcal{P}(\boldsymbol{X}, \Lambda, w)$. By using the canonical decomposition, we obtain that $\operatorname{Ker}\operatorname{Gr}^L_w v$ is a direct summand of $\operatorname{Gr}^{\tilde{L}}_w(\mathcal{T}', W, \boldsymbol{N}_{\Lambda_0})$. Hence, it is an object in $\mathcal{P}(\boldsymbol{X}, \Lambda_0, w)$. Thus, we obtain that $\operatorname{Gr}^L_w(\mathcal{T}', W, L, \boldsymbol{N}_\Lambda)$ is an object in $\mathcal{P}(\boldsymbol{X}, \Lambda, w)$.

The existence of relative monodromy filtrations of N_i $(i \in \Lambda)$ with respect to L follows from Proposition 8.3.14. Namely, we construct filtrations for N_i by using Deligne's formula. They give relative monodromy filtrations for N_i on $X \setminus D$ by Proposition 8.3.14. Hence, they are relative monodromy filtration on X. For the admissibility condition, we may assume that $X = \Delta^n$ and $D = \{z_1 = 0\}$, and that $\mathcal{I}$ is unramified. We obtain $\tilde{\psi}_{z_1,\mathfrak{a},u}(\mathcal{T}, W, L, \boldsymbol{N}_\Lambda)$ in $\mathcal{M}(D, \Lambda \sqcup \{1\})$ and $\tilde{\psi}_{z_1,\mathfrak{a},u}(\mathcal{T}', W, \tilde{L}, \boldsymbol{N}_{\Lambda_0})$ in $\mathcal{M}(D, \Lambda_0 \sqcup \{1\})$ with the induced morphisms

$$\Sigma^{1,0} \operatorname{res}^{\Lambda\sqcup\{1\}}_{\Lambda_0\sqcup\{1\}} \tilde{\psi}_{z_1,\mathfrak{a},u}(\mathcal{T}, W, L, \boldsymbol{N}_\Lambda) \longrightarrow \tilde{\psi}_{z_1,\mathfrak{a},u}(\mathcal{T}', W, \tilde{L}, \boldsymbol{N}_{\Lambda_0}) \longrightarrow \Sigma^{0,-1} \operatorname{res}^{\Lambda\sqcup\{1\}}_{\Lambda_0\sqcup\{1\}} \tilde{\psi}_{z_1,\mathfrak{a},u}(\mathcal{T}, W, L, \boldsymbol{N}_\Lambda). \tag{9.6}$$

We define $L_j\tilde{\psi}_{z_1,\mathfrak{a},u}(\mathcal{T}') := \tilde{\psi}_{z_1,\mathfrak{a},u}(L_j\mathcal{T}')$. Then, by the exactness in Corollary 9.1.9, we obtain that the filtration L on $\tilde{\psi}_{z_1,\mathfrak{a},u}(\mathcal{T}')$ is equal to the filtration obtained as the transfer of L on $\tilde{\psi}_{z_1,\mathfrak{a},u}(\mathcal{T})$ by the morphisms in (9.6). By Proposition 8.3.14, $\big(\tilde{\psi}_{z_1,\mathfrak{a},u}(\mathcal{T}'), W, L, \boldsymbol{N}_{\Lambda\sqcup\{1\}}\big)$ is an object in $\mathcal{M}(D, \Lambda \sqcup 1)$. Hence, we have a relative monodromy filtration of $(\tilde{\psi}_{z_1,\mathfrak{a},u}(\mathcal{T}'), L, N_1)$. Then, the claim of Lemma 9.3.6 follows, and the proof of Proposition 9.3.5 is finished. □

9.3.3 Some Operations

We use the notation in Sect. 9.2.5. The following lemma easily follows from Lemmas 9.1.12, 9.1.13, 9.1.14, 9.2.10 and 9.2.11.

Lemma 9.3.7

- *Let* $(\mathcal{T}, W, L, \boldsymbol{N}_\Lambda) \in \mathcal{M}(\boldsymbol{X}, \Lambda)$. *Then, we have*

$$(\mathcal{T}, W, L, \boldsymbol{N}_\Lambda)^* \in \mathcal{M}(\boldsymbol{X}, \Lambda), \quad (\mathcal{T}, W, L, \boldsymbol{N}_\Lambda)^\vee \in \mathcal{M}(\boldsymbol{X}^-, \Lambda),$$

$$j^*(\mathcal{T}, W, L, \boldsymbol{N}_\Lambda) \in \mathcal{M}(\boldsymbol{X}^-, \Lambda), \quad \tilde{\gamma}^*_{\mathrm{sm}}(\mathcal{T}, W, L, \boldsymbol{N}_\Lambda) \in \mathcal{M}(\boldsymbol{X}, \Lambda).$$

- *For* $(\mathcal{T}_1, W, L, \boldsymbol{N}_\Lambda) \in \mathcal{M}(\boldsymbol{X}, \Lambda)$ *and* $(\mathcal{T}_2, W, \boldsymbol{N}_\Lambda) \in \mathcal{M}((X, D, \boldsymbol{0}), \Lambda)$, *we have* $(\mathcal{T}_1, W, L, \boldsymbol{N}_\Lambda) \otimes (\mathcal{T}_2, W, L, \boldsymbol{N}_\Lambda) \in \mathcal{M}(\boldsymbol{X}, \Lambda)$. □

Let us consider the case $X = \Delta^n$, $D_i = \{z_i = 0\}$ and $D = \bigcup_{i=1}^{\ell} D_i$. We also obtain the following.

Lemma 9.3.8

- *Let* $(\mathcal{T}, W, L, \boldsymbol{N}_\Lambda) \in \mathcal{M}(\boldsymbol{X}, \Lambda)$. *Then, for any* $I \subset \underline{\ell}$, *we have the following natural isomorphisms:*

$${}^I\tilde{\psi}_{-\boldsymbol{u}}\big((\mathcal{T}, W, L, \boldsymbol{N}_\Lambda)^\vee\big) \simeq \big({}^I\tilde{\psi}_{\boldsymbol{u}}(\mathcal{T}, W, L, \boldsymbol{N}_\Lambda)\big)^\vee$$

$${}^I\tilde{\psi}_{\boldsymbol{u}}\big((\mathcal{T}, W, L, \boldsymbol{N}_\Lambda)^*\big) \simeq \big({}^I\tilde{\psi}_{\boldsymbol{u}}(\mathcal{T}, W, L, \boldsymbol{N}_\Lambda)\big)^*$$

$$
{}^{I}\widetilde{\psi}_{j^*(\boldsymbol{u})}\big(j^*(\mathcal{T},W,L,\boldsymbol{N}_\Lambda)\big) \simeq j^*\big({}^{I}\widetilde{\psi}_{\boldsymbol{u}}(\mathcal{T},W,L,\boldsymbol{N}_\Lambda)\big)
$$

$$
{}^{I}\widetilde{\psi}_{\widetilde{\gamma}^*(\boldsymbol{u})}\big(\widetilde{\gamma}_{sm}^*(\mathcal{T},W,L,\boldsymbol{N}_\Lambda)\big) \simeq \widetilde{\gamma}_{sm}^*\big({}^{I}\widetilde{\psi}_{\boldsymbol{u}}(\mathcal{T},W,L,\boldsymbol{N}_\Lambda)\big)
$$

- *For $(\mathcal{T}_1,W,\boldsymbol{N}_\Lambda) \in \mathcal{M}(\boldsymbol{X},\Lambda)$ and $(\mathcal{T}_2,W,L,\boldsymbol{N}_\Lambda) \in \mathcal{M}((X,D,\boldsymbol{0}),\Lambda,w_2)$, we have the following natural isomorphisms:*

$$
\begin{aligned}
&{}^{I}\widetilde{\psi}_{\boldsymbol{u}}\big((\mathcal{T}_1,W,L,\boldsymbol{N}_\Lambda)\otimes(\mathcal{T}_2,W,L,\boldsymbol{N}_\Lambda)\big)\\
&\quad\simeq \bigoplus_{(\boldsymbol{u}_1,\boldsymbol{u}_2)\in\mathcal{S}(\boldsymbol{u})} {}^{I}\widetilde{\psi}_{\boldsymbol{u}_1}(\mathcal{T}_1,W,L,\boldsymbol{N}_\Lambda)\otimes{}^{I}\widetilde{\psi}_{\boldsymbol{u}_2}(\mathcal{T}_2,W,L,\boldsymbol{N}_\Lambda)
\end{aligned}
\tag{9.7}
$$

Here, $\mathcal{S}(\boldsymbol{u})$ are as in Lemma 9.1.14. □

9.3.4 A Remark on Nearby Cycle Functors

Let us give a remark related with nearby cycle functors. Let $\mathbb{I}^{a,b}$ be the Beilinson IMTM. For $\boldsymbol{p} \in \mathbb{Z}_{\geq 0}^{\Lambda}$, we consider the induced $(\Lambda\sqcup\bullet)$-IMTM $\mathbb{I}^{a,b}_{\bullet,\boldsymbol{p}} := \big(\mathbb{I}^{a,b},\boldsymbol{p}N_{\mathbb{I}},N_{\mathbb{I}}\big)$. We obtain the Λ-IMTM $\widetilde{\mathbb{I}}^{a,b}_{\bullet,\boldsymbol{p}} := \mathrm{TNIL}_\bullet\big(\mathbb{I}^{a,b}_{\bullet,\boldsymbol{p}}\big)$ on $\mathbb{C}^*$ obtained as the twistor nilpotent orbit.

Set $X = \Delta^n$ and $D = \bigcup_{i=1}^{\ell}\{z_i = 0\}$. Let $K \subset \underline{\ell}$ and $\boldsymbol{m} \in \mathbb{Z}_{>0}^K$. We put $g := z^{\boldsymbol{m}}$, which gives $X \setminus \bigcup_{i\in K} D_i \longrightarrow \mathbb{C}^*$. We put $\widetilde{\mathbb{I}}^{a,b}_{\boldsymbol{m},\boldsymbol{p}} := g^*\widetilde{\mathbb{I}}^{a,b}_{\bullet,\boldsymbol{p}}$.

Let $(\mathcal{T},L,\boldsymbol{N}) \in \mathcal{M}(\boldsymbol{X},\Lambda)$. We obtain $\big(\Pi^{a,b}_{\boldsymbol{m},\boldsymbol{p}}\mathcal{T},L,\widetilde{\boldsymbol{N}}\big) := (\mathcal{T},L,\boldsymbol{N})\otimes\widetilde{\mathbb{I}}^{a,b}_{\boldsymbol{m},\boldsymbol{p}} \in \mathcal{M}(\boldsymbol{X},\Lambda)$. Take $I \subset \{i \in \Lambda \,|\, p_i > 0\}$, and we consider the following morphism in $\mathcal{M}(\boldsymbol{X},\Lambda)$:

$$
\widetilde{N}_I : \Sigma^{|I|,0}\big(\Pi^{0,N}_{\boldsymbol{m},\boldsymbol{p}}\mathcal{T},\widetilde{\boldsymbol{N}}_{I!}L,\widetilde{\boldsymbol{N}}_\Lambda\big) \longrightarrow \Sigma^{0,-|I|}\big(\Pi^{0,N}_{\boldsymbol{m},\boldsymbol{p}}\mathcal{T},\widetilde{\boldsymbol{N}}_{I*}L,\widetilde{\boldsymbol{N}}_\Lambda\big)
$$

The cokernel in $\mathcal{M}(\boldsymbol{X},\Lambda)$ is denoted by $\big(\psi^{(0)}_{\boldsymbol{m},\boldsymbol{p}}(\mathcal{T})_I,\hat{L},\widetilde{\boldsymbol{N}}_\Lambda\big)$. We define the naively induced filtration L on $\psi^{(0)}_{\boldsymbol{m},\boldsymbol{p}}(\mathcal{T})_I$ by $L_j\psi^{(0)}_{\boldsymbol{m},\boldsymbol{p}}(\mathcal{T})_I = \psi^{(0)}_{\boldsymbol{m},\boldsymbol{p}}(L_j\mathcal{T})_I$. We have the induced morphism $N_{\mathbb{I}} : \psi^{(0)}_{\boldsymbol{m},\boldsymbol{p}}(\mathcal{T})_I \longrightarrow \psi^{(0)}_{\boldsymbol{m},\boldsymbol{p}}(\mathcal{T})_I \otimes \boldsymbol{T}(-1)$ induced by $N_{\mathbb{I}}$ on $\widetilde{\mathbb{I}}^{a,b}_{\boldsymbol{m},\boldsymbol{p}}$. We immediately obtain the following lemma from Corollary 8.4.3.

Lemma 9.3.9 *We have $\hat{L} = M(N_{\mathbb{I}};L)[1]$.* □

9.4 Specialization of Admissible Mixed Twistor Structure

9.4.1 Statement

Let $X := \Delta^n$, $D_i := \{z_i = 0\}$ and $D := \bigcup_{i=1}^{\ell} D_i$. Let $(\mathcal{T}, W) \in \mathrm{MTS}^{\mathrm{adm}}(\boldsymbol{X})$. Assume that $\mathcal{I}$ is unramified. For $I \subset \underline{\ell}$, $\mathfrak{a} \in \mathcal{I}$ and $\boldsymbol{u} \in (\mathbb{R} \times \mathbb{C})^I$, we define the filtration L on ${}^I\tilde{\psi}_{\mathfrak{a},\boldsymbol{u}}(\mathcal{T})$ by

$$L_j {}^I\tilde{\psi}_{\mathfrak{a},\boldsymbol{u}}(\mathcal{T}) := {}^I\tilde{\psi}_{\mathfrak{a},\boldsymbol{u}}(W_j\mathcal{T}).$$

We obtain an object $\big({}^I\tilde{\psi}_{\mathfrak{a},\boldsymbol{u}}(\mathcal{T}), L, \boldsymbol{N}_I\big)$ in $\mathrm{TS}(\boldsymbol{D}_I(-\mathfrak{a}), I)^{\mathrm{fil}}$. (See Sect. 9.2.4 We have a relative monodromy filtration W on $\big({}^I\tilde{\psi}_{\mathfrak{a},\boldsymbol{u}}(\mathcal{T}), L, N(I)\big)_{|D_I^{\circ}}$, and the tuple $\big({}^I\tilde{\psi}_{\mathfrak{a},\boldsymbol{u}}(\mathcal{T}), W, L, \boldsymbol{N}_I\big)_{|D_I^{\circ}}$ is an object in $\mathcal{M}(D_I^{\circ}, I)$. Because W is given by Deligne's inductive formula, it is naturally extended to a filtration on ${}^I\tilde{\psi}_{\mathfrak{a},\boldsymbol{u}}(\mathcal{T})$ which is also denoted by W. Thus, we obtain an object $\big({}^I\tilde{\psi}_{\mathfrak{a},\boldsymbol{u}}(\mathcal{T}), W\big)$ in $\mathrm{MTS}(\boldsymbol{D}_I(-\mathfrak{a}))$. We will prove the following proposition in Sect. 9.4.3.

Proposition 9.4.1 *$\big({}^I\tilde{\psi}_{\mathfrak{a},\boldsymbol{u}}(\mathcal{T}), W\big)$ is admissible.*

Let us consider the case $\mathcal{I}$ is not necessarily unramified. For $(\mathcal{T}, W) \in \mathrm{MTS}^{\mathrm{adm}}(\boldsymbol{X})$, we have the objects $\big({}^I\tilde{\psi}_{\boldsymbol{u}}(\mathcal{T}), W\big) \in \mathrm{MTS}(\boldsymbol{D}_I)$ which is admissible by Proposition 9.4.1. We denote $\big({}^I\tilde{\psi}_{\boldsymbol{u}}(\mathcal{T}), W\big)$ by ${}^I\tilde{\psi}_{\boldsymbol{u}}(\mathcal{T}, W)$ if there is no risk of confusion. Thus, we obtain an exact functor

$${}^I\tilde{\psi}_{\boldsymbol{u}} : \mathrm{MTS}^{\mathrm{adm}}(\boldsymbol{X}) \longrightarrow \mathrm{MTS}^{\mathrm{adm}}(\boldsymbol{D}_I).$$

The following corollary is clear by Proposition 9.4.1 and the definition of $\mathcal{M}(\boldsymbol{X}, \Lambda)$.

Corollary 9.4.2 *Let $(\mathcal{T}, W, L, \boldsymbol{N}_\Lambda)$ be an object in $\mathcal{M}(\boldsymbol{X}, \Lambda)$.*

- *For any $\boldsymbol{u} \in (\mathbb{R} \times \mathbb{C})^I$, $({}^I\tilde{\psi}_{\boldsymbol{u}}(\mathcal{T}), L, \boldsymbol{N}_{\Lambda \sqcup I})$ are objects in $\mathcal{M}(\boldsymbol{D}_I, \Lambda \sqcup I)$. Thus, we obtain an exact functor ${}^I\tilde{\psi}_{\boldsymbol{u}} : \mathcal{M}(\boldsymbol{X}, \Lambda) \longrightarrow \mathcal{M}(\boldsymbol{D}_I, \Lambda \sqcup I)$.*
- *Assume that $\mathcal{T}$ is unramified. Then $({}^I\tilde{\psi}_{\mathfrak{a},\boldsymbol{u}}(\mathcal{T}), L, \boldsymbol{N}_{\Lambda \sqcup I})$ are objects in $\mathcal{M}(\boldsymbol{D}_I(-\mathfrak{a}), \Lambda \sqcup I)$ for any $I \subset \underline{\ell}$ and $\boldsymbol{u} \in (\mathbb{R} \times \mathbb{C})^I$. (See Sect. 9.2.4 for $\boldsymbol{D}_I(-\mathfrak{a})$.)* □

9.4.2 Some Notation

We introduce some notation which will be used in Chap. 10. We have the abelian category $\mathcal{A}_{D_I} := \mathrm{MTS}^{\mathrm{adm}}(\boldsymbol{D}_I)$. For any finite set Λ, we put

$$\mathcal{P}_{w,D_I}(\Lambda) := \mathcal{P}(\boldsymbol{D}_I, \Lambda, w), \quad \mathcal{M}_{D_I}(\Lambda) := \mathcal{M}(\boldsymbol{D}_I, \Lambda).$$

For any fixed I, the family $\{\mathcal{P}_{w,D_I}(\Lambda) \mid w, \Lambda\}$ satisfies the conditions **P0–P3**, and the family $\{\mathcal{M}_{D_I}(\Lambda) \mid \Lambda\}$ satisfies the conditions **M0–M3** in Sect. 6.3.1.

We obtain the categories $MLA_{D_I}(\Lambda_1, \Lambda_2)$ and $M'LA_{D_I}(\Lambda_1, \Lambda_2)$ as in Sects. 6.3.6 and 6.3.9. We have a natural equivalence $MLA_{D_I}(\Lambda_1, \Lambda_2) \longrightarrow M'LA_{D_I}(\Lambda_1, \Lambda_2)$ as in Theorem 6.3.18.

For $I \subset J$, we have a naturally defined exact functor

$$\psi_{J,I} : \mathcal{M}_{D_I}(\Lambda) \longrightarrow \mathcal{M}_{D_J}\big(\Lambda \sqcup (J \setminus I)\big)$$

given by ${}^{J\setminus I}\tilde{\psi}_{\boldsymbol{u}_0}$, where $\boldsymbol{u}_0 = (-1,0)^{J\setminus I} \in (\mathbb{R} \times \mathbb{C})^{J\setminus I}$. We have the following commutative diagram of the functors:

$$\begin{array}{ccc}
MLA_{D_I}(\Lambda_1, \Lambda_2) & \xrightarrow{\simeq} & M'LA_{D_I}(\Lambda_1, \Lambda_2) \\
\psi_{JI}\big\downarrow & & \psi_{JI}\big\downarrow \\
MLA_{D_J}(\Lambda_1, \Lambda_2 \sqcup (J\setminus I)) & \xrightarrow{\simeq} & M'LA_{D_J}(\Lambda_1, \Lambda_2 \sqcup (J\setminus I)) \\
\psi_{J\setminus I}\big\uparrow & & \psi_{J\setminus I}\big\uparrow \\
MLA_{D_J}(\Lambda_1 \sqcup (J\setminus I), \Lambda_2) & \xrightarrow{\simeq} & M'LA_{D_J}(\Lambda_1 \sqcup (J\setminus I), \Lambda_2)
\end{array} \tag{9.8}$$

The vertical arrows are exact functors.

9.4.3 Proof of Proposition 9.4.1

Proposition 9.4.3 *Suppose that D is smooth. Let $(\mathcal{T}, W, L, \boldsymbol{N}_\Lambda)$ be an object in* $\mathrm{MTS}(\boldsymbol{X})(\Lambda)^{\mathrm{fil}}$ *such that (i) the filtration L is compatible with the KMS-structure, (ii) $(\mathcal{T}, W, L, \boldsymbol{N}_\Lambda)$ satisfies* **R3** *and* **R4** *in* Sect. 9.3. *Then, it is an object in $\mathcal{M}(\boldsymbol{X}, \Lambda)$, and hence in* $\mathrm{MTS}^{\mathrm{adm}}(\boldsymbol{X}, \Lambda)^{\mathrm{fil}}$.

Proof We have only to prove that $(\mathcal{T}, W)$ is admissible. We may assume that $X = \Delta^n$ and $D = \{z_1 = 0\}$, and that $\mathcal{I}$ is unramified. By the condition **(R4)**, $(\tilde{\psi}_{\mathfrak{a},u}(\mathcal{T}), L, N_1)$ has a relative monodromy filtration $M(N_1; L)$. Let $\mathcal{M}_c$ $(c = 1, 2)$ be the underlying $\mathcal{R}_{X(*D)}$-modules of $\mathcal{T}$. For generic λ, we have $M(N_i; L)$ for $(\tilde{\psi}_{\mathfrak{a},u}(\mathcal{M}_c), L)_{|\lambda \times D}$. Hence, by Proposition 8.5.1, $\big(\tilde{\psi}_{\mathfrak{a},u}(\mathcal{T}), L, N(\Lambda \sqcup \{1\})\big)$ has a relative monodromy filtration $\tilde{W}$, and the tuple $\big(\tilde{\psi}_{\mathfrak{a},u}(\mathcal{T}), \tilde{W}, L, \boldsymbol{N}_{\Lambda\sqcup\{1\}}\big)$ is a $(\Lambda \sqcup \{1\})$-IMTM.

Take $\lambda_0 \in \mathbb{C}$. We have the good-KMS-family of filtered λ-flat bundles $\mathcal{Q}^{(\lambda_0)}_* \mathcal{M}_c$. We set $\mathcal{Q}^{(\lambda_0)}_b W_k(\mathcal{M}_c) := \mathcal{Q}^{(\lambda_0)}_b \mathcal{M}_c \cap W_k(\mathcal{M}_c)$ which give a filtration $\mathcal{Q}^{(\lambda_0)}_* W_k \mathcal{M}_c$ by $\mathcal{O}_{\mathcal{X}^{(\lambda_0)}}$-coherent subsheaves. We have the irregular decompositions

$\mathcal{M}_{c|\hat{\mathcal{D}}} = \bigoplus_{\mathfrak{a}\in\mathcal{I}} \mathcal{M}_{c,\mathfrak{a}}$ and $\mathcal{Q}_*^{(\lambda_0)}\mathcal{M}_{c|\hat{\mathcal{D}}} = \bigoplus_{\mathfrak{a}\in\mathcal{I}} \mathcal{Q}_*^{(\lambda_0)}\mathcal{M}_{c,\mathfrak{a}}$, and hence

$$\operatorname{Gr}_a^{\mathcal{Q}^{(\lambda_0)}} \mathcal{M}_c = \bigoplus_{\mathfrak{a}} \bigoplus_{\mathfrak{p}(\lambda_0,u)=a} \psi_{u,\mathfrak{a}}^{(\lambda_0)}\mathcal{M}_c. \tag{9.9}$$

We have the induced filtration L on $\operatorname{Gr}_a^{\mathcal{Q}^{(\lambda_0)}} \mathcal{M}_c$ which is compatible with (9.9). We have the nilpotent endomorphism $N(\Lambda)$ on $\operatorname{Gr}_a^{\mathcal{Q}^{(\lambda_0)}} \mathcal{M}_c$ which is also compatible with (9.9). Because each $\big(\psi_{u,\mathfrak{a}}^{(\lambda_0)}\mathcal{M}_c, L, N(\Lambda)\big)$ has a relative monodromy filtration, there exists a relative monodromy filtration $M\big(N(\Lambda);L\big)$ on $\operatorname{Gr}_a^{\mathcal{Q}^{(\lambda_0)}} \mathcal{M}_c$.

Lemma 9.4.4 *Each $\mathcal{Q}_*^{(\lambda_0)}W_m\mathcal{M}_c$ is a good-KMS family of filtered λ-flat bundles, and we have*

$$M\big(N(\Lambda);L\big)_m\big(\operatorname{Gr}_a^{\mathcal{Q}^{(\lambda_0)}} \mathcal{M}_c\big) = \operatorname{Gr}_a^{\mathcal{Q}^{(\lambda_0)}}\big(W_m\mathcal{M}_c\big). \tag{9.10}$$

Proof If L is pure, it follows from the assumption **R3**. We assume $L_k\mathcal{T} = \mathcal{T}$, and the claim holds for $(L_{k-1}\mathcal{T}, W, L, \boldsymbol{N}_\Lambda)$. The filtration $W(\mathcal{M}_c)$ is constructed by Deligne's formula:

$$W_{-i+k}\mathcal{M}_c = W_{-i+k}L_{k-1}\mathcal{M}_c + N(\Lambda)^i\big(W_{i+k}\mathcal{M}_c\big) \tag{9.11}$$

$$W_{i+k}\mathcal{M}_c = \operatorname{Ker}\Big(N(\Lambda)^{i+1} : \mathcal{M}_c \longrightarrow \mathcal{M}_c/W_{-i-2+k}\mathcal{M}_c\Big) \tag{9.12}$$

Let us prove that each $\mathcal{Q}_*^{(\lambda_0)}W_m\mathcal{M}_c$ is a good-KMS family of filtered λ-flat bundles. Assume that we know the claims for $W_{-i-2+k}\mathcal{M}_c$. Then, $\mathcal{M}_c/W_{-i-2+k}\mathcal{M}_c$ has the induced KMS-structure, and we have

$$\operatorname{Gr}_*^{\mathcal{Q}^{(\lambda_0)}}\big(\mathcal{M}_c/W_{-i-2+k}\mathcal{M}_c\big) \simeq \operatorname{Gr}_*^{\mathcal{Q}^{(\lambda_0)}} \mathcal{M}_c\big/\operatorname{Gr}_*^{\mathcal{Q}^{(\lambda_0)}} W_{-i-2+k}\mathcal{M}_c.$$

Because $(\tilde{\psi}_{z_1,\mathfrak{a},u}(\mathcal{T}), \tilde{W}, L, \boldsymbol{N}_{\Lambda\sqcup\{1\}})$ is a $(\Lambda\sqcup\{1\})$-IMTM, we obtain that the cokernel of the induced morphism

$$N(\Lambda)^{i+1} : \operatorname{Gr}_*^{\mathcal{Q}^{(\lambda_0)}} \mathcal{M}_c \longrightarrow \operatorname{Gr}_*^{\mathcal{Q}^{(\lambda_0)}}\big(\mathcal{M}_c/W_{-i-2+k}\mathcal{M}_c\big)$$

is strict. Hence, we obtain that the claim holds for $W_{i+k}\mathcal{M}_c$ by Lemma 9.4.5 below.

Assume the claim for $W_{i+k}\mathcal{M}_c$. Because $(\tilde{\psi}_{\mathfrak{a},u}(\mathcal{T}), \tilde{W}, L, \boldsymbol{N}_{\Lambda\sqcup\{1\}})$ is a $(\Lambda\sqcup\{1\})$-IMTM, we obtain that the cokernel of the following morphism is strict:

$$\operatorname{Gr}_*^{\mathcal{Q}^{(\lambda_0)}}\big(W_{-i+k}L_{k-1}\mathcal{M}_c\big) \oplus \operatorname{Gr}_*^{\mathcal{Q}^{(\lambda_0)}}\big(W_{i+k}\mathcal{M}_c\big) \xrightarrow{\text{inclusion}+N(\Lambda)^i} \operatorname{Gr}_*^{\mathcal{Q}^{(\lambda_0)}} \mathcal{M}_c$$

Hence, we obtain the claims for $W_{-i+k}\mathcal{M}_c$ by Lemma 9.4.5 below. Thus, the induction can proceed, and we obtain that $\mathcal{Q}_*^{(\lambda_0)}\mathcal{M}_c$ is a good-KMS family of filtered λ-flat bundles.

The above argument implies that the filtration $\mathcal{Q}_a^{(\lambda_0)}W_*\mathcal{M}_c$ of $\mathcal{Q}_a^{(\lambda_0)}(\mathcal{M}_c)$ is given by Deligne's formula for $N(\Lambda)$ and L on $\mathcal{Q}_a^{(\lambda_0)}(\mathcal{M}_c)$. Hence, we obtain (9.10). □

By Lemma 9.4.4, the filtration W is compatible with the KMS-structure. Because $(\tilde{\psi}_{\mathfrak{a},u}(\mathcal{T}), \tilde{W}, L, \boldsymbol{N}_{\Lambda\sqcup\{1\}})$ is a $(\Lambda \sqcup \{1\})$-IMTM, N_1 has a relative monodromy filtration with respect to the filtration the filtration $\mathcal{Q}_a^{(\lambda_0)}W_*\mathcal{M}_c$. Thus, we obtain that $(\mathcal{T}, W)$ is admissible, and the proof of Proposition 9.4.3 is finished. □

We have used the following standard lemma.

Lemma 9.4.5 *Let $X = \Delta^n$ and $D = \{z_1 = 0\}$. Suppose that $\mathcal{I}$ is unramified. Let $\mathcal{M}_i$ $(i = 1, 2)$ be $\mathcal{I}$-good-KMS smooth $\mathcal{R}_{X(*D)}$-modules. Let $F : \mathcal{M}_1 \longrightarrow \mathcal{M}_2$ be a morphism such that (i)* $\operatorname{Cok}\tilde{\psi}_{z_1,\mathfrak{a},u}(F)$ *are strict for any $\mathfrak{a}$ and u. Then, F is strictly compatible with the KMS-structure, and* $\operatorname{Ker} F$, $\operatorname{Im} F$ *and* $\operatorname{Cok} F$ *have naturally induced KMS-structure. Moreover, we have natural isomorphisms* $\tilde{\psi}_{z_1,\mathfrak{a},u}(\operatorname{Ker} F) \simeq \operatorname{Ker}\tilde{\psi}_{z_1,\mathfrak{a},u}(F)$, $\tilde{\psi}_{z_1,\mathfrak{a},u}(\operatorname{Im} F) \simeq \operatorname{Im}\tilde{\psi}_{z_1,\mathfrak{a},u}(F)$, *and* $\tilde{\psi}_{z_1,\mathfrak{a},u}(\operatorname{Cok} F) \simeq \operatorname{Cok}\tilde{\psi}_{z_1,\mathfrak{a},u}(F)$. □

Let us prove Proposition 9.4.1. Let $(\mathcal{T}, W) \in \mathrm{MTS}^{\mathrm{adm}}(\boldsymbol{X})$. We obtain **R3** for $({}^I\tilde{\psi}_{\boldsymbol{u}}(\mathcal{T}), W, L, \boldsymbol{N}_I)$ from Proposition 9.1.5 and Corollary 9.2.9. For the remaining conditions, we may assume that ∂D_I is smooth. We obtain **R4** from Proposition 9.3.3. Then, applying Proposition 9.4.3, we obtain the claim of Proposition 9.4.1. □

9.5 Integrable Case

9.5.1 Admissible Mixed Twistor Structure

An integrable mixed twistor structure on $\boldsymbol{X}$ is $(\mathcal{T}, W) \in \mathrm{TS}^{\mathrm{int}}(X,D)^{\mathrm{fil}}$ satisfying the conditions; (1) the underlying filtered smooth $\mathcal{R}_{X(*D)}$-triple is a mixed twistor structure on $\boldsymbol{X}$, (2) each $\mathrm{Gr}_w^W(\mathcal{T})$ has an integrable polarization. Let $\mathrm{MTS}^{\mathrm{int}}(\boldsymbol{X}) \subset \mathrm{TS}^{\mathrm{int}}(X,D)^{\mathrm{fil}}$ denote the corresponding full subcategory. It is an abelian category.

An object $(\mathcal{T}, W) \in \mathrm{MTS}^{\mathrm{int}}(\boldsymbol{X})$ is called admissible if the following holds:

- The underlying object in $\mathrm{MTS}(\boldsymbol{X})$ is admissible.
- For each w, $\mathrm{Gr}_w^W(\mathcal{T})$ is obtained as the canonical prolongation of a good wild integrable polarizable variation of pure twistor structure of weight w.

Let $\mathrm{MTS}^{\mathrm{int\,adm}}(\boldsymbol{X}) \subset \mathrm{MTS}^{\mathrm{int}}(\boldsymbol{X})$ be the corresponding full subcategory. We immediately obtain the following from Proposition 9.1.7.

Proposition 9.5.1 $\mathrm{MTS}^{\mathrm{int\,adm}}(\boldsymbol{X})$ *is an abelian subcategory.* □

The following lemma is an integrable analogue of Lemma 9.1.4.

Lemma 9.5.2 *Let $\mathcal{T} \in \mathrm{TS}(X, D)$ come from an integrable good wild polarizable variation of pure twistor structure of weight w. Then, it is naturally an admissible integrable mixed twistor structure on (X, D).* □

The following proposition can be reduced to Proposition 9.1.5.

Proposition 9.5.3 *Let $(\mathcal{T}, W) \in \mathrm{MTS}^{\mathrm{int\,adm}}(\boldsymbol{X})$. Let $P \in D_I^{\circ}$. Take a holomorphic coordinate neighbourhood $(X_P; z_1, \ldots, z_n)$ around P such that $\bigcup_{i=1}^{|I|}\{z_i = 0\}$.*

- ${}^I\tilde{\psi}_{\boldsymbol{u}}(\mathcal{T}, W)_P$ *is naturally an integrable I-IMTM for $\boldsymbol{u} \in (\mathbb{R} \times \{0\})^I$.*
- *If $\mathcal{I}$ is unramified around P, ${}^I\tilde{\psi}_{\mathfrak{a},\boldsymbol{u}}(\mathcal{T}, W)_P$ is naturally an integrable I-IMTM for any $\mathfrak{a} \in \mathrm{Irr}(\mathcal{T}, P)$ and for any $\boldsymbol{u} \in (\mathbb{R} \times \{0\})^I$.* □

Let us consider the specialization as in Sect. 9.4.1. Let $X = \Delta^n$, $D_i = \{z_i = 0\}$ and $D = \bigcup_{i=1}^{\ell} D_i$. The following is easy to see.

Proposition 9.5.4 *Suppose that $(\mathcal{T}, W) \in \mathrm{MTS}^{\mathrm{adm}}(\boldsymbol{X})$ is an integrable. Then, the induced object ${}^I\tilde{\psi}_{\boldsymbol{u}}(\mathcal{T}, W) \in \mathrm{MTS}^{\mathrm{adm}}(\boldsymbol{D}_I)$ is also naturally integrable. If $\mathcal{I}$ is unramified, ${}^I\tilde{\psi}_{\mathfrak{a},\boldsymbol{u}}(\mathcal{T}, W) \in \mathrm{MTS}^{\mathrm{adm}}(\boldsymbol{D}_I(-\mathfrak{a}))$ is also naturally integrable* □

9.5.2 Admissible Polarizable Mixed Twistor Structure

For any finite set Λ, we consider the category

$$\mathrm{MTS}^{\mathrm{int\,adm}}(\boldsymbol{X}, \Lambda) := \mathrm{MTS}^{\mathrm{int\,adm}}(\boldsymbol{X})(\Lambda).$$

An object $(\mathcal{T}, W, \boldsymbol{N}_\Lambda)$ in $\mathrm{MTS}^{\mathrm{int\,adm}}(\boldsymbol{X}, \Lambda)$ is called (w, Λ)-polarizable if the following holds:

- The underlying object in $\mathrm{MTS}^{\mathrm{adm}}(\boldsymbol{X}, \Lambda)$ is (w, Λ)-polarizable,
- It has an integrable polarization, i.e., there exists a Hermitian sesqui-linear duality $S : (\mathcal{T}, W, \boldsymbol{N}_\Lambda) \longrightarrow (\mathcal{T}, W, \boldsymbol{N}_\Lambda)^* \otimes \boldsymbol{T}(-w)$ of weight w such that the restriction $(\mathcal{T}, W, \boldsymbol{N}_\Lambda, S)_{|X \setminus D}$ is a polarized mixed twistor structure on $X \setminus D$.

Let $\mathcal{P}^{\mathrm{int}}(\boldsymbol{X}, \Lambda, w) \subset \mathrm{MTS}^{\mathrm{int\,adm}}(\boldsymbol{X}, \Lambda)$ denote the full subcategory of integrable admissible (w, Λ)-polarizable mixed twistor structure on $\boldsymbol{X}$. The following is an integrable analogue of Proposition 9.2.2.

Proposition 9.5.5 *The family of the categories $\{\mathcal{P}^{\mathrm{int}}(\boldsymbol{X}, \Lambda, w) \mid \Lambda, w\}$ has the property* **P0–3**. □

The following is an analogue of Lemma 9.2.10, and easy to see.

Lemma 9.5.6

- *Let* $(\mathcal{T}, W, \boldsymbol{N}_\Lambda) \in \mathcal{P}^{\mathrm{int}}(\boldsymbol{X}, \Lambda, w)$. *Then, we have*

$$(\mathcal{T}, W, \boldsymbol{N}_\Lambda)^\vee \in \mathcal{P}^{\mathrm{int}}(\boldsymbol{X}^-, \Lambda, -w), \quad (\mathcal{T}, W, \boldsymbol{N}_\Lambda)^* \in \mathcal{P}^{\mathrm{int}}(\boldsymbol{X}, \Lambda, -w).$$

$$j^*(\mathcal{T}, W, \boldsymbol{N}_\Lambda) \in \mathcal{P}^{\mathrm{int}}(\boldsymbol{X}^-, \Lambda, w), \quad \tilde{\gamma}_{sm}^*(\mathcal{T}, W, \boldsymbol{N}_\Lambda)^* \in \mathcal{P}^{\mathrm{int}}(\boldsymbol{X}, \Lambda, w).$$

- *For* $(\mathcal{T}_1, W, \boldsymbol{N}_\Lambda) \in \mathcal{P}^{\mathrm{int}}(\boldsymbol{X}, \Lambda, w_1)$ *and* $(\mathcal{T}_2, W, \boldsymbol{N}_\Lambda) \in \mathcal{P}^{\mathrm{int}}((X, D, \boldsymbol{0}), \Lambda, w_2)$, *the tensor product* $(\mathcal{T}_1, W, \boldsymbol{N}_\Lambda) \otimes (\mathcal{T}_2, W, \boldsymbol{N}_\Lambda)$ *is an object in* $\mathcal{P}^{\mathrm{int}}(\boldsymbol{X}, \Lambda, w_1 + w_2)$. □

Let us consider the specialization as in Sect. 9.2.4. Let $X = \Delta^n$, $D_i = \{z_i = 0\}$ and $D = \bigcup_{i=1}^{\ell} D_i$. Let $(\mathcal{T}, W, \boldsymbol{N}_\Lambda) \in \mathcal{P}^{\mathrm{int}}(\boldsymbol{X}, \Lambda, w)$. The following proposition is an integrable analogue of Proposition 9.2.8.

Proposition 9.5.7 *We have* ${}^I\tilde{\psi}_{\boldsymbol{u}}(\mathcal{T}, W, \boldsymbol{N}_\Lambda) \in \mathcal{P}^{\mathrm{int}}(\boldsymbol{D}_I, \Lambda \sqcup I, w)$. □

9.5.3 Admissible IMTM

Let $(\mathcal{T}, W, L, \boldsymbol{N}) \in \mathrm{MTS}^{\mathrm{int\,adm}}(\boldsymbol{X}, \Lambda)^{\mathrm{fil}}$. It is called an integrable admissible Λ-IMTM. if the following holds:

- It is an admissible Λ-IMTM.
- $\mathrm{Gr}_w^L(\mathcal{T}, W, \boldsymbol{N})$ has an integrable polarization.

Let $\mathcal{M}^{\mathrm{int}}(\boldsymbol{X}, \Lambda) \subset \mathrm{MTS}^{\mathrm{int\,adm}}(\boldsymbol{X}, \Lambda)$ denote the full subcategory of integrable admissible Λ-IMTM on $\boldsymbol{X}$. We can easily deduce the following proposition from Proposition 9.3.5.

Proposition 9.5.8 *The family of the subcategories* $\{\mathcal{M}^{\mathrm{int}}(\boldsymbol{X}, \Lambda) \mid \Lambda\}$ *has the property* **M0–3**. □

The following is an integrable analogue of Corollary 9.4.2. Let $X = \Delta^n$, $D_i = \{z_i = 0\}$ and $D = \bigcup_{i=1}^{\ell} D_i$.

Corollary 9.5.9 *Let* $(\mathcal{T}, W, L, \boldsymbol{N}_\Lambda) \in \mathcal{M}^{\mathrm{int}}(\boldsymbol{X}, \Lambda)$.

- *For any* $\boldsymbol{u} \in (\mathbb{R} \times \{0\})^I$, ${}^I\tilde{\psi}_{\boldsymbol{u}}(\mathcal{T}, W, L, \boldsymbol{N}_{\Lambda \sqcup I})$ *are objects in* $\mathcal{M}^{\mathrm{int}}(\boldsymbol{D}_I, \Lambda \sqcup I)$.
- *If* $\mathcal{I}$ *is unramified,* ${}^I\tilde{\psi}_{\mathfrak{a},\boldsymbol{u}}(\mathcal{T}, W, L, \boldsymbol{N}_{\Lambda \sqcup I})$ *are objects in* $\mathcal{M}^{\mathrm{int}}(\boldsymbol{D}_I(-\mathfrak{a}), \Lambda \sqcup I)$ *for any* $I \subset \underline{\ell}$ *and* $\boldsymbol{u} \in (\mathbb{R} \times \{0\})^I$.

Thus, we obtain an exact functor ${}^I\tilde{\psi}_{\boldsymbol{u}} : \mathcal{M}^{\mathrm{int}}(\boldsymbol{X}, \Lambda) \longrightarrow \mathcal{M}^{\mathrm{int}}(\boldsymbol{D}_I, \Lambda \sqcup I)$. □

Chapter 10
Good Mixed Twistor $\mathscr{D}$-Modules

We study filtered $\mathcal{R}$-triples locally expressed as the gluing of admissible mixed twistor structures given on the intersections of the irreducible components of normally crossing hypersurfaces. The main purpose in this chapter is to prove that such filtered $\mathcal{R}$-triples are mixed twistor $\mathscr{D}$-modules.

Let $X := \Delta^n$, $D_i := \{z_i = 0\}$ and $D := \bigcup_{i=1}^{\ell} D_i$. For any $I \subset \{1, \ldots, \ell\}$, we put $D_I := \bigcap_{i \in I} D_I$. We set $D_\emptyset := X$. We put $\partial D_I := \bigcup_{j \in \underline{\ell} \setminus I}(D_I \cap D_j)$. Let $\boldsymbol{\mathcal{V}} = (\mathcal{V}_I \,|\, I \subset \underline{\ell})$ be a tuple of admissible variations of mixed twistor structure on $(D_I, \partial D_I)$ $(I \subset \underline{\ell})$. Suppose that we are given morphisms $\psi_{z_i}^{(1)}(\mathcal{V}_I) \longrightarrow \mathcal{V}_{I \sqcup \{i\}} \longrightarrow \psi_{z_i}^{(0)}(\mathcal{V}_I)$ $(I \sqcup \{i\} \subset \underline{\ell})$ such that the composite is equal to the canonical morphism $\psi_{z_i}^{(1)}(\mathcal{V}_I) \longrightarrow \psi_{z_i}^{(0)}(\mathcal{V}_I)$ for each (I, i). We impose some commutativity conditions to the morphisms. Then, as explained in Sect. 5.6, we obtain an $\mathcal{R}_X$-triple $\Psi_X(\boldsymbol{\mathcal{V}})$ by using the formalism of Beilinson along z_i $(i \in \underline{\ell})$ successively. Our first issue is to construct the weight filtration on the $\mathcal{R}$-triple $\Psi_X(\boldsymbol{\mathcal{V}})$.

Let us recall the simplest case $n = \ell = 1$, which is already mentioned in the beginning of Chap. 6. Suppose that we are given admissible mixed twistor structure $\mathcal{V}_\emptyset$ on (X, D) and a mixed twistor structure $\mathcal{V}_1$ on D with morphisms $\psi_z^{(1)}(\mathcal{V}_\emptyset) \longrightarrow \mathcal{V}_1 \longrightarrow \psi_z^{(0)}(\mathcal{V}_\emptyset)$ such that the composite is equal to the canonical morphism $\psi_z^{(1)}(\mathcal{V}_\emptyset) \longrightarrow \psi_z^{(0)}(\mathcal{V}_\emptyset)$. The filtration W on $\mathcal{V}_\emptyset$ induces filtrations L on $\psi_z^{(a)}(\mathcal{V}_\emptyset)$ and $\Xi_z^{(0)}(\mathcal{V}_\emptyset)$. We have the weight filtration W on $\mathcal{V}_1$, and we replace the filtration W on $\mathcal{V}_1$ with the filtration obtained as the transfer of L with respect to the morphisms. Then, we obtain the correct weight filtration on $\Psi_X(\boldsymbol{\mathcal{V}})$ from the filtrations L on $\psi_z^{(a)}(\mathcal{V}_\emptyset)$, $\Xi_z^{(0)}(\mathcal{V}_\emptyset)$ and $\mathcal{V}_1$.

We will show that this procedure can be done for general n and ℓ (Proposition 10.1.1). We use Theorem 6.3.18 and Proposition 9.3.5.

Many issues for the filtered $\mathcal{R}$-triples $\Psi_X(\boldsymbol{\mathcal{V}})$ can be translated to the issues for the tuples $\boldsymbol{\mathcal{V}}$. For example, for any monomial function g, we have the operations $\Xi_g^{(a)}$, $\psi_g^{(a)}$ and $\phi_g^{(a)}$ for $\boldsymbol{\mathcal{V}}$, which are the counterparts for $\Xi_g^{(a)}$, $\psi_g^{(a)}$ and $\phi_g^{(a)}$ for $\Psi_X(\boldsymbol{\mathcal{V}})$. It admits us to prove that $\Psi_X(\boldsymbol{\mathcal{V}})$ are weakly admissibly specializable along

T. Mochizuki, *Mixed Twistor $\mathscr{D}$-modules*, Lecture Notes in Mathematics 2125,
DOI 10.1007/978-3-319-10088-3_10

any monomial function $\prod_{i=1}^{\ell} z_i^{m_i}$ on X in Proposition 10.2.5. (See Definition 10.2.1 for weakly admissible specializability.) Moreover, we prove that the canonical prolongations of admissible variations of mixed twistor structure are admissibly specializable along any holomorphic functions in Lemma 10.2.14.

After these preliminary, we shall prove that such $\Psi(\boldsymbol{\mathcal{V}})$ are indeed mixed twistor $\mathscr{D}$-modules in Sect. 10.3. In particular, from a given admissible variation of mixed twistor structure $\mathcal{V}$ on (X,D), we obtain various good mixed twistor $\mathscr{D}$-modules on X. For example, we obtain the mixed twistor $\mathscr{D}$-modules $\mathcal{V}[\star D]$ ($\star = *, !$). For any holomorphic function g on X such that $g^{-1}(0) \subset D$, we obtain the mixed twistor $\mathscr{D}$-modules $\Xi_g^{(a)}(\mathcal{V})$, $\psi_g^{(a)}(\mathcal{V})$ and $\phi_g^{(a)}(\mathcal{V})$. They play fundamental roles in the study of basic properties of general mixed twistor $\mathscr{D}$-modules in Chap. 11.

10.1 Good Gluing Data

10.1.1 An Equivalence

Set $X := \Delta^n$, $D_i = \{z_i = 0\}$ and $D := \bigcup_{i=1}^{\ell} D_i$. Let $\mathcal{I}$ be a good system of ramified irregular values, induced by a good set of ramified irregular values in $\tilde{\mathcal{O}}_X(*D)_O/\tilde{\mathcal{O}}_{X,O}$, where O is the origin of X. We set $\boldsymbol{X} := (X, D, \mathcal{I})$. The induced tuples $(D_I, \partial D_I, \mathcal{I}(I)_{|D_I})$ are denoted by $\boldsymbol{D}_I$. We omit to denote the weight filtration for objects in $\mathrm{MTS}^{\mathrm{adm}}(\boldsymbol{D}_I)$.

We shall use the notation in Sect. 9.4.2. In particular, we have the functors $\psi_{J,I} : \mathcal{M}_{D_I}(\Lambda) \longrightarrow \mathcal{M}_{D_J}(\Lambda \sqcup (J \setminus I))$. We naturally regard $\mathrm{MTS}^{\mathrm{adm}}(\boldsymbol{D}_I) \subset \mathcal{M}_{D_I}(\emptyset)$. Then, we have the following functors:

$$\mathrm{MTS}^{\mathrm{adm}}(\boldsymbol{D}_I) \longrightarrow \mathcal{M}_{D_I}(\emptyset) \xrightarrow{\psi_{J,I}} \mathcal{M}_{D_J}(J \setminus I) \xrightarrow{\mathrm{res}_{\emptyset}^{J\setminus I}} \mathrm{MTS}^{\mathrm{adm}}(\boldsymbol{D}_J)$$

The composite is denoted by $\overline{\psi}_{J,I}$.

Let $H = \bigcup_{i\in K}\{z_i = 0\}$ for some $K \subset \underline{\ell}$. We shall introduce categories $\mathfrak{G}_i(\boldsymbol{X}, *H)$ ($i = 0, 1$). (If $H = \emptyset$, they are denoted by $\mathfrak{G}_i(\boldsymbol{X})$.) Then, we will prove that $\mathfrak{G}_0(\boldsymbol{X}, *H)$ and $\mathfrak{G}_1(\boldsymbol{X}, *H)$ are equivalent. The argument in this subsection can work also in the integrable case.

10.1.1.1 Category $\mathfrak{G}_0(\boldsymbol{X}, *H)$

Objects in the category $\mathfrak{G}_0(\boldsymbol{X},*H)$ are tuples of objects $\mathcal{T}_I \in \mathrm{MTS}^{\mathrm{adm}}(\boldsymbol{D}_I)$ ($I \subset \underline{\ell}\setminus K$) with morphisms in $\mathrm{MTS}^{\mathrm{adm}}(\boldsymbol{D}_{Ii})$

$$\Sigma^{1,0}\overline{\psi}_i(\mathcal{T}_I) \xrightarrow{g_{I,i}} \mathcal{T}_{Ii} \xrightarrow{f_{I,i}} \Sigma^{0,-1}\overline{\psi}_i(\mathcal{T}_I)$$

for $i \in \underline{\ell} \setminus (I \cup K)$ such that $f_{I,i} \circ g_{I,i} = N_i$. For $j, k \in \underline{\ell} \setminus (I \cup K)$, we impose the commutativity of the following diagrams:

$$\begin{array}{ccc}
\Sigma^{1,0}\Big(\overline{\psi}_j(\mathcal{T}_{Ik})\Big) & \xrightarrow{\overline{\psi}_j(f_{I,k})} & \Sigma^{1,-1}\Big(\overline{\psi}_{jk}(\mathcal{T}_I)\Big) \\
{\scriptstyle g_{Ik,j}}\big\downarrow & & \big\downarrow{\scriptstyle \overline{\psi}_k(g_{I,j})} \\
\mathcal{T}_{Ijk} & \xrightarrow{f_{Ij,k}} & \Sigma^{0,-1}\Big(\overline{\psi}_k(\mathcal{T}_{Ij})\Big)
\end{array} \tag{10.1}$$

$$\begin{array}{ccc}
\Sigma^{2,0}\Big(\overline{\psi}_{jk}(\mathcal{T}_I)\Big) & \xrightarrow{\overline{\psi}_j(g_{I,k})} & \Sigma^{1,0}\Big(\overline{\psi}_j(\mathcal{T}_{Ik})\Big) \\
{\scriptstyle \overline{\psi}_k(g_{I,j})}\big\downarrow & & \big\downarrow{\scriptstyle g_{Ik,j}} \\
\Sigma^{1,0}\Big(\overline{\psi}_k(\mathcal{T}_{Ij})\Big) & \xrightarrow{g_{Ij,k}} & \mathcal{T}_{Ijk}
\end{array} \tag{10.2}$$

$$\begin{array}{ccc}
\mathcal{T}_{Ijk} & \xrightarrow{f_{Ik,j}} & \Sigma^{0,-1}\Big(\overline{\psi}_j(\mathcal{T}_{Ik})\Big) \\
{\scriptstyle f_{Ij,k}}\big\downarrow & & \big\downarrow{\scriptstyle \overline{\psi}_j(f_{I,k})} \\
\Sigma^{0,-1}\Big(\overline{\psi}_k(\mathcal{T}_{Ij})\Big) & \xrightarrow{\overline{\psi}_k(f_{I,j})} & \Sigma^{0,-2}\Big(\overline{\psi}_{jk}(\mathcal{T}_I)\Big)
\end{array} \tag{10.3}$$

For $\boldsymbol{\mathcal{T}}^{(i)} = (\mathcal{T}_I^{(i)}) \in \mathfrak{G}_0(\boldsymbol{X}, *H)$ $(i = 1, 2)$, a morphism $\boldsymbol{F} : \boldsymbol{\mathcal{T}}^{(1)} \longrightarrow \boldsymbol{\mathcal{T}}^{(2)}$ in $\mathfrak{G}_0(\boldsymbol{X}, *H)$ is a tuple of morphisms $F_I : \mathcal{T}_I^{(1)} \longrightarrow \mathcal{T}_I^{(2)}$ in $\mathrm{MTS}^{\mathrm{adm}}(\boldsymbol{D}_I)$ such that the following diagram is commutative:

$$\begin{array}{ccccc}
\Sigma^{1,0}\Big(\overline{\psi}_i(\mathcal{T}_I^{(1)})\Big) & \xrightarrow{g_{I,i}^{(1)}} & \mathcal{T}_{Ii}^{(1)} & \xrightarrow{f_{I,i}^{(1)}} & \Sigma^{0,-1}\Big(\overline{\psi}_i(\mathcal{T}_I^{(1)})\Big) \\
{\scriptstyle \overline{\psi}_i(F_I)}\big\downarrow & & {\scriptstyle F_{Ii}}\big\downarrow & & \big\downarrow{\scriptstyle \overline{\psi}_i(F_I)} \\
\Sigma^{1,0}\Big(\overline{\psi}_i(\mathcal{T}_I^{(2)})\Big) & \xrightarrow{g_{I,i}^{(2)}} & \mathcal{T}_{Ii}^{(2)} & \xrightarrow{f_{I,i}^{(2)}} & \Sigma^{0,-1}\Big(\overline{\psi}_i(\mathcal{T}_I^{(1)})\Big)
\end{array} \tag{10.4}$$

For an object $\boldsymbol{\mathcal{T}} = (\mathcal{T}_I)$ in $\mathfrak{G}_0(\boldsymbol{X}, *H)$, each $\mathcal{T}_I$ is equipped with a tuple of morphisms $\boldsymbol{N}_I = (N_i \,|\, i \in I)$, given by $N_i := g_{I\setminus i,i} \circ f_{I\setminus i,i}$. Moreover, for $I \subset J$, $\overline{\psi}_{J,I}(\mathcal{T}_I)$ is equipped with a tuple of the induced morphisms $\boldsymbol{N}_I = (N_i \,|\, i \in I)$ and a tuple of morphisms $\boldsymbol{N}_{J\setminus I} = (N_i \,|\, i \in J \setminus I)$ given by $N_i := f_{I,i} \circ g_{I,i}$. We denote the tuple $\boldsymbol{N}_I \sqcup \boldsymbol{N}_{J\setminus I}$ on $\overline{\psi}_{J,I}(\mathcal{T}_I)$ by $\boldsymbol{N}_J$.

10.1.1.2 Category $\mathfrak{G}_1(\boldsymbol{X}, *H)$

We consider objects $(\mathcal{T}, \mathcal{L})$ in $\mathfrak{G}_0(\boldsymbol{X}, *H)^{\mathrm{fil}}$ such that (i) $(\mathcal{T}_I, \mathcal{L}, \boldsymbol{N}_I) \in \mathcal{M}_{D_I}(I)$ for each $I \subset \underline{\ell} \setminus K$, (ii) the tuple $\big(\overline{\psi}_i(\mathcal{T}_I), \mathcal{T}_{Ii}; g_{I,i}, f_{I,i}; \mathcal{L}\big)$ is filtered S-decomposable. Let $\mathfrak{G}_1(\boldsymbol{X}, *H)$ be the full subcategory of such objects in $\mathfrak{G}_0(\boldsymbol{X}, *H)^{\mathrm{fil}}$.

10.1.1.3 An Equivalence

Proposition 10.1.1 *The forgetful functor $\Psi : \mathfrak{G}_1(\boldsymbol{X}, *H) \longrightarrow \mathfrak{G}_0(\boldsymbol{X}, *H)$ is an equivalence.*

Proof Let us prove the essential surjectivity. Let $\mathcal{T} \in \mathfrak{G}_0(\boldsymbol{X}, *H)$. For each $I \subset J$, we have $\mathcal{U}_I^{(J)} := \psi_{J\setminus I}(\mathcal{T}_I)$. The tuple $\big(\mathcal{U}_I^{(J)} \,\big|\, I \subset J\big)$ with the following induced morphisms for $I_2 \subset I_1$ in $\mathcal{M}_{D_J}(J \setminus I_1)$ give an object in $ML'\mathcal{A}_{D_J}(J)$:

$$\Sigma^{|I_1\setminus I_2|,0}\Big(\mathrm{res}^{J\setminus I_2}_{J\setminus I_1}\big(\mathcal{U}_{I_2}^{(J)}\big)\Big) \longrightarrow \mathcal{U}_{I_1}^{(J)} \longrightarrow \Sigma^{0,-|I_1\setminus I_2|}\Big(\mathrm{res}^{J\setminus I_2}_{J\setminus I_1}\big(\mathcal{U}_{I_2}^{(J)}\big)\Big)$$

By Theorem 6.3.18, we have the corresponding object in $MLA_{D_J}(J)$. Hence, we obtain the filtrations $L^{(J)}$ of $\overline{\psi}_{J\setminus I}(\mathcal{T}_I)$ in $\mathrm{MTS}^{\mathrm{adm}}(\boldsymbol{D}_J)$ for any $I \subset J$ such that $\big(\overline{\psi}_{J\setminus I}(\mathcal{T}_I), L^{(J)}, \boldsymbol{N}_J\big)$ is an object in $\mathcal{M}_{D_J}(J)$. In particular, we obtain a filtration $L^{(J)}$ of $\mathcal{T}_J$, and $(\mathcal{T}_J, L^{(J)}, \boldsymbol{N}_J) \in \mathcal{M}_{D_J}(J)$. By using the commutative diagram (9.8), we obtain that $\overline{\psi}_{J\setminus I}(L_j^{(I)}\mathcal{T}_I) = L_j^{(J)}\overline{\psi}_{J\setminus I}(\mathcal{T}_I)$. Hence, the tuple $\big(L^{(J)}(\mathcal{T}_J) \,\big|\, J \subset \underline{\ell}\big)$ gives a filtration $\mathcal{L}$ of $\mathcal{T}$ in the category $\mathfrak{C}_0(\boldsymbol{X}, *H)^{\mathrm{fil}}$, and $(\mathcal{T}, \mathcal{L})$ is an object in $\mathfrak{G}_1$. Thus, we obtain the essential surjectivity.

The full faithfulness of Ψ follows from the full faithfulness of the functors $MLA_{D_I}(I) \longrightarrow ML'\mathcal{A}_{D_I}(I)$ $(I \subset \underline{\ell})$. □

By definition, we have the forgetful functor $\mathfrak{G}_i(\boldsymbol{X}) \longrightarrow \mathfrak{G}_i(\boldsymbol{X}, *H)$ denoted by $\mathcal{T} \longmapsto \mathcal{T}(*H)$.

10.1.2 Canonical Prolongments

Let $K = K_1 \sqcup K_2$ be a decomposition. We have a functor $[*K_1!K_2]_0 : \mathfrak{G}_0(\boldsymbol{X}, *K) \longrightarrow \mathfrak{G}_0(\boldsymbol{X})$ given as follows. Let $\mathcal{T} = (\mathcal{T}_I, g_{I,i}, f_{I,i}) \in \mathfrak{G}_0(\boldsymbol{X}, *K)$. For $I \subset \underline{\ell}$, we put

$$\widetilde{\mathcal{T}}_I := \Sigma^{|I\cap K_2|,-|I\cap K_1|}\overline{\psi}_{I\cap K}(\mathcal{T}_{I\setminus K})$$

The tuple is equipped with the following morphisms in $\mathcal{A}_{D_{Ii}}$

$$\Sigma^{1,0}\overline{\psi}_i\widetilde{\mathcal{T}}_I \xrightarrow{\overline{\psi}_{K\cap I}(g_{I\setminus K,i})} \widetilde{\mathcal{T}}_{Ii} \xrightarrow{\overline{\psi}_{K\cap I}(f_{I\setminus K,i})} \Sigma^{0,-1}\overline{\psi}_i\widetilde{\mathcal{T}}_I \qquad (i \notin K)$$

$$\Sigma^{1,0}\overline{\psi}_i\widetilde{\mathcal{T}}_I \xrightarrow{\overline{\psi}_{I\cap K}(N_i)} \widetilde{\mathcal{T}}_{Ii} \xrightarrow{\mathrm{id}} \Sigma^{0,-1}\overline{\psi}_i\widetilde{\mathcal{T}}_I \qquad (i \in K_1)$$

$$\Sigma^{1,0}\overline{\psi}_i\widetilde{\mathcal{T}}_I \xrightarrow{\mathrm{id}} \widetilde{\mathcal{T}}_{Ii} \xrightarrow{\overline{\psi}_{K\cap I}(N_i)} \Sigma^{0,-1}\overline{\psi}_i\widetilde{\mathcal{T}}_I \qquad (i \in K_2)$$

We obtain an object $\mathcal{T}[*K_1!K_2]_0$ in $\mathfrak{G}_0(\boldsymbol{X})$. The procedure defines the functor $[*K_1!K_2]_0$. We obtain the induced functor

$$[*K_1!K_2]_0^{\mathrm{fil}} : \mathfrak{G}_0(\boldsymbol{X}, *K)^{\mathrm{fil}} \longrightarrow \mathfrak{G}_0(\boldsymbol{X})^{\mathrm{fil}}.$$

By Proposition 10.1.1, we also have the functor $[*K_1!K_2]_1 : \mathfrak{G}_1(\boldsymbol{X}, *K) \longrightarrow \mathfrak{G}_1(\boldsymbol{X})$ given as the composite of the following functors:

$$\mathfrak{G}_1(\boldsymbol{X}, *K) \xrightarrow[\simeq]{\Psi} \mathfrak{G}_0(\boldsymbol{X}, *K) \xrightarrow{[*K_1!K_2]_0} \mathfrak{G}_0(\boldsymbol{X}) \xleftarrow[\simeq]{\Psi} \mathfrak{G}_1(\boldsymbol{X})$$

Let $(\mathcal{T}, \mathcal{L}) \in \mathfrak{G}_1(\boldsymbol{X}, *K)$. We have $\mathcal{T}[*K_1!K_2]_0 \in \mathfrak{G}_0(\boldsymbol{X})$. It is equipped with two naturally induced filtrations. One is the filtration $\tilde{\mathcal{L}}$ of the object $(\mathcal{T}, \mathcal{L})[*K_1!K_2]_1 \in \mathfrak{G}_1(\boldsymbol{X})$. The other is the filtration $\mathcal{L}$ of the object $\mathcal{T}[*K_1!K_2]_0^{\mathrm{fil}} \in \mathfrak{G}_0(\boldsymbol{X})^{\mathrm{fil}}$. Note that they are not the same in general. The latter is called the naively induced filtration.

We obtain the induced functor $[*K_1!K_2]_0 : \mathfrak{G}_0(\boldsymbol{X}) \longrightarrow \mathfrak{G}_0(\boldsymbol{X})$ given by $\mathcal{T} \longmapsto \big(\mathcal{T}(*K)\big)[*K_1!K_2]_0$, and the corresponding functor $[*K_1!K_2]_1 : \mathfrak{G}_1(\boldsymbol{X}) \longrightarrow \mathfrak{G}_1(\boldsymbol{X})$.

Lemma 10.1.2 *We have the natural transformations* $[!K]_i \longrightarrow \mathrm{id} \longrightarrow [*K]_i$ *as functors on* $\mathfrak{G}_i(\boldsymbol{X})$ $(i = 0, 1)$.

Proof The claim is clear in the case $i = 0$ by construction, which implies the claim in the case $i = 1$. □

10.1.3 Beilinson Functors

We introduce Beilinson functors in the context of $\mathfrak{G}_0(\boldsymbol{X})$. We shall use the notation in Sect. 9.3.4.

Take a monomial $g = z^{\boldsymbol{m}}$, where $\boldsymbol{m} \in \mathbb{Z}^K_{>0}$. We define $\mathcal{T}[\star g]_0 := \mathcal{T}[\star K]_0$ for any $\mathcal{T} \in \mathfrak{G}_0(\boldsymbol{X})$ or $\mathcal{T} \in \mathfrak{G}_0(\boldsymbol{X}, *K)$.

Let $\mathcal{T} = (\mathcal{T}_I, f_{I,i}, g_{I,i}) \in \mathfrak{G}_0(\boldsymbol{X})$. We have $\mathcal{T}(*K) \in \mathfrak{G}_0(\boldsymbol{X}, *K)$. Let $\Pi_g^{a,b}\mathcal{T}(*K) = (\tilde{\mathcal{T}}_I, \tilde{f}_{I,i}, \tilde{g}_{I,i})$ be the object in $\mathfrak{G}_0(\boldsymbol{X}, *K)$ given as follows, for $I \subset \underline{\ell}\setminus K$ and $i \in \underline{\ell} \setminus (K \cup I)$:

$$\widetilde{\mathcal{T}}_I := \mathcal{T}_I \otimes \widetilde{\mathbb{I}}_{\boldsymbol{m}}^{a,b}, \qquad \Sigma^{1,0}\psi_i(\widetilde{\mathcal{T}}_I) \xrightarrow{g_{I,i}\otimes\mathrm{id}} \widetilde{\mathcal{T}}_{Ii} \xrightarrow{f_{I,i}\otimes\mathrm{id}} \Sigma^{0,-1}\psi_i(\widetilde{\mathcal{T}}_I)$$

(See Sect. 9.3.4 for $\widetilde{\mathbb{I}}_{\boldsymbol{m}}^{a,b}$ with $\Lambda = \emptyset$.) We set $\Pi_{g\star}^{a,b}\mathcal{T} := \big(\Pi_g^{a,b}\mathcal{T}\big)[\star g]_0 \in \mathfrak{G}_0(\boldsymbol{X})$ for $\star = *, !$.

Lemma 10.1.3 *Let $I \subset \underline{\ell}$. We set $I_0 := I \setminus K$ and $I_1 := I \cap K$. We also set $\boldsymbol{m}_J = (m_i \mid i \in J)$ for $J \subset K$. Then, we have the following natural isomorphism:*

$$\overline{\psi}_{I_1}(\mathcal{T}_{I_0} \otimes \widetilde{\mathbb{I}}^{a,b}_{\boldsymbol{m}}) \simeq \overline{\psi}_{I_1}(\mathcal{T}_{I_0}) \otimes \widetilde{\mathbb{I}}^{a,b}_{\boldsymbol{m}_{K\setminus I_1}, \boldsymbol{m}_{I_1}}$$

Moreover, for $a \geq a'$ and $b \geq b'$, the following diagram is commutative:

$$\begin{array}{ccc} \left(\Pi^{a,b}_{g!}\boldsymbol{\mathcal{T}}\right)_I & \longrightarrow & \Sigma^{|I_1|,0}\overline{\psi}_{I_1}(\mathcal{T}_{I_0}) \otimes \widetilde{\mathbb{I}}^{a,b}_{\boldsymbol{m}_{K\setminus I_1}, \boldsymbol{m}_{I_1}} \\ \big\downarrow & & \big\downarrow \widetilde{N}_{I_1} \\ \left(\Pi^{a',b'}_{g*}\boldsymbol{\mathcal{T}}\right)_I & \longrightarrow & \Sigma^{0,-|I_1|}\overline{\psi}_{I_1}(\mathcal{T}_{I_0}) \otimes \widetilde{\mathbb{I}}^{a',b'}_{\boldsymbol{m}_{K\setminus I_1}, \boldsymbol{m}_{I_1}} \end{array} \tag{10.5}$$

□

Lemma 10.1.4 *If M is sufficiently large for $\boldsymbol{\mathcal{T}}$, the natural morphisms*

$$\mathrm{Cok}\Big(\Pi^{a,M+1}_{g!}\boldsymbol{\mathcal{T}} \longrightarrow \Pi^{b,M+1}_{g*}\boldsymbol{\mathcal{T}}\Big) \longrightarrow \mathrm{Cok}\Big(\Pi^{a,M}_{g!}\boldsymbol{\mathcal{T}} \longrightarrow \Pi^{b,M}_{g*}\boldsymbol{\mathcal{T}}\Big) \tag{10.6}$$

$$\mathrm{Ker}\Big(\Pi^{-M,a}_{g!}\boldsymbol{\mathcal{T}} \longrightarrow \Pi^{-M,b}_{g*}\boldsymbol{\mathcal{T}}\Big) \longrightarrow \mathrm{Ker}\Big(\Pi^{-M-1,a}_{g!}\boldsymbol{\mathcal{T}} \longrightarrow \Pi^{-M-1,b}_{g*}\boldsymbol{\mathcal{T}}\Big) \tag{10.7}$$

are isomorphisms. We also have a natural isomorphism

$$\mathrm{Cok}\Big(\Pi^{a,M}_{g!}\boldsymbol{\mathcal{T}} \longrightarrow \Pi^{b,M}_{g*}\boldsymbol{\mathcal{T}}\Big) \simeq \mathrm{Ker}\Big(\Pi^{-M,a}_{g!}\boldsymbol{\mathcal{T}} \longrightarrow \Pi^{-M,b}_{g*}\boldsymbol{\mathcal{T}}\Big). \tag{10.8}$$

Proof We can easily deduce the claims of this lemma from the corresponding claims for vector spaces given in Sect. 10.1.3.1 below. □

So, for $a \leq b$, we define $\Pi^{a,b}_{g*!}\boldsymbol{\mathcal{T}} \in \mathfrak{S}_0(\boldsymbol{X})$ as the cokernel of $\Pi^{b,M}_{g!}\boldsymbol{\mathcal{T}} \longrightarrow \Pi^{a,M}_{g*}\boldsymbol{\mathcal{T}}$ for a sufficiently large M. It is isomorphic to the kernel of $\Pi^{-M,a}_{g!}\boldsymbol{\mathcal{T}} \longrightarrow \Pi^{-M,b}_{g!}\boldsymbol{\mathcal{T}}$ for a sufficiently large M.

10.1.3.1 Appendix

We recall a lemma for nilpotent maps on vector spaces. Let V be a finite dimensional $\mathbb{C}$-vector space. Let M be any non-negative integer. For $a \leq b$, we consider the vector space $V^{a,b} := V \otimes \big(t^a\mathbb{C}[t]/t^b\mathbb{C}[t]\big)$. Any element of $V^{a,b}$ is uniquely expressed as $\sum_{j=a}^{b-1} \beta_j t^j$ ($\beta_j \in V$). For any such pairs (a,b) and (a',b'), we define $\rho_{(a,b),(a',b')} : V^{a',b'} \longrightarrow V^{a,b}$ given by

$$\rho_{(a,b),(a',b')}\Big(\sum_{j=a}^{b-1} \beta_j t^j\Big) = \sum_{j=\max\{a,a'\}}^{\min\{b,b'\}-1} \beta_j t^j.$$

Let N_0 be any nilpotent endomorphism of V. We take a number k_0 such that $N_0^{k_0} = 0$. For any $\alpha \neq 0$, we consider the induced endomorphism $\widetilde{N}_0 = N_0 + \alpha t \operatorname{id}_V$ on $V^{a,b}$. The following is easy to check.

Lemma 10.1.5 *Suppose $b - a > k_0$.*

- *We have* $\operatorname{Ker} \tilde{N}_0 \subset \operatorname{Im} \rho_{(a,b),(b-k_0,b)}$. *We also have* $\operatorname{Im} \rho_{(a,b),(a+k_0,b)} \subset \operatorname{Im} \tilde{N}_0$.
- $\rho_{(b-1,b),(a,b)}$ *induces an isomorphism* $\operatorname{Ker} \tilde{N}_0 \simeq Vt^{b-1}$.
- *We have* $V^{a,b} = \operatorname{Im} \tilde{N}_0 \oplus \operatorname{Im} \rho_{(a,b),(a,a+1)}$. □

Let N_i $(i = 1, \ldots, c)$ be a commuting tuple of nilpotent endomorphisms of V. We take a positive integer k_0 such that $N_i^{k_0} = 0$ $(i = 1, \ldots, c)$. Take $\alpha_i \in \mathbb{C}^*$ $(i = 1, \ldots, c)$, and we consider the induced nilpotent maps $\tilde{N}_i = N_i + \alpha_i t \operatorname{id}_V$ on $V^{a,b}$. We set $\tilde{N}_{\underline{c}} := \prod_{i=1}^{c} \tilde{N}_i$.

Lemma 10.1.6 *Suppose $b - a > ck_0$.*

- *We have* $\operatorname{Im} \rho_{(a,b),(a+ck_0,b)} \subset \operatorname{Im} \tilde{N}_{\underline{c}}$.
- $\rho_{(b-c,b),(a,b)}$ *induces an isomorphism* $\operatorname{Ker} \tilde{N}_{\underline{c}} \simeq V_{(b-c,b)}$.
- *We have* $V^{a,b} = \operatorname{Im} \tilde{N}_{\underline{c}} \oplus \operatorname{Im} \rho_{(a,b),(a,a+c)}$. *In particular,* $\rho_{(a,b),(a,a+c)}$ *induces an isomorphism* $V^{a,a+c} \longrightarrow \operatorname{Cok} \tilde{N}_{\underline{c}}$.

Proof It is easy to observe the second claim directly. We use an induction on c. Suppose that the claim holds for $c - 1$. We can easily deduce $\operatorname{Im} \rho_{(a,b),(a+ck_0,b)} \subset \operatorname{Im} \tilde{N}_{\underline{c}}$ from $\operatorname{Im} \rho_{(a,b),(a+(c-1)k_0,b)} \subset \operatorname{Im} \tilde{N}_{\underline{c-1}}$. Because $\operatorname{Ker} \tilde{N}_c \subset \operatorname{Im} \rho_{(a,b),(b-k_0,b)} \subset \operatorname{Im} \rho_{(a,b),(a+(c-1)k_0,b)}$, we have $\operatorname{Ker} \tilde{N}_c \subset \operatorname{Im} \tilde{N}_{\underline{c-1}}$. Hence, we obtain

$$
\begin{aligned}
V^{a,b} &= \operatorname{Im} \tilde{N}_c \oplus \operatorname{Im} \rho_{(a,b),(a,a+1)} \\
&= \operatorname{Im} \tilde{N}_{\underline{c}} \oplus \tilde{N}_c\big(\operatorname{Im} \rho_{(a,b),(a,a+c-1)}\big) \oplus \operatorname{Im} \rho_{(a,b),(a,a+1)} \\
&= \operatorname{Im} \tilde{N}_{\underline{c}} \oplus \operatorname{Im} \rho_{(a,b),(a,a+c)} \qquad (10.9)
\end{aligned}
$$

Thus, we obtain Lemma 10.1.6. □

Corollary 10.1.7 *Let $a \leq b$. If $M > 0$ is sufficiently large, the following natural morphisms are isomorphisms:*

$$
\operatorname{Cok}\big(V^{b,M+1} \xrightarrow{\tilde{N}_c} V^{a,M+1}\big) \simeq \operatorname{Cok}\big(V^{b,M} \xrightarrow{\tilde{N}_c} V^{a,M}\big)
$$

$$
\operatorname{Ker}\big(V^{-M,b} \xrightarrow{\tilde{N}_c} V^{-M,a}\big) \simeq \operatorname{Ker}\big(V^{-M-1,b} \xrightarrow{\tilde{N}_c} V^{-M-1,a}\big)
$$

□

It is also easy to check the following.

Lemma 10.1.8 *Let $a \le b$. If $M > 0$ is sufficiently large, the following natural morphisms are isomorphisms.*

$$\operatorname{Cok}\bigl(V^{-M,M} \xrightarrow{\tilde{N}_{\underline{c}}} V^{-M,M}\bigr) \simeq \operatorname{Cok}\bigl(V^{-M,b} \xrightarrow{\tilde{N}_{\underline{c}}} V^{-M,a}\bigr)$$

$$\operatorname{Ker}\bigl(V^{b,M} \xrightarrow{\tilde{N}_{\underline{c}}} V^{a,M}\bigr) \simeq \operatorname{Ker}\bigl(V^{-M,M} \xrightarrow{\tilde{N}_{\underline{c}}} V^{-M,M}\bigr)$$

As a result, we have $\operatorname{Cok}\bigl(V^{b,M} \xrightarrow{\tilde{N}_{c}} V^{a,M}\bigr) \simeq \operatorname{Ker}\bigl(V^{-M,b} \xrightarrow{\tilde{N}_{c}} V^{-M,a}\bigr)$. □

10.1.4 Nearby Cycle Functors, Maximal Functors and Vanishing Cycle Functors

In particular, we obtain the following objects in $\mathfrak{G}_0(\boldsymbol{X})$:

$$\psi_g^{(a)}\mathcal{T} := \Pi_{g*!}^{a,a}\mathcal{T}, \qquad \Xi_g^{(a)}\mathcal{T} := \Pi_{g*!}^{a,a+1}\mathcal{T}.$$

We define $\phi_g^{(0)}\mathcal{T} \in \mathfrak{G}_0(\boldsymbol{X})$ as the cohomology of the complex in $\mathfrak{G}_0(\boldsymbol{X})$:

$$\mathcal{T}[!g]_0 \longrightarrow \mathcal{T} \oplus \Xi_g^{(0)}\mathcal{T} \longrightarrow \mathcal{T}[*g]_0$$

The induced morphisms $\psi_g^{(1)}(\mathcal{T}) \longrightarrow \phi_g^{(0)}(\mathcal{T})$ and $\phi_g^{(0)}(\mathcal{T}) \longrightarrow \psi_g^{(0)}(\mathcal{T})$ are denoted by can_g and var_g. We can reconstruct $\mathcal{T} \in \mathfrak{G}_0(\boldsymbol{X})$ as the cohomology of the complex in $\mathfrak{G}_0(\boldsymbol{X})$:

$$\psi_g^{(1)}\mathcal{T} \longrightarrow \phi_g^{(0)}\mathcal{T} \oplus \Xi_g^{(0)}\mathcal{T} \longrightarrow \psi_g^{(0)}\mathcal{T} \tag{10.10}$$

The following lemma can be checked by direct computations.

Lemma 10.1.9 *1. The functors $\psi_g^{(a)}$, $\Xi_g^{(a)}$ and $\phi_g^{(a)}$ are exact.*
2. If $\mathcal{T}[*g]_0 = \mathcal{T}$, then $\phi_g^{(0)}\mathcal{T} \simeq \psi_g^{(0)}\mathcal{T}$. If $\mathcal{T}[!g]_0 = \mathcal{T}$, then $\phi_g^{(0)}\mathcal{T} \simeq \psi_g^{(1)}\mathcal{T}$.

Proof We can easily deduce the exactness of $\psi_g^{(a)}$ and $\Xi_g^{(a)}$ from Lemma 10.1.6. Then, we easily obtain the exactness of $\phi_g^{(0)}$ by the construction. The second claim is also clear by the construction. □

We have the corresponding functors for $\mathfrak{G}_1(\boldsymbol{X})$, denoted by the same notation.

Lemma 10.1.10 *Let $(\mathcal{T}, L) \in \mathfrak{G}_1(\boldsymbol{X})$. If L is pure of weight w, then $\psi_g^{(1)}\mathcal{T} \longrightarrow \phi_g^{(0)}\mathcal{T} \longrightarrow \psi_g^{(0)}\mathcal{T}$ is S-decomposable.*

Proof It is enough to consider the case where $g_{I,i}$ are epimorphisms and $f_{I,i}$ are monomorphisms. We have only to consider the case where there exists $J \subset \underline{\ell}$ such that $\mathcal{T}_I = 0$ unless $J \subset I$. It is enough to consider the case $J = \emptyset$. We set $\mathcal{V} := \mathcal{T}_\emptyset$. By the above condition, the canonical morphism $\mathcal{V}[!g]_1 \longrightarrow \boldsymbol{\mathcal{T}}$ is an epimorphism, and the canonical morphism $\boldsymbol{\mathcal{T}} \longrightarrow \mathcal{V}[*g]_1$ is a monomorphism. By the claim 1 in the previous lemma, the induced morphism $\phi_g^{(0)}\mathcal{T}[!g] \longrightarrow \phi_g^{(0)}\boldsymbol{\mathcal{T}}$ is surjective, and $\phi_g^{(0)}\boldsymbol{\mathcal{T}} \longrightarrow \phi_g^{(0)}\mathcal{T}[*g]$ is injective. Then, the claim follows from the claim 2 of Lemma 10.1.9. □

For $\boldsymbol{\mathcal{T}} \in \mathfrak{G}_0(\boldsymbol{X})$, we define $\mathfrak{G}_0(\boldsymbol{X}) \ni \widetilde{\psi}_g(\boldsymbol{\mathcal{T}}) := \psi_g^{(1)}(\boldsymbol{\mathcal{T}}) \otimes \mathcal{U}(1,0) \simeq \psi_g^{(0)}(\boldsymbol{\mathcal{T}}) \otimes \mathcal{U}(0,1)$. More explicitly, it is given as follows. For any $I \subset \underline{\ell}$, we have $\psi_g^{(a)}(\boldsymbol{\mathcal{T}})_I \in \mathrm{MTS}^{\mathrm{adm}}(\boldsymbol{D}_I)$. We set

$$\tilde{\psi}_g(\boldsymbol{\mathcal{T}})_I := \psi_g^{(1)}(\boldsymbol{\mathcal{T}})_I \otimes \mathcal{U}(1,0) \simeq \psi_g^{(0)}(\boldsymbol{\mathcal{T}})_I \otimes \mathcal{U}(0,1).$$

Then, we have the naturally induced morphisms

$$\Sigma^{1,0}\overline{\psi}_i(\tilde{\psi}_g(\boldsymbol{\mathcal{T}})_I) \longrightarrow \tilde{\psi}_g(\boldsymbol{\mathcal{T}})_{Ii} \longrightarrow \Sigma^{0,-1}\overline{\psi}_i(\tilde{\psi}_g(\boldsymbol{\mathcal{T}})_I)$$

with which we obtain the object $\tilde{\psi}_g(\boldsymbol{\mathcal{T}})$. The procedure gives a functor $\tilde{\psi}_g : \mathfrak{G}_0(\boldsymbol{X}) \longrightarrow \mathfrak{G}_0(\boldsymbol{X})$. We also have the induced functors $\tilde{\psi}_g : \mathfrak{G}_1(\boldsymbol{X}) \longrightarrow \mathfrak{G}_1(\boldsymbol{X})$ and $\tilde{\psi}_g : \mathfrak{G}_0(\boldsymbol{X})^{\mathrm{fil}} \longrightarrow \mathfrak{G}_0(\boldsymbol{X})^{\mathrm{fil}}$.

Let $(\boldsymbol{\mathcal{T}}, L) \in \mathfrak{G}_1(\boldsymbol{X})$. We have two kinds of filtrations on $\tilde{\psi}_g(\boldsymbol{\mathcal{T}})$ and $\phi_g^{(0)}(\boldsymbol{\mathcal{T}})$. One is the naively induced filtration L, when we consider as functors on $\mathfrak{G}_0(\boldsymbol{X})^{\mathrm{fil}}$. The other is the weight filtration $\tilde{L}$ as the objects in $\mathfrak{G}_1(\boldsymbol{X})$. We have the naturally induced morphisms:

$$N : \phi_g^{(0)}(\boldsymbol{\mathcal{T}}) \longrightarrow \phi_g^{(-1)}(\boldsymbol{\mathcal{T}}) \simeq \phi_g^{(0)}(\boldsymbol{\mathcal{T}}) \otimes \boldsymbol{T}(-1)$$
$$N : \tilde{\psi}_g(\boldsymbol{\mathcal{T}}) = \psi_g^{(1)}(\boldsymbol{\mathcal{T}}) \otimes \mathcal{U}(1,0) \longrightarrow \psi_g^{(0)}(\boldsymbol{\mathcal{T}}) \otimes \mathcal{U}(1,0) \simeq \tilde{\psi}_g(\boldsymbol{\mathcal{T}}) \otimes \boldsymbol{T}(-1)$$

Proposition 10.1.11 *We have $\tilde{L} = M(N;L)$ on $\tilde{\psi}_g\boldsymbol{\mathcal{T}}$ and $\phi_g^{(0)}\boldsymbol{\mathcal{T}}$.*

Proof The claim for $\tilde{\psi}_g\boldsymbol{\mathcal{T}}$ follows from Lemma 9.3.9. Let us consider $\phi_g^{(0)}\boldsymbol{\mathcal{T}}$. We have only to consider the case L is pure. Then, we have the decomposition $\phi_g^{(0)}\boldsymbol{\mathcal{T}} = \mathrm{Im}\,\mathrm{can}_g \oplus \mathrm{Ker}\,\mathrm{var}_g$. We obtain the relation $\tilde{L} = M(N;L)$ on $\mathrm{Im}\,\mathrm{can}_g$ from the claim for $\tilde{\psi}_g\boldsymbol{\mathcal{T}}$. We have the decomposition $\boldsymbol{\mathcal{T}} = \boldsymbol{\mathcal{T}}_1 \oplus \mathrm{Ker}\,\mathrm{var}_g$, where $\boldsymbol{\mathcal{T}}_1$ has no subobject whose support is contained in $g^{-1}(0)$. Hence, we have $L = \tilde{L}$ on $\mathrm{Ker}\,\mathrm{var}_g$, and we obtain the claim for $\phi_g^{(0)}\boldsymbol{\mathcal{T}}$. □

Note that we can reconstruct the filtration L of $\boldsymbol{\mathcal{T}}$ from (10.10) with the naively induced filtrations L of $\psi_g^{(a)}\boldsymbol{\mathcal{T}}$ $(a = 0, 1)$, $\Xi_g^{(0)}\boldsymbol{\mathcal{T}}$ and $\phi_g^{(0)}\boldsymbol{\mathcal{T}}$. We can also reconstruct L of $\boldsymbol{\mathcal{T}}$ from (10.10) with the filtrations $\tilde{L}$ of $\psi_g^{(a)}\boldsymbol{\mathcal{T}}$ $(a = 0, 1)$, $\Xi_g^{(0)}\boldsymbol{\mathcal{T}}$ and $\phi_g^{(0)}\boldsymbol{\mathcal{T}}$ as objects in $\mathfrak{G}_1(\boldsymbol{X})$.

10.1.5 Gluing Along a Monomial Function

Let g be as above. Let $\boldsymbol{\mathcal{T}} \in \mathfrak{C}_1(\boldsymbol{X}, *K)$. We have $\Xi_g^{(0)}\boldsymbol{\mathcal{T}}$ and $\psi_g^{(a)}\boldsymbol{\mathcal{T}}$ in $\mathfrak{C}_1(\boldsymbol{X})$. Let $\boldsymbol{\mathcal{T}}' \in \mathfrak{C}_1(\boldsymbol{X})$ such that $\mathcal{T}_I' = 0$ if $I \cap K = \emptyset$. Suppose that we are given morphisms

$$\psi_g^{(1)}\boldsymbol{\mathcal{T}} \overset{u}{\longrightarrow} \boldsymbol{\mathcal{T}}' \overset{v}{\longrightarrow} \psi_g^{(0)}\boldsymbol{\mathcal{T}}$$

such that $v \circ u$ is equal to the natural morphism $\psi_g^{(1)}\boldsymbol{\mathcal{T}} \longrightarrow \psi_g^{(0)}\boldsymbol{\mathcal{T}}$. Then, we obtain $\mathrm{Glue}(\boldsymbol{\mathcal{T}}, \boldsymbol{\mathcal{T}}'; u, v) \in \mathfrak{C}_1(\boldsymbol{X})$ as the cohomology of $\psi_g^{(1)}\boldsymbol{\mathcal{T}} \longrightarrow \Xi_g^{(0)}\boldsymbol{\mathcal{T}} \oplus \boldsymbol{\mathcal{T}}' \longrightarrow \psi_g^{(0)}\boldsymbol{\mathcal{T}}$. Let $L^{(1)}$ denote the filtration of $\mathrm{Glue}(\boldsymbol{\mathcal{T}}, \boldsymbol{\mathcal{T}}'; u, v)$ as an object in $\mathfrak{C}_1(\boldsymbol{X})$.

We have the naively induced filtrations L on $\Xi_g^{(0)}\boldsymbol{\mathcal{T}}$ and $\psi_g^{(a)}\boldsymbol{\mathcal{T}}$. We have the filtration L of $\boldsymbol{\mathcal{T}}'$ obtained as the transfer of $L\big(\tilde{\psi}_g(\mathcal{T})\big)$. Then, we obtain a filtration $L^{(2)}$ of $\mathrm{Glue}(\boldsymbol{\mathcal{T}}, \boldsymbol{\mathcal{T}}'; u, v)$ with which $\mathrm{Glue}(\boldsymbol{\mathcal{T}}, \boldsymbol{\mathcal{T}}'; u, v) \in \mathfrak{C}_1(\boldsymbol{X})$.

Lemma 10.1.12 *We have $L^{(1)} = L^{(2)}$.*

Proof We apply the procedure in Sect. 10.1.3 to $\big(\mathrm{Glue}(\boldsymbol{\mathcal{T}}, \boldsymbol{\mathcal{T}}'; u, v), L^{(1)}\big)$. The naively induced filtration L on $\phi_g^{(0)}\,\mathrm{Glue}(\boldsymbol{\mathcal{T}}, \boldsymbol{\mathcal{T}}'; u, v) \simeq \boldsymbol{\mathcal{T}}'$ is the same the filtration obtained as the transfer, by Proposition 10.1.11. Then, the claim of the lemma follows. □

10.2 Good Pre-Mixed Twistor $\mathscr{D}$-Module

10.2.1 Weak Admissible Specializability

We introduce an auxiliary notion.

Definition 10.2.1 Let g be any holomorphic function on a complex manifold X. An object $(\mathcal{T}, W) \in \mathrm{MTW}(X)$ is called weakly filtered specializable along g if the following holds:

- $W_j\mathcal{T}$ and $\mathcal{T}/W_j\mathcal{T}$ $(j \in \mathbb{Z})$ are strictly specializable along g.
- For each $u \in \mathbb{R} \times \mathbb{C}$, the cokernel of $\tilde{\psi}_{g,u}(W_j\mathcal{T}) \longrightarrow \tilde{\psi}_{g,u}(\mathcal{T})$ and the kernel of $\tilde{\psi}_{g,u}(\mathcal{T}) \longrightarrow \tilde{\psi}_{g,u}(\mathcal{T}/W_j\mathcal{T})$ are strict. The cokernel of $\phi_g(W_j\mathcal{T}) \longrightarrow \phi_g(\mathcal{T})$ and the kernel of $\phi_g(\mathcal{T}) \longrightarrow \phi_g(\mathcal{T}/W_j\mathcal{T})$ are also strict. We set $L_j\tilde{\psi}_{g,u}(\mathcal{T}) := \tilde{\psi}_{g,u}(W_j\mathcal{T})$ and $L_j\phi_g(\mathcal{T}) := \phi_g(W_j\mathcal{T})$.

We say that $(\mathcal{T}, W)$ is weakly admissibly specializable if moreover the following holds:

- $\big(\tilde{\psi}_{g,u}(\mathcal{T}), L, \mathcal{N}\big)$ and $\big(\phi_g(\mathcal{T}), L, \mathcal{N}\big)$ have relative monodromy filtrations. □

Note that we do not consider ramified exponential twists in this definition.

10.2.2 Local Case

Let $X = \Delta^n$, $D_i = \{z_i = 0\}$ and $D = \bigcup_{i=1}^{\ell} D_i$. We have the category $\mathfrak{G}_1(\boldsymbol{X})$ in Sect. 10.1.1. We use a symbol $\boldsymbol{\mathcal{V}}$ to denote an object in $\mathfrak{G}_1(\boldsymbol{X})$. We may regard it as a filtered object in the category $\boldsymbol{C}(X, D)$ in Sect. 5.6.1. Then, we obtain a filtered $\mathcal{R}_X$-triple $\Psi_X(\boldsymbol{\mathcal{V}})$, as in Sect. 5.6.2.

Lemma 10.2.2 *The functor Ψ_X from $\mathfrak{C}_1(\boldsymbol{X})$ to $\mathcal{R}$-Tri(X) is exact.*

Proof We use an induction on $n = \dim X$. If $n = 0$, the claim is clear. Let $F : \boldsymbol{\mathcal{V}}_1 \longrightarrow \boldsymbol{\mathcal{V}}_2$ be any morphism in $\mathfrak{C}_1(\boldsymbol{X})$. By using Lemma 5.6.1 and the assumption of the induction, we obtain that $\Psi_X(F)$ is strictly specializable along z_1. Hence, we have the natural isomorphisms

$$\psi_{z_1}^{(a)} \operatorname{Ker} \Psi_X(F) \simeq \Psi_X \operatorname{Ker} \psi_{z_1}^{(a)} F, \quad \psi_{z_1}^{(a)} \operatorname{Im} \Psi_X F \simeq \Psi_X \operatorname{Im} \psi_{z_1}^{(a)} F. \tag{10.11}$$

$$\phi_{z_1}^{(a)} \operatorname{Ker} \Psi_X F \simeq \Psi_X \operatorname{Ker} \phi_{z_1}^{(a)} F, \quad \phi_{z_1}^{(a)} \operatorname{Im} \Psi_X F \simeq \Psi_X \operatorname{Im} \phi_{z_1}^{(a)} F. \tag{10.12}$$

We have similar isomorphisms for $\operatorname{Cok} \Psi_X(F)$. Let $\boldsymbol{\mathcal{V}}_1 \xrightarrow{F_1} \boldsymbol{\mathcal{V}}_2 \xrightarrow{F_2} \boldsymbol{\mathcal{V}}_3$ be an exact sequence in $\mathfrak{C}_1(\boldsymbol{X})$. Suppose the support S of $\operatorname{Ker} \Psi_X(F_2)/\operatorname{Im} \Psi_X(F_2)$ is non-empty. It is easy to see that S is contained in D. There exists $I \subset \underline{\ell}$ such that $S \cap D_I^\circ \neq \emptyset$ and $S \cap D_J^\circ = \emptyset$ for any $J \subsetneq I$. By shrinking X around $D_I^\circ \cap S$, we may assume that $S \subset D_{\underline{\ell}}$. We can reconstruct $\operatorname{Ker} \Psi_X(F_2)$ from $\Xi_{z_1}(\operatorname{Ker} \Psi(F_2)(*z_1))$, $\psi_{z_1}^{(a)}(\operatorname{Ker} \Psi(F_2))$ and $\phi_{z_1}^{(0)}(\operatorname{Ker} \Psi(F_2))$. We can also reconstruct $\operatorname{Im} \Psi_X(F_1)$ from $\Xi_{z_1}(\operatorname{Im} \Psi(F_1)(*z_1))$, $\psi_{z_1}^{(a)}(\operatorname{Im} \Psi(F_1))$ and $\phi_{z_1}^{(0)}(\operatorname{Im} \Psi(F_1))$. Hence, we obtain that the natural morphism $\operatorname{Im} \Psi_X(F_1) \longrightarrow \operatorname{Ker} \Psi_X(F_2)$ is an isomorphism from (10.11), (10.12) and the assumption of the induction. So, we obtain $S = \emptyset$. □

Lemma 10.2.3 *$\Psi_X(\boldsymbol{\mathcal{V}})$ is a pre-mixed twistor $\mathscr{D}$-module for any object $\boldsymbol{\mathcal{V}}$ in $\mathfrak{C}_1(\boldsymbol{X})$.*

Proof We have only to prove that, if $\boldsymbol{\mathcal{V}}$ is pure of weight w then $\Psi_X(\boldsymbol{\mathcal{V}})$ is a polarizable wild pure twistor $\mathscr{D}$-module of weight w. We may also assume that $g_{I,i}$ are epimorphisms and $f_{I,i}$ are monomorphisms. There exists $J \subset \underline{\ell}$ such that (i) $\mathcal{V}_I = 0$ unless $I \supset J$, (ii) $\mathcal{V}_J \neq 0$. By using Lemmas 5.6.1 and 10.2.2, we obtain that $\psi_{z_i}^{(1)} \Psi_X(\boldsymbol{\mathcal{V}}) \longrightarrow \phi_{z_i}^{(0)} \Psi_X(\boldsymbol{\mathcal{V}})$ are epimorphisms and that $\phi_{z_i}^{(0)} \Psi_X(\boldsymbol{\mathcal{V}}) \longrightarrow \psi_{z_i}^{(0)} \Psi_X(\boldsymbol{\mathcal{V}})$ are monomorphisms. Let $\iota_J : D_J \longrightarrow X$ be the inclusion. We have a wild pure twistor $\mathscr{D}$-module $\mathfrak{T}$ of weight w such that $\mathfrak{T}(*D(J^c)) = \iota_{J\dagger} \mathcal{V}_J$. If $\mathcal{M}$ is one of the underlying $\mathcal{R}_X$-modules of $\mathfrak{T}$ or $\Psi_X(\boldsymbol{\mathcal{V}})$, then it is the $\mathcal{R}_X$-submodule of $\mathcal{M}(*z_i)$ generated by $V_{<0}^{(\lambda_0)}$ on $\mathcal{X}^{(\lambda_0)}$, where $V^{(\lambda_0)}$ denotes the V-filtration along z_i. Hence, we obtain that $\mathfrak{T} = \Psi_X(\boldsymbol{\mathcal{V}})$. □

Definition 10.2.4 Let $\mathrm{MTW}^{\mathrm{good}}(\boldsymbol{X}) \subset \mathrm{MTW}(X)$ be the essential image of Ψ_X. Any object in $\mathrm{MTW}^{\mathrm{good}}(\boldsymbol{X})$ is called a good pre-mixed twistor $\mathscr{D}$-module on $\boldsymbol{X}$. □

It is independent of the choice of the coordinate system $(z_1,\dots,z_n)$ by the isomorphisms in Lemma 2.3.1. (See Lemma 5.6.5.)

By definition, Ψ_X is essentially surjective. It is also fully faithful according to Lemma 5.6.3. Hence, Ψ_X gives an equivalence $\mathfrak{G}_1(\boldsymbol{X}) \longrightarrow \mathrm{MTW}^{\mathrm{good}}(\boldsymbol{X})$.

Proposition 10.2.5 *Let $g = z^{\boldsymbol{m}}$ for some $\boldsymbol{m} \in \mathbb{Z}^K_{>0}$ where $K \subset \underline{\ell}$. Let $\boldsymbol{\mathcal{V}} \in \mathfrak{G}_1(\boldsymbol{X})$, and we set $\mathcal{T} := \Psi_X(\boldsymbol{\mathcal{V}})$. Then, $(\mathcal{T}, W) := \Psi_X(\boldsymbol{\mathcal{V}})$ is weakly admissibly specializable along g. Moreover, there exists $(\mathcal{T}, W)[\star g]$ in $\mathrm{MTW}^{\mathrm{good}}(\boldsymbol{X})$, and we have $(\mathcal{T}, W)[\star g] \simeq \Psi_X\big(\boldsymbol{\mathcal{V}}[\star g]\big)$ for which the following is commutative:*

$$\begin{array}{ccccc}
(\mathcal{T},W)[!g] & \longrightarrow & (\mathcal{T},W) & \longrightarrow & (\mathcal{T},W)[*g] \\
\simeq\big\downarrow & & \simeq\big\downarrow & & \simeq\big\downarrow \\
\Psi_X\big(\boldsymbol{\mathcal{V}}[!g]\big) & \longrightarrow & \Psi_X\big(\boldsymbol{\mathcal{V}}\big) & \longrightarrow & \Psi_X\big(\boldsymbol{\mathcal{V}}[*g]\big)
\end{array}$$

Proof Let us begin with the following lemma, forgetting the weight filtrations.

Lemma 10.2.6 *(i) $\mathcal{T} = \Psi_X(\boldsymbol{\mathcal{V}})$ is strictly specializable along g as an $\mathcal{R}_X$-triple, (ii) $\mathcal{T}[\star g]$ $(\star = *, !)$ exist as $\mathcal{R}_X$-triples, (iii) we have the following isomorphisms as $\mathcal{R}_X$-triples:*

$$\tilde{\psi}_{g,u}(\mathcal{T}) \simeq \Psi_X \circ \tilde{\psi}_{g,u}(\boldsymbol{\mathcal{V}}),\ \ \Xi_g(\mathcal{T}) \simeq \Psi_X \circ \Xi_g(\boldsymbol{\mathcal{V}}),\ \ \phi_g(\mathcal{T}) \simeq \Psi_X \circ \phi_g(\boldsymbol{\mathcal{V}}) \tag{10.13}$$

Proof Let us consider the case $\boldsymbol{\mathcal{V}} = \mathcal{V}[*I!J]_1$ in $\mathfrak{C}_1(\boldsymbol{X})$ for $\mathcal{V} \in \mathrm{MTS}^{\mathrm{adm}}(\boldsymbol{X})$, where $I \sqcup J = \underline{\ell}$. By Lemma 5.6.4, we naturally have $\Psi_X(\mathcal{V}[*I!J]_1) \simeq \mathcal{V}[*I!J]$ as an $\mathcal{R}_X$-triple. We set $I_0 := I \setminus K$ and $J_0 := J \setminus K$. By Proposition 5.4.1, the $\mathcal{R}_X(*g)$-triple $\Psi_X(\mathcal{V}[*I!J]_1)(*g)$ is strictly specializable along g, and

$$\Psi_X(\mathcal{V}[*I!J]_1)[\star g] \simeq \Psi_X(\mathcal{V}[*I_0!J_0 \star K]_1)$$

as an $\mathcal{R}_X$-triple. Moreover, we have

$$\Psi_X\Big(\big(\mathcal{V} \otimes \tilde{\mathbb{I}}_g^{a,b}\big)[*I!J]_1\Big)[\star g] = \Psi_X\Big(\big(\mathcal{V} \otimes \tilde{\mathbb{I}}_g^{a,b}\big)[*I_0!J_0 \star K]_1\Big)$$

$(\star = *, !)$ as $\mathcal{R}_X$-triples. Hence, $\Xi_g^{(0)}\big(\Psi_X(\mathcal{V}[*I!J]_1)\big)$ exists as an $\mathcal{R}_X$-triple, and it is given as follows:

$$\begin{aligned}
&\Xi_g^{(0)}\big(\Psi_X(\mathcal{V}[*I!J]_1)\big) \\
&= \mathrm{Ker}\Big(\Psi_X\big(\mathcal{V} \otimes \tilde{\mathbb{I}}_g^{-N,1}[*I_0!J_0!K]_1\big) \longrightarrow \Psi_X\big(\mathcal{V} \otimes \tilde{\mathbb{I}}_g^{-N,0}[*I_0!J_0 * K]_1\big)\Big) \\
&\simeq \Psi_X\big(\Xi_g^{(0)}(\mathcal{V}[*I!J]_1)\big).
\end{aligned} \tag{10.14}$$

Similarly, we have $\psi_g^{(a)}\big(\Psi_X(\mathcal{V}[*I!J]_1)\big) \simeq \Psi_X\psi_g^{(a)}(\mathcal{V}[*I!J]_1)$. Let $\mathcal{P}$ be obtained as the cohomology of the following:

$$\Psi_X\big(\mathcal{V}[*I_0!J_0!K]_1\big) \longrightarrow \Psi_X\big(\Xi_g^{(0)}\mathcal{V}[*I!J]_1\big) \oplus \Psi_X\big(\mathcal{V}[*I!J]_1\big) \longrightarrow \Psi_X\big(\mathcal{V}[*I_0!J_0 * K]_1\big) \tag{10.15}$$

It is naturally isomorphic to $\Psi_X(\phi_g^{(0)}\mathcal{V}[*I!J]_1)$, which is strict. Hence, we obtain that $\Psi_X(\mathcal{V}[*I!J]_1)$ is strictly specializable along g, and that $\mathcal{P}$ is isomorphic to $\phi_g^{(0)}\Psi_X(\mathcal{V}[*I!J]_1) \simeq \Psi_X\big(\phi_g^{(0)}(\mathcal{V}[*I!J]_1)\big)$. We can compute $\tilde{\psi}_{g,u}\big(\Psi_X(\mathcal{V}[*I!J]_1)\big)$ as above, and we obtain that $\tilde{\psi}_{g,u}\Psi_X(\mathcal{V}[*I!J]_1)$ is isomorphic to $\Psi_X\tilde{\psi}_{g,u}(\mathcal{V}[*I!J]_1)$.

By a similar argument, for $I \sqcup J \sqcup H = \underline{\ell}$, we can show the claims in the case $\boldsymbol{\mathcal{V}} = \mathcal{V}[*I!J]_1$ where $\mathcal{V} \in \mathrm{MTS}^{\mathrm{adm}}(\boldsymbol{D}_H)$.

A general object $\boldsymbol{\mathcal{V}} \in \mathfrak{G}_1(\boldsymbol{X})$ is expressed as the cohomology of a complex

$$\boldsymbol{\mathcal{V}}^0 \longrightarrow \boldsymbol{\mathcal{V}}^1 \longrightarrow \boldsymbol{\mathcal{V}}^2,$$

where $\boldsymbol{\mathcal{V}}^p = \bigoplus_{k_p} \mathcal{V}_{k_p}[*I_{k_p}!J_{k_p}]_1$ for some $\mathcal{V}_{k_p} \in \mathrm{MTS}^{\mathrm{adm}}(\boldsymbol{D}_{H_{k_p}})$ and $H_{k_p} = \underline{\ell} \setminus (I_{k_p} \sqcup J_{k_p})$. We have $\Psi_X(\boldsymbol{\mathcal{V}}) = H^1\big(\Psi_X(\boldsymbol{\mathcal{V}}^\bullet)\big)$. We have $\tilde{\psi}_g\Psi_X(\boldsymbol{\mathcal{V}}^i) \simeq \Psi_X\tilde{\psi}_g(\boldsymbol{\mathcal{V}}^i)$, and the cohomology of the complex $\Psi_X\tilde{\psi}_g(\boldsymbol{\mathcal{V}}^0) \longrightarrow \Psi_X\tilde{\psi}_g(\boldsymbol{\mathcal{V}}^1) \longrightarrow \Psi_X\tilde{\psi}_g(\boldsymbol{\mathcal{V}}^2)$ is strict. Similarly, we have $\phi_g^{(0)}\Psi_X(\boldsymbol{\mathcal{V}}^i) \simeq \Psi_X\phi_g^{(0)}(\boldsymbol{\mathcal{V}}^i)$, and the cohomology of the complex $\Psi_X\phi_g^{(0)}(\boldsymbol{\mathcal{V}}^0) \to \Psi_X\phi_g^{(0)}(\boldsymbol{\mathcal{V}}^1) \to \Psi_X\phi_g^{(0)}(\boldsymbol{\mathcal{V}}^2)$ is strict. Hence, we obtain that $\Psi_X(\boldsymbol{\mathcal{V}}^\bullet) = H^1\big(\Psi_X(\boldsymbol{\mathcal{V}}^\bullet)\big)$ is strictly specializable along g, and we have natural isomorphisms $\tilde{\psi}_{g,u}\Psi_X(\boldsymbol{\mathcal{V}}) \simeq \Psi_X\tilde{\psi}_{g,u}(\boldsymbol{\mathcal{V}})$ and $\phi_g^{(0)}\Psi_X(\boldsymbol{\mathcal{V}}) \simeq \Psi_X\phi_g^{(0)}(\boldsymbol{\mathcal{V}})$. In particular, the following morphisms are isomorphisms:

$$\mathrm{can} : \psi_g^{(1)}\Psi_X(\boldsymbol{\mathcal{V}}[!g]_1), \xrightarrow{\simeq} \phi_g^{(0)}\Psi_X(\boldsymbol{\mathcal{V}}[!g]_1)$$

$$\mathrm{var} : \phi_g^{(0)}\Psi_X(\boldsymbol{\mathcal{V}}[*g]_1) \xrightarrow{\simeq} \psi_g^{(0)}\Psi_X(\boldsymbol{\mathcal{V}}[*g]_1)$$

Hence, we obtain that $\Psi_X(\boldsymbol{\mathcal{V}})[\star g] \simeq \Psi_X(\boldsymbol{\mathcal{V}}[\star g]_1)$. We also obtain the isomorphism for $\Xi^{(0)}$. Thus, Lemma 10.2.6 is proved. □

Let $\boldsymbol{\mathcal{V}}_1 \longrightarrow \boldsymbol{\mathcal{V}}_2$ be a monomorphism in $\mathfrak{G}_1(\boldsymbol{X})$. The induced morphism $\tilde{\psi}_{g,u}\Psi_X(\boldsymbol{\mathcal{V}}_1) \longrightarrow \tilde{\psi}_{g,u}\Psi_X(\boldsymbol{\mathcal{V}}_2)$ is identified with $\Psi_X\tilde{\psi}_{g,u}(\boldsymbol{\mathcal{V}}_1) \longrightarrow \Psi_X\tilde{\psi}_{g,u}(\boldsymbol{\mathcal{V}}_2)$, and hence the cokernel is strict. Similarly, the cokernel of $\phi_g\Psi_X(\boldsymbol{\mathcal{V}}_1) \longrightarrow \phi_g\Psi_X(\boldsymbol{\mathcal{V}}_2)$ is strict. For an epimorphism $\boldsymbol{\mathcal{V}}_1 \longrightarrow \boldsymbol{\mathcal{V}}_2$ in $\mathfrak{G}_1(\boldsymbol{X})$, we obtain that the kernel of $\Psi_X\tilde{\psi}_{g,u}(\boldsymbol{\mathcal{V}}_1) \longrightarrow \Psi_X\tilde{\psi}_{g,u}(\boldsymbol{\mathcal{V}}_2)$ and $\phi_g\Psi_X(\boldsymbol{\mathcal{V}}_1) \longrightarrow \phi_g\Psi_X(\boldsymbol{\mathcal{V}}_2)$ are strict. Applying this strictness to the weight filtration of $\boldsymbol{\mathcal{V}}$, we obtain that $\Psi_X(\boldsymbol{\mathcal{V}})$ is filtered specializable along g. Then, the isomorphisms $\tilde{\psi}_{g,u}\Psi_X(\boldsymbol{\mathcal{V}}) \simeq \Psi_X\tilde{\psi}_{g,u}(\boldsymbol{\mathcal{V}})$ and $\phi_g\Psi_X(\boldsymbol{\mathcal{V}}) \simeq \Psi_X\phi_g(\boldsymbol{\mathcal{V}})$ are compatible with the naively induced filtrations. Hence, we obtain that $\Psi_X(\boldsymbol{\mathcal{V}})$ is weakly admissibly specializable by Proposition 10.1.11. We obtain $\Psi_X(\boldsymbol{\mathcal{V}})[\star g] \simeq \Psi_X(\boldsymbol{\mathcal{V}}[\star g])$ in $\mathrm{MTW}^{\mathrm{good}}(X, D)$ from the above results. □

We also obtained the following.

Corollary 10.2.7 *Let $\mathcal{V}$, $\mathcal{T}$ and g be as in Proposition 10.2.5.*

- *$\tilde{\psi}_{g,u}(\mathcal{T})$, $\Xi_g(\mathcal{T})$ and $\phi_g(\mathcal{T})$ are naturally equipped with the weight filtration W, with which they are good pre-mixed twistor $\mathscr{D}_X$-modules.*
- *We have $W = M(\mathcal{N};L)$ on $\phi_g\mathcal{T}$ and $\tilde{\psi}_{g,u}\mathcal{T}$, where L denote the naively induced filtrations, i.e., $L_j\phi_g(\mathcal{T}) = \phi_g(W_j\mathcal{T})$ and $L_j\tilde{\psi}_{g,u}(\mathcal{T}) = \tilde{\psi}_{g,u}(W_j\mathcal{T})$.*

Proof The first claim follows from the isomorphisms in (10.13). The second claim follows from Proposition 10.1.11. □

10.2.3 Global Case

Let X be a complex manifold with a normal crossing hypersurface $D = \bigcup_{i\in\Lambda} D_i$. Let $\mathcal{I}$ be a good set of ramified irregular values on (X,D). We set $\boldsymbol{X} := (X,D,\mathcal{I})$.

Definition 10.2.8 Let $\mathrm{MTW}^{\mathrm{good}}(\boldsymbol{X})$ be the full subcategory of $\mathrm{MTW}(X)$, whose objects $(\mathcal{T},W)$ satisfy the following:

- For any point $P\in X$, we take a coordinate neighbourhood $(X_P;z_1,\ldots,z_n)$ such that $D_P := X_P\cap D = \bigcup_{i=1}^{\ell}\{z_i=0\}$. We set $\boldsymbol{X}_P := (X_P,D_P,\mathcal{I}_{|X_P})$. Then, $(\mathcal{T},W)_{|X_P}\in\mathrm{MTW}^{\mathrm{good}}(\boldsymbol{X}_P)$.

An object in $\mathrm{MTW}^{\mathrm{good}}(\boldsymbol{X})$ is called a good pre-mixed twistor $\mathscr{D}$-module on $\boldsymbol{X}$. □

10.2.3.1 Weakly Admissible Specializability and Localizability

We obtain the following proposition from Proposition 10.2.5 and Corollary 10.2.7.

Proposition 10.2.9 *Let $(\mathcal{T},W)\in\mathrm{MTW}^{\mathrm{good}}(\boldsymbol{X})$. Let g be a holomorphic function on X such that $g^{-1}(0)\subset D$.*

- *$(\mathcal{T},W)$ is weakly admissibly specializable along g. There exist $(\mathcal{T},W)[\star g]$ in $\mathrm{MTW}^{\mathrm{good}}(\boldsymbol{X})$ for $\star = *,!$.*
- *We have $(\tilde{\psi}_{g,u}(\mathcal{T}),W)$, $(\Xi_g(\mathcal{T}),W)$ and $(\phi_g(\mathcal{T}),W)$ in $\mathrm{MTW}^{\mathrm{good}}(\boldsymbol{X})$.*
- *Let L be the filtrations of $\psi_{g,u}\mathcal{T}$ and $\phi_g\mathcal{T}$ naively induced by the weight filtration W of $\mathcal{T}$. (See Corollary 10.2.7.) Then, we have $W = M(\mathcal{N};L)$ on $\tilde{\psi}_{g,u}\mathcal{T}$ and $\phi_g\mathcal{T}$.* □

Let $I\subset\Lambda$. Let H be an effective divisor $\sum_{i\in I} m_iD_i$ $(m_i>0)$.

Proposition 10.2.10 *Let $\star = *$ or $!$. Take any $\mathcal{T}\in\mathrm{MTW}^{\mathrm{good}}(\boldsymbol{X})$. (We omit to denote the weight filtration W.)*

- *$\mathcal{T}$ is an object in* $\mathrm{MTW}^{\mathrm{loc}}(X,H)$.
- *We have* $\mathcal{T}[\star H] \in \mathrm{MTW}^{\mathrm{loc}}(X,H)$ *with an isomorphism* $\big(\mathcal{T}[\star H]\big)(*H) \simeq \mathcal{T}(*H)$ *satisfying the following:*
  - *Let P be any point of D. Let $(X_P, z_1, \dots, z_n)$ be a small coordinate neighbourhood around P such that $D = \bigcup_{i=1}^{\ell}\{z_i = 0\}$ and $H = \sum_{i\in I_P} m_i\{z_i = 0\}$. Then, we have an isomorphism $\mathcal{T}[\star H]_{|X_P} \simeq (\mathcal{T}_{|X_P})[\star I_P]$ which induces $(\mathcal{T}[\star H])(*H)_{|X_P} \simeq \mathcal{T}(*H)_{|X_P}$.*

  Such $\mathcal{T}[\star H]$ is unique up to canonical isomorphisms.
- *In particular,* $\mathcal{T}[\star H] \in \mathrm{MTW}^{\mathrm{good}}(\boldsymbol{X})$, *and it depends only on I.*

Proof The first claim follows from Proposition 10.2.9. By Proposition 7.1.51, $\mathcal{T}[\star H] \in \mathrm{MTW}^{\mathrm{loc}}(X,H)$. By Proposition 10.2.9, $\mathcal{T}[\star H]$ is determined by the filtered $\mathcal{R}$-triple satisfying the condition in the second claim. The third claim immediately follows from the second. □

10.2.3.2 Canonical Prolongments of Admissible Mixed Twistor Structure

Let $\mathcal{V} \in \mathrm{MTS}^{\mathrm{adm}}(\boldsymbol{X})$. Let $\Lambda = I \sqcup J$ be a decomposition. We have a filtered $\mathcal{R}_X$-triple $\mathcal{V}[*I!J]$ with an isomorphism $\big(\mathcal{V}[*I!J]\big)(*D) \simeq \mathcal{V}$ satisfying the following condition.

- Let P be any point of D. Let $(X_P, z_1, \dots, z_n)$ be a small coordinate neighbourhood around P such that $D_P := D \cap X_P = \bigcup_{i=1}^{\ell}\{z_i = 0\}$. We set $\boldsymbol{X}_P := (X_P, D_P, \boldsymbol{\mathcal{I}}_{|X_P})$. We have the decomposition $\underline{\ell} = I_P \sqcup J_P$ induced by $\Lambda = I \sqcup J$. We put $\mathcal{V}_P := \mathcal{V}_{|X_P}$, and we have the object $\mathcal{V}_P[*I_P!J_P]$ in $\mathfrak{G}_1(\boldsymbol{X}_P)$. Then, we have $\mathcal{V}[*I!J]_{|X_P} \simeq \Psi_{X_P}(\mathcal{V}_P[*I_P!J_P]_1)$ which induces $\big(\mathcal{V}[*I!J]\big)_{|X_P}(*D) \simeq \mathcal{V}_{|X_P}$.

Such $\mathcal{V}[*I!J]$ is unique up to canonical isomorphisms.

Proposition 10.2.11 *$\mathcal{V}[*I!J]$ is an object in* $\mathrm{MTW}^{\mathrm{good}}(\boldsymbol{X})$.

Proof By Proposition 10.2.10, we have only to consider the case $J = \emptyset$. We regard D as a reduced effective divisor. Because admissible specializability and localizability are checked locally, we obtain that the filtered $\mathcal{R}_X$-triple $\mathcal{V}[*D]$ is admissible and localizable along D by Proposition 10.2.5. As explained in Sect. 4.4.3.1, we have the $\mathcal{R}$-triples $(\mathrm{Gr}^W \psi_D^{(a)})(\mathcal{V}[*D])$ and $(\mathrm{Gr}^W \phi_D^{(a)})(\mathcal{V}[*D])$. The natural morphisms

$$(\mathrm{Gr}^W \phi_D^{(a)})(\mathcal{V}[*D]) \longrightarrow (\mathrm{Gr}^W \psi_D^{(a)})(\mathcal{V}[*D])$$

are isomorphisms. They are equipped with the naturally induced nilpotent maps $\mathcal{N}$.

Lemma 10.2.12 $\big(\mathrm{Gr}_k^W \psi_D^{(a)}\big)(\mathcal{V}[*D]) \in \mathrm{MT}(X,k)$.

Proof We have the canonical splitting (4.27). Note that $(\mathrm{Gr}_k^W \psi_D^{(a)})(\mathcal{V}[*D])$ and $(\mathrm{Gr}_k^W \psi_D^{(a)})(\mathrm{Gr}^W \mathcal{V}[*D])$ depend only on $\mathcal{V}$ and $\mathrm{Gr}^W(\mathcal{V})$. Hence, we have only to prove the claim of the lemma in the case that $\mathcal{V}$ is pure of weight 0. We may replace $\mathcal{V}[*D]$ with the polarized pure twistor $\mathscr{D}$-module associated to $\mathcal{V}$. Then, the claim follows from Proposition 7.1.29. □

For any $P \in D$, we take a small coordinate neighbourhood $(X_P, z_1, \ldots, z_n)$ around P such that $D_P := X_P \cap D = \bigcup_{i=1}^{\ell}\{z_i = 0\}$. Then, the formula (7.9) holds for $\mathrm{Gr}_k^W(\mathcal{V}[*D])_{|X_P}$. By varying P and by gluing the decompositions, we obtain that $\mathrm{Gr}_k^W(\mathcal{V}[*D])$ is isomorphic to the direct sum of the polarizable pure twistor $\mathscr{D}$-module associated to $\mathrm{Gr}_k^W(\mathcal{V})$, and a direct summand of $\big(\mathrm{Gr}_k^W \psi_D^{(0)}\big)(\mathcal{V}[*D])$. Hence, we obtain that $\mathrm{Gr}_k^W(\mathcal{V}[*D]) \in \mathrm{MT}(X,k)$. □

We obtain the following corollary from Proposition 10.2.5.

Corollary 10.2.13 *Let $K \subset \Lambda$ and $H = \sum_{i\in K} m_i D_i$ $(m_i > 0)$. Let $\mathcal{V} \in \mathrm{MTS}^{\mathrm{adm}}(\boldsymbol{X})$.*

- *$\mathcal{V}[*I!J]$ is weakly admissibly specializable along H.*
- *There exists $\big(\mathcal{V}[*I!J]\big)[\star H]$ in $\mathrm{MTW}^{\mathrm{good}}(\boldsymbol{X})$ and we have the following isomorphisms:*

$$\big(\mathcal{V}[*I!J]\big)[*H] \simeq \mathcal{V}[*(I\cup K)!(J\setminus K)], \quad \big(\mathcal{V}[*I!J]\big)[!H] = \mathcal{V}[*(I\setminus K)!(J\cup K)]$$

□

10.2.3.3 Admissible Specializability Along Any Monomial Functions

We consider $\mathcal{V} \in \mathrm{MTS}^{\mathrm{adm}}(\boldsymbol{X})$. Let $\Lambda = I \sqcup J$ be any decomposition. Let g be any holomorphic function on X such that $g^{-1}(0) \subset D$.

Lemma 10.2.14 *$\mathcal{V}[*I!J]$ is admissibly specializable along g.*

Proof We consider the ramified exponential twist by $\mathfrak{a} \in t^{-1/m}\mathbb{C}[t^{-1/m}]$. Let $\mathrm{Gal}(\varphi_m)$ be the Galois group of the ramified covering $\varphi_m : \mathbb{C}_{t_m} \longrightarrow \mathbb{C}_t$. We set $\mathcal{I}_{-\mathfrak{a}} := \big\{-\kappa^*\mathfrak{a} \,\big|\, \kappa \in \mathrm{Gal}(\varphi_m)\big\}$, which gives a good set of ramified irregular values on $(\mathbb{C}_t, 0)$. (See Sect. 15.1 below.) We obtain a good system of irregular values $\boldsymbol{\mathcal{I}}_{X,-\mathfrak{a}} := (g)^{-1}\mathcal{I}_{-\mathfrak{a}}$ on (X,D). For each $P \in D$, we set $\mathcal{I}_P^{(1)} := \big\{\mathfrak{b}+\mathfrak{c} \,\big|\, \mathfrak{b} \in \mathcal{I}_P,\ \mathfrak{c} \in \mathcal{I}_{X,-\mathfrak{a},P}\big\}$. Then, by applying Proposition 15.1.5 below to the system $\boldsymbol{\mathcal{I}}^{(1)} := \big(\mathcal{I}_P^{(1)} \,\big|\, P \in D\big)$, we obtain a projective morphism of complex manifolds $F : X' \longrightarrow X$ such that (i) $D' := F^{-1}(D)$ is simply normal crossing, (ii) $X' \setminus (g\circ F)^{-1}(0) \simeq X \setminus g^{-1}(0)$, (iii) $\boldsymbol{\mathcal{I}}' := F^{-1}\boldsymbol{\mathcal{I}}^{(1)}$ is a good system of irregular values on (X', D'). Let $\boldsymbol{X}' := (X', D', \boldsymbol{\mathcal{I}}')$. Then, we have $\mathcal{V}'_{-\mathfrak{a}} := F^*\big(\mathcal{V} \otimes g^*\varphi_{m\dagger}\mathcal{T}_{-\mathfrak{a}}\big) \in \mathrm{MTS}^{\mathrm{adm}}(\boldsymbol{X}')$.

Let $D' := \bigcup_{i\in\Lambda'} D'_i$ be the irreducible decomposition. Let $\Lambda' = I' \sqcup J'$ be any decomposition. Then, $\mathcal{V}'_{-\mathfrak{a}}[*I'!J']$ is weakly admissibly specializable along $g \circ F$, according to Proposition 10.2.9. By an argument in Lemma 7.1.31, we obtain that $F_{\dagger}(\mathcal{V}'_{-\mathfrak{a}}[*I'!J'])$ is weakly admissibly specializable along g. Then, we can easily deduce that $\mathcal{V}[*I!J] \otimes g^{*}\varphi_{m\dagger}\mathcal{T}_{-\mathfrak{a}}$ is weakly admissibly specializable along g. By applying it to any $\mathfrak{a}$, we obtain the admissible specializability of $\mathcal{V}[*I!J]$ along g. □

10.2.4 Gluing

Let g be a holomorphic function on X such that $D_1 := g^{-1}(0) \subset D$. Let $\mathcal{T} \in \mathrm{MTW}^{\mathrm{good}}(\boldsymbol{X})$. We have $\psi_g^{(a)}(\mathcal{T})$ and $\Xi_g^{(a)}(\mathcal{T})$ in $\mathrm{MTW}^{\mathrm{good}}(\boldsymbol{X})$. They are equipped with the weight filtrations W as objects in $\mathrm{MTW}^{\mathrm{good}}(\boldsymbol{X})$. They are also equipped with the naively induced filtrations L.

Let $\mathcal{T}' \in \mathrm{MTW}^{\mathrm{good}}(\boldsymbol{X})$ whose support is contained in D_1. Assume that we are given morphisms

$$\psi_g^{(1)}\mathcal{T} \xrightarrow{u} \mathcal{T}' \xrightarrow{v} \psi_g^{(0)}\mathcal{T}$$

in $\mathrm{MTW}^{\mathrm{good}}(\boldsymbol{X})$ such that $v \circ u$ is equal to the natural morphism $\psi_g^{(1)}\mathcal{T} \longrightarrow \psi_g^{(0)}\mathcal{T}$. Then, we obtain $\mathrm{Glue}(\mathcal{T}, \mathcal{T}', u, v)$ in $\mathrm{MTW}^{\mathrm{good}}(\boldsymbol{X})$ as the cohomology of the complex:

$$\psi_g^{(1)}\mathcal{T} \longrightarrow \Xi_g^{(0)}\mathcal{T} \oplus \mathcal{T}' \longrightarrow \psi_g^{(0)}\mathcal{T}. \tag{10.16}$$

We have the weight filtration W on $\mathrm{Glue}(\mathcal{T}, \mathcal{T}', u, v)$ induced by the filtration W of the objects in (10.16).

We have the filtration L of $\mathcal{T}'$ obtained as the transfer of L of $\tilde{\psi}_{g,-\boldsymbol{\delta}}(\mathcal{T})$. Then, we obtain a filtration $W^{(2)}$ of $\mathrm{Glue}(\mathcal{T}, \mathcal{T}', u, v)$ induced by the filtrations L on the objects in (10.16). We obtain the following lemma from Lemma 10.1.12.

Lemma 10.2.15 *We have* $W = W^{(2)}$. □

10.3 Good Mixed Twistor $\mathscr{D}$-Modules

10.3.1 Statement

Let X be a complex manifold. Let D be a normal crossing hypersurface. Let $\mathcal{I}$ be a good system of ramified irregular values on (X, D). We set $\boldsymbol{X} := (X, D, \mathcal{I})$. We shall prove the following theorem.

Theorem 10.3.1 *A good pre-mixed twistor* $\mathscr{D}_X$*-module is a mixed twistor* $\mathscr{D}_X$*-module.*

After the theorem is proved, good pre-mixed twistor $\mathscr{D}$-module is also called good mixed twistor $\mathscr{D}$-module, and $\mathrm{MTW}^{\mathrm{good}}(\boldsymbol{X})$ is also denoted by $\mathrm{MTM}^{\mathrm{good}}(\boldsymbol{X})$.

In particular, for a given admissible mixed twistor structure on $\boldsymbol{X}$, we have its canonical prolongation. Namely, let $\mathcal{V} \in \mathrm{MTS}^{\mathrm{adm}}(\boldsymbol{X})$, and let $D = \bigcup_{i \in \Lambda} D_i$ be the irreducible decomposition. For a decomposition $\Lambda = I \sqcup J$, we have the good mixed twistor $\mathscr{D}$-module $\mathcal{V}[*I!J]$.

We shall also prove the following in Sect. 10.3.4, which is a special case of Proposition 11.2.1 below.

Proposition 10.3.2 *Let $(\mathcal{T}, W) \in \mathrm{MTM}^{\mathrm{good}}(\boldsymbol{X})$. For any holomorphic function f on X, we have $(\mathcal{T}, W)[\star f]$ ($\star = *, !$) in $\mathrm{MTM}(X)$. Moreover, if $f^{-1}(0) = g^{-1}(0)$ for a holomorphic function g, we have a natural isomorphism $(\mathcal{T}, W)[\star f] \simeq (\mathcal{T}, W)[\star g]$.*

Corollary 10.3.3 *Let H be an effective divisor of X. For any object $(\mathcal{T}, W)$ in $\mathrm{MTM}^{\mathrm{good}}(\boldsymbol{X})$, we have $(\mathcal{T}, W)[\star H] \in \mathrm{MTM}(X)$. It depends only on the support of H.*

Proof It follows from Propositions 10.3.2 and 7.1.51. □

10.3.2 Preliminary

Let $\mathcal{V} \in \mathrm{MTS}^{\mathrm{adm}}(\boldsymbol{X})$. Let D_0 be a hypersurface of X. Let $\varphi : X' \longrightarrow X$ be a proper birational morphism such that $D' := \varphi^{-1}\big(D \cup D_0\big)$ is normal crossing. We set $\boldsymbol{X}' := (X', D', \varphi^{-1}\mathcal{I})$. We obtain an object $\mathcal{V}' := \varphi^*(\mathcal{V})(*D') \in \mathrm{MTS}^{\mathrm{adm}}(\boldsymbol{X}')$.

Let $D = D_1 \cup D_2$ be a decomposition into normal crossing hypersurfaces with $\operatorname{codim} D_1 \cap D_2 \geq 2$. We put $D'_2 := \varphi^{-1}(D_2)$, and let $D' = D'_1 \cup D'_2$ be the decomposition with $\operatorname{codim} D'_1 \cap D'_2 \geq 2$. We put $D''_1 := \varphi^{-1}(D_1)$, and let $D' = D''_1 \cup D''_2$ be the decomposition with $\operatorname{codim} D''_1 \cap D''_2 \geq 2$. We shall prove the following lemma in Sect. 10.3.5.

Lemma 10.3.4 *We have natural morphisms in* $\mathrm{MTW}(X)$

$$\varphi_\dagger \mathcal{V}'[*D''_1 !D''_2] \longrightarrow \mathcal{V}[*D_1 !D_2] \longrightarrow \varphi_\dagger \mathcal{V}'[*D'_1 !D'_2]$$

which induces natural isomorphisms $\varphi_\dagger \mathcal{V}' \simeq \mathcal{V}(*D_0) \simeq \varphi_\dagger \mathcal{V}'$.

If one of D_i is empty, $\mathcal{V}[*D_1 !D_2]$ is the image of the natural morphism $\varphi_\dagger \mathcal{V}'[*D''_1 !D''_2] \longrightarrow \varphi_\dagger \mathcal{V}'[*D'_1 !D'_2]$ in $\mathrm{MTW}(X)$.

10.3.3 Localizability of Good Pre-mixed Twistor $\mathscr{D}$-Modules

Let $\mathcal{V} \in \mathrm{MTS}^{\mathrm{adm}}(\boldsymbol{X})$. We have $\mathcal{V}[\star D] \in \mathrm{MTW}^{\mathrm{good}}(\boldsymbol{X})$ for $\star = *, !$.

Lemma 10.3.5 *For any holomorphic function f on X, $\mathcal{V}[\star D]$ ($\star = *, !$) are localizable along f.*

Proof We set $D_0 := f^{-1}(0)$, and we take $\varphi : X' \longrightarrow X$ as in Sect. 10.3.2. We set $D_1 := D$ and $D_2 := \emptyset$ if $\star = *$, and $D_1 = \emptyset$ and $D_2 := D$ if $\star =!$. We use the notation in Sect. 10.3.2. According to Proposition 10.2.9 and Lemma 10.2.14, $\mathcal{V}'[*D_1'!D_2']$ and $\mathcal{V}'[*D_1''!D_2'']$ are admissibly specializable along φ^*f. According to Lemma 7.1.31, we have $\varphi_\dagger\mathcal{V}'[*D_1'!D_2']$ and $\varphi_\dagger\mathcal{V}'[*D_1''!D_2'']$ are admissibly specializable along f. By Lemma 10.3.4 and Proposition 7.1.25, we obtain that $\mathcal{V}[\star D]$ is also admissibly specializable along f. It is easy to see that $\varphi_\dagger\big(\mathcal{V}'[*D_1'!D_2'][*\varphi^*f]\big)$ gives $(\mathcal{V}[\star D])[*f]$, and that $\varphi_\dagger\big(\mathcal{V}'[*D_1''!D_2''][!\varphi^*f]\big)$ gives $(\mathcal{V}[\star D])[!f]$. Thus, we obtain Lemma 10.3.5. □

Corollary 10.3.6 *Let g be a holomorphic function such that $g^{-1}(0) = D$. Then, $\psi_g^{(a)}\mathcal{V}[\star D]$ and $\Xi_g^{(a)}\mathcal{V}[\star D]$ are localizable along any holomorphic function f on X.* □

Lemma 10.3.7 *Let $\mathcal{T} \in \mathrm{MTW}^{\mathrm{good}}(\boldsymbol{X})$. Then, $\mathcal{T}$ is localizable along any holomorphic function f on X.*

Proof We have only to consider the case $X = \Delta^n$ and $D = \bigcup_{i=1}^{\ell}\{z_i = 0\}$. For $\mathcal{T} \in \mathrm{MTW}^{\mathrm{good}}(\boldsymbol{X})$, we put $\rho_1(\mathcal{T}) := \dim \mathrm{Supp}\,\mathcal{T}$, and let $\rho_2(\mathcal{T})$ denote the number of $I \subset \underline{\ell}$ such that $|I| + \dim \mathrm{Supp}\,\mathcal{T} = n$ and $\phi_I(\mathcal{T}) \neq 0$. We set $\rho(\mathcal{T}) := \big(\rho_1(\mathcal{T}), \rho_2(\mathcal{T})\big) \in \mathbb{Z}_{\geq 0} \times \mathbb{Z}_{\geq 0}$. We use the lexicographic order on $\mathbb{Z}_{\geq 0} \times \mathbb{Z}_{\geq 0}$.

We take $I \subset \underline{\ell}$ such that $|I| + \dim \mathrm{Supp}\,\mathcal{T} = n$ and $\phi_I(\mathcal{T}) \neq 0$. We put $h := \prod_{i \in I^c} z_i$. Then, $\mathcal{T}$ can be reconstructed as the cohomology of the complex:

$$\psi_h^{(1)}\mathcal{T} \longrightarrow \phi_h^{(0)}\mathcal{T} \oplus \Xi_h^{(0)}\mathcal{T} \longrightarrow \psi_h^{(0)}\mathcal{T}$$

Because $\rho(\psi_h^{(a)}\mathcal{T}) < \rho(\mathcal{T})$ and $\rho(\phi_h^{(a)}\mathcal{T}) < \rho(\mathcal{T})$, we can apply the hypothesis of the induction. Applying Corollary 10.3.6 to $\Xi_h^{(0)}\mathcal{T}$, we obtain $\mathcal{T} \in \mathrm{MTW}^{\mathrm{loc}}(X, f)$. Thus, the proof of Lemma 10.3.7 is finished. □

10.3.4 Proof of Theorem 10.3.1 and Proposition 10.3.2

Let X be a complex manifold. Let H be a normal crossing hypersurface of X. Let $\mathcal{I}$ be a good system of ramified irregular values on (X, D). We set $\boldsymbol{X} = (X, H, \mathcal{I})$. We consider the following.

P(n): The claim of the theorem holds for any objects $\mathcal{T} \in \mathrm{MTW}^{\mathrm{good}}(\boldsymbol{X})$ with $\dim \mathrm{Supp}\,\mathcal{T} \leq n$.

We prove $P(n)$ by an induction on n. Assume that $P(n-1)$ holds.

Let us consider the case that $X := \Delta^n$ and $H := \bigcup_{i=1}^{\ell}\{z_i = 0\}$. We take $\mathcal{V} \in \mathrm{MTS}^{\mathrm{adm}}(\boldsymbol{X})$, and let us prove that $\mathcal{T} = \mathcal{V}[*H]$ is a mixed twistor $\mathscr{D}$-module. We have already known that $\mathcal{T}$ is admissibly specializable. Let g be any holomorphic function on X. We take a projective birational morphism $\varphi : X' \longrightarrow X$ such that $H' := \varphi^{-1}(H \cup g^{-1}(0))$ is normal crossing. We obtain $\mathcal{V}' := \varphi^*(\mathcal{V})$ in $\mathrm{MTS}^{\mathrm{adm}}(\boldsymbol{X}')$. We put $H_1' := \varphi^{-1}(H)$, and let H_2' be the complement of H_1' in H'. We obtain $\mathcal{T}_1 := \mathcal{V}'[!H_2' * H_1']$ and $\mathcal{T}_2 := \mathcal{V}'[*H']$ in $\mathrm{MTW}^{\mathrm{good}}(\boldsymbol{X}')$. We put $g' := \varphi^*(g)$. By the hypothesis of the induction,

$$\phi_{g'}(\mathcal{T}_1), \quad \phi_{g'}(\mathcal{T}_2), \quad \tilde{\psi}_{g',u}(\mathcal{T}_1), \quad \tilde{\psi}_{g',u}(\mathcal{T}_2)$$

are mixed twistor $\mathscr{D}$-modules. By Lemma 10.3.4, we have $\mathcal{T} \simeq \mathrm{Im}\big(\varphi_\dagger(\mathcal{T}_1) \longrightarrow \varphi_\dagger(\mathcal{T}_2)\big)$. We obtain

$$\phi_g(\mathcal{T}) \simeq \mathrm{Im}\Big(\varphi_\dagger\phi_{g'}(\mathcal{T}_1) \longrightarrow \varphi_\dagger\phi_{g'}(\mathcal{T}_2)\Big),$$

$$\tilde{\psi}_{g,u}(\mathcal{T}) \simeq \mathrm{Im}\Big(\varphi_\dagger\tilde{\psi}_{g',u}(\mathcal{T}_1) \longrightarrow \varphi_\dagger\tilde{\psi}_{g',u}(\mathcal{T}_2)\Big).$$

Hence, we obtain $\phi_g(\mathcal{T})$ and $\tilde{\psi}_{g,u}(\mathcal{T})$ are mixed twistor $\mathscr{D}$-modules.

For any $\mathfrak{a} \in t^{-1/m}\mathbb{C}[t^{-1/m}]$, by applying the argument in Lemma 10.2.14, we obtain that $\tilde{\psi}_{g',\mathfrak{a},u}(\mathcal{T}_i)$ are mixed twistor $\mathscr{D}$-modules. We obtain that $\tilde{\psi}_{g,\mathfrak{a},u}(\mathcal{T})$ are mixed twistor $\mathscr{D}$-modules as above. Hence, we obtain $\mathcal{T} = \mathcal{V}[*H]$ is a mixed twistor $\mathscr{D}$-module. Similarly, we obtain that $\mathcal{V}[!H]$ is a mixed twistor $\mathscr{D}$-module.

Let us consider the case that $X = \Delta^N$ for $N \geq n$, and $H = \bigcup_{i=1}^{\ell}\{z_i = 0\}$. We set $H_{[n]} := \bigcup_{|I|=N-n} H_I$, where $H_I = \bigcap_{i\in I} H_i$. Let $\mathcal{T} \in \mathrm{MTW}^{\mathrm{good}}(\boldsymbol{X})$ with $\mathrm{Supp}\,\mathcal{T} \subset H_{[n]}$. Let us prove that $\mathcal{T} \in \mathrm{MTM}(X)$. We use an induction on the number $k(\mathcal{T})$ of the n-dimensional irreducible components of the support of $\mathcal{T}$. The case $k(\mathcal{T}) = 0$ follows from the hypothesis of the induction. We take $I \subset \underline{\ell}$ with $|I| = N-n$ such that $\phi_I\mathcal{T} \neq 0$. Let $g := \prod_{\underline{\ell}\setminus I} z_i$. By the hypothesis of the induction, we have $\phi_g^{(0)}(\mathcal{T}), \psi_g^{(a)}(\mathcal{T}) \in \mathrm{MTM}(X)$. By the result in the previous paragraph, we obtain that $\Xi_g^{(0)}(\mathcal{T}) \in \mathrm{MTM}(X)$. Because we can reconstruct $\mathcal{T}$ from $\phi_g^{(0)}(\mathcal{T}), \psi_g^{(a)}(\mathcal{T})$ $(a = 0, 1)$ and $\Xi_g^{(0)}(\mathcal{T})$, we obtain that $\mathcal{T}$ is a mixed twistor $\mathscr{D}$-module. Thus, the proof of Theorem 10.3.1 is finished.

We can observe that $(\mathcal{T}[\star D])[*f]$ and $(\mathcal{T}[\star D])[!f]$ in Lemma 10.3.5 are mixed twistor $\mathscr{D}$-modules, by their construction and Theorem 10.3.1. It follows that $\mathcal{T}[\star f]$ $(\star = *, !)$ in Lemma 10.3.7 are mixed twistor $\mathscr{D}$-modules. Thus, the proof of Proposition 10.3.2 is finished.

10.3.5 Proof of Lemma 10.3.4

Let us return to the situation in Sect. 10.3.2. Let $\mathcal{M}$ be one of the smooth $\mathcal{R}_X(*D)$-modules underlying $\mathcal{V}$. Let $\mathcal{M}' := \varphi^*\mathcal{M} \otimes \mathcal{O}_{\mathcal{X}}(*\mathcal{D}')$. We put $D_3' := \varphi^{-1}(D_0)$, and $D_4' := \varphi^{-1}(D)$. Let $\mathcal{M}_1' := \big(\mathcal{M}'[*D_3']\big)(*D_4')$ and $\mathcal{M}_2' := \big(\mathcal{M}'[!D_3']\big)(*D_4')$.

Lemma 10.3.8 *$\mathcal{M}$ is naturally isomorphic to the image of $\varphi_\dagger\mathcal{M}_2' \longrightarrow \varphi_\dagger\mathcal{M}_1'$.*

Proof By the construction of $\mathcal{M}_1'$, we have a naturally defined morphism $\mathcal{M} \longrightarrow \varphi_\dagger\mathcal{M}_1'$. For $\lambda \neq 0$, we have a naturally defined morphisms $\varphi_\dagger\mathcal{M}_2'^{\lambda} \longrightarrow \mathcal{M}^\lambda$. Hence, we obtain $\varphi_\dagger\mathcal{M}_2' \longrightarrow \mathcal{M}$.

Let $\lambda_0 \neq 0$ and $P \in D_0 \setminus D$. If we take a small neighbourhood $\mathcal{U}$ around (λ_0, P), we have an isomorphism of $\mathcal{M}$ with a natural $\mathcal{R}_X$-module $\mathcal{O}_{\mathcal{X}}$ on $\mathcal{U}$. Hence, it is easy to observe that $\mathcal{M}_{|\mathcal{U}}$ is the image of $\varphi_\dagger\mathcal{M}_2' \longrightarrow \varphi_\dagger\mathcal{M}_1'$ on $\mathcal{U}$. We obtain that $\mathcal{M}_{|\mathbb{C}_\lambda^*\times X}$ is the image of $\varphi_\dagger\mathcal{M}_2' \longrightarrow \varphi_\dagger\mathcal{M}_1'$ on $\mathbb{C}_\lambda^* \times X$. Because the cokernel of $\varphi_\dagger\mathcal{M}_2' \longrightarrow \varphi_\dagger\mathcal{M}_1'$ is strict, we obtain that $\mathcal{M}$ is the image of $\varphi_\dagger\mathcal{M}_2' \longrightarrow \varphi_\dagger\mathcal{M}_1'$. □

Let L denote the filtrations of $\mathcal{M}$ and $\mathcal{M}'$, induced by the weight filtration of the admissible variations of mixed twistor structure $\mathcal{V}$ and $\mathcal{V}'$. The naively induced filtrations of $\mathcal{M}_i'$ $(i = 1, 2)$ are denoted by L. Let $\tilde{L}$ denote the weight filtrations of $\mathcal{M}_i'$ induced by the weight filtration as objects of $\mathrm{MTW}^{\mathrm{good}}(\boldsymbol{X}')$. They induce the weight filtration of $\varphi_\dagger\mathcal{M}_i'$, which are also denoted by $\tilde{L}$.

We have the naturally induced morphisms $\varphi_\dagger(\mathcal{M}_2', L) \longrightarrow (\mathcal{M}, L) \longrightarrow \varphi_\dagger(\mathcal{M}_1', L)$. They are strictly compatible with the filtrations by Lemma 10.3.8.

Lemma 10.3.9 *The morphisms $\varphi_\dagger\mathcal{M}_2' \longrightarrow \mathcal{M} \longrightarrow \varphi_\dagger\mathcal{M}_1'$ give morphisms*

$$(\varphi_\dagger\mathcal{M}_2', \tilde{L}) \longrightarrow (\mathcal{M}, L) \longrightarrow (\varphi_\dagger\mathcal{M}_1', \tilde{L}),$$

which are strictly compatible with the filtrations.

Proof We have the induced morphisms $L_k\varphi_\dagger\mathcal{M}_2' \longrightarrow L_k\mathcal{M} \longrightarrow L_k\varphi_\dagger\mathcal{M}_1'$ for each k. We use an induction on k to prove that we have natural morphisms

$$(L_k\varphi_\dagger\mathcal{M}_2', \tilde{L}) \longrightarrow (L_k\mathcal{M}, L) \longrightarrow (L_k\varphi_\dagger\mathcal{M}_1', \tilde{L}),$$

and that they are strictly compatible with the filtrations. If k is sufficiently negative, the claim is trivial. Assume the claim in the case $k-1$. Let us look at the induced morphisms $\mathrm{Gr}_k^L\varphi_\dagger\mathcal{M}_2' \longrightarrow \mathrm{Gr}_k^L\mathcal{M}$. Note that $\mathrm{Gr}_k^L\varphi_\dagger\mathcal{M}_2' = \tilde{L}_k\,\mathrm{Gr}_k^L\varphi_\dagger\mathcal{M}_2'$, and the support of $\tilde{L}_{k-1}\,\mathrm{Gr}_k^L\varphi_\dagger\mathcal{M}_2'$ is contained in D_0. Hence, the induced morphism $\tilde{L}_{k-1}\,\mathrm{Gr}_k^L\varphi_\dagger\mathcal{M}_2' \longrightarrow \mathrm{Gr}_k^L\mathcal{M}$ is 0, it implies that the image of $\tilde{L}_{k-1}L_k\varphi_\dagger\mathcal{M}_2'$ is contained in $L_{k-1}\mathcal{M}$. Let us look at the following:

$$\tilde{L}_{k-1}L_k\varphi_\dagger\mathcal{M}_2' \xrightarrow{\ a\ } L_{k-1}\mathcal{M} \xrightarrow{\ b\ } L_{k-1}\varphi_\dagger\mathcal{M}_1'$$

The image of $\tilde{L}_m L_k \varphi_\dagger \mathcal{M}_2'$ $(m < k)$ via $b \circ a$ is contained in $\tilde{L}_m L_k \varphi_\dagger \mathcal{M}_1'$. Because b is injective and strict by the assumption of the induction, we obtain that the image of $\tilde{L}_m L_k \varphi_\dagger \mathcal{M}_2'$ $(m < k)$ via a is contained in $L_m \mathcal{M}$. Thus, we obtain $(L_k \varphi_\dagger \mathcal{M}_2', \tilde{L}) \longrightarrow (L_k \mathcal{M}, L)$. Moreover, it is strictly compatible with the filtrations.

Because $\varphi_\dagger \mathcal{M}_2' \longrightarrow \varphi_\dagger \mathcal{M}_1'$ is strictly compatible with $\tilde{L}$, we obtain that $(\mathcal{M}, L) \longrightarrow (\varphi_\dagger \mathcal{M}_1', \tilde{L})$ is strictly compatible. Hence, the induction can go on. □

Let $\mathcal{M}_3 := \varphi_\dagger \mathcal{M}'[*D_1'!D_2']$ and $\mathcal{M}_4 := \varphi_\dagger \mathcal{M}'[*D_1''!D_2'']$. Suppose $D_i = g_i^{-1}(0)$ for holomorphic functions g_i. We know that $\mathcal{M}[*D_1]$ and $\mathcal{M}_3(*g_2)$ are admissible specializable along g_1, and we have isomorphisms $\mathcal{M}[*D_1][*g_1] \simeq \mathcal{M}[*g_1]$ and $\mathcal{M}_3(*g_2) \simeq \mathcal{M}_3(*\mathcal{D})[*g_1](*g_2)$. Hence, we have a uniquely determined morphism $\mathcal{M}[*D_1] \longrightarrow \mathcal{M}_3(*g_2)$ induced by $\mathcal{M} \longrightarrow \varphi_\dagger \mathcal{M}_1' = \mathcal{M}_3(*\mathcal{D})$, which is compatible with the weight filtrations. We have that $\mathcal{M}[*D_1!D_2]$ and $\mathcal{M}_3$ are admissible specializable along g_2. We also have $\mathcal{M}[*D_1!D_2][!g_2] \simeq \mathcal{M}[*D_1!D_2]$ and $\mathcal{M}_3[!g_2] \simeq \mathcal{M}_3$. Hence, we have a uniquely determined morphism $\mathcal{M}[*D_1!D_2] \longrightarrow \mathcal{M}_3$. Similarly, we obtain $\mathcal{M}_4 \longrightarrow \mathcal{M}[*D_1!D_2]$. By the uniqueness, we obtain the global case. Thus, we obtain the first claim of Lemma 10.3.4.

Suppose $D_1 = \emptyset$. For any generic λ, the specialization $\mathcal{M}^\lambda[*D_1!D_2]$ is the image of the induced morphism $(\mathcal{M}_4)^\lambda \longrightarrow (\mathcal{M}_3)^\lambda$. Then, by using the strictness, we obtain that $\mathcal{M}[\star D] \longrightarrow \mathcal{M}_3$ is injective, and that $\mathcal{M}_4 \longrightarrow \mathcal{M}[\star D]$ is surjective, i.e., $\mathcal{M}[\star D]$ is the image of $\mathcal{M}_3 \longrightarrow \mathcal{M}[\star D] \longrightarrow \mathcal{M}_4$. Thus, the proof of Lemma 10.3.4 is finished. □

10.4 Integrable Case

Let X, D, $\boldsymbol{\mathcal{I}}$ and $\boldsymbol{X}$ be as in Sect. 10.3. Let $D = \bigcup_{i \in \Lambda} D_i$ be the irreducible decomposition. An integrable mixed twistor $\mathscr{D}$-module $(\mathcal{T}, W)$ is called good on $\boldsymbol{X}$, if the underlying mixed twistor $\mathscr{D}$-module is good on $\boldsymbol{X}$. Let $\mathrm{MTM}^{\mathrm{int\,good}}(\boldsymbol{X}) \subset \mathrm{MTM}^{\mathrm{int}}(\boldsymbol{X})$ be the full subcategory of good integrable mixed twistor $\mathscr{D}$-modules on $\boldsymbol{X}$.

Let $\mathcal{V} \in \mathrm{MTS}^{\mathrm{int\,adm}}(\boldsymbol{X})$. We have the good mixed twistor $\mathscr{D}$-module $\mathcal{V}[*I!J]$ constructed in Sect. 10.2.3.2.

Proposition 10.4.1 *We naturally have* $\mathcal{V}[*I!J] \in \mathrm{MTM}^{\mathrm{int\,good}}(\boldsymbol{X})$.

Proof By Lemma 7.1.57 with z_i $(i \in J)$, we have only to consider the case $\mathcal{V}[*D]$. By the construction, $\mathcal{V}[*D]$ and its weight filtration are integrable. To check that $\mathrm{Gr}_k^W(\mathcal{V}[*D]) \in \mathrm{MT}^{\mathrm{int}}(X, k)$, we have only to enhance the argument in the proof of Proposition 10.2.11 with integrability. □

Proposition 10.4.2 *Let* $\mathcal{T} \in \mathrm{MTM}^{\mathrm{int\,good}}(\boldsymbol{X})$. *Let g be a holomorphic function on X such that* $g^{-1}(0) \subset D$.

- *There exist* $\mathcal{T}[\star g]$ *in* $\mathrm{MTM}^{\mathrm{int\,good}}(\boldsymbol{X})$ *for* $\star = *, !$.
- $\tilde{\psi}_{g,\mathfrak{a},u}(\mathcal{T})$, $\Xi_g(\mathcal{T})$ *and* $\phi_g^{(a)}(\mathcal{T})$ *are objects in* $\mathrm{MTM}^{\mathrm{int\,good}}(\boldsymbol{X})$. □

Proposition 10.4.3 *Let H be an effective divisor of X. For any object $\mathcal{T} \in \mathrm{MTM}^{\mathrm{int\,good}}(\boldsymbol{X})$, we have $\mathcal{T}[\star H] \in \mathrm{MTM}^{\mathrm{int}}(X)$. It depends only on the support of H.*

Proof We only remark that the integrability of the polarization in Lemma 7.1.58. □

Chapter 11
Some Basic Property

In this chapter, we shall prove some basic properties of mixed twistor $\mathscr{D}$-modules. The most basic is the expression of mixed twistor $\mathscr{D}$-modules as the gluing of admissible variations of mixed twistor structure studied in Sect. 11.1.

Recall that any holonomic $\mathscr{D}$-modules are locally obtained as the gluing of some meromorphic flat bundles on subvarieties. Namely, let M be a holonomic $\mathscr{D}$-module on a complex manifold X. Let P be any point of X, and let X_P be a small neighbourhood of P in X. Let $\operatorname{Supp}(M_{|X_P}) = \bigcup_{i\in\Lambda} Z_i$ be the irreducible decomposition of the support of the restriction $M_{|X_P}$. Fix $i_0 \in \Lambda$. If X_P is sufficiently small, we can take a holomorphic function f such that $\bigcup_{i\neq i_0} Z_i \subset f^{-1}(0)$. Then, $M(*f)$ is obtained as the push-forward of a meromorphic flat bundle $V_{Z_{i_0}}$ given on a smooth open subset of Z_{i_0}. We can reconstruct M as the cohomology of the following:

$$\psi_f^{(1)}(M) \longrightarrow \Xi_f^{(0)}(M) \oplus \phi_f^{(0)}(M) \longrightarrow \psi_f^{(0)}(M) \tag{11.1}$$

Here, $\psi_f^{(a)}(M)$ and $\Xi_f^{(0)}(M)$ are induced by $V_{Z_{i_0}}$. The Noetherian induction admits us to describe M as the gluing of meromorphic flat bundles on subvarieties.

As an analogue, we see that any mixed twistor $\mathscr{D}$-modules are locally obtained as the gluing of admissible mixed twistor structures on subvarieties. Let $\mathcal{T}$ be a mixed twistor $\mathscr{D}$-module on X. Let P and X_P be as above. Let $\operatorname{Supp}(\mathcal{T})_{|X_P} = \bigcup_{i\in\Lambda} Z_i$ be the irreducible decomposition of the support of the restriction $\mathcal{T}_{|X_P}$. Take $i_0 \in \Lambda$ and a holomorphic function g such that $g^{-1}(0) \supset \bigcup_{i\neq i_0} Z_i$. We take a projective birational morphism $\varphi : Z^{(1)} \longrightarrow Z_{i_0}$ such that $Z^{(1)}$ is smooth and that $D^{(1)} := \varphi^{-1}(g^{-1}(0))$ is normal crossing. Then, it can be shown that we have a smooth $\mathcal{R}_{Z^{(1)}(*D^{(1)})}$-triple $\mathcal{T}_{Z^{(1)}}$ such that $\mathcal{T}(*g) = \varphi_\dagger \mathcal{T}_{Z^{(1)}}$. We can choose $(Z^{(1)}, \varphi)$ such that $\mathcal{T}_{Z^{(1)}}$ is an admissible mixed twistor structure on $(Z^{(1)}, D^{(1)})$. Once it is proved, the mixed twistor $\mathscr{D}$-module $\mathcal{T}_{|X_P}$ is reconstructed as the cohomology of a complex similar to (11.1) in the category of mixed twistor $\mathscr{D}$-modules.

T. Mochizuki, *Mixed Twistor $\mathscr{D}$-modules*, Lecture Notes in Mathematics 2125,
DOI 10.1007/978-3-319-10088-3_11

In Sect. 11.2, we study the localization functors for mixed twistor $\mathscr{D}$-modules. Let H be a hypersurface in a complex manifold X. For a holonomic $\mathscr{D}$-module M on X, we have the $\mathscr{D}$-modules $M[*H]$ and $M[!H]$ as mentioned in Chap. 3. The functors $[\star H]$ are fundamental in the operations for holonomic $\mathscr{D}$-modules along H. We have the analogue functors $[\star H]$ $(\star = *, !)$ for mixed twistor $\mathscr{D}$-modules. We may construct $\mathcal{T}[\star H]$ for a mixed twistor $\mathscr{D}$-module $\mathcal{T}$ as follows. Let $P \in X$, and let X_P be a small neighbourhood of P in X. Let $(Z^{(1)}, \varphi)$ and $\mathcal{T}_{Z^{(1)}}$ be as above. For the canonical prolongations of $\mathcal{T}_{Z^{(1)}}[\star D^{(1)}]$ on $Z^{(1)}$, according to results in Chap. 10, we have the localization with respect to any hypersurfaces contained in $D^{(1)}$. We may apply the hypothesis of the induction to the mixed twistor $\mathscr{D}$-modules with smaller support. So, we may obtain $\mathcal{T}[\star H]$ as their gluing. One remark is that we need to be careful with the existence of polarization on the graded pieces with respect to the weight filtration.

In Sect. 11.3, we study Beilinson functors for mixed twistor $\mathscr{D}$-modules along any holomorphic functions. Again we use the gluing expression for the construction of the functors with the same argument as that in the localizations.

In Sect. 11.4, we prove that the external tensor product of any mixed twistor $\mathscr{D}$-modules is also a mixed twistor $\mathscr{D}$-module.

11.1 Expression as Gluing of Admissible Mixed Twistor Structure

11.1.1 Cell

Let X be a complex manifold. An n-dimensional cell is a tuple $\mathcal{C} = (Z, U, \varphi, \mathcal{V})$ as follows:

(Cell 1) Z is an n-dimensional complex manifold, and $\varphi : Z \longrightarrow X$ is a morphism of complex manifolds. The image $\varphi(Z)$ is a locally closed subvariety of X, and $Z \longrightarrow \varphi(Z)$ is proper.

(Cell 2) $U \subset Z$ is the complement of a simply normal crossing hypersurface D_Z. The restriction $\varphi_{|U}$ is an immersion.

(Cell 3) $\mathcal{V} \in \mathrm{MTS}(Z, D_Z)$ satisfying **Adm0**.

If $\mathcal{V}$ is admissible, $\mathcal{C}$ is called an admissible cell. If $\mathcal{V}$ is integrable, $\mathcal{C}$ is called an integrable cell.

If the following additional condition is satisfied, $\mathcal{C}$ is called a cell at $P \in X$.

- P is contained in $\varphi(Z)$. There exists a neighbourhood X_P of P in X such that $\varphi : \varphi^{-1}(X_P) \longrightarrow X_P$ is projective. Moreover, there exists a hypersurface H of X_P such that $\varphi^{-1}(H) = D_Z \cap \varphi^{-1}(X_P)$.

A holomorphic function g on X is called a cell function of a cell $\mathcal{C} = (Z, U, \varphi, \mathcal{V})$, if $(\varphi^* g)^{-1}(0) = D_Z$.

11.1.2 Cell of Mixed Twistor $\mathscr{D}$-Modules

Let $\mathcal{T}$ be a mixed twistor $\mathscr{D}$-module on X. The weight filtration is denoted by W. Let Z be the support of $\mathcal{T}$. We have the graded pure twistor $\mathscr{D}$-module $\mathrm{Gr}^W(\mathcal{T}) = \bigoplus \mathrm{Gr}^W_w(\mathcal{T})$, whose support is Z. Suppose we have a hypersurface H of X satisfying the following.

- $Z \setminus H \neq \emptyset$ is smooth.
- $\mathcal{T}_{|X\setminus H}$ comes from a mixed twistor structure on $Z \setminus H$. (Note Proposition 7.2.8 and Lemma 8.1.9.) If $\mathcal{T} \in \mathrm{MTM}^{\mathrm{int}}(X)$, the mixed twistor structure is integrable.

For simplicity, we assume that X is a relative compact open subset in a complex manifold X_1, and H is the restriction of a hypersurface of X_1.

The following proposition implies the local existence of an admissible cell of $\mathcal{T}$, which we shall prove in Sect. 11.1.5.1.

Proposition 11.1.1 *There exists a projective morphism of complex manifolds $\varphi : Z_1 \longrightarrow X$ with the following property:*

- *$H_1 := \varphi^{-1}(H)$ is simply normal crossing, and φ induces an isomorphism $Z_1 \setminus H_1 \simeq Z \setminus H$.*
- *There exists an admissible mixed twistor structure $\mathcal{V}$ on (Z_1, H_1) such that $\mathcal{T}(*H) \simeq \varphi_\dagger \mathcal{V}$.*

The following proposition implies any cell of $\mathcal{T}$ is admissible, which we shall prove in Sect. 11.1.5.2.

Proposition 11.1.2 *Let $\varphi_{10} : Z_{10} \longrightarrow X$ be a projective morphism of complex manifolds such that (i) $H_{10} := \varphi_{10}^{-1}(H)$ is normal crossing (ii) φ_{10} induces $Z_{10} \setminus H_{10} \simeq Z \setminus H$, (iii) there exists $\mathcal{V}_{10} \in \mathrm{MTS}(Z_{10}, H_{10})$ which is a good smooth $\mathcal{R}_{Z_{10}(*H_{10})}$-triple, (iv) we have an isomorphism $\varphi_{10\dagger}\mathcal{V}_{10} \simeq \mathcal{T}(*H)$. Then, $\mathcal{V}$ is admissible, i.e., $(Z_{10}, Z_{10} \setminus H_{10}, \varphi_{10}, \mathcal{V}_{10})$ is an admissible cell.*

The following complementary claim is clear from the proof of the propositions.

Lemma 11.1.3 *In the setting of Propositions 11.1.1 or 11.1.2, the following holds:*

- *If $\mathcal{T} \in \mathrm{MTM}^{\mathrm{int}}(X)$, the admissible mixed twistor structure $\mathcal{V}$ is integrable.*
- *If $\mathcal{T} \in \mathrm{MTM}(X)_{\mathcal{A}}$, then the KMS-spectrum of $\mathcal{V}$ along any irreducible component of H_1 are contained in $\mathcal{A}$. (See Sects. 5.1.2 and 5.5 for the KMS-spectrum.)*

11.1.3 Expression as a Gluing

Let $\mathcal{T} \in \mathrm{MTM}(X)$. Let $P \in \mathrm{Supp}\,\mathcal{T}$. Let X_P denote a small neighbourhood of P in X. We will shrink X_P. We have an admissible cell $\mathcal{C} = (Z, U, \varphi, \mathcal{V})$ at P and a cell function g such that $\mathcal{T}(*g) = \varphi_\dagger(\mathcal{V})$. By Propositions 7.2.7 and 10.3.2, we have $\varphi_\dagger(\mathcal{V})[\star g]$ in $\mathrm{MTM}(X_P)$. According to Lemma 7.1.44, we have the morphisms $\varphi_\dagger(\mathcal{V})[!g] \longrightarrow \mathcal{T} \longrightarrow \varphi_\dagger(\mathcal{V})[*g]$ in $\mathrm{MTM}(X_P)$. We also have $\Xi_g \varphi_\dagger(\mathcal{V})$ in $\mathrm{MTM}(X_P)$, which is isomorphic to $\varphi_\dagger \Xi_{g\circ\varphi}(\mathcal{V})$ in $\mathrm{MTM}(X_P)$. So, we obtain $\big(\phi_g^{(0)}(\mathcal{T}), W\big)$ as the cohomology of the following complex in $\mathrm{MTM}(X_P)$:

$$\varphi_\dagger(\mathcal{V})[!g] \longrightarrow \mathcal{T} \oplus \Xi_g \varphi_\dagger(\mathcal{V}) \longrightarrow \varphi_\dagger(\mathcal{V})[*g]$$

We give a remark on the weight filtration on $\phi_g^{(0)}(\mathcal{T})$. We have the naively induced filtration $L\phi_g^{(0)}(\mathcal{T})$ defined by $L_j\phi_g^{(0)}(\mathcal{T}) = \phi_g^{(0)}(W_j\mathcal{T})$. It is a filtration of $\big(\phi_g^{(0)}(\mathcal{T}), W\big)$ in the category $\mathrm{MTM}(X_P)$. By definition of mixed twistor $\mathscr{D}$-module, we have the relative monodromy filtration $\tilde{L} := M(\mathcal{N}, L\phi_g^{(0)}\mathcal{T})$, with which $\big(\phi_g^{(0)}(\mathcal{T}), \tilde{L}\big) \in \mathrm{MTM}(X_P)$.

Lemma 11.1.4 *We have $\tilde{L} = W$ on $\phi_g^{(0)}(\mathcal{T})$.*

Proof Let us begin with a preliminary. Set $g_1 := g \circ \varphi$. We have $(\psi_{g_1}^{(a)}(\mathcal{V}), W)$ in MTM(Z). Let L denote the filtration on $\psi_{g_1}^{(a)}(\mathcal{V})$ given by $L_j\psi_{g_1}^{(a)}(\mathcal{V}) = \psi_{g_1}^{(a)}(W_j\mathcal{V})$. We have $W = M(\mathcal{N}; L)[-2a+1]$ on $\psi_{g_1}^{(a)}(\mathcal{V})$ by Proposition 10.2.9. We obtain the mixed twistor $\mathscr{D}_{X_P}$-module $(\varphi_\dagger\psi_{g_1}^{(a)}(\mathcal{V}), W)$. We have natural isomorphisms of $\mathcal{R}_{X_P}$-triples $\varphi_\dagger\psi_{g_1}^{(a)}(\mathcal{V}) \simeq \psi_g^{(a)}(\mathcal{T})$. Let L be the filtration on $\psi_g^{(a)}(\mathcal{T})$ given by $L_j\psi_g^{(a)}(\mathcal{T}) = \psi_g^{(a)}(W_j\mathcal{T})$. By Corollary 7.1.32, the weight filtration W on $\varphi_\dagger\psi_{g_1}^{(a)}(\mathcal{V})$ is equal to $M(\mathcal{N}; L)[-2a+1]$ on $\psi_g^{(a)}(\mathcal{T})$ under the isomorphism.

Let us return to the proof of Lemma 11.1.4. By a standard induction on the length of the filtration W, we have only to consider the case that $\mathcal{T}$ and $\mathcal{V}$ are pure of weight 0. Let us consider the morphisms $\psi_g^{(1)}(\mathcal{T}) \xrightarrow{\text{can}} \phi_g^{(0)}(\mathcal{T}) \xrightarrow{\text{var}} \psi_g^{(0)}(\mathcal{T})$ of $\mathcal{R}_{X_P}$-triples. The morphisms can and var are morphisms in $\mathrm{MTM}(X_P)$ for the both filtrations $\tilde{L}$ and W on $\phi_g^{(0)}(\mathcal{T})$. We have the decomposition $\phi_g^{(0)}(\mathcal{T}) = \mathrm{Im}\,\mathrm{can} \oplus \mathrm{Ker}\,\mathrm{var}$, which is compatible with both $\tilde{L}$ and W. The restriction of $\tilde{L}$ to Ker var is pure of weight 0 by definition. Because $\mathcal{T} = \mathcal{T}_0 \oplus \mathrm{Ker}\,\mathrm{var}$ as pure twistor $\mathscr{D}$-module, W is also pure of weight 0 on Ker var. Because can and var are strict with respect to both $\tilde{L}$ and W, we have $\tilde{L} = W$ on Im can. Thus, we obtain Lemma 11.1.4. □

We obtain the local expression of $\mathcal{T}$ as the cohomology of the following complex in $\mathrm{MTM}(X_P)$:

$$\psi_g^{(1)}(\varphi_\dagger\mathcal{V}) \longrightarrow \Xi_g^{(0)}(\varphi_\dagger\mathcal{V}) \oplus \phi_g^{(0)}(\mathcal{T}) \longrightarrow \psi_g^{(0)}(\varphi_\dagger\mathcal{V}) \tag{11.2}$$

11.1.3.1 Expression as a Sub-quotient

We obtain the following proposition.

Proposition 11.1.5 *Let $(\mathcal{T}, W) \in \mathrm{MTM}(X)$. For any $P \in X$, there exist a neighbourhood X_P, an admissible cell $\mathcal{C}_0 = (Z_0, U_0, \varphi_0, \mathcal{V}_0)$ with a cell function g for $(\mathcal{T}, W)_{|X_P}$, and a mixed twistor $\mathscr{D}$-module $(\mathcal{T}_0, W)$ with*

$$\mathrm{Supp}(\mathcal{T}_0) \subset g^{-1}(0) \cap \mathrm{Supp}(\mathcal{T}),$$

such that $(\mathcal{T}, W)_{|X_P}$ is a sub-quotient of $(\mathcal{T}_0, W) \oplus \varphi_{0\dagger}(\mathcal{V}_0)[!g]$ in $\mathrm{MTM}(X_P)$.

Proof We take $\mathcal{C} = (Z, U, \varphi, \mathcal{V})$ and g as in the beginning of Sect. 11.1.3. Let $D_Z := Z \setminus U$. Note that $\Xi_g \varphi_\dagger \mathcal{V}$ is expressed as the kernel of $\varphi_\dagger \mathcal{V}'[!g] \longrightarrow \varphi_\dagger \mathcal{V}''[*g]$ for some admissible variations of mixed twistor structure $\mathcal{V}'$ and $\mathcal{V}''$ on (Z, D_Z). Hence, the claim of the proposition follows. □

Remark 11.1.6 M. Saito emphasized the convenience of a description of mixed Hodge modules as in Proposition 11.1.5. □

11.1.4 Gluing

Let $\mathcal{C} = (Z, U, \varphi, \mathcal{V})$ be an admissible cell with a cell function g at P. We will shrink X around P. Let $\mathcal{T}' \in \mathrm{MTM}(X)$ such that $\mathrm{Supp}\,\mathcal{T}' \subset g^{-1}(0)$. Assume that we are given morphisms

$$\psi_g^{(1)} \varphi_\dagger \mathcal{V} \xrightarrow{u} \mathcal{T}' \xrightarrow{v} \psi_g^{(0)} \varphi_\dagger \mathcal{V} \tag{11.3}$$

in $\mathrm{MTM}(X)$ such that $v \circ u$ is equal to the canonical morphism $\psi_g^{(1)} \varphi_\dagger \mathcal{V} \longrightarrow \psi_g^{(0)} \varphi_\dagger \mathcal{V}$. Then, we obtain the $\mathcal{R}$-triple $(\mathrm{Glue}(\mathcal{C}, \mathcal{T}', u, v), W) \in \mathrm{MTM}(X)$ as the cohomology of the following complex in $\mathrm{MTM}(X)$:

$$\psi_g^{(1)} \varphi_\dagger \mathcal{V} \longrightarrow \Xi_g^{(0)} \varphi_\dagger \mathcal{V} \oplus \mathcal{T}' \longrightarrow \psi_g^{(0)} \varphi_\dagger \mathcal{V} \tag{11.4}$$

We can reconstruct $\mathcal{T}'$ as the cohomology of the complex in $\mathrm{MTM}(X)$:

$$\varphi_\dagger \mathcal{V}[!g] \longrightarrow \Xi_g \varphi_\dagger \mathcal{V} \oplus \mathrm{Glue}(\mathcal{C}, \mathcal{T}', u, v) \longrightarrow \varphi_\dagger \mathcal{V}[*g] \tag{11.5}$$

We have a similar expression for an integrable mixed twistor $\mathscr{D}$-module as the gluing of an integrable admissible cell and the integrable mixed twistor $\mathscr{D}$-module obtained as the vanishing cycle sheaf.

Let $L\psi_g^{(a)} \varphi_\dagger(\mathcal{V})$ and $L\Xi_g^{(0)} \varphi_\dagger(\mathcal{V})$ be the filtrations naively induced by the weight filtration of $\mathcal{V}$, i.e., $L_j \psi_g^{(a)} \varphi_\dagger(\mathcal{V}) = \psi_g^{(a)} \varphi_\dagger(W_j \mathcal{V})$ and $L_j \Xi_g^{(0)} \varphi_\dagger(\mathcal{V}) = \Xi_g^{(0)} \varphi_\dagger(W_j \mathcal{V})$. Let L be the filtration on $\mathcal{T}'$ obtained as the transfer of L on $\psi_g^{(a)} \varphi_\dagger(\mathcal{V})$ with respect to (11.3). We obtain a filtration $W^{(2)}$ on $\mathrm{Glue}(\mathcal{C}, \mathcal{T}', u, v)$ from (11.4) with the filtrations L on the objects.

Lemma 11.1.7 *We have* $W = W^{(2)}$ *on* $\mathrm{Glue}(\mathcal{C}, \mathcal{T}', u, v)$.

Proof We have the filtration $L^{(1)}$ on $\mathcal{T}'$ naively induced by the filtrations W on $\mathrm{Glue}(\mathcal{C}, \mathcal{T}', u, v)$. By using Lemma 11.1.4 and the uniqueness of the transfer, we obtain $L^{(1)} = L$ on $\mathcal{T}'$. Then, we obtain $W = W^{(2)}$ on $\mathrm{Glue}(\mathcal{C}, \mathcal{T}', u, v)$. □

11.1.5 Admissibility of Cells

11.1.5.1 Proof of Proposition 11.1.1

We use the notation in Sect. 11.1.2. Let us consider the case where (i) we have $X = Z = \Delta$ and $H := \{O\}$, (ii) $(\mathcal{T}, W)$ is a mixed twistor $\mathscr{D}$-module such that $(\mathcal{T}, W)_{|X\setminus H}$ is a mixed twistor structure. Let $\mathcal{M}_i$ $(i = 1, 2)$ be the underlying $\mathcal{R}_X$-modules. Let L be the filtrations on $\mathcal{M}_i$ induced by W on $\mathcal{T}$. Because $\mathrm{Gr}^W_w \mathcal{T}(*H)$ is obtained as the canonical meromorphic prolongation of a wild variation of polarizable pure twistor structure of weight w, $\mathrm{Gr}^L_w(\mathcal{M}_i)(*H)$ are good-KMS smooth $\mathcal{R}_{X(*H)}$-modules. Hence, $\mathcal{M}_i$ are also good-KMS smooth $\mathcal{R}_{X(*H)}$-modules. If we take an appropriate ramified covering $\varphi : (X_2, H_2) \longrightarrow (X, H)$, we have the irregular decomposition $\varphi^*(\mathcal{M}_i, L)_{|\hat{H}_2} = \bigoplus(\hat{\mathcal{M}}_{i,\mathfrak{a}}, L)$. Because $(\mathcal{M}_i, L)$ are admissibly specializable, each $\mathcal{M}_i$ has KMS-structure compatible with the filtration L, and the condition **(Adm2)** in Sect. 9.1.3 is satisfied. Hence, $(\mathcal{T}, L)(*H)$ comes from an admissible mixed twistor structure. The claim of Lemma 11.1.3 is clear in this situation.

Let us consider the general case. Let Z_0 be the closure of $Z \setminus H$ in X. We take a projective birational morphism $\varphi : X' \longrightarrow X$ such that (i) the proper transform Z_0' of Z_0 is smooth, (ii) Z_0' intersects with $H' := \varphi^{-1}(H)$ in a normal crossing way, (iii) $Z_0' \setminus H' \simeq Z_0 \setminus H$. (See [91].) We set $H_0' := Z_0' \cap H'$. We have the filtered $\mathcal{R}_{X'(*H')}$-triple $(\mathcal{T}', W)$ obtained as the lift of $(\mathcal{T}, W)(*H)$, as in Lemma 2.1.11. (We use Lemma 2.1.13 in the integrable case.) We have the graded $\mathcal{R}_{X'(*H')}$-triple $\mathrm{Gr}^W(\mathcal{T}')$.

Lemma 11.1.8 *There exists a projective birational morphism* $\varphi_1 : Z_1 \longrightarrow Z_0'$ *such that (i)* $H_1 := \varphi_1^{-1}(H_0')$ *is normal crossing, (ii)* $Z_1 \setminus H_1 \simeq Z_0' \setminus H_0'$, *(iii) there exist filtered* $\mathcal{R}_{Z_1}(*H_1)$*-triples* $\mathcal{T}_w^{(1)}$ $(w \in \mathbb{Z})$ *satisfying the following:*

- $\mathcal{T}_w^{(1)}$ *is obtained as the canonical prolongment of a graded good wild variation of pure twistor structure of weight* w *on* (Z_1, H_1).
- $\bigoplus_w \mathcal{T}_w^{(1)}$ *is a good smooth* $\mathcal{R}_{Z_1}(*H_1)$*-triple.*
- $\mathrm{Gr}^W_w(\mathcal{T}') \simeq (\iota_{Z_0'} \circ \varphi_1)_\dagger \mathcal{T}_w^{(1)}$, *where* $\iota_{Z_0'} : Z_0' \longrightarrow X'$ *denotes the inclusion.*

Proof Let $\mathrm{Gr}^W(\mathcal{T})_0$ be the sum of the direct summands of $\mathrm{Gr}^W(\mathcal{T})$ whose strict supports are Z_0. According to [55], there exists a projective birational morphism $\varphi_2 : Z_2 \longrightarrow Z_0$ such that the following holds:

- There exist a normal crossing hypersurface $H_2 \subset Z_2$ satisfying $H_2 \supset \varphi_2^{-1}(H \cap Z_0)$, and good wild polarizable variations of pure twistor structure $\mathcal{V}_w^{(2)}$ of weight w on (Z_2, H_2). Let $\mathfrak{T}_w^{(2)}$ be the polarizable pure twistor $\mathscr{D}$-module of weight w on Z_2 obtained as the minimal extension of $\mathcal{V}_w^{(2)}$.
- Let $\iota_{Z_0} : Z_0 \longrightarrow X$ denote the inclusion. Then, $\mathrm{Gr}_w^W(\mathcal{T})(*H)$ is a direct summand of $(\iota_{Z_0} \circ \varphi_2)_\dagger^0(\mathfrak{T}_w^{(2)})(*H)$, and the support of $(\iota_{Z_0} \circ \varphi_2)_\dagger^0(\mathfrak{T}_w^{(2)})(*H) / \mathrm{Gr}_w^W(\mathcal{T})(*H)$ is strictly smaller than Z_0.

We may assume that φ_2 factors through Z_0'.

We have the polarizable variation of pure twistor structure $\mathcal{V}_w^{(0)\prime}$ of weight w on $Z_0' \setminus H_0'$ corresponding to $\mathrm{Gr}_w^W(\mathcal{T}')_{|X' \setminus H'}$. Applying Corollary 15.2.9 below with the existence of φ_2 and $\mathcal{V}_w^{(2)}$ as above, we can take a projective birational morphism $\varphi_1 : Z_1 \longrightarrow Z_0'$ such that (i) $H_1 := \varphi_1^{-1}(H_0')$ is normal crossing, (ii) each $\mathcal{V}_w^{(1)} := \varphi_1^{-1}\mathcal{V}_w^{(0)\prime}$ comes from a good wild harmonic bundle $(E_w^{(1)}, \overline{\partial}_{E_w^{(1)}}, \theta_w^{(1)}, h_w^{(1)})$ on (Z_1, H_1) up to shift of the weight, (iii) $\bigoplus(E_w^{(1)}, \overline{\partial}_{E_w^{(1)}}, \theta_w^{(1)}, h_w^{(1)})$ is good wild. We may assume that φ_2 factors through Z_1, i.e., we may have $\psi : Z_2 \longrightarrow Z_1$ such that $\varphi_2 = \varphi_1 \circ \psi$. Let $\mathcal{T}_w^{(1)}$ is the $\mathcal{R}_{Z_1}(*H_1)$-triple obtained as the canonical prolongment of $\mathcal{V}_w^{(1)}$. Let $\mathcal{T}_w^{(2)}$ be the $\mathcal{R}_{Z_2}(*H_2)$-triple obtained as the canonical prolongment of $\mathcal{V}_w^{(2)}$. We have a natural isomorphism $\psi^*\mathcal{T}_w^{(1)}(*H_2) \simeq \mathcal{T}_w^{(2)}$ by the construction. We obtain $\psi^*\mathcal{T}_w^{(1)} \simeq \mathfrak{T}_w^{(2)}(*\psi^{-1}H_1)$. Hence, we obtain a morphism $(\iota_{Z_0'} \circ \varphi_1)_\dagger \mathcal{T}_w^{(1)} \longrightarrow \mathrm{Gr}_w^L(\mathcal{T}')(*H_0')$. Because its restriction to $X' \setminus H_0'$ is an isomorphism by construction, it is an isomorphism on X'. Thus, we obtain Lemma 11.1.8. □

Lemma 11.1.9 *There exists a filtered $\mathcal{R}_{Z_0'(*H_0')}$-triple $(\mathcal{T}_0', W)$ such that $(\mathcal{T}', W)$ is the push-forward of $(\mathcal{T}_0', W)$ by $(Z_0', H_0') \longrightarrow (X', H')$.*

Proof We have only to prove the underlying filtered $\mathcal{R}_{X'(*H')}$-modules $\mathcal{M}_i$ of $\mathcal{T}'$ are the push-forward of $\mathcal{R}_{Z_0'(*H_0')}$-modules $\tilde{\mathcal{K}}_i$. Such $\tilde{\mathcal{K}}_i$ are unique if they exist. So, it is enough to construct them locally around any point of H_0'.

Let L denote the filtration on $\mathcal{M}_i$ induced by W on $\mathcal{T}'$. By Lemma 11.1.8, $\mathrm{Gr}^L(\mathcal{M}_i)$ are the push-forward of graded $\mathcal{R}_{Z_0'(*H_0')}$-modules $\bigoplus_i \mathcal{K}_{iw}$. By the admissible specializability of mixed twistor $\mathscr{D}$-modules, the restriction $(\mathcal{M}_i, L)_{|X' \setminus H'}$ are the push-forward of smooth filtered $\mathcal{R}_{Z_0' \setminus H_0'}$-modules $(\mathcal{K}_i, L)$. We have $\mathrm{Gr}^L(\mathcal{K}_i) = \bigoplus_i \mathcal{K}_{iw|Z_0' \setminus H_0'}$. Then, by applying Corollary 11.1.12 below, we obtain Lemma 11.1.9. □

By using Lemma 2.1.11, we obtain a filtered $\mathcal{R}_{Z_1(*H_1)}$-triple $(\mathcal{T}^{(1)}, W)$ such that (i) $\mathrm{Gr}^W(\mathcal{T}^{(1)}) = \bigoplus \mathcal{T}_w^{(1)}$, (ii) $\varphi_{1\dagger}(\mathcal{T}^{(1)}, W) \simeq (\mathcal{T}_0', W)$. Because the underlying $\mathcal{R}_{Z_1(*H_1)}$-modules are good and smooth, $\mathcal{T}^{(1)}$ is a good smooth $\mathcal{R}_{Z_1(*H_1)}$-triple by Proposition 5.5.7. Then, we obtain that $(\mathcal{T}^{(1)}, W)$ is an object in $\mathrm{MTS}(Z_1, H_1)$. The condition **(Adm0)** is satisfied. Proposition 11.1.1 is reduced to the following.

Lemma 11.1.10 $(\mathcal{T}^{(1)}, W)$ *is admissible.*

Proof We omit to denote the weight filtration W. First, let us consider the case that $\dim Z_0 = 1$. We may assume $X = \Delta^n$ and $H = \{z_1 = 0\}$. Let $q_1 : X \longrightarrow \Delta^1$ be the projection onto the first component. We may assume that the induced morphism $F : Z_1 \longrightarrow \Delta^1$ is a covering with a ramification along 0. We obtain $\mathcal{T}_1 := q_{1\dagger}(\mathcal{T}) \in \mathrm{MTM}(\Delta^1)$. By the result in the one dimensional case, $\mathcal{T}_2 := \mathcal{T}_1(*0)$ is an object in $\mathrm{MTS}^{\mathrm{adm}}(\Delta^1, 0)$. We obtain $F^*\mathcal{T}_2 \in \mathrm{MTS}^{\mathrm{adm}}(Z_1, H_1)$. Because $\mathcal{T}_2 = F_\dagger(\mathcal{T}^{(1)})$, $\mathcal{T}^{(1)}$ is a direct summand of $F^*\mathcal{T}_2$ as a filtered smooth $\mathcal{R}_{Z_1(*H_1)}$-triple. Hence, $\mathcal{T}^{(1)} \in \mathrm{MTS}^{\mathrm{adm}}(Z_1, H_1)$.

Let us consider the general case. Let C be a smooth curve in Z_1, which intersects with the smooth part of H_1 transversally. By using the nearby cycle functor and the vanishing cycle functor to $\mathcal{T}$ successively, we obtain a mixed twistor $\mathscr{D}$-module $\mathcal{T}_3$ such that (i) the support of $\mathcal{T}_3(*H)$ is $\varphi_1(C)$, (ii) the lift of $\mathcal{T}_3(*H)$ to C is $\mathcal{T}^{(1)}_{|C}$. Then, by using the result in the case $\dim Z_0 = 1$, we obtain that $\mathcal{T}^{(1)}_{|C}$ is admissible. By Proposition 9.1.15, we obtain the admissibility of $\mathcal{T}^{(1)}$. Thus, we obtain Lemma 11.1.10, and the proof of Proposition 11.1.1 is finished. □

If $(\mathcal{T}, W)$ is integrable, then $(\mathcal{T}^{(1)}, W)$ is also integrable by construction. If $(\mathcal{T}, W) \in \mathrm{MTM}(X)_{\mathcal{A}}$ we can check that the KMS-spectra of $(\mathcal{T}^{(1)}, W)$ are contained in $\mathcal{A}$ by considering the restrictions to curves as in the proof of Lemma 11.1.10. Thus, we obtain the claim of Lemma 11.1.3 in the setting of Proposition 11.1.1. □

11.1.5.2 Proof of Proposition 11.1.2

By Proposition 11.1.1, there exists a commutative diagram of projective morphisms of complex manifolds:

$$\begin{array}{ccc} Z_{11} & \xrightarrow{\kappa_1} & Z_{10} \\ {\scriptstyle\kappa_2}\downarrow & & \downarrow{\scriptstyle\varphi_{10}} \\ Z_1 & \xrightarrow{\varphi} & X \end{array}$$

We have an isomorphism $\kappa_1^*\mathcal{V}_{10} \simeq \kappa_2^*\mathcal{V}$. Then, we can easily check the condition **(Adm0)** for $\mathcal{V}_{10}$. By using Proposition 9.1.15, we can easily check the admissibility of $\mathcal{V}_{10}$. Thus, we obtain Proposition 11.1.2. The claim of Lemma 11.1.3 also follows from the comparison above. □

11.1.5.3 Strict Specializability

We give a proof of a claim used in the proof of Proposition 11.1.1. Let $X := \Delta^n$ and $D = \bigcup_{i=1}^{\ell}\{z_i = 0\}$. Let $Z = \bigcap_{i=\ell+1}^{m}\{z_i = 0\}$ and $D_Z := D \cap Z$. Let $(\mathcal{M}, L)$ be a filtered $\mathcal{R}_X(*D)$-module satisfying the following conditions:

- $(\mathcal{M}, L)_{|\mathbb{C}_\lambda \times (X \setminus D)}$ is the push-forward of a smooth $\mathcal{R}_{Z \setminus D_Z}$-module.
- Each $\mathrm{Gr}^L_w(\mathcal{M})$ is the push-forward of a strict coherent $\mathcal{R}_{Z(*D_Z)}$-module.

Let $X_1 := \{z_{\ell+1} = 0\}$ and $D_1 := X_1 \cap D$. Let $i : X_1 \longrightarrow X$ be the inclusion. Let $f : \mathcal{M} \longrightarrow \mathcal{M}$ be given by the multiplication of $z_{\ell+1}$. We obtain an $\mathcal{R}_{X_1(*D_1)}$-module $\mathrm{Ker} f$, which is naturally equipped with a filtration L.

Lemma 11.1.11 *We have a natural isomorphism* $\iota_\dagger(\mathrm{Ker} f, L) \simeq (\mathcal{M}, L)$. *We also have* $\mathrm{Gr}^L(\mathrm{Ker} f) \simeq \mathrm{Ker}\,\mathrm{Gr}^L(f)$.

Proof If L is pure, the claim is trivial. We use an induction on the length of L. Assume that $L_{-1}\mathcal{M} = 0$ and $L_0\mathcal{M} \neq 0$. By the assumption, $L_0\mathcal{M}$ comes from a strict coherent $\mathcal{R}_{Z(*D_Z)}$-module. By a direct computation, we can check that the cokernel $\mathrm{Cok}(f_{|L_0\mathcal{M}})$ is naturally isomorphic to $L_0\mathcal{M}/\lambda L_0\mathcal{M}$. Hence, a section of $g \in \mathrm{Cok}\big(f_{|L_0\mathcal{M}}\big)$ is 0, if its restriction to $X \setminus D$ is 0. We have the following commutative diagram:

$$\begin{array}{ccccc}
L_0\mathcal{M} & \longrightarrow & \mathcal{M} & \longrightarrow & \mathcal{M}/L_0\mathcal{M} \\
{\scriptstyle f}\big\downarrow & & {\scriptstyle f}\big\downarrow & & {\scriptstyle f}\big\downarrow \\
L_0\mathcal{M} & \longrightarrow & \mathcal{M} & \longrightarrow & \mathcal{M}/L_0\mathcal{M}
\end{array}$$

We may apply the assumption of the induction to $\mathcal{M}/L_0\mathcal{M}$. Then, we have only to show that the induced morphism

$$\mathrm{Ker}\big(f : \mathcal{M}/L_0\mathcal{M} \longrightarrow \mathcal{M}/L_0\mathcal{M}\big) \longrightarrow \mathrm{Cok}\big(f : L_0\mathcal{M} \longrightarrow L_0\mathcal{M}\big) \tag{11.6}$$

is 0. By the assumption, the restriction of (11.6) to $X \setminus D$ vanish. Hence, (11.6) vanishes because of the previous consideration. □

Corollary 11.1.12 *Under the assumption,* $(\mathcal{M}, L)$ *comes from a filtered strict coherent* $\mathcal{R}_{Z(*D_Z)}$*-module.* □

11.2 Localization

11.2.1 Localization Along Functions

Let $\mathcal{T} \in \mathrm{MTM}(X)$. Let f be a holomorphic function on X. Let $\iota_f : X \longrightarrow X \times \mathbb{C}_t$ be the graph. Recall that we have constructed $(\iota_{f\dagger}\mathcal{T})[\star t] \in \mathrm{MTW}(X \times \mathbb{C}_t)$ in Sect. 7.1.6.

Proposition 11.2.1

- *For* $\star = *, !$, *we have* $\mathcal{T}[\star f] \in \mathrm{MTM}(X)$ *such that* $\iota_{f\dagger}(\mathcal{T}[\star f]) \simeq (\iota_{f\dagger}\mathcal{T})[\star t]$.
- *If* $f_1^{-1}(0) = f_2^{-1}(0)$, *we have* $\mathcal{T}[\star f_1] \simeq \mathcal{T}[\star f_2]$ *naturally.*
- *If* $\mathcal{T} \in \mathrm{MTM}^{\mathrm{int}}(X)$, *we naturally have* $\mathcal{T}[\star f] \in \mathrm{MTM}^{\mathrm{int}}(X)$.

- *Let $\mathcal{A}$ be a $\mathbb{Q}$-vector subspace of $\mathbb{R}\times\mathbb{C}$ such that $\mathbb{Q}\times\{0\}\subset\mathcal{A}$. If $\mathcal{T}\in \mathrm{MTM}(X)_{\mathcal{A}}$ then $\mathcal{T}[\star f]\in \mathrm{MTM}(X)_{\mathcal{A}}$.*

Proof As remarked in Lemma 7.1.46, we have only to consider the issues locally. Let $P\in \operatorname{Supp}\mathcal{T}$. We use the Noetherian induction on the support around P. We will shrink X around P in the following argument. Let $(Z,U,\varphi,\mathcal{V})$ be a cell of $\mathcal{T}$ with a cell function g at P. We have the expression of $\mathcal{T}$ as the cohomology of the following complex in $\mathrm{MTM}(X)$:

$$\psi_g^{(1)}\varphi_\dagger(\mathcal{V})\longrightarrow \Xi_g^{(0)}\varphi_\dagger(\mathcal{V})\oplus\phi_g^{(0)}(\mathcal{T})\longrightarrow\psi_g^{(0)}\varphi_\dagger(\mathcal{V})$$

We have already known the claims for good mixed twistor $\mathscr{D}$-modules. Hence, we obtain the claims for $\psi_g^{(a)}\varphi_\dagger(\mathcal{V})$ and $\Xi_g^{(0)}\varphi_\dagger(\mathcal{V})$. We can apply the hypothesis of the induction to $\phi_g^{(0)}(\mathcal{T})$. Then, we obtain the claims for $\mathcal{T}$. The claim in the integrable case follows from Lemma 7.1.60. The claim for $\mathrm{MTM}(X)_{\mathcal{A}}$ follows from Lemma 7.1.66. □

We have naturally defined morphisms $\mathcal{T}[!f]\longrightarrow\mathcal{T}\longrightarrow\mathcal{T}[*f]$, as remarked in Lemma 7.1.44.

11.2.2 Localization Along Hypersurfaces

Let H be a hypersurface of X. Let $\mathcal{T}\in\mathrm{MTM}(X)$. By Propositions 11.2.1 and 7.1.51, we have $\mathcal{T}[\star H]$ in $\mathrm{MTM}(X)$ with an isomorphism $\mathcal{T}[\star H](*H)\simeq\mathcal{T}(*H)$ satisfying the following.

- Let P be any point of X. Let X_P be a small neighbourhood around P. We have an expression $H=\{f=0\}$ on X_P. Then, $\mathcal{T}[\star H]_{|X_P}\simeq\mathcal{T}_{|X_P}[\star f]$ which induces $\mathcal{T}[\star H](*H)_{|X_P}\simeq\mathcal{T}(*H)_{|X_P}$.

We have naturally defined morphisms $\mathcal{T}[!H]\longrightarrow\mathcal{T}\longrightarrow\mathcal{T}[*H]$. If $\mathcal{T}$ is integrable, $\mathcal{T}[\star H]$ are also integrable.

In particular, we have the following.

Proposition 11.2.2 *Let $\mathcal{T}$ be a mixed twistor $\mathscr{D}$-module on X.*

- *$\mathcal{T}$ is localizable along any effective divisor D of X.*
- *If the support of effective divisors D_i $(i=1,2)$ are the same, we naturally have $\mathcal{T}[\star D_1]\simeq\mathcal{T}[\star D_2]$.*
- *If $\mathcal{T}$ is integrable, $\mathcal{T}[\star H]$ is also integrable.*
- *Let $\mathcal{A}\subset\mathbb{R}\times\mathbb{C}$ be a $\mathbb{Q}$-vector subspace such that $\mathbb{Q}\times\{0\}\subset\mathcal{A}$. If $\mathcal{T}\in\mathrm{MTM}(X)_{\mathcal{A}}$ then $\mathcal{T}[\star H]\in\mathrm{MTM}(X)_{\mathcal{A}}$.* □

For $\star=*,!$, we obtain the full subcategory $\mathrm{MTM}(X,[\star H])$ of $\mathcal{T}\in\mathrm{MTM}(X)$ such that $\mathcal{T}=\mathcal{T}[\star H]$. We use the symbols $\mathrm{MTM}^{\mathrm{int}}(X,[\star H])$ and $\mathrm{MTM}(X,[\star H])_{\mathcal{A}}$

with similar meanings. We have naturally defined functors:

$$[\star H] : \mathrm{MTM}(X) \longrightarrow \mathrm{MTM}(X, [\star H]) \; (\star = *, !) \tag{11.7}$$

We obtain the following from Lemma 7.1.49.

Proposition 11.2.3 MTM$(X, [\star H])$ *are abelian subcategories in* MTM(X)*, and the functors (11.7) are exact. Similar claims also hold for the abelian subcategories* $\mathrm{MTM}^{\mathrm{int}}(X, [\star H])$ *and* $\mathrm{MTM}(X, [\star H])_{\mathcal{A}}$. □

We obtain the following from Lemma 7.1.43.

Lemma 11.2.4 *Let* $\star = *$ *or* $!$. *Let* $\mathcal{T}_i \in \mathrm{MTM}(X)$ $(i = 1, 2)$ *such that* $\mathcal{T}_i[\star H] = \mathcal{T}_i$. *We have a natural bijective correspondence between morphisms* $\mathcal{T}_1(*H) \longrightarrow \mathcal{T}_2(*H)$ *as filtered* $\mathcal{R}_X(*H)$*-triples, and morphisms* $\mathcal{T}_1 \longrightarrow \mathcal{T}_2$ *in* MTM(X). *If* $\mathcal{T}_i$ *are integrable, we have the bijection of integrable morphisms.* □

Let $\mathrm{MTM}_H(X) \subset \mathrm{MTM}(X)$ be the full subcategory of mixed twistor $\mathscr{D}$-modules whose supports are contained in H.

Lemma 11.2.5 *For* $\mathcal{T} \in \mathrm{MTM}(X)$*, we have* $\mathcal{T}[\star H] = 0$ *if and only if* $\mathcal{T} \in \mathrm{MTM}_H(X)$.

Proof The if part is clear. If $\mathcal{T}[\star H] = 0$, we have $\mathcal{T}(*H) = 0$, and hence $\mathcal{T} \in \mathrm{MTM}_H(X)$. □

We obtain the following from Corollary 7.1.45

Lemma 11.2.6 *Let* $\mathcal{T}_i \in \mathrm{MTM}(X)$ $(i = 1, 2)$. *We have natural bijections:*

$$\mathrm{Hom}_{\mathrm{MTM}(X)}\big(\mathcal{T}_1[*H],\, \mathcal{T}_2[*H]\big) \simeq \mathrm{Hom}_{\mathrm{MTM}(X)}\big(\mathcal{T}_1,\, \mathcal{T}_2[*H]\big)$$

$$\mathrm{Hom}_{\mathrm{MTM}(X)}\big(\mathcal{T}_1[!H],\, \mathcal{T}_2[!H]\big) \simeq \mathrm{Hom}_{\mathrm{MTM}(X)}\big(\mathcal{T}_1[!H],\, \mathcal{T}_2\big)$$

Similarly, we have the bijections for integrable morphisms, if $\mathcal{T}_i$ *are integrable.* □

Proposition 11.2.7

- *Let* $F : X \longrightarrow Y$ *be a projective morphism. Let* H_Y *be a hypersurface of* Y. *We put* $H_X := F^{-1}(H_Y)$. *For* $\mathcal{T} \in \mathrm{MTM}(X)$*, we have a natural isomorphism* $\big(F^j_\dagger(\mathcal{T})\big)[\star H_Y] \simeq F^j_\dagger\big(\mathcal{T}[\star H_X]\big)$.
- *For* $\mathcal{T} \in \mathrm{MTM}(X)$*, we have natural isomorphisms* $\big(\mathcal{T}[*H]\big)^* \simeq \big(\mathcal{T}^*\big)[!H]$ *and* $j^*(\mathcal{T}[\star H]) \simeq (j^*\mathcal{T})[\star H]$.

Proof The second claim is trivial. Let us prove the first claim. We have a natural isomorphism $\big(F^j_\dagger\mathcal{T}\big)[\star H_Y](*H_Y) \simeq F^j_\dagger\big(\mathcal{T}[\star H_X]\big)(*H_Y)$. Because $F^j_\dagger\big(\mathcal{T}[\star H_X]\big)$ satisfies the characterization of the localization, we obtain the desired isomorphism. □

11.2.3 The Underlying $\mathscr{D}$-Modules

Let $\mathcal{A}$ be a $\mathbb{Q}$-vector subspace in $\mathbb{R} \times \mathbb{C}$ such that $\mathbb{Q} \times \{0\} \subset \mathcal{A}$. We say that a non-zero complex number λ_0 is generic with respect to $\mathcal{A}$ if the map $\mathfrak{e}(\lambda_0) : \mathcal{A} \longrightarrow \mathbb{C}$ is injective.

Example 11.2.8 If $\mathcal{A}$ is contained in $\mathbb{R} \times (\sqrt{-1}\mathbb{R})$, then $\lambda_0 = 1$ is generic with respect to $\mathcal{A}$. □

Let X be a complex manifold with an effective divisor H. Let $\mathcal{T} \in \mathrm{MTM}(X)_{\mathcal{A}}$. Let $(\mathcal{M}_1, \mathcal{M}_2, C)$ be the underlying $\mathcal{R}_X$-triple of $\mathcal{T}$. For any non-zero complex number λ_0, we have the naturally defined morphisms of $\mathscr{D}_X$-modules:

$$\mathcal{M}_i[*H]^{\lambda_0} \longrightarrow \mathcal{M}_i^{\lambda_0}[*H] \tag{11.8}$$

$$\mathcal{M}_i^{\lambda_0}[!H] \longrightarrow \mathcal{M}_i[!H]^{\lambda_0} \tag{11.9}$$

We obtain the following proposition from Lemma 3.3.4.

Proposition 11.2.9 *If λ_0 is generic with respect to $\mathcal{A}$, the morphisms (11.8) and (11.9) are isomorphisms.* □

Corollary 11.2.10 *For any $\mathcal{T} \in \mathrm{MTM}(X, \mathbb{R} \times \sqrt{-1}\mathbb{R})$, we have the following natural isomorphisms*

$$\mathcal{M}_i^1[\star H] \simeq \big(\mathcal{M}_i[\star H]\big)^1 \tag{11.10}$$

Here, $\mathcal{M}_i$ are the underlying $\mathcal{R}_X$-modules of $\mathcal{T}$. In particular, for any integrable mixed twistor $\mathscr{D}$-module $\mathcal{T}$, we have the isomorphisms (11.10). □

We have the following consequence on the localization in which we do not impose the restriction on KMS-spectra.

Corollary 11.2.11 *Suppose $H_1 \subset H$ be hypersurfaces in X. Let $\mathcal{T} \in \mathrm{MTM}(X)$. Then, the natural morphisms*

$$\mathcal{T}[*H] \longrightarrow (\mathcal{T}[*H])[*H_1] \quad \textit{and} \quad (\mathcal{T}[!H])[!H_1] \longrightarrow \mathcal{T}[!H]$$

are isomorphisms.

Proof It is enough to check the claims locally at any point P of X. Let X_P be a small neighbourhood of P in X such that the closure of X_P is compact. By using the description as a gluing, we can easily observe that there exists a finite dimensional $\mathbb{Q}$-vector subspace $\mathcal{A} \subset \mathbb{R} \times \mathbb{C}$ such that (i) $\mathbb{Q} \times \{0\} \subset \mathcal{A}$, (ii) $\mathcal{T}_{|X_P} \in \mathrm{MTM}(X_P)_{\mathcal{A}}$. We take a non-zero complex number λ_0 which is generic with respect to $\mathcal{A}$.

Let $(\mathcal{M}_1, \mathcal{M}_2, C)$ be the underlying $\mathcal{R}_X$-triple of $\mathcal{T}$. By Proposition 11.2.9, we obtain that the induced morphism

$$\left(\mathcal{M}_2[*H]\right)^{\lambda_0} \longrightarrow \left(\left(\mathcal{M}_2[*H]\right)[*H_1]\right)^{\lambda_0}$$

is an isomorphism. Then, we obtain that $\mathcal{T}_1[*H] \longrightarrow (\mathcal{T}_1[*H])[*H_1]$ is an isomorphism. We can argue the other in a similar way. □

11.2.4 Independence from Compactification

Let X be a complex manifold with a hypersurface H. Let $F : X' \longrightarrow X$ be a projective birational morphism such that $X' \setminus H' \simeq X \setminus H$, where $H' := \varphi^{-1}(H)$.

Proposition 11.2.12 *For $\star = *, !$, the push-forward induces an equivalence of the categories*

$$F_\dagger : \mathrm{MTM}(X', [\star H']) \longrightarrow \mathrm{MTM}(X, [\star H]).$$

Similar claims hold for $\mathrm{MTM}^{\mathrm{int}}(X, [\star H])$ *and* $\mathrm{MTM}(X, [\star H])_{\mathcal{A}}$.

Proof We prove only the ordinary case for $\star = *$. The other cases can be argued similarly. We obtain the fully faithfulness from Lemma 11.2.4. Let us prove the essential surjectivity. Let $\mathcal{T} \in \mathrm{MTM}(X, [*H])$. We have the filtered $\mathcal{R}_X(*H)$-triple $\mathcal{T}(*H)$. We have the corresponding $\mathcal{R}_{X'}(*H')$-triple $\mathcal{T}_1$, as remarked in Lemma 2.1.11. We have only to show that there exists $\mathcal{T}' \in \mathrm{MTM}(X')$ such that $\mathcal{T}'(*H') = \mathcal{T}_1$ as filtered $\mathcal{R}_{X'(*H')}$-triples.

First, let us consider the local problem. Let P be a point of X. We take a small neighbourhood X_P of P. We set $H_P := H \cap X_P$, $X'_P := F^{-1}(X_P)$ and $H'_P := F^{-1}(H_P)$. We set $\mathcal{T}_P := \mathcal{T}_{|X_P}$. Let $F_P := F_{|X'_P}$.

Lemma 11.2.13 *If we shrink X_P, there exists $\mathcal{T}'_P \in \mathrm{MTM}(X'_P, [*H'_P])$ such that $F_{P\dagger}\mathcal{T}'_P \simeq \mathcal{T}_P$.*

Proof In the proof, we shrink X instead of considering X_P. So, we omit the subscript P to simplify the notation. We use the Noetherian induction. We take a cell $\mathcal{C} = (Z, U, \varphi, \mathcal{V})$ of $\mathcal{T}$ at P. We may assume that $\varphi : Z \longrightarrow X$ factors through X', i.e., φ is the composition of $\varphi' : Z \longrightarrow X'$ and $F : X' \longrightarrow X$. Let g be a cell function for $\mathcal{C}$ with $H \subset g^{-1}(0)$. We have the expression of $\mathcal{T}$ as the cohomology of $\psi_g^{(1)}\varphi_\dagger\mathcal{V} \longrightarrow \Xi_g^{(0)}\varphi_\dagger\mathcal{V} \oplus \phi_g^{(0)}(\mathcal{T}) \longrightarrow \psi_g^{(0)}\varphi_\dagger\mathcal{V}$. Let $g' := g \circ F$. We have the mixed twistor $\mathscr{D}$-modules $\psi_{g'}^{(a)}\varphi'_\dagger\mathcal{V}[*H']$ and $\Xi_{g'}^{(a)}\varphi'_\dagger\mathcal{V}[*H']$ on X'. By the assumption of the induction, we have $\mathcal{Q} \in \mathrm{MTM}(X')$ such that $\mathcal{Q}[*H'] = \mathcal{Q}$ and $F_\dagger\mathcal{Q} = \phi_g^{(0)}\mathcal{T}[*H]$.

By the fully faithfulness, we have the morphisms

$$\psi_{g'}^{(1)}\varphi'_{\dagger}\mathcal{V}[*H'] \longrightarrow \mathcal{Q} \longrightarrow \psi_{g'}^{(0)}\varphi'_{\dagger}\mathcal{V}[*H']$$

corresponding to $\psi_g^{(1)}\varphi_{\dagger}\mathcal{V}[*H] \longrightarrow \phi_g^{(0)}\mathcal{T}[*H] \longrightarrow \psi_g^{(0)}\varphi_{\dagger}\mathcal{V}[*H]$. We obtain $\mathcal{T}' \in \mathrm{MTM}(X')$ as the cohomology of

$$\psi_{g'}^{(1)}\varphi_{\dagger}\mathcal{V}[*H'] \longrightarrow \Xi_{g'}^{(0)}\varphi_{\dagger}\mathcal{V}[*H'] \oplus \mathcal{Q} \longrightarrow \psi_{g'}^{(0)}\varphi_{\dagger}\mathcal{V}[*H'].$$

It satisfies $\mathcal{T}'(*H') = \mathcal{T}_1$. Thus, we obtain Lemma 11.2.13. □

Let W denote the weight filtration of $\mathcal{T}'_P$ as an object in $\mathrm{MTM}(X'_P)$. We have the decomposition of the polarizable pure twistor $\mathscr{D}$-modules:

$$\mathrm{Gr}_w^W(\mathcal{T}'_P) = \mathcal{P}'_{P,w} \oplus \mathrm{Ker}\Big(\mathrm{Gr}_w^{\hat{N}_*L}\mathrm{Gr}_w^W \psi_{H'_P}^{(0)}(\mathcal{T}'_P) \longrightarrow \mathrm{Gr}_w^L \mathrm{Gr}_w^W \psi_{H'_P}^{(0)}(\mathcal{T}'_P)\Big) \tag{11.11}$$

Here, $\mathcal{P}'_{P,w}$ denotes the sum of the direct summand of $\mathrm{Gr}_w^W(\mathcal{T}'_P)$ whose strict supports are not contained in H'_P.

Let us consider the global problem. By gluing $(\mathcal{T}'_P, W)$ for varied P, we obtain a filtered $\mathcal{R}_{X'}$-triple $(\mathcal{T}', W)$ on X'. It is localizable along H', and we have $\mathcal{T}'[*H'] \simeq \mathcal{T}'$. To establish $\mathcal{T}'[*H'] \in \mathrm{MTM}(X', [*H'])$, we have only to prove that $\mathrm{Gr}_w^W(\mathcal{T}')$ is polarizable pure twistor $\mathscr{D}$-module of weight w. By gluing (11.11), we have the following decomposition:

$$\mathrm{Gr}_w^W(\mathcal{T}') = \mathcal{P}'_w \oplus \mathrm{Ker}\Big(\mathrm{Gr}_w^{\hat{N}_*L}\mathrm{Gr}_w^W \psi_{H'}^{(0)}(\mathcal{T}') \longrightarrow \mathrm{Gr}_w^L \mathrm{Gr}_w^W \psi_{H'}^{(0)}(\mathcal{T}')\Big) \tag{11.12}$$

Let W denote the weight filtration of $\mathcal{T}$ as an object in $\mathrm{MTM}(X)$. Let $\mathrm{Gr}_w^W(\mathcal{T}) = \bigoplus \mathcal{P}_{Z,w}$ be the decomposition by the strict supports, where Z runs through closed irreducible subvariety of X. By using the correspondence between wild harmonic bundles and wild pure twistor $\mathscr{D}$-modules, for $Z \not\subset H$, we take a projective birational morphism $\kappa_Z : Z_1 \longrightarrow Z$ satisfying the following conditions:

- Z_1 is smooth with a normal crossing hypersurface H_1.
- We have a polarizable pure twistor $\mathscr{D}$-module $\mathfrak{P}_{Z,w}$ on Z_1 whose strict support is Z_1, such that $\mathcal{P}_{Z,w}$ is the component of $\kappa_{Z\dagger}^0(\mathfrak{P}_{Z,w})$ whose strict support is Z.
- κ_Z factors through X', i.e., $\iota_Z \circ \kappa_Z = F \circ \kappa'_Z$ for some $\kappa'_Z : Z_1 \longrightarrow X'$, where $\iota_Z : Z \longrightarrow X$ denote the inclusion.

Let $\mathcal{P}'_{Z,w}$ be the polarizable pure twistor $\mathscr{D}$-module on X' obtained as the component of $\kappa'_{Z\dagger}\mathfrak{P}_{Z,w}$ whose strict support is $\kappa'_Z(Z_1)$. We naturally have $F_{\dagger}\big(\mathcal{P}'_{Z,w}(*H')\big) \simeq \mathcal{P}_{Z,w}(*H)$. By the uniqueness in the local construction, we have $\bigoplus_Z \mathcal{P}'_{Z,w|X'_P} \simeq \mathcal{P}'_{P,w}$. Hence, we obtain that $\mathcal{P}'_w = \bigoplus_Z \mathcal{P}'_{Z,w}$, i.e., $\mathcal{P}'_w$ is a polarizable pure twistor $\mathscr{D}$-module of weight w.

We have the canonical decomposition:

$$\mathrm{Gr}^W_w \psi^{(0)}_{H'}(\mathcal{T}') \simeq \mathrm{Gr}^W_w \psi^{(0)}_{H'}(\mathrm{Gr}^W(\mathcal{T}')) \simeq \mathrm{Gr}^W_w \psi^{(0)}_{H'}\Big(\bigoplus_{w'} \mathcal{P}'_{w'}\Big)$$

Because $\mathcal{P}'_{w'}$ are polarizable pure twistor $\mathscr{D}$-modules of weight w', we obtain that $\mathrm{Gr}^W_w \psi^{(0)}_{H'}(\mathcal{T}')$ is a polarizable pure twistor $\mathscr{D}$-module of weight w by Proposition 7.1.29. Thus, the proof of Proposition 11.2.12 is finished. □

11.3 Twist by Admissible Twistor Structure and Beilinson Functors

11.3.1 Smooth Case

Let $\mathcal{T} \in \mathrm{MTM}(X)$ and $\mathcal{V} \in \mathrm{MTS}(X)$. We have a naturally defined filtered $\mathcal{R}_X$-triple $\mathcal{T} \otimes \mathcal{V}$.

Lemma 11.3.1 *We have $\mathcal{T} \otimes \mathcal{V} \in \mathrm{MTM}(X)$. If $\mathcal{T}$ and $\mathcal{V}$ are integrable, then we naturally have $\mathcal{T} \otimes \mathcal{V} \in \mathrm{MTM}^{\mathrm{int}}(X)$.*

Proof We have only to use the Noetherian induction. □

Lemma 11.3.2 *Let $f : X \longrightarrow Y$ be a projective morphism. For $\mathcal{T} \in \mathrm{MTM}(X)$ and $\mathcal{V} \in \mathrm{MTS}(Y)$, we have natural isomorphisms*

$$f^i_\dagger(\mathcal{T} \otimes f^*\mathcal{V}) \simeq f^i_\dagger(\mathcal{T}) \otimes \mathcal{V} \tag{11.13}$$

in $\mathrm{MTM}(Y)$. We have similar isomorphisms in the integrable case.

Proof It follows from Lemma 2.1.16. □

11.3.2 Admissible Case

Let H be a hypersurface of X which is not necessarily normal crossing. Suppose that we are given a projective morphism of complex manifolds $F : X' \longrightarrow X$ such that (i) $H' := F^{-1}(H)$ is normal crossing, (ii) F induces an isomorphism $X' \setminus H' \simeq X \setminus H$. For any $\mathcal{T}_0 \in \mathrm{MTS}^{\mathrm{adm}}(X', H')$, the push-forward $F_\dagger \mathcal{T}_0$ is also called an admissible variation of mixed twistor structure on (X, H).

Let $\mathcal{T} \in \mathrm{MTM}(X)$ and $\mathcal{T}_0 \in \mathrm{MTS}^{\mathrm{adm}}(X', H')$. We naturally obtain a filtered $\mathcal{R}_{X(*H)}$-triple $\mathcal{T} \otimes F^0_\dagger \mathcal{T}_0$.

Proposition 11.3.3 *The filtered $\mathcal{R}_X(*H)$-triple $\mathcal{T}\otimes F_\dagger^0\mathcal{T}_0$ is naturally extended to a mixed twistor $\mathscr{D}_X$-module $(\mathcal{T}\otimes F_\dagger^0\mathcal{T}_0)[\star H]$ with isomorphisms:*

$$\Big((\mathcal{T}\otimes F_\dagger^0\mathcal{T}_0)[\star H]\Big)(*H)\simeq\mathcal{T}\otimes F_\dagger^0\mathcal{T}_0,$$
$$((\mathcal{T}\otimes F_\dagger^0\mathcal{T}_0)[\star H])[\star H]\simeq(\mathcal{T}\otimes F_\dagger^0\mathcal{T}_0)[\star H] \tag{11.14}$$

Such a mixed twistor $\mathscr{D}$-module $(\mathcal{T}\otimes F_\dagger^0\mathcal{T}_0)[\star H]$ is unique up to canonical isomorphisms. Moreover, the following holds.

- *If $\mathcal{T}$ and $\mathcal{T}_0$ are integrable, then $(\mathcal{T}\otimes F_\dagger^0\mathcal{T}_0)[\star H]$ is also naturally integrable.*
- *Let $\mathcal{A}\subset\mathbb{R}\times\mathbb{C}$ be a $\mathbb{Q}$-vector subspace such that $\mathbb{Q}\times\{0\}\subset\mathcal{A}$. Suppose that $\mathcal{T}\in\mathrm{MTM}(X)_{\mathcal{A}}$ and that the KMS-spectra of $\mathcal{T}_0$ along any irreducible component of H are contained in $\mathcal{A}$. Then, $(\mathcal{T}\otimes F_\dagger^0\mathcal{T}_0)[\star H]\in\mathrm{MTM}(X)_{\mathcal{A}}$.*

Proof The uniqueness is clear. It is enough to construct such a mixed twistor $\mathscr{D}$-module.

Let $(\varphi,Z,U,\mathcal{V})$ be an admissible cell with a cell function g satisfying the following conditions.

- φ factors through X', i.e., there exists $\varphi_1:Z\longrightarrow X'$ such that $\varphi=F\circ\varphi_1$.
- $D'_Z:=\varphi_1^{-1}(H')\cup D_Z$ is normal crossing.

We obtain admissible mixed twistor structures $\varphi_1^*(\mathcal{T}_0)(*D'_Z)$ and $\mathcal{V}(*D'_Z)$ on (Z,D'_Z). By Proposition 9.1.16, we may assume that $\mathcal{V}\otimes\varphi_1^*\mathcal{T}_0$ is also admissible on (Z,D'_Z).

If $\mathcal{T}=\varphi_\dagger(\mathcal{V})[\star_1 g]$ ($\star_1=*,!$) we set

$$(\mathcal{T}\otimes F_\dagger^0\mathcal{T}_0)[\star H]:=\Big(\varphi_\dagger\big(\mathcal{V}\otimes\varphi_1^*\mathcal{T}_0\big)[\star_1 D'_Z]\Big)[\star H].$$

Then, it satisfies the condition in this case. Thus, we obtain the claim for $\mathcal{T}=\varphi_\dagger(\mathcal{V})[\star_1 g]$ ($\star_1=*,!$). We easily obtain the claims in the cases that $\mathcal{T}$ is $\Xi_g^{(a)}(\varphi_\dagger\mathcal{V})$ or $\psi_g^{(a)}(\varphi_\dagger\mathcal{V})$.

Let P be any point of X. On a small neighbourhood X_P of P in X, by using the above consideration together with a Noetherian induction on the support, we can construct a mixed twistor $\mathscr{D}$-module $\big(\mathcal{T}\otimes F_\dagger^0\mathcal{T}_0\big)[\star H]_P$ on X_P with isomorphisms:

$$\Big((\mathcal{T}\otimes F_\dagger^0\mathcal{T}_0)[\star H]\Big)_P(*H_P)\simeq\big(\mathcal{T}\otimes F_\dagger^0\mathcal{T}_0\big)_{|X_P}$$
$$((\mathcal{T}\otimes F_\dagger^0\mathcal{T}_0)[\star H])_P[\star H_P]\simeq((\mathcal{T}\otimes F_\dagger^0\mathcal{T}_0)[\star H])_P$$

By varying $P\in X$ and gluing $((\mathcal{T}\otimes F_\dagger^0\mathcal{T}_0)[\star H])_P$, we construct a filtered $\mathcal{R}$-triple $\mathcal{T}\otimes\mathcal{T}_0[\star H]$. By the construction, we have the isomorphisms in (11.14). By the construction, it is admissibly specializable and localizable along H. We can check that it is an object in $\mathrm{MTW}(X)$ by the argument in the proof of

Proposition 11.2.12. Then, we obtain that it is an object in MTM(X) by the construction. The complementary claims are easy to check by construction. □

Lemma 11.3.4 *Let X, H, $F : X' \longrightarrow X$ and $\mathcal{T}_0$ be as above. Let $f : Y \longrightarrow X$ be any projective morphism of complex manifolds. We set $H_Y := f^{-1}(H)$. Let $\mathcal{T}_Y \in$* MTM(Y). *Then, we have natural isomorphisms:*

$$f_\dagger^i\big((\mathcal{T}_Y \otimes f^* F_\dagger^0 \mathcal{T}_0)[\star H_Y]\big) \simeq \big(f_\dagger^i(\mathcal{T}_Y) \otimes F_\dagger^0(\mathcal{T}_0)\big)[\star H_Y].$$

in MTM(X). *We have similar isomorphisms in the integrable case.*

Proof It follows from Lemma 11.3.2. □

11.3.3 Beilinson Functors

As a consequence of Proposition 11.3.3, for any holomorphic function g on X and for any integers $a \leq b$, we define the functors $\Pi_{g\star}^{a,b}$ on MTM(X) as follows:

$$\Pi_{g\star}^{a,b}(\mathcal{T}) := \big(\mathcal{T} \otimes g^* \tilde{\mathbb{I}}^{a,b}\big)[\star g]$$

We obtain Beilinson functors $\Pi_{g*!}^{a,b}$ on MTM(X) defined as follows:

$$\Pi_{g*!}^{a,b}(\mathcal{T}) := \varinjlim_{N \to \infty} \operatorname{Ker}\Big(\Pi_{g!}^{-N,b}(\mathcal{T}) \longrightarrow \Pi_{g*}^{-N,a}(\mathcal{T})\Big)$$

Note that the right hand side is locally independent of any sufficiently large N. In particular, we obtain the maximal functors $\Xi_g^{(a)}$ and the nearby cycle functors $\psi_g^{(a)}$.

$$\Xi_g^{(a)} := \Pi_{g,*!}^{a,a+1}, \qquad \psi_g^{(a)} := \Pi_{g,*!}^{a,a}$$

We obtain the vanishing cycle functors $\phi_g^{(a)}$ on MTM(X) as in Sect. 4.2.3. Namely, for any $\mathcal{T} \in$ MTM(X), the mixed twistor $\mathscr{D}$-modules $\phi_g^{(a)}(\mathcal{T})$ are defined as the cohomology of the natural complexes in MTM(X):

$$\mathcal{T}[!g] \otimes \boldsymbol{T}(a) \longrightarrow \Xi_g^{(a)}(\mathcal{T}) \oplus \mathcal{T} \otimes \boldsymbol{T}(a) \longrightarrow \mathcal{T}[*g] \otimes \boldsymbol{T}(a)$$

We can reconstruct $\mathcal{T}$ as the cohomology of the natural complex in MTM(X):

$$\psi_g^{(1)}(\mathcal{T}) \longrightarrow \Xi_g^{(0)}(\mathcal{T}) \oplus \phi_g^{(0)}(\mathcal{T}) \longrightarrow \psi^{(0)}(\mathcal{T}).$$

We also have similar functors on $\mathrm{MTM}^{\mathrm{int}}(X)$ and $\mathrm{MTM}(X)_{\mathcal{A}}$ defined in the same ways. We obtain the following from Lemma 11.3.4, Proposition 4.2.10 and Corollary 4.2.11.

Lemma 11.3.5 *Let X and g be as above. Let $f : Y \longrightarrow X$ be a projective morphism. We set $g_Y := g \circ f$. Let $\mathcal{T}_Y \in \mathrm{MTM}(Y)$. We have the natural isomorphisms*

$$f_\dagger^j\bigl(\Pi^{a,b}_{g_Y*!}(\mathcal{T}_Y)\bigr) \simeq \Pi^{a,b}_{g,*!}\bigl(f_\dagger^j(\mathcal{T}_Y)\bigr)$$

In particular, we have the natural isomorphisms

$$f_\dagger^j \circ \psi^{(a)}_{g_Y}(\mathcal{T}_Y) \simeq \psi^{(a)}_g \circ f_\dagger^j(\mathcal{T}_Y), \quad f_\dagger^j \circ \Xi^{(a)}_{g_Y}(\mathcal{T}_Y) \simeq \Xi^{(a)}_g \circ f_\dagger^j(\mathcal{T}_Y).$$

We also have $f_\dagger^j \circ \phi^{(a)}_{g_Y}(\mathcal{T}_Y) \simeq \phi^{(a)}_g \circ f_\dagger^j(\mathcal{T}_Y)$. We have similar isomorphisms in the integrable case. □

We give a remark on the weight filtration on $\phi^{(a)}_g(\mathcal{T})$ and $\psi^{(a)}_g(\mathcal{T})$. Let L be the filtration on $\phi^{(a)}_g(\mathcal{T})$ and $\psi^{(a)}_g(\mathcal{T})$ given by $L_j\phi^{(a)}_g(\mathcal{T}) = \phi^{(a)}_g(W_j\mathcal{T})$ and $L_j\psi^{(a)}_g(\mathcal{T}) = \psi^{(a)}_g(W_j\mathcal{T})$.

Proposition 11.3.6 *Let $M(\mathcal{N};L)$ denote the relative monodromy filtration of $\mathcal{N}$ with respect to L, and let W denote the weight filtration on $\phi^{(a)}_g(\mathcal{T})$ and $\psi^{(a)}_g(\mathcal{T})$ obtained as the above construction. Then, we have $W = M(\mathcal{N};L)[-2a]$ on $\phi^{(a)}_g(\mathcal{T})$, and $W = M(\mathcal{N};L)[-2a+1]$ on $\psi^{(a)}_g(\mathcal{T})$.*

Proof We use an induction on $\dim \operatorname{Supp} \mathcal{T}$. If $\dim \operatorname{Supp}(\mathcal{T}) = 0$, then the claim is clear.

It is enough to check the claim locally around any point P of X. We take a small neighbourhood X_P, an admissible cell $\mathcal{C}_1 = (Z_1, U_1, \varphi_1, \mathcal{V}_1)$ and a cell function g_1 for $\mathcal{T}_{|X_P}$. We impose the condition that the dimension of $g_1^{-1}(0) \cap \operatorname{Supp}(\mathcal{T})$ is strictly smaller than $\dim \operatorname{Supp}(\mathcal{T})$. We express $\mathcal{T}_{|X_P}$ as the cohomology of the following complex:

$$\psi^{(1)}_{g_1}(\mathcal{T}) \longrightarrow \Xi^{(0)}_{g_1}\varphi_{1\dagger}(\mathcal{V}) \oplus \phi^{(0)}_{g_1}(\mathcal{T}) \longrightarrow \psi^{(0)}_{g_1}(\mathcal{T})$$

Then, $\phi^{(a)}_g(\mathcal{T})$ is expressed as the cohomology of the following complex:

$$\phi^{(a)}_g\psi^{(1)}_{g_1}(\mathcal{T}) \longrightarrow \phi^{(a)}_g\Xi^{(0)}_{g_1}\varphi_{1\dagger}(\mathcal{V}) \oplus \phi^{(a)}_g\phi^{(0)}_{g_1}(\mathcal{T}) \longrightarrow \phi^{(a)}_g\psi^{(0)}_{g_1}(\mathcal{T})$$

By the assumption the induction, we have $W = M(\mathcal{N};L)[-2a]$ on $\phi^{(a)}_g\psi^{(b)}_{g_1}(\mathcal{T})$ and $\phi^{(a)}_g\phi^{(0)}_{g_1}(\mathcal{T})$. We put $g' := g \circ \varphi_1$ and $g_1' := g_1 \circ \varphi_1$. By Proposition 10.2.9, we have $W = M(\mathcal{N};L)[-2a]$ on $\phi^{(a)}_{g'}\Xi^{(0)}_{g_1'}\mathcal{V}_1$. By Corollary 7.1.32, we have $W = M(\mathcal{N};L)[-2a]$ on $\phi^{(a)}_g\Xi^{(0)}_{g_1}\varphi_{1\dagger}(\mathcal{V}_1)$. Then, we can easily obtain that $W = M(\mathcal{N};L)[-2a]$ on $\phi^{(a)}_g(\mathcal{T})$. (See Sect. 7.1.1.1, for example.) We obtain the claim for $\psi^{(a)}_g(\mathcal{T})$ similarly. □

11.4 External Tensor Product

11.4.1 Preliminary

11.4.1.1 $\mathcal{O}$-Modules

Let X_i $(i = 1, 2)$ be complex manifolds. We set $X := X_1 \times X_2$. Let $p_i : X \longrightarrow X_i$ be the projections. We set $\mathcal{X} := \mathbb{C}_\lambda \times X$ and $\mathcal{X}_i := \mathbb{C}_\lambda \times X_i$. The induced morphisms $\mathcal{X} \longrightarrow \mathcal{X}_i$ are also denoted by p_i. The pull back of $\mathcal{O}_{\mathbb{C}_\lambda}$ by the projections $\mathcal{X} \longrightarrow \mathbb{C}_\lambda$ and $\mathcal{X}_i \longrightarrow \mathbb{C}_\lambda$ are also denoted by $\mathcal{O}_{\mathbb{C}_\lambda}$.

Let $\mathcal{M}_i$ $(i = 1, 2)$ be $\mathcal{O}_{\mathcal{X}_i}$-modules. We have $\mathcal{O}_{\mathcal{X}}$-modules $p_i^* \mathcal{M}_i$ obtained as the pull back.

Lemma 11.4.1 *Suppose that $\mathcal{M}_1$ is flat over $\mathcal{O}_{\mathbb{C}_\lambda}$. Then, $p_1^* \mathcal{M}_1$ is flat over $p_2^{-1} \mathcal{O}_{\mathcal{X}_2}$.*

Proof We set $\mathcal{A} := p_1^{-1} \mathcal{O}_{\mathcal{X}_1} \otimes_{\mathcal{O}_{\mathbb{C}_\lambda}} p_2^{-1} \mathcal{O}_{\mathcal{X}_2}$. Because $\mathcal{M}_1$ is flat over $\mathcal{O}_{\mathbb{C}_\lambda}$, the $\mathcal{A}$-module $\mathcal{A} \otimes_{p_1^{-1} \mathcal{O}_{\mathcal{X}_1}} p_1^{-1} \mathcal{M}_1 \simeq p_2^{-1} \mathcal{O}_{\mathcal{X}_2} \otimes_{\mathcal{O}_{\mathbb{C}_\lambda}} p_1^{-1} \mathcal{M}_1$ is flat over $p_2^{-1} \mathcal{O}_{\mathcal{X}_2}$. For any $\mathcal{O}_{\mathcal{X}_2}$-module N, we obtain $\mathrm{Tor}_k^{\mathcal{A}}\big(\mathcal{A} \otimes_{p_1^{-1} \mathcal{O}_{\mathcal{X}_1}} p_1^{-1} \mathcal{M}_1, \mathcal{A} \otimes_{p_2^{-1} \mathcal{O}_{\mathcal{X}_2}} p_2^{-1} N\big) = 0$ $(k > 0)$. Note that $\mathcal{A} \longrightarrow \mathcal{O}_{\mathcal{X}}$ is fully faithful which follows from Proposition 4.10 in [41], for example. Then, we obtain $\mathrm{Tor}_k^{\mathcal{O}_{\mathcal{X}}}(p_1^* \mathcal{M}_1, p_2^* N) = 0$ $(k > 0)$. □

Lemma 11.4.2 *If $\mathcal{M}_i$ $(i = 1, 2)$ are flat over $\mathcal{O}_{\mathbb{C}_\lambda}$, then $p_1^* \mathcal{M}_1 \otimes_{\mathcal{O}_{\mathcal{X}}} p_2^* \mathcal{M}_2$ is flat over $\mathcal{O}_{\mathbb{C}_\lambda}$, and $\mathrm{Tor}_k^{\mathcal{O}_{\mathcal{X}}}(p_1^* \mathcal{M}_1, p_2^* \mathcal{M}_2) = 0$ $(k > 0)$.*

Proof We directly obtain $\mathrm{Tor}_k^{\mathcal{O}_{\mathcal{X}}}(p_1^* \mathcal{M}_1, p_2^* \mathcal{M}_2) = 0$ $(k > 0)$ from Lemma 11.4.1. For any $\mathcal{O}_{\mathbb{C}_\lambda}$-module J, we have $\mathcal{M}_2 \otimes^L_{\mathcal{O}_{\mathbb{C}_\lambda}} J \simeq \mathcal{M}_2 \otimes_{\mathcal{O}_{\mathbb{C}_\lambda}} J$ and $\mathrm{Tor}_k^{\mathcal{O}_{\mathcal{X}}}(p_1^* \mathcal{M}_1, p_2^* \mathcal{M}_2 \otimes J) = 0$ $(k > 0)$. Hence, we obtain $(\mathcal{M}_1 \otimes_{\mathcal{O}_{\mathcal{X}}} \mathcal{M}_2) \otimes^L_{\mathcal{O}_{\mathbb{C}_\lambda}} J \simeq (\mathcal{M}_1 \otimes_{\mathcal{O}_{\mathcal{X}}} \mathcal{M}_2) \otimes_{\mathcal{O}_{\mathbb{C}_\lambda}} J$, i.e., $\mathcal{M}_1 \otimes_{\mathcal{O}_{\mathcal{X}}} \mathcal{M}_2$ is flat over $\mathcal{O}_{\mathbb{C}_\lambda}$. □

Let $L\mathcal{M}_i$ be an increasing filtration on $\mathcal{M}_i$ such that (i) $L_k(\mathcal{M}_i) = 0$ $(k << 0)$, and $L_k(\mathcal{M}_i) = \mathcal{M}_i$ $(k >> 0)$, (ii) $\mathrm{Gr}^L(\mathcal{M}_i)$ are $\mathcal{O}_{\mathbb{C}_\lambda}$-flat. We naturally regard $L_a(\mathcal{M}_1) \boxtimes L_b(\mathcal{M}_2) \subset \mathcal{M}_1 \boxtimes \mathcal{M}_2$, and we set

$$L_k(\mathcal{M}_1 \boxtimes \mathcal{M}_2) := \sum_{a+b \leq k} L_a(\mathcal{M}_1) \boxtimes L_b(\mathcal{M}_2). \tag{11.15}$$

Thus, we obtain a filtration L on $\mathcal{M}_1 \boxtimes \mathcal{M}_2$. We have naturally induced morphisms

$$\bigoplus_{a+b=k} \mathrm{Gr}^L_a(\mathcal{M}_1) \boxtimes \mathrm{Gr}^L_b(\mathcal{M}_2) \longrightarrow \mathrm{Gr}^L_k(\mathcal{M}_1 \boxtimes \mathcal{M}_2) \tag{11.16}$$

Lemma 11.4.3 *The morphisms (11.16) are isomorphisms.*

Proof It is enough to check the claim locally around any point of $X_1 \times X_2$. So, we may assume X_i are small neighborhood of $P_i \in X_i$. If $\mathrm{Gr}^L(\mathcal{M}_i)$ are locally free, we can easily check the claim by taking splittings. For general case, we can take free resolutions

$$\mathcal{P}_i^\bullet = \left(\cdots \to \mathcal{P}_i^{-2} \xrightarrow{\alpha_{-2}} \mathcal{P}_i^{-1} \xrightarrow{\alpha_{-1}} \mathcal{P}_i^0\right) \xrightarrow{\alpha_0} \mathcal{M}_i.$$

with filtration L on $\mathcal{P}_i^\bullet$ such that (i) α_p are strictly compatible with L, i.e., $\alpha_p(L_j\mathcal{P}_i^p) = L_j\mathcal{P}_i^{p+1} \cap \alpha_p(\mathcal{P}_i^p)$ for $p < 0$ and $\alpha_0(L_j\mathcal{P}_i^0) = L_j\mathcal{M}_i \cap \alpha_0(\mathcal{P}_i^0)$, (ii) the induced complex $\mathrm{Gr}_j^L \mathcal{P}_i^\bullet$ is naturally a resolution of $\mathrm{Gr}_j^L(\mathcal{M}_i)$. We have the filtrations $L(\mathcal{P}_1^p \boxtimes \mathcal{P}_2^q)$, which induces the total complex of the double complex $\mathcal{P}^\bullet := \mathrm{Tot}(\mathcal{P}_1^\bullet \boxtimes \mathcal{P}_2^\bullet)$ with a filtration L. Then, by Lemma 11.4.2, $\mathcal{P}^\bullet$ is a resolution of $\mathcal{M}_1 \boxtimes \mathcal{M}_2$. By construction of L, we have $\big(\mathcal{H}^0(\mathcal{P}^\bullet), L\big) \simeq (\mathcal{M}_1 \boxtimes \mathcal{M}_2, L)$. We also have that $\mathrm{Tot}\big(\mathrm{Gr}_a^L(\mathcal{P}_1^\bullet) \boxtimes \mathrm{Gr}_b^L(\mathcal{P}_2^\bullet)\big)$ are resolutions of $\mathrm{Gr}_a^L(\mathcal{M}_1) \boxtimes \mathrm{Gr}_b^L(\mathcal{M}_2)$. Because $\mathrm{Gr}_k^L(\mathcal{P}^\bullet) = \bigoplus_{a+b=k} \mathrm{Tot}\big(\mathrm{Gr}_a^L(\mathcal{P}_1^\bullet) \boxtimes \mathrm{Gr}_b^L(\mathcal{P}_2^\bullet)\big)$, the latter implies that $\mathcal{H}^j \mathrm{Gr}_k^L(\mathcal{P}^\bullet) = 0$ for any $j < 0$. Hence, we obtain $\mathcal{H}^0 \mathrm{Gr}^L(\mathcal{P}^\bullet) \simeq \mathrm{Gr}^L \mathcal{H}^0(\mathcal{P}^\bullet)$. □

11.4.1.2 $\mathcal{R}$-Modules

Let $\mathcal{M}_i$ $(i = 1,2)$ be $\mathcal{R}_{X_i}$-modules. We obtain $\mathcal{R}_X$-modules $p_i^*\mathcal{M}_i := \mathcal{O}_{\mathcal{X}} \otimes_{p_i^{-1}\mathcal{O}_{\mathcal{X}_i}} \mathcal{M}_i$, and we set

$$\mathcal{M}_1 \boxtimes \mathcal{M}_2 := p_1^*\mathcal{M}_1 \otimes_{\mathcal{O}_{\mathcal{X}}} p_2^*\mathcal{M}_2.$$

It is called the external tensor product of $\mathcal{R}_{X_i}$-modules. If $\mathcal{M}_i$ are strict, i.e., flat over $\mathcal{O}_{\mathbb{C}_\lambda}$, $\mathcal{M}_1 \boxtimes \mathcal{M}_2$ is also strict. For filtrations $L_\bullet\mathcal{M}_i$ on $\mathcal{M}_i$ by $\mathcal{R}_{X_i}$-modules such that $\mathrm{Gr}^L(\mathcal{M}_i)$ are also strict, we define a filtration $L(\mathcal{M}_1 \boxtimes \mathcal{M}_2)$ by the formula (11.15).

11.4.1.3 Distributions

Let U_i $(i = 1,2)$ be open subsets in X_i. Let $I \subset \boldsymbol{S}$ be an open interval. Let Φ_i be sections of $\mathfrak{Db}_{\boldsymbol{S}\times X_i/\boldsymbol{S}}$ on $I \times U_i$. For each $\theta \in I$, we have the distributions Φ_i^θ on U_i. By using expressions of distributions as derivatives and the continuity of expressions with respect to the parameter (see [49]), we obtain the naturally defined distribution $\Phi_1^\theta \boxtimes \Phi_2^\theta$ on $U_1 \times U_2$, and they depend on θ continuously. Hence, they give a section $\Phi_1 \boxtimes \Phi_2$ of $\mathfrak{Db}_{\boldsymbol{S}\times X/\boldsymbol{S}}$ on $I \times U_1 \times U_2$.

11.4.1.4 $\mathcal{R}$-Triples

Let $\mathcal{T}_i = (\mathcal{M}_i', \mathcal{M}_i'', C_i)$ $(i = 1,2)$ be strict $\mathcal{R}_{X_i}$-triples. By using the external product of distributions mentioned above, we have the induced sesqui-linear pairing $C_1 \boxtimes C_2$ of $\mathcal{M}_1' \boxtimes \mathcal{M}_2'$ and $\mathcal{M}_1'' \boxtimes \mathcal{M}_2''$. Thus, we obtain an $\mathcal{R}_X$-triple

$$\mathcal{T}_1 \boxtimes \mathcal{T}_2 = \big(\mathcal{M}_1' \boxtimes \mathcal{M}_2', \mathcal{M}_1'' \boxtimes \mathcal{M}_2'', C_1 \boxtimes C_2\big).$$

Suppose that $\mathcal{T}_i$ are equipped with increasing filtrations L such that (i) $L_k(\mathcal{T}_i) = 0$ ($k << 0$) and $L_k(\mathcal{T}_i) = \mathcal{T}_i$ ($k >> 0$), (ii) $\mathrm{Gr}^L(\mathcal{T}_i)$ are strict. We can naturally regard $L_k\mathcal{T}_i$ as $\mathcal{R}_X$-sub-triples of $\mathcal{T}_i$. We set

$$L_k(\mathcal{T}_1 \boxtimes \mathcal{T}_2) = \sum_{a+b\leq k} L_a(\mathcal{T}_1) \boxtimes L_b(\mathcal{T}_2). \tag{11.17}$$

We have the naturally defined morphisms:

$$\bigoplus_{a+b=k} \mathrm{Gr}^L_a(\mathcal{T}_1) \boxtimes \mathrm{Gr}^L_b(\mathcal{T}_2) \longrightarrow \mathrm{Gr}^L_k(\mathcal{T}_1 \boxtimes \mathcal{T}_2) \tag{11.18}$$

We obtain the following lemma from Lemma 11.4.3.

Lemma 11.4.4 *The morphisms (11.18) are isomorphisms.* □

The following is clear by construction.

Lemma 11.4.5 *Let $\mathcal{T}_i$ ($i = 1,2$) be strict $\mathcal{R}_{X_i}$-triples with a finite increasing filtration L such that $\mathrm{Gr}^L(\mathcal{T}_i)$ are also strict. We have the following natural isomorphisms*

$$j^*\big((\mathcal{T}_1, L) \boxtimes (\mathcal{T}_2, L)\big) \simeq j^*(\mathcal{T}_1, L) \boxtimes j^*(\mathcal{T}_2, L)$$

$$\big((\mathcal{T}_1, L) \boxtimes (\mathcal{T}_2, L)\big)^* \simeq (\mathcal{T}_1, L)^* \boxtimes (\mathcal{T}_2, L)^*$$

□

11.4.2 External Tensor Product of Mixed Twistor $\mathscr{D}$-Modules

For any $(\mathcal{T}_i, W) \in \mathrm{MTM}(X_i)$ ($i = 1,2$), we obtain a naturally defined filtered $\mathcal{R}_X$-triple $\mathcal{T} = \mathcal{T}_1 \boxtimes \mathcal{T}_2$. The weight filtration W on $\mathcal{T}$ is given by (11.17). If $(\mathcal{T}_i, W)$ are integrable, $(\mathcal{T}, W)$ is also integrable. We shall prove the following proposition in Sects. 11.4.2.1–11.4.2.4.

Proposition 11.4.6 *$(\mathcal{T}, W)$ is a mixed twistor $\mathscr{D}$-module on X. Moreover, the following holds.*

- *If $(\mathcal{T}_i, W)$ are integrable, then $(\mathcal{T}, W)$ is also naturally integrable.*
- *Let $\mathcal{A} \subset \mathbb{R} \times \mathbb{C}$ be a $\mathbb{Q}$-vector subspace such that $\mathbb{Q} \times \{0\} \subset \mathcal{A}$. If $(\mathcal{T}_i, W) \in \mathrm{MTM}(X_i)_{\mathcal{A}}$, then $(\mathcal{T}, W) \in \mathrm{MTM}(X)_{\mathcal{A}}$.*

11.4.2.1 Admissibly Specializability and Localizability

Let $(\mathcal{T}_i, W)$ be mixed twistor $\mathscr{D}$-modules on X_i. Let f_1 be a holomorphic function on X_1.

Lemma 11.4.7 *The filtered $\mathcal{R}_X$-triple $(\mathcal{T},W)=(\mathcal{T}_1,W)\boxtimes(\mathcal{T}_2,W)$ is admissibly specializable along $f_1\circ p_1$.*

Proof We may assume that $X_1=X_{10}\times\mathbb{C}_t$ and $f_1=t$. Let $\mathcal{M}$ be one of the underlying $\mathcal{R}_{X_1}$-module of $\mathcal{T}_1$. It is equipped with the induced filtration W. Let $\lambda_0\in\mathbb{C}$. Let $V^{(\lambda_0)}(L\mathcal{M})$ be the V-filtration of $\mathcal{M}$ on $\mathcal{X}_1^{(\lambda_0)}$. By using Lemma 11.4.1, we obtain that $V^{(\lambda_0)}(\mathcal{M}_1)\boxtimes\mathcal{M}_2$ is the V-filtration of $\mathcal{M}_1\boxtimes\mathcal{M}_2$ on $\mathcal{X}^{(\lambda_0)}$. Then, the claim can be checked easily. □

Let H_1 be a hypersurface in X_1. We set $\tilde{H}_1:=p_1^{-1}(H_1)$.

Lemma 11.4.8 *The filtered $\mathcal{R}_X$-triple $(\mathcal{T}_1,W)\boxtimes(\mathcal{T}_2,W)$ is admissibly specializable along $\tilde{H}_1$. There exist $\big((\mathcal{T}_1,W)\boxtimes(\mathcal{T}_2,W)\big)[\star\tilde{H}_1]$, and we have natural isomorphisms $\big((\mathcal{T}_1,W)\boxtimes(\mathcal{T}_2,W)\big)[\star\tilde{H}_1]\simeq\big((\mathcal{T}_1,W)[\star H_1]\big)\boxtimes(\mathcal{T}_2,W)$.*

Proof The first claim is a consequence of Lemma 11.4.7. We can also check the second claim by using Lemma 11.4.1. □

11.4.2.2 Pure Case

Let us consider the pure case.

Lemma 11.4.9 *Suppose that $\mathcal{T}_i$ are polarizable pure twistor $\mathscr{D}$-modules of weight w_i on X_i. Then, $\mathcal{T}_1\boxtimes\mathcal{T}_2$ is a polarizable pure twistor $\mathscr{D}$-module of weight w_1+w_2 on $X_1\times X_2$.*

Proof We can take complex manifolds Z_i, projective morphisms $\varphi_i:Z_i\longrightarrow X_i$, and polarizable pure twistor $\mathscr{D}$-modules $\mathfrak{T}_i$ whose strict supports are Z_i, such that $\mathcal{T}_i$ are direct summands of $\varphi_{i\dagger}^0(\mathfrak{T}_i)$. Then, $\mathcal{T}_1\boxtimes\mathcal{T}_2$ is a direct summand of $\varphi_{1\dagger}^0(\mathfrak{T}_1)\boxtimes\varphi_{2\dagger}^0(\mathfrak{T}_2)\simeq(\varphi_1\times\varphi_2)_\dagger^0(\mathfrak{T}_1\boxtimes\mathfrak{T}_2)$. Hence, it is enough to prove that $\mathfrak{T}_1\boxtimes\mathfrak{T}_2$ is a polarizable pure twistor $\mathscr{D}$-module on $Z_1\times Z_2$ of weight w_1+w_2. Hence, we have only to consider the case where the strict supports of $\mathcal{T}_i$ are X_i. We may also assume that there exist normal crossing hypersurface $H_i\subset X_i$ $(i=1,2)$, good sets of ramified irregular values $\boldsymbol{\mathcal{I}}_i$ on (X_i,H_i), and good wild variation of pure twistor structure $\mathcal{V}_i$ of weight w_i on $(X_i,H_i,\boldsymbol{\mathcal{I}}_i)$ such that $\mathcal{T}_i$ are obtained as the image of $\mathcal{V}_i[!H_i]\longrightarrow\mathcal{V}_i[*H_i]$ as $\mathcal{R}_{X_i}$-triples.

We set $H:=(H_1\times X_2)\cup(X_1\times H_2)$. We naturally have a smooth $\mathcal{R}_{X(*H)}$-triple $\mathcal{V}:=\mathcal{V}_1\boxtimes\mathcal{V}_2$ as in the case of the external tensor product of $\mathcal{R}_{X_i}$-triples. By the construction, we have natural isomorphisms

$$\big(\mathcal{V}_1[\star H_1]\boxtimes\mathcal{V}_2[\star H_2]\big)(*H)\simeq\mathcal{V}. \tag{11.19}$$

Let $\boldsymbol{\mathcal{I}}_1\boxtimes\boldsymbol{\mathcal{I}}_2$ denote the system on (X,H) given by $(\boldsymbol{\mathcal{I}}_1\boxtimes\boldsymbol{\mathcal{I}}_2)_{(P_1,P_2)}=\big\{p_1^*\mathfrak{a}_1+p_2^*\mathfrak{a}_2\,\big|\,\mathfrak{a}_i\in\mathcal{I}_{P_i}\big\}$ for $(P_1,P_2)\in X$. By Proposition 15.1.5, we take a projective morphism $F:X'\longrightarrow X$ such that (i) $H'=F^{-1}(H)$ is normal crossing, (ii) F induces an isomorphism $X'\setminus H'\simeq X\setminus H$, (iii) $F^{-1}(\boldsymbol{\mathcal{I}}_1\boxtimes\boldsymbol{\mathcal{I}}_2)$ is a good set of ramified irregular values. Then, $\mathcal{V}' = F^*\mathcal{V}$ is a good wild polarizable variation of pure twistor structure of weight $w_1 + w_2$ on (X', H'). We have the mixed twistor $\mathscr{D}$-modules $F_\dagger^0\mathcal{V}'[\star H']$ on X.

Lemma 11.4.10 *We have isomorphisms of* $\mathcal{R}_X$*-triples*

$$\mathcal{V}_1[\star H_1] \boxtimes \mathcal{V}_2[\star H_2] \simeq F_\dagger^0(\mathcal{V}'[\star H']) \tag{11.20}$$

which induce (11.19). We also have the following commutative diagram of the natural morphisms:

$$\begin{array}{ccc} \mathcal{V}_1[!H_1] \boxtimes \mathcal{V}_2[!H_2] & \longrightarrow & F_\dagger^0(\mathcal{V}'[!H']) \\ \downarrow & & \downarrow \\ \mathcal{V}_1[*H_1] \boxtimes \mathcal{V}_2[*H_2] & \longrightarrow & F_\dagger^0(\mathcal{V}'[*H']) \end{array}$$

Proof Let $\tilde{H}_i := p_i^{-1}(H_i)$. By Lemma 11.4.8, $\mathcal{V}_1[\star H_1] \boxtimes \mathcal{V}_2[\star H_2]$ is localizable along $\tilde{H}_i$, and we have the following natural isomorphisms:

$$\begin{aligned} \Big(\mathcal{V}_1[\star H_1] \boxtimes \mathcal{V}_2[\star H_2]\Big) &\simeq \Big(\mathcal{V}_1[\star H_1] \boxtimes \mathcal{V}_2[\star H_2]\Big)[\star \tilde{H}_1] \\ &\simeq \Big(\mathcal{V}_1[\star H_1] \boxtimes \mathcal{V}_2[\star H_2]\Big)[\star \tilde{H}_2] \end{aligned} \tag{11.21}$$

By the property of $F_\dagger^0(\mathcal{V}'[\star H'])$, we also have the following natural isomorphisms:

$$F_\dagger^0(\mathcal{V}'[\star H']) \simeq F_\dagger^0(\mathcal{V}'[\star H'])[\star \tilde{H}_1] \simeq F_\dagger^0(\mathcal{V}'[\star H'])[\star \tilde{H}_2].$$

Hence, we have the isomorphism (11.20), and we obtain Lemma 11.4.10. □

Hence, the $\mathcal{R}_{X_1 \times X_2}$-triple $\mathcal{T}_1 \boxtimes \mathcal{T}_2$ is isomorphic to the image of the natural morphism $F_\dagger^0(\mathcal{V}'[!H']) \longrightarrow F_\dagger^0(\mathcal{V}'[*H'])$. We have $\mathrm{Gr}_w^W F_\dagger^0(\mathcal{V}'[*H']) = 0$ for $w < w_1 + w_2$ and $\mathrm{Gr}_w^W F_\dagger^0(\mathcal{V}'[!H']) = 0$ for $w > w_1 + w_2$. Hence, we obtain that $\mathcal{T}_1 \boxtimes \mathcal{T}_2$ is isomorphic to the image of $\mathrm{Gr}_w^W(F_\dagger^0\mathcal{V}'[!H']) \longrightarrow \mathrm{Gr}_w^W(F_\dagger^0\mathcal{V}'[*H'])$. Hence, it is a polarizable pure twistor $\mathscr{D}$-module of weight $w_1 + w_2$, and we obtain Lemma 11.4.9. □

Corollary 11.4.11 *For* $(\mathcal{T}_i, W) \in \mathrm{MTM}(X_i)$, *the induced filtered* $\mathcal{R}_{X_1 \times X_2}$*-triple* $(\mathcal{T}_1, W) \boxtimes (\mathcal{T}_2, W)$ *is an object in* $\mathrm{MTW}(X_1 \times X_2)$. □

11.4.2.3 Admissible Mixed Twistor Structure

Suppose that X_i are equipped with simple normal crossing hypersurfaces H_i. Let $H_i = H_{i1} \cup H_{i2}$ be decompositions such that $\mathrm{codim}_{X_i} H_{i1} \cap H_{i2} \geq 2$. Let $\mathcal{V}_i$

be admissible mixed twistor structure on (X_i, H_i). We have the mixed twistor $\mathscr{D}$-modules $(\mathcal{T}_i, W) = (\mathcal{V}_i[!H_{i1} * H_{i2}], W)$ on X_i.

We consider the following special case of Proposition 11.4.6.

Lemma 11.4.12 *The filtered $\mathcal{R}_X$-triple $(\mathcal{T}, W) = (\mathcal{T}_1, W) \boxtimes (\mathcal{T}_2, W)$ is a mixed twistor $\mathscr{D}$-module on X.*

Proof By Corollary 11.4.11, we have already known $\mathcal{T} \in \mathrm{MTW}(X)$. The conditions in Definition 7.2.1 are local. By Proposition 9.1.16, there exists a projective morphism $F : X' \longrightarrow X$ such that (i) $H' = F^{-1}(H)$ is normal crossing, (ii) F induces an isomorphism $X' \setminus H' \simeq X \setminus H$, (iii) $\mathcal{V}' := F^*\mathcal{V}$ is an admissible mixed twistor structure on (X', H'). We obtain the mixed twistor $\mathscr{D}$-module $F^0_\dagger(\mathcal{V}'[*H'], W)$.

We set $K_a := p_1^{-1}(H_{1a}) \cup p_2^{-1}(H_{2a})$ $(a = 1, 2)$. We have the mixed twistor $\mathscr{D}$-module:

$$(\mathcal{T}_{10}, W) := \big(F^0_\dagger(\mathcal{V}'[*H'], W)\big)[!K_2]$$

We have natural isomorphisms $(\mathcal{T}_{10}, W)[!H_{j2}] \simeq (\mathcal{T}_{10}, W)$ $(j = 1, 2)$. We also have $(\mathcal{T}_{10}, W)[*H_{j1}](*K_2) \simeq (\mathcal{T}_{10}, W)(*K_2)$ $(j = 1, 2)$. We have similar isomorphisms for $(\mathcal{T}, W)$. Hence, we have $(\mathcal{T}_{10}, W) \simeq (\mathcal{T}, W)$ by using Lemma 7.1.35 successively, and we obtain Lemma 11.4.12. □

11.4.2.4 End of the Proof of Proposition 11.4.6

Let us return to the situation in Proposition 11.4.6. We shall often omit to denote the weight filtrations. We have already known $(\mathcal{T}_1, W) \boxtimes (\mathcal{T}_2, W) \in \mathrm{MTW}(X)$. Because the conditions in Definition 7.2.1 are local, we may and will shrink X_i.

Let $\mathcal{C} = (Z, U, \varphi, \mathcal{V})$ be any admissible cell with a cell function g on X_1.

Lemma 11.4.13 $\Pi^{a,b}_{g*!}(\varphi_\dagger \mathcal{V}) \boxtimes \mathcal{T}_2 \in \mathrm{MTM}(X)$. *In particular,* $\Xi^{(a)}_g(\varphi_\dagger \mathcal{V}) \boxtimes \mathcal{T}_2 \in \mathrm{MTM}(X)$ *and* $\psi^{(a)}_g(\varphi_\dagger \mathcal{V}) \boxtimes \mathcal{T}_2 \in \mathrm{MTM}(X)$

Proof We use a Noetherian induction on the support of $\mathcal{T}_2$. Shrinking X_2, we can take an admissible cell $\mathcal{C}_2 = (Z_2, U_2, \varphi_2, \mathcal{V}_2)$ with a cell function g_2 for $\mathcal{T}_2$. Then, $\Pi^{a,b}_{g*!}(\varphi_\dagger \mathcal{V}) \boxtimes \mathcal{T}_2$ is isomorphic to the cohomology of the following complex of filtered $\mathcal{R}_{X_1 \times X_2}$-triples:

$$\Pi^{a,b}_{g*!}(\varphi_\dagger \mathcal{V}) \boxtimes \Big(\psi^{(1)}_{g_2}(\varphi_{2\dagger}\mathcal{V}_2) \longrightarrow \Xi^{(0)}_{g_2}(\varphi_{2\dagger}\mathcal{V}_2) \oplus \phi^{(0)}_{g_2}(\mathcal{T}_2) \longrightarrow \psi^{(1)}_{g_2}(\varphi_{2\dagger}\mathcal{V}_2)\Big)$$

By the hypothesis of the induction, we know that $\Pi^{a,b}_{g*!}(\varphi_\dagger \mathcal{V}) \boxtimes \psi^{(a)}_{g_2}(\varphi_{2\dagger}\mathcal{V}_2)$ and $\Pi^{a,b}_{g*!}(\varphi_\dagger \mathcal{V}) \boxtimes \phi^{(a)}_{g_2}(\varphi_{2\dagger}\mathcal{V}_2)$ are mixed twistor $\mathscr{D}$-modules. By using Lemma 11.4.12, we obtain that $\Pi^{a,b}_{g*!}(\varphi_\dagger \mathcal{V}) \boxtimes \Xi^{(0)}_{g_2}(\varphi_{2\dagger}\mathcal{V}_2) \in \mathrm{MTM}(X)$. Then, the claim of Lemma 11.4.13 follows. □

Let us finish the proof of Proposition 11.4.6. We use a Noetherian induction on the support of $\mathcal{T}_1$. We take an admissible cell $\mathcal{C}_1 = (Z_1, U_1, \varphi_1, \mathcal{V}_1)$ with a cell function g_1 for $\mathcal{T}_1$. Then, the filtered $\mathcal{R}_X$-triple $\mathcal{T}_1 \boxtimes \mathcal{T}_2$ is isomorphic to the cohomology of the following complex:

$$\Big(\psi^{(1)}_{g_1}(\varphi_{1\dagger}\mathcal{V}_1) \longrightarrow \Xi^{(0)}_{g_1}(\varphi_{1\dagger}\mathcal{V}_1) \oplus \phi^{(0)}_{g_1}(\mathcal{T}_1) \longrightarrow \psi^{(0)}_{g_1}(\varphi_{1\dagger}\mathcal{V}_1)\Big) \boxtimes \mathcal{T}_2$$

By Lemma 11.4.13, $\psi^{(a)}_{g_1}(\varphi_{1\dagger}\mathcal{V}_1) \boxtimes \mathcal{T}_2$ and $\Xi^{(0)}_{g_1}(\varphi_{1\dagger}\mathcal{V}_1) \boxtimes \mathcal{T}_2$ are mixed twistor $\mathscr{D}$-modules on X. By the hypothesis of the induction, $\phi^{(0)}_{g_1}(\varphi_{1\dagger}\mathcal{V}_1) \boxtimes \mathcal{T}_2$ is a mixed twistor $\mathscr{D}$-module. Hence, we obtain that $\mathcal{T}_1 \boxtimes \mathcal{T}_2$ is a mixed twistor $\mathscr{D}$-module on $X_1 \times X_2$.

The claim for the integrability is clear by the construction of $\mathcal{T}_1 \boxtimes \mathcal{T}_2$ with the weight filtration. We can easily check that, in the proof of Lemma 11.4.9, if $\mathcal{T}_i \in \mathrm{MTM}(X_i)_{\mathcal{A}}$, then the KMS-spectra of $\mathcal{V}'$ along any irreducible component of H' are contained in $\mathcal{A}$. Then, we can obtain the restriction of the KMS-spectra. Thus, the proof of Proposition 11.4.6 is finished. □

11.4.3 Compatibility

Let $\mathcal{T}_i \in \mathrm{MTM}(X_i)$ $(i = 1, 2)$.

Lemma 11.4.14 *Let $f_i : X_i \longrightarrow Y_i$ be projective morphisms. Let $f : X_1 \times X_2 \longrightarrow Y_1 \times Y_2$ be the induced morphism. We have a natural isomorphism:*

$$f^m_\dagger(\mathcal{T}_1 \boxtimes \mathcal{T}_2) \simeq \bigoplus_{k+\ell=m} f^k_\dagger(\mathcal{T}_1) \boxtimes f^\ell_\dagger(\mathcal{T}_2)$$

Proof By Lemma 11.4.1, we have natural isomorphisms of the underlying filtered $\mathcal{R}_{Y_1 \times Y_2}$-modules. We can check that the isomorphisms are compatible with the sesqui-linear pairings by elementary computations. □

Let H_i $(i = 1, 2)$ be hypersurfaces of X_i. Let $\varphi_i : X_i' \longrightarrow X_i$ be projective morphism of complex manifolds such that $H_i' := \varphi_i^{-1}(H_i)$ is normal crossing. Let $\mathcal{V}_i \in \mathrm{MTS}^{\mathrm{adm}}(X_i', H_i')$.

Lemma 11.4.15 *Let $\star_i \in \{*, !\}$. We have the natural isomorphisms*

$$\Big((\mathcal{T}_1 \otimes \varphi_{1\dagger}(\mathcal{V}_1))[\star_1 H_1]\Big) \boxtimes \Big((\mathcal{T}_2 \otimes \varphi_{2\dagger}\mathcal{V}_2)[\star_2 H_2]\Big)$$
$$\simeq \Big((\mathcal{T}_1 \boxtimes \mathcal{T}_2) \otimes (\varphi_{1\dagger}\mathcal{V}_1 \boxtimes \varphi_{2\dagger}\mathcal{V}_2)\Big)[\star_1(H_1 \times X_2)][\star_2(X_1 \times H_2)]. \tag{11.22}$$

Proof Let $\mathcal{T}_3$ and $\mathcal{T}_4$ denote the left hand side and the right hand side, respectively. We set $\tilde{H} := (H_1 \times X_2) \cup (X_1 \times H_2)$. We also set $\tilde{H}_1 := H_1 \times X_2$ and $\tilde{H}_2 := X_1 \times H_2$. We have $\mathcal{T}_3(*H) = \mathcal{T}_4(*H)$. By using Lemma 11.4.1, we can check that $\mathcal{T}_i(*\tilde{H}_2)[\star_1 \tilde{H}_1] = \mathcal{T}_i(*\tilde{H}_2)$ for $i = 3, 4$. Hence, we have $\mathcal{T}_3(*\tilde{H}_2) = \mathcal{T}_4(*\tilde{H}_2)$. We can also check that $\mathcal{T}_i[*\tilde{H}_2] = \mathcal{T}_i$ $(i = 3, 4)$, and hence we have $\mathcal{T}_3 = \mathcal{T}_4$. □

Corollary 11.4.16 *Let g_1 be any holomorphic function on X_1. Let g be the induced function on X. Then, we have $\Pi^{a,b}_{g_1*!}(\mathcal{T}_1) \boxtimes \mathcal{T}_2 \simeq \Pi^{a,b}_{g*!}(\mathcal{T}_1 \boxtimes \mathcal{T}_2)$. In particular, we have $\psi^{(a)}_{g_1}(\mathcal{T}_1) \boxtimes \mathcal{T}_2 \simeq \psi^{(a)}_{g}(\mathcal{T}_1 \boxtimes \mathcal{T}_2)$ and $\Xi^{(a)}_{g_1}(\mathcal{T}_1) \boxtimes \mathcal{T}_2 \simeq \Xi^{(a)}_{g}(\mathcal{T}_1 \boxtimes \mathcal{T}_2)$. We also have $\phi^{(a)}_{g_1}(\mathcal{T}_1) \boxtimes \mathcal{T}_2 \simeq \phi^{(a)}_{g}(\mathcal{T}_1 \boxtimes \mathcal{T}_2)$.* □

Chapter 12
$\mathscr{D}$-Triples and Their Functoriality

Let M_i $(i = 1, 2)$ be holonomic $\mathscr{D}$-modules on a complex manifold X. Let $\overline{X}$ denote the complex manifold obtained as the conjugate of X. We naturally have the $\mathscr{D}_{\overline{X}}$-module $\overline{M}_2$. Let $\mathfrak{Db}_X$ be the sheaf of distributions on X. It is naturally a $\mathscr{D}_X \otimes_{\mathbb{C}} \mathscr{D}_{\overline{X}}$-module. A $\mathscr{D}_X \otimes \mathscr{D}_{\overline{X}}$-homomorphism $C : M_1 \otimes_{\mathbb{C}} \overline{M}_2 \longrightarrow \mathfrak{Db}_X$ is called a sesquilinear pairing of M_1 and M_2, and we call such a tuple (M_1, M_2, C) by a $\mathscr{D}$-triple. In this chapter, we study the basic functorial properties of $\mathscr{D}$-triples.

We may naturally regard a $\mathscr{D}$-triple as an underlying object of an $\mathcal{R}$-triple. Indeed, if we choose $\lambda_0 \in \mathbb{C}$ with $|\lambda_0| = 1$, we obtain a $\mathscr{D}$-triple $(\mathcal{M}_1^{-\lambda_0}, \mathcal{M}_2^{\lambda_0}, C^{\lambda_0})$ from any $\mathcal{R}$-triple $(\mathcal{M}_1, \mathcal{M}_2, C)$.

The main purpose is to establish the duality of $\mathscr{D}$-triples. (See Sect. 12.4.) It will be useful in the study of the duality of mixed twistor $\mathscr{D}$-modules (Theorem 13.3.1). We also introduce the concept of real structure of $\mathscr{D}$-triples, and we compare it with the real structures of the underlying perverse sheaves or the $\mathbb{R}$-Betti structures of the underlying holonomic $\mathscr{D}$-modules.

This type of objects were first studied by Kashiwara [33]. He introduced the Hermitian dual of $\mathscr{D}$-modules. Namely, for any $\mathscr{D}$-module on X, we set $\mathcal{C}_X(M) := R\mathcal{H}om_{\mathscr{D}_X}(M, \mathfrak{Db}_X)$ in $D^b(\mathscr{D}_{\overline{X}})$. He proved that if M is regular singular then the j-th cohomology sheaf $\mathcal{H}^j\mathcal{C}_X(M)$ of $\mathcal{C}_X(M)$ vanishes unless $j = 0$, and $\mathcal{H}^0\mathcal{C}_X(M)$ is a regular holonomic $\mathscr{D}_{\overline{X}}$-module. He conjectured that similar claims hold even for holonomic $\mathscr{D}$-modules, which was proved in [58, 65] on the basis of the development in the theory of meromorphic flat bundles.

We also have the duality functor of $\mathscr{D}$-modules. Namely, let ω_X denote the canonical bundle of X. Then, for any coherent $\mathscr{D}$-module M on X, we have $\boldsymbol{D}_X(M) := R\mathcal{H}om_{\mathscr{D}_X}(M, \mathscr{D}_X \otimes \omega_X^{-1})[\dim X]$ in $D^b(\mathscr{D}_X)$. If M is holonomic, then it is classically known that the j-th cohomology sheaf of $\boldsymbol{D}_X(M)$ vanishes unless $j = 0$.

Let DR_X denote the de Rham functor for $\mathscr{D}$-modules on X. We have a natural isomorphism $\mathrm{DR}_X \boldsymbol{D}_X(M) \simeq \mathrm{DR}_{\overline{X}} \mathcal{C}_X(M)$ for any holonomic $\mathscr{D}$-module M on X,

T. Mochizuki, *Mixed Twistor $\mathscr{D}$-modules*, Lecture Notes in Mathematics 2125,
DOI 10.1007/978-3-319-10088-3_12

and both of them are isomorphic to the dual of $\mathrm{DR}_X(M)$ in the category of perverse sheaves on X.

Because we have a natural isomorphism $\mathrm{DR}_{\overline{X}} \boldsymbol{D}_{\overline{X}} \mathcal{C}_X(M) \simeq \mathrm{DR}_X(M)$, it is natural to define a real structure of M as an isomorphism of $\mathscr{D}$-modules $M \simeq \overline{\boldsymbol{D}_{\overline{X}} \mathcal{C}_X(M)}$ on X satisfying the involutivity property. But, for the involutivity, we should have an equivalence $\boldsymbol{D}_{\overline{X}} \circ \mathcal{C}_X \simeq \mathcal{C}_X \circ \boldsymbol{D}_X$ in a functorial way. Because we naturally have $\mathrm{DR}_{\overline{X}} \boldsymbol{D}_{\overline{X}} \mathcal{C}_X(M) \simeq \mathrm{DR}_{\overline{X}} \mathcal{C}_X \boldsymbol{D}_X(M)$, we obtain $\boldsymbol{D}_{\overline{X}} \circ \mathcal{C}_X(M) \simeq \mathcal{C}_X \circ \boldsymbol{D}_X(M)$ for any regular holonomic $\mathscr{D}$-modules by the Riemann-Hilbert correspondence [28, 30, 46–48]. In the non-regular case, we need some additional work. It is equivalent to the above mentioned issue on the duality of $\mathscr{D}$-triples.

The rough idea is simple. We can directly construct such an isomorphism $\boldsymbol{D}_{\overline{X}} \circ \mathcal{C}_X(M) \simeq \mathcal{C}_X \circ \boldsymbol{D}_X(M)$ if M is a good meromorphic flat bundle. Because any holonomic $\mathscr{D}$-module is locally expressed as gluing of meromorphic flat bundles, we can obtain the desired isomorphism by gluing the isomorphisms for good meromorphic flat bundles.

This chapter is a preparation for Chap. 13.

12.1 $\mathscr{D}$-Triples and Their Push-Forward

12.1.1 $\mathscr{D}$-triples and $\mathscr{D}$-Complex-Triples

We introduce the notions of $\mathscr{D}$-triples and $\mathscr{D}$-complex-triples, which are variants of $\mathcal{R}$-triples in [66]. Let X be a complex manifold. The conjugate complex manifold is denoted by $\overline{X}$. We set $\mathscr{D}_{X,\overline{X}} := \mathscr{D}_X \otimes_{\mathbb{C}} \mathscr{D}_{\overline{X}}$, which is naturally a sheaf of algebras. Let $\mathfrak{Db}_X$ denote the sheaf of distributions on X. It is naturally a $\mathscr{D}_{X,\overline{X}}$-module.

Let M_i $(i = 1,2)$ be $\mathscr{D}_X$-modules. A sesqui-linear pairing of M_1 and M_2 is a $\mathscr{D}_{X,\overline{X}}$-homomorphism $C : M_1 \otimes_{\mathbb{C}} \overline{M_2} \longrightarrow \mathfrak{Db}_X$. Such a tuple (M_1, M_2, C) is called a $\mathscr{D}_X$-triple. A morphism of $\mathscr{D}_X$-triples $(M_1', M_2', C') \longrightarrow (M_1, M_2, C)$ is a pair of morphisms $\varphi_1 : M_1 \longrightarrow M_1'$ and $\varphi_2 : M_2' \longrightarrow M_2$ such that $C'\bigl(\varphi_1(m_1), \overline{m_2'}\bigr) = C\bigl(m_1, \overline{\varphi_2(m_2')}\bigr)$. A $\mathscr{D}_X$-triple (M_1, M_2, C) is called coherent (good, holonomic, etc.), if the underlying $\mathscr{D}_X$-modules are coherent (good, holonomic, etc.).

Let $\mathscr{D}\text{-Tri}(X)$ denote the category of $\mathscr{D}_X$-triples. It is an abelian category.

12.1.1.1 $\mathscr{D}$-Complex-Triples

Let $M_i^\bullet$ $(i = 1,2)$ be bounded complexes of $\mathscr{D}_X$-modules. A sesqui-linear pairing of $M_1^\bullet$ and $M_2^\bullet$ is a morphism of $\mathscr{D}_{X,\overline{X}}$-complexes $C : \mathrm{Tot}\bigl(M_1^\bullet \otimes \overline{M_2^\bullet}\bigr) \longrightarrow \mathfrak{Db}_X$. Namely, a tuple of morphisms $C^p : M_1^{-p} \otimes \overline{M_2^p} \longrightarrow \mathfrak{Db}_X$ such that

$$C^p(dx^{-p-1}, y^p) + (-1)^{p+1} C^{p+1}(x^{-p-1}, dy^p) = 0.$$

Such $(M_1^\bullet, M_2^\bullet, C)$ is called a $\mathscr{D}_X$-complex-triple. A morphism of $\mathscr{D}_X$-complex-triples

$$(M_1^\bullet, M_2^\bullet, C_M) \longrightarrow (N_1^\bullet, N_2^\bullet, C_N)$$

is a pair of morphisms of $\mathscr{D}_X$-complexes $\varphi_1 : N_1^\bullet \longrightarrow M_1^\bullet$ and $\varphi_2 : M_2^\bullet \longrightarrow N_2^\bullet$ such that $C_M(\varphi_1(x), y) = C_N(x, \varphi_2(y))$.

Let $\mathcal{C}(\mathscr{D})$-Tri(X) denote the category of $\mathscr{D}_X$-complex-triples. It is an abelian category. A morphism in $\mathcal{C}(\mathscr{D})$-Tri(X) is called a quasi-isomorphism if the underlying morphisms of $\mathscr{D}_X$-complexes are quasi-isomorphisms.

Let $\mathfrak{T} = (M_1^\bullet, M_2^\bullet, C) \in \mathcal{C}(\mathscr{D})$-Tri$(X)$. We put $\epsilon(\ell) := (-1)^{\ell(\ell-1)/2}$ for any integer ℓ. We set

$$\mathcal{S}_\ell(\mathfrak{T}) := (M_1^\bullet[-\ell], M_2^\bullet[\ell], C[\ell])$$

where we define

$$C[\ell]^p(x^{-p-\ell}, y^{p+\ell}) := (-1)^{\ell p} \epsilon(\ell)\, C^{\ell+p}(x^{-p-\ell}, y^{p+\ell})$$

for $x^{-p-\ell} \in M_1^{-p-\ell} = M_1[-\ell]^{-p}$ and $y^{p+\ell} \in M_2^{p+\ell} = M_2[\ell]^p$. It is called the shift functor. We naturally have $\mathcal{S}_k \circ \mathcal{S}_\ell(\mathfrak{T}) = \mathcal{S}_{k+\ell}(\mathfrak{T})$.

Let $H^j(M_i^\bullet)$ be the j-th cohomology of the complexes $M_i^\bullet$. We have the induced sesqui-linear pairing $H^j(C)$ of $H^{-j}(M_1^\bullet)$ and $H^j(M_2^\bullet)$. We have the induced $\mathscr{D}_X$-complex-triple

$$\big(H^{-j}(M_1^\bullet)[j], H^j(M_2^\bullet)[-j], H^j(C)\big)$$

denoted by $\boldsymbol{H}^j(\mathfrak{T})$. We set

$$H^j(\mathfrak{T}) := \mathcal{S}_j \boldsymbol{H}^j(\mathfrak{T}) = \big(H^{-j}(M_1^\bullet), H^j(M_2^\bullet), \epsilon(j) H^j(C)\big).$$

Let $(M_1^\bullet, M_2^\bullet, C_M)$ and $(N_1^\bullet, N_2^\bullet, C_N)$ be $\mathscr{D}$-complex-triples. Let $\varphi_i : M_i^\bullet \longrightarrow N_i^\bullet$ be quasi-isomorphisms. We say that C_M and C_N are the same under the quasi-isomorphisms, if $C_M(m_1, \overline{m}_2) = C_N\big(\varphi_1(m_1), \overline{\varphi_2(m_2)}\big)$. In this case, we have the following natural quasi-isomorphisms of $\mathscr{D}$-complex-triples:

$$(M_1^\bullet, M_2^\bullet, C_M) \longrightarrow (M_1^\bullet, N_2^\bullet, C') \longleftarrow (N_1^\bullet, N_2^\bullet, C_N)$$

Here, $C'(m_1, \overline{n}_2) = C_N(\varphi_1(m_1), n_2)$.

12.1.1.2 Complex of $\mathscr{D}$-Triples

A complex of $\mathscr{D}_X$-triples consists of a tuple of $\mathscr{D}_X$-triples $\mathcal{T}^p = (M_1^{-p}, M_2^p, C^p)$ ($p \in \mathbb{Z}$) and morphisms $\delta^p : \mathcal{T}^p \longrightarrow \mathcal{T}^{p+1}$ such that $\delta^{p+1} \circ \delta^p = 0$. The morphisms are described as $\delta^p = (\delta_1^p, \delta_2^p)$, where $\delta_1^p : M_1^{-p-1} \longrightarrow M_1^{-p}$ and $\delta_2^p : M_2^p \longrightarrow M_2^{p+1}$. They give complexes $M_i^\bullet$, and satisfy

$$C^{p+1}(x^{-p-1}, \delta_2^p y^p) = C^p(\delta_1^p x^{-p-1}, y^p) \tag{12.1}$$

Let $\mathcal{C}\big(\mathscr{D}\text{-Tri}(X)\big)$ denote the category of bounded complexes of $\mathscr{D}_X$-triples. For $\mathcal{T}^\bullet \in \mathcal{C}(\mathscr{D}\text{-Tri}(X))$, its j-th cohomology is denoted by $H^j(\mathcal{T}^\bullet)$. For any integer ℓ, we define the shift $\mathcal{T}^\bullet[\ell]$ in the standard way. Namely, we set $\mathcal{T}[q]^p := \mathcal{T}^{p+q}$, and the differentials $\tilde{\delta}^p : \mathcal{T}[q]^p \longrightarrow \mathcal{T}[q]^{p+1}$ are given by $(-1)^q \delta^{p+q}$.

We put $\epsilon(p) := (-1)^{p(p-1)/2}$ for integers p. For $\mathcal{T}^\bullet \in \mathcal{C}\big(\mathscr{D}\text{-Tri}(X)\big)$ with the above description, we define an object $\Psi_1(\mathcal{T}^\bullet) := \big(M_1^\bullet, M_2^\bullet, \tilde{C}\big)$ in $\mathcal{C}(\mathscr{D})\text{-Tri}(X)$, where $\tilde{C}^p := \epsilon(p)\, C^p$. Thus, we obtain a functor

$$\Psi_1 : \mathcal{C}(\mathscr{D}\text{-Tri}(X)) \longrightarrow \mathcal{C}(\mathscr{D})\text{-Tri}(X).$$

It is easy to check the following.

Proposition 12.1.1 *The functor Ψ_1 is an equivalence. We have $\Psi_1(\mathcal{T}^\bullet[\ell]) \simeq \mathcal{S}_\ell(\Psi_1(\mathcal{T}^\bullet))$ which is given by the identity of the underlying $\mathscr{D}$-complexes. We also have $\boldsymbol{H}^j\Psi_1(\mathcal{T}^\bullet) \simeq \Psi_1\big(H^j(\mathcal{T}^\bullet)[-j]\big)$ and $H^j\Psi_1(\mathcal{T}^\bullet) \simeq \Psi_1 H^j(\mathcal{T}^\bullet)$.* □

12.1.1.3 $\mathscr{D}$-Double-Complex-Triples and Their Total Complexes

A $\mathscr{D}_X$-double-complex-triple is a tuple $\big(M_1^{\bullet,\bullet}, M_2^{\bullet,\bullet}, C\big)$:

- $M_i^{\bullet,\bullet}$ ($i = 1, 2$) are double complexes of $\mathscr{D}_X$-modules, i.e., they are $\mathbb{Z}^2$-graded $\mathscr{D}$-modules $\{M_i^{\boldsymbol{p}} \mid \boldsymbol{p} \in \mathbb{Z}\}$ with morphisms $d_j : M_i^{\boldsymbol{p}} \longrightarrow M_i^{\boldsymbol{p}+\boldsymbol{\delta}_j}$ ($i = 1, 2, j = 1, 2$) such that $d_j \circ d_j = 0$ and $d_j \circ d_m = d_m \circ d_j$. Here, $\boldsymbol{\delta}_1 = (1, 0)$ and $\boldsymbol{\delta}_2 = (0, 1)$. For simplicity, we assume the boundedness.
- $C : M_1^{\bullet,\bullet} \otimes \overline{M}_2^{\bullet,\bullet} \longrightarrow \mathfrak{Db}_X$ be a morphism of $\mathscr{D}_{X,\overline{X}}$-double-complexes, i.e., a tuple of $\mathscr{D}_{X,\overline{X}}$-morphisms $C^{\boldsymbol{p}} : M_1^{-\boldsymbol{p}} \otimes M_2^{\boldsymbol{p}} \longrightarrow \mathfrak{Db}_X$ such that

$$C^{\boldsymbol{p}}(d_1 x^{-\boldsymbol{p}-\boldsymbol{\delta}_1}, y^{\boldsymbol{p}}) + (-1)^{p_1+1} C^{\boldsymbol{p}+\boldsymbol{\delta}_1}(x^{-\boldsymbol{p}-\boldsymbol{\delta}_1}, d_1 y^{\boldsymbol{p}}) = 0$$

$$C^{\boldsymbol{p}}(d_2 x^{-\boldsymbol{p}-\boldsymbol{\delta}_2}, y^{\boldsymbol{p}}) + (-1)^{p_2+1} C^{\boldsymbol{p}+\boldsymbol{\delta}_2}(x^{-\boldsymbol{p}-\boldsymbol{\delta}_2}, d_2 y^{\boldsymbol{p}}) = 0$$

Let $\mathcal{C}^{(2)}(\mathscr{D})\text{-Tri}(X)$ denote the category of $\mathscr{D}_X$-double-complex-triples.

Let $(M_1^{\bullet,\bullet}, M_2^{\bullet,\bullet}, C)$ be an object in $\mathcal{C}^{(2)}(\mathscr{D})$-Tri$(X)$. We define the total complex object Tot$(M_1^{\bullet,\bullet}, M_2^{\bullet,\bullet}, C)$ in $\mathcal{C}(\mathscr{D})$-Tri(X). The underlying $\mathscr{D}$-complexes are the total complexes Tot$(M_i^{\bullet,\bullet})$, i.e.,

$$\mathrm{Tot}(M_i^{\bullet,\bullet})^p = \bigoplus_{p_1+p_2=p} M_i^{\boldsymbol{p}},$$

with the differential $d\boldsymbol{x}^{\boldsymbol{p}} = d_1 x^{\boldsymbol{p}} + (-1)^{p_1} d_2 x^{\boldsymbol{p}}$. The pairings are given by

$$\tilde{C}^p = \bigoplus_{p_1+p_2=p} (-1)^{p_1 p_2} C^{\boldsymbol{p}}. \tag{12.2}$$

Let us show that they give a $\mathscr{D}_X$-complex-triple. We have only to check (12.1). We have

$$\begin{aligned}
&\tilde{C}^{\boldsymbol{p}}(dx^{-\boldsymbol{p}-\boldsymbol{\delta}_1}, y^{\boldsymbol{p}}) + (-1)^{p_1+p_2+1}\tilde{C}^{\boldsymbol{p}+\boldsymbol{\delta}_1}(x^{-\boldsymbol{p}-\boldsymbol{\delta}_1}, dy^{\boldsymbol{p}}) \\
&\quad = (-1)^{p_1p_2} C^{\boldsymbol{p}}(d_1 x^{-\boldsymbol{p}-\boldsymbol{\delta}_1}, y^{\boldsymbol{p}}) \\
&\qquad + (-1)^{p_1+p_2+1}(-1)^{(p_1+1)p_2} C^{\boldsymbol{p}+\boldsymbol{\delta}_1}(x^{-\boldsymbol{p}-\boldsymbol{\delta}_1}, d_1 y^{\boldsymbol{p}}) \\
&\quad = (-1)^{p_1p_2}\big(C^{\boldsymbol{p}}(d_1 x^{-\boldsymbol{p}-\boldsymbol{\delta}_1}, y^{\boldsymbol{p}}) + (-1)^{p_1+1} C^{\boldsymbol{p}+\boldsymbol{\delta}_1}(x^{-\boldsymbol{p}-\boldsymbol{\delta}_1}, d_1 y^{\boldsymbol{p}})\big) = 0
\end{aligned} \tag{12.3}$$

We also have

$$\begin{aligned}
&\tilde{C}^{\boldsymbol{p}}(dx^{-\boldsymbol{p}-\boldsymbol{\delta}_2}, y^{\boldsymbol{p}}) + (-1)^{p_1+p_2+1}\tilde{C}^{\boldsymbol{p}+\boldsymbol{\delta}_2}(x^{-\boldsymbol{p}-\boldsymbol{\delta}_2}, dy^{\boldsymbol{p}}) \\
&\quad = (-1)^{p_1p_2+p_1} C^{\boldsymbol{p}}(d_2 x^{-\boldsymbol{p}-\boldsymbol{\delta}_2}, y^{\boldsymbol{p}}) \\
&\qquad + (-1)^{p_1+p_2+1+p_1(p_2+1)+p_1} C^{\boldsymbol{p}+\boldsymbol{\delta}_2}(x^{-\boldsymbol{p}-\boldsymbol{\delta}_2}, d_2 y^{\boldsymbol{p}}) \\
&\quad = (-1)^{p_1p_2+p_1}\big(C^{\boldsymbol{p}}(d_2 x^{-\boldsymbol{p}-\boldsymbol{\delta}_2}, y^{\boldsymbol{p}}) \\
&\qquad + (-1)^{p_2+1} C^{\boldsymbol{p}+\boldsymbol{\delta}_2}(x^{-\boldsymbol{p}-\boldsymbol{\delta}_2}, d_2 y^{\boldsymbol{p}})\big) = 0
\end{aligned} \tag{12.4}$$

We can easily deduce (12.1) from (12.3) and (12.4). Thus, we obtain a natural functor

$$\mathrm{Tot} : \mathcal{C}^{(2)}(\mathscr{D})\text{-Tri}(X) \longrightarrow \mathcal{C}(\mathscr{D})\text{-Tri}(X).$$

12.1.2 The Push-Forward

Let $\mathcal{T} = (M_1, M_2, C)$ be a $\mathscr{D}_X$-triple. Let $f : X \longrightarrow Y$ be a morphism of complex manifolds such that the restriction of f to the support of $\mathcal{T}$ is proper. As in the case of

$\mathcal{R}$-triples [66], we shall construct $f_{\dagger}^{(0)}\mathcal{T} = (f_{\dagger}M_1, f_{\dagger}M_2, f_{\dagger}^{(0)}C)$ in $\mathcal{C}(\mathscr{D})$-Tri(Y) and correspondingly $f_{\dagger}\mathcal{T} = (f_{\dagger}M_1, f_{\dagger}M_2, f_{\dagger}C)$ in $\mathcal{C}(\mathscr{D}\text{-Tri}(Y))$ i.e., $\Psi_1\big(f_{\dagger}\mathcal{T}\big) = f_{\dagger}^{(0)}\mathcal{T}$. We remark that this is a specialization of the push-forward for $\mathcal{R}$-triples given in [66]. We shall compare it with a more naive construction in Sect. 12.1.4.

12.1.2.1 Closed Immersion

If f is a closed immersion, we have $f_{\dagger}M_i = \omega_X \otimes f^{-1}(\mathscr{D}_Y \otimes \omega_Y^{-1}) \otimes_{\mathscr{D}_X} M_i$. Let η_X and η_Y denote local generators of ω_X and ω_Y, respectively. We put $d_X := \dim X$ and $d_Y := \dim Y$. We set

$$f_{\dagger}^{(0)}C\Big((\eta_X/\eta_Y)\cdot m_1, \overline{(\eta_X/\eta_Y)\cdot m_2}\Big) := \frac{1}{\eta_Y\overline{\eta}_Y} f_*\big(\eta_X\overline{\eta}_X\cdot C(m_1,\overline{m}_2)\big)\left(\frac{1}{2\pi\sqrt{-1}}\right)^{d_X-d_Y}\epsilon(d_X)\,\epsilon(d_Y) \tag{12.5}$$

Namely, for a test form $\varphi = \phi\cdot\eta_Y\overline{\eta}_Y$, we define

$$\Big\langle f_{\dagger}^{(0)}C\Big((\eta_X/\eta_Y)\cdot m_1, \overline{(\eta_X/\eta_Y)\cdot m_2}\Big), \varphi\Big\rangle := \Big\langle C(m_1,\overline{m}_2), f^*\phi\,\eta_X\overline{\eta}_X\Big\rangle\left(\frac{1}{2\pi\sqrt{-1}}\right)^{d_X-d_Y}\epsilon(d_X)\,\epsilon(d_Y) \tag{12.6}$$

In this case, we have $f_{\dagger}C = f_{\dagger}^{(0)}C$.

Let $f : X \longrightarrow Y$ and $g : Y \longrightarrow Z$ be closed immersions. We have natural isomorphisms $(g\circ f)_{\dagger}M_i \simeq g_{\dagger}\big(f_{\dagger}M_i\big)$ $(i = 1,2)$. The following lemma is easy to see.

Lemma 12.1.2 *We have $(g\circ f)_{\dagger}^{(0)}(C) = g_{\dagger}^{(0)}(f_{\dagger}^{(0)}C)$ under the isomorphisms.* □

Remark 12.1.3 If $Y = \Delta^n$, $X = \{z_{\ell+1} = \cdots z_n = 0\}$, $\eta_X = dz_1\cdots dz_\ell$ and $\eta_Y = dz_1\cdots dz_n$, we have

$$\eta_X\,\overline{\eta}_X = \eta_Y\,\overline{\eta}_Y\Big/\prod_{i=\ell+1}^{n} dz_i\,d\overline{z}_i \times \epsilon(d_X)\,\epsilon(d_Y)$$

Hence, for a test form φ on Y, we have

$$\Big\langle f_{\dagger}^{(0)}C\big((\eta_X/\eta_Y)\,m_1, \overline{(\eta_X/\eta_Y)\,m_2}\big), \varphi\Big\rangle = \Big\langle C(m_1,\overline{m}_2), f^*\varphi\Big/\prod_{i=\ell+1}^{n} dz_i d\overline{z}_i\Big\rangle\left(\frac{1}{2\pi\sqrt{-1}}\right)^{d_X-d_Y} \tag{12.7}$$

When we consider $\mathcal{R}$-triples, η_X/η_Y is replaced with $(\lambda^{-d_X}\eta_X)/(\lambda^{-d_Y}\eta_Y)$. Then, the signature $(-1)^{d_X-d_Y}$ appears in the right hand side of *(12.7)*. The formula is the same as that in *§1.6.d [66]*, which we recalled in *(2.2)*. □

12.1.2.2 Projection

Let us consider the case $f : X = Z \times Y \longrightarrow Y$ is the projection. Let $\mathcal{E}^p_{X/Y}$ denote the sheaf of C^∞ relative p-forms on X over Y which are holomorphic in the Y-direction. Set $d_Z = \dim Z$. We naturally have $f_\dagger M_i = f_!\big(\mathcal{E}^\bullet_{X/Y}[d_Z] \otimes M_i\big)$. Let $\eta^{d_Z+j} \otimes m_i$ denote sections of $\mathcal{E}^{d_Z+j}_{X/Y} \otimes M_i$. We set

$$f_\dagger^{(0)} C\big(\eta^{d_Z-p}\, m_1,\ \overline{\eta^{d_Z+p} m_2}\big)$$
$$:= \int \eta^{d_Z-p} \wedge \overline{\eta}^{d_Z+p} C(m_1, \overline{m}_2) \left(\frac{1}{2\pi\sqrt{-1}}\right)^{d_Z} \epsilon(d_Z)\,(-1)^{pd_Z} \qquad (12.8)$$

We have the following:

$$f_\dagger C\big(\eta^{d_Z-p}\, m_1,\ \overline{\eta^{d_Z+p} m_2}\big)$$
$$= \int \eta^{d_Z-p} \wedge \overline{\eta}^{d_Z+p} C(m_1, \overline{m}_2) \left(\frac{1}{2\pi\sqrt{-1}}\right)^{d_Z} \epsilon(p + d_Z) \qquad (12.9)$$

Note that $\epsilon(d_Z)(-1)^{pd_Z}\epsilon(p) = \epsilon(p + d_Z)$. The formula (12.9) is the same as the formula in §1.6.d [66] for the push-forward of $\mathcal{R}$-triples by projections, which we recalled in (2.3).

12.1.2.3 Some Compatibility

For our later purpose, we give some lemmas on compatibility of the push-forward for closed immersion and projection (Lemmas 12.1.4 and 12.1.5). Let $\ell \leq k$. We put $Y := \Delta^n$. Let X and Z be the submanifold of Y given as $X := \{z_{\ell+1} = \cdots = z_n = 0\}$ and $Z := \{z_{k+1} = \cdots = z_n = 0\}$. Let $f : X \longrightarrow Y$ be the natural inclusion. Let $g : Y \longrightarrow Z$ be the projection forgetting the last components. The composite $g \circ f : X \longrightarrow Z$ is the natural inclusion.

For $p = 1, \ldots, n$, we set $\eta_p := dz_1 \cdots dz_p$. For $p_1 \leq p_2$, we set $\eta_{p_1,p_2} := \eta_{p_2}/\eta_{p_1}$. Let $(M_1, M_2, C) \in \mathscr{D}\text{-Tri}(X)$. We put $d_{Y/Z} := d_Y - d_Z$. Let $\Omega^\bullet_{Y/Z}$ denote the relative holomorphic de Rham complex on Y over Z. We have

$$g_\dagger\big(f_\dagger M_i\big) \simeq g_*\Big(\Omega^\bullet_{Y/Z}[d_{Y/Z}] \otimes f_\dagger M_i\Big)$$

We have the quasi-isomorphism $g_{\dagger}\big(f_{\dagger}M_i\big) \simeq (g\circ f)_{\dagger}M_i$ induced by

$$\eta_{k,n}\otimes\big(m_i/\eta_{\ell,n}\big)\longmapsto m_i/\eta_{\ell,k}.$$

Lemma 12.1.4 *We have $g_{\dagger}^{(0)}(f_{\dagger}^{(0)}C) = (g\circ f)_{\dagger}^{(0)}C$ under the above quasi-isomorphisms.*

Proof We have the following:

$$\begin{aligned}
&\Big\langle g_{\dagger}^{(0)}\big(f_{\dagger}^{(0)}C\big)\Big(\eta_{k,n}\,(m_1/\eta_{\ell,n}),\ \overline{\eta_{k,n}\,(m_2/\eta_{\ell,n})}\Big),\ \varphi\Big\rangle\\
&=\Big(\frac{1}{2\pi\sqrt{-1}}\Big)^{n-k}\epsilon(n-k)\int\eta_{k,n}\overline{\eta}_{k,n}f_{\dagger}^{(0)}C\Big(m_1/\eta_{\ell,n},\ \overline{m_2/\eta_{\ell,n}}\Big)g^{*}\varphi\\
&=\Big(\frac{1}{2\pi\sqrt{-1}}\Big)^{n-k}\epsilon(n-k)\Big\langle f_{\dagger}^{(0)}C\big(m_1/\eta_{\ell,n},\overline{m_2/\eta_{\ell,n}}\big),\ \eta_{k,n}\overline{\eta}_{k,n}g^{*}\varphi\Big\rangle\\
&=\Big(\frac{1}{2\pi\sqrt{-1}}\Big)^{\ell-k}\epsilon(n-k)\,\epsilon(\ell)\,\epsilon(n)\Big\langle(\eta_n\overline{\eta}_n)^{-1}f_{*}\big(\eta_{\ell}\overline{\eta}_{\ell}C(m_1,\overline{m}_2)\big),\ g^{*}\varphi\,\eta_{k,n}\overline{\eta}_{k,n}\Big\rangle\\
&=\Big(\frac{1}{2\pi\sqrt{-1}}\Big)^{\ell-k}\epsilon(n-k)\,\epsilon(\ell)\,\epsilon(n)\Big\langle\eta_{\ell}\overline{\eta}_{\ell}C(m_1,\overline{m}_2),\ f^{*}\Big(\frac{g^{*}\varphi\,\eta_{k,n}\overline{\eta}_{k,n}}{\eta_n\overline{\eta}_n}\Big)\Big\rangle
\end{aligned}\tag{12.10}$$

We also have the following:

$$\begin{aligned}
&\Big\langle(g\circ f)_{\dagger}^{(0)}C\big(m_1/\eta_{\ell,k},\ \overline{m_2/\eta_{\ell,k}}\big),\ \varphi\Big\rangle\\
&=\Big(\frac{1}{2\pi\sqrt{-1}}\Big)^{\ell-k}\Big\langle(\eta_k\overline{\eta}_k)^{-1}(g\circ f)_{*}\big(\eta_{\ell}\overline{\eta}_{\ell}C(m_1,\overline{m}_2)\big),\ \varphi\Big\rangle\times\epsilon(\ell)\,\epsilon(k)
\end{aligned}\tag{12.11}$$

Let $\varphi=\eta_k\overline{\eta}_k\,\phi$. Then, we have

$$(g\circ f)^{*}\Big(\frac{\varphi}{\eta_k\overline{\eta}_k}\Big)=(g\circ f)^{*}\phi=(-1)^{k(n-k)}f^{*}\Big(\frac{g^{*}(\eta_k\overline{\eta}_k\phi)\,\eta_{k,n}\overline{\eta}_{k,n}}{\eta_n\overline{\eta}_n}\Big)$$

Because $(n-k)(n-k-1)/2-n(n-1)/2-k(k-1)/2+k(n-k)\equiv 0$ modulo 2, we obtain $(g\circ f)_{\dagger}^{(0)}C=g_{\dagger}^{(0)}f_{\dagger}^{(0)}C$. Thus, Lemma 12.1.4 is proved. □

We consider another compatibility. Let Z be a complex manifold. Let us consider the following diagram:

$$\begin{array}{ccc} Z\times\Delta^k & \xrightarrow{g_1} & Z\times\Delta^{k+\ell} \\ f_1\downarrow & & f_2\downarrow \\ \Delta^k & \xrightarrow{g_2} & \Delta^{k+\ell} \end{array}$$

Here, g_2 is the inclusion into the first k-components, g_1 is the induced morphism, and f_i are the projections. Let $(M_1, M_2, C) \in \mathscr{D}\text{-Tri}(Z\times\Delta^k)$ such that the support of M_i are proper with respect to f_1. We have natural isomorphisms $g_{2\dagger}\big(f_{1\dagger}M_i\big) \simeq f_{2\dagger}g_{1\dagger}M_i$.

Lemma 12.1.5 *We have $g^{(0)}_{2\dagger}\circ f^{(0)}_{1\dagger}(C) = f^{(0)}_{2\dagger}\circ g^{(0)}_{1\dagger}(C)$ under the isomorphisms.*

Proof We put $n = \dim Z$. We have the following:

$$\begin{aligned}
&g^{(0)}_{2\dagger}\big(f^{(0)}_{1\dagger}(C)\big)\Big(\big((\eta_k/\eta_{k+\ell})\cdot\xi^{n-p}m_1,\ \overline{(\eta_k/\eta_{k+\ell})\,\xi^{n+p}m_2}\big)\Big)\\
&=\Big(\frac{1}{2\pi\sqrt{-1}}\Big)^{-\ell}(\eta_{k+\ell}\overline{\eta}_{k+\ell})^{-1}g_{2*}\Big(\eta_k\overline{\eta}_k f^{(0)}_{1\dagger}(C)(\xi^{n-p}m_1,\overline{\xi^{n+p}m_2})\Big)\\
&\quad\times\epsilon(k+\ell)\,\epsilon(k)\\
&=\Big(\frac{1}{2\pi\sqrt{-1}}\Big)^{n-\ell}(\eta_{k+\ell}\overline{\eta}_{k+\ell})^{-1}g_{2*}\Big(\eta_k\overline{\eta}_k\int_{f_1}\xi^{n-p}\overline{\xi}^{n+p}C(m_1,\overline{m}_2)\Big)\\
&\quad\times\epsilon(n)\,(-1)^{np}\,\epsilon(k+\ell)\,\epsilon(k)
\end{aligned}\tag{12.12}$$

Suppose that the supports $\mathrm{Supp}(\xi^{n-p})$ and $\mathrm{Supp}(\xi^{n+p})$ are sufficiently small so that there exists a C^∞-local generator η_Z of ω_Z on a neighbourhood $\mathcal{U}$ of $\mathrm{Supp}(\xi^{n-p})\cup\mathrm{Supp}(\xi^{n+p})$. We set $\eta'_m := \eta_Z\cdot\eta_m$ on $\mathcal{U}$. We also have the following:

$$\begin{aligned}
&f^{(0)}_{2\dagger}\big(g^{(0)}_{1\dagger}C\big)\Big(\xi^{n-p}(\eta'_k/\eta'_{k+\ell})\,m_1,\ \overline{\xi^{n+p}(\eta'_k/\eta'_{k+\ell})m_2}\Big)\\
&=\Big(\frac{1}{2\pi\sqrt{-1}}\Big)^{n}\int_{f_2}\xi^{n-p}\overline{\xi}^{n+p}g_{1\dagger}\big((\eta'_k/\eta'_{k+\ell})\,m_1,\ \overline{(\eta'_k/\eta'_{k+\ell})\,m_2}\big)\cdot\epsilon(n)\,(-1)^{-np}\\
&=\Big(\frac{1}{2\pi\sqrt{-1}}\Big)^{n-\ell}\epsilon(n)(-1)^{-np}\epsilon(n+k+\ell)\epsilon(n+k)\\
&\quad\times\int_{f_2}\xi^{n-p}\overline{\xi}^{n+p}(\eta'_{k+\ell}\overline{\eta}'_{k+\ell})^{-1}g_{1*}\big(\eta'_k\overline{\eta}'_kC(m_1,\overline{m}_2)\big)
\end{aligned}\tag{12.13}$$

We have the following equality modulo 2:

$$
\begin{aligned}
&(k+\ell)(k+\ell-1)/2 - k(k-1)/2 - (n+k+\ell)(n+k+\ell-1)/2 \\
&\quad -(n+k)(n+k-1)/2 \\
&\quad \equiv k\ell + (n+k)\ell = n\ell
\end{aligned}
\tag{12.14}
$$

We have the following:

$$
\begin{aligned}
&\Big\langle (\eta_{k+\ell}\overline{\eta}_{k+\ell})^{-1} g_{2*}\Big(\eta_k\overline{\eta}_k \int_{f_1} \xi^{n-p}\overline{\xi}^{n+p} C(m_1,\overline{m}_2)\Big),\, \phi\eta_{k+\ell}\overline{\eta}_{k+\ell}\Big\rangle \\
&\quad = \Big\langle \int_{f_1} \xi^{n-p}\overline{\xi}^{n+p} C(m_1,\overline{m}_2),\, g_2^*(\phi)\eta_k\overline{\eta}_k\Big\rangle
\end{aligned}
\tag{12.15}
$$

We have the following:

$$
\begin{aligned}
&\Big\langle \int_{f_2} \xi^{n-p}\overline{\xi}^{n+p} \Big((\eta'_{k+\ell}\overline{\eta}'_{k+\ell})^{-1} g_{1*}\big(\eta'_k\overline{\eta}'_k \cdot C(m_1,\overline{m}_2)\big)\Big),\, \phi\cdot\eta_{k+\ell}\overline{\eta}_{k+\ell}\Big\rangle \\
&\quad = \Big\langle g_{1*}\big(\eta'_k\overline{\eta}'_k C(m_1,\overline{m}_2)\big),\, \frac{f_2^*\phi\, \xi^{n-p}\overline{\xi}^{n+p}\, \eta_{k+\ell}\overline{\eta}_{k+\ell}}{\eta'_{k+\ell}\overline{\eta}'_{k+\ell}}\Big\rangle
\end{aligned}
\tag{12.16}
$$

We have the description $\xi^{n-p}\overline{\xi}^{n+p} = A\eta_Z\overline{\eta}_Z$. Then, we have

$$
\frac{f_2^*\phi\, \xi^{n-p}\overline{\xi}^{n+p}\, \eta_{k+\ell}\overline{\eta}_{k+\ell}}{\eta'_{k+\ell}\overline{\eta}'_{k+\ell}} = f_2^*\phi\, A\,(-1)^{n(k+\ell)}
$$

Then, (12.16) is rewritten as follows:

$$
\begin{aligned}
&\Big\langle C(m_1,\overline{m}_2),\, g_1^*(f_2^*\phi^* A(-1)^{n(k+\ell)})\eta'_k\overline{\eta}'_k\Big\rangle \\
&\quad = \Big\langle C(m_1,\overline{m}_2),\, g_1^*(f_2^*\phi^* A(-1)^{n\ell})\eta_Z\overline{\eta}_Z\eta_k\overline{\eta}_k\Big\rangle \\
&\quad = (-1)^{n\ell}\Big\langle C(m_1,\overline{m}_2),\, \xi^{n-p}\overline{\xi}^{n+p}(g_2\circ f_1)^*\phi\, \eta_k\overline{\eta}_k\Big\rangle
\end{aligned}
\tag{12.17}
$$

By comparing (12.15) and (12.17), we obtain the desired equality in the case that $\mathrm{Supp}(\xi^{n-p}) \cup \mathrm{Supp}(\xi^{n+p})$ is sufficiently small. We obtain the general case by using the partition of the unity, and thus Lemma 12.1.5 is proved. □

12.1.2.4 Construction of the Push-Forward in the General Case

In the general case, we factor f into the closed immersion $f_1 : X \longrightarrow X \times Y$ and the projection $f_2 : X \times Y \longrightarrow Y$. We obtain a $\mathscr{D}$-complex-triple $f_{2\dagger}^{(0)}\big(f_{1\dagger}^{(0)}(M_1, M_2, C)\big)$ on Y. We obtain the following lemma from Lemmas 12.1.4 and 12.1.5.

Lemma 12.1.6 *If f is a closed immersion or a projection, then the object $f_{2\dagger}^{(0)}\big(f_{1\dagger}^{(0)}(M_1, M_2, C)\big)$ is naturally isomorphic to $f_{\dagger}^{(0)}(M_1, M_2, C)$ given previously.* □

12.1.2.5 Push-Forward of $\mathscr{D}$-Complex-Triples

Let $(M_1^\bullet, M_2^\bullet, C) \in \mathcal{C}(\mathscr{D})\text{-Tri}(X)$ such that the restriction of f to the support of M_i are proper. We have the following object in $\mathcal{C}^{(2)}(\mathscr{D})\text{-Tri}(Y)$:

$$\Big(f_\dagger(M_1^\bullet)^\bullet, f_\dagger(M_2^\bullet)^\bullet, f_\dagger^{(0)}C\Big)$$

By taking the total complex, we obtain an object in $\mathcal{C}(\mathscr{D})\text{-Tri}(Y)$, which we denote by $f_\dagger^{(0)}(M_1^\bullet, M_2^\bullet, C)$. Correspondingly, we have the push-forward for complexes of $\mathscr{D}$-triples.

12.1.2.6 Composition

Let $f : X \longrightarrow Y$ and $g : Y \longrightarrow Z$ be morphisms of complex manifolds. Let $(M_i^\bullet, M_2^\bullet, C)$ be a $\mathscr{D}_X$-complex-triple such that the restriction of f and $g \circ f$ to the supports of M_i are proper. We have natural quasi-isomorphisms $(g \circ f)_\dagger M_i \simeq g_\dagger\big(f_\dagger M_i\big)$. The following lemma is implied in [66].

Lemma 12.1.7 *We have $(g \circ f)_\dagger^{(0)}(C) = g_\dagger^{(0)}(f_\dagger^{(0)}C)$ under the natural quasi-isomorphisms.*

Proof By using Lemmas 12.1.4 and 12.1.5, we can reduce the issue to the case that $f : W \times Z \times Y \longrightarrow Z \times Y$ and $g : Z \times Y \longrightarrow Y$ are the projections. We set $m := \dim W$ and $n := \dim Z$. We have the following:

$$\begin{aligned}
&(g \circ f)_\dagger^{(0)} C\big(\xi^{n-q}\eta^{m-p}m_1, \overline{\xi^{n+q}\eta^{m+p}m_2}\big) \\
&= \Big(\frac{1}{2\pi\sqrt{-1}}\Big)^{m+n} \int \xi^{n-q}\eta^{m-p}\overline{\xi^{n+q}\eta^{m+p}} C(m_1, \overline{m}_2)\, \epsilon(m+n)\, (-1)^{-(m+n)(p+q)}
\end{aligned} \tag{12.18}$$

We also have the following (recall (12.2)):

$$
\begin{aligned}
& g_{\dagger}^{(0)}(f_{\dagger}^{(0)}C)\Big(\xi^{n-q}\eta^{m-p}m_1,\ \overline{\xi^{n+q}\eta^{m+p}m_2}\Big)\times(-1)^{pq} \\
&= \left(\frac{1}{2\pi\sqrt{-1}}\right)^{n}\int \xi^{n-q}\overline{\xi}^{n+q}f_{\dagger}^{(0)}C(\eta^{m-p}m_1,\ \overline{\eta^{m+p}m_2})\,\epsilon(n)\,(-1)^{-nq+pq} \\
&= \left(\frac{1}{2\pi\sqrt{-1}}\right)^{m+n}\int \xi^{n-q}\eta^{m-p}\overline{\xi}^{n+q}\overline{\eta}^{m+p}C(m_1,\overline{m}_2) \\
&\quad\times\epsilon(m)\epsilon(n)(-1)^{-mp-nq+pq+(n+q)(m-p)}
\end{aligned}
\tag{12.19}
$$

We have the following equality modulo 2:

$$
\begin{aligned}
& m(m-1)/2+n(n-1)/2+mp+nq+pq+(n+q)(m-p) \\
&\quad\equiv (m+n)(m+n-1)/2+(m+n)(p+q)
\end{aligned}
\tag{12.20}
$$

Thus, we obtain Lemma 12.1.7. □

12.1.2.7 Correspondence of Left and Right Triples (Appendix)

The correspondence between sesqui-linear pairings for left and right $\mathscr{D}$-modules are given as follows. Let $\mathfrak{Db}_X^{p,q}$ denote the sheaf of (p,q)-currents. Set $d_X := \dim X$.

Let C be a sesqui-linear pairing $M_1\times\overline{M_2}\longrightarrow\mathfrak{Db}_X$. The corresponding pairing of right $\mathscr{D}$-modules $(\omega_X\otimes M_1)\otimes_{\mathbb{C}}\overline{(\omega_X\otimes M_2)}\longrightarrow\mathfrak{Db}_X^{d_X,d_X}$ is given as follows:

$$
C^r\big(\eta_1\otimes m_1,\ \overline{\eta_2\otimes m_2}\big):=\eta_1\overline{\eta}_2\,C(m_1,\overline{m}_2)\,\epsilon(d_X)\left(\frac{1}{2\pi\sqrt{-1}}\right)^{d_X}
$$

For a closed immersion $f: X\longrightarrow Y$, we should have

$$
f_*C^r(\eta_X\otimes m_1,\ \overline{\eta_X\otimes m_2})=(f_{\dagger}^{(0)}C)^r\big(\eta_X\otimes m_1,\ \overline{\eta_X\otimes m_2}\big).
$$

It implies the formula in the closed immersion, as follows.

$$
\begin{aligned}
& \Big\langle C(m_1,\overline{m}_2),\,f^*\phi\ \eta_X\overline{\eta}_X\Big\rangle\,\epsilon(d_X)\left(\frac{1}{2\pi\sqrt{-1}}\right)^{d_X} \\
&= \Big\langle f_*C^r(\eta_X\otimes m_1,\overline{\eta_X\otimes m_2}),\,\phi\Big\rangle = \Big\langle (f_{\dagger}^{(0)}C)^r(\eta_X\otimes m_1,\overline{\eta_X\otimes m_2}),\,\phi\Big\rangle \\
&= \Big\langle (f_{\dagger}^{(0)}C)\big((\eta_X/\eta_Y)m_1,\overline{(\eta_X/\eta_Y)m_2}\big),\,f^*\phi\ \eta_Y\overline{\eta}_Y\Big\rangle\,\epsilon(d_Y)\left(\frac{1}{2\pi\sqrt{-1}}\right)^{d_Y}
\end{aligned}
\tag{12.21}
$$

12.1.3 Hermitian Adjoint of $\mathscr{D}$-Complex-Triples

For $\mathcal{T}^\bullet \in \mathcal{C}(\mathscr{D}\text{-Tri}(X))$, we define $\boldsymbol{D}^{\mathrm{herm}}(\mathcal{T}^\bullet)$ in $\mathcal{C}(\mathscr{D}\text{-Tri}(X))$ as follows. The p-th member $\boldsymbol{D}^{\mathrm{herm}}(\mathcal{T}^\bullet)^p$ is $(\mathcal{T}^{-p})^*$. The differential $\boldsymbol{D}^{\mathrm{herm}}(\mathcal{T}^\bullet)^p \longrightarrow \boldsymbol{D}^{\mathrm{herm}}(\mathcal{T}^\bullet)^{p+1}$ is defined by $\boldsymbol{D}^{\mathrm{herm}}(\delta_{-p-1}) = \delta^*_{-p-1}$ as in §1.6.c [66]. Then, $\boldsymbol{D}^{\mathrm{herm}}$ gives a contravariant auto-equivalence of $\mathcal{C}(\mathscr{D}\text{-Tri}(X))$. We have a natural isomorphism $\boldsymbol{D}^{\mathrm{herm}}(\mathcal{T}^\bullet[\ell]) = \boldsymbol{D}^{\mathrm{herm}}(\mathcal{T}^\bullet)[-\ell]$.

By the equivalence Ψ_1, we obtain a contravariant auto-equivalence $\boldsymbol{D}^{(0)\,\mathrm{herm}}$ on $\mathcal{C}(\mathscr{D})\text{-Tri}(X)$:

$$\boldsymbol{D}^{(0)\,\mathrm{herm}}(M_1^\bullet, M_2^\bullet, C) = (M_2^\bullet, M_1^\bullet, C^*)$$

Here, $(C^*)^p = (-1)^p\big(C^p\big)^*$. We have a natural isomorphism $\mathcal{S}_{-\ell} \circ \boldsymbol{D}^{\mathrm{herm}} \simeq \boldsymbol{D}^{\mathrm{herm}} \circ \mathcal{S}_\ell$.

Lemma 12.1.8 *Let $f : X \longrightarrow Y$ be a morphism of complex manifolds. Let $\mathfrak{T}$ be an object in $\mathcal{C}(\mathscr{D})$-Tri(X) such that the restriction of f to the support of $\mathfrak{T}$ is proper. We have $f_\dagger^{(0)} \circ \boldsymbol{D}^{(0)\,\mathrm{herm}}(\mathfrak{T}) = \boldsymbol{D}^{(0)\,\mathrm{herm}} \circ f_\dagger^{(0)}(\mathfrak{T})$ in $\mathcal{C}(\mathscr{D})$-Tri(Y). As a consequence, for $\mathcal{T}^\bullet \in \mathcal{C}(\mathscr{D}\text{-Tri}(X))$, we have $f_\dagger \circ \boldsymbol{D}^{\mathrm{herm}}(\mathcal{T}^\bullet) = \boldsymbol{D}^{\mathrm{herm}} \circ f_\dagger(\mathcal{T}^\bullet)$ in $\mathcal{C}(\mathscr{D}\text{-Tri}(Y))$.*

Proof We use the notation in the construction of push-forward in Sect. 12.1.2. Let us consider the case that f is a closed immersion. We have the following:

$$\begin{aligned}
&f_\dagger^{(0)}(C^*)\big((\eta_X/\eta_Y)\,m_2,\ \overline{(\eta_X/\eta_Y)\,m_1}\big)\\
&= (\eta_Y\overline{\eta}_Y)^{-1} f_*\big(\eta_X\,\overline{\eta}_X\,C^*(m_2,\overline{m}_1)\big)\left(\frac{1}{2\pi\sqrt{-1}}\right)^{d_X-d_Y}\epsilon(d_X)\,\epsilon(d_Y)\\
&= (\eta_Y\overline{\eta}_Y)^{-1} f_*\big(\eta_X\overline{\eta}_X\,\overline{C(m_1,\overline{m}_2)}\big)\left(\frac{1}{2\pi\sqrt{-1}}\right)^{d_X-d_Y}\epsilon(d_X)\,\epsilon(d_Y)\\
&= \overline{(\overline{\eta}_Y\eta_Y)^{-1} f_*\big(\overline{\eta}_X\eta_X C(m_1,\overline{m}_2)\big)\left(\frac{1}{2\pi\sqrt{-1}}\right)^{d_X-d_Y}}(-1)^{d_X-d_Y}\epsilon(d_X)\,\epsilon(d_Y)\\
&= \overline{(\eta_Y\overline{\eta}_Y)^{-1} f_*\big(\eta_X\overline{\eta}_X C(m_1,\overline{m}_2)\big)\left(\frac{1}{2\pi\sqrt{-1}}\right)^{d_X-d_Y}}\epsilon(d_X)\,\epsilon(d_Y)\\
&= (f_\dagger^{(0)}C)^*\big((\eta_X/\eta_Y)\,m_1,\overline{(\eta_X/\eta_Y)\,m_2}\big)
\end{aligned}\tag{12.22}$$

Let us consider the case that $f : Z \times Y \longrightarrow Y$. We put $m := \dim Z$. We have the following:

$$\begin{aligned}
&(f_\dagger^{(0)}C^*)(\eta^{m+p}m_2,\ \overline{\eta^{m-p}m_1})\\
&= \left(\frac{1}{2\pi\sqrt{-1}}\right)^m \int \eta^{m+p}\overline{\eta}^{m-p}\,C^*(m_2,\overline{m}_1)\,\epsilon(m)\,(-1)^{mp}
\end{aligned}$$

$$= \overline{\left(\frac{1}{2\pi\sqrt{-1}}\right)^m (-1)^m \int \eta^{m-p}\overline{\eta}^{m+p}(-1)^{m-p} C(m_1, \overline{m_2})\, \epsilon(m)\, (-1)^{mp}}$$

$$= (-1)^p (f_\dagger^{(0)} C)^*(\eta^{m+p} m_2, \overline{\eta^{m-p} m_1}). \tag{12.23}$$

Thus, we are done. □

12.1.4 Comparison with the Naive Push-Forward

12.1.4.1 Trace Morphism

Let us observe that we have a natural morphism, called the trace morphism:

$$\mathrm{tr} : f_! \Big(\big(\mathscr{D}_{Y\leftarrow X} \otimes_{\mathbb{C}} \mathscr{D}_{\overline{Y}\leftarrow\overline{X}} \big) \otimes^L_{\mathscr{D}_{X,\overline{X}}} \mathfrak{Db}_X \Big) \longrightarrow \mathfrak{Db}_Y \tag{12.24}$$

We have the following:

$$\begin{aligned}
&\Big(\omega_X \otimes f^{-1}(\mathscr{D}_Y \otimes \omega_Y^{-1}) \otimes_{\mathbb{C}} \omega_{\overline{X}} \otimes f^{-1}(\mathscr{D}_{\overline{Y}} \otimes \omega_{\overline{Y}}^{-1}) \Big) \otimes^L_{\mathscr{D}_{X,\overline{X}}} \mathfrak{Db}_X \\
&\longrightarrow f^{-1}(\mathscr{D}_{Y,\overline{Y}} \otimes \omega_{Y,\overline{Y}}^{-1}) \otimes \mathrm{Tot}\big(\Omega^\bullet_X[d_X] \otimes \Omega^\bullet_{\overline{X}}[d_X]\big) \otimes_{\mathcal{O}_{X,\overline{X}}} \mathfrak{Db}_X \\
&\longrightarrow f^{-1}(\mathscr{D}_{Y,\overline{Y}} \otimes \omega_{Y,\overline{Y}}^{-1}) \otimes \mathfrak{Db}^\bullet_X[2d_X]
\end{aligned} \tag{12.25}$$

Here, we use the pairing $\mathrm{Tot}\big(\Omega^\bullet_X[d_X] \otimes \Omega^\bullet_{\overline{X}}[d_X]\big) \longrightarrow \mathfrak{Db}^\bullet_X[2d_X]$ given as follows. We have the identification $\mathrm{Tot}\big(\Omega^\bullet_X[d_X] \otimes \Omega^\bullet_{\overline{X}}[d_X]\big) \simeq \mathrm{Tot}\big(\Omega^\bullet_X \otimes \Omega^\bullet_{\overline{X}}\big)[2d_X]$ which is given by the multiplication of $(-1)^{pd_X}\epsilon(d_X)$ on $\Omega^{p+d_X}_X \otimes \Omega^{q+d_X}_{\overline{X}}$. We have the natural isomorphism $\mathrm{Tot}\big(\Omega^\bullet_X \otimes \Omega^\bullet_{\overline{X}}\big) \otimes \mathfrak{Db}_X \longrightarrow \mathfrak{Db}^\bullet_X$. The composition gives the desired pairing. Then, we obtain

$$\begin{aligned}
f_!\Big(\big(\mathscr{D}_{Y\leftarrow X} \otimes_{\mathbb{C}} \mathscr{D}_{\overline{Y}\leftarrow\overline{X}} \big) \otimes^L_{\mathscr{D}_{X,\overline{X}}} \mathfrak{Db}_X \Big) &\longrightarrow \big(\mathscr{D}_{Y,\overline{Y}} \otimes \omega_{Y,\overline{Y}}^{-1}\big) \otimes f_!(\mathfrak{Db}^\bullet_X[2d_X]) \\
&\xrightarrow{A} \big(\mathscr{D}_{Y,\overline{Y}} \otimes \omega_{Y,\overline{Y}}^{-1}\big) \otimes \mathfrak{Db}^\bullet_Y[2d_Y] \\
&\overset{B}{\simeq} \mathrm{Tot}\Big(\big(\omega_Y^{-1} \otimes \mathscr{D}_Y \otimes \Omega^\bullet_Y[d_Y]\big) \\
&\otimes \big(\omega_{\overline{Y}}^{-1} \otimes \mathscr{D}_{\overline{Y}} \otimes \Omega^\bullet_{\overline{Y}}[d_Y]\big)\Big) \otimes \mathfrak{Db}_Y \simeq \mathfrak{Db}_Y
\end{aligned} \tag{12.26}$$

The morphism A is given by the integration $f_! \mathfrak{Db}^{p+d_X,q+d_X}_X \longrightarrow \mathfrak{Db}^{p+d_Y,q+d_Y}_Y$ multiplied with $(2\pi\sqrt{-1})^{-d_X+d_Y}$. For B, we use the identification as in the case of (12.25).

12.1.4.2 Naive Push-Forward of Pairings

Let $C : M_1 \times \overline{M}_2 \longrightarrow \mathfrak{Db}_X$. We have the following induced pairing:

$$\left(\mathscr{D}_{Y\leftarrow X} \otimes^L_{\mathscr{D}_X} M_1\right) \otimes_{\mathbb{C}} \overline{\left(\mathscr{D}_{Y\leftarrow X} \otimes^L_{\mathscr{D}_X} M_2\right)} \longrightarrow \left(\mathscr{D}_{Y\leftarrow X} \otimes_{\mathbb{C}} \mathscr{D}_{\overline{Y}\leftarrow\overline{X}}\right) \otimes^L_{\mathscr{D}_{X,\overline{X}}} \mathfrak{Db}_X$$

Hence, we obtain the following pairing

$$f_\dagger M_1 \otimes f_\dagger \overline{M}_2 \longrightarrow f_!\left(\left(\mathscr{D}_{Y\leftarrow X} \otimes \mathscr{D}_{\overline{Y}\leftarrow\overline{X}}\right) \otimes^L_{\mathscr{D}_{X,\overline{X}}} \mathfrak{Db}^\bullet_X[2d_X]\right) \xrightarrow{\mathrm{tr}} \mathfrak{Db}_Y$$

It is denoted by $f_\dagger^{(1)} C$. We have the following comparison.

Lemma 12.1.9 *We have* $f_\dagger^{(1)}(C) = f_\dagger^{(0)}(C)$.

Proof Let us consider the case that f is a closed immersion. We have the following:

$$\begin{aligned} & f_\dagger^{(1)}(C)\left((\eta_X/\eta_Y)\, m_1,\ \overline{(\eta_X/\eta_Y)\, m_2}\right) \\ &\quad = (\eta_Y\, \overline{\eta}_Y)^{-1} f_*\left(\eta_X \overline{\eta}_X\, C(m_1, \overline{m}_2)\right) \left(\frac{1}{2\pi\sqrt{-1}}\right)^{d_X - d_Y} \epsilon(d_X)\, \epsilon(d_Y) \end{aligned} \tag{12.27}$$

Here, $\epsilon(d_X)$ and $\epsilon(d_Y)$ appear by the identifications $\Omega^\bullet_X[d_X] \otimes \Omega^\bullet_{\overline{X}}[d_X] \simeq \left(\Omega^\bullet_X \otimes \Omega^\bullet_{\overline{X}}\right)[2d_X]$ and $\Omega^\bullet_Y[d_Y] \otimes \Omega^\bullet_{\overline{Y}}[d_Y] \simeq \left(\Omega^\bullet_Y \otimes \Omega^\bullet_{\overline{Y}}\right)[2d_Y]$. By definition, we have the following:

$$\begin{aligned} & f_\dagger^{(0)}(C)\left((\eta_X/\eta_Y)\, m_1,\ \overline{(\eta_X/\eta_Y)\, m_2}\right) \\ &\quad = (\eta_Y \overline{\eta}_Y)^{-1} f_*\left(\eta_X \overline{\eta}_X\, C(m_1, \overline{m}_2)\right) \cdot \left(\frac{1}{2\pi\sqrt{-1}}\right)^{d_X - d_Y} \epsilon(d_X)\, \epsilon(d_Y) \end{aligned} \tag{12.28}$$

Hence, we are done in this case.

Let us consider the case $f : X = Z \times Y \longrightarrow Y$. We have the identification

$$\begin{aligned} f_\dagger M &\simeq f_\dagger(M \otimes \mathcal{E}^\bullet_{X/Y}[d_Z]) \\ &\simeq f_\dagger\left(M \otimes \mathcal{E}^\bullet_{X/Y}[d_Z] \otimes_{f^{-1}\mathcal{O}_Y} f^{-1}\left(\Omega^\bullet_Y[d_Y] \otimes \mathscr{D}_Y \otimes \omega_Y^{-1}\right)\right) \end{aligned} \tag{12.29}$$

Let $m = \dim Z$ and $n = \dim Y$. We have

$$\begin{aligned} & f_\dagger^{(1)} C\left(\eta_Y^{-1} \otimes (m_1 \cdot \xi^{m-p}\, \eta_Y),\ \overline{\eta_Y^{-1} \otimes (m_2 \cdot \xi^{m+p} \eta_Y)}\right) \\ &\quad = (\eta_Y \overline{\eta}_Y)^{-1} \int C(m_1, \overline{m}_2)\, \xi^{m-p} \eta_Y\, \overline{\xi^{m+p} \eta_Y}\, (-1)^{pd_X} \left(\frac{1}{2\pi\sqrt{-1}}\right)^m \epsilon(d_X)\, \epsilon(d_Y) \end{aligned}$$

$$= (\eta_Y \overline{\eta}_Y)^{-1} \int C(m_1, \overline{m}_2) \, \xi^{m-p} \overline{\xi}^{m+p} \eta_Y \overline{\eta}_Y$$

$$\times (-1)^{n(m+p)+pd_X} \left(\frac{1}{2\pi\sqrt{-1}} \right)^m \epsilon(d_X) \, \epsilon(d_Y) \tag{12.30}$$

We also have the following:

$$f_{\dagger}^{(0)} C(m_1 \, \xi^{m-p}, \overline{m_2 \, \xi^{m+p}}) = \int C(m_1, \overline{m}_2) \, \xi^{m-p} \overline{\xi}^{m+p} (-1)^{mp} \left(\frac{1}{2\pi\sqrt{-1}} \right)^m \epsilon(m) \tag{12.31}$$

It is easy to check that (12.30) and (12.31) are equal. Hence, we obtain $f_{\dagger}^{(1)} C = f_{\dagger}^{(0)} C$. □

Remark 12.1.10 We will not distinguish $f_{\dagger}^{(0)}$ and $f_{\dagger}^{(1)}$ in the following. □

12.1.5 *Rules for Signature (Appendix)*

12.1.5.1 Contravariant Functors and Shifts of Degree

Let $A^{\bullet}$ be a $\mathbb{C}_X$-complex with a differential ∂. Recall that, for any integer ℓ, the shift $A^{\bullet}[\ell]$ is the complex such that $(A^{\bullet}[\ell])^p = A^{\ell+p}$ with the differential $(-1)^{\ell}\partial$. We naturally identify $(A^{\bullet}[\ell])[k] = A^{\bullet}[\ell + k]$.

We recall the rule in [13] on the signature for contravariant functor. Let F be a contravariant functor from the category of $\mathbb{C}_X$-modules to a $\mathbb{C}$-linear abelian category $\boldsymbol{C}$. Then, we obtain a complex $\mathcal{B}$ in $\boldsymbol{C}$ given by $\mathcal{B}^k := F(A^{-k})$. The differential $\mathcal{B}^k \longrightarrow \mathcal{B}^{k+1}$ is given by $(-1)^{k+1}F(\partial) : F(A^{-k}) \longrightarrow F(A^{-k-1})$. The complex is denoted by $F(A^{\bullet})$. For any integer ℓ, we have a natural isomorphism $F(A^{\bullet})[-\ell] \simeq F(A^{\bullet}[\ell])$ given by the multiplication of $\epsilon(\ell)(-1)^{p\ell}$ on $F(A^{\bullet})[-\ell]^p = F(A^{-p+\ell})$, where $\epsilon(\ell) := (-1)^{\ell(\ell-1)/2}$. The composite of the isomorphisms $F(A^{\bullet})[-\ell - k] \simeq F(A^{\bullet}[\ell])[-k] \simeq F(A^{\bullet}[\ell + k])$ is equal to the direct one $F(A^{\bullet})[-\ell - k] \simeq F(A^{\bullet}[\ell + k])$.

12.1.5.2 Naive *n*-Complexes and the Total Complexes

Let $\mathcal{A}$ be a naive n-complex of $\mathbb{C}_X$-modules, i.e., it is a $\mathbb{Z}^n$-graded $\mathbb{C}_X$-module $\mathcal{A} = \bigoplus_{\boldsymbol{k}\in\mathbb{Z}^n} \mathcal{A}^{\boldsymbol{k}}$ equipped with differentials $\partial_i : \mathcal{A}^{\boldsymbol{k}} \longrightarrow \mathcal{A}^{\boldsymbol{k}+\boldsymbol{\delta}_i}$ $(i = 1, \ldots, n)$ such that $\partial_i \circ \partial_i = 0$ and $[\partial_i, \partial_j] = 0$. Here, the j-th component of $\boldsymbol{\delta}_i$ is 1 if $i = j$, and 0 if $i \neq j$. We put $|\boldsymbol{k}| = \sum k_i$ for $\boldsymbol{k} = (k_i) \in \mathbb{Z}^n$. Then, the total complex $\mathrm{Tot}(\mathcal{A})$ is defined as a $\mathbb{Z}$-graded $\mathbb{C}_X$-module $\mathrm{Tot}(\mathcal{A})^k = \bigoplus_{|\boldsymbol{k}|=k} \mathcal{A}^{\boldsymbol{k}}$ with the differential $\partial : \mathrm{Tot}(\mathcal{A})^k \longrightarrow \mathrm{Tot}(\mathcal{A})^{k+1}$ given by $\partial(a^{\boldsymbol{k}}) = \sum_i (-1)^{\sum_{j<i} k_i} \partial_i(a^{\boldsymbol{k}})$.

Let $\mathcal{A}$ be an n-complex. Let $\mathcal{A}[\ell\boldsymbol{\delta}_i]$ be the n-complex given by $\mathcal{A}[\ell\boldsymbol{\delta}_i]^{\boldsymbol{k}} = \mathcal{A}^{\boldsymbol{k}+\ell\boldsymbol{\delta}_i}$ with the differentials ∂_j $(j \neq i)$ and $(-1)^\ell\partial_i$. We have an isomorphism $\mathrm{Tot}(\mathcal{A}[\ell\boldsymbol{\delta}_i]) \simeq \mathrm{Tot}(\mathcal{A})[\ell]$ given by the multiplication of $(-1)^{\sum_{j<i}\ell k_i}$ on $\mathcal{A}^{\boldsymbol{k}}$.

12.1.5.3 Inner Homomorphism

Let $A_i^\bullet$ $(i = 1, 2)$ be $\mathbb{C}_X$-complexes. We have the $\mathbb{C}_X$-double complex $\mathcal{B}^{\bullet,\bullet}$ given by $\mathcal{B}^{p,q} := \mathcal{H}om(A_1^{-q}, A_2^p)$. We often denote the double complex just by $\mathcal{H}om(A_1^\bullet, A_2^\bullet)$. By the standard rule on signatures, the differential of the total complex is given by $\partial(f) = \partial \circ f - (-1)^{p-q} f \circ \partial$ for $f \in \mathcal{H}om(A_1^q, A_2^p)$, i.e., $\partial(f)(a) = \partial\big(f(a)\big) - (-1)^{|f|} f(\partial a)$.

We have the isomorphism

$$\mathcal{H}om(A_1^\bullet[\ell], A_2^\bullet) \simeq \mathcal{H}om(A_1^\bullet, A_2^\bullet)[-\ell\boldsymbol{\delta}_2]$$

in Sect. 12.1.5.1. The isomorphism in Sect. 12.1.5.2 induces the following isomorphism

$$\mathrm{Tot}\,\mathcal{H}om(A_1^\bullet[\ell], A_2^\bullet) \simeq \mathrm{Tot}\,\mathcal{H}om(A_1^\bullet, A_2^\bullet)[-\ell]$$

which is given by the multiplication of $\epsilon(\ell)(-1)^{r\ell}$ on $\mathrm{Tot}\,\mathcal{H}om(A_1^\bullet[\ell], A_2^\bullet)^r$.

We can check the following lemma by a direct computation.

Lemma 12.1.11 *The following natural diagram of complexes is commutative:*

$$\begin{array}{ccc}
A_1^\bullet[\ell] & \longrightarrow & \mathcal{H}om\Big(\mathcal{H}om\big(A_1^\bullet[\ell], A_2^\bullet\big), A_2^\bullet\Big) \\
\downarrow & & \simeq\downarrow \\
\mathcal{H}om\Big(\mathcal{H}om\big(A_1^\bullet, A_2^\bullet\big), A_2^\bullet\Big)[\ell] & \xrightarrow{\simeq} & \mathcal{H}om\Big(\mathcal{H}om\big(A_1^\bullet, A_2^\bullet\big)[-\ell], A_2^\bullet\Big)
\end{array}$$

Here, we omit to denote Tot. □

12.1.5.4 Pairing

For any integer ℓ, we have an isomorphism of complexes

$$\mathrm{Tot}(A_1^\bullet \otimes A_2^\bullet) \simeq \mathrm{Tot}(A_1^\bullet[-\ell] \otimes A_2^\bullet[\ell])$$

given by the multiplication of $\epsilon(\ell)(-1)^{\ell p}$ on $\big(A_1^\bullet[-\ell] \otimes A_2^\bullet[\ell]\big)^{p,q}$. The composite of the isomorphisms $\mathrm{Tot}(A_1^\bullet \otimes A_2^\bullet) \simeq \mathrm{Tot}(A_1^\bullet[-\ell] \otimes A_2^\bullet[\ell]) \simeq \mathrm{Tot}(A_1^\bullet[-\ell-k] \otimes A_2^\bullet[\ell+k])$ is equal to the direct one $\mathrm{Tot}(A_1^\bullet \otimes A_2^\bullet) \simeq \mathrm{Tot}(A_1^\bullet[-\ell-k] \otimes A_2^\bullet[\ell+k])$.

Let $I^\bullet$ be a $\mathbb{C}_X$-complex. Let $C : \mathrm{Tot}(A_1^\bullet \otimes A_2^\bullet) \longrightarrow I^\bullet$ be a morphism. We have the induced morphism

$$C[\ell] : \mathrm{Tot}\big(A_1^\bullet[-\ell] \otimes A_2^\bullet[\ell]\big) \longrightarrow I^\bullet.$$

A morphism $C : \mathrm{Tot}(A_1^\bullet \otimes A_2^\bullet) \longrightarrow I^\bullet$ is equivalent to a morphism $\Psi_C : A_1^\bullet \longrightarrow \mathrm{Tot}\,\mathcal{H}om(A_2^\bullet, I^\bullet)$. The correspondence is given by $\Psi_C(a)(b) = C(a,b)$. The following lemma can be checked in an elementary way.

Lemma 12.1.12 *We have $\Psi_{C[\ell]} = \Psi_C$ under the identification*

$$\mathrm{Tot}\,\mathcal{H}om(A_2^\bullet[\ell], I^\bullet) \simeq \mathrm{Tot}\,\mathcal{H}om(A_2^\bullet, I^\bullet)[-\ell].$$

□

12.2 Some Basic Functors for Non-degenerate $\mathscr{D}$-Triples

12.2.1 Category of Non-degenerate $\mathscr{D}$-Triples

For any $\mathscr{D}_X$-module M, we have the object $\mathcal{C}_X M := R\mathcal{H}om_{\mathscr{D}_X}(M, \mathfrak{Db}_X)$ in the derived category of $\mathscr{D}_{\overline{X}}$-complexes. If M is holonomic, we have the natural identification

$$\mathcal{C}_X M \simeq \mathcal{H}^0\mathcal{C}_X M = \mathcal{H}om_{\mathscr{D}_X}(M, \mathfrak{Db}_X),$$

and it is holonomic. (See [33, 58, 65, 68].)

A $\mathscr{D}_X$-triple (M_1, M_2, C) is called holonomic if M_i are holonomic. It is called non-degenerate if moreover the induced morphism $\varphi_{C,1,2} : \overline{M_2} \longrightarrow \mathcal{C}_X(M_1)$ is an isomorphism. The condition is equivalent to that the induced morphism $\varphi_{C,2,1} : M_1 \longrightarrow \mathcal{C}_{\overline{X}}(\overline{M_2})$ is an isomorphism. Let $\mathscr{D}\text{-Tri}^{nd}(X) \subset \mathscr{D}\text{-Tri}(X)$ denote the full subcategory of non-degenerate $\mathscr{D}_X$-triples. It is easy to check that $\mathscr{D}\text{-Tri}^{nd}(X)$ is an abelian subcategory of $\mathscr{D}\text{-Tri}(X)$.

12.2.2 Localization

Let g be any holomorphic function on a complex manifold X. Let $\mathcal{T} = (M_1, M_2, C)$ be a holonomic $\mathscr{D}_X$-triple. As in the case of $\mathcal{R}$-triples, we have the uniquely induced pairings $C[!g] : M_1[*g] \times \overline{M_2[!g]} \longrightarrow \mathfrak{Db}_X$ and $C[*g] : M_1[!g] \times \overline{M_2[*g]} \longrightarrow \mathfrak{Db}_X$. (See Sect. 3.3.) Thus, we have $\mathcal{T}[!g] = \big(M_1[*g], M_2[!g], C[!g]\big)$ and $\mathcal{T}[*g] = \big(M_1[!g], M_2[*g], C[*g]\big)$.

Lemma 12.2.1 *If $\mathcal{T}$ is non-degenerate, then $\mathcal{T}[!g]$ and $\mathcal{T}[*g]$ are also non-degenerate.*

Proof Let N be a holonomic $\mathscr{D}_X$-module with an isomorphism $N(*g) \simeq C_{\overline{X}}(\overline{M}_2[*g])(*g)$. We have an isomorphism $C_X(N)(*g) \simeq \overline{M}_2(*g)$. Hence, we have a unique morphism $\overline{M}_2[!g] \longrightarrow C_X(N)$. It induces $N \longrightarrow C_{\overline{X}}(\overline{M}_2[!g])$. By this universal property, we obtain that $C_{\overline{X}}(\overline{M}_2[!g]) \simeq C_{\overline{X}}(\overline{M}_2[!g])[*g]$. Therefore, we obtain that the induced morphism $M_1[*g] \longrightarrow C_{\overline{X}}(\overline{M}_2[!g])$ is an isomorphism. Namely, $\mathcal{T}[!g]$ is non-degenerate. Similarly, we obtain that $\mathcal{T}[*g]$ is non-degenerate. □

Lemma 12.2.2 *If $g_1^{-1}(0) = g_2^{-1}(0)$, we have $\mathcal{T}[\star g_1] = \mathcal{T}[\star g_2]$.*

Proof We have $M[*g_1] = M[*g_2]$ for any holonomic $\mathscr{D}$-module M, which implies $M[!g_1] = M[!g_2]$. Hence, we have the coincidence of the underlying $\mathscr{D}_X$-modules of $\mathcal{T}[\star g_i]$ $(i = 1, 2)$. Then, it is easy to deduce the coincidence of the pairings on X from the coincidence on $X \setminus g_i^{-1}(0)$. □

Let H be a hypersurface of X. By Lemma 12.2.2, we obtain sesqui-linear pairings:

$$C[*H] : M_1[!H] \times \overline{M_2[*H]} \longrightarrow \mathfrak{Db}_X, \qquad C[!H] : M_1[*H] \times \overline{M_2[!H]} \longrightarrow \mathfrak{Db}_X.$$

Thus, we obtain the induced $\mathscr{D}$-triples $\mathcal{T}[*H] = \big(M_1[!H], M_2[*H], C[*H]\big)$ and $\mathcal{T}[!H] = \big(M_1[*H], M_2[!H], C[!H]\big)$ for holonomic $\mathcal{T} \in \mathscr{D}\text{-Tri}(X)$. If $\mathcal{T}$ is non-degenerate, $\mathcal{T}[\star H]$ are also non-degenerate. It is easy to see the following.

Lemma 12.2.3 *We naturally have*

$$\boldsymbol{D}^{\mathrm{herm}}(\mathcal{T}[*H]) = (\boldsymbol{D}^{\mathrm{herm}}\mathcal{T})[!H], \quad \boldsymbol{D}^{\mathrm{herm}}(\mathcal{T}[!H]) = (\boldsymbol{D}^{\mathrm{herm}}\mathcal{T})[*H].$$

□

12.2.3 Tensor Product with Smooth $\mathscr{D}$-Triples

A $\mathscr{D}_X$-module is called smooth if it is coherent as an $\mathcal{O}_X$-module, which implies that it is a locally free $\mathcal{O}_X$-module. A $\mathscr{D}_X$-triple is called smooth if the underlying $\mathscr{D}_X$-modules are smooth. The values of the sesqui-linear pairing is contained in the sheaf of C^∞-functions on X. Let $\mathcal{T}_i = (\mathcal{M}_{i,1}, \mathcal{M}_{i,2}, C_i) \in \mathscr{D}\text{-Tri}(X)$ $(i = 1, 2)$. If $\mathcal{T}_2$ is smooth, we naturally have an object $\mathcal{T}_1 \otimes \mathcal{T}_2 := \big(\mathcal{M}_{1,1} \otimes \mathcal{M}_{2,1}, \mathcal{M}_{2,1} \otimes \mathcal{M}_{2,2}, C_1 \otimes C_2\big)$ in $\mathscr{D}\text{-Tri}(X)$. We clearly have $\boldsymbol{D}^{\mathrm{herm}}(\mathcal{T}_1 \otimes \mathcal{T}_2) = \boldsymbol{D}^{\mathrm{herm}}(\mathcal{T}_1) \otimes \boldsymbol{D}^{\mathrm{herm}}(\mathcal{T}_2)$.

Lemma 12.2.4 *If $\mathcal{T}_i$ are non-degenerate, $\mathcal{T}_1 \otimes \mathcal{T}_2$ is also non-degenerate.*

Proof We have only to consider the case that $\mathcal{T}_2 = (\mathcal{O}, \mathcal{O}, C_0)$. Then, the claim is clear. □

Let us consider a more general case. Let H be a hypersurface of X. A $\mathscr{D}_X(*H)$-module M is called smooth if $M_{|X\setminus H}$ is a locally free $\mathcal{O}_{X\setminus H}$-module. A smooth $\mathscr{D}_X(*H)$-triple is a tuple of smooth $\mathscr{D}_X(*H)$-modules V_i with a pairing $C_V : V_1 \otimes \overline{V_2} \longrightarrow \mathcal{C}_X^{\infty \bmod H}$. It is called non-degenerate, if $(V_1, V_2, C_V)_{|X\setminus H}$ is non-degenerate.

Let $\mathcal{T}_V = (V_1, V_2, C_V)$ be a non-degenerate smooth $\mathscr{D}_X(*H)$-triple. Let $\mathcal{T} = (M_1, M_2, C) \in \mathscr{D}\text{-Tri}^{nd}(X)$. We have the induced pairing

$$C \otimes C_V : (M_1 \otimes_{\mathcal{O}_X} V_1) \otimes \overline{(M_2 \otimes_{\mathcal{O}_X} V_2)} \longrightarrow \mathfrak{Db}_X^{\bmod H},$$

where $\mathfrak{Db}_X^{\bmod H}$ denotes the sheaf of distributions on X with moderate growth along H. We obtain the following sesqui-linear pairings, as in Sect. 12.2.2:

$$(C \otimes C_V)[*H] : (M_1 \otimes_{\mathcal{O}_X} V_1)[!H] \otimes \overline{(M_2 \otimes_{\mathcal{O}_X} V_2)[*H]} \longrightarrow \mathfrak{Db}_X$$

$$(C \otimes C_V)[!H] : (M_1 \otimes_{\mathcal{O}_X} V_1)[*H] \otimes \overline{(M_2 \otimes_{\mathcal{O}_X} V_2)[!H]} \longrightarrow \mathfrak{Db}_X$$

Thus, we obtain sesqui-linear pairings:

$$(\mathcal{T} \otimes \mathcal{T}_V)[*H] := \big((M_1 \otimes_{\mathcal{O}_X} V_1)[!H], (M_2 \otimes_{\mathcal{O}_X} V_2)[*H], (C \otimes C_V)[*H]\big)$$
$$(\mathcal{T} \otimes \mathcal{T}_V)[!H] := \big((M_1 \otimes_{\mathcal{O}_X} V_1)[*H], (M_2 \otimes_{\mathcal{O}_X} V_2)[!H], (C \otimes C_V)[!H]\big)$$

We clearly have

$$\boldsymbol{D}^{\text{herm}}\big((\mathcal{T} \otimes \mathcal{T}_V)[!H]\big) = \Big(\boldsymbol{D}^{\text{herm}}(\mathcal{T}) \otimes \boldsymbol{D}^{\text{herm}}(\mathcal{T}_V)\Big)[*H]$$

$$\boldsymbol{D}^{\text{herm}}\big((\mathcal{T} \otimes \mathcal{T}_V)[*H]\big) = \Big(\boldsymbol{D}^{\text{herm}}(\mathcal{T}) \otimes \boldsymbol{D}^{\text{herm}}(\mathcal{T}_V)\Big)[!H].$$

Lemma 12.2.5 $(\mathcal{T} \otimes \mathcal{T}_V)[\star H]$ $(\star = *, !)$ *are also non-degenerate.*

Proof Let us consider the case $\star = *$. By the previous lemma, the restriction of the induced morphism $(M_1 \otimes V_1)[!H] \longrightarrow \mathcal{C}_{\overline{X}}\big((M_2 \otimes V_2)[*H]\big)$ to $X \setminus H$ is an isomorphism. Hence, $(M_1 \otimes V_1)[!H] \longrightarrow \mathcal{C}_{\overline{X}}\big((M_2 \otimes V_2)[*H]\big)$ is an isomorphism. The other case can be shown similarly. □

12.2.4 Beilinson Functors for $\mathscr{D}$-Triples

Let us consider a meromorphic flat bundle $\tilde{\mathbb{I}}^{a,b} := \bigoplus_{a\leq j<b} \mathcal{O}_{\mathbb{C}_z}(*z)\, s^j$ on $(\mathbb{C}_z, 0)$ with a connection $z\partial_z s^j = s^{j+1}$. We have the non-degenerate sesqui-linear pairing $\tilde{\mathbb{I}}^{-b+1,-a+1}$ and $\tilde{\mathbb{I}}^{a,b}$ with values in $\mathcal{C}_{\mathbb{C}_z}^{\infty \bmod 0}$ given as follows:

$$\tilde{C}(s^i, \overline{s}^j) = \frac{(\log|z|^2)^{-i-j}}{(-i-j)!}\chi_{i+j\leq 0}.$$

Here, $\chi_{i+j\leq 0}$ is 1 if $i+j\leq 0$, or 0 otherwise. For a holomorphic function g on X, we set $\tilde{\mathbb{I}}_g^{a,b} := g^*\tilde{\mathbb{I}}^{a,b}$. For a $\mathscr{D}_X$-module M, we have a $\mathscr{D}_X(*g)$-module $\Pi_g^{a,b}M := M\otimes\tilde{\mathbb{I}}_g^{a,b}$. We set $\Pi_{g\star}^{a,b}M := (\Pi_g^{a,b}M)[\star g]$ for $\star = *,!$.

Let $\mathcal{T} = (M_1, M_2, C) \in \mathscr{D}\text{-Tri}(X)$. We obtain objects

$$\Pi_{g!}^{a,b}\mathcal{T} = \big(\Pi_{g*}^{-b+1,-a+1}M_1, \Pi_{g!}^{a,b}M_2, \Pi_{g!}^{a,b}C\big),$$
$$\Pi_{g*}^{a,b}\mathcal{T} = \big(\Pi_{g!}^{-b+1,-a+1}M_1, \Pi_{g*}^{a,b}M_2, \Pi_{g*}^{a,b}C\big),$$

in $\mathscr{D}\text{-Tri}(X)$. As in the case of $\mathcal{R}$-triples, we set

$$\Pi_{g*!}^{a,b}(\mathcal{T}) := \varprojlim_{N\to\infty} \mathrm{Cok}\Big(\Pi_{g!}^{b,N}(\mathcal{T}) \longrightarrow \Pi_{g*}^{a,N}(\mathcal{T})\Big)$$
$$\simeq \varinjlim_{N\to\infty} \mathrm{Ker}\Big(\Pi_{g!}^{-N,b}(\mathcal{T}) \longrightarrow \Pi_{g*}^{-N,a}(\mathcal{T})\Big) \qquad (12.32)$$

If $\mathcal{T}$ is non-degenerate, $\Pi_{g\star}^{a,b}(\mathcal{T})$ $(\star = *,!)$ and $\Pi_{g*!}^{a,b}(\mathcal{T})$ are also non-degenerate.

In particular, we define $\psi_g^{(a)}(\mathcal{T}) := \Pi_{g*!}^{a,a}(\mathcal{T})$ and $\Xi_g^{(a)}(\mathcal{T}) := \Pi_{g*!}^{a,a+1}(\mathcal{T})$. As in the case of $\mathcal{R}$-triples, we define $\phi_g^{(0)}(\mathcal{T})$ as the cohomology of the naturally obtained complex:

$$\mathcal{T}[!g] \longrightarrow \Xi_g^{(0)}(\mathcal{T}) \oplus \mathcal{T} \longrightarrow \mathcal{T}[*g]$$

We can recover $\mathcal{T}$ as the cohomology of the following complex:

$$\psi_g^{(1)}(\mathcal{T}) \longrightarrow \Xi_g^{(0)}(\mathcal{T}) \oplus \phi_g^{(0)}(\mathcal{T}) \longrightarrow \psi_g^{(0)}(\mathcal{T})$$

If $\mathcal{T}$ is non-degenerate, $\psi_g^{(a)}(\mathcal{T})$, $\Xi_g^{(a)}(\mathcal{T})$ and $\phi_g^{(0)}(\mathcal{T})$ are also non-degenerate. It is easy to see the following.

Lemma 12.2.6 *We also have* $\boldsymbol{D}^{\mathrm{herm}} \circ \prod_{g*!}^{a,b} = \prod_{g*!}^{-b+1,-a+1} \circ \boldsymbol{D}^{\mathrm{herm}}$. *In particular, we have* $\boldsymbol{D}^{\mathrm{herm}} \circ \Xi_g^{(a)} = \Xi_g^{(-a)} \circ \boldsymbol{D}^{\mathrm{herm}}$ *and* $\boldsymbol{D}^{\mathrm{herm}} \circ \psi_g^{(a)} = \psi_g^{(-a+1)} \circ \boldsymbol{D}^{\mathrm{herm}}$. *We also have* $\boldsymbol{D}^{\mathrm{herm}} \circ \phi_g^{(a)} = \phi_g^{(-a)} \circ \boldsymbol{D}^{\mathrm{herm}}$. □

Remark 12.2.7 For regular holonomic $\mathscr{D}$-modules, the compatibilities of the functors in *Sect. 12.2* are essentially contained in *[64]*. □

12.2.5 External Tensor Product

Let X_i $(i = 1,2)$ be complex manifolds. We set $X := X_1 \times X_2$. For any $\mathcal{O}_{X_i}$-modules M_i $(i = 1,2)$, we set $M_1 \otimes M_2 := q_1^*M_1 \otimes q_2^*M_2$. We have $\mathscr{D}_{X_1} \boxtimes \mathscr{D}_{X_2} = \mathscr{D}_X$. For $\mathscr{D}_{X_i}$-modules M_i $(i = 1,2)$, we naturally obtain the $\mathscr{D}_X$-module $M_1 \boxtimes M_2$.

Let $\mathcal{T}_1 = (M_1, M_2, C_M) \in \mathscr{D}\text{-Tri}(X_1)$ and $\mathcal{T}_2 = (N_1, N_2, C_N) \in \mathscr{D}\text{-Tri}(X_2)$. We have the $\mathscr{D}_X$-modules $M_i \boxtimes N_i$. We have the induced sesqui-linear pairing $C_M \boxtimes C_N$ of $M_1 \boxtimes N_1$ and $M_2 \boxtimes N_2$. In this way, we obtain $\mathcal{T}_1 \boxtimes \mathcal{T}_2 = (M_1 \boxtimes N_1, M_2 \boxtimes N_2, C_M \boxtimes C_N)$.

Lemma 12.2.8 *If $\mathcal{T}_i$ are non-degenerate, then $\mathcal{T}_1 \boxtimes \mathcal{T}_2$ is non-degenerate.*

Proof We have a natural morphism

$$\rho : \mathcal{C}_{X_1}(M_1) \boxtimes \mathcal{C}_{X_2}(N_1) \longrightarrow \mathcal{C}_{X_1 \times X_2}(M_1 \boxtimes N_1).$$

Applying the de Rham functor, it is easy to see that the induced morphism $\mathrm{DR}(\rho)$: $\mathrm{DR}_{\overline{X}_1} \mathcal{C}_{X_1}(M_1) \boxtimes \mathrm{DR}_{\overline{X}_2} \mathcal{C}_{X_2}(N_1) \longrightarrow \mathrm{DR}_{\overline{X}_1 \times \overline{X}_2} \mathcal{C}_{X_1 \times X_2}(M_1 \boxtimes N_1)$ is an isomorphism. Hence, ρ is also an isomorphism. Then, the claim of the lemma is clear. □

Let $\mathfrak{T}_1^\bullet = (M_1^\bullet, M_2^\bullet, C_{M^\bullet}) \in \mathcal{C}(\mathscr{D})\text{-Tri}(X_1)$ and $\mathfrak{T}_2^\bullet = (N_1^\bullet, N_2^\bullet, C_{N^\bullet}) \in \mathcal{C}(\mathscr{D})\text{-Tri}(X_2)$. We have the sesqui-linear pairings $C_{M^\bullet}^p$ of M_1^{-p} and M_2^p, and $C_{N^\bullet}^q$ of N_1^{-q} and N_2^q. Applying the procedure in Sect. 12.1.1.3, we define

$$\mathrm{Tot}\big(\mathfrak{T}_1^\bullet \boxtimes \mathfrak{T}_2^\bullet\big) = \big(\mathrm{Tot}(M_1^\bullet \boxtimes N_1^\bullet), \mathrm{Tot}(M_2^\bullet \boxtimes N_2^\bullet), C_{M^\bullet} \boxtimes C_{N^\bullet}\big)$$

by the formula $(C_{M^\bullet} \boxtimes C_{N^\bullet})^m = \bigoplus_{p+q=m} (-1)^{pq} C_{M^\bullet}^p \boxtimes C_{N^\bullet}^q$. We can easily check that $\mathcal{S}_\ell \,\mathrm{Tot}(\mathfrak{T}_1^\bullet \boxtimes \mathfrak{T}_2^\bullet) = \mathrm{Tot}\big(\mathcal{S}_\ell(\mathfrak{T}_1^\bullet) \boxtimes \mathfrak{T}_2^\bullet\big) = \mathrm{Tot}\big(\mathfrak{T}_1^\bullet \boxtimes \mathcal{S}_\ell(\mathfrak{T}_2^\bullet)\big)$.

Let $\mathcal{T}_i^\bullet \in \mathcal{C}\big(\mathscr{D}\text{-Tri}(X_i)\big)$ $(i = 1, 2)$. We have $\mathcal{T}_1^p \boxtimes \mathcal{T}_2^q \in \mathscr{D}\text{-Tri}(X)$. Let $\mathcal{T}_1^p = (M_1^{-p}, M_2^p, C_M^p)$ and $\mathcal{T}_2^q = (N_1^{-q}, N_2^q, C_N^q)$. We set $\tilde{\mathcal{T}}^m = \bigoplus_{p+q=m} \mathcal{T}_1^p \boxtimes \mathcal{T}_2^q$. We have the complexes of $\mathscr{D}_X$-modules $\mathrm{Tot}(M_i^\bullet \boxtimes N_i^\bullet)$. In particular, we have the morphisms $\bigoplus_{p+q=m+1} M_1^{-p} \boxtimes N_1^{-q} \longrightarrow \bigoplus_{p+q=m} M_1^{-p} \boxtimes N_1^{-q}$ and $\bigoplus_{p+q=m} M_2^p \boxtimes N_2^q \longrightarrow \bigoplus_{p+q=m+1} M_2^p \boxtimes N_2^q$. They give a morphism $\delta^m : \tilde{\mathcal{T}}^m \longrightarrow \tilde{\mathcal{T}}^{m+1}$, and we obtain a complex $\mathrm{Tot}(\mathcal{T}_1 \boxtimes \mathcal{T}_2) := \tilde{\mathcal{T}}^\bullet$. We have $\mathrm{Tot}(\mathcal{T}_1 \boxtimes \mathcal{T}_2)[\ell] = \mathrm{Tot}(\mathcal{T}_1[\ell] \boxtimes \mathcal{T}_2) = \mathrm{Tot}(\mathcal{T}_1 \boxtimes \mathcal{T}_2[\ell])$. It is easy to check that $\Psi_{1X_1}(\mathcal{T}_1) \boxtimes \Psi_{1X_2}(\mathcal{T}_2) = \Psi_{1X_1 \times X_2}(\mathcal{T}_1 \boxtimes \mathcal{T}_2)$.

12.2.5.1 Push-Forward

Let $f_i : X_i \longrightarrow Y_i$ be morphisms of complex manifolds. Let $f : X_1 \times X_2 \longrightarrow Y_1 \times Y_2$ be the induced morphism. Let $\mathcal{T}_i = (M_{i1}, M_{i2}, C_i) \in \mathscr{D}\text{-Tri}(X_i)$ such that the support of $\mathcal{T}_i$ is proper with respect to f_i. The following lemma can be checked by direct computations.

Lemma 12.2.9 *The natural isomorphisms $f_\dagger(M_{1j} \boxtimes M_{2j}) \simeq f_{1\dagger}(M_{1j}) \boxtimes f_{2\dagger}(M_{2j})$ induce $f_\dagger^{(0)}(\mathcal{T}_1 \boxtimes \mathcal{T}_2) \simeq \mathrm{Tot}\big(f_{1\dagger}^{(0)}(\mathcal{T}_1) \boxtimes f_{2\dagger}^{(0)}(\mathcal{T}_2)\big)$ and $f_\dagger(\mathcal{T}_1 \boxtimes \mathcal{T}_2) \simeq \mathrm{Tot}\big(f_{1\dagger}(\mathcal{T}_1) \boxtimes f_{2\dagger}(\mathcal{T}_2)\big)$.* □

We have the naturally induced isomorphisms of $\mathscr{D}_Y$-modules:

$$f_\dagger^m(M_{1j} \boxtimes M_{2j}) \simeq \bigoplus_{k+\ell=m} f_{1\dagger}^k(M_{1j}) \boxtimes f_{2\dagger}^\ell(M_{2j}) \tag{12.33}$$

Corollary 12.2.10 *The isomorphisms* (12.33) *induce the isomorphisms of $\mathscr{D}_Y$-triples $f_{\dagger}^{m}(\mathcal{T}_1 \boxtimes \mathcal{T}_2) \simeq \bigoplus_{k+\ell=m} f_{1\dagger}^{k}(\mathcal{T}_1) \boxtimes f_{2\dagger}^{\ell}(\mathcal{T}_2)$.* □

12.2.5.2 Twist

Let H_i be hypersurfaces of X_i $(i = 1, 2)$. Let $\mathcal{V}_i$ be smooth $\mathscr{D}_{X_i}(*H_i)$-triples. Let $\mathcal{T}_i \in \mathscr{D}\text{-Tri}(X_i)$. Let $\star_i \in \{*, !\}$.

Lemma 12.2.11 *We have the natural isomorphisms*

$$
\begin{aligned}
&(\mathcal{T}_1 \otimes \mathcal{V}_1)[\star_1 H_1] \boxtimes (\mathcal{T}_2 \otimes \mathcal{V}_2)[\star_2 H_2] \\
&\quad \simeq \big((\mathcal{T}_1 \boxtimes \mathcal{T}_2) \otimes (\mathcal{V}_1 \boxtimes \mathcal{V}_2)\big)[\star_1(H_1 \times X_2)][\star_2(X_1 \times H_2)]
\end{aligned}
\tag{12.34}
$$

Proof Let $\mathcal{T}_3$ denote the left hand side, and $\mathcal{T}_4$ denote the right hand side. Let $H := (H_1 \times X_2) \cup (X_1 \times H_2)$. We have $\mathcal{T}_3(*H) = \mathcal{T}_4(*H)$. It induces the isomorphisms of the underlying $\mathscr{D}$-modules of $\mathcal{T}_3$ and $\mathcal{T}_4$. It is easy to see that the isomorphisms are compatible with the sesqui-linear pairings. □

Corollary 12.2.12 *Let g_1 be any holomorphic function on X_1. Let g be the induced holomorphic function on X. Let $\mathcal{T}_i \in \mathscr{D}$-Tri$(X_i)$. We have natural isomorphisms $\Pi^{a,b}_{g_1 *,!}(\mathcal{T}_1) \boxtimes \mathcal{T}_2 \simeq \Pi^{a,b}_{g*!}(\mathcal{T}_1 \boxtimes \mathcal{T}_2)$. In particular, we have $\psi^{(a)}_{g_1}(\mathcal{T}_1) \boxtimes \mathcal{T}_2 \simeq \psi^{(a)}_{g}(\mathcal{T}_1 \boxtimes \mathcal{T}_2)$ and $\Xi^{(a)}_{g_1}(\mathcal{T}_1) \boxtimes \mathcal{T}_2 \simeq \Xi^{(a)}_{g}(\mathcal{T}_1 \boxtimes \mathcal{T}_2)$. We also have $\phi^{(a)}_{g_1}(\mathcal{T}_1) \boxtimes \mathcal{T}_2 \simeq \phi^{(a)}_{g}(\mathcal{T}_1 \boxtimes \mathcal{T}_2)$.* □

12.3 De Rham Functor

12.3.1 $\mathbb{C}_X$-Complex-Triples

For any $\mathbb{C}_X$-sheaf $\mathcal{F}$, let $\overline{\mathcal{F}}$ denote the complex conjugate of $\mathcal{F}$. Namely, we have $\overline{\mathcal{F}} = \mathcal{F}$ as a sheaf, and $\alpha \cdot a = \overline{\alpha} a$ for local sections $\alpha \in \mathbb{C}_X$ and $a \in \overline{\mathcal{F}} = \mathcal{F}$.

Let $I_X^{\bullet}$ be a $\mathbb{C}_X$-complex. A $(\mathbb{C}_X, I_X^{\bullet})$-complex-triple is a tuple $(\mathcal{F}_1^{\bullet}, \mathcal{F}_2^{\bullet}, C)$ which consists of $\mathbb{C}_X$-complexes $\mathcal{F}_i^{\bullet}$ and a morphism of complexes $C : \mathcal{F}_1^{\bullet} \otimes \overline{\mathcal{F}}_2^{\bullet} \longrightarrow I_X^{\bullet}$, i.e., such that $dC(x^i \otimes y^j) = C(dx^i \otimes y^j) + (-1)^i C(x^i \otimes dy^j)$.

Let $I_X^{\bullet,\bullet}$ be a $\mathbb{C}_X$-double-complex. A $(\mathbb{C}_X, I_X^{\bullet,\bullet})$-double-complex-triple is a tuple $(\mathcal{F}_1^{\bullet,\bullet}, \mathcal{F}_2^{\bullet,\bullet}, C)$ which consists of $\mathbb{C}_X$-double complexes $\mathcal{F}_i^{\bullet,\bullet}$ $(i = 1, 2)$ and a morphism of double complexes $C : \mathcal{F}_1^{\bullet,\bullet} \otimes \overline{\mathcal{F}}_2^{\bullet,\bullet} \longrightarrow I_X^{\bullet,\bullet}$:

$$d_1 C(x^{\boldsymbol{p}}, y^{\boldsymbol{q}}) = C(d_1 x^{\boldsymbol{p}}, y^{\boldsymbol{q}}) + (-1)^{p_1} C(x^{\boldsymbol{p}}, d_1 y^{\boldsymbol{q}})$$

$$d_2 C(x^{\boldsymbol{p}}, y^{\boldsymbol{q}}) = C(d_2 x^{\boldsymbol{p}}, y^{\boldsymbol{q}}) + (-1)^{p_2} C(x^{\boldsymbol{p}}, d_2 y^{\boldsymbol{q}})$$

Let $J_X^\bullet$ be the total complex of $I_X^{\bullet,\bullet}$. For any $(\mathbb{C}_X, I_X^{\bullet,\bullet})$-double-complex-triple $(\mathcal{F}_1^{\bullet,\bullet}, \mathcal{F}_2^{\bullet,\bullet}, C)$, we have the associated $(\mathbb{C}_X, J_X^\bullet)$-complex-triple given as follows. The underlying $\mathbb{C}_X$-complexes are the total complexes of $\mathcal{F}_i^{\bullet,\bullet}$. We set $\tilde{C}(x^{\boldsymbol{p}}, y^{\boldsymbol{q}}) := (-1)^{p_2 q_1} C(x^{\boldsymbol{p}}, y^{\boldsymbol{q}})$. We can check that they give a $(\mathbb{C}_X, J_X^\bullet)$-complex-triple by a direct computation, as follows:

$$
\begin{aligned}
d\tilde{C}(x^{\boldsymbol{p}}, y^{\boldsymbol{q}}) &= d_1\tilde{C}(x^{\boldsymbol{p}}, y^{\boldsymbol{q}}) + (-1)^{p_1+q_1} d_2\tilde{C}(x^{\boldsymbol{p}}, y^{\boldsymbol{q}}) \\
&= (-1)^{p_2 q_1} d_1 C(x^{\boldsymbol{p}}, y^{\boldsymbol{q}}) + (-1)^{p_1+q_1+p_2 q_1} d_2 C(x^{\boldsymbol{p}}, y^{\boldsymbol{q}}) \\
&= (-1)^{p_2 q_1}\Big(C(d_1 x^{\boldsymbol{p}}, y^{\boldsymbol{q}}) + (-1)^{p_1} C(x^{\boldsymbol{p}}, d_1 y^{\boldsymbol{q}})\Big) \\
&\quad + (-1)^{p_1+q_1+p_2 q_1}\Big(C(d_2 x^{\boldsymbol{p}}, y^{\boldsymbol{q}}) + (-1)^{p_2} C(x^{\boldsymbol{p}}, d_2 y^{\boldsymbol{q}})\Big) \\
&= (-1)^{p_2 q_1} C(d_1 x^{\boldsymbol{p}}, y^{\boldsymbol{q}}) + (-1)^{p_1+p_2+(q_1+1)p_2} C(x^{\boldsymbol{p}}, d_1 y^{\boldsymbol{q}}) \\
&\quad + (-1)^{(p_2+1)q_1} C\big((-1)^{p_1} d_2 x^{\boldsymbol{p}}, y^{\boldsymbol{q}}\big) + (-1)^{p_1+p_2+p_2 q_1} C(x^{\boldsymbol{p}}, (-1)^{q_1} d_2 y^{\boldsymbol{q}}) \\
&= \tilde{C}(d_1 x^{\boldsymbol{p}}, y^{\boldsymbol{q}}) + \tilde{C}((-1)^{p_1} d_2 x^{\boldsymbol{p}}, y^{\boldsymbol{q}}) \\
&\quad + (-1)^{p_1+p_2}\Big(\tilde{C}(x^{\boldsymbol{p}}, d_1 y^{\boldsymbol{q}}) + \tilde{C}(x^{\boldsymbol{p}}, (-1)^{q_1} d_2 y^{\boldsymbol{q}})\Big) \\
&= \tilde{C}(dx^{\boldsymbol{p}}, y^{\boldsymbol{q}}) + (-1)^{p_1+p_2}\tilde{C}(x^{\boldsymbol{p}}, dy^{\boldsymbol{q}}) = 0 \qquad (12.35)
\end{aligned}
$$

For any $(\mathbb{C}_X, I_X^\bullet)$-triple $(\mathcal{F}_1^\bullet, \mathcal{F}_2^\bullet, C)$ and any $\ell \in \mathbb{Z}$, we define the shift $(\mathcal{F}_1^\bullet[-\ell], \mathcal{F}_2^\bullet[\ell], C[\ell])$ as in Sect. 12.1.5.4, which is denoted by $\mathcal{S}_\ell(\mathcal{F}_1^\bullet, \mathcal{F}_2^\bullet, C)$.

12.3.2 De Rham Functor for $\mathscr{D}_X$-Complex-Triples

Put $\omega_X^{top} := \mathfrak{Db}_X^\bullet[2d_X]$. We shall use the identification $\mathrm{Tot}\Big(\Omega_X^\bullet[d_X] \otimes \Omega_{\overline{X}}^\bullet[d_X] \otimes \mathfrak{Db}_X\Big) \simeq \omega_X^{top}$ given as in Sect. 12.1.4.1. For a $\mathscr{D}_X$-module M, we have the naturally defined $\mathscr{D}_{\overline{X}}$-module $\overline{M}$. We have the complex $\mathrm{DR}_{\overline{X}}(\overline{M})$. We have a natural anti-$\mathbb{C}$-linear isomorphism $\mathrm{DR}_X(M) \simeq \mathrm{DR}_{\overline{X}}(\overline{M})$ with which we often identify $\mathrm{DR}_{\overline{X}}(\overline{M})$ with the conjugate $\overline{\mathrm{DR}_X(M)}$.

For a $\mathscr{D}_X$-triple (M_1, M_2, C), we define the pairing

$$\mathrm{DR}(C) : \mathrm{DR}_X(M_1) \otimes \mathrm{DR}_{\overline{X}}(\overline{M}_2) \longrightarrow \omega_X^{top}$$

by $\mathrm{DR}(C)(\eta^{d_X+p}\, m_1, \overline{\eta^{d_X+q}\, m_2}) := \eta^{d_X+p}\overline{\eta}^{d_X+q}\, C(m_1, \overline{m}_2)\,\epsilon(d_X)(-1)^{pd_X}$. We obtain a $(\mathbb{C}_X, \omega_X^{top})$-complex-triple

$$\mathrm{DR}_X(M_1, M_2, C) := \big(\mathrm{DR}_X(M_1), \mathrm{DR}_X(M_2), \mathrm{DR}(C)\big),$$

where we use $\overline{\mathrm{DR}_X(M_2)} \simeq \mathrm{DR}_{\overline{X}}(\overline{M}_2)$.

Let $(M_1^\bullet, M_2^\bullet, C)$ be a $\mathscr{D}_X$-complex-triple. The de Rham functor induces the $(\mathbb{C}_X, \omega_X^{top})$-double-complex-triple $\big(\mathrm{DR}_X(M_1^\bullet), \mathrm{DR}_X(M_2^\bullet), \mathrm{DR}(C)\big)$. Thus, we obtain a $(\mathbb{C}_X, \omega_X^{top})$-complex-triple

$$\mathrm{DR}(M_1^\bullet, M_2^\bullet, C) := \mathrm{Tot}\big(\mathrm{DR}_X(M_1^\bullet), \mathrm{DR}_X(M_2^\bullet), \mathrm{DR}(C)\big).$$

12.3.3 Compatibility with the Shift

We can check the following lemma by a direct computation.

Lemma 12.3.1 *For any $\ell \in \mathbb{Z}$, we have $\mathcal{S}_\ell \,\mathrm{DR} \simeq \mathrm{DR} \circ \mathcal{S}_\ell$ induced by the natural identifications* $\mathrm{DR}(M_1[-\ell]) \simeq \mathrm{DR}(M_1)[-\ell]$ *and* $\mathrm{DR}(M_2[\ell]) \simeq \mathrm{DR}(M_2)[\ell]$.

Proof We have

$$\begin{aligned}
&\mathrm{Tot}\,\mathrm{DR}(C[\ell])\big(\eta_1^{d_X+p} m_1^{-\ell+a}, \overline{\eta_2^{d_X+q} m_2^{\ell+b}}\big)\\
&\quad = (-1)^{aq+pd_X+\ell a}\epsilon(d_X)\epsilon(\ell)\eta_1^{d_X+p}\overline{\eta_2^{d_X+q}}C(m_1^{-\ell+a}, \overline{m_2^{\ell+b}})
\end{aligned} \tag{12.36}$$

We also have

$$\begin{aligned}
&\big(\mathrm{Tot}\,\mathrm{DR}\, C\big)[\ell]\big(\eta_1^{d_X+p} m_1^{-\ell+a}, \overline{\eta_2^{d_X+q} m_2^{\ell+b}}\big)\\
&\quad = (-1)^{(\ell+a)q+pd_X+\ell(a+p)}\epsilon(d_X)\epsilon(\ell)\eta_1^{d_X+p}\overline{\eta_2^{d_X+q}}C(m_1^{-\ell+a}, \overline{m_2^{\ell+b}})
\end{aligned} \tag{12.37}$$

In the natural isomorphisms $\mathrm{DR}(M_1^\bullet[-\ell]) \simeq \mathrm{DR}(M_1^\bullet)[-\ell]$ and $\mathrm{DR}(M_2^\bullet[\ell]) \simeq \mathrm{DR}(M_2^\bullet)[\ell]$, the signatures $(-1)^{\ell p}$ and $(-1)^{\ell q}$ appear. Hence, the sesqui-linear pairings are equal under the identifications. □

12.3.4 Compatibility with the Hermitian Adjoint

For complexes $A^\bullet$ and $B^\bullet$, let $\mathrm{ex} : \mathrm{Tot}(A^\bullet \otimes B^\bullet) \longrightarrow \mathrm{Tot}(B^\bullet \otimes A^\bullet)$ be given by $A^p \otimes B^q \longrightarrow B^q \otimes A^p$; $a \otimes b \longmapsto (-1)^{pq} b \otimes a$. For a $(\mathbb{C}_X, \omega_X^{top})$-triple $(\mathcal{F}_1^\bullet, \mathcal{F}_2^\bullet, C)$, we have the $(\mathbb{C}_X, \omega_X^{top})$-triple

$$\boldsymbol{D}^{\mathrm{herm}}(\mathcal{F}_1^\bullet, \mathcal{F}_2^\bullet, C) := (\mathcal{F}_2^\bullet, \mathcal{F}_1^\bullet, C^*),$$

where $C^* := (-1)^{d_X}\overline{C} \circ \mathrm{ex}$. Here, the complex conjugate on $\mathfrak{Db}_X^\bullet[2d_X]$ is the ordinary one. It is called the Hermitian adjoint of $(\mathcal{F}_1, \mathcal{F}_2, C)$.

Lemma 12.3.2 *We have $\boldsymbol{D}^{\mathrm{herm}}\,\mathrm{DR} = \mathrm{DR}\,\boldsymbol{D}^{\mathrm{herm}}$. Namely, for any $\mathscr{D}_X$-complex-triple $(M_1^\bullet, M_2^\bullet, C)$, we have $(-1)^{d_X}\overline{\mathrm{Tot}\,\mathrm{DR}(C)} \circ \mathrm{ex} = \mathrm{Tot}\,\mathrm{DR}(C^*)$.*

Proof Let η^a be local sections of Ω^a, and m_i^ℓ be local sections of M_i^ℓ. We have the following equalities:

$$\begin{aligned}
&\mathrm{Tot}\big(\mathrm{DR}(C^*)\big)\big(\eta^{d_X+p} m_2^\ell,\ \overline{\eta^{d_X+q} m_1^{-\ell}}\big) \\
&= (-1)^{\ell q}\, \mathrm{DR}(C^*)\big(\eta^{d_X+p} m_2^\ell,\ \overline{\eta^{d_X+q} m_1^{-\ell}}\big) \\
&= \eta^{d_X+p}\overline{\eta}^{d_X+q}\, C^*\big(m_2^\ell, \overline{m}_1^{-\ell}\big)\, \epsilon(d_X)\, (-1)^{pd_X+\ell q} \\
&= \overline{\eta}^{d_X+q}\eta^{d_X+p}\,\overline{C(m_1^{-\ell}, \overline{m}_2^\ell)}\epsilon(d_X)\, (-1)^{(p+d_X)(q+d_X)+pd_X+\ell q+\ell} \\
&= \overline{\mathrm{Tot}\,\mathrm{DR}(C)(\eta^{d_X+q} m_1^{-\ell}, \overline{\eta^{d_X+p} m_2^\ell})}\, (-1)^{qd_X+pd_X+(p+d_X)(q+d_X)+\ell p+\ell q+\ell} \\
&= \overline{\mathrm{Tot}\,\mathrm{DR}(C)(\eta^{q+d_X} m_1^{-\ell}, \overline{\eta^{p+d_X} m_2^\ell})}\, (-1)^{d_X+(\ell+p)(-\ell+q)}
\end{aligned} \tag{12.38}$$

Thus, we are done. □

12.3.5 Compatibility with the Push-Forward

Let $f : X \longrightarrow Y$ be a morphism of complex manifolds. We have the trace morphism $\mathrm{tr} : f_! \omega_X^{top} \longrightarrow \omega_Y^{top}$, which is given by the natural integral $f_! \mathfrak{Db}_X^\bullet[2d_X] \longrightarrow \mathfrak{Db}_Y^\bullet[2d_Y]$ multiplied with $(2\pi\sqrt{-1})^{-d_X+d_Y}$. Let $(\mathcal{F}_1, \mathcal{F}_2, C)$ be a c-soft $(\mathbb{C}_X, \omega_X^{top})$-complex-triple. The following naturally induced pairing is denoted by $f_* C$:

$$f_! \mathcal{F}_1 \otimes f_! \overline{\mathcal{F}_2} \longrightarrow f_! \omega_X^{top} \xrightarrow{\ \mathrm{tr}\ } \omega_Y^{top}$$

We obtain a $(\mathbb{C}_Y, \omega_Y^{top})$-complex triple $f_!(\mathcal{F}_1, \mathcal{F}_2, C) := (f_! \mathcal{F}_1, f_! \mathcal{F}_2, f_* C)$.

Proposition 12.3.3 *Take $\mathfrak{T} = (M_1^\bullet, M_2^\bullet, C) \in C(\mathscr{D})$-Tri$(X)$ such that the restriction of f to the cohomological support of $\mathfrak{T}$ is proper. We have a natural quasi-isomorphism*

$$\mathrm{DR}_Y f_\dagger^{(0)}(M_1^\bullet, M_2^\bullet, C) \simeq f_! \circ \mathrm{DR}_X(M_1^\bullet, M_2^\bullet, C).$$

Proof We have the standard quasi-isomorphisms of the underlying complexes. We have only to compare the pairings. We have only to consider the issue for a $\mathscr{D}_X$-triple (M_1, M_2, C). Let us argue the case that f is a closed immersion. Let η_N be a local generator of ω_Y/ω_X. The quasi-isomorphism $f_* \mathrm{DR}_X(M) \longrightarrow \mathrm{DR}_Y(f_\dagger M)$ is given by $\xi \otimes m \longmapsto \eta_N \cdot \xi \otimes (m/\eta_N)$. Let $\ell = d_Y - d_X$. Then, we have the following:

$$\begin{aligned}
&\mathrm{DR} f_\dagger^{(0)} C\big(\eta_N\, \xi^p \otimes (m_1/\eta_N),\ \overline{\eta_N \otimes \xi^q (m_2/\eta_N)}\big) \\
&= \eta_N \overline{\eta}_N \xi^p \overline{\xi}^q f_\dagger^{(0)}\big(C(m_1/\eta_N, \overline{m_2/\eta_N})\big)\epsilon(d_Y)(-1)^{(p+\ell-d_Y)d_Y+\ell p}
\end{aligned}$$

$$
\begin{aligned}
&= f_*\Big(\xi^p\overline{\xi}^q C(m_1,\overline{m}_2)\Big)\epsilon(d_Y)\epsilon(d_Y-d_X)(-1)^{(p+\ell-d_Y)d_Y+\ell p}(2\pi\sqrt{-1})^\ell \\
&= f_*\,\mathrm{DR}(C)\big(\xi^p\,m_1,\overline{\xi^q\,m_2}\big)\epsilon(d_X)(-1)^{(p-d_X)d_X}\epsilon(d_Y)(-1)^{(p+\ell-d_Y)d_Y+\ell p}\epsilon(d_Y-d_X)
\end{aligned}
\tag{12.39}
$$

It is easy to see $\epsilon(d_X)\epsilon(d_Y)\epsilon(d_Y-d_X)(-1)^{(p-d_X)d_X+(p+\ell-d_Y)d_Y+\ell p}=1$. Hence, we are done in this case.

Let us consider the case $f: X=Z\times Y\longrightarrow Y$. Let $\dim Z=m$ and $\dim Y=n$. We have the following:

$$
\begin{aligned}
&\mathrm{DR}_Y(f_\dagger^{(0)}C)\big([m_1\eta^{m-p}]\,\omega^r,\ \overline{[m_2\eta^{m+p}]\,\omega^s}\big)\,(-1)^{p(n-r)} \\
&= f_\dagger^{(0)}C\big([m_1\eta^{m-p}],\ \overline{[m_2\eta^{m+p}]}\big)\cdot\omega^r\overline{\omega}^s\epsilon(n)(-1)^{(r-n)n+p(n-r)} \\
&= \int \eta^{m-p}\overline{\eta}^{m+p}C(m_1,\overline{m}_2)\,\omega^r\overline{\omega}^s\epsilon(m)\epsilon(n)(-1)^{pm+(r-n)n+p(n-r)}\left(\frac{1}{2\pi\sqrt{-1}}\right)^m \\
&= \int \eta^{m-p}\omega^r\overline{\eta}^{m+p}\overline{\omega}^s C(m_1,\overline{m}_2) \\
&\qquad\times\epsilon(m)\epsilon(n)(-1)^{r(m+p)-pm+n(r-n)+p(n-r)}\left(\frac{1}{2\pi\sqrt{-1}}\right)^m
\end{aligned}
\tag{12.40}
$$

We also have the following:

$$
\begin{aligned}
&f_*\,\mathrm{DR}(C)\big(m_1\eta^{m-p}\omega^r,\ \overline{m_2\eta^{m+p}\omega^s}\big) \\
&= \int_f\big(\eta^{m-p}\omega^r\overline{\eta^{m+p}\omega^s}C(m_1,\overline{m}_2)\big)\epsilon(m+n) \\
&\qquad(-1)^{(m+n)(r+m-p-m-n)}\left(\frac{1}{2\pi\sqrt{-1}}\right)^m
\end{aligned}
\tag{12.41}
$$

It is easy to check that (12.40) and (12.41) are equal. Hence, we have $\mathrm{DR}(F_\dagger C)=F_*\,\mathrm{DR}(C)$. □

12.3.6 Compatibility with the External Tensor Product

Let X_i $(i=1,2)$ be complex manifolds. We set $X:=X_1\times X_2$. We have the isomorphism $\Omega^\bullet_{X_1}[d_{X_1}]\boxtimes\Omega^\bullet_{X_2}[d_{X_2}]\simeq\Omega^\bullet_X[d_X]$ given by $\eta^{d_{X_1}+p}\otimes\xi^{d_{X_2}+q}\longmapsto(-1)^{d_{X_2}(p+d_{X_1})}\eta^{d_{X_1}+p}\wedge\xi^{d_{X_2}+q}$. It induces an isomorphism $\mathrm{DR}_X(M_1\boxtimes M_2)\simeq\mathrm{DR}_{X_1}(M_1)\boxtimes\mathrm{DR}_{X_2}(M_2)$ for any $\mathscr{D}_{X_i}$-modules M_i. We can check the following lemma by elementary computations.

Lemma 12.3.4 *Let $\mathcal{T}_i \in \mathscr{D}$-Tri$(X_i)$ $(i = 1, 2)$. We have the isomorphism*

$$\mathrm{DR}_X(\mathcal{T}_1 \boxtimes \mathcal{T}_2) \simeq \mathrm{DR}_{X_1}(\mathcal{T}_1) \boxtimes \mathrm{DR}_{X_2}(\mathcal{T}_2)$$

induced by the above isomorphisms of the underlying $\mathscr{D}_X$-modules and the natural morphism $\mathfrak{Db}^{\bullet}_{X_1}[2d_{X_1}] \boxtimes \mathfrak{Db}^{\bullet}_{X_2}[2d_{X_2}] \longrightarrow \mathfrak{Db}^{\bullet}_X[2d_X]$. □

12.4 Duality of $\mathscr{D}$-Triples

12.4.1 Duality for Non-degenerate $\mathscr{D}$-Triples

We set $d_X := \dim X$. For a $\mathscr{D}_X$-module M, we have the object

$$\boldsymbol{D}_X M := R\mathcal{H}om_{\mathscr{D}_X}(M, \mathscr{D}_X \otimes \omega_X^{-1})[d_X]$$

in the derived category $D^b(\mathscr{D}_X)$. If M is holonomic, we have a natural identification $\boldsymbol{D}_X M \simeq \mathcal{H}^0 \boldsymbol{D}_X M$, and it is holonomic. Because we have the natural isomorphisms

$$\mathrm{DR}_{\overline{X}} \mathcal{C}_X(M) \simeq \mathcal{H}om_{\mathscr{D}_X}(M, \mathfrak{Db}_X^{0,\bullet})[d_X],$$
$$\mathrm{DR}_X \boldsymbol{D}_X M \simeq R\mathcal{H}om_{\mathscr{D}_X}(M, \mathcal{O}_X)[d_X],$$

we have a natural isomorphism $\nu_M : \mathrm{DR}_{\overline{X}} \mathcal{C}_X(M) \simeq \mathrm{DR}_X \boldsymbol{D}_X M$ in $D^b_c(\mathbb{C}_X)$, induced by the canonical isomorphism $\mathfrak{Db}_X^{0,\bullet} \simeq \mathcal{O}_X$ in $D^b(\mathscr{D}_X)$.

Let $(M_1, M_2, C) \in \mathscr{D}$-Tri$^{nd}(X)$. We have the induced isomorphism $\varphi_{C,1,2} : \overline{M}_2 \simeq \mathcal{C}_X(M_1)$. Hence, we have an isomorphism

$$\nu_{M_1} \circ \mathrm{DR}\,\varphi_{C,1,2} : \mathrm{DR}_{\overline{X}}(\overline{M}_2) \simeq \mathrm{DR}_X \boldsymbol{D}_X(M_1).$$

Similarly, we also have an induced isomorphism

$$\nu_{\overline{M}_2} \circ \mathrm{DR}\,\varphi_{C,2,1} : \mathrm{DR}_X(M_1) \simeq \mathrm{DR}_{\overline{X}} \boldsymbol{D}_{\overline{X}}(\overline{M}_2).$$

We will prove the following proposition in Sect. 12.5.

Theorem 12.4.1 *There uniquely exists a non-degenerate sesqui-linear pairing $\boldsymbol{D}C : \boldsymbol{D}_X M_1 \times \boldsymbol{D}_{\overline{X}} \overline{M}_2 \longrightarrow \mathfrak{Db}_X$ such that the following diagram is commutative.*

$$\begin{array}{ccc} \mathrm{DR}_X \boldsymbol{D}_X M_1 \otimes \mathrm{DR}_{\overline{X}} \boldsymbol{D}_{\overline{X}} \overline{M}_2 & \xrightarrow{\mathrm{DR}(\boldsymbol{D}C)} & \omega_X^{top} \\ \simeq \big\downarrow \kappa & & = \big\downarrow \\ \mathrm{DR}_{\overline{X}}(\overline{M}_2) \otimes \mathrm{DR}_X(M_1) & \xrightarrow{\mathrm{DR}(C) \circ \mathrm{ex}} & \omega_X^{top} \end{array} \tag{12.42}$$

Here, κ is induced by $\nu_{M_1} \circ \mathrm{DR}_X(\varphi_{C,1,2})$ and $\nu_{\overline{M}_2} \circ \mathrm{DR}_X(\varphi_{C,2,1})$, and ex *denotes the isomorphism $A^\bullet \otimes B^\bullet \simeq B^\bullet \otimes A^\bullet$ given by $x \otimes y \longmapsto (-1)^{\deg(x)\deg(y)} y \otimes x$.*

The correspondence $\mathcal{T} = (M_1, M_2, C) \longmapsto \boldsymbol{D}\mathcal{T} := (\boldsymbol{D}M_1, \boldsymbol{D}M_2, \boldsymbol{D}C)$ gives a contravariant exact functor on $\mathscr{D}$-$\mathrm{Tri}^{nd}(X)$. We have $\boldsymbol{D}(\boldsymbol{D}C) = C$ under the natural identifications $\boldsymbol{D}(\boldsymbol{D}M_i) = M_i$.

Note that we have only to show the existence of a sesqui-linear pairing $\boldsymbol{D}C$ such that the diagram (12.42) is commutative. The other claims immediately follow.

Example 12.4.2 Let us consider the case that M_i $(i = 1,2)$ are smooth $\mathscr{D}$-modules, i.e., the case in which M_i are locally free $\mathcal{O}_X$-modules. Let L_i $(i = 1,2)$ be the corresponding local systems. Then, we have a natural quasi-isomorphism $L_i[d_X] \simeq \mathrm{DR}_X(M_i)$. The non-degenerate sesqui-linear pairing $C : M_1 \times \overline{M}_2 \longrightarrow \mathfrak{Db}_X$ corresponds to a non-degenerate sesqui-linear pairing $C_0 : L_1 \otimes \overline{L}_2 \longrightarrow \mathbb{C}_X$. We have the dual $C_0^\vee : L_1^\vee \otimes \overline{L}_2^\vee \longrightarrow \mathbb{C}_X$.

Let us observe that the dual $\boldsymbol{D}C$ corresponds to $(-1)^{d_X} C_0^\vee$. The pairing $\mathrm{DR}(C)$ on $L_1[d_X] \otimes \overline{L}_2[d_X]$ is given by $\epsilon(d_X)(-1)^{d_X} C_0$. The signature comes from that of $\Omega_X^\bullet[d_X] \otimes \Omega_{\overline{X}}^\bullet[d_X] \longrightarrow \mathfrak{Db}_X^\bullet[2d_X]$. Hence, $\mathrm{DR}_X(C) \circ \mathrm{ex}$ on $\overline{L}_2[d_X] \otimes L_1[d_X]$ is given by $(\overline{b}, a) \longmapsto \epsilon(d_X) C_0(a, \overline{b})$. Let $\Psi_{C_0 21} : L_1^\vee \simeq \overline{L}_2$ and $\Psi_{C_0 12} : L_2^\vee \simeq \overline{L}_1$ denote the morphisms induced by the non-degenerate pairing C_0. The isomorphism $\mathrm{DR}_{\overline{X}}(\overline{M}_2) \simeq \mathrm{DR}_X \boldsymbol{D}_X M_1$ is the composite of

$$\mathrm{DR}_{\overline{X}}(\overline{M}_2) \simeq R\mathcal{H}om_{\mathscr{D}_X}(M_1, \Omega_{\overline{X}}^\bullet[d_X] \otimes_{\mathcal{O}_{\overline{X}}} \mathfrak{Db}_X) \simeq R\mathcal{H}om_{\mathscr{D}_X}(M_1, \mathcal{O}_X[d_X]).$$

The induced map $\overline{L}_2[d_X] \simeq L_1^\vee[d_X]$ is the shift of $\Psi_{C_0 21}$. Similarly, the isomorphism $\mathrm{DR}_X(M_1) \simeq \mathrm{DR}_{\overline{X}}(\boldsymbol{D}M_2)$ is the composite of

$$\mathrm{DR}_X(M_1) \simeq R\mathcal{H}om_{\mathscr{D}_{\overline{X}}}(\overline{M}_2, \Omega_X^\bullet[d_X] \otimes_{\mathcal{O}_X} \mathfrak{Db}_X) \simeq R\mathcal{H}om_{\mathscr{D}_{\overline{X}}}(\overline{M}_2, \mathcal{O}_{\overline{X}}[d_X]).$$

The induced map $L_1[d_X] \simeq \overline{L}_2^\vee[d_X]$ is the shift of $\Psi_{C_0 12}$. Then, we obtain that $(-1)^{d_X} C_0^\vee$ gives $\boldsymbol{D}C_0$. □

Remark 12.4.3 We have the induced pairing $C \circ \mathrm{ex} : \overline{M}_2 \otimes M_1 \longrightarrow \mathfrak{Db}_X$. Note that $\mathrm{DR}(C \circ \mathrm{ex}) = \mathrm{DR}(C) \circ \mathrm{ex}$ as pairings $\mathrm{DR}_{\overline{X}}(\overline{M}_2) \times \mathrm{DR}_X(M_1) \longrightarrow \omega^{top}$, if we use the above identification $\mathrm{Tot}\big(\Omega_X^\bullet[d_X] \otimes \Omega_{\overline{X}}^\bullet[d_X] \otimes \mathfrak{Db}_X\big) \simeq \mathfrak{Db}_X[2d_X]$ given by $\xi^{d_X+p} \otimes \overline{\xi}^{d_X+q} \otimes \tau \longleftrightarrow \epsilon(d_X)(-1)^{p d_X} \xi^{d_X+p} \wedge \overline{\xi}^{d_X+q} \wedge \tau$. Note that the induced identification $\mathrm{Tot}\big(\Omega_{\overline{X}}^\bullet[d_X] \otimes \Omega_X^\bullet[d_X] \otimes \mathfrak{Db}_X\big) \simeq \mathfrak{Db}_X[2d_X]$ is given by

$$\overline{\xi}^{d_X+q} \otimes \xi^{d_X+p} \otimes \tau \longleftrightarrow \epsilon(d_X)(-1)^{q d_X + d_X} \overline{\xi}^{d_X+q} \wedge \xi^{d_X+p} \wedge \tau.$$ □

12.4.1.1 Another Formulation

Let $(M_1, M_2, C) \in \mathscr{D}$-$\mathrm{Tri}^{nd}(X)$. We have the induced isomorphism $\varphi_{C,1,2} : \overline{M}_2 \simeq \mathcal{C}_X(M_1)$, which induces $\boldsymbol{D}\varphi_{C,1,2} : \boldsymbol{D}_{\overline{X}}\mathcal{C}_X(M_1) \simeq \boldsymbol{D}_{\overline{X}}\overline{M}_2$. According to Theorem 12.4.1, we have a dual $(\boldsymbol{D}_X M_1, \boldsymbol{D}_X M_2, \boldsymbol{D}C)$, from which we obtain the isomorphism $\varphi_{\boldsymbol{D}C,1,2} : \boldsymbol{D}_{\overline{X}}\overline{M}_2 \simeq C_X(\boldsymbol{D}_X M_1)$. By the composition, we obtain an isomorphism

$$\Xi_M : C_X(\boldsymbol{D}_X M_1) \simeq \boldsymbol{D}_{\overline{X}} C_X(M_1).$$

Proposition 12.4.4 *The following diagram is commutative:*

$$\begin{array}{ccc} \mathrm{DR}_{\overline{X}}\, C_X \boldsymbol{D}_X M_1 & \xrightarrow{\mathrm{DR}\,\Xi_M} & \mathrm{DR}_{\overline{X}}\, \boldsymbol{D}_{\overline{X}} C_X M_1 \\ {\scriptstyle a_1}\big\downarrow & & {\scriptstyle a_2}\big\downarrow \\ \boldsymbol{D}\,\mathrm{DR}_X\, \boldsymbol{D}_X M_1 & \xrightarrow{a_3} & \boldsymbol{D}\,\mathrm{DR}_{\overline{X}}\, C_X M_1 \end{array}$$

Here, a_i $(i = 1, 2, 3)$ are naturally induced morphisms.

Proof Let us consider the composite of the following isomorphisms:

$$\mathrm{DR}_{\overline{X}}(\boldsymbol{D}\overline{M}_2) \xrightarrow{\mathrm{DR}\,\varphi_{\boldsymbol{D}C,1,2}} \mathrm{DR}_X(C_X \boldsymbol{D}_X M_1) \simeq \boldsymbol{D}\,\mathrm{DR}_{\overline{X}}(\boldsymbol{D}_X M_1) \simeq \boldsymbol{D}\,\mathrm{DR}_X\, C_X M_1$$
$$\xrightarrow{\boldsymbol{D}\,\mathrm{DR}\,\varphi_{C,1,2}} \boldsymbol{D}\,\mathrm{DR}\,\overline{M}_2 \tag{12.43}$$

By the condition of $\boldsymbol{D}C$, it is equal to the composite of the following isomorphisms:

$$\mathrm{DR}_{\overline{X}}(\boldsymbol{D}\overline{M}_2) \simeq \mathrm{DR}_X(C_X \overline{M}_2) \xrightarrow{\mathrm{DR}\,\varphi^{-1}_{C,2,1}} \mathrm{DR}_X(M_1) \xrightarrow{b} \boldsymbol{D}\,\mathrm{DR}_{\overline{X}}(\overline{M}_2) \tag{12.44}$$

Here, b is the morphism induced by $\mathrm{DR}(C) \circ \mathrm{ex}$. We can check that the composite of (12.44) is equal to the naturally induced morphism. Moreover, $\mathrm{DR}\,\Xi_M$ is identified with the composite of the following morphisms:

$$\boldsymbol{D}\,\mathrm{DR}_X\, \boldsymbol{D} M_1 \simeq \mathrm{DR}_X\, C_X \boldsymbol{D}_X M_1 \xrightarrow{\mathrm{DR}\,\varphi^{-1}_{\boldsymbol{D}C,1,2}} \mathrm{DR}_{\overline{X}}\, \boldsymbol{D}\overline{M}_2 \simeq \boldsymbol{D}\,\mathrm{DR}_{\overline{X}}\, \overline{M}_2$$
$$\xrightarrow{\boldsymbol{D}\,\mathrm{DR}\,\varphi^{-1}_{C,1,2}} \boldsymbol{D}\,\mathrm{DR}_X\, C_X M_1$$

Then, the claim of the proposition follows. □

12.4.1.2 Appendix

Let us recall the duality functor for $\mathscr{D}$-modules, very briefly. See [34, 69–71] for more details. Let X be a complex manifold, and let H be a hypersurface.

Let N be a left-$\mathscr{D}_{X(*H)}$-bi-module, i.e., it is equipped with mutually commuting two $\mathscr{D}_{X(*H)}$-actions ρ_i $(i = 1, 2)$. The left $\mathscr{D}_{X(*H)}$-module by ρ_i is denoted by (N, ρ_i). For a $\mathscr{D}_{X(*H)}$-module L, let $\mathcal{H}om_{\mathscr{D}_{X(*H)}}(L^\bullet, N^{\rho_1,\rho_2})$ denote the sheaf of $\mathscr{D}_{X(*H)}$-homomorphisms $L \longrightarrow (N, \rho_1)$. It is equipped with a $\mathscr{D}_{X(*H)}$-action induced by ρ_2. Note that, for a $\mathscr{D}_{X(*H)}$-complex L, we have the naturally defined $\mathscr{D}_{X(*H)}$-homomorphism

$$L^\bullet \longrightarrow \mathcal{H}om_{\mathscr{D}_{X(*H)}}\Big(\mathcal{H}om_{\mathscr{D}_{X(*H)}}(L^\bullet, N^{\rho_2,\rho_1}), N^{\rho_1,\rho_2}\Big) \tag{12.45}$$

given by $x \longmapsto \big(F \longmapsto (-1)^{|x|\,|F|}F(x)\big)$.

We have the natural two $\mathscr{D}_{X(*H)}$-action on $\mathscr{D}_{X(*H)} \otimes \omega_X^{-1}$. The left multiplication is denoted by ℓ. Let r denote the action induced by the right multiplication. More generally, for a left $\mathscr{D}_{X(*H)}$-module N, we have two induced left $\mathscr{D}_{X(*H)}$-module structure on $N \otimes \mathscr{D}_{X(*H)} \otimes \omega_X^{-1}$. One is given by ℓ and the left $\mathscr{D}$-action on N, which is denoted by ℓ. The other is induced by the right multiplication, denoted by r. We have the automorphism of $\Phi_N : N \otimes \mathscr{D}_{X(*H)} \otimes \omega_X^{-1}$, which exchanges ℓ and r, as in [69].

Let $\mathcal{G}^\bullet$ be a $\mathscr{D}_{X(*H)} \otimes_{\mathbb{C}} \mathscr{D}_{X(*H)}$-injective resolution of $\mathscr{D}_{X(*H)} \otimes \omega_X^{-1}$. Note that $\mathcal{G}^\bullet$ is also a $\mathscr{D}_{X(*H)}$-injective resolution with respect to each $\mathscr{D}_{X(*H)}$-action. The isomorphism $\Phi_{\mathcal{O}_{X(*H)}} : \mathscr{D}_{X(*H)} \otimes \omega_X^{-1} \simeq \mathscr{D}_{X(*H)} \otimes \omega_X^{-1}$ is extended to a $\mathbb{C}$-linear quasi-isomorphism $\Phi_{\mathcal{G}^\bullet} : \mathcal{G}^\bullet \longrightarrow \mathcal{G}^\bullet$ which exchanges the two $\mathscr{D}_{X(*H)}$-actions. For any $\mathscr{D}_{X(*H)}$-module, we define

$$\boldsymbol{D}M := \mathcal{H}om_{\mathscr{D}_{X(*H)}}\Big(M, \mathcal{G}^{\bullet\,\ell,r}\Big).$$

The morphism as in (12.45) and the morphism $\Phi_{\mathcal{G}^\bullet}$ induces an isomorphism $M \longrightarrow \boldsymbol{D}_{X(*H)} \circ \boldsymbol{D}_{X(*H)}(M)$ in the derived category.

Let us consider the case $H = \emptyset$. For any perverse $\mathbb{C}_X$-sheaf $\mathcal{F}$, put $\boldsymbol{D}_{\mathbb{C}_X}(\mathcal{F}) := R\mathcal{H}om_{\mathbb{C}_X}(\mathcal{F}, \omega_X^{top})$. If M is holonomic, we have the naturally defined isomorphism $\mathrm{DR}_X \boldsymbol{D}_X(M) \simeq \boldsymbol{D}_{\mathbb{C}_X} \mathrm{DR}_X(M)$. See [71] for more details on this isomorphism. In particular, the following diagram is commutative:

$$\begin{array}{ccc}
\mathrm{DR}_X \boldsymbol{D}_X \circ \boldsymbol{D}_X M & \xrightarrow{\simeq} & \boldsymbol{D}_{\mathbb{C}_X} \circ \boldsymbol{D}_{\mathbb{C}_X} \mathrm{DR}_X M \\
\uparrow & & \uparrow \\
\mathrm{DR}_X M & \xrightarrow{=} & \mathrm{DR}_X M
\end{array}$$

The vertical arrows are induced by $\boldsymbol{D} \circ \boldsymbol{D} \simeq \mathrm{id}$, and the upper horizontal arrow is induced by the exchange of the functors DR and $\boldsymbol{D}$.

12.4.2 Duality of Complexes of Non-degenerate $\mathscr{D}_X$-Triples

By assuming Theorem 12.4.1, for any $\mathcal{T}^\bullet \in \mathcal{C}(\mathscr{D}\text{-Tri}^{nd}(X))$, we define the object $\boldsymbol{D}(\mathcal{T}^\bullet)$ in $\mathcal{C}(\mathscr{D}\text{-Tri}^{nd}(X))$ as follows. The p-th member is given by

$$\boldsymbol{D}(\mathcal{T}^\bullet)^p = \big(\boldsymbol{D}M_1^p, \boldsymbol{D}M_2^{-p}, (-1)^p \boldsymbol{D}C^{-p}\big).$$

The differential $\tilde{\delta}^p : \boldsymbol{D}(\mathcal{T}^\bullet)^p \longrightarrow \boldsymbol{D}(\mathcal{T}^\bullet)^{p+1}$ is given as the pair of the following morphisms

$$\boldsymbol{D}(M_1^p) \xleftarrow{(-1)^p \boldsymbol{D}(\delta_{-p-1,1})} \boldsymbol{D}(M_1^{p+1}),$$

$$\boldsymbol{D}(M_2^{-p}) \xrightarrow{(-1)^{p+1} \boldsymbol{D}(\delta_{-p-1,2})} \boldsymbol{D}(M_2^{-p-1}).$$

Note that the signature is adjusted so that the underlying $\mathscr{D}$-complexes are $\boldsymbol{D}(M_i^\bullet)$.

We have the saturated full subcategory $\mathcal{C}(\mathscr{D})\text{-Tri}_0^{nd}(X) \subset \mathcal{C}(\mathscr{D})\text{-Tri}(X)$ corresponding to $\mathcal{C}\big(\mathscr{D}\text{-Tri}^{nd}(X)\big) \subset \mathcal{C}\big(\mathscr{D}\text{-Tri}(X)\big)$ by Ψ_1. We obtain a contravariant auto-equivalence $\boldsymbol{D}$ on $\mathcal{C}(\mathscr{D})\text{-Tri}_0^{nd}(X)$. We have

$$\boldsymbol{D}(M_1^\bullet, M_2^\bullet, C) = (\boldsymbol{D}M_1^\bullet, \boldsymbol{D}M_2^\bullet, \boldsymbol{D}C),$$

where $(\boldsymbol{D}C)^p = \boldsymbol{D}(C^{-p})$.

12.4.3 *Compatibility of the Push-Forward and the Duality*

To state the stability for push-forward, we introduce a notion of duality for $\mathscr{D}$-complex-triples, which is more general than that in Sect. 12.4.2. Let $\mathfrak{T} = (M_1^\bullet, M_2^\bullet, C) \in \mathcal{C}(\mathscr{D})\text{-Tri}(X)$ such that each $H^p(\mathfrak{T}) \in \mathscr{D}\text{-Tri}^{nd}(X)$. We say that $\mathfrak{T}$ has a dual object if the following holds:

- We have $\mathscr{D}$-complexes $Q_i^\bullet$ with an isomorphism $Q_i^\bullet \simeq \boldsymbol{D}M_i^\bullet$ in the derived category.
- We have a sesqui-linear pairing C_Q of Q_1 and Q_2 such that the following diagram is commutative in $D_c^b(\mathbb{C}_X)$:

$$\begin{array}{ccc} \operatorname{Tot} \mathrm{DR}_X(Q_1^\bullet) \otimes \operatorname{Tot} \mathrm{DR}_{\overline{X}}(\overline{Q_2^\bullet}) & \xrightarrow{\mathrm{DR}\, C_Q} & \omega_X^{top} \\ \simeq\downarrow & & =\downarrow \\ \operatorname{Tot} \mathrm{DR}_{\overline{X}}(\overline{M_2^\bullet}) \otimes \operatorname{Tot} \mathrm{DR}_X(M_1^\bullet) & \xrightarrow{\mathrm{DR}(C)\circ \mathrm{ex}} & \omega^{top} \end{array}$$

Here, the left vertical arrow is induced as the composite of the following morphisms, where a_i are given as in the diagram (12.42):

$$\operatorname{Tot} \mathrm{DR}_X(Q_1^\bullet) \longrightarrow \operatorname{Tot} \mathrm{DR}_X(\boldsymbol{D}_X M_1^\bullet) \xrightarrow{a_1} \operatorname{Tot} \mathrm{DR}_{\overline{X}}(\overline{M_2^\bullet})$$

$$\operatorname{Tot} \mathrm{DR}_{\overline{X}}(\overline{Q_2^\bullet}) \longrightarrow \operatorname{Tot} \mathrm{DR}_{\overline{X}}(\boldsymbol{D}_{\overline{X}} \overline{M_2^\bullet}) \xrightarrow{a_2} \operatorname{Tot} \mathrm{DR}_X(M_1^\bullet)$$

In that case, $(Q_1^\bullet, Q_2^\bullet, C_Q)$ is called a dual object of $\mathfrak{T}$, and denoted by $\boldsymbol{D}\mathfrak{T}$.

The full subcategory of such $\mathfrak{T}$ in $\mathcal{C}(\mathscr{D})$-Tri(X) is denoted by $\mathcal{C}(\mathscr{D})\text{-Tri}_1^{nd}(X)$.

We have the following stability of the category of the non-degenerate pairings with respect to the push-forward, and the compatibility of the duality and the push-forward, which we will prove in Sect. 12.5.

Theorem 12.4.5 *Let $F : X \longrightarrow Y$ be a morphism of complex manifolds. Let $\mathcal{T} = (M_1, M_2, C) \in \mathscr{D}\text{-Tri}^{nd}(X)$. Assume that the restriction of F to the support of $\mathcal{T}$ is proper. Then, we have $F_\dagger^{(0)}\mathcal{T} \in \mathcal{C}(\mathscr{D})\text{-Tri}_1^{nd}(Y)$, and $F_\dagger^{(0)}\boldsymbol{D}\mathcal{T}$ gives a dual object of $F_\dagger^{(0)}\mathcal{T}$ in the above sense under the natural quasi-isomorphism of the $\mathscr{D}$-complexes. In particular, $\boldsymbol{H}^j F_\dagger^{(0)}\boldsymbol{D}\mathcal{T}$ is a dual object of $\boldsymbol{D}\boldsymbol{H}^{-j}F_\dagger^{(0)}\mathcal{T}$ under the natural isomorphism of the $\mathscr{D}$-modules.*

(See Sect. 12.1 for $F_\dagger^{(0)}$.) As a consequence, we obtain the following for $\mathscr{D}$-triples.

Theorem 12.4.6 *Let F and $\mathcal{T}$ be as in Theorem 12.4.5. Then, we have $F_\dagger^j\mathcal{T} \in \mathscr{D}\text{-Tri}^{nd}(Y)$ for any j. Moreover, we have $F_\dagger^j\boldsymbol{D}\mathcal{T}[-j] \simeq \boldsymbol{D}\big(F_\dagger^{-j}\mathcal{T}[j]\big)$, namely we have $F_\dagger^{-j}\boldsymbol{D}C = (-1)^j\boldsymbol{D}F_\dagger^j C$ under the natural isomorphisms $F_\dagger^j\boldsymbol{D}M_1 \simeq \boldsymbol{D}F_\dagger^{-j}M_1$ and $F_\dagger^{-j}\boldsymbol{D}M_2 \simeq \boldsymbol{D}F_\dagger^j M_2$.*

Corollary 12.4.7 *Suppose $\mathcal{T}^\bullet \in \mathcal{C}(\mathscr{D})\text{-Tri}_1^{nd}(X)$. Then, $F_\dagger^{(0)}(\mathcal{T}^\bullet)$ is an object in $\mathcal{C}(\mathscr{D})\text{-Tri}_1^{nd}(Y)$, and we have $\boldsymbol{D}F_\dagger^{(0)}\mathcal{T}^\bullet \simeq F_\dagger^{(0)}\boldsymbol{D}\mathcal{T}^\bullet$ under the natural identifications of the underlying $\mathscr{D}_Y$-complexes.* □

12.4.4 Compatibility with Other Functors

12.4.4.1 Shift

We give the compatibility with the shift functor.

Lemma 12.4.8 *We have natural transforms $\mathcal{S}_{-\ell} \circ \boldsymbol{D} \simeq \boldsymbol{D} \circ \mathcal{S}_\ell$ on the categories $\mathcal{C}\big(\mathscr{D}\text{-Tri}^{nd}(X)\big)$ and $\mathcal{C}(\mathscr{D})\text{-Tri}_0^{nd}(X)$. The isomorphisms of the underlying complexes of $\mathscr{D}_X$-modules are given as in* Sect. 12.1.5.1.

Proof Let $\mathfrak{T} = (M_1^\bullet, M_2^\bullet, C) \in \mathcal{C}(\mathscr{D})\text{-Tri}_0^{nd}(X)$. We have

$$\boldsymbol{D}\mathcal{S}_\ell(\mathfrak{T}) = \big(\boldsymbol{D}(M_1^\bullet[-\ell]), \boldsymbol{D}(M_2^\bullet[\ell]), \boldsymbol{D}(C[\ell])\big),$$
$$\mathcal{S}_{-\ell}\boldsymbol{D}(\mathfrak{T}) = \big(\boldsymbol{D}(M_1^\bullet)[\ell], \boldsymbol{D}(M_2^\bullet)[-\ell], \boldsymbol{D}(C)[-\ell]\big).$$

We have natural identifications

$$\boldsymbol{D}(M_1^\bullet[-\ell])^{-p} = \boldsymbol{D}(M_1^\bullet[-\ell]^p) = \boldsymbol{D}(M_1^{p-\ell}),$$
$$\boldsymbol{D}(M_1^\bullet)[\ell]^{-p} = \boldsymbol{D}(M_1^\bullet)^{-p+\ell} = \boldsymbol{D}(M_1^{p-\ell}).$$

We also have $\boldsymbol{D}(M_2^\bullet[\ell])^p = \boldsymbol{D}(M_2^{-p+\ell})$ and $\boldsymbol{D}(M_2^\bullet)[-\ell]^p = \boldsymbol{D}(M_2^{-p+\ell})$. Under the identifications, we have $\boldsymbol{D}(C[\ell])^p = (-1)^\ell\big(\boldsymbol{D}(C)[-\ell]\big)^p$. Indeed, we have the following equalities:

$$\boldsymbol{D}(C[\ell])^p = \boldsymbol{D}\big((C[\ell])^{-p}\big) = (-1)^{\ell(\ell-1)/2+\ell p}\boldsymbol{D}(C^{-p+\ell}) \tag{12.46}$$

$$\big(\boldsymbol{D}(C)[-\ell]\big)^p = (-1)^{\ell(\ell+1)/2+\ell p}(\boldsymbol{D}C)^{p-\ell} = (-1)^{\ell(\ell+1)/2+\ell p}\boldsymbol{D}(C^{-p+\ell}) \tag{12.47}$$

As mentioned in Sect. 12.1.5.1, the isomorphisms of the $\mathscr{D}_X$-complexes are given as follows:

$$\begin{aligned} \boldsymbol{D}(M_1^\bullet)[\ell]^{-p} &\xleftarrow{(-1)^{\ell p}\epsilon(-\ell)} \boldsymbol{D}(M_1^\bullet[-\ell])^{-p} \\ \boldsymbol{D}(M_2^\bullet)[-\ell]^{p} &\xrightarrow{(-1)^{\ell p}\epsilon(\ell)} \boldsymbol{D}(M_2^\bullet[\ell])^{p} \end{aligned} \tag{12.48}$$

It is easy to check that they are compatible with the sesqui-linear pairings, i.e., $\mathcal{S}_{-\ell}\boldsymbol{D}\mathfrak{T} \simeq \boldsymbol{D}\mathcal{S}_\ell\mathfrak{T}$.

By using Ψ_1, we also obtain that the isomorphisms (12.48) of $\mathscr{D}_X$-complexes induce $\mathcal{S}_{-\ell}\boldsymbol{D}(\mathcal{T}) \simeq \boldsymbol{D}\mathcal{S}_\ell\mathcal{T}$ for any $\mathcal{T} \in \mathcal{C}(\mathscr{D}\text{-Tri}^{nd}(X))$. □

12.4.4.2 Hermitian Adjoint

For any $\mathcal{T} = (M_1, M_2, C) \in \mathscr{D}\text{-Tri}(X)$, let $\mathcal{T}^* := (M_2, M_1, C^*)$, where $C^*(m_2, \overline{m_1}) = \overline{C(m_1, \overline{m_2})}$. Once we know Theorem 12.4.1, it is easy to obtain the compatibility with the Hermitian adjoint and the duality.

Proposition 12.4.9 *For any $\mathcal{T} \in \mathscr{D}\text{-Tri}^{nd}(X)$, we have $\boldsymbol{D}(C^*) = (\boldsymbol{D}C)^*$, i.e., $\boldsymbol{D}(\mathcal{T}^*) \simeq (\boldsymbol{D}\mathcal{T})^*$ under the natural identification of the underlying $\mathscr{D}_X$-modules.* □

12.4.4.3 Localization

Once we know Theorem 12.4.1, it is easy to obtain the compatibility with the localization in Sect. 12.2.2.

Proposition 12.4.10 *Let H be a hypersurface of X. For $\mathcal{T} \in \mathscr{D}\text{-Tri}^{nd}(X)$, we have natural isomorphisms*

$$\boldsymbol{D}\big(\mathcal{T}[!H]\big) \simeq \big(\boldsymbol{D}\mathcal{T}\big)[*H], \qquad \boldsymbol{D}\big(\mathcal{T}[*H]\big) \simeq \big(\boldsymbol{D}\mathcal{T}\big)[!H].$$

Proof Let us show $\boldsymbol{D}\big(\mathcal{T}[!H]\big) \simeq \big(\boldsymbol{D}\mathcal{T}\big)[*H]$. We have only to compare the pairings $\boldsymbol{D}\big(C[!H]\big)$ and $\big(\boldsymbol{D}C\big)[*H]$ under the natural identifications $\boldsymbol{D}\big(M_1[*H]\big) \simeq \big(\boldsymbol{D}M_1\big)[!H]$

and $\boldsymbol{D}\big(M_2[!H]\big) \simeq \big(\boldsymbol{D}M_2\big)[*H]$. It is easy to compare their restrictions to $X \setminus H$. We obtain the comparison on X by the uniqueness of the extension. □

12.4.4.4 Tensor Product with Smooth $\mathscr{D}$-Triples

Let us consider the compatibility with the tensor product of a non-degenerate smooth $\mathscr{D}$-triples in Sect. 12.2.3. Let $\mathcal{T}_V = (V_1, V_2, C)$ be a smooth $\mathscr{D}_X$-triple. If C is non-degenerate, the tuple $\mathcal{T}_V$ is locally isomorphic to a direct sum of the trivial pairing $(\mathcal{O}, \mathcal{O}, C_0)$, where $C_0(f, \overline{g}) = f\overline{g}$. We put $V_i^{\vee} := \mathcal{H}om_{\mathcal{O}_X}\big(V_i, \mathcal{O}_X\big)$. If C is non-degenerate, we have the induced pairing $C^{\vee} : V_1^{\vee} \otimes \overline{V_2^{\vee}} \longrightarrow \mathcal{C}_X^{\infty} \longrightarrow \mathfrak{Db}_X$. In this case, for $\mathcal{T} = (V_1, V_2, C)$, the dual $(V_1^{\vee}, V_2^{\vee}, C^{\vee})$ is denoted by $\mathcal{T}_V^{\vee}$. Once we know Theorem 12.4.1, the following lemma is obvious.

Lemma 12.4.11 *Let $\mathcal{T}_i \in \mathscr{D}\text{-}Tri^{nd}(X)$ $(i = 1, 2)$. Assume that $\mathcal{T}_2$ is smooth. Then, we naturally have*

$$\boldsymbol{D}(\mathcal{T}_1 \otimes \mathcal{T}_2) \simeq \boldsymbol{D}(\mathcal{T}_1) \otimes \mathcal{T}_2^{\vee}.$$

Proof We have the natural isomorphisms of the underlying holonomic $\mathscr{D}$-modules, which is reviewed in Remark 12.4.13 below. For the comparison of the pairings, we have only to consider the case $\mathcal{T}_2 = (\mathcal{O}, \mathcal{O}, C_0)$, and then it is clear. □

Let us consider a more general case. Let H be a hypersurface of X. Let $\mathcal{T}_V = (V_1, V_2, C)$ be a $\mathscr{D}_X$-triple such that V_i are smooth $\mathscr{D}_{X(*H)}$-modules. Such $\mathcal{T}_V$ is called a smooth $\mathscr{D}_{X(*H)}$-triple. We define the dual triple $\mathcal{T}_V^{\vee}$ as in the case of smooth $\mathscr{D}_X$-triples. Once we know Theorem 12.4.1, the following lemma is easy to see.

Lemma 12.4.12 *Let $\mathcal{T} \in \mathscr{D}\text{-}Tri^{nd}(X)$. Let $\mathcal{T}_V$ be a smooth non-degenerate $\mathscr{D}_{X(*H)}$-triple. Then, we have natural isomorphisms:*

$$\boldsymbol{D}\big((\mathcal{T} \otimes \mathcal{T}_V)[!H]\big) \simeq \big(\boldsymbol{D}(\mathcal{T}) \otimes \mathcal{T}_V^{\vee}\big)[*H], \quad \boldsymbol{D}\big((\mathcal{T} \otimes \mathcal{T}_V)[*H]\big) \simeq \big(\boldsymbol{D}(\mathcal{T}) \otimes \mathcal{T}_V^{\vee}\big)[!H].$$

Proof The underlying $\mathscr{D}$-modules of $\boldsymbol{D}\big((\mathcal{T} \otimes \mathcal{T}_V)[*H]\big)$ and $\big(\boldsymbol{D}(\mathcal{T}) \otimes \mathcal{T}_V^{\vee}\big)[!H]$ are naturally isomorphic. (See Remark 12.4.13 below.) For comparison of the pairings, we have only to compare their restrictions to $X \setminus H$, which follows from Lemma 12.4.11. The other case can be checked similarly. □

Remark 12.4.13 Let X and H be as above. Let M be a holonomic $\mathscr{D}_X(*H)$-module. Let V be a smooth $\mathscr{D}_X(*H)$-module. We have the natural isomorphism $\boldsymbol{D}_{X(*H)}(M \otimes_{\mathcal{O}_X(*H)} V) \simeq \boldsymbol{D}_{X(*H)}(M) \otimes_{\mathcal{O}_X(*H)} V^{\vee}$, given as follows. We have two naturally induced left $\mathscr{D}$-module structure ℓ and r on $V^{\vee} \otimes_{\mathcal{O}_X} (\mathscr{D}_{X(*H)} \otimes \omega_X^{-1})$, as explained in the appendix of *Sect. 12.4.1*. We have

$$\begin{aligned} \boldsymbol{D}_{X(*H)}(M \otimes V) &\simeq R\mathcal{H}om_{\mathscr{D}_X(*H)}\big(M \otimes V, (\mathscr{D}_{X(*H)} \otimes \omega_X^{-1})^{\ell,r}\big)[d_X] \\ &\simeq R\mathcal{H}om_{\mathscr{D}_X(*H)}\big(M, (V^{\vee} \otimes \mathscr{D}_{X(*H)} \otimes \omega_X^{-1})^{\ell,r}\big)[d_X] \end{aligned} \qquad (12.49)$$

By $\Phi_{V^\vee}$, it is isomorphic to

$$\begin{aligned} &R\mathcal{H}om_{\mathscr{D}_{X(*H)}}\big(M, (V^\vee \otimes \mathscr{D}_{X(*H)} \otimes \omega_X^{-1})^{r,\ell}\big)[d_X] \\ &\simeq R\mathcal{H}om_{\mathscr{D}_{X(*H)}}\big(M, (\mathscr{D}_{X(*H)} \otimes \omega_X^{-1})^{r,\ell}\big) \otimes V^\vee[d_X], \end{aligned} \tag{12.50}$$

which is isomorphic to $R\mathcal{H}om_{\mathscr{D}_{X(*H)}}\big(M, (\mathscr{D}_{X(*H)} \otimes \omega_X^{-1})^{\ell,r}\big) \otimes V^\vee[d_X]$ by $\Phi_{\mathcal{O}_{X(*H)}}$. Thus, we obtain the desired isomorphism

$$\boldsymbol{D}_{X(*H)}(M \otimes V) \simeq \boldsymbol{D}_{X(*H)}(M) \otimes V^\vee.$$

We can deduce $\boldsymbol{D}_X(M\otimes V)\simeq(\boldsymbol{D}_X(M)\otimes V^\vee)[!H]$ and $\boldsymbol{D}_X((M\otimes V)[!H])\simeq\boldsymbol{D}_X(M)\otimes V^\vee$ easily. □

12.4.4.5 Beilinson Functors

We have the perfect pairing $\tilde{\mathbb{I}}^{a,b} \times \tilde{\mathbb{I}}^{-b+1,-a+1} \longrightarrow \mathcal{O}_{\mathbb{C}_z}(*z)$ given as follows:

$$\big\langle f(s), g(s)\big\rangle = \operatorname*{Res}_{s=0}\Big(f(s)g(-s)\frac{ds}{s}\Big)$$

It gives an identification of the dual of $\tilde{\mathbb{I}}^{a,b}$ with $\tilde{\mathbb{I}}^{-b+1,-a+1}$. As in Sect. 2.3, we can check

$$\big(\tilde{\mathbb{I}}^{-b-1,-a-1}, \tilde{\mathbb{I}}^{a,b}, \tilde{C}_{\mathbb{I}}\big)^\vee \simeq \big(\tilde{\mathbb{I}}^{a,b}, \tilde{\mathbb{I}}^{-b+1,-a+1}, \tilde{C}_{\mathbb{I}}\big).$$

Then, once we know Theorem 12.4.1, we obtain the compatibility of the duality and Beilinson functors in Sect. 12.2.4.

Proposition 12.4.14 *Let $\mathcal{T} \in \mathscr{D}\text{-}Tri^{nd}(X)$. Let g be a holomorphic function on X.*

- *We have $\boldsymbol{D}\big(\Pi^{a,b}_{g*!}(\mathcal{T})\big) \simeq \Pi^{-b+1,-a+1}_{g*!}(\boldsymbol{D}\mathcal{T})$.*
- *In particular, $\psi^{(a)}_g(\mathcal{T})$, $\Xi^{(a)}_g(\mathcal{T})$ and $\phi^{(0)}_g(\mathcal{T})$ are non-degenerate, and we have natural isomorphisms:*

$$\boldsymbol{D}\psi^{(a)}_g(\mathcal{T}) \simeq \psi^{(-a+1)}_g(\boldsymbol{D}\mathcal{T}),\ \ \boldsymbol{D}\Xi^{(a)}_g(\mathcal{T}) \simeq \Xi^{(-a)}_g(\boldsymbol{D}\mathcal{T}),\ \ \boldsymbol{D}\phi^{(0)}_g(\mathcal{T}) \simeq \phi^{(0)}_g(\boldsymbol{D}\mathcal{T}).$$

Proof By Lemma 12.4.12, $\Pi^{a,N}_{g\star}(\mathcal{T})$ are non-degenerate. Hence, we obtain that $\Pi^{a,b}_{g*!}\mathcal{T}$ are non-degenerate. We have the following diagram:

$$\begin{array}{ccccc} \Pi^{-N+1,-b+1}_{g!}(\boldsymbol{D}\mathcal{T}) & \longrightarrow & \Pi^{-N+1,-a+1}_{g*}(\boldsymbol{D}\mathcal{T}) & \longrightarrow & \Pi^{-b+1,-a+1}_{g*!}(\boldsymbol{D}\mathcal{T}) \\ \simeq\big\downarrow & & \simeq\big\downarrow & & \big\downarrow \\ \boldsymbol{D}\Pi^{b,N}_{g*}(\mathcal{T}) & \longrightarrow & \boldsymbol{D}\Pi^{a,N}_{g!}(\mathcal{T}) & \longrightarrow & \boldsymbol{D}\Pi^{a,b}_{g*!}(\mathcal{T}) \end{array}$$

Hence, by the exactness of $\boldsymbol{D}$, we obtain $\Pi_{g*!}^{-b+1,-a+1}(\boldsymbol{D}\mathcal{T}) \simeq \boldsymbol{D}\Pi_{g*!}^{a,b}(\mathcal{T})$. We also obtain $\boldsymbol{D}\phi_g^{(0)} \simeq \phi_g^{(0)}\boldsymbol{D}$ by using the exactness of $\boldsymbol{D}$. □

12.4.4.6 External Tensor Product

Let X_i $(i = 1, 2)$ be complex manifolds. We set $X = X_1 \times X_2$. Let $q_i : X \longrightarrow X_i$ be the projection. For coherent $\mathscr{D}_{X_i}$-modules M_i, we have a naturally defined isomorphism

$$\boldsymbol{D}_{X_1}(M_1) \boxtimes \boldsymbol{D}_{X_2}(M_2) \simeq \boldsymbol{D}_X(M_1 \boxtimes M_2) \tag{12.51}$$

in the derived category of $\mathscr{D}_X$-modules. Indeed, let $\mathcal{I}_i^\bullet$ be a $\mathscr{D}_{X_i} \otimes \mathscr{D}_{X_i}$-injective resolution of $\mathscr{D}_{X_i} \otimes \omega_{X_i}^{-1}$. We take an injective $\mathscr{D}_X \otimes \mathscr{D}_X$-injective resolution $\mathcal{I}^\bullet$ of $\mathscr{D}_X \otimes \omega_X^{-1}$ with a quasi-isomorphism $\mathrm{Tot}(\mathcal{I}_1^\bullet \boxtimes \mathcal{I}_2^\bullet) \longrightarrow \mathcal{I}^\bullet$ which is an extension of $(\mathscr{D}_{X_1} \otimes \omega_{X_1}^{-1}) \boxtimes (\mathscr{D}_{X_2} \otimes \omega_{X_2}^{-1}) \simeq \mathscr{D}_X \otimes \omega_X^{-1}$. Then, we have a naturally defined morphism

$$\mathcal{H}om_{\mathscr{D}_{X_1}}\big(M_1, (\mathcal{I}_1^\bullet)^{\ell,r}\big) \boxtimes \mathcal{H}om_{\mathscr{D}_{X_2}}\big(M_2, (\mathcal{I}_2^\bullet)^{\ell,r}\big) \longrightarrow \mathcal{H}om_{\mathscr{D}_X}\big(M_1 \boxtimes M_2, (\mathcal{I}^\bullet)^{\ell,r}\big) \tag{12.52}$$

By taking free resolutions of M_i locally, we can check that (12.52) is a quasi-isomorphism. Thus, we obtain (12.51). In particular, if M_i are holonomic, (12.51) is an isomorphism of holonomic $\mathscr{D}_X$-modules.

Let $\mathcal{T}_i = (M_{i1}, M_{i2}, C_i) \in \mathscr{D}\text{-}\mathrm{Tri}^{nd}(X_i)$ $(i = 1, 2)$. We have the $\mathscr{D}_X$-modules $M_{1j} \boxtimes M_{2j}$ $(j = 1, 2)$. We also have the sesqui-linear pairing $C_1 \boxtimes C_2$. We set $\mathcal{T}_1 \boxtimes \mathcal{T}_2 := \big(M_{11} \boxtimes M_{21}, M_{12} \boxtimes M_{22}, C_1 \boxtimes C_2\big)$.

Proposition 12.4.15 *$\mathcal{T}_1 \boxtimes \mathcal{T}_2$ is an object in $\mathscr{D}$-Tri$^{nd}(X)$, and we have a natural isomorphism*

$$\boldsymbol{D}(\mathcal{T}_1 \boxtimes \mathcal{T}_2) \simeq \boldsymbol{D}(\mathcal{T}_1) \boxtimes \boldsymbol{D}(\mathcal{T}_2)$$

The isomorphisms of the underlying $\mathscr{D}_X$-modules are given by (12.51).

Proof Set $d_i := \dim X_i$ and $d := \dim X$. We have the isomorphism of complexes $\mathrm{Tot}\big(\Omega_{X_1}^\bullet[d_1] \boxtimes \Omega_{X_2}^\bullet[d_2]\big) \simeq \Omega_X^\bullet[d]$ given by

$$\eta^{d_1+p} \otimes \xi^{d_2+q} \longmapsto (-1)^{d_2(p+d_1)} \eta^{d_1+p} \wedge \xi^{d_2+q}.$$

It induces an isomorphism $\mathrm{DR}_{X_1}(M_1) \boxtimes \mathrm{DR}_{X_2}(M_2) \simeq \mathrm{DR}_X(M_1 \boxtimes M_2)$. Similarly, we have an isomorphism $\mathrm{DR}_{\overline{X}_1}(\overline{M}_1) \boxtimes \mathrm{DR}_{\overline{X}_2}(\overline{M}_2) \simeq \mathrm{DR}_{\overline{X}}(\overline{M}_1 \boxtimes \overline{M}_2)$. By an elementary computation, we can check $\mathrm{DR}_X(C_1 \boxtimes C_2) = \mathrm{DR}_{X_1}(C_1) \boxtimes \mathrm{DR}_{X_2}(C_2)$. Then, we can easily check that $\boldsymbol{D}C_1 \boxtimes \boldsymbol{D}C_2$ satisfies the condition for $\boldsymbol{D}(C_1 \boxtimes C_2)$. □

12.4.5 Functor $\tilde{\gamma}^*$

We obtain the following auto-equivalence $\tilde{\gamma}^*$ on the category $\mathcal{C}(\mathscr{D}\text{-Tri}^{nd}(X))$ as follows:

$$\tilde{\gamma}^* := \boldsymbol{D}^{\mathrm{herm}} \circ \boldsymbol{D} = \boldsymbol{D} \circ \boldsymbol{D}^{\mathrm{herm}}$$

We also define $\tilde{\gamma}^{(0)*} := \boldsymbol{D}^{(0)\,\mathrm{herm}} \circ \boldsymbol{D}^{(0)} = \boldsymbol{D}^{(0)} \circ \boldsymbol{D}^{(0)\,\mathrm{herm}}$ on $\mathcal{C}(\mathscr{D})\text{-Tri}_0^{nd}(X)$. We have $\Psi_1 \circ \tilde{\gamma}^{(0)*} = \tilde{\gamma}^* \circ \Psi_1$. In the following, we will often omit to denote "(0)".

We obtain the following proposition from Lemma 12.1.8 and Theorem 12.4.1.

Proposition 12.4.16 *Let $F : X \longrightarrow Y$ be a morphism of complex manifolds.*

- *Let $\mathcal{T}^\bullet \in \mathcal{C}(\mathscr{D}\text{-Tri}^{nd}(X))$ such that the restriction of f to the support of $\mathcal{T}^\bullet$ is proper. Then, we have the following natural isomorphisms in $\mathcal{C}(\mathscr{D}\text{-Tri}^{nd}(Y))$:*

$$H^j\big(F_\dagger(\tilde{\gamma}^*\mathcal{T}^\bullet)\big)[-j] \simeq \tilde{\gamma}^*\Big(H^jF_\dagger(\mathcal{T}^\bullet)[-j]\Big)$$

- *Let $\mathfrak{T} \in \mathcal{C}(\mathscr{D})\text{-Tri}_0^{nd}(X)$ such that the restriction of f to the cohomological support of $\mathfrak{T}$ is proper. Then, we have the following natural isomorphisms in $\mathcal{C}(\mathscr{D})\text{-Tri}_0^{nd}(Y)$:*

$$\boldsymbol{H}^j\big(F_\dagger^{(0)}(\tilde{\gamma}^{(0)*}\mathfrak{T})\big) \simeq \tilde{\gamma}^{(0)*}\boldsymbol{H}^jF_\dagger^{(0)}(\mathfrak{T})$$

In the both cases, the underlying isomorphisms of $\mathscr{D}_X$-modules are the natural isomorphisms $\boldsymbol{D}F_\dagger^jM \simeq F_\dagger^{-j}\boldsymbol{D}M$ for any $\mathscr{D}_X$-module M whose support is proper over Y with respect to f. □

Lemma 12.4.17 *We have $\kappa_{1,\ell} : \mathcal{S}_{-\ell} \circ \boldsymbol{D}^{\mathrm{herm}} \simeq \boldsymbol{D}^{\mathrm{herm}} \circ \mathcal{S}_\ell$ and $\kappa_{2,\ell} : \mathcal{S}_{-\ell} \circ \boldsymbol{D} \simeq \boldsymbol{D} \circ \mathcal{S}_\ell$ on $\mathcal{C}(\mathscr{D}\text{-Tri}^{nd}(X))$ and $\mathcal{C}(\mathscr{D})\text{-Tri}_0^{nd}(X)$. As a consequence, we have $\kappa_{3,\ell} : \mathcal{S}_{-\ell} \circ \tilde{\gamma}^* \simeq \tilde{\gamma}^* \circ \mathcal{S}_\ell$ on the categories. They satisfy $\kappa_{a,\ell+m} = \kappa_{a,\ell} \circ \kappa_{a,m}$.*

Proof As mentioned in Sect. 12.1.3 and Lemma 12.4.8, the natural isomorphisms of the $\mathscr{D}_X$-complexes induce the natural transforms $\kappa_{i,\ell}$ $(i = 1, 2)$. Hence, we obtain the natural transforms $\kappa_{3,\ell}$. □

By the construction, $\kappa_{3,\ell}$ are given by the natural isomorphisms of $\mathscr{D}_X$-complexes.

Let H be a hypersurface of X. Let $V = (V_1, V_2, C)$ be any non-degenerate smooth $\mathscr{D}_{X(*H)}$-triple. We set $\tilde{\gamma}_{sm}^*(V) := (V_2^\vee, V_1^\vee, (C^*)^\vee)$. We can easily deduce the following lemma from Lemma 12.4.12.

Proposition 12.4.18 *For any non-degenerate smooth $\mathscr{D}_{X(*H)}$-triple V and any $\mathcal{T} \in \mathscr{D}\text{-Tri}^{nd}(X)$, we have natural isomorphisms $\tilde{\gamma}^*\big((\mathcal{T} \otimes V)[\star H]\big) \simeq \big(\tilde{\gamma}^*(\mathcal{T}) \otimes \tilde{\gamma}_{sm}^*(V)\big)[\star H]$ $(\star = *, !)$.* □

Because Beilinson triples are equipped with the natural real structure, we can easily deduce the following from Proposition 12.4.14.

Proposition 12.4.19 *Let $\mathcal{T} \in \mathscr{D}\text{-Tri}^{nd}(X)$. Let g be a holomorphic function on X. We have natural isomorphisms $\tilde{\gamma}^*\Pi^{a,b}_{g*!}(\mathcal{T}) \simeq \Pi^{a,b}_{g*!}(\tilde{\gamma}^*\mathcal{T})$. In particular, We have the following natural isomorphisms:*

$$\tilde{\gamma}^*\psi_g^{(a)}(\mathcal{T}) \simeq \psi_g^{(a)}(\tilde{\gamma}^*\mathcal{T}), \quad \tilde{\gamma}^*\Xi_g^{(a)}(\mathcal{T}) \simeq \Xi_g^{(a)}(\tilde{\gamma}^*\mathcal{T}), \quad \tilde{\gamma}^*\phi_g^{(0)}(\mathcal{T}) \simeq \phi_g^{(0)}(\tilde{\gamma}^*\mathcal{T}).$$

□

We obtain the following from Proposition 12.4.15.

Proposition 12.4.20 *Let X_i $(i = 1, 2)$ be complex manifolds. We set $X := X_1 \times X_2$. For $\mathcal{T}_i \in \mathscr{D}\text{-Tri}^{nd}(X_i)$ $(i = 1, 2)$, we have a natural isomorphism $\tilde{\gamma}^*(\mathcal{T}_1 \boxtimes \mathcal{T}_2) \simeq \tilde{\gamma}^*(\mathcal{T}_1) \boxtimes \tilde{\gamma}^*(\mathcal{T}_2)$.* □

12.4.6 Push-Forward and Duality of $\mathscr{D}$-Modules (Appendix)

We recall some details on the compatibility of the push-forward and the duality. (See [34, 71].)

12.4.6.1 Preliminary

Let Y be a complex manifold. Let ℓ and r denote the left and right multiplication of $\mathscr{D}_Y$ on itself. The induced actions on $\mathscr{D}_Y \otimes \omega_Y^{-1}$ are denoted by the same symbols.

We set $\mathcal{M}_{0,Y} := (\mathscr{D}_Y \otimes \omega_Y^{-1})({}^{\ell}\otimes^{\ell}_{\mathcal{O}_Y})(\mathscr{D}_Y \otimes \omega_Y^{-1})$. Here, "${}^{\ell}\otimes^{\ell}_{\mathcal{O}_Y}$" means that we use the $\mathcal{O}$-actions induced by ℓ for the tensor product. We have the $\mathscr{D}_Y$-action $\ell_1 \otimes \ell_2$ on $\mathcal{M}_{0,Y}$ induced by the actions ℓ on the first and second components. We also have the $\mathscr{D}_Y$-actions r_i $(i = 1, 2)$ on $\mathcal{M}_{0,Y}$ induced by the $\mathscr{D}_Y$-action r on the i-th component.

We set $\mathcal{M}_{1,Y} := (\mathscr{D}_Y \otimes \omega_Y^{-1})({}^{r}\otimes^{\ell}_{\mathcal{O}_Y})(\mathscr{D}_Y \otimes \omega_Y^{-1})$. Here, "${}^{r}\otimes^{\ell}_{\mathcal{O}_Y}$" means that we use the $\mathcal{O}_Y$-action induced by r on the first component, and the $\mathcal{O}_Y$-action induced by ℓ on the second component. We have the $\mathscr{D}_Y$-action $r_1 \otimes \ell_2$ on $\mathcal{M}_{1,Y}$ induced by the $\mathscr{D}_Y$-action r on the first component and the $\mathscr{D}_Y$-action ℓ on the second component. We have the $\mathscr{D}_Y$-action ℓ_1 on $\mathcal{M}_{1,Y}$ induced by the $\mathscr{D}_Y$-action ℓ on the first component. We have the $\mathscr{D}_Y$-action r_2 on $\mathcal{M}_{1,Y}$ induced by the $\mathscr{D}_Y$-action r on the second component.

We have a unique $\mathbb{C}$-linear isomorphism $\Psi : \mathcal{M}_{1,Y} \longrightarrow \mathcal{M}_{0,Y}$ such that (1) $\Psi \circ r_2 = r_2 \circ \Psi$, (2) $\Psi \circ \ell_1 = (\ell_1 \otimes \ell_2) \circ \Psi$, (3) Ψ induces the identity on $\omega_Y^{-1} \otimes \omega_Y^{-1}$. We can prove the following lemma by a direct computation. (See the proof of Lemma 13.1.3 below.)

Lemma 12.4.21 *We have $\Psi \circ (r_1 \otimes \ell_2) = \ell_1 \circ \Psi$.* □

Remark 12.4.22 We consider $\omega_Y \otimes^L_{\mathscr{D}_Y} \mathcal{M}_{i,Y}$ $(i = 1, 2)$, where we use the $\mathscr{D}_Y$-action r_2 on $\mathcal{M}_{i,Y}$. They are naturally isomorphic to $\mathscr{D}_Y \otimes \omega_Y^{-1}$, and the morphism Ψ induces the identity on $\mathscr{D}_Y \otimes \omega_Y^{-1}$. □

12.4.6.2 A Morphism

Let $f : X \longrightarrow Y$ be a morphism of complex manifolds. We consider

$$\mathscr{D}_{Y\leftarrow X} \otimes^L_{\mathscr{D}_X} \big(\mathcal{O}_X[dx] \otimes^{\ell}_{f^{-1}\mathcal{O}_Y} f^{-1}(\mathscr{D}_Y \otimes \omega_Y^{-1})\big), \tag{12.53}$$

where "$\otimes^{\ell}_{f^{-1}\mathcal{O}_Y}$" means that we use the $f^{-1}\mathcal{O}_Y$-action on $f^{-1}(\mathscr{D}_Y \otimes \omega_Y^{-1})$ induced by ℓ. We have two $f^{-1}\mathscr{D}_Y$-actions on (12.53). One is induced by the $f^{-1}(\mathscr{D}_Y)$-action on $\mathscr{D}_{Y\leftarrow X}$, denoted by κ_1. The other is induced by the $f^{-1}\mathscr{D}_Y$-action r on $f^{-1}(\mathscr{D}_Y \otimes \omega_Y^{-1})$, denoted by κ_2.

We consider

$$\big(\mathscr{D}_{Y\leftarrow X} \otimes^L_{\mathscr{D}_X} \mathcal{O}_X[dx]\big) \otimes^{\ell}_{f^{-1}\mathcal{O}_Y} f^{-1}(\mathscr{D}_Y \otimes \omega_Y^{-1}), \tag{12.54}$$

where "$\otimes^{\ell}_{f^{-1}\mathcal{O}_Y}$" means that we use the $f^{-1}\mathcal{O}_Y$-action ℓ on $f^{-1}(\mathscr{D}_Y \otimes \omega_Y^{-1})$ for the tensor product. We have two $f^{-1}\mathscr{D}_Y$-actions on (12.54). One is induced by the $f^{-1}\mathscr{D}_Y$-action on $\mathscr{D}_{Y\leftarrow X} \otimes^L_{\mathscr{D}_X} \mathcal{O}_X[dx]$ and the $f^{-1}\mathscr{D}_Y$-action ℓ on $f^{-1}(\mathscr{D}_Y \otimes \omega_Y^{-1})$, denoted by κ_1'. The other is induced by that on $f^{-1}(\mathscr{D}_Y \otimes \omega_Y^{-1})$, denoted by κ_2'.

Lemma 12.4.23 *The isomorphism Ψ^{-1} in* Sect. 12.4.6.1 *induces a $\mathbb{C}$-linear isomorphism*

$$\begin{aligned} F : \mathscr{D}_{Y\leftarrow X} &\otimes^L_{\mathscr{D}_X} \big(\mathcal{O}_X[dx] \otimes_{f^{-1}\mathcal{O}_Y} f^{-1}(\mathscr{D}_Y \otimes \omega_Y^{-1})\big) \\ &\simeq \big(\mathscr{D}_{Y\leftarrow X} \otimes^L_{\mathscr{D}_X} \mathcal{O}_X[dx]\big) \otimes_{f^{-1}\mathcal{O}_Y} f^{-1}(\mathscr{D}_Y \otimes \omega_Y^{-1}) \end{aligned} \tag{12.55}$$

such that $F \circ \kappa_i = \kappa_i' \circ F$ $(i = 1, 2)$.

Proof We have $\mathscr{D}_{Y\leftarrow X} = \omega_X \otimes f^{-1}(\mathscr{D}_Y \otimes \omega_Y^{-1}) \simeq \big(\Omega^{\bullet}_X[dx] \otimes \mathscr{D}_X\big) \otimes_{f^{-1}\mathcal{O}_Y} f^{-1}(\mathscr{D}_Y \otimes \omega_Y^{-1})$. We have the following:

$$\begin{aligned} &\mathscr{D}_{Y\leftarrow X} \otimes^L_{\mathscr{D}_X} \big(\mathcal{O}_X[dx] \otimes^{\ell}_{f^{-1}\mathcal{O}_Y} f^{-1}(\mathscr{D}_Y \otimes \omega_Y^{-1})\big) \\ &\simeq \Big(\Omega^{\bullet}_X[dx] \otimes \mathscr{D}_X \otimes^{\ell}_{f^{-1}\mathcal{O}_Y} f^{-1}(\mathscr{D}_Y \otimes \omega_Y^{-1})\Big) \\ &\otimes_{\mathscr{D}_X} \Big(\mathcal{O}_X[dx] \otimes^{\ell}_{f^{-1}\mathcal{O}_Y} f^{-1}(\mathscr{D}_Y \otimes \omega_Y^{-1})\Big) \\ &\simeq \Omega^{\bullet}_X[dx] \otimes^{\ell_1 \otimes \ell_2}_{f^{-1}\mathcal{O}_Y} f^{-1}\big(\mathcal{M}_{0,Y}\big)[dx] \end{aligned} \tag{12.56}$$

In the last term, "$\otimes^{\ell_1\otimes\ell_2}_{f^{-1}\mathcal{O}_Y}$ means that we use the $f^{-1}\mathcal{O}_Y$-action induced by $\ell_1\otimes\ell_2$ on $f^{-1}\mathcal{M}_{0,Y}$ for the tensor product. We also have the following:

$$\begin{aligned}&\bigl(\mathscr{D}_{Y\leftarrow X}\otimes^L_{\mathscr{D}_X}\mathcal{O}_X[d_X]\bigr)\otimes^{\ell}_{f^{-1}\mathcal{O}_Y}f^{-1}(\mathscr{D}_Y\otimes\omega_Y^{-1})\\&\simeq\Bigl(\bigl(\Omega^\bullet_X[d_X]\otimes\mathscr{D}_X\otimes^{\ell}_{f^{-1}\mathcal{O}_Y}f^{-1}(\mathscr{D}_Y\otimes\omega_Y^{-1})\bigr)\otimes_{\mathscr{D}_X}\mathcal{O}_X[d_X]\Bigr)\\&\otimes^{\ell}_{f^{-1}\mathcal{O}_Y}f^{-1}(\mathscr{D}_Y\otimes\omega_Y^{-1})\\&\simeq\Omega^\bullet_X[d_X]\otimes^{\ell_1}_{f^{-1}\mathcal{O}_Y}f^{-1}\Bigl(\mathcal{M}_{1,Y}\Bigr)[d_X]\end{aligned}\tag{12.57}$$

In the last term, "$\otimes^{\ell_1}_{f^{-1}\mathcal{O}_Y}$" means that we use the $f^{-1}(\mathcal{O}_Y)$-action ℓ_1 on $f^{-1}\mathcal{M}_{1,Y}$. Then, we obtain Lemma 12.4.23 from Lemma 12.4.21. □

12.4.6.3 Compatibility of the Push-Forward and the Duality

Recall that we have the trace morphism $f_\dagger(\mathcal{O}_X[d_X])\longrightarrow\mathcal{O}_Y[d_Y]$, induced by the trace $f_!\mathfrak{D}\mathfrak{b}^\bullet_X[2d_X]\longrightarrow\mathfrak{D}\mathfrak{b}^\bullet_Y[2d_Y]$. Indeed,

$$\begin{aligned}f_\dagger\bigl(\mathcal{O}_X[d_X]\bigr)&\longrightarrow f_!\bigl(\mathfrak{D}\mathfrak{b}^\bullet_X[2d_X]\bigr)\otimes_{f^{-1}\mathcal{O}_Y}f^{-1}(\mathscr{D}_Y\otimes\omega_Y^{-1})\\&\longrightarrow\mathfrak{D}\mathfrak{b}^\bullet_Y[2d_Y]\otimes\mathscr{D}_Y\otimes\omega_Y^{-1}\simeq\mathcal{O}_Y[d_Y]\end{aligned}\tag{12.58}$$

From Lemma 12.4.23 and (12.58), we obtain the following trace morphism

$$\begin{aligned}&f_!\bigl(\mathscr{D}_{Y\leftarrow X}\otimes^L_{\mathscr{D}_X}(\mathcal{O}_X[d_X]\otimes_{f^{-1}\mathcal{O}_Y}f^{-1}(\mathscr{D}_Y\otimes\omega_Y^{-1}))\bigr)\\&\simeq f_!\Bigl((\mathscr{D}_{Y\leftarrow X}\otimes^L_{\mathscr{D}_X}\mathcal{O}_X[d_X])\Bigr)\otimes_{\mathcal{O}_Y}(\mathscr{D}_Y\otimes\omega_Y^{-1})\\&\longrightarrow\mathcal{O}_Y[d_Y]\otimes\mathscr{D}_Y\otimes\omega_Y^{-1}=\mathscr{D}_Y\otimes\omega_Y^{-1}[d_Y]\end{aligned}\tag{12.59}$$

Suppose that M is a $\mathscr{D}_X$-module whose support is proper relative to f. By using this trace morphism (12.59), we obtain

$$\begin{aligned}&f_\dagger\circ\boldsymbol{D}M\simeq f_!\Bigl(R\mathcal{H}om_{\mathscr{D}_X}\bigl(M,\mathcal{O}_X[d_X]\otimes_{f^{-1}\mathcal{O}_Y}f^{-1}(\mathscr{D}_Y\otimes\omega_Y^{-1})\bigr)\Bigr)\\&\longrightarrow f_!\Bigl(R\mathcal{H}om_{f^{-1}(\mathscr{D}_Y)}\Bigl(\mathscr{D}_{Y\leftarrow X}\otimes^L_{\mathscr{D}_X}M,\,\mathscr{D}_{Y\leftarrow X}\otimes^L_{\mathscr{D}_X}\bigl(\mathcal{O}_X[d_X]\otimes_{f^{-1}\mathcal{O}_Y}f^{-1}(\mathscr{D}_Y\otimes\omega_Y^{-1})\bigr)\Bigr)\Bigr)\\&\longrightarrow R\mathcal{H}om_{\mathscr{D}_Y}\bigl(f_\dagger M,f_!(\mathscr{D}_{Y\leftarrow X}\otimes^L_{\mathscr{D}_X}(\mathcal{O}_X[d_X]\otimes_{f^{-1}\mathcal{O}_Y}f^{-1}(\mathscr{D}_Y\otimes\omega_Y^{-1})))\bigr)\\&\longrightarrow R\mathcal{H}om_{\mathscr{D}_Y}\bigl(f_\dagger M,\mathscr{D}_Y\otimes\omega_Y^{-1}[d_Y]\bigr)\simeq\boldsymbol{D}f_\dagger M\end{aligned}\tag{12.60}$$

The composite is an isomorphism, if M is good relatively to f. See [34] for example.

Remark 12.4.24 We obtain the following morphism by applying "$\omega_Y \otimes^L_{\mathscr{D}_Y}$" to *(12.59)* with the $\mathscr{D}_Y$-action r_2:

$$\omega_Y \otimes^L_{\mathscr{D}_Y} \Big(f_! \big(\mathscr{D}_{Y \leftarrow X} \otimes^L_{\mathscr{D}_X} \mathcal{O}_X[d_X] \big) \otimes \mathscr{D}_Y \otimes \omega_Y^{-1} \Big)$$
$$\longrightarrow \omega_Y \otimes^L_{\mathscr{D}_Y} \big(\mathcal{O}_Y[d_Y] \otimes \mathscr{D}_Y \otimes \omega_Y^{-1} \big) = \mathcal{O}_Y[d_Y] \tag{12.61}$$

It is the same as the trace morphism $f_\dagger \mathcal{O}_X[d_X] \longrightarrow \mathcal{O}_Y[d_Y]$. □

12.4.6.4 Compatibility of the de Rham Functor, the Push-Forward and the Duality

Let $f : X \longrightarrow Y$ be a morphism of complex manifolds. For simplicity, we assume that f is proper. We have the following diagram of the functors from the category of holonomic $\mathscr{D}_X$-modules to the derived category of the cohomologically constructible complexes on Y:

$$\begin{array}{ccc} \mathrm{DR}_Y \circ f_\dagger \circ \boldsymbol{D} & \xrightarrow{\simeq} & \mathrm{DR}_Y \circ \boldsymbol{D} \circ f_\dagger \\ \simeq \downarrow & & \simeq \downarrow \\ f_* \circ \boldsymbol{D} \circ \mathrm{DR}_X & \xrightarrow{\simeq} & \boldsymbol{D} \circ f_* \circ \mathrm{DR}_X \end{array} \tag{12.62}$$

Here, the left vertical arrow is given by $\mathrm{DR}_Y \circ f_\dagger \circ \boldsymbol{D} \simeq f_* \circ \mathrm{DR}_X \circ \boldsymbol{D} \simeq f_* \circ \boldsymbol{D} \circ \mathrm{DR}_X$, and right vertical arrow is given by $\mathrm{DR}_Y \circ \boldsymbol{D} \circ f_\dagger \simeq \boldsymbol{D} \circ \mathrm{DR}_X \circ f_\dagger \simeq \boldsymbol{D} \circ f_* \circ \mathrm{DR}_X$.

Proposition 12.4.25 *The diagram* (12.62) *is commutative.*

Proof Let $N = \mathcal{O}_X \otimes_{f^{-1}\mathcal{O}_Y} f^{-1}(\mathscr{D}_Y \otimes \omega_Y^{-1})[d_X]$. We have $\mathrm{DR}_Y f_\dagger \boldsymbol{D} M = \omega_Y \otimes^L_{\mathscr{D}_Y} f_* R\mathcal{H}om_{\mathscr{D}_X}(M, N)$. The morphism $\mathrm{DR}_Y f_\dagger \boldsymbol{D} M \longrightarrow \mathrm{DR}_Y \boldsymbol{D} f_\dagger M \longrightarrow \boldsymbol{D} f_* \mathrm{DR}_X M$ is expressed as follows, where $R\mathcal{H}om_{\mathcal{A}}(Q_1, Q_1)$ are denoted by $(Q_1, Q_2)_{\mathcal{A}}$:

$$\begin{array}{ccccc} \omega_Y \otimes^L_{\mathscr{D}_Y} f_\dagger(M,N)_{\mathscr{D}_X} \longrightarrow & \omega_Y \otimes^L_{\mathscr{D}_Y} (f_\dagger M, f_\dagger N)_{\mathscr{D}_Y} & \longrightarrow \omega_Y \otimes^L_{\mathscr{D}_Y} (\mathrm{DR}_Y f_\dagger M, \mathrm{DR}_Y f_\dagger N)_{\mathbb{C}_Y} \\ & \downarrow & \downarrow \\ & \omega_Y \otimes^L_{\mathscr{D}_Y} (f_\dagger M, \mathscr{D}_Y \otimes \omega_Y^{-1}[d_Y])_{\mathscr{D}_Y} \longrightarrow & (\mathrm{DR}_Y f_\dagger M, \mathrm{DR}_Y \mathcal{O}_Y[d_Y])_{\mathbb{C}_Y} \\ & & \downarrow \\ & & (f_* \mathrm{DR}_X M, \mathrm{DR}_Y \mathcal{O}_Y[d_Y])_{\mathbb{C}_Y} \end{array}$$

The morphism $\mathrm{DR}_Y f_\dagger \boldsymbol{D} M \longrightarrow f_* \boldsymbol{D} \mathrm{DR}_X M \longrightarrow \boldsymbol{D} f_* \mathrm{DR}_X M$ is expressed as follows:

$$\begin{array}{ccc} \omega_Y \otimes^L_{\mathscr{D}_Y} f_\dagger(M,N)_{\mathscr{D}_X} \rightarrow & \omega_Y \otimes^L_{\mathscr{D}_Y} f_*(\mathrm{DR}\, M, \mathrm{DR}\, N)_{\mathbb{C}_X} & \rightarrow \omega_Y \otimes^L_{\mathscr{D}_Y} (f_* \mathrm{DR}\, M, f_* \mathrm{DR}\, N)_{\mathbb{C}_Y} \\ & \downarrow & \downarrow \\ & f_*(\mathrm{DR}\, M, f^{-1}\omega_Y \otimes^L_{f^{-1}\mathscr{D}_Y} \mathrm{DR}\, N)_{\mathbb{C}_X} \rightarrow & (f_* \mathrm{DR}\, M, \omega_Y \otimes^L_{\mathscr{D}_Y} f_* \mathrm{DR}\, N)_{\mathbb{C}_Y} \\ & & \downarrow \\ & & (f_* \mathrm{DR}\, M, \mathrm{DR}_Y \mathcal{O}_Y[d_Y])_{\mathbb{C}_Y} \end{array}$$

Then, we can deduce the claim of the proposition. □

Corollary 12.4.26 *Let $\mathcal{T} \in \mathscr{D}\text{-Tri}^{nd}(X)$. The following diagram of natural isomorphisms is commutative:*

$$\begin{array}{ccc} \mathrm{DR}_Y \circ H^j f_\dagger \circ \boldsymbol{D}(\mathcal{T}) & \longrightarrow & \mathrm{DR}_Y \boldsymbol{D} \circ H^{-j} f_\dagger(\mathcal{T}) \\ \downarrow & & \downarrow \\ H^j f_* \circ \boldsymbol{D} \circ \mathrm{DR}_X(\mathcal{T}) & \longrightarrow & \boldsymbol{D} H^{-j} f_* \circ \mathrm{DR}_X(\mathcal{T}) \end{array}$$

Here, the cohomology for $\mathbb{C}$-complexes are taken with respect to the middle perversity.

Proof We have only to check it for the underlying $\mathscr{D}$-complexes and $\mathbb{C}$-complexes, which have already been done in Proposition 12.4.25. □

12.5 Proof of Theorems 12.4.1 and 12.4.5

12.5.1 Preliminary

We recall the following lemma, which will be used implicitly.

Lemma 12.5.1 *Let $\mathcal{A}$ and $\mathcal{B}$ be sheaves of $\mathbb{C}$-algebras on X. Let $\mathcal{L}_1$ be an $\mathcal{A}$-module, and let $\mathcal{L}_2$ be an $(\mathcal{A} \otimes \mathcal{B})$-injective module. Then, $\mathcal{H}om_{\mathcal{A}}(\mathcal{L}_1, \mathcal{L}_2)$ is $\mathcal{B}$-injective.*

Proof Let $J_1 \to J_2$ be a monomorphism of $\mathcal{B}$-modules, and let $J_1 \to \mathcal{H}om_{\mathcal{A}}(\mathcal{L}_1, \mathcal{L}_2)$ be any $\mathcal{B}$-morphism. Note that $J_1 \otimes_{\mathbb{C}} \mathcal{L}_1 \longrightarrow J_2 \otimes_{\mathbb{C}} \mathcal{L}_1$ is a monomorphism. Hence, we have a morphism $J_2 \otimes \mathcal{L}_1 \longrightarrow \mathcal{L}_2$ whose restriction to $J_1 \otimes \mathcal{L}_1$ is equal to the given morphism. It means we obtain $J_2 \longrightarrow \mathcal{H}om_{\mathcal{A}}(\mathcal{L}_1, \mathcal{L}_2)$ whose restriction to J_1 is equal to the given morphism. □

12.5.2 Push Forward and the Functor $\mathcal{C}_X$

Let $F : X \longrightarrow Y$ be a morphism of complex manifolds. Let M be a holonomic $\mathscr{D}_X$-module such that the restriction of F to $\operatorname{Supp} M$ is proper. We have the natural morphism $F_\dagger \mathcal{C}_X(M) \longrightarrow \mathcal{C}_Y F_\dagger(M)$ given as follows:

$$RF_! R\mathcal{H}om_{\mathscr{D}_X}\Big(M,\ \mathscr{D}_{\overline{Y} \leftarrow \overline{X}} \otimes^L_{\mathscr{D}_{\overline{X}}} \mathfrak{Db}_X\Big)$$

$$\longrightarrow RF_! R\mathcal{H}om_{F^{-1}\mathscr{D}_Y}\Big(\mathscr{D}_{Y \leftarrow X} \otimes^L_{\mathscr{D}_X} M,\ \mathscr{D}_{Y \leftarrow X, \overline{Y} \leftarrow \overline{X}} \otimes^L_{\mathscr{D}_{X,\overline{X}}} \mathfrak{Db}_X\Big)$$

$$\longrightarrow R\mathcal{H}om_{\mathscr{D}_Y}\Big(F_\dagger M,\, RF_!\big(\mathscr{D}_{Y\leftarrow X,\overline{Y}\leftarrow\overline{X}}\otimes^L_{\mathscr{D}_{X,\overline{X}}}\mathfrak{Db}_X\big)\Big)$$
$$\longrightarrow R\mathcal{H}om_{\mathscr{D}_Y}\big(F_\dagger M,\mathfrak{Db}_Y\big) \tag{12.63}$$

Here, we have used the trace map (12.24) in the last map. Recall the compatibility of the push-forward and the de Rham functor $\mathrm{DR}_Y\circ F_\dagger\simeq F_!\circ\mathrm{DR}_X$ given as follows:

$$\omega_Y\otimes^L_{\mathscr{D}_Y}F_!(\mathscr{D}_{Y\leftarrow X}\otimes^L_{\mathscr{D}_X}N)\simeq F_!\big(F^{-1}(\omega_Y)\otimes^L_{F^{-1}(\mathscr{D}_Y)}\mathscr{D}_{Y\leftarrow X}\otimes^L_{\mathscr{D}_X}N\big)$$
$$\simeq F_!\big(\omega_X\otimes^L_{\mathscr{D}_X}N\big) \tag{12.64}$$

Hence, we have the induced morphism

$$F_!\,\mathrm{DR}_{\overline{X}}\mathcal{C}_XM\simeq\mathrm{DR}_{\overline{Y}}F_\dagger\mathcal{C}_XM\longrightarrow\mathrm{DR}_{\overline{Y}}\mathcal{C}_YF_\dagger M. \tag{12.65}$$

We also have the following isomorphism:

$$F_!\,\mathrm{DR}_X\boldsymbol{D}_XM\simeq\mathrm{DR}_Y\,F_\dagger\boldsymbol{D}_XM\longrightarrow\mathrm{DR}_Y\boldsymbol{D}_YF_\dagger M. \tag{12.66}$$

(See [34] for the compatibility $\boldsymbol{D}_X\circ\mathrm{DR}_X\simeq\mathrm{DR}_X\circ\boldsymbol{D}_X$.) We shall show the following lemma.

Lemma 12.5.2 *The following diagram in $D^b_c(\mathbb{C}_Y)$ is commutative:*

$$\begin{array}{ccc}F_!\,\mathrm{DR}_X\boldsymbol{D}_XM & \xrightarrow{\simeq} & F_!\,\mathrm{DR}_{\overline{X}}\mathcal{C}_XM\\ \downarrow & & \downarrow\\ \mathrm{DR}_Y\boldsymbol{D}_YF_\dagger M & \xrightarrow{\simeq} & \mathrm{DR}_{\overline{Y}}\mathcal{C}_YF_\dagger M\end{array} \tag{12.67}$$

The vertical arrows are given by (12.65) *and* (12.66)*. In particular, the induced morphism $F_\dagger\mathcal{C}_X(M)\longrightarrow\mathcal{C}_YF_\dagger(M)$ is an isomorphism.*

Proof We consider objects $C_1^X:=\mathcal{O}_X[d_X]$ and $C_2^X:=\omega_{\overline{X}}\otimes^L_{\mathscr{D}_{\overline{X}}}\mathfrak{Db}_X$ in $D^b(\mathscr{D}_X)$. We have the following expressions:

$$\mathrm{DR}_X\boldsymbol{D}_X(M)=R\mathcal{H}om_{\mathscr{D}_X}\big(M,\,C_1^X\big),\quad \mathrm{DR}_{\overline{X}}\mathcal{C}_X(M)=R\mathcal{H}om_{\mathscr{D}_X}\big(M,\,C_2^X\big).$$

We have the natural isomorphism $\eta_X : C_1^X\simeq C_2^X$ in $D^b(\mathscr{D}_X)$ which induces the isomorphism $\mathrm{DR}_X\boldsymbol{D}_X(M)\simeq\mathrm{DR}_{\overline{X}}\mathcal{C}_X(M)$. We have the trace morphisms tr : $F_\dagger C_i^X\longrightarrow C_i^Y$ in $D^b(\mathscr{D}_Y)$, and the following diagram is commutative in $D^b(\mathscr{D}_Y)$:

$$\begin{array}{ccc}F_\dagger C_1^X & \xrightarrow{F_\dagger\eta_X} & F_\dagger C_2^X\\ {\scriptstyle\mathrm{tr}}\downarrow & & {\scriptstyle\mathrm{tr}}\downarrow\\ C_1^Y & \xrightarrow{\eta_Y} & C_2^Y\end{array} \tag{12.68}$$

Lemma 12.5.3 *The natural isomorphisms*

$$F_! \,\mathrm{DR}_{\overline{X}}\, \mathcal{C}_X(M) \to \mathrm{DR}_{\overline{Y}}\, \mathcal{C}_Y F_\dagger(M) \quad \text{and} \quad F_! \,\mathrm{DR}_X \boldsymbol{D}_X(M) \to \mathrm{DR}_Y \boldsymbol{D}_Y F_\dagger(M)$$

are given as the composite of the following natural morphisms:

$$\begin{aligned} F_! R\mathcal{H}om_{\mathscr{D}_X}\big(M, C_i^X\big) &\longrightarrow F_! R\mathcal{H}om_{F^{-1}(\mathscr{D}_Y)}\big(\mathscr{D}_{Y\leftarrow X} \otimes^L_{\mathscr{D}_X} M,\ \mathscr{D}_{Y\leftarrow X} \otimes^L_{\mathscr{D}_X} C_i^X\big) \\ &\longrightarrow R\mathcal{H}om_{\mathscr{D}_Y}\big(F_\dagger(M),\ F_\dagger(C_i^X)\big) \longrightarrow R\mathcal{H}om_{\mathscr{D}_Y}\big(F_\dagger(M),\ C_i^Y\big) \end{aligned} \tag{12.69}$$

Proof Let us consider the claim for $\mathcal{C}$. The claim for $\boldsymbol{D}$ can be argued similarly. Let $\mathcal{I}_1^\bullet$ be a $\mathscr{D}_X \otimes F^{-1}(\mathscr{D}_{\overline{Y}})$-injective resolution of $\mathscr{D}_{\overline{Y}\leftarrow\overline{X}} \otimes^L_{\mathscr{D}_{\overline{X}}} \mathfrak{Db}_X$ obtained by the Godement construction. Then, $F_! \mathcal{H}om_{\mathscr{D}_X}\big(M, \mathcal{I}_1^\bullet\big)$ represents

$$F_! R\mathcal{H}om_{\mathscr{D}_X}(M, \mathscr{D}_{\overline{Y}\leftarrow\overline{X}} \otimes^L_{\mathscr{D}_{\overline{X}}} \mathfrak{Db}_X).$$

We remark that $\omega_{\overline{Y}}$ is represented by $\omega_{\overline{Y}}^\sim := \Omega_{\overline{Y}}^\bullet \otimes_{\mathcal{O}_{\overline{Y}}} \mathscr{D}_{\overline{Y}}[d_Y]$, and $\Omega_{\overline{Y}}^j \otimes_{\mathcal{O}_{\overline{Y}}} \mathscr{D}_{\overline{Y}}$ are right locally $\mathscr{D}_{\overline{Y}}$-free modules. Then, $F^{-1}(\omega_{\overline{Y}}^\sim) \otimes_{F^{-1}\mathscr{D}_{\overline{Y}}} \mathcal{I}_1^\bullet$ is a $\mathscr{D}_X$-injective resolution of $F^{-1}\omega_{\overline{Y}}^\sim \otimes_{F^{-1}\mathscr{D}_{\overline{Y}}} \mathscr{D}_{\overline{Y}\leftarrow\overline{X}} \otimes^L_{\mathscr{D}_{\overline{X}}} \mathfrak{Db}_X \simeq \omega_{\overline{X}} \otimes^L_{\mathscr{D}_{\overline{X}}} \mathfrak{Db}_X$. Hence, the isomorphism $\mathrm{DR}_{\overline{Y}} F_\dagger \mathcal{C}_X M \simeq F_! \,\mathrm{DR}_{\overline{X}}\, \mathcal{C}_X M$ is expressed as follows:

$$\begin{aligned} &\omega_{\overline{Y}}^\sim \otimes_{\mathscr{D}_{\overline{Y}}} F_! R\mathcal{H}om_{\mathscr{D}_X}\big(M,\ \mathscr{D}_{\overline{Y}\leftarrow\overline{X}} \otimes^L_{\mathscr{D}_{\overline{X}}} \mathfrak{Db}_X\big) \\ &\quad \simeq F_! R\mathcal{H}om_{\mathscr{D}_X}\big(M,\ F^{-1}\omega_{\overline{Y}}^\sim \otimes_{F^{-1}\mathscr{D}_{\overline{Y}}} \mathscr{D}_{\overline{Y}\leftarrow\overline{X}} \otimes^L_{\mathscr{D}_{\overline{X}}} \mathfrak{Db}_X\big) \\ &\quad \simeq F_! R\mathcal{H}om_{\mathscr{D}_X}\big(M, \omega_{\overline{X}} \otimes^L_{\mathscr{D}_{\overline{X}}} \mathfrak{Db}_X\big) \end{aligned} \tag{12.70}$$

The morphism $\mathrm{DR}_{\overline{Y}} F_\dagger \mathcal{C}_X(M) \longrightarrow \mathrm{DR}_{\overline{Y}}\, \mathcal{C}_Y F_\dagger(M)$ is the composite of the following:

$$\begin{aligned} &\omega_{\overline{Y}}^\sim \otimes_{\mathscr{D}_{\overline{Y}}} F_! R\mathcal{H}om_{\mathscr{D}_X}\big(M,\ \mathscr{D}_{\overline{Y}\leftarrow\overline{X}} \otimes^L_{\mathscr{D}_{\overline{X}}} \mathfrak{Db}_X\big) \\ &\quad \longrightarrow \omega_{\overline{Y}}^\sim \otimes_{\mathscr{D}_{\overline{Y}}} F_! R\mathcal{H}om_{F^{-1}\mathscr{D}_Y}\big(\mathscr{D}_{Y\leftarrow X} \otimes^L_{\mathscr{D}_X} M,\ \mathscr{D}_{Y\leftarrow X, \overline{Y}\leftarrow\overline{X}} \otimes^L_{\mathscr{D}_{X,\overline{X}}} \mathfrak{Db}_X\big) \\ &\quad \longrightarrow \omega_{\overline{Y}} \otimes^L_{\mathscr{D}_{\overline{Y}}} R\mathcal{H}om_{\mathscr{D}_Y}\Big(F_*(\mathscr{D}_{Y\leftarrow X} \otimes^L_{\mathscr{D}_X} M),\ F_!\big(\mathscr{D}_{Y\leftarrow X, \overline{Y}\leftarrow X} \otimes^L_{\mathscr{D}_{X,\overline{X}}} \mathfrak{Db}_X\big)\Big) \\ &\quad \longrightarrow \omega_{\overline{Y}} \otimes^L_{\mathscr{D}_{\overline{Y}}} R\mathcal{H}om_{\mathscr{D}_Y}\big(F_\dagger M,\ \mathfrak{Db}_Y\big) \end{aligned} \tag{12.71}$$

Let A_i denote the i-th term in (12.69). Let B_i denote the i-th term in (12.71). For each $i = 1, 2, 3$, we have a natural morphism $B_i \longrightarrow A_i$ given as in (12.70). We also have a natural isomorphism $B_4 \simeq A_4$. We have only to show the commutativity of the following diagrams for $i = 1,2,3$:

$$\begin{array}{ccc} B_i & \longrightarrow & B_{i+1} \\ \downarrow & & \downarrow \\ A_i & \longrightarrow & A_{i+1} \end{array} \tag{12.72}$$

Let us consider the diagram (12.72) with $i = 1,2$. Let $\mathcal{I}_1^\bullet$ and $\omega_{\widetilde{Y}}$ be as above. We take a resolution $\mathscr{D}^f_{Y\leftarrow X}$ of $\mathscr{D}_{Y\leftarrow X}$, which is $\mathscr{D}_X$-flat and c-soft with respect to F. We take a $F^{-1}(\mathscr{D}_{Y,\overline{Y}})$-injective resolution $\mathcal{I}_2^\bullet$ of $\mathscr{D}_{Y\leftarrow X,\overline{Y}\leftarrow\overline{X}} \otimes^L_{\mathscr{D}_{X,\overline{X}}} \mathfrak{Db}_X$ (see the proof of Lemma 12.5.4, for example), and a morphism $\mathscr{D}^f_{Y\leftarrow X} \otimes_{\mathscr{D}_X} \mathcal{I}_1^\bullet \longrightarrow \mathcal{I}_2^\bullet$, which represents the natural isomorphism

$$\mathscr{D}_{Y\leftarrow X} \otimes^L_{\mathscr{D}_X} \big(\mathscr{D}_{\overline{Y}\leftarrow\overline{X}} \otimes^L_{\mathscr{D}_{\overline{X}}} \mathfrak{Db}\big) \longrightarrow \mathscr{D}_{Y\leftarrow X,\overline{Y}\leftarrow\overline{X}} \otimes^L_{\mathscr{D}_{X,\overline{X}}} \mathfrak{Db}_X.$$

Because $F^{-1}(\omega_{\widetilde{Y}}) \otimes_{F^{-1}\mathscr{D}_{\overline{Y}}} \mathcal{I}_2^\bullet$ is a $F^{-1}(\mathscr{D}_Y)$-injective resolution of $\mathscr{D}_{Y\leftarrow X} \otimes^L_{\mathscr{D}_X} \big(\omega_{\overline{X}} \otimes^L_{\mathscr{D}_{\overline{X}}} \mathfrak{Db}_X\big)$, the diagram (12.72) with $i = 1$ is represented by the following diagram, which is commutative:

$$\begin{array}{ccc} \omega_{\widetilde{Y}} \otimes_{\mathscr{D}_{\overline{Y}}} F_!\mathcal{H}om_{\mathscr{D}_X}(M, \mathcal{I}_1^\bullet) & \longrightarrow & \omega_{\widetilde{Y}} \otimes_{\mathscr{D}_{\overline{Y}}} F_!\mathcal{H}om_{F^{-1}(\mathscr{D}_Y)}(\mathscr{D}^f_{Y\leftarrow X} \otimes M, \mathcal{I}_2^\bullet) \\ \downarrow & & \downarrow \\ F_!\mathcal{H}om_{\mathscr{D}_X}(M, F^{-1}(\omega_{\widetilde{Y}}) \otimes_{F^{-1}\mathscr{D}_{\overline{Y}}} \mathcal{I}_1^\bullet) & \longrightarrow & F_!\mathcal{H}om_{F^{-1}\mathscr{D}_Y}(\mathscr{D}^{\widetilde{f}}_{Y\leftarrow X} \otimes_{\mathscr{D}_X} M, F^{-1}(\omega_{\widetilde{Y}}) \otimes_{F^{-1}\mathscr{D}_{\overline{Y}}} \mathcal{I}_2^\bullet) \end{array}$$

The diagram (12.72) with $i = 2$ is expressed by the following diagram, which is commutative:

$$\begin{array}{ccc} \omega_{\widetilde{Y}} \otimes_{\mathscr{D}_{\overline{Y}}} F_!\mathcal{H}om_{F^{-1}(\mathscr{D}_Y)}(\mathscr{D}^f_{Y\leftarrow X} \otimes M, \mathcal{I}_2^\bullet) & \to & \omega_{\widetilde{Y}} \otimes_{\mathscr{D}_{\overline{Y}}} \mathcal{H}om_{\mathscr{D}_Y}(F_\dagger(M), F_!\mathcal{I}_2^\bullet) \\ \downarrow & & \downarrow \\ F_!\mathcal{H}om_{F^{-1}\mathscr{D}_Y}(\mathscr{D}^f_{Y\leftarrow X} \otimes_{\mathscr{D}_X} M, F^{-1}(\omega_{\widetilde{Y}}) \otimes_{F^{-1}\mathscr{D}_{\overline{Y}}} \mathcal{I}_2^\bullet) & \to & \mathcal{H}om_{\mathscr{D}_Y}\Big(F_\dagger M, F_!\big(F^{-1}(\omega_{\widetilde{Y}}) \otimes_{F^{-1}\mathscr{D}_{\overline{Y}}} \mathcal{I}_2^\bullet\big)\Big) \end{array}$$

The diagram (12.72) with $i = 3$ is commutative because of the construction of the trace morphism. Thus, the proof of Lemma 12.5.3 is finished. □

Now, Lemma 12.5.2 follows from Lemma 12.5.3 and the commutativity of (12.68).

12.5.3 Pairing on the Push-Forward

Let $(M_1, M_2, C) \in \mathscr{D}\text{-Tri}^{nd}(X)$. Let $F : X \longrightarrow Y$ be a morphism such that the restriction of F to $\operatorname{Supp} M_i$ are proper. We have $F_\dagger\varphi_{C,1,2} : F_\dagger\overline{M}_2 \longrightarrow F_\dagger C_X M_1$, and the morphism $b : F_\dagger C_X M_1 \simeq C_Y F_\dagger M_1$ in Sect. 12.5.2. We also have $\varphi_{F_\dagger^{(0)}C,1,2} :$ $F_\dagger\overline{M}_2 \longrightarrow C_Y F_\dagger M_1$.

Lemma 12.5.4 *We have* $b \circ F_{\dagger}\varphi_{C,1,2} = \varphi_{F_{\dagger}^{(0)}C,1,2}$.

Proof We set $\mathscr{D}^{f}_{\overline{Y}\leftarrow\overline{X}} := F^{-1}(\mathscr{D}_{\overline{Y}}\otimes\omega_{\overline{Y}}^{-1})\otimes_{F^{-1}\mathcal{O}_{\overline{Y}}}\mathcal{E}^{\bullet}_{\overline{X}}\otimes_{\mathcal{O}_{\overline{X}}}\mathscr{D}_{\overline{X}}[d_X]$. We use the notation $\mathscr{D}^{f}_{Y\leftarrow X}$ in a similar meaning. It naturally gives a resolution of $\mathscr{D}_{\overline{Y}\leftarrow\overline{X}}$, which is $\mathscr{D}_{\overline{X}}$-flat and c-soft with respect to F. Then, $\varphi_{F_{\dagger}^{(0)}C,1,2}$ is represented as the composition of the following morphisms:

$$\begin{aligned}
&F_!\big(\mathscr{D}^{f}_{\overline{Y}\leftarrow\overline{X}}\otimes_{\mathscr{D}_{\overline{X}}}\overline{M}_2\big)\\
&\longrightarrow F_!\mathcal{H}om_{F^{-1}\mathscr{D}_Y}\big(\mathscr{D}^{f}_{Y\leftarrow X}\otimes_{\mathscr{D}_X}M_1, F^{-1}(\mathscr{D}_{Y,\overline{Y}}\otimes\omega_{Y,\overline{Y}}^{-1})\otimes_{F^{-1}(\mathcal{O}_{Y,\overline{Y}})}\mathfrak{Db}^{\bullet}_X\big)[2d_X]\\
&\longrightarrow \mathcal{H}om_{\mathscr{D}_Y}\Big(F_*\big(\mathscr{D}^{f}_{Y\leftarrow X}\otimes_{\mathscr{D}_X}M_1\big), \mathscr{D}_{Y,\overline{Y}}\otimes\omega_{Y,\overline{Y}}^{-1}\otimes_{\mathcal{O}_{Y,\overline{Y}}}F_!\mathfrak{Db}^{\bullet}_X\Big)[2d_X]\\
&\longrightarrow \mathcal{H}om_{\mathscr{D}_Y}\Big(F_*\big(\mathscr{D}^{f}_{Y\leftarrow X}\otimes_{\mathscr{D}_X}M_1\big), \mathfrak{Db}_Y\Big)
\end{aligned}\tag{12.73}$$

Take a $F^{-1}(\mathscr{D}_{Y,\overline{Y}})$-injective resolution $\mathcal{J}_1^{\bullet}$ of $F^{-1}(\mathscr{D}_{Y,\overline{Y}}\otimes\omega_{Y,\overline{Y}}^{-1})\otimes_{F^{-1}\mathcal{O}_{Y,\overline{Y}}}\mathfrak{Db}^{\bullet}_X[2d_X]$ which is isomorphic to $\mathscr{D}_{Y\leftarrow X,\overline{Y}\leftarrow\overline{X}}\otimes^{L}_{\mathscr{D}_{X,\overline{X}}}\mathfrak{Db}_X$ in $D^b(F^{-1}(\mathscr{D}_{Y,\overline{Y}}))$. We take a $\mathscr{D}_{Y,\overline{Y}}$-injective resolution $\mathcal{J}_2^{\bullet}$ of $\mathfrak{Db}_Y$, and a morphism $F_!\mathcal{J}_1^{\bullet}\longrightarrow\mathcal{J}_2^{\bullet}$ such that the following diagram in $D^b(\mathscr{D}_{Y,\overline{Y}})$ is commutative:

$$\begin{array}{ccc}
F_!\Big(F^{-1}(\mathscr{D}_{Y,\overline{Y}}\otimes\omega_{Y,\overline{Y}}^{-1})\otimes_{F^{-1}\mathcal{O}_{Y,\overline{Y}}}\mathfrak{Db}^{\bullet}_X[2d_X]\Big) & \longrightarrow & \mathfrak{Db}_Y\\
\downarrow & & \downarrow\\
F_!\mathcal{J}_1^{\bullet} & \longrightarrow & \mathcal{J}_2^{\bullet}
\end{array}$$

Then, $\varphi_{F_{\dagger}^{(0)}C,1,2}$ is represented by the composition of the following morphisms:

$$\begin{aligned}
F_!(\mathscr{D}^{f}_{\overline{Y}\leftarrow\overline{X}}\otimes_{\mathscr{D}_{\overline{X}}}\overline{M}_2) &\longrightarrow F_!\mathcal{H}om_{F^{-1}(\mathscr{D}_Y)}\Big(\mathscr{D}^{f}_{Y\leftarrow X}\otimes_{\mathscr{D}_X}M_1, \mathcal{J}_1^{\bullet}\Big)\\
&\longrightarrow \mathcal{H}om_{\mathscr{D}_Y}\Big(F_!\big(\mathscr{D}^{f}_{Y\leftarrow X}\otimes_{\mathscr{D}_X}M_1\big), F_!\mathcal{J}_1^{\bullet}\Big)\\
&\longrightarrow \mathcal{H}om_{\mathscr{D}_Y}\Big(F_!\big(\mathscr{D}^{f}_{Y\leftarrow X}\otimes_{\mathscr{D}_X}M_1\big), \mathcal{J}_2^{\bullet}\Big)
\end{aligned}\tag{12.74}$$

Hence, we obtain the following factorization of $\varphi_{F_{\dagger}^{(0)}C,1,2}$:

$$\begin{aligned}
F_{\dagger}\overline{M}_2 &\xrightarrow{A} F_!R\mathcal{H}om_{F^{-1}(\mathscr{D}_Y)}\big(\mathscr{D}_{Y\leftarrow X}\otimes^{L}_{\mathscr{D}_X}M_1,\ \mathscr{D}_{Y\leftarrow X,\overline{Y}\leftarrow\overline{X}}\otimes^{L}_{\mathscr{D}_{X,\overline{X}}}\mathfrak{Db}_X\big)\\
&\xrightarrow{A_1} R\mathcal{H}om_{\mathscr{D}_Y}\Big(F_{\dagger}M_1,\ F_!\big(\mathscr{D}_{Y\leftarrow X,\overline{Y},\leftarrow\overline{X}}\otimes^{L}_{\mathscr{D}_{X,\overline{X}}}\mathfrak{Db}_X\big)\Big)\\
&\xrightarrow{A_2} R\mathcal{H}om_{\mathscr{D}_Y}\big(F_{\dagger}M_1,\mathfrak{Db}_Y\big)
\end{aligned}\tag{12.75}$$

Let us look at the morphism A. It is obtained as the push-forward of the composition A' of the following morphisms:

$$\begin{aligned}
&\mathscr{D}^f_{\overline{Y}\leftarrow\overline{X}} \otimes_{\mathscr{D}_{\overline{Y}}} \overline{M}_2 \\
&\longrightarrow \mathcal{H}om_{F^{-1}(\mathscr{D}_Y)}\Big(\mathscr{D}^f_{Y\leftarrow X}\otimes_{\mathscr{D}_X} M_1,\, F^{-1}\big(\mathscr{D}_{Y,\overline{Y}}\otimes\omega^{-1}_{Y,\overline{Y}}\big)\otimes_{F^{-1}\mathcal{O}_{Y,\overline{Y}}}\mathfrak{Db}^{\bullet}_X\Big)[2d_X] \\
&\longrightarrow \mathcal{H}om_{F^{-1}(\mathscr{D}_Y)}\big(\mathscr{D}^f_{Y\leftarrow X}\otimes_{\mathscr{D}_X} M_1, \mathcal{J}_1^{\bullet}\big)
\end{aligned}\tag{12.76}$$

We set $\mathscr{D}^g_{\overline{Y}\leftarrow\overline{X}} := F^{-1}(\mathscr{D}_{\overline{Y}}\otimes\omega^{-1}_{\overline{Y}})\otimes_{F^{-1}\mathcal{O}_{\overline{Y}}}\Omega^{\bullet}_{\overline{X}}\otimes\mathscr{D}_{\overline{X}}[d_X]$ and $\mathscr{D}^g_{Y\leftarrow X} := F^{-1}(\mathscr{D}_Y\otimes\omega^{-1}_Y)\otimes_{F^{-1}\mathcal{O}_Y}\Omega^{\bullet}_X\otimes\mathscr{D}_X[d_X]$. We have $\big(\mathscr{D}^g_{Y\leftarrow X}\otimes_{\mathbb{C}}\mathscr{D}^g_{\overline{Y}\leftarrow\overline{X}}\big)\otimes_{\mathscr{D}_{X,\overline{X}}}\mathfrak{Db}_X \simeq F^{-1}(\mathscr{D}_{Y,\overline{Y}}\otimes\omega^{-1}_{Y,\overline{Y}})\otimes_{F^{-1}\mathcal{O}_{Y,\overline{Y}}}\mathfrak{Db}^{\bullet}_X[2d_X]$, where we use the identification of $\Omega^{\bullet}_X[d_X]\otimes\Omega^{\bullet}_{\overline{X}}[d_X]\otimes\mathfrak{Db}_X\simeq\mathfrak{Db}^{\bullet}[2d_X]$. In $D^b(F^{-1}\mathscr{D}_{\overline{Y}})$, the morphism A' is represented by the composition A'' of the following morphisms:

$$\begin{aligned}
&\mathscr{D}^g_{\overline{Y}\leftarrow\overline{X}} \otimes_{\mathscr{D}_{\overline{Y}}} \overline{M}_2 \\
&\xrightarrow{B_0} \mathcal{H}om_{F^{-1}(\mathscr{D}_Y)}\Big(\mathscr{D}^g_{Y\leftarrow X}\otimes_{\mathscr{D}_X} M_1,\, F^{-1}\big(\mathscr{D}_{Y,\overline{Y}}\otimes\omega_{Y,\overline{Y}}\big)\otimes_{F^{-1}\mathcal{O}_{Y,\overline{Y}}}\mathfrak{Db}^{\bullet}_X\Big)[2d_X] \\
&\longrightarrow \mathcal{H}om_{F^{-1}(\mathscr{D}_Y)}\big(\mathscr{D}^g_{Y\leftarrow X}\otimes_{\mathscr{D}_X} M_1, \mathcal{J}_1^{\bullet}\big)
\end{aligned}\tag{12.77}$$

We take a $\mathscr{D}_X\otimes F^{-1}(\mathscr{D}_{\overline{Y}})$-injective resolution $\mathcal{J}_3^{\bullet}$ of $\mathscr{D}^g_{\overline{Y}\leftarrow\overline{X}}\otimes_{\mathscr{D}_{\overline{X}}}\mathfrak{Db}_X$. We have a morphism $\mathscr{D}^g_{Y\leftarrow X}\otimes_{\mathscr{D}_X}\mathcal{J}_3^{\bullet}\longrightarrow\mathcal{J}_1^{\bullet}$, which represents the natural isomorphism

$$\mathscr{D}^g_{Y\leftarrow X}\otimes_{\mathscr{D}_X}\big(\mathscr{D}^g_{\overline{Y}\leftarrow\overline{X}}\otimes_{\mathscr{D}_{\overline{X}}}\mathfrak{Db}\big)\simeq F^{-1}(\mathscr{D}_{Y,\overline{Y}}\otimes\omega^{-1}_{Y,\overline{Y}})\otimes_{F^{-1}\mathcal{O}_{Y,\overline{Y}}}\mathfrak{Db}^{\bullet}_X[2d_X].$$

The composition of the following morphisms is denoted by B_1:

$$\mathscr{D}^g_{\overline{Y}\leftarrow\overline{X}}\otimes_{\mathscr{D}_{\overline{X}}}\overline{M}_2\longrightarrow\mathcal{H}om_{\mathscr{D}_X}\big(M_1,\mathscr{D}^g_{\overline{Y}\leftarrow\overline{X}}\otimes\mathfrak{Db}_X\big)\longrightarrow\mathcal{H}om_{\mathscr{D}_X}(M_1,\mathcal{J}_3^{\bullet})$$

Here, the first one is induced by $\varphi_{C,1,2}:\overline{M}_2\longrightarrow\mathcal{H}om_{\mathscr{D}_X}(M_1,\mathfrak{Db}_X)$. The composite of the following morphisms is denoted by B_2:

$$\begin{aligned}
\mathcal{H}om_{\mathscr{D}_X}(M_1,\mathcal{J}_3^{\bullet})&\longrightarrow\mathcal{H}om_{F^{-1}\mathscr{D}_Y}\big(\mathscr{D}^g_{Y\leftarrow X}\otimes_{\mathscr{D}_X}M_1,\,\mathscr{D}^g_{Y\leftarrow X}\otimes_{\mathscr{D}_X}\mathcal{J}_3^{\bullet}\big)\\
&\longrightarrow\mathcal{H}om_{F^{-1}\mathscr{D}_Y}\big(\mathscr{D}^g_{Y\leftarrow X}\otimes_{\mathscr{D}_X}M_1,\,\mathcal{J}_1^{\bullet}\big)
\end{aligned}\tag{12.78}$$

The composite $B_2\circ B_1$ factors through B_0. Hence, $B_2\circ B_1 = A''$ in $D^b(F^{-1}\mathscr{D}_Y)$. It means that the morphism A is factorized into

$$\begin{aligned}
&F_{\dagger}\overline{M}_2\xrightarrow{F_{\dagger}\varphi_{C,1,2}}F_{\dagger}\mathcal{C}_XM_1\xrightarrow{B_3}\\
&\qquad F_!R\mathcal{H}om_{F^{-1}(\mathscr{D}_Y)}\Big(\mathscr{D}_{Y\leftarrow X}\otimes^L_{\mathscr{D}_X}M_1,\,\mathscr{D}_{Y\leftarrow X,\overline{Y}\leftarrow\overline{X}}\otimes^L_{\mathscr{D}_{X,\overline{X}}}\mathfrak{Db}_X\Big)
\end{aligned}\tag{12.79}$$

We have $A_2 \circ A_1 \circ B_3 = b$. Thus, the proof of Lemma 12.5.4 is finished. □

We have the following commutative diagram as the compatibility of the de Rham functor and the push-forward:

$$\begin{array}{ccc} F_! \, \mathrm{DR}_{\overline{X}} \overline{M}_2 & \xrightarrow[\simeq]{F_! \, \mathrm{DR} \, \varphi_{C,1,2}} & F_! \, \mathrm{DR}_{\overline{X}} \mathcal{C}_X M_1 \\ \simeq \downarrow & & \simeq \downarrow \\ \mathrm{DR}_{\overline{Y}} \, F_\dagger \overline{M}_2 & \xrightarrow[\simeq]{\mathrm{DR} \, F_\dagger \varphi_{C,1,2}} & \mathrm{DR}_{\overline{Y}} \, F_\dagger \mathcal{C}_X M_1 \end{array} \tag{12.80}$$

Corollary 12.5.5 *We have $F_\dagger^j(M_1, M_2, C) \in \mathscr{D}\text{-}Tri^{nd}(Y)$.*

Proof By Lemmas 12.5.2, 12.5.4 and the commutative diagram (12.80), we obtain that the induced morphism $\mathrm{DR}\,\varphi_{F_\dagger^{(0)} C,1,2} : \mathrm{DR}_{\overline{Y}} F_\dagger \overline{M}_2 \longrightarrow \mathrm{DR}_{\overline{Y}} \mathcal{C}_Y F_\dagger M_1$ is an isomorphism. Because $F_\dagger^{(0)} C$ and $F_\dagger C$ are equal up to the signature, we obtain the corollary. □

12.5.4 Duality and Push-Forward

Let $\mathcal{T} = (M_1, M_2, C) \in \mathscr{D}\text{-Tri}^{nd}(X)$. Let $F : X \longrightarrow Y$ be a morphism such that its restriction to $\mathrm{Supp}\, M_i$ are proper. Assume that there exists a non-degenerate sesquilinear pairing $\boldsymbol{D}_X C : \boldsymbol{D}_X M_1 \times \boldsymbol{D}_{\overline{X}} \overline{M}_2 \longrightarrow \mathfrak{Db}_X$, which is a dual pairing of C. We have the induced pairings:

$$F_\dagger^{(0)} C : F_\dagger M_1 \times F_\dagger \overline{M}_2 \longrightarrow \mathfrak{Db}_Y$$

$$F_\dagger^{(0)} \boldsymbol{D}_X C : F_\dagger \boldsymbol{D}_X M_1 \times F_\dagger \boldsymbol{D}_{\overline{X}} \overline{M}_2 \longrightarrow \mathfrak{Db}_Y$$

Recall that we have a natural equivalence $F_\dagger \boldsymbol{D} M_i \simeq \boldsymbol{D} F_\dagger M_i$ under which we have the induced morphism

$$F_\dagger^{(0)} \boldsymbol{D}_X C : \boldsymbol{D}_Y F_\dagger M_1 \times \boldsymbol{D}_{\overline{Y}} F_\dagger \overline{M}_2 \longrightarrow \mathfrak{Db}_Y$$

in the derived category of $\mathscr{D}_{Y,\overline{Y}}$-complexes.

Lemma 12.5.6 *The dual $\boldsymbol{D}_Y F_\dagger^{(0)} C$ exists, and it is equal to $F_\dagger^{(0)} \boldsymbol{D}_X C$.*

Proof We shall compare the following morphisms

$$\mathrm{DR}\, F_\dagger^{(0)} \boldsymbol{D} C : \mathrm{DR}_Y \boldsymbol{D}_Y F_\dagger M_1 \otimes \mathrm{DR}_{\overline{Y}} \boldsymbol{D}_{\overline{Y}} F_\dagger \overline{M}_2 \longrightarrow \omega_Y^{top} \tag{12.81}$$

$$\big(\mathrm{DR}\, F_\dagger^{(0)} C\big) \circ \mathrm{ex} : \mathrm{DR}_{\overline{Y}}\big(F_\dagger \overline{M}_2\big) \otimes \mathrm{DR}_Y\big(F_\dagger M_1\big) \longrightarrow \omega_Y^{top} \tag{12.82}$$

We have the isomorphism

$$\Upsilon_1 : \mathrm{DR}_Y \boldsymbol{D}_Y F_\dagger M_1 \otimes \mathrm{DR}_{\overline{Y}} \boldsymbol{D}_{\overline{Y}} F_\dagger \overline{M}_2 \simeq \mathrm{DR}_{\overline{Y}}\big(F_\dagger \overline{M}_2\big) \otimes \mathrm{DR}_Y\big(F_\dagger M_1\big)$$

induced by the following isomorphisms:

$$\nu_{F_\dagger M_1} \circ \mathrm{DR}(\varphi_{F_\dagger^{(0)} C,1,2}) : \mathrm{DR}_{\overline{Y}} F_\dagger \overline{M}_2 \simeq \mathrm{DR}_Y \boldsymbol{D}_Y F_\dagger M_1$$

$$\nu_{F_\dagger \overline{M}_2} \circ \mathrm{DR}(\varphi_{F_\dagger^{(0)} C,2,1}) : \mathrm{DR}_Y F_\dagger M_1 \simeq \mathrm{DR}_{\overline{Y}} \boldsymbol{D}_{\overline{Y}} F_\dagger \overline{M}_2$$

We have only to show that (12.81) and (12.82) are equal under the identification Υ_1.

We have another identification

$$\Upsilon_2 : \mathrm{DR}_Y \boldsymbol{D}_Y F_\dagger M_1 \otimes \mathrm{DR}_{\overline{Y}} \boldsymbol{D}_{\overline{Y}} F_\dagger \overline{M}_2 \simeq \mathrm{DR}_{\overline{Y}}\big(F_\dagger \overline{M}_2\big) \otimes \mathrm{DR}_Y\big(F_\dagger M_1\big).$$

Indeed, we have the isomorphism $\kappa_{1,2} : \mathrm{DR}_{\overline{Y}} F_\dagger \overline{M}_2 \simeq \mathrm{DR}_Y \boldsymbol{D}_Y F_\dagger M_1$ induced by $\nu_{M_1} \circ \mathrm{DR}(\varphi_{C,1,2})$ given as follows:

$$\begin{aligned} \mathrm{DR}_{\overline{Y}} F_\dagger \overline{M}_2 \simeq F_! \mathrm{DR}_{\overline{X}} \overline{M}_2 \simeq F_! \mathrm{DR}_{\overline{X}} \mathcal{C}_X M_1 \simeq F_! \mathrm{DR}_X \boldsymbol{D}_X M_1 \\ \simeq \mathrm{DR}_Y \boldsymbol{D}_Y F_\dagger M_1 \end{aligned} \tag{12.83}$$

Similarly, we have the isomorphism $\kappa_{2,1} : \mathrm{DR}_Y F_\dagger M_1 \simeq \mathrm{DR}_{\overline{Y}} \boldsymbol{D}_{\overline{Y}} F_\dagger \overline{M}_2$. The identification Υ_2 is induced by $\kappa_{1,2}$ and $\kappa_{2,1}$. We can check that (12.81) and (12.82) are equal under the identification Υ_2 by construction. Hence, we have only to show $\nu_{F_\dagger M_1} \circ \mathrm{DR}\,\varphi_{F_\dagger^{(0)} C,1,2} = \kappa_{1,2}$ and $\nu_{F_\dagger \overline{M}_2} \circ \mathrm{DR}\,\varphi_{F_\dagger^{(0)} C,2,1} = \kappa_{2,1}$.

As for the first, we have only to show the commutativity of the following diagrams:

$$\begin{array}{ccccc} F_! \mathrm{DR}_{\overline{X}} \overline{M}_2 & \xrightarrow{F_! \mathrm{DR}\,\varphi_C} & F_! \mathrm{DR}_{\overline{X}} \mathcal{C}_X M_1 & \xrightarrow{F_! \nu_{M_1}} & F_! \mathrm{DR}_X \boldsymbol{D}_X M_1 \\ \simeq\downarrow & & \simeq\downarrow & & \simeq\downarrow \\ \mathrm{DR}_{\overline{Y}} F_\dagger \overline{M}_2 & \xrightarrow{\mathrm{DR}\,\varphi_{F_\dagger^{(0)} C}} & \mathrm{DR}_{\overline{Y}} \mathcal{C}_Y F_\dagger M_1 & \xrightarrow{\nu_{F_\dagger M_1}} & \mathrm{DR}_{\overline{X}} \boldsymbol{D}_{\overline{X}} F_\dagger M_1 \end{array} \tag{12.84}$$

The commutativity of the left square follows from Lemma 12.5.4 and the commutativity of (12.80). The commutativity of the right square is given in Lemma 12.5.2.

We can prove the second one by considering the non-degenerate $\mathscr{D}$-triple $C' : \overline{M}_2 \times M_1 \longrightarrow \mathfrak{Db}_X$ on the conjugate $\overline{X}$ given by $C'(\overline{m}_2, m_1) = C(m_1, \overline{m}_2)$. Note that the signature $(-1)^{d_X - d_Y}$ appears twice. One is caused by the effect of the change of the orientation on the trace map. The other is the difference of the identifications $\mathrm{Tot}\big(\Omega_X^\bullet[d_X] \otimes \Omega_{\overline{X}}^\bullet[d_X] \otimes \mathfrak{Db}_X\big) \simeq \mathfrak{Db}_X^\bullet[2d_X]$ and $\mathrm{Tot}\big(\Omega_Y^\bullet[d_Y] \otimes$

$\Omega^{\bullet}_{\overline{Y}}[d_Y] \otimes \mathfrak{Db}_Y) \simeq \mathfrak{Db}^{\bullet}_Y[2d_Y]$. Then, we can apply the argument for the first. Thus, we obtain Lemma 12.5.6. □

12.5.5 Canonical Prolongation of Good Meromorphic Flat Bundles

Let V_1 be a good meromorphic flat bundle on (X, D). We obtain a good meromorphic flat bundle $\mathcal{C}_X(V_1)(*D)$ on $(\overline{X}, \overline{D})$, and a good meromorphic flat bundle $V_2 := \overline{\mathcal{C}_X(V_1)(*D)}$ on (X, D).

Lemma 12.5.7 *The natural isomorphism $\overline{\mathcal{C}_X(V_1)_{|X\setminus D}} \simeq V_{2|X\setminus D}$ is uniquely extended to isomorphisms*

$$\overline{\mathcal{C}_X(V_1)} \simeq V_2[!D], \qquad \overline{\mathcal{C}_X(V_1[!D])} \simeq V_2.$$

Proof We set $\mathcal{M} := \overline{\mathcal{C}_X(V_1)}$. We have $\mathcal{M}(*D) = V_2$. Let $\mathcal{M}'$ be any holonomic $\mathscr{D}_X$-module such that $\mathcal{M}'(*D) = V_2$. We have a unique morphism $\varphi_1 : \overline{\mathcal{C}_X(\mathcal{M}')} \longrightarrow V_1$ such that $\varphi_{1|X\setminus D} = \mathrm{id}$. Hence, we have a morphism $\varphi_2 : \mathcal{M} \longrightarrow \mathcal{M}'$ such that $\varphi_{2|X\setminus D} = \mathrm{id}$. It means $\mathcal{M} = \mathcal{M}[!D] = V_2[!D]$ by the universal property of $V_2[!D]$. We obtain the other isomorphism similarly. □

Hence, for integers m, we obtain non-degenerate holonomic sesqui-linear pairings

$$\mathcal{T}_!(V_1, V_2, m) := \big(V_1[*D], V_2[!D], (-1)^m C_1\big),$$

$$\mathcal{T}_*(V_1, V_2, m) := \big(V_1[!D], V_2[*D], (-1)^m C_2\big).$$

Here C_i are non-degenerate sesqui-linear pairings corresponding to the isomorphisms in Lemma 12.5.7.

Let us consider their dual triple. By using the description of the Stokes structure, we have $V_2^{\vee} = \overline{\mathcal{C}_X(V_1^{\vee})(*D)}$. Hence, we have $\mathcal{T}_{\star}(V_1^{\vee}, V_2^{\vee}, m)$ for $\star = *, !$. We have the canonical isomorphisms $\boldsymbol{D}_X(V_i) \simeq V_i^{\vee}[!D]$ and $\boldsymbol{D}_X(V_i[!D]) \simeq V_i^{\vee}$.

Lemma 12.5.8 *$\mathcal{T}_{\star}(V_1, V_2, m)$ have duals, and we have*

$$\boldsymbol{D}\mathcal{T}_*(V_1, V_2, m) = \mathcal{T}_!(V_1^{\vee}, V_2^{\vee}, m + d_X), \quad \boldsymbol{D}\mathcal{T}_!(V_1, V_2, m) = \mathcal{T}_*(V_1^{\vee}, V_2^{\vee}, m + d_X).$$

Proof The case $D = \emptyset$ has already been proved in Example 12.4.2. Let $\pi : \tilde{X}(D) \longrightarrow X$ be the real blow up. Let $\mathcal{L}_i$ be the local system on $\tilde{X}(D)$ with the Stokes structure associated to V_i. Let $\mathcal{L}_i^{\vee}$ be the local system on $\tilde{X}(D)$ with the Stokes structure associated to $V_i^{\vee}$.

We have the associated constructible sheaves $\mathcal{L}_i^{<D}$ and $\mathcal{L}_i^{\leq D}$ from $\mathcal{L}_i$ with the Stokes structure. (See [57] for the notation.) Similarly, we have $\mathcal{L}_i^{\vee\,<D}$ and $\mathcal{L}_i^{\vee\,\leq D}$,

from $\mathcal{L}_i^{\vee}$ with the Stokes structure. Let $k : X \setminus D \longrightarrow \tilde{X}(D)$ be the inclusion. We set $\omega_{\tilde{X}(D)} := k_! \mathbb{C}_{X\setminus D}[2d_X]$, which is a dualizing complex. We set $\boldsymbol{D}_{\tilde{X}(D)}(\mathcal{F}) := R\mathcal{H}om_{\mathbb{C}_{\tilde{X}(D)}}(\mathcal{F}, \omega_{\tilde{X}(D)})$. We have natural identifications $\boldsymbol{D}_{\tilde{X}(D)}\mathcal{L}_i^{<D}[d_X] \simeq \mathcal{L}_i^{\vee \leq D}[d_X]$ and $\boldsymbol{D}_{\tilde{X}(D)}\mathcal{L}_i^{\leq D}[d_X] \simeq \mathcal{L}_i^{\vee <D}[d_X]$. We have naturally defined non-degenerate pairings $\mathcal{L}_1^{<D}[d_X] \times \overline{\mathcal{L}_2}^{\leq D}[d_X] \longrightarrow \omega_{\tilde{X}(D)}$ and $\mathcal{L}_1^{\leq D}[d_X] \times \overline{\mathcal{L}_2}^{<D}[d_X] \longrightarrow \omega_{\tilde{X}(D)}$. We also have $\mathcal{L}_1^{\vee <D}[d_X] \times \overline{\mathcal{L}}_2^{\vee \leq D}[d_X] \longrightarrow \omega_{\tilde{X}(D)}$ and $\mathcal{L}_1^{\vee \leq D}[d_X] \times \overline{\mathcal{L}}_2^{\vee <D}[d_X] \longrightarrow \omega_{\tilde{X}(D)}$.

Look at $\mathrm{DR}_X V_1[!D] \longrightarrow \boldsymbol{D}\,\mathrm{DR}_{\overline{X}} \overline{V}_2$ and $\mathrm{DR}_{\overline{X}} \overline{V}_2 \longrightarrow \boldsymbol{D}\,\mathrm{DR}_X V_1[!D]$ induced by the pairing $V_1[!D] \times \overline{V}_2 \longrightarrow \mathfrak{Db}_X$. The morphism $\mathrm{DR}_X V_1[!D] \longrightarrow \boldsymbol{D}\,\mathrm{DR}_{\overline{X}} \overline{V}_2$ is factorized as follows:

$$\mathrm{DR}_X V_1[!D] \longrightarrow \mathcal{H}om_{\mathbb{C}_X}\big(\mathrm{DR}_{\overline{X}} \overline{V}_2, \mathfrak{Db}_X^{\bullet}\big)[2d_X] \longrightarrow \boldsymbol{D}\,\mathrm{DR}_{\overline{X}} \overline{V}_2$$

We have the following commutative diagram:

$$\begin{array}{ccc}
\mathrm{DR}_X^{<D} V_1 & \longrightarrow & \mathcal{H}om_{\mathbb{C}_X}\big(\mathrm{DR}_{\overline{X}}^{\leq D} \overline{V}_2, \Omega_X^{\bullet <D}\big)[2d_X] \\
\downarrow & & \downarrow \\
\mathrm{DR}_X V_1[!D] & \longrightarrow & \mathcal{H}om_{\mathbb{C}_X}\big(\mathrm{DR}_{\overline{X}}^{\leq D} \overline{V}_2, \mathfrak{Db}_X^{\bullet}\big)[2d_X]
\end{array}$$

Hence, the isomorphism $\mathrm{DR}_X V_1[!D] \longrightarrow \boldsymbol{D}\,\mathrm{DR}_{\overline{X}} \overline{V}_2$ is obtained as the push-forward of the unique isomorphism $\mathcal{L}_1^{<D}[d_X] \simeq \boldsymbol{D}_{\tilde{X}(D)} \overline{\mathcal{L}_2}^{\leq D}[d_X]$ whose restriction to $X \setminus D$ is the identity.

The morphism $\mathrm{DR}_{\overline{X}} \overline{V}_2 \longrightarrow \boldsymbol{D}\,\mathrm{DR}_{\overline{X}} V_1[!D]$ is represented as the composite of the following morphisms

$$\begin{aligned}
\mathrm{DR}_{\overline{X}} \overline{V}_2 &\longrightarrow \mathcal{H}om_{\mathbb{C}_X}\big(\mathrm{DR}_X^{<D} V_1, \mathcal{E}_X^{\bullet}\big)[2d_X] \\
&\longrightarrow \mathcal{H}om_{\mathbb{C}_X}\big(\mathrm{DR}_X^{<D} V_1, \mathfrak{Db}_X^{\bullet}\big)[2d_X] \longrightarrow \boldsymbol{D}\,\mathrm{DR}_X^{<D} V_1
\end{aligned} \tag{12.85}$$

Hence, the isomorphism $\mathrm{DR}_{\overline{X}} \overline{V}_2 \longrightarrow \boldsymbol{D}\,\mathrm{DR}_X V_1[!D]$ is obtained as the push-forward of the unique isomorphism $\overline{\mathcal{L}_2}^{\leq D}[d_X] \simeq \boldsymbol{D}_{\tilde{X}(D)}\mathcal{L}_1^{<D}[d_X] = \mathcal{L}_1^{\vee \leq D}[d_X]$ whose restriction to $X \setminus D$ is the identity.

Let us consider the diagram:

$$\begin{array}{ccc}
\mathrm{DR}_X(V_1^{\vee}) \times \mathrm{DR}_{\overline{X}}(\overline{V}_2[!D]) & \xrightarrow{A_2} & \omega_X^{top} \\
{\scriptstyle A_1}\downarrow & & \downarrow{\scriptstyle =} \\
\mathrm{DR}_{\overline{X}}(\overline{V}_2) \times \mathrm{DR}_X(V_1[!D]) & \xrightarrow{(-1)^{d_X} A_3} & \omega_X^{top}
\end{array} \tag{12.86}$$

The identification A_1 and the pairings A_i $(i = 2, 3)$ are obtained as the push forward of the identification and the pairings on the real blow up. Hence, the diagram (12.86) is commutative. □

Let g be a holomorphic function on X such that $g^{-1}(0) = D$. Combined with Beilinson construction, we obtain the following non-degenerate sesqui-linear pairings, which are mutually dual:

$$\Pi_{g!}^{a,b} C_1 : \Pi_{g*}^{-b+1,-a+1} V_1 \times \overline{\Pi_{g!}^{a,b} V_2} \longrightarrow \mathfrak{Db}_X$$

$$(-1)^{d_X} \Pi_{g*}^{-b+1,-a+1} C_2 : \Pi_{g!}^{a,b} V_1^{\vee} \times \overline{\Pi_{g*}^{-b+1,-a+1} V_2^{\vee}} \longrightarrow \mathfrak{Db}_X$$

We obtain the following non-degenerate sesqui-linear pairings, which are mutually dual:

$$\Xi_g^{(a)} C_1 : \Xi_g^{(-a)} V_1 \times \overline{\Xi_g^{(a)} V_2} \longrightarrow \mathfrak{Db}_X$$

$$(-1)^{d_X} \Xi_g^{(-a)} C_2 : \Xi_g^{(a)} V_1^{\vee} \times \overline{\Xi_g^{(-a)} V_2^{\vee}} \longrightarrow \mathfrak{Db}_X$$

We also obtain the following sesqui-linear pairings which are mutually dual:

$$\psi_g^{(a)} C_1 : \psi_g^{(-a+1)} V_1 \times \overline{\psi_g^{(a)} V_2} \longrightarrow \mathfrak{Db}_X$$

$$(-1)^{d_X} \psi_g^{(-a+1)} C_2 : \psi_g^{(a)} V_1^{\vee} \times \overline{\psi_g^{(-a+1)} V_2^{\vee}} \longrightarrow \mathfrak{Db}_X$$

Namely, there exist $\boldsymbol{D}\psi_g^{(a)} \mathcal{T}_{\star}(V_1, V_2)$ and $\boldsymbol{D}\Xi_g^{(a)} \mathcal{T}_{\star}(V_1, V_2)$.

12.5.6 Special Case

Let $\mathcal{T} = (M_1, M_2, C) \in \mathscr{D}\text{-Tri}^{nd}(X)$. Assume that there exists a good cell $\mathcal{C} = (Z, U, \varphi, V_1)$ for M_1 at $P \in \operatorname{Supp} \mathcal{T}$. (See [57] for the notion of good cells.) Let us show that, on a small neighbourhood X_P of P, there exists a dual pairing $\boldsymbol{D}C$: $\boldsymbol{D}M_{1|X_P} \times \overline{\boldsymbol{D}M_{2|X_P}} \longrightarrow \mathfrak{Db}_{X_P}$.

Let $V_2 := \overline{\mathcal{C}_Z(V_1)}(*D_Z)$. Because $M_2 \longrightarrow \overline{\mathcal{C}_X(M_1)}$ is an isomorphism, the tuple (Z, U, φ, V_2) is a good cell of M_2. We have the objects $\mathcal{T}_{V!} := (V_{1*}, V_{2!}, C_{1!}) \in \mathscr{D}\text{-Tri}^{nd}(Z)$ and $\mathcal{T}_{V*} := (V_{1!}, V_{2*}, C_{1*}) \in \mathscr{D}\text{-Tri}^{nd}(Z)$. We have $\varphi_{\dagger} V_{1!} \longrightarrow \mathcal{M}_1 \longrightarrow \varphi_{\dagger} V_{1*}$. By applying the functor $\overline{\mathcal{C}_X}$, we obtain $\varphi_{\dagger} V_{2!} \longrightarrow \mathcal{M}_2 \longrightarrow \varphi_{\dagger} V_{2*}$. By construction of the pairings, we obtain the morphisms $\varphi_{\dagger} \mathcal{T}_{V!} \longrightarrow \mathcal{T} \longrightarrow \varphi_{\dagger} \mathcal{T}_{V*}$. Let g be a cell function for (Z, U, φ, V_1), and we put $g_Z := g \circ \varphi$. Recall that we obtain $\phi_g^{(0)} \mathcal{T} \in \mathscr{D}\text{-Tri}^{nd}(X)$ as the cohomology of the complex $\varphi_{\dagger} \mathcal{T}_{V!} \longrightarrow \varphi_{\dagger} \Xi_{g_Z}^{(0)} \mathcal{T}_V \oplus \mathcal{T} \longrightarrow \varphi_{\dagger} \mathcal{T}_{V*}$.

Suppose that the claim of Theorem 12.4.1 holds for any non-degenerate pairings such that the dimensions of the supports are strictly less than $\dim \operatorname{Supp} \mathcal{T}$. Assume also that $\dim(g^{-1}(0) \cap \operatorname{Supp}(\mathcal{T})) < \dim \operatorname{Supp}(\mathcal{T})$. Then, we have the dual $\boldsymbol{D}\phi_g^{(0)}\mathcal{T}$ and $\boldsymbol{D}\psi_g^{(a)}\mathcal{T}$ in $\mathscr{D}\text{-Tri}^{nd}(X)$ of $\phi_g^{(0)}\mathcal{T}$ and $\psi_g^{(a)}\mathcal{T}$, respectively. We have the naturally induced morphisms $\boldsymbol{D}\psi_g^{(0)}\mathcal{T} \longrightarrow \boldsymbol{D}\phi_g^{(0)}\mathcal{T} \longrightarrow \boldsymbol{D}\psi_g^{(1)}\mathcal{T}$. We also have the duals $\boldsymbol{D}\varphi_\dagger\psi_{gz}^{(a)}\mathcal{T}$ and $\boldsymbol{D}\varphi_\dagger\Xi_{gz}^{(a)}\mathcal{T}$ (Sect. 12.5.5). We have natural isomorphisms $\boldsymbol{D}\psi_g^{(a)}\mathcal{T} \simeq \boldsymbol{D}\varphi_\dagger\psi_{gz}^{(a)}\mathcal{T}$ (Sect. 12.5.4). We define a non-degenerate sesqui-linear pairing $\boldsymbol{D}\mathcal{T}$ as the cohomology of the complex:

$$\boldsymbol{D}\varphi_\dagger\psi_{gz}^{(0)}\mathcal{T}_V \longrightarrow \boldsymbol{D}\varphi_\dagger\Xi_{gz}^{(0)}\mathcal{T}_V \oplus \boldsymbol{D}\phi_g^{(0)}\mathcal{T} \longrightarrow \boldsymbol{D}\varphi_\dagger\psi_{gz}^{(1)}\mathcal{T}_V$$

It is easy to check that $\boldsymbol{D}\mathcal{T}$ gives a dual of $\mathcal{T}$.

Corollary 12.5.9 *Suppose that the claim of Theorem* 12.4.1 *holds for any non-degenerate pairings such that the dimensions of the support are less than n.*

Let $\mathcal{T} \in \mathscr{D}\text{-Tri}^{nd}(X)$ with $\dim \operatorname{Supp} \mathcal{T} = n$. *If there exists a good cell at any point of $P \in \mathcal{T}$, there exists a dual of $\mathcal{T}$.* □

12.5.7 End of the Proof

Let V_1 be a meromorphic flat bundle on (X, D), which is not necessarily good. We have a non-degenerate pairing $\mathcal{T} = \big(V_1, \overline{\mathcal{C}_X(V_1)}, C\big)$, where C is the naturally defined pairing. By applying Corollary 12.5.9 and resolution of turning points ([38, 39], see also [54, 55]), we obtain the existence of a dual of $\mathcal{T}$. Similarly, we have a dual of $(V_{1!}, \overline{\mathcal{C}_X(V_{1!})}, C')$. Then, by using the argument in Sect. 12.5.6 for cells which are not necessarily good, we can finish the proof of Theorem 12.4.1. Once we obtain Theorem 12.4.1, we obtain Theorem 12.4.5 from Corollary 12.5.5 and Lemma 12.5.6. □

12.6 Real Structure

12.6.1 Real Structure of Non-degenerate $\mathscr{D}$-Triple

12.6.1.1 Real Structure

A real structure of $\mathcal{T}^\bullet \in \mathcal{C}\big(\mathscr{D}\text{-Tri}^{nd}(X)\big)$ is defined to be an isomorphism $\kappa : \tilde{\gamma}^*\mathcal{T}^\bullet \simeq \mathcal{T}^\bullet$ such that $\kappa \circ \tilde{\gamma}^*\kappa = \mathrm{id}$. Similarly, a real structure of $\mathfrak{T} \in \mathcal{C}(\mathscr{D})\text{-Tri}_0^{nd}(X)$ is defined to be an isomorphism $\kappa : \tilde{\gamma}^*\mathfrak{T} \simeq \mathfrak{T}$ such that $\kappa \circ \tilde{\gamma}^*\kappa = \mathrm{id}$. A real structure of $\mathcal{T}^\bullet \in \mathcal{C}\big(\mathscr{D}\text{-Tri}^{nd}(X)\big)$ naturally corresponds to a real structure of $\Psi_1\big(\mathcal{T}^\bullet\big) \in \mathcal{C}(\mathscr{D})\text{-Tri}_0^{nd}(X)$. We obtain the following lemma from Lemma 12.4.17.

Lemma 12.6.1 *A real structure of $\mathfrak{T} \in \mathcal{C}(\mathscr{D})\text{-}Tri_0^{nd}(X)$ naturally induces a real structure of $\mathcal{S}_\ell(\mathfrak{T})$. A real structure of $\mathcal{T}^\bullet \in \mathcal{C}\big(\mathscr{D}\text{-}Tri^{nd}(X)\big)$ naturally induces a real structure of $\mathcal{S}_\ell(\mathcal{T}^\bullet)$.* □

12.6.1.2 Functoriality of the Real Structure

Let $\mathscr{D}\text{-Tri}^{nd}(X, \mathbb{R})$ be the category of non-degenerate $\mathscr{D}$-triples with real structures.

We obtain the following functoriality from the results in Sect. 12.4.5.

Proposition 12.6.2

- *$\mathscr{D}\text{-}Tri^{nd}(X, \mathbb{R})$ is equipped with the functors $\mathbf{D}$ and $\mathbf{D}^{\text{herm}}$.*
- *For any hypersurface $H \subset X$, we have the exact functors $[\star H]$ $(\star = *, !)$ on $\mathscr{D}\text{-}Tri^{nd}(X, \mathbb{R})$, which are compatible with the corresponding functors on $\mathscr{D}\text{-}Tri(X)$.*
- *For a proper morphism $f : X \longrightarrow Y$, we have the cohomological functor $H^j f_\dagger : \mathscr{D}\text{-}Tri^{nd}(X, \mathbb{R}) \longrightarrow \mathscr{D}\text{-}Tri^{nd}(Y, \mathbb{R})$ compatible with $H^j f_\dagger : \mathscr{D}\text{-}Tri(X) \longrightarrow \mathscr{D}\text{-}Tri(Y)$.*
- *Let X_i $(i = 1, 2)$ be complex manifolds. The external tensor product induces a bi-functor $\mathscr{D}\text{-}Tri^{nd}(X_1, \mathbb{R}) \times \mathscr{D}\text{-}Tri^{nd}(X_2, \mathbb{R}) \longrightarrow \mathscr{D}\text{-}Tri^{nd}(X_1 \times X_2, \mathbb{R})$.* □

Let H be any hypersurface of X. Let $\mathcal{T}_1 = (V_1, V_2, C)$ be a non-degenerate smooth $\mathscr{D}_{X(*H)}$-triple. Recall we have $\mathcal{T}_1^\vee = (V_1^\vee, V_2^\vee, C^\vee)$ and $\tilde{\gamma}_{sm}^*(\mathcal{T}_1) = (\mathcal{T}_1^\vee)^*$. A real structure of $\mathcal{T}_1$ as a smooth $\mathscr{D}_{X(*H)}$-triple is an isomorphism $\kappa : \tilde{\gamma}_{sm}^*(\mathcal{T}_1) \simeq \mathcal{T}_1$ such that $\kappa \circ \tilde{\gamma}_{sm}^*(\kappa) = \text{id}$. We obtain the following from Proposition 12.4.18.

Proposition 12.6.3 *Let $\mathcal{T} \in \mathscr{D}\text{-}Tri^{nd}(X, \mathbb{R})$. Let $\mathcal{T}_1$ be a smooth $\mathscr{D}_{X(*H)}$-triple with real structure. Then, $(\mathcal{T} \otimes \mathcal{T}_1)[\star H] \in \mathscr{D}\text{-}Tri^{nd}(X)$ are equipped with naturally induced real structure.* □

We obtain the following from Proposition 12.4.19.

Proposition 12.6.4 *For any holomorphic function g on X, we have Beilinson's functors $\Pi_{g,*!}^{a,b}$ on $\mathscr{D}\text{-}Tri^{nd}(X, \mathbb{R})$, compatible with those on $\mathscr{D}\text{-}Tri(X)$. In particular, we have the nearby cycle functor $\psi_g^{(a)}$ and the maximal functor $\Xi_g^{(a)}$. Moreover, we have the vanishing cycle functor $\phi_g^{(a)}$.* □

12.6.2 Descriptions of Real Perverse Sheaves

12.6.2.1 Preliminary

Let X be a d_X-dimensional complex manifold. Let $D_c^b(\mathbb{C}_X)$ denote the derived category of cohomologically bounded constructible $\mathbb{C}_X$-complexes. We set $a_X^!\mathbb{C} := \mathbb{C}_X[2d_X]$ in $D_c^b(\mathbb{C}_X)$ with a naturally defined real structure. For any $\mathcal{F} \in D_c^b(\mathbb{C}_X)$, we set $\boldsymbol{D}(\mathcal{F}) := R\mathcal{H}om_{\mathbb{C}_X}(\mathcal{F}, a_X^!\mathbb{C})$. The real structure of $a_X^!\mathbb{C}$ induces an identification $\boldsymbol{D}(\overline{\mathcal{F}}) \simeq \overline{\boldsymbol{D}\mathcal{F}}$.

We have the quasi-isomorphism $a_X^!\mathbb{C} \longrightarrow \mathfrak{Db}_X^{\bullet}[2d_X]$ given by $s \longmapsto (2\pi\sqrt{-1})^{d_X}\iota(s)$ at the degree $-2d_X$, where $\iota : \mathbb{C}_X \longrightarrow \mathfrak{Db}_X$ is the inclusion. We shall identify them in $D_c^b(\mathbb{C}_X)$. We have the compatible real structure on $\mathfrak{Db}_X^{\bullet}[2d_X]$, which is twisted by $(2\pi\sqrt{-1})^{d_X}$ from the ordinary one.

Remark 12.6.5 The real structure is also induced by the identification $\Omega_X^{\bullet}[d_X] \otimes \Omega_{\overline{X}}^{\bullet}[d_X] \otimes \mathfrak{Db}_X \simeq \mathfrak{Db}_X^{\bullet}[2d_X]$. Namely, after the composite of the following identifications

$$\begin{aligned}\overline{\mathfrak{Db}_X^{\bullet}[2d_X]} &\simeq \overline{\Omega_X^{\bullet}[d_X] \otimes \Omega_{\overline{X}}^{\bullet}[d_X] \otimes \mathfrak{Db}_X} \simeq \Omega_{\overline{X}}^{\bullet}[d_X] \otimes \Omega_X^{\bullet}[d_X] \otimes \mathfrak{Db}_X \\ &\simeq \Omega_X^{\bullet}[d_X] \otimes \Omega_{\overline{X}}^{\bullet}[d_X] \otimes \mathfrak{Db}_X \simeq \mathfrak{Db}_X^{\bullet}[2d_X],\end{aligned} \tag{12.87}$$

the conjugate of ω is identified with $(-1)^{d_X}\overline{\omega}$. □

Let $F : X \longrightarrow Y$ be a proper morphism. We use the trace morphism $F_*(a_X^!\mathbb{C}) \longrightarrow a_Y^!\mathbb{C}$, induced by the identification $a_X^!\mathbb{C} \simeq \mathfrak{Db}_X^{\bullet}[2d_X]$. The trace morphism is compatible with the real structures of $a_X^!\mathbb{C}$ and $a_Y^!\mathbb{C}$. We have the Verdier duality $F_*\boldsymbol{D}(\mathcal{F}) \simeq \boldsymbol{D}F_*\mathcal{F}$ for any $\mathcal{F}$ in $D_c^b(\mathbb{C}_Y)$, and the following natural diagram is commutative:

$$\begin{array}{ccc} F_*\boldsymbol{D}\overline{\mathcal{F}} & \xrightarrow{\simeq} & \boldsymbol{D}F_*\overline{\mathcal{F}} \\ \simeq\downarrow & & \simeq\downarrow \\ \overline{F_*\boldsymbol{D}\mathcal{F}} & \xrightarrow{\simeq} & \overline{\boldsymbol{D}F_*\mathcal{F}} \end{array}$$

Let $\mathcal{F}_i$ $(i = 1, 2)$ be objects in $D_c^b(\mathbb{C}_X)$. A morphism $C : \mathcal{F}_1 \otimes \mathcal{F}_2 \longrightarrow a_X^!\mathbb{C}$ is equivalent to a morphism $\Psi_C : \mathcal{F}_1 \longrightarrow \boldsymbol{D}\mathcal{F}_2$. We have the induced pairings:

$$\overline{C} : \overline{\mathcal{F}}_1 \otimes \overline{\mathcal{F}}_2 \longrightarrow a_X^!\mathbb{C}, \quad C \circ \mathrm{ex} : \mathcal{F}_2 \otimes \mathcal{F}_1 \longrightarrow a_X^!\mathbb{C}$$

We have $\Psi_{\overline{C}} = \overline{\Psi_C}$ under the natural isomorphism $\boldsymbol{D}\overline{\mathcal{F}}_2 \simeq \overline{\boldsymbol{D}\mathcal{F}_2}$. We have $\Psi_{C\circ\mathrm{ex}} = \boldsymbol{D}\Psi_C$ under the natural isomorphism $\mathcal{F}_2 \simeq \boldsymbol{D}\big(\boldsymbol{D}\mathcal{F}_2\big)$. For any integer ℓ, the shift $C[\ell] : \mathcal{F}_1[-\ell] \otimes \mathcal{F}_2[\ell] \longrightarrow a_X^!\mathbb{C}$ is induced. (See Sect. 12.1.5.) We have $\Psi_{C[\ell]} = \Psi_C$ under the natural identification $\boldsymbol{D}(\mathcal{F}_2[\ell]) \simeq \boldsymbol{D}(\mathcal{F}_2)[-\ell]$, as remarked in Lemma 12.1.12.

We say that C is non-degenerate if Ψ_C is a quasi-isomorphism. In that case, we have the induced non-degenerate pairing $\boldsymbol{D}'C : \boldsymbol{D}\mathcal{F}_2 \otimes \boldsymbol{D}\mathcal{F}_1 \longrightarrow a_X^!\mathbb{C}$ in $D_c^b(\mathbb{C}_X)$ given by $\boldsymbol{D}\mathcal{F}_2 \otimes \boldsymbol{D}\mathcal{F}_1 \longrightarrow \mathcal{F}_1 \otimes \mathcal{F}_2 \longrightarrow a_X^!\mathbb{C}$. By construction, we have $\Psi_{\boldsymbol{D}'C} = \Psi_C^{-1}$.

12.6.2.2 Category of Non-degenerate Pairings

Let $\mathcal{C}(\mathbb{C})\text{-Tri}(X)$ denote the category of tuples of $\mathcal{F}_i \in D^b_c(\mathbb{C}_X)$ $(i = 1, 2)$ with a pairing $C : \mathcal{F}_1 \otimes \overline{\mathcal{F}_2} \to a_X^! \mathbb{C}$ in $D^b_c(\mathbb{C}_X)$. A morphism $(\mathcal{F}_1, \mathcal{F}_2, C) \to (\mathcal{F}_1', \mathcal{F}_2', C')$ is a pair of morphisms $\varphi_1 : \mathcal{F}_1' \to \mathcal{F}_1$ and $\varphi_2 : \mathcal{F}_2 \to \mathcal{F}_2'$ such that $C' \circ (\mathrm{id} \otimes \varphi_2) = C \circ (\varphi_1 \otimes \mathrm{id})$. Let $\mathcal{C}(\mathbb{C})\text{-Tri}^{nd}(X) \subset \mathcal{C}(\mathbb{C})\text{-Tri}(X)$ be the full subcategory of the objects $(\mathcal{F}_1, \mathcal{F}_2, C)$ such that C is non-degenerate.

We define contravariant auto-equivalences $\boldsymbol{D}^{\mathrm{herm}}$ and $\boldsymbol{D}$ on $\mathcal{C}(\mathbb{C})\text{-Tri}^{nd}(X)$ as follows:

$$\boldsymbol{D}^{\mathrm{herm}}(\mathcal{F}_1, \mathcal{F}_2, C) := (\mathcal{F}_2, \mathcal{F}_1, \overline{C \circ \mathrm{ex}}),\ \boldsymbol{D}(\mathcal{F}_1, \mathcal{F}_2, C) := (\boldsymbol{D}\mathcal{F}_1, \boldsymbol{D}\mathcal{F}_2, \boldsymbol{D}'C \circ \mathrm{ex})$$

We define an auto-equivalence $\tilde{\gamma}^*$ on $\mathcal{C}(\mathbb{C})\text{-Tri}^{nd}(X)$ by $\tilde{\gamma}^* := \boldsymbol{D} \circ \boldsymbol{D}^{\mathrm{herm}} = \boldsymbol{D}^{\mathrm{herm}} \circ \boldsymbol{D}$. We have

$$\tilde{\gamma}^*(\mathcal{F}_1, \mathcal{F}_2, C) = (\boldsymbol{D}\mathcal{F}_2, \boldsymbol{D}\mathcal{F}_1, \overline{\boldsymbol{D}'C}).$$

For any integer ℓ, the shift is defined by $\mathcal{S}_\ell(\mathcal{F}_1, \mathcal{F}_2, C) = (\mathcal{F}_1[-\ell], \mathcal{F}_2[\ell], C[\ell])$.

Lemma 12.6.6 *We have the natural transforms $\nu_{1,\ell} : \mathcal{S}_{-\ell} \circ \boldsymbol{D}^{\mathrm{herm}} \simeq \boldsymbol{D}^{\mathrm{herm}} \circ \mathcal{S}_\ell$ and $\nu_{2,\ell} : \mathcal{S}_{-\ell} \circ \boldsymbol{D} \simeq \boldsymbol{D} \circ \mathcal{S}_\ell$. As a consequence, we have $\nu_{3,\ell} : \mathcal{S}_{-\ell} \circ \tilde{\gamma}^* \simeq \tilde{\gamma}^* \circ \mathcal{S}_\ell$. They satisfy $\nu_{a,\ell+m} = \nu_{a,\ell} \circ \nu_{a,m}$. The isomorphisms of the underlying $\mathbb{C}_X$-complexes are the natural isomorphisms.*

Proof We have

$$\boldsymbol{D}^{\mathrm{herm}} \mathcal{S}_\ell(\mathfrak{F}) = (\mathcal{F}_2[\ell], \mathcal{F}_1[-\ell], \overline{C[\ell] \circ \mathrm{ex}}),$$

$$\mathcal{S}_{-\ell} \boldsymbol{D}^{\mathrm{herm}}(\mathfrak{F}) = (\mathcal{F}_2[\ell], \mathcal{F}_1[-\ell], \overline{C \circ \mathrm{ex}}[-\ell]).$$

The isomorphism $\nu_{1,\ell}$ is given by the natural identifications of the underlying $\mathbb{C}_X$-complexes. We have the following commutative diagram of isomorphisms:

$$\begin{array}{ccccc}
\mathcal{F}_2[\ell] & \longrightarrow & (\boldsymbol{D}\boldsymbol{D}\mathcal{F}_2)[\ell] & \xrightarrow{\overline{\boldsymbol{D}\Psi_C}} & \boldsymbol{D}(\overline{\mathcal{F}_1})[\ell] \\
\downarrow & & \downarrow & & \downarrow \\
\boldsymbol{D}(\boldsymbol{D}(\mathcal{F}_2[\ell])) & \longrightarrow & \boldsymbol{D}((\boldsymbol{D}\mathcal{F}_2)[-\ell]) & \xrightarrow{\overline{\boldsymbol{D}\Psi_{C[\ell]}}} & \boldsymbol{D}(\overline{\mathcal{F}_1}[-\ell])
\end{array}$$

In this sense, we identify $\overline{\boldsymbol{D}\Psi_{C[\ell]}}$ and $\overline{\boldsymbol{D}\Psi_C}$. Because $\Psi_{\overline{C[\ell] \circ \mathrm{ex}}} = \overline{\boldsymbol{D}\Psi_{C[\ell]}} = \overline{\boldsymbol{D}\Psi_C} = \Psi_{\overline{C \circ \mathrm{ex}}}$, the pairings are equal.

Let us consider the case of $\boldsymbol{D}$. We have the following:

$$\mathcal{S}_{-\ell}\boldsymbol{D}(\mathfrak{F}) = \big((\boldsymbol{D}\mathcal{F}_1)[\ell], (\boldsymbol{D}\mathcal{F}_2)[-\ell], (\boldsymbol{D}'C \circ \mathrm{ex})[-\ell]\big)$$

$$\boldsymbol{D}\mathcal{S}_\ell(\mathfrak{F}) = \big(\boldsymbol{D}(\mathcal{F}_1[-\ell]), \boldsymbol{D}(\mathcal{F}_2[\ell]), \boldsymbol{D}'\big(C[\ell]\big) \circ \mathrm{ex}\big)$$

We have natural isomorphisms $\boldsymbol{D}(\mathcal{F}_1)[\ell] \simeq \boldsymbol{D}(\mathcal{F}_1[-\ell])$ and $\boldsymbol{D}(\mathcal{F}_2)[-\ell] \simeq \boldsymbol{D}(\mathcal{F}_2[\ell])$, as in Sect. 12.1.5. Under the identification, we obtain the equality of the pairings from $\Psi_{(\boldsymbol{D}'C\circ\mathrm{ex})[-\ell]} = \Psi_{\boldsymbol{D}'(C[\ell])\circ\mathrm{ex}}$, which can be deduced easily. □

12.6.2.3 Real Structure

A real structure of an object $\mathfrak{F} \in \mathcal{C}(\mathbb{C})\text{-Tri}^{nd}(X)$ is an isomorphism $\kappa : \tilde{\gamma}^*\mathfrak{F} \simeq \mathfrak{F}$ such that $\kappa \circ \tilde{\gamma}^*\kappa = \mathrm{id}$. We have the following corollary.

Corollary 12.6.7 *A real structure of $\mathfrak{F} \in \mathcal{C}(\mathbb{C})\text{-Tri}^{nd}(X)$ naturally induces a real structure of $\mathcal{S}_\ell(\mathfrak{F})$.* □

Let $F : X \longrightarrow Y$ be a proper map. For any $\mathfrak{F} \in \mathcal{C}(\mathbb{C})\text{-Tri}^{nd}(X)$ we have a naturally induced object $F_*\mathfrak{F} := (F_*\mathcal{F}_1, F_*\mathcal{F}_2, F_*C)$. It gives a functor

$$\mathcal{C}(\mathbb{C})\text{-Tri}^{nd}(X) \longrightarrow \mathcal{C}(\mathbb{C})\text{-Tri}^{nd}(Y).$$

The following lemma is easy to see.

Lemma 12.6.8 *We have natural transformations $F_*\boldsymbol{D}^{\mathrm{herm}} \simeq \boldsymbol{D}^{\mathrm{herm}}F_*$, $F_*\boldsymbol{D} \simeq \boldsymbol{D}F_*$ and $F_*\widetilde{\gamma}^* \simeq \widetilde{\gamma}^*F_*$. The isomorphisms of the underlying $\mathbb{C}_Y$-complexes are the natural isomorphisms.*

As a result, if an object $\mathfrak{F}$ in $\mathcal{C}(\mathbb{C})\text{-Tri}^{nd}(X)$ is equipped with a real structure κ, then $F_*\mathfrak{F}$ is equipped with a naturally induced real structure $F_*\kappa$.

12.6.2.4 Pairs of Perverse Sheaves with a Pairing

Let $\mathbb{C}\text{-Tri}^{nd}(X)$ denote the full subcategory of objects $(\mathcal{F}_1, \mathcal{F}_2, C)$ such that $\mathcal{F}_i$ are perverse in $\mathcal{C}(\mathbb{C})\text{-Tri}^{nd}(X)$. We have the natural contravariant auto-equivalences $\boldsymbol{D}$ and $\boldsymbol{D}^{\mathrm{herm}}$ on $\mathbb{C}\text{-Tri}^{nd}(X)$. We also have $\tilde{\gamma}^*$ on $\mathbb{C}\text{-Tri}^{nd}(X)$. A real structure of an object in $\mathbb{C}\text{-Tri}^{nd}(X)$ is defined to be a real structure as an object in $\mathcal{C}(\mathbb{C})\text{-Tri}^{nd}(X)$.

Let $\mathfrak{F} = (\mathcal{F}_1, \mathcal{F}_2, C) \in \mathcal{C}(\mathbb{C})\text{-Tri}^{nd}(X)$. Let $H^j(\mathcal{F}_i)$ $(i = 1, 2)$ denote the j-th cohomology with respect to the t-structure associated to the middle perversity. We have the induced pairing $H^j(C) : H^{-j}(\mathcal{F}_1) \otimes \overline{H^j(\mathcal{F}_2)} \longrightarrow a_X^!\mathbb{C}$. We set

$$H^j\big(\mathfrak{F}\big) := \big(H^{-j}(\mathcal{F}_1), H^j(\mathcal{F}_2), H^j(C)\big).$$

Thus, we obtain a functor $H^j : \mathcal{C}(\mathbb{C})\text{-Tri}^{nd}(X) \longrightarrow \mathbb{C}\text{-Tri}^{nd}(X)$. We have natural isomorphisms $H^j \circ \boldsymbol{D} \simeq \boldsymbol{D} \circ H^{-j}$, $H^j \circ \boldsymbol{D}^{\text{herm}} \simeq \boldsymbol{D}^{\text{herm}} \circ H^{-j}$ and $H^j \circ \tilde{\gamma}^* \simeq \tilde{\gamma}^* \circ H^{-j}$. Therefore, if $\mathfrak{F}$ is equipped with a real structure, $H^j(\mathfrak{F})$ are equipped with induced real structures.

Let $\mathcal{C}(\mathbb{C})\text{-Tri}^{nd}(X,\mathbb{R})$ denote the category of objects $\mathfrak{F} \in \mathcal{C}(\mathbb{C})\text{-Tri}^{nd}(X)$ with a real structure κ. Morphisms are defined in a natural way. Let $\mathbb{C}\text{-Tri}^{nd}(X,\mathbb{R}) \subset \mathcal{C}(\mathbb{C})\text{-Tri}^{nd}(X,\mathbb{R})$ denote the full subcategory of the objects $(\mathfrak{F},\kappa)$ such that $\mathfrak{F} \in \mathbb{C}\text{-Tri}^{nd}(X)$.

12.6.2.5 Perverse Sheaves with Real Structure

For a field K, let $\text{Per}(K_X)$ be the category of K_X-perverse sheaves. Let $\text{Per}(\mathbb{C}_X,\mathbb{R})$ be the category of $\mathbb{C}_X$-perverse sheaves $\mathcal{F}$ with real structure ρ, i.e., $\rho : \overline{\mathcal{F}} \simeq \mathcal{F}$ such that $\rho \circ \overline{\rho} = \text{id}$. A morphism $(\mathcal{F}_1,\rho_1) \longrightarrow (\mathcal{F}_2,\rho_2)$ is a morphism $\varphi : \mathcal{F}_1 \longrightarrow \mathcal{F}_2$ in $\text{Per}(\mathbb{C}_X)$ such that $\varphi \circ \rho_1 = \rho_2 \circ \varphi$. We have a natural equivalence $\text{Per}(\mathbb{R}_X) \simeq \text{Per}(\mathbb{C}_X,\mathbb{R})$ given by $\mathcal{F}_{\mathbb{R}} \longmapsto \mathcal{F}_{\mathbb{R}} \otimes \mathbb{C}$ with $\rho(x \otimes \alpha) = x \otimes \overline{\alpha}$.

We have a natural functor $\Lambda : \text{Per}(\mathbb{C}_X,\mathbb{R}) \longrightarrow \mathbb{C}\text{-Tri}^{nd}(X,\mathbb{R})$. For any $(\mathcal{F},\rho)$ in $\text{Per}(\mathbb{C}_X,\mathbb{R})$, we have the induced non-degenerate pairing $C_\rho : \boldsymbol{D}\mathcal{F} \otimes \overline{\mathcal{F}} \longrightarrow a_X^!\mathbb{C}$:

$$\boldsymbol{D}\mathcal{F} \otimes \overline{\mathcal{F}} \xrightarrow{\boldsymbol{D}\rho\otimes\text{id}} \boldsymbol{D}\overline{\mathcal{F}} \otimes \overline{\mathcal{F}} \xrightarrow{\ b\ } a_X^!\mathbb{C}$$

The morphism b is naturally given one. We have $\Psi_{C_\rho} = \boldsymbol{D}\rho$. It is equal to the composite of $\boldsymbol{D}\mathcal{F} \otimes \overline{\mathcal{F}} \xrightarrow{\text{id}\otimes\rho} \boldsymbol{D}\mathcal{F} \otimes \mathcal{F} \longrightarrow a_X^!\mathbb{C}$. It is easy to check that $(\boldsymbol{D}\mathcal{F},\mathcal{F},C_\rho)$ with $\kappa = (\text{id},\text{id})$ is an object in $\mathbb{C}\text{-Tri}^{nd}(X,\mathbb{R})$, which is denoted by $\Lambda(\mathcal{F},\rho)$. The following lemma is easy to check.

Lemma 12.6.9 *The functor Λ is an equivalence.* □

12.6.2.6 Example

Let us look at an example. Let $\mathcal{F}_i = \mathbb{C}[d_X]$ $(i = 1,2)$. Let e_i be the section of $\mathbb{C}[d_X]$ corresponding to 1. Let u be the section of $a_X^!\mathbb{C} = \mathbb{C}[2d_X]$ corresponding to 1. We consider a pairing $C : \mathcal{F}_1 \otimes \overline{\mathcal{F}_2} \longrightarrow a_X^!\mathbb{C}$ given by $C(e_1,\overline{e}_2) = u$. We have $\boldsymbol{D}\mathcal{F}_i = \mathbb{C}\, e_i^\vee$, where $e_i^\vee$ is determined by the condition $e_i^\vee \otimes e_i \longmapsto u$. Then, $\overline{\boldsymbol{D}'C} : \boldsymbol{D}\mathcal{F}_2 \otimes \overline{\boldsymbol{D}\mathcal{F}_1} \longrightarrow a_X^!\mathbb{C}$ is given by $\overline{\boldsymbol{D}'C}(e_2^\vee,e_1^\vee) = (-1)^{d_X}u$. Note that the induced morphisms $\mathcal{F}_1 \longrightarrow \boldsymbol{D}\overline{\mathcal{F}}_2$ and $\overline{\mathcal{F}}_2 \longrightarrow \boldsymbol{D}\mathcal{F}_1$ are given by $e_1 \longmapsto \overline{e}_2^\vee$ and $\overline{e}_2 \longmapsto (-1)^{d_X}e_1^\vee$, respectively. Let $\kappa = (\kappa_1,\kappa_2) : (\boldsymbol{D}\mathcal{F}_2,\boldsymbol{D}\mathcal{F}_1,\overline{\boldsymbol{D}'C}) \longrightarrow (\mathcal{F}_1,\mathcal{F}_2,C)$ be given by $\kappa_1(e_1) = e_2^\vee$ and $\kappa_2(e_1^\vee) = (-1)^{d_X}e_2$. The induced map

$$\widetilde{\gamma}^*\kappa : \Big(\boldsymbol{D}(\boldsymbol{D}\mathcal{F}_1),\boldsymbol{D}(\boldsymbol{D}\mathcal{F}_2),\overline{(\boldsymbol{D}'(\overline{\boldsymbol{D}'C}))}\Big) \longrightarrow (\boldsymbol{D}\mathcal{F}_2,\boldsymbol{D}\mathcal{F}_1,\overline{\boldsymbol{D}'C})$$

given by $(\gamma^*\kappa)_1(e_2^\vee) = (-1)^{d_X}(e_1^\vee)^\vee$ and $(\gamma^*\kappa)_2(e_2^\vee)^\vee = e_1^\vee$. Note the natural identification $(e_i^\vee)^\vee = (-1)^{d_X} e_i$. Hence, we have $\kappa \circ \widetilde{\gamma}^*\kappa = \mathrm{id}$, i.e., κ is a real structure.

12.6.3 The de Rham Functor

For any object $\mathcal{T} \in \mathcal{C}(\mathscr{D})$-Tri$(X)$, we define $\mathrm{DR}_X(\mathcal{T}) \in \mathcal{C}(\mathbb{C})$-Tri$(X)$ as in Sect. 12.3.2. We obtain a functor $\mathrm{DR}_X : \mathcal{C}(\mathscr{D})\text{-Tri}(X) \longrightarrow \mathcal{C}(\mathbb{C})\text{-Tri}(X)$.

Lemma 12.6.10 *By the rule of the signatures, we have natural transforms $\mathcal{S}_\ell \circ \mathrm{DR} \simeq \mathrm{DR} \circ \mathcal{S}_\ell$ of the functors from $\mathcal{C}(\mathscr{D})$-Tri(X) to $\mathcal{C}(\mathbb{C})$-Tri(X).*

Proof We have the natural isomorphism

$$\mathrm{Tot}\,\mathrm{DR}_X(M_i^\bullet[\pm\ell]) \simeq \mathrm{Tot}\,\mathrm{DR}_X(M_i^\bullet)[\pm\ell],$$

which are given by the multiplication of $(-1)^{p\ell}$ on $\Omega^{d_X+p} \otimes M_i^{r\pm\ell}$. We can compare the pairings by a direct computation. □

12.6.3.1 Compatibility with the Dualities in the Perverse Case

We have the induced functor $\mathrm{DR}_X : \mathscr{D}\text{-Tri}^{nd}(X) \longrightarrow \mathbb{C}\text{-Tri}^{nd}(X)$.

Proposition 12.6.11 *For $\mathcal{T} \in \mathscr{D}$-Tri$^{nd}(X)$, we have natural isomorphisms*

$$\boldsymbol{D} \circ \mathrm{DR}_X(\mathcal{T}) \simeq \mathrm{DR}_X \circ \boldsymbol{D}(\mathcal{T}), \qquad \boldsymbol{D}^{\mathrm{herm}} \circ \mathrm{DR}_X(\mathcal{T}) \simeq \mathrm{DR}_X \circ \boldsymbol{D}^{\mathrm{herm}}(\mathcal{T}),$$

which are given by the natural identifications of the underlying complexes.

Proof Let us consider the case of $\boldsymbol{D}^{\mathrm{herm}}$. We remark that the conjugate $\overline{\mathrm{DR}_X(C)}$ in Lemma 12.3.2 is taken for the natural real structure of $\mathfrak{Db}_X^\bullet[2d_X]$. Because the quasi-isomorphism $a_X^!\mathbb{C} \simeq \mathfrak{Db}_X^\bullet[2d_X]$ is given by the multiplication of $(2\pi\sqrt{-1})^{d_X}$, the conjugate with respect to the real structure of $a_X^!\mathbb{C}$ is $(-1)^{d_X}\overline{\mathrm{DR}_X(C)}$. Then, by using Lemma 12.3.2, we can compare the pairings.

Let us consider the case of $\boldsymbol{D}$. We can check that the following diagram is commutative:

$$\begin{array}{ccc}
\mathrm{DR}_{\overline{X}}(\overline{M}_2) & \xrightarrow{\simeq} & \overline{\mathrm{DR}_X(M_2)} \\
\big\downarrow & & \big\downarrow{\scriptstyle \Psi_{DR(C)\circ \mathrm{ex}}} \\
\mathrm{DR}_X(\boldsymbol{D}M_1) & \longrightarrow & \boldsymbol{D}\,\mathrm{DR}_X(M_1)
\end{array}$$

Here, the upper horizontal arrow is given by the conjugate, and the left vertical arrow is the morphism used in Theorem 12.4.1. We can check that the following natural diagram is commutative:

$$\begin{array}{ccc}
\mathrm{DR}_X(M_1) & \xrightarrow{=} & \mathrm{DR}_X(M_1) \\
{\scriptstyle a_1}\downarrow & & \downarrow{\scriptstyle \Psi_{\mathrm{DR}(C)}} \\
\mathrm{DR}_{\overline{X}}\,\boldsymbol{D}_{\overline{X}}\overline{M}_2 & \xrightarrow{a_2} & \boldsymbol{D}\overline{\mathrm{DR}_X\,M_2} \\
\simeq\downarrow & & \downarrow{\scriptstyle a_3} \\
\overline{\mathrm{DR}_X\,\boldsymbol{D}_X M_2} & \xrightarrow{a_4} & \overline{\boldsymbol{D}_X\,\mathrm{DR}_X(M_2)}
\end{array}$$

Here, a_1 is the morphism used in Theorem 12.4.1, and a_2 is given as follows:

$$\mathcal{H}om_{\mathscr{D}_{\overline{X}}}(\overline{M}_2, \mathfrak{Db}_X^{\bullet,0}[d_X]) \longrightarrow R\mathcal{H}om_{\mathbb{C}_X}\big(\mathrm{DR}_{\overline{X}}\overline{M}_2, \Omega_X^{\bullet}[d_X] \otimes \Omega_{\overline{X}}^{\bullet}[d_X] \otimes \mathfrak{Db}_X\big)$$
$$\longrightarrow R\mathcal{H}om_{\mathbb{C}_X}\big(\mathrm{DR}_{\overline{X}}\overline{M}_2, \mathfrak{Db}_X^{\bullet}[2d_X]\big) \tag{12.88}$$

The morphism a_3 is induced by the conjugate of $a_X^!\mathbb{C}_X$. The morphism a_4 is induced by the natural isomorphism, i.e.,

$$\overline{\mathcal{H}om_{\mathscr{D}_X}(M_2, \mathfrak{Db}_X^{0,\bullet}[d_X])} \longrightarrow \overline{R\mathcal{H}om_{\mathbb{C}_X}\big(\mathrm{DR}_X\,M_2, \Omega_X^{\bullet}[d_X] \otimes \Omega_{\overline{X}}^{\bullet}[d_X] \otimes \mathfrak{Db}_X\big)}$$
$$\longrightarrow \overline{R\mathcal{H}om_{\mathbb{C}_X}\big(\mathrm{DR}_X\,M_2, \mathfrak{Db}_X^{\bullet}[2d_X]\big)} \tag{12.89}$$

Note that, in (12.89), we use the fixed identification $\Omega_X^{\bullet}[d_X] \otimes \Omega_{\overline{X}}^{\bullet}[d_X] \otimes \mathfrak{Db}_X \simeq \mathfrak{Db}^{\bullet}[2d_X]$, which means that we use the real structure of $a_X^!\mathbb{C}_X$. We can easily compare the pairings of $\boldsymbol{D}\circ\mathrm{DR}(\mathcal{T})$ and $\mathrm{DR}\,\boldsymbol{D}\mathcal{T}$ by using the commutative diagrams. □

Corollary 12.6.12 *We have $\tilde{\gamma}^*\,\mathrm{DR}_X(\mathcal{T}) \simeq \mathrm{DR}_X\,\tilde{\gamma}^*\mathcal{T}$ given by the natural isomorphisms of the underlying complexes. In particular, the de Rham functor gives a functor $\mathscr{D}$-$Tri^{nd}(X,\mathbb{R}) \longrightarrow \mathbb{C}$-$Tri^{nd}(X,\mathbb{R})$.* □

12.6.3.2 Compatibility with the Dualities in the Case of Complexes

We consider the full subcategory $\mathcal{C}(\mathscr{D})$-$\mathrm{Tri}_0^{nd}(X) \subset \mathcal{C}(\mathscr{D})$-$\mathrm{Tri}(X)$ in Sect. 12.4.2. The functor DR gives a functor $\mathcal{C}(\mathscr{D})$-$\mathrm{Tri}_0^{nd}(X) \longrightarrow \mathcal{C}(\mathbb{C})$-$\mathrm{Tri}^{nd}(X)$.

Proposition 12.6.13 *We have natural equivalences $\mathrm{DR}\,\boldsymbol{D}^{\mathrm{herm}} \simeq \boldsymbol{D}^{\mathrm{herm}}\,\mathrm{DR}$ and $\mathrm{DR}\,\boldsymbol{D} \simeq \boldsymbol{D}\,\mathrm{DR}$, which are given by the natural identifications of the underlying complexes. As a consequence, we have $\mathrm{DR}\circ\tilde{\gamma}^* \simeq \tilde{\gamma}^*\circ\mathrm{DR}$.*

Proof We have the transforms $\mathrm{DR}\,\boldsymbol{D}^{\mathrm{herm}} \simeq \boldsymbol{D}^{\mathrm{herm}}\,\mathrm{DR}$ and $\mathrm{DR}\,\boldsymbol{D} \simeq \boldsymbol{D}\,\mathrm{DR}$ by Proposition 12.6.11 by Lemmas 12.4.17, 12.6.6 and 12.6.10. It is enough to check that the isomorphisms of the underlying complexes are given by the natural identifications.

In the case of $\boldsymbol{D}^{\text{herm}}$, we note that the change of the signature can happen only for the exchange $\mathcal{S}_\ell \circ \mathrm{DR} \simeq \mathrm{DR} \circ \mathcal{S}_\ell$, which are canceled out. Let us consider the case of $\boldsymbol{D}$. We take a $\mathscr{D}_X$-injective resolution $\mathcal{I}^\bullet$ of $\mathcal{O}_X[d_X]$. We can observe that the composite of the isomorphisms $\mathrm{DR}\,\boldsymbol{D}(M^\bullet[\ell]) \simeq \mathrm{DR}\big(\boldsymbol{D}(M^\bullet)[-\ell]\big) \simeq \mathrm{DR}\big(\boldsymbol{D}(M^\bullet)\big)[-\ell]$ is equal to the natural isomorphism

$$\operatorname{Tot}\mathcal{H}om_{\mathscr{D}_X}(M^\bullet[\ell],\mathcal{I}^\bullet) \simeq \operatorname{Tot}\mathcal{H}om_{\mathscr{D}_X}(M^\bullet,\mathcal{I}^\bullet)[-\ell].$$

Let $\mathcal{J}^\bullet$ be a $\mathbb{C}_X$-injective resolution of $a_X^!\mathbb{C}$ with an isomorphism $\mathrm{DR}_X\,\mathcal{I}^\bullet \longrightarrow \mathcal{J}^\bullet$. We can observe that the composite of the isomorphisms

$$\boldsymbol{D}\,\mathrm{DR}(M^\bullet[\ell]) \simeq \boldsymbol{D}\big(\mathrm{DR}(M^\bullet)[\ell]\big) \simeq \boldsymbol{D}\big(\mathrm{DR}(M^\bullet)\big)[-\ell]$$

is given by the natural isomorphism

$$\begin{aligned}&\operatorname{Tot} R\mathcal{H}om_{\mathbb{C}_X}\big(\Omega_X^\bullet[d_X]\otimes(M^\bullet[\ell]),\mathcal{J}^\bullet\big)\\ &\longrightarrow \operatorname{Tot} R\mathcal{H}om_{\mathbb{C}_X}\big(\Omega_X^\bullet[d_X]\otimes M^\bullet,\mathcal{J}^\bullet\big)[-\ell].\end{aligned}\tag{12.90}$$

We can check that the following diagram of the natural morphisms is commutative:

$$\begin{array}{ccc}\operatorname{Tot}\mathcal{H}om_{\mathscr{D}_X}(M^\bullet[\ell],\mathcal{I}^\bullet) & \longrightarrow & \operatorname{Tot}\mathcal{H}om_{\mathscr{D}_X}(M^\bullet,\mathcal{I}^\bullet)[-\ell]\\ \downarrow & & \downarrow\\ \operatorname{Tot} R\mathcal{H}om_{\mathbb{C}_X}\big(\Omega_X^\bullet[d_X]\otimes(M^\bullet[\ell]),\mathcal{J}^\bullet\big) & \longrightarrow & \operatorname{Tot} R\mathcal{H}om_{\mathbb{C}_X}\big(\Omega_X^\bullet[d_X]\otimes M^\bullet,\mathcal{J}^\bullet\big)[-\ell].\end{array}$$

Then, the claim for the case $\boldsymbol{D}$ follows. □

12.6.3.3 Complement

We remark the commutativity of the following diagrams of the natural isomorphisms:

$$\begin{array}{ccc}\mathcal{S}_{-\ell}\boldsymbol{D}\,\mathrm{DR}(\mathcal{T}) & \longrightarrow & \boldsymbol{D}\mathcal{S}_\ell\,\mathrm{DR}(\mathcal{T})\\ \downarrow & & \downarrow\\ \mathrm{DR}\,\mathcal{S}_{-\ell}\boldsymbol{D}(\mathcal{T}) & \longrightarrow & \mathrm{DR}\,\boldsymbol{D}\mathcal{S}_\ell(\mathcal{T})\end{array}$$

We have similar diagrams for $\boldsymbol{D}^{\text{herm}}$ and $\tilde{\gamma}^*$. In particular, the shift functor of real structures are compatible with the de Rham functor.

We have the commutativity of the following diagram:

$$\begin{array}{ccc} \boldsymbol{D}\widetilde{\gamma}^* \,\mathrm{DR}\,\mathcal{T} & \longrightarrow & \widetilde{\gamma}^* \boldsymbol{D}\,\mathrm{DR}\,\mathcal{T} \\ \downarrow & & \downarrow \\ \mathrm{DR}\,\boldsymbol{D}\widetilde{\gamma}^*\mathcal{T} & \longrightarrow & \mathrm{DR}\,\widetilde{\gamma}^*\boldsymbol{D}\mathcal{T} \end{array}$$

Hence, the duality of real structures is compatible with the de Rham functor. Similarly, the functors $\boldsymbol{D}^{\mathrm{herm}}$ and $\tilde{\gamma}^*$ of real structures are compatible with the de Rham functor.

Let $\mathcal{T} \in \mathcal{C}(\mathscr{D})\text{-Tri}^{nd}(X)$. Let $F : X \longrightarrow Y$ be a morphism of complex manifolds such that $Supp(\mathcal{T})$ is proper with respect to F. By Proposition 12.4.25, the following is commutative:

$$\begin{array}{ccc} \mathrm{DR}_Y \circ F_\dagger^j \circ \widetilde{\gamma}^*(\mathcal{T}) & \longrightarrow & \mathrm{DR}_Y \circ \widetilde{\gamma}^* \circ F_\dagger^j(\mathcal{T}) \\ \downarrow & & \downarrow \\ F_*^j \circ \widetilde{\gamma}^* \circ \mathrm{DR}_X(\mathcal{T}) & \longrightarrow & \widetilde{\gamma}^* \circ F_*^j \circ \mathrm{DR}_X(\mathcal{T}) \end{array}$$

In particular, the push-forward functor of real structures is compatible with the de Rham functor.

Let X_i $(i = 1, 2)$ be complex manifolds. Let $\mathcal{T}_i \in \mathscr{D}\text{-Tri}(X_i)$. As remarked in the proof of Proposition 12.4.15, we have a natural isomorphism $\mathrm{DR}_{X_1\times X_2}(\mathcal{T}_1 \boxtimes \mathcal{T}_2) \simeq \mathrm{DR}_{X_1}(\mathcal{T}_1) \boxtimes \mathrm{DR}_{X_2}(\mathcal{T}_2)$. The following induced diagram is commutative:

$$\begin{array}{ccc} \mathrm{DR}\,\widetilde{\gamma}^*(\mathcal{T}_1 \boxtimes \mathcal{T}_2) & \stackrel{\simeq}{\longrightarrow} & \widetilde{\gamma}^*(\mathrm{DR}\,\mathcal{T}_1 \boxtimes \mathrm{DR}\,\mathcal{T}_2) \\ \simeq\downarrow & & \simeq\downarrow \\ \mathrm{DR}(\widetilde{\gamma}^*\mathcal{T}_1 \boxtimes \widetilde{\gamma}^*\mathcal{T}_2) & \stackrel{\simeq}{\longrightarrow} & \widetilde{\gamma}^*\,\mathrm{DR}\,\mathcal{T}_1 \boxtimes \widetilde{\gamma}^*\,\mathrm{DR}\,\mathcal{T}_2 \end{array}$$

Indeed, it is enough to check the commutativity of the diagram of the morphisms of the underlying $\mathbb{C}_X$-complexes, which is easy to see.

12.6.4 Regular Case

Let $\mathscr{D}\text{-Tri}^{nd}_{reg}(X, \mathbb{R}) \subset \mathscr{D}\text{-Tri}^{nd}(X, \mathbb{R})$ denote the full subcategory of the objects whose underlying $\mathscr{D}$-modules are regular holonomic. By the Riemann-Hilbert correspondence, the de Rham functor induces an equivalence

$$\mathrm{DR} : \mathscr{D}\text{-Tri}^{nd}_{reg}(X, \mathbb{R}) \longrightarrow \mathbb{C}\text{-Tri}^{nd}(X, \mathbb{R}) \simeq \mathrm{Per}(\mathbb{R}_X).$$

According to the results in Sect. 12.6.3, it is compatible with the push-forward by proper morphisms. It is also compatible with the duality.

12.6.4.1 Description

Let M be any regular holonomic $\mathscr{D}_X$-module such that $\mathrm{DR}_X(M)$ has a real structure, i.e., $\iota : \overline{\mathrm{DR}_X(M)} \simeq \mathrm{DR}_X(M)$ such that $\iota \circ \overline{\iota} = \mathrm{id}$. It induces a pairing $C_\iota : \boldsymbol{D}_X M \otimes \overline{M} \longrightarrow \mathfrak{D}\mathfrak{b}_X$ corresponding to

$$\mathrm{DR}_X(\boldsymbol{D}_X M) \otimes \mathrm{DR}_{\overline{X}}(\overline{M}) \longrightarrow \boldsymbol{D}_X \,\mathrm{DR}_X(M) \otimes \mathrm{DR}_X(M) \longrightarrow \omega^{top}.$$

We put $\mathcal{T} := (\boldsymbol{D}M, M, C_\iota)$.

Proposition 12.6.14 *$\mathcal{T}$ has a real structure given by $\kappa = (\mathrm{id}, \mathrm{id})$, i.e., $(\mathcal{T}, \kappa) \in \mathscr{D}\text{-}Tri^{nd}_{reg}(X, \mathbb{R})$. Conversely, any object in $\mathscr{D}\text{-}Tri^{nd}_{reg}(X, \mathbb{R})$ has such a description up to isomorphisms.*

Proof Let us prove the first claim. We have only to check that $(\mathrm{id}, \mathrm{id})$ gives a morphism $\tilde{\gamma}^*\mathcal{T} \longrightarrow \mathcal{T}$. Namely, we have only to check that $(\boldsymbol{D}C_\iota)^* = C_\iota$ under the natural isomorphism $\boldsymbol{D}_X\boldsymbol{D}_X(M) \simeq M$. The morphism $\overline{M} \longrightarrow \mathcal{C}_X(\boldsymbol{D}M)$ given by C_ι induces $\iota : \mathrm{DR}_{\overline{X}}(\overline{M}) \simeq \mathrm{DR}_X(M)$ if we use the identification $\Omega^\bullet_X[d_X] \otimes \Omega^\bullet_{\overline{X}}[d_X] \otimes \mathfrak{D}\mathfrak{b}_X \simeq \mathfrak{D}\mathfrak{b}^\bullet_X[2d_X]$. See Remark 12.4.3 for the signature. The morphism $\mathrm{DR}_X(\boldsymbol{D}_X M) \simeq \mathrm{DR}_{\overline{X}}(\boldsymbol{D}_{\overline{X}}\overline{M})$ induced by C_ι is identified with the morphism induced by $\iota : \mathrm{DR}_{\overline{X}}\overline{M} \simeq \mathrm{DR}_X M$. Then, we obtain the commutativity of the following diagram:

$$\begin{array}{ccc}
\mathrm{DR}\, M \otimes \mathrm{DR}\, \boldsymbol{D}\overline{M} & \xrightarrow{\mathrm{DR}(\boldsymbol{D}C_\iota)} & a_X^!\mathbb{C} \\
\big\downarrow{\scriptstyle \iota\otimes\iota^*} & & \big\downarrow{\scriptstyle \mathrm{id}} \\
\mathrm{DR}_{\overline{X}}(\overline{M}) \otimes \boldsymbol{D}\,\mathrm{DR}_X\, M & \xrightarrow{\mathrm{DR}(C_\iota)\circ \mathrm{ex}} & a_X^!\mathbb{C}
\end{array}$$

Then, we obtain the commutativity of the following diagram:

$$\begin{array}{ccc}
\mathrm{DR}\, \boldsymbol{D}M \otimes \mathrm{DR}\, \overline{M} & \xrightarrow{\overline{\mathrm{DR}(\boldsymbol{D}C_\iota)\circ \mathrm{ex}}} & a_X^!\mathbb{C} \\
\big\downarrow{\scriptstyle \mathrm{id}\otimes\mathrm{id}} & & \big\downarrow{\scriptstyle \mathrm{id}} \\
\mathrm{DR}_X(\boldsymbol{D}M) \otimes \mathrm{DR}_{\overline{X}}\, \overline{M} & \xrightarrow{\mathrm{DR}(C_\iota)} & a_X^!\mathbb{C}
\end{array}$$

It means $\boldsymbol{D}(C_\iota)^* = C_\iota$. As for the second claim, we have only to care the signatures which can be checked as in the case of first claim. □

12.6.4.2 Localizations

Let H be a hypersurface of X. Let $j : X \setminus H \longrightarrow X$ be the inclusion. We have the functor $[*H]$ on $\mathscr{D}\text{-Tri}^{nd}_{reg}(X,\mathbb{R})$. It is equivalent to j_*j^* on $\text{Per}(\mathbb{R}_X)$. Similarly, the functor $[!H]$ on $\mathscr{D}\text{-Tri}^{nd}_{\text{reg}}(X,\mathbb{R})$ is equivalent to $j_!j^*$ on $\text{Per}(\mathbb{R}_X)$.

12.6.4.3 Beilinson Functors

For any integers $a \leq b$, let $\mathcal{I}^{a,b}$ be the $\mathbb{R}$-local system on $\mathbb{C}^*$ underlying the smooth $\mathcal{R}_{\mathbb{C}(*0)}$-triple $\mathbb{I}^{a,b}$ with the real structure. Let g be a holomorphic function on X. By using $g^*\mathcal{I}^{a,b}$, we obtain the nearby cycle functors $\psi_g^{(a)}$, the maximal functors $\Xi_g^{(a)}$ and the vanishing cycle functors $\phi_g^{(a)}$ on $\text{Per}(\mathbb{R}_X)$. By the compatibility of the localizations with the de Rham functor $\text{DR}_X : \mathscr{D}\text{-Tri}^{nd}_{reg}(X,\mathbb{R}) \longrightarrow \text{Per}(\mathbb{R}_X)$, we obtain that DR_X is compatible with the functors $\psi_g^{(a)}$, $\Xi_g^{(a)}$ and $\phi_g^{(a)}$.

12.6.5 $\mathbb{R}$-Betti Structure

Let $\mathcal{T} = (M_1, M_2, C) \in \mathscr{D}\text{-Tri}^{nd}(X,\mathbb{R})$. By the construction above, we have a real structure of $\text{DR}_X(M_2)$, i.e., a pre-$\mathbb{R}$-Betti structure of M_2 in the sense of [57].

Proposition 12.6.15 *The pre-$\mathbb{R}$-Betti structure is an $\mathbb{R}$-Betti structure of M_2 in the sense of* §7.2 *in* [57].

Proof We have only to check the claim locally around any points P of the support of M_2. We shall shrink X without mention. We use a Noetherian induction on the support. If the support of M_2 is a point, then the claim is clear. We take a cell $\mathcal{C} = (Z, U, \varphi, V)$ of M_2 at P with a cell function g. We have the $\mathbb{R}$-structure of the meromorphic flat bundle V in the sense of §6.4 of [57].

Lemma 12.6.16 *The $\mathbb{R}$-structure is good in the sense of* §6.4 *of* [57].

Proof We consider the special case $X = Z = \Delta = \{|z| < 1\}$ with $\dim X = 1$. We may assume $M_i(*P) = V_i$. We have the pairing $M_1(*P) \times \overline{M_2(*P)} \longrightarrow \mathfrak{Db}_X(*P)$. By Lemma 12.6.10 of [55], we can check that it comes from a pairing on the real blow up. Hence, the real structure is compatible with the Stokes structure. Moreover, we may replace $\mathfrak{Db}_X(*P)$ with $C_X^{\infty \bmod P}$.

Let us consider the case $\dim Z = 1$. We may assume that $X = \Delta_t \times \Delta^m$ with $g = t$. Let $h : X \longrightarrow \Delta_t$ denote the projection. We obtain $h_\dagger(M_1, M_2, C)$ with an induced real structure. Then, V is a direct summand of $(h \circ \varphi)^* h_\dagger M_2(*g)$. Hence, by the result in the previous paragraph, we obtain that the $\mathbb{R}$-structure of V is good.

Let us consider the case in which $\dim Z$ is arbitrary. For any irreducible smooth curve $C \subset Z$ with $C \not\subset Z \setminus U$, we can check that the $\mathbb{R}$-structure of $V_{|C}$ is good, by using the result in the previous paragraph.

Let Q be any point of Z. Let (Z_Q, ψ_Q) be a local resolution of Z around Q. (See §6.4 of [57].) We can easily deduce that the $\mathbb{R}$-structure of $\psi_Q^* V$ is good, by using the result in the previous paragraph. Hence, the $\mathbb{R}$-structure of V is good. □

We have the induced $\mathbb{R}$-structures of $\mathcal{T}[\star g]$, and the morphisms $\mathcal{T}[!g] \longrightarrow \mathcal{T} \longrightarrow \mathcal{T}[*g]$ are compatible with the $\mathbb{R}$-structures. We have two pre-$\mathbb{R}$-Betti structures on $M_2[\star g] = \varphi_\star(V)$. One is induced by the $\mathbb{R}$-structure of $\mathcal{T}[\star g]$. The other is induced by the good $\mathbb{R}$-structure of V.

Lemma 12.6.17 *The two pre-$\mathbb{R}$-Betti structures are the same.*

Proof Let $V_2 := V$. Let (Z, U, φ, V_1) be a cell of M_1. We have the naturally defined sesqui-linear pairing $C_V^U : V_{1|U} \times \overline{V}_{2|U} \longrightarrow \mathcal{C}_U^\infty$. For any irreducible smooth curve $C \subset Z$ with $C \not\subset D := Z \setminus U$, the pairing of $V_{i|U\cap C}$ is extended to $V_{1|C} \times \overline{V}_{2|C} \longrightarrow \mathcal{C}_Z^{\infty \bmod D}$, which can be shown by an argument in the proof of Lemma 12.6.16. For any $Q \in Z$, we take a local resolution (Z_Q, ψ_Q) of V_i. By considering the Stokes filtrations of V_i, and by using Lemma 12.6.10 of [55], we obtain a unique extension of the pairing $\psi_Q^* V_1 \times \overline{\psi_Q^* V_2} \longrightarrow \mathcal{C}_{Z_Q}^{\infty \bmod D_Q}$. Then, we obtain a unique extension of the pairing $C_V^{\mathrm{mod}} : V_1 \times \overline{V_2} \longrightarrow \mathcal{C}_Z^{\infty \bmod D}$. Hence, we have $C_{V*} : V_{1!} \times \overline{V_{2*}} \longrightarrow \mathfrak{Db}_Z$ and $C_{V!} : V_{1*} \times \overline{V_{2!}} \longrightarrow \mathfrak{Db}_Z$. We have the $\mathbb{R}$-structure of $\mathcal{T}_V^U := (V_{1|U}, V_{2|U}, C_V^U)$. Let $\mathcal{T}_{V*} := (V_{1!}, V_{2*}, C_{V*})$ and $\mathcal{T}_{V!} := (V_{1*}, V_{2!}, C_{V!})$. For any irreducible curve $C \subset Z$, the $\mathbb{R}$-structure of $\mathcal{T}_{V|C\cap U}^U$ is extended to those of $\mathcal{T}_{V\star|C}$, which can be shown by an argument in the proof of Lemma 12.6.16. It implies that the morphisms $V_{1|U} \longrightarrow \boldsymbol{D}V_{2|U}$ and $V_{2|U} \longrightarrow \boldsymbol{D}V_{1|U}$ of the real structure of $\mathcal{T}_V^U$ have meromorphic extensions on Z. Hence, the $\mathbb{R}$-structure of $\mathcal{T}_V^U$ is extended to those of $\mathcal{T}_{V\star}$. The induced pre-$\mathbb{R}$-Betti structures of $V_{2\star}$ are the $\mathbb{R}$-Betti structures. Indeed, by using local resolutions, we have only to check it in the case V_i are good meromorphic flat bundles, which is easy. Because the $\mathbb{R}$-structures of $\mathcal{T}[\star g]$ are obtained as the push-forward of the $\mathbb{R}$-structures of $\mathcal{T}_{V\star}$, we obtain the claim of Lemma 12.6.17. □

By Lemmas 12.6.16 and 12.6.17, the cell $\mathcal{C}$ is compatible with the pre-$\mathbb{R}$-Betti structure of M_2 in the sense of §7.1.3 in [57]. We have the induced $\mathbb{R}$-structure on $\phi_g^{(0)}(\mathcal{T})$. The induced pre-$\mathbb{R}$-Betti structure of $\phi_g^{(0)}(M_2)$ is a $\mathbb{R}$-Betti structure, by the hypothesis of the induction. Thus, the pre-$\mathbb{R}$-Betti structure of M_2 is a $\mathbb{R}$-Betti-structure. □

Let $\mathrm{Hol}(X, \mathbb{R})$ denote the category of holonomic $\mathscr{D}_X$-modules with $\mathbb{R}$-Betti structure. For any morphism $\mathcal{T}^{(1)} \longrightarrow \mathcal{T}^{(2)}$ in $\mathscr{D}\text{-Tri}^{nd}(X, \mathbb{R})$, the underlying morphism $M_2^{(1)} \longrightarrow M_2^{(2)}$ is compatible with the pre-$\mathbb{R}$-Betti structure, where $\mathcal{T}^{(i)} = (M_1^{(i)}, M_2^{(i)}, C^{(i)})$. Hence, by Proposition 12.6.15, we obtain a functor $\Upsilon : \mathscr{D}\text{-Tri}^{nd}(X, \mathbb{R}) \longrightarrow \mathrm{Hol}(X, \mathbb{R})$.

Proposition 12.6.18 *The functor Υ is an equivalence.*

Proof It is clearly faithful. Let us observe that the functor is full. Let $\mathcal{T}^{(i)} = (M_1^{(i)}, M_2^{(i)}, C^{(i)})$ with $\kappa^{(i)}$ be objects in $\mathscr{D}\text{-Tri}^{nd}(X, \mathbb{R})$. Let $f : M_2^{(1)} \longrightarrow M_2^{(2)}$ be a morphism in $\mathrm{Hol}(X, \mathbb{R})$. The following induced morphism is commutative:

$$\begin{array}{ccc} \mathrm{DR}(\boldsymbol{D}M_2^{(2)}) & \xrightarrow{\mathrm{DR}(\boldsymbol{D}f)} & \mathrm{DR}(\boldsymbol{D}M_2^{(1)}) \\ \downarrow & & \downarrow \\ \overline{\mathrm{DR}(\boldsymbol{D}M_2^{(2)})} & \xrightarrow{\overline{\mathrm{DR}(\boldsymbol{D}f)}} & \overline{\mathrm{DR}(\boldsymbol{D}M_2^{(1)})} \end{array} \tag{12.91}$$

Here, the vertical arrows are induced by the real structure of $\mathrm{DR}\,\boldsymbol{D}(M_2^{(i)})$.

We set $f_2 := f$, and let $f_1 : M_1^{(2)} \longrightarrow M_1^{(1)}$ be the morphism induced by $\boldsymbol{D}(f)$: $\boldsymbol{D}M_2^{(2)} \longrightarrow \boldsymbol{D}M_2^{(1)}$ with isomorphisms $M_1^{(i)} \simeq \boldsymbol{D}M_2^{(i)}$ underlying $\kappa^{(i)}$. Let us consider the following diagram:

$$\begin{array}{ccc} \mathrm{DR}\,\boldsymbol{D}M_2^{(2)} \simeq \mathrm{DR}\,M_1^{(2)} & \simeq & \mathrm{DR}\,C_{\overline{X}}\overline{M}_2^{(2)} \\ \downarrow & & \downarrow \\ \mathrm{DR}\,\boldsymbol{D}M_2^{(1)} \simeq \mathrm{DR}\,M_1^{(1)} & \simeq & \mathrm{DR}\,C_{\overline{X}}\overline{M}_2^{(1)} \end{array} \tag{12.92}$$

Here, the vertical arrows are given as $\mathrm{DR}\,\boldsymbol{D}f$ and $\mathrm{DR}\,C_{\overline{X}}\overline{f}$, respectively. Under the natural isomorphism $\mathrm{DR}_X\,C_{\overline{X}}\overline{M}_2^{(i)} \simeq \mathrm{DR}_{\overline{X}}\,\boldsymbol{D}_{\overline{X}}\overline{M}_2^{(i)}$, we have $\mathrm{DR}\,C_{\overline{X}}(\overline{f}) = \overline{\mathrm{DR}\,\boldsymbol{D}f}$. The horizontal arrows are induced by the real structures of $\mathrm{DR}\,\boldsymbol{D}M_2^{(i)}$. Hence, (12.92) is commutative. Then, we obtain that (f_1, f_2) gives a morphism $\mathcal{T}^{(1)} \longrightarrow \mathcal{T}^{(2)}$ such that $\Upsilon(f_1, f_2) = f$. Hence, Υ is full.

Let us observe that Υ is essentially surjective. By the full faithfulness, we have only to check it locally around any point P of X. We may shrink X without mention. We use a Noetherian induction on the support of M. Let $M \in \mathrm{Hol}(X, \mathbb{R})$. We take a cell $\mathcal{C} = (Z, U, \varphi, V)$ of M at P with a cell function g. We have a description of M as the cohomology of

$$\psi_g^{(1)}(\varphi_\dagger V) \longrightarrow \Xi_g^{(0)}(\varphi_\dagger V) \oplus \phi_g^{(0)}(M) \longrightarrow \psi_g^{(0)}(\varphi_\dagger V) \tag{12.93}$$

The $\mathbb{R}$-structure of V is good. As in the proof of Lemma 12.6.17, we obtain $\mathcal{T}_{V\star} \in \mathscr{D}\text{-Tri}^{nd}(Z, \mathbb{R})$ $(\star = *, !)$. By using the Beilinson construction, we obtain $\Xi_g^{(0)}(\varphi_\dagger V) \in \mathscr{D}\text{-Tri}^{nd}(X, \mathbb{R})$. By the hypothesis of the induction, we have $\mathcal{T}_i^{(a)} \in \mathscr{D}\text{-Tri}^{nd}(X, \mathbb{R})$ $(i = 1, 2,\ a = 0, 1)$ such that $\Upsilon(\mathcal{T}_1^{(a)}) \simeq \psi_g^{(a)}(\varphi_\dagger V)$ and $\Upsilon(\mathcal{T}_2^{(a)}) \simeq \phi_g^{(a)}(\varphi_\dagger V)$. By the fully faithfulness of Υ, (12.93) comes from a complex in $\mathscr{D}\text{-Tri}^{nd}(X, \mathbb{R})$. Hence, M comes from an object in $\mathscr{D}\text{-Tri}^{nd}(X, \mathbb{R})$, i.e., we obtain the essential surjectivity of Υ. □

Proposition 12.6.19 *The functor Υ is compatible with the following functors:*

- *The push-forward by any proper morphisms.*
- *The dual $\boldsymbol{D}$.*

Proof It follows from the compatibilities in Sect. 12.6.3. □

For any hypersurface H on X, we have the localization functors $[\star H]$ ($\star = *, !$) on $\mathrm{Hol}(X, \mathbb{R})$.

Proposition 12.6.20 *The localizations are compatible with Υ, i.e., for any $M \in \mathscr{D}\text{-}\mathit{Tri}^{nd}(X, \mathbb{R})$, we naturally have $\Upsilon(M[\star H]) \simeq \Upsilon(M)[\star H]$.*

Proof It follows from the characterization of the localization as in Theorem 8.1.4 of [57]. □

Let H be any hypersurface of X. Let $\mathcal{T}_1 = (V_1, V_2, C)$ be a smooth $\mathscr{D}_{X(*H)}$-triple with a real structure. Then, V_2 is a meromorphic flat bundle on (X, H) with a good real structure.

Proposition 12.6.21 *Let $\mathcal{T} \in \mathscr{D}\text{-}\mathit{Tri}^{nd}(X, \mathbb{R})$. We have $\Upsilon(\mathcal{T} \otimes \mathcal{T}_1[\star H]) \simeq \big(\Upsilon(\mathcal{T}) \otimes V_2\big)[\star H]$ in $\mathrm{Hol}(X, \mathbb{R})$.*

Proof If $H = \emptyset$, the claim is clear. The general case easily follows. □

Corollary 12.6.22 *We naturally have $\Upsilon \circ \Xi_g^{(a)} \simeq \Xi_g^{(a)} \circ \Upsilon$, $\Upsilon \circ \psi_g^{(a)} \simeq \psi_g^{(a)} \circ \Upsilon$ and $\Upsilon \circ \phi_g^{(a)} \simeq \phi_g^{(a)} \circ \Upsilon$ for any holomorphic function g.* □

12.6.6 Basic Examples

12.6.6.1 A Duality Isomorphism

Let Z be a complex manifold. We have the natural identification of the cohomology sheaf $H^{-d_Z}\big(\mathrm{DR}\,\mathcal{O}_Z\big)$ and the sheaf of flat sections of $\mathcal{O}_Z$. The natural isomorphism $\mathrm{DR}\,\boldsymbol{D}\mathcal{O}_Z \simeq R\mathcal{H}om_{\mathscr{D}_Z}(\mathcal{O}_Z, \mathcal{O}_Z[d_Z])$ induces the following:

$$H^{-d_Z}\big(\mathrm{DR}\,\boldsymbol{D}\mathcal{O}_Z\big) \simeq \mathcal{H}om_{\mathscr{D}_Z}(\mathcal{O}_Z, \mathcal{O}_Z) \simeq \mathcal{H}om_{\mathbb{C}_Z}\big(\mathbb{C}_Z, \mathbb{C}_Z\big) \tag{12.94}$$

Let $\nu : \boldsymbol{D}\mathcal{O}_Z \simeq \mathcal{O}_Z$ be the unique isomorphism which induces

$$H^{-d_Z}(\mathrm{DR}\,\boldsymbol{D}\mathcal{O}_Z) \simeq H^{-d_Z}(\mathrm{DR}\,\mathcal{O}_Z)$$

given by $\mathrm{id} \longmapsto (-1)^{d_Z}$. We have another expression of the isomorphism ν. Let ℓ and r denote the $\mathscr{D}_Z$-actions on $\mathscr{D}_Z \otimes \omega_Z^{-1}$ induced by the left and right multiplications. Let Θ_Z be the tangent sheaf of Z. We set $\Theta_Z^p := \bigwedge^{-p} \Theta_Z$. We have the isomorphism of complexes given as follows, which induces an isomorphism $\rho : \mathcal{O}_Z \simeq \boldsymbol{D}\mathcal{O}_Z$:

$$\mathscr{D}_Z \otimes \Theta_Z^{\bullet} \longrightarrow \mathcal{H}om_{\mathscr{D}_Z}\Big(\mathscr{D}_Z \otimes \Theta_Z^{\bullet},\ \big(\mathscr{D}_Z \otimes \omega_Z^{-1}[d_Z]\big)^{\ell,r}\Big) \tag{12.95}$$

$$P \otimes \tau \longmapsto \Big(Q \otimes \omega \longmapsto (-1)^{|\tau|\,|\omega|} C(Q \otimes \omega, P \otimes \tau)\Big) \tag{12.96}$$

Here, $C(Q \otimes \omega, P \otimes \tau) = \ell(Q) r(P)(\omega \wedge \tau)\,(-1)^{|\tau|}$. (See also Sect. 13.1.5 below.)

Lemma 12.6.23 *We have* $\nu = \rho^{-1}$.

Proof By taking tensor product $\omega_Z \otimes_{\mathscr{D}_Z}$, we obtain

$$\omega_Z \otimes \Theta_Z^{\bullet} \longrightarrow \mathcal{H}om_{\mathscr{D}_Z}\big(\mathscr{D}_Z \otimes \Theta_Z^{\bullet}, \mathcal{O}_Z[d_Z]\big) \longrightarrow \mathcal{H}om_{\mathscr{D}_Z}\big(\mathcal{O}_Z, \mathcal{O}_Z[d_Z]\big)$$

Let η be a local generator of ω_Z. Then, $1 = \eta \otimes \eta^{-1} \in \omega_Z \otimes \Theta_Z^{d_Z}$ is sent to $(-1)^{d_Z}\,\mathrm{id} \in \mathcal{H}om_{\mathscr{D}_Z}(\mathcal{O}_Z, \mathcal{O}_Z)$. Then, we obtain the desired equality by the construction of ν. □

We remark that the composite of the natural isomorphisms

$$\mathcal{O}_Z \longrightarrow \boldsymbol{D}(\boldsymbol{D}\mathcal{O}_Z) \xleftarrow{\boldsymbol{D}\nu} \boldsymbol{D}\mathcal{O}_Z \xrightarrow{\nu} \mathcal{O}_Z$$

is $(-1)^{d_Z}$. We can check it by using the argument in the proof of Lemma 13.1.6 below. In other words, $(-1)^{d_Z}\nu$ can be identified with $\boldsymbol{D}\nu$.

The following lemma is easy to see by construction.

Lemma 12.6.24 *Let* Z_i $(i = 1, 2)$ *be complex manifolds. Under the natural isomorphisms* $\mathcal{O}_{Z_1 \times Z_2} \simeq \mathcal{O}_{Z_1} \boxtimes \mathcal{O}_{Z_2}$ *and* $\boldsymbol{D}_{Z_1 \times Z_2}(\mathcal{O}_{Z_1 \times Z_2}) \simeq \boldsymbol{D}_{Z_1}(\mathcal{O}_{Z_1}) \boxtimes \boldsymbol{D}_{Z_2}(\mathcal{O}_{Z_2})$, *we have* $\nu_{Z_1 \times Z_2} = \nu_{Z_1} \boxtimes \nu_{Z_2}$. □

12.6.6.2 Real Structures on the Simplest $\mathscr{D}$-Triple

We consider the $\mathscr{D}$-triple $\mathcal{U}_Z = (\mathcal{O}_Z, \mathcal{O}_Z, C_0)$, where $C_0(f, \overline{g}) = f \cdot \overline{g}$. We have $\tilde{\gamma}^* \mathcal{U}_Z = \big(\boldsymbol{D}\mathcal{O}_Z, \boldsymbol{D}\mathcal{O}_Z, \boldsymbol{D}C_0\big)$. The following proposition is clear.

Proposition 12.6.25 *Let a be a complex number such that* $|a| = 1$. *Then,*

$$(a\,\nu^{-1}, (-1)^{d_Z}\overline{a}\,\nu) : \gamma^* \mathcal{U}_Z \simeq \mathcal{U}_Z$$

gives a real structure. □

12.6.6.3 The $\mathscr{D}$-Triple Corresponding to the Simplest $\mathbb{R}$-Perverse Sheaf

We have the $\mathscr{D}$-triple $(\boldsymbol{D}\mathcal{O}_Z, \mathcal{O}_Z, C_\iota)$ corresponding to the simplest $\mathbb{R}$-perverse sheaf $\mathbb{R}[d_Z]$, as in Sect. 12.6.4.1. Here, C_ι is the pairing induced by the natural real structure of $\mathrm{DR}_Z(\mathcal{O}_Z) \simeq \mathbb{C}[d_Z]$. The $(\mathrm{id}, \mathrm{id})$ gives the real structure of $(\boldsymbol{D}\mathcal{O}_Z, \mathcal{O}_Z, C_\iota)$ corresponding to $\mathbb{R}[d_Z]$.

Proposition 12.6.26 *We have an isomorphism* $(\nu^{-1}, 1) : (\boldsymbol{D}\mathcal{O}_Z, \mathcal{O}_Z, C_\iota) \simeq \mathcal{U}_Z$. *The corresponding real structure of* $\mathcal{U}_Z$ *is given by* $(\nu^{-1}, (-1)^{d_Z}\nu)$.

Proof The second claim follows from the first. Let us prove the first. We have the quasi-isomorphism $\mathbb{C}[d_Z] \simeq \mathrm{DR}(\mathcal{O}_Z)$. Let e denote the section of $(\mathbb{C}[d_Z])^{-d_Z}$ corresponding to $1 \in \mathbb{C}$. We have $\mathrm{DR}(C_0)(e,\overline{e}) = (-1)^{d_Z}\epsilon(d_Z)$.

Let us consider the composite of the following morphisms:

$$\begin{aligned} \mathrm{DR}(\mathcal{O}_Z) &\xrightarrow{\mathrm{DR}\,\nu^{-1}} \mathrm{DR}(\boldsymbol{D}\mathcal{O}_Z) \longrightarrow \mathcal{H}om_{\mathscr{D}_Z}(\mathcal{O}_Z, \mathcal{O}_Z[d_Z]) \\ &\longrightarrow R\mathcal{H}om_{\mathbb{C}_Z}(\mathrm{DR}\,\mathcal{O}_Z, \mathrm{DR}_Z\,\mathcal{O}_Z[d_Z]) \\ &\longrightarrow R\mathcal{H}om_{\mathbb{C}_Z}(\mathrm{DR}\,\mathcal{O}_Z, \mathfrak{Db}^{\bullet}_Z[2d_Z]) \\ &\simeq R\mathcal{H}om_{\mathbb{C}_Z}(\mathbb{C}[d_Z], \mathbb{C}[2d_Z]) \end{aligned} \tag{12.97}$$

By the definition, e is mapped to $(-1)^{d_Z}\,\mathrm{id} \in \big(\mathcal{H}om_{\mathscr{D}_Z}(\mathcal{O}_Z, \mathcal{O}_Z[d_Z])\big)^{-d_Z}$. Let g_0 be the image in $R\mathcal{H}om_{\mathbb{C}_Z}(\mathrm{DR}\,\mathcal{O}_Z, \mathrm{DR}_Z\,\mathcal{O}_Z[d_Z])^{-d_Z}$. We have

$$g_0(e) = (-1)^{d_Z}(-1)^{d_Z} = 1 \in \mathcal{O}_Z = \mathrm{DR}_Z(\mathcal{O}_Z[d_Z])^{-2d_Z}.$$

Let g_1 be the image in $R\mathcal{H}om_{\mathbb{C}_Z}(\mathrm{DR}\,\mathcal{O}_Z, \mathfrak{Db}^{\bullet}_Z[2d_Z])^{-d_Z}$. Recall that our identification $\Omega^{\bullet}_Z[d_Z] \otimes \Omega^{\bullet}_{\overline{Z}}[d_Z] \otimes \mathfrak{Db}_Z \simeq \mathfrak{Db}^{\bullet}_Z[2d_Z]$ is given by

$$\xi^{d_Z+p} \otimes \overline{\xi}^{d_Z+q} \longmapsto \epsilon(d_Z)(-1)^{pd_Z}\xi^{d_Z+p}\overline{\xi}^{d_Z+q}.$$

Hence, we have $g_1(e) = (-1)^{d_Z}\epsilon(d_Z)$. Then, it follows that the pair $(\nu^{-1}, 1)$ gives a morphism $(\boldsymbol{D}\mathcal{O}_Z, \mathcal{O}_Z, C_\iota) \longrightarrow \mathcal{U}_Z$, which is clearly an isomorphism. □

Corollary 12.6.27 *The real structure* $((-1)^a\nu^{-1}, (-1)^{d_Z+a}\nu)$ *of* $\mathcal{U}_Z$ *corresponds to the real structure of* $\mathbb{C}[d_Z]$ *given by* $(2\pi\sqrt{-1})^a\mathbb{R}[d_Z] \subset \mathbb{C}[d_Z]$. □

12.6.6.4 Natural Real Structures of $\mathcal{U}_Z[d_Z]$ and $\mathcal{U}_Z[-d_Z]$

Let us observe that $\mathcal{U}_Z[d_Z]$ and $\mathcal{U}_Z[-d_Z]$ in $\mathcal{C}(\mathscr{D}\text{-Tri}(Z))$, or equivalently in $\mathcal{C}(\mathscr{D})\text{-Tri}(Z)$, have natural real structures.

Let μ be the unique isomorphism $\mu : \boldsymbol{D}(\mathcal{O}_Z[d_Z]) \simeq \mathcal{O}_Z[-d_Z]$ such that the following diagram is commutative:

$$\begin{array}{ccc} \boldsymbol{D}(\mathcal{O}_Z[d_Z]) & \xrightarrow{\mu} & \mathcal{O}_Z[-d_Z] \\ \downarrow & & \downarrow = \\ \boldsymbol{D}(\mathcal{O}_Z)[-d_Z] & \xrightarrow{(-1)^{d_Z}\cdot\nu} & \mathcal{O}_Z[-d_Z] \end{array} \tag{12.98}$$

Here, the left vertical arrow is given by the exchange of the dual and the shift. (It is given by the multiplication of $(-1)^{pd_Z}\epsilon(d_Z)$ on the degree p-part.)

Proposition 12.6.28 *The pair of the morphisms $\big(\mu^{-1}, \boldsymbol{D}\mu\big)$ gives a real structure of the object $\mathcal{U}_Z[d_Z]$ in $\mathcal{C}(\mathscr{D})$-Tri(Z) obtained as the shift of the real structure $\big((-1)^{d_Z}\nu^{-1}, \nu\big)$ of $\mathcal{U}_Z$. The pair $\big((\boldsymbol{D}\mu)^{-1}, \mu\big)$ is a real structure of $\mathcal{U}_Z[-d_Z]$ in $\mathcal{C}(\mathscr{D})$-Tri(Z) obtained from the real structure $\big(\nu^{-1}, (-1)^{d_Z}\nu\big)$ of $\mathcal{U}_Z$ by the shift.*

Proof We obtain the commutativity of the following diagram from the commutativity of (12.98):

$$\begin{array}{ccc} \boldsymbol{D}(\mathcal{O}_Z[-d_Z]) & \xrightarrow{\boldsymbol{D}\mu} & \mathcal{O}_Z[d_Z] \\ \big\downarrow & & =\big\downarrow \\ \boldsymbol{D}(\mathcal{O}_Z)[d_Z] & \xrightarrow{\nu} & \mathcal{O}_Z[d_Z] \end{array} \tag{12.99}$$

Here, the left vertical arrow is given by the exchange of the dual and the shift. The commutativity of the diagrams (12.98) and (12.99) implies the claims of the proposition. □

Corollary 12.6.29 *We have the corresponding real structures on the objects $\mathcal{U}_Z[d_Z]$ and $\mathcal{U}_Z[-d_Z]$ in $\mathcal{C}(\mathscr{D}$-Tri(Z)). They correspond to the real structures of $\mathbb{C}[2d_Z]$ and $\mathbb{C}[0]$ given by $(2\pi\sqrt{-1})^{d_Z}\mathbb{R}[2d_Z]$ and $\mathbb{R}[0]$, respectively.* □

12.6.6.5 A Trace Morphism

We assume that Z is compact. Let $a_Z : Z \longrightarrow \mathrm{pt}$ be the canonical morphism of Z to a point. We have $a_{Z\dagger}\mathcal{O}_Z = Ra_{Z*}\mathfrak{Db}^{\bullet}_Z[d_Z]$. Let $\mathrm{tr}_1 : a^{d_Z}_{Z\dagger}\mathcal{O}_Z \longrightarrow \mathbb{C}$ be given by

$$\mathrm{tr}_1(\eta^{2d_Z}) := \left(\frac{1}{2\pi\sqrt{-1}}\right)^{d_Z} \int_Z \eta^{2d_Z}.$$

Let us look at the push-forward $a_{Z\dagger}\mathcal{U}_Z = (a_{Z\dagger}\mathcal{O}_Z, a_{Z\dagger}\mathcal{O}_Z, a_{Z\dagger}C_0)$ as the complex of $\mathscr{D}$-triples. We have $\mathcal{U}_{\mathrm{pt}} = \big(\mathbb{C}, \mathbb{C}, C_0\big)$ on pt. Let us observe that we have natural morphisms $a^0_{Z\dagger}\big(\mathcal{U}_Z[d_Z]\big) \longrightarrow \mathcal{U}_{\mathrm{pt}}$ and $\mathcal{U}_{\mathrm{pt}} \longrightarrow a^0_{Z\dagger}\big(\mathcal{U}_Z[-d_Z]\big)$. We have

$$a_{Z\dagger}C_0(\eta^{d_Z-p}, \overline{\eta^{d_Z+p}}) := \left(\frac{1}{2\pi\sqrt{-1}}\right)^{d_Z} \int \eta^{d_Z-p}\overline{\eta}^{d_Z+p}\,\epsilon(p+d_Z).$$

In particular, $a^{d_Z}_{Z\dagger}\mathcal{U}_Z$ is given as $\big(a^{-d_Z}_{Z\dagger}\mathcal{O}_Z,\, a^{d_Z}_{Z\dagger}\mathcal{O}_Z,\, C_1\big)$ where

$$\begin{aligned} C_1(1, \overline{\eta}^{2d_Z}) &= \left(\frac{1}{2\pi\sqrt{-1}}\right)^{d_Z}\int \overline{\eta}^{2d_Z}(-1)^{d_Z} = \overline{\left(\frac{1}{2\pi\sqrt{-1}}\right)^{d_Z}\int \eta^{2d_Z}} \\ &= C_0(1, \mathrm{tr}_1(\overline{\eta}^{2d_Z})). \end{aligned} \tag{12.100}$$

We also have $a_{Z\dagger}^{-d_Z}\mathcal{U}_Z = \bigl(a_{Z\dagger}^{d_Z}\mathcal{O}_Z,\ a_{Z\dagger}^{-d_Z}\mathcal{O}_Z,\ C_2\bigr)$, where

$$C_2(\eta^{2d_Z}, 1) = \left(\frac{1}{2\pi\sqrt{-1}}\right)^{d_Z} \int \eta^{2d_Z} = C_0\bigl(\mathrm{tr}_1(\eta^{2d_Z}), 1\bigr). \tag{12.101}$$

The equalities (12.100) and (12.101) mean that we have the following natural morphisms:

$$(a_Z^{-1}, \mathrm{tr}_1) : a_{Z\dagger}^0\bigl(\mathcal{U}_Z[d_Z]\bigr) \longrightarrow \mathcal{U}_{\mathrm{pt}}, \qquad (\mathrm{tr}_1, a_Z^{-1}) : \mathcal{U}_{\mathrm{pt}} \longrightarrow a_{Z\dagger}^0\bigl(\mathcal{U}_Z[-d_Z]\bigr) \tag{12.102}$$

The natural real structure of $\mathcal{U}_{\mathrm{pt}}$ is given by (id, id). The real structures of $\mathcal{U}_Z[d_Z]$ and $\mathcal{U}_Z[-d_Z]$ are natural in the following sense.

Proposition 12.6.30 *The morphisms in* (12.102) *are compatible with the real structures.*

Proof Let us consider the dual of $\mathrm{tr}_1 : a_{Z\dagger}^{d_Z}\mathcal{O}_Z \longrightarrow \mathbb{C}$. We consider Φ_1 induced as follows:

$$\begin{array}{ccc} \bigl(a_{Z\dagger}^0(\mathcal{O}_Z[d_Z])\bigr)^\vee & \xleftarrow{\mathrm{tr}_1^\vee} & \mathbb{C}^\vee \\ {\scriptstyle B}\downarrow\simeq & & {\scriptstyle A}\downarrow\simeq \\ a_{Z\dagger}^0\bigl(\boldsymbol{D}(\mathcal{O}_Z[d_Z])\bigr) & \xleftarrow{\Phi_1} & \mathbb{C} \end{array}$$

Here, A is given by $\mathrm{id} \longmapsto 1$, and B is given by the compatibility of the dual and the push-forward. We have the identification:

$$a_{Z\dagger}^0\bigl(\boldsymbol{D}(\mathcal{O}_Z[d_Z])\bigr) \simeq a_{Z*}^0\mathcal{H}om_{\mathscr{D}_Z}\bigl(\mathcal{O}_Z[d_Z], \mathcal{O}_Z[d_Z]\bigr) \tag{12.103}$$

Lemma 12.6.31 *We have* $\Phi_1(1) = \epsilon(d_Z)(-1)^{d_Z}\,\mathrm{id}$ *under the isomorphism* (12.103).

Proof Let us consider the following morphisms:

$$\begin{aligned} a_{Z*}^0 R\mathcal{H}om_{\mathscr{D}_Z}(\mathcal{O}_Z[d_Z], \mathcal{O}_Z[d_Z]) &\longrightarrow a_{Z*}^0 R\mathcal{H}om_{\mathbb{C}_Z}\bigl(\mathrm{DR}\,\mathcal{O}_Z[d_Z], \mathrm{DR}\,\mathcal{O}_Z[d_Z]\bigr) \\ &\xrightarrow{b} \bigl[R\mathcal{H}om_{\mathbb{C}}\bigl(a_{Z*}\,\mathrm{DR}(\mathcal{O}_Z)[d_Z],\ a_{Z*}\mathfrak{Db}_Z^\bullet[2d_Z]\bigr)\bigr]^0 \\ &\longrightarrow R\mathcal{H}om_{\mathbb{C}}\bigl(a_{Z*}^{d_Z}\,\mathrm{DR}(\mathcal{O}_Z), \mathbb{C}\bigr) \end{aligned} \tag{12.104}$$

Note that, for the morphism b, we use the identification $\Omega_Z^\bullet[d_Z] \otimes \Omega_{\overline{Z}}^\bullet[d_Z] \otimes \mathfrak{Db}_Z \simeq \mathfrak{Db}_Z^\bullet[2d_Z]$ given by $\xi^{d_Z+p} \otimes \overline{\xi}^{d_Z+q} \longmapsto \epsilon(d_Z)(-1)^{pd_Z}$. Then, the claim of the lemma follows. □

Hence, the following diagram is commutative, by the commutativity of (12.98):

$$\begin{array}{ccccc} a^0_{Z\dagger}(\mathcal{O}_Z[d_Z])^\vee & \xrightarrow{B} & a^0_{Z\dagger}\boldsymbol{D}(\mathcal{O}_Z[d_Z]) & \xrightarrow{a_{Z\dagger}\mu} & a^0_{Z\dagger}\mathcal{O}_Z[-d_Z] \\ \uparrow{\scriptstyle \mathrm{tr}_1^\vee} & & \uparrow{\scriptstyle \Phi_1} & & \uparrow{\scriptstyle a_Z^{-1}} \\ \mathbb{C}^\vee & \xrightarrow{A} & \mathbb{C} & \xrightarrow{\mathrm{id}} & \mathbb{C} \end{array} \tag{12.105}$$

Let us consider the dual of $a_Z^{-1} : \mathbb{C} \longrightarrow a^0_{Z\dagger}(\mathcal{O}_Z[-d_Z])$. It is enough to prove the commutativity of the following diagram:

$$\begin{array}{ccccc} a^0_{Z\dagger}(\mathcal{O}_Z[-d_Z])^\vee & \xrightarrow{B'} & a^0_{Z\dagger}(\boldsymbol{D}(\mathcal{O}_Z[-d_Z])) & \xrightarrow{a_{Z\dagger}\boldsymbol{D}\mu} & a^0_{Z\dagger}(\mathcal{O}_Z[d_Z]) \\ \downarrow{\scriptstyle (a_Z^{-1})^\vee} & & & & \downarrow{\scriptstyle \mathrm{tr}_1} \\ \mathbb{C}^\vee & \xrightarrow{A'} & \mathbb{C} & \xrightarrow{=} & \mathbb{C} \end{array} \tag{12.106}$$

Here, A' is given by $\mathrm{id} \longmapsto 1$, and B' is induced by the exchange of the push-forward and the dual. For that purpose, we have only to prove $(a_{Z\dagger}\mu \circ B)^\vee = a_{Z\dagger}\boldsymbol{D}\mu \circ B'$ under the natural identification $(a^0_{Z\dagger}\mathcal{O}_Z[d_Z])^{\vee\vee} \simeq a^0_{Z\dagger}\mathcal{O}_Z[d_Z]$. By a simple diagram chasing, it can be reduced to the following, which can be directly checked:

- $\mu \circ (\boldsymbol{D}\mu)^{-1}$ is the natural isomorphism $\boldsymbol{D}\boldsymbol{D}(\mathcal{O}_Z[d_Z]) \simeq \mathcal{O}_Z[d_Z]$.
- The natural isomorphisms $a_{Z\dagger} \simeq a_{Z\dagger}\boldsymbol{D} \circ \boldsymbol{D}$ and $a_{Z\dagger} \simeq \boldsymbol{D} \circ \boldsymbol{D}a_{Z\dagger}$ are compatible with the identification $a_{Z\dagger}\boldsymbol{D} \circ \boldsymbol{D} \simeq \boldsymbol{D} \circ a_{Z\dagger} \circ \boldsymbol{D} \simeq \boldsymbol{D} \circ \boldsymbol{D}a_{Z\dagger}$.

Then, we obtain the claim of Proposition 12.6.30. □

Chapter 13
Duality and Real Structure of Mixed Twistor $\mathscr{D}$-Modules

The first goal in this chapter is to introduce the duality functor in the context of mixed twistor $\mathscr{D}$-modules. As in the case of holonomic $\mathscr{D}$-modules, it should be fundamental in the theory of mixed twistor $\mathscr{D}$-modules.

In Sects. 13.1–13.2, we study the duality functor for $\mathcal{R}$-modules. The definition is given as in the case of $\mathscr{D}$-modules. One remark is that we need to twist the naively defined one as in (13.2), where $\boldsymbol{D}'_X$ is the naive one and $\boldsymbol{D}_X$ is the correct one. The twist admits us to have the compatibility of the duality and the push-forward, as we will see in Sect. 13.1.2. The definition is consistent with the duality of filtered $\mathscr{D}$-modules of Saito in [69]. We also study various compatibilities.

Then, we introduce the duality of mixed twistor $\mathscr{D}$-modules in Sect. 13.3. For the proof of the existence of the dual with the desired property, we use the argument in the case of $\mathscr{D}$-triples. Namely, we directly construct the dual for cells, and we construct the dual of a mixed twistor $\mathscr{D}$-module by gluing the dual of its cells.

The second goal is to introduce the concept of real structure for mixed twistor $\mathscr{D}$-modules as in Sect. 13.4. To explain the idea, let us recall the concept of real structure in the context of mixed twistor structure [82].

A mixed twistor structure is a holomorphic vector bundle V equipped with an increasing filtration W indexed by integers such that (a) $W_j(V) = 0$ ($j << 0$), $W_j(V) = V$ ($j >> 0$), (b) $\mathrm{Gr}^W_n(V)$ is isomorphic to a direct sum of $\mathcal{O}_{\mathbb{P}^1}(n)$. Let σ and γ be the anti-holomorphic involutions on $\mathbb{P}^1$ given by $\sigma(\lambda) = -\overline{\lambda}^{-1}$ and $\gamma(\lambda) = \overline{\lambda}^{-1}$. We also set $j(\lambda) := -\lambda$ on $\mathbb{P}^1$. For a mixed twistor structure (V, W), the pull back $\gamma^*(V, W)$, $\sigma^*(V, W)$ and $j^*(V, W)$ are naturally mixed twistor structures. Note that $\gamma = j \circ \sigma$. Then, a real structure of a mixed twistor structure (V, W) is defined to be an isomorphism $\kappa : \gamma^*(V, W) \simeq (V, W)$ such that $\gamma^*(\kappa) \circ \kappa = \mathrm{id}$.

In the context of mixed twistor $\mathscr{D}$-modules, the counterpart of γ^* is given by the functor $\mathcal{T} \longmapsto \boldsymbol{D}(j^*\mathcal{T}^*) =: \tilde{\gamma}^*\mathcal{T}$. So, a real structure of mixed twistor structure $\mathcal{T}$ is given as an isomorphism $\kappa : \tilde{\gamma}^*\mathcal{T} \simeq \mathcal{T}$ such that $\tilde{\gamma}^*\kappa \circ \kappa = id$. We shall observe the functoriality of real structures. We also observe that the underlying holonomic $\mathscr{D}$-modules are naturally equipped with an $\mathbb{R}$-Betti structure in the sense of [57]. In the

T. Mochizuki, *Mixed Twistor $\mathscr{D}$-modules*, Lecture Notes in Mathematics 2125,
DOI 10.1007/978-3-319-10088-3_13

integrable case, it is interesting and natural to impose the compatibility condition of the real structure and the Stokes structure along $\lambda = 0$. The integrable mixed twistor $\mathscr{D}$-modules equipped with such a good $\mathbb{R}$-structure (or more generally a good K-structure) would be one of the basic ingredients in the study of non-commutative Hodge theory.

In Sect. 13.5, we shall observe that the category of mixed $\mathbb{R}$-Hodge modules is naturally embedded to the category of integrable mixed twistor $\mathscr{D}$-modules with good $\mathbb{R}$-structure.

13.1 Duality of $\mathcal{R}$-Modules

13.1.1 Duality

Let X be any complex manifold, and let H be any hypersurface. Set $d_X := \dim X$. Let N be any left-$\mathcal{R}_{X(*H)}$-bi-module, i.e., it is equipped with mutually commuting two $\mathcal{R}_{X(*H)}$-actions ρ_i $(i = 1,2)$. Let (N, ρ_i) denote the left $\mathcal{R}_{X(*H)}$-modules by ρ_i. For any $\mathcal{R}_{X(*H)}$-module L, let $\mathcal{H}om_{\mathcal{R}_{X(*H)}}(L, N^{\rho_1,\rho_2})$ denote the sheaf of $\mathcal{R}_{X(*H)}$-homomorphisms $L \longrightarrow (N, \rho_1)$. The sheaf is equipped with an $\mathcal{R}_{X(*H)}$-action induced by ρ_2. For any $\mathcal{R}_{X(*H)}$-complex $L^\bullet$, we obtain an $\mathcal{R}_{X(*H)}$-complex $\mathcal{H}om_{\mathcal{R}_{X(*H)}}(L^\bullet, N^{\rho_1,\rho_2})$. We have the naturally defined $\mathcal{R}_{X(*H)}$-homomorphism

$$L^\bullet \longrightarrow \mathcal{H}om_{\mathcal{R}_{X(*H)}}\Big(\mathcal{H}om_{\mathcal{R}_{X(*H)}}(L^\bullet, N^{\rho_2,\rho_1}), N^{\rho_1,\rho_2}\Big) \tag{13.1}$$

given by $x \longmapsto \big(F \longmapsto (-1)^{|x|\,|F|}F(x)\big)$.

Let Θ_X denote the tangent sheaf of X, and let $\Omega^1_X := \mathcal{H}om_{\mathcal{O}_X}\big(\Theta_X, \mathcal{O}_X\big)$. Let $p_\lambda : \mathcal{X} \longrightarrow X$ denote the projection. Let $\Theta_{\mathcal{X}} := \lambda \cdot p_\lambda^*\Theta_X$, and let $\Omega^1_{\mathcal{X}} := \mathcal{H}om_{\mathcal{O}_{\mathcal{X}}}\big(\Theta_{\mathcal{X}}, \mathcal{O}_{\mathcal{X}}\big)$. We have $\Omega^1_{\mathcal{X}} = \lambda^{-1}p_\lambda^*\Omega^1_X$. Its j-th exterior product is denoted by $\Omega^j_{\mathcal{X}}$. In particular, we set $\omega_{\mathcal{X}} := \Omega^{d_X}_{\mathcal{X}}$.

Recall that we have the natural two $\mathcal{R}_{X(*H)}$-actions on $\mathcal{R}_{X(*H)} \otimes \omega_{\mathcal{X}}^{-1}$. The left multiplication is denoted by ℓ. Let r denote the action induced by the right multiplication. More generally, for a left $\mathcal{R}_{X(*H)}$-module N, we have two induced left $\mathcal{R}_{X(*H)}$-module structures on $N \otimes \mathcal{R}_{X(*H)} \otimes \omega_{\mathcal{X}}^{-1}$. One is given by ℓ and the left $\mathcal{R}$-action on N, which is denoted by ℓ. The other is induced by the right multiplication, denoted by r. As in [66, 69], we have the unique $\mathbb{C}$-linear isomorphism $\Phi_N : N \otimes \mathcal{R}_{X(*H)} \otimes \omega_{\mathcal{X}}^{-1} \simeq N \otimes \mathcal{R}_{X(*H)} \otimes \omega_{\mathcal{X}}^{-1}$, such that (a) $\Phi_N \circ r = \ell \circ \Phi_N$, (b) $\Phi_N \circ \ell = r \circ \Phi_N$, (c) Φ_N induces the identity on $N \otimes \omega_{\mathcal{X}}^{-1}$.

Let $\mathcal{G}^\bullet$ be an injective $\mathcal{R}_{X(*H)}$-bi-module resolution of $\mathcal{R}_{X(*H)} \otimes \omega_{\mathcal{X}}^{-1}[d_X]$. It also gives an injective $\mathcal{R}_{X(*H)}$-module resolution with respect to both ℓ and r. The isomorphism of $\mathcal{R}_{X(*H)}$-bi-modules $\Phi_{\mathcal{O}} : \big(\mathcal{R}_{X(*H)} \otimes \omega_{\mathcal{X}}^{-1}, \ell, r\big) \longrightarrow \big(\mathcal{R}_{X(*H)} \otimes \omega_{\mathcal{X}}^{-1}, r, \ell\big)$ is extended to a quasi-isomorphism $\Phi_0 : (\mathcal{G}^\bullet, \ell, r) \longrightarrow (\mathcal{G}^\bullet, r, \ell)$. Let $\mathcal{M}$

be any coherent $\mathcal{R}_{X(*H)}$-module. We define

$$\boldsymbol{D}'_{X(*H)}\mathcal{M} := \mathcal{H}om_{\mathcal{R}_{X(*H)}}\big(\mathcal{M}, \mathcal{G}^{\bullet\,\ell,r}\big), \quad \boldsymbol{D}_{X(*H)}\mathcal{M} := \lambda^{d_X}\cdot\boldsymbol{D}'_{X(*H)}\mathcal{M}. \tag{13.2}$$

The morphism (13.1) with Φ_0 induces an isomorphism $\mathcal{M} \simeq \boldsymbol{D}_{X(*H)} \circ \boldsymbol{D}_{X(*H)}(\mathcal{M})$.

Remark 13.1.1 Because $\mathcal{O}_{\mathbb{C}_\lambda}$ is the center of $\mathcal{R}_{X(*H)}$, we can naturally regard $\mathcal{R}_{X(*H)} \otimes \omega_{\mathcal{X}}^{-1}$ as an $\mathcal{R}_{X(*H)} \otimes_{\mathcal{O}_{\mathbb{C}_\lambda}} \mathcal{R}_{X(*H)}$-module. For any injective $\mathcal{R}_{X(*H)} \otimes_{\mathcal{O}_{\mathbb{C}_\lambda}} \mathcal{R}_{X(*H)}$-resolution $\mathcal{G}_1^\bullet$ of $\mathcal{R}_{X(*H)} \otimes \omega_{\mathcal{X}}^{-1}[d_X]$, we have a natural isomorphism in the derived category of $\mathcal{R}_{X(*H)}$-modules:

$$\lambda^{d_X}\mathcal{H}om_{\mathcal{R}_{X(*H)}}(\mathcal{M}, \mathcal{G}_1^{\bullet\,\ell,r}) \simeq \boldsymbol{D}_{X(*H)}\mathcal{M}$$

Indeed, we may assume to have an isomorphism $\mathcal{G}_1^\bullet \longrightarrow \mathcal{G}^\bullet$ of complexes of $\mathcal{R}_{X(*H)}$-bi-modules in the derived category, which induces the desired isomorphism. □

13.1.2 Compatibility with Push-Forward

Let $F : X \longrightarrow Y$ be a morphism of complex manifolds. We shall construct a trace morphism

$$\lambda^{d_X}\cdot F_\dagger(\mathcal{O}_{\mathcal{X}})[d_X] \longrightarrow \lambda^{d_Y}\cdot \mathcal{O}_{\mathcal{Y}}[d_Y] \tag{13.3}$$

in $D^b(\mathcal{R}_Y)$ by a standard method. Let $\mathfrak{Db}_{\mathcal{X}/\mathbb{C}_\lambda}$ denote the sheaf of distributions on $\mathcal{X}$ which are holomorphic in λ. We set

$$\mathfrak{Db}^\bullet_{\mathcal{X}/\mathbb{C}_\lambda} := \mathrm{Tot}\Big(p_\lambda^{-1}\Omega_X^{0,\bullet} \otimes_{p_\lambda^{-1}C^\infty(X)} \big(\Omega^\bullet_{\mathcal{X}} \otimes_{\mathcal{O}_{\mathcal{X}}} \mathfrak{Db}_{\mathcal{X}/\mathbb{C}_\lambda}\big)\Big)$$

We have a natural quasi-isomorphism $\Omega^\bullet_{\mathcal{X}} \longrightarrow \mathfrak{Db}^\bullet_{\mathcal{X}/\mathbb{C}_\lambda}$, and hence an isomorphism $\mathcal{R}_X \otimes_{\mathcal{O}_{\mathcal{X}}} \big(\mathfrak{Db}^\bullet_{\mathcal{X}/\mathbb{C}_\lambda}\big)[d_X] \simeq \omega_{\mathcal{X}}$ in $D^b(\mathcal{R}_X)$. We obtain the following morphism of complexes, by the integration multiplied with $(2\pi\sqrt{-1})^{-d_X+d_Y}$:

$$\lambda^{d_X}\cdot F_!\mathfrak{Db}^\bullet_{\mathcal{X}/\mathbb{C}_\lambda}[2d_X] \longrightarrow \lambda^{d_Y}\cdot\mathfrak{Db}^\bullet_{\mathcal{Y}/\mathbb{C}_\lambda}[2d_Y]$$

Hence, we obtain the trace morphism (13.3) as follows:

$$\begin{aligned} F_\dagger\mathcal{O}_{\mathcal{X}}[d_X] &\longrightarrow F_!\big(\mathcal{R}_{Y\leftarrow X} \otimes^L_{\mathcal{R}_X} \mathcal{O}_{\mathcal{X}}[d_X]\big) \\ &= F_!\big(\mathfrak{Db}^\bullet_{\mathcal{X}/\mathbb{C}_\lambda} \otimes_{F^{-1}\mathcal{O}_{\mathcal{Y}}} F^{-1}(\mathcal{R}_Y \otimes \omega_{\mathcal{Y}}^{-1})[2d_X]\big) \\ &\longrightarrow \lambda^{d_Y-d_X}\cdot\mathfrak{Db}^\bullet_{\mathcal{Y}/\mathbb{C}_\lambda} \otimes \mathcal{R}_Y \otimes \omega_{\mathcal{Y}}^{-1}[2d_Y] \simeq \lambda^{d_Y-d_X}\cdot\mathcal{O}_{\mathcal{Y}}[d_Y] \end{aligned} \tag{13.4}$$

By using the isomorphism in Sect. 13.1.2.1, we obtain the following morphism in the derived category of $\mathcal{R}_Y$-bi-complexes:

$$\begin{aligned} F_\dagger\bigl(\mathcal{O}_{\mathcal{X}}[d_X]\otimes_{F^{-1}\mathcal{O}_{\mathcal{Y}}}F^{-1}(\mathcal{R}_Y\otimes\omega_{\mathcal{Y}}^{-1})\bigr) &\simeq RF_*\bigl(\mathcal{O}_{\mathcal{X}}[d_X]\otimes_{F^{-1}\mathcal{O}_{\mathcal{Y}}}F^{-1}\mathcal{M}_{Y0}\bigr)\\ &\simeq RF_*\bigl(\mathcal{O}_{\mathcal{X}}[d_X]\otimes_{F^{-1}\mathcal{O}_{\mathcal{Y}}}F^{-1}\mathcal{M}_{Y1}\bigr)\\ &\simeq F_\dagger\mathcal{O}_{\mathcal{X}}[d_X]\otimes_{\mathcal{O}_{\mathcal{Y}}}(\mathcal{R}_Y\otimes\omega_{\mathcal{Y}}^{-1})\\ &\longrightarrow \lambda^{d_Y-d_X}\mathcal{O}_{\mathcal{Y}}[d_Y]\otimes_{\mathcal{O}_{\mathcal{Y}}}(\mathcal{R}_Y\otimes\omega_{\mathcal{Y}}^{-1})\\ &\simeq \lambda^{d_Y-d_X}\mathcal{R}_Y\otimes\omega_{\mathcal{Y}}^{-1}[d_Y] \end{aligned} \tag{13.5}$$

Let $\mathcal{M}$ be a coherent $\mathcal{R}_X$-module. We have a natural morphism $\varphi: F_\dagger \boldsymbol{D}_X\mathcal{M}\longrightarrow \boldsymbol{D}_Y F_\dagger\mathcal{M}$ induced by the composite of the following morphisms:

$$\begin{aligned} &F_\dagger R\mathcal{H}om_{\mathcal{R}_X}(\mathcal{M},\mathcal{R}_X\otimes\omega_{\mathcal{X}}^{-1})[d_X]\\ &\quad\longrightarrow F_! R\mathcal{H}om_{\mathcal{R}_X}\Bigl(\mathcal{M},\mathcal{O}_{\mathcal{X}}\otimes_{F^{-1}\mathcal{O}_{\mathcal{Y}}}F^{-1}(\mathcal{R}_Y\otimes\omega_{\mathcal{Y}}^{-1})\Bigr)[d_X]\\ &\quad\longrightarrow F_! R\mathcal{H}om_{F^{-1}\mathcal{R}_Y}\Bigl(\mathcal{R}_{Y\leftarrow X}\otimes^L_{\mathcal{R}_Y}\mathcal{M},\,\mathcal{R}_{Y\leftarrow X}\otimes^L_{\mathcal{R}_Y}\mathcal{O}_{\mathcal{X}}\otimes_{F^{-1}\mathcal{O}_{\mathcal{Y}}}F^{-1}(\mathcal{R}_Y\otimes\omega_{\mathcal{Y}}^{-1})\Bigr)[d_X]\\ &\quad\longrightarrow R\mathcal{H}om_{\mathcal{R}_Y}(F_\dagger\mathcal{M},\,F_\dagger(\mathcal{O}_{\mathcal{X}}\otimes_{F^{-1}\mathcal{O}_{\mathcal{Y}}}F^{-1}(\mathcal{R}_Y\otimes\omega_{\mathcal{Y}}^{-1})))[d_X]\\ &\quad\longrightarrow R\mathcal{H}om_{\mathcal{R}_Y}(F_\dagger\mathcal{M},\,(\mathcal{R}_Y\otimes\omega_{\mathcal{Y}}^{-1}))[d_Y]\cdot\lambda^{d_Y-d_X} \end{aligned} \tag{13.6}$$

The following lemma can be shown by an argument in the proof of Proposition 4.39 of [34].

Lemma 13.1.2 *Assume that (i) the restriction of F to* Supp $\mathcal{M}$ *is proper, (ii) $\mathcal{M}$ is good over Y. Then, φ is an isomorphism.* □

13.1.2.1 Appendix

Let Y be a complex manifold. We set $\mathcal{M}_{Y0}:=\bigl(\mathcal{R}_Y\otimes\omega_{\mathcal{Y}}^{-1}\bigr)^{\ell}\otimes^{\ell}_{\mathcal{O}_{\mathcal{Y}}}\bigl(\mathcal{R}_Y\otimes\omega_{\mathcal{Y}}^{-1}\bigr)$, where "${}^{\ell}\otimes^{\ell}_{\mathcal{O}_{\mathcal{Y}}}$" means that we use the $\mathcal{O}_{\mathcal{Y}}$-actions induced by ℓ on the both first and second components. Let r_i $(i=1,2)$ denote the $\mathcal{R}_Y$-actions induced by the $\mathcal{R}_Y$-action r on the i-th component. We also have the $\mathcal{R}_Y$-action $\ell_1\otimes\ell_2$ induced by the $\mathcal{R}_Y$-actions ℓ on the first and the second.

We set $\mathcal{M}_{Y1}:=(\mathcal{R}_Y\otimes\omega_{\mathcal{Y}}^{-1})^r\otimes^{\ell}_{\mathcal{O}_{\mathcal{Y}}}(\mathcal{R}_Y\otimes\omega_{\mathcal{Y}}^{-1})$, where "${}^r\otimes^{\ell}_{\mathcal{O}_{\mathcal{Y}}}$" means that for the tensor product we use the $\mathcal{O}_{\mathcal{Y}}$-actions r and ℓ on the first factor and the second factor, respectively. Let ℓ_1 (resp. r_2) denote the $\mathcal{R}_Y$-action induced by the $\mathcal{R}_Y$-action ℓ (resp. r) on the first (resp. second) component. We also have the $\mathcal{R}_Y$-action $r_1\otimes\ell_2$ induced by the $\mathcal{R}_Y$-actions r and ℓ on the first and the second, respectively.

We have a unique $\mathbb{C}$-linear isomorphism $\Psi:\mathcal{M}_{Y1}\longrightarrow\mathcal{M}_{Y0}$ such that (a) $\Psi\circ r_2=r_2\circ\Psi$, (b) $\Psi\circ\ell_1=(\ell_1\otimes\ell_2)\circ\Psi$, (c) Ψ induces the identity on $\omega_{\mathcal{Y}}^{-1}\otimes_{\mathcal{O}_{\mathcal{Y}}}\omega_{\mathcal{Y}}^{-1}$. Here, $\Psi\circ r_2=r_2\circ\Psi$ means that $\Psi(r_2(a)m_1\otimes m_2)=r_2\Psi(m_1\otimes m_2)$ for local sections

$a \in \mathcal{R}_Y$ and $m_i \in \mathcal{R}_Y \otimes \omega_{\mathcal{Y}}^{-1}$. Similar for the condition $\Psi \circ \ell_1 = (\ell_1 \otimes \ell_2) \circ \Psi$. For a local frame τ of $\omega_{\mathcal{Y}}^{-1}$, and local sections $P \in \mathcal{R}_Y$ and $m \in \mathcal{R}_Y \otimes \omega_{\mathcal{Y}}^{-1}$, we have

$$\Psi\big((P \otimes \tau) \otimes m\big) = (\ell_1 \otimes \ell_2)(P)\big((1 \otimes \tau) \otimes m\big).$$

Lemma 13.1.3 *We have $\Psi \circ (r_1 \otimes \ell_2) = r_1 \circ \Psi$.*

Proof Let $f \in \mathcal{O}_{\mathcal{Y}}$. Then, we have

$$\begin{aligned} \Psi\big((r_1 \otimes \ell_2)(f)(P \otimes \tau) \otimes m\big) &= \Psi\big((P \otimes \tau) \otimes fm\big) = (\ell_1 \otimes \ell_2)(P)\big((1 \otimes \tau) \otimes fm\big) \\ &= (\ell_1 \otimes \ell_2)(P) r_1(f)\big((1 \otimes \tau) \otimes m\big) \\ &= r_1(f)\Psi\big((P \otimes \tau) \otimes m\big). \end{aligned} \tag{13.7}$$

Suppose that $\tau = \lambda^{-n} dx_1 \wedge \cdots \wedge dx_n$ ($n = \dim Y$) for a holomorphic coordinate system $(x_1, \ldots, x_n)$. For a local section v of $\lambda p_\lambda^* \Theta_Y$, we have the following:

$$\begin{aligned} \Psi\big((r_1 \otimes \ell_2)(v)(P \otimes \tau) \otimes m\big) &= \Psi\big((-Pv \otimes \tau) \otimes m\big) + \Psi\big((P \otimes \tau) \otimes vm\big) \\ &= (\ell_1 \otimes \ell_2)(-Pv)\big((1 \otimes \tau) \otimes m\big) \\ &\quad + (\ell_1 \otimes \ell_2)(P)\big((1 \otimes \tau) \otimes vm\big) \\ &= (\ell_1 \otimes \ell_2)(P)\big((-v \otimes \tau) \otimes m\big) \\ &= (\ell_1 \otimes \ell_2)(P) \circ r_1(v)\big((1 \otimes \tau) \otimes m\big) \\ &= r_1(v)\Psi\big((P \otimes \tau) \otimes m\big) \end{aligned} \tag{13.8}$$

Then, we obtain the claim of Lemma 13.1.3. □

13.1.3 Specialization Along $\mathcal{X}^{\lambda_0}$

Let X be a complex manifold with a hypersurface H. Let $\omega_{X(*H)} := \Omega_X^{d_X}(*H)$. Take $\lambda_0 \in \mathbb{C}$. We set

$$\mathcal{R}_{X(*H)}^{\lambda_0} := \mathcal{R}_{X(*H)} \big/ (\lambda - \lambda_0)\mathcal{R}_{X(*H)}.$$

We can also naturally regard it as a sheaf of algebras on $\mathcal{X}^{\lambda_0} := \{\lambda_0\} \times X$. If $\lambda_0 \neq 0$, it is naturally isomorphic to $\mathscr{D}_{X(*H)}$. If $\lambda_0 = 0$, $\mathcal{R}_{X(*H)}^0$ is naturally isomorphic to $\operatorname{Sym}^\bullet \Theta_X(*H)$.

Let $i_{\lambda_0} : \mathcal{X}^{\lambda_0} \longrightarrow \mathcal{X}$ be the inclusion. For any $\mathcal{R}_{X(*H)}$-module $\mathcal{M}$, we set $i_{\lambda_0}^* \mathcal{M} := \mathcal{R}_{X(*H)}^{\lambda_0} \otimes_{i_{\lambda_0}^{-1} \mathcal{R}_{X(*H)}} i_{\lambda_0}^{-1} \mathcal{M}$. For any $\mathcal{R}_{X(*H)}^{\lambda_0}$-module $\mathcal{N}$, we may naturally regard $i_{\lambda_0 *} \mathcal{N}$ as an $\mathcal{R}_{X(*H)}$-module.

Proposition 13.1.4 *Let $\mathcal{M}$ be any coherent strict $\mathcal{R}_{X(*H)}$-module. We have a natural isomorphism*

$$i_{\lambda_0 *}\mathcal{O}_X \otimes^L_{\mathcal{O}_{\mathcal{X}}} \boldsymbol{D}_{X(*H)}\mathcal{M} \simeq i_{\lambda_0 *} R\mathcal{H}om_{\mathcal{R}^{\lambda_0}_{X(*H)}}\big(i^*_{\lambda_0}\mathcal{M}, \mathcal{R}^{\lambda_0}_{X(*H)} \otimes \omega^{-1}_{X(*H)}\big)[d_X]$$

Proof We omit to denote $(*H)$ in the proof. Note $\lambda^{d_X}\omega_{\mathcal{X}}^{-1} = \lambda^{2d_X} p_\lambda^* \omega_X^{-1}$. We have an isomorphism $i^*_{\lambda_0}(\mathcal{R}_X \otimes \lambda^{d_X}\omega_{\mathcal{X}}^{-1}) \simeq \mathcal{R}_X^{\lambda_0} \otimes \omega_X^{-1}$. We take an injective $\mathcal{R}_X^{\lambda_0} \otimes_{\mathbb{C}} \mathcal{R}_X^{\lambda_0}$-resolution $\mathcal{G}_2^\bullet$ of $\mathcal{R}_X^{\lambda_0} \otimes \omega_X^{-1}[d_X]$. We have

$$\mathcal{H}om_{\mathcal{R}_X}\big(\mathcal{M}, i_{\lambda_0 *}\mathcal{G}_2^\bullet\big) \simeq i_{\lambda_0 *}\mathcal{H}om_{\mathcal{R}_X^{\lambda_0}}\big(i^*_{\lambda_0}\mathcal{M}, \mathcal{G}_2^\bullet\big)$$

We take an injective $\mathcal{R}_X \otimes_{\mathcal{O}_{\mathbb{C}_\lambda}} \mathcal{R}_X$-resolution $\mathcal{G}_3^\bullet$ of $i_{\lambda_0 *}\big(\mathcal{R}_X^{\lambda_0} \otimes \omega_X^{-1}\big)[d_X]$. We may assume to have a morphism $i_{\lambda_0}\mathcal{G}_2^\bullet \longrightarrow \mathcal{G}_3^\bullet$ of $\mathcal{R}_X \otimes_{\mathcal{O}_{\mathbb{C}_\lambda}} \mathcal{R}_X$-complexes compatible with the identity on $i_{\lambda_0 *}(\mathcal{R}_X^{\lambda_0} \otimes \omega_X^{-1})[d_X]$. Because $\mathcal{M}$ is strict, we can easily check that the morphism $\mathcal{H}om_{\mathcal{R}_X}(\mathcal{M}, i_{\lambda_0 *}\mathcal{G}_2^\bullet) \longrightarrow \mathcal{H}om_{\mathcal{R}_X}(\mathcal{M}, \mathcal{G}_3^\bullet)$ is a quasi-isomorphism. Thus, we obtain the following isomorphisms:

$$i_{\lambda_0 *}\mathcal{H}om_{\mathcal{R}_X^{\lambda_0}}(i^*_{\lambda_0}\mathcal{M}, \mathcal{G}_2^\bullet) \simeq \mathcal{H}om_{\mathcal{R}_X}(\mathcal{M}, \mathcal{G}_3^\bullet) \simeq i_{\lambda_0 *}\mathcal{O}_X \otimes^L_{\mathcal{X}} \boldsymbol{D}_X(\mathcal{M}) \tag{13.9}$$

The isomorphism (13.9) gives the desired one. □

13.1.4 Twist by Smooth $\mathcal{R}$-Modules

Let X and H be as above. Let $\mathcal{M}$ be a coherent $\mathcal{R}_{X(*H)}$-module. Let V be a smooth $\mathcal{R}_{X(*H)}$-module. We have the natural isomorphism $\boldsymbol{D}_{X(*H)}(\mathcal{M} \otimes V) \simeq \boldsymbol{D}_{X(*H)}(\mathcal{M}) \otimes V^\vee$, given as follows. We have two naturally induced left $\mathcal{R}$-module structures ℓ and r on $V^\vee \otimes_{\mathcal{O}_{\mathcal{X}}} (\mathcal{R}_{X(*H)} \otimes \omega_{\mathcal{X}}^{-1})$. Then, we obtain

$$\begin{aligned} \boldsymbol{D}_{X(*H)}(\mathcal{M} \otimes V) &\simeq \lambda^{d_X} R\mathcal{H}om_{\mathcal{R}_{X(*H)}}\big(\mathcal{M} \otimes V, (\mathcal{R}_{X(*H)} \otimes \omega_{\mathcal{X}}^{-1})^{\ell,r}\big)[d_X] \\ &\simeq \lambda^{d_X} R\mathcal{H}om_{\mathcal{R}_{X(*H)}}\big(\mathcal{M}, (V^\vee \otimes \mathcal{R}_{X(*H)} \otimes \omega_{\mathcal{X}}^{-1})^{\ell,r}\big)[d_X]. \end{aligned} \tag{13.10}$$

By $\Phi_{V^\vee}$ in Sect. 13.1.1, it is isomorphic to

$$\begin{aligned} &\lambda^{d_X} R\mathcal{H}om_{\mathcal{R}_{X(*H)}}\big(\mathcal{M}, (V^\vee \otimes \mathcal{R}_{X(*H)} \otimes \omega_{\mathcal{X}}^{-1})^{r,\ell}\big)[d_X] \\ &\simeq \lambda^{d_X} R\mathcal{H}om_{\mathcal{R}_{X(*H)}}\big(\mathcal{M}, (\mathcal{R}_{X(*H)} \otimes \omega_{\mathcal{X}}^{-1})^{r,\ell}\big) \otimes V^\vee[d_X]. \end{aligned} \tag{13.11}$$

By $\Phi_{\mathcal{O}_{\mathcal{X}}(*\mathcal{H})}$, it is isomorphic to $\lambda^{d_X} R\mathcal{H}om_{\mathcal{R}_{X(*H)}}\big(\mathcal{M}, (\mathcal{R}_{X(*H)} \otimes \omega_{\mathcal{X}}^{-1})^{\ell,r}\big) \otimes V^\vee[d_X]$. We obtain the desired isomorphism $\boldsymbol{D}_{X(*H)}(\mathcal{M} \otimes V) \simeq \boldsymbol{D}_{X(*H)}(\mathcal{M}) \otimes V^\vee$.

13.1.5 Duality of Smooth $\mathcal{R}$-Modules

Let $\mathcal{M}$ be a smooth $\mathcal{R}_{X(*H)}$-module. Set $\mathcal{M}^\vee := \mathcal{H}om_{\mathcal{O}_{\mathcal{X}(*\mathcal{H})}}(\mathcal{M}, \mathcal{O}_\mathcal{X}(*\mathcal{H}))$, which is also naturally a smooth $\mathcal{R}_{X(*H)}$-module. Although it is easy to see that $\mathcal{M}^\vee$ is isomorphic to $\boldsymbol{D}'_{X(*H)}\mathcal{M}$, let us look at the isomorphism more closely, and check the signature of the isomorphism $\mathcal{M} \simeq \boldsymbol{D}' \circ \boldsymbol{D}'(\mathcal{M}) \simeq \boldsymbol{D}'(\mathcal{M}^\vee) \simeq \mathcal{M}$. For simplicity of the description, we omit to denote $(*H)$.

We have two natural $\mathcal{R}_X$-actions ℓ and r on $\mathcal{R}_X \otimes \omega_\mathcal{X}^{-1}$. For $p \in \mathbb{Z}$, let $\Theta^p := \bigwedge^{-p} \Theta_\mathcal{X}$. We have the Spencer resolution $\Theta^\bullet \otimes \mathcal{R}_X$ of $\mathcal{O}_\mathcal{X}$ as in [66]. We have an $\mathcal{R}_X \otimes_{\mathcal{O}_{\mathbb{C}_\lambda}} \mathcal{R}_X$-homomorphism

$$C : \big(\mathcal{R}_X \otimes \Theta_\mathcal{X}^\bullet\big) \otimes \big(\mathcal{R}_X \otimes \Theta_\mathcal{X}^\bullet\big) \longrightarrow \mathcal{R}_X \otimes \omega_\mathcal{X}^{-1}[d_X]$$

given as follows:

$$C(P_1 \otimes \tau_1, P_2 \otimes \tau_2) = \begin{cases} (-1)^{|\tau_2|}\ell(P_1)\, r(P_2)\tau_1 \wedge \tau_2 & (|\tau_1| + |\tau_2| = d_X) \\ 0 & \text{otherwise} \end{cases}$$

Here, ℓ and r denote the $\mathcal{R}_X$-actions on $\mathcal{R}_X \otimes \omega_\mathcal{X}^{-1}$. Suppose that there exists a holomorphic coordinate $(z_1, \dots, z_n)$, and that $\tau_j = \bigwedge_{j \in I_i} \eth_j$ such that $I_1 \sqcup I_2 = \{1, \dots, n\}$, where $\eth_j = \lambda \partial_j$. Then, we have $C(P_1 \otimes \tau_1, P_2 \otimes \tau_2) = (-1)^{|\tau_2|} P_1\, {}^t\!P_2\, \tau_1 \wedge \tau_2$.

Lemma 13.1.5 *C is a morphism of complexes.*

Proof We have only to show that $C \circ \partial(P_1 \otimes \tau_1 \otimes P_2 \otimes \tau_2) = 0$ if $|\tau_1| + |\tau_2| = d + 1$. We may assume that there exists a holomorphic coordinate $(z_1, \dots, z_n)$, and we have only to consider the case $\tau_i = \bigwedge_{j \in I_i} \eth_j$. We have

$$\begin{aligned} \partial(P_1 \otimes \tau_1 \otimes P_2 \otimes \tau_2) = &\sum_{i=1}^n P_1 \eth_i \otimes \big(\iota(\lambda^{-1} dz_i)\tau_1\big) \otimes (P_2 \otimes \tau_2) \\ &+ \sum_{i=1}^n (-1)^{|\tau_1|} P_1 \otimes \tau_1 \otimes (P_2 \eth_i \otimes \iota(\lambda^{-1} dz_i)\tau_2) \end{aligned} \tag{13.12}$$

Hence, we have the following:

$$\begin{aligned} &C \circ \partial\big(P_1 \otimes \tau_1 \otimes P_2 \otimes \tau_2\big) \\ &= \sum_{i=1}^n P_1 \eth_i {}^t\!P_2 \otimes \iota(\lambda^{-1} dz_i)\tau_1 \wedge \tau_2 (-1)^{|\tau_2|} \end{aligned}$$

$$+\sum_{i=1}^{n} P_1 {}^t(P_2 \eth_i) \otimes \tau_1 \wedge \iota(\lambda^{-1} dz_i)\tau_2 (-1)^{|\tau_2|+|\tau_1|-1}$$

$$= \sum_{i=1}^{n} P_1 \eth_i {}^t P_2 \otimes \Big(\iota(\lambda^{-1} dz_i)\tau_1 \wedge \tau_2 + (-1)^{|\tau_1|} \tau_1 \wedge \iota(\lambda^{-1} dz_i)\tau_2 \Big) (-1)^{|\tau_2|} = 0 \tag{13.13}$$

We have used $\tau_1 \wedge \tau_2 = 0$. Thus, we are done. □

We obtain the induced isomorphism of $\mathcal{R}_X$-complexes

$$\Psi : \mathcal{R}_X \otimes \Theta^\bullet \longrightarrow \mathcal{H}om_{\mathcal{R}_X}\big(\mathcal{R}_X \otimes \Theta^\bullet, (\mathcal{R}_X \otimes \omega_{\mathcal{X}}^{-1}[d_X])^{\ell,r}\big),$$

given by $\Psi(P \otimes \tau)(Q \otimes \omega) = (-1)^{|\tau||\omega|} C(Q \otimes \omega, P \otimes \tau)$.

Let $C_{\mathcal{M}} : \mathcal{M}^\vee \otimes \mathcal{M} \longrightarrow \mathcal{O}_{\mathcal{X}}$ be the natural perfect pairing of the $\mathcal{O}_{\mathcal{X}}$-modules. We obtain the $\mathcal{R}_X \otimes \mathcal{R}_X$-homomorphism

$$C : \big(\mathcal{R}_X \otimes \Theta^\bullet \otimes \mathcal{M}^\vee\big) \otimes \big(\mathcal{R}_X \otimes \Theta^\bullet \otimes \mathcal{M}\big) \longrightarrow \mathcal{R}_X \otimes \omega_{\mathcal{X}}^{-1}[d_X]$$

given as follows:

$$C(P_1 \otimes \tau_1 \otimes x_1, P_2 \otimes \tau_2 \otimes x_2) = \begin{cases} (-1)^{|\tau_2|} \ell(P_1)\, r(P_2) \big(\tau_1 \wedge \tau_2\, C_{\mathcal{M}}(x_1, x_2)\big) & (|\tau_1| + |\tau_2| = d_X) \\ 0 & \text{(otherwise)} \end{cases} \tag{13.14}$$

It is a morphism of complexes. Hence, we obtain the induced isomorphism of $\mathcal{R}_X$-complexes:

$$\Psi_{\mathcal{M}} : \mathcal{R}_X \otimes \Theta^\bullet \otimes \mathcal{M}^\vee \longrightarrow \mathcal{H}om_{\mathcal{R}_X}\big(\mathcal{R}_X \otimes \Theta^\bullet \otimes \mathcal{M}, (\mathcal{R}_X \otimes \omega_{\mathcal{X}}^{-1}[d_X])^{\ell,r}\big)$$

By the natural quasi isomorphism $\mathcal{R}_X \otimes \Theta^\bullet \otimes \mathcal{M} \simeq \mathcal{M}$, we obtain an isomorphism $\mathcal{M}^\vee \simeq \boldsymbol{D}'\mathcal{M}$ in the derived category which is also denoted by $\Psi_{\mathcal{M}}$.

We obtain the isomorphism $(\Psi_{\mathcal{M}}^*)^{-1} \circ \Psi_{\mathcal{M}^\vee} : \mathcal{M} \simeq \boldsymbol{D}' \circ \boldsymbol{D}'(\mathcal{M})$ in the derived category given as follows:

$$\mathcal{R}_X \otimes \Theta^\bullet \otimes \mathcal{M} \xrightarrow{\Psi_{\mathcal{M}^\vee}} \mathcal{H}om_{\mathcal{R}_X}\big(\mathcal{R}_X \otimes \Theta^\bullet \otimes \mathcal{M}^\vee, (\mathcal{R}_X \otimes \omega_{\mathcal{X}}^{-1})^{\ell,r}[d_X]\big)$$
$$\xleftarrow{\Psi_{\mathcal{M}}^*} \mathcal{H}om_{\mathcal{R}_X}\Big(\mathcal{H}om_{\mathcal{R}_X}\big(\mathcal{R}_X \otimes \Theta^\bullet \otimes \mathcal{M}, (\mathcal{R}_X \otimes \omega_{\mathcal{X}}^{-1})^{\ell,r}\big), (\mathcal{R}_X \otimes \omega_{\mathcal{X}}^{-1})^{\ell,r}[d_X]\Big) \tag{13.15}$$

Here, $\Psi^*_{\mathcal{M}}(F) := F \circ \Psi_{\mathcal{M}}$. Briefly, the isomorphism is obtained as the composite of $\boldsymbol{D}'(\boldsymbol{D}'\mathcal{M}) \simeq \boldsymbol{D}'(\mathcal{M}^\vee) \simeq (\mathcal{M}^\vee)^\vee \simeq \mathcal{M}$.

Let $\Phi : \mathcal{R}_X \otimes \omega_{\mathcal{X}}^{-1}[d_X] \simeq \mathcal{R}_X \otimes \omega_{\mathcal{X}}^{-1}[d_X]$ be the automorphism, which exchanges the actions ℓ and r. It induces

$$\Phi_* : \mathcal{H}om_{\mathcal{R}_X}\big(\mathcal{R}_X \otimes \Theta^{-\bullet} \otimes \mathcal{M}, (\mathcal{R}_X \otimes \omega_{\mathcal{X}}^{-1})^{\ell,r}[d_X]\big) \\ \simeq \mathcal{H}om_{\mathcal{R}_X}\big(\mathcal{R}_X \otimes \Theta^{-\bullet} \otimes \mathcal{M}, (\mathcal{R}_X \otimes \omega_{\mathcal{X}}^{-1})^{r,\ell}[d_X]\big). \tag{13.16}$$

It induces the following isomorphism:

$$\Phi^* : \mathcal{H}om_{\mathcal{R}_X}\Big(\mathcal{H}om_{\mathcal{R}_X}\big(\mathcal{R}_X \otimes \Theta^{-\bullet} \otimes \mathcal{M}, (\mathcal{R}_X \otimes \omega_{\mathcal{X}}^{-1})^{r,\ell}\big), (\mathcal{R}_X \otimes \omega_{\mathcal{X}}^{-1})^{\ell,r}[d_X]\Big) \\ \simeq \mathcal{H}om_{\mathcal{R}_X}\Big(\mathcal{H}om_{\mathcal{R}_X}\big(\mathcal{R}_X \otimes \Theta^{-\bullet} \otimes \mathcal{M}, (\mathcal{R}_X \otimes \omega_{\mathcal{X}}^{-1})^{\ell,r}\big), (\mathcal{R}_X \otimes \omega_{\mathcal{X}}^{-1})^{\ell,r}[d_X]\Big) \\ = \boldsymbol{D}' \circ \boldsymbol{D}'(\mathcal{M}) \tag{13.17}$$

We have the quasi-isomorphism $\Lambda : \mathcal{R}_X \otimes \Theta^{-\bullet} \otimes \mathcal{M} \longrightarrow \boldsymbol{D}' \circ \boldsymbol{D}'(\mathcal{M})$ induced by the morphisms (13.1) and (13.17).

Lemma 13.1.6 *We have $\Lambda = (-1)^{d_X}(\Psi^*_{\mathcal{M}})^{-1} \circ \Psi_{\mathcal{M}^\vee}$ in the derived category.*

Proof For local sections $P \otimes \tau \otimes x \in \mathcal{R}_X \otimes \omega_{\mathcal{X}} \otimes \mathcal{M}$ and $Q \otimes \omega \otimes y \in \mathcal{R}_X \otimes \omega_{\mathcal{X}} \otimes \mathcal{M}^\vee$, we have the following:

$$\begin{aligned} \big(\Psi^*_{\mathcal{M}}\Lambda(P \otimes \tau \otimes x)\big)(Q \otimes \omega \otimes y) &= \Lambda(P \otimes \tau \otimes x)\big(\Psi_{\mathcal{M}}(Q \otimes \omega \otimes y)\big) \\ &= \Phi\big((-1)^{|\tau|\,|\omega|}\Psi_{\mathcal{M}}(Q \otimes \omega \otimes y)(P \otimes \tau \otimes x)\big) \\ &= \Phi\big(C(P \otimes \tau \otimes x, Q \otimes \omega \otimes y)\big) \\ &= (-1)^{d_X+|\tau|\,|\omega|}C(Q \otimes \omega \otimes y, P \otimes \tau \otimes x) \\ &= \big((-1)^{d_X}\Psi_{\mathcal{M}^\vee}(P \otimes \tau \otimes x)\big)(Q \otimes \omega \otimes y) \end{aligned} \tag{13.18}$$

Thus, we are done. □

13.1.6 Duality of Integrable $\mathcal{R}_X$-Modules

We can naturally regard that $\mathcal{R}_{X(*H)}$ is equipped with a left $\tilde{\mathcal{R}}_{X(*H)}$-action and a right $\mathcal{R}_{X(*H)}$-action. We naturally regard $\mathcal{R}_{X(*H)} \otimes \omega_{\mathcal{X}}^{-1}$ as an $\tilde{\mathcal{R}}_{X(*H)} \otimes_{\mathcal{O}_{\mathbb{C}_\lambda}} \mathcal{R}_{X(*H)}$-module. Note that the actions of $\lambda^2\partial_\lambda \otimes 1$ and $1 \otimes \mathcal{R}_{X(*H)}$ also gives an $\tilde{\mathcal{R}}_{X(*H)}$-action. In other words, it can be regarded as an $\mathcal{R}_{X(*H)} \otimes_{\mathcal{O}_{\mathbb{C}_\lambda}} \tilde{\mathcal{R}}_{X(*H)}$-module. The isomorphism of $\mathcal{R}_{X(*H)} \otimes_{\mathcal{O}_{\mathbb{C}_\lambda}} \mathcal{R}_{X(*H)}$-modules $\Psi_{\mathcal{O}} : (\mathcal{R}_{X(*H)} \otimes \omega_{\mathcal{X}}^{-1}, \ell, r) \longrightarrow$

$(\mathcal{R}_{X(*H)} \otimes \omega_{\mathcal{X}}^{-1}, r, \ell)$ naturally gives an isomorphism of $\tilde{\mathcal{R}}_{X(*H)} \otimes_{\mathcal{O}_{\mathbb{C}_\lambda}} \mathcal{R}_{X(*H)}$-modules.

We take an injective $\tilde{\mathcal{R}}_{X(*H)} \otimes_{\mathcal{O}_{\mathbb{C}_\lambda}} \mathcal{R}_{X(*H)}$-resolution $\mathcal{G}_0^\bullet$ of $\mathcal{R}_{X(*H)} \otimes \omega_{\mathcal{X}}^{-1}[d_X]$. Note that $\mathcal{G}_0^\bullet$ is also an injective $\mathcal{R}_{X(*H)} \otimes_{\mathcal{O}_{\mathbb{C}_\lambda}} \mathcal{R}_{X(*H)}$-resolution $\mathcal{G}_0^\bullet$ of $\mathcal{R}_{X(*H)} \otimes \omega_{\mathcal{X}}^{-1}[d_X]$. The isomorphism $\Psi_{\mathcal{O}}$ is extended to an isomorphism of $\tilde{\mathcal{R}}_{X(*H)} \otimes_{\mathcal{O}_{\mathbb{C}_\lambda}} \mathcal{R}_{X(*H)}$-complexes $\mathcal{G}_0^\bullet \longrightarrow \mathcal{G}_0^\bullet$.

Let $\mathcal{M}$ be an integrable $\mathcal{R}_{X(*H)}$-module. Then, we have an isomorphism in the derived category of $\mathcal{R}_{X(*H)}$-modules:

$$\boldsymbol{D}_{X(*H)}\mathcal{M} \simeq \mathcal{H}om_{\mathcal{R}_{X(*H)}}\big(\mathcal{M}, \lambda^{d_X}\mathcal{G}_0^{\bullet\,\ell,r}\big).$$

Note that the right hand side is naturally an $\tilde{\mathcal{R}}_{X(*H)}$-complex. The action of $\lambda^2\partial_\lambda$ is given by $\lambda^2\partial_\lambda(f)(s) = \lambda^2\partial_\lambda(f(s)) - f(\lambda^2\partial_\lambda s)$ for $f \in \mathcal{H}om_{\mathcal{R}_{X(*H)}}\big(\mathcal{M}, \mathcal{G}_0^i\big)$ and $s \in \mathcal{M}$. It determines an object in the derived category of $\tilde{\mathcal{R}}_{X(*H)}$-modules up to natural isomorphisms.

The following lemma is clear by the construction.

Lemma 13.1.7

- *In Lemma 13.1.2, we assume moreover that $\mathcal{M}$ is integrable. Then, φ is given as an isomorphism in the derived category of integrable $\mathcal{R}$-modules.*
- *In Sect. 13.1.4, the isomorphism $\boldsymbol{D}_{X(*H)}(\mathcal{M} \otimes V) \simeq \boldsymbol{D}_{X(*H)}(\mathcal{M}) \otimes V^\vee$ is given in the derived category of integrable $\mathcal{R}$-modules for any integrable coherent $\mathcal{R}_X$-module $\mathcal{M}$ and smooth integrable $\mathcal{R}_X$-module V.*
- *In Sect. 13.1.5, the isomorphism $\mathcal{M}^\vee \simeq \boldsymbol{D}'_{X(*H)}(\mathcal{M})$ is given in the derived category of integrable $\mathcal{R}$-modules for any smooth integrable $\mathcal{R}_X$-module $\mathcal{M}$.*

□

See [60] for more details on the duality of integrable $\mathcal{R}$-modules.

13.2 Duality and Strict Specializability of $\mathcal{R}$-Modules

13.2.1 Statement

Let $X = X_0 \times \mathbb{C}_t$. Let $\mathcal{M}$ be a coherent $\mathcal{R}_X$-module which is strictly specializable along t. Assume the following:

(P0) We have either (i) $\mathcal{M} = \mathcal{M}[*t]$, or (ii) $\mathcal{M} = \mathcal{M}[!t]$.

(P1) We have $\mathcal{H}^j(\boldsymbol{D}_{X_0}\operatorname{Gr}_a^{V^{(\lambda_0)}}\mathcal{M}) = 0$ for any $j \neq 0$, any $a \in \mathbb{R}$ and any $\lambda_0 \in \mathbb{C}$. Moreover, $\mathcal{H}^0\boldsymbol{D}_{X_0}\operatorname{Gr}_a^{V^{(\lambda_0)}}\mathcal{M}$ are strict.

(P2) We have $\mathcal{H}^j\boldsymbol{D}\mathcal{M}(*t) = 0$ for any $j \neq 0$, and $\mathcal{H}^0(\boldsymbol{D}\mathcal{M}(*t))$ is strict.

We shall prove the following proposition in the rest of this subsection.

Proposition 13.2.1 *Under the assumptions, the following holds:*

- *We have* $\mathcal{H}^j(\boldsymbol{D}\mathcal{M}) = 0$ *for any* $j \neq 0$.
- $\mathcal{H}^0(\boldsymbol{D}\mathcal{M})$ *is strict, and strictly specializable along t.*
- $\mathcal{H}^0\boldsymbol{D}\mathcal{M} = (\mathcal{H}^0\boldsymbol{D}\mathcal{M})[!t]$ *in the case (i), and* $\mathcal{H}^0\boldsymbol{D}\mathcal{M} = (\mathcal{H}^0\boldsymbol{D}\mathcal{M})[*t]$ *in the case (ii).*
- *We have* $\mathrm{KMS}(\mathcal{H}^0\boldsymbol{D}\mathcal{M}, t) = \{-u \mid u \in \mathrm{KMS}(\mathcal{M}, t)\}$.

We follow the argument due to Saito in [70].

13.2.2 Preliminary

We prepare some general lemmas. Recall that $V_0\mathcal{R}_X \subset \mathcal{R}_X$ is the sheaf of subalgebras generated by $\mathcal{R}_{X_0}$ and $t\eth_t = \lambda t\partial_t$ over $\mathcal{O}_{\mathcal{X}}$.

Lemma 13.2.2 *Let $\mathcal{M}$ be a coherent $\mathcal{R}_X$-module. Assume the following.*

- $\mathcal{M}(*t)$ *is strict*
- $\mathcal{M}$ *is equipped with a $V_0\mathcal{R}_X$-coherent filtration $V^{(\lambda_0)}(\mathcal{M})$ such that (i) monodromic, (ii)* $\mathrm{Gr}^{V^{(\lambda_0)}}$ *is strict.*

Then, $\mathcal{M}$ is strict.

Proof Let us show that $V^{(\lambda_0)}_{<0}\mathcal{M} \longrightarrow \mathcal{M}(*t)$ is injective. Let $f \in V^{(\lambda_0)}_{<0}\mathcal{M}$ be mapped to 0 in $\mathcal{M}(*t)$. We may assume that $tf = 0$. Assume that there exists $a \in \mathbb{R}$ such that $f \in V^{(\lambda_0)}_a \setminus V^{(\lambda_0)}_{<a}$. Then, we obtain a non-zero element $[f]$ in $\mathrm{Gr}^{V^{(\lambda_0)}}_a$. By using the monodromic property, we can easily deduce that $[f] = 0$, which is a contradiction. Hence, we obtain $f \in V^{(\lambda_0)}_a$ for any $a \in \mathbb{R}$. We obtain $\mathcal{R}_X \cdot f \in V^{(\lambda_0)}_{<0}$. We have the decomposition $\mathcal{R}_X \cdot f = \bigoplus_{j\geq 0} \eth_t^j \mathcal{R}_{X_0} f$ as an $\mathcal{R}_{X_0}$-module. Then, it is easy to check that $\mathcal{R}_X \cdot f$ is not finitely generated as $V_0\mathcal{R}_X$-module. It contradicts with the coherence of $V^{(\lambda_0)}_{<0}\mathcal{M}$. Hence, $V^{(\lambda_0)}_{<0}\mathcal{M} \longrightarrow \mathcal{M}(*t)$ is injective. It implies that $V^{(\lambda_0)}_{<0}\mathcal{M}$ is strict. Then, we can easily deduce that $\mathcal{M}$ is also strict. □

We can show the following lemma by a similar argument.

Lemma 13.2.3 *Let $\mathcal{M}$ be a coherent $\mathcal{R}_X$-module. Assume that (i) it is also $V_0\mathcal{R}_X$-coherent, (ii)* $\mathrm{Supp}\,\mathcal{M} \subset X_0$. *Then, we have $\mathcal{M} = 0$.* □

13.2.3 $\mathcal{R}_{X_0[t]}$-Modules

Let X_0 be a complex manifold. Let t be a formal variable. On $\mathcal{X}_0 = \mathbb{C}_\lambda \times X_0$, we set $\omega_{\mathcal{X}_0[t]} := \omega_{\mathcal{X}_0}[t] \cdot dt/\lambda$ and $\mathcal{R}_{X_0[t]} := \mathcal{R}_{X_0}[t]\langle\eth_t\rangle$. For a coherent $\mathcal{R}_{X_0[t]}$-module $\mathcal{M}$, we set

$$\boldsymbol{D}'_{X_0[t]}\mathcal{M} := R\mathcal{H}om_{\mathcal{R}_{X_0[t]}}\big(\mathcal{M}, \mathcal{R}_{X_0[t]} \otimes \omega^{-1}_{\mathcal{X}_0[t]}\big)[d_{X_0} + 1].$$

Let us compute some specific examples. Let $\mathcal{M}_0$ be a coherent strict $\mathcal{R}_{X_0}$-module. We set $\mathcal{M}_0\langle t, \eth_t\rangle := \mathcal{R}_{X_0[t]} \otimes_{\mathcal{R}_{X_0}} \mathcal{M}_0$. For $P(t, \eth_t) \in \mathcal{O}_{\mathbb{C}_\lambda}\langle t, \eth_t\rangle$, let $R\big(P(t, \eth_t)\big)$ denote the right multiplication of $P(t, \eth_t)$ on $\mathcal{M}_0\langle t, \eth_t\rangle$. Let $\mathcal{N}$ be a nilpotent $\mathcal{R}_{X_0[t]}$-endomorphism on $\mathcal{M}_0$. Take $\lambda_0 \in \mathbb{C}$. Let $\mathcal{X}_0^{(\lambda_0)}$ denote neighbourhoods of $\{\lambda_0\}\times X_0$ in $\mathcal{X}_0$. Let $u \in \mathbb{R}\times\mathbb{C}$ with $-1 \leq \mathfrak{p}(\lambda_0, u) \leq 0$. On $\mathcal{X}_0^{(\lambda_0)}$, we define $\mathcal{B}(\mathcal{M}_0, u, \mathcal{N})$ as the cokernel of the injective $\mathcal{R}_{X_0[t]}$-homomorphism:

$$R\big(-\eth_t t + \mathfrak{e}(\lambda, u)\big) + \mathcal{N} : \mathcal{M}_0\langle t, \eth_t\rangle \longrightarrow \mathcal{M}_0\langle t, \eth_t\rangle.$$

The V-filtration $V^{(\lambda_0)}$ is given as follows. For $n \in \mathbb{Z}$, we put

$$V^{(\lambda_0)}_{\mathfrak{p}(\lambda_0,u)+n}\mathcal{B}(\mathcal{M}_0, u, \mathcal{N}) := \begin{cases} \operatorname{Cok}\big(t^{-n}\mathcal{M}_0[t] \longrightarrow t^{-n}\mathcal{M}_0[t]\big) & (n \leq 0) \\ \sum_{\substack{i+j\leq n \\ j<n}} \eth_t^i V^{(\lambda_0)}_{\mathfrak{p}(\lambda_0,u)+j} & (n > 0) \end{cases}$$

For any $b \in \mathbb{R}$, we have $b_0 := \max\{\mathfrak{p}(\lambda_0, u) + n \leq b \,\big|\, n \in \mathbb{Z}\}$, and we put

$$V_b^{(\lambda_0)}\mathcal{B}(\mathcal{M}_0, u, \mathcal{N}) := V_{b_0}^{(\lambda_0)}\mathcal{B}(\mathcal{M}_0, u, \mathcal{N}).$$

By construction, we have $t \cdot V_a^{(\lambda_0)} \subset V_{a-1}^{(\lambda_0)}$ and $\eth_t \cdot V_a^{(\lambda_0)} \subset V_{a+1}^{(\lambda_0)}$. If $a \leq \mathfrak{p}(\lambda_0, u)$, we have $t : V_a^{(\lambda_0)} \simeq V_{a-1}^{(\lambda_0)}$. If $a \geq \mathfrak{p}(\lambda_0, u)$, we have $\eth_t : \operatorname{Gr}_a^{V^{(\lambda_0)}} \simeq \operatorname{Gr}_{a+1}^{V^{(\lambda_0)}}$.

Let us compute $\boldsymbol{D}'\mathcal{B}(\mathcal{M}_0, u, \mathcal{N})$. We assume that $\boldsymbol{D}'_{X_0}\mathcal{M}_0 \simeq \mathcal{H}^0\boldsymbol{D}'_{X_0}\mathcal{M}_0$, and that it is strict, for simplicity. We consider it locally, and we take an $\mathcal{R}_{X_0}$-free resolution $\mathcal{P}^\bullet$ of $\mathcal{M}_0$. We have $\mathcal{N}^\bullet : \mathcal{P}^\bullet \longrightarrow \mathcal{P}^\bullet$ which induces $\mathcal{N}$. Then, $\mathcal{B}(\mathcal{M}_0, u, \mathcal{N})$ is naturally quasi-isomorphic to the complex associated to the following double complex:

$$R\big(-\eth_t t + \mathfrak{e}(\lambda, u)\big) + \mathcal{N}^\bullet : \mathcal{P}^\bullet\langle t, \eth_t\rangle \longrightarrow \mathcal{P}^\bullet\langle t, \eth_t\rangle$$

As a right $\mathcal{R}_{X_0[t]}$-complex, $\boldsymbol{D}'\mathcal{B}(\mathcal{M}_0, u, \mathcal{N})$ is quasi-isomorphic to the complex associated to the following double complex:

$$\mathcal{H}om_{\mathcal{R}_{X_0}}\big(\mathcal{P}^\bullet(\mathcal{M}_0), \mathcal{R}_{X_0}\big)\langle t, \eth_t\rangle \xrightarrow{-\eth_t t+\mathfrak{e}(\lambda,u)+\mathcal{N}^\vee} \mathcal{H}om_{\mathcal{R}_{X_0}}\big(\mathcal{P}^\bullet(\mathcal{M}_0), \mathcal{R}_{X_0}\big)\langle t, \eth_t\rangle$$

As a left $\mathcal{R}_{X_0}\langle t, \eth_T\rangle$-module, it is quasi isomorphic to

$$t\eth_t + \mathfrak{e}(\lambda, u) + \mathcal{N}^\vee : \boldsymbol{D}'_{X_0}\mathcal{M}_0\langle t, \eth_t\rangle \longrightarrow \boldsymbol{D}'_{X_0}\mathcal{M}_0\langle t, \eth_t\rangle.$$

Note $t\eth_t + \mathfrak{e}(\lambda, u) = \eth_t t + \mathfrak{e}(\lambda, u + \boldsymbol{\delta})$. Hence, we obtain

$$\boldsymbol{D}'\mathcal{B}\big(\mathcal{M}_0, u, \mathcal{N}\big) \simeq \mathcal{B}\big(\boldsymbol{D}'_{X_0}\mathcal{M}_0, -u - \boldsymbol{\delta}, -\mathcal{N}^\vee\big)$$

13.2.4 Filtered Free Module

Let $X = X_0 \times \mathbb{C}_t$. Let $\mathcal{P}$ be a free $\mathcal{R}_X$-module with a generator e. Let a be a real number. We have the filtration V given as follows. For $n \leq 0$, we put $V_{a+n}\mathcal{P} := t^{-n} V_0 \mathcal{R}_X \cdot e$. For $n > 0$, we put $V_{a+n}\mathcal{P} := \sum_{i \leq n} \eth_t^i V_a \mathcal{P}$. For $b \notin a + \mathbb{Z}$, we put $V_b\mathcal{P} := V_c\mathcal{P}$, where $c := \max\{a + n \mid a + n \leq b, n \in \mathbb{Z}\}$. Such a filtered $\mathcal{R}$-module is denoted by $\mathcal{R}_X(e, a)$. A filtered $\mathcal{R}_X$-module is called $(\mathcal{R}, V)$-free, if it is isomorphic to a direct sum $\bigoplus_{i=1}^m \mathcal{R}_X(e_i, a_i)$.

We have similar notions for $\mathcal{R}_{X_0[t]}$-modules. Let $\mathcal{Q}$ be a free $\mathcal{R}_{X_0[t]}$-module with a generator e. Let a be a real number. We have the filtration V given as follows. For $n \leq 0$, we put $V_{a+n}\mathcal{Q} := t^{-n} V_0 \mathcal{R}_{X_0[t]} \cdot e$. For $n > 0$, we put $V_{a+n}\mathcal{Q} := \sum_{i \leq n} \eth_t^i V_a \mathcal{Q}$. For $b \notin a + \mathbb{Z}$, we put $V_b\mathcal{Q} := V_c\mathcal{Q}$, where $c := \max\{a + n \mid a + n \leq b, n \in \mathbb{Z}\}$. Such a filtered $\mathcal{R}_{X_0[t]}$-module is denoted by $\mathcal{R}_{X_0[t]}(e, a)$. A filtered $\mathcal{R}_{X_0[t]}$-module is called $(\mathcal{R}_{X_0[t]}, V)$-free, if it is isomorphic to a direct sum $\bigoplus_{i=1}^m \mathcal{R}_{X_0[t]}(e_i, a_i)$.

By the identification $\mathcal{R}_X = \mathcal{R}_X(1, 0)$, $\mathcal{R}_X$ is equipped with a filtration V, and $(\mathcal{R}_X, V)$ is a filtered ring. Note that $\mathrm{Gr}^V \mathcal{R}_X$ is naturally isomorphic to $\mathcal{R}_{X_0[t]}$, and we have a natural isomorphism $\mathrm{Gr}^V \mathcal{R}_X(e, a) \simeq \mathcal{R}_{X_0[t]}(e, a)$. If a filtered $\mathcal{R}_X$-module $(\mathcal{L}, V)$ is $(\mathcal{R}_X, V)$-free, then $\mathrm{Gr}^V(\mathcal{L}, V)$ is $(\mathcal{R}_{X_0[t]}, V)$-free.

13.2.4.1 Strictness

Let $(\mathcal{P}_1, V) \xrightarrow{\varphi_1} (\mathcal{P}_2, V) \xrightarrow{\varphi_2} (\mathcal{P}_3, V)$ be a complex of coherent $(\mathcal{R}_X, V)$-free modules.

Lemma 13.2.4 *Assume that the induced complex*

$$\mathrm{Gr}^V \mathcal{P}_1 \longrightarrow \mathrm{Gr}^V \mathcal{P}_2 \longrightarrow \mathrm{Gr}^V \mathcal{P}_3$$

is exact. Then, φ_i are strict with respect to V.

Proof The strictness of φ_2 is easy. Let us argue the strictness of φ_1. We fix a sufficiently small $b < 0$. Let $a \in \mathbb{R}$. Let us observe that there exists $N(a) > 0$ such that $\mathrm{Im}\, \varphi_1 \cap V_{b-N(a)}\mathcal{P}_2 \subset \varphi_1\big(V_{a-1}\mathcal{P}_1\big)$. We obtain a $V_0\mathcal{R}_X$-coherent submodule $\varphi_1^{-1}(V_b\mathcal{P}_2)$. If we take sufficiently large $N(a)$, we have $\varphi_1^{-1}(V_b\mathcal{P}_2) \subset V_{a+N(a)-1}\mathcal{P}_1$. Then, if b is sufficiently small, we have $t^{N(a)} V_b \mathcal{P}_2 = V_{b-N(a)}\mathcal{P}_2$. Hence, we have $\varphi_1^{-1}(V_{b-N(a)}\mathcal{P}_2) \subset V_{a-1}\mathcal{P}_1$.

Take $f \in V_a\mathcal{P}_1$ such that $\varphi_1(f) \in V_b\mathcal{P}_2$ for some $b < a$. By using the exactness of $\mathrm{Gr}(\mathcal{P}_\bullet)$, we can find $g \in V_{<a}\mathcal{P}_1$ such that $\varphi_1(f - g) \in V_{b-N(a)-1}\mathcal{P}_2$. Then, we can find $h \in V_{a-1}\mathcal{P}_1$ such that $\varphi_1(f - g - h) = 0$, i.e., $\varphi_1(f) = \varphi_1(g + h)$. By an easy inductive argument, we obtain that $\varphi_1(f) \in \varphi_1\big(V_b(\mathcal{P}_1)\big)$. □

13.2.4.2 Duality

We consider the dual $\mathcal{R}$-module of $\mathcal{R}_X(e,a)$, i.e.,

$$\boldsymbol{D}\mathcal{R}_X(e,a) := \mathcal{H}om_{\mathcal{R}_X}(\mathcal{R}_X(e,a), \mathcal{R}_X \otimes \omega_{\mathcal{X}}).$$

Let $e^\vee \in \boldsymbol{D}\mathcal{R}_X(e,a)$ be given by $e^\vee(e) = 1$. Then, $\boldsymbol{D}\mathcal{R}_X(e,a) \simeq \mathcal{R}_X \cdot e^\vee$. A filtration of $\boldsymbol{D}\mathcal{R}_X(e,a)$ is induced by the filtration of $\mathcal{R}_X(e^\vee, -a-1)$. Similarly, we have the filtration of $\boldsymbol{D}\mathcal{R}_{X_0[t]}(e,a)$ induced by $\boldsymbol{D}\mathcal{R}_{X_0[t]}(e,a) \simeq \mathcal{R}_{X_0[t]}(e^\vee, -a-1)$. Hence, we have naturally induced filtrations on the dual of free $(\mathcal{R}_X, V)$-modules or $(\mathcal{R}_{X_0[t]}, V)$.

13.2.5 A Filtered Free Resolution

Let $X = X_0 \times \mathbb{C}_t$. Let $\mathcal{M}$ be a coherent strictly specializable $\mathcal{R}_X$-module along t. For simplicity, we assume either (i) $\mathcal{M} = \mathcal{M}[*t]$, or (ii) $\mathcal{M} = \mathcal{M}[!t]$. Locally, we shall construct a complex of coherent $(\mathcal{R}_X, V)$-free modules

$$\cdots \longrightarrow (\mathcal{P}_{\ell+1}(\mathcal{M}), V) \longrightarrow (\mathcal{P}_\ell(\mathcal{M}), V) \longrightarrow \cdots\cdots \longrightarrow (\mathcal{P}_1(\mathcal{M}), V) \longrightarrow (\mathcal{P}_0(\mathcal{M}), V)$$

with the following property:

- $\mathcal{P}_\bullet(\mathcal{M})$ is a free resolution of $\mathcal{M}$.
- The morphisms $\mathcal{P}_{\ell+1}(\mathcal{M}) \longrightarrow \mathcal{P}_\ell(\mathcal{M})$ and $\mathcal{P}_0(\mathcal{M}) \longrightarrow \mathcal{M}$ are strict with respect to the filtrations.
- In particular, $\mathrm{Gr}^V(\mathcal{P}_\bullet(\mathcal{M}))$ gives a free $\mathcal{R}_{X_0[t]}$-resolution of $\mathrm{Gr}^V(\mathcal{M})$.

First, we construct a $(\mathcal{R}_X, V)$-free module $\mathcal{P}_0(\mathcal{M})$ with a surjection $\mathcal{P}_0(\mathcal{M}) \longrightarrow \mathcal{M}$.

Let us consider the case $\mathcal{M} = \mathcal{M}[*t]$. We put $a_0 := \min\bigl\{-1 < a \leq 0 \bigm| \mathrm{Gr}_a^{V^{(\lambda_0)}} \mathcal{M} \neq 0\bigr\}$. We take a generator $\tilde{\boldsymbol{e}}_{a_0} = (\tilde{e}_{a_0,i})$ of $V_{a_0}^{(\lambda_0)}\mathcal{M}$. We take generators $\boldsymbol{e}_a$ of $\mathrm{Gr}_a^{V^{(\lambda_0)}}(\mathcal{M})$ for $a_0 < a \leq 0$, and we take lifts $\tilde{\boldsymbol{e}}_a = (\tilde{e}_{a,i})$ to $V_a^{(\lambda_0)}(\mathcal{M})$. Let $\mathcal{P}_0\mathcal{M}$ be the $(\mathcal{R}_X, V)$-free module generated by $\tilde{\boldsymbol{e}}_a$ $(a_0 \leq a \leq 0)$, where a is associated to each $\tilde{e}_{a,i}$. We have a naturally defined filtered morphism $(\mathcal{P}_0, V) \longrightarrow (\mathcal{M}, V^{(\lambda_0)})$.

Let us consider the case $\mathcal{M} = \mathcal{M}[!t]$. We take a generator $\tilde{\boldsymbol{e}}_{-1} = (\tilde{e}_{-1,i})$ of $V_{-1}^{(\lambda_0)}\mathcal{M}$. We take generators $\boldsymbol{e}_a$ of $\mathrm{Gr}_a^{V^{(\lambda_0)}}(\mathcal{M})$ for $-1 < a < 0$, and we take lifts $\tilde{\boldsymbol{e}}_a = (\tilde{e}_{a,i})$ to $V_a^{(\lambda_0)}(\mathcal{M})$. Let $\mathcal{P}_0\mathcal{M}$ be the $(\mathcal{R}_X, V)$-free module generated by $\tilde{\boldsymbol{e}}_a$ $(-1 \leq a < 0)$, where a is associated to each $\tilde{e}_{a,i}$. We have the naturally defined filtered morphism $(\mathcal{P}_0, V) \longrightarrow (\mathcal{M}, V^{(\lambda_0)})$.

By construction, for each $a \in \mathbb{R}$, the morphism $V_a\mathcal{P}_0 \longrightarrow V_a^{(\lambda_0)}\mathcal{M}$ is surjective. By construction, the induced morphism $\mathrm{Gr}^V \mathcal{P}_0 \longrightarrow \mathrm{Gr}^{V^{(\lambda_0)}} \mathcal{M}$ is surjective. Let $\mathcal{K}_0(\mathcal{M})$ denote the kernel of $\mathcal{P}_0(\mathcal{M}) \longrightarrow \mathcal{M}$. It is equipped with a naturally induced filtration V. By construction, we have $t \cdot V_a\mathcal{K}_0(\mathcal{M}) \subset V_{a-1}\mathcal{K}_0(\mathcal{M})$ and $\eth_t \cdot V_a\mathcal{K}_0(\mathcal{M}) \subset V_{a+1}\mathcal{K}_0(\mathcal{M})$.

- We have $t : V_b(\mathcal{K}_0(\mathcal{M})) \simeq V_{b-1}(\mathcal{K}_0(\mathcal{M}))$ for $b \leq 0$ in the case (i), or for $b < 0$ in the case (ii).
- We have the exact sequence $0 \longrightarrow \mathrm{Gr}^V \mathcal{K}_0 \longrightarrow \mathrm{Gr}^V \mathcal{P}_0 \longrightarrow \mathrm{Gr}^{V^{(\lambda_0)}} \mathcal{M} \longrightarrow 0$. In particular, we have $\eth_t : \mathrm{Gr}_b^V \mathcal{K}_0 \simeq \mathrm{Gr}_{b+1}^V \mathcal{K}_0$ for $b > -1$ in the case (i), or for $b \geq -1$ in the case (ii).

Inductively, for any $\ell \geq 0$, we construct filtered $\mathcal{R}_X$-modules $(\mathcal{P}_\ell(\mathcal{M}), V)$ and $(\mathcal{K}_\ell(\mathcal{M}), V)$ with exact sequences $0 \longrightarrow \mathcal{K}_{\ell+1} \longrightarrow \mathcal{P}_{\ell+1} \longrightarrow \mathcal{K}_\ell \longrightarrow 0$, such that

- $(\mathcal{P}_\ell(\mathcal{M}), V)$ are free $(\mathcal{R}_X, V)$-modules.
- For both $\mathcal{P}_\ell$ and $\mathcal{K}_\ell$, the morphisms $t : V_b \simeq V_{b-1}$ $(b \leq 0)$ and $\eth_t : \mathrm{Gr}_b^V \simeq \mathrm{Gr}_{b+1}^V$ $(b > -1)$ are isomorphisms in the case (i). The morphisms $t : V_b \simeq V_{b-1}$ $(b < 0)$ and $\eth_t : \mathrm{Gr}_b^V \simeq \mathrm{Gr}_{b+1}^V$ $(b \geq -1)$ are isomorphisms in the case (ii).
- The morphisms $\mathcal{K}_{\ell+1} \longrightarrow \mathcal{P}_{\ell+1}$ and $\mathcal{P}_{\ell+1} \longrightarrow \mathcal{K}_\ell$ are strict with respect to the filtrations V. In particular, the induced sequence $0 \longrightarrow \mathrm{Gr}^V \mathcal{K}_{\ell+1} \longrightarrow \mathrm{Gr}^V \mathcal{P}_{\ell+1} \longrightarrow \mathrm{Gr}^V \mathcal{K}_\ell \longrightarrow 0$ is exact.

We obtain a $(\mathcal{R}_X, V)$-free resolution $\mathcal{P}_\bullet(\mathcal{M})$ of $\mathcal{M}$. We have a natural isomorphism

$$\mathrm{Gr}^V \mathcal{H}om_{\mathcal{R}_X}\big(\mathcal{P}_\bullet, \mathcal{R}_X\big) \simeq \mathcal{H}om_{\mathrm{Gr}^V \mathcal{R}_X}\big(\mathrm{Gr}^V \mathcal{P}_\bullet, \mathrm{Gr}^V \mathcal{R}_X\big).$$

It naturally gives a $(\mathcal{R}_{X_0[t]}, V)$-free resolution of $\mathrm{Gr}^V \mathcal{M}$.

13.2.6 Proof of Proposition 13.2.1

Let $\mathcal{M}$ be as in Sect. 13.2.1. Let $\mathcal{P}_\bullet(\mathcal{M})$ be a $(\mathcal{R}_X, V)$-free resolution of $\mathcal{M}$ as in Sect. 13.2.5. Let us study the dual complex of $\mathcal{M}$ by using $\mathcal{P}_\bullet(\mathcal{M})$. The complexes

$$\mathcal{C} := \mathcal{H}om_{\mathcal{R}_X}(\mathcal{P}_\bullet(\mathcal{M}), \mathcal{R}_X \otimes \omega_{\mathcal{X}}^{-1})[d_X]$$

and

$$\mathrm{Gr}^V(\mathcal{C}) = \mathcal{H}om_{\mathrm{Gr}^V \mathcal{R}_X}(\mathrm{Gr}^V \mathcal{P}_\bullet(\mathcal{M}), \mathrm{Gr}^V \mathcal{R}_X \otimes \omega_{\mathcal{X}_0[t]}^{-1})[d_{X_0} + 1]$$

express $\boldsymbol{D}'\mathcal{M}$ and $\boldsymbol{D}'_{X_0[t]} \mathrm{Gr}^V \mathcal{M}$, respectively.

Lemma 13.2.5 *Assume (P0) and (P1). Then, the following holds.*

- *We have $\mathcal{H}^j\boldsymbol{D}'_{X_0[t]}\operatorname{Gr}^V(\mathcal{M}) = 0$ $(j \neq 0)$, and $\mathcal{H}^0\boldsymbol{D}'_{X_0[t]}\operatorname{Gr}^V(\mathcal{M})$ is strict.*
- *The induced filtration $V^{(\lambda_0)}$ on $\mathcal{H}^0\boldsymbol{D}'_{X_0[t]}\operatorname{Gr}^V(\mathcal{M})$ is monodromic. Moreover, we have $t\cdot V^{(\lambda_0)}_a = V^{(\lambda_0)}_{a-1}$ for $a<0$ and $\eth_t : \operatorname{Gr}^{V^{(\lambda_0)}}_a \simeq \operatorname{Gr}^{V^{(\lambda_0)}}_{a+1}$ for $a>-1$.*
- *In the case (i) the morphism $\eth_t : \operatorname{Gr}^{V^{(\lambda_0)}}_{-1}\boldsymbol{D}'_{X_0[t]}\operatorname{Gr}^V(\mathcal{M}) \to \operatorname{Gr}^{V^{(\lambda_0)}}_0\boldsymbol{D}'_{X_0[t]}\operatorname{Gr}^V(\mathcal{M})$ is an isomorphism. In the case (ii) the morphism $t : \operatorname{Gr}^{V^{(\lambda_0)}}_0\boldsymbol{D}'_{X_0[t]}\operatorname{Gr}^V(\mathcal{M}) \to \operatorname{Gr}^{V^{(\lambda_0)}}_{-1}\boldsymbol{D}'_{X_0[t]}\operatorname{Gr}^V(\mathcal{M})$ is an isomorphism.*

Proof It follows from the computation in Sect. 13.2.3. □

Let us finish the proof of Proposition 13.2.1. By Lemmas 13.2.4 and 13.2.5, the morphisms in the complex $\mathcal{C}$ is strict with respect to the filtration V. Hence, we have the commutativity $\operatorname{Gr}^V\mathcal{H}^i\mathcal{C} \simeq \mathcal{H}^i\operatorname{Gr}^V\mathcal{C}$. According to Lemma 13.2.5, we have $\operatorname{Gr}^V\mathcal{H}^i\mathcal{C} = 0$ unless $i=0$. We also have $\big(\mathcal{H}^i\mathcal{C}\big)(*t) = 0$ unless $i=0$. Hence, we obtain $\mathcal{H}^i\boldsymbol{D}\mathcal{M} = \mathcal{H}^i\mathcal{C} = 0$ unless $i=0$, by Lemma 13.2.3.

By Lemma 13.2.5, $\operatorname{Gr}^V\mathcal{H}^0\boldsymbol{D}\mathcal{M}$ is strict and monodromic. Hence, by using Lemma 13.2.2, we obtain that $\mathcal{H}^0\boldsymbol{D}\mathcal{M}$ is strict and strictly specializable along t. Moreover the induced filtration $V^{(\lambda_0)}$ gives a V-filtration. By the second claim of Lemma 13.2.5, we obtain $(\boldsymbol{D}\mathcal{M})[!t] = \boldsymbol{D}\mathcal{M}$ in the case (i), or $(\boldsymbol{D}\mathcal{M})[*t] = \boldsymbol{D}\mathcal{M}$ in the case (ii). We obtain the last claim in Proposition 13.2.1 from the computation in Sect. 13.2.3. □

13.3 Duality of Mixed Twistor $\mathscr{D}$-Modules

13.3.1 Statements

Let us consider the duality for mixed twistor $\mathscr{D}$-modules. We will prove the following theorem in Sects. 13.3.2–13.3.8.

Theorem 13.3.1 *Let $(\mathcal{T}, W)$ be a mixed twistor $\mathscr{D}$-module on X with the underlying $\mathcal{R}_X$-triple $(\mathcal{M}_1, \mathcal{M}_2, C)$.*

- *We have $\mathcal{H}^j(\boldsymbol{D}\mathcal{M}_i) = 0$ $(j\neq 0)$, and $\mathcal{H}^0(\boldsymbol{D}\mathcal{M}_i)$ are strict.*
- *We have a unique sesqui-linear pairing $\boldsymbol{D}C$ of $\mathcal{H}^0\boldsymbol{D}\mathcal{M}_1$ and $\mathcal{H}^0\boldsymbol{D}\mathcal{M}_2$ with the following property: Let X_1 be any open subset in X such that $\mathcal{T}_{|X_1} \in \operatorname{MTM}(X_1)_{\mathcal{A}}$ for a finite dimensional $\mathbb{Q}$-vector subspace $\mathcal{A} \subset \mathbb{R}\times\mathbb{C}$ such that $\mathbb{Q}\times\{0\} \subset \mathcal{A}$. (See Proposition 7.1.67.) Then, for any $\lambda \in \boldsymbol{S}$ which is generic to $\mathcal{A}$, we have*

$$(\boldsymbol{D}C)_{|\mathcal{X}_1^\lambda} = \boldsymbol{D}\big(C_{|\mathcal{X}_1^\lambda}\big).$$

Here, the latter is defined for non-degenerate sesqui-linear pairings of holonomic $\mathscr{D}$-modules in Theorem 12.4.1. (Note that $C_{|\mathcal{X}_1^\lambda}$ is non-degenerate. In the pure case, it follows from the description of pure twistor $\mathscr{D}$-modules as the minimal extension of a polarizable wild variation of pure twistor structure. The mixed case is easily reduced to the pure case.)

- *The $\mathcal{R}$-triple $\boldsymbol{D}\mathcal{T} := \big(\boldsymbol{D}\mathcal{M}_1, \boldsymbol{D}\mathcal{M}_2, \boldsymbol{D}C\big)$ with the naturally induced filtration W is a mixed twistor $\mathscr{D}$-module. Here, W on $\boldsymbol{D}\mathcal{T}$ is given as follows:*

$$W_j(\boldsymbol{D}\mathcal{T}) := \mathrm{Ker}\big(\boldsymbol{D}\mathcal{T} \longrightarrow \boldsymbol{D}(W_{-j-1}\mathcal{T})\big)$$

- *The duality $\boldsymbol{D}$ gives a contravariant functor on* $\mathrm{MTM}(X)$. *We have a natural transform $\boldsymbol{D} \circ \boldsymbol{D} \simeq \mathrm{id}$.*
- *If $\mathcal{T}$ is integrable, then $\boldsymbol{D}\mathcal{T}$ is also naturally integrable.*

Before going to the proof of Theorem 13.3.1, we give some consequences. We shall not distinguish $\mathcal{H}^0\boldsymbol{D}\mathcal{M}$ and $\boldsymbol{D}\mathcal{M}$ for the underlying $\mathcal{R}_X$-modules $\mathcal{M}$ of mixed twistor $\mathscr{D}$-modules.

For any $\mathcal{R}_X$-triple $\mathcal{T} = (\mathcal{M}_1, \mathcal{M}_2, C)$ and $\ell \in \mathbb{Z}$, we set

$$\mu_\ell(\mathcal{T}) := (\mathcal{M}_1, \mathcal{M}_2, (-1)^\ell C).$$

Once we know Theorem 13.3.1, we obtain the following compatibility of the duality and the push-forward, from Theorem 12.4.6.

Theorem 13.3.2 *Let $\mathcal{T} \in$* $\mathrm{MTM}(X)$, *and let $F : X \longrightarrow Y$ be a projective morphism. Then, the natural isomorphisms of the underlying $\mathcal{R}_Y$-modules (Lemma 13.1.2) give a natural isomorphism $\boldsymbol{D}F_\dagger^j\mathcal{T} \simeq \mu_j F_\dagger^{-j}(\boldsymbol{D}\mathcal{T})$ in* $\mathrm{MTM}(Y)$. □

In other words, we obtain the following.

Corollary 13.3.3 $\big(\epsilon(j)\varphi_1^{-1}, \epsilon(-j)\varphi_2\big)$ *gives an isomorphism*

$$F_\dagger^j(\boldsymbol{D}\mathcal{T}) \simeq \boldsymbol{D}F_\dagger^{-j}\mathcal{T},$$

where $\varphi_1 : F_\dagger^{-j}\boldsymbol{D}_X\mathcal{M}_1 \simeq \boldsymbol{D}_Y F_\dagger^j\mathcal{M}_1$ and $\varphi_2 : F_\dagger^j\boldsymbol{D}_X\mathcal{M}_2 \simeq \boldsymbol{D}_Y F_\dagger^{-j}\mathcal{M}_2$ are the isomorphisms in Lemma 13.1.2, and $\epsilon(m) := (-1)^{m(m-1)/2}$ for integers m. □

The signature comes from the commutation of the shift of the duality functor, i.e., $\boldsymbol{D}(F_\dagger\mathcal{M})[k] \simeq \boldsymbol{D}(F_\dagger\mathcal{M}[-k])$.

Proposition 13.3.4 *We have natural isomorphisms*

$$\boldsymbol{D}(\mathcal{T}^*) \simeq (\boldsymbol{D}\mathcal{T})^*, \quad \boldsymbol{D}j^*\mathcal{T} \simeq j^*\boldsymbol{D}\mathcal{T}$$

in $\mathrm{MTM}(X)$.

Proof We have natural isomorphisms for the underlying filtered $\mathcal{R}$-modules. We have only to show the compatibility of the pairings $\boldsymbol{D}(C^*) = (\boldsymbol{D}C)^*$ and $j^*\boldsymbol{D}C = \boldsymbol{D}j^*C$. Both of them can be reduced to the claims for the $\mathscr{D}$-modules, and easy to check. (See Sect. 12.4.3.) □

Proposition 13.3.5 *Let H be a hypersurface of X. We have natural isomorphisms $\boldsymbol{D}(\mathcal{T}[*H]) \simeq (\boldsymbol{D}\mathcal{T})[!H]$ and $\boldsymbol{D}(\mathcal{T}[!H]) \simeq (\boldsymbol{D}\mathcal{T})[*H]$ in* $\mathrm{MTM}(X)$.

Proof Let us observe that $\boldsymbol{D}(\mathcal{T}[*H]) \simeq \boldsymbol{D}(\mathcal{T})[!H]$. We have the isomorphisms of the underlying $\mathcal{R}$-modules (Proposition 13.2.1). We have the compatibility of the sesqui-linear pairings (see Proposition 12.4.10). Hence, we obtain the isomorphism $\boldsymbol{D}(\mathcal{T}[*H]) \simeq \boldsymbol{D}(\mathcal{T})[!H]$ as $\mathcal{R}$-triples. Then, the claim follows from Lemma 7.1.52. □

Proposition 13.3.6 *Let $\mathcal{V} \in \mathrm{MTS}^{\mathrm{adm}}(X,H)$. Let $\mathcal{T} \in \mathrm{MTM}(X)$. Then, we have natural isomorphisms $\boldsymbol{D}\big((\mathcal{T}\otimes\mathcal{V})[!H]\big) \simeq \big(\boldsymbol{D}\mathcal{T}\otimes\mathcal{V}^{\vee}\big)[*H]$ and $\boldsymbol{D}\big((\mathcal{T}\otimes\mathcal{V})[*H]\big) \simeq \big(\boldsymbol{D}\mathcal{T}\otimes\mathcal{V}^{\vee}\big)[!H]$.*

Proof We have only to consider the case $H=\emptyset$. (See Proposition 13.3.5.) We have natural isomorphisms as in Sect. 13.1.4. We have the compatibility of the sesqui-linear pairings in Lemma 12.4.12. □

Corollary 13.3.7 *We naturally have $\boldsymbol{D}(\Pi^{a,b}_{*!}\mathcal{T}) \simeq \Pi^{-b+1,-a+1}_{*!}(\boldsymbol{D}\mathcal{T})$. In particular, we have*

$$\boldsymbol{D}\psi_g^{(a)}(\mathcal{T}) \simeq \psi_g^{(-a+1)}\boldsymbol{D}\mathcal{T}, \qquad \boldsymbol{D}\Xi_g^{(a)}(\mathcal{T}) \simeq \Xi_g^{(-a)}\boldsymbol{D}\mathcal{T}.$$

We also obtain $\boldsymbol{D}\phi_g^{(a)}(\mathcal{T}) \simeq \phi_g^{(-a)}\boldsymbol{D}\mathcal{T}$. □

Proposition 13.3.8 *Let $\mathcal{A} \subset \mathbb{R}\times\mathbb{C}$ be a $\mathbb{Q}$-vector subspace such that $\mathbb{Q}\times\{0\} \subset \mathcal{A}$. If $\mathcal{T} \in \mathrm{MTM}(X)_{\mathcal{A}}$, then $\boldsymbol{D}\mathcal{T} \in \mathrm{MTM}(X)_{\mathcal{A}}$.*

Proof It follows from Proposition 13.2.1 and the above propositions. □

Let X_i ($i=1,2$) be complex manifolds. For any strict coherent $\mathcal{R}_{X_i}$-modules $\mathcal{M}_i$, we have a naturally induced isomorphism

$$\boldsymbol{D}_{X_1}(\mathcal{M}_1) \boxtimes \boldsymbol{D}_{X_2}(\mathcal{M}_2) \longrightarrow \boldsymbol{D}_{X_1\times X_2}(\mathcal{M}_1\boxtimes\mathcal{M}_2). \tag{13.19}$$

We obtain the following from Proposition 12.4.15.

Proposition 13.3.9 *For $\mathcal{T}_i \in \mathrm{MTM}(X_i)$ ($i=1,2$) we have an isomorphism $(\boldsymbol{D}_{X_1}\mathcal{T}_1)\boxtimes(\boldsymbol{D}_{X_2}\mathcal{T}_2) \simeq \boldsymbol{D}_{X_1\times X_2}(\mathcal{T}_1\boxtimes\mathcal{T}_2)$. The isomorphisms of the underlying $\mathcal{R}_X$-modules are given by (13.19).* □

13.3.2 Relative Monodromy Filtrations

For the proof of the theorems, we shall use an induction on the support of the dimensions. We consider the following

A(n) If $\dim \operatorname{Supp} \mathcal{T} \leq n$, the claim of Theorem 13.3.1 holds.

Assume $A(n)$. Let $\mathcal{T} \in \mathrm{MTM}(X)$ with $\dim \operatorname{Supp} \mathcal{T} \leq n$. Let W be the weight filtration of $\mathcal{T}$, and let L be a filtration of $\mathcal{T}$ in $\mathrm{MTM}(X)$. Let $\mathcal{N} : (\mathcal{T}, W, L) \longrightarrow \big((\mathcal{T}, W) \otimes \boldsymbol{T}(-1), L\big)$ be a morphism such that $W = M(\mathcal{N}; L)$. We have $\boldsymbol{D}(\mathcal{T}, W) = (\boldsymbol{D}\mathcal{T}, \boldsymbol{D}W)$ in $\mathrm{MTM}(X)$. It is equipped with the induced filtration $\boldsymbol{D}L$ and the induced morphism $\boldsymbol{D}\mathcal{N} : \boldsymbol{D}(\mathcal{T}, W, L) \longrightarrow \big(\boldsymbol{D}(\mathcal{T}, W) \otimes \boldsymbol{T}(-1), \boldsymbol{D}L\big)$. We can show the following lemma by a standard argument.

Lemma 13.3.10 *We have* $\boldsymbol{D}W = M\big(\boldsymbol{D}\mathcal{N}; \boldsymbol{D}L\big)$.

Proof Let us consider the case $\mathrm{Gr}^L_j(\mathcal{T}) = 0$ unless $j = w$. We may assume that $w = 0$. We naturally have $\mathrm{Gr}^{\boldsymbol{D}W}_j \boldsymbol{D}\mathcal{T} \simeq \boldsymbol{D}\,\mathrm{Gr}^W_{-j}\mathcal{T}$, and the morphism $\boldsymbol{D}\mathcal{N}^j : \mathrm{Gr}^{\boldsymbol{D}W}_j \boldsymbol{D}\mathcal{T} \longrightarrow \mathrm{Gr}^{\boldsymbol{D}W}_{-j} \boldsymbol{D}\mathcal{T}$ is the dual of $\mathcal{N}^j : \mathrm{Gr}^W_j \mathcal{T} \longrightarrow \mathrm{Gr}^W_{-j}\mathcal{T}$. Hence, it is an isomorphism, and $\boldsymbol{D}W$ on $\boldsymbol{D}\mathcal{T}$ is the monodromy filtration of $\boldsymbol{D}\mathcal{N}$.

The induced morphism $L_j\boldsymbol{D}\mathcal{T} \longrightarrow \mathrm{Gr}^L_j \boldsymbol{D}\mathcal{T}$ is strict with respect to $\boldsymbol{D}W$. Hence, we can deduce that $\boldsymbol{D}W$ is the relative monodromy filtration with respect to L. □

13.3.3 Duality of Smooth $\mathcal{R}$-Triples

We give a remark on the signature. Let $\mathcal{V} = (\mathcal{V}_1, \mathcal{V}_2, C) \in \mathrm{TS}(X)$. Let $\mathcal{V}^\vee = (\mathcal{V}_1^\vee, \mathcal{V}_2^\vee, C^\vee)$ be the dual in $\mathrm{TS}(X)$. We have the isomorphisms $\Phi_i : \mathcal{V}_i^\vee \simeq \boldsymbol{D}\mathcal{V}_i$ ($i = 1, 2$) given by $\Phi_i(v_i) = \lambda^{d_X}\, \Psi_{\mathcal{V}_i} v_i$. (See Sect. 13.1.5.) We have the induced pairing C_1 of $\boldsymbol{D}\mathcal{V}_1$ and $\boldsymbol{D}\mathcal{V}_2$.

Lemma 13.3.11 *We have* $C_1^\lambda = \boldsymbol{D}(C^\lambda)$ *for each* $\lambda \in \boldsymbol{S}$, *where* $\boldsymbol{D}(C^\lambda)$ *denotes the duality functor for non-degenerate* $\mathscr{D}$*-triples. (See Chap. 12.)*

Proof Let $\lambda_0 \in \boldsymbol{S}$. We have $C_1\big(\lambda^{d_X} f, \overline{\lambda^{d_X} g}\big) = C^\vee(f, \overline{g})$ by construction. On the other side, we have

$$\begin{aligned} \boldsymbol{D}(C^{\lambda_0})\big((\lambda^{d_X} f)_{\lambda_0}, (\overline{\lambda^{d_X} g})_{-\lambda_0}\big) &= (-1)^{d_X} \lambda_0^{d_X} (-\lambda_0)^{-d_X} (C^{\lambda_0})^\vee \big(f_{\lambda_0}, \overline{g}_{-\lambda_0}\big) \\ &= C^\vee(f, \overline{g})_{\lambda_0} \end{aligned} \tag{13.20}$$

(See Example 12.4.2.) Then, the claim of the lemma follows. □

So we have an isomorphism of $\mathcal{R}$-triples $\boldsymbol{D}\mathcal{V} \simeq \mathcal{V}^\vee$. If $\mathcal{V}$ is integrable, we have an isomorphism of integrable $\mathcal{R}_X$-triples $\boldsymbol{D}\mathcal{V} \simeq \big(\lambda^{d_X}\mathcal{V}_1^\vee, \lambda^{d_X}\mathcal{V}_2^\vee, (-1)^{d_X} C^\vee\big)$ as observed in the proof of the lemma.

13.3.4 Duality of Canonical Prolongation as $\mathcal{R}$-Triples

Let X be any n-dimensional complex manifold with a normal crossing hypersurface D. Let $\mathcal{V} \in \mathrm{MTS}^{\mathrm{adm}}(X,D)$. Let $\mathcal{V}^{\vee} \in \mathrm{MTS}^{\mathrm{adm}}(X,D)$ be its dual. We have the underlying smooth $\mathcal{R}_{X(*D)}$-triples $\mathcal{V} = (\mathcal{V}_1, \mathcal{V}_2, C)$ and $\mathcal{V}^{\vee} = (\mathcal{V}_1^{\vee}, \mathcal{V}_2^{\vee}, C^{\vee})$. We put $\mathcal{V}_{i\star} := \mathcal{V}_i[\star D]$ for $\star = *, !$. We use the symbol $\mathcal{V}_{i\star}^{\vee}$ in similar meanings. We have the $\mathcal{R}_X$-triples $\mathcal{V}_* = (\mathcal{V}_{1!}, \mathcal{V}_{2*}, C_*)$ and $\mathcal{V}_! = (\mathcal{V}_{1*}, \mathcal{V}_{2!}, C_!)$. We also have the $\mathcal{R}_X$-triples $\mathcal{V}_*^{\vee} = (\mathcal{V}_{1!}^{\vee}, \mathcal{V}_{2*}^{\vee}, C_*^{\vee})$ and $\mathcal{V}_!^{\vee} = (\mathcal{V}_{1*}^{\vee}, \mathcal{V}_{2!}^{\vee}, C_!^{\vee})$.

Lemma 13.3.12 *Assume $A(n-1)$. Then, we have natural isomorphisms of $\mathcal{R}_X$-modules $\boldsymbol{D}\mathcal{V}_{i!} \simeq \lambda^{d_X}\mathcal{V}_{i*}^{\vee}$ and $\boldsymbol{D}\mathcal{V}_{i*} \simeq \lambda^{d_X}\mathcal{V}_{i!}^{\vee}$.*

Proof We may assume that there exists a holomorphic function g on X such that $g^{-1}(0) = D$. Let $\iota_g : X \longrightarrow X \times \mathbb{C}_t$ be the graph. By the assumption $A(n-1)$, the condition $(P1)$ in Sect. 13.2.1 is satisfied. The other conditions $(P0)$ and $(P2)$ are also satisfied. Hence, we obtain that $\boldsymbol{D}\mathcal{V}_{i*} = \mathcal{H}^0\boldsymbol{D}\mathcal{V}_{i*}$ by Proposition 13.2.1. Moreover, it is strictly specializable along g and we have $\boldsymbol{D}(\mathcal{V}_{i*})[!g] = \boldsymbol{D}\mathcal{V}_{i*}$. Because $\boldsymbol{D}(\mathcal{V}_{i*})(*g) \simeq \lambda^{d_X}\mathcal{V}_i^{\vee}$ naturally, we obtain $\boldsymbol{D}(\mathcal{V}_{i*}) \simeq \lambda^{d_X}\mathcal{V}_{i!}^{\vee}$. We obtain the other isomorphism in a similar way. □

Lemma 13.3.13 *$\mathcal{V}_\star$ $(\star = *, !)$ have their duals as $\mathcal{R}$-triples, and the isomorphisms in Lemma 13.3.12 induce $\boldsymbol{D}\mathcal{V}_! \simeq \mathcal{V}_*^{\vee}$ and $\boldsymbol{D}\mathcal{V}_* \simeq \mathcal{V}_!^{\vee}$.*

Proof By Lemma 13.3.11, we have the coincidence of pairings on $\{\lambda\}\times(X\setminus D)$ $(\lambda \in \boldsymbol{S})$ under the isomorphisms of the underlying $\mathcal{R}$-modules. Then, it is extended to $\boldsymbol{D}C_{!|\mathcal{X}^\lambda} \simeq C^{\vee}_{*|\mathcal{X}^\lambda}$ for generic λ. It means $\boldsymbol{D}\mathcal{V}_!$ exists and $\boldsymbol{D}\mathcal{V}_! \simeq \mathcal{V}_*^{\vee}$. We can show the claim for the other in a similar way. □

We shall argue the comparison of the weight filtrations later.

13.3.5 Duality of Minimal Extensions in the Pure Case

Let (X,D) and $\mathcal{V}$ be as in Sect. 13.3.4. Let us assume that $\mathcal{V}$ is pure. We set $\mathcal{V}_{!*} := \mathrm{Im}(\mathcal{V}_! \longrightarrow \mathcal{V}_*)$, which is a polarizable wild pure twistor $\mathscr{D}$-module. The underlying $\mathcal{R}_X$-modules are denoted by $\mathcal{V}_{i!*}$ $(i = 1,2)$. Similarly, we obtain a polarizable wild pure twistor $\mathscr{D}$-module $\mathcal{V}_{!*}^{\vee}$ with the underlying $\mathcal{R}_X$-modules $\mathcal{V}_{i!*}^{\vee}$.

Lemma 13.3.14 *We have natural isomorphisms $\boldsymbol{D}\mathcal{V}_{i!*} \simeq \mathcal{V}_{i!*}^{\vee}$. The dual of $\mathcal{V}_{!*}$ as an $\mathcal{R}_X$-triple exists, and it is isomorphic to the $\mathcal{R}_X$-triple $\mathcal{V}_{!*}^{\vee}$.*

Proof Let K_i denote the kernel of $\mathcal{V}_{i!} \longrightarrow \mathcal{V}_{i!*}$, and let C_i denote the cokernel of $\mathcal{V}_{i!*} \longrightarrow \mathcal{V}_{i*}$. By the assumption $A(n-1)$, we have $\mathcal{H}^0\boldsymbol{D}K_i = \boldsymbol{D}K_i$ and $\mathcal{H}^0\boldsymbol{D}C_i = \boldsymbol{D}C_i$.

From the exact sequence $0 \longrightarrow K_i \longrightarrow \mathcal{V}_{i!} \longrightarrow \mathcal{V}_{i!*} \longrightarrow 0$, we obtain $\mathcal{H}^j\boldsymbol{D}\mathcal{V}_{i!*} = 0$ unless $j = 0, -1$. From the exact sequence $0 \longrightarrow \mathcal{V}_{i!} \longrightarrow \mathcal{V}_{i!*} \longrightarrow C_i \longrightarrow 0$, we obtain $\mathcal{H}^j\boldsymbol{D}\mathcal{V}_{i!*} = 0$ unless $j = 0, 1$. Hence, we obtain $\mathcal{H}^j\boldsymbol{D}\mathcal{V}_{i!*} = 0$ unless $j = 0$. We also obtain that $\mathcal{H}^0\boldsymbol{D}\mathcal{V}_{i!*}$ is the image of $\mathcal{H}^0\boldsymbol{D}\mathcal{V}_{i*} \longrightarrow \mathcal{H}^0\boldsymbol{D}\mathcal{V}_{i!}$. Therefore, $\boldsymbol{D}\mathcal{V}_{i!*} \simeq \mathcal{V}_{i!*}^{\vee}$. By using the uniqueness of the extension of sesqui-linear pairings, we obtain $\boldsymbol{D}\mathcal{V}_{!*} \simeq \mathcal{V}_{!*}^{\vee}$. □

We have immediate consequences on the dual of the filtered $\mathcal{R}$-modules underlying good mixed twistor $\mathscr{D}$-modules.

Corollary 13.3.15 *Let $\mathcal{T} \in \mathrm{MTM}^{\mathrm{good}}(X, D)$. Let $\mathcal{M}_i$ $(i = 1, 2)$ be the underlying $\mathcal{R}_X$-modules of $\mathcal{T}$. Then, we have $\mathcal{H}^0\boldsymbol{D}\mathcal{M}_i \simeq \boldsymbol{D}\mathcal{M}_i$, and they are strict. Moreover, they are equipped with induced filtrations.* □

Let $\varphi : \mathcal{T}_1 \longrightarrow \mathcal{T}_2$ be a morphism in $\mathrm{MTM}^{\mathrm{good}}(X, D)$. Let $\mathcal{M}_i'$ and $\mathcal{M}_i''$ be the $\mathcal{R}_X$-modules underlying $\mathcal{T}_i$. We have the underlying morphisms $\varphi' : \mathcal{M}_2' \longrightarrow \mathcal{M}_1'$ and $\varphi'' : \mathcal{M}_1'' \longrightarrow \mathcal{M}_2''$.

Corollary 13.3.16 *We have natural isomorphisms*

$$\operatorname{Ker}\boldsymbol{D}\varphi' = \boldsymbol{D}\operatorname{Cok}\varphi', \quad \operatorname{Im}\boldsymbol{D}\varphi' = \boldsymbol{D}\operatorname{Im}\varphi', \quad \operatorname{Cok}\boldsymbol{D}\varphi' = \boldsymbol{D}\operatorname{Ker}\varphi'.$$

We have similar isomorphism for φ''. □

We shall use it in the special case. Let g be a holomorphic function such that $g^{-1}(0) = D$. Let $\mathcal{V} \in \mathrm{MTS}^{\mathrm{adm}}(X, D)$ with the underlying $\mathcal{R}_{X(*D)}$-modules. For any a, b, we have $\left(\Pi_g^{a,b}\mathcal{V}\right)^{\vee} \simeq \Pi_g^{-b+1,-a+1}\mathcal{V}^{\vee}$ in $\mathrm{MTS}^{\mathrm{adm}}(X, D)$.

Corollary 13.3.17 *We have natural isomorphisms of $\mathcal{R}_X$-modules*

$$\boldsymbol{D}\Pi_{g!*}^{a,b}\mathcal{V}_i \simeq \Pi_{g!*}^{-b+1,-a+1}\mathcal{V}_i^{\vee}.$$

In particular, we have $\boldsymbol{D}\psi_g^{(a)}\mathcal{V}_i \simeq \psi_g^{(-a+1)}\mathcal{V}_i^{\vee}$ and $\boldsymbol{D}\Xi_g^{(a)}\mathcal{V}_i \simeq \Xi_g^{(-a)}\mathcal{V}_i^{\vee}$.

Proof Let us consider the natural morphism $\Pi_{g!}^{a,N}\mathcal{V}_i \longrightarrow \Pi_{g*}^{b,N}\mathcal{V}_i$ for a sufficiently large N. According to Lemma 13.3.12, its dual is naturally identified with

$$\Pi_{g!}^{-N+1,-b+1}\mathcal{V}_i^{\vee} \longrightarrow \Pi_{g*}^{-N+1,-a+1}\mathcal{V}_i^{\vee}.$$

Hence, the claim follows from Corollary 13.3.16. □

Corollary 13.3.18 *$\Pi_{g!*}^{-b+1,-a+1}\mathcal{V}^{\vee}$ is a dual of $\Pi_{g!*}^{a,b}\mathcal{V}$ as an $\mathcal{R}_X$-triple. In particular, $\psi_g^{(a)}(\mathcal{V})$ and $\Xi_g^{(a)}(\mathcal{V})$ have duals as $\mathcal{R}_X$-triples, and naturally $\boldsymbol{D}\psi_g^{(a)}(\mathcal{V}) \simeq \psi_g^{(-a+1)}(\mathcal{V}^{\vee})$ and $\boldsymbol{D}\Xi_g^{(a)}(\mathcal{V}) \simeq \Xi_g^{(-a)}(\mathcal{V}^{\vee})$.* □

13.3.6 Duality of the Canonical Prolongations in MTM

Let X and D be as above. Let $\mathcal{V} \in \mathrm{MTS}^{\mathrm{adm}}(X, D)$. We have already obtained an isomorphism $\boldsymbol{D}(\mathcal{V}_!) \simeq \mathcal{V}_*^\vee$ as $\mathcal{R}_X$-triples. We have two natural filtrations on $\boldsymbol{D}(\mathcal{V}_!)$. One is the filtration $\boldsymbol{D}W$ obtained as the dual of W on $\mathcal{V}_!$, and the other is the filtration W of $\mathcal{V}_*^\vee$.

Lemma 13.3.19 *They are the same.*

Proof Recall that $\mathcal{V}_!$ is obtained as the cohomology of the complex of $\mathcal{R}_X$-triples

$$\psi_g^{(1)}\mathcal{V} \longrightarrow \Xi_g^{(0)}\mathcal{V} \oplus \phi_g^{(0)}(\mathcal{V}_!) \longrightarrow \psi_g^{(0)}\mathcal{V},$$

where $\phi_g^{(0)}(\mathcal{V}_!) \simeq \psi_g^{(1)}(\mathcal{V})$. The weight filtration of $\mathcal{V}_!$ is obtained from the naively induced filtrations L on $\psi_g^{(a)}(\mathcal{V})$ and $\Xi_g^{(0)}(\mathcal{V})$, and the filtration L of $\phi_g^{(0)}(\mathcal{V}_!)$ obtained as the transfer of L on $\psi_g^{(a)}(\mathcal{V})$.

Hence, $\boldsymbol{D}(\mathcal{V}_!)$ is obtained as the cohomology of the complex of $\mathcal{R}$-triples

$$\boldsymbol{D}\psi_g^{(0)}(\mathcal{V}) \longrightarrow \boldsymbol{D}\Xi_g^{(0)}(\mathcal{V}) \oplus \boldsymbol{D}\phi_g^{(0)}(\mathcal{V}_!) \longrightarrow \boldsymbol{D}\psi_g^{(1)}(\mathcal{V}),$$

and the filtration $\boldsymbol{D}W$ is induced by $\boldsymbol{D}L$ on $\boldsymbol{D}\psi_g^{(a)}(\mathcal{V})$, $\boldsymbol{D}\Xi_g^{(0)}(\mathcal{V})$, and $\boldsymbol{D}\phi_g^{(0)}(\mathcal{V}_!)$. It is easy to check that $\boldsymbol{D}L$ on $\boldsymbol{D}\psi_g^{(a)}(\mathcal{V})$ and $\boldsymbol{D}\Xi_g^{(0)}(\mathcal{V})$ are the same as the naively induced filtrations under the isomorphisms $\boldsymbol{D}\psi_g^{(a)}(\mathcal{V}) \simeq \psi_g^{(-a+1)}(\mathcal{V}^\vee)$ and $\boldsymbol{D}\Xi_g^{(0)}(\mathcal{V}) \simeq \Xi_g^{(0)}(\mathcal{V}^\vee)$. Because $\boldsymbol{D}W$ on $\boldsymbol{D}\psi_g^{(a)}(\mathcal{V})$ is $M(\boldsymbol{D}\mathcal{N}; \boldsymbol{D}L)[1-2a]$, we obtain $\boldsymbol{D}W$ is the canonical weight filtration of $\psi_g^{(-a+1)}(\mathcal{V}^\vee)$. Hence, $\boldsymbol{D}W$ on $\boldsymbol{D}\phi_g^{(0)}(\mathcal{V}_!)$ is the same as the canonical weight filtration W of $\boldsymbol{D}\phi_g^{(0)}(\mathcal{V}_!) \simeq \phi_g^{(0)}(\mathcal{V}_*^\vee)$. We obtain that $\boldsymbol{D}L$ on $\boldsymbol{D}\phi_g^{(0)}(\mathcal{V}_!)$ is the transfer of L on $\psi_g^{(a)}(\mathcal{V}^\vee)$. Hence, $\boldsymbol{D}W$ on $\boldsymbol{D}\mathcal{V}_!$ is the same as the canonical weight filtration of $\mathcal{V}_*^\vee$. □

Corollary 13.3.20 *The induced filtrations of $\boldsymbol{D}\psi_g^{(a)}\mathcal{V}$ and $\boldsymbol{D}\Xi_g^{(a)}\mathcal{V}$ are the canonical weight filtrations under the identifications $\boldsymbol{D}\psi_g^{(a)}\mathcal{V} \simeq \psi_g^{(-a+1)}\mathcal{V}^\vee$ and $\boldsymbol{D}\Xi_g^{(a)}\mathcal{V} \simeq \Xi_g^{(-a)}\mathcal{V}^\vee$.* □

Remark 13.3.21 In the case of $\boldsymbol{D}\psi_g^{(a)}\mathcal{V}$, we can also deduce it by using the characterization as the relative monodromy filtration. □

13.3.7 Local Construction of the Pairing $\boldsymbol{D}\mathcal{T}$

Let $\mathcal{T} \in \mathrm{MTM}(X)$ such that $\dim \mathrm{Supp}(\mathcal{T}) = n$. Let P be any point of X. We take a small neighbourhood X_P and a finite dimensional $\mathbb{Q}$-vector subspace $\mathcal{A} \subset \mathbb{R} \times \mathbb{C}$ such that $\mathbb{Q} \times \{0\} \subset \mathcal{A}$ and $\mathcal{T}_{|X_P} \in \mathrm{MTM}(X_P)_{\mathcal{A}}$, as in Proposition 7.1.67. Suppose that we have a cell $\mathcal{C} = (Z, U, \varphi, \mathcal{V})$ of $\mathcal{T}$ at P with a cell function g. We may assume that φ gives a proper morphism $Z \longrightarrow X_P$. We have the expression of $\mathcal{T}_{|X_P}$ as the cohomology of the complex in $\mathrm{MTM}(X_P)_{\mathcal{A}}$:

$$\psi_g^{(1)}\varphi_\dagger(\mathcal{V}) \longrightarrow \Xi_g^{(0)}\varphi_\dagger(\mathcal{V}) \oplus \phi_g^{(0)}\mathcal{T}_{|X_P} \longrightarrow \psi_g^{(0)}\varphi_\dagger(\mathcal{V})$$

By the assumption $A(n-1)$, Corollary 13.3.20 and the result in Sect. 13.3.2, we obtain the following complex in $\mathrm{MTM}(X_P)_{\mathcal{A}}$:

$$\boldsymbol{D}\psi_g^{(0)}\varphi_\dagger(\mathcal{V}) \longrightarrow \boldsymbol{D}\Xi_g^{(0)}\varphi_\dagger(\mathcal{V}) \oplus \boldsymbol{D}\phi_g^{(0)}\mathcal{T}_{|X_P} \longrightarrow \boldsymbol{D}\psi_g^{(1)}\varphi_\dagger(\mathcal{V}). \tag{13.21}$$

We obtain a mixed twistor $\mathscr{D}$-module $\boldsymbol{D}\mathcal{T}_{|X_P}$ on X_P as the cohomology of (13.21). The underlying filtered $\mathcal{R}_{X_P}$-modules are the dual of the underlying filtered $\mathcal{R}_{X_P}$-modules of $\mathcal{T}_{|X_P}$. Let C_1 and C_2 denote the pairings of $\mathcal{T}_{|X_P}$ and $\boldsymbol{D}\mathcal{T}_{|X_P}$, respectively. By construction, $C_{2|\mathcal{X}_P^\lambda} = \boldsymbol{D}C_{1|\mathcal{X}_P^\lambda}$ for $\lambda \in \boldsymbol{S}$ which is generic with respect to $\mathcal{A}$. Hence, we have $C_2 = \boldsymbol{D}C_1$ in the sense of Theorem 13.3.1.

As a result, we obtain the following.

Lemma 13.3.22 *For any point $P \in X$, there exists a neighbourhood X_P of P in X such that there exists $\boldsymbol{D}(\mathcal{T}_{|X_P})$ in* MTM(X_P).

If $\mathcal{T}$ is integrable then $\boldsymbol{D}(\mathcal{T}_{|X_P})$ is also integrable.

Proof We have already proved the first claim. For the second claim, if $\mathcal{T} \in \mathrm{MTM}^{\mathrm{int}}(X)$, then $\mathcal{V}$ and $\mathcal{V}^\vee$ are integrable. Hence, by construction, $\boldsymbol{D}\psi_g^{(a)}\varphi_\dagger(\mathcal{V})$ and $\boldsymbol{D}\Xi_g^{(0)}\varphi_\dagger(\mathcal{V})$ are also naturally integrable. By the hypothesis of the induction, $\boldsymbol{D}\phi_g^{(0)}\mathcal{T}$ is integrable. Hence, $\boldsymbol{D}\mathcal{T}$ is integrable. Thus, the induction can go further. □

Lemma 13.3.23 *Let $\mathcal{M}_i$ $(i = 1, 2)$ be the underlying $\mathcal{R}$-module of the mixed twistor $\mathscr{D}$-module $\mathcal{T}$. Then, we have $\mathcal{H}^j\boldsymbol{D}\mathcal{M}_i = 0$ unless $j = 0$.*

Proof We have only to check the claim locally at any point of X. Then, the claim follows from the existence of $\boldsymbol{D}\mathcal{T}_{|X_P}$ on a small neighbourhood X_P of P in X (Lemma 13.3.22). □

13.3.8 End of the Proof of Theorem 13.3.1

Let $\mathcal{T} \in \mathrm{MTM}(X)$ with $\dim \mathrm{Supp}(\mathcal{T}) = n$. We construct a filtered $\mathcal{R}_X$-triple $\boldsymbol{D}\mathcal{T}$ determined by the following conditions. Let $\mathcal{M}_i$ $(i = 1, 2)$ be the underlying $\mathcal{R}_X$-modules. As remarked in Lemma 13.3.23, we have already known $\mathcal{H}^j\boldsymbol{D}\mathcal{M}_i = 0$

unless $j = 0$. We have the filtered $\mathcal{R}_X$-triples $\boldsymbol{D}\mathcal{T}$ determined by the following conditions:

- The underlying $\mathcal{R}_X$-modules are $\mathcal{H}^0\boldsymbol{D}\mathcal{M}_1$ and $\mathcal{H}^0\boldsymbol{D}\mathcal{M}_2$.
- For any point $P \in X$, we take a small neighbourhood X_P of P in X as in Lemma 13.3.22. Then, we have $\boldsymbol{D}\mathcal{T}_{|X_P} \simeq \boldsymbol{D}(\mathcal{T}_{|X_P})$ where the morphism of the underlying $\mathcal{R}_{|X_P}$-modules are the natural isomorphisms $\boldsymbol{D}(\mathcal{M}_i)_{|X_P} \simeq \boldsymbol{D}(\mathcal{M}_{i|X_P})$.

Indeed, it is enough to glue $\boldsymbol{D}(\mathcal{T}_{|X_P})$ by using the uniqueness of the dual pairing.

Let W be the filtration of $\boldsymbol{D}\mathcal{T}$. Let us check that $\mathrm{Gr}^W_w \boldsymbol{D}\mathcal{T} \in \mathrm{MT}(X, w)$. We have to check that $\mathrm{Gr}^W_w \boldsymbol{D}\mathcal{T}$ is equipped with a polarization. By the exactness of $\boldsymbol{D}$ for the $\mathcal{R}_X$-modules underlying mixed twistor $\mathscr{D}$-modules, we have $\mathrm{Gr}^W_w \boldsymbol{D}\mathcal{T} = \boldsymbol{D}\,\mathrm{Gr}^W_{-w}\mathcal{T}$ as an $\mathcal{R}_X$-triple. Hence, it is enough to consider the case where $\mathcal{T}$ is pure of weight 0. We have already known that $\boldsymbol{D}\mathcal{T}$ is strictly S-decomposable. It is enough to consider the case that the support Z of $\mathcal{T}$ is irreducible. We have a closed complex analytic subset $Z_1 \subset Z$ and a wild variation of pure twistor structure $\mathcal{V}$ of weight 0 on (Z, Z_1) such that $\mathcal{T}$ is the polarizable pure twistor $\mathscr{D}$-module associated to $\mathcal{V}$. Then, by using the S-decomposability of $\boldsymbol{D}\mathcal{T}$, we can check that $\boldsymbol{D}\mathcal{T}$ is isomorphic to the $\mathcal{R}$-triple obtained as the polarizable pure twistor $\mathscr{D}$-module associated to $\mathcal{V}^{\vee}$.

The other condition for $(\boldsymbol{D}\mathcal{T}, W)$ to be a mixed twistor $\mathscr{D}$-module are clearly satisfied by construction. If $\mathcal{T}$ is integrable $\boldsymbol{D}\mathcal{T}$ is also integrable by Lemma 13.3.22. Thus, the proof of Theorem 13.3.1 is finished. □

13.4 Real Structure of Mixed Twistor $\mathscr{D}$-Modules

13.4.1 Some Functors

Let $\boldsymbol{D}^{\mathrm{herm}}$ denote the Hermitian adjoint, i.e., $\boldsymbol{D}^{\mathrm{herm}}(\mathcal{T}) := \mathcal{T}^*$. Formally, we set $\tilde{\sigma}^* := \boldsymbol{D} \circ \boldsymbol{D}^{\mathrm{herm}} = \boldsymbol{D}^{\mathrm{herm}} \circ \boldsymbol{D}$. It gives an involution on $\mathrm{MTM}(X)$. We have a natural transformation $\tilde{\sigma}^* \circ \tilde{\sigma}^* \simeq \mathrm{id}$. We define $\tilde{\gamma}^* := j^* \circ \tilde{\sigma}^* = \tilde{\sigma}^* \circ j^*$. We naturally have $\tilde{\gamma}^* \circ \tilde{\gamma}^* \simeq \mathrm{id}$.

More explicitly, for $\mathcal{T} = (\mathcal{M}', \mathcal{M}'', C) \in \mathrm{MTM}(X)$, we have

$$\tilde{\sigma}^*\mathcal{T} = \boldsymbol{D}(\mathcal{T}^*) = \big(\boldsymbol{D}\mathcal{M}'', \boldsymbol{D}\mathcal{M}', \boldsymbol{D}C^*\big), \quad \tilde{\gamma}^*\mathcal{T} = \Big(j^*\boldsymbol{D}\mathcal{M}'', j^*\boldsymbol{D}\mathcal{M}', j^*\boldsymbol{D}C^*\Big)$$

with the induced weight filtrations.

Remark 13.4.1 Let σ, γ and j be the involutions on $\mathbb{P}^1$ given by $\sigma(\lambda) = -\overline{\lambda}^{-1}$, $\gamma(\lambda) = \overline{\lambda}^{-1}$ and $j(\lambda) = -\lambda$. They induce functors on the category of holomorphic vector bundles on $\mathbb{P}^1$, i.e., the category of twistor structures. The above functors $\tilde{\sigma}^*$, $\tilde{\gamma}^*$ and j^* are their counterparts on the category of mixed twistor $\mathscr{D}$-modules. □

The following lemma is clear by construction.

Lemma 13.4.2

- *We have* $\tilde{\sigma}^* \circ \tilde{\gamma}^* = \tilde{\gamma}^* \circ \tilde{\sigma}^* = j^*$. *We also have* $\tilde{\sigma}^* \circ j^* = j^* \circ \tilde{\sigma}^* = \tilde{\gamma}^*$ *and* $j^* \circ \tilde{\gamma}^* = \tilde{\gamma}^* \circ j^* = \tilde{\sigma}^*$.
- *Let* $\tilde{\upsilon}^*$ *be one of* $\tilde{\sigma}^*$, $\tilde{\gamma}^*$ *or* j^*. *We have natural commutativity* $\tilde{\upsilon}^* \circ \boldsymbol{D} \simeq \boldsymbol{D} \circ \tilde{\upsilon}^*$ *and* $\tilde{\upsilon}^* \circ \boldsymbol{D}^{\mathrm{herm}} \simeq \boldsymbol{D}^{\mathrm{herm}} \circ \tilde{\upsilon}^*$ *by the natural isomorphisms of the underlying filtered* $\mathcal{R}$*-modules.*
- *Let* F *be a projective morphism. The natural transform* $\boldsymbol{D}F_{\dagger} \simeq F_{\dagger}\boldsymbol{D}$ *for* $\mathcal{R}$*-modules induces* $F_{\dagger}^p \tilde{\gamma}^*(\mathcal{T}) \simeq \mu_p \circ \tilde{\gamma}^* F_{\dagger}^p(\mathcal{T})$ *and* $F_{\dagger}^p \tilde{\sigma}^*(\mathcal{T}) \simeq \mu_p \circ \tilde{\sigma}^* F_{\dagger}^p(\mathcal{T})$. *We also have* $j^* \circ F_{\dagger} = F_{\dagger} \circ j^*$ *naturally. (See Sect. 13.3.1 for* μ_p*.)* □

We consider $\mathcal{T} = (\mathcal{M}_1, \mathcal{M}_2, C) \in \mathrm{MTM}(X)$. Let $\varphi_1 : F_{\dagger}^{-p}\boldsymbol{D}\mathcal{M}_1 \simeq \boldsymbol{D}F_{\dagger}^p\mathcal{M}_1$ and $\varphi_2 : F_{\dagger}^p\boldsymbol{D}\mathcal{M}_2 \simeq \boldsymbol{D}F_{\dagger}^{-p}\mathcal{M}_2$ be the natural isomorphisms. Then, the pair $\big(\epsilon(p)\varphi_2, \epsilon(-p)\varphi_1^{-1}\big)$ gives an isomorphism $F_{\dagger}^p\tilde{\sigma}^*(\mathcal{T}) \simeq \tilde{\sigma}^*F_{\dagger}^p(\mathcal{T})$, and the pair $\big(\epsilon(p)j^*\varphi_2, \epsilon(-p)j^*\varphi_1^{-1}\big)$ gives an isomorphism $F_{\dagger}^p\tilde{\gamma}^*(\mathcal{T}) \simeq \tilde{\gamma}^*F_{\dagger}^p(\mathcal{T})$.

The following lemma is obvious by construction.

Lemma 13.4.3 *Let* $\mathcal{T} \in \mathrm{MTM}(X)$ *and* $\mathcal{V} \in \mathrm{MTS}^{\mathrm{adm}}(X, H)$. *Then, we have the relation* $\tilde{\gamma}^*(\mathcal{T} \otimes \mathcal{V}[\star H]) \simeq \big(\tilde{\gamma}^*(\mathcal{T}) \otimes \tilde{\gamma}^*_{\mathrm{sm}}(\mathcal{V})\big)[\star H]$.

Proof We naturally have $j^*\big((\mathcal{T} \otimes \mathcal{V})[*H]\big) \simeq \big(j^*\mathcal{T} \otimes j^*\mathcal{V}\big)[*H]$, $\Big((\mathcal{T} \otimes \mathcal{V})[*H]\Big)^* \simeq (\mathcal{T}^* \otimes \mathcal{V}^*)[!H]$ and $\boldsymbol{D}\Big((\mathcal{T} \otimes \mathcal{V})[!H]\Big) \simeq \big(\boldsymbol{D}(\mathcal{T}) \otimes \mathcal{V}^{\vee}\big)[*H]$, which induce the desired isomorphisms. □

The following lemma is clear from Proposition 13.3.9.

Lemma 13.4.4 *Let* X_i $(i = 1, 2)$ *be complex manifolds. For* $\mathcal{T}_i \in \mathrm{MTM}(X_i)$ $(i = 1, 2)$, *we have a natural isomorphism* $(\tilde{\gamma}^*\mathcal{T}_1) \boxtimes (\tilde{\gamma}^*\mathcal{T}_2) \simeq \tilde{\gamma}^*(\mathcal{T}_1 \boxtimes \mathcal{T}_2)$. *The isomorphisms of the underlying* $\mathcal{R}_X$*-modules are induced by (13.19).* □

13.4.1.1 Complex of Mixed Twistor $\mathscr{D}$-Modules

Let $\mathcal{C}(\mathrm{MTM}(X))$ denote the category of bounded complexes of mixed twistor $\mathscr{D}$-modules. Let $(\mathcal{T}^{\bullet}, \delta^{\bullet}) \in \mathcal{C}(\mathrm{MTM}(X))$.

- Let $\boldsymbol{D}^{\mathrm{herm}}(\mathcal{T}^{\bullet}) \in \mathcal{C}(\mathrm{MTM}(X))$ be given by $\boldsymbol{D}^{\mathrm{herm}}(\mathcal{T}^{\bullet})^p = \boldsymbol{D}^{\mathrm{herm}}(\mathcal{T}^{-p})$ with the differentials $\boldsymbol{D}^{\mathrm{herm}}(\delta)^p = \boldsymbol{D}^{\mathrm{herm}}(\delta^{-p-1})$.
- Let $\boldsymbol{D}(\mathcal{T}^{\bullet}) \in \mathcal{C}(\mathrm{MTM}(X))$ be given by $\boldsymbol{D}(\mathcal{T}^{\bullet})^p = \mu_p\boldsymbol{D}(\mathcal{T}^{-p})$ with the differentials

$$\boldsymbol{D}(\delta)^p = \big((-1)^p\boldsymbol{D}\delta_1^p, (-1)^{p+1}\boldsymbol{D}\delta_2^{-p-1}\big).$$

- Let $j^*(\mathcal{T}^{\bullet}) \in \mathcal{C}(\mathrm{MTM}(X))$ be given by $j^*(\mathcal{T}^{\bullet})^p = j^*(\mathcal{T}^p)$ with the differential $j^*(\delta)^p = j^*(\delta^p)$.

- We define $\mathcal{S}_\ell(\mathcal{T}^\bullet)^p \in \mathcal{C}(\mathrm{MTM}(X))$ by $\mathcal{S}_\ell(\mathcal{T}^\bullet)^p = \mathcal{T}^{p+\ell}$ with $\mathcal{S}_\ell(\delta)^p = (-1)^\ell \delta^{p+\ell}$.

The natural quasi-isomorphisms of the underlying complexes of $\mathcal{R}$-modules give the commutativity of the functors $\boldsymbol{D}^{\mathrm{herm}}$, $\boldsymbol{D}$ and j^*. We also have $\mathcal{S}_\ell \circ \boldsymbol{D}^{\mathrm{herm}} \simeq \boldsymbol{D}^{\mathrm{herm}} \circ \mathcal{S}_{-\ell}$, $\mathcal{S}_\ell \circ \boldsymbol{D} \simeq \boldsymbol{D} \circ \mathcal{S}_{-\ell}$ and $j^* \circ \mathcal{S}_\ell = \mathcal{S}_\ell \circ j^*$. (See Sect. 12.6.1 for the signature.)

We define the functors $\tilde{\sigma}^*$ and $\tilde{\gamma}^*$ on $\mathcal{C}(\mathrm{MTM}(X))$ by $\tilde{\sigma}^* := \boldsymbol{D} \circ \boldsymbol{D}^{\mathrm{herm}}$ and $\tilde{\gamma}^* := j^* \circ \tilde{\sigma}^*$. The natural quasi-isomorphisms of the underlying $\mathcal{R}$-modules give $\mathcal{S}_\ell \circ \tilde{\nu}^* \simeq \tilde{\nu}^* \circ \mathcal{S}_\ell$, where $\tilde{\nu}^*$ is one of j^*, $\tilde{\gamma}^*$ or $\tilde{\sigma}^*$. The compatibility with the push-forward is reformulated as follows.

Lemma 13.4.5 *Let F be a projective morphism. We have natural isomorphisms $\mathcal{S}_{-p} \circ F_\dagger^p(\boldsymbol{D}\mathcal{T}) \simeq \boldsymbol{D} \circ \mathcal{S}_p \circ F_\dagger^{-p}(\mathcal{T})$, $\mathcal{S}_{-p} \circ F_\dagger^p(\tilde{\sigma}^*\mathcal{T}) \simeq \tilde{\sigma}^* \circ \mathcal{S}_{-p} \circ F_\dagger^p(\mathcal{T})$ and $\mathcal{S}_{-p} \circ F_\dagger^p(\tilde{\gamma}^*\mathcal{T}) \simeq \tilde{\gamma}^* \circ \mathcal{S}_{-p} \circ F_\dagger^p(\mathcal{T})$.* □

13.4.2 Real Structure of Mixed Twistor $\mathscr{D}$-Modules

Let $\mathcal{T}$ be an (integrable) mixed twistor $\mathscr{D}$-module. A real structure of $\mathcal{T}$ is an (integrable) isomorphism $\kappa : \tilde{\gamma}^*\mathcal{T} \simeq \mathcal{T}$ such that $\tilde{\gamma}^*\kappa \circ \kappa = \mathrm{id}$. A morphism of (integrable) mixed twistor $\mathscr{D}$-modules with real structure is an (integrable) morphism compatible with the real structure. Let $\mathrm{MTM}(X,\mathbb{R})$ (resp. $\mathrm{MTM}^{\mathrm{int}}(X,\mathbb{R})$) denote the category of mixed twistor $\mathscr{D}$-modules (resp. integrable mixed twistor $\mathscr{D}$-modules) equipped with a real structure (resp. integrable real structure). We obtain the following proposition.

Proposition 13.4.6 *Let $(\mathcal{T},\kappa)$ be an (integrable) mixed twistor $\mathscr{D}$-module with (integrable) real structure on X.*

- *$\boldsymbol{D}(\mathcal{T})$, $\boldsymbol{D}^{\mathrm{herm}}(\mathcal{T})$ and $j^*(\mathcal{T})$ are equipped with naturally induced (integrable) real structures.*
- *Let $F : X \longrightarrow Y$ be a projective morphism. Then, $F_\dagger^i(\mathcal{T})$ are equipped with induced (integrable) real structures.*
- *Let H be a hypersurface of X. Then, $\mathcal{T}[\star H]$ ($\star = *, !$) are equipped with naturally induced (integrable) real structures.*
- *Let $\mathcal{V}$ be an (integrable) admissible mixed twistor structure on (X,H). If $\mathcal{V}$ is equipped with a real structure $\kappa_{\mathcal{V}}$, $(\mathcal{T} \otimes \mathcal{V})[\star H]$ is also equipped with an induced (integrable) real structure for $\star = *, !$. (See Sect. 2.1.7.2 for the concept of real structure of smooth $\mathcal{R}$-triples.)*

In the first statement, the (integrable) mixed twistor $\mathscr{D}$-modules with the (integrable) real structures are denoted by $\boldsymbol{D}(\mathcal{T},\kappa)$, $\boldsymbol{D}^{\mathrm{herm}}(\mathcal{T},\kappa)$ and $j^*(\mathcal{T},\kappa)$. We use the symbols $F_\dagger^j(\mathcal{T},\kappa)$, $(\mathcal{T},\kappa)[\star H]$, $\big((\mathcal{T},\kappa) \otimes (\mathcal{V},\kappa_{\mathcal{V}})\big)[\star H]$. □

Corollary 13.4.7 *Let $(\mathcal{T},\kappa)$ be an (integrable) mixed twistor $\mathscr{D}$-module with an (integrable) real structure. Let g be any holomorphic function. Then, $\psi_g^{(a)}(\mathcal{T})$,*

$\Xi_g^{(a)}\mathcal{T}$ *and* $\phi_g^{(a)}\mathcal{T}$ *are naturally equipped with induced (integrable) real structures. The induced objects in* $\mathrm{MTM}(X,\mathbb{R})$ *or* $\mathrm{MTM}^{\mathrm{int}}(X,\mathbb{R})$ *are denoted by* $\psi_g^{(a)}(\mathcal{T},\kappa)$, $\Xi_g^{(a)}(\mathcal{T},\kappa)$ *and* $\phi_g^{(a)}(\mathcal{T},\kappa)$. □

Let X be a complex manifold with a hypersurface H. We have the full subcategories $\mathrm{MTM}(X,[\star H],\mathbb{R}) \subset \mathrm{MTM}(X,\mathbb{R})$ ($\star = *,!$) of the objects $(\mathcal{T},\kappa) \in \mathrm{MTM}(X,\mathbb{R})$ satisfying $\mathcal{T}[\star H] = \mathcal{T}$. We use the notation $\mathrm{MTM}^{\mathrm{int}}(X,[\star H],\mathbb{R})$ with a similar meaning.

Proposition 13.4.8 *Let* $F : (X',H') \longrightarrow (X,H)$ *be a projective birational morphism such that* $X' \setminus H' \simeq X \setminus H$. *Then,* $F_\dagger$ *induces equivalences* $\mathrm{MTM}(X',[\star H'],\mathbb{R}) \simeq \mathrm{MTM}(X,[\star H],\mathbb{R})$ *and* $\mathrm{MTM}^{\mathrm{int}}(X',[\star H'],\mathbb{R}) \simeq \mathrm{MTM}^{\mathrm{int}}(X,[\star H],\mathbb{R})$. □

Proposition 13.4.9 *Let* X_i $(i = 1,2)$ *be complex manifolds. Let* $(\mathcal{T}_i,\kappa_i)$ *be (integrable) mixed twistor* $\mathscr{D}$*-modules on* X_i. *Then,* $\mathcal{T}_1 \boxtimes \mathcal{T}_2$ *is naturally equipped with the real structure* $\kappa_1 \boxtimes \kappa_2$. *In other words, the external tensor product naturally induces bi-functors* $\mathrm{MTM}(X_1,\mathbb{R}) \times \mathrm{MTM}(X_2,\mathbb{R}) \longrightarrow \mathrm{MTM}(X_1 \times X_2,\mathbb{R})$ *and* $\mathrm{MTM}^{\mathrm{int}}(X_1,\mathbb{R}) \times \mathrm{MTM}^{\mathrm{int}}(X_2,\mathbb{R}) \longrightarrow \mathrm{MTM}^{\mathrm{int}}(X_1 \times X_2,\mathbb{R})$. □

13.4.2.1 Example

Let us look at the smooth $\mathcal{R}$-triple $\mathcal{U}_X(d_X,0)$. It is a mixed twistor $\mathscr{D}$-module according to Theorem 10.3.1. It is naturally equipped with an integrable structure. We naturally identify $j^*\mathcal{O}_{\mathcal{X}}$ with $\mathcal{O}_{\mathcal{X}}$. We have the following isomorphism:

$$\begin{aligned}\tilde{\gamma}^*\mathcal{U}(d_X,0) &= \boldsymbol{D}j^*\big(\mathcal{O}_{\mathcal{X}},\, \mathcal{O}_{\mathcal{X}}\,\lambda^{d_X},\, C_0\big) = \boldsymbol{D}\big(\mathcal{O}_{\mathcal{X}},\, \mathcal{O}_{\mathcal{X}}\,\lambda^{d_X}, C_0\big)\\ &\simeq \big(\mathcal{O}_{\mathcal{X}}\lambda^{d_X},\, \mathcal{O}_{\mathcal{X}},\, (-1)^{d_X}C_0\big) \qquad (13.22)\end{aligned}$$

Here, we fix the integrable isomorphism $\boldsymbol{D}\mathcal{O}_{\mathcal{X}} \simeq \lambda^{d_X}\,\mathcal{O}_{\mathcal{X}}$ determined by the condition that its restriction to $\lambda = 1$ induces the isomorphism ν in Sect. 12.6.6. Hence, for any complex number a with $|a| = 1$, $\big((-1)^{d_X}\overline{a}, a\big)$ gives a real structure on $\mathcal{U}_X(d_X,0)$. (We remark the signature in Lemma 13.1.6.)

13.4.3 $\mathbb{R}$-Betti Structure of the Underlying $\mathscr{D}$-Modules

Let $\mathcal{A} \subset \mathbb{R} \times \mathbb{C}$ be a $\mathbb{Q}$-vector subspace such that $\mathbb{Q} \times \{0\} \subset \mathcal{A}$. We also impose that $(a,0)$ and $(0,\alpha)$ are elements of $\mathcal{A}$ if $u = (a,\alpha) \in \mathcal{A}$. Suppose that there exists $\lambda_0 \in \boldsymbol{S}$ which is generic with respect to $\mathcal{A}$. Let $\mathrm{MTM}(X,\mathbb{R})_{\mathcal{A}} \subset \mathrm{MTM}(X,\mathbb{R})$ denote the full subcategory of the objects whose KMS-spectra are contained in $\mathcal{A}$.

Let us consider $(\mathcal{T},\kappa) \in \mathrm{MTM}(X,\mathbb{R})_{\mathcal{A}}$. Let $\mathcal{M}_i$ $(i = 1,2)$ be the underlying $\mathcal{R}_X$-modules. The real structure κ of $\mathcal{T}$ induces a real structure of the non-degenerate holonomic $\mathscr{D}$-triple $(\mathcal{M}_1^{-\lambda_0}, \mathcal{M}_2^{\lambda_0}, C^{\lambda_0})$. (See Sect. 12.6.1.) As explained in Sects. 12.6.2, 12.6.3 and 12.6.5, $\mathcal{M}_2^{\lambda_0}$ is equipped with a naturally induced $\mathbb{R}$-Betti structure in the sense of Sect. 7.2 of [57]. Thus, we obtain a functor $\Upsilon^{\lambda_0} : \mathrm{MTM}(X,\mathbb{R})_{\mathcal{A}} \longrightarrow \mathrm{Hol}(X,\mathbb{R})$, where $\mathrm{Hol}(X,\mathbb{R})$ denotes the category of holonomic $\mathscr{D}$-modules on X with $\mathbb{R}$-Betti structure. We obtain the following from Proposition 12.6.19.

Proposition 13.4.10 *The functor Υ^{λ_0} is compatible with the duality. It is also compatible with the push-forward by any projective morphisms.* □

We obtain the following from Propositions 11.2.9 and 12.6.20.

Proposition 13.4.11 *For any hypersurface H, the localizations $[*H]$ and $[!H]$ are compatible with the functor Υ^{λ_0}.* □

Let $\mathcal{V} = (\mathcal{V}_1, \mathcal{V}_2, C) \in \mathrm{MTS}^{\mathrm{adm}}(X,H)$ with a real structure as a smooth $\mathcal{R}_{X(*H)}$-triple. Suppose that the KMS-spectra along any irreducible component of H are contained in $\mathcal{A}$. Then, the meromorphic flat bundle $\mathcal{V}_2^{\lambda_0}$ has a good $\mathbb{R}$-structure. We obtain the following from Proposition 12.6.21.

Proposition 13.4.12 *Let $\mathcal{T} \in \mathrm{MTM}(X,\mathbb{R})_{\mathcal{A}}$. Then, we naturally have*

$$\Upsilon^{\lambda_0}\big((\mathcal{T} \otimes \mathcal{V})[\star H]\big) \simeq \big(\Upsilon^{\lambda_0}(\mathcal{T}) \otimes \mathcal{V}_2^{\lambda_0}\big)[\star H]$$

in $\mathrm{Hol}(X,\mathbb{R})$. □

Corollary 13.4.13 *Let g be any holomorphic function on X. We naturally have $\Upsilon^{\lambda_0} \circ \Xi_g^{(a)} \simeq \Xi_g^{(a)} \circ \Upsilon^{\lambda_0}$, $\Upsilon^{\lambda_0} \circ \psi_g^{(a)} \simeq \psi_g^{(a)} \circ \Upsilon^{\lambda_0}$ and $\Upsilon^{\lambda_0} \circ \phi_g^{(a)} \simeq \phi_g^{(a)} \circ \Upsilon^{\lambda_0}$.* □

We obtain the following from Lemma 13.4.4.

Proposition 13.4.14 *Let X_i $(i = 1,2)$ be complex manifolds. For any $\mathcal{T}_i \in \mathrm{MTM}(X_i,\mathbb{R})_{\mathcal{A}}$, we have a natural isomorphism $\Upsilon^{\lambda_0}(\mathcal{T}_1 \boxtimes \mathcal{T}_2) \simeq \Upsilon^{\lambda_0}(\mathcal{T}_1) \boxtimes \Upsilon^{\lambda_0}(\mathcal{T}_2)$ in* $\mathrm{Hol}(X,\mathbb{R})$. □

13.4.4 Real Structure in the Integrable Case

13.4.4.1 Preliminary

Note that if a mixed twistor $\mathscr{D}$-module on X is integrable then it is also an object in $\mathrm{MTM}(X)_{\mathbb{R}\times\{0\}}$. We note that any non-zero complex number λ is generic with respect to $\mathbb{R} \times \{0\}$.

Let $\mathcal{T}$ be an integrable mixed twistor $\mathscr{D}$-module on X. Let $\mathcal{M}_i$ $(i = 1,2)$ be the underlying $\tilde{\mathcal{R}}_X$-modules. Let $\pi_0 : \mathcal{H} \longrightarrow \mathbb{C}_\lambda^*$ be a universal covering. The induced morphism $\mathcal{H} \times X \longrightarrow \mathbb{C}_\lambda^* \times X$ is also denoted by π_0. Let $p_{\mathcal{H}} : \mathcal{H} \times X \longrightarrow X$ be the projection. Fix any $\lambda_0 \neq 0$, and $\tilde{\lambda}_0 \in \mathcal{H}$ such that $\pi_0(\tilde{\lambda}_0) = \lambda_0$. We set

$\mathcal{M}_i^{\lambda_0} := \mathcal{O}_{\{\lambda_0\}\times X} \otimes_{\mathcal{O}_{\mathbb{C}_\lambda \times X}} \mathcal{M}_i$. Let $i_{\lambda_0} : \{\tilde{\lambda}_0\} \times X \longrightarrow \mathcal{H} \times X$ denote the inclusion. We have a natural morphism

$$\Omega^\bullet_{\mathcal{H}\times X} \otimes_{\mathcal{O}_{\mathcal{H}\times X}} \pi_0^{-1}(\mathcal{M}_{i|\mathbb{C}^*_\lambda \times X}) \longrightarrow i_{\lambda_0 *}\big(\Omega^\bullet_X \otimes_{\mathcal{O}_X} \mathcal{M}_i^{\lambda_0}\big) \tag{13.23}$$

Proposition 13.4.15 *We have a unique isomorphism in $D^b_c(\mathcal{H} \times X)$*

$$\Phi_i^{\lambda_0} : \Omega^\bullet_{\mathcal{H}\times X} \otimes_{\mathcal{O}_{\mathcal{H}\times X}} \pi_0^{-1}\mathcal{M}_{i|\mathbb{C}^*_\lambda \times X} \simeq p_{\mathcal{H}}^{-1}\big(\Omega^\bullet_X \otimes_{\mathcal{O}_X} \mathcal{M}_i^{\lambda_0}\big)$$

such that the composite of $\Phi_i^{\lambda_0}$ and the natural morphism $p_{\mathcal{H}}^{-1}\big(\Omega^\bullet_X \otimes_{\mathcal{O}_X} \mathcal{M}_i^{\lambda_0}\big) \longrightarrow i_{\lambda_0 *}\big(\Omega^\bullet_X \otimes_{\mathcal{O}_X} \mathcal{M}_i^{\lambda_0}\big)$ is equal to (13.23).

Proof We begin with a special case. Let H be a simply normal crossing hypersurface of X. Let $(\mathcal{V}_1, \mathcal{V}_2, C)$ be an integrable good admissible mixed twistor structure on (X, H). Note that $\mathcal{V}_{i|\mathbb{C}^* \times X}$ are good meromorphic flat bundles on $\mathbb{C}^* \times (X, H)$. Recall Lemma 5.3.5. By Theorem 11.3.3 of [36], the micro-supports of $\Omega^\bullet_{\mathcal{H}\times X} \otimes \pi_0^{-1}\mathcal{V}_i[\star H]$ is contained in $\mathcal{H} \times \mathcal{S}$ for a Lagrangian cone $\mathcal{S}$ in the cotangent bundle of X. By Proposition 5.4.5 of [36], there exists an object $G_{i\star} \in D^b_c(X)$ such that $\Omega^\bullet_{\mathcal{H}\times X} \otimes \pi_0^{-1}\mathcal{V}_i[\star H]$ is isomorphic to $p_{\mathcal{H}}^{-1} G_{i\star}$ in $D^b_c(\mathcal{H} \times X)$. By using the description of the de Rham complex of $\mathcal{V}_i[\star H]$ in Sect. 5.1.1 of [57], we can observe that

$$G_{i\star} \simeq i_{\lambda_0}^{-1}\big(\Omega^\bullet_{\mathcal{H}\times X} \otimes \pi_0^{-1}\mathcal{V}_i[\star H]\big) \simeq \Omega^\bullet_X \otimes \mathcal{V}_i^{\lambda_0}[\star H].$$

Let g be any holomorphic function such that $g^{-1}(0) = H$. We have unique isomorphisms

$$\Omega^\bullet_{\mathcal{H}\times X} \otimes \pi_0^{-1}\big(\Pi^{a,b}_{g\star}(\mathcal{V}_i)\big) \simeq p_{\mathcal{H}}^{-1}\big(\Omega^\bullet_X \otimes \Pi^{a,b}_{g\star}(\mathcal{V}_i)^{\lambda_0}\big)$$

in $D^b_c(\mathcal{H} \times X)$ with the desired property. By Proposition 2.7.8 of [36], the natural morphism $\Omega^\bullet_{\mathcal{H}\times X} \otimes \pi_0^{-1}\big(\Pi^{a,b}_{g!}(\mathcal{V}_i)\big) \longrightarrow \Omega^\bullet_{\mathcal{H}\times X} \otimes \pi_0^{-1}\big(\Pi^{a,b}_{g*}(\mathcal{V}_i)\big)$ is obtained as the pull back of the natural morphism $\Omega^\bullet_X \otimes \Pi^{a,b}_{g!}(\mathcal{V}_i)^{\lambda_0} \longrightarrow \Omega^\bullet_X \otimes \Pi^{a,b}_{g*}(\mathcal{V}_i)^{\lambda_0}$. Hence, we have a unique isomorphism

$$\Omega^\bullet_{\mathcal{H}\times X} \otimes \pi_0^{-1}\big(\Pi^{a,b}_{g!*}(\mathcal{V}_i)\big) \simeq p_{\mathcal{H}}^{-1}\big(\Omega^\bullet_X \otimes \Pi^{a,b}_{g!*}(\mathcal{V}_i)^{\lambda_0}\big)$$

with the desired property. In particular, we have a desired isomorphism for $\Xi^{(a)}_g(\mathcal{V}_i)$. Then, we obtain the desired isomorphisms for the $\tilde{\mathcal{R}}_X$-modules underlying integrable mixed twistor $\mathscr{D}$-modules, by using a standard Noetherian induction on the supports. □

Let $\mathcal{H}_{\mathbb{R}} := \pi_0^{-1}(\boldsymbol{S})$. Let $\pi_1 : \mathcal{H}_{\mathbb{R}} \longrightarrow \boldsymbol{S}$ be the restriction of π_0. For any C^∞-manifold Z, let $\mathcal{E}^\bullet_Z$ denote the C^∞-de Rham complex of Z. Let $p_{\mathcal{H}_{\mathbb{R}}} : \mathcal{H}_{\mathbb{R}} \times X \longrightarrow X$ be the projection. Fix any $\lambda_0 \in \boldsymbol{S}$. Let $\iota : \mathcal{H}_{\mathbb{R}} \times X \longrightarrow \mathcal{H} \times X$ denote the inclusion.

The inclusion $i_{1,\lambda_0} : \{\lambda_0\} \times X \longrightarrow \mathcal{H}_{\mathbb{R}} \times X$ induces

$$\mathcal{E}^{\bullet}_{\mathcal{H}_{\mathbb{R}}\times X} \otimes_{\iota^{-1}\mathcal{O}_{\mathcal{H}\times X}} \iota^{-1}\pi_0^{-1}\big(\mathcal{M}_{i|\mathbb{C}^*\times X}\big) \longrightarrow i_{1,\lambda_0 *}\big(\mathcal{E}^{\bullet}_X \otimes_{\mathcal{O}_X} \mathcal{M}_i^{\lambda_0}\big). \tag{13.24}$$

Proposition 13.4.16 *We have unique isomorphisms in* $D^b_c(\mathcal{H}_{\mathbb{R}} \times X)$

$$\Phi^{\lambda_0}_{1,i} : \mathcal{E}^{\bullet}_{\mathcal{H}_{\mathbb{R}}\times X} \otimes_{\iota^{-1}\mathcal{O}_{\mathcal{H}\times X}} \iota^{-1}\pi_0^{-1}\big(\mathcal{M}_{i|\mathbb{C}^*\times X}\big) \simeq p^{-1}_{\mathcal{H}_{\mathbb{R}}}\big(\mathcal{E}^{\bullet}_X \otimes_{\mathcal{O}_X} \mathcal{M}_i^{\lambda_0}\big)$$

such that the composite of $\Phi^{\lambda_0}_{1,i}$ *and the natural* $p^{-1}_{\mathcal{H}_{\mathbb{R}}}\big(\mathcal{E}^{\bullet}_X \otimes_{\mathcal{O}_X} \mathcal{M}_i^{\lambda_0}\big) \longrightarrow i_{1,\lambda_0 *}\big(\mathcal{E}^{\bullet}_X \otimes_{\mathcal{O}_X} \mathcal{M}_i^{\lambda_0}\big)$ *is equal to (13.24).*

Proof Let $q : \mathcal{H} \longrightarrow \mathcal{H}_{\mathbb{R}}$ be the morphism corresponding to the projection $\mathbb{C}^* \simeq \boldsymbol{S}\times\mathbb{R}_{>0} \longrightarrow \boldsymbol{S}$. By using the argument in the proof of Proposition 13.4.15, we obtain isomorphisms in $D^b_c(\mathcal{H} \times X)$

$$\mathcal{E}^{\bullet}_{\mathcal{H}\times X} \otimes_{\mathcal{O}_{\mathcal{H}\times X}} \pi_0^{-1}(\mathcal{M}_{i|\mathbb{C}^*_\lambda\times X}) \simeq q^{-1}\Big(\mathcal{E}^{\bullet}_{\mathcal{H}_{\mathbb{R}}\times X} \otimes_{\iota^{-1}\mathcal{O}_{\mathcal{H}\times X}} \iota^{-1}\pi_0^{-1}(\mathcal{M}_{i|\mathbb{C}^*_\lambda\times X})\Big)$$

such that its restriction to $\mathcal{H}_{\mathbb{R}} \times X$ is the natural one. Then, the claim of Proposition 13.4.16 is obtained as the specialization of Proposition 13.4.15. □

13.4.4.2 Real Structure of the Underlying Integrable $\mathcal{R}_X$-Modules

Let $(\mathcal{T}, \kappa) \in \mathrm{MTM}^{\mathrm{int}}(X, \mathbb{R})$. Let $\mathcal{M}_i$ $(i = 1, 2)$ be the underlying $\tilde{\mathcal{R}}_X$-modules. Take any $\lambda_0 \in \boldsymbol{S}$. We have the induced real structure of $\mathcal{M}_2^{\lambda_0}$. The isomorphism in Proposition 13.4.15 induces a real structure of the perverse sheaf $\mathrm{DR}_{\mathcal{H}\times X}\big(\pi_0^{-1}\mathcal{M}_{2|\mathbb{C}^*_\lambda\times X}\big)$.

Lemma 13.4.17 *The real structure is independent of the choices of* λ_0 *and* $\tilde{\lambda}_0$.

Proof By Proposition 13.4.16, the pairings $\mathrm{DR}(\mathcal{M}_1^{-\lambda}) \otimes \overline{\mathrm{DR}(\mathcal{M}_2^{\lambda})} \longrightarrow \omega^{top}$ are independent of $\lambda \in \boldsymbol{S}$. The isomorphisms $\mathrm{DR}(\mathcal{M}_2^{\lambda}) \simeq \boldsymbol{D}\mathrm{DR}(\mathcal{M}_1^{-\lambda})$ are also independent of λ. Then, the claim follows. □

Hence, we obtain a pre-$\mathbb{R}$-Betti structure of $\mathcal{M}_{i|\mathbb{C}^*\times X}$.

Lemma 13.4.18 *The pre-$\mathbb{R}$-Betti structure of* $\mathcal{M}_{i|\mathbb{C}^*\times X}$ *is a $\mathbb{R}$-Betti structure.*

Proof We have only to check the claim locally around any point P of X. We shall shrink X without mention. We take an admissible cell $(Z, U, \varphi, \mathcal{V})$ of $\mathcal{T}$ at P with a cell function g. We have $\varphi_\dagger(\mathcal{V})[\star g] \simeq \mathcal{T}[\star g]$. By the standard argument with a Noetherian induction on the support, we have only to prove the claim for $\varphi_\dagger(\mathcal{V})[\star g]$. Let $\mathcal{V} = (\mathcal{V}_1, \mathcal{V}_2, C_{\mathcal{V}})$. We set $D_Z := Z\setminus U$ and $g_Z := \varphi^*(g)$. We can observe that $\mathcal{V}_2$ is a good meromorphic flat bundle on $\mathbb{C}^*_\lambda \times (Z, D_Z)$ with a good $\mathbb{R}$-structure. Hence, $\mathcal{V}_2[\star g_Z]$ has an $\mathbb{R}$-Betti structure. The real structure of $\pi_0^{-1}\big(\Omega^{\bullet}_{\mathbb{C}^*\times Z} \otimes \mathcal{V}_2[\star g_Z]\big)$ is equal to the one induced by the isomorphism with $p^{-1}_{\mathcal{H}}\big(\Omega^{\bullet}_Z \otimes \mathcal{V}_2^{\lambda_0}[\star g_Z]\big)$. Then, the claim for $\varphi_\dagger(\mathcal{V})[\star g]$ easily follows. □

Thus, we obtain a functor $\tilde{\Upsilon} : \mathrm{MTM}^{\mathrm{int}}(X,\mathbb{R}) \longrightarrow \mathrm{Hol}(\mathbb{C}^* \times X, \mathbb{R})$.

Proposition 13.4.19 *The functor $\tilde{\Upsilon}$ is compatible with the following functors.*

- *The push-forward by any projective morphisms.*
- *The duality $\boldsymbol{D}$.*

Proof Let us consider the first claim. Let $F : X \longrightarrow Y$ be any projective morphism. Let $(\mathcal{T},\kappa) \in \mathrm{MTM}^{\mathrm{int}}(X,\mathbb{R})$, and let $\mathcal{M}_i$ be the underlying $\tilde{\mathcal{R}}_X$-modules. Set $S(\lambda_0) := \{\lambda \in \mathbb{C}^* \,\big|\, |\arg(\lambda) - \arg(\lambda_0)| < \epsilon\}$ for a small ϵ. We have only to compare the $\mathbb{R}$-structures on $F^i_\dagger \,\mathrm{DR}_{\mathbb{C}^*\times X}(\mathcal{M}_2)_{|S(\lambda_0)\times X} \simeq \mathrm{DR}_{\mathbb{C}^*\times Y} F^i_\dagger(\mathcal{M}_2)_{|S(\lambda_0)\times Y}$. The specializations at $\lambda_0 \in \boldsymbol{S}$ are the same by Proposition 13.4.10. Let $p_{S(\lambda_0)}$ denote the projections $S(\lambda_0) \times X \longrightarrow X$ and $S(\lambda_0) \times Y \longrightarrow Y$. Because the $\mathbb{R}$-structures on $S(\lambda_0) \times Y$ are obtained by the isomorphisms

$$\mathrm{DR}_{\mathbb{C}^*\times X}(\mathcal{M}_2)_{|S(\lambda_0)\times X} \simeq p^{-1}_{S(\lambda_0)} \,\mathrm{DR}_X(\mathcal{M}_2^{\lambda_0})[1],$$

$$\mathrm{DR}_{\mathbb{C}^*\times Y} F^i_\dagger(\mathcal{M}_2)_{|S(\lambda_0)\times Y} \simeq p^{-1}_{S(\lambda_0)} \,\mathrm{DR}_Y F^i_\dagger(\mathcal{M}_2^{\lambda_0})[1],$$

we obtain the desired coincidence. The second claim can be shown similarly.

We obtain the following from the characterization of $[\star H]$ as in Theorem 8.1.4 of [57].

Proposition 13.4.20 *For any hypersurface H, the localizations $[\star H]$ ($\star = *, !$) are compatible with $\tilde{\Upsilon}$.* □

Let $\mathcal{V} = (\mathcal{V}_1, \mathcal{V}_2, C) \in \mathrm{MTS}^{\mathrm{int\,adm}}(X,H)$ with an integrable real structure $\kappa_{\mathcal{V}}$ as a smooth integrable $\mathcal{R}_{X(*H)}$-triple. Then, $\mathcal{V}_{2|\mathbb{C}^*\times X}$ has a good $\mathbb{R}$-structure.

Proposition 13.4.21 *Let $\mathcal{T} \in \mathrm{MTM}^{\mathrm{int}}(X,\mathbb{R})$. Let $\mathcal{V} \in \mathrm{MTS}^{\mathrm{int\,adm}}(X,H)$ with an integrable real structure. Then, we have a natural isomorphism $\tilde{\Upsilon}\big((\mathcal{T}\otimes\mathcal{V})[\star H]\big) \simeq \big(\tilde{\Upsilon}(\mathcal{T}) \otimes \mathcal{V}_2\big)[\star H]$ in* $\mathrm{Hol}(\mathbb{C}^* \times X, \mathbb{R})$. □

Corollary 13.4.22 *Let g be any holomorphic function on X. We naturally have $\tilde{\Upsilon} \circ \Xi_g^{(a)} \simeq \Xi_g^{(a)} \circ \tilde{\Upsilon}$, $\tilde{\Upsilon} \circ \psi_g^{(a)} \simeq \psi_g^{(a)} \circ \tilde{\Upsilon}$ and $\tilde{\Upsilon} \circ \phi_g^{(a)} \simeq \phi_g^{(a)} \circ \tilde{\Upsilon}$.* □

We have the following for external products.

Proposition 13.4.23 *Let X_i ($i = 1,2$) be complex manifolds. For any $\mathcal{T}_i \in \mathrm{MTM}^{\mathrm{int}}(X_i,\mathbb{R})$, we have a natural isomorphism $\tilde{\Upsilon}(\mathcal{T}_1 \boxtimes \mathcal{T}_2) \simeq \tilde{\Upsilon}(\mathcal{T}_1) \boxtimes \tilde{\Upsilon}(\mathcal{T}_2)$ in* $\mathrm{Hol}(\mathbb{C}^* \times X, \mathbb{R})$. □

13.4.4.3 Good Real Structure of Integrable Mixed Twistor $\mathscr{D}$-Modules

Let $(\mathcal{T},\kappa) \in \mathrm{MTM}^{\mathrm{int}}(X,\mathbb{R})$. Let $\mathcal{M}_i$ ($i = 1,2$) be the underlying integrable $\mathcal{R}_X$-modules. We obtain $\mathscr{D}_{\mathbb{C}\times X}$-modules $\mathcal{M}_i(*\mathcal{X}^0)$, where $\mathcal{X}^0 := \{0\} \times X$. We say that the real structure κ of $\mathcal{T}$ is good if $\mathcal{M}_2(*\mathcal{X}^0)$ has a $\mathbb{R}$-Betti structure whose restriction to $\mathcal{M}_{2|\mathbb{C}^*\times X}$ is equal to the one induced by the real structure of $\mathcal{T}$. Here, a $\mathbb{R}$-Betti structure of $\mathcal{M}_2(*\mathcal{X}^0)$ is a perverse sheaf $P_{\mathbb{R}}$ with an isomorphism $P_{\mathbb{R}} \otimes \mathbb{C} \simeq \mathrm{DR}_{\mathcal{X}}(\mathcal{M}_2(*\mathcal{X}^0))$ satisfying a compatibility condition with the Stokes structure. (See [57] for more details.) Let $\mathrm{MTM}^{\mathrm{int}}_{\mathrm{good}}(X,\mathbb{R}) \subset \mathrm{MTM}^{\mathrm{int}}(X,\mathbb{R})$ denote the full subcategory of integrable mixed twistor $\mathscr{D}$-modules with good real structure. By using the results in [57], we obtain the following.

Proposition 13.4.24 *Let* $(\mathcal{T},\kappa) \in \mathrm{MTM}^{\mathrm{int}}_{\mathrm{good}}(X,\mathbb{R})$.

- *The push-forward* $F^i_\dagger(\mathcal{T},\kappa)$ *is an object in* $\mathrm{MTM}^{\mathrm{int}}_{\mathrm{good}}(Y,\mathbb{R})$ *for any projective morphism* $F : X \longrightarrow Y$.
- $\boldsymbol{D}(\mathcal{T},\kappa) \in \mathrm{MTM}^{\mathrm{int}}_{\mathrm{good}}(X,\mathbb{R})$.
- *For any hypersurface* H *of* X, *we have* $(\mathcal{T},\kappa)[\star H] \in \mathrm{MTM}^{\mathrm{int}}_{\mathrm{good}}(X,\mathbb{R})$.
- *For any holomorphic function* g *on* X, *we have* $\Xi^{(a)}_g(\mathcal{T},\kappa)$, $\psi^{(a)}_g(\mathcal{T},\kappa)$ *and* $\phi^{(a)}_g(\mathcal{T},\kappa)$ *in* $\mathrm{MTM}^{\mathrm{int}}_{\mathrm{good}}(X,\mathbb{R})$. □

13.4.4.4 Good K-Structure of Integrable Mixed Twistor $\mathscr{D}$-Modules

Let K be a subfield of $\mathbb{R}$. Let $(\mathcal{T},\kappa) \in \mathrm{MTM}^{\mathrm{int}}(X,\mathbb{R})$. A good K-structure of $(\mathcal{T},\kappa)$ is a K-Betti structure P_K of $\mathcal{M}_2(*\mathcal{X}^0)$ whose restriction to $\mathbb{C}^* \times X$ induces the $\mathbb{R}$-Betti structure of $\mathcal{M}_{2|\mathbb{C}^*\times X}$ by the extension $\mathbb{R}/K$. Here, a K-Betti structure of $\mathcal{M}_2(*\mathcal{X}^0)$ is a K-perverse sheaf P_K with an isomorphism $P_K \otimes \mathbb{C} \simeq \mathrm{DR}(\mathcal{M}_2(*\mathcal{X}^0))$ satisfying a compatibility condition with the Stokes structure. (See [57] for more details.) If $(\mathcal{T},\kappa)$ has a good K-structure, we have the naturally induced object $(\mathcal{T},\kappa) \in \mathrm{MTM}^{\mathrm{int}}_{\mathrm{good}}(X,\mathbb{R})$.

Let $\mathrm{MTM}^{\mathrm{int}}_{\mathrm{good}}(X,K)$ denote the category of integrable mixed twistor $\mathscr{D}$-modules with good K-structure. By using the results in [57], we obtain the following.

Proposition 13.4.25 *Let* $(\mathcal{T},\kappa,P_K) \in \mathrm{MTM}^{\mathrm{int}}_{\mathrm{good}}(X,K)$.

- *We have the push-forward* $F^i_\dagger(\mathcal{T},\kappa,P_K)$ *in* $\mathrm{MTM}^{\mathrm{int}}_{\mathrm{good}}(Y,K)$ *for any projective morphism* $F : X \longrightarrow Y$.
- *We have* $\boldsymbol{D}(\mathcal{T},\kappa,P_K) \in \mathrm{MTM}^{\mathrm{int}}_{\mathrm{good}}(X,K)$.
- *For any hypersurface* H *of* X, *we have* $(\mathcal{T},\kappa,P_K)[\star H] \in \mathrm{MTM}^{\mathrm{int}}_{\mathrm{good}}(X,K)$.
- *For any holomorphic function* g *on* X, $\Xi^{(a)}_g(\mathcal{T},\kappa,P_K)$, $\psi^{(a)}_g(\mathcal{T},\kappa,P_K)$ *and* $\phi^{(a)}_g(\mathcal{T},\kappa,P_K)$ *are objects in* $\mathrm{MTM}^{\mathrm{int}}_{\mathrm{good}}(X,K)$. □

We also have the following.

Proposition 13.4.26 *Let* X_i *be complex manifolds. The external product induces* $\mathrm{MTM}^{\mathrm{int}}_{\mathrm{good}}(X_1,K) \times \mathrm{MTM}^{\mathrm{int}}_{\mathrm{good}}(X_2,K) \longrightarrow \mathrm{MTM}^{\mathrm{int}}_{\mathrm{good}}(X_1 \times X_2,K)$. □

13.4.4.5 Full Faithfulness

We introduce a category $\text{Mod}(\tilde{\mathcal{R}}_X, \mathbb{R})$. Objects are tuples $(\mathcal{M}, P)$ of a holonomic strict $\tilde{\mathcal{R}}_X$-module $\mathcal{M}$ and a $\mathbb{R}$-Betti structure P of $\mathcal{M}_{|\mathbb{C}^*_\lambda \times X}$. A morphism $(\mathcal{M}_1, P_1) \longrightarrow (\mathcal{M}_2, P_2)$ is a morphism of $\tilde{\mathcal{R}}_X$-modules $\varphi : \mathcal{M}_1 \longrightarrow \mathcal{M}_2$ such that $\varphi_{|\mathbb{C}^* \times X}$ is compatible with the $\mathbb{R}$-Betti structures. Let $\text{Mod}(\tilde{\mathcal{R}}_X, \mathbb{R})^{\text{fil}}$ be the category of filtered objects in $\text{Mod}(\tilde{\mathcal{R}}_X, \mathbb{R})$. We have the induced functor $\hat{\Upsilon} : \text{MTM}^{\text{int}}(X, \mathbb{R}) \longrightarrow \text{Mod}(\tilde{\mathcal{R}}_X, \mathbb{R})^{\text{fil}}$, i.e., for any $(\mathcal{T}, W, \kappa)$, let $\hat{\Upsilon}(\mathcal{T}, W, \kappa) = (\mathcal{M}_2, W, \tilde{\Upsilon}(\mathcal{T}))$.

Proposition 13.4.27 *The functor $\hat{\Upsilon}$ is fully faithful.*

Proof The functor $\hat{\Upsilon}$ is clearly faithful. Let us observe that it is full. Let

$$(\mathcal{T}^{(i)}, W, \kappa^{(i)})(i = 1, 2)$$

be objects in $\text{MTM}^{\text{int}}(X, \mathbb{R})$ with a morphism $\hat{\Upsilon}(\mathcal{T}^{(1)}, W, \kappa^{(1)}) \longrightarrow \hat{\Upsilon}(\mathcal{T}^{(2)}, W, \kappa^{(2)})$. We may assume that the underlying $\mathcal{R}$-triples of $\mathcal{T}^{(i)}$ are of the form $(j^*\boldsymbol{D}\mathcal{M}^{(i)}, \mathcal{M}^{(i)}, C^{(i)})$, and that $\kappa^{(i)}$ are given by (id, id). Let $\hat{\Upsilon}(\mathcal{T}^{(i)}, W, \kappa) = (\mathcal{M}^{(i)}, P^{(i)})$. The morphism $f_2 : \mathcal{M}^{(1)} \longrightarrow \mathcal{M}^{(2)}$ induces $f_1 : j^*\boldsymbol{D}\mathcal{M}^{(2)} \longrightarrow j^*\boldsymbol{D}\mathcal{M}^{(1)}$. It is enough to prove that (f_1, f_2) gives a morphism of $\mathcal{R}$-triples. We have only to check the claim locally around any point P of X. We use the Noetherian induction for the supports of $\mathcal{T}^{(i)}$. Let $(Z, U, \varphi, \mathcal{V}^{(i)}, \kappa_0^{(i)})$ be integrable cells for $\mathcal{T}^{(i)}$ with a cell function g. Let $\mathcal{V}^{(i)} = (\mathcal{V}_1^{(i)}, \mathcal{V}_2^{(i)}, C_0^{(i)})$. We have the induced morphisms $h_1 : \mathcal{V}_1^{(2)} \longrightarrow \mathcal{V}_1^{(1)}$ and $h_2 : \mathcal{V}_2^{(1)} \longrightarrow \mathcal{V}_2^{(2)}$. We have the isomorphism $\mathcal{V}_1^{(i)} = j^*\boldsymbol{D}\mathcal{V}_2^{(i)}(*D)$, where $D := Z \setminus U$. Then, the sesqui-linear pairings $C_0^{(i)}$ are induced by the real structures. Because h_i preserve the real structure, we obtain that (h_1, h_2) gives a morphism of smooth $\mathcal{R}_{Z(*D)}$-triples $h : \mathcal{V}^{(1)} \longrightarrow \mathcal{V}^{(2)}$. Then, it is standard to obtain that (f_1, f_2) gives a morphism of $\mathcal{R}_X$-triples $f : \mathcal{T}_1 \longrightarrow \mathcal{T}_2$. □

We introduce a category $\text{Mod}^{\text{good}}(\tilde{\mathcal{R}}_X, K)$. Objects are pairs $(\mathcal{M}, P)$ of a strict holonomic $\tilde{\mathcal{R}}_X$-module and a K-Betti structure of $\mathcal{M}(*\mathcal{X}^0)$. A morphism $(\mathcal{M}_1, P) \longrightarrow (\mathcal{M}_2, P)$ is defined to be a morphism of $\tilde{\mathcal{R}}_X$-modules $\varphi : \mathcal{M}_1 \longrightarrow \mathcal{M}_2$ such that the induced morphism $\varphi : \mathcal{M}_1(*\mathcal{X}^0) \longrightarrow \mathcal{M}_2(*\mathcal{X}^0)$ is compatible with the K-Betti structure. Let $\text{Mod}^{\text{good}}(\tilde{\mathcal{R}}_X, K)^{\text{fil}}$ be the category of filtered objects in $\text{Mod}^{\text{good}}(\tilde{\mathcal{R}}_X, K)$. We have the induced functor $\hat{\Upsilon}_K^{\text{good}} : \text{MTM}_{\text{good}}^{\text{int}}(X, K) \longrightarrow \text{Mod}^{\text{good}}(\tilde{\mathcal{R}}_X, K)^{\text{fil}}$. The following proposition is similar to Proposition 13.4.27.

Proposition 13.4.28 *The functor $\hat{\Upsilon}_K^{\text{good}}$ is fully faithful.* □

13.5 Relation with Mixed Hodge Modules

Let X be a complex manifold. Let K be a subfield of $\mathbb{R}$. Let P be a K-perverse sheaf on X. We have a regular holonomic $\mathscr{D}_X$-module M with an isomorphism κ : $\mathrm{DR}(M) \simeq P \otimes_K \mathbb{C}$. A Hodge filtration on P is a filtration F of M by coherent $\mathcal{O}_X$-submodules indexed by integers such that $F_j(\mathscr{D}_X)\, F_i(M) \subset F_{i+j}(M)$, where $F_*(\mathscr{D}_X)$ is the filtration by the order of differential operators. The category $MF_h(\mathscr{D}_X, K)$ of K-perverse sheaves with a Hodge filtration is naturally defined. Objects are tuples (M, F, P, κ) as above. We often omit to denote κ. Morphisms are naturally defined. Recall that filtered objects in $\mathrm{MF}_h(\mathscr{D}_X, K)$ are ingredients of mixed Hodge modules.

Let (M, F, P) with a weight filtration W be an object in $\mathrm{MF}_h(\mathscr{D}_X, K)^{\mathrm{fil}}$ which gives a mixed Hodge module [73]. Then, we have the naturally associated filtered $\tilde{\mathcal{R}}_X$-triple. Indeed, let $\mathcal{M}$ be the $\tilde{\mathcal{R}}_X$-module obtained as the analytification of the Rees module of (M, F). By construction, $\boldsymbol{D}\mathcal{M}$ is the analytification of the Rees module of the dual of (M, F) in the category of filtered $\mathscr{D}_X$-modules [70]. Hence, we have $\mathcal{H}^j\boldsymbol{D}\mathcal{M} = 0$ unless $j = 0$. In the following, we shall not distinguish $\boldsymbol{D}\mathcal{M}$ and $\mathcal{H}^0(\boldsymbol{D}\mathcal{M})$. We have the naturally defined sesqui-linear pairing $C : j^*\boldsymbol{D}\mathcal{M}_{|\boldsymbol{S}\times X} \times \sigma^*\mathcal{M}_{|\boldsymbol{S}\times X} \longrightarrow \mathfrak{Db}_{\boldsymbol{S}\times X/X}$ induced by the real structure of $\mathrm{DR}(M)$. In this way, we obtain an $\mathcal{R}$-triple $\mathcal{T} = (j^*\boldsymbol{D}\mathcal{M}, \mathcal{M}, C)$ with a naturally induced filtration W. We set $\Phi_X(M, F, P, W) := (\mathcal{T}, W)$.

Lemma 13.5.1 *For any projective morphism $f : X \longrightarrow Y$ and any mixed Hodge module (M, F, P, W) on X, we have a natural isomorphism*

$$\Phi_Y \mathcal{H}^i f_*(M, F, P, W) \simeq f_\dagger^i \Phi_X(M, F, P, W).$$

Proof It follows from the construction of the functors, and the compatibility of the push-forward and the duality. □

Lemma 13.5.2 *Let $a : (M_1, F, P, W) \longrightarrow (M_2, F, P, W)$ be a morphism of mixed Hodge modules on X. Then, we have $\Phi_X(\mathrm{Ker}(a)) = \mathrm{Ker}\,\Phi_X(a)$, $\Phi_X(\mathrm{Im}(a)) = \mathrm{Im}\,\Phi_X(a)$ and $\Phi_X(\mathrm{Cok}(a)) = \mathrm{Cok}\,\Phi_X(a)$.*

Proof It follows from that $a : (M_1, F, W) \longrightarrow (M_2, F, W)$ is bi-strict. □

Lemma 13.5.3 *Let (M, F, P, W) be a mixed Hodge module on X.*

- *If (M, F, P, W) is pure, $\Phi_X(M, F, P, W)$ is strictly S-decomposable.*
- *In general, $\Phi_X(M, F, P, W)$ is admissibly specializable.*
- *Let g be any holomorphic function on X. Let $j_g : X \setminus g^{-1}(0) \longrightarrow X$ denote the inclusion. For $\star = *, !$, if $j_{g\star} j_g^{-1}(M, F, P, W) \simeq (M, F, P, W)$, then we have $\Phi_X(M, F, P, W)[\star g] \simeq \Phi_X(M, F, P, W)$.*

Proof Let $R^F(M, F)$ denote the Rees module associated to (M, F). Let U be any open subset of X with a holomorphic function g. Let $i_g : U \longrightarrow U \times \mathbb{C}_t$ denote the graph of g. The V-filtration of $i_{g\dagger}M$ naturally induces a V-filtration of $R^F i_{g\dagger}(M, F)$. Because $i_{g\dagger}(M, F)$ is quasi-unipotent and regular along $\{t = 0\}$ in the sense of

Sect. 3.2 of [69], we obtain that $t : V_aR^F i_{g\dagger}(M,F) \longrightarrow V_{a-1}R^F i_{g\dagger}(M,F)$ are isomorphisms for any $a < 0$, and that $\eth_t : \mathrm{Gr}^V_a R^F i_{g\dagger}(M,F[1]) \longrightarrow \mathrm{Gr}^V_{a+1} R^F i_{g\dagger}(M,F)$ are isomorphisms for any $a > -1$. By the construction, $\mathrm{Gr}^V_a R^F i_{g\dagger}(M,F)$ are flat over $\mathbb{C}[\lambda]$. Hence, we obtain that $\mathcal{M}$ is strictly specializable along g. By the construction, $\boldsymbol{D}\mathcal{M}$ is the analytification of the Rees module associated to the dual $(\boldsymbol{D}M,F)$ of the filtered module (M,F). Hence, $\boldsymbol{D}\mathcal{M}$ is also strictly specializable along g. Moreover, if (M,F,P,W) is pure, we have $\mathrm{Gr}^V_0(R^F(M)) = \mathrm{Im}\,\mathrm{can} \oplus \mathrm{Ker}\,\mathrm{var}$, where $\mathrm{can} : \mathrm{Gr}^V_{-1} R^F i_{g\dagger}(M,F[1]) \longrightarrow \mathrm{Gr}^V_0 R^F i_{g\dagger}(M,F)$ and $\mathrm{var} : \mathrm{Gr}^V_0 R^F i_{g\dagger}(M,F) \longrightarrow \mathrm{Gr}^V_{-1} R^F i_{g\dagger}(M,F)$ are induced by the action of $\eth_t$ and t, respectively. Hence, $\mathcal{M}$ is strictly S-decomposable along g. Similarly $\boldsymbol{D}\mathcal{M}$ is also strictly S-decomposable along g. Because M is regular, it is easy to see that $\tilde{\psi}_{g,\mathfrak{a},u}(\mathcal{M}) = \tilde{\psi}_{g,\mathfrak{a},u}(\boldsymbol{D}\mathcal{M}) = 0$ if $\mathfrak{a} \neq 0$.

For any mixed Hodge module (M,F,P,W), the filtrations F, V and W on $i_{g\dagger}M$ are compatible. It means that

$$V_a W_j R^F i_{g\dagger}(M,F) \longrightarrow V_a \mathrm{Gr}^W_j R^F i_{g\dagger}(M,F)$$

are surjective for any a and j. It implies that we obtain the filtered strictly specializability of $\mathcal{M}$ along g. We set $\big(\psi_g(M),F\big) := \bigoplus_{-1\leq a<0} \mathrm{Gr}^V_a(M,F[1])$ and $\big(\phi_{g,1}(M),F\big) := \mathrm{Gr}^V_0(M,F)$. We also set $L_j\psi_g(M) := \psi(W_{j+1}M)$ and $L_j\phi_{g,1}(M) := \phi_{g,1}(W_jM)$. Let N denote the nilpotent part of the action of $-\eth_t t$. By the condition for mixed Hodge modules, there exists a relative monodromy filtration W on $(\psi_g(M),L)$ and $(\phi_{g,1}(M),L)$, and $(\psi_g(M,F),W)$ and $(\phi_{g,1}(M,F),W)$ are mixed Hodge modules. It implies that the induced actions of $-\lambda\eth_t t$ on $\bigoplus_{-1\leq a<0} \mathrm{Gr}^V_a R^F i_{g\dagger}(M,F)$ and $\mathrm{Gr}^V_0 R^F i_{g\dagger}(M,F)$ with L have relative monodromy filtrations. Together with the first claim, we obtain that $(\mathcal{T},W)$ is admissibly specializable along g.

The third claim follows from the characterization of $\Phi_X(M,F,P,W)[\star g]$. □

Proposition 13.5.4 *$\Phi_X(M,F,P,W)$ is a mixed twistor $\mathscr{D}_X$-module for any mixed Hodge module (M,F,P,W) on X.*

Proof Let us consider the case (M,F,P,W) is a polarizable pure Hodge module of weight w. We may assume that it is irreducible. There exists a polarizable variation of pure Hodge structure $(\mathcal{H},F)$ of weight $w-n$ on a Zariski open subset $U \subset \mathrm{Supp}(M)$, where $n = \dim U$, and that (M,F,P,W) is obtained as the minimal extension of $(\mathcal{H},F)$. Let $\mathcal{M}_0$ be the $\mathcal{R}_U$-module obtained as the analytification of $R^F(\mathcal{H},F)$. We have the polarizable variation of pure twistor structure $(\boldsymbol{D}\mathcal{M}_0,\mathcal{M}_0,C)$ of weight w on U, which is tame on $(\mathrm{Supp}(M),U)$. It is uniquely extended to a polarizable pure twistor $\mathscr{D}$-module $\mathcal{T}_1$ of weight w, whose strict support is $\mathrm{Supp}(M)$. Let $Z := \mathrm{Supp}(M) \setminus U$. We have a natural isomorphism $\Upsilon : \mathcal{T}_{1|X\setminus Z} \simeq \Phi_X(M,F,P,W)_{|X\setminus Z}$. We shall prove that the isomorphism Υ is uniquely extended to $\mathcal{T}_1 \simeq \Phi_X(M,F,P,W)$. By the uniqueness, we have only to check the claim locally around any point of $\mathrm{Supp}(M)$. We may assume that there exists a holomorphic function g such that $g^{-1}(0) \cap \mathrm{Supp}(M) = Z$. By the strict

S-decomposability of $\mathcal{T}_1$ and $\Phi_X(M,F,P,W)$, we have only to prove that $\mathcal{T}_1(*g) = \Phi_X(M,F,P,W)(*g)$. We may assume that there exists a complex manifold Y with a projective birational morphism $\varphi : Y \longrightarrow \mathrm{Supp}(M)$ such that (i) $H_Y := \varphi^{-1}(Z)$ is normal crossing hypersurface, (ii) $Y \setminus H_Y = U$. Let $\tilde{\mathcal{H}}$ denote the $\mathscr{D}_Y(*H_Y)$-module obtained as the regular extension of $\mathcal{H}$. It is equipped with the filtration by locally free $\mathcal{O}_Y(*H_Y)$-modules, induced by F, denoted by $\tilde{F}$. As the analytification of the Rees module $R^F(\tilde{\mathcal{H}},\tilde{F})$, we have an $\mathcal{R}_{Y(*H_Y)}$-module $\tilde{\mathcal{M}}$. Then, we can observe that the underlying $\mathcal{R}_X(*g)$-modules of $\mathcal{T}_1(*g)$ and $\Phi_X(M,F,P,W)(*g)$ are naturally isomorphic, by the construction of the prolongations of variations of polarizable pure Hodge (twistor) structure to polarizable pure Hodge (twistor) modules. Thus, we are done in the pure case.

We consider the mixed case. We use an induction on $\dim \mathrm{Supp} M$. The case $\dim \mathrm{Supp} M = 0$ is clear. Let us consider the case $\dim \mathrm{Supp} M > 0$. We have already known $\Phi_X(M,F,P,W) \in \mathrm{MTW}(X)$. To check $\Phi_X(M,F,P,W) \in \mathrm{MTM}(X)$, we have only to prove it locally around any point of $P \in \mathrm{Supp}(M)$. We may assume that there exists a holomorphic function g on X such that (i) $\dim \mathrm{Supp}(M) \cap g^{-1}(0) < \dim \mathrm{Supp}(M)$, (ii) $\mathrm{Supp}(M) \setminus g^{-1}(0)$ is a complex submanifold of $X \setminus g^{-1}(0)$, (iii) $(M,F,P,W)_{|X\setminus g^{-1}(0)}$ comes from an admissible variation of mixed Hodge structure $(\mathcal{H},F,W)$ on $\mathrm{Supp}(M) \setminus g^{-1}(0)$. We may assume to have a complex manifold Y with a projective birational morphism $\varphi : Y \longrightarrow \mathrm{Supp}(M)$ such that (i) $H_Y := \varphi^{-1}(g^{-1}(0))$ is normal crossing, (ii) $Y \setminus H_Y \simeq \mathrm{Supp}(M) \setminus g^{-1}(0)$. We have an admissible variation of mixed Hodge structure on (Y,H_Y) given by $\varphi^*(\mathcal{H},F,W)$. It induces an admissible variation of mixed twistor structure $(\mathcal{V},W)$ on (Y,H_Y). As in the pure case, we obtain that $\Phi_X(M,F,P,W)(*g) \simeq \varphi_\dagger(\mathcal{V},W)$. Hence, by using the admissible specializability along g and the characterization as in Lemma 7.1.43, we have $j_{g\star}j_g^{-1}\Phi_X(M,F,P,W) \simeq \varphi_\star(\mathcal{V},W)$. By using Beilinson's functor for mixed Hodge modules as in [76], we have a description of (M,F,P,W) as the cohomology of the complex in the category of mixed Hodge modules:

$$\psi_g(M,F,P,W) \to \Xi_g(M,F,P,W) \oplus \phi_g(M,F,P,W) \to \psi_g(M,F,P,W)(-1)$$

By the construction of $\Xi_g(M,F,P,W)$, there exist admissible variations of mixed Hodge structure $(\mathcal{V}_i,W)$ $(i = 1,2)$ on (Y,H_Y) such that $\Xi_g(M,F,P,W)$ is isomorphic to the kernel of a morphism $\varphi_!(\mathcal{V}_1,W) \longrightarrow \varphi_*(\mathcal{V}_2,W)$. Hence, $\Phi_X\Xi_g(M,F,P,W)$ is a mixed twistor $\mathscr{D}$-module on X. Similarly, we obtain that $\Phi_X\psi_g(M,F,P,W)$ is a mixed twistor $\mathscr{D}$-module. By the hypothesis of the induction, $\Phi_X\phi_g(M,F,P,W)$ is a mixed twistor $\mathscr{D}$-module on X. Then, we obtain that $\Phi_X(M,F,P,W) \in \mathrm{MTM}(X)$. □

For a mixed Hodge module (M,F,P,W), the mixed twistor $\mathscr{D}$-module $(\mathcal{T},W) = \Phi_X(M,F,P,W)$ has a natural real structure $\kappa = (\mathrm{id},\mathrm{id})$, Moreover, $(\mathcal{T},W)$ is clearly integrable and equipped with a good K-structure. Thus, we obtain a functor Φ_X from the category of mixed Hodge modules $\mathrm{MHM}(X,K)$ with K-structure to the category $\mathrm{MTM}^{\mathrm{int}}_{\mathrm{good}}(X,K)$.

13.5.1 Some Compatibilities

Let X be a complex manifold. Let K be a subfield of $\mathbb{R}$.

Proposition 13.5.5 *The functor* $\Phi_X : \mathrm{MHM}(X, K) \longrightarrow \mathrm{MTM}^{\mathrm{int}}_{\mathrm{good}}(X, K)$ *is exact and fully faithful.*

Proof By Lemma 13.5.2, Φ_X is exact. The functor is clearly faithful. Let us prove that it is full. Let $(M_i, F, P_i, W) \in \mathrm{MHM}(X, K)$ $(i = 1, 2)$. Suppose that we are given a morphism $f : \Phi_X(M_1, F, P_1, W) \longrightarrow \Phi_X(M_2, F, P_2, W)$ in $\mathrm{MTM}^{\mathrm{int}}_{\mathrm{good}}(X, K)$. By taking the specialization to $\lambda = 1$, we have the morphism $f_0 : (M_1, P_1, W) \longrightarrow (M_2, P_2, W)$. It is enough to check that the morphism is compatible with the Hodge filtrations.

Let $\mathcal{M}_i$ be the $\tilde{\mathcal{R}}_X$-modules obtained as the analytification of the Rees module of (M_i, F). We have the given morphism of $\tilde{\mathcal{R}}_X$-modules $f : \mathcal{M}_1 \longrightarrow \mathcal{M}_2$.

We set $\mathfrak{X} := \mathbb{P}^1 \times X$ and $\mathfrak{X}^\kappa := \{\kappa\} \times X$ $(\kappa \in \mathbb{P}^1)$. Let $\tilde{\mathfrak{R}}_X$ be the sheaf of algebras on $\mathfrak{X}$ determined by the conditions $\tilde{\mathfrak{R}}_{X|\mathfrak{X}\setminus\mathfrak{X}^\infty} = \tilde{\mathcal{R}}_X$ and $\tilde{\mathfrak{R}}_{X|\mathfrak{X}\setminus\mathfrak{X}^0} = \mathscr{D}_{\mathfrak{X}\setminus\mathfrak{X}^0}(*\mathfrak{X}^\infty)$. Let $p : \mathfrak{X} \longrightarrow X$ be the projection. We have the $\tilde{\mathfrak{R}}_X$-module $\mathfrak{M}_i$ determined by the condition $\mathfrak{M}_{i|\mathfrak{X}\setminus\mathfrak{X}^\infty} = \mathcal{M}_i$ and $\mathfrak{M}_{i|\mathfrak{X}\setminus\mathfrak{X}^0} = p^*(M_i)(*\mathfrak{X}^\infty)_{|\mathfrak{X}\setminus\mathfrak{X}^0}$. Note that $p_*\mathfrak{M}_i$ is naturally isomorphic to the Rees module $R^F M_i$. Note that $R^F M_i$ is equipped with the natural action of $\lambda\partial_\lambda$ which induces actions of $\lambda\partial_\lambda$ on $\mathfrak{M}_i$. The isomorphisms $p_*\mathfrak{M}_i \simeq R^F M_i$ are compatible with the actions of $\lambda\partial_\lambda$.

We have the induced morphism of $\tilde{\mathfrak{R}}_X$-modules $\overline{f} : \mathfrak{M}_1 \longrightarrow \mathfrak{M}_2$, which induces a morphism of $R^F\mathscr{D}_X$-modules $R^F M_1 \longrightarrow R^F M_2$. It is also compatible with the actions of $\lambda\partial_\lambda$. Hence, we can easily observe that $\overline{f}$ is induced by the restriction f_0, and that f_0 preserves the Hodge filtrations. □

Proposition 13.5.6 *The functors Φ_X for complex manifolds X are compatible with the push-forward by projective morphisms, the duality and the localizations.*

Proof We obtain the compatibility with the projective push-forward by Lemma 13.5.1. We obtain the compatibility with the localizations by Lemma 13.5.3. The compatibility with the duality is clear by construction. □

Let H be a hypersurface of X. Let $\mathcal{V}_0 = (V, F, L_K, W)$ be a graded polarizable variation of mixed Hodge structure on $X \setminus H$ which is admissible along H. Here, L_K is a K-local system on $X \setminus H$, V is a flat bundle on $X \setminus H$ equipped with an isomorphism $\mathrm{DR}_X(V) \simeq L_K[d_X] \otimes \mathbb{C}$, and F and W denote a Hodge filtration and a weight filtration, respectively.

We have the associated admissible variation of mixed twistor structure on (X, H) given as follows. We take a projective morphism $F : X' \longrightarrow X$ such that (a) $H' := F^{-1}(H)$ is normal crossing, (b) F induces $X'\setminus H' \simeq X\setminus H$. The flat bundle $V' := F^*V$ on $X' \setminus H'$ is extended to a regular singular meromorphic flat bundle $\tilde{V}'$ on (X', H') which is equipped with the induced Hodge filtration $\tilde{F}$ and the weight filtration $\tilde{W}$. By applying the Rees construction to $(\tilde{V}', \tilde{F})$, and taking the analytification, we obtain a smooth $\tilde{\mathcal{R}}_{X'}(*H')$-module $\mathcal{M}'$. We also obtain a smooth $\tilde{\mathcal{R}}_{X'}(*H')$-module

$(\mathcal{M}')^{\vee}$. The real structure of the flat bundle V' induces a smooth $\mathcal{R}_{X'(*H')}$-triple $\mathcal{T}'_{\mathcal{V}_0} := \big(j^*(\mathcal{M}')^{\vee}, \mathcal{M}', C_{\mathcal{M}'}\big)$ with a real structure $(\mathrm{id}, \mathrm{id}) : \tilde{\gamma}^*_{sm}\mathcal{T}'_{\mathcal{V}_0} \simeq \mathcal{T}'_{\mathcal{V}_0}$. It is an admissible variation of mixed twistor structure on (X', H'). By the push-forward, we obtain the desired admissible variation of twistor structure $\mathcal{T}_{\mathcal{V}_0} := F_{\dagger}\mathcal{T}'_{\mathcal{V}_0}$ on (X, H).

Let $(M, F, P, W) \in \mathrm{MTM}(X, K)$. Let $\mathcal{V}_0$ be a graded polarizable admissible variation of mixed Hodge structure on (X, H). We have the mixed Hodge module $(M, F, P, W)_{|X\setminus H} \otimes \mathcal{V}_0$ on $X \setminus H$. It is extended to a mixed Hodge module $\big((M, F, P, W) \otimes \mathcal{V}_0\big)[\star H]$ on X by the admissibility of $\mathcal{V}_0$ along H. It is naturally equipped with K-structure. We obtain the following from Lemma 13.5.3.

Proposition 13.5.7 *We have a natural isomorphism*

$$\Phi_X\Big(\big((M, F, P, W) \otimes \mathcal{V}_0\big)[\star H]\Big) \simeq \big(\Phi_X(M, F, P, W) \otimes \mathcal{T}_{\mathcal{V}_0}\big)[\star H]$$

in $\mathrm{MTM}^{\mathrm{int}}_{\mathrm{good}}(X, K)$. □

Because Beilinson triples come from admissible variation of mixed Hodge structure with $\mathbb{Q}$-structure, we may define the Beilinson functors $\Pi^{a,b}_{g\star}$, $\Pi^{a,b}_{g*!}$, $\Xi^{(a)}_g$, $\psi^{(a)}$ and $\phi^{(a)}_g$ on the category of mixed Hodge modules in the way explained in Sect. 11.3.3. We have the following.

Corollary 13.5.8 *The functor Φ_X is compatible with $\Pi^{a,b}_{g\star}$, $\Pi^{a,b}_{g*!}$, $\Xi^{(a)}_g$, $\psi^{(a)}_g$ and $\phi^{(a)}_g$.* □

13.5.2 Polarization

13.5.2.1 Basic Examples

Let X be a complex manifold. Set $d_X := \dim X$. We have $(\mathcal{O}_X, F, \mathbb{R}_X[d_X]) \in \mathrm{MF}_h(\mathscr{D}_X, K)$, where F is given by $F_j(\mathcal{O}_X) = 0$ $(j < 0)$ and $F_j(\mathcal{O}_X) = \mathcal{O}_X$ $(j \geq 0)$. Let W be the filtration on $(\mathcal{O}_X, F, \mathbb{R}_X[d_X])$ which is pure of weight d_X. Then, $(\mathcal{O}_X, F, \mathbb{R}_X[d_X], W)$ is a mixed Hodge module on X which is pure of weight d_X.

The analytification of the Rees module of $(\mathcal{O}_X, F)$ is naturally isomorphic to $\mathcal{O}_{\mathcal{X}}$. The underlying $\mathcal{R}_X$-triple of $\Phi_X(\mathcal{O}_X, F, \mathbb{R}_X[d_X], W)$ is $\mathcal{T}_0 := \big(j^*\boldsymbol{D}\mathcal{O}_{\mathcal{X}}, \mathcal{O}_{\mathcal{X}}, C_\iota\big)$, where C_ι is the sesqui-linear pairing induced by the real structure. We have the $\tilde{\mathcal{R}}_X$-isomorphism $\nu : j^*\boldsymbol{D}\mathcal{O}_{\mathcal{X}} \simeq \lambda^{d_X}\mathcal{O}_{\mathcal{X}}$ induced by $\nu : \boldsymbol{D}\mathcal{O}_X \simeq \mathcal{O}_X$ given in Sect. 12.6.6. According to Proposition 12.6.26, we have the isomorphism $\varphi = (\nu^{-1}, 1) : \mathcal{T}_0 \simeq \mathcal{U}(d_X, 0)$ of $\mathcal{R}_X$-triples. Let $\mathcal{S}_0 : \mathcal{T}_0 \longrightarrow \mathcal{T}_0^* \otimes \boldsymbol{T}(-d_X)$ be given by

$\big((-1)^{d_X}\nu, (-1)^{d_X}\nu\big)$. The following diagram is commutative:

$$\begin{array}{ccc} \mathcal{T}_0 & \longrightarrow & \mathcal{T}_0^* \otimes \boldsymbol{T}(-d_X) \\ \varphi\big\downarrow & & \varphi^*\big\uparrow \\ \mathcal{U}(d_X,0) & \xrightarrow{((-1)^{d_X},(-1)^{d_X})} & \mathcal{U}(d_X,0)^* \otimes \boldsymbol{T}(-d_X) \end{array}$$

Hence, $\mathcal{S}_0$ is a polarization of the pure twistor $\mathscr{D}$-module $\mathcal{T}_0$ of weight d_X.

Note that $(-1)^{d_X}\,\mathrm{DR}(\nu) : \mathrm{DR}_X \boldsymbol{D}\mathcal{O}_X \simeq \mathrm{DR}_X\,\mathcal{O}_X$ is given as follows:

$$\begin{aligned} H^{-d_X}\,\mathrm{DR}_X \boldsymbol{D}\mathcal{O}_X = \mathrm{Hom}(\mathbb{C}_X, \mathbb{C}_X) &\longrightarrow H^{-d_X}\,\mathrm{DR}_X\,\mathcal{O}_X = \mathbb{C}_X \\ \mathrm{id} &\longmapsto (-1)^{d_X(d_X-1)/2} \end{aligned}$$

Hence, the polarization $\mathcal{S}_0$ corresponds to the bi-linear form

$$\mathbb{C}_X[d_X] \times \mathbb{C}_X[d_X] \longrightarrow \mathbb{C}_X[2d_X]$$

given by $(a,b) \longmapsto (-1)^{d_X(d_X-1)/2}ab$:

$$\begin{array}{ccc} H^{-d_X}(\mathbb{C}_X[d_X]) \times H^{-d_X}(\mathbb{C}_X[d_X]) & \longrightarrow & H^{-2d_X}(\mathbb{C}_X[2d_X]) \\ \| & & \| \\ \mathbb{C} \times \mathbb{C} & \longrightarrow & \mathbb{C} \end{array}$$

13.5.2.2 Polarized Variation of Pure Hodge Structure

For any integer n, we set $T_K(n) := (\mathcal{O}_X, F^{(n)}, K_X(n))$. Here, $F_j^{(n)}\mathcal{O}_X = 0$ $(j < n)$ and $F_j^{(n)}\mathcal{O}_X = \mathcal{O}_X$ $(j \geq n)$, and $K_X(n)$ denotes the K-local subsystem $(2\pi\sqrt{-1})^n K_X \subset \mathbb{C}_X$.

Let (V, F, L_K) be a polarizable variation of pure Hodge structure of weight $w-d_X$ on X. Here, V denotes a flat bundle on X, F is a Hodge filtration of V, and L_K is a K-local system on X with an isomorphism $L_K[d_X] \otimes \mathbb{C} \simeq \mathrm{DR}_X(V)$. Let S be a polarization of (V, F, L_K). In particular, it is a morphism of $(V, F, L_K) \otimes (V, F, L_K) \longrightarrow T_K(-w + d_X)$. We have the object $\mathcal{H}_1 := (V, F, L_K[d_X])$ in $\mathrm{MF}_h(\mathscr{D}_X, K)$. We have the induced bi-linear morphism $S_1 : L_K[d_X] \otimes_{K_X} L_K[d_X] \longrightarrow K_X(-w + d_X)[2d_X]$ such that the induced morphism

$$H^{-2d_X}(S_1) : L_K \otimes L_K \longrightarrow K_X(-w + d_X)$$

is given by $(-1)^{d_X(d_X-1)/2}S$. According to [69], $\mathcal{H}_1$ is a pure Hodge module of weight w, and S_1 is a polarization of $\mathcal{H}_1$.

Let $\mathcal{T}_1 = \Phi_X(\mathcal{H}_1)$ which is a polarizable pure twistor $\mathscr{D}$-module of weight w. Let $\mathcal{M}$ be the $\mathcal{R}_X$-module obtained as the analytification of the Rees module of (V, F). By the compatibility of S_1 and the Hodge filtration, we have the isomorphism a_{S_1} : $\mathcal{M} \simeq \lambda^{-w} j^* \boldsymbol{D}\mathcal{M}$ induced by S_1, where we use the natural isomorphism $j^*\mathcal{M} \simeq \mathcal{M}$.

Lemma 13.5.9 *The pair (a_{S_1}, a_{S_1}) gives a morphism $\Psi_1 : \mathcal{T}_1 \longrightarrow \mathcal{T}_1^* \otimes \boldsymbol{T}(-w)$, and it is a polarization of $\mathcal{T}_1$.*

Proof We have already checked the claim in the case $T_K(0)$ with the natural polarization. (See Sect. 13.5.2.1) Let $\mathcal{T}_0$ and $\mathcal{S}_0$ be the corresponding pure twistor $\mathscr{D}$-module of weight d_X.

Let us consider the general case. From $\mathcal{H}_1$, we obtain the smooth $\mathcal{R}_X$-triple $\mathcal{T}_2 = (j^*\mathcal{M}^\vee, \mathcal{M}, C_{2\iota})$. It is equipped with the isomorphism $\Psi_2 : \mathcal{T}_2 \simeq \mathcal{T}_2^* \otimes \boldsymbol{T}(-w)$ induced by S which is a polarization of variation of pure twistor structure of weight w. Then, we have $\mathcal{T}_1 = \mathcal{T}_0 \otimes \mathcal{T}_2$, and $\Psi_1 = \Psi_0 \otimes \Psi_2$. Hence, Ψ_1 is a polarization. □

13.5.2.3 General Case

Let $(M, F, P) \in \mathrm{MF}_h(\mathscr{D}_X, K)$ be a pure Hodge module of weight w on X with a polarization S. (See [69].) Let $\mathcal{T} = \Phi_X(M, F, P)$ be the associated pure twistor $\mathscr{D}$-module of weight w. Let $\mathcal{M}$ be the $\mathcal{R}_X$-module obtained as the analytification of the Rees module of (M, F). Note that S induces an isomorphism $a_S : \mathcal{M} \simeq \lambda^{-w} j^* \boldsymbol{D}\mathcal{M}$, by the compatibility of the Hodge filtration and S.

Proposition 13.5.10 *The pair (a_S, a_S) gives a morphism $\mathcal{T} \longrightarrow \mathcal{T}^* \otimes \boldsymbol{T}(-w)$, denoted by $\mathcal{S}$, and it is a polarization of $\mathcal{T}$.*

Proof Because a pure Hodge module with a polarization is the minimal extension of a variation of pure Hodge structure with a polarization given on a Zariski open subset of a closed subvariety. Hence, the claim is reduced to Lemma 13.5.9. □

13.5.2.4 Push-Forward

We continue to use the notation in Sect. 13.5.2.3. Let us check that the compatibility of the induced polarizations on the push-forward. Let $f : X \longrightarrow Y$ be any projective morphism of complex manifolds. Let $f_\dagger^i P$ be the i-th cohomology perverse sheaf of the push-forward $Rf_* P$. We have the naturally induced Hodge filtration $f_\dagger^i F$ on $f_\dagger M$, and $(f_\dagger^i M, f_\dagger^i F, f_\dagger^i P)$ is a polarizable pure Hodge module of weight $w + i$ [69]. Let $\mathcal{L}$ be a line bundle on X which is relatively ample to f. We take any Hermitian metric of $\mathcal{L}$, and let τ be the curvature form of the Chern connection. Note also that the multiplication of τ induces morphisms $\ell : (f_\dagger^i M, f_\dagger^i F, f_\dagger^i P) \longrightarrow (f_\dagger^{i+2} M, f_\dagger^{i+2} F, f_\dagger^{i+2} P)(1)$. The Hard Lefschetz Theorem [69] implies that ℓ^i : $(f_\dagger^{-i} M, f_\dagger^{-i} F, f_\dagger^{-i} P) \longrightarrow (f_\dagger^i M, f_\dagger^i F, f_\dagger^i P)(i)$ are isomorphisms for any $i \geq 0$. The kernel $\mathcal{P}_i$ of $\ell^{i+1} : (f_\dagger^{-i}M, f_\dagger^{-i}F, f_\dagger^{-i}P) \longrightarrow (f_\dagger^{i+2}M, f_\dagger^{i+2}F, f_\dagger^{i+2}P)(i+1)$ is called the primitive part. The Hard Lefschetz Theorem [69] implies that a polarization of the primitive part $\mathcal{P}_i$ is induced by

$$S_i := (-1)^{i(i-1)/2}\{f_\dagger S\}_i \circ (\mathrm{id} \otimes \ell^i) : f_\dagger^{-i}P \otimes f_\dagger^{-i}P \longrightarrow K_Y(-w+i+d_Y)[2d_Y].$$

Here, for any $j \in \mathbb{Z}$, $\{f_\dagger S\}_j : f_\dagger^{-j}P \otimes f_\dagger^{j}P \longrightarrow K_Y(-w-d_Y)[2d_Y]$ is induced as follows. From the trace morphism $f_* K_X(-w+d_X)[2d_X] \longrightarrow K_Y(-w+d_Y)[2d_Y]$, we obtain $Rf_*P \otimes Rf_*P \longrightarrow K_Y(-w+d_Y)[2d_Y]$. It naturally induces $f_\dagger^{-j}P[j] \otimes f_\dagger^{j}P[-j] \longrightarrow K_Y(-w+d_Y)[2d_Y]$. Then, $\{f_\dagger S\}_j$ is obtained as the composite of the following:

$$\begin{aligned} f_\dagger^{-j}P \otimes f_\dagger^{j}P \simeq \big(\mathbb{Z}[j] \otimes \mathbb{Z}[-j]\big) \otimes f_\dagger^{-j}P \otimes f_\dagger^{j}P \simeq f_\dagger^{-j}P[j] \otimes f_\dagger^{j}P[-j] \\ \longrightarrow K_Y(-w+d_Y)[2d_Y] \qquad (13.25) \end{aligned}$$

Here, $H^{-j}(\mathbb{Z}[j]) \otimes H^j(\mathbb{Z}[-j]) = \mathbb{Z} \otimes \mathbb{Z} \simeq \mathbb{Z}$ is given by the multiplication. We have $S_i = (-1)^{i(i-1)/2}\{f_\dagger S\}_{-i} \circ (\ell^i \otimes \mathrm{id})$.

We also have the morphism $(f_\dagger S)_{-i} : f_\dagger^{i}P \otimes f_\dagger^{-i}P \longrightarrow K_Y(-w+d_Y)$ induced by $f_\dagger^{i}P[-i] \otimes f_\dagger^{-i}P[i] \longrightarrow K_Y(-w+d_Y)[2d_Y]$ and the signature rule in Sect. 12.1.5.4. Then, we have $S_i = (f_\dagger S)_{-i} \circ ((-\ell)^i \otimes \mathrm{id})$.

Let ℓ also denote the morphism $f_\dagger^{i}\mathcal{M} \longrightarrow \lambda f_\dagger^{i+2}\mathcal{M}$ induced by the multiplication of τ. We have the morphism $L : f_\dagger^{i}\mathcal{T} \longrightarrow f_\dagger^{i+2}(\mathcal{T}) \otimes \boldsymbol{T}(1)$ given by $(-\ell, -\ell)$. According to the Hard Lefschetz Theorem for pure twistor $\mathscr{D}$-modules [52, 55, 66], the induced morphisms $L^i : f_\dagger^{-i}\mathcal{T} \longrightarrow f_\dagger^{i}\mathcal{T} \otimes \boldsymbol{T}(i)$ are isomorphisms. Let $Pf_\dagger^{-i}\mathcal{T}$ be the kernel of $L^{i+1} : f_\dagger^{-i}\mathcal{T} \longrightarrow f_\dagger^{i+2}\mathcal{T} \otimes \boldsymbol{T}(i+1)$. Let $\mathcal{S} : \mathcal{T} \longrightarrow \mathcal{T}^* \otimes \boldsymbol{T}(-w)$ be a polarization on $\mathcal{T}$ induced by S. Then, the Hard Lefschetz Theorem implies that $f_\dagger^{i}\mathcal{S} \circ L^i$ induces a polarization on $Pf_\dagger^{-i}\mathcal{T}$. (See Remark 2.1.27 for the signature. Note that $(2\pi\sqrt{-1})^{-1}\tau$ is the real $(1,1)$-form and that its cohomology class is the first Chern class in the ordinary sense.)

The analytification of the Rees module of $(f_\dagger^{-i}M, f_\dagger^{-i}F)$ is $f_\dagger^{-i}\mathcal{M}$. Note that the morphism $f_\dagger^{-i}\mathcal{M} \longrightarrow \lambda^{-w+i} \cdot j^*\boldsymbol{D}(f_\dagger^{-i}\mathcal{M})$ induced by $(f_\dagger S)_{-i} \circ \big((-\ell)^i \otimes \mathrm{id}\big)$ is equal to the composite of the following, i.e., the underlying morphism of $f_\dagger^{i}\mathcal{S} \circ L^i$:

$$f_\dagger^{-i}\mathcal{M} \xrightarrow{(-\ell)^i} \lambda^i \cdot f_\dagger^{i}\mathcal{M} \xrightarrow{f_\dagger^{i}as} \lambda^{i-w} \cdot f_\dagger^{i}\boldsymbol{D}\mathcal{M} \overset{\beta}{\simeq} \lambda^{i-w} \cdot \boldsymbol{D}f_\dagger^{-i}\mathcal{M} \qquad (13.26)$$

Here, β is induced by $f_\dagger \boldsymbol{D}(\mathcal{M})[i] \simeq \boldsymbol{D}\big(f_\dagger\mathcal{M}\big)[i] \simeq \boldsymbol{D}\big(f_\dagger\mathcal{M}[-i]\big)$. More explicitly, it equals the natural isomorphism $f_\dagger^{i}\boldsymbol{D}\mathcal{M} \simeq \lambda^{i-w} \cdot \boldsymbol{D}f_\dagger^{-i}\mathcal{M}$ multiplied with $(-1)^{i(i+1)/2}$. Hence, we have the coincidence of the polarizations in the Hodge setting and the twistor setting.

Chapter 14
Algebraic Mixed Twistor $\mathscr{D}$-Modules and Their Derived Category

We study algebraic mixed twistor $\mathscr{D}$-modules and their derived categories. In particular, we obtain the six operations on the derived categories.

Let X be a smooth complex algebraic varieties. We can take a smooth complete variety $\overline{X}$ with an open immersion $X \subset \overline{X}$ such that $D = \overline{X} \setminus X$ is a hypersurface. Algebraic mixed twistor $\mathscr{D}$-modules on X are just filtered $\mathcal{R}_{\overline{X}(*D)}$-triples $(\mathcal{T}, W)$ obtained as $(\overline{\mathcal{T}}, W)(*D)$ from a mixed twistor $\mathscr{D}$-module $(\overline{\mathcal{T}}, W)$ on $\overline{X}$.

Obviously, functors for mixed twistor $\mathscr{D}$-modules have their counterparts in the context of algebraic mixed twistor $\mathscr{D}$-modules. Moreover, passing to derived categories, we may also have functors which are not given for mixed twistor $\mathscr{D}$-modules in a general analytic setting. Namely, we have the push-forward and the pull back for any algebraic morphisms. So, we have six operations on the derived categories on algebraic mixed twistor $\mathscr{D}$-modules.

To deal with the derived category, we closely follow the methods developed in [2, 73]. It might be instructive to recall a part of the work of Beilinson briefly. There are two useful ways to consider "the derived category of holonomic $\mathscr{D}$-modules". One naive way is to consider the derived category $D^b \operatorname{Hol}(X)$ of the bounded complexes in the abelian category $\operatorname{Hol}(X)$ of holonomic $\mathscr{D}$-modules on X. The other way is given as follows. We have the category $D^b(\mathscr{D}_X)$ of cohomologically bounded complexes of $\mathscr{D}_X$-modules. Then, we have the full subcategory $D^b_h(\mathscr{D}_X) \subset D^b(\mathscr{D}_X)$ of the cohomologically bounded complexes of $\mathscr{D}_X$-modules whose cohomology sheaves are holonomic. In the general analytic context, $D^b_h(\mathscr{D}_X)$ is much more useful than $D^b \operatorname{Hol}(X)$. For example, the push-forward functor for any proper morphism $f : X \longrightarrow Y$ is naturally defined as $f_\dagger : D^b_h(\mathscr{D}_X) \longrightarrow D^b_h(\mathscr{D}_Y)$.

The classical theorem of Beilinson [2] ensures that, in the algebraic context, the natural functor $D^b \operatorname{Hol}(X) \longrightarrow D^b_h(\mathscr{D}_X)$ is an equivalence. Moreover, he gave a direct way to construct the push-forward functors $f_\star : D^b \operatorname{Hol}(X) \longrightarrow D^b \operatorname{Hol}(Y)$ ($\star = *, !$) for any algebraic morphism $f : X \longrightarrow Y$.

We mention one of the basic ideas due to Beilinson for the construction of the push-forward. For simplicity, suppose that X is projective, and we consider the

T. Mochizuki, *Mixed Twistor $\mathscr{D}$-modules*, Lecture Notes in Mathematics 2125,
DOI 10.1007/978-3-319-10088-3_14

push-forward for the morphism a_X from X to a point. Let M be any holonomic $\mathscr{D}$-module on X. If $M(*D) = M$ for an ample hypersurface D, we have $a^j_{X\dagger}(M) = 0$ for $j > 0$. Indeed, $a^j_{X\dagger}(M)$ is the j-th hypercohomology group of the de Rham complex $\Omega^\bullet_X \otimes M[\dim X]$. Because $Ra^j_{X*}(\Omega^i_X \otimes M) = 0$ $(j > 0)$, we can easily obtain $a^j_{X\dagger}(M) = 0$ for $j > 0$. By using the duality, if $M(!D) = D$ for an ample hypersurface D, we have $a^j_{X\dagger}(M) = 0$ for $j < 0$. For general holonomic $\mathscr{D}$-module M on X, we can find generic ample hypersurfaces D_i $(i = 1, 2)$ such that $M(!D_1)(*D_2) = M(*D_2)(!D_1)$. Hence, we can easily find a complex $\tilde{M}^\bullet$ such that (a) $\tilde{M}^\bullet$ is isomorphic to M in $D^b\operatorname{Hol}(X)$, (b) $a^j_{X\dagger}(\tilde{M}^i) = 0$ $(j \neq 0)$. So, we set $a_{X\dagger}(M) := a^0_{X\dagger}(\tilde{M}^\bullet)$. Similarly, for any algebraic morphism $f : X \longrightarrow Y$, we can find a complex $\tilde{M}^\bullet$ quasi-isomorphic to M satisfying $f^j_\dagger\tilde{M}^i = 0$ $(j \neq 0)$. So, it is natural to define $f_\dagger(M)$ by $f^0_\dagger(\tilde{M}^\bullet)$. See [2] for more details and precision.

The method of Beilinson was adopted in the context of derived category of algebraic mixed Hodge modules by Saito [73]. He also studied the pull back functors. Their method can be also applied in the context of mixed twistor $\mathscr{D}$-modules which we shall explain in Sect. 14.3.

Finally in this introductory part, we mention that for a given algebraic mixed twistor $\mathscr{D}$-module it is not expected that the underlying $\mathcal{R}_X$-modules are algebraic. But, we shall prove that their specializations at any general complex numbers are algebraic. (See Sect. 14.1.5.2 for a more precise statement.) We also prove that if an algebraic mixed twistor $\mathscr{D}$-module is integrable then the underlying $\tilde{\mathcal{R}}_X$-modules are algebraic (Theorem 14.4.8).

We consider only algebraic varieties of finite type over the complex number field unless otherwise specified.

14.1 Algebraic Mixed Twistor $\mathscr{D}$-Modules

14.1.1 Definition

Let X be a smooth algebraic variety. In the following, a principal completion of $(\overline{X}, D)$ means a complete smooth complex variety $\overline{X}$ with an open immersion $X \subset \overline{X}$ such that $D = \overline{X} \setminus X$ is a hypersurface. For any principal completion $(\overline{X}, D)$ of X, we have the natural functor $\Psi_{(\overline{X},D)} : \operatorname{MTM}(\overline{X}) \longrightarrow \mathcal{R}\text{-Tri}(\overline{X}, D)^{\mathrm{fil}}$, where $\Psi_{(\overline{X},D)}(\mathcal{T}, W) = (\mathcal{T}, W)(*D)$.

Definition 14.1.1 Let $\operatorname{MTM}(X^{\mathrm{alg}}) \subset \mathcal{R}\text{-Tri}(\overline{X}, D)^{\mathrm{fil}}$ denote the essential image of $\Psi_{(\overline{X},D)}$. Any object in $\operatorname{MTM}(X^{\mathrm{alg}})$ is called an algebraic mixed twistor $\mathscr{D}$-module on X. □

In other words, an object $(\mathcal{T}, W) \in \mathcal{R}\text{-Tri}(\overline{X}, D)^{\mathrm{fil}}$ is called an algebraic mixed twistor $\mathscr{D}$-module on X if there exists a mixed twistor $\mathscr{D}$-module $(\overline{\mathcal{T}}, W)$ on $\overline{X}$ such that $(\mathcal{T}, W) \simeq \Psi_{(\overline{X},D)}(\overline{\mathcal{T}}, W)$ in $\mathcal{R}\text{-Tri}(\overline{X}, D)^{\text{fil}}$. A morphism $(\mathcal{T}_1, W) \longrightarrow (\mathcal{T}_2, W)$ in $\text{MTM}(X^{\text{alg}})$ is a morphism as filtered $\mathcal{R}_{\overline{X}(*D)}$-triples.

Lemma 14.1.2 $\Psi_{(\overline{X},D)}$ *induces an equivalence*

$$\text{MTM}(\overline{X}, [*D]) \longrightarrow \text{MTM}(X^{\text{alg}}).$$

It also gives an equivalence $\text{MTM}(\overline{X}, [!D]) \longrightarrow \text{MTM}(X^{\text{alg}})$.

Proof It follows from Lemma 7.1.59. □

Let $(\overline{X}', D')$ be another principal completion of X. Let $\text{MTM}(X^{\text{alg}})'$ be the essential image of $\Psi_{(\overline{X}',D')} : \text{MTM}(\overline{X}', [*D']) \longrightarrow \mathcal{R}\text{-Tri}(\overline{X}', D')^{\text{fil}}$.

Lemma 14.1.3 *We have a natural equivalence* $\text{MTM}(X^{\text{alg}})' \simeq \text{MTM}(X^{\text{alg}})$. *In this sense, the category* $\text{MTM}(X^{\text{alg}})$ *is independent of the choice of a projective completion* $\overline{X}$ *as above.*

Proof We can take projective morphisms $\rho_1 : X'' \longrightarrow X$ and $\rho_2 : X'' \longrightarrow X'$ such that (i) $\rho_1^{-1}(H) = \rho_2^{-1}(H') =: H''$ is a hypersurface, (ii) we have $X''\setminus H'' \simeq X$, and ρ_i induces the identity on X. Then, we obtain the desired equivalence by Lemma 2.1.11 and Proposition 11.2.12. □

14.1.2 Restriction of KMS-Spectrum

Suppose that we are given a $\mathbb{Q}$-vector space $\mathcal{A} \subset \mathbb{R} \times \mathbb{C}$ such that $\mathbb{Q} \times \{0\} \subset \mathcal{A}$. We take a principal completion $(\overline{X}, D)$ of X.

We define the subcategory $\text{MTM}(X^{\text{alg}})_{\mathcal{A}} \subset \text{MTM}(X^{\text{alg}})$ as the essential image of $\text{MTM}(\overline{X})_{\mathcal{A}}$ by $\Psi_{(\overline{X},D)}$. It is equivalent to the category $\text{MTM}(\overline{X}, [*D])_{\mathcal{A}}$. As in Lemma 14.1.3, it is essentially independent of the choice of $(\overline{X}, D)$.

Lemma 14.1.4 *For any* $(\mathcal{T}, W) \in \text{MTM}(X^{\text{alg}})$, *there exists a finite dimensional* $\mathbb{Q}$*-vector space* $\mathcal{A}$ *such that* $(\mathcal{T}, W) \in \text{MTM}(X^{\text{alg}})_{\mathcal{A}}$.

Proof It is enough to prove the claim in the case where X is complete. Then, the claim follows from Proposition 7.1.67. □

14.1.3 Some Functors for Algebraic Mixed Twistor $\mathscr{D}$-Modules

14.1.3.1 Localizations

Let H be an algebraic hypersurface of X. We take a principal completion $(\overline{X}, D)$ of X. Let $\overline{H}$ be the closure of H in $\overline{X}$. We have the functors $[\star\overline{H}] : \text{MTM}(\overline{X}) \longrightarrow$

$\mathrm{MTM}(\overline{X})$ ($\star = *, !$). We have the induced functors $\mathrm{MTM}(X^{\mathrm{alg}}) \longrightarrow \mathrm{MTM}(X^{\mathrm{alg}})$ denoted by $[\star H]$. Equivalently, we define the functors $[\star \overline{H}]_{*D} : \mathrm{MTM}(\overline{X}, [*D]) \longrightarrow \mathrm{MTM}(\overline{X}, [*D])$ ($\star = *, !$) as the composite of the following:

$$\mathrm{MTM}(\overline{X}, [*D]) \xrightarrow{[\star \overline{H}]} \mathrm{MTM}(\overline{X}) \xrightarrow{[*D]} \mathrm{MTM}(\overline{X}, [*D]) \tag{14.1}$$

They give the functors $[\star H]$ on $\mathrm{MTM}(X^{\mathrm{alg}}) \simeq \mathrm{MTM}(\overline{X}, [*D])$.

Lemma 14.1.5 *The functors $[\star H]$ are independent of the choice of a principal completion.*

Proof Suppose that we are given principal completions $(\overline{X}, D)$ and $(\overline{X}', D')$ with a morphism $F : \overline{X}' \longrightarrow \overline{X}$ such that F induces the identity on X. It is enough to compare the functors $[\star H]$ determined by $(\overline{X}, D)$ and $(\overline{X}', D')$.

We have the equivalence $\mathrm{MTM}(\overline{X}', [*D']) \longrightarrow \mathrm{MTM}(\overline{X}, [*D])$. We have $F^{-1}(D) = D'$. Let $\overline{H}$ and $\overline{H}'$ be the closure of H in $\overline{X}$ and $\overline{X}'$, respectively. We set $\overline{H}'_1 := F^{-1}(\overline{H})$. We have $\overline{H}'_1 \cap X = H$. Let $(\mathcal{T}', W) \in \mathrm{MTM}(\overline{X}', [\star D'])$. We have $F^j_\dagger(\mathcal{T}, W) = 0$ unless $j = 0$. Set $(\mathcal{T}, W) := F^0_\dagger(\mathcal{T}', W)$. We have $F^0_\dagger((\mathcal{T}', W)[\star \overline{H}'_1]) \simeq (\mathcal{T}, W)[\star \overline{H}]$. Hence, we have

$$F^0_\dagger\big((\mathcal{T}', W[\star \overline{H}'_1][*D'])\big) \simeq F^0_\dagger\big((\mathcal{T}', W)[\star \overline{H}'_1]\big)[*D] \simeq (\mathcal{T}, W)[\star \overline{H}][*D]$$

Hence, we obtain the comparison of the functors determined by $(\overline{X}, D)$ and $(\overline{X}', D')$. □

The following claim is clear by the description as the composite of (14.1).

Lemma 14.1.6

- *The functors $[\star H]$ are exact.*
- *For any $\mathcal{T} \in \mathrm{MTM}(X^{\mathrm{alg}})$, we have natural morphisms $\mathcal{T}[!H] \longrightarrow \mathcal{T} \longrightarrow \mathcal{T}[*H]$. We also have natural isomorphisms $(\mathcal{T}[\star H])[\star H] \simeq \mathcal{T}[\star H]$.* □

Let $\mathrm{MTM}(X^{\mathrm{alg}}, [\star H]) \subset \mathrm{MTM}(X^{\mathrm{alg}})$ denote the full subcategory of the objects $\mathcal{T} \in \mathrm{MTM}(X^{\mathrm{alg}})$ such that $\mathcal{T}[\star H] = \mathcal{T}$.

14.1.3.2 Duality and Hermitian Adjoint

We have the naturally defined duality functors $\boldsymbol{D}_X$ and the Hermitian adjoint $\boldsymbol{D}^{\mathrm{herm}}_X$ on $\mathrm{MTM}(X^{\mathrm{alg}})$. We take a principal completion $(\overline{X}, D)$ of X. For any $\mathcal{T} \in \mathrm{MTM}(X^{\mathrm{alg}})$, we take $\overline{\mathcal{T}} \in \mathrm{MTM}(\overline{X})$ such that $\Psi_{(\overline{X},D)}(\overline{\mathcal{T}}) = \mathcal{T}$. We set

$$\boldsymbol{D}_X(\mathcal{T}) := \Psi_{(\overline{X},D)}\big(\boldsymbol{D}_{\overline{X}}(\overline{\mathcal{T}})\big), \qquad \boldsymbol{D}^{\mathrm{herm}}_X(\mathcal{T}) := \Psi_{(\overline{X},D)}\big(\boldsymbol{D}^{\mathrm{herm}}_{\overline{X}}(\overline{\mathcal{T}})\big).$$

They are independent of the choice of $\overline{\mathcal{T}}$. We can check the independence from the choice of $\overline{X}$ by using the argument in the proof of Lemma 14.1.5. The following is clear by construction.

Lemma 14.1.7 *The functors $\boldsymbol{D}_X$ and $\boldsymbol{D}_X^{\mathrm{herm}}$ are exact.* □

Lemma 14.1.8 *Set $\overline{*} =!$ and $\overline{!} = *$. For any $\mathcal{T} \in \mathrm{MTM}(X^{\mathrm{alg}})$, we have natural isomorphisms*

$$\boldsymbol{D}_X(\mathcal{T}[\star H]) \simeq \boldsymbol{D}_X(\mathcal{T})[\overline{\star} H], \quad \boldsymbol{D}_X^{\mathrm{herm}}(\mathcal{T}[\star H]) \simeq \boldsymbol{D}_X^{\mathrm{herm}}(\mathcal{T})[\overline{\star} H].$$

Proof It follows from Proposition 13.3.5. □

14.1.3.3 Functors j^* and $\tilde{\gamma}^*$

Take a principal completion $(\overline{X}, D)$ of X. The functor j^* on $\mathcal{R}\text{-Tri}(\overline{X}, D)^{\mathrm{fil}}$ clearly preserves the full subcategory $\mathrm{MTM}(X^{\mathrm{alg}})$. We set $\tilde{\sigma}^* := \boldsymbol{D} \circ \boldsymbol{D}^{\mathrm{herm}} = \boldsymbol{D}^{\mathrm{herm}} \circ \boldsymbol{D}$ on $\mathrm{MTM}(X^{\mathrm{alg}})$. We also set $\tilde{\gamma}^* := j^* \circ \tilde{\sigma}^*$. The functors j^*, $\tilde{\sigma}^*$ and $\tilde{\gamma}^*$ are auto-equivalences on $\mathrm{MTM}(X^{\mathrm{alg}})$.

For $\rho^* = j^*, \tilde{\gamma}^*, \tilde{\sigma}^*$, we have $\rho^* \circ \boldsymbol{D} \simeq \boldsymbol{D} \circ \rho^*$ and $\rho^* \circ \boldsymbol{D}^{\mathrm{herm}} \simeq \boldsymbol{D}^{\mathrm{herm}} \circ \rho^*$. We also have natural isomorphisms $\rho^*(\mathcal{T}[\star H]) \simeq (\rho^*\mathcal{T})[\star H]$ for any $\mathcal{T} \in \mathrm{MTM}(X^{\mathrm{alg}})$ and any algebraic hypersurface H in X.

14.1.3.4 Push-Forward by Quasi-projective Morphisms

Let $f : X \longrightarrow Y$ be a quasi-projective morphism of smooth algebraic varieties. We have cohomological functors $f_*^i, f_!^i : \mathrm{MTM}(X^{\mathrm{alg}}) \longrightarrow \mathrm{MTM}(Y^{\mathrm{alg}})$ $(i \in \mathbb{Z})$. Indeed, we take principal completions $X \subset \overline{X}$ and $Y \subset \overline{Y}$ such that (a) $D_X := \overline{X} \setminus X$ and $D_Y := \overline{Y} \setminus Y$, (b) we have a projective morphism $\overline{f} : \overline{X} \longrightarrow \overline{Y}$ which induces $f : X \longrightarrow Y$. We have $f^{-1}(D_Y) \subset D_X$. Then, for $\mathcal{T} \in \mathrm{MTM}(X^{\mathrm{alg}})$, we take $\overline{\mathcal{T}}$ in $\mathrm{MTM}(\overline{X})$ such that $\Psi_{(\overline{X}, D_X)}(\overline{\mathcal{T}}) = \mathcal{T}$. We set

$$f_\star^i(\mathcal{T}) := \Psi_{(\overline{Y}, D_Y)}\big(\overline{f}_\dagger^i \overline{\mathcal{T}}[\star D_X]\big)$$

They are independent of the choice of $\overline{\mathcal{T}}$. By Corollary 7.1.20, we can also deduce that they are also independent of the choice of principal completions. The following is clear by the results in Chaps. 11 and 13.

Lemma 14.1.9 *Let $f : X \longrightarrow Y$ be a quasi-projective morphism of smooth algebraic varieties.*

- *Let H_Y be any algebraic hypersurface in Y. We put $H_X := f^{-1}(H_Y)$. Then, we have natural isomorphisms $f_\star^i(\mathcal{T}[\star H_X]) \simeq f_\star^i(\mathcal{T})[\star H_Y]$ for $\star = *, !$.*

- *We have natural isomorphisms* $f^i_\star \circ \boldsymbol{D}_X \simeq \boldsymbol{D}_Y \circ f^{-i}_{\overline{\star}}$. *Here,* $\star$ *and* $\overline{\star}$ *are as in Lemma 14.1.8. (See Corollary 13.3.3.)*
- *For* $\rho^* = j^*, \tilde{\gamma}^*, \tilde{\sigma}^*$, *we have natural isomorphisms* $\rho^* \circ f^i_\star \simeq f^i_\star \circ \rho^*$.

Later in Sect. 14.3.2, we shall study the push-forward by the morphisms which are not necessarily quasi-projective.

14.1.3.5 External Tensor Product

Let X_i $(i = 1, 2)$ be smooth algebraic varieties. Let us observe that the external tensor product induces a bi-functor $\mathrm{MTM}(X_1^{\mathrm{alg}}) \times \mathrm{MTM}(X_2^{\mathrm{alg}}) \longrightarrow \mathrm{MTM}(X_1^{\mathrm{alg}} \times X_2^{\mathrm{alg}})$. We take principal completions $(\overline{X}_i, D_i)$ of X_i. We set $\overline{X} := \overline{X}_1 \times \overline{X}_2$ and $D := (D_1 \times X_2) \cup (X_1 \times D_2)$. Then, $(\overline{X}, \overline{D})$ is a principal completion of $X_1 \times X_2$. For $\mathcal{T}_i \in \mathrm{MTM}(X_i^{\mathrm{alg}})$, we take $\overline{\mathcal{T}}_i \in \mathrm{MTM}(\overline{X}_i^{\mathrm{alg}})$, and we set $\mathcal{T}_1 \boxtimes \mathcal{T}_2 := \Psi_{(\overline{X},D)}\big(\overline{\mathcal{T}}_1 \boxtimes \overline{\mathcal{T}}_2\big)$. The following lemma is clear by the construction and by the results in Chaps. 11 and 13.

Lemma 14.1.10 *Let* $\mathcal{T}_i \in \mathrm{MTM}(X_i^{\mathrm{alg}})$ $(i = 1, 2)$.

- *For* $i = 1, 2$, *let* H_i *be any algebraic hypersurfaces in* X_i, *and let* $\star_i \in \{*, !\}$. *We have natural isomorphisms*

$$(\mathcal{T}_1[\star_1 H_1]) \boxtimes (\mathcal{T}_2[\star_2 H_2]) \simeq (\mathcal{T}_1 \boxtimes \mathcal{T}_2)\big[\star_1(H_1 \times X_2)\big]\big[\star_2(X_1 \times H_2)\big]$$

- *We have natural isomorphisms* $\boldsymbol{D}_{X_1 \times X_2}(\mathcal{T}_1 \boxtimes \mathcal{T}_2) \simeq (\boldsymbol{D}_{X_1}\mathcal{T}_1) \boxtimes (\boldsymbol{D}_{X_2}\mathcal{T}_2)$.
- *For* $\rho^* = j^*, \tilde{\gamma}^*, \tilde{\sigma}^*$, *we naturally have* $\rho^*(\mathcal{T}_1 \boxtimes \mathcal{T}_2) \simeq (\rho^*\mathcal{T}_1) \boxtimes (\rho^*\mathcal{T}_2)$.
- *Let* $f_i : X_i \longrightarrow Y_i$ $(i = 1, 2)$ *be quasi-projective morphisms of smooth algebraic varieties. Let* $f : X_1 \times X_2 \longrightarrow Y_1 \times Y_2$ *be the induced morphism. We have natural isomorphisms* $f^m_\star(\mathcal{T}_1 \boxtimes \mathcal{T}_2) \simeq \bigoplus_{k+\ell=m} f^k_{1\star}(\mathcal{T}_1) \boxtimes f^\ell_{2\star}(\mathcal{T}_2)$. □

14.1.3.6 Twist by Admissible Mixed Twistor Structure

Let X be a smooth algebraic variety with an algebraic hypersurface H. We take a principal completion $(\overline{X}, D)$ of X. Let $\overline{H}$ be the closure of H in $\overline{X}$. We set $\overline{H}_1 := \overline{H} \cup D$. Let $F : \overline{X}' \longrightarrow \overline{X}$ be a projective morphism such that (a) $\overline{H}'_1$ is a simple normal crossing hypersurface where $\overline{H}'_1 := \varphi^{-1}(\overline{H}_1)$, (b) F induces $\overline{X}' \setminus \overline{H}'_1 \simeq \overline{X} \setminus \overline{H}_1$. Let $\mathcal{V}' \in \mathrm{MTS}^{\mathrm{adm}}(\overline{X}', \overline{H}'_1)$. We obtain a filtered smooth $\mathcal{R}_{\overline{X}(*\overline{H}_1)}$-triple $\mathcal{V} := F^0_\dagger \mathcal{V}'$.

Let $\mathcal{T} \in \mathrm{MTM}(X^{\mathrm{alg}})$. Take $\overline{\mathcal{T}} \in \mathrm{MTM}(\overline{X})$ such that $\Psi_{(\overline{X},D)}(\overline{\mathcal{T}}) = \mathcal{T}$. We obtain a mixed twistor $\mathscr{D}$-module $\big(\overline{\mathcal{T}} \otimes \mathcal{V}\big)[\star \overline{H}_1]$ on $\overline{X}$. We set

$$\big(\mathcal{T} \otimes \mathcal{V}\big)[\star H] := \Psi_{(\overline{X},D)}\Big(\big(\overline{\mathcal{T}} \otimes \mathcal{V}\big)[\star \overline{H}_1]\Big).$$

The following lemma is clear by the construction and by the results in Chaps. 11 and 13.

Lemma 14.1.11 *Let X, H and $\mathcal{V}$ be as above. Let $\mathcal{T} \in \mathrm{MTM}(X^{\mathrm{alg}})$.*

- *We have natural isomorphisms:*

$$\boldsymbol{D}(\mathcal{T} \otimes \mathcal{V}[\star H]) \simeq \big(\boldsymbol{D}\mathcal{T} \otimes \mathcal{V}^{\vee}\big)[\overline{\star} H],\ \boldsymbol{D}^{\mathrm{herm}}(\mathcal{T} \otimes \mathcal{V}[\star H]) \simeq \big(\boldsymbol{D}^{\mathrm{herm}}\mathcal{T} \otimes \mathcal{V}^{*}\big)[\overline{\star} H]$$

- *We have natural isomorphisms:*

$$j^{*}\big((\mathcal{T} \otimes \mathcal{V})[\star H]\big) \simeq \big(j^{*}\mathcal{T} \otimes j^{*}\mathcal{V}\big)[\star H]$$
$$\tilde{\gamma}^{*}\big((\mathcal{T} \otimes \mathcal{V})[\star H]\big) \simeq \big(\tilde{\gamma}^{*}\mathcal{T} \otimes \tilde{\gamma}^{*}_{sm}\mathcal{V}\big)[\star H]$$
$$\tilde{\sigma}^{*}\big((\mathcal{T} \otimes \mathcal{V})[\star H]\big) \simeq \big(\tilde{\sigma}^{*}\mathcal{T} \otimes (\mathcal{V}^{\vee})^{*}\big)[\star H]$$

□

Lemma 14.1.12 *Let X, H and $\mathcal{V}$ be as above. Let $f : Y \longrightarrow X$ be any quasi-projective morphism of smooth algebraic varieties. We set $H_Y := f^{-1}(H)$. Let $\mathcal{T}_Y \in$ $\mathrm{MTM}(Y^{\mathrm{alg}})$. Then, we have natural isomorphisms*

$$f^{i}_{\dagger}\big((\mathcal{T}_Y \otimes f^{*}\mathcal{V})[\star H_Y]\big) \simeq \big(f^{i}_{\dagger}(\mathcal{T}_Y) \otimes \mathcal{V}\big)[\star H].$$

□

Let X_i $(i = 1, 2)$ be smooth algebraic varieties. Let H_i be algebraic hypersurfaces in X_i. Let $\varphi_i : X'_i \longrightarrow X_i$ be projective morphisms of smooth algebraic varieties such that $H'_i := \varphi_i^{-1}(H_i)$ are simply normal crossing. We take principal completions $(\overline{X}_i, D_i)$ of X_i and $(\overline{X}'_i, D'_i)$ of X'_i. We may assume to have projective morphisms $\overline{X}'_i \longrightarrow \overline{X}_i$ which induce $X'_i \longrightarrow X_i$. We may also assume that $D'_i \cup H'_i$ are simply normal crossing.

Lemma 14.1.13 *Let $\mathcal{V}_i \in \mathrm{MTS}^{\mathrm{adm}}(\overline{X}'_i, D'_i \cup H'_i)$. Let $\mathcal{T}_i \in \mathrm{MTM}(X_i^{\prime\,\mathrm{alg}})$. Let $\star_i \in \{*, !\}$. Then, we have the following natural isomorphism in $\mathrm{MTM}\big((X_1 \times X_2)^{\mathrm{alg}}\big)$:*

$$\Big((\mathcal{T}_1 \otimes \varphi_{1\dagger}\mathcal{V}_1)[\star_1 H_1]\Big) \boxtimes \Big((\mathcal{T}_2 \otimes \varphi_{2\dagger}\mathcal{V}_1)[\star_2 H_2]\Big)$$
$$\simeq \Big((\mathcal{T}_1 \boxtimes \mathcal{T}_2) \otimes \big(\varphi_{1\dagger}\mathcal{V}_1 \boxtimes \varphi_{2\dagger}\mathcal{V}_2\big)\Big)[\star_1(H_1 \times X_2)][\star_2(X_1 \times H_2)] \qquad (14.2)$$

□

14.1.3.7 Beilinson Functors

Let g be an algebraic function on X. Set $H := g^{-1}(0)$. We take a principal completion $(\overline{X}, D)$ of X. We take $F : \overline{X}' \longrightarrow X$ as in Sect. 14.1.3.6. We have a meromorphic function $g' := F^*(g)$ on $(\overline{X}', \overline{H}'_1)$. For simplicity, we assume that g' gives a morphism $\overline{X}' \longrightarrow \mathbb{P}^1$ so that $(g')^{-1}(\{0, \infty\}) \subset \overline{H}'_1$.

By the pull back of the Beilinson triples, we obtain admissible mixed twistor structures $\tilde{\mathbb{I}}^{a,b}_{g'} := (g')^* \tilde{\mathbb{I}}^{a,b}(*\overline{H}'_1)$. By taking the push-forward, we obtain filtered $\mathcal{R}_{\overline{X}(*\overline{H}_1)}$-triples $\tilde{\mathbb{I}}^{a,b}_g := F_\dagger(\tilde{\mathbb{I}}^{a,b}_{g'})$.

Let $\mathcal{T} \in \mathrm{MTM}(X^{\mathrm{alg}})$. We take $\overline{\mathcal{T}} \in \mathrm{MTM}(\overline{X})$ such that $\Psi_{(\overline{X},D)}(\overline{\mathcal{T}}) = \mathcal{T}$. As in the ordinary case, for $\star = *, !$, we define

$$\Pi^{a,b}_{g\star}(\mathcal{T}) := \Psi_{(\overline{X},D)}\Big(\big(\overline{\mathcal{T}} \otimes \mathbb{I}^{a,b}_g\big)[\star \overline{H}_1]\Big).$$

We also define

$$\Pi^{a,b}_{g*!}(\mathcal{T}) := \varprojlim_{N\to\infty} \mathrm{Cok}\big(\Pi^{b,N}_{g!}(\mathcal{T}) \longrightarrow \Pi^{a,N}_{g*}(\mathcal{T})\big).$$

Lemma 14.1.14 *Fix $\mathcal{T}$. If N_0 is sufficiently large, for $N_0 \leq N_1 \leq N_2$, the natural morphisms*

$$\mathrm{Cok}\big(\Pi^{b,N_2}_{g!}(\mathcal{T}) \longrightarrow \Pi^{a,N_2}_{g*}(\mathcal{T})\big) \longrightarrow \mathrm{Cok}\big(\Pi^{b,N_1}_{g!}(\mathcal{T}) \longrightarrow \Pi^{a,N_1}_{g*}(\mathcal{T})\big)$$

are isomorphisms. In particular, we have $\Pi^{a,b}_{g*!}(\mathcal{T}) \simeq \mathrm{Cok}\big(\Pi^{b,N_0}_{g!}(\mathcal{T}) \longrightarrow \Pi^{a,N_0}_{g*}(\mathcal{T})\big)$.

Moreover, if N_0 is sufficiently large, for $N_0 \leq N_1 \leq N_2$, the following natural morphisms are also isomorphisms

$$\mathrm{Ker}\big(\Pi^{-N_1,b}_{g!}(\mathcal{T}) \longrightarrow \Pi^{-N_1,a}_{g*}(\mathcal{T})\big) \longrightarrow \mathrm{Ker}\big(\Pi^{-N_2,b}_{g!}(\mathcal{T}) \longrightarrow \Pi^{-N_2,a}_{g*}(\mathcal{T})\big).$$

Proof It is enough to check the claims locally around any point of X. Then, they are reduced to Lemmas 4.1.1 and 4.1.5. □

Similarly, by using the results in Sect. 4.1, we obtain the following natural isomorphisms:

$$\Pi^{a,b}_{g*!}(\mathcal{T}) \simeq \varinjlim_{N\to\infty} \mathrm{Ker}\big(\Pi^{-N,b}_{g!}(\mathcal{T}) \longrightarrow \Pi^{-N,a}_{g*}(\mathcal{T})\big)$$

The right hand side is isomorphic to $\mathrm{Ker}\big(\Pi^{-N_0,b}_{g!}(\mathcal{T}) \longrightarrow \Pi^{-N_0,a}_{g*}(\mathcal{T})\big)$ for a sufficiently large N_0. By the construction, the functors $\Pi^{a,b}_{g\star}$ and $\Pi^{a,b}_{g*!}$ are exact on

$\mathrm{MTM}(X^{\mathrm{alg}})$. In particular, we set

$$\Xi_g^{(a)}(\mathcal{T}) := \Pi_{g*!}^{a,a+1}(\mathcal{T}), \quad \psi_g^{(a)}(\mathcal{T}) := \Pi_{g*!}^{a,a}(\mathcal{T}).$$

Let $\mathcal{T}[*g] := \mathcal{T}[*H]$. We have the exact sequences:

$$0 \longrightarrow \psi_g^{(a+1)}(\mathcal{T}) \longrightarrow \Xi_g^{(a)}(\mathcal{T}) \longrightarrow \mathcal{T}[*g] \otimes \boldsymbol{T}(a) \longrightarrow 0$$

$$0 \longrightarrow \mathcal{T}[!g] \otimes \boldsymbol{T}(a) \longrightarrow \Xi_g^{(a)}(\mathcal{T}) \longrightarrow \psi_g^{(a)}(\mathcal{T}) \longrightarrow 0$$

As in the ordinary case, we define $\phi_g^{(a)}(\mathcal{T})$ as the cohomology of the following:

$$\mathcal{T}[!g] \otimes \boldsymbol{T}(a) \longrightarrow \Xi_g^{(a)}(\mathcal{T}) \oplus \big(\mathcal{T} \otimes \boldsymbol{T}(a)\big) \longrightarrow \mathcal{T}[*g] \otimes \boldsymbol{T}(a)$$

We have the naturally defined morphisms $\psi_g^{(a+1)}(\mathcal{T}) \longrightarrow \phi_g^{(a)}(\mathcal{T}) \longrightarrow \psi_g^{(a)}(\mathcal{T})$. We can reconstruct $\mathcal{T}$ as the cohomology of the following complex:

$$\psi_g^{(1)}(\mathcal{T}) \longrightarrow \Xi_g^{(0)}(\mathcal{T}) \oplus \phi_g^{(0)}(\mathcal{T}) \longrightarrow \psi_g^{(0)}(\mathcal{T})$$

The functor $\phi_g^{(a)}$ is also exact on $\mathrm{MTM}(X^{\mathrm{alg}})$.

The following lemmas are clear by the construction and by the results in Chaps. 11 and 13.

Lemma 14.1.15 *Let $f : X \longrightarrow Y$ be any projective morphism of smooth algebraic varieties. Let g_Y be an algebraic function on Y. We set $g_X := g_Y \circ f$. We have natural isomorphisms $\psi_{g_Y}^{(a)} \circ f_\dagger^j \simeq f_\dagger^j \circ \psi_{g_X}^{(a)}$, $\Xi_{g_Y}^{(a)} \circ f_\dagger^j \simeq f_\dagger^j \circ \Xi_{g_X}^{(a)}$ and $\phi_{g_Y}^{(a)} \circ f_\dagger^j \simeq f_\dagger^j \circ \phi_{g_X}^{(a)}$.* □

Lemma 14.1.16 *Let g be any algebraic function on X. We have natural isomorphisms $\boldsymbol{D} \circ \psi_g^{(a)} \simeq \psi_g^{(1-a)} \circ \boldsymbol{D}$ and $\boldsymbol{D}^{\mathrm{herm}} \circ \psi_g^{(a)} \simeq \psi_g^{(1-a)} \circ \boldsymbol{D}^{\mathrm{herm}}$. We also have $\boldsymbol{D} \circ \Xi_g^{(a)} \simeq \Xi_g^{(-a)} \circ \boldsymbol{D}$, $\boldsymbol{D} \circ \phi_g^{(a)} \simeq \phi_g^{(-a)} \circ \boldsymbol{D}$, $\boldsymbol{D}^{\mathrm{herm}} \circ \Xi_g^{(a)} \simeq \Xi_g^{(-a)} \circ \boldsymbol{D}^{\mathrm{herm}}$, and $\boldsymbol{D}^{\mathrm{herm}} \circ \phi_g^{(a)} \simeq \phi_g^{(-a)} \circ \boldsymbol{D}^{\mathrm{herm}}$.* □

Lemma 14.1.17 *Let g be any algebraic function on X. For $\rho^* = j^*, \tilde{\gamma}^*, \tilde{\sigma}^*$ and for $\upsilon_g^{(a)} = \Xi_g^{(a)}, \psi_g^{(a)}, \phi_g^{(a)}$, we have natural isomorphisms $\rho^* \circ \upsilon_g^{(a)} \simeq \upsilon_g^{(a)} \circ \rho^*$.* □

14.1.3.8 Gluing

We use the notation in Sect. 14.1.3.7. Let $X_2 := X \setminus H$. Let $\mathcal{T}_2 \in \mathrm{MTM}(X_2^{\mathrm{alg}})$. By definition, we have $\mathcal{T} \in \mathrm{MTM}(X^{\mathrm{alg}})$ such that $\mathcal{T}(*H) = \mathcal{T}_2$. We can easily observe that

$$\psi_g^{(a)}(\mathcal{T}_2) := \psi_g^{(a)}(\mathcal{T}), \quad \Xi_g^{(a)}(\mathcal{T}_2) := \Xi_g^{(a)}(\mathcal{T})$$

are independent of the choice of $\mathcal{T}$.

Suppose that we are given $\mathcal{Q} \in \mathrm{MTM}(X^{\mathrm{alg}})$ whose support is contained in H, and morphisms

$$\psi_g^{(1)}(\mathcal{T}_2) \xrightarrow{u} \mathcal{Q} \xrightarrow{v} \psi_g^{(0)}(\mathcal{T}_2)$$

in $\mathrm{MTM}(X^{\mathrm{alg}})$ such that $v \circ u$ is equal to the canonical morphism $\psi_g^{(1)}(\mathcal{T}_2) \longrightarrow \psi_g^{(0)}(\mathcal{T}_2)$. As in the ordinary case, we obtain an object $\mathrm{Glue}(\mathcal{T}_2, \mathcal{Q}, u, v)$ in $\mathrm{MTM}(X^{\mathrm{alg}})$ as the cohomology of the following:

$$\psi_g^{(1)}\mathcal{T}_2 \longrightarrow \Xi_g^{(0)}(\mathcal{T}_2) \oplus \mathcal{Q} \longrightarrow \psi_g^{(0)}(\mathcal{T}_2)$$

14.1.4 Cech Resolutions

Let X be a smooth algebraic variety. Let $\boldsymbol{H} = (H_i \mid i \in \Lambda)$ be a finite tuple of algebraic hypersurfaces. For any non-empty subset $I \subset \Lambda$, we put $H_I := \bigcup_{i \in I} H_i$.

For $I \subset J$, we have the naturally defined morphisms $a_{*,J,I} : \mathcal{T}[*H_I] \longrightarrow \mathcal{T}[*H_J]$ and $a_{!,I,J} : \mathcal{T}[!H_J] \longrightarrow \mathcal{T}[!H_I]$. We have $a_{*,K,J} \circ a_{*,J,I} = a_{*,K,I}$ and $a_{!,I,J} \circ a_{!,J,K} = a_{!,I,K}$ for $I \subset J \subset K$ in Λ.

Let $\mathbb{C}\boldsymbol{e} = \bigoplus_{i \in \Lambda} \mathbb{C}e_i$ be a vector space with a Hermitian metric such that $\boldsymbol{e} = (e_i \mid i \in \Lambda)$ is an orthogonal basis. Let V_I be the one-dimensional subspace in $\bigwedge^{|I|} \mathbb{C}\boldsymbol{e}$ generated by $\bigwedge_{i \in I} e_i$. We have the smooth $\mathcal{R}$-triple $E_I := \bigl(V_I \otimes \mathcal{O}_{\mathbb{C}_\lambda}, V_I \otimes \mathcal{O}_{\mathbb{C}_\lambda}, C_I\bigr)$, where C_I is the naturally induced sesqui-linear pairing. For $I \subset \Lambda$ and $i \in \Lambda \setminus I$, we set $Ii := I \sqcup \{i\}$. We have the morphism $b_{*,i} = (b_{i1}, b_{i2}) : E_I \longrightarrow E_{Ii}$, where b_{i1} is given by the inner product of e_i, and b_{i2} is given by the exterior product of e_i. By taking the Hermitian adjoint, we also have $b_{!,i} : E_{Ii} \longrightarrow E_I$.

For $k \geq 0$, we set

$$\mathcal{C}^k(\mathcal{T}, *\boldsymbol{H}) := \bigoplus_{|I|=k+1} \mathcal{T}[*H_I] \otimes E_I$$

We have the morphisms $\mathcal{T}[*H_I] \otimes E_I \longrightarrow \mathcal{T}[*H_{Ii}] \otimes E_{Ii}$ induced by $a_{*,Ii,I}$ and $b_{*,i}$. They give $\partial^k : \mathcal{C}^k(\mathcal{T}, *\boldsymbol{H}) \longrightarrow \mathcal{C}^{k+1}(\mathcal{T}, *\boldsymbol{H})$. We have $\partial^{k+1} \circ \partial^k = 0$. Thus, we obtain a complex $\mathcal{C}^\bullet(\mathcal{T}, *\boldsymbol{H})$ in $\mathrm{MTM}(X^{\mathrm{alg}})$. The morphisms $\mathcal{T} \longrightarrow \mathcal{T}[*H_i]$ $(i \in \Lambda)$ induce a morphism of complexes $\mathcal{T} \longrightarrow \mathcal{C}^\bullet(\mathcal{T}, *\boldsymbol{H})$.

Similarly, for $k \geq 0$, we set

$$\mathcal{C}^{-k}(\mathcal{T}, !\boldsymbol{H}) := \bigoplus_{|I|=k+1} \mathcal{T}[!H_I] \otimes E_I$$

We have the morphisms $\mathcal{T}_{Ii} \otimes E_{Ii} \longrightarrow \mathcal{T}_I \otimes E_I$ induced by $a_{!,I,Ii}$ and $b_{!,i}$. They give $\partial^{-k-1} : \mathcal{C}^{-k-1}(\mathcal{T}, !\boldsymbol{H}) \longrightarrow \mathcal{C}^{-k}(\mathcal{T}, !\boldsymbol{H})$. Thus, we obtain a complex $\mathcal{C}^\bullet(\mathcal{T}, !\boldsymbol{H})$ in

$\mathrm{MTM}(X^{\mathrm{alg}})$. The natural morphisms $\mathcal{T}[!H_i] \longrightarrow \mathcal{T}$ $(i \in \Lambda)$ induce a morphism of complexes $\mathcal{C}^\bullet(\mathcal{T}, !\boldsymbol{H}) \longrightarrow \mathcal{T}$.

Lemma 14.1.18 *If there exists a hypersurface H such that $\mathcal{T}[*H] = \mathcal{T}$ and $\bigcap_{i\in\Lambda} H_i \subset H$, then the morphism $\mathcal{T} \longrightarrow \mathcal{C}^\bullet(\mathcal{T}, *\boldsymbol{H})$ is a quasi-isomorphism.*

If there exists a hypersurface H such that $\mathcal{T}[!H] = \mathcal{T}$ and $\bigcap_{i\in\Lambda} H_i \subset H$, then the morphism $\mathcal{C}^\bullet(\mathcal{T}, !\boldsymbol{H}) \longrightarrow \mathcal{T}$ is a quasi-isomorphism.

In particular, if $\bigcap_{i\in\Lambda} H_i = \emptyset$, the morphisms $\mathcal{T} \longrightarrow \mathcal{C}^\bullet(\mathcal{T}, *\boldsymbol{H})$ and $\mathcal{C}^\bullet(\mathcal{T}, !\boldsymbol{H}) \longrightarrow \mathcal{T}$ are quasi-isomorphisms.

Proof By construction, if $\mathcal{T}[\star H] = \mathcal{T}$, then we have $\mathcal{C}^\bullet(\mathcal{T}, \star\boldsymbol{H})[\star H] = \mathcal{C}^\bullet(\mathcal{T}, \star\boldsymbol{H})$. Hence, it is enough to consider the case $\bigcap_{i\in\Lambda} H_i = \emptyset$.

We have a finite dimensional $\mathbb{Q}$-vector subspace $\mathcal{A} \subset \mathbb{R} \times \mathbb{C}$ such that (i) $\mathbb{Q} \times \{0\} \subset \mathcal{A}$, (ii) $\mathcal{T} \in \mathrm{MTM}(X^{\mathrm{alg}})_{\mathcal{A}}$. We take a non-zero complex number λ_0 which is generic with respect to $\mathcal{A}$. It is a standard fact that $\Upsilon^{\lambda_0}(\mathcal{T}) \longrightarrow \Upsilon^{\lambda_0}\mathcal{C}^\bullet(\mathcal{T}, *\boldsymbol{H})$ and $\Upsilon^{\lambda_0}\mathcal{C}^\bullet(\mathcal{T}, !\boldsymbol{H}) \longrightarrow \Upsilon^{\lambda_0}(\mathcal{T})$ are quasi-isomorphisms. Hence, we obtain the claim of the lemma. □

14.1.5 The Underlying $\mathcal{R}^{\lambda_0}$-Modules

14.1.5.1 Algebraic $\mathcal{R}^{\lambda_0}$-Modules

Let Y be any smooth algebraic variety. Let $\mathscr{D}_Y^{\mathrm{alg}}$ be the sheaf of algebraic differential operators on Y with the Zariski topology. It is equipped with the filtration F by the order of differential operators. We have the Rees algebra $R^F\mathscr{D}_Y^{\mathrm{alg}}$. We put

$$\mathcal{R}_Y^{\lambda_0,\mathrm{alg}} := R^F\mathscr{D}_Y^{\mathrm{alg}}/(\lambda - \lambda_0)R^F\mathscr{D}_Y^{\mathrm{alg}}.$$

Let D be an algebraic hypersurface of Y. We set $\mathcal{R}_{Y(*D)}^{\lambda_0,\mathrm{alg}} := \mathcal{R}_Y^{\lambda_0,\mathrm{alg}} \otimes_{\mathcal{O}_Y} \mathcal{O}_Y(*D)$. If $\lambda_0 \neq 0$, it is isomorphic to $\mathscr{D}_{Y(*D)}^{\mathrm{alg}}$. If $\lambda_0 = 0$, it is $\mathrm{Sym}^\bullet\, \Theta_Y(*D)$.

An $\mathcal{R}_{Y(*D)}^{\lambda_0}$-module is called algebraic if it is the analytification of an $\mathcal{R}_{Y(*D)}^{\lambda_0,\mathrm{alg}}$-module.

14.1.5.2 Algebraicity of the Underlying $\mathcal{R}^{\lambda_0}$-Modules

Let X be an algebraic variety with a principal completion $(\overline{X}, D)$. Let $\mathcal{T} \in \mathrm{MTM}(X^{\mathrm{alg}})$. Let $(\mathcal{M}_1, \mathcal{M}_2, C)$ be the underlying $\mathcal{R}_{\overline{X}(*D)}$-triple. Let λ_0 be any complex number. We have the $\mathcal{R}_{\overline{X}(*D)}^{\lambda_0}$-module $\Upsilon^{\lambda_0}(\mathcal{T}) := \mathcal{M}_2^{\lambda_0}$ on $\overline{X}$.

Proposition 14.1.19 *$\Upsilon^{\lambda_0}(\mathcal{T})$ is algebraic.*

Proof First, let us consider the case where X is projective. We take a small neighbourhood $U(\lambda_0)$ of λ_0, and set $\mathcal{X}^{(\lambda_0)} := U(\lambda_0) \times X$. Note that $\mathcal{M}_{2|\mathcal{X}^{(\lambda_0)}}$ is coherent over $\mathcal{R}_{X|\mathcal{X}^{(\lambda_0)}}$ and good over $\mathcal{O}_{\mathcal{X}^{(\lambda_0)}}$. Hence, we have coherent $\mathcal{O}_{\mathcal{X}^{(\lambda_0)}}$-modules Q_i $(i = 1, 2)$ and an exact sequence

$$\mathcal{R}_{X|\mathcal{X}^{(\lambda_0)}} \otimes_{\mathcal{O}_{\mathcal{X}^{(\lambda_0)}}} Q_2 \xrightarrow{\varphi_2} \mathcal{R}_{X|\mathcal{X}^{(\lambda_0)}} \otimes_{\mathcal{O}_{\mathcal{X}^{(\lambda_0)}}} Q_1 \xrightarrow{\varphi_1} \mathcal{M}_{2|\mathcal{X}^{(\lambda_0)}} \longrightarrow 0.$$

Set $Q_i^{\lambda_0} := \mathcal{O}_{\mathcal{X}^{\lambda_0}} \otimes_{\mathcal{O}_{\mathcal{X}^{(\lambda_0)}}} Q_i$, which are $\mathcal{O}_X$-coherent. We obtain the following exact sequence of $\mathcal{R}_X^{\lambda_0}$-modules:

$$\mathcal{R}_X^{\lambda_0} \otimes_{\mathcal{O}_X} Q_2^{\lambda_0} \xrightarrow{\varphi_2^{\lambda_0}} \mathcal{R}_X^{\lambda_0} \otimes_{\mathcal{O}_X} Q_1^{\lambda_0} \xrightarrow{\varphi_1^{\lambda_0}} \mathcal{M}_2^{\lambda_0} \longrightarrow 0$$

Because X is assumed to be projective, $Q_i^{\lambda_0}$ are algebraic. The morphism $\varphi_2^{\lambda_0}$ is determined by $Q_2^{\lambda_0} \longrightarrow \mathcal{R}_X^{\lambda_0} \otimes_{\mathcal{O}_X} Q_1^{\lambda_0}$, and hence it is also algebraic. Thus, we obtain the algebraicity of $\mathcal{M}_2^{\lambda_0}$ in the case where X is projective. We also obtain the algebraicity in the case where X is quasi-projective, or in particular, affine.

Let us consider the general case. We may assume that X is complete and irreducible. We take a tuple of hypersurfaces $\boldsymbol{H} = (H_i \,|\, i \in \Lambda)$ of X such that (i) $X \setminus H_i$ are affine, (ii) $\bigcap_{i\in\Lambda} H_i = \emptyset$. We have the Cech resolution $\mathcal{C}^{\bullet}(\mathcal{T}, *\boldsymbol{H})$ of $\mathcal{T}$ in MTM(X^{alg}). Then, $\Upsilon^{\lambda_0}\mathcal{C}^{\bullet}(\mathcal{T}, \boldsymbol{H})$ is a resolution of $\Upsilon^{\lambda_0}(\mathcal{T})$ in the category of $\mathscr{D}_X$-modules. We have already known that each $\Upsilon^{\lambda_0}\mathcal{C}^k(\mathcal{T}, *\boldsymbol{H})$ is algebraic. We can also obtain that the morphisms $\Upsilon^{\lambda_0}\mathcal{C}^k(\mathcal{T}, *\boldsymbol{H}) \longrightarrow \Upsilon^{\lambda_0}\mathcal{C}^{k+1}(\mathcal{T}, *\boldsymbol{H})$ are algebraic. Hence, we obtain that $\Upsilon^{\lambda_0}(\mathcal{T})$ is algebraic. □

We may naturally regard the $\mathcal{R}_{\overline{X}(*D)}^{\lambda_0\,\mathrm{alg}}$-module $\Upsilon^{\lambda_0}(\mathcal{T})$ as an $\mathcal{R}_X^{\lambda_0\,\mathrm{alg}}$-module.

14.1.5.3 Compatibility of the Functors

Let Hol(X^{alg}) denote the category of algebraic holonomic $\mathscr{D}_X$-modules. We have the duality functor $\boldsymbol{D}$ on Hol(X^{alg}). For any algebraic hypersurface H, we have the localization functors $[\star H]$ on Hol(X^{alg}). For any algebraic function g on X, we have the Beilinson functors $\Pi^{a,b}_{g*!}$. In particular, we have the maximal functors $\Xi_g^{(a)}$ and the nearby cycle functors $\psi_g^{(a)}$. The vanishing cycle functors $\phi_g^{(a)}$ are also induced. For an algebraic projective morphism $X \longrightarrow Y$, we have the cohomological functors $f_{\dagger}^i : \mathrm{Hol}(X^{\mathrm{alg}}) \longrightarrow \mathrm{Hol}(Y^{\mathrm{alg}})$ $(i \in \mathbb{Z})$. For algebraic varieties X_i $(i = 1, 2)$, we have the external tensor products $\boxtimes$ of algebraic holonomic $\mathscr{D}_{X_i}$-modules.

For simplicity, we fix a $\mathbb{Q}$-vector subspace $\mathcal{A} \subset \mathbb{R} \times \mathbb{C}$ such that $\mathbb{Q} \times \{0\} \subset \mathcal{A}$. For any non-zero complex number λ_0, we have the functor

$$\Upsilon^{\lambda_0} : \mathrm{MTM}(X^{\mathrm{alg}})_{\mathcal{A}} \longrightarrow \mathrm{Hol}(X^{\mathrm{alg}}) \tag{14.3}$$

for any smooth algebraic variety X. The following proposition is easy to see by construction.

Proposition 14.1.20 *The functor Υ^{λ_0} is compatible with the duality, the push-forward by projective morphisms and the external tensor products. If λ_0 is non-zero and generic with respect to $\mathcal{A}$, Υ^{λ_0} is also compatible with the localizations, the maximal functors, the nearby cycle functors, the vanishing cycle functors, the push-forward by quasi-projective morphisms, and the external tensor products.*

Proof The claim follows from the compatibility in Chaps. 11 and 13. (See [27] for the compatibility of the functors for algebraic $\mathscr{D}$-modules and the associated analytic $\mathscr{D}$-modules.) □

14.1.6 Real Structure

Let X be a smooth algebraic variety. A real structure on $\mathcal{T} \in \mathrm{MTM}(X^{\mathrm{alg}})$ is an isomorphism $\kappa : \tilde{\gamma}^*\mathcal{T} \simeq \mathcal{T}$ such that $\tilde{\gamma}^*\kappa \circ \kappa = \mathrm{id}$. Let $\mathrm{MTM}(X^{\mathrm{alg}}, \mathbb{R})$ denote the category of algebraic mixed twistor $\mathscr{D}$-modules with real structure on X. The functor $\Psi_{(\overline{X},D)}$ induces equivalences $\mathrm{MTM}(\overline{X}, [\star D], \mathbb{R}) \simeq \mathrm{MTM}(X^{\mathrm{alg}}, \mathbb{R})$ $(\star = *, !)$. We have the following as in the case of Proposition 13.4.6 and Corollary 13.4.7.

Proposition 14.1.21 *The functors in Sect. 14.1.3 are enhanced to the functors for algebraic mixed twistor $\mathscr{D}$-modules with real structure.* □

Let $\mathcal{A} \subset \mathbb{R} \times \mathbb{C}$ be a $\mathbb{Q}$-vector subspace such that $\mathbb{Q} \times \{0\} \subset \mathcal{A}$. We also impose that $(a, 0)$ and $(0, \alpha)$ are elements of $\mathcal{A}$ if $u = (a, \alpha) \in \mathcal{A}$. We have the full subcategory $\mathrm{MTM}(X^{\mathrm{alg}}, \mathbb{R})_{\mathcal{A}}$ in $\mathrm{MTM}(X^{\mathrm{alg}}, \mathbb{R})$ of the objects whose KMS-spectra are contained in $\mathcal{A}$. Suppose that a non-zero complex number λ_0 is generic with respect to $\mathcal{A}$. Let $\mathrm{Hol}(X^{\mathrm{alg}}, \mathbb{R})$ denote the category of algebraic $\mathscr{D}_X$-modules with $\mathbb{R}$-Betti structure. We obtain the following.

Lemma 14.1.22 *We have the functor* $\mathrm{MTM}(X^{\mathrm{alg}}, \mathbb{R})_{\mathcal{A}} \longrightarrow \mathrm{Hol}(X^{\mathrm{alg}}, \mathbb{R})$ *induced by Υ^{λ_0}. It is compatible with the functors as in Proposition 14.1.20.* □

14.1.7 Algebraic Integrable Mixed Twistor $\mathscr{D}$-Modules

14.1.7.1 Definition

Let X be a smooth algebraic variety. We take a principal completion $(\overline{X}, D)$ of X. We have the functor $\Psi^{\mathrm{int}}_{(\overline{X},D)} : \mathrm{MTM}^{\mathrm{int}}(\overline{X}) \longrightarrow \mathcal{R}\text{-}\mathrm{Tri}^{\mathrm{int}}(\overline{X}, D)^{\mathrm{fil}}$ given by $\Psi^{\mathrm{int}}_{\overline{X}}(\mathcal{T}, W) = (\mathcal{T}, W)(*D)$.

Definition 14.1.23 Let $\mathrm{MTM}^{\mathrm{int}}(X^{\mathrm{alg}})$ denote the essential image of $\Psi^{\mathrm{int}}_{(\overline{X},D)}$ in $\mathcal{R}\text{-Tri}^{\mathrm{int}}(\overline{X},D)^{\mathrm{fil}}$. Any object in $\mathrm{MTM}^{\mathrm{int}}(X^{\mathrm{alg}})$ is called an algebraic integrable mixed twistor $\mathscr{D}$-module. □

We have the equivalences $\mathrm{MTM}^{\mathrm{int}}(\overline{X},[\star D]) \simeq \mathrm{MTM}^{\mathrm{int}}(X^{\mathrm{alg}})$ ($\star = *,!$) given by $\Psi^{\mathrm{int}}_{(\overline{X},D)}$. For another principal completion $(\overline{X}',D')$, let $\mathrm{MTM}^{\mathrm{int}}(X^{\mathrm{alg}})'$ denote the essential image of $\Psi^{\mathrm{int}}_{(\overline{X}',D')}:\mathrm{MTM}^{\mathrm{int}}(\overline{X}',D')\longrightarrow\mathcal{R}\text{-Tri}^{\mathrm{int}}(\overline{X}',D')^{\mathrm{fil}}$. Then, we have a natural equivalence $\mathrm{MTM}^{\mathrm{int}}(X^{\mathrm{alg}})\simeq\mathrm{MTM}^{\mathrm{int}}(X^{\mathrm{alg}})'$. In this sense, $\mathrm{MTM}^{\mathrm{int}}(X^{\mathrm{alg}})$ is well defined. The following proposition is clear by the construction of the functors.

Proposition 14.1.24 *The functors in Sect. 14.1.3 are enhanced to the functors for algebraic integrable mixed twistor $\mathscr{D}$-modules.* □

Note that the KMS-spectra of any integrable mixed twistor $\mathscr{D}$-modules are contained in $\mathbb{R}\times\{0\}$. So, the functor $\Upsilon^1:\mathrm{MTM}^{\mathrm{int}}(X^{\mathrm{alg}})\longrightarrow\mathrm{Hol}(X^{\mathrm{alg}})$ is compatible with the functors in Sect. 14.1.3.

Later, we shall study the algebraicity of the underlying $\tilde{\mathcal{R}}$-modules of algebraic integrable mixed twistor $\mathscr{D}$-modules (see Sect. 14.4).

14.1.7.2 Real Structure

Let X be any smooth algebraic variety with a principal completion $(\overline{X},D)$. A real structure of $\mathcal{T}\in\mathrm{MTM}^{\mathrm{int}}(X^{\mathrm{alg}})$ is an isomorphism $\kappa:\tilde{\gamma}^*\mathcal{T}\simeq\mathcal{T}$ in $\mathrm{MTM}^{\mathrm{int}}(X^{\mathrm{alg}})$ such that $\tilde{\gamma}^*\kappa\circ\kappa=\mathrm{id}$. Let $\mathrm{MTM}(X^{\mathrm{alg}},\mathbb{R})$ be the category of algebraic integrable mixed twistor $\mathscr{D}$-modules with real structure on X.

Let $(\mathcal{T},\kappa)\in\mathrm{MTM}^{\mathrm{int}}(X,\mathbb{R})$. Let $\overline{\mathcal{T}}$ be the object in $\mathrm{MTM}^{\mathrm{int}}(\overline{X})$ such that $\Psi_{(\overline{X},D)}(\overline{\mathcal{T}})=\mathcal{T}$ and $\overline{\mathcal{T}}[*D]=\overline{\mathcal{T}}$. Then, κ naturally induces a real structure $\overline{\kappa}$ on $\overline{\mathcal{T}}$.

Let $(\mathcal{M}_1,\mathcal{M}_2,C)$ be the $\tilde{\mathcal{R}}_{\overline{X}(*D)}$-triple underlying $\mathcal{T}$. Let $(\overline{\mathcal{M}}_1,\overline{\mathcal{M}}_2,\overline{C})$ be the $\tilde{\mathcal{R}}_{\overline{X}}$-triple underlying $\overline{\mathcal{T}}$. Then, we have $\mathcal{M}_{2|\mathbb{C}_\lambda^*\times\overline{X}}=\overline{\mathcal{M}}_{2|\mathbb{C}_\lambda^*\times\overline{X}}$. Recall that $\overline{\kappa}$ determines an $\mathbb{R}$-Betti structure of $\overline{\mathcal{M}}_{2|\mathbb{C}_\lambda^*\times\overline{X}}$. Hence, we have the $\mathbb{R}$-Betti structure on $\mathcal{M}_{2|\mathbb{C}_\lambda^*\times\overline{X}}$ induced by κ.

14.1.7.3 Good K-Structure

Let X and $(\overline{X},D)$ be as above. Let K be a subfield of $\mathbb{R}$. Let $(\mathcal{T},\kappa)\in\mathrm{MTM}^{\mathrm{int}}(X^{\mathrm{alg}},\mathbb{R})$. Let $(\mathcal{M}_1,\mathcal{M}_2,C)$ be the $\tilde{\mathcal{R}}_{\overline{X}(*D)}$-triple underlying $\mathcal{T}$. Set $\overline{\mathcal{X}}^0:=\{0\}\times\overline{X}\subset\overline{\mathcal{X}}=\mathbb{C}_\lambda\times\overline{X}$. We have the $\mathscr{D}_{\mathbb{C}_\lambda\times\overline{X}}$-module $\mathcal{M}_2(*\overline{\mathcal{X}}^0)$. A good K-structure of $(\mathcal{T},\kappa)$ is a K-Betti structure P_K of $\mathcal{M}_2(*\overline{\mathcal{X}}^0)$ such that $P_K\otimes_K\mathbb{R}_{|\mathbb{C}_\lambda^*\times\overline{X}}$ is equal to the $\mathbb{R}$-Betti structure induced by κ. (See [57] for the concept of K-Betti structure of holonomic $\mathscr{D}$-modules in the context of irregular singularities. We also refer [57] for their basic functorial property.) We denote the category of integrable algebraic mixed twistor $\mathscr{D}_X$-module with a good K-structure by $\mathrm{MTM}^{\mathrm{int}}_{\mathrm{good}}(X^{\mathrm{alg}}, K)$. We have the following as in Proposition 13.4.25.

Proposition 14.1.25 *The functors in Sect. 14.1.3 are enhanced to the functors for algebraic mixed twistor $\mathscr{D}$-modules with good K-structures.* □

14.2 Derived Category of Algebraic Mixed Twistor $\mathscr{D}$-Modules

For any abelian category A, let D^bA denote the derived category of bounded complexes in A. For objects $C_i \in A$ $(i = 1, 2)$, let $\mathrm{Ext}^j_A(C_1, C_2) = \mathrm{Hom}_{D^bA}(C_1, C_2[j])$.

14.2.1 Some Exact Functors

Let X be any smooth algebraic variety. We have some triangulated functors induced by the exact functors on $\mathrm{MTM}(X^{\mathrm{alg}})$.

14.2.1.1 Localization

Let H be any algebraic hypersurface in X. The localization functors $[\star H]$ $(\star = *, !)$ on $\mathrm{MTM}(X^{\mathrm{alg}})$ induce exact functors on $D^b\,\mathrm{MTM}(X^{\mathrm{alg}})$, which are also denoted by the same notation. By comparing the Yoneda extensions, for $\mathcal{T}_i \in \mathrm{MTM}(X^{\mathrm{alg}})$ $(i = 1, 2)$, we have the following natural isomorphisms:

$$\mathrm{Ext}^i_{\mathrm{MTM}(X^{\mathrm{alg}})}\big(\mathcal{T}_1, \mathcal{T}_2[*H]\big) \simeq \mathrm{Ext}^i_{\mathrm{MTM}(X^{\mathrm{alg}})}\big(\mathcal{T}_1[*H], \mathcal{T}_2[*H]\big)$$

$$\mathrm{Ext}^i_{\mathrm{MTM}(X^{\mathrm{alg}})}\big(\mathcal{T}_1[!H], \mathcal{T}_2\big) \simeq \mathrm{Ext}^i_{\mathrm{MTM}(X^{\mathrm{alg}})}\big(\mathcal{T}_1[!H], \mathcal{T}_2[!H]\big)$$

14.2.1.2 Beilinson Functors

Let g be any algebraic function on X. The exact functors $\Pi^{a,b}_{g\star}$ on $\mathrm{MTM}(X^{\mathrm{alg}})$ induce exact functors on $D^b\,\mathrm{MTM}(X^{\mathrm{alg}})$ where $\star = *, !$ and $a, b \in \mathbb{Z}$. They are denoted by the same notation. In particular, we naturally obtain the exact functors $\Xi^{(a)}_g$, $\psi^{(a)}_g$ and $\phi^{(a)}_g$ on $D^b\,\mathrm{MTM}(X^{\mathrm{alg}})$.

14.2.1.3 Duality and Hermitian Adjoint

We extend $\boldsymbol{D}$ and $\boldsymbol{D}^{\rm herm}$ on $\mathrm{MTM}(X^{\rm alg})$ to the functors on $D^b\,\mathrm{MTM}(X^{\rm alg})$ as in Sect. 13.4.1.1.

Proposition 14.2.1

- *We have natural isomorphisms* $\boldsymbol{D}\circ\boldsymbol{D}\simeq\mathrm{id}$ *and* $\boldsymbol{D}^{\rm herm}\circ\boldsymbol{D}^{\rm herm}=\mathrm{id}$. *We also have* $\boldsymbol{D}\circ\boldsymbol{D}^{\rm herm}=\boldsymbol{D}^{\rm herm}\circ\boldsymbol{D}$. *The isomorphism of the underlying* $\mathcal{R}$*-modules are the natural isomorphisms.*
- *We have natural isomorphisms:*

$$\boldsymbol{D}(\mathcal{T}^{\bullet}[*H])\simeq\boldsymbol{D}(\mathcal{T}^{\bullet})[!H],\quad \boldsymbol{D}^{\rm herm}(\mathcal{T}^{\bullet}[*H])\simeq\boldsymbol{D}^{\rm herm}(\mathcal{T}^{\bullet})[!H].$$

- *Let g be any algebraic function on X. We have natural isomorphisms*

$$\Xi^{a,b}_{g*!}\boldsymbol{D}\mathcal{T}^{\bullet}\simeq\boldsymbol{D}\,\Xi^{-b+1,-a+1}_{g*!}\mathcal{T}^{\bullet},\quad \Xi^{a,b}_{g*!}\boldsymbol{D}^{\rm herm}\mathcal{T}^{\bullet}\simeq\boldsymbol{D}^{\rm herm}\,\Xi^{-b+1,-a+1}_{g*!}\mathcal{T}^{\bullet}.$$

In particular, the duality and the Hermitian adjoint are compatible with the nearby cycle functor, the vanishing cycle functor, and the maximal functor. □

By comparing Yoneda extensions, we have the following natural isomorphisms:

$$\mathrm{Ext}^i_{\mathrm{MTM}(X^{\rm alg})}(\mathcal{T}_1,\mathcal{T}_2)\simeq\mathrm{Ext}^i_{\mathrm{MTM}(X^{\rm alg})}\big(\boldsymbol{D}_X\mathcal{T}_2,\boldsymbol{D}_X\mathcal{T}_1\big)$$
$$\mathrm{Ext}^i_{\mathrm{MTM}(X^{\rm alg})}(\mathcal{T}_1,\mathcal{T}_2)\simeq\mathrm{Ext}^i_{\mathrm{MTM}(X^{\rm alg})}\big(\boldsymbol{D}^{\rm herm}_X\mathcal{T}_2,\boldsymbol{D}^{\rm herm}_X\mathcal{T}_1\big).$$

14.2.1.4 Functors $\tilde{\gamma}^*$ and j^*

The functor j^* on $\mathrm{MTM}(X^{\rm alg})$ naturally induces that on $D^b\,\mathrm{MTM}(X^{\rm alg})$, which is denoted by the same notation. We define $\tilde{\gamma}^*:=j^*\circ\boldsymbol{D}^{\rm herm}\circ\boldsymbol{D}$ on $D^b\,\mathrm{MTM}(X^{\rm alg})$ as in Sect. 13.4.1. We have a natural isomorphism $\tilde{\gamma}^*\circ\tilde{\gamma}^*=\mathrm{id}$.

- We have natural equivalences $\tilde{\gamma}^*\circ\boldsymbol{D}\simeq\boldsymbol{D}\circ\tilde{\gamma}^*$ and $\tilde{\gamma}^*\circ\boldsymbol{D}^{\rm herm}\simeq\boldsymbol{D}^{\rm herm}\circ\tilde{\gamma}^*$.
- We have natural isomorphisms $\tilde{\gamma}^*(\mathcal{T}^{\bullet}[*H])\simeq\tilde{\gamma}^*(\mathcal{T}^{\bullet})[*H]$.
- Let g be an algebraic function on X. We have natural isomorphisms

$$\Xi^{a,b}_{g\star}(\tilde{\gamma}^*\mathcal{T}^{\bullet})\simeq\tilde{\gamma}^*\,\Xi^{a,b}_{g\star}(\mathcal{T}^{\bullet}),\qquad \Xi^{a,b}_{g*!}(\tilde{\gamma}^*\mathcal{T}^{\bullet})\simeq\tilde{\gamma}^*\,\Xi^{a,b}_{g*!}(\mathcal{T}^{\bullet}).$$

In particular, we have the compatibility of $\tilde{\gamma}^*$ with the nearby cycle functor, the vanishing cycle functor, and the maximal functor.

14.2.1.5 External Tensor Product

First, let us consider the complex analytic case. Let X_i $(i = 1,2)$ be complex manifolds. Let $\mathcal{T}_i^\bullet \in C(\mathrm{MTM}(X_i))$. We set $\tilde{\mathcal{T}}^m := \bigoplus_{p+q=m} \mathcal{T}_1^p \boxtimes \mathcal{T}_2^q \in \mathrm{MTM}(X_1 \times X_2)$. Let $(\mathcal{M}_{i1}^{-p}, \mathcal{M}_{i2}^p, C_i^p)$ denote the underlying $\mathcal{R}_{X_i}$-triples of $\mathcal{T}_i^p$. We have the complexes $\mathcal{M}_{ij}^\bullet$ of $\mathcal{R}_{X_i}$-modules. They induce the complexes of $\mathcal{R}_{X_1\times X_2}$-modules $\mathrm{Tot}(\mathcal{M}_{1j}^\bullet \boxtimes \mathcal{M}_{2j}^\bullet)$. The morphisms induce $\tilde{\mathcal{T}}^m \longrightarrow \tilde{\mathcal{T}}^{m+1}$. In this way, we obtain the complex $\mathrm{Tot}(\mathcal{T}_1^\bullet \boxtimes \mathcal{T}_2^\bullet) := \tilde{\mathcal{T}}^\bullet$ in $\mathrm{MTM}(X_1 \times X_2)$. It is easy to check that $\mathcal{S}_p \mathrm{Tot}(\mathcal{T}_1^\bullet \boxtimes \mathcal{T}_2^\bullet) = \mathrm{Tot}\big(\mathcal{S}_p(\mathcal{T}_1^\bullet) \boxtimes \mathcal{T}_2^\bullet\big) = \mathrm{Tot}\big(\mathcal{T}_1^\bullet \boxtimes \mathcal{S}_p(\mathcal{T}_2^\bullet)\big)$ for any $p \in \mathbb{Z}$.

Let us consider the algebraic situation. Let X_i $(i = 1,2)$ be smooth algebraic varieties. Set $X := X_1 \times X_2$. We have the following functor induced by the above external tensor product of complexes of mixed twistor $\mathscr{D}$-modules:

$$\boxtimes : D^b\,\mathrm{MTM}(X_1^{\mathrm{alg}}) \times D^b\,\mathrm{MTM}(X_2^{\mathrm{alg}}) \longrightarrow D^b\,\mathrm{MTM}(X).$$

14.2.1.6 Real Structure

A real structure of $\mathcal{T}^\bullet \in D^b\,\mathrm{MTM}(X^{\mathrm{alg}})$ is defined to be an isomorphism $\kappa : \tilde{\gamma}^*\mathcal{T}^\bullet \simeq \mathcal{T}^\bullet$ in $D^b\,\mathrm{MTM}(X^{\mathrm{alg}})$ such that $\tilde{\gamma}^*\kappa \circ \kappa = \mathrm{id}$. Let $D^b(\mathrm{MTM}(X^{\mathrm{alg}}), \mathbb{R})$ be the category of objects in $D^b\,\mathrm{MTM}(X^{\mathrm{alg}})$ equipped with a real structure. A morphism $(\mathcal{T}_1^\bullet, \kappa_1) \longrightarrow (\mathcal{T}_2^\bullet, \kappa_2)$ in $D^b(\mathrm{MTM}(X^{\mathrm{alg}}), \mathbb{R})$ is a morphism $\varphi : \mathcal{T}_1^\bullet \longrightarrow \mathcal{T}_2^\bullet$ in $D^b(\mathrm{MTM}(X^{\mathrm{alg}}))$ such that $\varphi \circ \kappa_1 = \kappa_2 \circ \tilde{\gamma}^*\varphi$. The category $D^b(\mathrm{MTM}(X^{\mathrm{alg}}), \mathbb{R})$ is naturally equipped with localizations, Beilinson's functors, $\boldsymbol{D}$ and $\boldsymbol{D}^{\mathrm{herm}}$.

Let us look at basic examples. We have the objects $\mathcal{U}_X(0, d_X)[d_X]$ and $\mathcal{U}_X(d_X, 0)[-d_X]$ in $D^b\,\mathrm{MTM}(X^{\mathrm{alg}})$. We naturally identify $j^*\mathcal{O}_{\mathcal{X}}$ and $\mathcal{O}_{\mathcal{X}}$. We fix the integrable isomorphism $\mu_X : j^*\boldsymbol{D}\big(\lambda^{d_X}\mathcal{O}_{\mathcal{X}}[d_X]\big) \simeq \mathcal{O}_{\mathcal{X}}[-d_X]$ whose restriction to $\{\lambda\} \times X$ $(\lambda \neq 0)$ is given as in Sect. 12.6.6. Then, we have natural real structures $\big(\mu_X^{-1}, j^*\boldsymbol{D}(\mu_X)\big)$ of $\mathcal{U}_X(0, d_X)[d_X]$, and $\big(j^*\boldsymbol{D}(\mu_X)^{-1}, \mu_X\big)$ of $\mathcal{U}_X(d_X, 0)[-d_X]$. As mentioned in Proposition 12.6.28, they are obtained as the shift of the real structure $((-1)^{d_X}\nu_X^{-1}, \nu_X)$ of $\mathcal{U}_X(0, d_X)$, and the real structure $(\nu_X^{-1}, (-1)^{d_X}\nu_X)$ of $\mathcal{U}_X(d_X, 0)$. Here, $\nu_X : j^*\boldsymbol{D}\mathcal{O}_{\mathcal{X}} \simeq \lambda^{d_X}\mathcal{O}_{\mathcal{X}}$ is the isomorphism whose restriction to $\{\lambda\} \times X$ $(\lambda \neq 0)$ is equal to the isomorphism ν in Sect. 12.6.6.1.

14.2.2 A Version of Kashiwara's Equivalence

Let X be a smooth algebraic variety. Let A be any algebraic subset of X. Let $\mathrm{MTM}_A(X^{\mathrm{alg}}) \subset \mathrm{MTM}(X^{\mathrm{alg}})$ be the full subcategory of algebraic mixed twistor $\mathscr{D}$-modules on X whose supports are contained in A. Let $D_A^b\,\mathrm{MTM}(X^{\mathrm{alg}}) \subset D^b\,\mathrm{MTM}(X^{\mathrm{alg}})$ be the full subcategory of the objects $\mathcal{T}^\bullet$ such that the supports of the cohomology objects of $\mathcal{T}^\bullet$ are contained in A.

Proposition 14.2.2 *The natural functor*

$$D^b\,\mathrm{MTM}_A(X^{\mathrm{alg}}) \longrightarrow D^b_A\,\mathrm{MTM}(X^{\mathrm{alg}})$$

is an equivalence.

Proof According to [4], we have only to check the following effaceability:

- Let $\mathcal{T}_i \in \mathrm{MTM}_A(X^{\mathrm{alg}})$ $(i = 1, 2)$. For any $f \in \mathrm{Ext}^i_{\mathrm{MTM}(X^{\mathrm{alg}})}(\mathcal{T}_1, \mathcal{T}_2)$, there exists a monomorphism $\mathcal{T}_2 \longrightarrow \mathcal{T}_2'$ in $\mathrm{MTM}_A(X^{\mathrm{alg}})$ such that the image of f in $\mathrm{Ext}^i_{\mathrm{MTM}(X^{\mathrm{alg}})}(\mathcal{T}_1, \mathcal{T}_2')$ is 0.

We can show it by using the arguments in Sects. 2.2 and 2.2.1 in [2]. □

14.2.3 Enhancement

Let X be a smooth algebraic variety. Let K be a subfield of $\mathbb{R}$. The following proposition is clear by the construction.

Proposition 14.2.3 *The functors in Sect. 14.2.1 are naturally enhanced to the functors for*

$$D^b\,\mathrm{MTM}^{\mathrm{int}}(X^{\mathrm{alg}}),\quad D^b\,\mathrm{MTM}^{\mathrm{int}}(X^{\mathrm{alg}}, \mathbb{R}),\quad D^b\,\mathrm{MTM}^{\mathrm{int}}_{\mathrm{good}}(X^{\mathrm{alg}}, K).$$

□

Let A be any closed algebraic subset in X. Let $\mathrm{MTM}^{\mathrm{int}}_{A,\mathrm{good}}(X^{\mathrm{alg}}, K)$ denote the full subcategory in $\mathrm{MTM}^{\mathrm{int}}_{\mathrm{good}}(X^{\mathrm{alg}}, K)$ of the objects whose supports are contained in A. Let $D^b_A\,\mathrm{MTM}^{\mathrm{int}}_{\mathrm{good}}(X^{\mathrm{alg}}, K)$ denote the full subcategory in $D^b\,\mathrm{MTM}^{\mathrm{int}}_{\mathrm{good}}(X^{\mathrm{alg}}, K)$ of the objects whose cohomological supports are contained in A. The following is the enhancement of Proposition 14.2.2.

Proposition 14.2.4 *The natural functor*

$$D^b\,\mathrm{MTM}^{\mathrm{int}}_{A,\mathrm{good}}(X^{\mathrm{alg}}, K) \longrightarrow D^b_A\,\mathrm{MTM}^{\mathrm{int}}_{\mathrm{good}}(X^{\mathrm{alg}}, K)$$

is an equivalence. We have similar claims for $\mathrm{MTM}^{\mathrm{int}}(X)$ *and* $\mathrm{MTM}^{\mathrm{int}}(X, \mathbb{R})$. □

14.3 Push-Forward and Pull Back

14.3.1 Push-Forward of Algebraic Holonomic D-Modules

For any smooth algebraic variety X, let $D^b_h(\mathscr{D}_X^{\rm alg})$ denote the derived category of cohomologically bounded holonomic algebraic $\mathscr{D}_X$-complexes. Let $f : X \longrightarrow Y$ be a morphism of smooth algebraic varieties. If f is proper, we have the push-forward $f_\dagger : D^b_h(\mathscr{D}_X^{\rm alg}) \longrightarrow D^b_h(\mathscr{D}_Y^{\rm alg})$ obtained as $f_\dagger(M^\bullet) = Rf_*\big(\mathscr{D}^{\rm alg}_{Y\leftarrow X} \otimes^L_{\mathscr{D}_X^{\rm alg}} M^\bullet\big)$. If f is an open immersion such that $Y \setminus f(X)$ is a hypersurface, we have the natural push-forward ${}^D\!f_* : D^b_h(\mathscr{D}_X^{\rm alg}) \longrightarrow D^b_h(\mathscr{D}_Y^{\rm alg})$ and ${}^D\!f_! := \boldsymbol{D}_Y \circ {}^D\!f_* \circ \boldsymbol{D}_X$. In general, we take a smooth algebraic variety X_1 with morphisms $X \stackrel{\iota}{\longrightarrow} X_1 \stackrel{\overline{f}}{\longrightarrow} Y$ such that (a) ι is an open immersion and $X_1 \setminus \iota(X)$ is a hypersurface, (b) $\overline{f}$ is proper, (c) $f = \overline{f} \circ \iota$. Then, we obtain triangulated functors ${}^D\!f_\star := \bar{f}_\dagger \circ {}^D\iota_\star : D^b_h(\mathscr{D}_X^{\rm alg}) \longrightarrow D^b_h(\mathscr{D}_Y^{\rm alg})$ for $\star = *, !$. In particular, we have the cohomological functors ${}^D\!f^i_\star : \mathrm{Hol}(X^{\rm alg}) \longrightarrow \mathrm{Hol}(Y^{\rm alg})$ $(\star = *, !)$

Beilinson [2] proved that the natural functor $D^b\,\mathrm{Hol}(X^{\rm alg}) \longrightarrow D^b_h(\mathscr{D}_X^{\rm alg})$ is an equivalence. He also gave the direct method to construct the push-forward functors ${}^h\!f_\star : D^b\,\mathrm{Hol}(X^{\rm alg}) \longrightarrow D^b\,\mathrm{Hol}(Y^{\rm alg})$ $(\star = *, !)$ from ${}^D\!f^i_\star$. The method was applied by Saito [73] in the construction of the push-forward functors for algebraic mixed Hodge modules.

Proposition 14.3.1 (Beilinson) *Let $f : X \longrightarrow Y$ be any morphism of smooth algebraic varieties. We have triangulated functors*

$$ {}^h\!f_\star : D^b\,\mathrm{Hol}(X^{\rm alg}) \longrightarrow D^b\,\mathrm{Hol}(Y^{\rm alg}) \quad (\star = *, !), $$

with an equivalence of cohomological functors ${}^h\!f^i_\star \simeq {}^D\!f^i_\star$ $(i \in \mathbb{Z})$. Here $\{{}^h\!f^i_\star \mid i \in \mathbb{Z}\}$ denotes the cohomological functor induced by ${}^h\!f_\star$.

- *For any such triangulated functors ${}^h\!f^{(0)}_\star$, we naturally have equivalences ${}^h\!f_\star \simeq {}^h\!f^{(0)}_\star$.*
- *Let $g : Y \longrightarrow Z$ be a morphism of smooth algebraic varieties. Then, we naturally have ${}^h(g \circ f)_\star \simeq {}^h g_\star \circ {}^h\!f_\star$ $(\star = *, !)$.*

We shall recall the proof of Proposition 14.3.1 in a way convenient to us because it can be applied in the construction of the push-forward functors for mixed twistor $\mathscr{D}$-modules. We shall give some complement in Sect. 14.3.1.11. We refer [2, 73] for more details and precisions.

14.3.1.1 Preliminary

Let M be any holonomic $\mathscr{D}$-module on a complex manifold Y. We have the stratification $\mathrm{Supp}(M) = \coprod_{i\in\Gamma} S_i$ such that (a) S_i are smooth locally closed complex analytic subvarieties of Y, (b) the characteristic variety of M is $\coprod_{i\in\Gamma} T^*_{S_i}Y$. Let H be a smooth hypersurface of Y. Let $\iota_H : H \longrightarrow Y$ be the inclusion. We say that H is transversal with the stratification $\coprod_{i\in\Gamma} T^*_{S_i}Y$ if H is transversal with S_i $(i \in \Gamma)$. If H is transversal with the stratification for M, then H is non-characteristic for M. It implies that $\iota_{H\dagger}\mathcal{O}_H$ and $\mathcal{O}_Y(*H)$ are also non-characteristic to M. Hence, we have the following exact sequence:

$$0 \longrightarrow M \longrightarrow M(*H) \longrightarrow \iota_{H\dagger}\iota_H^*M \longrightarrow 0 \tag{14.4}$$

Note that $\iota_{H\dagger}(\iota_H^*M) \simeq \iota_{H\dagger}(\mathcal{O}_H) \otimes M$. By using the duality, we obtain

$$0 \longrightarrow \iota_{H\dagger}\iota_H^*M \longrightarrow M(!H) \longrightarrow M \longrightarrow 0. \tag{14.5}$$

Set $S_{iH} := S_i \cap H$ and $S_i^\circ := S_i \setminus H$. Then, we have the stratification $Y = \coprod_{i\in\Gamma} S_{iH} \sqcup \coprod_{i\in\Gamma} S_i^\circ$, and the characteristic variety of $M(\star H)$ is $\coprod_{i\in\Gamma} T^*_{S_{iH}}Y \sqcup \coprod_{i\in\Gamma} T^*_{S_i^\circ}Y$.

Lemma 14.3.2 *If moreover we have a hypersurface H_1 such that $M(*H_1) = M$, then we have $\big(M(!H)\big)(*H_1) = M(!H)$. Dually, if $M(!H_1) = M$, then we have $\big(M(*H)(!H_1)\big) = M(*H)$.*

Proof We may assume $H_1 = \{f = 0\}$ for a holomorphic function f. The multiplication of f on M and $\iota_{H\dagger}\iota_H^*M$ are invertible. Hence, we obtain that the multiplication of f is invertible on $M(!H)$ by (14.5). □

14.3.1.2 Acyclic Algebraic Holonomic $\mathscr{D}$-Modules

Let Y be any smooth complete algebraic variety. For $\star = *, !$, let $\mathrm{Hol}(Y^{\mathrm{alg}})^{ac}$ denote the full subcategories in $\mathrm{Hol}(Y^{\mathrm{alg}})$ of the objects M with the following property.

- Let $f : Y \longrightarrow Z$ be any morphism of smooth algebraic varieties. Then, $f_\dagger^j M = 0$ unless $j = 0$.

Note that $f_\dagger^0 M \in \mathrm{Hol}(Z^{\mathrm{alg}})^{ac}$ in that case.

14.3.1.3 Generic Hypersurfaces

Let Y be a projective complex manifold with a very ample line bundle $\mathcal{O}_Y(1)$. Let $M_1, \ldots, M_m$ be any algebraic holonomic $\mathscr{D}$-modules on Y.

Lemma 14.3.3 *We can take a finite tuple of hyperplane sections H_j $(j \in \mathcal{S})$ of $\mathcal{O}_Y(1)$ such that (i) H_j are smooth, (ii) $\bigcap_{j\in\mathcal{S}} H_j = \emptyset$, (iii) for any i and for any $J \subset \mathcal{S}$ and $k \in \mathcal{S} \setminus J$, H_k is non-characteristic for $M_i(\star H_J)$ $(\star = *, !)$, where $H_J := \bigcup_{j\in J} H_j$.*

Proof By setting $M = \bigoplus M_i$, it is enough to consider the case $m = 1$. Let $Y = \coprod_{\ell \in \Gamma} S_\ell$ be the stratification for M as above. By Bertini's theorem, we can take a smooth hyperplane H_1 which is transversal with the stratification. Inductively, we take smooth hyperplanes H_k such that (i) H_k is transversal with the stratifications of $M(\star H_J)$ for $J \subset \{1, \ldots, k-1\}$, (ii) $\bigcup_{j=1}^{k} H_k$ is normal crossing. If k is sufficiently large, we have $\bigcap_{j=1}^{k} H_j = \emptyset$. By using the description of the stratification for the $\mathscr{D}$-modules, we can check that the tuple H_j satisfies the last condition. □

Let H_0 be an ample hypersurface of Y.

Lemma 14.3.4 *If $M_i(*H_0) = M_i$, then $M_i(!H_I) \in \mathrm{Hol}(Y^{\mathrm{alg}})^{ac}$ for any non-empty subset $I \subset \mathcal{S}$. If $M_i(!H_0)$, then $M_i(*H_I) \in \mathrm{Hol}(Y^{\mathrm{alg}})^{ac}$ for any non-empty subset $I \subset \mathcal{S}$.*

Proof Let us consider the first claim. Note $M_i(!H_I) = M_i(!H_I)(*H_0)$ under the assumption. Let $F : Y \longrightarrow Y'$ be any morphism of smooth algebraic varieties. Because H_I is ample, we have $F_\dagger^j M_i(!H_I) = 0$ $(j < 0)$. Because $M_i(!H_I)(*H_0) = M_i(!H_I)$ for the ample hypersurface H_0, we have $F_\dagger^j M_i(!H_I) = 0$ $(j > 0)$. Thus, we obtain the first claim. The second claim can be proved similarly. □

14.3.1.4 Cut of $\mathscr{D}$-Modules on Affine Open Subsets

Let X be any complete algebraic variety. Let U be any affine open subset in X. Set $D := X \setminus U$. Let H be any hypersurface in X.

We take a smooth projective completion X' of U with a morphism $\varphi : X' \longrightarrow X$ which induces the identity on U. Let $D' := X' \setminus U = \varphi^{-1}(D)$. We set $H' := \varphi^{-1}(H)$.

Let M be any holonomic $\mathscr{D}$-module on X such that $M(!D)(*H) = M$. We have the $\mathscr{D}$-module M' on X' such that $\varphi_\dagger(M') = M$ and $M'(!D')(*H') = M'$. Indeed, we have the $\mathscr{D}$-module M'' on X' such that $\varphi_\dagger M'' = M(!D)$ and $M''(!D') = M''$. Then, $M' := M''(*H')$ has the desired property.

Let H_1' be an ample hypersurface in X_1 which is non-characteristic for M'. Then, we have

$$M'(!H_1') = M'(!H_1')(*H') = M'(!D')(*H')(!H_1')(*H') = M'(!D')(!H_1')(*H').$$

We set $H_1 := \varphi(H_1')$.

Lemma 14.3.5 *We have $M(!D)(!H_1)(*H) = \varphi_\dagger\big(M'(!H_1')\big)$.*

Proof Note that $\varphi^{-1}(H_1 \cup D) = H_1' \cup D'$. Hence, we have

$$\begin{aligned} \varphi_\dagger\big(M'(!H_1')\big) &\simeq \varphi_\dagger\big(M'(!D')(!H_1')(*H')\big) \simeq \varphi_\dagger(M')(!D)(!H_1)(*H) \\ &= M(!D)(!H_1)(*H) \end{aligned} \tag{14.6}$$

Thus, we obtain the claim of the lemma. □

Lemma 14.3.6 *Suppose that $X \setminus H$ is affine. Then, $M(!D)(!H_1)(*H)$ is an object in* $\mathrm{Hol}(X^{\mathrm{alg}})^{ac}$.

Proof It is enough to prove that $f_\dagger^i M(!D)(!H_1)(*H) = 0$ $(i < 0)$ for any morphism $f : X \longrightarrow Y$. Because H_1' is ample, we have $g_\dagger^i\big(M'(!H_1')\big) = 0$ for any morphism $g : X' \longrightarrow Y$ and for $i < 0$. We also have $\varphi_\dagger^i\big(M'(!H_1')\big) = 0$ for $i \neq 0$. Then, the claim of the lemma follows. □

14.3.1.5 Cech Resolutions

Let X be any smooth algebraic variety. Let $\boldsymbol{H} = (H_i \,|\, i \in \Lambda)$ be a tuple of hypersurfaces. For any subset $I \subset \Lambda$, we set $H_I := \bigcup_{i \in I} H_i$.

Let $M \in \mathrm{Hol}(X^{\mathrm{alg}})$. For $I \subset J$, we have the naturally defined morphisms $a_{*,J,I} : M(*H_I) \longrightarrow M(*H_J)$ and $a_{!,I,J} : M(!H_J) \longrightarrow M(!H_I)$.

Let $\mathbb{C}(\boldsymbol{e})$ be the $\mathbb{C}$-vector space with a base $\boldsymbol{e} = (e_i \,|\, i \in \Lambda)$. For any $I \subset \Lambda$ with $|I| = k$, let $V_I \subset \bigwedge^k \mathbb{C}(\boldsymbol{e})$ be the one-dimensional subspace generated by $\bigwedge_{i \in I} e_i$.

For $k \geq 0$, we set

$$\mathcal{C}^k(M, *\boldsymbol{H}) = \bigoplus_{|I|=k+1} M(*H_I) \otimes V_I.$$

The morphisms $a_{*,Ii,I}$ and the multiplication of e_i induce $M(*H_I) \otimes V_I \longrightarrow M(*H_{Ii}) \otimes V_{Ii}$. They give morphisms $\partial^k : \mathcal{C}^k(M, *\boldsymbol{H}) \longrightarrow \mathcal{C}^{k+1}(M, *\boldsymbol{H})$ satisfying $\partial^{k+1} \circ \partial^k = 0$. Thus, we obtain a complex $\mathcal{C}^\bullet(M, *\boldsymbol{H})$ in $\mathrm{Hol}(X^{\mathrm{alg}})$.

For $k \geq 0$, we set

$$\mathcal{C}^{-k}(M, !\boldsymbol{H}) := \bigoplus_{|I|=k+1} M(!I) \otimes V_I^\vee.$$

The natural morphism $a_{!I,Ii}$ and the inner product of e_i induce $M(!Ii) \otimes V_{Ii}^\vee \longrightarrow M(!I) \otimes V_I^\vee$. They give a morphism $\partial^{-k-1} : \mathcal{C}^{-k-1}(M, !\boldsymbol{H}) \longrightarrow \mathcal{C}^{-k}(M, !\boldsymbol{H})$ satisfying $\partial^{-k} \circ \partial^{-k-1} = 0$. Thus, we obtain a complex $\mathcal{C}^\bullet(M, !\boldsymbol{H})$.

For any complex $M^\bullet$ in $\mathrm{Hol}(X^{\mathrm{alg}})$, we obtain the double complexes $\mathcal{C}^\bullet(M, \star\boldsymbol{H})$ $(\star = *, !)$, and we have the naturally defined morphisms of complexes:

$$M^\bullet \longrightarrow \mathrm{Tot}\, \mathcal{C}^\bullet(M^\bullet, *\boldsymbol{H}) \tag{14.7}$$

$$\mathrm{Tot}\, \mathcal{C}^\bullet(M^\bullet, !\boldsymbol{H}) \longrightarrow M^\bullet \tag{14.8}$$

Lemma 14.3.7 *Suppose that there exists a hypersurface H such that (i) $\bigcap_{i \in \Lambda} H_i \subset H$, (ii) $M^\bullet(*H) = M^\bullet$. Then, (14.7) is a quasi-isomorphism. If we have $M^\bullet(!H) = M^\bullet$ instead of (ii), then (14.8) is a quasi-isomorphism.*

Proof It is standard that (14.7) and (14.8) are quasi-isomorphisms on $X \setminus \bigcap_{i \in \Lambda} H_i$. Then, the claim is clear. □

14.3.1.6 Filtered Complexes of Acyclic $\mathscr{D}$-Modules

Let X be any smooth complete algebraic variety. Let $\mathfrak{U} = (U_k \mid k \in \Lambda)$ be a family of affine open subsets in X. Let $H_k := X \setminus U_k$. Let $\mathrm{Cpx}_*(\mathfrak{U})$ denote the category of bounded complexes $C^\bullet$ in $\mathrm{Hol}(X^{\mathrm{alg}})$ equipped with decompositions $C^n = \bigoplus_{k \in \Lambda} C_k^n$ $(n \in \mathbb{Z})$ satisfying the following conditions.

- $C_k^n(*H_k) = C_k^n$.
- Let $\partial_{k_1,k_2}^n : C_{k_2}^n \longrightarrow C_{k_1}^{n+1}$ be the component of the differential $\partial^n : C^n \longrightarrow C^{n+1}$ with respect to the decompositions. Then, we have $\partial_{k_1,k_2}^n = 0$ unless $U_{k_1} \subset U_{k_2}$, i.e., $H_{k_1} \supset H_{k_2}$.

A morphism of $C_1^\bullet \longrightarrow C_2^\bullet$ in $\mathrm{Cpx}_*(\mathfrak{U})$ is a morphism of complexes $\varphi : C_1^\bullet \longrightarrow C_2^\bullet$ such that the components $\varphi_{k_1,k_2}^n : C_{1,k_2}^n \longrightarrow C_{2,k_1}^n$ are 0 unless $U_{k_1} \subset U_{k_2}$. Let $\mathrm{Cpx}_*(\mathfrak{U})^{ac}$ be the full subcategory in $\mathrm{Cpx}_*(\mathfrak{U})$ of the objects $C^\bullet$ such that $C^n \in \mathrm{Hol}(X^{\mathrm{alg}})^{ac}$ for any n.

We have the dual concept. Let $\mathrm{Cpx}_!(\mathfrak{U})$ be the category of bounded complexes $C^\bullet$ in $\mathrm{Hol}(X^{\mathrm{alg}})$ equipped with decompositions $C^n = \bigoplus_{k \in \Lambda} C_k^n$ $(n \in \mathbb{Z})$ satisfying the following conditions:

- $C_k^n(!H_k) = C_k^n$.
- $\partial_{k_1,k_2}^n = 0$ unless $U_{k_1} \supset U_{k_2}$, i.e., $H_{k_1} \subset H_{k_2}$.

A morphism of $C_1^\bullet \longrightarrow C_2^\bullet$ in $\mathrm{Cpx}_!(\mathfrak{U})$ is a morphism of complexes $\varphi : C_1^\bullet \longrightarrow C_2^\bullet$ such that the components $\varphi_{k_1,k_2}^n : C_{1,k_2}^n \longrightarrow C_{2,k_1}^n$ are 0 unless $U_{k_1} \supset U_{k_2}$. Let $\mathrm{Cpx}_!(\mathfrak{U})^{ac}$ be the full subcategory in $\mathrm{Cpx}_!(\mathfrak{U})$ of the objects $C^\bullet$ such that $C^n \in \mathrm{Hol}(X^{\mathrm{alg}})^{ac}$ for any n.

14.3.1.7 Quasi Isomorphisms

Let X and $\mathfrak{U}$ be as in Sect. 14.3.1.6. Let Γ be a category which consists of finite objects and finite morphisms.

Proposition 14.3.8 *Suppose that we are given a functor $\varphi : \Gamma \longrightarrow \mathrm{Cpx}_*(\mathfrak{U})$. Then, we have a functor $\tilde{\varphi} : \Gamma \longrightarrow \mathrm{Cpx}_*(\mathfrak{U})^{ac}$ and a natural transform $\tilde{\varphi} \longrightarrow \varphi$ such that $\tilde{\varphi}(b) \longrightarrow \varphi(b)$ $(b \in \Gamma)$ are quasi-isomorphisms.*

Similarly, if we are given a functor $\varphi : \Gamma \longrightarrow \mathrm{Cpx}_!(\mathfrak{U})$, we have a functor $\tilde{\varphi} : \Gamma \longrightarrow \mathrm{Cpx}_!(\mathfrak{U})^{ac}$ and a natural transform $\varphi \longrightarrow \tilde{\varphi}$ such that $\varphi(b) \longrightarrow \tilde{\varphi}(b)$ $(b \in \Gamma)$ are quasi-isomorphisms.

For example, if we are given morphisms $C_1 \longrightarrow C_2 \longleftarrow C_3$ in $\mathrm{Cpx}_*(\mathfrak{U})$, then we can take morphisms $\tilde{C}_1 \longrightarrow \tilde{C}_2 \longleftarrow \tilde{C}_3$ in $\mathrm{Cpx}_*(\mathfrak{U})^{ac}$ with quasi-isomorphisms $\varphi_i : \tilde{C}_i \longrightarrow C_i$ $(i = 1, 2, 3)$ in $\mathrm{Cpx}_*(\mathfrak{U})$ such that the following is commutative

$$\begin{array}{ccccc} \widetilde{C}_1 & \longrightarrow & \widetilde{C}_2 & \longleftarrow & \widetilde{C}_3 \\ \downarrow & & \downarrow & & \downarrow \\ C_1 & \longrightarrow & C_2 & \longleftarrow & C_3 \end{array}$$

14.3.1.8 Proof of Proposition 14.3.8

We give a proof for $\mathrm{Cpx}_*(\mathfrak{U})$. The claim for $\mathrm{Cpx}_!(\mathfrak{U})$ can be proved in a dual way.

For $b \in \Gamma$, we denote the complex $\varphi(b)$ by $C_b^\bullet$. We shall use an induction on $\max_{b\in\Gamma} \dim \mathrm{Supp}\, C_b^\bullet$. We take a finite affine covering $\mathfrak{U}^{(1)} = (U_j^{(1)} \mid j \in \Lambda^{(1)})$ of X. We set $H_j^{(1)} := X \setminus U_j^{(1)}$. Let $\boldsymbol{H}^{(1)}$ denote the tuple $(H_j^{(1)} \mid j \in \Lambda^{(1)})$. In the following, for a given tuple of hypersurfaces $(K_i \mid i \in \mathcal{S})$, we set $K_I := \bigcup_{i\in I} K_i$ for any $I \subset \mathcal{S}$.

For $b \in \Gamma$, we have the decompositions $C_b^n = \bigoplus_{k\in\Lambda} C_{b,k}^n$, where each $C_{b,k}^n$ satisfies $C_{b,k}^n(*H_k) = C_{b,k}^n$. We have the Cech resolutions of $C_{b,k}^n$:

$$\mathcal{C}^\bullet(C_{b,k}^n, !\boldsymbol{H}^{(1)}) \longrightarrow C_{b,k}^n.$$

Because $C_{b.k}^n(*H_k) = C_{b,k}^n$, we obtain the following resolution:

$$\mathcal{C}^\bullet(C_{b,k}^n, !\boldsymbol{H}^{(1)})(*H_k) \longrightarrow C_{b,k}^n.$$

By construction, we have the decompositions

$$\mathcal{C}^{-\ell}(C_{b,k}^n, !\boldsymbol{H}^{(1)})(*H_k) = \bigoplus_{|I|=\ell+1} C_{b,k}^n(!H_I^{(1)})(*H_k)$$

By the condition for objects in $\mathrm{Cpx}_*(\mathfrak{U})$, the morphisms $C_{b,k_2}^n \longrightarrow C_{b,k_1}^{n+1}$ induce

$$\mathcal{C}^\bullet(C_{b,k_2}^n, !\boldsymbol{H}^{(1)})(*H_{k_2}) \longrightarrow \mathcal{C}^\bullet(C_{b,k_1}^{n+1}, !\boldsymbol{H}^{(1)})(*H_{k_1}). \tag{14.9}$$

We set $\overline{C}_b^{\bullet,n} := \bigoplus_k \mathcal{C}^\bullet(C_{b,k}^n, !\boldsymbol{H}^{(1)})(*H_k)$. We have the morphisms $\overline{C}_b^{\bullet,n} \longrightarrow \overline{C}_b^{\bullet,n+1}$ induced by (14.9) with which $\overline{C}_b^{\bullet,\bullet}$ are double complexes. We set $\overline{\varphi}(b) := \mathrm{Tot}\, \overline{C}_b^{\bullet,\bullet}$. Then, we obtain a functor $\overline{\varphi} : \Gamma \longrightarrow \mathrm{Cpx}_*(\mathfrak{U})$. By construction, we have a natural transform $\overline{\varphi} \longrightarrow \varphi$ such that each $\overline{\varphi}(b) \longrightarrow \varphi(b)$ $(b \in \Gamma)$ is a quasi-isomorphism.

For each $j \in \Lambda^{(1)}$, we take a smooth projective completion X_j of $U_j^{(1)}$ with a morphism $\varphi_j : X_j \longrightarrow X$ which induces the identity of $U_j^{(1)}$. We set $H_{j,\ell}^{(1)} := \varphi_j^{-1}(H_\ell^{(1)})$ and $H_{j,\ell} := \varphi_j^{-1}(H_\ell)$. For $j \in I$, we take the $\mathscr{D}$-modules $M_{b,j,k,I}^n$ on X_j such that $\varphi_{j\dagger}(M_{b,j,k,I}^n) \simeq C_{b,k}^n(!H_I^{(1)})(*H_k)$ and $M_{b,j,k,I}^n(!H_{j,I}^{(1)})(*H_{j,k}) = M_{b,j,k,I}^n$.

We take a total order on $\Lambda^{(1)}$, and we identify $\Lambda^{(1)} = \{1, \dots, |\Lambda^{(1)}|\}$. Inductively, we take smooth ample hypersurfaces $H_{j,j}^{(2)}$ on X_j with the following property.

- For $\ell < j$, we set $H^{(2)}_{j,\ell} := \varphi_j^{-1}\varphi_\ell(H^{(2)}_{\ell,\ell})$. For $I \ni j$, we set $I_{<j} := \{\ell \in I \,|\, \ell < j\}$. Then, $H^{(2)}_{j,j}$ is non-characteristic to $M^n_{b,j,k,I}(!H^{(2)}_{j,I_{<j}})(*H_{j,k})$. Here, $H^{(2)}_{j,I_{<j}} := \bigcup_{\ell\in I_{<j}} H^{(2)}_{j,\ell}$.

We set $H^{(2)}_j = \varphi_j(H^{(2)}_{j,j})$.

Lemma 14.3.9 *For any non-empty subset $I \subset \Lambda^{(1)}$, $C^n_{b,k}(!H^{(1)}_I)(!H^{(2)}_I)(*H_k)$ is an object in* $\mathrm{Hol}(X^{\mathrm{alg}})^{ac}$.

Proof For $j \in I$, we set $I_{\le j} := I_{<j} \cup \{j\}$. We have

$$\varphi_{j\dagger}\Big(M^n_{b,j,k,I}(!H^{(2)}_{j,I_{<j}})(*H_{j,k})(!H^{(2)}_{j,j})\Big) \simeq \varphi_{j\dagger}\Big(M^n_{b,j,k,I}(!H^{(2)}_{j,I_{\le j}})(*H_{j,k})\Big)$$
$$\simeq \varphi_{j\dagger}\Big(M^n_{b,j,k,I}(!H^{(2)}_{j,j})\Big)(!H^{(2)}_{I_{<j}})(*H_k) \qquad (14.10)$$

Because $M^n_{b,j,k,I}(!H^{(2)}_{j,j}) \simeq M^n_{b,j,k,I}(!H^{(1)}_{j,j})(!H^{(2)}_{j,j})(*H_{j,k})$, the right hand side of (14.10) is isomorphic to

$$\varphi_{j\dagger}\Big(M^n_{b,j,k,I}\Big)(!H^{(1)}_j)(!H^{(2)}_j)(!H^{(2)}_{I_{<j}})(*H_k) \simeq C^n_{b,k}(!H^{(1)}_I)(!H^{(2)}_{I_{\le j}})(*H_k).$$

We obtain $f^i_\dagger C^n_{b,k}(!H^{(1)}_I)(!H^{(2)}_I)(*H_k) = 0$ $(i \ne 0)$ for any morphism $f : X \longrightarrow Y$ by using the isomorphisms in the case $j = \max I$. □

We set $H^{(3)}_i := H^{(1)}_i \cup H^{(2)}_i$ $(i \in \Lambda^{(1)})$. For $b \in \Gamma$, we put

$$\tilde{C}^{-\ell,n}_{1,b} := \bigoplus_k \mathcal{C}^{-\ell}(C^n_{b,k}, !\boldsymbol{H}^{(3)})(*H_k) = \bigoplus_k \bigoplus_{|I|=\ell+1} C^n_{b,k}(!H^{(1)}_I)(!H^{(2)}_I)(*H_k).$$

Because $C^\bullet_b \in \mathrm{Cpx}_*(\mathfrak{U})$, we have the morphism of complexes

$$\mathcal{C}^\bullet(C^n_{b,k_2}, !\boldsymbol{H}^{(3)})(*H_{k_2}) \longrightarrow \mathcal{C}^\bullet(C^{n+1}_{b,k_1}, !\boldsymbol{H}^{(3)})(*H_{k_1}) \qquad (14.11)$$

induced by $C^n_{b,k_2} \longrightarrow C^{n+1}_{b,k_1}$. They give morphisms $\tilde{C}^{\bullet,n}_{1,b} \longrightarrow \tilde{C}^{\bullet,n+1}_{1,b}$ with which $\tilde{C}^{\bullet,\bullet}_{1,b}$ are double complexes. We set $\varphi_1(b) := \mathrm{Tot}\,\tilde{C}^{\bullet,\bullet}_{1,b}$. By construction, they give a functor $\varphi_1 : \Gamma \longrightarrow \mathrm{Cpx}_*(\mathfrak{U})^{ac}$, and we have a natural transform $a : \varphi_1 \longrightarrow \overline{\varphi}$. For each $b \in \Gamma$, the morphisms $a_b : \varphi_1(b) \longrightarrow \overline{\varphi}(b)$ are epimorphism, and $\mathrm{Ker}(a_b) \in \mathrm{Cpx}_*(\mathfrak{U})$. We obtain a functor $\mathrm{Ker}(a) : \Gamma \longrightarrow \mathrm{Cpx}_*(\mathfrak{U})$ which satisfies $\max_\Gamma \dim \mathrm{Supp}\,\mathrm{Ker}(a_b) < \max_\Gamma \dim \mathrm{Supp}\,\varphi(b)$. By applying the hypothesis of the induction, we can take a functor $\varphi_2 : \Gamma \longrightarrow \mathrm{Cpx}_*(\mathfrak{U})^{ac}$ and a natural transform $\varphi_2 \longrightarrow \mathrm{Ker}(a)$ such that $\varphi_2(b) \longrightarrow \mathrm{Ker}(a_b)$ $(b \in \Gamma)$ are quasi-isomorphisms.

Let $\tilde{\varphi}(b)$ be the cone of $\varphi_2(b) \longrightarrow \varphi_1(b)$ $(b \in \Gamma)$. Then, we obtain a functor $\tilde{\varphi} : \Gamma \longrightarrow \mathrm{Cpx}(\mathfrak{U})^{ac}$ and a natural transform $\tilde{\varphi} \longrightarrow \varphi$ such that $\tilde{\varphi}(b) \longrightarrow \varphi(b)$ $(b \in \Gamma)$ are quasi-isomorphisms. □

14.3.1.9 Morphisms in the Derived Category

Let X be any smooth algebraic variety. Let $\overline{X}$ be a principal completion of X. Let $\mathfrak{U}_0 = (U_{0,k} \mid k \in \Lambda_0)$ be a finite affine open covering of X. Let $\mathfrak{U}$ be the family of the affine open subsets $\bigcap_{k \in I} U_{0,k}$ in $\overline{X}$ for non-empty subsets $I \subset \Lambda_0$.

Lemma 14.3.10 *Let $\star$ be $*$ or $!$. Let C_i $(i = 1, 2)$ be objects in $\mathrm{Cpx}_\star(\mathfrak{U})^{ac}$.*

- *For any morphism $\varphi : C_1 \longrightarrow C_2$ in $D^b(\mathscr{D}_{\overline{X}})$, we have morphisms $C_1 \xleftarrow{a_1} C_3 \xrightarrow{a_2} C_2$ in $\mathrm{Cpx}_\star(\mathfrak{U})^{ac}$ which expresses φ. Here, a_1 is a quasi-isomorphism.*
- *Suppose that we are given morphisms $C_1 \xleftarrow{a_1} C_3 \xrightarrow{a_2} C_2$ in $\mathrm{Cpx}_\star(\mathfrak{U})^{ac}$, where a_1 is a quasi-isomorphism. If it is 0 as a morphism $C_1 \longrightarrow C_2$ in $D^b(\mathscr{D}_{\overline{X}})$, we have C_4 in $\mathrm{Cpx}_\star(\mathfrak{U})^{ac}$ with a quasi-isomorphism $C_4 \longrightarrow C_3$ such that the composite $C_4 \longrightarrow C_3 \longrightarrow C_2$ is chain homotopy equivalent to 0.*

Proof We give a proof in the case $\star = *$. The other case can be proved in a dual way. Let $\varphi : C_1 \longrightarrow C_2$ be a morphism in $D^b(\mathscr{D}_{\overline{X}})$. We have morphisms of complexes of holonomic $\mathscr{D}_{\overline{X}}$-modules $C_1 \longleftarrow M^\bullet \longrightarrow C_2$. We may assume that $M^\bullet(*D) = M^\bullet$, where $D = \overline{X} \setminus X$. Let $H_k := \overline{X} \setminus U_k$ for $k \in \Lambda$.

Let Λ be the set of non-empty subsets of Λ_0. For any $J \subset \Lambda_0$ and $k \in \Lambda$, we have the element $Jk := k \cup J \in \Lambda$. We have the decompositions $C_i^n = \bigoplus_{k \in \Lambda} C_{i,k}^n$ such that $C_{i,k}^n(*H_k) = C_{i,k}^n$. For any non-empty $J \subset \Lambda_0$, we have $C_i^n(*H_J) = \bigoplus_{k \in \Lambda} C_{i,k}^n(*H_J)$, and $C_{i,k}^n(*H_J)(*H_{Jk}) = C_{i,k}^n(*H_J)$. In this way, we regard $\mathcal{C}^\bullet(C_i, *\boldsymbol{H})$ as an object in $\mathrm{Cpx}_*(\mathfrak{U})$. We can also naturally regard $\mathcal{C}^\bullet(M^\bullet, *\boldsymbol{H})$ as an object in $\mathrm{Cpx}_*(\mathfrak{U})$. We have the following commutative diagram:

$$\begin{array}{ccccc} C_1 & \longleftarrow & M^\bullet & \longrightarrow & C_2 \\ \downarrow & & \downarrow & & \downarrow \\ \mathcal{C}^\bullet(C_1, *\boldsymbol{H}) & \longleftarrow & \mathcal{C}^\bullet(M^\bullet, *\boldsymbol{H}) & \longrightarrow & \mathcal{C}^\bullet(C_2, *\boldsymbol{H}) \end{array}$$

Thus, we can find morphisms $C_1 \xleftarrow{a_1'} C_3' \xrightarrow{a_2'} C_2$ in $\mathrm{Cpx}_*(\mathfrak{U})$ which represents φ. We can take the following commutative diagram in $\mathrm{Cpx}_*(\mathfrak{U})$:

$$\begin{array}{ccccc} \widetilde{C}_1 & \longleftarrow & \widetilde{C}_3 & \longrightarrow & \widetilde{C}_2 \\ \downarrow & & \downarrow & & \downarrow \\ C_1 & \longleftarrow & C_3' & \longrightarrow & C_2 \end{array}$$

Here, the vertical arrows are quasi-isomorphisms, and $\widetilde{C}_i$ are objects in $\mathrm{Cpx}_*(\mathfrak{U})^{ac}$. Then, we can find morphisms $C_1 \longleftarrow C_3 \longrightarrow C_2$ in $\mathrm{Cpx}_*(\mathfrak{U})^{ac}$ which expresses φ. Thus, we obtain the first claim.

Let us consider the second claim. It is enough to consider the case $C_3 = C_1$. Suppose we are given a quasi-isomorphism of holonomic $\mathscr{D}_{\overline{X}}$-complexes $M^\bullet \longrightarrow C_1$

such that the induced morphism $M^{\bullet} \longrightarrow C_2$ is chain homotopy equivalent to 0. We may assume $M^{\bullet}(*D) = M^{\bullet}$. By using the Cech resolution, we can take a quasi-isomorphism $C_5 \longrightarrow C_1$ in $\mathrm{Cpx}_*(\mathfrak{U})$ such that the composite

$$C_5 \longrightarrow C_1 \longrightarrow C_2 \longrightarrow C^{\bullet}(C_2, *\boldsymbol{H})$$

is chain homotopy equivalent to 0. It is standard that there exists a morphism $C_6 \longrightarrow C_5$ in $\mathrm{Cpx}_*(\mathfrak{U})$ such that the induced morphism $C_6 \longrightarrow C_2$ is chain homotopy equivalent to 0. Then, we can find the following diagram in $\mathrm{Cpx}_*(\mathfrak{U})^{ac}$:

$$\begin{array}{ccccc} \widetilde{C}_6 & \longrightarrow & \widetilde{C}_1 & \longrightarrow & \widetilde{C}_2 \\ \downarrow & & \downarrow & & \downarrow \\ C_6 & \longrightarrow & C_1 & \longrightarrow & C_2 \end{array}$$

Here, $\tilde{C}_i \in \mathrm{Cpx}_*(\mathfrak{U})^{ac}$, and the vertical arrows are quasi-isomorphisms. Then, we can find desired morphism $C_4 \longrightarrow C_1$. $\square$

14.3.1.10 Proof of Proposition 14.3.1

We take principal completions $\iota_X : X \longrightarrow \overline{X}$ and $\iota_Y : Y \longrightarrow \overline{Y}$ with a morphism $\overline{f} : \overline{X} \longrightarrow \overline{Y}$ such that $\overline{f} \circ \iota_X = \iota_Y \circ f$.

Take a finite affine covering $\mathfrak{U}_0 = (U_{0k} \mid k \in \Lambda_0)$ of X. Let $\mathfrak{U}$ be the family of affine open subsets $\bigcap_{k \in I} \iota_X(U_{0k})$ in $\overline{X}$ for $\emptyset \neq I \subset \Lambda_0$. Let $H_{0i} := \overline{X} \setminus \iota_X(U_{0i})$ for $i \in \Lambda_0$.

Let us consider the case $\star = *$. For any complex $M^{\bullet}$ in $\mathrm{Hol}(X^{\mathrm{alg}})$, we fix a complex $\tilde{M}^{\bullet}_*$ in $\mathrm{Hol}(X^{\mathrm{alg}})$ with a quasi-isomorphism $\iota_{X*}\tilde{M}^{\bullet}_* \longrightarrow C(\iota_{X*}M^{\bullet}, *\boldsymbol{H}_0)$ such that $\iota_{X*}\tilde{M}^{\bullet}_* \in \mathrm{Cpx}_*(\mathfrak{U})^{ac}$. Note that any object in $\mathrm{Cpx}_*(\mathfrak{U})^{ac}$ is obtained as the push-forward of a complex in $\mathrm{Hol}(X^{\mathrm{alg}})$ via ι_{X*}. We impose $\tilde{M}^{\bullet}_*[\ell] = \widetilde{(M^{\bullet}[\ell])}_*$. We note that the natural morphism $\iota_{X*}M^{\bullet} \longrightarrow C^{\bullet}(\iota_{X*}M^{\bullet}, *\boldsymbol{H}_0)$ is a quasi-isomorphism. We obtain the following complexes in $\mathrm{Hol}(Y^{\mathrm{alg}})$:

$${}^{h}f_*(M^{\bullet}) := {}^{D}f^{0}_*(\tilde{M}^{\bullet}_*).$$

We can check that ${}^{h}f_*$ gives a functor $D^b\,\mathrm{Hol}(X^{\mathrm{alg}}) \longrightarrow D^b\,\mathrm{Hol}(Y^{\mathrm{alg}})$ by using Lemma 14.3.10. It is compatible with the translation by construction. It is easy to see that the image of any distinguished triangle is isomorphic to a distinguished triangle in $D^b\,\mathrm{Hol}(Y^{\mathrm{alg}})$.

We have the following natural isomorphisms in $D^b_h(\mathscr{D}^{\mathrm{alg}}_Y)$:

$${}^{h}f_*(M^{\bullet}) = {}^{D}f^{0}_*(\tilde{M}^{\bullet}_*) \simeq {}^{D}f_*(\tilde{M}^{\bullet}_*) \simeq {}^{D}f_*(M^{\bullet})$$

They give the equivalence ${}^{h}f_{*} \simeq {}^{D}f_{*}$ as functors $D^{b}\operatorname{Hol}(X^{\mathrm{alg}}) \longrightarrow D_{h}^{b}(\mathscr{D}_{Y}^{\mathrm{alg}})$. In particular, we have the equivalence of the cohomological functors ${}^{h}f_{*}^{i} \simeq {}^{D}f_{*}^{i}$ $(i \in \mathbb{Z})$.

Let ${}^{h}f_{*}^{(0)} : D^{b}\operatorname{Hol}(X^{\mathrm{alg}}) \longrightarrow D^{b}\operatorname{Hol}(Y^{\mathrm{alg}})$ be another triangulated functor with an equivalence $({}^{h}f_{*}^{(0)})^{i} \simeq {}^{D}f_{*}^{i}$ $(i \in \mathbb{Z})$, where $({}^{h}f_{*}^{(0)})^{i}$ denote the induced cohomological functor. We have the following isomorphisms in $D^{b}\operatorname{Hol}(Y^{\mathrm{alg}})$:

$$
{}^{h}f_{*}^{(0)}(M^{\bullet}) \simeq {}^{h}f_{*}^{(0)}(\tilde{M}_{*}^{\bullet})
$$

Because $({}^{h}f_{*}^{(0)})^{j}(\tilde{M}_{*}^{i}) = 0$ unless $j = 0$, we have the following natural isomorphisms in $D^{b}\operatorname{Hol}(Y^{\mathrm{alg}})$:

$$
{}^{h}f_{*}^{(0)}(\tilde{M}_{*}^{\bullet}) \simeq ({}^{h}f_{*}^{(0)})^{0}(\tilde{M}_{*}^{\bullet}) \simeq {}^{D}f_{*}^{0}(\tilde{M}_{*}^{\bullet}) = {}^{h}f_{*}(M^{\bullet})
$$

Hence, we obtain isomorphisms ${}^{h}f_{*}^{(0)}(M^{\bullet}) \simeq {}^{h}f_{*}(M^{\bullet})$ in $D^{b}\operatorname{Hol}(Y^{\mathrm{alg}})$, which give an equivalence of ${}^{h}f_{*}^{(0)}$ and ${}^{h}f_{*}$.

Let $g : Y \longrightarrow Z$ be any morphism of smooth algebraic varieties. We may assume to have a morphism $\overline{g} : \overline{Y} \longrightarrow \overline{Z}$ for a principal completion $(\overline{Z}, D_{Z})$ of Z.

We take a finite affine coverings $\mathfrak{U}_{0}^{Y} = (U_{0j}^{Y} \mid j \in \Lambda_{0}^{Y})$ of Y. We may assume that for any $U_{0i} \in \mathfrak{U}_{0}$, there exists $U_{0j}^{Y} \in \mathfrak{U}_{0}^{Y}$ such that $f(U_{0i}) \subset U_{0j}^{Y}$. Let $\mathfrak{U}^{Y}$ be the set of $\bigcap_{j \in J} U_{0,j}^{Y}$ for non-empty subsets $J \subset \Lambda_{0}^{Y}$. Then, for any complex $M^{\bullet}$ in $\operatorname{Hol}(X^{\mathrm{alg}})$, we have ${}^{h}f_{*}(M^{\bullet}) \in \operatorname{Cpx}_{*}(\mathfrak{U}^{Y})^{ac}$. Hence, we have the following morphisms in $D^{b}\operatorname{Hol}(Z^{\mathrm{alg}})$:

$$
\begin{aligned}
{}^{h}g_{*}({}^{h}f_{*}(M^{\bullet})) &= {}^{D}g_{*}^{0}\Big(\widetilde{({}^{h}f_{*}(M^{\bullet}))}_{*}\Big) \simeq {}^{D}g_{*}^{0}(({}^{h}f_{*}(M^{\bullet}))) = {}^{D}g_{*}^{0} \circ {}^{D}f_{*}^{0}(\tilde{M}_{*}^{\bullet}) \\
&= {}^{D}(g \circ f)_{*}^{0}(\tilde{M}_{*}^{\bullet}) = {}^{h}(g \circ f)_{*}(M^{\bullet}) \qquad (14.12)
\end{aligned}
$$

Hence, we obtain ${}^{h}(g \circ f)_{*} \simeq {}^{h}g_{*} \circ {}^{h}f_{*}$.

Let us consider the case $\star = !$. For each complex of algebraic holonomic $\mathscr{D}_{X}$-module $M^{\bullet}$, take a complex $\tilde{M}_{!}^{\bullet}$ in $\operatorname{Hol}(X^{\mathrm{alg}})$ with a quasi-isomorphism $C^{\bullet}(\iota_{X!}M^{\bullet}, !\boldsymbol{H}_{0}) \longrightarrow \iota_{X!}\tilde{M}_{!}^{\bullet}$ such that $\iota_{X!}\tilde{M}_{!}^{\bullet} \in \operatorname{Cpx}_{!}(\mathfrak{U})^{ac}$. We impose $\tilde{M}_{!}^{\bullet}[\ell] = \widetilde{(M^{\bullet}[\ell])}_{!}$. We set

$$
{}^{h}f_{!}(M^{\bullet}) := {}^{D}f_{!}^{0}(\tilde{M}_{!}^{\bullet}).
$$

By using Lemma 14.3.10, we can check that ${}^{h}f_{!}$ gives a triangulated functor $D^{b}\operatorname{Hol}(X^{\mathrm{alg}}) \longrightarrow D^{b}\operatorname{Hol}(Y^{\mathrm{alg}})$. As in the case of ${}^{h}f_{*}$, we have the natural equivalence of the cohomological functors ${}^{h}f_{!}^{i} \simeq {}^{D}f_{!}^{i}$.

Let ${}^{h}f_{!}^{(0)} : D^{b}\operatorname{Hol}(X^{\mathrm{alg}}) \longrightarrow D^{b}\operatorname{Hol}(Y^{\mathrm{alg}})$ be another triangulated functor with an equivalence $({}^{h}f_{!}^{(0)})^{i} \simeq {}^{D}f_{!}^{i}$ $(i \in \mathbb{Z})$, where $({}^{h}f_{!}^{(0)})^{i}$ denote the associated cohomological functor. We have the following isomorphisms in $D^b\,\mathrm{Hol}(Y^{\mathrm{alg}})$:

$$ {}^h f_!^{(0)}(M^\bullet) \simeq {}^h f_!^{(0)}\big(C^\bullet(\iota_{X!}M^\bullet, !\boldsymbol{H}_0)_{|X}\big) \simeq {}^h f_!^{(0)}\big(\tilde{M}_!^\bullet\big) $$

Because $({}^h f_!^{(0)})^j(\tilde{M}_!^i) = 0$ $(j \neq 0)$, we have

$$ {}^h f_!^{(0)}\big(\tilde{M}_!^\bullet\big) \simeq ({}^h f_!^{(0)})^0(\tilde{M}_!^\bullet) \simeq {}^D f_!^0(\tilde{M}_!^\bullet) = {}^h f_!(M^\bullet). $$

We obtain ${}^h(g \circ f)_! \simeq {}^h g_! \circ {}^h f_!$ as in the case of $\star = *$. Thus, the proof of Proposition 14.3.1 is finished. □

14.3.1.11 Basic Property

Proposition 14.3.11 *If f is proper, then we have a natural equivalence ${}^h f_! \simeq {}^h f_*$.*

Proof In this case, we have $f_\dagger^i = {}^D f_*^i = {}^D f_!^i$. We use the notation in the proof of Proposition 14.3.1. We have the following morphisms in $D^b\,\mathrm{Hol}(Y^{\mathrm{alg}})$:

$$ {}^h f_*(\tilde{M}_!^\bullet) \simeq {}^h f_*(M^\bullet) $$

Because ${}^h f_*^j(\tilde{M}_!^i) \simeq {}^D f_*^j(\tilde{M}_!^i) \simeq f_\dagger^j(\tilde{M}_!^i)$, it is 0 unless $j = 0$. Hence, we have the natural isomorphisms ${}^h f_*(\tilde{M}_!^\bullet) \simeq {}^D f_!^0(\tilde{M}_!^\bullet) \simeq {}^h f_!(M^\bullet)$. The isomorphisms give the equivalence ${}^h f_* \simeq {}^h f_!$. □

Proposition 14.3.12 *Let $f : X \longrightarrow Y$ be any morphism of smooth algebraic varieties.*

- *We have a natural transform ${}^h f_! \longrightarrow {}^h f_*$.*
- *We have an equivalence $\boldsymbol{D}_Y \circ {}^h f_* \circ \boldsymbol{D}_X \simeq {}^h f_!$.*

Proof We take morphisms $X \xrightarrow{\iota} X_1 \xrightarrow{\overline{f}} Y$ such that (i) $f = \overline{f} \circ \iota$, (ii) ι is an open immersion, and $X_1 \setminus \iota(X)$ is a hypersurface, (iii) $\overline{f}$ is proper. Then, we have ${}^h f_\star = {}^h\overline{f}_\star \circ \iota_\star$. We have the equivalence ${}^h\overline{f}_! \simeq {}^h\overline{f}_*$ and the natural transform $\iota_! \longrightarrow \iota_*$. They induce $f_! \longrightarrow f_*$. It is easy to see that the natural transform is independent of the choice of the factorization $X \longrightarrow X_1 \longrightarrow Y$. Thus, we obtain the first claim.

For the second claim, it is enough to consider the case where f is proper. We have a natural isomorphism $\boldsymbol{D}\big(\tilde{M}_*^\bullet\big) \simeq (\widetilde{\boldsymbol{D}M^\bullet})_!$. We have

$$ {}^h f_!(\boldsymbol{D}_X M^\bullet) \simeq f_\dagger^0(\boldsymbol{D}_X \tilde{M}_*^\bullet) \simeq \boldsymbol{D}_Y f_\dagger^0(\tilde{M}_*^\bullet) \simeq \boldsymbol{D}_Y {}^h f_*(\tilde{M}) $$

Thus, we obtain the second claim. □

Proposition 14.3.13 *Let $f : X \longrightarrow Y$ be any morphism of smooth algebraic varieties. Let H_Y be any algebraic hypersurface in Y. We set $H_X := f^{-1}(H_Y)$. Then, we naturally have ${}^h f_\star\big(M^\bullet(\star H_X)\big) \simeq {}^h f_\star(M^\bullet)(\star H_Y)$.*

Proof Note that if $M^\bullet(\star H_X) = M^\bullet$, we can impose $\tilde{M}^\bullet_\star(\star H_X) \simeq \tilde{M}^\bullet_\star$. Then, the claim is clear. □

14.3.2 Push-Forward of Algebraic Mixed Twistor $\mathscr{D}$-Modules

14.3.2.1 Preliminary

Let X be any smooth algebraic variety. For $\star = *, !$, let $\mathrm{MTM}^\circ_\star(X^{\mathrm{alg}})$ be the full subcategory in $\mathrm{MTM}(X^{\mathrm{alg}})$ which consists of the objects $(\mathcal{T}, W)$ satisfying the following condition.

- There exists a hypersurface $H \subset X$ such that $X \setminus H$ is quasi-projective and that $(\mathcal{T}, W)[\star H] = (\mathcal{T}, W)$.

Let $f : X \longrightarrow Y$ be any morphism of smooth algebraic varieties. We take principal completions $(\overline{X}, D_X)$ and $(\overline{Y}, D_Y)$ of X and Y, respectively, with a morphism $\overline{f} : \overline{X} \longrightarrow \overline{Y}$ which induces $f : X \longrightarrow Y$. Let $\iota_X : X \longrightarrow \overline{X}$ and $\iota_Y : Y \longrightarrow \overline{Y}$ denote the inclusions.

For any $(\overline{\mathcal{T}}, W) \in \mathrm{MTW}(\overline{X})$, we define a filtration W on $\overline{f}^i_\dagger\overline{\mathcal{T}}$ as in Sect. 7.1.2. Namely, $W_m\overline{f}^j_\dagger(\overline{\mathcal{T}})$ are the image of $\overline{f}^j_\dagger W_{m-j}\overline{\mathcal{T}} \longrightarrow \overline{f}^j_\dagger\overline{\mathcal{T}}$.

Lemma 14.3.14 *The filtered $\mathcal{R}_{\overline{Y}}$-triples $\overline{f}^i_\dagger(\overline{\mathcal{T}}, W) := (\overline{f}^i_\dagger\overline{\mathcal{T}}, W)$ are pre-mixed twistor $\mathscr{D}$-modules on $\overline{Y}$.*

Proof Suppose that $\overline{\mathcal{T}}$ is pure of weight w. By using Chow's lemma and the description of pure twistor $\mathscr{D}$-modules as the minimal extension of wild variation of pure twistor structure, we can take a smooth projective variety X_1 with a morphism $\varphi : X_1 \longrightarrow X$ and a pure twistor $\mathscr{D}$-module $\mathcal{T}_1$ of weight w on X_1 such that $\overline{\mathcal{T}}$ is a direct summand of $\varphi^0_\dagger\mathcal{T}_1$. Then, $\overline{f}^j_\dagger(\overline{\mathcal{T}})$ are direct summands of $(\overline{f} \circ \varphi)^j_\dagger(\mathcal{T}_1)$. Hence, we obtain that $\overline{f}^j_\dagger\overline{\mathcal{T}}$ are pure of weight $w + j$. Then, by the argument in Proposition 7.1.6, we obtain that $(\overline{f}^i_\dagger(\overline{\mathcal{T}}), W) \in \mathrm{MTW}(\overline{Y})$ in the general case. □

Let $(\mathcal{T}_\star, W) \in \mathrm{MTM}(X^{\mathrm{alg}})$. We take a mixed twistor $\mathscr{D}$-module $(\overline{\mathcal{T}}_\star, W)$ on $\overline{X}$ such that $\Psi_{(\overline{X}, D_X)}(\overline{\mathcal{T}}_\star, W) = (\mathcal{T}_\star, W)$ and that $(\overline{\mathcal{T}}_\star, W)[\star D_X] = (\overline{\mathcal{T}}_\star, W)$. We obtain filtered $\mathcal{R}_{Y(*D_Y)}$-triples $f^j_\star(\mathcal{T}_\star, W) := \Psi_{(\overline{Y}, D_Y)}\overline{f}^j_\dagger(\overline{\mathcal{T}}_\star, W)$.

Lemma 14.3.15 *If $(\mathcal{T}_\star, W)$ is an object in $\mathrm{MTM}^\circ_\star(X^{\mathrm{alg}})$, then $(f^j_\star(\mathcal{T}_\star), W) \in \mathrm{MTM}(Y^{\mathrm{alg}})$.*

Proof We have a hypersurface H in X such that $U := X \setminus H$ is quasi-projective and that $\mathcal{T}_\star[\star H] = \mathcal{T}_\star$. We take a principal projective completion $U \subset X_1$ with a morphism $\varphi : X_1 \longrightarrow \overline{X}$ which induces the identity on U. We set $D_U := X_1 \setminus U = \varphi^{-1}(D_X \cup H)$. We have $(\mathcal{T}_1, W) \in \mathrm{MTM}(X_1)$ such that $(\mathcal{T}_1, W)[\star D_U] = (\mathcal{T}_1, W)$ and that $\varphi_\dagger^0(\mathcal{T}_1, W) = (\overline{\mathcal{T}_\star}, W)$. Because $\varphi_\dagger^j \mathcal{T}_1 = 0$ $(j \neq 0)$, we have natural isomorphisms of $\mathcal{R}$-triples $(\overline{f} \circ \varphi)_\dagger^i \mathcal{T}_1 \simeq \overline{f}_\dagger^i \overline{\mathcal{T}}_\star$. By the argument in Proposition 7.1.11, we obtain that they are isomorphisms of mixed twistor $\mathscr{D}$-modules. Then, the claim is clear. □

Hence, we obtain functors $f_\star^i : \mathrm{MTM}_\star^\circ(X^{\mathrm{alg}}) \longrightarrow \mathrm{MTM}(Y^{\mathrm{alg}})$. For any algebraic hypersurface H of X such that $X \setminus H$ is quasi-projective, the restriction of $f_\star^i$ to $\mathrm{MTM}(X^{\mathrm{alg}}, [\star H])$ is cohomological.

14.3.2.2 Statement

Let $f : X \longrightarrow Y$ be any morphism of smooth algebraic varieties. We take principal completions $(\overline{X}, D_X)$ and $(\overline{Y}, D_Y)$ of X and Y, respectively, with a morphism $\overline{f} : \overline{X} \longrightarrow \overline{Y}$ which induces $f : X \longrightarrow Y$. We shall prove the following theorem in Sect. 14.3.2.7.

Theorem 14.3.16 *For* $\star = *, !$, *we have triangulated functors*

$$^{T}\!f_\star : D^b\,\mathrm{MTM}(X^{\mathrm{alg}}) \longrightarrow D^b\,\mathrm{MTM}(Y^{\mathrm{alg}})$$

with equivalences of the functors $^{T}\!f_\star^i \simeq f_\star^i$ $(i \in \mathbb{Z})$ *on* $\mathrm{MTM}^\circ(X^{\mathrm{alg}})$, *where* $^{T}\!f_\star^i : \mathrm{MTM}(X^{\mathrm{alg}}) \longrightarrow \mathrm{MTM}(Y^{\mathrm{alg}})$ $(i \in \mathbb{Z})$ *denote the cohomological functor induced by* $^{T}\!f_\star$. *For any hypersurface* H *such that* $X \setminus H$ *is quasi-projective, the equivalences* $^{T}\!f_\star^i \simeq f_\star^i$ *on* $\mathrm{MTM}(X^{\mathrm{alg}}, [\star H])$ *gives an equivalence of the cohomological functors.*

For any triangulated functors $^{T}\!f_\star^{(0)} : D^b\,\mathrm{MTM}(X^{\mathrm{alg}}) \longrightarrow D^b\,\mathrm{MTM}(Y^{\mathrm{alg}})$ *with the same equivalences, we have natural equivalences of functors* $^{T}\!f_\star \simeq {}^{T}\!f_\star^{(0)}$.

By construction, the following will be clear.

Proposition 14.3.17 *Let* $f : X \longrightarrow Y$ *be a morphism of smooth algebraic varieties.*

- *If* f *is a closed immersion, then we naturally have* $^{T}\!f_* \simeq {}^{T}\!f_! \simeq f_\dagger$. *We also have* $^{T}\!f_\star^i = 0$ $(i \neq 0)$.
- *If* f *is an open immersion such that* $Y \setminus f(X)$ *is a hypersurface, then we naturally have* $^{T}\!f_\star = f_\star$. *We also have* $^{T}\!f_\star^i = 0$ $(i \neq 0)$. □

We mention some basic properties which can be proved as in the case of $\mathscr{D}$-modules (Sect. 14.3.1.11) once the construction is given.

Proposition 14.3.18 *Let* $f : X \longrightarrow Y$ *be any morphism of smooth algebraic varieties.*

- *We have natural equivalences* $\boldsymbol{D}_X \circ {}^T\!f_* \simeq {}^T\!f_! \circ \boldsymbol{D}_Y$ *and* $\boldsymbol{D}_X^{\mathrm{herm}} \circ {}^T\!f_* \simeq {}^T\!f_! \circ \boldsymbol{D}^{\mathrm{herm}}$.
- *Let H be a hypersurface in Y. We set $H_X := f^{-1}(H)$. Then, we naturally have* ${}^T\!f_\star(\mathcal{T}^\bullet[\star H]) \simeq {}^T\!f_\star(\mathcal{T}^\bullet)[\star H]$ *for any* $\mathcal{T} \in \mathrm{MTM}(X^{\mathrm{alg}})$.
- *If f is proper, we have a natural equivalence* ${}^T\!f_* \simeq {}^T\!f_!$. *In general, we have a natural transform* ${}^T\!f_! \longrightarrow {}^T\!f_*$.
- *Let $g : Y \longrightarrow Z$ be any morphism of smooth algebraic varieties. Then, we have natural equivalences* ${}^T(g \circ f)_\star \simeq {}^T g_\star \circ {}^T\!f_\star$. □

Let $\mathcal{A} \subset \mathbb{R} \times \mathbb{C}$ be a $\mathbb{Q}$-vector subspace with $\mathbb{Q} \times \{0\} \subset \mathcal{A}$. Let λ_0 be a non-zero complex number which is generic with respect to $\mathcal{A}$. Recall that we have Υ^{λ_0} : $\mathrm{MTM}(X^{\mathrm{alg}})_{\mathcal{A}} \longrightarrow \mathrm{Hol}(X^{\mathrm{alg}})$ for any smooth algebraic varieties. The following is clear by construction.

Proposition 14.3.19 *Let $f : X \longrightarrow Y$ be any morphism of smooth algebraic varieties. We have natural equivalences* $\Upsilon^{\lambda_0} \circ {}^T\!f_\star \simeq {}^h\!f_\star \circ \Upsilon^{\lambda_0}$. □

14.3.2.3 Acyclic Algebraic Mixed Twistor $\mathscr{D}$-Modules

Let X be any smooth complete algebraic variety.

Definition 14.3.20 A mixed twistor $\mathscr{D}$-module $(\mathcal{T}, W)$ on X is called acyclic if we have $f_\dagger^i \mathcal{T} = 0$ $(i \neq 0)$ for any morphism $f : X \longrightarrow Y$ of smooth algebraic varieties. Here, $f_\dagger^i \mathcal{T}$ denotes the push-forward as $\mathcal{R}$-triple. □

Let $\mathrm{MTM}(X^{\mathrm{alg}})^{ac} \subset \mathrm{MTM}(X^{\mathrm{alg}})$ denote the full subcategory of acyclic algebraic mixed twistor $\mathscr{D}$-modules on X.

Lemma 14.3.21 *Suppose that X is projective. Let λ_0 be any non-zero complex number. Then, the condition in Definition 14.3.20 is equivalent to that $\Upsilon^{\lambda_0}(\mathcal{T})$ is acyclic in the sense of Sect. 14.3.1.2.*

Proof Let $f : X \longrightarrow Y$ be any morphism to a smooth algebraic variety Y. Let $(\mathcal{M}_1, \mathcal{M}_2, C)$ be the $\mathcal{R}_X$-triple underlying $(\mathcal{T}, W)$. Because $f_\dagger^j \mathcal{M}_2$ are strict, we have $f_\dagger^j(\mathcal{M}_2^{\lambda_0}) = f_\dagger^j(\mathcal{M}_2)^{\lambda_0}$. By the twistor property, we have $f_\dagger^j(\mathcal{M}_2) = 0$ if and only if $f_\dagger^j(\mathcal{M}_2)^{\lambda_0} = 0$. Then, the claim of the lemma follows. □

14.3.2.4 Filtered Complexes of Acyclic Mixed Twistor $\mathscr{D}$-Modules

Let X be any smooth complete algebraic variety. Let $\mathfrak{U} = (U_k \mid k \in \Lambda)$ be a finite family of affine open subsets in X. Let $H_k := X \setminus U_k$. Let $\mathrm{Cpx}_*^{\mathrm{MTM}}(\mathfrak{U})$ be the category of bounded complexes $C^\bullet$ in $\mathrm{MTM}(X^{\mathrm{alg}})$ with decompositions $C^n = \bigoplus_{k \in \Lambda} C_k^n$ satisfying the following conditions.

- $C_k^n[*H_k] = C_k^n$ for $k \in \Lambda$.

- Let $\partial^n_{k_1,k_2} : C^n_{k_2} \longrightarrow C^{n+1}_{k_1}$ be the component of $\partial^n : C^n \longrightarrow C^{n+1}$ with respect to the decompositions. Then, we have $\partial^n_{k_1,k_2} = 0$ unless $U_{k_1} \subset U_{k_2}$.

A morphism $C_1^\bullet \longrightarrow C_2^\bullet$ in $\mathrm{Cpx}_*^{\mathrm{MTM}}(\mathfrak{U})$ is a morphism of complexes $\varphi : C_1^\bullet \longrightarrow C_2^\bullet$ in $\mathrm{MTM}(X^{\mathrm{alg}})$ such that $\varphi^n_{k_1,k_2} = 0$ unless $U_{k_1} \subset U_{k_2}$. Here $\varphi^n_{k_1,k_2} : C^n_{1,k_2} \longrightarrow C^n_{1,k_1}$ are the components of $\varphi^n : C_1^n \longrightarrow C_2^n$ with respect to the decompositions. Let $\mathrm{Cpx}_*^{\mathrm{MTM}}(\mathfrak{U})^{ac}$ denote the full subcategory in $\mathrm{Cpx}_*^{\mathrm{MTM}}(\mathfrak{U})$ of the objects $\mathcal{T}^\bullet$ such that $\mathcal{T}^n \in \mathrm{MTM}(X^{\mathrm{alg}})^{ac}$.

Let $\mathrm{Cpx}_!^{\mathrm{MTM}}(\mathfrak{U})$ be the category of bounded complexes $C^\bullet$ in $\mathrm{MTM}(X^{\mathrm{alg}})$ with decompositions $C^n = \bigoplus_{k\in\Lambda} C^n_k$ satisfying the following conditions.

- $C^n_k[!H_k] = C^n_k$ for $k \in \Lambda$.
- Let $\partial^n_{k_1,k_2} : C^n_{k_2} \longrightarrow C^{n+1}_{k_1}$ be the component of $\partial^n : C^n \longrightarrow C^{n+1}$ with respect to the decompositions. Then, we have $\partial^n_{k_1,k_2} = 0$ unless $U_{k_1} \supset U_{k_2}$.

A morphism $C_1^\bullet \longrightarrow C_2^\bullet$ in $\mathrm{Cpx}_!^{\mathrm{MTM}}(\mathfrak{U})$ is a morphism of complexes $\varphi : C_1^\bullet \longrightarrow C_2^\bullet$ in $\mathrm{MTM}(X^{\mathrm{alg}})$ such that $\varphi^n_{k_1,k_2} = 0$ unless $U_{k_1} \supset U_{k_2}$. Here, $\varphi^n_{k_1,k_2} : C^n_{1,k_2} \longrightarrow C^n_{2,k_1}$ are the components of $\varphi^n : C_1^n \longrightarrow C_2^n$ with respect to the decompositions. Let $\mathrm{Cpx}_!^{\mathrm{MTM}}(\mathfrak{U})^{ac}$ denote the full subcategory in $\mathrm{Cpx}_!^{\mathrm{MTM}}(\mathfrak{U})$ of the objects $\mathcal{T}^\bullet$ such that $\mathcal{T}^n \in \mathrm{MTM}(X^{\mathrm{alg}})^{ac}$.

14.3.2.5 Quasi-Isomorphisms with Complexes of Acyclic Algebraic Mixed Twistor $\mathscr{D}$-Modules

Let Γ be a category which consists of finite objects and finite morphisms.

Proposition 14.3.22 *For any functor $\varphi : \Gamma \longrightarrow \mathrm{Cpx}_*^{\mathrm{MTM}}(\mathfrak{U})$, we can take a functor $\tilde{\varphi} : \Gamma \longrightarrow \mathrm{Cpx}_*^{\mathrm{MTM}}(\mathfrak{U})^{ac}$ with a natural transform $\tilde{\varphi} \longrightarrow \varphi$ such that $\tilde{\varphi}(b) \longrightarrow \varphi(b)$ $(b \in \Gamma)$ are quasi-isomorphisms.*

Similarly, for any functor $\varphi : \Gamma \longrightarrow \mathrm{Cpx}_!^{\mathrm{MTM}}(\mathfrak{U})$, we can take a functor $\tilde{\varphi} : \Gamma \longrightarrow \mathrm{Cpx}_!^{\mathrm{MTM}}(\mathfrak{U})^{ac}$ with a natural transform $\varphi \longrightarrow \tilde{\varphi}$ such that $\varphi(b) \longrightarrow \tilde{\varphi}(b)$ $(b \in \Gamma)$ are quasi-isomorphisms.

Proof We have only to repeat the construction in the proof of Proposition 14.3.8. □

14.3.2.6 Morphisms in the Derived Category

Let X be a smooth algebraic variety. Let $\overline{X}$ be a principal completion of X. Let $\mathfrak{U}_0 = (U_{0,k} \,|\, k \in \Lambda_0)$ be a finite affine open covering of X. Let $\mathfrak{U}$ be the family of affine open subsets $\bigcap_{k\in I} U_{0,k}$ for non-empty subsets $I \subset \Lambda_0$. The following lemma is similar to Lemma 14.3.10.

Lemma 14.3.23 *Let $\star = *, !$. Let C_i $(i = 1, 2)$ be objects in $\mathrm{Cpx}_\star^{\mathrm{MTM}}(\mathfrak{U})^{ac}$.*

- *For any morphism $\varphi : C_1 \longrightarrow C_2$ in $D^b\,\mathrm{MTM}(\overline{X}^{\mathrm{alg}})$, we have morphisms $C_1 \xleftarrow{a_1} C_3 \longrightarrow C_2$ in $\mathrm{Cpx}_{\star}^{\mathrm{MTM}}(\mathfrak{U})$ which expresses φ. Here, a_1 is a quasi-isomorphism.*
- *Suppose that we are given morphisms $C_1 \xleftarrow{a_1} C_3 \longrightarrow C_2$ in $\mathrm{Cpx}_{\star}^{\mathrm{MTM}}(\mathfrak{U})$, where a_1 is a quasi-isomorphism. If it is 0 as a morphism $C_1 \longrightarrow C_2$ in $D^b\,\mathrm{MTM}(\overline{X}^{\mathrm{alg}})$, we have C_4 in $\mathrm{Cpx}_{\star}^{\mathrm{MTM}}(\mathfrak{U})^{ac}$ with a quasi-isomorphism $C_4 \longrightarrow C_3$ such that the composite $C_4 \longrightarrow C_3 \longrightarrow C_2$ is chain homotopy equivalent to 0.* □

14.3.2.7 Proof of Theorem 14.3.16

We take principal completions $\iota_X : X \subset \overline{X}$ and $\iota_Y : Y \subset \overline{Y}$ with a morphism $\overline{f} : \overline{X} \longrightarrow \overline{Y}$ such that $\overline{f} \circ \iota_X = \iota_Y \circ f$. We take a finite affine open covering $\mathfrak{U}_0 = (U_{0,k} \mid k \in \Lambda_0)$ of X. Let $\mathfrak{U}$ be the family of affine open subsets $\bigcap_{k\in I} \iota_X(U_{0,k})$ in $\overline{X}$ for $\emptyset \neq I \subset \Lambda_0$. Let $H_{0,i} := \overline{X} \setminus \iota_X(U_{0,i})$ for $i \in \Lambda_0$.

Let us consider the case $\star = *$. For any complex $(\mathcal{T}^{\bullet}, W)$ in $\mathrm{MTM}(X^{\mathrm{alg}})$, we fix a complex $(\tilde{\mathcal{T}}_*^{\bullet}, W)$ in $\mathrm{MTM}(X^{\mathrm{alg}})$ with a quasi-isomorphism

$$\iota_{X*}(\tilde{\mathcal{T}}_*^{\bullet}, W) \longrightarrow \mathcal{C}^{\bullet}(\iota_{X*}(\mathcal{T}^{\bullet}, W), *\boldsymbol{H}_0)$$

in $\mathrm{Cpx}_*^{\mathrm{MTM}}(\mathfrak{U})$ such that $\iota_{X*}(\tilde{\mathcal{T}}_*^{\bullet}, W) \in \mathrm{Cpx}_*^{\mathrm{MTM}}(\mathfrak{U})^{ac}$. Note that any object in $\mathrm{Cpx}_*^{\mathrm{MTM}}(\mathfrak{U})$ is obtained as the push-forward of an object in $\mathrm{MTM}(X^{\mathrm{alg}})$ via ι_{X*}. We impose $\tilde{\mathcal{T}}_*^{\bullet}[\ell] = \widetilde{\mathcal{T}^{\bullet}[\ell]}_*$. We note that the natural morphism $\iota_{X*}(\mathcal{T}^{\bullet}, W) \longrightarrow \mathcal{C}^{\bullet}(\iota_{X*}(\mathcal{T}^{\bullet}, W), *\boldsymbol{H}_0)$ is a quasi-isomorphism. We also remark that we have $f_*^j(\tilde{\mathcal{T}}_*^i) = 0$ $(j \neq 0)$. We obtain the following complex in $\mathrm{MTM}(Y^{\mathrm{alg}})$:

$${}^{T}\!f_*(\mathcal{T}^{\bullet}, W) := f_*^0(\tilde{\mathcal{T}}_*^{\bullet}, W).$$

We can check that ${}^{T}\!f_*$ gives a functor $D^b\,\mathrm{MTM}(X^{\mathrm{alg}}) \longrightarrow D^b\,\mathrm{MTM}(Y^{\mathrm{alg}})$ by using Lemma 14.3.23. The functor is compatible with the translations. It is easy to check that the image of any distinguished triangle is isomorphic to a distinguished triangle in $D^b\,\mathrm{MTM}(Y^{\mathrm{alg}})$.

Let us compare ${}^{T}\!f_*^i$ and f_*^i on $\mathrm{MTM}_*^{\circ}(X^{\mathrm{alg}})$. Let $U \subset X$ be any Zariski open subset such that (i) U is quasi-projective, (ii) $H := X \setminus U$ is a hypersurface. Let $\iota_U : U \longrightarrow X$ denote the inclusion. Let $(\mathcal{Q}, W) \in \mathrm{MTM}(U^{\mathrm{alg}})$. We have $(\tilde{\mathcal{Q}}_*^{\bullet}, W)$ in $\mathrm{MTM}(U^{\mathrm{alg}})$ with a quasi-isomorphism $\iota_{X*}\iota_{U*}(\tilde{\mathcal{Q}}_*^{\bullet}, W) \longrightarrow \mathcal{C}^{\bullet}(\iota_{X*}\iota_{U*}(\mathcal{Q}, W), *\boldsymbol{H}_0)$ such that $\iota_{X*}\iota_{U*}(\tilde{\mathcal{Q}}_*^{\bullet}, W) \in \mathrm{Cpx}_*^{\mathrm{MTM}}(\mathfrak{U})^{ac}$. We have the following quasi-isomorphisms of complexes in $\mathrm{MTM}(\overline{X}, [*H_1])$:

$$\iota_{X*}\iota_{U*}(\mathcal{Q}, W) \longrightarrow \mathcal{C}^{\bullet}\big(\iota_{X*}\iota_{U*}(\mathcal{Q}, W), *\boldsymbol{H}_0\big) \longleftarrow \iota_{X*}\iota_{U*}(\tilde{\mathcal{Q}}^{\bullet}, W)$$

Here, $H_1 := \overline{X} \setminus U$. Thus, we obtain isomorphisms of $\mathcal{R}_Y$-triples $f_*^i\mathcal{Q} \longrightarrow {}^{T}\!f_*^i\mathcal{Q}$ $(i \in \mathbb{Z})$.

Lemma 14.3.24 *The isomorphisms preserve the weight filtrations, i.e., we have the isomorphisms* $f_*^i(\mathcal{Q}, W) \longrightarrow {}^T\!f_*^i(\mathcal{Q}, W)$ *in* $\mathrm{MTM}(Y^{\mathrm{alg}})$*. Hence,* ${}^T\!f_*$ *has the desired property.*

Proof It is enough to consider the case $\overline{X} = X = U$ is projective. First, let us consider $(\mathcal{T}, W) \in \mathrm{MTM}(X^{\mathrm{alg}})$ such that $(\mathcal{T}, W)[*H_2] = (\mathcal{T}, W)$ for an ample hypersurface H_2. We can take a complex $(C^\bullet, W)$ in $\mathrm{MTM}(X^{\mathrm{alg}})$ with a quasi-isomorphism $(C^\bullet, W) \longrightarrow (\mathcal{T}, W)$ such that (i) $C^j = 0$ $(j > 0)$, (ii) $(C^j, W) \in \mathrm{MTM}(X^{\mathrm{alg}})^{ac}$. We may assume to have a quasi-isomorphism $(C^\bullet, W) \longrightarrow (\tilde{\mathcal{T}}^\bullet, W)$ of complexes in $\mathrm{MTM}(X^{\mathrm{alg}})$, which induces the following quasi-isomorphism of the complexes in $\mathrm{MTM}(Y^{\mathrm{alg}})$:

$$f_\dagger^0(C^\bullet, W) \longrightarrow {}^T\!f_*(\mathcal{T}, W)$$

The morphism $(C^0, W) \longrightarrow (\mathcal{T}, W)$ in $\mathrm{MTM}(X^{\mathrm{alg}})$ induces $f_\dagger^0(C^0, W) \longrightarrow f_\dagger^0(\mathcal{T}, W)$ in $\mathrm{MTM}(Y^{\mathrm{alg}})$. It induces the isomorphism between $f_\dagger^0(\mathcal{T}, W)$ and the 0-th cohomology object of the complex $f_\dagger^0(C^\bullet, W)$ in $\mathrm{MTM}(Y^{\mathrm{alg}})$. Hence, the isomorphism of the $\mathcal{R}$-triple ${}^T\!f_*^0(\mathcal{T}) \simeq f_\dagger^0(\mathcal{T})$ gives an isomorphism in $\mathrm{MTM}(Y^{\mathrm{alg}})$.

Let $(\mathcal{T}, W)$ be a polarizable pure twistor $\mathscr{D}$-module of weight w on X. We prove that ${}^T\!f_*^i(\mathcal{T}, W)$ are pure of weight $w + i$, which implies that the isomorphism ${}^T\!f_*^i(\mathcal{T}) \simeq f_\dagger^i(\mathcal{T})$ as $\mathcal{R}_Y$-triples are isomorphisms in $\mathrm{MTM}(Y^{\mathrm{alg}})$. We shall use an induction on the dimension of the support of $\mathcal{T}$. We take an ample hypersurface H_3 of X which is non-characteristic to $\mathcal{T}$. Then, the cokernel $(\mathcal{T}_1, W)$ of $(\mathcal{T}, W) \longrightarrow (\mathcal{T}, W)[*H_3]$ is pure of weight $w + 1$. We have the exact sequence

$$\cdots \longrightarrow {}^T\!f_*^i(\mathcal{T}, W) \longrightarrow {}^T\!f_*^i((\mathcal{T}, W)[*H_3]) \longrightarrow {}^T\!f_*^i(\mathcal{T}_1, W) \longrightarrow {}^T\!f_*^{i+1}(\mathcal{T}, W) \longrightarrow \cdots \tag{14.13}$$

By the hypothesis of the induction, ${}^T\!f_*^i(\mathcal{T}_1, W)$ are pure of weight $w + 1 + i$. We have already known that ${}^T\!f_*^0((\mathcal{T}, W)[*H_3]) \simeq f_\dagger^0\big((\mathcal{T}, W)[*H_3]\big)$ in $\mathrm{MTM}(Y^{\mathrm{alg}})$, which implies $\mathrm{Gr}_j^W\, {}^T\!f_*^0\big((\mathcal{T}, W)[*H_3]\big) = 0$ $(j < w)$. Hence, we obtain that ${}^T\!f_*^i(\mathcal{T}, W)$ $(i > 0)$ are pure of weight $w + i$, and that $\mathrm{Gr}_j^W\, {}^T\!f_*^i(\mathcal{T}, W) = 0$ for $j < w$. By using the Hermitian adjoint, we also obtain that ${}^T\!f_*^i(\mathcal{T}, W)$ $(i < 0)$ are pure of weight $w + i$, and that $\mathrm{Gr}_j^W({}^T\!f_*^0\mathcal{T}) = 0$ for $j > w$. Thus, the claim is proved in the pure case.

Let us prove the claim in the general case. We use an induction on the length of the weight filtration. Suppose that $W_w\mathcal{T} = \mathcal{T}$ and that we have already known that the isomorphisms ${}^T\!f_*^i(W_{w-1}\mathcal{T}) \simeq f_\dagger^i(W_{w-1}\mathcal{T})$ preserve the weight filtrations. We have the following commutative diagram of $\mathcal{R}_Y$-triples:

$$\begin{array}{ccccc}
f_\dagger^i(W_{w-1}\mathcal{T}) & \xrightarrow{a_1} & f_\dagger^i\mathcal{T} & \xrightarrow{a_2} & f_\dagger^i\,\mathrm{Gr}_w^W\mathcal{T} \\
\simeq\downarrow c_1 & & \simeq\downarrow c_2 & & \simeq\downarrow c_3 \\
{}^T\!f_*^i(W_{w-1}\mathcal{T}) & \xrightarrow{b_1} & {}^T\!f_*^i(\mathcal{T}) & \xrightarrow{b_2} & {}^T\!f_*^i\,\mathrm{Gr}_w^W\mathcal{T}
\end{array}$$

Here, the horizontal arrows give exact sequences in $\mathrm{MTM}(Y^{\mathrm{alg}})$, and the vertical arrows are isomorphisms of $\mathcal{R}_Y$-triples. We know that $\mathrm{Cok}(a_1)$ and $\mathrm{Cok}(b_1)$ are pure of weight $w+i$. We have already known that c_1 preserves the weight filtrations. Moreover, $\mathrm{Gr}_m^W f_\dagger^i(W_{w-1}\mathcal{T}) = 0$ for $m \geq w+i$. Then, we obtain that c_2 also preserves the weight filtrations. □

Let ${}^T\!f_*^{(0)} : D^b\,\mathrm{MTM}(X^{\mathrm{alg}}) \longrightarrow D^b\,\mathrm{MTM}(Y^{\mathrm{alg}})$ be another triangulated functor with the desired equivalences. Then, for any complex $(\mathcal{T}^\bullet, W)$ in $\mathrm{MTM}(X^{\mathrm{alg}})$, we have the following isomorphisms in $D^b\,\mathrm{MTM}(Y^{\mathrm{alg}})$:

$$\begin{aligned} {}^T\!f_*^{(0)}(\mathcal{T}^\bullet, W) &\simeq {}^T\!f_*^{(0)}(\tilde{\mathcal{T}}^\bullet, W) \simeq \left({}^T\!f_*^{(0)}\right)^0(\tilde{\mathcal{T}}^\bullet, W) \\ &\simeq f_*^0(\tilde{\mathcal{T}}^\bullet, W) = {}^T\!f_*(\mathcal{T}^\bullet, W). \end{aligned} \tag{14.14}$$

Thus, the proof for the claims in the case $\star = *$ is finished.

Although the case $\star = !$ is given by $\boldsymbol{D}_Y^{\mathrm{herm}} \circ f_! \circ \boldsymbol{D}_X^{\mathrm{herm}}$, we outline the construction given in a parallel way. For any complex $(\mathcal{T}^\bullet, W)$ in $\mathrm{MTM}(X^{\mathrm{alg}})$, we fix a complex $(\tilde{\mathcal{T}}_!^\bullet, W)$ in $\mathrm{MTM}(X^{\mathrm{alg}})$ with a quasi-isomorphism

$$\mathcal{C}^\bullet(\iota_{X!}(\mathcal{T}^\bullet, W), !\boldsymbol{H}_0) \longrightarrow \iota_{X!}(\tilde{\mathcal{T}}_!^\bullet, W)$$

such that $\iota_{X!}(\tilde{\mathcal{T}}_!^\bullet, W) \in \mathrm{Cpx}_!^{\mathrm{MTM}}(\mathfrak{U})$. We impose $\tilde{\mathcal{T}}_![\ell] = (\widetilde{\mathcal{T}[\ell]})_!$. We set

$${}^T\!f_!(\mathcal{T}^\bullet, W) := f_!^0(\tilde{\mathcal{T}}_!^\bullet, W).$$

Then, we can check that ${}^T\!f_!$ gives a triangulated functor $D^b\,\mathrm{MTM}(X^{\mathrm{alg}}) \longrightarrow D^b\,\mathrm{MTM}(Y^{\mathrm{alg}})$ with the desired property. Uniqueness can be argued as in the case of $\star = *$. Thus, we finish the proof of Theorem 14.3.16. □

14.3.3 Pull Back of Algebraic Mixed Twistor $\mathscr{D}$-Modules

As in [73], the pull back functors on the derived categories of algebraic mixed twistor $\mathscr{D}$-modules are defined to be the adjoint of the push-forward functors.

Proposition 14.3.25 *${}^T\!f_!$ has the right adjoint ${}^T\!f^!$, and ${}^T\!f_*$ has the left adjoint ${}^T\!f^*$. Thus, we obtain the following triangulated functors:*

$${}^T\!f^\star : D^b\,\mathrm{MTM}(Y^{\mathrm{alg}}) \longrightarrow D^b\,\mathrm{MTM}(X^{\mathrm{alg}}) \qquad (\star = !, *)$$

Similarly, we have the pull back ${}^h\!f^\star : D^b\,\mathrm{Hol}(Y^{\mathrm{alg}}) \longrightarrow D^b\,\mathrm{Hol}(X^{\mathrm{alg}})$ $(\star = *, !)$. Let $\mathcal{A} \subset \mathbb{R} \times \mathbb{C}$ be a $\mathbb{Q}$-vector space with $\mathbb{Q} \times \{0\} \subset \mathcal{A}$. Let λ_0 be a non-zero complex number which is generic with respect to λ_0. We have natural equivalences $\Upsilon^{\lambda_0} \circ {}^T\!f^\star \simeq {}^h\!f^\star \circ \Upsilon^{\lambda_0}$.

We have only to follow the argument in [73]. It is enough to construct adjoint functors in the cases (i) f is a closed immersion, (ii) f is a projection $X \times Y \longrightarrow Y$. We give only an indication for the constructions in Sects. 14.3.3.1 and 14.3.3.3.

14.3.3.1 The Case (i)

Let $f : X \longrightarrow Y$ be a closed immersion. The open immersion $Y \setminus X \longrightarrow Y$ is denoted by j. Let $\mathcal{T}^\bullet$ be a complex in $\mathrm{MTM}(Y^{\mathrm{alg}})$. We take hypersurfaces H_i $(i = 1, \ldots, N)$ in Y such that $\bigcap_{i=1}^N H_i = X$. For any $J \subset \{1, \ldots, N\}$, we set $H_J := \bigcup_{j \in J} H_j$. For any subset $I = (i_1, \ldots, i_m) \subset \{1, \ldots, N\}$, let $\mathbb{C}_I$ be the subspace of $\bigwedge^m \mathbb{C}^N$ generated by $e_{i_1} \wedge \cdots \wedge e_{i_m}$, where $e_i \in \mathbb{C}^N$ denotes an element whose j-th entry is 1 $(j = i)$ or 0 $(j \neq i)$. We use a natural inner product of $\mathbb{C}^N$ given by $(e_i, e_j) = 0$ $(i \neq j)$ and $(e_i, e_i) = 1$. For $I = I_0 \sqcup \{i\}$, the canonical morphism $\mathcal{T}^p[*H_{I_0}] \longrightarrow \mathcal{T}^p[*H_I]$, the multiplication and the inner product of e_i induce $\mathcal{T}^p[*H_{I_0}] \otimes \mathbb{C}_{I_0} \longrightarrow \mathcal{T}^p[*H_I] \otimes \mathbb{C}_I$. For $m \geq 0$, we put $\mathcal{C}^m(\mathcal{T}^p, *\boldsymbol{H}) := \bigoplus_{|I|=m} \mathcal{T}^p[*H_I] \otimes \mathbb{C}_I$, and we obtain the double complex $\mathcal{C}^\bullet(\mathcal{T}^\bullet, *\boldsymbol{H})$. The total complex is denoted by $\mathrm{Tot}\, \mathcal{C}^\bullet(\mathcal{T}^\bullet, *\boldsymbol{H})$. It is easy to observe that the support of the cohomology of $\mathrm{Tot}\, \mathcal{C}^\bullet(\mathcal{T}^\bullet, *\boldsymbol{H})$ is contained in X. According to Proposition 14.2.2, we have an object ${}^T\!f^! \mathcal{T}^\bullet$ in $D^b\, \mathrm{MTM}(X^{\mathrm{alg}})$ with an isomorphism ${}^T\!f_! {}^T\!f^! \mathcal{T}^\bullet \simeq \mathrm{Tot}\, \mathcal{C}^\bullet(\mathcal{T}^\bullet, *\boldsymbol{H})$. Thus, we obtain a functor ${}^T\!f^! : D^b\, \mathrm{MTM}(Y^{\mathrm{alg}}) \longrightarrow D^b\, \mathrm{MTM}(X^{\mathrm{alg}})$. We have the naturally defined morphism ${}^T\!f_! {}^T\!f^! \mathcal{T}^\bullet \longrightarrow \mathcal{T}^\bullet$. It induces

$$\mathrm{Hom}_{D^b\, \mathrm{MTM}(X^{\mathrm{alg}})}(\mathcal{T}_1^\bullet, {}^T\!f^! \mathcal{T}_2^\bullet) \longrightarrow \mathrm{Hom}_{D^b\, \mathrm{MTM}(Y^{\mathrm{alg}})}({}^T\!f_! \mathcal{T}_1^\bullet, \mathcal{T}_2^\bullet) \tag{14.15}$$

Let us check that (14.15) is bijective. Let $c_1 : {}^T\!f_! \mathcal{T}_1^\bullet \longrightarrow \mathcal{T}_2^\bullet$ be a morphism in $D^b\, \mathrm{MTM}(Y^{\mathrm{alg}})$ expressed by morphisms ${}^T\!f_! \mathcal{T}_1^\bullet \overset{a_1}{\longleftarrow} \mathcal{T}_3^\bullet \longrightarrow \mathcal{T}_2^\bullet$, where a_1 is a quasi-isomorphism. Then, $\mathrm{Tot}\, \mathcal{C}^\bullet(\mathcal{T}_3^\bullet, *\boldsymbol{H}) \longrightarrow \mathcal{T}_3^\bullet$ is a quasi-isomorphism because $\mathcal{T}_3^\bullet[*H_J]$ is acyclic for any non-empty $J \subset \{1, \ldots, N\}$. Hence, we obtain a morphism $c_2 : {}^T\!f_! \mathcal{T}_1^\bullet \longrightarrow \mathrm{Tot}\, \mathcal{C}^\bullet(\mathcal{T}_2^\bullet, *\boldsymbol{H})$ in $D^b\, \mathrm{MTM}(Y^{\mathrm{alg}})$ expressed by morphisms ${}^T\!f_! \mathcal{T}_1^\bullet \overset{a_2}{\longleftarrow} \mathrm{Tot}\, \mathcal{C}^\bullet(\mathcal{T}_3^\bullet, *\boldsymbol{H}) \longrightarrow \mathrm{Tot}\, \mathcal{C}^\bullet(\mathcal{T}_2^\bullet, *\boldsymbol{H})$, where a_2 is a quasi-isomorphism. Then, c_1 is the image of c_2. Let $c_3 : {}^T\!f_! \mathcal{T}_4^\bullet \longrightarrow \mathrm{Tot}\, \mathcal{C}^\bullet(\mathcal{T}_5^\bullet, *\boldsymbol{H})$ be a morphism in $D^b\, \mathrm{MTM}(Y^{\mathrm{alg}})$ such that the induced morphism ${}^T\!f_! \mathcal{T}_4^\bullet \longrightarrow \mathcal{T}_5^\bullet$ is 0 as a morphism in $D^b\, \mathrm{MTM}(Y^{\mathrm{alg}})$. Then, c_3 is induced by a morphism $c_4 : {}^T\!f_! \mathcal{T}_4^\bullet \longrightarrow \mathrm{Ker}\big(\mathrm{Tot}\, \mathcal{C}^\bullet(\mathcal{T}_5^\bullet, *\boldsymbol{H}) \longrightarrow \mathcal{T}_5^\bullet\big)$. We can easily observe that c_4 is 0 by using $X \subset H_i$ $(i = 1, \ldots, N)$. Hence, we have $c_3 = 0$.

We may obtain ${}^T\!f^*$ as $\boldsymbol{D}_Y \circ {}^T\!f^! \circ \boldsymbol{D}_X$. But, we outline the construction of ${}^T\!f^*$ given in a parallel way. Let $\mathbb{C}_I^\vee$ denote the dual of $\mathbb{C}_I$. For $I = I_0 \sqcup \{i\}$, we have a natural morphism $\mathcal{T}[!H_I] \otimes \mathbb{C}_I^\vee \longrightarrow \mathcal{T}[!H_{I_0}] \otimes \mathbb{C}_{I_0}^\vee$. For $m \geq 0$, we set

$$\mathcal{C}^{-m}(\mathcal{T}^\bullet, !\boldsymbol{H}) := \bigoplus_{|I|=m} \mathcal{T}^p[!H_I] \otimes \mathbb{C}_I,$$

and we obtain the double complex $\mathcal{C}^\bullet(\mathcal{T}^\bullet, !\boldsymbol{H})$. As in the previous case, the support of the cohomology of $\operatorname{Tot}\mathcal{C}^\bullet(\mathcal{T}^\bullet, !\boldsymbol{H})$ is contained in X. We have ${}^{T}\!f^*(\mathcal{T}^\bullet) \in D^b\operatorname{MTM}(X^{\mathrm{alg}})$ such that ${}^{T}\!f_*{}^{T}\!f^*(\mathcal{T}^\bullet) \simeq \operatorname{Tot}\mathcal{C}^\bullet(\mathcal{T}^\bullet, !\boldsymbol{H})$. We obtain a functor ${}^{T}\!f^* : D^b\operatorname{MTM}(Y^{\mathrm{alg}}) \longrightarrow D^b\operatorname{MTM}(X^{\mathrm{alg}})$. We have a natural morphism $\mathcal{T}^\bullet \longrightarrow {}^{T}\!f_*{}^{T}\!f^*\mathcal{T}^\bullet$. It induces

$$\operatorname{Hom}_{D^b\operatorname{MTM}(X^{\mathrm{alg}})}({}^{T}\!f^*\mathcal{T}_1^\bullet, \mathcal{T}_2^\bullet) \longrightarrow \operatorname{Hom}_{D^b\operatorname{MTM}(Y^{\mathrm{alg}})}(\mathcal{T}_1^\bullet, {}^{T}\!f_*\mathcal{T}_2^\bullet) \tag{14.16}$$

It is easy to check that (14.16) is bijective.

14.3.3.2 Basic Examples

Let $f : X \longrightarrow Y$ be a closed immersion. We put $d := \dim Y - \dim X$.

Lemma 14.3.26 *We have natural isomorphisms*

$$f^*\mathcal{U}_Y(p,q) \simeq \mathcal{U}_X(p-d,q)[d], \quad f^!\mathcal{U}_Y(p,q) \simeq \mathcal{U}_X(p,q-d)[-d].$$

Proof Let $L^if^*\mathcal{U}(p,q)$ denote the i-th cohomology triple of $f^*\mathcal{U}(p,q)$. We have $L^if^!\mathcal{U}_Y(p,q) = 0$ unless $i = d$, and $L^if^*\mathcal{U}_Y(p,q) = 0$ unless $i = -d$. We set $\mathcal{T}_0 := L^df^!\mathcal{U}_Y(p,q)$. We shall construct $\mathcal{T}_0 \simeq \mathcal{U}_X(p,q-d)$. We obtain isomorphisms for the other as the Hermitian adjoint.

Let us consider the case that X is a smooth hypersurface of Y, i.e., $d = 1$. In this case, we have $\mathcal{T}_0 \simeq \operatorname{Cok}(\mathcal{U}(p,q) \longrightarrow \mathcal{U}(p,q)[*X])$. Let Y_1 be an open subset of Y in the classical topology with a holomorphic coordinate $(z_1,\ldots,z_n)$ such that $X_1 := X \cap Y_1 = \{z_1 = 0\}$. We have the isomorphism $\mathcal{O}_{\mathcal{Y}_1}[*\mathcal{X}_1]/\mathcal{O}_{\mathcal{Y}_1} \simeq i_\dagger\lambda^{-1}\mathcal{O}_{\mathcal{X}_1}$ given by $z_1^{-1} \longleftrightarrow \lambda^{-1}1\cdot(dz_1/\lambda)^{-1}$. We also have the isomorphism $\operatorname{Ker}\bigl(\mathcal{O}_{\mathcal{Y}_1}[!\mathcal{X}_1] \longrightarrow \mathcal{O}_{\mathcal{Y}_1}\bigr) \simeq i_\dagger\mathcal{O}_{\mathcal{X}_1}$ given by $-\eth_{z_1}\otimes 1 \longleftrightarrow 1\cdot(dz_1/\lambda)^{-1}$. They give an isomorphism of $\mathcal{R}$-triples. (See Sect. 4.3.3.) It is independent of the choice of such a coordinate system. Hence, we can glue the isomorphisms for varied coordinate neighbourhoods, and we obtain the global isomorphism $\mathcal{T}_0 \simeq \mathcal{U}_X(p,q-1)$.

Let us consider the general case. We take a Zariski open subset $Y_2 \subset Y$ with hypersurfaces $H_1,\ldots,H_d$ such that $\bigcup H_i$ is a normal crossing, and that $\bigcap_{i=1}^d H_i = Y_2 \cap X =: X_2$. By using the result in the previous paragraph successively, we obtain an isomorphism $\mathcal{T}_{0|X_2} \simeq \mathcal{U}_X(p,q-d)_{|X_2}$. Because it is given on an Zariski open subset of X, it is extended to an isomorphism $\mathcal{T}_0 \simeq \mathcal{U}_X(p,q-d)$, denoted by $\varphi_{Y_2,H_1,\ldots,H_d}$.

Let us observe that it is independent of the choice of Y_2 and $H_1,\ldots,H_d$. Note that the isomorphism is already determined up to constant multiplications. First, we shall observe the independence of the order of $H_1,\dots,H_d$. We set $\underline{d} := \{1,\dots,d\}$. We set

$$\mathcal{C}^j := \bigoplus_{\substack{I\subset \underline{d}\\ |I|=j}} \mathcal{O}_{\mathcal{Y}_2}(*H_I)\otimes \mathbb{C}_I$$

The inclusion $\mathcal{O}_{\mathcal{Y}_2}[*H_I] \longrightarrow \mathcal{O}_{\mathcal{Y}_2}[*H_{I\sqcup i}]$ and the multiplication of e_i induce a map $\delta : \mathcal{C}^j \longrightarrow \mathcal{C}^{j+1}$. Thus, we obtain a complex $(\mathcal{C}^\bullet,\delta)$. It is quasi-isomorphic to the d-th cohomology sheaf $\mathcal{H}^d(\mathcal{C}^\bullet)$.

We set $\mathcal{M}_0 := \mathcal{O}_{\mathcal{Y}_2}$. Inductively, we set $\mathcal{M}_i := \mathcal{M}_{i-1}[*H_i]$. We have $\mathcal{O}_{\mathcal{Y}_2}[*H_{\underline{d}}] = \mathcal{M}_d$. By the construction, $\mathcal{O}_{\mathcal{Y}_2}[*H_{\underline{d}}]$ is the $\mathcal{R}_{Y_2}$-submodule in $\mathcal{O}_{\mathcal{Y}_2}(*H_{\underline{d}})$ generated by $\mathcal{O}_{\mathcal{Y}_2}(H_{\underline{d}})$. We have the unique $\mathcal{R}_{Y_2}$-homomorphism

$$\mathcal{O}_{\mathcal{Y}_2}[*H_{\underline{d}}]\otimes \bigwedge^d \mathbb{C}^d \longrightarrow \iota_\dagger\lambda^{-d}\mathcal{O}_{\mathcal{X}_2}$$

given by $(z_1\cdots z_d)^{-1}\otimes e_1\wedge\cdots\wedge e_d \longmapsto \lambda^d\cdot (dz_1/\lambda)^{-1}\cdots(dz_d/\lambda)^{-1}$. It induces an isomorphism $\mathcal{H}^d(\mathcal{C}^\bullet)\simeq \iota_\dagger\lambda^{-d}\mathcal{O}_{\mathcal{X}_2}$. By using this isomorphism, we can observe that $\varphi_{Y_2,H_1,\dots,H_d}$ is independent of the order of $H_1,\dots,H_d$. Next, if $H_i = H_i'$ $(i = 1,\dots,d-1)$, then we have $\varphi_{Y_2,H_1,\dots,H_d} = \varphi_{Y_2,H_1',\dots,H_d'}$ by the construction. We can also observe that, if $Y_3\subset Y_2$ and $H_i' = H_i\cap Y_3$, then we have $\varphi_{Y_2,H_1,\dots,H_d} = \varphi_{Y_3,H_1',\dots,H_d'}$. Then, we can easily obtain that $\varphi_{Y_2,H_1,\dots,H_d}$ is independent of Y_2 and $H_1,\dots,H_d$. □

14.3.3.3 The Case (ii)

Let us consider the case that f is the projection of $X = Y\times Z$ to Y. We put $d_Z := \dim Z$. We set

$${}^{T}\!f^*\mathcal{T} := \mathcal{T}\boxtimes\mathcal{U}_Z(d_Z,0)[-d_Z],\quad {}^{T}\!f^!\mathcal{T} := \mathcal{T}\boxtimes\mathcal{U}_Z(0,d_Z)[d_Z].$$

Let us prove that ${}^{T}\!f^*$ is the left adjoint of ${}^{T}\!f_*$. We can reduce the claim for ${}^{T}\!f^!$ and ${}^{T}\!f_!$ to the claim for ${}^{T}\!f^*$ and ${}^{T}\!f_*$ by using the Hermitian adjoint. To simplify the description, we omit the superscript "T".

We have only to construct natural transforms $\alpha : \mathrm{id}\longrightarrow f_*f^*$ and $\beta : f^*f_*\longrightarrow \mathrm{id}$ such that

$$\beta\circ f^*\alpha : f^*\mathcal{T}^\bullet \longrightarrow f^*f_*f^*\mathcal{T}^\bullet \longrightarrow f^*\mathcal{T}^\bullet \tag{14.17}$$

$$f_*\beta\circ\alpha : f_*\mathcal{N}^\bullet \longrightarrow f_*f^*f_*\mathcal{N}^\bullet \longrightarrow {}^{T}\!f_*\mathcal{N}^\bullet \tag{14.18}$$

are the identities. We have the following natural morphisms:

$$(\mathrm{tr}, a_Z^{-1}) : \mathcal{U}_{\mathrm{pt}}(0,0) \longrightarrow a_{Z*}\mathcal{U}_Z(d_Z,0)[-d_Z],$$
$$(a_Z^{-1}, \mathrm{tr}) : a_{Z!}\mathcal{U}_Z(0,d_Z)[d_Z] \longrightarrow \mathcal{U}_{\mathrm{pt}}(0,0).$$

Here, the morphism $\mathrm{tr} : a_{Z!}\mathcal{O}_{\mathcal{Z}}[d_Z]\lambda^{d_Z} \longrightarrow \mathcal{O}_{\mathbb{C}_\lambda}$ is given by the trace map, and $a_Z^{-1} : \mathcal{O}_{\mathbb{C}_\lambda} \longrightarrow a_{Z*}\mathcal{O}_{\mathcal{Z}}[-d_Z]$ is given by the pull back. In particular, we obtain a natural transform $\alpha : \mathrm{id} \longrightarrow f_*f^*$. For the construction of β, the following diagram is used as in [73]:

$$\begin{array}{ccccc} Z \times Y & \xrightarrow{i} & Z \times Z \times Y & \xrightarrow{q_1} & Z \times Y \\ & & \big\downarrow q_2 & & \big\downarrow p_1 \\ & & Z \times Y & \xrightarrow{p_2} & Y \end{array}$$

Here, i is induced by the diagonal $Z \longrightarrow Z \times Z$, q_j are induced by the projection $Z \times Z \longrightarrow Z$ onto the j-th component, and p_j are the projections. We have the following morphisms of complexes of mixed twistor $\mathscr{D}$-modules:

$$f^*f_*\mathcal{T}^\bullet = p_2^*p_{1*}\mathcal{T}^\bullet \simeq q_{2*}q_1^*\mathcal{T}^\bullet \longrightarrow q_{2*}\bigl(i_*i^*q_1^*\mathcal{T}^\bullet\bigr) \simeq i^*q_1^*\mathcal{T}^\bullet \tag{14.19}$$

We have a natural isomorphism $i^*\mathcal{U}_{Z\times Z}(d_Z,0)[-d_Z] \simeq \mathcal{U}_Z(0,0)$ according to Lemma 14.3.26. It induces $i^*q_1^*\mathcal{T}^\bullet \simeq \mathcal{T}^\bullet$ in $D^b\,\mathrm{MTM}\bigl((Z\times Y)^{\mathrm{alg}}\bigr)$. We define β as the composite of (14.19) with the isomorphism. Note that this construction is compatible with that in the case of $\mathscr{D}$-modules. Then, we obtain that the transforms in (14.17) and (14.18) are the identities, because the transforms for the underlying $\mathscr{D}$-modules are the identity. (See Sect. 9.2.5.2 of [57], for example.) Thus, the proof of Proposition 14.3.25 is finished. □

14.3.4 Base Change

Let $f : X \longrightarrow Y$ and $g : Y_1 \longrightarrow Y$ be morphisms of smooth algebraic varieties. We assume that either f or g is smooth. We set $X_1 := X \times_Y Y_1$ which is also a smooth algebraic variety. We have the induced morphisms $f_1 : X_1 \longrightarrow Y_1$ and $g_1 : X_1 \longrightarrow X$.

Proposition 14.3.27 *For any $\mathcal{T}^\bullet \in D^b\,\mathrm{MTM}(X)$, we have natural isomorphisms:*

$$^{T}\!f_{1*} \circ {}^{T}\!g_1^!(\mathcal{T}^\bullet) \simeq {}^{T}\!g^! \circ {}^{T}\!f_*(\mathcal{T}^\bullet), \qquad {}^{T}\!g^* \circ {}^{T}\!f_!(\mathcal{T}^\bullet) \simeq {}^{T}\!f_{1!} \circ {}^{T}\!g_1^*(\mathcal{T}^\bullet)$$

Proof It is enough to consider the case that f is smooth. The other case is obtained by using the duality. As for g, we have only to consider the case that (i) g is a closed immersion, (ii) g is the projection $Y_1 = Y_2 \times Y \longrightarrow Y$. The case (ii) is easy to see. Let us consider the case (i). For simplicity of the description, we omit to denote the superscript "T". We take hypersurfaces H_{Yi} $(i = 1, \dots, m)$ in Y such that $Y_1 = \bigcap_{i=1}^m H_{Yi}$. We put $H_{Xi} := f^{-1}(H_{Yi})$. We have $X_1 = \bigcap_{i=1}^m H_{Xi}$. We take $\tilde{\mathcal{T}}_!^\bullet$ for $\mathcal{T}^\bullet$ as in Sect. 14.3.2.7. Then, we have

$$g_* g^* f_! \mathcal{T}^\bullet \simeq \mathcal{C}^\bullet(f_! \mathcal{T}^\bullet, !\boldsymbol{H}_Y) \simeq \mathcal{C}^\bullet\big(f_!^0 \tilde{\mathcal{T}}_!^\bullet, !\boldsymbol{H}_Y\big) \simeq f_!^0 \mathcal{C}^\bullet(\tilde{\mathcal{T}}_!^\bullet, !\boldsymbol{H}_X)$$

We take $\widetilde{\mathcal{C}^\bullet(\mathcal{T}^\bullet, !\boldsymbol{H}_X)_!}$ for $\mathcal{C}^\bullet(\mathcal{T}^\bullet, !\boldsymbol{H}_X)$ as in Sect. 14.3.2.7. We have

$$g_* f_{1!} g_1^*(\mathcal{T}^\bullet) = g_! f_{1!} g_1^*(\mathcal{T}^\bullet) \simeq f_! g_{1!} g_1^*(\mathcal{T}^\bullet) \simeq f_! \mathcal{C}^\bullet(\mathcal{T}^\bullet, !\boldsymbol{H}_X) \simeq f_!^0 \widetilde{\mathcal{C}^\bullet(\mathcal{T}^\bullet, !\boldsymbol{H}_X)_!}$$

By the construction, we may assume to have $\mathcal{C}^\bullet(\tilde{\mathcal{T}}_!^\bullet, !\boldsymbol{H}_X) \simeq \widetilde{\mathcal{C}^\bullet(\mathcal{T}^\bullet, !\boldsymbol{H}_X)_!}$. We obtain the isomorphism ${}^T g^* \circ {}^T f_!(\mathcal{T}^\bullet) \simeq {}^T f_{1!} \circ {}^T g_1^*(\mathcal{T}^\bullet)$. We obtain the other isomorphism similarly. □

14.3.5 Tensor and Inner Homomorphism

Let X be a smooth algebraic variety. Let $\delta_X : X \longrightarrow X \times X$ be the diagonal morphism. As in [73], we obtain the functors $\otimes$ and $R\mathcal{H}om$ on $D^b\,\mathrm{MTM}(X^{\mathrm{alg}})$ in the standard ways:

$$\mathcal{T}_1 \otimes \mathcal{T}_2 := \delta_X^*\big(\mathcal{T}_1 \boxtimes \mathcal{T}_2\big), \quad R\mathcal{H}om(\mathcal{T}_1, \mathcal{T}_2) := \delta_X^!\big(\boldsymbol{D}_X \mathcal{T}_1 \boxtimes \mathcal{T}_2\big)$$

They are compatible with the corresponding functors on $D^b\,\mathrm{Hol}(X^{\mathrm{alg}})$.

14.3.6 Enhancement

Let K be a subfield of $\mathbb{R}$. For any smooth algebraic variety X, let $\mathscr{C}(X^{\mathrm{alg}})$ denote $\mathrm{MTM}^{\mathrm{int}}(X^{\mathrm{alg}})$, $\mathrm{MTM}^{\mathrm{int}}(X^{\mathrm{alg}}, \mathbb{R})$ or $\mathrm{MTM}^{\mathrm{int}}_{\mathrm{good}}(X^{\mathrm{alg}}, K)$.

Let $f : X \longrightarrow Y$ be any morphism of smooth algebraic varieties. The following proposition is the natural enhancement of Theorem 14.3.16 and Proposition 14.3.25. The proof is given in the same way.

Proposition 14.3.28 *We have the triangulated functors*

$${}^T f_\star : D^b\mathscr{C}(X^{\mathrm{alg}}) \longrightarrow D^b\mathscr{C}(Y^{\mathrm{alg}}) \quad (\star = *, !)$$

with the property as in Theorem 14.3.16. We also have the right adjoint ${}^T f^!$ of ${}^T f_!$, and the left adjoint ${}^T f^*$ of ${}^T f_*$. Hence, the categories $D^b\mathscr{C}(X^{\mathrm{alg}})$ are equipped with the six operations. □

14.3.7 Mixed Hodge Modules

Let K be a subfield of $\mathbb{R}$. Let $\mathrm{MHM}(X^{\mathrm{alg}}, K)$ denote the abelian category of algebraic mixed Hodge modules with K-structure. As in Sect. 13.5, we have the functor $\mathrm{MHM}(X^{\mathrm{alg}}, K) \longrightarrow \mathrm{MTM}^{\mathrm{int}}_{\mathrm{good}}(X^{\mathrm{alg}}, K)$ which induces

$$\Phi_X : D^b\,\mathrm{MHM}(X^{\mathrm{alg}}, K) \longrightarrow D^b\,\mathrm{MTM}^{\mathrm{int}}_{\mathrm{good}}(X^{\mathrm{alg}}, K).$$

The following proposition is clear by the construction of the functors.

Proposition 14.3.29 *The functor Φ_X is compatible with the six operations.* □

14.4 Algebraicity of the $\tilde{\mathcal{R}}$-Modules in the Integrable Case

14.4.1 Preliminary

14.4.1.1 $\tilde{\mathfrak{R}}$-Modules

Let Y be any complex manifold. Let H be a hypersurface of Y. We set $\mathfrak{Y} := \mathbb{P}^1 \times Y$. We set $\mathfrak{Y}^\lambda := \{\lambda\} \times Y$ for $\lambda \in \mathbb{P}^1$. We use the notation $\mathfrak{H}$ and $\mathfrak{H}^\lambda$ with similar meaning.

We have the sheaf of differential operators $\mathscr{D}_{\mathfrak{Y}}$ on $\mathfrak{Y}$. Let $p : \mathfrak{Y} \longrightarrow Y$ be the projection. Let $\tilde{\mathfrak{R}}_{Y(*H)}$ be the sheaf of subalgebras $\mathscr{D}_{\mathfrak{Y}}\big(*(\mathfrak{Y}^\infty \cup \mathfrak{H})\big)$ generated by $\lambda p^*\Theta_Y(*H)$ and $\lambda^2\partial_\lambda$ over $\mathcal{O}_{\mathfrak{Y}}\big(*(\mathfrak{Y}^\infty \cup \mathfrak{H})\big)$.

If Y and H are algebraic, we can consider the algebraic version $\tilde{\mathfrak{R}}^{\mathrm{alg}}_{Y(*H)}$ of $\tilde{\mathfrak{R}}_{Y(*H)}$. If an $\tilde{\mathfrak{R}}_{Y(*H)}$-module $\mathfrak{M}$ is isomorphic to the analytification of an $\tilde{\mathfrak{R}}^{\mathrm{alg}}_{Y(*H)}$-module, $\mathfrak{M}$ is called an algebraic $\tilde{\mathfrak{R}}_{Y(*H)}$-module.

Lemma 14.4.1 *Let $\mathfrak{M}$ be a coherent $\tilde{\mathfrak{R}}_Y$-module which is good as an $\mathcal{O}_{\mathfrak{Y}}$-module. Suppose that Y is projective. Then, $\mathfrak{M}$ is an algebraic coherent $\tilde{\mathfrak{R}}_Y$-module.*

Proof Because $\mathfrak{M}$ is good, we can take a coherent $\mathcal{O}_{\mathfrak{Y}}$-module N_1 with a surjection $\varphi : \tilde{\mathfrak{R}}_Y \otimes_{\mathcal{O}_{\mathfrak{Y}}} N_1 \longrightarrow \mathfrak{M}$. Because $\operatorname{Ker}\varphi$ is also good as an $\mathcal{O}_{\mathfrak{Y}}$-module, we have a coherent $\mathcal{O}_{\mathfrak{Y}}$-module N_2 with a morphism $\varphi_2 : \tilde{\mathfrak{R}}_Y \otimes_{\mathcal{O}_{\mathfrak{Y}}} N_2 \longrightarrow \tilde{\mathfrak{R}}_Y \otimes_{\mathcal{O}_{\mathfrak{Y}}} N_1$ such that $\mathrm{Cok}(\varphi_2)$ is isomorphic to $\mathfrak{M}$. Because $\mathbb{P}^1 \times Y$ is projective, N_i are algebraic. The morphism φ_2 is determined by an $\mathcal{O}_{\mathfrak{Y}}$-homomorphism $N_2 \longrightarrow \tilde{\mathfrak{R}}_Y \otimes N_1$, which is also algebraic. Hence, we have the algebraicity of φ_2 and $\mathrm{Cok}(\varphi_2)$. □

Let $\mathfrak{R}_Y$ denote the sheaf of subalgebras in $\mathscr{D}_{\mathfrak{Y}}(*\mathfrak{Y}^\infty)$ generated by $\lambda p^*\Theta_Y$ over $\mathcal{O}_{\mathfrak{Y}}(*\mathfrak{Y}^\infty)$. The algebraic version is denoted by $\mathfrak{R}^{\mathrm{alg}}_Y$.

We have the push-forward of $\mathfrak{R}$-modules and $\tilde{\mathfrak{R}}$-modules as in the cases of $\mathcal{R}$-modules and $\tilde{\mathcal{R}}$-modules. For simplicity, let $f : X \longrightarrow Y$ be a proper morphism of complex manifolds. The induced morphism $\mathfrak{X} \longrightarrow \mathfrak{Y}$ is also denoted by f. We put

$\omega_{\mathfrak{X}} := \lambda^{-\dim X} p^* \omega_X$. For any $\mathfrak{R}_X$-module $\mathfrak{M}$, we set

$$f_\dagger(\mathfrak{M}) := Rf_*\Big(\big(\omega_{\mathfrak{X}} \otimes_{f^{-1}\mathcal{O}_{\mathfrak{Y}}} f^{-1}(\mathfrak{R}_Y \otimes \omega_{\mathfrak{Y}}^{-1})\big) \otimes^L_{\mathfrak{R}_X} \mathfrak{M}\Big).$$

Let $f_\dagger^i \mathfrak{M}$ denote the $\mathfrak{R}_Y$-module obtained as the i-th cohomology sheaf of $f_\dagger \mathfrak{M}$. If $\mathfrak{M}$ is an $\tilde{\mathfrak{R}}_X$-module, they are naturally $\tilde{\mathfrak{R}}_Y$-modules. The construction is compatible with the push-forward of $\mathcal{R}$-modules and $\tilde{\mathcal{R}}$-modules. The following lemma can be proved as in the case of $\mathscr{D}$-modules [34].

Lemma 14.4.2 *If $\mathfrak{M}$ is a coherent $\tilde{\mathfrak{R}}_X$-module which is good as an $\mathcal{O}_{\mathfrak{X}}$-module, then $f_\dagger^j \mathfrak{M}$ are also coherent $\tilde{\mathfrak{R}}_Y$-module which are good as $\mathcal{O}_{\mathfrak{Y}}$-modules.* □

14.4.1.2 Algebraic $\tilde{\mathcal{R}}$-Modules

Let Y be any algebraic variety. Let $\tilde{\mathcal{R}}_Y^{\mathrm{alg}}$ denote the sheaf of subalgebras in $\mathscr{D}^{\mathrm{alg}}_{\mathbb{C}_\lambda \times Y}$ generated by $\lambda^2 \partial_\lambda$ and $\mathcal{R}_Y^{\mathrm{alg}}$ over $\mathcal{O}^{\mathrm{alg}}_{\mathbb{C}_\lambda \times Y}$. An $\tilde{\mathcal{R}}_Y^{\mathrm{alg}}$-module means a sheaf of left $\tilde{\mathcal{R}}_Y^{\mathrm{alg}}$-modules.

An $\tilde{\mathcal{R}}_Y$-module $\mathcal{M}$ is called algebraic if it is the analytification of an $\tilde{\mathcal{R}}_Y^{\mathrm{alg}}$-module $\mathcal{M}^{\mathrm{alg}}$. If moreover $\mathcal{M}^{\mathrm{alg}}$ is good and coherent, $\mathcal{M}$ is called an algebraic good coherent $\tilde{\mathcal{R}}_Y$-module.

Let $\mathcal{M}$ be an algebraic $\tilde{\mathcal{R}}_Y$-module. An algebraization of $\mathcal{M}$ naturally induces an $\tilde{\mathfrak{R}}_Y$-module $\mathfrak{M}$ such that $\mathfrak{M}_{|\mathcal{Y}} = \mathcal{M}$. Conversely, we have the following which is clear by Lemma 14.4.1.

Lemma 14.4.3 *Suppose that we are given a coherent $\tilde{\mathfrak{R}}_Y$-module $\mathfrak{M}$ which is good as an $\mathcal{O}_{\mathfrak{Y}}$-module such that $\mathfrak{M}_{|\mathcal{Y}} = \mathcal{M}$. Suppose that Y is projective. Then, $\mathcal{M}$ is algebraic.* □

14.4.1.3 Regularity Along ∞

Let Y be any complex manifold. Note that $\tilde{\mathfrak{R}}_{Y|\mathfrak{Y}\setminus\mathfrak{Y}^0}$ is equal to the restriction of $\mathscr{D}_{\mathfrak{Y}}(*\mathfrak{Y}^\infty)$. So, we have the concept of V-filtrations of $\tilde{\mathfrak{R}}_Y$-modules along $\mathfrak{Y}^\infty$. We briefly recall it.

We fix a section $\sigma : \mathbb{C}/\mathbb{Z} \longrightarrow \mathbb{C}$. Any complex number is uniquely expressed as $n + \sigma(a)$ by $n \in \mathbb{Z}$ and $a \in \mathbb{C}/\mathbb{Z}$. We fix a total order on $\mathbb{C}/\mathbb{Z}$. The lexicographic order on $\mathbb{Z} \times \mathbb{C}/\mathbb{Z}$ induces a total order $\leq_{\mathbb{C}}$ on $\mathbb{C}$.

Let $V^\infty \tilde{\mathfrak{R}}_Y$ be the sheaf of subalgebras in $\tilde{\mathfrak{R}}_Y$ on $\mathfrak{Y} \setminus \mathfrak{Y}^0$ generated by $p^*\mathscr{D}_Y$ and $\mu\partial_\mu$ over $\mathcal{O}_{\mathfrak{Y}\setminus\mathfrak{Y}^0}$.

A V-filtration of an $\tilde{\mathfrak{R}}_Y$-module $\mathfrak{M}$ along $\mathfrak{Y}^\infty$ is a filtration $V^\infty\mathfrak{M}$ by coherent $V^\infty\tilde{\mathfrak{R}}_Y$-modules indexed by $(\mathbb{C}, \leq)$ such that (a) $\bigcup_{\alpha\in\mathbb{C}} V^\infty_\alpha \mathfrak{M} = \mathfrak{M}$ and $V^\infty_\alpha \mathfrak{M} = \bigcap_{\beta>\alpha} V^\infty_\beta \mathfrak{M}$, (b) $\mu \cdot V^\infty_\alpha(\mathfrak{M}) = V^\infty_{\alpha-1}(\mathfrak{M})$ and $\partial_\mu \cdot V^\infty_\alpha(\mathfrak{M}) \subset V^\infty_{\alpha+1}(\mathfrak{M})$, (c) the induced actions of $\partial_{\mu}\mu + \alpha$ on $\mathrm{Gr}_{\alpha}^{V^{\infty}}(\mathfrak{M})$ are nilpotent. If such a V-filtration exists, it is unique. Recall the following standard lemma.

Lemma 14.4.4 *Let $f : \mathfrak{M}_1 \longrightarrow \mathfrak{M}_2$ be a morphism of $\tilde{\mathfrak{R}}_Y$-modules. Suppose that $\mathfrak{M}_i$ are equipped with a V-filtration along ∞.*

- *f is strictly compatible with the V-filtrations.*
- *The induced filtrations on* Ker(f), Cok(f) *and* Im(f) *are V-filtrations.* □

Definition 14.4.5 An $\tilde{\mathfrak{R}}_Y$-module $\mathfrak{M}$ is called regular along ∞ if the following holds.

- $\mathfrak{M}$ has a V-filtration along ∞. Moreover, each $V_{\alpha}^{\infty}(\mathfrak{M})$ is coherent over $p^{*}\mathscr{D}_Y$.

An algebraic $\tilde{\mathcal{R}}_Y$-module $\mathcal{M}$ is called regular along ∞ if the associated $\tilde{\mathfrak{R}}_Y$-module is regular along ∞. □

We obtain the following from Lemma 14.4.4.

Lemma 14.4.6 *Let $f : \mathfrak{M}_1 \longrightarrow \mathfrak{M}_2$ be a morphism of $\tilde{\mathfrak{R}}_Y$-modules. Suppose that $\mathfrak{M}_i$ are regular along ∞. Then,* Ker(f), Im(f) *and* Cok(f) *are also regular along ∞.* □

The following is well-known, and a consequence of the functoriality of the V-filtrations for $\mathscr{D}$-modules.

Lemma 14.4.7 *Let $f : X \longrightarrow Y$ be a proper morphism of complex manifolds. Let $\mathfrak{M}$ be a coherent $\tilde{\mathfrak{R}}_X$-module which is good as an $\mathcal{O}_{\mathfrak{X}}$-module. Suppose that $\mathfrak{M}$ is regular along ∞. Then, $f_{\dagger}^{j}\mathfrak{M}$ are also regular along ∞.* □

14.4.2 Statement

The following theorem implies the algebraicity of the $\tilde{\mathcal{R}}$-modules underlying algebraic mixed twistor $\mathscr{D}$-modules.

Theorem 14.4.8 *Let X be any smooth algebraic variety.*

- *Let $(\mathcal{T}, W)$ be an algebraic integrable mixed twistor $\mathscr{D}$-module on X. Let $\mathcal{M}_i$ $(i = 1, 2)$ be the underlying $\tilde{\mathcal{R}}_X$-modules. Then, we have algebraic $\tilde{\mathfrak{R}}_X$-modules $\mathfrak{M}_i$ with an isomorphism $\mathfrak{M}_{i|\mathcal{X}} \simeq \mathcal{M}_i$ such that $\mathfrak{M}_i$ are regular along ∞. Such $\tilde{\mathfrak{R}}_X$-modules are unique up to canonical isomorphisms.*
- *For any morphism of $\mathcal{T}_1 \longrightarrow \mathcal{T}_2$ in* $\mathrm{MTM}^{\mathrm{int}}(X^{\mathrm{alg}})$, *the morphisms of the underlying $\tilde{\mathcal{R}}_X$-modules are uniquely extended to morphisms of the $\tilde{\mathfrak{R}}_X$-modules.*

14.4.3 Preliminary

Set $X := \{(z_1,\ldots,z_n) \in \mathbb{C}^n \,\big|\, |z_i| < \epsilon_i\}$. We set $D_i := \{z_1 = 0\} \cap X$ $(i = 1,2)$ and $D = D_1 \cup D_2$. Let $\mathcal{I}$ be a good set of irregular values on (X,D). We assume that $\mathrm{ord}(\mathfrak{a}) \in \mathbb{Z}^2_{<0} \cup (\mathbb{Z}_{<0} \times \{0\})$ for non-zero element $\mathfrak{a} \in \mathcal{I}$ and that $\mathrm{ord}(\mathfrak{a}-\mathfrak{b}) \in \mathbb{Z}^2_{<0} \cup (\mathbb{Z}_{<0} \times \{0\})$ for $(\mathfrak{a},\mathfrak{b}) \in \mathcal{I}^2$ with $\mathfrak{a} \neq \mathfrak{b}$.

We set $X' := X \setminus D_2$ and $D_1' := D_1 \setminus D_2$. Let (V',∇) be an unramifiedly good meromorphic flat bundle on (X',D_1') whose good set of irregular values is contained in $\mathcal{I}$.

Proposition 14.4.9 *We have an unramifiedly good meromorphic flat bundle (V,∇) on (X,D) with an isomorphism $(V,\nabla)_{|X'} \simeq (V',\nabla)$. Such (V,∇) is uniquely determined up to canonical isomorphisms.*

Proof This proposition can be proved with the argument in Sect. 4.4 of [55]. We give only an outline.

We fix a lift $\tilde{\mathfrak{a}} \in \mathcal{O}_X(*D)$ of each $\mathfrak{a} \in \mathcal{I}$. Let $\pi : \tilde{X}(D) \longrightarrow X$ be the real blow up of X along D, i.e., it is the fiber product of the real blow up $\tilde{X}(D_i)$ of X along D_i over X. For $\boldsymbol{m} = (m_1,m_2) \in \mathbb{Z}^2$, we set $\boldsymbol{z}^{\boldsymbol{m}} = z_1^{m_1} z_2^{m_2}$. For each $(\mathfrak{a},\mathfrak{b}) \in \mathcal{I}$ with $\mathfrak{a} \neq \mathfrak{b}$, we have the function $F_{\mathfrak{a},\mathfrak{b}} := -|\boldsymbol{z}^{-\mathrm{ord}(\mathfrak{a}-\mathfrak{b})}|\,\mathrm{Re}(\tilde{\mathfrak{a}}-\tilde{\mathfrak{b}})$ on $\tilde{X}(D)$. For any $Q \in \pi^{-1}(D_1)$, we have the partial order $\leq_Q$ on $\mathcal{I}$ given as follows. We define $\mathfrak{a} <_Q \mathfrak{b}$ if $F_{\mathfrak{a},\mathfrak{b}}(Q) < 0$. We define $\mathfrak{a} \leq_Q \mathfrak{b}$ if $\mathfrak{a} = \mathfrak{b}$ or $\mathfrak{a} <_Q \mathfrak{b}$ holds.

Set $O := (0,\ldots,0) \in X$. We use the coordinate system

$$(\theta_1,\theta_2,r_1,r_2,z_3,\ldots,z_n)$$

on $\tilde{X}(D)$, where $z_i = r_i e^{\sqrt{-1}\theta_i}$ $(i = 1,2)$. Take $Q_0 = (\theta_{01},\theta_{02},0,0,0,\ldots,0) \in \pi^{-1}(O)$. For $(\theta_2,r_2,z_3,\ldots,z_n)$ and for $\delta_1 > 0$, we put

$$I_{\theta_{01},\delta_1}(\theta_2,r_2,z_3,\ldots,z_n) := \{(\theta_1,\theta_2,0,r_2,z_3,\ldots,z_n) \,\big|\, |\theta_1-\theta_{01}| < \delta_1\}.$$

If $\delta_1 > 0$ is sufficiently small, we have the following.

- For any $(\mathfrak{a},\mathfrak{b}) \in \mathcal{I}$ and for any $Q \in I_{\theta_{01},\delta_1}(\theta_{02},0,0,\ldots,0)$, the condition $\mathfrak{a} \leq_{Q_0} \mathfrak{b}$ implies $\mathfrak{a} \leq_Q \mathfrak{b}$.

The following lemma is elementary.

Lemma 14.4.10 *Take Q_0 and $\delta_1 > 0$ as above. If $\delta_2 > 0$ and $\epsilon > 0$ are sufficiently small, the following holds.*

- *Take $(\theta_2,r_2,z_3,\ldots,z_n)$ such that $|\theta_2-\theta_{02}| < \delta_2$, $r_2 \leq \epsilon$ and $|z_i| \leq \epsilon$. Take any $\mathfrak{a},\mathfrak{b} \in \mathcal{I}$. Then, $\mathfrak{a} \leq_{Q_0} \mathfrak{b}$ holds if and only if $\mathfrak{a} \leq_Q \mathfrak{b}$ holds for any $Q \in I_{\theta_{01},\delta_1}(\theta_2,r_2,z_3,\ldots,z_n)$.* □

Let L_1 be the local system on $X \setminus D = X' \setminus D_1'$ associated to the flat bundle $(V',\nabla)_{|X\setminus D}$. It is naturally extended to a local system L on $\tilde{X}(D)$. For any point

$Q \in \pi^{-1}(D_1')$, we have the Stokes filtration $\tilde{\mathcal{F}}^Q(L_Q)$ indexed by $(\mathcal{I}, \leq_Q)$, associated to the unramifiedly good meromorphic flat bundle (V', ∇) on (X', D_1'). Let $Q_0 \in \pi^{-1}(O)$. We take δ_1, δ_2 and ϵ as in Lemma 14.4.10. Take $(\theta_2, r_2, z_3, \ldots, z_n)$ such that $|\theta_2 - \theta_{02}| < \delta_2$, $r_2 \leq \epsilon$, $|z_i| \leq \epsilon$. According to Proposition 4.1.5 of [55], there exists a filtration $\tilde{\mathcal{F}}^{\theta_2, r_2, z_3, \ldots, z_n}$ indexed by $(\mathcal{I}, \leq_{Q_0})$ on the space of sections of L on $I_{\theta_{01}, \delta_1}(\theta_2, r_2, z_3, \ldots, z_n)$ such that $\tilde{\mathcal{F}}^{\theta_2, r_2, z_3, \ldots, z_n}$ is compatible with the filtrations $\tilde{\mathcal{F}}^Q$ over $(\mathcal{I}, \leq_{Q_0}) \longrightarrow (\mathcal{I}, \leq_Q)$ for any $Q \in I_{\theta_{01}, \delta_1}(\theta_2, r_2, z_3, \ldots, z_n)$. (See [55] for the compatibility of the filtrations in this situation.) It is easy to see that the filtrations $\tilde{\mathcal{F}}^{\theta_2, r_2, z_3, \ldots, z_n}$ are independent of $(\theta_2, r_2, z_3, \ldots, z_n)$ under the parallel transport. We define the filtration $\tilde{\mathcal{F}}^{Q_0}$ on L_{Q_0} by the parallel transport of $\tilde{\mathcal{F}}^{\theta_2, r_2, z_3, \ldots, z_n}$. Then, it is easy to check that the family of filtrations $\tilde{\mathcal{F}}^Q$ $(Q \in \pi^{-1}(O))$ gives a Stokes structure of L along $\pi^{-1}(O)$. Hence, by the Riemann-Hilbert-Birkhoff correspondence in [55], we obtain a germ of meromorphic flat bundle (V_O, ∇) on a neighbourhood X_O. Its restriction to $X_O \cap X'$ is equal to the restriction of (V', ∇) by construction. So, they give a meromorphic flat bundle on $(X' \cup X_O, D \cap (X' \cup X_O))$. According to Proposition 4.4.6 of [55], it is extended to a meromorphic flat bundle on (X, D). □

14.4.4 Extension of $\tilde{\mathcal{R}}$-Modules with Good-KMS Structure

Let Y be any complex manifold with a normal crossing hypersurface H. Let $\mathcal{V}$ be a smooth $\tilde{\mathcal{R}}_{Y(*H)}$-module which is good-KMS as a smooth $\mathcal{R}_{Y(*H)}$-module. (See Chap. 5.)

We set $Y_1 := Y \setminus H$. The vector bundle $\mathcal{V}_{|\mathcal{Y}_1}$ with the meromorphic flat connection is uniquely extended to a locally free $\mathcal{O}_{\mathfrak{Y}_1}(*\mathfrak{Y}_1^\infty)$-module $\widetilde{(\mathcal{V}_{|\mathcal{Y}_1})}$ with a meromorphic flat connection which is regular singular along $\mathfrak{Y}_1^\infty$. By gluing $\mathcal{V}$ and $\widetilde{(\mathcal{V}_{|\mathcal{Y}_1})}$, we obtain a locally free $\mathcal{O}_{\mathfrak{Y}}(*\mathfrak{Y}^\infty)$-module $\mathfrak{V}_1$ with a meromorphic flat connection given on the open subset $\mathfrak{Y} \setminus \mathfrak{H}^\infty$. Note that the codimension of $\mathfrak{H}^\infty$ in $\mathfrak{Y}$ is 2.

Proposition 14.4.11 *$\mathfrak{V}_1$ is extended to a coherent reflexive $\mathcal{O}_{\mathfrak{Y}}\big(*(\mathfrak{Y}^\infty \cup \mathfrak{H})\big)$-module on $\mathfrak{Y}$ uniquely.*

Proof The uniqueness is clear. It is enough to consider the case where Y is a neighbourhood of $(0, \ldots, 0)$ in $\mathbb{C}^n$, and $H = \bigcup_{i=1}^{\ell} \{z_i = 0\} \cap Y$. We have only to consider the case where $\mathcal{V}$ is unramifiedly good. According to a theorem of Malgrange [45], any meromorphic flat connection given outside a codimension 3 set is extended to a meromorphic flat connection on the whole space. So, it is enough to consider the case where H is given by $\{z_1 = 0\}$. We may assume that Y is the product of H and a disk.

Let $\mathcal{I} \subset \mathcal{O}_Y(*H)/\mathcal{O}_Y$ be the set of irregular values of $\mathcal{V}$, and we set $m_0 := \max\{-\operatorname{ord}(\mathfrak{a} - \mathfrak{b}) \mid \mathfrak{a}, \mathfrak{b} \in \mathcal{I}\}$.

We set $\mathfrak{Y}^{(0)} := \mathfrak{Y}$ and $Z^{(0)} := \mathfrak{H}^{\infty}$. We also set $\mathfrak{Y}^{\infty(0)} := \mathfrak{Y}^{\infty}$. Let $\mathfrak{Y}^{(1)}$ be the blow up $\mathfrak{Y}^{(0)}$ along $Z^{(0)}$. The exceptional divisor $E^{(1)}$ is isomorphic to $\mathbb{P}^1 \times Z^{(0)}$. Let $\mathfrak{Y}^{\infty(1)} \subset \mathfrak{Y}^{(1)}$ denote the strict transform of $\mathfrak{Y}^{\infty}$. Let $Z^{(1)} := E^{(1)} \cap \mathfrak{Y}^{\infty(1)}$. We define $\mathfrak{Y}^{(j)}$ and $Z^{(j)}$ inductively. Namely, let $\mathfrak{Y}^{(j+1)}$ be the blow up of $\mathfrak{Y}^{(j)}$ along $Z^{(j)}$. Let $\mathfrak{Y}^{\infty(j+1)} \subset \mathfrak{Y}^{(j+1)}$ be the strict transform of $\mathfrak{Y}^{\infty}$. Let $E^{(j+1)} \subset \mathfrak{Y}^{(j+1)}$ be the exceptional divisor of $\mathfrak{Y}^{(j+1)} \longrightarrow \mathfrak{Y}^{(j)}$. Let $Z^{(j+1)} := E^{(j+1)} \cap \mathfrak{Y}^{\infty(j+1)}$. It is easy to see that each Z^j is naturally isomorphic to $\mathfrak{H}^{\infty}$, and each $E^{(j)}$ is isomorphic to $\mathfrak{H}^{\infty} \times \mathbb{P}^1$.

Let $\pi_k : \mathfrak{Y}^{(k)} \longrightarrow \mathfrak{Y}$ be the induced morphism. It is easy to see that if $k > m_0$ $\pi_k^{-1}(\mathcal{I})$ determines a good system of unramifiedly irregular values on $(\mathfrak{Y}^{(k)}, D^{(k)})$. Then, by using Proposition 4.4.6 of [55] and Proposition 14.4.9 successively, we obtain the extension of $\mathfrak{V}_1$ on $\mathfrak{Y}^{(k)}$. □

Let $H = \bigcup_{i \in \Lambda} H_i$ be the irreducible decomposition. Because $\mathcal{V}$ is integrable, the KMS-spectra along any H_i are contained in $\mathbb{R} \times \{0\}$. So, we have a filtered bundle $\mathcal{Q}_*\mathcal{V}$ over $\mathcal{V}$ on $(\mathcal{X}, \mathcal{H})$ indexed by $(\mathbb{R}^{\Lambda}, \leq)$. Here, $\boldsymbol{a} \leq \boldsymbol{b}$ is defined to be $a_i \leq b_i$ $(i \in \Lambda)$. Note that $\mathcal{Q}_*\mathcal{V}_{|\mathbb{C}_\lambda^* \times X}$ is the Deligne-Malgrange filtered bundle. (See [44, 55, 61].)

We have the irreducible decomposition $\mathfrak{Y}^{\infty} \cup \mathfrak{H} = \mathfrak{Y}^{\infty} \cup \bigcup_{i \in \Lambda} \mathfrak{H}_i$. We have the Deligne-Malgrange filtered sheaf $\mathcal{Q}_{*,*}\mathfrak{V}_{|\mathfrak{Y} \setminus \mathfrak{Y}^0}$ over $\mathfrak{V}_{|\mathfrak{Y} \setminus \mathfrak{Y}^0}$ which is indexed by $(\mathbb{C} \times \mathbb{R}^{\Lambda}, \leq)$. Here, $(\alpha, \boldsymbol{a}) \leq (\beta, \boldsymbol{b})$ is defined to be $\alpha \leq_{\mathbb{C}} \beta$ and $a_i \leq b_i$ $(\forall i \in \Lambda)$. Each $\mathcal{Q}_{\alpha, \boldsymbol{a}}\mathfrak{V}_{|\mathfrak{Y} \setminus \mathfrak{Y}^0}$ is a coherent reflexive $\mathcal{O}_{\mathfrak{Y} \setminus \mathfrak{Y}^0}$-module.

By gluing them, we obtain a filtered sheaf $\mathcal{Q}_{*,*}\mathfrak{V}$ over $\mathfrak{V}$ on $(\mathfrak{Y}, \mathfrak{Y}^{\infty} \cup \mathfrak{H})$. Any $\mathcal{Q}_{\alpha, \boldsymbol{a}}\mathfrak{V}$ are reflexive coherent $\mathcal{O}_{\mathfrak{Y}}$-modules.

Lemma 14.4.12 *If the base space Y is projective, then $\mathcal{Q}_{\alpha, \boldsymbol{a}}\mathfrak{V}$ and $\mathcal{Q}_{\alpha, <\boldsymbol{a}}\mathfrak{V}$ are algebraic. In particular, $\mathcal{V}$ is an algebraic $\tilde{\mathcal{R}}_Y$-module.* □

14.4.5 The Extension of Admissible Mixed Twistor Structure

Let X be a complex manifold with a simple normal crossing hypersurface H. Let $H = \bigcup_{i \in \Lambda} H_i$ be the irreducible decomposition. For simplicity, we assume that Λ is finite. Let $(\mathcal{T}_0, W)$ be an admissible integrable mixed twistor structure on (X, H). Let $(\mathcal{M}_1, \mathcal{M}_2, C)$ be the underlying good-KMS smooth $\tilde{\mathcal{R}}_{X(*H)}$-triple. We have already known that $\mathcal{M}_i$ are extended to $\tilde{\mathfrak{R}}_{X(*H)}$-modules $\mathfrak{M}_i$ on $\mathfrak{X}$. They are equipped with lattices $\mathcal{Q}_{\alpha, \boldsymbol{a}}\mathfrak{M}_i$ which are coherent $\mathcal{O}_{\mathfrak{X}}$-modules.

Let $V_0\mathfrak{R}_X \subset \mathfrak{R}_X$ be the sheaf of subalgebras generated by $\lambda p^*\Theta_X(\log H)$ over $\mathcal{O}_{\mathfrak{X}}(*\mathfrak{X}^{\infty})$. For any subset $J \subset \Lambda$, let $\boldsymbol{\delta}_J$ be the element of $\mathbb{R}^{\Lambda}$ whose i-th component is 1 $(i \in J)$ or 0 $(i \notin J)$.

Let $I \sqcup J = \Lambda$ be a decomposition. Take a sufficiently small $\epsilon > 0$. We set $\boldsymbol{a}(I,J) := \boldsymbol{\delta}_I + (1-\epsilon)\boldsymbol{\delta}_J$. Let $H_I := \bigcup_{i\in I} H_i$ and $H_J := \bigcup_{i\in J} H_i$. We have the following $\tilde{\mathfrak{R}}_X$-modules:

$$\mathfrak{M}_i[*H_I!H_J] := \mathfrak{R}_X \otimes_{V_0\mathfrak{R}_X} \mathcal{Q}_{\alpha,\boldsymbol{a}(I,J)}\mathfrak{M}_i$$

They are independent of the choice of $\alpha \in \mathbb{C}$. By construction, $\mathcal{M}_i[*H_I!H_J]$ are the restrictions of $\mathfrak{M}_i[*H_I!H_J]$ to $\mathcal{X}$.

Lemma 14.4.13 *The $\tilde{\mathcal{R}}_X$-modules $\mathfrak{M}_i[*H_I!H_J]$ are regular along ∞. They are good $\mathcal{O}_{\mathfrak{X}}$-modules.*

Proof Because they are generated by $\mathcal{Q}_{\alpha,\boldsymbol{a}(I,J)}\mathfrak{M}_i$ over $\tilde{\mathfrak{R}}_X$, they are good as $\mathcal{O}_{\mathfrak{X}}$-modules. Thus, we obtain the second claim.

To consider the first claim, we may restrict ourselves on $\mathfrak{X} \setminus \mathfrak{X}^0$. Let $V\mathscr{D}_X \subset \mathscr{D}_X$ be the sheaf of subalgebras generated by $\Theta_X(\log H)$ over $\mathcal{O}_X$. On $\mathfrak{X} \setminus \mathfrak{X}^0$, let $V^{\infty}_{\alpha}(\mathfrak{M}_i[*H_I!H_J])$ be the image of $p^*\mathscr{D}_X \otimes_{p^*V\mathscr{D}_X} \mathcal{Q}_{\alpha+1,\boldsymbol{a}(I,J)}\mathfrak{M}_i \longrightarrow \mathfrak{M}_i[*H_I!H_J]$. Then, the filtration satisfies the condition of V-filtration along ∞ for $\mathfrak{M}_i[*H_I!H_J]$. Moreover, they are $p^*\mathscr{D}_X$-coherent. So, $\mathfrak{M}_i[*H_I!H_J]$ are regular along ∞. □

Let g be a meromorphic function on (X,H) such that $g^{-1}(0) \subset H$. Let $I \sqcup J = \Lambda$ be a decomposition such that the pole of g is contained in H_I and that $H_J \subset g^{-1}(0)$.

We have the integrable admissible mixed twistor structure $\Pi^{a,b}_g\mathcal{T}_0$ on (X,H), and the associated integrable mixed twistor $\mathscr{D}$-modules $\Pi^{a,b}_g\mathcal{T}_0[*H]$ and $\Pi^{a,b}_g\mathcal{T}_0[*H_I!H_J]$ on X. According to Lemma 14.4.13, the underlying $\tilde{\mathcal{R}}_X$-modules of $\Pi^{a,b}_g\mathcal{T}_0[*H]$ and $\Pi^{a,b}_g\mathcal{T}_0[*H_I!H_J]$ are extended to $\tilde{\mathfrak{R}}_X$-modules which are regular along ∞ and good as $\mathcal{O}_{\mathfrak{X}}$-modules.

We set

$$\Pi^{a,b}_{g*!}(\mathcal{T}_0)[*H_I] := \varprojlim_{N\to\infty} \operatorname{Cok}\bigl(\Pi^{b,b+N}_g(\mathcal{T}_0)[*H_I!H_J] \longrightarrow \Pi^{a,a+N}_g(\mathcal{T}_0)[*H]\bigr)$$

In particular, we set $\psi^{(a)}_g(\mathcal{T}_0)[*H_I] := \Pi^{a,a}_{g,*!}(\mathcal{T}_0)[*H_I]$ and $\Xi^{(a)}_g(\mathcal{T}_0)[*H_I] := \Pi^{a,a+1}_{g,*!}(\mathcal{T}_0)[*H_I]$. We obtain the following lemma from Lemma 14.4.6.

Lemma 14.4.14 *The underlying $\tilde{\mathcal{R}}_X$-modules of the algebraic integrable mixed twistor $\mathscr{D}$-modules $\psi^{(a)}_g\mathcal{T}_0[*H_I]$ and $\Xi^{(a)}_g(\mathcal{T}_0)[*H_I]$ are extended to $\tilde{\mathfrak{R}}_X$-modules which are regular along ∞ and good as $\mathcal{O}_{\mathfrak{X}}$-modules.* □

14.4.6 Affine Case

Let X be a smooth affine variety. Let $\overline{X}$ be any principal completion of X. Let $D := \overline{X} \setminus X$.

Lemma 14.4.15 *Let $\mathcal{T}$ be an integrable mixed twistor $\mathscr{D}$-module on $\overline{X}$ such that $\mathcal{T}[*D] = \mathcal{T}$. Then, the underlying $\tilde{\mathcal{R}}_{\overline{X}}$-modules of $\mathcal{T}$ are uniquely extended to $\tilde{\mathfrak{R}}_{\overline{X}}$-modules which are regular along ∞ and good as $\mathcal{O}_{\mathfrak{X}}$-modules. They are algebraic.*

Proof We use an induction on the dimension of the support. If the support is 0-dimensional, the claim is clear. We use an induction on the dimension of the support. If the support is 0-dimensional, the claim is clear.

We take an algebraic function g on X such that (i) $\mathcal{T}\big(*(g^{-1}(0) \cup D)\big)$ is the push-forward of admissible mixed twistor structure, (ii) $\dim\big(\mathrm{Supp}(\mathcal{T}) \cap g^{-1}(0)\big) < \dim \mathrm{Supp}(\mathcal{T})$. We put $H := g^{-1}(0) \cup D$. We take a smooth projective variety Z with a morphism $\varphi : Z \longrightarrow \overline{X}$ such that (i) $H_Z = \varphi^{-1}(H)$ is a normal crossing hypersurface, (ii) we have an admissible integrable mixed twistor structure $\mathcal{V}$ on (Z, H_Z) with an isomorphism $\varphi_\dagger(\mathcal{V}) \simeq \mathcal{T}(*H)$. We set $g_Z := g \circ \varphi$ and $D_Z := \varphi^{-1}(D)$. We have $H_Z = D_Z \cup g_Z^{-1}(0)$. We can reconstruct $\mathcal{T}$ as the cohomology of

$$\varphi_\dagger\big(\psi_{g_Z}^{(1)}(\mathcal{V})[*D_Z]\big) \to \phi_g^{(0)}(\mathcal{T})[*D] \oplus \varphi_\dagger\big(\Xi_{g_Z}^{(0)}(\mathcal{V})[*D_Z]\big) \to \varphi_\dagger\big(\psi_{g_Z}^{(0)}(\mathcal{V})[*D_Z]\big).$$

By Lemmas 14.4.7 and 14.4.14, we obtain that the underlying $\tilde{\mathcal{R}}_{\overline{X}}$-modules of $\varphi_\dagger\big(\psi_{g_Z}^{(a)}(\mathcal{V})[*D_Z]\big)$ and $\varphi_\dagger\big(\Xi_{g_Z}^{(a)}(\mathcal{V})[*D_Z]\big)$ are extended to $\tilde{\mathfrak{R}}_{\overline{X}}$-modules which are regular along ∞ and good as $\mathcal{O}_{\mathfrak{X}}$-modules. By the assumption of the induction, the underlying $\tilde{\mathcal{R}}_{\overline{X}}$-modules of $\phi_g^{(0)}(\mathcal{T})[*D]$ are also extended to $\tilde{\mathfrak{R}}_{\overline{X}}$-modules which are regular along ∞ and good as $\mathcal{O}_{\overline{\mathfrak{X}}}$-modules. Moreover the morphisms $\varphi_\dagger\big(\psi_{g_Z}^{(1)}(\mathcal{V})[*D_Z]\big) \longrightarrow \phi_g^{(0)}(\mathcal{T})[*D] \longrightarrow \varphi_\dagger\big(\psi_{g_Z}^{(0)}(\mathcal{V})[*D_Z]\big)$ are extended to $\tilde{\mathfrak{R}}$-homomorphisms. Hence, the underlying $\tilde{\mathcal{R}}$-modules of $\mathcal{T}$ are also extended to $\tilde{\mathfrak{R}}$-modules which are regular along ∞ and good as $\mathcal{O}_{\overline{\mathfrak{X}}}$-modules.

Let us study the uniqueness of the extension. Let $\mathcal{M}_i$ $(i = 1, 2)$ be the $\tilde{\mathcal{R}}_{\overline{X}}$-modules underlying $\mathcal{T}$. We have already obtained $\tilde{\mathfrak{R}}_{\overline{X}}$-modules $\mathfrak{M}_i$ with an isomorphism $\mathfrak{M}_{i|\overline{\mathcal{X}}} \simeq \mathcal{M}_i$ which are regular along ∞. For $i = 1, 2$, let $\overline{\mathfrak{M}}_i$ be another $\tilde{\mathfrak{R}}_{\overline{X}}$-module with an isomorphism $\overline{\mathfrak{M}}_{i|\overline{\mathcal{X}}} \simeq \mathcal{M}_i$, which is regular along ∞. Let us prove that the isomorphism $\mathfrak{M}_{i|\overline{\mathcal{X}}} \simeq \overline{\mathfrak{M}}_{i|\overline{\mathcal{X}}}$ is extended to an isomorphism $\mathfrak{M}_i \simeq \overline{\mathfrak{M}}_i$. It is enough to compare the $\mathscr{D}_{\overline{\mathfrak{X}}}$-modules $\mathfrak{M}'_i = \mathfrak{M}_i(*\overline{\mathfrak{X}}^0)$ and $\overline{\mathfrak{M}}'_i = \overline{\mathfrak{M}}_i(*\overline{\mathfrak{X}}^0)$.

We set $Z_1 := \varphi(Z)$. We take a projective morphism $F : \overline{X}_2 \longrightarrow \overline{X}$ such that (i) $H_2 := F^{-1}(H)$ is normal crossing, (ii) the strict transform Z_2 of Z_1 in $\overline{X}_2$ is smooth, (iii) Z_2 intersects with H_2 in a normal crossing way, (iv) $Z_2 \setminus H_2 \simeq Z_1 \setminus H$. We have the $\mathscr{D}_{\overline{\mathfrak{X}}_2}(*H_2)$-module $\overline{\mathfrak{M}}'_{i2}$ such that $F_\dagger(\overline{\mathfrak{M}}'_{i2})$ are isomorphic to $\overline{\mathfrak{M}}'_i(*\mathfrak{H})$ and that $\mathrm{Supp}(\overline{\mathfrak{M}}'_{i2}) = Z_2$. Set $H_{Z_2} := Z_2 \cap H_2$. By using Kashiwara's equivalence, we have the $\mathscr{D}_{\mathfrak{Z}_2}(*\mathfrak{H}_{Z_2})$-module $\overline{\mathfrak{M}}'_{i3}$ whose push-forward via the inclusion is isomorphic to $\overline{\mathfrak{M}}'_{i2}$. We may assume that φ factors through $Z \xrightarrow{\varphi_2} Z_2 \longrightarrow \overline{X}$, such that $\varphi_2^{-1}(H_{Z_2}) = H_Z$ and that φ_2 induces $Z \setminus H_Z \simeq Z_2 \setminus H_{Z_2}$. Then, we have $\mathscr{D}_3(*H_{Z_2})$-modules $\overline{\mathfrak{M}}'_{i4}$ with an isomorphism $\varphi_{2\dagger}\overline{\mathfrak{M}}'_{i4} \simeq \overline{\mathfrak{M}}'_{i3}$.

Let $\mathcal{V} = (\mathcal{V}_1, \mathcal{V}_2, C)$ be the admissible mixed twistor structure considered above. We have the extension $\mathfrak{V}_i$ of $\mathcal{V}_i$ to $\tilde{\mathfrak{R}}_{Z(*H_Z)}$-modules. We set $\mathfrak{V}'_i := \mathfrak{V}_i(*\mathfrak{Z}^0)$. By the construction, we have a natural isomorphism $\overline{\mathfrak{M}}'_{i4|\mathcal{Z}} \simeq \mathfrak{V}'_{i|\mathcal{Z}}$. Both of the meromorphic flat bundles $\overline{\mathfrak{M}}'_{i4|\mathfrak{Z}\setminus\mathfrak{H}_Z}$ and $\overline{\mathfrak{V}}'_{i|\mathfrak{Z}\setminus\mathfrak{H}_Z}$ are regular along ∞, and hence the isomorphism on $\mathfrak{Z}\setminus\mathfrak{H}_Z$ is extended on $\mathfrak{Z}\setminus\mathfrak{H}_Z^{\infty}$. By using the Deligne-Malgrange lattices, we obtain that the isomorphism is extended on $\mathfrak{Z}$. Hence, we obtain $\overline{\mathfrak{M}}_i(*\mathfrak{H}) \simeq \varphi_{\dagger}(\mathfrak{V}_i)$.

We take the hypersurface H_{Z1} in Z such that $H_{Z1}\cup D_Z = H_Z$ and that the codimension of $H_{Z1}\cap D_Z$ is larger than 2. We have $\overline{\mathfrak{M}}'_i[\star_1 g][\star_2\mathfrak{D}] \simeq \varphi_{\dagger}\big(\mathfrak{V}'_i[\star_1 H_{Z_1}][\star_2 D_Z]\big)$ for $\star_1, \star_2 \in \{*, !\}$. In particular, $\overline{\mathfrak{M}}'_i[\star_1 g][\star_2\mathfrak{D}]$ are regular along ∞.

Let us look at $\overline{\mathfrak{M}}'_2$. We consider the following complex:

$$\varphi_{\dagger}\mathfrak{V}'_2[!\mathfrak{H}_{Z1} * \mathfrak{D}_Z] \longrightarrow \overline{\mathfrak{M}}'_2 \oplus \varphi_{\dagger}\Xi^{(0)}_{gz}\mathfrak{V}'_2[*\mathfrak{D}_Z] \longrightarrow \varphi_{\dagger}\mathfrak{V}'_2[*\mathfrak{H}_Z]$$

The cohomology C is the extension of $\phi_g^{(0)}(\mathcal{M})(*\mathfrak{X}^0)$ to a $\mathscr{D}_{\mathfrak{X}}(*\mathfrak{X}^{\infty})$-module which is regular along ∞. We can reconstruct $\overline{\mathfrak{M}}'_2$ as the cohomology of

$$\varphi_{\dagger}\psi^{(1)}_{gz}(\mathfrak{V}'_2) \longrightarrow C \oplus \varphi_{\dagger}\Xi^{(0)}_{gz}(\mathfrak{V}'_2) \longrightarrow \varphi_{\dagger}\psi^{(0)}_{gz}(\mathfrak{V}'_2).$$

By applying the uniqueness of the extension to $\phi_g^{(0)}(\mathcal{M})$, we obtain that $\overline{\mathfrak{M}}_2 \simeq \mathfrak{M}_2$. We obtain $\overline{\mathfrak{M}}_1 \simeq \mathfrak{M}_1$ similarly.

Let us prove the algebraicity. We take a smooth projective completion $\overline{X}^{(1)}$ of X with a morphism $F : \overline{X}^{(1)} \longrightarrow \overline{X}$ which induces the identity on X. Let $D^{(1)} := X^{(1)} \setminus X$. We have $\mathcal{T}^{(1)} \in \mathrm{MTM}^{\mathrm{int}}(\overline{X}^{(1)}, [*D^{(1)}])$ such that $F_{\dagger}\mathcal{T}^{(1)} = \mathcal{T}$. Let $(\mathcal{M}_1^{(1)}, \mathcal{M}_2^{(1)}, C^{(1)})$ be the underlying $\tilde{\mathcal{R}}_{\overline{X}^{(1)}}$-triple of $\mathcal{T}^{(1)}$. We have the extensions of $\mathcal{M}_i^{(1)}$ to $\tilde{\mathfrak{R}}_{X^{(1)}}$-modules $\mathfrak{M}_i^{(1)}$ which are regular along ∞ and good as $\mathcal{O}_{\mathfrak{X}^{(1)}}$-modules. Because $X^{(1)}$ is projective, they are algebraic. By using the uniqueness, we obtain $F_{\dagger}\mathfrak{M}_i^{(1)} \simeq \mathfrak{M}_i$. Hence, $\mathfrak{M}_i$ are also algebraic. □

Let $f : \mathcal{T}_1 \longrightarrow \mathcal{T}_2$ be a morphism in $\mathrm{MTM}^{\mathrm{int}}(\overline{X}, [*D])$. Let $(\mathcal{M}_{i1}, \mathcal{M}_{i2}, C_i)$ be the underlying $\tilde{\mathcal{R}}_{\overline{X}}$-triples. We have the extensions $\mathfrak{M}_{i1}$.

Lemma 14.4.16 *The morphisms of $\tilde{\mathcal{R}}_{\overline{X}}$-modules $f_1 : \mathcal{M}_{21} \longrightarrow \mathcal{M}_{11}$ and $f_2 : \mathcal{M}_{12} \longrightarrow \mathcal{M}_{22}$ are uniquely extended to $\tilde{\mathfrak{R}}_{\overline{X}}$-homomorphism $\mathfrak{M}_{21} \longrightarrow \mathfrak{M}_{11}$ and $\mathfrak{M}_{12} \longrightarrow \mathfrak{M}_{22}$. The morphism is algebraic.*

Proof The uniqueness is clear. The existence is proved by an induction on the dimension of the support as in the case of Lemma 14.4.15. The algebraicity is reduced to the case where $\overline{X}$ is projective as in the proof of Lemma 14.4.15. □

14.4.7 Proof of Theorem 14.4.8

Let X be any algebraic variety. Let $\mathcal{T} \in \mathrm{MTM}^{\mathrm{int}}(X^{\mathrm{alg}})$.

We take a finite affine covering $\mathfrak{U} = (U_i \mid i \in \Lambda)$ of X. We set $H_i := X \setminus U_i$ for $i \in \Lambda$. We have the quasi-isomorphism $\mathcal{T} \longrightarrow \mathcal{C}^\bullet(\mathcal{T}, *\boldsymbol{H})$. By Lemma 14.4.15, we obtain that the underlying $\tilde{\mathcal{R}}_X$-modules of $\mathcal{T}[*H_I]$ $(I \subset \Lambda, I \neq \emptyset)$ are extended to algebraic $\tilde{\mathfrak{R}}$-modules which are regular along ∞. By Lemma 14.4.16, the morphisms are also uniquely extended to algebraic $\tilde{\mathfrak{R}}$-homomorphisms. Hence, we obtain that the underlying $\tilde{\mathcal{R}}_X$-modules of $\mathcal{T}$ are extended to $\tilde{\mathfrak{R}}$-modules which are good as an $\mathcal{O}_{\mathfrak{X}}$-module and regular along ∞.

Let $(\mathcal{M}_1, \mathcal{M}_2, C)$ be the $\tilde{\mathcal{R}}_X$-triple underlying $\mathcal{T}$. We have already obtained the extensions of $\mathcal{M}_i$ to $\tilde{\mathfrak{R}}_X$-modules $\mathfrak{M}_i$ which are regular along ∞ and good as $\mathcal{O}_{\mathfrak{X}}$-modules. Let $\overline{\mathfrak{M}}_i$ be other such extensions. Let us observe that $\overline{\mathfrak{M}}_2$ is naturally isomorphic to $\mathfrak{M}_2$.

Let H be any hypersurface of X such that $X \setminus H$ is affine. Then, we can prove that $\overline{\mathfrak{M}}_2(*H) \simeq \mathfrak{M}_2(*H)$ by using the argument for the uniqueness in Lemma 14.4.15. By varying such H, we obtain $\overline{\mathfrak{M}}_2 \simeq \mathfrak{M}_2$.

The extension of morphisms is also obtained by using any finite affine Cech covering. The uniqueness of the extension is clear. Thus, we obtain Theorem 14.4.8.

□

Chapter 15
Good Systems of Ramified Irregular Values

This chapter is an appendix. We have two purposes. One is to introduce the concept of good system of ramified irregular values, which would be convenient for the description of the irregularity of meromorphic flat bundles and Higgs bundles. The other is to prove the existence of a resolution for meromorphic Lagrangian covers.

Let $X := \{z \in \mathbb{C} \mid |z| < 1\}$ and $D := \{0\}$. Let (V, ∇) be a meromorphic flat bundle on (X, D), i.e., V is a locally free $\mathcal{O}_X(*D)$-module with a connection ∇. For any $e \in \mathbb{Z}_{>0}$, we consider $X^{(e)} := \{z_e \in \mathbb{C} \mid |z_e| < 1\}$ with a morphism $\varphi_e : X^{(e)} \longrightarrow X$ given by $\varphi_e(z_e) = z_e^e$. According to the classical Hukuhara-Levelt-Turrittin theorem, there exists a positive integer e and a decomposition

$$\varphi_e^*(V, \nabla)_{|\hat{0}} = \bigoplus_{\mathfrak{a} \in \mathbb{C}((z_e))/\mathbb{C}[\![z_e]\!]} (V_{\mathfrak{a}}, \nabla_{\mathfrak{a}}) \tag{15.1}$$

such that $\nabla_{\mathfrak{a}} - d\tilde{\mathfrak{a}}$ are regular singular, where "$\hat{0}$" means the formal completion at $0 \in X$, and $\tilde{\mathfrak{a}} \in \mathbb{C}((z_e))$ denote lifts of $\mathfrak{a}$. We set $\mathrm{Irr}(V, \nabla) := \{\mathfrak{a} \mid V_{\mathfrak{a}} \neq 0\}$. We can naturally regard it as a finite subset in $\varinjlim_e \mathbb{C}((z^{1/e}))/\mathbb{C}[\![z^{1/e}]\!]$ which is invariant under the natural action of the Galois group.

The higher dimensional case was studied by several people including, André, Kedlaya, Majima, Malgrange, Sabbah and the author, and we now have an appropriate generalization. Namely, let X be a complex manifold with a simple normal crossing hypersurface D. Let (V, ∇) be a meromorphic flat bundle on (X, D). Then, there exists a projective birational morphism $F : X' \longrightarrow X$ such that $D' := F^{-1}(D)$ is a simple normal crossing hypersurface and that $F^*(V, \nabla)$ is a good meromorphic flat bundle. Namely, for any point $P \in D'$, the pull back of the formal completion $F^*(V, \nabla)_{|\hat{P}}$ via an appropriate covering ramified over D' has a decomposition as in (15.1). Such a birational morphism is called "a resolution of turning points".

In the above situation, we have a family of the index sets of the decompositions for $P \in D'$. That is a good system ramified irregular values formulated in

T. Mochizuki, *Mixed Twistor $\mathscr{D}$-modules*, Lecture Notes in Mathematics 2125,
DOI 10.1007/978-3-319-10088-3_15

Sect. 15.1. In the case where any ramification never happens, such families are already formulated in [55]. It seems more convenient to give the terminology in the ramified case.

The good property is not preserved by various easy operations. For example, even if (V_i, ∇) $(i = 1, 2)$ are good meromorphic flat bundle, $(V_1, \nabla) \otimes (V_2, \nabla)$ and $(V_1, \nabla) \oplus (V_2, \nabla)$ are not good. It would be useful to have a general result to ensure that the good property can be recovered after an appropriate blow up (Proposition 15.1.5).

In our study of the existence of a resolution of turning points in the algebraic case [55], one of the main ideas is to replace the issue for meromorphic flat bundles with a similar issue for Higgs bundles in a special case which was solved in [55]. We study the issue for Higgs bundles more systematically. We consider the case that the spectral covers given by the Higgs fields are Lagrangian. Such covers are called Lagrangian covers, and the condition is satisfied if the Higgs bundles come from harmonic bundles. Then, we prove the existence of resolutions of turning points for such Higgs bundles under the assumption that the spectral varieties are meromorphic (Theorem 15.2.7).

It gives us a simplification of the definition of wild harmonic bundles. Previously, unramifiedly good wild harmonic bundles are defined in a rather complicated way, then good wild harmonic bundles are defined as the local descent of unramifiedly good harmonic bundles, and then wild harmonic bundles are defined as the push-forward of good wild harmonic bundles. But, after Theorem 15.2.7, wild harmonic bundles can be defined directly in terms of their spectral varieties.

15.1 Good System of Ramified Irregular Values

15.1.1 Good Set of Irregular Values

Let X be a complex manifold with a normal crossing hypersurface D. Let P be any point of D. We take a holomorphic coordinate neighbourhood $(X_P; z_1, \dots, z_n)$ around P such that $X_P \cap D = \bigcup_{i=1}^{\ell} \{z_i = 0\}$. For any $\boldsymbol{m} = (m_i) \in \mathbb{Z}^n$, we set $\boldsymbol{z}^{\boldsymbol{m}} = \prod_{i=1}^{n} z_i^{m_i}$. We naturally regard $\mathbb{Z}^{\ell} \subset \mathbb{Z}^n$ by $\boldsymbol{m} \longmapsto (\boldsymbol{m}, 0, \dots, 0)$.

Definition 15.1.1 A finite subset $\mathcal{I} \subset \mathcal{O}_X(*D)_P / \mathcal{O}_{X,P}$ is called a good set of irregular values at P, if the following holds ([55, 65]):

(i) For any $\mathfrak{a} \in \mathcal{I} \setminus \{0\}$, we take a lift $\tilde{\mathfrak{a}} \in \mathcal{O}_X(*D)_P$. Then, there exists $\operatorname{ord}(\mathfrak{a}) \in \mathbb{Z}^{\ell}_{\leq 0}$ such that $\tilde{\mathfrak{a}} \boldsymbol{z}^{-\operatorname{ord}(\mathfrak{a})}$ is an invertible element of $\mathcal{O}_{X,P}$.

(ii) For any pair $\mathfrak{a}_i \in \mathcal{I}$ $(i = 1, 2)$ with $\mathfrak{a}_1 \neq \mathfrak{a}_2$, we take lifts $\tilde{\mathfrak{a}}_i \in \mathcal{O}_X(*D)_P$. Then, there exists $\operatorname{ord}(\mathfrak{a}_1 - \mathfrak{a}_2) \in \mathbb{Z}^{\ell}_{\leq 0}$ such that $(\tilde{\mathfrak{a}}_1 - \tilde{\mathfrak{a}}_2) \boldsymbol{z}^{-\operatorname{ord}(\mathfrak{a}_1 - \mathfrak{a}_2)}$ is an invertible element of $\mathcal{O}_{X,P}$.

(iii) Let $(\mathfrak{a}_1, \mathfrak{a}_2)$ and $(\mathfrak{b}_1, \mathfrak{b}_2)$ be such pairs in $\mathcal{I}$. Then, we have either $\boldsymbol{z}^{\operatorname{ord}(\mathfrak{a}_1 - \mathfrak{a}_2) - \operatorname{ord}(\mathfrak{b}_1 - \mathfrak{b}_2)} \in \mathcal{O}_{X,P}$ or $\boldsymbol{z}^{-\operatorname{ord}(\mathfrak{a}_1 - \mathfrak{a}_2) + \operatorname{ord}(\mathfrak{b}_1 - \mathfrak{b}_2)} \in \mathcal{O}_{X,P}$.

The condition is independent of the choices of lifts. □

15.1.2 Good System of Ramified Irregular Values

Let X be a complex manifold with a normal crossing hypersurface D. We generalize the notion of good system of irregular values in Sect. 2.4.1 of [55].

Let P be any point of D. We introduce a category $C_P(X,D)$ as follows. Objects in $C_P(X,D)$ are holomorphic maps $\varphi : (Z,Q) \longrightarrow (X,P)$ of smooth complex manifolds which are coverings with ramification along D. We set $D_Z := \varphi^{-1}(D)$. Morphisms $F : \big(Z,Q,\varphi\big) \longrightarrow \big(Z',Q',\varphi'\big)$ are holomorphic maps $F : (Z,Q) \longrightarrow (Z',Q')$ such that $\varphi' \circ F = \varphi$. Such morphisms induce the morphisms $\mathcal{O}_{Z'}(*D_{Z'})_{Q'} \longrightarrow \mathcal{O}_Z(*D_Z)_Q$ over $\mathcal{O}_X(*D)_Q$. Let $\tilde{\mathcal{O}}_X(*D)_P$ denote an inductive limit of $\mathcal{O}_Z(*D_Z)_Q$. Similarly, let $\tilde{\mathcal{O}}_{X,P}$ denote the inductive limit of $\mathcal{O}_{Z,Q}$.

We have another description of the rings. Let $\mathbb{C}\{z_1,\ldots,z_n\}$ denote the ring of convergent power series. Let $\mathbb{C}\{z_1,\ldots,z_n\}_{z_1\ldots z_\ell}$ denote its localization with respect to $z_1\cdots z_\ell$. For a coordinate $(z_1,\ldots,z_n)$ such that $D = \bigcup_{i=1}^{\ell}\{z_i = 0\}$, we have natural isomorphisms

$$\tilde{\mathcal{O}}_{X,P} \simeq \varinjlim_{e} \mathbb{C}\big\{z_1^{1/e},\ldots,z_\ell^{1/e},z_{\ell+1},\ldots,z_n\big\},$$

$$\tilde{\mathcal{O}}_X(*D)_P \simeq \varinjlim_{e} \mathbb{C}\big\{z_1^{1/e},\ldots,z_\ell^{1/e},z_{\ell+1},\ldots,z_n\big\}_{z_1^{1/e}\cdots z_\ell^{1/e}}.$$

Definition 15.1.2 A finite subset $\mathcal{I} \subset \tilde{\mathcal{O}}_X(*D)_P\big/\tilde{\mathcal{O}}_{X,P}$ can be regarded as $\mathcal{I} \subset \mathcal{O}_Z(*D_Z)_Q\big/\mathcal{O}_{Z,Q}$ for some $\big(Z,Q,\varphi\big) \in C_P(X,D)$. It is called a good set of ramified irregular values at P, if (i) it is a good set of irregular values on (Z,D_Z), (ii) it is stable under the action of the Galois group of φ. □

Let $\mathcal{I}_P$ be a good set of ramified irregular values at P. Take $(Z,Q,\varphi) \in C_P(X,D)$ as in $\mathcal{I}_P \subset \mathcal{O}_Z(*D_Z)_Q\big/\mathcal{O}_{Z,Q}$. When P_1 is close to P, we choose $Q_1 \in \varphi^{-1}(P_1)$, and then we have the naturally defined maps $\mathcal{I}_P \longrightarrow \mathcal{O}_Z(*D_Z)_{Q_1}\big/\mathcal{O}_{Z,Q_1} \longrightarrow \tilde{\mathcal{O}}_X(*D)_{P_1}\big/\tilde{\mathcal{O}}_{X,P_1}$. Because $\mathcal{I}_P$ is invariant under action of the Galois group of φ, the image of the composite of the maps is well defined, and it gives a good set of ramified irregular values at P_1.

Definition 15.1.3 A good system of ramified irregular values on (X,D) is a family of good sets of ramified irregular values $\boldsymbol{\mathcal{I}} = \big\{\mathcal{I}_P \bigm| P \in D\big\}$ satisfying the following condition.

(A) If P_1 is sufficiently close to P, the image of $\mathcal{I}_P$ in $\tilde{\mathcal{O}}_X(*D)_{P_1}\big/\tilde{\mathcal{O}}_{X,P_1}$ is equal to $\mathcal{I}_{P_1}$. □

The easiest example of good system of ramified irregular values is given by setting $\mathcal{I}_P = \{0\}$ for any $P \in D$. It is called the trivial good system of ramified irregular values, and denoted by $\mathbf{0}$.

Remark 15.1.4 Let $\boldsymbol{\mathcal{I}} = (\mathcal{I}_P \mid P \in D)$ be a good system of ramified irregular values on (X,D).

- We set $-\mathcal{I} := (-\mathcal{I}_P \mid P \in D)$, where $-\mathcal{I}_P := \{-\mathfrak{a} \mid \mathfrak{a} \in \mathcal{I}_P\}$, which is a good system of irregular values on (X, D).
- Let $F : X' \longrightarrow X$ be a morphism of complex manifolds such that $D' := F^{-1}(D)$ is normal crossing. For any $P' \in D'$, we have a naturally defined morphism $\tilde{\mathcal{O}}_X(*D)_{F(P')}/\tilde{\mathcal{O}}_{X,F(P')} \longrightarrow \tilde{\mathcal{O}}_{X'}(*D')_{P'}/\tilde{\mathcal{O}}_{X',P'}$. The image of $\mathcal{I}_{F(P')}$ is denoted by $F^{-1}(\mathcal{I})_{P'}$, which is a good set of irregular values at P'. Thus, we obtain a good system of irregular values $F^{-1}(\mathcal{I})$ on (X', D').
- In particular, for any open subset $X' \subset X$, we have the naturally defined good system $\mathcal{I}_{|X'} := (\mathcal{I}_P \mid P \in X' \cap D)$ on $(X', D \cap X')$. □

15.1.3 Specialization of Good Set of Ramified Irregular Values

Let X and D be as above. Let $D = \bigcup_{i\in\Lambda} D_i$ be the irreducible decomposition. Let $I \subset \Lambda$ be any subset. We set $I^c := \Lambda \setminus I$. We put $D_I := \bigcap_{i\in I} D_i$, $D(I) := \bigcup_{i\in I} D_i$, $\partial D_I := D_I \cap D(I^c)$ and $D_I^\circ := D_I \setminus \partial D_I$. For any $P \in D_I$, we have the naturally defined maps:

$$\varphi_{I1} : \tilde{\mathcal{O}}_X(*D)_P/\tilde{\mathcal{O}}_{X,P} \longrightarrow \tilde{\mathcal{O}}_X(*D)_P/\tilde{\mathcal{O}}_X(*D(I^c))_P$$

$$\varphi_{I2} : \tilde{\mathcal{O}}_X(*D(I^c))_P/\tilde{\mathcal{O}}_{X,P} \longrightarrow \tilde{\mathcal{O}}_{D_I}(*\partial D_I)_P/\tilde{\mathcal{O}}_{D_I,P}$$

For any subset $\mathcal{J}_P \subset \tilde{\mathcal{O}}_X(*D)_P/\tilde{\mathcal{O}}_{X,P}$, we set $\mathcal{J}_P(I)_{|D_I} := \varphi_{I2}\big(\varphi_{I1}^{-1}(0)\big)$.

Let $\mathcal{I}$ be a good system of ramified irregular values on (X, D). By applying the above procedure to any $P \in D_I$, we obtain a good system of ramified irregular values on $(D_I, \partial D_I)$, denoted by $\mathcal{I}(I)_{|D_I}$.

In the local case, we may also use the following procedure. Let $\mathcal{J}$ be any finite subset of $\tilde{\mathcal{O}}_X(*D)_P/\tilde{\mathcal{O}}_{X,P}$. For any $\mathfrak{a} \in \mathcal{O}_X(*D)_P/\mathcal{O}_{X,P}$, we define $\mathcal{J}(-\mathfrak{a}) := \{\mathfrak{b} - \mathfrak{a} \mid \mathfrak{b} \in \mathcal{J}\}$. We obtain the set $\mathcal{J}(-\mathfrak{a}, I)_{|D_I} := \mathcal{J}(-\mathfrak{a})(I)_{|D_I} \subset \tilde{\mathcal{O}}_{D_I}(*\partial D_I)_P/\tilde{\mathcal{O}}_{D_I,P}$ in a similar way. If $\mathcal{J} \subset \mathcal{O}_X(*D)_P/\mathcal{O}_X$ is a good set of irregular values at $P \in X$, then $\mathcal{J}(-\mathfrak{a}, I)_{|D_I} \subset \mathcal{O}_{D_I}(*\partial D_I)_P/\mathcal{O}_{D_I,P}$ is a good set of irregular values at $P \in D_I$ for each $\mathfrak{a} \in \mathcal{J}$.

15.1.4 Resolution

Let X be a complex manifold with a simple normal crossing hypersurface D. For simplicity, we assume that there exists a complex manifold X_1 with a simple normal crossing hypersurface D_1 such that (X, D) is contained in (X_1, D_1) as a relatively compact set. Let $\mathcal{I} = (\mathcal{I}_P \mid P \in D)$ be a family of finite subsets $\mathcal{I}_P \subset \tilde{\mathcal{O}}_X(*D)_P/\tilde{\mathcal{O}}_{X,P}$ such that (i) each $\mathcal{I}_P$ is Galois invariant, (ii) it satisfies the condition **(A)** in Definition 15.1.3. The following proposition is essentially proved in [55]. We will give a proof in Sects. 15.1.4.1–15.1.4.2.

Proposition 15.1.5 *There exists a projective morphism of complex manifolds $F : X' \longrightarrow X$ such that (i) $D' := F^{-1}(D)$ is simply normal crossing, (ii) $X' \setminus D' \simeq X \setminus D$, (iii) $F^{-1}\boldsymbol{\mathcal{I}}$ is a good set of irregular values on (X', D').*

Remark 15.1.6 Proposition 15.1.5 is applied typically as follows. Let $\boldsymbol{\mathcal{I}}_i$ $(i = 1, 2)$ be good systems of ramified irregular values. We set $(\mathcal{I}_1 \oplus \mathcal{I}_2)_P := \mathcal{I}_{1P} \cup \mathcal{I}_{2P}$ and $(\mathcal{I}_1 \otimes \mathcal{I}_2)_P := \{\mathfrak{a}_1 + \mathfrak{a}_2 \,|\, \mathfrak{a}_i \in \mathcal{I}_{iP}\}$. Then, the systems $\boldsymbol{\mathcal{I}}_1 \oplus \boldsymbol{\mathcal{I}}_2$ and $\boldsymbol{\mathcal{I}}_1 \otimes \boldsymbol{\mathcal{I}}_2$ are not good systems of ramified irregular values, in general. We use Proposition 15.1.5 to eliminate the points at which they are not good. □

15.1.4.1 Systems of Finite Sets of Ramified Ideals

Let X and D be as above. Let $D = \bigcup_{i \in \Lambda} D_i$ be the irreducible decomposition.

Let $P \in D$. Suppose that we are given a finite set of finitely generated ideals $\mathfrak{I} = \{\tilde{\mathcal{O}}_{X,P}, \mathfrak{I}_1, \ldots, \mathfrak{I}_m\}$ of $\tilde{\mathcal{O}}_{X,P}$ for $P \in D$ such that $\mathfrak{I}$ is invariant under the action of Galois group of $\tilde{\mathcal{O}}_{X,P}/\mathcal{O}_{X,P}$. We can take $(Z, Q, \varphi) \in C_P(X, D)$ such that $\mathfrak{I}_i$ $(i = 1, \ldots, m)$ are induced by ideals $\mathfrak{I}_{Z,i}$ in $\mathcal{O}_{Z,Q}$. If P_1 is sufficiently close to P, we choose $Q_1 \in \varphi^{-1}(P_1)$, and we have the ideals $\mathfrak{I}^{(Q_1)}_{Z,i}$ of $\mathcal{O}_{Z,Q_1}$ induced by $\mathfrak{I}_{Z,i}$. They give a finite set of ideals $\{\tilde{\mathcal{O}}_{Z,Q}, \mathfrak{I}^{(Q_1)}_{Z,1}, \ldots, \mathfrak{I}^{(Q_1)}_{Z,m}\}$ in $\mathcal{O}_{Z,Q_1}$, which induces a finite set of ideals $\mathfrak{I}^{(P_1)}$ in $\tilde{\mathcal{O}}_{X,P_1}$. Because $\mathfrak{I}$ is assumed to be Galois invariant, $\mathfrak{I}^{(P_1)}$ is well defined and Galois invariant.

Suppose that for each $P \in D$ we are given finite sets of finitely generated ideals $\mathfrak{I}_P = \{\tilde{\mathcal{O}}_{X,P}, \mathfrak{I}_{P,1}, \ldots, \mathfrak{I}_{P,s(P)}\}$ of $\tilde{\mathcal{O}}_{X,P}$. Such a family $\boldsymbol{\mathfrak{I}} = \{\mathfrak{I}_P \,|\, P \in D\}$ is called a system of finite sets of ramified ideals on (X, D) if the following conditions are satisfied.

- Each $\mathfrak{I}_P$ is Galois invariant.
- If P_1 is sufficiently close to P, we have $\mathfrak{I}^{(P_1)}_P = \mathfrak{I}_{P_1}$. Here $\mathfrak{I}^{(P_1)}_P$ denotes the set of ideals in $\tilde{\mathcal{O}}_{X,P_1}$ induced by $\mathfrak{I}_P$ as above.

The system of sets of ramified ideals on (X, D) is called principal, if the following holds:

- For any P, $\mathfrak{I}_{P,i}$ are principal ideals, and $\{\mathfrak{I}_{P,i}\}$ is totally ordered with respect to the inclusion.

The following lemma is clear.

Lemma 15.1.7 *For a given system of finite sets of ramified ideals on (X, D), there exists $(e_i \,|\, i \in \Lambda) \in \mathbb{Z}^{\Lambda}_{>0}$ such that the following holds:*

- *Let P be any point of D_I. We take a holomorphic coordinate system $(z_1, \ldots, z_n)$ around P such that D_j $(j \in I)$ are expressed as $\{z_{\alpha(j)} = 0\}$. Then, $\mathfrak{I}_{Pi}$ are contained in the extension $\mathcal{O}_{X,P}[z^{1/e_j}_{\alpha(j)} \,|\, j \in I]$.* □

Let $F : X' \longrightarrow X$ be any morphism of complex manifolds such that $D' := F^{-1}(D)$ is also normal crossing. Let $\boldsymbol{\mathfrak{I}} = (\mathfrak{I}_P \,|\, P \in D)$ be a system of finite sets of ramified ideals. For each $P' \in D'$, the set of ideals $\mathfrak{I}_{F(P')}$ in $\tilde{\mathcal{O}}_{X,F(P')}$ and the morphism $F^{\sharp}_{P'} : \tilde{\mathcal{O}}_{X,F(P')} \longrightarrow \tilde{\mathcal{O}}_{X',P'}$ induce the set of ideals $\mathfrak{I}_{P'}$ in $\tilde{\mathcal{O}}_{X',P'}$. Thus, we obtain a system of finite sets of ramified ideals $(\mathfrak{I}_{P'})$ on (X', D') which is denoted by $F^{-1}\boldsymbol{\mathfrak{I}}$.

Lemma 15.1.8 *Let $\boldsymbol{\mathfrak{I}}$ be any family of finite sets of ramified ideals on (X, D). There exists a projective morphism $F : X' \longrightarrow X$ such that (i) $D' := F^{-1}(D)$ is simply normal crossing, (ii) F induces an isomorphism $X' \setminus D' \simeq X \setminus D$, (iii) $F^{-1}\boldsymbol{\mathfrak{I}}$ is principal.*

Proof For each $P \in D$, we take a small neighbourhood X_P of P, and a ramified covering $\varphi_P : Z_P \longrightarrow X_P$ whose ramification indexes along D_i are e_i. We have the set of the ideals $\boldsymbol{\mathfrak{I}}_{Z_P} := \big(\mathfrak{I}_{Z_P,i} \,\big|\, i = 1, \ldots, s(P)\big)$ of $\mathcal{O}_{Z_P}$ which induces $\boldsymbol{\mathfrak{I}}_P$, and it is invariant under the action of the Galois group $G(\varphi_P)$ of φ_P. By applying the construction in Sect. 15.1 of [55], we canonically obtain a normal complex variety Z'_P equipped with a $G(\varphi)$-action, and a $G(\varphi)$-equivariant projective birational morphism $G_P : Z'_P \longrightarrow Z_P$ such that (i) $Z'_P \setminus (\varphi_P \circ G_P)^{-1}(D) \simeq Z_P \setminus \varphi_P^{-1}(D)$, (ii) each ideal $G_P^{-1}(\mathfrak{I}_{Z_P,i})\mathcal{O}_{Z'_P}$ is principal, (iii) the ideals are totally ordered with respect to the inclusion. In particular, we obtain $X_P^{(1)} = Z'_P/G(\varphi)$ with an induced birational morphism $F_P^{(1)} : X_P^{(1)} \longrightarrow X_P$ which satisfies $X_P^{(1)} \setminus (F_P^{(1)})^{-1}(D) \simeq X_P \setminus D$. We can glue $(X_P^{(1)}, F_P^1)$ $(P \in D)$, and we obtain a complex variety $X^{(1)}$ with a morphism $F^{(1)} : X^{(1)} \longrightarrow X$. Then, by taking an appropriate projective birational morphism $X' \longrightarrow X^{(1)}$, we obtain the desired one. □

15.1.4.2 Proof of Proposition 15.1.5

The following lemma is clear.

Lemma 15.1.9 *There exists $(e_i \,|\, i \in \Lambda) \in \mathbb{Z}_{>0}^{\Lambda}$ such that the following holds:*

- *Let P be any point of D_I. We take a holomorphic coordinate system $(z_1, \ldots, z_n)$ around P such that D_j $(j \in I)$ are expressed as $\{z_{\alpha(j)} = 0\}$. Then, $\mathcal{I}_P$ is contained in $\mathcal{O}_X(*D)_P[z_{\alpha(j)}^{1/e_j} \,|\, j \in I]\big/\mathcal{O}_{X,P}[z_{\alpha(j)}^{1/e_j} \,|\, j \in I]$.* □

For each $P \in D$, we take a small neighbourhood X_P and a ramified covering $\varphi_P : (Z_P, Q) \longrightarrow (X_P, P)$ whose ramification indexes along D_i are e_i. We set $D_{Z_P} := \varphi_P^{-1}(D)$. We have $\mathcal{I}_P \subset \mathcal{O}_{Z_P}(*D_{Z_P})_Q\big/\mathcal{O}_{Z_P,Q}$. For each $\mathfrak{a} \in \mathcal{I}_P$, we take lift $\tilde{\mathfrak{a}} \in \mathcal{O}_{Z_P}(*D_{Z_P})_Q$. We take a coordinate $(\xi_1, \ldots, \xi_n)$ of Z_P such that $D_{Z_P} = \bigcup_{i=1}^{\ell}\{\xi_i = 0\}$. For $\mathfrak{a} \in \mathcal{I}_P$, let $m_i(\mathfrak{a}) \geq 0$ be the pole order of $\mathfrak{a}$ along $\{\xi_i = 0\}$. We put $m_i(P) := \max_{\mathfrak{a} \in \mathcal{I}_P} m_i(\mathfrak{a})$. We set $\boldsymbol{\xi}^{\boldsymbol{m}(P)} := \prod_{i=1}^{\ell} \xi_i^{m_i(P)}$. We consider the ideals $\mathfrak{I}_{P,\mathfrak{a}}$ generated by $\boldsymbol{\xi}^{\boldsymbol{m}(P)}$ and $\tilde{\mathfrak{a}}\boldsymbol{\xi}^{\boldsymbol{m}(P)}$. It is independent of the choice of $\tilde{\mathfrak{a}}$. Note that the ideal $\mathfrak{I}_{P,\mathfrak{a}}$ is principal means either one of the following holds; (i) $\mathfrak{a} = 0$ in $\tilde{\mathcal{O}}_X(*D)_P\big/\tilde{\mathcal{O}}_{X,P}$, (ii) there exists $\mathrm{ord}(\tilde{\mathfrak{a}})$ in $\mathbb{Z}_{\leq 0}^{\ell}$.

They induce the set of finitely generated ideals $\mathfrak{I}_P = \{\tilde{\mathcal{O}}_{X,P}\} \cup \{\mathfrak{I}_{P,\mathfrak{a}} \,|\, \mathfrak{a} \in \mathcal{I}_P\}$ in $\tilde{\mathcal{O}}_{X,P}$, which is Galois invariant. The family $\boldsymbol{\mathfrak{I}} = (\mathfrak{I}_P \,|\, P \in D)$ is a system of tuples of ramified ideals. Note that $\mathfrak{I}$ is principal implies that for any $P \in D$ and for each pair $(\mathfrak{a}, \mathfrak{b})$ in $\mathcal{I}_P$, we have either $\xi^{m(\mathfrak{a})-m(\mathfrak{b})} \in \mathcal{O}_{Z_P,Q}$ or $\xi^{-m(\mathfrak{a})+m(\mathfrak{b})} \in \mathcal{O}_{Z_P,Q}$. (We may use it below to deduce the condition (iii) in Definition 15.1.1 at each P.)

By applying Lemma 15.1.8, we obtain a projective birational morphism $F_1 : X_1 \longrightarrow X$ such that (i) $D_1 := F_1^{-1}(D)$ is simply normal crossing, (ii) F_1 induces an isomorphism $X_1 \setminus D_1 \simeq X \setminus D$, (iii) $F^{-1}\mathcal{I}$ satisfies the condition (i) in Definition 15.1.1 at each $P_1 \in D_1$.

By applying similar arguments to tuples $\mathcal{J}_P := \{\mathfrak{a} - \mathfrak{b} \mid \mathfrak{a}, \mathfrak{b} \in \mathcal{I}_P\}$, we obtain a projective birational morphism $F_2 : X_2 \longrightarrow X$ such that (i) $D_2 := F_2^{-1}(D)$ is simply normal crossing, (ii) F_2 induces an isomorphism $X_2 \setminus D_2 \simeq X \setminus D$, (iii) $F^{-1}\mathcal{I}$ satisfies the condition (i) and (ii) in Definition 15.1.1 at each $P_2 \in D_2$. The condition (iii) in Definition 15.1.1 at each $P_2 \in D_2$ is also satisfied. Note that the above remark. □

15.2 Resolution of Turning Points for Lagrangian Covers

15.2.1 Lagrangian Cover

Let Y be any complex manifold. We consider a reduced complex analytic closed subset $\Sigma \subset T^*Y$ such that (i) it is finite over Y, (ii) the smooth part of Σ is Lagrangian with respect to the natural symplectic structure of T^*Y. We call such Σ by a Lagrangian cover of Y. The following is well known and easy to check.

Lemma 15.2.1 *Let ω be a holomorphic section of $\omega \in \Omega_Y^1$. Then,* $\operatorname{Im}\omega \subset T^*Y$ *is Lagrangian cover of Y if and only if $d\omega = 0$.* □

If we are given a Lagrangian cover Σ and any holomorphic closed one form ω, then $\omega + \Sigma \subset T^*Y$ denotes the translation of Σ by ω, and it is also a Lagrangian cover.

15.2.1.1 Meromorphic Lagrangian Cover

Let X be an n dimensional complex manifold with a simple normal crossing hypersurface H. For any point $P \in X$, let X_P denote a small neighbourhood of P in X, and $H_P := X_P \cap H$. We have the irreducible decomposition $H_P = \bigcup_{i \in \Lambda_P} H_{P,i}$. For any $\boldsymbol{N} \in \mathbb{Z}^{\Lambda_P}$, let $\boldsymbol{N}H_P := \sum_{i \in \Lambda_P} N_i H_{P,i}$.

Definition 15.2.2 A Lagrangian cover Σ is called meromorphic on (X, H) if the following holds for any P.

- Take a small neighbourhood X_P of P in X. Then, the closure of $\Sigma_{|X_P}$ in $T^*X_P(\log H_P) \otimes \mathcal{O}_{X_P}(\boldsymbol{N}H_P)$ is a complex analytic subset which is finite over X_P for some $\boldsymbol{N} \in \mathbb{Z}_{>0}^{\Lambda_P}$. □

We introduce some conditions for the behaviour of meromorphic Lagrangian cover around H.

Definition 15.2.3 A meromorphic Lagrangian cover is called logarithmic if the closure in $T^*X(\log H)$ is finite over X. □

We obtain the following from Lemma 15.2.1.

Lemma 15.2.4 *Let Σ be a logarithmic Lagrangian cover on (X, H). Take $P \in H$ such that Σ is unramified on a neighbourhood of P, i.e., there exists a neighbourhood X_P of P in X such that $\Sigma_{|X_P} = \coprod_{i \in \Lambda_P} \Sigma_{P,i}$ and that the induced morphisms $\Sigma_{P,i} \longrightarrow X_P$ are isomorphic. Then, we have closed meromorphic one forms ω_i on X_P which are logarithmic along $D \cap X_P$ such that $\Sigma_{P,i}$ are the image of ω_i.* □

Note that on a holomorphic coordinate neighbourhood $(X_P, z_1, \dots, z_n)$, if ω is a closed meromorphic one form which is logarithmic along $\bigcup_{i=1}^{\ell}\{z_i = 0\}$, then ω is the sum of $\sum_{i=1}^{\ell} \alpha_i dz_i/z_i$ $(\alpha_i \in \mathbb{C})$ and a holomorphic one form.

Definition 15.2.5 A meromorphic Lagrangian cover Σ is called unramifiedly good at $P \in H$ if there exist a good set of irregular values $\mathcal{I}_P \subset M(X_P, H_P)/H(X_P)$ and logarithmic Lagrangian covers $\Sigma_{P,\mathfrak{a}}$ $(\mathfrak{a} \in \mathcal{I}_P)$ on (X_P, H_P) such that

$$\Sigma_{|X_P \setminus H_P} = \coprod_{\mathfrak{a} \in \mathcal{I}_P} \big(d\tilde{\mathfrak{a}} + \Sigma_{P,\mathfrak{a}}\big). \tag{15.2}$$

Here, $\tilde{\mathfrak{a}} \in M(X_P, H_P)$ are lifts of $\mathfrak{a}$. The meromorphic Lagrangian cover is called unramifiedly good on (X, H) if it is unramifiedly good at any $P \in H$. □

Definition 15.2.6 A meromorphic Lagrangian cover Σ is called good at $P \in H$ if there exists a ramified covering $\psi_P : (X'_P, H'_P) \longrightarrow (X_P, H_P)$ such that $\psi_P^{-1}(\Sigma)$ is unramifiedly good on (X'_P, H'_P). It is called good if it is good at any $P \in H$. □

15.2.2 Statement

Let X be an n dimensional complex manifold with a simple normal crossing hypersurface H. For simplicity, we assume that there exists a complex manifold $X^{(1)}$ with a simple normal crossing hypersurface $H^{(1)}$ such that (X, H) is contained in $(X^{(1)}, H^{(1)})$ as a relatively compact open subset. We will prove the following theorem in Sects. 15.2.3–15.2.5.

Theorem 15.2.7 *For any meromorphic Lagrangian cover Σ, there exists a projective morphism $F : X' \longrightarrow X$ such that (i) $H' := F^{-1}(H)$ is normal crossing, (ii) F induces an isomorphism $X' \setminus H' \simeq X \setminus H$, (iii) $F^{-1}(\Sigma)$ is good on (X', H').*

15.2.2.1 Wild Harmonic Bundles

Before going to the proof, we give consequences on wildness of harmonic bundles. Let X and H be as above. Let $(E, \overline{\partial}_E, \theta, h)$ be a harmonic bundle on $X \setminus H$. Let $\Sigma(\theta) \subset T^*(X \setminus H)$ denote the spectral variety of θ. It is easy to observe that $\Sigma(\theta)$ is a Lagrangian cover of $X \setminus H$, by using Gabber's theorem. (See Sect. A:III.3 of [6], for example.) The following corollary is an easy characterization of wildness of harmonic bundles.

Corollary 15.2.8 *$\Sigma(\theta)$ is meromorphic if and only if there exists a projective morphism $F : X' \longrightarrow X$ such that (i) $H' := F^{-1}(H)$ is simply normal crossing, (ii) $X' \setminus H' \simeq X \setminus H$, (iii) $F^{-1}(E, \overline{\partial}_E, \theta, h)$ be a good wild harmonic bundle on (X', H').*

Proof The if part is clear. The only if part follows from Theorem 15.2.7. □

We call F as in Corollary 15.2.8 by a resolution for the wild harmonic bundle $(E, \overline{\partial}_E, \theta, h)$ on (X, H). The following corollary means that we do not have to care with the choice of H.

Corollary 15.2.9 *There exists a resolution for $(E, \overline{\partial}_E, \theta, h)$ on (X, H) if and only if there exists a resolution for $(E, \overline{\partial}_E, \theta, h)$ on (X, H_1) for some $H_1 \supset H$.* □

Corollary 15.2.10 *There exists a resolution for $(E, \overline{\partial}_E, \theta, h)$ on (X, H) if and only if $(E, \overline{\partial}_E, \theta)$ is extended to a meromorphic Higgs sheaf on (X, H).*

Proof If $(E, \overline{\partial}_E, \theta)$ is extended to a meromorphic Higgs sheaf, $\Sigma(\theta)$ is meromorphic. Hence, there exists a resolution. Conversely, if we have a resolution $F : (X', H') \longrightarrow (X, H)$, According to [55], $F^{-1}(E, \overline{\partial}_E, \theta)$ is extended to a good filtered Higgs bundle on (X', H'). By taking the push-forward, $(E, \overline{\partial}_E, \theta)$ is extended to a meromorphic Higgs sheaf on (X, H). □

15.2.3 Separation of Ramification and Polar Part

15.2.3.1 Separation of the Zeroes in the Interior Part

Let Z be any irreducible normal complex analytic space with a hypersurface $H = \bigcup_{i \in \Lambda} H_i$. Suppose that (Z, H) is equipped with an action of a finite group G. For any $\boldsymbol{N} \in \mathbb{Z}^{\Lambda}$, let $\boldsymbol{N}H$ denote the divisor $\sum_{i \in \Lambda} N_i H_i$.

Let $\mathcal{E}$ be any G-equivariant locally free $\mathcal{O}_Z$-module on Z. Let $f_1, \dots, f_m$ be sections of $\mathcal{E}(\boldsymbol{N}H)$ for some $\boldsymbol{N} \in \mathbb{Z}^{\Lambda}_{>0}$, such that the G-action induces a permutation on $\{f_1, \dots, f_m\}$. Let $\mathfrak{I}_{f_i}$ denote the ideal sheaf of the 0-set of f_i regarded as a section of $\mathcal{E}(\boldsymbol{N}H)$. We set $\mathfrak{I}^{(0)}_{f_i} := \mathfrak{I}_{f_i} + \mathcal{O}(-\boldsymbol{N}H)$ in $\mathcal{O}_Z$. By applying the construction in Sect. 15.1 of [55], we obtain the following lemma.

Lemma 15.2.11 *There exist an irreducible normal complex analytic space $Z^{(1)}$ with a G-action, and a G-equivariant projective morphism $F : Z^{(1)} \longrightarrow Z$ such that the following holds:*

- *F induces an isomorphism $Z^{(1)} \setminus F^{-1}(H) \simeq Z \setminus H$. (It implies that there exists a closed analytic subset $A \subset H$ such that $Z^{(1)} \setminus F^{-1}(A) \simeq Z \setminus A$.)*
- *The ideals $\mathfrak{I}_{f_i}^{(1)} := \mathcal{O}_{Z^{(1)}} F^{-1}(\mathfrak{I}_{f_i}^{(0)})$ are principal.* □

We have the induced sections $F^*(f_i)$ of $F^*\mathcal{E}(*F^{-1}(H))$. Let $\mathcal{Z}(F^*(f_i)) \subset Z^{(1)} \setminus F^{-1}(H)$ be the set of the points $P \in Z^{(1)} \setminus F^{-1}(H)$ such that $F^*(f_i)(P) = 0$. The second condition in Lemma 15.2.11 implies the following.

Lemma 15.2.12 *For any $Q \in F^{-1}(H)$ and for any $i = 1, \ldots, m$, one of the following holds:*

(A1) *Q is not contained in the closure of $\mathcal{Z}(F^*(f_i))$.*
(A2) *$F^*(f_i)$ is a section of $F^*\mathcal{E}$ around Q.*

Proof Fix $Q \in F^{-1}(H)$ and f_i. We set $P := F(Q)$. We take a frame $e_1, \ldots, e_r$ of $\mathcal{E}$. We have the expression $f_i = \sum_{p=1}^{r} A_p e_p$. Let h_j be defining functions of D_j around P. If $P \notin D_j$, we put $h_j := 1$. We set $\alpha := \prod h_j^{N_j}$. Then, the ideal $\mathfrak{I}_{f_i}$ is generated by αA_p $(p = 1, \ldots, r)$ and α. By the construction, $\mathfrak{I}_{f_i}^{(1)}$ is principal. If $F^*\alpha$ is a generator, we have that $F^*(A_p) \in \mathcal{O}_{Z^{(1)},Q}$ for any p, which means that **(A2)** holds. If one of αA_p, say αA_{p_0}, is a generator, A_{p_0} is invertible outside $F^{-1}(H)$, which implies **(A1)**. Thus, Lemma 15.2.12 is proved. □

15.2.3.2 A General Construction

Let W be any irreducible normal complex analytic space. Let D be an effective divisor on W. The support of D is denoted by $|D|$. Let $\pi : E \longrightarrow W$ be a holomorphic vector bundle on W. Let $\Upsilon \subset E(D) := E \otimes \mathcal{O}_W(D)$ be a reduced irreducible closed analytic subset of $E(D)$ such that (i) the induced map $\pi : \Upsilon \longrightarrow W$ is finite and flat, (ii) $\dim \Upsilon = \dim W$. Let d be the degree of $\Upsilon \longrightarrow W$. We set

$$A(\Upsilon) := \Big\{(x_1, \ldots, x_d) \in \overbrace{\Upsilon \times_W \cdots \times_W \Upsilon}^{d} \;\Big|\; \pi(x_i) \notin D,\ x_i \neq x_j\ (i \neq j)\Big\}.$$

Let $\overline{A}(\Upsilon)$ denote the closure of $A(\Upsilon)$ in $E(D) \times_W \cdots \times_W E(D)$ with the reduced structure. Let $Y(\Upsilon)$ denote the normalization of $\overline{A}(\Upsilon)$. Let $\pi_{Y(\Upsilon)}$ denote the naturally induced morphism $Y(\Upsilon) \longrightarrow W$.

Let $\mathfrak{S}_d$ denote the d-th symmetric group. We have a naturally induced action on $\mathfrak{S}_d$ on $Y(\Upsilon)$.

Lemma 15.2.13 *The natural morphism $[\pi_{Y(\Upsilon)}] : Y(\Upsilon)/\mathfrak{S}_d \longrightarrow W$ is an isomorphism of complex analytic spaces.*

Proof Let $W_0(\Upsilon)$ denote the set of the points $P \in W \setminus D$ such that the number of the points in $\Upsilon_W \times \{P\}$ is d. Then, $A(\Upsilon)$ is an $\mathfrak{S}_d$-principal bundle over $W_0(\Upsilon)$. We naturally have $A(\Upsilon)/\mathfrak{S}_d \simeq W_0(\Upsilon)$.

Because $\overline{A}(\Upsilon)$ is contained in $\Upsilon \times_W \cdots \times_W \Upsilon$, it is finite over W. Hence, $\pi_{Y(\Upsilon)}$ is finite. By the construction, the dimension of each irreducible component $Y(\Upsilon)$ is $\dim W$. Hence, we obtain that $\pi_{Y(\Upsilon)}$ is an open map (see [20]). Then, it is easy to check that $[\pi_{Y(\Upsilon)}]$ gives a homeomorphism of topological spaces. Both of $Y(\Upsilon)/\mathfrak{S}_d$ and W are normal. They are isomorphic on the open subset which is the complement of a nowhere dense closed analytic subset. Hence, we obtain that they are isomorphic as complex analytic spaces. □

Let $D_{Y(\Upsilon)}$ be the pull back of D by $Y(\Upsilon) \longrightarrow W$. The projection of $E(D) \times_W \times \cdots \times_W E(D)$ onto the i-th component induces a section of $\pi^*_{Y(\Upsilon)}E(D)$ on $Y(\Upsilon)$, which is denoted by s_i. By applying the construction in Lemma 15.2.11 to $\pi^*_{Y(\Upsilon)}E$ with the tuple of the sections $\{s_i \,|\, i = 1, \dots, m\} \cup \{s_i - s_j \,|\, i,j = 1, \dots, m\}$, we obtain a reduced irreducible normal complex analytic space Y_1 with an $\mathfrak{S}_d$-action, and an $\mathfrak{S}_d$-equivariant projective morphism $F_Y : Y_1 \longrightarrow Y(\Upsilon)$.

We set $W_1 := Y_1/\mathfrak{S}_d$ which is an irreducible normal complex analytic space. We have a naturally induced morphism $F_W : W_1 \longrightarrow W$. It satisfies $W_1 \setminus |D_1| \simeq W \setminus |D|$. We set $D_1 := F_W^*(D)$, $E_1 := F_W^*E$, and $\Upsilon_1 := F_W^*\Upsilon$ over W_1.

Lemma 15.2.14 *Y_1 is naturally isomorphic to $Y(\Upsilon_1)$.*

Proof By the construction, we naturally have $Y_1 \longrightarrow E_1(D_1) \times_{W_1} \cdots \times_{W_1} E_1(D_1)$. The image is contained in $\overline{A}(\Upsilon_1)$. Hence, we have the naturally defined morphism $\psi : Y_1 \longrightarrow Y(\Upsilon_1)$. The restriction of ψ over $W_1 \setminus |D_1|$ is an isomorphism. Because $Y_1 \longrightarrow W_1$ is finite, ψ is also finite. Both of Y_1 and $Y(\Upsilon)$ is normal. Then, we obtain that ψ is an isomorphism. □

Let Q be any point of $|D_{Y_1}|$. For each i, one of the following holds:

(A$_i$1) Q is not contained in the closure of $\mathcal{Z}(F_Y^*(s_i))$.
(A$_i$2) $F_Y^*(s_i)$ are sections of $\pi^*_{Y(\Upsilon_1)}E_1$.

For each pair (i,k), one of the following holds:

(A$_{i,k}$1) Q is not contained in the closure of $\mathcal{Z}(F_Y^*(s_i - s_k))$.
(A$_{i,k}$2) $F_Y^*(s_i - s_k)$ are sections of $\pi^*_{Y(\Upsilon_1)}E_1$.

15.2.4 Separation of Cover

Let $X = \Delta^n$ and $H = \bigcup_{i=1}^{\ell}\{z_i = 0\}$. Let Σ be a meromorphic Lagrangian cover on (X,H). Let $\overline{\Sigma} \subset T^*X(\log H) \otimes \mathcal{O}(NH)$ be the closure which is finite over X. We obtain a normal complex space $Y(\overline{\Sigma})$ with a morphism $\pi_{Y(\overline{\Sigma})} : Y(\overline{\Sigma}) \longrightarrow X$ as in Sect. 15.2.3.2. We also have the sections s_i $(i = 1, \dots, d)$ of $\pi^*_{Y(\overline{\Sigma})}T^*X(\log D) \otimes \mathcal{O}(NH)$. Let $O = (0, \dots, 0)$.

Proposition 15.2.15 *Fix $Q \in \pi^{-1}_{Y(\overline{\Sigma})}(O)$. Suppose that for any i, at least one of the following holds:*

(A$_i$1) *Q is not contained in the closure of $\mathcal{Z}(s_i)$.*
(A$_i$2) *s_i is a section of $\pi^*_{Y(\overline{\Sigma})}T^*X(\log D)$*

Moreover, suppose that for any (i,k), one of the following holds:

(A$_{ik}$1) *Q is not contained in the closure of $\mathcal{Z}(s_i - s_k)$.*
(A$_{ik}$2) *$s_i - s_k$ is a section of $\pi^*_{Y(\overline{\Sigma})}T^*X(\log D)$*

Then, there exists a finite subset $\mathcal{I}_O \subset \tilde{\mathcal{O}}_X(*H)_P/\tilde{\mathcal{O}}_{X,P}$ for which we have the decomposition (15.2) *around O.*

Proof By taking an appropriate ramified covering, we have only to consider the case that $Y(\overline{\Sigma}) \longrightarrow X$ is not ramified along $\{z_i = 0\}$ $(i = 1, \ldots, \ell)$. We will shrink X around O without mention in the following argument.

We define an equivalence relation on $\{s_1, \ldots, s_d\}$ as follows: $s_i \sim s_j$ if $s_i - s_j$ is a section of $\pi^*_{Y(\overline{\Sigma})}T^*X(\log D)$ around Q. We obtain a decomposition by the equivalence relation:

$$\{s_1, \ldots, s_d\} = \bigsqcup_{\kappa \in T} B_\kappa \tag{15.3}$$

Let $G_Q \subset \mathfrak{S}_d$ be the stabilizer of Q. Then, the action of G_Q on $\{s_1, \ldots, s_d\}$ preserves the decomposition (15.3) Indeed, if $g^*s_i = s_j$ for some $g \in G_Q$, Q is contained in the closure of $\mathcal{Z}(s_i - s_j)$. So, we have $s_i, s_j \in B_\kappa$ for some κ.

For each κ, we have a section ω_κ of $\pi^*_{Y(\overline{\Sigma})}\Big(T^*X(\log D) \otimes \big(\mathcal{O}_X(NH)/\mathcal{O}_X\big)\Big)$ induced by $s_i \in B_\kappa$, which is independent of the choice of s_i. By the consideration in the previous paragraph, ω_κ is invariant under the action of G_Q. Hence, it is the pull back of a section $\omega_{\kappa,0}$ of $T^*X(\log H) \otimes \mathcal{O}_X(NH)/\mathcal{O}_X$. Note that we have the induced exterior derivative d on $\Omega^\bullet_X(\log H) \otimes \mathcal{O}_X(NH)/\mathcal{O}_X$. Because we have $d\omega_\kappa = 0$ around general points of H, we have $d\omega_\kappa = 0$. Hence, we can take a meromorphic function $\mathfrak{a}_i \in M(X, H)$ such that $d\mathfrak{a}_i$ induces $\omega_{\kappa,0}$. We set $t_i := s_i - d\mathfrak{a}_\kappa$, which are sections of $\pi^*_{Y(\overline{\Sigma})}T^*X(\log D)$. We obtain decompositions $s_i = \pi^*_{Y(\overline{\Sigma})}d\mathfrak{a}_\kappa + t_i$. Then, we have the Lagrangian covers given by $\{t_i \mid s_i \in B_\kappa\}$, denoted by Σ_κ, and we have $\Sigma = \bigsqcup \big(d\mathfrak{a}_\kappa + \Sigma_\kappa\big)$. □

15.2.5 Proof of Theorem 15.2.7

Let us return to the situation in Sect. 15.2.2. By applying the construction in Sect. 15.2.3.2 to $T^*X(\log H) \otimes \mathcal{O}(NH)$ with the closure $\overline{\Sigma}$ of Σ, and by applying the resolution of singularity, we obtain a projective morphism $F : X_1 \longrightarrow X$ of complex manifolds, satisfying the following conditions:

- $H_1 := F^{-1}(H)$ is normal crossing.
- F induces $X_1 \setminus H_1 \simeq X \setminus H$,
- We have $F^*\overline{\Sigma}$ in $F^*\big(T_X^*(\log H) \otimes \mathcal{O}(\boldsymbol{N}H)\big)$. For any $Q \in Y(F^*\overline{\Sigma})$ and for any i (resp. (i,k)) we have **(A_i1)** or **(A_i2)** (resp. **(A_{ik}1)** or **(A_{ik}2)**).

We obtain an effective divisor $\boldsymbol{N}_1 H_1 = F^*(\boldsymbol{N}H)$. We have a naturally defined morphism $F^*T^*X(\log H) \otimes \mathcal{O}_X(\boldsymbol{N}H) \longrightarrow T^*X_1(\log H_1) \otimes \mathcal{O}_{X_1}(\boldsymbol{N}_1 H_1)$. Let $\overline{\Sigma}_1$ denote the image of $F^*\overline{\Sigma}$. Because $X_1 \setminus H_1 \simeq X \setminus H$, for any $Q \in Y(\overline{\Sigma}_1)$, and for any i (resp. (i,k)) we have **(A_i1)** or **(A_i2)** (resp. **(A_{ik}1)** or **(A_{ik}2)**). Then, by applying Proposition 15.2.15, for each $P \in H_1$, we obtain a finite subset $\mathcal{I}_P \subset \tilde{\mathcal{O}}_{X_1}(*H_1)_P\big/\tilde{\mathcal{O}}_{X_1,P}$ such that we have the decomposition (15.2) around P. It is easy to observe that the system $\boldsymbol{\mathcal{I}} = \big\{\mathcal{I}_P \,|\, P \in H_1\big\}$ satisfies the assumption in Proposition 15.1.5. Hence, we can take a projective morphism of complex manifolds $F_1 : X' \longrightarrow X_1$ such that (i) $H' := F_1^{-1}(H_1)$ is simply normal crossing, (ii) $X' \setminus H' \simeq X_1 \setminus H_1$, (iii) $F^{-1}\boldsymbol{\mathcal{I}}$ is a good set of irregular values on (X', H'). Then, the induced projective morphism $X' \longrightarrow X$ has the desired property. Thus, the proof of Theorem 15.2.7 is finished. □

References

1. C. Banica, Le complété formel d'un espace analytique le long d'un sous-espace: Un théoréme de comparaison. Manuscripta Math. **6**, 207–244 (1972)
2. A. Beilinson, On the derived category of perverse sheaves, in *K-Theory, Arithmetic and Geometry (Moscow, 1984–1986)*. Lecture Notes in Mathematics, vol. 1289 (Springer, Berlin, 1987), pp. 27–41
3. A. Beilinson, How to glue perverse sheaves, in *K-Theory, Arithmetic and Geometry (Moscow, 1984–1986)*. Lecture Notes in Mathematics, vol. 1289 (Springer, Berlin, 1987), pp. 42–51
4. A. Beilinson, J. Bernstein, P. Deligne, Faisceaux pervers, in *Analysis and topology on singular spaces, I (Luminy, 1981)*. Astérisque, vol. 100 (1982), pp. 5–171
5. J. Bingener, Über Formale Komplexe Räume. Manuscripta Math. **24**, 253–293 (1978)
6. J.-E. Björk, *Analytic D-Modules and Applications* (Kluwer Academic, Dordrecht, 1993)
7. E. Cattani, A. Kaplan, Polarized mixed Hodge structures and the local monodromy of variation of Hodge structure. Invent. Math. **67**, 101–115 (1982)
8. E. Cattani, A. Kaplan, W. Schmid, Degeneration of Hodge structures. Ann. Math. **123**, 457–535 (1986)
9. E. Cattani, A. Kaplan, W. Schmid, L^2 and intersection cohomologies for a polarized variation of Hodge structure. Invent. Math. **87**, 217–252 (1987)
10. M.A. de Cataldo, L. Migliorini, The Hodge theory of algebraic maps. Ann. Sci. École Norm. Sup. (4) **38**, 693–750 (2005)
11. P. Deligne, Théorème de Lefschetz et critères de dégénérescence de suites spectrales. Inst. Hautes Études Sci. Publ. Math. **35**, 259–278 (1968)
12. P. Deligne, *Équation Différentielles à Points Singuliers Réguliers*. Lectures Notes in Mathematics, vol. 163 (Springer, Berlin, 1970)
13. P. Deligne, *Cohomologie á supports propres*, Théorie des Topos et Cohomologie Etale des Schémas, Springer Lecture Notes in Mathematics **305** (1973), 250–480
14. P. Deligne, Théorie de Hodge. I, in *Actes du Congrès International des Mathématiciens (Nice, 1970)*, Tome 1 (Gauthier-Villars, Paris, 1971), pp. 425–430
15. P. Deligne, Théorie de Hodge, II. Inst. Hautes Études Sci. Publ. Math. **40**, 5–57 (1971)
16. P. Deligne, Théorie de Hodge, III. Inst. Hautes Études Sci. Publ. Math. **44**, 5–77 (1974)
17. P. Deligne, B. Malgrange, J-P. Ramis, Singularités irrégulières, in *Documents Mathématiques*, vol. 5 (Société Mathématique de France, Paris, 2007)
18. A. Douady, Prolongement de faisceaux analytique cohérents (Travaux de Trautmann, Frisch-Guenot, Siu), in *Seminaure Bourbaki 22e année*, no. 366 (1969/1970)
19. H. Esnault, C. Sabbah, J.-D. Yu, E_1-Degeneration of the irregular Hodge filtration [arXiv:1302.4537] (2013)

T. Mochizuki, *Mixed Twistor $\mathscr{D}$-modules*, Lecture Notes in Mathematics 2125,
DOI 10.1007/978-3-319-10088-3

20. H. Grauert, R. Remmert, *Coherent Analytic Sheaves* (Springer, Berlin, 1984)
21. P.A. Griffiths, Hodge theory and geometry. Bull. Lond. Math. Soc. **36**, 721–757 (2004)
22. C. Hertling, *tt** geometry, Frobenius manifolds, their connections, and the construction for singularities. J. Reine Angew. Math. **555**, 77–161 (2003)
23. C. Hertling, C. Sevenheck, Nilpotent orbits of a generalization of Hodge structures. J. Reine Angew. Math. **609**, 23–80 (2007)
24. C. Hertling, C. Sevenheck, Limits of families of Brieskorn lattices and compactified classifying spaces. Adv. Math. **223**, 1155–1224 (2010)
25. C. Hertling, C. Sevenheck, Twistor structures, tt*-geometry and singularity theory, in *From Hodge Theory to Integrability and TQFT tt*-Geometry.* Proceedings of Symposia in Pure Mathematics, vol.78 (American Mathematical Society, Providence, 2008), pp. 49–73
26. N. Hitchin, The self-duality equations on a Riemann surface. Proc. Lond. Math. Soc. **55**, 59–126 (1987)
27. R. Hotta, K. Takeuchi, T. Tanisaki, *D-Modules, Perverse Sheaves, and Representation Theory*. Progress in Mathematics, vol. 236 (Birkhäuser, Boston, 2008)
28. M. Kashiwara, On the maximally overdetermined system of linear differential equations, I. Publ. Res. Inst. Math. Sci. **10**, 563–579 (1974/1975)
29. M. Kashiwara, Vanishing cycle sheaves and holonomic systems of differential equations, in *Algebraic Geometry (Tokyo/Kyoto, 1982)*. Lecture Notes in Mathematics, vol. 1016 (Springer, Berlin, 1983), pp. 134–142
30. M. Kashiwara, The Riemann-Hilbert problem for holonomic systems. Publ. Res. Inst. Math. Sci. **20**, 319–365 (1984)
31. M. Kashiwara, The asymptotic behavior of a variation of polarized Hodge structure. Publ. Res. Inst. Math. Sci. **21**, 853–875 (1985)
32. M. Kashiwara, A study of variation of mixed Hodge structure. Publ. Res. Inst. Math. Sci. **22**, 991–1024 (1986)
33. M. Kashiwara, Regular holonomic D-modules and distributions on complex manifolds. in *Complex Analytic Singularities*. Advanced Studies in Pure Mathematics, vol. 8 (North-Holland, Amsterdam, 1987), pp. 199–206
34. M. Kashiwara, *D-Modules and Microlocal Calculus*. Translations of Mathematical Monographs. Iwanami Series in Modern Mathematics, vol. 217 (American Mathematical Society, Providence, 2003)
35. M. Kashiwara, T. Kawai, The Poincaré lemma for variations of polarized Hodge structure. Publ. Res. Inst. Math. Sci. **23**, 345–407 (1987)
36. M. Kashiwara, P. Schapira, *Sheaves on Manifolds* (Springer, Berlin, 1990)
37. L. Katzarkov, M. Kontsevich, T. Pantev, Hodge theoretic aspects of mirror symmetry, in *From Hodge Theory to Integrability and TQFT tt*-Geometry.* Proceedings of Symposia in Pure Mathematics, vol. 78 (American Mathematical Society, Providence, 2008), pp. 87–174
38. K. Kedlaya, Good formal structures for flat meromorphic connections, I: surfaces. Duke Math. J. **154**, 343–418 (2010)
39. K. Kedlaya, Good formal structures for flat meromorphic connections, II: excellent schemes. J. Am. Math. Soc. **24**, 183–229 (2011)
40. R. MacPherson, K. Vilonen, Elementary construction of perverse sheaves. Invent. Math. **84**, 403–435 (1986)
41. B. Malgrange, *Ideals of differentiable functions*, in *Tata Institute of Fundamental Research Studies in Mathematics*, vol. 3 (Tata Institute of Fundamental Research/Oxford University Press, Bombay/London, 1967)
42. B. Malgrange, Polynômes de Bernstein-Sato et cohomologie évanescente, in *Analysis and Topology on Singular Spaces, II, III (Luminy, 1981)*. Astérisque, vol. 101–102 (1983), pp. 243–267
43. B. Malgrange, *Équations Différentielles à Coefficients Polynomiaux.* Progress in Mathematics, vol. 96 (Birkhäuser, Boston, 1991)
44. B. Malgrange, Connexions méromorphes, II. Le réseau canonique. Invent. Math. **124**, 367–387 (1996)

45. B. Malgrange, On irregular holonomic D-modules, in *Éléments de la Théorie des Systèmes Différentiels Géométriques*. Sémin. Congr., vol. 8 (Société Mathématique de France, Paris, 2004), pp. 391–410
46. Z. Mebkhout, Une équivalence de catégories. Compos. Math. **51**, 51–62 (1984)
47. Z. Mebkhout, Une autre équivalence de catégories. Compos. Math. **51**, 63–88 (1984)
48. Z. Mebkhout, *Le formalisme des six opérations de Grothendieck pour les D_X-modules cohérents*. With supplementary material by the author and L. Narváez Macarro. Travaux en Cours [Works in Progress], vol. 35 (Hermann, Paris, 1989), pp. x+254
49. S. Mizohata, *The Theory of Partial Differential Equations* (Translated from the Japanese by Katsumi Miyahara) (Cambridge University Press, New York, 1973)
50. T. Mochizuki, Asymptotic behaviour of tame nilpotent harmonic bundles with trivial parabolic structure. J. Diff. Geom. **62**, 351–559 (2002)
51. T. Mochizuki, Kobayashi-Hitchin correspondence for Tame harmoinc bundles and an application. Astérisque **309** viii+117 pp (2006)
52. T. Mochizuki, Asymptotic behaviour of tame harmonic bundles and an application to pure twistor D-modules I, II. Mem. Am. Math. Soc. **185** vii+565 pp (2007)
53. T. Mochizuki, Kobayashi-Hitchin correspondence for tame harmonic bundles II. Geom. Topol. **13**, 359–455 (2009)
54. T. Mochizuki, Good formal structure for meromorphic flat connections on smooth projective surfaces, in *Algebraic Analysis and Around*. Advanced Studies in Pure Mathematics, Math. Soc. Japan, Tokyo, vol. 54 (2009), pp. 223–253
55. T. Mochizuki, Wild harmonic bundles and wild pure twistor D-modules. Astérisque **340** x+607 pp (2011)
56. T. Mochizuki, Asymptotic behaviour of variation of pure polarized TERP structures. Publ. Res. Inst. Math. Sci. **47**, 419–534 (2011)
57. T. Mochizuki, *Holonomic $\mathcal{D}$-Module with Betti Structure*. Mémoire de la SMF, vol. 138–139 (Société Mathématique de France, Paris, 2014)
58. T. Mochizuki, The Stokes structure of good meromorphic flat bundle. J. Inst. Math. Jussieu **10**, 675–712 (2011)
59. T. Mochizuki, A twistor approach to the Kontsevich complexes [arXiv:1501.04145] (2015)
60. T. Mochizuki, Twistor property of GKZ-hypergeometric systems [arXiv:1501.04146] (2015)
61. T. Mochizuki, *On Deligne-Malgrange lattices, resolution of turning points and harmonic bundles*, Ann. Inst. Fourier (Grenoble) **59**, 2819–2837 (2009)
62. C. Peters, J. Steenbrink, *Mixed Hodge Structure* (Springer, Berlin, 2008)
63. W. Rudin, Functional analysis, in *International Series in Pure and Applied Mathematics*, 2nd edn. (McGraw-Hill, New York, 1991)
64. C. Sabbah, Vanishing cycles and Hermitian duality. Proc. Steklov Inst. Math. **238**, 194–214 (2002)
65. C. Sabbah, Équations différentielles à points singuliers irréguliers et phénomène de Stokes en dimension 2. Astérisque **263** viii+190 pp (2000)
66. C. Sabbah, Polarizable twistor D-modules. Astérisque **300** vi+208 pp (2005)
67. C. Sabbah, Wild twistor D-modules, in *Algebraic Analysis and Around*. Advanced Studies in Pure Mathematics, vol. 54 (Mathematical Society of Japan, Tokyo, 2009), pp. 293–353
68. C. Sabbah, *Introduction to Stokes Structures*. Lecture Notes in Mathematics, vol. 2060 (Springer, Heidelberg, 2013)
69. M. Saito, Modules de Hodge polarisables. Publ. Res. Inst. Math. Sci. **24**, 849–995 (1988)
70. M. Saito, Duality for vanishing cycle functors. Publ. Res. Inst. Math. Sci. **25**, 889–921 (1989)
71. M. Saito, Induced D-modules and differential complexes. Bull. Soc. Math. Fr. **117**, 361–387 (1989)
72. M. Saito, Introduction to mixed Hodge modules, Actes du Colloque de Théorie de Hodge (Luminy, 1987). Astérisque No. 179–180 **10**, 145–162 (1989)
73. M. Saito, Mixed Hodge modules. Publ. Res. Inst. Math. Sci. **26**, 221–333 (1990)
74. M. Saito, Mixed Hodge modules and applications, Proceedings of the International Congress of Mathematicians, Vol. I, II (Kyoto, 1990), 725–734, Math. Soc. Japan, Tokyo, 1991

75. M. Saito, On the theory of mixed Hodge modules, Selected papers on number theory, algebraic geometry, and differential geometry, 47–61, Amer. Math. Soc. Transl. Ser. 2, 160, Amer. Math. Soc., Providence, RI, 1994
76. M. Saito, Appendix to H. Esnault, C. Sabbah, J.-D. Yu, E_1-Degeneration of the irregular Hodge filtration [arXiv:1302.4537]
77. W. Schmid, Variation of Hodge structure: the singularities of the period mapping. Invent. Math. **22**, 211–319 (1973)
78. C. Simpson, Constructing variations of Hodge structure using Yang-Mills theory and applications to uniformization. J. Am. Math. Soc. **1**, 867–918 (1988)
79. C. Simpson, Harmonic bundles on noncompact curves. J. Am. Math. Soc. **3**(3), 713–770 (1990)
80. C. Simpson, Higgs bundles and local systems. Publ. I.H.E.S. **75**, 5–95 (1992)
81. C. Simpson, Some families of local systems over smooth projective varieties. Ann. Math. **138**(2), 337–425 (1993)
82. C. Simpson, Mixed twistor structures [math.AG/9705006] (1997)
83. C. Simpson, The Hodge filtration on nonabelian cohomology, in *Algebraic geometry—Santa Cruz 1995*. Proceedings of the Symposia Pure Mathematics, vol. 62, Part 2 (American Mathematical Society, Providence, 1997), pp. 217–281
84. C. Simpson, Local systems on proper algebraic V-manifolds. Pure Appl. Math. Q. **7**, 1675–1759 (2011) [Special Issue: In memory of Eckart Viehweg]
85. Y. T. Siu, *Techniques of Extension of Analytic Objects*, vol. 8 (Marcel Dekker, New York, 1974)
86. J. Steenbrink, Limits of Hodge structures. Invent. Math. **31**, 229–257 (1975/1976)
87. J. Steenbrink, S. Zucker, Variation of mixed Hodge structure, I. Invent. Math. **80**, 489–542 (1985)
88. S. Zucker, Hodge theory with degenerating coefficients: L^2 cohomology in the Poincaré metric. Ann. Math. (2) **109**, 415–476 (1979)
89. S. Zucker, Variation of mixed Hodge structure, II. Invent. Math. **80**, 543–565 (1985)
90. J-L. Verdier, Extension of a perverse sheaf over a closed subspace, in *Differential Systems and Singularities (Luminy, 1983).* Astérisque, vol. 130 (1985), pp. 210–217
91. J. Włodarczyk, Resolution of singularities of analytic spaces, in *Proceedings of Gökova Geometry-Topology Conference 2008, Gökova Geometry/Topology Conference (GGT)*, Gökova (2009), pp. 31–63

Index

T. Mochizuki, *Mixed Twistor $\mathscr{D}$-modules*, Lecture Notes in Mathematics 2125,
DOI 10.1007/978-3-319-10088-3

LECTURE NOTES IN MATHEMATICS

Edited by J.-M. Morel, B. Teissier; P.K. Maini

Editorial Policy (for the publication of monographs)

1. Lecture Notes aim to report new developments in all areas of mathematics and their applications - quickly, informally and at a high level. Mathematical texts analysing new developments in modelling and numerical simulation are welcome.
   Monograph manuscripts should be reasonably self-contained and rounded off. Thus they may, and often will, present not only results of the author but also related work by other people. They may be based on specialised lecture courses. Furthermore, the manuscripts should provide sufficient motivation, examples and applications. This clearly distinguishes Lecture Notes from journal articles or technical reports which normally are very concise. Articles intended for a journal but too long to be accepted by most journals, usually do not have this "lecture notes" character. For similar reasons it is unusual for doctoral theses to be accepted for the Lecture Notes series, though habilitation theses may be appropriate.

2. Manuscripts should be submitted either online at www.editorialmanager.com/lnm to Springer's mathematics editorial in Heidelberg, or to one of the series editors. In general, manuscripts will be sent out to 2 external referees for evaluation. If a decision cannot yet be reached on the basis of the first 2 reports, further referees may be contacted: The author will be informed of this. A final decision to publish can be made only on the basis of the complete manuscript, however a refereeing process leading to a preliminary decision can be based on a pre-final or incomplete manuscript. The strict minimum amount of material that will be considered should include a detailed outline describing the planned contents of each chapter, a bibliography and several sample chapters.
   Authors should be aware that incomplete or insufficiently close to final manuscripts almost always result in longer refereeing times and nevertheless unclear referees' recommendations, making further refereeing of a final draft necessary.
   Authors should also be aware that parallel submission of their manuscript to another publisher while under consideration for LNM will in general lead to immediate rejection.

3. Manuscripts should in general be submitted in English. Final manuscripts should contain at least 100 pages of mathematical text and should always include

   – a table of contents;
   – an informative introduction, with adequate motivation and perhaps some historical remarks: it should be accessible to a reader not intimately familiar with the topic treated;
   – a subject index: as a rule this is genuinely helpful for the reader.

   For evaluation purposes, manuscripts may be submitted in print or electronic form (print form is still preferred by most referees), in the latter case preferably as pdf- or zipped ps-files. Lecture Notes volumes are, as a rule, printed digitally from the authors' files. To ensure best results, authors are asked to use the LaTeX2e style files available from Springer's web-server at:

   ftp://ftp.springer.de/pub/tex/latex/svmonot1/ (for monographs) and
   ftp://ftp.springer.de/pub/tex/latex/svmultt1/ (for summer schools/tutorials).

Additional technical instructions, if necessary, are available on request from lnm@springer.com.

4. Careful preparation of the manuscripts will help keep production time short besides ensuring satisfactory appearance of the finished book in print and online. After acceptance of the manuscript authors will be asked to prepare the final LaTeX source files and also the corresponding dvi-, pdf- or zipped ps-file. The LaTeX source files are essential for producing the full-text online version of the book (see http://www.springerlink.com/openurl.asp?genre=journal&issn=0075-8434 for the existing online volumes of LNM). The actual production of a Lecture Notes volume takes approximately 12 weeks.

5. Authors receive a total of 50 free copies of their volume, but no royalties. They are entitled to a discount of 33.3 % on the price of Springer books purchased for their personal use, if ordering directly from Springer.

6. Commitment to publish is made by letter of intent rather than by signing a formal contract. Springer-Verlag secures the copyright for each volume. Authors are free to reuse material contained in their LNM volumes in later publications: a brief written (or e-mail) request for formal permission is sufficient.

Addresses:

Professor J.-M. Morel, CMLA,
École Normale Supérieure de Cachan,
61 Avenue du Président Wilson, 94235 Cachan Cedex, France
E-mail: morel@cmla.ens-cachan.fr

Professor B. Teissier, Institut Mathématique de Jussieu,
UMR 7586 du CNRS, Équipe "Géométrie et Dynamique",
175 rue du Chevaleret
75013 Paris, France
E-mail: teissier@math.jussieu.fr

For the "Mathematical Biosciences Subseries" of LNM:

Professor P. K. Maini, Center for Mathematical Biology,
Mathematical Institute, 24-29 St Giles,
Oxford OX1 3LP, UK
E-mail: maini@maths.ox.ac.uk

Springer, Mathematics Editorial, Tiergartenstr. 17,
69121 Heidelberg, Germany,
Tel.: +49 (6221) 4876-8259

Fax: +49 (6221) 4876-8259
E-mail: lnm@springer.com

Printed in the United States
By Bookmasters